College Algebra:

Building Concepts and Connections

College Algebra:
Building Concepts and Connections

Revathi Narasimhan
Kean University

HOUGHTON MIFFLIN COMPANY   Boston   New York

Publisher: Richard Stratton
Senior Sponsoring Editor: Molly Taylor
Senior Marketing Manager: Jennifer Jones
Senior Development Editor: Erin Brown
Senior Project Editor: Tamela Ambush
Art and Design Manager: Jill Haber
Cover Design Director: Tony Saizon
Senior Photo Editor: Jennifer Meyer Dare
Senior Composition Buyer: Chuck Dutton
New Title Project Manager: James Lonergan
Editorial Associate: Andrew Lipsett
Editorial Assistant: Joanna Carter-O'Connell

Cover image: © Ralph Mercer Photography

Printed in the U.S.A.

Library of Congress Control Number: 2007938673

Instructor's Annotated Edition:
  ISBN-10: 0-618-26036-6
  ISBN-13: 978-0-618-26036-2

For orders, use student text ISBNs:
  ISBN-10: 0-618-26035-8
  ISBN-13: 978-0-618-26035-5

1 2 3 4 5 6 7 8 9-CRK-12 11 10 09 08

Contents

Chapter 3 Quadratic Functions   213

Chapter 4 Polynomial and Rational Functions   283

Chapter 5 Exponential and Logarithmic Functions   363

ABOUT THE AUTHOR

Revathi Narasimhan received her Ph.D. in Applied Mathematics from the University of Maryland at College Park. She grew up in Mesa, Arizona and received her undergraduate degree at Arizona State University. She is currently on the faculty of the Mathematics Department at Kean University in Union, New Jersey.

Professionally trained to apply the principles of analysis and algebra, she is keen on helping her students understand the "big picture" concepts in mathematics, whether at the graduate or undergraduate level. In addition to this textbook, she has written scholarly articles for academic journals and technology supplements for other textbooks.

She and her husband, a research microbiologist, have two sons. Reva likes to garden and sew and is an avid reader.

PREFACE

Our Mission

The goal of *College Algebra: Building Concepts and Connections* is to teach students to more clearly see how mathematical concepts connect and relate. We set out to accomplish this goal in two fundamental ways.

First, we created a pedagogy that "recalls" previous topics and skills by way of linked examples and Just in Time exercises and references. Through these devices, students receive consistent prompts that enable them to better remember and apply what they have learned.

We also considered the order in which functions should be presented relative to their corresponding equations. Accordingly, rather than present a comprehensive review of equations and equation solving in Chapter 1, we introduce functions in Chapter 1. We then present related equations and techniques for solving those equations in the context of their associated functions. When equations are presented in conjunction with their "functional" counterparts in this way, students come away with a more coherent picture of the mathematics.

Ultimately, our hope is that through *College Algebra: Building Concepts and Connections*, students will develop a better conceptual understanding of the subject and achieve greater preparedness for future math courses.

Instruction and Pedagogy

The instruction and pedagogy have been designed to help students make greater sense of the mathematics and to condition good study habits. We endeavor to keep students engaged, to help them make stronger connections between concepts, and to encourage exploration and review on a regular basis.

Engage

Contemporary and Classical Applications Applications are derived from a wide variety of topics in business, economics, and the social and natural sciences. While modern data is well represented, classical applications are also infused in various exercise sets and examples. Integrating applications throughout the text improves the accessibility of the writing by providing a firm context. It also helps students to develop a stronger sense of how mathematics is used to analyze problems in a variety of disciplines, to draw comparisons between discrete sets of data, and to make more informed decisions.

Writing Style We make every effort to write in an "open" and friendly manner to reduce the intimidation sometimes experienced by students when reading a mathematics textbook. We provide patient explanations while maintaining the mathematical rigor expected at this level. We also reference previously-introduced topics when appropriate, to help students draw stronger links between concepts. In this way, we hope to keep students more engaged and promote their success when working outside the classroom.

Connect

Functions and Equations: Gradual Progression   In each of the first five chapters, a different function is presented. Along with each unique function, the corresponding equation is given. For example, in Sections 1.3–1.5, linear equations and inequalities are discussed in the context of linear functions. This allows for a smoother transition given that students are likely familiar with linear equations and therefore can make a connection between linear equations and linear functions.

After Chapter 1, the discussion of functions builds gradually, with more abstract concepts appearing in Chapter 2. This presentation is unique in that two full chapters covering the fundamental properties and concepts of functions are presented before quadratic functions are presented in Chapter 3. All topics related to quadratics appear separately in Chapter 3, which also distinguishes the presentation. The discussion of functions then progresses to more complex topics. By organizing the contents in this way, particularly the introductory chapters on functions, we hope to offer students an opportunity to master core concepts before they advance to higher-level topics.

Just in Time References   These references are found in the margins throughout the textbook, where appropriate. They point to specific pages within the textbook where the referenced topics were first introduced and thus enable students to quickly turn back to the original discussions of the cited topics.

Just in Time Exercises   These exercises are included as the first set of exercises at the end of many sections. These exercises correlate to the Just in Time references that appear within the section. They are provided to help students recall what they have previously learned for direct application to new concepts presented in the current section.

Repeated Themes   We frequently revisit examples and exercises to illustrate how ideas may be advanced and extended. In particular, certain examples, called **Linked Examples,** have been labeled with ⋯⟡ icons so that instructors and students can connect them with other examples in the book. Through these devices, students can synthesize various concepts and skills associated with a specific example or exercise topic.

Explore

Keystroke Appendix   A Keystroke Appendix for the TI-83/84 family of calculators is included at the end of the book for quick reference. The appendix contents parallel the order of topics covered in the textbook and offer detailed instruction on keystrokes, commands, and menus.

Technology Notes   Technology Notes appear in the margins to support the optional use of graphing calculator technology and reference the Keystroke Appendix when appropriate. The screen shots and instructions found within the Technology Notes have been carefully prepared to illustrate and support some of the more subtle details of graphing calculator use that can often be overlooked.

Discover and Learn   These instructor-guided exercises appear within the discussions of selected topics. They are designed as short, in-class activities and are meant to encourage further exploration of the topic at hand.

Review and Reinforce

Chapter P   Chapter P has been developed for students or instructors who want to review prerequisite skills for the course. Topics include the real number system; exponents and scientific notation; roots, radicals, and rational exponents; polynomials; factoring; rational expressions; geometry; and rudimentary equation-solving.

Check It Out   A Check It Out exercise follows every example. These exercises provide students with an opportunity to try a problem similar to that given in the example. The answers to each Check It Out are provided in an appendix at the back of the textbook so that students can immediately check their work and self-assess.

Observations   Observations appear as short, bulleted lists that directly follow the graphs of functions. Typically, the Observations highlight key features of the graphs of functions, but they may also illustrate patterns that can help students organize their thinking. Since observations are repeated throughout the textbook, students will get into the habit of analyzing key features of functions. In this way, the Observations will condition students to better interpret and analyze what they see.

Notes to the Student   Placed within the exposition where appropriate, the Notes provide tips on avoiding common errors or offer further information on the topic under discussion.

Key Points   At the end of every section, the Key Points summarize major themes from the section. They are presented in bullet form for ease of use.

Three-Column Chapter Summary   A detailed Summary appears at the end of every chapter. It is organized by section and illustrates the main concepts presented in each section. Examples are provided to accompany the concepts, along with references to examples or exercises within the chapter. This format helps students quickly identify key problems to practice and review, ultimately leading to more efficient study sessions.

Additional Resources

INSTRUCTOR RESOURCES	STUDENT RESOURCES
Instructor's Annotated Edition (IAE)—a replica of the student textbook with answers to all exercises either embedded within the text pages or given in the Instructor Answer Appendix at the back of the textbook.	**Student Solutions Manual**—a manual containing complete solutions to all odd-numbered exercises and all of the solutions to the Chapter Tests.
HM Testing (Powered by Diploma™)—a computerized test bank that offers a wide array of algorithms. Instructors can create, author/edit algorithmic questions, customize, and deliver multiple types of tests.	

 Instructional DVDs—Hosted by Dana Mosley, these DVDs cover all sections of the text and provide explanations of key concepts in a lecture-based format. DVDs are closed-captioned for the hearing-impaired.

HM [math] SPACE **HM MathSPACE®** encompasses the interactive online products and services integrated with Houghton Mifflin textbook programs. HM MathSPACE is available through text-specific student and instructor websites and via Houghton Mifflin's online course management system. HM MathSPACE includes homework powered by **WebAssign®**; a **Multimedia eBook;** self-assessment and remediation tools; videos, tutorials, and **SMARTHINKING®.**

▶ **WebAssign®**—Developed by teachers, for teachers, WebAssign allows instructors to create assignments from an abundant ready-to-use database of algorithmic questions, or write and customize their own exercises. With WebAssign, instructors can create, post, and review assignments 24 hours a day, 7 days a week; deliver, collect, grade, and record assignments instantly; offer more practice exercises, quizzes, and homework; assess student performance to keep abreast of individual progress; and capture the attention of online or distance learning students.

▶ **Online Multimedia eBook**—Integrates numerous assets such as video explanations and tutorials to expand upon and reinforce concepts as they appear in the text.

▶ **SMARTHINKING® Live, Online Tutoring**—Provides an easy-to-use and effective online, text-specific tutoring service. A dynamic *Whiteboard* and a *Graphing Calculator* function enable students and e-structors to collaborate easily.

▶ **Student Website**—Students can continue their learning here with a multimedia eBook, glossary flash cards, and more.

▶ **Instructor Website**—Instructors can download solutions to textbook exercises via the Online Instructor's Solutions Manual, digital art and figures, and more.

Online Course Management Content for Blackboard®, WebCT®, and eCollege®—Deliver program- or text-specific Houghton Mifflin content online using your institution's local course management system. Houghton Mifflin offers homework, tutorials, videos, and other resources formatted for Blackboard, WebCT, eCollege, and other course management systems. Add to an existing online course or create a new one by selecting from a wide range of powerful learning and instructional materials.

For more information, visit **college.hmco.com/pic/narasimhanCA1e** or contact your local Houghton Mifflin sales representative.

Acknowledgments

We would like to thank the following instructors and students who participated in the development of this textbook. We are very grateful for your insightful comments and detailed review of the manuscript.

Manuscript Reviewers and Other Pre-Publication Contributors

April Allen
Baruch College

Carolyn Allred-Winnett
Columbia State Community College

Jann Avery
Monroe Community College

Rich Avery
Dakota State University

Robin Ayers
Western Kentucky University

Donna J. Bailey
Truman State University

Andrew Balas
University of Wisconsin, Eau Claire

Michelle Benedict
Augusta State University

Marcelle Bessman
Jacksonville University

Therese Blyn
Wichita State University

Bill Bonnell
Glendale Community College

Beverly Broomell
Suffolk County Community College

Bruce Burdick
Roger Williams University

Veena Chadha
University of Wisconsin, Eau Claire

Jodi Cotten
Westchester Community College

Anne Darke
Bowling Green State University

Amit Dave
Dekalb Technical Institute

Luz De Alba
Drake University

Kamal Demian
Cerritos College

Tristan Denley
University of Mississippi

Richard T. Driver
Washburn University

Douglas Dunbar
Okaloosa Walton Community College

Royetta Ealba
Henry Ford Community College

Carolyn Edmond
University of Arizona

Donna Fatheree
University of Louisiana, Lafayette

Kevin A. Fox
Shasta College

Jodie Fry
Broward Community College

Cathy Gardner
Grand Valley State University

Don Gibbons
Roxbury Community College

Dauhrice K. Gibson
Gulf Coast Community College

Gregory Gibson
North Carolina A & T State University

Irie Glajar
Austin Community College

Sara Goldammer
University of South Dakota

Patricia Gramling
Trident Technical College

Michael Greenwich
Community College of Southern Nevada

Robert Griffiths
Miami Dade College, Kendall

David Gross
University of Connecticut

Margaret Gruenwald
University of Southern Indiana

Shirley Hagewood
Austin Peay State University

Shawna Haider
Salt Lake Community College

Andrea Hendricks
Georgia Perimeter College

Jada Hill
Richland Community College

Gangadhar Hiremath
University of North Carolina, Pembroke

Lori Holden
Manatee Community College

Eric Hsu
San Francisco State University

Charlie Huang
McHenry County College

Rebecca Hubiak
Tidewater Community College, Virginia Beach

Jennifer Jameson
Coconino County Community College

Larry Odell Johnson
Dutchess Community College

Tina Johnson
Midwestern State University

Michael J. Kantor
University of Wisconsin, Madison

Mushtaq Khan
Norfolk State University

Helen Kolman
Central Piedmont Community College

Tamela Kostos
McHenry County College

Marc Lamberth
North Carolina A & T State University

Charles G. Laws
Cleveland State Community College

Matt Lunsford
Union University

Jo Major
Fayetteville Technical Community College

Kenneth Mann
Catawba Valley Community College

Mary Barone Martin
Middle Tennessee State University

Dave Matthews
Minnesota West Community and Technical College

Marcel Maupin
Oklahoma State University, Oklahoma City

Melissa E. McDermid
Eastern Washington University

Mikal McDowell
Cedar Valley College

Terri Miller
Arizona State University

Ferne Mizell
Austin Community College, Rio Grande

Shahram Nazari
Bauder College

Katherine Nichols
University of Alabama

Lyn Noble
Florida Community College, South

Tanya O' Keefe
Darton College, Albany

Susan Paddock
San Antonio College

Dennis Pence
Western Michigan University

Nancy Pevey
Pellissippi State Technical Community College

Jane Pinnow
University of Wisconsin, Parkside

David Platt
Front Range Community College

Margaret Poitevint
North Georgia College

Julia Polk
Okaloosa Walton Community College

Jennifer Powers
Michigan State University

Anthony Precella
Del Mar College

Ken Prevot
Metro State College, Denver

Laura Pyzdrowski
West Virginia University

Bala Rahman
Fayetteville Technical Community College

Jignasa Rami
Community College Baltimore County, Catonsville

Margaret Ramsey
Chattanooga State Technical Community College

Richard Rehberger
Montana State University

David Roach
Murray State University

Dan Rothe
Alpena Community College

Cynthia Schultz
Illinois Valley Community College

Shannon Schumann
University of Colorado, Colorado Springs

Edith Silver
Mercer County Community College

Randy Smith
Miami Dade College

Jed Soifer
Atlantic Cape Community College

Donald Solomon
University of Wisconsin, Milwaukee

Dina Spain
Horry-Georgetown Technical College

Carolyn Spillman
Georgia Perimeter College

Peter Staab
Fitchburg State College

Robin Steinberg
Pima Community College

Jacqui Stone
University of Maryland

Clifford Story
Middle Tennessee State University

Scott R. Sykes
State University of West Georgia

Fereja Tahir
Illinois Central College

Willie Taylor
Texas Southern University

Jo Ann Temple
Texas Tech University

Peter Thielman
University of Wisconsin, Stout

J. Rene Torres
University of Texas-Pan American

Craig Turner
Georgia College & State University

Clen Vance
Houston Community College, Central

Susan A. Walker
Montana Tech, The University of Montana

Barrett Walls
Georgia Perimeter College

Fred Warnke
University of Texas, Brownsville

Carolyn Warren
University of Mississippi

Jan Wehr
University of Arizona

Richard West
Francis Marion University

Beth White
Trident Technical College

Jerry Williams
University of Southern Indiana

Susan Williford
Columbia State Community College

Class Test Participants

Irina Andreeva
Western Illinois University

Richard Andrews
Florida A&M University

Mathai Augustine
Cleveland State Community College

Laurie Battle
Georgia College & State University

Sam Bazzi
Henry Ford Community College

Chad Bemis
Riverside Community College District

Rajeed Carriman
Miami Dade College, North

Martha M. Chalhoub
Collin County Community College

Tim Chappell
Penn Valley Community College

Oiyin Pauline Chow
Harrisburg Area Community College

Allan Danuff
Central Florida Community College

Ann Darke
Bowling Green State University

Steven M. Davis
Macon State College

Jeff Dodd
Jacksonville State University

Jennifer Duncan
Manatee Community College

Abid Elkhader
Northern State University

Nicki Feldman
Pulaski Technical College

Perry Gillespie
Fayetteville State University

Susan Grody
Broward Community College

Don Groninger
Middlesex County College

Martha Haehl
Penn Valley Community College

Katherine Hall
Roger Williams University

Allen C. Hamlin
Palm Beach Community College, Lake Worth

Celeste Hernandez
Richland College

Lynda Hollingsworth
Northwest Missouri State University

Sharon Holmes
Tarrant County College

David Hope
Palo Alto College

Jay Jahangiri
Kent State University

Susan Jordan
Arkansas Technical University

Rahim G. Karimpour
Southern Illinois University, Edwardsville

William Keigher
Rutgers University, Newark

Jerome Krakowiak
Jackson Community College

Anahipa Lorestani
San Antonio College

Cyrus Malek
Collin County Community College

Jerry Mayfield
Northlake College

M. Scott McClendon
University of Central Oklahoma

Francis Miller
Rappahannock Community College

Sharon Morrison
St. Petersburg College

Adelaida Quesada
Miami Dade College, Kendall

Sondra Roddy
Nashville State Community College

Randy K. Ross
Morehead State University

Susan W. Sabrio
Texas A&M University, Kingsville

Manuel Sanders
University of South Carolina

Michael Schroeder
Savannah State University

Mark Sigfrids
Kalamazoo Valley Community College

Mark Stevenson
Oakland Community College

Pam Stogsdill
Bossier Parish Community College

Denise Szecsei
Daytona Beach Community College

Katalin Szucs
East Carolina University

Mahbobeh Vezvaei
Kent State University

Lewis J. Walston
Methodist University

Jane-Marie Wright
Suffolk County Community College

Tzu-Yi Alan Yang
Columbus State Community College

Marti Zimmerman
University of Louisville

Focus Group Attendees

Dean Barchers
Red Rocks Community College

Steven Castillo
Los Angeles Valley College

Diedra Collins
Glendale Community College

Rohan Dalpatadu
University of Nevada, Las Vegas

Mahmoud El-Hashash
Bridgewater State College

Angela Everett
Chattanooga State Technical Community College

Brad Feldser
Kennesaw State University

Eduardo Garcia
Santa Monica Community College District

Lee Graubner
Valencia Community College

Barry Griffiths
University of Central Florida

Dan Harned
Lansing Community College

Brian Hons
San Antonio College

Grant Karamyan
University of California, Los Angeles

Paul Wayne Lee
St. Philips College

Richard Allen Leedy
Polk Community College

Aaron Levin
Holyoke Community College

Austin Lovenstein
Pulaski Technical College

Janice Lyon
Tallahassee Community College

Jane Mays
Grand Valley State University

Barry Monk
Macon State College

Sanjay Mundkur
Kennesaw State University

Kenneth Pothoven
University of South Florida

Jeff Rushall
Northern Arizona University

Stephanie Sibley
Roxbury Community College

Jane Smith
University of Florida

Joyce Smith
Chattanooga State Technical Community College

Jean Thorton
Western Kentucky University

Razvan Verzeanu
Fullerton College

Thomas Welter
Bethune Cookman College

Steve White
Jacksonville State University

Bonnie Lou Wicklund
Mount Wachusett Community College

Don Williamson
Chadron State College

Mary D. Wolfe
Macon State College

Maureen Woolhouse
Quinsigamond Community College

Student Class Test Participants

Olutokumbo Adebusuyi
Florida A&M University

Jeremiah Aduei
Florida A&M University

Jennifer Albornoz
Broward Community College, North

Steph Allison
Bowling Green State University

Denise Anderson
Daytona Beach Community College

Aaron Anderson
Daytona Beach Community College

India Yvette Anderson
Jacksonville State University

Sharon Auguste
Broward Community College, North

Danielle Ault
Jacksonville State University

Genisa Autin-Holliday
Florida A&M University

Dylan Baker
Saint Petersburg College

Mandie Baldwin
Bowling Green State University

Heather Balsamo
Collin County Community College, Spring Creek

Emmanuel Barker
Daytona Beach Community College

Nataliya V. Battles
Arkansas Tech University

Jason Beardslee
North Lake College

Barbara Belizaire
Broward Community College, North

Akira Benison
Jacksonville State University

Derek S. Bent
Kalamazoo Valley Community College

Janice Berbridge
Broward Community College, North

Corey Bieber
Jackson Community College

Michael Biegler
Northern State University

Robert Bogle
Florida A&M University

Skyy Bond
Florida A&M University

Brittany Bradley
Arkansas Tech University

Josh Braun
Northern State University

Marcus Brewer
Manatee Community College

Jeniece Brock
Bowling Green State University

Channing Brooks
Collin County Community College, Spring Creek

Renetta Brooks
Florida A&M University

Jonathan Brown
Arkansas Tech University

Jawad Brown
Florida A&M University

Ray Brown
Manatee Community College

Jill Bunge
Bowling Green State University

Melissa Buss
Northern State University

Kimberly Calhoun
Jacksonville State University

Joshua Camper
Northern State University

Andrew Capone
Franklin and Marshall

Bobby Caraway
Daytona Beach Community College

Brad Carper
Morehead State University

Megan Champion
Jacksonville State University

Kristen Chapman
Jacksonville State University

Shaina Chesser
Arkansas Tech University

Alisa Chirochanapanich
Saint Petersburg College

Holly Cobb
Jacksonville State University

Travis Coleman
Saint Petersburg College

Jazmin Colon
Miami Dade College, North

Stephen Coluccio
Suffolk County Community College

Cynthia Y. Corbett
Miami Dade College, North

Maggie Coyle
Northern State University

Theresa Craig
Broward Community College, North

Elle Crofton
Rutgers University

Shanteen Daley
Florida A&M University

Joann DeLucia
Suffolk County Community College

Christopher Deneen
Suffolk County Community College

William Deng
Northern State University

Erica Derreberry
Manatee Community College

Brendan DiFerdinand
Daytona Beach Community College

Rathmony Dok
North Lake College

Julie Eaton
Palm Beach Community College

Courtnee Eddington
Florida A&M University

Shaheen Edison
Florida A&M University

Jessica Ellis
Jacksonville State University

Deborah J. Ellis
Morehead State University

Victoria Enos
Kalamazoo Valley Community College

Amber Evangelista
Saint Petersburg College

Ruby Exantus
Florida A&M University

Staci Farnan
Arkansas Tech University

Falon R. Fentress
Tarrant County College, Southeast

Kevin Finan
Broward Community College, North

Daniela Flinner
Saint Petersburg College

Shawn Flora
Morehead State University

Lisa Forrest
Northern State University

Ryan Frankart
Bowling Green State University

Ashley Frystak
Bowling Green State University

Desiree Garcia
Jacksonville State University

Benjamin Garcia
North Lake College

Josie Garcia
Palo Alto College

Jahmal Garrett
Bowling Green State University

Robyn Geiger
Kalamazoo Valley Community College

Melissa Gentner
Kalamazoo Valley Community College

James Gillespie
University of Louisville

Jeanette Glass
Collin County Community College, Spring Creek

Holly Gonzalez
Northern State University

Jennifer Gorsuch
Jackson Community College

Josh Govan
Arkansas Tech University

Lindsey Graft
Jacksonville State University

Sydia Graham
Broward Community College, North

Donald R. Gray III
Morehead State University

Melissa Greene
Bowling Green State University

Stacy Haenig
Saint Petersburg College

Mitchell Haley
Morehead State University

Seehee Han
San Antonio College

Kimberly Harrison
Jacksonville State University

Emily Diane Harrison
Morehead State University

Joshua Hayes
San Antonio College

Leeza Heaven
Miami Dade College, North

David Heinzen
Arkansas Tech University

Katrina Henderson
Jacksonville State University

Ashley Hendry
Saint Petersburg College

Johnathan Hentschel
Arkansas Tech University

Amber Hicks
Jacksonville State University

Ryan Hilgemann
Northern State University

Maurice Hillman
Bowling Green State University

Matt Hobe
Jackson Community College

Neda Hosseiny
North Lake College

Laura Hyden
Bowling Green State University

Blake Jackson
Jacksonville State University

David Jaen
North Lake College

K. C. Jansson
Collin County Community College, Spring Creek

Wesley Jennings
Northern State University

Racel Johnson
Collin County Community College, Spring Creek

Kevin Jones
Miami Dade College, North

Matt Kalkbrenner
Pulaski Technical College

Brenda Kohlman
Northern State University

Tanya Kons
Daytona Beach Community College

Kevin LaRose-Renner
Daytona Beach Community College

Alex David Lasurdo
Suffolk County Community College

Amber Lee
Jacksonville State University

Amanda Lipinski
Northern State University

Ryan Lipsley
Pulaski Technical College

Amber Logan
Florida A&M University

Chris Lundgren
Miami Dade College, North

Lindsay Lvens
Kalamazoo Valley Community College

Lisa Mainz
Palo Alto College

Patricia Mantooth
Jacksonville State University

Jaclyn Margolis
Suffolk County Community College

Summer Martin
Jacksonville State University

Miguel Martinez
San Antonio College

Ali Masumi
Collin County Community College, Spring Creek

Somer Dawn Matter
Saint Petersburg College

Tiffany Mauriquez
Collin County Community College, Spring Creek

Tania Maxwell
Daytona Beach Community College

Durya McDonald
Florida A&M University

Bekah McCarley
Arkansas Tech University

W. McLeod
Jackson Community College

Shannon McNeal
Morehead State University

Carolina Medina
Tarrant County College, Southeast

Lilliam Mercado
Miami Dade College, North

Chelsea Metcalf
North Lake College

John H. Meyer
University of Louisville

Christina Michling
Collin County Community College, Spring Creek

Chris Migge
Northern State University

Della Mitchell
Florida A&M University

Debra Mogro
Miami Dade College, North

Emily Mohney
Kalamazoo Valley Community College

Demaris Moncada
Miami Dade College, North

Kathleen Monk
Jackson Community College

Virginia Mora
Palo Alto College

James Morales
San Antonio College

Amber Morgan
Jacksonville State University

Justin Murray
Bowling Green State University

Ernesto Noguera García
Jacksonville State University

Courtney Null
Jacksonville State University

Gladys Okoli
North Lake College

Ashley Olivier
San Antonio College

Jonathan Orjuela
Daytona Beach Community College

Lucia Orozco
Miami Dade College, North

Elizabeth Patchak
Kalamazoo Valley Community College

Natasha Patel
San Antonio College

Braden Peterson
Collin County Community College, Spring Creek

Jenny Phillips
Northern State University

Karina Pierce
Fayetteville State University

Joe Pietrafesa
Suffolk County Community College

Lacie Pine
Kalamazoo Valley Community College

Brandon Pisacrita
Jacksonville State University

Andrea Prempel
Tarrant County College, Southeast

Kaylin Purcell
Daytona Beach Community College

Mary Michelle Quillian
Jacksonville State University

Elizabeth Quinlisk
Saint Petersburg College

Kristina Randolph
Florida A&M University

Ian Rawls
Florida A&M University

Heather Rayburg
Manatee Community College

Samantha Reno
Arkansas Tech University

Marcus Revilla
San Antonio College

Kyle Rosenberger
Bowling Green State University

Cassie Rowland
Kalamazoo Valley Community College

Jason Russell
Jacksonville State University

Brian P. Rzepa
Manatee Community College

Matt Sanderson
Arkansas Tech University

Ana Santos
Daytona Beach Community College

Allison Schacht
Bowling Green State University

Dwayne Scheuneman
Saint Petersburg College

Jacqueline Schmidt
Northern State University

Danielle Serra
Suffolk County Community College

Kelly LeAnn Shelton
Morehead State University

Naomi Shoemaker
Palo Alto College

Chelsey Siebrands
Northern State University

Justin Silvia
Tarrant County College, Southeast

Bethany Singrey
Northern State University

Eron Smith
Arkansas Tech University

Klye Smith
Jackson Community College

Nicholas Solozano
Collin County Community College, Spring Creek

Joslyn Sorensen
Manatee Community College

Hailey Stimpson
Palo Alto College

Yanti Sunggono
Pulaski Technical College

Sharne Sweeney
Fayetteville State University

Katherine Sweigart
Arkansas Tech University

Amanda Tewksbury
Northern State University

Jenna Thomson
Manatee Community College

Diego F. Torres
San Antonio College

Tiffany Truman
Manatee Community College

Alice Turnbo
Morehead State University

Anselma Valcin-Greer
Broward Community College, North

A'Donna Wafer
Bowling Green State University

Christy Ward
Ohio University

Portia Wells
Daytona Beach Community College

Larissa Wess
Bowling Green State University

Ben White
Florida A&M University

Theresa Williams
Suffolk County Community College

Amy Wisler
Bowling Green State University

Aikaterini Xenaki
Daytona Beach Community College

Kristen Yates
Morehead State University

Amanda Young
Manatee Community College, Bradenton

Stephanie Zinter
Northern State University

Kristen Zook
Collin County Community College, Spring Creek

In addition, many thanks to Georgia Martin, who provided valuable feedback on the entire manuscript. Also, thanks to Brenda Burns, Noel Kamm, Joan McCarter, Rekha Natarajan, Danielle Potvin, George Pasles, Sally Snelson, and Douglas Yates for their assistance with the manuscript and exercise sets; Carrie Green, Lauri Semarne, and Christi Verity for accuracy reviews; Mark Stevenson for writing the solutions manuals; and Dana Mosely for the videos.

At Houghton Mifflin, I wish to thank Erin Brown and Molly Taylor for taking special care in guiding the book from its manuscript stages to production; Jennifer Jones for her creative marketing ideas; Tamela Ambush for superbly managing the production process; and Richard Stratton for his support of this project.

Special thanks to my husband, Prem Sreenivasan, our children, and our parents for their loving support throughout.

TEXTBOOK FEATURES

Chapter Opener ▶

Each **Chapter Opener** includes an applied example of content that will be introduced in the chapter. An outline of the section provides a clear picture of the topics being presented.

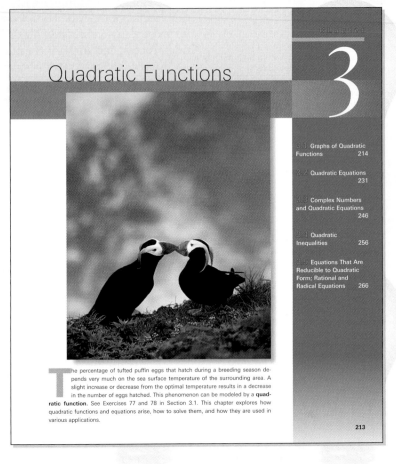

Quadratic Functions

Chapter

3

The percentage of tufted puffin eggs that hatch during a breeding season depends very much on the sea surface temperature of the surrounding area. A slight increase or decrease from the optimal temperature results in a decrease in the number of eggs hatched. This phenomenon can be modeled by a **quadratic function.** See Exercises 77 and 78 in Section 3.1. This chapter explores how quadratic functions and equations arise, how to solve them, and how they are used in various applications.

213

▼ Section Objectives

Each section begins with a list of bulleted objectives offering an at-a-glance overview of what will be covered.

5.1 Inverse Functions

Objectives

▶ Define the inverse of a function

▶ Verify that two functions are inverses of each other

▶ Define a one-to-one function

▶ Define the conditions for the existence of an inverse function

▶ Find the inverse of a function

Inverse Functions

Section 2.2 treated the composition of functions, which entails using the output of one function as the input for another. Using this idea, we can sometimes find a function that will undo the action of another function—a function that will use the output of the original function as input, and will in turn output the number that was input to the original function. A function that undoes the action of a function f is called the **inverse** of f.

As a concrete example of undoing the action of a function, Example 1 presents a function that converts a quantity of fuel in gallons to an equivalent quantity of that same fuel in liters.

Example 2 Algebraic Solution of a Quadratic Inequality

Solve the inequality $-x^2 + 5x - 4 \le 0$ algebraically.

▶Solution

STEPS	EXAMPLE
1. The inequality should be written so that one side consists only of zero.	$-x^2 + 5x - 4 \le 0$
2. Factor the expression on the nonzero side of the inequality; this will transform it into a product of two linear factors.	$(-x + 4)(x - 1) \le 0$
3. Find the zeros of the expression on the nonzero side of the inequality—that is, the zeros of $(-x + 4)(x - 1)$. These are the only values of x at which the expression on the nonzero side can change sign. To find the zeros, set each of the factors found in the previous step equal to zero, and solve for x.	$-x + 4 = 0 \implies x = 4$ $x - 1 = 0 \implies x = 1$
4. If the zeros found in the previous step are distinct, use them to break up the number line into three disjoint intervals. Otherwise, break it up into just two disjoint intervals. Indicate these intervals on the number line.	

◀ Examples

Well-marked and with descriptive titles, the text **Examples** further illustrate the subject matter being discussed. In cases where the solution to an example may involve multiple steps, the steps are presented in tabular format for better organization.

◀ Linked Examples

Where appropriate, some examples are linked throughout a section or chapter to promote in-depth understanding and to build stronger connections between concepts. While each example can be taught on its own, it's suggested that the student review examples from previous sections when they have a bearing on the problem under discussion.

Linked Examples are clearly marked with an icon. •••⁝

Example 1 Modeling Bacterial Growth

Example 1 in Section 5.3 builds upon this example. ···⁞

Suppose a bacterium splits into two bacteria every hour.
(a) Fill in Table 5.2.1, which shows the number of bacteria present, $P(t)$, after t hours.

Table 5.2.1

t (hours)	0	1	2	3	4	5	6	7	8
$P(t)$ (number of bacteria)									

(b) Find an expression for $P(t)$.

▶Solution

(a) Since we start with one bacterium and each bacterium splits into two bacteria every hour, the population is *doubled* every hour. This gives us Table 5.2.2.

Table 5.2.2

t (hours)	0	1	2	3	4	5	6	7	8
$P(t)$ (number of bacteria)	1	2	4	8	16	32	64	128	256

(b) To find an expression for $P(t)$, note that the number of bacteria present after t hours will be double the number of bacteria present an hour earlier. This gives

$$P(1) = 2(1) = 2^1; \quad P(2) = 2(P(1)) = 2(2) = 4 = 2^2;$$
$$P(3) = 2(P(2)) = 2(2^2) = 2^3; \quad P(4) = 2(P(3)) = 2(2^3) = 2^4;$$

Following this pattern, we find that $P(t) = 2^t$. Here, the variable, t, is in the *exponent*. This is quite different from the examined in the previous chapters, where the independent variable w *fixed power*. The function $P(t)$ is an example of an exponential functio

☑ *Check It Out 1:* In Example 1, evaluate $P(9)$ and interpret your result.

Example 1 Bacterial Growth

⁝··· *This example builds on Example 1 of Section 5.2.*

A bacterium splits into two bacteria every hour. How many hours will it take for the bacterial population to reach 128?

▶Solution Note that in this example, we are given the ending population and must figure out how long it takes to reach that population. Table 5.3.1 gives the population for various values of the time t, in hours. (See Example 1 from Section 5.2 for details.)

Table 5.3.1

t (hours)	0	1	2	3	4	5	6	7	8
$P(t) = 2^t$ (number of bacteria)	1	2	4	8	16	32	64	128	256

From the table, we see that the bacterial population reaches 128 after 7 hours. Put another way, we are asked to find the *exponent* t such that $2^t = 128$. The answer is $t = 7$.

☑ *Check It Out 1:* Use the table in Example 1 to determine when the bacterial population will reach 64. ■

WHAT REVIEWERS SAY ABOUT JUST IN TIME

"I struggle every semester: Do I spend a week doing the review chapter? With the **Just in Time** feature I don't have to; it gives me more time to teach!"

Dean Barchers, Red Rocks Community College

▼ Check It Out

Following every example, these exercises provide the student with an opportunity to try a problem similar to that presented in the example.

The answers to each **Check It Out** are provided in an appendix at the back of the book so that students will receive immediate feedback.

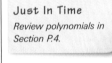 *Check It Out 3:* Use the model in Example 3 to project the national debt in the year 2012. ■

▼ Just in Time

Just in Time references, found in the margin of the text, are helpful in that they reduce the amount of time needed to review prerequisite skills. They refer to content previously introduced for "on-the-spot" review.

Just In Time

Review polynomials in Section P.4.

Some of the constants that appear in the definition of a polynomial function have specific names associated with them:

▶ The nonnegative integer n is called the **degree** of the polynomial. Polynomials are usually written in **descending order,** with the exponents decreasing from left to right.

▶ The constants $a_0, a_1, \ldots, a_n$ are called **coefficients.**

▶ The term $a_n x^n$ is called the **leading term,** and the coefficient a_n is called the **leading coefficient.**

▶ A function of the form $f(x) = a_0$ is called a **constant polynomial** or a **constant function.**

WHAT REVIEWERS SAY ABOUT OBSERVATIONS AND DISCOVER AND LEARN

"...[What **Observations**] helps us help students do is to analyze what's happening in a particular problem...it helps you pick it apart in a way that can be challenging sometimes...to pick out and observe some of those details and some of those characteristics that you want to come out...it helps you enter into that conversation with the students.

"The **Discover and Learn**...some of those kinds of problems push you to go beyond a service understanding of what it is you're talking about."

Stephanie Sibley, Roxbury Community College

▼ Observations

Observations are integrated throughout various sections. They often follow graphs and help to highlight and analyze important features of the graphs shown.

Presented as bulleted lists, they help students focus on what is most important when they look at similar graphs. By studying Observations, students can learn to better interpret and analyze what they see.

Observations:

▶ The y-intercept is $(0, -1)$.

▶ The domain of h is the set of all real numbers.

▶ From the sketch of the graph, we see that the *range of h is the set of*
 numbers, or $(-\infty, 0)$ in interval notation.

▶ As $x \to +\infty$, $h(x) \to -\infty$.

▶ As $x \to -\infty$, $h(x) \to 0$. Thus, the *horizontal asymptote* is the line $y =$

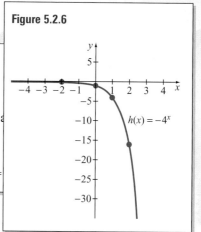

Figure 5.2.6

$h(x) = -4^x$

◀ Discover and Learn

These instructor-guided exercises are placed closest to the discussion of the topic to which they apply and encourage further exploration of the concepts at hand.

They facilitate student interaction and participation and can be used by the instructor for in-class discussions or group exercises.

Discover *and* Learn

In Example 3b, verify that the order of the vertical and horizontal translations does not matter by first shifting the graph of $f(x) = |x|$ down by 2 units and then shifting the resulting graph horizontally to the left by 3 units.

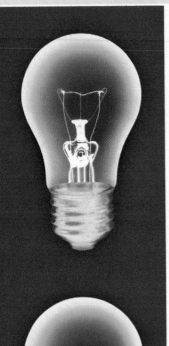

WHAT REVIEWERS SAY ABOUT TECHNOLOGY NOTES

Technology Note

Use a table of values to find a suitable window to graph $Y_1(x) = 10000(0.92)^x$. One possible window size is [0, 30](5) by [0, 11000](1000). See Figure 5.2.9.

Keystroke Appendix:
Sections 6 and 7

Figure 5.2.9

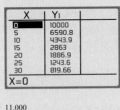

X	Y₁
0	10000
5	6590.8
10	4343.9
15	2863
20	1886.9
25	1243.6
30	819.66

X=0

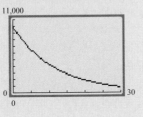

11,000

0
0   30

◀ Technology Notes

Technology Notes appear in the margins to support the optional use of graphing calculator technology. Look for the

graphing calculator icon .

Sometimes the Notes acknowledge the limitations of graphing calculator technology, and they often provide tips on ways to work through those limitations.

▼ Keystroke Appendix

A **Keystroke Guide** at the end of the book orients students to specific keystrokes for the TI-83/84 series of calculators.

Example 3 Graphing a System of Inequalities

Graph the following system of inequalities.

$$\begin{cases} y \geq x \\ y \leq -x \end{cases}$$

▶Solution To satisfy this system of inequalities, we must shade the area above $y = x$ and below $y = -x$.

1. In the Y= Editor, enter $\boxed{X, T, \theta, n}$ in Y₁ and then use the $\boxed{\triangleleft}$ key to move to the leftmost end of the screen. Press $\boxed{\text{ENTER}}$ $\boxed{\text{ENTER}}$ to activate the "shade above" command. See Figure A.8.7.

2. In the Y= Editor, enter $\boxed{(-)}$ $\boxed{X, T, \theta, n}$ in Y₂ and then use the $\boxed{\triangleleft}$ key to move to the leftmost end of the screen. Press $\boxed{\text{ENTER}}$ $\boxed{\text{ENTER}}$ $\boxed{\text{ENTER}}$ to activate the "shade below" command. See Figure A.8.7.

▼ Notes to the Student

Placed within the exposition where appropriate, these **Notes** speak to the reader in a conversational, one-on-one tone. Notes may be cautionary or informative, providing tips on avoiding common errors or further information on the topic at hand.

> **Note** Example 7 shows how an amount of money invested can grow over a period of time. Note that the amount in the account at the beginning of each year is multiplied by a constant of 1.05 to obtain the amount in the account at the end of that year.

> **Note** The symbol for infinity, ∞, is not a number. Therefore, it cannot be followed by the bracket symbol in interval notation. Any interval extending infinitely is denoted by the infinity symbol followed by a parenthesis. Similarly, the $-\infty$ symbol is preceded by a parenthesis.

Key Points ▶

Key Points are presented in bulleted format at the end of each section. These easy-to-read summaries review the topics that have just been covered.

5.4 Key Points

In the following statements, let $x, y > 0$, let k be any real number, and let $a > 0$, $a \neq 1$.

▶ From the definition of logarithm,

$$k = \log_a x \quad \text{implies} \quad a^k = x.$$

Also,

$$a^k = x \quad \text{implies} \quad k = \log_a x.$$

▶ Using the definition of logarithm,

$$a^{\log_a x} = x.$$

▶ $\log_a a = 1, \quad \log_a 1 = 0$

▶ **The product, power, and quotient properties**

$$\log_a(xy) = \log_a x + \log_a y \qquad \text{Product property of logarithms}$$
$$\log_a x^k = k \log_a x \qquad \text{Power property of logarithms}$$

Section-Ending Exercises

The section-ending exercises are organized as follows: **Just In Time** Exercises (where appropriate), **Skills, Applications,** and **Concepts.** Exercises that encourage use of a graphing calculator are denoted with an icon.

5.3 Exercises

▶**Just in Time Exercises** These exercises correspond to the Just in Time references in this section. Complete them to review topics relevant to the remaining exercises.

In Exercises 1–4, rewrite using rational exponents.

1. $\sqrt{3}$ 2. $\sqrt{5}$

3. $\sqrt[3]{10}$ 4. $\sqrt[3]{12}$

In Exercises 5–8, f and g are inverses of each other.

5. True or False? $(f \circ g)(x) = x$

6. True or False? The domain of f equals the domain of g.

7. True or False? The domain of f equals the range of g.

8. True or False? f is a one-to-one function.

◀ **Just in Time Exercises**

These exercises correspond to the Just in Time references that appear in the section. By completing these exercises, students review topics relevant to the **Skills, Applications,** and **Concepts** exercises that follow.

Skills ▶

These exercises reinforce the skills illustrated in the section.

▶**Skills** This set of exercises will reinforce the skills illustrated in this section.

In Exercises 5–34, solve the exponential equation. Round to three decimal places, when needed.

5. $5^x = 125$ 6. $7^{2x} = 49$

7. $10^x = 1000$ 8. $10^x = 0.0001$

9. $4^x = \dfrac{1}{16}$ 10. $6^x = \dfrac{1}{216}$

11. $4e^x = 36$ 12. $5e^x = 60$

13. $2^x = 5$ 14. $3^x = 7$

WHAT REVIEWERS SAY ABOUT THE EXERCISES

"The quality of exercises is outstanding. I found myself applauding the author for her varied applications problems—they are excellent and representative of the subject matter."

Kevin Fox, Shasta College

"The one feature that I most appreciate is the **'Concepts'** problems incorporated in the homework problems of most sections. I feel that these problems provide a great opportunity to encourage students to think and to challenge their understanding."

Bethany Seto, Horry-Georgetown Technical College

▼ Applications

A wide range of **Applications** are provided, emphasizing how the math is applied in the real world.

55. **Environment** Sulfur dioxide (SO_2) is emitted by power-generating plants and is one of the primary sources of acid rain. The following table gives the total annual SO_2 emissions from the 263 highest-emitting sources for selected years. (*Source:* Environmental Protection Agency)

Year	Annual SO₂ Emissions (millions of tons)
1980	9.4
1985	9.3
1990	8.7
1994	7.4
1996	4.8
1998	4.7
2000	4

(a) Let t denote the number of years since 1980. Make a scatter plot of sulfur dioxide emissions versus t.
(b) Find an expression for the cubic curve of best fit for this data.
(c) Plot the cubic model for the years 1980–2005. Remember that for the years 2001–2005, the curve gives only a *projection*.
(d) Forecast the amount of SO_2 emissions for the year 2005 using the cubic function from part (b).
(e) Do you think the projection found in part (d) is attainable? Why or why not?
(f) The Clean Air Act was passed in 1990, in part to implement measures to reduce the amount of sulfur dioxide emissions. According to the model presented here, have these measures been successful? Explain.

▼ Concepts

These exercises appear toward the end of the section-ending exercise sets. They are designed to help students think critically about the content in the existing section.

▶**Concepts** This set of exercises will draw on the ideas presented in this section and your general math background.

90. Do the equations $\ln x^2 = 1$ and $2 \ln x = 1$ have the same solutions? Explain.

91. Explain why the equation $2e^x = -1$ has no solution.

92. What is wrong with the following step?
$$\log x + \log(x + 1) = 0 \Rightarrow x(x + 1) = 0$$

93. What is wrong with the following step?
$$2^{x+5} = 3^{4x} \Rightarrow x + 5 = 4x$$

In Exercises 94–97, solve using any method, and eliminate extraneous solutions.

94. $\ln(\log x) = 1$

95. $e^{\log x} = e$

96. $\log_5 |x - 2| = 2$

97. $\ln |2x - 3| = 1$

WHAT REVIEWERS SAY ABOUT THE CHAPTER SUMMARY

▼ Chapter Summary

This unique, three-column format, broken down by section, provides the ultimate study guide.

Chapter 5		Summary

Section 5.1 Inverse Functions

Concept	Illustration	Study and Review
Definition of an inverse function Let f be a function. A function g is said to be the **inverse function** of f if the domain of g is equal to the range of f and, for every x in the domain of f and every y in the domain of g, $g(y) = x$ if and only if $f(x) = y$. The notation for the inverse function of f is f^{-1}.	The inverse of $f(x) = 4x + 1$ is $f^{-1}(x) = \frac{x-1}{4}$.	Examples 1, 2 Chapter 5 Review, Exercises 1–12
Composition of a function and its inverse If f is a function with an inverse function f^{-1}, then • for every x in the domain of f, $f^{-1}(f(x))$ is defined and $f^{-1}(f(x)) = x$. • for every x in the domain of f^{-1}, $f(f^{-1}(x))$ is defined and $f(f^{-1}(x)) = x$.	Let $f(x) = 4x + 1$ and $f^{-1}(x) = \frac{x-1}{4}$. Note that $f^{-1}(f(x)) = \frac{(4x+1)-1}{4} = x$. Similarly, $f(f^{-1}(x)) = 4\left(\frac{x-1}{4}\right) + 1 = x$.	Examples 2, 3 Chapter 5 Review, Exercises 1–4
One-to-one function A function f is **one-to-one** if $f(a) = f(b)$ implies $a = b$. For a function to have an inverse, it must be one-to-one.	The function $f(x) = x^3$ is one-to-one, whereas the function $g(x) = x^2$ is not.	Example 4 Chapter 5 Review, Exercises 5–12
Graph of a function and its inverse The graphs of a function f and its inverse function f^{-1} are symmetric with respect to the line $y = x$.	The graphs of $f(x) = 4x + 1$ and $f^{-1}(x) = \frac{x-1}{4}$ are pictured. Note the symmetry about the line $y = x$. 	Examples 5, 6 Chapter 5 Review, Exercises 13–16

Continued

The first column, "Concept," describes the mathematical topic in words.

The second column, "Illustration," shows this concept being performed mathematically.

The third column, "Study and Review," provides suggested examples and chapter review exercises that should be completed to review each concept.

▼ Chapter Review Exercises

Each chapter concludes with an extensive exercise set, broken down by section, so that students can easily identify which sections of the chapter they have mastered and which sections might require more attention.

Chapter 5   Review Exercises

Section 5.1

In Exercises 1–4, verify that the functions are inverses of each other.

1. $f(x) = 2x + 7; g(x) = \dfrac{x - 7}{2}$

2. $f(x) = -x + 3; g(x) = -x + 3$

3. $f(x) = 8x^3; g(x) = \dfrac{\sqrt[3]{x}}{2}$

4. $f(x) = -x^2 + 1, x \geq 0; g(x) = \sqrt{1 - x}$

In Exercises 5–12, find the inverse of each one-to-one function.

5. $f(x) = -\dfrac{4}{5}x$ 6. $g(x) = \dfrac{2}{3}x$

7. $f(x) = -3x + 6$

8. $f(x) = -2x - \dfrac{5}{3}$

9. $f(x) = x^3 + 8$

10. $f(x) = -2x^3 + 4$

11. $g(x) = -x^2 + 8, x \geq 0$

12. $g(x) = 3x^2 - 5, x \geq 0$

In Exercises 13–16, find the inverse of each one-to-one function. Graph the function and its inverse on the same set of axes, making sure the scales on both axes are the same.

13. $f(x) = -x - 7$ 14. $f(x) = 2x + 1$

15. $f(x) = -x^3 + 1$ 16. $f(x) = x^2 - 3, x \geq 0$

Section 5.3

25. Complete the table by filling in the exponential statements that are equivalent to the given logarithmic statements.

Logarithmic Statement	Exponential Statement
$\log_3 9 = 2$	
$\log 0.1 = -1$	
$\log_5 \dfrac{1}{25} = -2$	

26. Complete the table by filling in the logarithmic statements that are equivalent to the given exponential statements.

Exponential Statement	Logarithmic Statement
$3^5 = 243$	
$4^{1/5} = \sqrt[5]{4}$	
$8^{-1} = \dfrac{1}{8}$	

In Exercises 27–36, evaluate each expression without using a calculator.

27. $\log_5 625$

29. $\log_9 81$

31. $\log \sqrt{10}$

33. $\ln \sqrt[3]{e}$

Chapter Test ▶

Each chapter ends with a test that includes questions based on each section of the chapter.

Chapter 5   Test

1. Verify that the functions $f(x) = 3x - 1$ and $g(x) = \dfrac{x + 1}{3}$ are inverses of each other.

2. Find the inverse of the one-to-one function
$$f(x) = 4x^3 - 1.$$

3. Find $f^{-1}(x)$ given $f(x) = x^2 - 2, x \geq 0$. Graph f and f^{-1} on the same set of axes.

In Exercises 4–6, sketch a graph of the function and describe its behavior as $x \to \pm\infty$.

4. $f(x) = -3^x + 1$

5. $f(x) = 2^{-x} - 3$

6. $f(x) = e^{-2x}$

7. Write in exponential form: $\log_6 \dfrac{1}{216} = -3$.

8. Write in logarithmic form: $2^5 = 32$.

In Exercises 9 and 10, evaluate the expression without using a calculator.

9. $\log_8 \dfrac{1}{64}$ 10. $\ln e^{3.2}$

11. Use a calculator to evaluate $\log_7 4.91$ to four decimal places.

12. Sketch the graph of $f(x) = \ln(x + 2)$. Find all asymptotes and intercepts.

In Exercises 13 and 14, write the expression as a sum or difference of logarithmic expressions. Eliminate exponents and radicals when possible.

13. $\log \sqrt[3]{x^3 y^4}$ 14. $\ln(e^2 x^2 y)$

In Exercises 17–22, solve.

17. $6^{2x} = 36^{3x-1}$ 18. $4^x = 7.1$

19. $4e^{x+2} - 6 = 10$ 20. $200e^{0.2t} = 800$

21. $\ln(4x + 1) = 0$

22. $\log x + \log(x + 3) = 1$

23. For an initial deposit of \$3000, find the total amount in a bank account after 6 years if the interest rate is 5%, compounded quarterly.

24. Find the value in 3 years of an initial investment of \$4000 at an interest rate of 7%, compounded continuously.

25. The depreciation rate of a laptop computer is about 40% per year. If a new laptop computer was purchased for \$900, find a function that gives its value t years after purchase.

26. The magnitude of an earthquake is measured on the Richter scale using the formula $R(I) = \log\left(\dfrac{I}{I_0}\right)$, where I represents the actual intensity of the earthquake and I_0 is a baseline intensity used for comparison. If an earthquake registers 6.2 on the Richter scale, express its intensity in terms of I_0.

27. The number of college students infected with a cold virus in a dormitory can be modeled by the logistic function $N(t) = \dfrac{120}{1 + 3e^{-0.8t}}$, where t is the number of days after the breakout of the infection.
   (a) How many students were initially infected?
   (b) Approximately how many students will be infected after 10 days?

College Algebra:
Building Concepts and Connections

Algebra and Geometry Review

Yachts that sail in the America's Cup competition must meet strict design rules. One of these rules involves finding square and cube roots to determine the appropriate dimensions of the yacht. See Exercise 71 in Section P.3. This chapter will review algebra and geometry topics that are prerequisite to the material in the main text, including how to find square and cube roots. It will also serve as a reference as you work through the problems in Chapters 1–8.

P.1 The Real Number System

Objectives

▶ Understand basic properties of real numbers

▶ Understand interval notation

▶ Represent an interval of real numbers on the number line

▶ Find the absolute value of a real number

▶ Simplify expressions using order of operations

The number system that you are familiar with is formally called the **real number system.** The real numbers evolved over time as the need arose to numerically represent various types of values. When people first started using numbers, they counted using the numbers 1, 2, 3, and so on. It is customary to enclose a list of numbers in braces and to formally call the list a **set** of numbers. The counting numbers are called the set of **natural numbers:**

$$\{1, 2, 3, 4, 5, 6, \ldots\}.$$

Soon a number was needed to represent nothing, and so zero was added to the set of natural numbers to form the set of **whole numbers:**

$$\{0, 1, 2, 3, 4, 5, \ldots\}.$$

As commerce intensified, the concept of debt led to the introduction of negative numbers. The set consisting of the natural numbers, their negatives, and zero is called the set of **integers:**

$$\{\ldots, -6, -5, -4, -3, -2, -1, 0, 1, 2, 3, 4, 5, 6, \ldots\}.$$

The next step in the evolution of numbers was the introduction of **rational numbers.** Using rational numbers made it easier to describe things, such as a loaf of bread cut into two or more pieces or subdivisions of time and money. Rational numbers are represented by the division of two integers $\frac{r}{s}$, with $s \neq 0$. They can be described by either terminating decimals or nonterminating, repeating decimals. The following are examples of rational numbers.

$$\frac{1}{3} = 0.3333\ldots, \quad -0.25 = -\frac{1}{4}, \quad 4\frac{2}{5} = \frac{22}{5}$$

When the ancient Greeks used triangles and circles in designing buildings, they discovered that some measurements could not be represented by rational numbers. This led to the introduction of a new set of numbers called the **irrational numbers.** The irrational numbers can be expressed as nonterminating, nonrepeating decimals. New symbols were introduced to represent the exact values of these numbers. Examples of irrational numbers include

$$\sqrt{2}, \quad \pi, \quad \text{and} \quad \frac{\sqrt{5}}{2}.$$

Note A calculator display of 3.141592654 is just an approximation of π. Calculators and computers can only approximate irrational numbers because they can store only a finite number of digits.

Properties of Real Numbers

The **associative property** of real numbers states that numbers can be grouped in any order when adding or multiplying, and the result will be the same.

Associative Properties

Associative property of addition	$a + (b + c) = (a + b) + c$
Associative property of multiplication	$a(bc) = (ab)c$

The **commutative property** states that numbers can be added or multiplied in any order, and the result will be the same.

> **Commutative Properties**
>
> Commutative property of addition $\qquad a + b = b + a$
>
> Commutative property of multiplication $\quad ab = ba$

The **distributive property** of multiplication over addition changes sums to products or products to sums.

> **Distributive Property**
>
> $$ab + ac = a(b + c),$$ where a, b, and c are any real numbers.

We also have the following definitions for additive and multiplicative identities.

> **Additive and Multiplicative Identities**
>
> There exists a unique real number 0 such that, for any real number a, $a + 0 = 0 + a = a$. The number 0 is called the **additive identity.**
>
> There exists a unique real number 1 such that, for any real number a, $a \cdot 1 = 1 \cdot a = a$. The number 1 is called the **multiplicative identity.**

Finally, we have the following definitions for additive and multiplicative inverses.

> **Additive and Multiplicative Inverses**
>
> The **additive inverse** of a real number a is $-a$, since $a + (-a) = 0$.
>
> The **multiplicative inverse** of a real number a, $a \neq 0$, is $\dfrac{1}{a}$, since $a \cdot \dfrac{1}{a} = 1$.

Discover *and* **Learn**

Give an example to show that subtraction is not commutative.

Example 1 Properties of Real Numbers

What property does each of the following equations illustrate?

(a) $4 + 6 = 6 + 4$ (b) $3(5 - 8) = 3(5) - 3(8)$

(c) $2 \cdot (3 \cdot 5) = (2 \cdot 3) \cdot 5$ (d) $4 \cdot 1 = 4$

▶Solution

(a) The equation $4 + 6 = 6 + 4$ illustrates the commutative property of addition.

(b) The equation $3(5 - 8) = 3(5) - 3(8)$ illustrates the distributive property.

(c) The equation $2 \cdot (3 \cdot 5) = (2 \cdot 3) \cdot 5$ illustrates the associative property of multiplication.

(d) The equation $4 \cdot 1 = 4$ illustrates the multiplicative identity.

✔ *Check It Out 1:* What property does the equation $7(8 + 5) = 7 \cdot 8 + 7 \cdot 5$ illustrate?

Ordering of Real Numbers

The real numbers can be represented on a **number line.** Each real number corresponds to exactly one point on the number line. The number 0 corresponds to the **origin** of the number line. The positive numbers are to the right of the origin and the negative numbers are to the left of the origin. Figure P.1.1 shows a number line.

Figure P.1.1

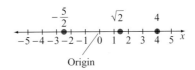

Origin

If we consider any two real numbers a and b, we can put them in a relative order. For example, $a < b$ is read as "a is less than b." It can also be read as "b is greater than a," or $b > a$. Both expressions mean that a is to the left of b on the number line.

The statement $a \le b$ is read as "a is less than or equal to b." It means that *either* $a < b$ *or* $a = b$. Only one of these conditions needs to be satisfied for the entire statement to be true.

To express the set of real numbers that lie between two real numbers a and b, including a and b, we write an **inequality** of the form $a \le x \le b$. This inequality can also be expressed in interval notation as $[a, b]$. The interval $[a, b]$ is called a **closed interval.** If the endpoints of the interval, a and b, are not included in the set, we write the inequality as an **open interval** (a, b). The symbol ∞, or **positive infinity,** is used to show that an interval extends forever in the positive direction. The symbol $-\infty$, or **negative infinity,** is used to show that an interval extends forever in the negative direction.

On a number line, when an endpoint of an interval is included, it is indicated by a filled-in, or closed, circle. If an endpoint is not included, it is indicated by an open circle.

Note   The symbol for infinity, ∞, is not a number. Therefore, it cannot be followed by the bracket symbol in interval notation. Any interval extending infinitely is denoted by the infinity symbol followed by a parenthesis. Similarly, the $-\infty$ symbol is preceded by a parenthesis.

Example 2 Graphing an Interval

Graph each of the following intervals on the number line, and give a verbal description of each.

(a) $(-2, 3]$

(b) $(-\infty, 4]$

▶Solution

Figure P.1.2

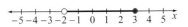

(a) The interval $(-2, 3]$ is graphed in Figure P.1.2. Because -2 is not included in the set, it is represented by an open circle. Because 3 *is* included, it is represented by a closed circle. The interval $(-2, 3]$ consists of all real numbers greater than -2 and less than or equal to 3.

Figure P.1.3

(b) The interval $(-\infty, 4]$ is graphed in Figure P.1.3. Because 4 is included in the set, it is represented by a closed circle. The interval $(-\infty, 4]$ consists of all real numbers less than or equal to 4.

✔ *Check It Out 2:* Graph the interval $[3, 5]$ on the number line, and give a verbal description of the interval. ■

Table P.1.1 lists different types of inequalities, their corresponding interval notations, and their graphs on the number line.

Table P.1.1 Interval Notation and Graphs

Inequality	Interval Notation	Graph
$a \le x \le b$	$[a, b]$	 •———————• a … b … x
$a < x < b$	(a, b)	 ○———————○ a … b … x
$a \le x < b$	$[a, b)$	 •———————○ a … b … x
$a < x \le b$	$(a, b]$	 ○———————• a … b … x
$x \ge a$	$[a, \infty)$	 •—————→ a … x
$x > a$	(a, ∞)	 ○—————→ a … x
$x \le a$	$(-\infty, a]$	 ←—————• a … x
$x < a$	$(-\infty, a)$	 ←—————○ a … x
All real numbers	$(-\infty, \infty)$	 ←—————————→ x

Figure P.1.4

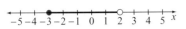

Figure P.1.5

Example 3 Writing a Set in Interval Notation

Write the interval graphed in Figure P.1.4 in interval notation.

▶Solution The interval consists of all numbers greater than or equal to -3 and less than 2. The number 2 is not included, since it is represented by an open circle. Thus, in interval notation, the set of points is written as $[-3, 2)$.

☑ *Check It Out 3:* Write the interval graphed in Figure P.1.5 in interval notation. ▪

Absolute Value

The distance from the origin to a real number c is defined as the **absolute value** of c, denoted by $|c|$. The absolute value of a number is also known as its **magnitude.** The algebraic definition of absolute value is given next.

Definition of Absolute Value

$$|x| = \begin{cases} x, & \text{if } x \geq 0 \\ -x, & \text{if } x < 0 \end{cases}$$

Example 4 Evaluating Absolute Value Expressions

Evaluate the following.

(a) $|-3|$ (b) $-4 - |4.5|$ (c) $|-4 + 9| + 3$

▶Solution

(a) Because $-3 < 0, |-3| = -(-3) = 3$.

(b) Because $4.5 > 0, |4.5| = 4.5$. Thus

$$-4 - |4.5| = -4 - 4.5 = -8.5.$$

(c) Perform the addition first and then evaluate the absolute value.

$$|-4 + 9| + 3 = |5| + 3 \qquad \text{Evaluate } -4 + 9 = 5 \text{ first}$$
$$= 5 + 3 = 8 \qquad \text{Because } 5 > 0, |5| = 5$$

☑ *Check It Out 4:* Evaluate $|4 - 6| - 3$. ▪

Next we list some basic properties of absolute value that can be derived from the definition of absolute value.

Properties of Absolute Value

For any real numbers a and b, we have

1. $|a| \geq 0$ 2. $|-a| = |a|$ 3. $|ab| = |a||b|$ 4. $\left|\dfrac{a}{b}\right| = \dfrac{|a|}{|b|}, b \neq 0$

Using absolute value, we can determine the distance between two points on the number line. For instance, the distance between 7 and 12 is 5. We can write this using absolute value notation as follows:

$$|12 - 7| = 5 \quad or \quad |7 - 12| = 5.$$

The distance is the same regardless of the order of subtraction.

Distance Between Two Points on the Real Number Line

Let a and b be two points on the real number line. Then the distance between a and b is given by

$$|b - a| \quad \text{or} \quad |a - b|.$$

Example **5** **Distance on the Real Number Line**

Find the distance between -6 and 4 on the real number line.

▶ **Solution** Letting $a = -6$ and $b = 4$, we have

$$|a - b| = |-6 - 4| = |-10| = 10.$$

Figure P.1.6

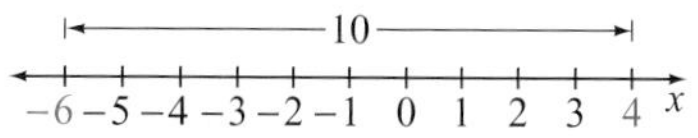

Thus the distance between -6 and 4 is 10 as shown in Figure P.1.6. We will obtain the same result if we compute $|b - a|$ instead:

$$|b - a| = |4 - (-6)| = |10| = 10.$$

✔ *Check It Out 5:* Find the distance between -4 and 5 on the real number line. ■

Order of Operations

When evaluating an arithmetic expression, the result must be the same regardless of who performs the operations. To ensure that this is the case, certain conventions for combining numbers must be followed. These conventions are outlined next.

Rules for Order of Operations

When evaluating a mathematical expression, perform the operations in the following order, beginning with the innermost parentheses and working outward.

Step 1 Simplify all numbers with exponents, working from left to right.

Step 2 Perform all multiplications and divisions, working from left to right.

Step 3 Perform all additions and subtractions, working from left to right.

When an expression is written as the quotient of two other expressions, the numerator and denominator are evaluated separately, and then the division is performed.

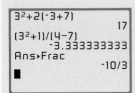

Example 6 **Simplifying Using Order of Operations**

Simplify each expression.

(a) $3^2 + 2(-3 + 7)$

(b) $\dfrac{3^2 + 1}{4 - 7}$

(c) $6 + 3(5 - (-3^2 + 2))$

▶Solution

(a) Follow the order of operations.

$$
\begin{aligned}
3^2 + 2(-3 + 7) &= 3^2 + 2(4) && \text{Evaluate within parentheses} \\
&= 9 + 2(4) && \text{Simplify number with exponent: } 3^2 = 9 \\
&= 9 + 8 && \text{Multiply } 2(4) \\
&= 17 && \text{Add}
\end{aligned}
$$

(b) Because the expression $\dfrac{3^2 + 1}{4 - 7}$ is a quotient of two other expressions, evaluate the numerator and denominator separately, and then divide.

$$
\begin{aligned}
3^2 + 1 &= 10 && \text{Evaluate numerator} \\
4 - 7 &= -3 && \text{Evaluate denominator} \\
\frac{3^2 + 1}{4 - 7} &= -\frac{10}{3} && \text{Divide}
\end{aligned}
$$

(c) $6 + 3(5 - (-3^2 + 2)) = 6 + 3(5 - (-7))$ Evaluate within innermost parentheses. Note that $-3^2 + 2 = -9 + 2 = -7$.

$= 6 + 3(12) = 6 + 36$ Evaluate within parentheses; then multiply

$= 42$ Add

✔ *Check It Out 6:* Simplify the expression $5^2 - 3(15 - 12) + 4 \div 2$. ▪

Understanding order of operations is extremely important when entering and evaluating expressions using a calculator.

Example 7 **Calculator Use and Order of Operations**

Evaluate the following expressions on your calculator. Check the calculator output against a hand calculation.

(a) $4 + 10 \div 5 + 2$

(b) $\dfrac{4 + 10}{5 + 2}$

▶Solution

(a) Enter $4 + 10 \div 5 + 2$ in the calculator as

$$4 \; + \; 10 \; \div \; 5 \; + \; 2.$$

The result is 8. This result checks with a hand calculation of

$$4 + 10 \div 5 + 2 = 4 + 2 + 2 = 8.$$

(b) To calculate $\frac{4+10}{5+2}$, we need to proceed with caution. The additions in the numerator and denominator must be performed first, before the division. Enter the expression into the calculator as

$$(\ 4 \ + \ 10 \ ) \ \div \ (\ 5 \ + \ 2 \ ).$$

The result is 2, which can be checked quickly by hand. The parentheses are a *must* in entering this expression. Omitting them will result in a wrong answer, because the calculator will perform the operations in a different order.

✔ *Check It Out 7:* Evaluate the expression $\frac{3-10}{4+3} - 5$ using a calculator. Check your result by hand. ▪

P.1 Key Points

▶ The **real number system** consists of the **rational numbers** and the **irrational numbers.**

▶ The **associative property** states that numbers can be grouped in any order when adding or multiplying, and the result will be the same.

▶ The **commutative property** states that numbers can be added or multiplied in any order, and the result will be the same.

▶ The **distributive property** of multiplication states that
$$ab + ac = a(b + c), \text{ where } a, b, \text{ and } c \text{ are any real numbers.}$$

▶ The **additive identity** and **multiplicative identity** for a real number a are 0 and 1, respectively.

▶ The **additive inverse** of a real number a is $-a$.

▶ The **multiplicative inverse** of a real number a, $a \neq 0$, is $\frac{1}{a}$.

▶ An **inequality** of the form $a \leq x \leq b$ can be expressed in interval notation as $[a, b]$. This interval is called a **closed interval.**

▶ If the endpoints of the interval, a and b, are not included, we write the inequality as the **open interval** (a, b).

▶ The **absolute value** of a number a is denoted by $|a|$, and represents the distance of a from the origin.

▶ The **distance between two points** a and b on the real number line is given by
$$|b - a| \quad \text{or} \quad |a - b|.$$

▶ When evaluating a numerical expression, we use the convention called **order of operations.** When following order of operations, we (1) remove parentheses; (2) simplify numbers with exponents; (3) perform multiplications and divisions from left to right; and (4) perform additions and subtractions from left to right.

P.1 Exercises

▶**Skills** This set of exercises will reinforce the skills illustrated in this section.

In Exercises 1–8, consider the following numbers.

$$\sqrt{2}, 0.5, \frac{4}{5}, -1, 0, 40, \pi, 10, -1.67$$

1. Which are integers?

2. Which are natural numbers?

3. Which are rational numbers?

4. Which are irrational numbers?

5. Which are whole numbers?

6. Which are rational numbers that are not integers?

7. Which are integers that are not positive?

8. Which are real numbers?

In Exercises 9–14, name the property illustrated by each equality.

9. $3(8 \cdot 9) = (3 \cdot 8)9$

10. $(5 + x) + z = 5 + (x + z)$

11. $-4(x - 2) = -4x + 8$

12. $3x + y = y + 3x$

13. $9a = a9$

14. $b(x + 2) = bx + 2b$

In Exercises 15–24, graph each interval on the real number line.

15. $[-2, 4]$ 16. $[-3, -1]$

17. $[-5, 0)$ 18. $(-2, 4)$

19. $(-\infty, 3)$ 20. $[2, \infty)$

21. $\left[\dfrac{1}{2}, 4\right]$ 22. $\left[-\dfrac{3}{2}, 2\right]$

23. $(-2.5, 3)$ 24. $[3.5, \infty)$

In Exercises 25–32, describe the graph using interval notation.

25. 26.

27. 28.

29. 30.

31. 32.

In Exercises 33–38, fill in the table.

	Inequality	Interval Notation	Graph
33.	$4 \le x \le 10$		
34.		$(-\infty, 6)$	
35.	$-3 \le x < 0$		
36.		$(-5, 10]$	
37.			
38.			

In Exercises 39–54, evaluate each expression without using a calculator.

39. $|-3.2|$ 40. $|-45.5|$

41. $|253|$ 42. $|-37|$

43. $\left|\dfrac{5}{4}\right|$ 44. $\left|-\dfrac{1}{2}\right|$

45. $\left|-\dfrac{4}{3}\right|$ 46. $\left|\dfrac{7}{2}\right|$

47. $|-2| + 4$ 48. $-|5|$

49. $-|-4.5|$ 50. $-|3.2|$

51. $|5 - 2| + 4$ 52. $4 + |6 - 7|$

53. $5 - |12 - 4|$ 54. $-3 - |6 - 10|$

In Exercises 55–66, find the distance between the numbers on the real number line.

55. $-2, 4$ 56. $-5, 7$

57. $0, 9$ 58. $-12, 0$

59. $-12, -7.5$ 60. $-5.5, 6$

61. $-4.3, 7.9$ 62. $6.7, 13.4$

63. $-\dfrac{1}{2}, \dfrac{5}{2}$ 64. $-\dfrac{4}{3}, \dfrac{7}{3}$

65. $-\dfrac{4}{5}, \dfrac{1}{3}$ 66. $-\dfrac{2}{3}, \dfrac{1}{4}$

In Exercises 67–76, evaluate each expression without using a calculator.

67. $9 - 3(2) - 8$ 68. $6 + 4(3) - 14$

69. $(3 - 5)^2$ 70. $8 - (7 - 9)^2$

71. $3^2 + 3(5) - 10$ 72. $10 + 2^3 \div 4$

73. $(-3)^5 + 3$ 74. $-6(1 + 2^2)$

75. $\dfrac{10 - 4^2}{2 + 3^2}$ 76. $\dfrac{4(2) - 6^2}{2^3 - 1}$

In Exercises 77–84, evaluate each expression using a calculator. Check your solution by hand.

77. -3^2

78. $(-3)^2$

79. $1 + \dfrac{2}{3} - 4^2$

80. $\left(-\dfrac{3}{5}\right)\left(\dfrac{25}{6}\right)$

81. $\dfrac{2}{3+5}$

82. $\dfrac{5+7}{3}$

83. $\dfrac{4^2+3}{5}$

84. $\dfrac{-6^2+3(7)}{5-2^3}$

▶**Applications** In this set of exercises, you will use real numbers and interval notation to study real-world problems.

85. **Shoe Sizes** The available shoe sizes at a store are whole numbers from 5 to 9, inclusive. List all the possible shoe sizes in the store.

86. **Salary** The annual starting salary for an administrative assistant at a university can range from $30,000 to $40,000, inclusive. Write the salary range in interval notation.

87. **Temperature** On a particular winter day in Chicago, the temperature ranged from a low of 25°F to a high of 36°F. Write this range of temperatures in interval notation.

88. **Distance** On the Garden State Parkway in New Jersey, the distance between Exit A and Exit B, in miles, is given by $|A - B|$. How many miles are traveled between Exit 88 and Exit 127 on the Garden State Parkway?

89. **Elevation** The highest point in the United States is Mount McKinley, Alaska, with an elevation of 20,320 feet. The lowest point in the U.S. is Death Valley, California, with an elevation of -282 feet (282 feet below sea level). Find the absolute value of the difference in elevation between the lowest and highest points in the U.S. (*Source:* U.S. Geological Survey)

▶**Concepts** This set of exercises will draw on the ideas presented in this section and your general math background.

90. Find two numbers a and b such that $|a + b| \neq |a| + |b|$. (Answers may vary.)

91. If a number is nonnegative, must it be positive? Explain.

92. Find two points on the number line that are a distance of 5 units from 1.

93. Find two points on the number line that are a distance of 4 units from -3.

94. Find a rational number less than π. (Answers may vary.)

P.2 Integer Exponents and Scientific Notation

Objectives

▶ Evaluate algebraic expressions

▶ Define positive and negative exponents

▶ Simplify expressions involving exponents

▶ Write numbers using scientific notation

▶ Determine significant figures for a given number

In this section, we will discuss integers as exponents and the general rules for evaluating and simplifying expressions containing exponents. We will also discuss scientific notation, a method used for writing very large and very small numbers.

Algebraic Expressions

In algebra, letters such as x and y are known as **variables.** You can use a variable to represent an unknown quantity. For example, if you earn x dollars per hour and your friend earns $2 more than you per hour, then $x + 2$ represents the amount in dollars per hour earned by your friend. When you combine variables and numbers using multiplication, division, addition, and subtraction, as well as powers and roots, you get an **algebraic expression.**

We are often interested in finding the value of an algebraic expression for a given value of a variable. This is known as *evaluating an algebraic expression* and is illustrated in the next example.

Example **1** **Evaluating an Expression**

The number of students at an elementary school is given by $300 + 20x$, where x is the number of years since 2003. Evaluate this expression for $x = 4$, and describe what the result means.

▶Solution We substitute $x = 4$ in the algebraic expression to obtain

$$300 + 20x = 300 + 20(4)$$
$$= 300 + 80 \qquad \text{Multiply first—recall order of operations}$$
$$= 380.$$

The answer tells us that there will be 380 students in the elementary school in 2007, which is 4 years after 2003.

✔ *Check It Out 1:* Evaluate the expression in Example 1 for $x = 8$, and describe what the result means. ■

Integer Exponents

If we want to multiply the same number by itself many times, writing out the multiplication becomes tedious, and so a new notation is needed. For example, $6 \cdot 6 \cdot 6$ can be written compactly as 6^3. The number 3 is called the **exponent** and the number 6 is called the **base.** The exponent tells you how many 6's are multiplied together. In general, we have the following definition.

> **Definition of Positive Integer Exponents**
>
> For any positive integer n,
>
> $$a^n = \underbrace{a \cdot a \cdot a \cdots a}_{n \text{ factors}}.$$
>
> The number a is the **base** and the number n is the **exponent.**

Negative integer exponents are defined in terms of positive integer exponents as follows.

> **Definition of Negative Integer Exponents**
>
> Let a be any nonzero real number and let m be a positive integer. Then
>
> $$a^{-m} = \frac{1}{a^m}.$$

Technology Note

To evaluate an expression involving exponents, use the x^y key found on most scientific calculators. On a graphing calculator, use the $\wedge$ (hat) key.

Keystroke Appendix: Section 4

Example **2** **Writing an Expression with Positive Exponents**

Write each expression using positive exponents.

(a) 4^{-2}

(b) 1.45^{-3}

(c) $\dfrac{x^{-4}}{y^5}$, $x, y \neq 0$

▶Solution

(a) $4^{-2} = \dfrac{1}{4^2}$, using the rule for negative exponents.

(b) $1.45^{-3} = \dfrac{1}{1.45^3}$

(c) $\dfrac{x^{-4}}{y^5} = x^{-4} \cdot \dfrac{1}{y^5}$

$\qquad = \dfrac{1}{x^4} \cdot \dfrac{1}{y^5} = \dfrac{1}{x^4 y^5}$

✔ *Check It Out 2:* Write $x^{-2}y^3$ using positive exponents. ■

The following properties of exponents are used to simplify expressions containing exponents.

Properties of Integer Exponents

Let a and b be real numbers and let m and n be integers. Then the following properties hold.

Table P.2.1 Properties of Integer Exponents

PROPERTY	ILLUSTRATION
1. $a^m \cdot a^n = a^{m+n}$	$3^4 \cdot 3^{-2} = 3^{4+(-2)} = 3^2$
2. $(a^m)^n = a^{mn}$	$(5^4)^3 = 5^{4 \cdot 3} = 5^{12}$
3. $(ab)^m = a^m b^m$	$(4x)^3 = 4^3 x^3 = 64x^3$
4. $\left(\dfrac{a}{b}\right)^m = \dfrac{a^m}{b^m}, b \neq 0$	$\left(\dfrac{3}{7}\right)^2 = \dfrac{3^2}{7^2} = \dfrac{9}{49}$
5. $\dfrac{a^r}{a^s} = a^{r-s}, a \neq 0$	$\dfrac{5^4}{5^8} = 5^{4-8} = 5^{-4}$ or $\dfrac{1}{5^4}$
6. $a^1 = a$	$7^1 = 7$
7. $a^0 = 1, a \neq 0$	$10^0 = 1$

Discover *and* **Learn**

Verify that $4^2 \cdot 4^3 = 4^5$ by expanding the left-hand side and multiplying.

Example 3 shows how the properties of exponents can be combined to simplify expressions.

Example **3** **Simplifying Expressions Containing Exponents**

Simplify each expression and write it using positive exponents. Assume that variables represent nonzero real numbers.

(a) $\dfrac{16x^{10}y^4}{4x^{10}y^8}$

(b) $\dfrac{(3x^{-2})^3}{x^{-5}}$

(c) $\left(\dfrac{54s^2 t^{-3}}{6zt}\right)^{-2}$

▶Solution

(a) $\dfrac{16x^{10}y^4}{4x^{10}y^8} = \dfrac{16}{4}\,x^{10-10}y^{4-8} = 4x^0y^{-4} = \dfrac{4}{y^4}$

The property $x^0 = 1$ was used in the final step. It is usually a good idea to wait until the last step to write the expression using only positive exponents.

(b) $\dfrac{(3x^{-2})^3}{x^{-5}} = \dfrac{3^3x^{-2\cdot3}}{x^{-5}} = \dfrac{27x^{-6}}{x^{-5}} = 27x^{-6-(-5)} = 27x^{-1} = \dfrac{27}{x}$

(c) $\left(\dfrac{54s^2t^{-3}}{6zt}\right)^{-2} = \left(\dfrac{9s^2t^{-3-1}}{z}\right)^{-2}$ Simplify within parentheses

$= \left(\dfrac{9s^2t^{-4}}{z}\right)^{-2}$

$= \dfrac{9^{-2}s^{-4}t^8}{z^{-2}}$ Use Property (4)

$= \dfrac{z^2t^8}{81s^4}$ Note that $9^{-2} = \dfrac{1}{81}$

✔ *Check It Out 3:* Simplify the expression $\dfrac{(4y^{-3})^4}{(xy)^{-2}}$ and write it using positive exponents. ■

Scientific Notation

There may be times when you wish to write a nonzero number in a way that enables you to easily get a rough idea of how large or how small the number is, or to make a rough comparison of two or more numbers. The system called **scientific notation** is ideal for both of these purposes.

Definition of Scientific Notation

A nonzero number x is written in scientific notation as

$$a \times 10^b$$

where $1 \le a < 10$ if $x > 0$ and $-10 < a \le -1$ if $x < 0$, and b is an integer.
    The number b is sometimes called the **order of magnitude** of x.

If $x = a \times 10^b$, then the number b indicates the number of places the decimal point in a has to be shifted in order to "get back" to x.

▶ If $b > 0$, the decimal point has to be shifted b places to the right.

▶ If $b = 0$, the decimal point is not shifted at all.

▶ If $b < 0$, the decimal point has to be shifted b places to the left.

Technology Note

On most calculators, you can enter a number in scientific notation by using the EE key.

Keystroke Appendix: Section 4

Example 4 **Expressing a Number in Scientific Notation**

Express each of the following numbers in scientific notation.

(a) 328.5

(b) 4.69

(c) 0.00712

▶Solution

(a) Because 328.5 is greater than zero, $1 \le a < 10$. In this case, $a = 3.285$. In going from 328.5 to 3.285, we shifted the decimal point two places to the left. In order to *start* from 3.285 and "get back" to 328.5, we have to shift the decimal point two places to the right. Thus $b = 2$. In scientific notation, we have

$$328.5 = 3.285 \times 10^2.$$

(b) Since $1 \le 4.69 < 10$, there is no need to shift the decimal point, and so $a = 4.69$ and $b = 0$. In scientific notation, 4.69 is written as 4.69×10^0.

(c) The first nonzero digit of 0.00712 is the 7. In scientific notation, the 7 will be the digit to the left of the decimal point. Thus $a = 7.12$. To go from 0.00712 to 7.12, we had to shift the decimal point three places to the right. To get back to 0.00712 from 7.12, we have to shift the decimal point three places to the left. Thus $b = -3$. In scientific notation,

$$0.00712 = 7.12 \times 10^{-3}.$$

✔ *Check It Out 4:* Express 0.0315 in scientific notation. ▨

Sometimes we may need to convert a number in scientific notation to decimal form.

Converting from Scientific Notation to Decimal Form

To convert a number in the form $a \times 10^b$ into decimal form, proceed as follows.
- ▶ If $b > 0$, the decimal point is shifted b places *to the right*.
- ▶ If $b = 0$, the decimal point is not shifted at all.
- ▶ If $b < 0$, the decimal point is shifted $|b|$ places *to the left*.

Example 5 Converting from Scientific Notation to Decimal Form

Write the following numbers in decimal form.
(a) 2.1×10^5
(b) 3.47×10^{-3}

▶Solution

(a) To express 2.1×10^5 in decimal form, move the decimal point in 2.1 five places to the right. You will need to append four zeros.

$$2.1 \times 10^5 = 210{,}000$$

(b) To express 3.47×10^{-3} in decimal form, move the decimal point in 3.47 three places to the left. You will need to attach two zeros to the right of the decimal point in the final answer.

$$3.47 \times 10^{-3} = 0.00347$$

✔ *Check It Out 5:* Write 7.05×10^{-4} in decimal form. ▨

To multiply and divide numbers using scientific notation, we apply the rules of exponents, as illustrated in Example 6.

Example 6 **Determining Population**

In 2005, Japan had a land area of 3.75×10^5 square kilometers and a population of 1.27×10^8. What is the population density of Japan? Here, population density is defined as the number of people per square kilometer of land. Express your answer in scientific notation. (*Source:* CIA World Factbook)

▶Solution To compute the number of people per square kilometer of land, divide the total population by the total land area.

$$\frac{\text{Total population}}{\text{Total land area}} = \frac{1.27 \times 10^8}{3.75 \times 10^5}$$

$$= \frac{1.27}{3.75} \cdot \frac{10^8}{10^5}$$

$$\approx 0.339 \times 10^3 \qquad \textit{Use properties of exponents}$$

$$\approx 3.39 \times 10^2 \qquad \textit{Write in scientific notation}$$

Thus the population density is approximately 339 people per square kilometer.

✔ *Check It Out 6:* Find the population density of Italy, with a population of 5.81×10^7 and a land area of 2.94×10^5 square kilometers (2005 estimates). ■

Significant Figures

When you carry out a numerical computation, you may be unsure about how to express the result: whether to round off the answer, how many digits to write down, and so on. Before demonstrating how we make such decisions, we need the following definition.

Properties of Significant Figures

A digit of a nonzero number x is a **significant figure** if it satisfies one of the following conditions.

▶ The digit is the **first nonzero digit** of x, going from left to right.

▶ The digit lies **to the right of the first nonzero digit** of x.

The significant figures in a number range from the **most significant figure** (at the far left) to the **least significant figure** (at the far right).

It is important to distinguish between a significant figure and a decimal place. A digit to the right of the decimal point is not necessarily a significant figure, whereas any digit to the right of the decimal point occupies a decimal place.

Example 7 **Determining Significant Figures**

For each of the following numbers, list all of its significant figures in decreasing order of significance. Also give the number of decimal places in each number.

(a) 52.074

(b) 0.018

▶Solution

(a) The leftmost digit of 52.074 is 5, which is nonzero. Thus 5 is the most significant figure. There are five significant figures—5, 2, 0, 7, 4.

   There are five digits in 52.074, all of which are significant figures, but there are only three decimal places, occupied by 0, 7, and 4.

(b) The number 0.018 has no nonzero digit to the left of the decimal point, so its most significant figure lies to the right of the decimal point. The digit in the first decimal place is zero, which is not significant. So the 1 is the most significant figure, followed by the 8. Thus 0.018 has two significant figures, 1 and 8, but three decimal places, occupied by 0, 1, and 8.

☑ *Check It Out 7:* Determine the number of significant figures in 1.012. ▪

To determine the appropriate number of significant figures in the answer when we carry out a numerical computation, we note the number of significant figures in each of the quantities that enter into the computation, and then take the *minimum*. Quantities that are known exactly do not have any "uncertainty" and are not included in reckoning the number of significant figures.

Example 8 Using Significant Figures in Computations

Compute the following. Express your answers using the appropriate number of significant figures.

(a) The distance traversed by someone who rides a bike for a total of 1.26 hours at a speed of 8.95 miles per hour

(b) The cost of 15 oranges at $0.39 per orange

▶Solution

(a) Since the bicyclist is traveling at a constant speed, the distance traversed is the product of the speed and the time. When we multiply 8.95 by 1.26, we get 11.2770. Because the values of speed and time can be measured with limited precision, each of these two quantities has *some* uncertainty.

   The speed and the time have three significant figures each. The *minimum* of 3 and 3 is 3, so we round off the answer to three significant figures to obtain 11.3 miles as the distance traversed.

(b) The cost of 15 oranges is the product of the price per orange, $0.39, and the number of oranges purchased, 15.

   When we perform the multiplication, we get $5.85. Since the price per orange and the number of oranges are known *exactly*, neither of these quantities is uncertain, so we keep all three significant figures.

☑ *Check It Out 8:* Compute the distance traveled by a car driven at 43.5 miles per hour for 2.5 hours. Use the proper number of significant figures in your answer. ▪

Note In many applications using real-world data, computational results are rounded to the number of significant digits in the problem, and not necessarily the number of decimal places.

P.2 Key Points

▶ **Positive integer exponents**

For any positive integer n,

$$a^n = \underbrace{a \cdot a \cdot a \cdots a}_{n \text{ factors}}.$$

The number a is the **base** and the number n is the **exponent.**

▶ **Negative integer exponents**

Let a be any nonzero real number and let m be a positive integer. Then

$$a^{-m} = \frac{1}{a^m}.$$

▶ **Zero as an exponent**

If $a \neq 0$, then $a^0 = 1$.

▶ A nonzero number x is written in **scientific notation** as

$$a \times 10^b$$

where $1 \le a < 10$ if $x > 0$ and $-10 < a \le -1$ if $x < 0$, and b is an integer.

▶ **Significant figures** are used in computations to determine how many digits should be retained in the final answer.

P.2 Exercises

▶**Skills** This set of exercises will reinforce the skills illustrated in this section.

In Exercises 1–8, evaluate each expression for the given value of the variable.

1. $4x + 5, x = 2$
2. $3x - 1, x = 3$

3. $-2x + 4, x = -3$
4. $-x + 5, x = 0$

5. $3(a + 4) - 2, a = 5$
6. $2(a - 2) + 7, \ a = 1$

7. $-(2x + 1), x = 2$
8. $-(-3x + 4), x = -4$

In Exercises 9–16, simplify each expression without using a calculator.

9. -3^2
10. $(-3)^2$

11. 4^{-3}
12. 6^{-2}

13. -2^0
14. $(-3)^0$

15. $\left(\frac{3}{4}\right)^{-2}$
16. $\left(\frac{5}{2}\right)^{-2}$

In Exercises 17–38, simplify each expression and write it using positive exponents. Assume that all variables represent nonzero numbers.

17. $-(4x^2y^4)^2$
18. $(-4xy^3)^2$

19. $(2x^5)^{-2}$
20. $(4y^2)^{-3}$

21. $(-3a^2b^3)^2$
22. $(-4ab^4)^3$

23. $-2(a^2b^5)^2$
24. $-3(a^4b^2)^2$

25. $(-4xy^2)^{-2}$
26. $(-3x^2y^2)^{-3}$

27. $\left(\frac{3x}{2x^2y}\right)^{-2}$
28. $\left(\frac{2y^2}{5yx^2}\right)^{-2}$

29. $\frac{7x^6y^2}{21x^3}$
30. $\frac{6x^3y^4}{24x^2y}$

31. $\frac{(2y^{-3})^2}{y^{-5}}$
32. $\frac{(-4x^{-1})^2}{x^{-3}}$

33. $\left(\frac{2x^{-2}y^3}{xy^4}\right)^2$
34. $\left(\frac{3x^3y}{xy^2}\right)^3$

35. $\left(\dfrac{4x^3y^2z^3}{16x^{-2}yz}\right)^{-1}$

36. $\left(\dfrac{9x^3y^4z^7}{3x^{-4}y^3z^{-1}}\right)^2$

37. $\left(\dfrac{-4s^5t^{-2}}{12s^3t}\right)^{-2}$

38. $\left(\dfrac{27s^3z^5t^{-2}}{3z^2t}\right)^{-3}$

In Exercises 39–50, write each number in scientific notation.

39. 0.0051

40. 23.37

41. 5600

42. 497

43. 0.0000567

44. 0.0000032

45. 1,760,000

46. 5,341,200

47. 31.605

48. 457.31

49. 280,000,000

50. 62,000,000,000

In Exercises 51–62, write each number in decimal form.

51. 3.71×10^2

52. 4.26×10^4

53. 2.8×10^{-2}

54. 6.25×10^{-3}

55. 5.96×10^5

56. 2.5×10^3

57. 4.367×10^7

58. 3.105×10^{-2}

59. 8.673×10^{-3}

60. 7.105×10^4

61. 4.65×10^{-6}

62. 1.37×10^{-9}

In Exercises 63–70, simplify and write the answer in scientific notation.

63. $(2 \times 10^{-2})(3 \times 10^4)$

64. $(5 \times 10^4)(6 \times 10^{-5})$

65. $(2.1 \times 10^3)(4.3 \times 10^4)$

66. $(3.7 \times 10^{-1})(5.1 \times 10^3)$

67. $\dfrac{8 \times 10^4}{4 \times 10^2}$

68. $\dfrac{9 \times 10^{-2}}{3 \times 10^{-3}}$

69. $\dfrac{9.4 \times 10^2}{4.7 \times 10^3}$

70. $\dfrac{1.3 \times 10^{-4}}{3.9 \times 10^{-2}}$

In Exercises 71–76, find the number of significant figures for each number.

71. 1.42

72. 0.0134

73. 3.901

74. 4.00

75. 3.005

76. 2.0

In Exercises 77–82, calculate and round your answer to the correct number of significant figures.

77. $2.31 \cdot 5.2$

78. $4.06 \cdot 3.0$

79. $12.5 \div 0.5$

80. $30.2 \div 0.01$

81. $14.3 \cdot (2.4 \div 1.2)$

82. $(3.001 \cdot 4.00) \div 0.500$

▶**Applications** In this set of exercises, you will use exponents and scientific notation to study real-world problems.

In Exercises 83–86, express each number using scientific notation.

83. **Astronomy** The moon orbits the earth at 36,800 kilometers per hour.

84. **Astronomy** The diameter of Saturn is 74,978 miles.

85. **Biology** The length of a large amoeba is 0.005 millimeter.

86. **Land Area** The total land area of the United States is approximately 3,540,000 square miles.

87. **Population** The United States population estimate for 2004 is 2.93×10^8 people. If the total land area of the United States is approximately 3,540,000 square miles, how many people are there per square mile of land in the U.S.? Express your answer in decimal form.

88. **National Debt** The national debt of the United States as of January 2006 was 8.16×10^{12} dollars. In January 2006, one dollar was worth 0.828 euros. Express the U.S. national debt in euros, using scientific notation.

89. **Economics** The gross domestic product (GDP) is the total annual value of goods and services produced by a country. In 2004, Canada had a GDP of 1.02×10^{12} (in U.S. dollars) and a population of 32.8 million. Find the per capita GDP (that is, the GDP per person) of Canada in 2004.

90. **Carpentry** A small piece of wood measures 6.5 inches. If there are 2.54 centimeters in an inch, how long is the piece of wood in centimeters? Round your answer to the correct number of significant figures.

91. **Area** A rectangular room is 12.5 feet wide and 10.3 feet long. If the area of the room is the product of its length and width, find the area of the room in square feet. Round your answer to the correct number of significant figures.

▶**Concepts** This set of exercises will draw on the ideas presented in this section and your general math background.

92. Does simplifying the expressions $10 - (4 + 3)$ and $10 - 4 + 3$ produce the same result? Explain.

93. Does simplifying the expressions $2 \cdot 5 \div 10 \cdot 2$ and $2 \cdot 5 \div (10 \cdot 2)$ produce the same result? Explain.

94. For what value(s) of x is the expression $\dfrac{3x^{-2}}{5}$ defined?

95. For what values of x and y is the expression $\dfrac{3x^2 y^{-3}}{y}$ defined?

P.3 Roots, Radicals, and Rational Exponents

Objectives

▶ Find the nth root of a number

▶ Understand and use rules for radicals

▶ Simplify radical expressions

▶ Add and subtract radical expressions

▶ Understand and use rules for rational exponents

▶ Simplify expressions containing rational exponents

Roots and Radicals

The square root of 25 is defined to be the positive number 5, since $5^2 = 25$. The square root of 25 is indicated by $\sqrt{25}$. The fourth root of 25 is denoted by $\sqrt[4]{25}$. Note that in the real number system, we can take even roots of only nonnegative numbers.

Similarly, the cube root of -8 is indicated by $\sqrt[3]{-8} = -2$, since $(-2)^3 = -8$. We can take odd roots of all real numbers, negative and nonnegative. This leads us to the following definition.

> **The nth root of a**
>
> Let n be an even positive integer and let a be a nonnegative real number. Then
>
> $\sqrt[n]{a}$ denotes the nonnegative number whose nth power is a.
>
> Let n be an odd positive integer and let a be any real number. Then
>
> $\sqrt[n]{a}$ denotes the number whose nth power is a.
>
> The number $\sqrt[n]{a}$ is called the **nth root of a**. The $\sqrt[n]{}$ symbol is called a **radical.** If the n is omitted in the radical symbol, the root is assumed to be a square root.

Example **1** **Evaluating the nth Root of a Number**

Determine $\sqrt[3]{64}$, $\sqrt[4]{16}$, and $\sqrt[5]{-32}$.

▶**Solution** First, $\sqrt[3]{64} = 4$, since $4^3 = 64$. Next, $\sqrt[4]{16} = 2$, since $2^4 = 16$. Finally, $\sqrt[5]{-32} = -2$, since $(-2)^5 = -32$.

✔ *Check It Out 1:* Determine $\sqrt[4]{81}$. ■

We now introduce the following rules for radicals.

Discover *and* Learn

Give an example to show that $\sqrt{x + y} \neq \sqrt{x} + \sqrt{y}$.

> **Rules for Radicals**
>
> Suppose a and b are real numbers such that their nth roots are defined.
>
> **Product Rule:** $\sqrt[n]{a} \cdot \sqrt[n]{b} = \sqrt[n]{ab}$
>
> **Quotient Rule:** $\dfrac{\sqrt[n]{a}}{\sqrt[n]{b}} = \sqrt[n]{\dfrac{a}{b}}, b \neq 0$

It is conventional to leave radicals only in the numerator of a fraction. This is known as **rationalizing the denominator,** a technique that is illustrated in the following three examples.

Example 2 Rationalizing the Denominator

Simplify and rationalize the denominator of each expression.

(a) $\sqrt{\dfrac{18}{5}}$ (b) $\dfrac{\sqrt[3]{10}}{\sqrt[3]{3}}$

▶ Solution

(a) $\sqrt{\dfrac{18}{5}} = \dfrac{\sqrt{18}}{\sqrt{5}} \cdot \dfrac{\sqrt{5}}{\sqrt{5}} = \dfrac{\sqrt{90}}{5} = \dfrac{3\sqrt{10}}{5}$ $\sqrt{90} = \sqrt{9} \cdot \sqrt{10} = 3\sqrt{10}$

(b) $\dfrac{\sqrt[3]{10}}{\sqrt[3]{3}} = \dfrac{\sqrt[3]{10}}{\sqrt[3]{3}} \cdot \dfrac{\sqrt[3]{9}}{\sqrt[3]{9}}$ Multiply numerator and denominator by $\sqrt[3]{9}$ to obtain a perfect cube in the denominator

$\quad = \dfrac{\sqrt[3]{90}}{\sqrt[3]{27}}$

$\quad = \dfrac{\sqrt[3]{90}}{3}$ $\sqrt[3]{27} = 3$

Note that $\sqrt[3]{90}$ cannot be simplified further, since there are no perfect cubes that are factors of 90.

✔ *Check It Out 2:* Rationalize the denominator of $\dfrac{5}{\sqrt{2}}$. ▪

The product and quotient rules for radicals can be used to simplify expressions involving radicals. It is conventional to write radical expressions such that, for the nth root, there are no powers under the radical greater than or equal to n. We also make the following rule for finding the nth root of a^n, $a \neq 0$.

Finding the nth root of a^n

If n is odd, then $\sqrt[n]{a^n} = a$, $a \neq 0$.

If n is even, then $\sqrt[n]{a^n} = |a|$, $a \neq 0$.

For instance, $\sqrt[3]{(-4)^3} = -4$, whereas $\sqrt{(-4)^2} = |-4| = 4$.

Note Unless otherwise stated, variables are assumed to be positive to avoid the issue of taking absolute values.

Example 3 Simplifying Radicals

Simplify the following.

(a) $\sqrt{75}$ (b) $\sqrt[3]{\dfrac{125}{108}}$ (c) $\sqrt{x^5 y^7}$, $x, y > 0$

▶Solution

(a) Because $75 = 25 \cdot 3$ and 25 is a perfect square, we can use the product rule for radicals to obtain

$$\sqrt{75} = \sqrt{25 \cdot 3} = \sqrt{25}\sqrt{3} = 5\sqrt{3}.$$

(b) Apply the quotient rule for radicals. Because $125 = 5^3$ and $108 = 27 \cdot 4$, we can write

$$\sqrt[3]{\frac{125}{108}} = \frac{\sqrt[3]{125}}{\sqrt[3]{108}} = \frac{5}{\sqrt[3]{27 \cdot 4}} = \frac{5}{\sqrt[3]{27}\sqrt[3]{4}} = \frac{5}{3\sqrt[3]{4}}.$$

Next, we clear the radical $\sqrt[3]{4}$ in the denominator by multiplying the numerator and denominator by $\sqrt[3]{2}$.

$$\frac{5}{3\sqrt[3]{4}} = \frac{5}{3\sqrt[3]{4}} \cdot \frac{\sqrt[3]{2}}{\sqrt[3]{2}} = \frac{5\sqrt[3]{2}}{3\sqrt[3]{8}} = \frac{5\sqrt[3]{2}}{6}$$

(c) Because $x^5 = x \cdot x^4 = x(x^2)^2$ and $y^7 = y \cdot y^6 = y(y^3)^2$, we have

$$\sqrt{x^5 y^7} = \sqrt{x(x^2)^2 y(y^3)^2} = x^2 y^3 \sqrt{xy}.$$

Because we were given that $x, y > 0$, the square root of the quantity under the radical is always defined.

☑ *Check It Out 3:* Simplify $\sqrt{45}$ and $\sqrt[3]{x^5 y^6}$. ▪

If the denominator of an expression is of the form $\sqrt{a} + \sqrt{b}$, we can eliminate the radicals by multiplying the numerator and denominator by $\sqrt{a} - \sqrt{b}$, $a, b > 0$. Note that

$$(\sqrt{a} + \sqrt{b})(\sqrt{a} - \sqrt{b}) = (\sqrt{a} + \sqrt{b})(\sqrt{a}) - (\sqrt{a} + \sqrt{b})(\sqrt{b})$$
$$= \sqrt{a} \cdot \sqrt{a} + \sqrt{b}\sqrt{a} - \sqrt{a}\sqrt{b} - \sqrt{b}\sqrt{b}$$
$$= \sqrt{a} \cdot \sqrt{a} - \sqrt{b}\sqrt{b} = a - b.$$

Therefore, the radicals in the denominator have been eliminated. If the denominator is of the form $\sqrt{a} - \sqrt{b}$, then multiply both numerator and denominator by $\sqrt{a} + \sqrt{b}$. The same approach can be used for denominators of the form $a \pm \sqrt{b}$ and $\sqrt{a} \pm b$.

Example 4 Rationalizing a Denominator Containing Two Terms

Rationalize the denominator.

$$\frac{5}{4 - \sqrt{2}}$$

▶Solution To remove the radical in the denominator, multiply the numerator and denominator by $4 + \sqrt{2}$.

$$\frac{5}{4 - \sqrt{2}} = \frac{5}{4 - \sqrt{2}} \cdot \frac{4 + \sqrt{2}}{4 + \sqrt{2}}$$
$$= \frac{5(4 + \sqrt{2})}{4^2 - (\sqrt{2})^2} = \frac{5(4 + \sqrt{2})}{14}$$

☑ *Check It Out 4:* Rationalize the denominator of $\dfrac{4}{1 + \sqrt{3}}$. ▪

Adding and Subtracting Radical Expressions

Two or more radicals of the form $\sqrt[n]{\ }$ can be combined provided they all have the same expression under the radical and they all have the same value of n. For instance, $3\sqrt{2} - 2\sqrt{2} = \sqrt{2}$ and $\sqrt[3]{x} + 2\sqrt[3]{x} = 3\sqrt[3]{x}$.

Example 5 Combining Radical Expressions

Simplify the following radical expressions. Assume $x \geq 0$.
(a) $\sqrt{48} + \sqrt{27} - \sqrt{12}$ (b) $(3 + \sqrt{x})(4 - 2\sqrt{x})$

▶Solution

(a) $\sqrt{48} + \sqrt{27} - \sqrt{12} = \sqrt{16 \cdot 3} + \sqrt{9 \cdot 3} - \sqrt{4 \cdot 3}$
$$= 4\sqrt{3} + 3\sqrt{3} - 2\sqrt{3} = 5\sqrt{3}$$
(b) $(3 + \sqrt{x})(4 - 2\sqrt{x}) = 12 - 6\sqrt{x} + 4\sqrt{x} - 2\sqrt{x}\sqrt{x}$
$$= 12 - 2\sqrt{x} - 2x$$

✔ *Check It Out 5:* Simplify $\sqrt{16x} - \sqrt{9x} + \sqrt{x^3}$. Assume $x > 0$. ▪

Rational Exponents

We have already discussed integer exponents in Section P.2. We can also define exponents that are rational.

Definition of $a^{1/n}$

If a is a real number and n is a positive integer greater than 1, then
$$a^{1/n} = \sqrt[n]{a}, \quad \text{where } a \geq 0 \text{ when } n \text{ is even.}$$
The quantity $a^{1/n}$ is called the **nth root of a.**

Example 6 Expressions Involving nth Roots

Evaluate the following.
(a) $27^{1/3}$ (b) $-36^{1/2}$

▶Solution

(a) Applying the definition of $a^{1/n}$ with $n = 3$,
$$27^{1/3} = \sqrt[3]{27} = 3.$$

(b) Applying the definition of $a^{1/n}$ with $n = 2$, $-36^{1/2} = -(\sqrt{36}) = -6$. Note that only the number 36 is raised to the $1/2$ power, and the result is multiplied by -1.

✔ *Check It Out 6:* Evaluate $(-8)^{1/3}$. ▪

Next we give the definition of $a^{m/n}$, where m/n is in lowest terms and $n \geq 2$. The definition is given in such a way that the laws of exponents from Section P.2 also hold for rational exponents.

> **Definition of $a^{m/n}$**
>
> Let a be a positive real number and let m and n be integers such that m/n is in lowest terms and $n \geq 2$. We then have
> $$a^{m/n} = \sqrt[n]{a^m} = (\sqrt[n]{a})^m.$$

For example, $64^{2/3} = (\sqrt[3]{64})^2 = 4^2 = 16$ and $64^{2/3} = \sqrt[3]{64^2} = \sqrt[3]{4096} = 16$.

The rules for integer exponents given earlier also hold for rational exponents, with some restrictions. These rules are summarized below.

> **Properties of Rational Exponents**
>
> Assume that a and b are real numbers and s and t are rational numbers. Whenever s and t indicate even roots, assume that a and b are *nonnegative* real numbers. Then the following properties hold.
>
> 1. $a^s \cdot a^t = a^{s+t}$ 2. $(a^s)^t = a^{st}$ 3. $(ab)^s = a^s b^s$
>
> 4. $a^{-s} = \dfrac{1}{a^s}, a \neq 0$ 5. $\left(\dfrac{a}{b}\right)^s = \dfrac{a^s}{b^s}, b \neq 0$ 6. $\dfrac{a^s}{a^t} = a^{s-t}, a \neq 0$

Example 7 Expressions Containing Rational Exponents

Evaluate the following expressions without using a calculator.

(a) $(4)^{3/2}$

(b) $\left(\dfrac{1}{3}\right)^{-2}$

(c) $(23.1)^0$

▶Solution

(a) Using the rules for rational exponents,
$$(4)^{3/2} = (4^{1/2})^3 = 2^3 = 8.$$

(b) Using the rules for exponents,
$$\left(\frac{1}{3}\right)^{-2} = \frac{1^{-2}}{3^{-2}} = \frac{1}{\frac{1^2}{3^2}} = \frac{3^2}{1^2} = 9. \qquad \text{Recall that } 3^{-2} = \frac{1}{3^2}$$

(c) Note that $(23.1)^0 = 1$, since any nonzero number raised to the zero power is defined to be equal to 1.

✔ *Check It Out 7:* Evaluate the following expressions without using a calculator.

(a) $(8)^{2/3}$

(b) $\left(\dfrac{1}{4}\right)^{-3}$

(c) $(16)^{-1/2}$ ■

Example 8 Simplifying Expressions Containing Rational Exponents

Simplify and write with positive exponents.

(a) $(25)^{3/2}$

(b) $(16s^{4/3}t^{-3})^{3/2}$, $s, t > 0$

(c) $\dfrac{(x^{-7/3}y^{5/2})^3}{y^{-1/2}x^2}$, $x, y > 0$

▶**Solution**

(a) $(25)^{3/2} = (25^{1/2})^3 = (\sqrt{25})^3 = 5^3 = 125$

(b) First use Property (3) from the Properties of Rational Exponents box on page 24 to rewrite the power of a product.

$$
\begin{aligned}
(16s^{4/3}t^{-3})^{3/2} &= 16^{3/2}(s^{4/3})^{3/2}(t^{-3})^{3/2} && \text{Use Property (3)} \\
&= 16^{3/2}s^{(4/3)(3/2)}t^{-3(3/2)} && \text{Use Property (2)} \\
&= 16^{3/2}s^2t^{-9/2} && \frac{4}{3} \cdot \frac{3}{2} = 2 \text{ and } (-3)\left(\frac{3}{2}\right) = -\frac{9}{2} \\
&= \frac{64s^2}{t^{9/2}} && 16^{3/2} = (16^{1/2})^3 = 64
\end{aligned}
$$

(c) Use Property (3) to simplify the numerator. This gives

$$
\begin{aligned}
\frac{(x^{-7/3}y^{5/2})^3}{y^{-1/2}x^2} &= \frac{x^{(-7/3)3}y^{(5/2)3}}{y^{-1/2}x^2} && \text{Use Property (3)} \\
&= \frac{x^{-7} \cdot y^{15/2}}{y^{-1/2}x^2} && \frac{-7}{3} \cdot 3 = -7 \text{ and } \frac{5}{2} \cdot 3 = \frac{15}{2} \\
&= x^{-7-2}y^{(15/2)-(-1/2)} = x^{-9}y^8 && \text{Use Property (6)} \\
&= \frac{y^8}{x^9}. && \text{Use Property (4) to write with} \\
& && \text{positive exponents}
\end{aligned}
$$

✔ *Check It Out 8:* Simplify the following.

(a) $(16)^{5/4}$

(b) $(8s^{5/3}t^{-1/6})^3$ ■

P.3 Key Points

▶ Let n be an integer. The number $\sqrt[n]{a}$ is called the **nth root of a**. If n is even, then $a \geq 0$. If n is odd, then a can be any real number.

▶ Rules for radicals:

Product Rule: $\sqrt[n]{a} \cdot \sqrt[n]{b} = \sqrt[n]{ab}$

Quotient Rule: $\dfrac{\sqrt[n]{a}}{\sqrt[n]{b}} = \sqrt[n]{\dfrac{a}{b}}, b \neq 0$

▶ Radicals usually are not left in the denominator when simplifying. Removing the radical from the denominator is called **rationalizing the denominator.**

▶ If a is a real number and n is a positive integer greater than 1, then

$$a^{1/n} = \sqrt[n]{a}, \quad \text{where } a \geq 0 \text{ when } n \text{ is even.}$$

▶ Let a be a positive real number and let m and n be integers. Then

$$a^{m/n} = \sqrt[n]{a^m} = (\sqrt[n]{a})^m.$$

P.3 Exercises

▶**Skills** This set of exercises will reinforce the skills illustrated in this section.

In Exercises 1–8, evaluate without using a calculator.

1. $\sqrt{49}$

2. $\sqrt[3]{64}$

3. $\sqrt[3]{\dfrac{1}{8}}$

4. $-\sqrt{\dfrac{9}{4}}$

5. $(49)^{3/2}$

6. $(27)^{2/3}$

7. $\left(\dfrac{16}{625}\right)^{1/4}$

8. $\left(\dfrac{-8}{125}\right)^{1/3}$

In Exercises 9–50, simplify the radical expression. Assume that all variables represent positive real numbers.

9. $\sqrt{32}$

10. $\sqrt{75}$

11. $\sqrt[3]{250}$

12. $\sqrt[3]{80}$

13. $\sqrt{32} \cdot \sqrt{8}$

14. $\sqrt{27} \cdot \sqrt{12}$

15. $\sqrt{30} \cdot \sqrt{15}$

16. $\sqrt{40} \cdot \sqrt{20}$

17. $\sqrt[3]{\dfrac{-16}{125}}$

18. $\sqrt[3]{\dfrac{32}{125}}$

19. $\sqrt{\dfrac{50}{147}}$

20. $\sqrt{\dfrac{32}{125}}$

21. $\sqrt{\dfrac{3}{5}}$

22. $\sqrt{\dfrac{2}{7}}$

23. $\sqrt[3]{\dfrac{7}{9}}$

24. $\sqrt[3]{\dfrac{5}{2}}$

25. $\sqrt[3]{\dfrac{-48}{81}}$

26. $\sqrt[3]{\dfrac{-375}{32}}$

27. $\sqrt{x^3 y^4}$

28. $\sqrt{s^6 y^3}$

29. $\sqrt[3]{x^8 y^4}$

30. $\sqrt[3]{s^{10} t^7}$

31. $\sqrt{3x^2 y} \cdot \sqrt{15xy^3}$

32. $\sqrt{5yz^3} \cdot \sqrt{8y^2 z^2}$

33. $\sqrt{12yz} \cdot \sqrt{3y^3 z^5}$

34. $\sqrt{6x^3 y^2} \cdot \sqrt{3xy^4}$

35. $\sqrt[3]{6x^3 y^2} \cdot \sqrt[3]{4x^2 y}$

36. $\sqrt[3]{9xy^4} \cdot \sqrt[3]{6x^2 y^2}$

37. $\sqrt{98} + 3\sqrt{32}$

38. $2\sqrt{200} - \sqrt{72}$

39. $\sqrt{216} - 4\sqrt{24} + \sqrt{3}$

40. $-\sqrt{125} + \sqrt{20} - \sqrt{50}$

41. $(1 + \sqrt{5})(1 - \sqrt{5})$

42. $(3 - \sqrt{2})(3 + \sqrt{2})$

43. $(-2\sqrt{3} + 1)(1 + \sqrt{2})$

44. $(-\sqrt{2} + 1)(2 - \sqrt{3})$

45. $(\sqrt{6} + \sqrt{3})(\sqrt{5} - \sqrt{2})$

46. $(\sqrt{7} - \sqrt{3})(\sqrt{6} - \sqrt{5})$

47. $\dfrac{4}{1 - \sqrt{5}}$

48. $\dfrac{3}{2 + \sqrt{6}}$

49. $\dfrac{1}{\sqrt{3} - \sqrt{2}}$

50. $\dfrac{2}{\sqrt{5} + \sqrt{2}}$

In Exercises 51–66, simplify and write with positive exponents. Assume that all variables represent positive real numbers.

51. $3^{2/3} \cdot 3^{-4/3}$

52. $2^{1/2} \cdot 2^{-1/3}$

53. $5^{-1/2} \cdot 5^{1/2}$

54. $3^{2/3} \cdot 3^{1/3}$

55. $\dfrac{7^{-1/4}}{7^{1/2}}$

56. $\dfrac{5^{1/2}}{5^{1/3}}$

57. $\dfrac{4^{-1/3}}{4^{1/4}}$

58. $\dfrac{2^{3/2}}{2^{1/4}}$

59. $(x^2 y^3)^{-1/2}$

60. $(s^4 y^5)^{-1/3}$

61. $\dfrac{x^{1/3} \cdot x^{1/2}}{x^2}$

62. $\dfrac{y^{2/3} \cdot y^{3/2}}{y^3}$

63. $(8r^{3/2} s^{-2})^{2/3}$

64. $(27rs^{-3})^{1/3}$

65. $\dfrac{(x^{2/3} y^{3/2})^2}{y^{-1/3} x^2}$

66. $\dfrac{(x^{4/3} y^{1/4})^3}{xy^{1/2}}$

▶**Applications** In this set of exercises, you will use radicals and rational exponents to study real-world problems.

67. **Geometry** The length of a diagonal of a square with a side of length s is $s\sqrt{2}$. Find the sum of the lengths of the two diagonals of a square whose side is 5 inches long.

68. **Physics** If an object is dropped from a height of h meters, it will take $\sqrt{\dfrac{h}{4.9}}$ seconds to hit the ground. How long will

it take a ball dropped from a height of 30 meters to hit the ground?

69. **Ecology** The number of tree species in a forested area of Malaysia is given approximately by the expression $386a^{1/4}$, where a is the area of the forested region in square kilometers. Determine the number of tree species in an area of 20 square kilometers. (*Source:* Plotkin et al., *Proceedings of the National Academy of Sciences*)

70. **Learning Theory** Researchers discovered that the time it took for a 13-year-old student to solve the equation $7x + 1 = 29$ depended on the number of days, d, that the student was exposed to the equation. They found that the time, in seconds, to solve the equation was given by $0.63d^{-0.28}$. Using a scientific calculator, determine how long it would take the student to solve the equation if she was exposed to it for 3 days. (*Source:* Qin et al., *Proceedings of the National Academy of Sciences*)

71. **Sailing** Racing yachts must meet strict design requirements to ensure some level of fairness in the race. In 2006, the International America's Cup Class (IACC) rules included the following requirement for the dimensions of a racing sailboat.

$$L + 1.25\sqrt{S} - 9.8\sqrt[3]{D} \le 16.464 \text{ meters}$$

where L is the length of the yacht in meters, S is the area of the sail in square meters, and D is the amount of water displaced in cubic meters. If you design a yacht that is 18 meters long, has 250 square meters of sail area, and displaces 20 cubic meters of water, will it meet the IACC requirements? (*Source:* America's Cup Properties, Inc.)

▶ **Concepts** This set of exercises will draw on the ideas presented in this section and your general math background.

72. Find numbers a and b such that $\sqrt[3]{ax^b} = 4x^2$.

73. Find numbers a, b, and c such that $\sqrt{ax^b y^c} = 6x^4 y^2$.

74. For what values of b is the expression $\sqrt{-b}$ a real number?

75. For what values of b is the expression $\sqrt[3]{-b}$ a real number?

76. Show with a numerical example that $\sqrt{x^2 + y^2} \ne x + y$. (Answers may vary.)

77. Without using a calculator, explain why $\sqrt{10}$ must be greater than 3.

P.4 Polynomials

Objectives

▶ Define a polynomial

▶ Write a polynomial in descending order

▶ Add and subtract polynomials

▶ Multiply polynomials

In this section, we discuss a specific type of algebraic expression known as a **polynomial.** Some examples of polynomials are

$$2x + 7, \quad 3y^2 + 8y - \frac{3}{2}, \quad 10x^7 + x\sqrt{5}, \quad \text{and} \quad 5.$$

Polynomials consist of sums of individual expressions, where each expression is the product of a real number and a variable raised to a nonnegative power.

> **A Polynomial Expression in One Variable**
>
> A **polynomial in one variable** is an algebraic expression of the form
>
> $$a_n x^n + a_{n-1}x^{n-1} + a_{n-2}x^{n-2} + \cdots + a_1 x + a_0$$
>
> where n is a nonnegative integer, $a_n, a_{n-1}, \ldots, a_0$ are real numbers, and $a_n \ne 0$.

All polynomials discussed in this section are in one variable only. They will form the basis for our work with quadratic and polynomial functions in Chapters 3 and 4.

Polynomial Terminology

The following is a list of important definitions related to polynomials.

▶ The **degree** of a polynomial is n, the highest power to which a variable is raised.

▶ The parts of a polynomial separated by plus signs are called **terms.**

▶ The numbers $a_n, a_{n-1}, \ldots, a_0$ are called **coefficients,** and the **leading coefficient** is a_n.

▶ The **constant term** is a_0.

▶ Polynomials are usually written in **descending order,** with the exponents decreasing from left to right.

Note   *The variable x in the definition of a polynomial can be consistently replaced by any other variable.*

Example **1** **Identifying Features of a Polynomial**

Write the following polynomial in descending order and find its degree, terms, coefficients, and constant term.

$$-4x^2 + 3x^5 - 2x - 7$$

▶**Solution**   Writing the polynomial in descending order, with the exponents decreasing from left to right, we have

$$3x^5 - 4x^2 - 2x - 7.$$

The degree of the polynomial is 5 because that is the highest power to which a variable is raised. The terms of the polynomial are

$$3x^5, \quad -4x^2, \quad -2x, \quad \text{and} \quad -7.$$

The coefficients of the polynomial are

$$a_5 = 3, \quad a_2 = -4, \quad a_1 = -2, \quad \text{and} \quad a_0 = -7 \qquad \text{Note that } a_4 = a_3 = 0$$

The constant term is $a_0 = -7$.

☑ *Check It Out 1:* Write the following polynomial in descending order and find its degree, terms, coefficients, and constant term.

$$-3x^2 + 7 + 4x^5 ■$$

Special Names for Polynomials

Polynomials with one, two, or three terms have specific names.

▶ A polynomial with one term, such as $2y^3$, is called a **monomial.**

▶ A polynomial with two terms, such as $-3z^4 + \frac{1}{5}z^2$, is called a **binomial.**

▶ A polynomial with three terms, such as $4t^6 - \frac{3}{4}t^2 - \sqrt{3}$, is called a **trinomial.**

Addition and Subtraction of Polynomials

Terms of an expression that have the same variable raised to the same power are called **like terms.** To add or subtract polynomials, we *combine* or *collect like terms* by adding their respective coefficients using the distributive property. This is illustrated in the following example.

Example 2 Adding and Subtracting Polynomials

Add or subtract each of the following.

(a) $(3x^3 + 2x^2 - 5x + 7) + (x^3 - x^2 + 5x - 2)$

(b) $\left(s^4 + \dfrac{3}{4}s^2\right) - (s^4 - s^2)$

▶ Solution

(a) Rearranging terms so that the terms with the same power are grouped together gives

$$(3x^3 + 2x^2 - 5x + 7) + (x^3 - x^2 + 5x - 2)$$
$$= (3x^3 + x^3) + (2x^2 - x^2) + (-5x + 5x) + (7 - 2).$$

Using the distributive property, we can write

$$= (3 + 1)x^3 + (2 - 1)x^2 + (-5 + 5)x + 7 - 2$$
$$= 4x^3 + x^2 + 5. \qquad\qquad \text{Simplify}$$

(b) We must distribute the minus sign throughout the second polynomial.

$$\left(s^4 + \frac{3}{4}s^2\right) - (s^4 - s^2) = s^4 + \frac{3}{4}s^2 - s^4 + s^2$$

$$= (s^4 - s^4) + \left(\frac{3}{4}s^2 + s^2\right) \qquad \text{Collect like terms}$$

$$= (1 - 1)s^4 + \left(\frac{3}{4} + 1\right)s^2 \qquad \text{Use distributive property}$$

$$= \frac{7}{4}s^2$$

✔ *Check It Out 2:* Add the following.

$$(5x^4 - 3x^2 + x - 4) + (-4x^4 + x^3 + 6x^2 + 4x) \; ■$$

Multiplication of Polynomials

We will first explain the multiplication of monomials, since these are the simplest of the polynomials. To multiply two monomials, we multiply the coefficients of the monomials and then multiply the variable expressions.

Example 3 Multiplication of Monomials

Multiply $(-2x^2)(4x^7)$.

▶ Solution

$$(-2x^2)(4x^7) = (-2)(4)(x^2 x^7) \quad \text{Commutative and associative properties of multiplication}$$
$$= -8x^9 \qquad\qquad \text{Multiply coefficients and add exponents of same base}$$

✔ *Check It Out 3:* Multiply $(4x^4)(-3x^5)$. ■

To multiply binomials, apply the distributive property twice and then apply the rules for multiplying monomials.

Example 4 **Multiplying Binomials**

Multiply $(3x + 2)(-2x - 3)$.

▶Solution

$$(3x + 2)(-2x - 3) = 3x(-2x - 3) + 2(-2x - 3) \qquad \textit{Apply distributive property}$$

Note that the terms $3x$ and 2 are *each* multiplied by $(-2x - 3)$.

$$= -6x^2 - 9x - 4x - 6 \qquad \textit{Remove parentheses—}$$
$$\textit{apply distributive property}$$

$$= -6x^2 - 13x - 6 \qquad \textit{Combine like terms}$$

✔ *Check It Out 4:* Multiply $(x + 4)(2x - 3)$. ▧

Examining our work, we can see that to multiply binomials, each term in the first polynomial is multiplied by each term in the second polynomial. To make sure you have multiplied all combinations of the terms, use the memory aid **FOIL**: multiply the First terms, then the Outer terms, then the Inner terms, and then the Last terms. Collect like terms and simplify if possible.

Example 5 **Using FOIL to Multiply Binomials**

Multiply $(-7x + 4)(5x - 1)$.

▶Solution Since we are multiplying two binomials, we apply FOIL to get

$$\begin{array}{ccc} & \text{Outer} & \text{Last} \\ (-7x + 4)(5x - 1) = \underbrace{(-7x)(5x)} + \overbrace{(-7x)(-1)} + \underbrace{(4)(5x)} + \overbrace{(4)(-1)} \\ \text{First} & \text{Inner} \end{array}$$

$$= -35x^2 + 7x + 20x - 4$$
$$= -35x^2 + 27x - 4.$$

✔ *Check It Out 5:* Use FOIL to multiply $(2x - 5)(3x + 1)$. ▧

To multiply general polynomials, use the distributive property repeatedly, as illustrated in Example 6.

Example 6 **Multiplying General Polynomials**

Multiply $(4y^3 - 3y + 1)(y - 2)$.

▶Solution We have

$$(4y^3 - 3y + 1)(y - 2)$$
$$= (4y^3 - 3y + 1)(y) - (4y^3 - 3y + 1)(2). \qquad \textit{Use distributive property}$$

Making sure to distribute the second negative sign throughout, we have

$$= 4y^4 - 3y^2 + y - 8y^3 + 6y - 2$$ Use distributive property again

$$= 4y^4 - 8y^3 - 3y^2 + 7y - 2.$$ Combine like terms

✔ *Check It Out 6:* Multiply $(3y^2 - 6y + 5)(y + 3)$. ■

The products of binomials given in Table P.4.1 occur often enough that they are worth committing to memory. They will be used in the next section to help us factor polynomial expressions.

Table P.4.1 Special Products of Binomials

SPECIAL PRODUCT	ILLUSTRATION
Square of a sum $(A + B)^2 = A^2 + 2AB + B^2$	$(3y + 4)^2 = (3y)^2 + 2(3y)(4) + 4^2$ $= 9y^2 + 24y + 16$
Square of a difference $(A - B)^2 = A^2 - 2AB + B^2$	$(4x^2 - 5)^2 = (4x^2)^2 - 2(4x^2)(5) + (5)^2$ $= 16x^4 - 40x^2 + 25$
Product of a sum and a difference $(A + B)(A - B) = A^2 - B^2$	$(7y + 3)(7y - 3) = (7y)^2 - (3)^2$ $= 49y^2 - 9$

P.4 Key Points

▸ A **polynomial in one variable** is an algebraic expression of the form

$$a_n x^n + a_{n-1} x^{n-1} + a_{n-2} x^{n-2} + \cdots + a_1 x + a_0$$

where n is a nonnegative integer, $a_n, a_{n-1}, \ldots, a_0$ are real numbers, and $a_n \neq 0$.

▸ To **add or subtract polynomials,** *combine* or *collect like terms* by adding their respective coefficients.

▸ To **multiply polynomials,** apply the distributive property and then apply the rules for multiplying monomials.

▸ **Special products of polynomials**

$$(A + B)^2 = A^2 + 2AB + B^2$$
$$(A - B)^2 = A^2 - 2AB + B^2$$
$$(A + B)(A - B) = A^2 - B^2$$

P.4 Exercises

▸**Skills** This set of exercises will reinforce the skills illustrated in this section.

In Exercises 1–10, collect like terms and arrange the polynomial in descending order. Give the degree of the polynomial.

1. $3y + 16 + 2y + 10$

2. $5z + 3 - 6z + 2$

3. $2t^2 - 2t + 5 + t^2$

4. $v^2 + v + v^2 - 1$

5. $-7s^2 - 6s + 3 + 3s^2 + 4$

6. $3s^3 + 4s - 2 + 5s^2 - 6s + 18$

7. $-v^3 - 3 - 3v^3$

8. $-5t^2 + t^3 - 2t + 4 + 4t + 5t^2 - 10t^3$

9. $1 + z^3 - 10z^5 - 3z^4 + 6z^3$

10. $-2u + 3u^2 + 4u^3 + 5u^5 - u^2 - 6u$

In Exercises 11–78, perform the given operations. Express your answer as a single polynomial in descending order.

11. $(z + 6) + (5z + 8)$

12. $(2x - 3) - (x + 6)$

13. $(-5y^2 + 6y) - (y - 3)$

14. $(-9x^2 - 32x + 14) + (-2x^2 + 15x - 6)$

15. $(-9x^3 + 6x^2 - 20x + 3) + (x^2 - 5x - 6)$

16. $(5t^3 + 16t - 12) - (2t^3 - 4t^2 + t) + (-5t^2 - 3t + 7)$

17. $(3x^4 - x^3 + x^4 - 4) + (x^5 + 7x^3 - 3x^2 + 5)$

18. $(z^5 - 4z^4 + 7) - (-z^3 + 15z^2 - 8z)$

19. $(x^5 - 3x^4 + 7x) - (4x^5 - 3x^3 + 8x + 1)$

20. $(3v^4 - 6v + 5) + (25v^3 - 16v^3 + 3v^2)$

21. $(2t^4 + 3t^3 - 1) + (-9t^5 + 4t^2) + (4t^3 - 7t^2 - 3)$

22. $(-2x^3 + 4x^2 - 2x) - (7x^5 + 4x^3 - 1)$

23. $(4v + 6) - (9v^2 - 13v)$

24. $(-16s^2 - 3s + 5) + (9s^2 + 3s + 5)$

25. $(-21t^2 + 21t + 21) + (9t^3 + 2)$

26. $(x^2 + 3) - (3x^2 + x - 10) + (2x^3 - 3x)$

27. $s(2s + 1)$

28. $-v(3v - 4)$

29. $3z(-6z^2 - 5)$

30. $5u(4u^2 - 10)$

31. $-t(-7t^2 + 3t + 9)$

32. $-5z(9z^2 - 2z - 4)$

33. $7z^2(z^2 + 9z - 8)$

34. $-6v^2(5v^2 - 3v + 7)$

35. $(y + 6)(y + 5)$

36. $(x - 4)(x + 7)$

37. $(-v - 12)(v - 3)$

38. $(x + 8)(x - 3)$

39. $(t + 8)(7t - 4)$

40. $(2x + 5)(x + 3)$

41. $(5 + 4v)(-7v - 6)$

42. $(-12 + 6z)(3z - 7)$

43. $(u^2 - 9)(u + 3)$

44. $(s^2 + 5)(s - 1)$

45. $(x + 4)^2$

46. $(t - 5)^2$

47. $(s + 6)^2$

48. $(-v + 3)^2$

49. $(5t + 4)^2$

50. $(7x + 1)^2$

51. $(6v - 3)^2$

52. $(4x + 5)^2$

53. $(3z - 1)^2$

54. $(2y + 1)^2$

55. $(6 - 5t)^2$

56. $(-9 + 2u)^2$

57. $(v + 9)(v - 9)$

58. $(z - 7)(z + 7)$

59. $(9s + 7)(7 - 9s)$

60. $(-6 + 5t)(-5t - 6)$

61. $(v^2 + 3)(v^2 - 3)$

62. $(7 - z^2)(7 + z^2)$

63. $(5y^2 - 4)(5y^2 + 4)$

64. $(2x^2 + 3)(2x^2 - 3)$

65. $(4z^2 + 5)(4z^2 - 5)$

66. $(6 - 7u^2)(6 + 7u^2)$

67. $(x + 2z)(x - 2z)$

68. $(u - 3v)(u + 3v)$

69. $(-5s - 4t)(4t - 5s)$

70. $(6y - 7z)(6y + 7z)$

71. $(-t^2 - 5t + 1)(t + 6)$

72. $(v^2 + 3v - 7)(-v - 2)$

73. $(4z^2 + 3z + 5)(7z - 4)$

74. $(7u^2 + 6u - 11)(-3u + 2)$

75. $(x - 2)(x^2 + 3x - 7)$

76. $(u + 3)(6u^2 - 4u + 5)$

77. $(5u^3 - 6u^2 - 7u + 9)(-4u + 7)$

78. $(-8v^3 + 7v^2 + 5v - 4)(3v - 9)$

▶**Applications** In this set of exercises, you will use polynomials to study real-world problems.

79. **Home Improvement** The amount of paint needed to cover the walls of a bedroom is $132x$, where x is the thickness of the coat of paint. The amount of paint of the same thickness that is needed to cover the walls of the den is $108x$. How much more paint is needed for the bedroom than for the den? Express your answer as a monomial in x.

80. **Geometry** Two circles have a common center. Let r denote the radius of the smaller circle. What is the area of the region between the two circles if the area of the larger circle is $9\pi r^2$ and the area of the smaller circle is πr^2? Express your answer as a monomial in r.

81. **Geometry** The *perimeter* of a square is the sum of the lengths of all four sides.
    (a) If one side is of length s, find the perimeter of the square in terms of s.
    (b) If each side of the square in part (a) is doubled, find the perimeter of the new square.

82. **Shopping** At the Jolly Ox, a gallon of milk sells for $3.30 and apples go for $0.49 per pound. Suppose Tania bought x gallons of milk and y pounds of apples.
    (a) How much did she spend altogether (in dollars)? Express your answer as a binomial in x and y.
    (b) If Tania gave the cashier a $20 bill, how much would she receive in change? Express your answer in terms of x and y, and assume Tania's purchases do not exceed $20.

83. **Investment** Suppose an investment of $1000 is worth $1000(1 + r)^2$ after 2 years, where r is the interest rate. Assume that no additional deposits or withdrawals are made.
    (a) Write $1000(1 + r)^2$ as a polynomial in descending order.
    (b) If the interest rate is 5%, use a calculator to determine how much the $1000 investment is worth after 2 years. (In the formula $1000(1 + r)^2$, r is assumed to be in decimal form.)

84. **Investment** Suppose an investment of $500 is worth $500(1 + r)^3$ after 3 years, where r is the interest rate. Assume that no additional deposits or withdrawals are made.
    (a) Write $500(1 + r)^3$ as a polynomial in descending order.
    (b) If the interest rate is 4%, use a calculator to determine how much the $500 investment is worth after 3 years. (In the formula $500(1 + r)^3$, r is assumed to be in decimal form.)

▶ **Concepts** This set of exercises will draw on the ideas presented in this section and your general math background.

85. Think of an integer x. Subtract 3 from it, and then square the result. Subtract 9 from *that* number, and then add six times your original integer to the result. Show algebraically that your answer is x^2.

86. What is the coefficient of the y^2 term in the sum of the polynomials $6y^4 - 2y^3 + 4y^2 - 7$ and $5y^3 - 4y^2 + 3$?

87. If two polynomials of degree 3 are added, is their sum necessarily a polynomial of degree 3? Explain.

88. A student writes the following on an exam: $(x + 2)^2 = x^2 + 4$. Explain the student's error and give the correct answer for the simplification of $(x + 2)^2$.

89. What is the constant term in the product of $5x^2 - 3x + 2$ and $6x^2 - 9x$?

90. If a polynomial of degree 2 is multiplied by a polynomial of degree 3, what is the degree of their product?

91. For what value(s) of a is $-8x^3 + 5x^2 + ax$ a binomial?

P.5 Factoring

Objectives

▶ Factor by grouping

▶ Factor trinomials

▶ Factor differences of squares and perfect square trinomials

▶ Factor sums and differences of cubes

The process of factoring a polynomial reverses the process of multiplication. That is, we find *factors* that can be multiplied together to produce the original polynomial expression. Factoring skills are of great importance in understanding the quadratic, polynomial, and rational functions discussed in Chapters 3 and 4. This section reviews important factoring strategies.

Common Factors

When factoring, the first step is to look for the *greatest common factor* in all the terms of the polynomial and then factor it out using the distributive property.

The **greatest common factor** is a monomial whose constant part is an integer with the largest absolute value common to all terms. Its variable part is the variable with the largest exponent common to all terms.

Example 1 Factoring the Greatest Common Factor

Factor the greatest common factor from each of the following.

(a) $3x^4 + 9x^3 + 18x^2$

(b) $-8y^2 - 6y + 4$

▶Solution

(a) Because $3x^2$ is common to all the terms, we can use the distributive property to write

$$3x^4 + 9x^3 + 18x^2 = 3x^2(x^2 + 3x + 6).$$

Inside the parentheses, there are no further factors common to all the terms. Therefore, $3x^2$ is the greatest common factor of all the terms in the polynomial.

To check the factoring, multiply $3x^2(x^2 + 3x + 6)$ to see that it gives the original polynomial expression.

(b) Because 2 is common to all the terms, we have

$$-8y^2 - 6y + 4 = 2(-4y^2 - 3y + 2) \quad \text{or} \quad -2(4y^2 + 3y - 2).$$

There are no variable terms to factor out. You can check that the factoring is correct by multiplying. Also note there can be more than one way to factor.

☑ *Check It Out 1:* Factor the greatest common factor from $5y + 10y^2 - 25y^3$. ■

Factoring by Grouping

Suppose we have an expression of the form $pA + qA$, where p, q, and A can be any expression. Using the distributive property, we can write

$$pA + qA = (p + q)A \quad \text{or} \quad A(p + q).$$

This is the key to a technique called **factoring by grouping.**

Example 2 Factoring by Grouping

Factor $x^3 - x^2 + 2x - 2$ by grouping.

▶Solution Group the terms as follows.

$$x^3 - x^2 + 2x - 2 = (x^3 - x^2) + (2x - 2)$$
$$= x^2(x - 1) + 2(x - 1) \qquad \text{Common factor in both groups is } (x - 1)$$

Using the distributive property to factor out the term $(x - 1)$, we have

$$= (x - 1)(x^2 + 2).$$

☑ *Check It Out 2:* Factor $x^2 + 3x + 4x + 12$ by grouping. ■

When using factoring by grouping, it is essential to group together terms with the *same* common factor. Not all polynomials can be factored by grouping.

Note *Factoring does not change an expression; it simply puts it in a different form. The factored form of a polynomial is quite useful when solving equations and when graphing quadratic and other polynomial functions.*

Factoring Trinomials of the Form $ax^2 + bx + c$

One of the most common factoring problems involves trinomials of the form $ax^2 + bx + c$. In such problems, we assume that a, b, and c have no common factors other than 1 or -1. If they do, simply factor out the greatest common factor first. We will consider two methods for factoring these types of trinomials; use whichever method you are comfortable with.

Method 1: The FOIL Method   The first method simply reverses the FOIL method for multiplying polynomials. We try to find integers P, Q, R, and S such that

$$(Px + Q)(Rx + S) = \underbrace{PR}_{\text{First}} x^2 + \overbrace{(PS + QR)}^{\text{Outer + Inner}}x + \underbrace{QS}_{\text{Last}} = ax^2 + bx + c.$$

We see that $PR = a$, $PS + QR = b$, and $QS = c$. That is, we find factors of a and factors of c and choose only those factor combinations for which the sum of the inner and outer terms adds to bx. This method is illustrated in Example 3.

Example 3 Factoring Trinomials

Factor each of the following.

(a) $x^2 - 2x - 8$

(b) $8x^3 - 10x^2 - 12x$

▶Solution

(a) First, note that there is no common factor to factor out. The factors of 1 are ± 1. The factors of -8 are ± 1, ± 2, ± 4, ± 8. Since $a = 1$, we must have a factorization of the form

$$x^2 - 2x - 8 = (x + \square)(x + \square)$$

where the numbers in the boxes are yet to be determined. The factors of $c = -8$ must be chosen so that the coefficient of x in the product is -2. Since 2 and -4 satisfy this condition, we have

$$x^2 - 2x - 8 = (x + 2)(x + (-4)) = (x + 2)(x - 4).$$

You can multiply the factors to check your answer.

(b) Factor out the greatest common factor $2x$ to get

$$8x^3 - 10x^2 - 12x = 2x(4x^2 - 5x - 6).$$

The expression in parentheses is a trinomial of the form $ax^2 + bx + c$. We wish to factor it as follows:

$$4x^2 - 5x - 6 = (\square x + \triangle)(\square x + \triangle).$$

The factors of $a = 4$ are placed in the boxes and the factors of $c = -6$ are placed in the triangles.

$a = 4$	Factors: $\pm 1, \pm 2, \pm 4$
$c = -6$	Factors: $\pm 1, \pm 2, \pm 3, \pm 6$

Find a pair of factors each for a and c such that the middle term of the trinomial is $-5x$. Note also that the two factors of -6 must be opposite in sign. We try different possibilities until we get the correct result.

Binomial Factors	ax^2	bx	c
$(2x + 3)(2x - 2)$	$4x^2$	$2x$	-6
$(4x + 1)(x - 6)$	$4x^2$	$-23x$	-6
$(4x - 3)(x + 2)$	$4x^2$	$5x$	-6
$(4x + 3)(x - 2)$	$4x^2$	$-5x$	-6

The last factorization is the correct one. Thus,

$$8x^3 - 10x^2 - 12x = 2x(4x^2 - 5x - 6) = 2x(4x + 3)(x - 2).$$

You should multiply out the factors to check your answer.

✔ *Check It Out 3:* Factor: $2x^2 + 8x - 10$. ■

Discover *and* Learn

Can $x^2 - 1$ be factored as $(x - 1)(x - 1)$? Explain.

Method 2: Factoring by Grouping Another method for factoring trinomials of the form $ax^2 + bx + c$ uses the technique of grouping. This method is also known as the *ac method*. The idea is to rewrite the original expression in a form suitable for factoring by grouping, discussed in the earlier part of this section. The procedure is as follows.

The Grouping Method for Factoring $ax^2 + bx + c$

In the following steps, we assume that the only common factors among a, b, and c are ± 1. If that's not the case, simply factor out the greatest common factor first.

Step 1 Form the product ac, where a is the coefficient of x^2 and c is the constant term.

Step 2 Find two integers p and q such that $pq = ac$ and $p + q = b$.

Step 3 Using the numbers p and q found in step 2, split the middle term bx into a sum of two like terms, using p and q as coefficients: $bx = px + qx$.

Step 4 Factor the expression from step 3, $ax^2 + px + qx + c$, by grouping.

Example 4 illustrates this method.

Example 4 Factoring Trinomials by Grouping

Factor $6x^3 - 3x^2 - 45x$.

▶Solution First factor out the greatest common factor of $3x$ to get

$$6x^3 - 3x^2 - 45x = 3x(2x^2 - x - 15).$$

Now factor the expression $2x^2 - x - 15$ with $a = 2$, $b = -1$, and $c = -15$. There are no common factors among a, b, and c other than ± 1.

Step 1 The product ac is $ac = (2)(-15) = -30$.

Step 2 Make a list of pairs of integers whose product is $ac = -30$, and find their sum. Then select the pair that adds to $b = -1$.

Factors of -30	Sum
$1, -30$	-29
$-1, 30$	29
$2, -15$	-13
$-2, 15$	13
$3, -10$	-7
$-3, 10$	7
$5, -6$	-1
$-5, 6$	1

We see that $p = 5$ and $q = -6$ satisfy $pq = -30$ and $p + q = -1$.

Step 3 Split the middle term, $-x$, into a sum of two like terms using $p = 5$ and $q = -6$.

$$-x = 5x - 6x$$

Step 4 Factor the resulting expression by grouping.

$$2x^2 - x - 15 = 2x^2 + 5x - 6x - 15$$
$$= x(2x + 5) - 3(2x + 5)$$
$$= (2x + 5)(x - 3)$$

Thus, the factorization of the original trinomial is

$$6x^3 - 3x^2 - 45x = 3x(2x^2 - x - 15) = 3x(2x + 5)(x - 3).$$

You should check the factorization by multiplying out the factors.

✔ *Check It Out 4:* Factor using the grouping method: $2x^2 + x - 6$. ■

Special Factorization Patterns

One of the most efficient ways to factor is to remember the special factorization patterns that occur frequently. We have categorized them into two groups—quadratic

Just in Time

Review special products in Section P.4.

factoring patterns and cubic factoring patterns. The quadratic factoring patterns follow directly from the special products of binomials mentioned in Section P.4. Tables P.5.1 and P.5.2 list the quadratic and cubic factoring patterns, respectively.

Table P.5.1 Quadratic Factoring Patterns

QUADRATIC FACTORING PATTERN	ILLUSTRATION
Difference of squares $A^2 - B^2 = (A + B)(A - B)$	$9x^2 - 5 = (3x)^2 - (\sqrt{5})^2$ $\quad = (3x + \sqrt{5})(3x - \sqrt{5})$ where $A = 3x$ and $B = \sqrt{5}$.
Perfect square trinomial $A^2 + 2AB + B^2 = (A + B)^2$	$25t^2 + 30t + 9 = (5t)^2 + 2(5t)(3) + 3^2$ $\quad = (5t + 3)^2$ where $A = 5t$ and $B = 3$.
Perfect square trinomial $A^2 - 2AB + B^2 = (A - B)^2$	$16s^2 - 8s + 1 = (4s)^2 - 2(4s)(1) + 1^2$ $\quad = (4s - 1)^2$ where $A = 4s$ and $B = 1$.

Table P.5.2 Cubic Factoring Patterns

CUBIC FACTORING PATTERN	ILLUSTRATION
Difference of cubes $A^3 - B^3 = (A - B)(A^2 + AB + B^2)$	$y^3 - 27 = y^3 - 3^3$ $\quad = (y - 3)(y^2 + 3y + 3^2)$ $\quad = (y - 3)(y^2 + 3y + 9)$ where $A = y$ and $B = 3$.
Sum of cubes $A^3 + B^3 = (A + B)(A^2 - AB + B^2)$	$8x^3 + 125 = (2x)^3 + 5^3$ $\quad = (2x + 5)((2x)^2 - 2x(5) + 5^2)$ $\quad = (2x + 5)(4x^2 - 10x + 25)$ where $A = 2x$ and $B = 5$.

Example 5 Special Factorization Patterns

Factor using one of the special factorization patterns.

(a) $27x^3 - 64$

(b) $8x^2 + 32x + 32$

▶**Solution**

(a) Because $27x^3 - 64 = (3x)^3 - 4^3$, we can use the formula for the difference of cubes.

$$27x^3 - 64 = (3x)^3 - 4^3 \qquad \text{Use } A = 3x \text{ and } B = 4$$
$$= (3x - 4)((3x)^2 + (3x)(4) + 4^2)$$
$$= (3x - 4)(9x^2 + 12x + 16)$$

(b) $8x^2 + 32x + 32 = 8(x^2 + 4x + 4)$ Factor out 8

$$= 8(x + 2)^2 \qquad \text{Perfect square trinomial with } A = x \text{ and } B = 2$$

✔ *Check It Out 5:* Factor $4y^2 - 100$ using a special factorization pattern. ■

Note Not all polynomial expressions can be factored using the techniques covered in this section. A more detailed study of the factorization of polynomials is given in Chapter 4.

P.5 Key Points

▶ A trinominal can be factored either by reversing the FOIL method of multiplication or by the grouping method. If they are of a special form, polynomials also can be factored using quadratic and cubic factoring patterns.

▶ **Quadratic factoring patterns**
$$A^2 - B^2 = (A + B)(A - B)$$
$$A^2 + 2AB + B^2 = (A + B)^2$$
$$A^2 - 2AB + B^2 = (A - B)^2$$

▶ **Cubic factoring patterns**
$$A^3 - B^3 = (A - B)(A^2 + AB + B^2)$$
$$A^3 + B^3 = (A + B)(A^2 - AB + B^2)$$

P.5 Exercises

▶**Just in Time Exercises** These exercises correspond to the Just in Time references in this section. Complete them to review topics relevant to the remaining exercises.

In Exercises 1–6, simplify the expression.

1. $(x + 6)^2$

2. $(3x - 2)^2$

3. $2(6x + 7)^2$

4. $(u + 7)(u - 7)$

5. $(3y + 10)(3y - 10)$

6. $3(6t + 5)(6t - 5)$

▶**Skills** This set of exercises will reinforce the skills illustrated in this section.

In Exercises 7–14, factor the greatest common factor from each expression.

7. $2x^3 + 6x^2 - 8x$

8. $4x^4 - 8x^3 + 12$

9. $-3y^3 + 6y - 9$

10. $y^4 + 2y^2 + 5y$

11. $-2t^6 - 4t^5 + 10t^2$

12. $12x^5 - 6x^3 - 18x^2$

13. $-5x^7 + 10x^5 - 15x^3$

14. $-14z^5 + 7z^3 + 28$

In Exercises 15–20, factor each expression by grouping.

15. $3(x + 1) + x(x + 1)$

16. $x(x - 2) + 4(x - 2)$

17. $s^3 - 5s^2 - 9s + 45$

18. $-27v^3 - 36v^2 + 3v + 4$

19. $12u^3 + 4u^2 - 3u - 1$

20. $75t^3 + 25t^2 - 12t - 4$

In Exercises 21–34, factor each trinomial.

21. $x^2 + 4x + 3$

22. $x^2 + 2x - 35$

23. $x^2 - 6x - 16$

24. $x^2 - 10x + 24$

25. $3s^2 + 15s + 12$

26. $4y^2 - 20y + 24$

27. $-6t^2 + 24t + 72$

28. $9u^2 - 27u + 18$

29. $-5z^2 - 20z + 60$

30. $2x^2 - 4x + 6$

31. $3x^2 - 5x - 12$

32. $2x^2 - 7x + 6$

33. $4z^2 - 23z - 6$

34. $6z^2 + 11z - 10$

In Exercises 35–60, factor each polynomial using one of the special factorization patterns.

35. $x^2 - 16$

36. $t^2 - 25$

37. $9x^2 - 4$

38. $4y^2 - 25$

39. $y^2 - 3$

40. $16x^2 - 7$

41. $3x^2 - 12$

42. $5x^2 - 5$

43. $x^2 + 6x + 9$

44. $x^2 + 24x + 144$

45. $y^2 - 14y + 49$

46. $y^2 - 26y + 169$

47. $4x^2 + 4x + 1$

48. $9x^2 + 12x + 4$

49. $9x^2 - 6x + 1$

50. $4x^2 - 20x + 25$

51. $12x^2 + 12x + 3$

52. $50x^2 - 60x + 18$

53. $y^3 + 64$

54. $t^3 + 1$

55. $u^3 - 125$

56. $8x^3 - 27$

57. $2x^3 - 16$

58. $24t^3 + 3$

59. $8y^3 + 1$

60. $64x^3 - 8$

In Exercises 61–114, factor each expression completely, using any of the methods from this section.

61. $z^2 + 13z + 42$

62. $z^2 + z - 30$

63. $x^2 + 12x + 36$

64. $x^2 + 8x + 16$

65. $-y^2 + 4y - 4$

66. $-y^2 - 6y - 9$

67. $z^2 - 16z + 64$

68. $z^2 - 8z + 16$

69. $-2y^2 + 7y - 3$

70. $-3y^2 - 2y + 8$

71. $9y^2 + 12y + 4$

72. $4x^2 + 12x + 9$

73. $9z^2 - 6z + 1$

74. $-12x^2 + 5x + 2$

75. $4z^2 - 20z + 25$

76. $18y^2 + 43y - 5$

77. $6z^2 - 3z - 18$

78. $8v^2 + 20v - 12$

79. $-15t^2 - 70t + 25$

80. $14y^2 - 7y - 21$

81. $-10u^2 - 45u - 20$

82. $-6x^2 + 27x - 30$

83. $-s^2 + 49$

84. $y^2 - 9$

85. $v^2 - 4$

86. $-u^2 + 36$

87. $-25t^2 + 4$

88. $-49v^2 + 16$

89. $9z^2 - 1$

90. $-16s^2 + 9$

91. $t^3 - 16t^2$

92. $x^3 - 9x^2$

93. $12u^3 + 4u^2 - 40u$

94. $6x^3 - 15x^2 + 9x$

95. $-10t^3 + 5t^2 + 15t$

96. $-8y^3 - 44y^2 - 20y$

97. $-15z^3 - 5z^2 + 20z$

98. $6x^3 + 14x^2 - 12x$

99. $2y^3 + 3y^2 - 8y - 12$

100. $-18z^3 + 27z^2 + 32z - 48$

101. $4x^4 + 20x^3 + 24x^2$

102. $-10s^4 - 25s^3 + 15s^2$

103. $3y^4 + 18y^3 + 24y^2$

104. $21v^4 - 28v^3 + 7v^2$

105. $-x^4 + x^3 + 6x^2$

106. $2y^4 + 4y^3 - 16y^2$

107. $7x^5 - 63x^3$

108. $-6s^5 - 30s^3$

109. $5y^5 - 20y^3$

110. $15u^5 + 18u^3$

111. $8x^3 + 64$

112. $27x^3 + 1$

113. $-8y^3 + 1$

114. $-64z^3 + 27$

▶**Concepts** This set of exercises will draw on the ideas presented in this section and your general math background.

115. Give an example of a monomial that can be factored into two polynomials, each of degree 2. Then factor the monomial accordingly.

116. Is $(x^2 - 4)(x + 5)$ completely factored? Explain.

117. Give an example of a polynomial of degree 2 that can be expressed as the square of a binomial, and then express it as such.

118. Can $y^2 + a^2$ be factored as $(y + a)^2$? Explain.

119. Express $16x^4 - 81$ as the product of three binomials.

120. Find one value of a such that the expressions $x^3 - a^3$ and $(x - a)^3$ are equal (for all real numbers x).

P.6 Rational Expressions

Objectives

▶ Simplify a rational expression

▶ Multiply and divide rational expressions

▶ Add and subtract rational expressions

▶ Simplify complex fractions

A quotient of two polynomial expressions is called a **rational expression.** A rational expression is defined whenever the denominator is not equal to zero.

Example 1 Values for Which a Rational Expression is Defined

For what values of x is the following rational expression defined?

$$\frac{x+1}{(x-3)(x-5)}$$

▶Solution The rational expression is defined only when the denominator is *not* zero. This happens whenever

$$x - 3 \neq 0 \implies x \neq 3 \quad \text{or} \quad x - 5 \neq 0 \implies x \neq 5.$$

Thus, the rational expression is defined whenever x is *not* equal to 3 or 5. We can also say that 3 and 5 are *excluded values* of x.

✔ *Check It Out 1:* For what values of x is the rational expression $\frac{x}{x^2-1}$ defined? ▪

> **Note** Throughout this section, we will assume that any rational expression is meaningful only for values of the variables that are not excluded. Unless they are specifically stated in a given example or problem, we will not list excluded values.

Simplifying Rational Expressions

Recall that if you have a fraction such as $\frac{4}{12}$, you simplify it by first factoring the numerator and denominator and then dividing out the common factors:

$$\frac{4}{12} = \frac{2 \cdot 2}{2 \cdot 2 \cdot 3} = \frac{1}{3}.$$

When simplifying rational expressions containing variables, you factor polynomials instead of numbers. Familiarity with the many factoring techniques is the most important tool in manipulating rational expressions.

Just In Time

Review factoring in Section P.5.

Example 2 Simplifying a Rational Expression

Simplify: $\dfrac{x^2 - 2x + 1}{1 - 4x + 3x^2}$.

▶Solution

$$\frac{x^2 - 2x + 1}{1 - 4x + 3x^2} = \frac{(x-1)(x-1)}{(1-x)(1-3x)} \qquad \text{Factor completely}$$

$$= \frac{(x-1)(x-1)}{-(x-1)(1-3x)} \qquad 1-x = -(x-1)$$

$$= \frac{x-1}{-(1-3x)} \qquad \text{Divide out } x-1, \text{ a common factor}$$

$$= \frac{x-1}{3x-1} \qquad -(1-3x) = 3x-1$$

✔ *Check It Out 2:* Simplify: $\dfrac{x^2 - 4}{x^2 + 5x + 6}$. ▪

Next we discuss the arithmetic of rational expressions, which is very similar to the arithmetic of rational numbers.

Multiplication and Division of Rational Expressions

Multiplication of rational expressions is straightforward. You multiply the numerators, multiply the denominators, and then simplify your answer.

Example 3 Multiplication of Rational Expressions

Multiply the following rational expressions and express your answers in lowest terms. For what values of the variable is the expression meaningful?

(a) $\dfrac{3a}{8} \cdot \dfrac{24}{6a^3}$

(b) $\dfrac{x^2 + x - 6}{x^2 - 4} \cdot \dfrac{(x + 2)^2}{x^2 + 9}$

▶Solution

(a) $\dfrac{3a}{8} \cdot \dfrac{24}{6a^3} = \dfrac{(3a)(24)}{(8)(6a^3)}$

$= \dfrac{3 \cdot a \cdot 6 \cdot 4}{4 \cdot 2 \cdot 6 \cdot a^3}$ Factor and divide out common factors

$= \dfrac{3}{2a^2}$

The expression is meaningful for $a \neq 0$.

(b) $\dfrac{x^2 + x - 6}{x^2 - 4} \cdot \dfrac{(x + 2)^2}{x^2 + 9} = \dfrac{(x^2 + x - 6)(x + 2)^2}{(x^2 - 4)(x^2 + 9)}$

$= \dfrac{(x + 3)(x - 2)(x + 2)^2}{(x + 2)(x - 2)(x^2 + 9)}$ Factor

$= \dfrac{(x + 3)(x + 2)}{x^2 + 9}$ Divide out common factors

The expression is meaningful for $x \neq 2, -2$ because $x^2 - 4 = 0$ for $x = 2, -2$. Observe that $x^2 + 9$ cannot be factored further using real numbers, and is never equal to zero.

✔ *Check It Out 3:* Multiply and simplify: $\dfrac{x^2 + 2x + 1}{x^2 - 4} \cdot \dfrac{x^2 + 4x + 4}{x + 1}$. ■

When dividing two rational expressions, the expression following the division symbol is called the **divisor.** To divide rational expressions, multiply the first expression by the reciprocal of the divisor.

Example 4 Dividing Rational Expressions

Divide and simplify: $\dfrac{3x^2 - 5x - 2}{x^2 - 4x + 4} \div \dfrac{9x^2 - 1}{x + 5}$.

▶Solution Taking the reciprocal of the divisor and multiplying, we have

$$\frac{3x^2 - 5x - 2}{x^2 - 4x + 4} \div \frac{9x^2 - 1}{x + 5} = \frac{3x^2 - 5x - 2}{x^2 - 4x + 4} \cdot \frac{x + 5}{9x^2 - 1}.$$

Factor, divide out common factors, and multiply to get

$$= \frac{(3x + 1)(x - 2)}{(x - 2)^2} \cdot \frac{x + 5}{(3x + 1)(3x - 1)}$$

$$= \frac{x + 5}{(x - 2)(3x - 1)}.$$

☑ *Check It Out 4:* Divide and simplify: $\dfrac{7x + 14}{x^2 - 4} \div \dfrac{7x}{x^2 + x - 6}$. ■

Addition and Subtraction of Rational Expressions

To add and subtract rational expressions, follow the same procedure used for adding and subtracting rational numbers. Before we can add rational expressions, we must write them in terms of the same denominator, known as the **least common denominator.** For instance, to compute $\dfrac{1}{4} + \dfrac{1}{6}$, we find the least common multiple of 4 and 6, which is 12. The number $12 = 2 \cdot 2 \cdot 3$ is the smallest number whose factors include the factors of 4, which are 2 and 2, *and* the factors of 6, which are 2 and 3.

> **Definition of the Least Common Denominator**
>
> The **least common denominator (LCD)** of a set of rational expressions is the simplest expression that includes all the factors of each of the denominators.

Example 5 Adding and Subtracting Rational Expressions

Add or subtract the following expressions. Express your answers in lowest terms.

(a) $\dfrac{x + 4}{3x + 6} + \dfrac{2x + 1}{x^2 + 7x + 10}$ (b) $\dfrac{3x}{4 - 2x} - \dfrac{x + 5}{x^2 - 4}$

▶Solution

(a) Factor the denominators and find the least common denominator.

$$\frac{x + 4}{3x + 6} + \frac{2x + 1}{x^2 + 7x + 10} = \frac{x + 4}{3(x + 2)} + \frac{2x + 1}{(x + 5)(x + 2)}$$

The LCD is $3(x + 2)(x + 5)$. Write both expressions as equivalent rational expressions using the LCD.

$$= \frac{x + 4}{3(x + 2)} \cdot \frac{x + 5}{x + 5} + \frac{2x + 1}{(x + 5)(x + 2)} \cdot \frac{3}{3}$$

Simplify the numerators and add the two fractions.

$$= \frac{x^2 + 9x + 20}{3(x + 2)(x + 5)} + \frac{6x + 3}{3(x + 2)(x + 5)} = \frac{x^2 + 15x + 23}{3(x + 2)(x + 5)}$$

The expression cannot be simplified further.

(b) Factor the denominators and find the least common denominator.

$$\frac{3x}{4-2x} - \frac{x+5}{x^2-4} = \frac{3x}{2(2-x)} - \frac{x+5}{(x+2)(x-2)} = \frac{3x}{-2(x-2)} - \frac{x+5}{(x+2)(x-2)}$$

The LCD is $-2(x-2)(x+2)$. Note that $2(2-x) = -2(x-2)$. Write both expressions as equivalent rational expressions using the LCD.

$$= \frac{3x}{-2(x-2)} \cdot \frac{x+2}{x+2} - \frac{x+5}{(x+2)(x-2)} \cdot \frac{-2}{-2}$$

$$= \frac{3x^2+6x}{-2(x-2)(x+2)} - \frac{-2x-10}{-2(x-2)(x+2)} \qquad \text{Simplify the numerators}$$

$$= \frac{(3x^2+6x)-(-2x-10)}{-2(x-2)(x+2)} = \frac{3x^2+8x+10}{-2(x-2)(x+2)} \qquad \text{Subtract, taking care to distribute the minus sign}$$

The expression cannot be simplified further.

✔ *Check It Out 5:* Subtract and express your answer in lowest terms: $\dfrac{-2}{x+2} - \dfrac{3}{x^2-4}$.

> **Note** When adding or subtracting rational expressions, you factor polynomials instead of numbers to find the least common denominator.

Complex Fractions

A **complex fraction** is one in which the numerator and/or denominator of the fraction contains a rational expression. Complex fractions are also commonly referred to as **complex rational expressions.**

Example 6 **Simplifying a Complex Fraction**

Simplify: $\dfrac{\dfrac{x^2-4}{2x+1}}{\dfrac{x^2+x-6}{x-1}}$.

▶Solution Because we have a quotient of two rational expressions, we can write

$$\frac{\dfrac{x^2-4}{2x+1}}{\dfrac{x^2+x-6}{x-1}} = \frac{x^2-4}{2x+1} \div \frac{x^2+x-6}{x-1}$$

$$= \frac{x^2-4}{2x+1} \cdot \frac{x-1}{x^2+x-6}$$

$$= \frac{(x+2)(x-2)}{2x+1} \cdot \frac{x-1}{(x+3)(x-2)} \qquad \text{Factor}$$

$$= \frac{(x+2)(x-1)}{(2x+1)(x+3)}. \qquad \text{Cancel } (x-2) \text{ term}$$

There are no more common factors, so the expression is simplified.

✔ *Check It Out 6:* Simplify: $\dfrac{\dfrac{x+y}{y}}{\dfrac{x^2-y^2}{x}}$. ■

Another way to simplify a complex fraction is to multiply the numerator and denominator by the least common denominator of all the denominators.

Example 7 Simplifying a Complex Fraction

Simplify: $\dfrac{\dfrac{1}{x} + \dfrac{1}{xy}}{\dfrac{3}{y^2} + \dfrac{1}{y}}$.

▶**Solution** First find the LCD of the four rational expressions. The denominators are

$$x, \ xy, \ y^2, \text{ and } y.$$

Thus, the LCD is xy^2. We then can write

$$\dfrac{\dfrac{1}{x} + \dfrac{1}{xy}}{\dfrac{3}{y^2} + \dfrac{1}{y}} = \dfrac{\dfrac{1}{x} + \dfrac{1}{xy}}{\dfrac{3}{y^2} + \dfrac{1}{y}} \cdot \dfrac{xy^2}{xy^2}$$

$$= \dfrac{\left(\dfrac{1}{x} + \dfrac{1}{xy}\right)xy^2}{\left(\dfrac{3}{y^2} + \dfrac{1}{y}\right)xy^2}$$

$$= \dfrac{\dfrac{1}{x}(xy^2) + \dfrac{1}{xy}(xy^2)}{\dfrac{3}{y^2}(xy^2) + \dfrac{1}{y}(xy^2)} \qquad \text{Distribute } xy^2$$

$$= \dfrac{y^2 + y}{3x + xy} \qquad \text{Simplify each term}$$

$$= \dfrac{y(y+1)}{x(3+y)}. \qquad \text{Factor to see if any common factors can be removed}$$

There are no common factors, so the expression is simplified.

✔ *Check It Out 7:* Simplify: $\dfrac{\dfrac{2}{x} + \dfrac{1}{xy}}{\dfrac{1}{y} - \dfrac{2}{x}}$. ■

P.6 Key Points

▶ A quotient of two polynomial expressions is called a **rational expression.** A rational expression is defined whenever the denominator is not equal to zero.

▶ To find the **product of two rational expressions,** multiply the numerators, multiply the denominators, and then simplify the answer.

▶ To add and subtract rational expressions, first write them in terms of the same denominator, known as the **least common denominator (LCD).**

▶ To simplify a complex fraction, multiply the numerator and denominator by the LCD of all the denominators in the expression.

P.6 Exercises

▶**Just in Time Exercises** These exercises correspond to the Just in Time references in this section. Complete them to review topics relevant to the remaining exercises.

In Exercises 1–6, factor.

1. $2x^2 - 14x$

2. $y^2 + y - 2$

3. $x^2 - 81$

4. $4y^2 - 400$

5. $-x^2 + 6x - 9$

6. $v^3 - 27$

▶**Skills** This set of exercises will reinforce the skills illustrated in this section.

In Exercises 7–16, simplify each rational expression and indicate the values of the variable for which the expression is defined.

7. $\dfrac{57}{24}$

8. $\dfrac{56}{49}$

9. $\dfrac{x^2 - 4}{6(x + 2)}$

10. $\dfrac{3(x - 3)}{x^2 - 9}$

11. $\dfrac{x^2 - x - 6}{x^2 - 9}$

12. $\dfrac{z^2 - 1}{z^2 + 2z + 1}$

13. $\dfrac{x^4 - x^2}{x + 1}$

14. $\dfrac{y^3 - y}{y - 1}$

15. $\dfrac{x^3 - 1}{x^2 - 1}$

16. $\dfrac{y^3 + 8}{y^2 - 4}$

In Exercises 17–32, multiply or divide. Express your answer in lowest terms.

17. $\dfrac{3x}{6y^2} \cdot \dfrac{2xy}{x^3}$

18. $\dfrac{x^2y}{2y^2} \cdot \dfrac{4y^4}{x^3}$

19. $\dfrac{x + 2}{x^2 - 9} \cdot \dfrac{x + 3}{x^2 + 4x + 4}$

20. $\dfrac{x - 3}{x^2 - 2x + 1} \cdot \dfrac{x - 1}{2x - 6}$

21. $\dfrac{3x + 9}{x^2 + x - 6} \cdot \dfrac{2x - 4}{x + 6}$

22. $\dfrac{x - 3}{4x + 16} \cdot \dfrac{3x + 12}{x^2 - 5x + 6}$

23. $\dfrac{6x - 12}{3x^3 - 12x} \cdot \dfrac{x^2 - 4x + 4}{x^2 + 3x - 10}$

24. $\dfrac{4x^4 - 36x^2}{8x - 8} \cdot \dfrac{x^2 - 2x + 1}{x^2 + 2x - 15}$

25. $\dfrac{x^3 + 1}{x^2 - 1} \cdot \dfrac{2x^2 - x - 1}{x + 2}$

26. $\dfrac{a^3 - 1}{a^2 - 2a + 1} \cdot \dfrac{(a - 1)^2}{a^2 + a + 1}$

27. $\dfrac{5x - 20}{x^2 - 4x - 5} \div \dfrac{x^2 - 8x + 16}{x - 5}$

28. $\dfrac{3x - 6}{2x + 2} \div \dfrac{3x^2 - 5x - 2}{x^2 - 5x + 6}$

29. $\dfrac{6x^3 - 24x}{3x^2 - 3} \div \dfrac{2x^2 + 4x}{x^2 - 2x + 1}$

30. $\dfrac{5x^4 - 45x^2}{7x - 14} \div \dfrac{3x^2 + 9x}{x^2 + 3x - 10}$

31. $\dfrac{x^3 - 8}{2x^2 - 3x - 2} \div \dfrac{x^2 - 4}{2x + 1}$

32. $\dfrac{a^3 + 27}{a^2 - 1} \div \dfrac{a^2 + 6a + 9}{a^2 + 2a + 1}$

In Exercises 33–56, add or subtract. Express your answer in lowest terms.

33. $\dfrac{2}{x} + \dfrac{3}{x^2}$

34. $\dfrac{-3}{y^2} + \dfrac{4}{y}$

35. $\dfrac{3}{x} - \dfrac{4}{x^2}$

36. $\dfrac{-7}{x^2} - \dfrac{1}{x}$

37. $\dfrac{1}{x + 1} + \dfrac{4}{x - 1}$

38. $\dfrac{5}{x - 3} + \dfrac{6}{x + 2}$

39. $\dfrac{3}{x + 4} - \dfrac{3}{2x - 1}$

40. $\dfrac{-1}{x + 1} - \dfrac{3}{2x - 1}$

41. $\dfrac{4x}{x^2 - 9} + \dfrac{2x^2}{3x + 9}$

42. $\dfrac{3y}{2y + 4} + \dfrac{y^2}{y^2 - 4}$

43. $\dfrac{2z}{5z - 10} + \dfrac{z + 1}{z^2 - 4z + 4}$

44. $\dfrac{x + 2}{3x + 9} + \dfrac{3x}{x^2 - x - 12}$

45. $\dfrac{x}{x + 1} - \dfrac{x - 4}{x - 1}$

46. $\dfrac{x + 1}{x - 3} - \dfrac{6}{x + 2}$

47. $\dfrac{-3x}{x^2 - 16} - \dfrac{3x^2}{3x + 12}$

48. $\dfrac{5x}{x^2 - 16} - \dfrac{3x^2}{2x - 8}$

49. $\dfrac{z}{3z - 15} - \dfrac{z - 1}{z^2 - 10z + 25}$

50. $\dfrac{3x + 1}{2x + 4} - \dfrac{x - 1}{x^2 - x - 6}$

51. $\dfrac{3}{x - 1} + \dfrac{4}{1 - x}$

52. $\dfrac{6}{2x - 1} + \dfrac{4}{1 - 2x}$

53. $\dfrac{4}{x + 2} - \dfrac{2}{x - 2} + \dfrac{1}{x^2 - 4}$

54. $\dfrac{-1}{x - 1} + \dfrac{2}{x + 1} - \dfrac{3}{x^2 - 1}$

55. $\dfrac{7}{3 - x} - \dfrac{1}{x + 2} + \dfrac{4}{x^2 - x - 6}$

56. $\dfrac{3}{y - 4} + \dfrac{2}{y^2 - 5y + 4} + \dfrac{2}{1 - y}$

In Exercises 57–72, simplify each complex fraction.

57. $\dfrac{\dfrac{x + 1}{x}}{\dfrac{x^2 - 1}{x^2}}$

58. $\dfrac{\dfrac{a^2 - 1}{a}}{\dfrac{a - 1}{a^3}}$

59. $\dfrac{\dfrac{1}{x} + \dfrac{1}{y}}{\dfrac{1}{y^2} - \dfrac{2}{x}}$

60. $\dfrac{\dfrac{1}{y} - \dfrac{1}{x^2}}{\dfrac{1}{x} + \dfrac{2}{y}}$

61. $\dfrac{1}{\dfrac{1}{r} + \dfrac{1}{s} + \dfrac{1}{t}}$

62. $\dfrac{2}{\dfrac{1}{x^2} + \dfrac{1}{xy} + \dfrac{1}{y^2}}$

63. $\dfrac{1 + x^{-1}}{x^{-2} - 1}$

64. $\dfrac{a^{-1} + b^{-1}}{a + b}$

65. $\dfrac{\dfrac{1}{x - 1} - \dfrac{1}{x - 3}}{\dfrac{2}{x - 1} + \dfrac{3}{x + 1}}$

66. $\dfrac{\dfrac{2}{x - 2} + \dfrac{1}{x - 1}}{\dfrac{3}{x + 3} - \dfrac{1}{x - 2}}$

67. $\dfrac{\dfrac{1}{x + h} - \dfrac{1}{x}}{h}$

68. $\dfrac{\dfrac{1}{x} - \dfrac{1}{a}}{x - a}$

69. $\dfrac{\dfrac{2}{x^2 - 4} + \dfrac{1}{x - 2}}{\dfrac{4}{x + 2}}$

70. $\dfrac{\dfrac{3}{x^2 - 9} - \dfrac{1}{x + 3}}{\dfrac{2}{x - 3}}$

71. $\dfrac{\dfrac{a}{a^2 - b^2} + \dfrac{b}{a + b}}{\dfrac{1}{a - b}}$

72. $\dfrac{\dfrac{1}{a + b} + \dfrac{3b}{a^2 + 2ab + b^2}}{\dfrac{a}{a + b}}$

▶**Applications** In this set of exercises, you will use rational expressions to study real-world problems.

73. **Average Cost** The average cost per book for printing x booklets is $\dfrac{300 + 0.5x}{x}$. Evaluate this expression for $x = 100$, and interpret the result.

74. **Driving Speed** If it takes t hours to drive a distance of 400 miles, then the average driving speed is given by $\dfrac{400}{t}$. Evaluate this expression for $t = 8$, and interpret the result.

75. **Work Rate** One pump can fill a pool in 4 hours, and another can fill it in 3 hours. Working together, it takes the pumps $t = \dfrac{1}{\frac{1}{4} + \frac{1}{3}}$ hours to fill the pool. Find t.

76. **Physics** In an electrical circuit, if three resistors are connected in parallel, then their total resistance is given by

$$R = \dfrac{1}{\dfrac{1}{R_1} + \dfrac{1}{R_2} + \dfrac{1}{R_3}}.$$

Simplify the expression for R.

▶**Concepts** This set of exercises will draw on the ideas presented in this section and your general math background.

77. Find two numbers x and y such that $\dfrac{1}{x} + \dfrac{1}{y} \neq \dfrac{2}{x + y}$. (Answers may vary.)

78. The expression $\dfrac{x^2 - 1}{x + 1}$ simplifies to $x - 1$. What value(s) of x must be excluded when performing the simplification?

79. Does $\dfrac{x^2}{x} = x$ for all values of x? Explain.

80. In an answer to an exam question, $\dfrac{x^2 + 4}{x + 2}$ is simplified as $x + 2$. Is this correct? Explain.

P.7 Geometry Review

Objectives

▶ Know and apply area and perimeter formulas

▶ Know and apply volume and surface area formulas

▶ Know and apply the Pythagorean Theorem

In this section, we will review formulas for the perimeter, area, and volume of common figures that will be used throughout this textbook. We will also discuss the Pythagorean Theorem.

Formulas for Two-Dimensional Figures

Table P.7.1 gives formulas for the perimeter and area of common two-dimensional figures.

Table P.7.1. Formulas for Two-Dimensional Figures

Rectangle	Square	Circle	Triangle	Parallelogram
Length: l Width: w	Length: s Width: s	Radius: r	Base: b Height: h	Base: b Height: h Side: s
Perimeter $P = 2l + 2w$	Perimeter $P = 4s$	Circumference $C = 2\pi r$	Perimeter $P = a + b + c$	Perimeter $P = 2b + 2s$
Area $A = lw$	Area $A = s^2$	Area $A = \pi r^2$	Area $A = \dfrac{1}{2}bh$	Area $A = bh$

Example 1 gives an application of these formulas.

Example 1 Using Area Formulas

Figure P.7.1

|←——— 4 in. ———→|

Find the area of the figure shown in Figure P.7.1, which consists of a semicircle mounted on top of a square.

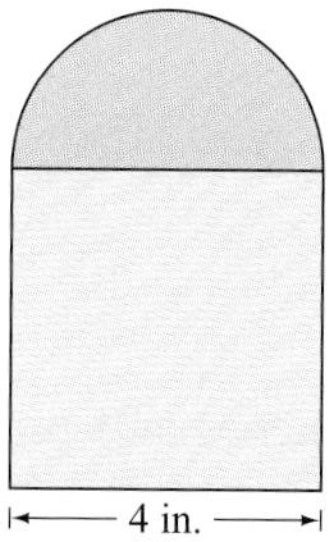

▶**Solution** The figure consists of two shapes, a semicircle and a square. The diameter of the semicircle is 4 inches, and its radius is 2 inches. Thus we have

Area = area of square + area of semicircle

$$= s^2 + \frac{1}{2}\pi r^2 \qquad \text{Area of semicircle is half area of circle}$$

$$= (4)^2 + \frac{1}{2}\pi(2)^2 \qquad \text{Substitute } s = 4 \text{ and } r = 2$$

$$= 16 + 2\pi \approx 22.283 \text{ square inches.}$$

The area of the figure is about 22.283 square inches.

☑ *Check It Out 1:* Rework Example 1 if the side of the square is 6 inches. ▨

Formulas for Three-Dimensional Figures

Table P.7.2 gives formulas for the surface area and volume of common three-dimensional figures.

Table P.7.2. Formulas for Three-Dimensional Figures

Rectangular Solid	Right Circular Cylinder	Sphere	Right Circular Cone
Length: l Width: w Height: h	Radius: r Height: h	Radius: r	Radius: r Height: h
Surface Area $S = 2(wh + lw + lh)$	Surface Area $S = 2\pi rh + 2\pi r^2$	Surface Area $S = 4\pi r^2$	Surface Area $S = \pi r(r^2 + h^2)^{1/2} + \pi r^2$
Volume $V = lwh$	Volume $V = \pi r^2 h$	Volume $V = \frac{4}{3}\pi r^3$	Volume $V = \frac{1}{3}\pi r^2 h$

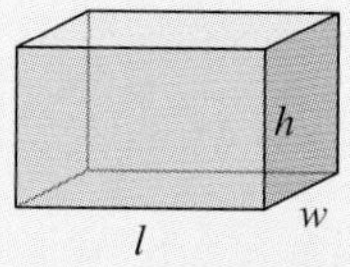

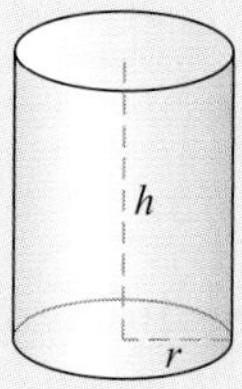

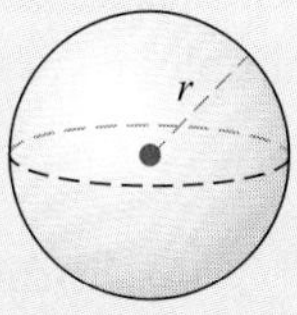

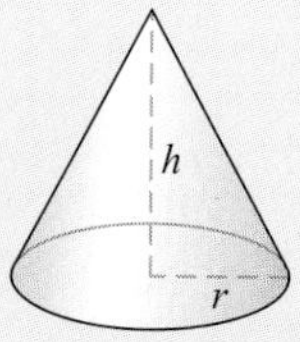

Example **2** **Finding the Volume of a Cone**

Find the volume of an ice cream cone in the shape of a right circular cone with a radius of 1 inch and a height of 4 inches.

▶**Solution** Using the formula for the volume of a right circular cone gives

$$V = \frac{1}{3}\pi r^2 h \qquad\qquad \text{Volume formula}$$

$$V = \frac{1}{3}\pi(1)^2 4 \qquad\qquad \text{Substitute } r = 1 \text{ and } h = 4$$

$$V = \frac{\pi}{3}(1)(4) \qquad\qquad \text{Simplify}$$

$$V = \frac{4\pi}{3} \approx 4.189 \text{ cubic inches.}$$

The volume of the ice cream cone is about 4.189 cubic inches.

✔ *Check It Out 2:* Rework Example 2 if the radius of the cone is 1.5 inches and the height is 5 inches. ■

The Pythagorean Theorem

When two sides of a triangle intersect at a right angle, the triangle is called a **right triangle.** For right triangles, there exists a relationship among the lengths of the three sides known as the **Pythagorean Theorem.**

The Pythagorean Theorem

In a right triangle, the side opposite the 90° angle is called the **hypotenuse.** The other sides are called **legs.** If the legs have lengths a and b and the hypotenuse has length c, then

$$c^2 = a^2 + b^2.$$

See Figure P.7.2.

Figure P.7.2

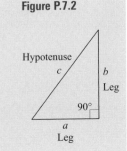

Examples 3 and 4 illustrate the use of the Pythagorean Theorem.

Example **3** **Finding the Hypotenuse of a Right Triangle**

If a right triangle has legs of lengths 5 and 12, what is the length of the hypotenuse?

▶**Solution** Because this is a right triangle, we can apply the Pythagorean Theorem with $a = 5$ and $b = 12$ to find the length c of the hypotenuse. We have

$$a^2 + b^2 = c^2 \qquad\qquad \text{The Pythagorean Theorem}$$
$$c^2 = 5^2 + 12^2 \qquad\qquad \text{Use } a = 5 \text{ and } b = 12$$
$$c^2 = 25 + 144 \qquad\qquad \text{Simplify}$$
$$c^2 = 169 \implies c = \sqrt{169} = 13. \qquad \text{Solve for } c$$

The hypotenuse has length 13.

✔ *Check It Out 3:* If a right triangle has legs of lengths 1 and 1, what is the length of the hypotenuse? ▪

Example 4 Application of the Pythagorean Theorem

Figure P.7.3

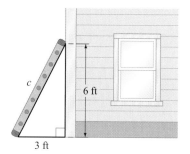

6 ft

c

3 ft

A ladder leans against a wall as shown in Figure P.7.3. The top of the ladder is 6 feet above the ground and the bottom of the ladder is 3 feet away from the wall. How long is the ladder?

▶Solution Because the wall and the floor make a right angle, we can apply the Pythagorean Theorem. Denote the length of the ladder by c. The lengths of the legs are 3 feet and 6 feet.

$$c^2 = a^2 + b^2$$ The Pythagorean Theorem

$$c^2 = 3^2 + 6^2$$ Use $a = 3$ and $b = 6$

$$c^2 = 9 + 36$$ Simplify

$$c^2 = 45 \implies c = \sqrt{45} = 3\sqrt{5}$$ Solve for c

Thus the ladder is $3\sqrt{5} \approx 6.71$ feet long.

✔ *Check It Out 4:* A small garden plot is in the shape of a right triangle. The lengths of the legs of this triangle are 4 feet and 6 feet. What is the length of the third side of the triangular plot? ▪

P.7 Key Points

▶ **Area and perimeter formulas** for specific two-dimensional shapes are given in Table P.7.1.

▶ **Volume and surface area formulas** for specific three-dimensional shapes are given in Table P.7.2.

▶ **Pythagorean Theorem:** In a right triangle, if the legs have lengths a and b and the hypotenuse has length c, then

$$c^2 = a^2 + b^2.$$

P.7 Exercises

▶**Skills** This set of exercises will reinforce the skills illustrated in this section.

In Exercises 1–22, compute the given quantity. Round your answer to three decimal places.

1. Perimeter of a rectangle with length 5 inches and width 7 inches

2. Perimeter of a rectangle with length 14 centimeters and width 10 centimeters

3. Circumference of a circle with radius 6 inches

4. Circumference of a circle with radius 4 centimeters

5. Perimeter of a parallelogram with side lengths of 8 centimeters and 3 centimeters

6. Perimeter of a parallelogram with side lengths of 5 inches and 2 inches

7. Area of a parallelogram with base 3 centimeters and height 5 centimeters

8. Area of a parallelogram with base 5 inches and height 6 inches

9. Area of a circle with radius 3 feet

10. Area of a circle with radius 5 inches

11. Volume of a right circular cylinder with radius 3 inches and height 7 inches

12. Volume of a right circular cylinder with radius 6 centimeters and height 7 centimeters

13. Volume of a sphere with radius 4 inches

14. Volume of a sphere with radius 6 inches

15. Volume of a right circular cylinder with radius 3 centimeters and height 8 centimeters

16. Volume of a right circular cylinder with radius 5 inches and height 6 inches

17. Surface area of a sphere with radius 3 inches

18. Surface area of a sphere with radius 5 centimeters

19. Surface area of a right circular cylinder with radius 2 inches and height 3 inches

20. Surface area of a right circular cylinder with radius 3 centimeters and height 5 centimeters

21. Surface area of a right circular cone with radius 4 centimeters and height 6 centimeters

22. Surface area of a right circular cone with radius 2 feet and height 5 feet

In Exercises 23–26, find the area of the figure.

23.

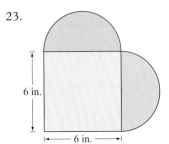

24.

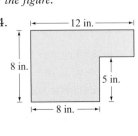

25.

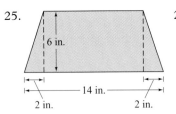

26.

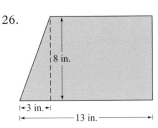

In Exercises 27–34, find the hypotenuse of the right triangle, given the lengths of its legs.

27. $a = 3, b = 4$

28. $a = 6, b = 8$

29. $a = 10, b = 24$

30. $a = 8, b = 15$

31. $a = 20, b = 21$

32. $a = 7, b = 24$

33. $a = 3, b = 5$

34. $a = 7, b = 3$

▶**Applications** In this set of exercises, you will use area formulas, volume formulas, and the Pythagorean Theorem to study real-world problems. Round all answers to three decimal places, unless otherwise noted.

35. **Construction** A rectangular fence has a length of 10 feet. Its width is half its length. Find the perimeter of the fence and the area of the rectangle it encloses.

36. **Geometry** A square fence has a perimeter of 100 feet. Find the area enclosed by the fence.

37. **Manufacturing** The height of a right cylindrical drum is equal to its radius. If the radius is 2 feet, find the volume of the drum.

38. **Manufacturing** The diameter of a beach ball is 10 inches. Find the amount of material necessary to manufacture one beach ball.

39. **Design** The cross-section of a paperweight is in the shape of a right triangle. The lengths of the legs of this triangle are 2 inches and 3 inches. What is the length of the third side of the triangular cross-section?

40. **Geometry** The perimeter of an equilateral triangle is 36 inches. Find the length of each side. An equilateral triangle is one in which all sides are equal.

41. **Landscaping** The diameter of a sundial in a school's courtyard is 6 feet. The garden club wants to put a thin border around the sundial. Find the circumference of the sundial.

42. Commerce If an ice cream cone is in the shape of a right circular cone with a diameter of 12 centimeters and a height of 15 centimeters, find the volume of ice cream the cone will hold. Round to the nearest cubic centimeter.

43. Carpentry A solid piece of wood is in the form of a right circular cylinder with a radius of 2 inches and a height of 6 inches. If a hole of radius 1 inch is drilled through the center of the cylinder, find the volume of the resulting piece of wood.

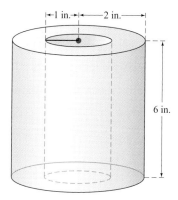

44. Manufacturing A portable refrigerator in the shape of a rectangular solid is 2 feet long, 28 inches deep, and 2 feet high. What is the volume of the refrigerator, in cubic feet?

▶ **Concepts** This set of exercises will draw on the ideas presented in this section and your general math background.

45. What values of r are meaningful in the formula for the circumference of a circle?

46. Suppose there are two circles with radii r and R, where $R > r$. Find and factor the expression that gives the difference in their areas.

47. If the radius of a circle is doubled, by what factor does the circle's area increase?

48. If the length of each side of a cube is doubled, by what factor does the cube's volume increase?

P.8 Solving Basic Equations

Objectives

▶ Solve simple equations

▶ Solve equations involving fractions

▶ Solve equations involving decimals

▶ Solve equations for one variable in terms of another

In this section, we will review some basic equation-solving skills that you learned in your previous algebra courses. When you set two algebraic expressions equal to each other, you form an **equation.** If you can find a value of the variable that makes the equation true, you have **solved the equation.** The following strategies can help you solve an equation.

Equation-Solving Strategies

When solving an equation, you must isolate the variable on one side of the equation using one or more of the following steps.

Step 1 Simplify an expression by removing parentheses. Then combine **like terms**—that is, combine real numbers or expressions with the same variable names.

Step 2 Add or subtract the *same* real number or expression to (from) *both sides* of the equation.

$$a = b \text{ is equivalent to } a + c = b + c.$$

Step 3 Multiply or divide *both sides* of the equation by the *same nonzero* real number.

$$a = b \text{ is equivalent to } ac = bc, c \neq 0.$$

Example 1 Solving an Equation

Solve the following equation for x.

$$3(x + 2) - 2 = 4x$$

▶Solution Proceed as follows.

$$
\begin{aligned}
3(x + 2) - 2 &= 4x && \text{Given equation}\\
3x + 6 - 2 &= 4x && \text{Remove parentheses}\\
3x + 6 - 2 - 4x &= 4x - 4x && \text{Subtract 4x from both sides}\\
-x + 4 &= 0 && \text{Combine like terms}\\
-x &= -4 && \text{Isolate term containing x}\\
x &= 4 && \text{Multiply both sides by } -1
\end{aligned}
$$

Thus $x = 4$ is the solution to the given equation. Check the solution by substituting $x = 4$ in the original equation:

$$3(4 + 2) - 2 = 4(4) \implies 16 = 16.$$

✔ *Check It Out 1:* Solve the equation $2(x + 4) = 3x + 2$ for x. ■

When an equation involves fractions, it is easier to solve if the denominators are cleared first, as illustrated in the next example.

Example 2 Solving an Equation Involving Fractions

Solve the equation.

$$\frac{x + 5}{2} + \frac{2x - 1}{5} = 5$$

▶Solution Clear the denominators by multiplying *both* sides of the equation by the least common denominator, which is 10.

$$
\begin{aligned}
10\left(\frac{x + 5}{2} + \frac{2x - 1}{5}\right) &= 5 \cdot 10 && \text{Multiply both sides by LCD}\\
5(x + 5) + 2(2x - 1) &= 50 && \text{Simplify each term}\\
5x + 25 + 4x - 2 &= 50 && \text{Remove parentheses}\\
9x + 23 &= 50 && \text{Combine like terms}\\
9x &= 27 && \text{Subtract 23 from both sides}\\
x &= 3 && \text{Divide both sides by 9 to solve for x}
\end{aligned}
$$

You can check the answer in the original equation.

✔ *Check It Out 2:* Solve the equation.

$$\frac{x + 3}{4} + \frac{x + 5}{2} = 7$$ ■

We can also clear decimals in an equation to make it easier to work with. This procedure is illustrated in Example 3.

Example 3 Solving an Equation Involving Decimals

Solve the equation.

$$0.3(x + 2) - 0.02x = 0.5$$

▶**Solution** There are two decimal coefficients, 0.3 and 0.02. Multiply both sides of the equation by the smallest power of 10 that will eliminate the decimals. In this case, multiply both sides by 100.

$0.3(x + 2) - 0.02x = 0.5$	Given equation
$100(0.3(x + 2) - 0.02x) = 0.5(100)$	Multiply both sides by 100
$30(x + 2) - 2x = 50$	$100(0.3) = 30$ and $100(0.02) = 2$
$30x + 60 - 2x = 50$	Remove parentheses
$28x + 60 = 50$	Combine like terms
$28x = -10$	Subtract 60 from both sides
$x = -\dfrac{10}{28} = -\dfrac{5}{14}$	Divide each side by 28 and reduce the fraction

✔ *Check It Out 3:* Solve the equation $0.06(2x + 1) - 0.03(x - 1) = 0.15$. ■

In Example 4, we solve for one variable in terms of another. In this case, the solution is not just a number.

Example 4 Solving for One Variable in Terms of Another

The perimeter of a rectangular fence is 15 feet. Write the width of the fence in terms of the length.

▶**Solution** The perimeter formula for a rectangle is $P = 2l + 2w$. Thus we have

$2l + 2w = 15$	$P = 15$
$2w = 15 - 2l$	Isolate w term
$w = \dfrac{1}{2}(15 - 2l).$	Divide by 2 to solve for w

✔ *Check It Out 4:* Rework Example 4 if the perimeter is 20 feet. ■

P.8 Key Points

▶ To solve an equation, simplify it using basic operations until you arrive at the form $x = c$ for some number c.

▶ If an equation contains fractions, multiply the equation by the LCD to clear the fractions. This makes the equation easier to work with.

▶ If an equation contains decimals, multiply the equation by the smallest power of 10 that will eliminate the decimals. This makes the equation easier to work with.

▶ If an equation contains two variables, you can solve for one variable in terms of the other.

P.8 Exercises

▶ **Skills** This set of exercises will reinforce the skills illustrated in this section.

In Exercises 1–30, solve the equation.

1. $3x + 5 = 8$

2. $4x + 1 = 17$

3. $-2x - 5 = 3x + 10$

4. $4x - 2 = 2x + 8$

5. $-3(x - 1) = 12$

6. $5(x + 2) = 20$

7. $-2(x + 4) - 3 = 7$

8. $5(x - 2) + 4 = 19$

9. $-3(x - 4) = -(x + 1) - 6$

10. $6(2x + 1) = 3(x - 3) + 7$

11. $-2(5 + x) - (x - 2) = 10(x + 1)$

12. $3(4 + x) + 2(x + 2) = 2(2x - 1)$

13. $\dfrac{1}{2} + \dfrac{x}{3} = \dfrac{7}{6}$

14. $-\dfrac{1}{3} + \dfrac{x}{5} = \dfrac{2}{3}$

15. $\dfrac{x + 3}{4} + \dfrac{x}{3} = 6$

16. $\dfrac{x - 1}{5} + \dfrac{x}{2} = 4$

17. $\dfrac{2x - 3}{3} - \dfrac{x}{2} = -\dfrac{2}{3}$

18. $\dfrac{3x + 1}{2} - \dfrac{2x}{3} = \dfrac{3}{2}$

19. $\dfrac{3x + 4}{2} + x = 4$

20. $\dfrac{7x - 1}{3} - x = 1$

21. $0.4(x - 1) + 1 = 0.5x$

22. $-0.3(2x + 1) - 3 = 0.2x$

23. $1.2(x + 5) = 3.1x$

24. $2.6(x - 1) = 4.5x$

25. $0.01(x - 3) - 0.02 = 0.05$

26. $-0.03(x + 4) + 0.05 = 0.03$

27. $0.5(2x - 1) - 0.02x = 0.3$

28. $0.4(x - 2) - 0.05x = 0.7$

29. $\pi x + 3 = 4\pi x$

30. $\sqrt{2}(x + 1) - 1 = 3\sqrt{2}$

In Exercises 31–38, solve each equation for y in terms of x.

31. $x + y = 5$

32. $-x + y = 3$

33. $-4x + 2y = 6$

34. $6x + 3y = 12$

35. $5x + 4y = 10$

36. $3x + 2y = 12$

37. $4x + y - 5 = 0$

38. $-5x + y + 4 = 0$

▶ **Applications** In this set of exercises, you will use basic equations to study real-world problems.

39. **Commerce** The profit in dollars from selling x DVD players is given by $40x - 200$. Set up and solve an equation to find out how many DVD players must be sold to obtain a profit of $800.

40. **Commerce** The profit in dollars from selling x plasma televisions is given by $200x - 500$. Set up and solve an equation to determine how many plasma televisions must be sold to obtain a profit of $3500.

41. **Geometry** The circumference of a circular hoop is 14π inches. Find the radius of the hoop.

42. **Geometry** The perimeter of a right triangle is 12 inches. If the hypotenuse is 5 inches long and one of the legs is 3 inches long, find the length of the third side of the triangle.

43. **Construction** A contractor builds a square fence with 50 feet of fencing material. Find the length of a side of the square.

44. **Construction** A contractor is enclosing a rectangular courtyard with 100 feet of fence. If the width of the courtyard is 10 feet, find the length of the courtyard.

45. **Art** A rectangular frame for a painting has a perimeter of 96 inches. If the length of the frame is 30 inches, find the width of the frame.

46. **Art** The surface area of a rectangular crate used to ship a sculpture is 250 square inches. If the base of the crate is a square with sides of length 5 inches, find the height of the crate.

47. **Manufacturing** The volume of a small drum in the shape of a right circular cylinder is 10π cubic feet. The radius of the drum is 2 feet. Find the height of the drum.

48. **Manufacturing** The volume of a rectangular fish tank is 720 cubic inches. The base of the fish tank has dimensions 6 inches by 12 inches. Find the height of the tank.

▶**Concepts** This set of exercises will draw on the ideas presented in this section and your general math background.

49. Can the equation $x + 2 = x$ be solved for x? Explain.

50. In Example 4, the width of the fence is given by $w = \frac{1}{2}(15 - 2l)$. Evaluate w when $l = 2.5$ feet. If you try to evaluate w for $l = 10$ feet, do you get a realistic value for w? Explain.

51. Find the mistake in the following "solution" of the equation $\frac{x+1}{4} + 1 = 4$.

$$\frac{x+1}{4} + 1 = 4$$
$$(x + 1) + 1 = 4 \qquad \text{Multiply by 4}$$
$$x = 2$$

52. For what values of a does the equation $ax + x = 5$ have a solution?

Chapter P   Summary

Section P.1 The Real Number System

Concept	Illustration	Study and Review
Properties of real numbers Any real number is either a rational or an irrational number. All real numbers satisfy the following properties. • The **associative properties** of addition and multiplication: $$a + (b + c) = (a + b) + c$$ and $a(bc) = (ab)c$ • The **commutative properties** of addition and multiplication: $$a + b = b + a \quad \text{and} \quad ab = ba$$ • The **distributive property** of multiplication: $$ab + ac = a(b + c)$$ where a, b, and c are any real numbers.	• Associative property: $$3 + (4 + 5) = (3 + 4) + 5$$ and $3(4 \cdot 5) = (3 \cdot 4)5$ • Commutative property: $$5 + 8 = 8 + 5 \quad \text{and} \quad 5 \cdot 8 = 8 \cdot 5$$ • Distributive property: $$3(4 + 5) = 3 \cdot 4 + 3 \cdot 5$$	Example 1 Chapter P Review, Exercises 1–6
Ordering of real numbers • An **inequality** of the form $a \le x \le b$ can be expressed in interval notation as $[a, b]$. This interval is called a **closed interval.** • If the endpoints of an interval, a and b, are not included, we write the inequality as an **open interval** (a, b).	The inequality $0 \le x \le 5$ is written as $[0, 5]$ in interval form. The inequality $0 < x < 5$ is written as $(0, 5)$ in interval form.	Examples 2, 3 Chapter P Review, Exercises 7–10
Absolute value of a number $$\lvert x \rvert = \begin{cases} x & \text{if } x \ge 0 \\ -x & \text{if } x < 0 \end{cases}$$ The distance between two points a and b is given by $\lvert b - a \rvert$ or $\lvert a - b \rvert$.	Using the definition, $\lvert -3 \rvert = -(-3) = 3$ and $\lvert 7 \rvert = 7$. The distance between -3 and 7 is $\lvert -3 - 7 \rvert = 10$.	Examples 4, 5 Chapter P Review, Exercises 11–13

Continued

Section P.1 The Real Number System

Concept	Illustration	Study and Review
Rules for order of operations When evaluating a numerical expression, the proper **order of operations** is to (1) remove parentheses, (2) simplify expressions containing exponents, (3) perform multiplications and divisions from left to right; and (4) perform additions and subtractions from left to right.	$$\dfrac{3^2 + \dfrac{12}{4}}{(8-6)3} = \dfrac{9+3}{(2)3}$$ $$= \dfrac{12}{6}$$ $$= 2$$	Examples 6, 7 Chapter P Review, Exercises 13–18

Section P.2 Integer Exponents and Scientific Notation

Concept	Illustration	Study and Review
Algebraic expressions An **algebraic expression** is a combination of numbers and variables using mathematical operations. We can evaluate an expression containing a variable for a given value of the variable.	Evaluating $3x + 5$ for $x = 2$, we have $3(2) + 5 = 11$.	Example 1 Chapter P Review, Exercises 19, 20
Positive and negative integer exponents **Positive integer exponents** For any positive integer n, $$a^n = \underbrace{a \cdot a \cdot a \cdots a}_{n \text{ factors}}.$$ The number a is the **base** and the number n is the **exponent.**	We can write $$4 \cdot 4 \cdot 4 = 4^3$$ because 4 is a factor three times.	Examples 2, 3 Chapter P Review, Exercises 21–26
Negative integer exponents Let a be any nonzero real number and let m be a positive integer. Then $$a^{-m} = \frac{1}{a^m}.$$	Using the definition of negative exponents, $$4^{-3} = \frac{1}{4^3} = \frac{1}{64}.$$	
Properties of integer exponents 1. $a^m \cdot a^n = a^{m+n}$ 2. $(a^m)^n = a^{mn}$ 3. $(ab)^m = a^m b^m$ 4. $\left(\dfrac{a}{b}\right)^m = \dfrac{a^m}{b^m}, b \neq 0$ 5. $\dfrac{a^r}{a^s} = a^{r-s}, a \neq 0$ 6. $a^1 = a$ 7. $a^0 = 1, a \neq 0$	1. $3^2 \cdot 3^4 = 3^6$ 2. $(3^2)^4 = 3^8$ 3. $(3x)^3 = 3^3 x^3$ 4. $\left(\dfrac{2}{3}\right)^2 = \dfrac{2^2}{3^2}$ 5. $\dfrac{3^4}{3^5} = 3^{-1}$ 6. $3^1 = 3$ 7. $4^0 = 1$	Examples 2, 3 Chapter P Review, Exercises 21–26

Continued

Section P.2 Integer Exponents and Scientific Notation

Concept	Illustration	Study and Review
Scientific notation A nonzero number x is written in **scientific notation** as $$a \times 10^b$$ where $1 \le a < 10$ if $x > 0$ and $-10 < a \le -1$ if $x < 0$, and b is an integer.	In scientific notation, $$0.00245 = 2.45 \times 10^{-3}$$	Examples 4–6 Chapter P Review, Exercises 27–32
Significant figures A digit of a nonzero number x is a **significant figure** if it satisfies one of the following conditions. • The digit is the **first nonzero digit** of x, going from left to right. • The digit lies **to the right of the first nonzero digit** of x.	The number 12.341 has five significant figures, whereas 0.341 has only three significant figures.	Examples 7, 8 Chapter P Review, Exercises 33, 34

Section P.3 Roots, Radicals, and Rational Exponents

Concept	Illustration	Study and Review
The nth root of a number For n an integer, $\sqrt[n]{a}$ is called the **nth root of a.** It denotes the number whose nth power is a. If n is even, then $a \ge 0$. If n is odd, then a can be any real number.	The square root of 64 is $\sqrt{64} = 8$ because $8^2 = 64$. Likewise, $\sqrt[3]{8} = 2$ because $2^3 = 8$.	Example 1 Chapter P Review, Exercises 35, 36
Rules for radicals Suppose a and b are real numbers such that their nth roots are defined. Product Rule: $\sqrt[n]{a} \cdot \sqrt[n]{b} = \sqrt[n]{ab}$ Quotient Rule: $\dfrac{\sqrt[n]{a}}{\sqrt[n]{b}} = \sqrt[n]{\dfrac{a}{b}}, b \ne 0$	Product rule: $\sqrt[3]{2} \cdot \sqrt[3]{4} = \sqrt[3]{8} = 2$ Quotient rule: $$\dfrac{\sqrt[3]{81}}{\sqrt[3]{3}} = \sqrt[3]{\dfrac{81}{3}} = \sqrt[3]{27} = 3$$	Examples 2–5 Chapter P Review, Exercises 37–44
Rational exponents • If a is a real number and n is a positive integer greater than 1, then $a^{1/n} = \sqrt[n]{a}$, where $a \ge 0$ when n is even. • Let a be a positive real number and let m and n be integers. Then $a^{m/n} = \sqrt[n]{a^m} = (\sqrt[n]{a})^m.$	$\sqrt{5} = 5^{1/2}$ and $8^{1/3} = \sqrt[3]{8}$ $\sqrt[3]{5^2} = (5^2)^{1/3} = 5^{2/3}$	Examples 6–8 Chapter P Review, Exercises 45–52

Section P.4 **Polynomials**

Concept	Illustration	Study and Review
Definition of polynomial A **polynomial in one variable** is an algebraic expression of the form $a_n x^n + a_{n-1} x^{n-1} + a_{n-2} x^{n-2} + \cdots + a_1 x + a_0$ where n is a nonnegative integer and a_n, $a_{n-1}, \ldots, a_0$ are real numbers, $a_n \neq 0$. The **degree** of the polynomial is n, the highest power to which a variable is raised.	The expression $3x^4 + 5x^2 + 1$ is a polynomial of degree 4.	Example 1 Chapter P Review, Exercises 53–56
Addition of polynomials To add or subtract polynomials, *combine* or *collect like terms* by adding their respective coefficients.	Adding $3x^3 + 6x^2 - 2$ and $-x^3 + 3x + 4$ gives $2x^3 + 6x^2 + 3x + 2$.	Example 2 Chapter P Review, Exercises 53–56
Products of polynomials To multiply polynomials, apply the distributive property and then apply the rules for multiplying monomials. Special products of polynomials are listed below. $(A + B)^2 = A^2 + 2AB + B^2$ $(A - B)^2 = A^2 - 2AB + B^2$ $(A + B)(A - B) = A^2 - B^2$	Using FOIL, $$(x + 4)(x - 3) = x^2 - 3x + 4x - 12$$ $$= x^2 + x - 12.$$ An example of a special product is $(x - 2)(x + 2) = x^2 - 4$.	Examples 3–6 Chapter P Review, Exercises 57–66

Section P.5 **Factoring**

Concept	Illustration	Study and Review
General factoring techniques A trinomial can be factored either by reversing the FOIL method of multiplication or by the grouping method.	By working backward and trying various factors, we obtain $$x^2 - 2x - 8 = (x - 4)(x + 2).$$	Examples 1–4 Chapter P Review, Exercises 67–74
Quadratic factoring patterns $A^2 - B^2 = (A + B)(A - B)$ $A^2 + 2AB + B^2 = (A + B)^2$ $A^2 - 2AB + B^2 = (A - B)^2$	 $x^2 - 1 = (x + 1)(x - 1)$ $x^2 + 2x + 1 = (x + 1)^2$ $x^2 - 2x + 1 = (x - 1)^2$	Table P.5.1 Chapter P Review, Exercises 75–80
Cubic factoring patterns $A^3 - B^3 = (A - B)(A^2 + AB + B^2)$ $A^3 + B^3 = (A + B)(A^2 - AB + B^2)$	 $x^3 - 1 = (x - 1)(x^2 + x + 1)$ $x^3 + 1 = (x + 1)(x^2 - x + 1)$	Table P.5.2 Chapter P Review, Exercises 81, 82

Section P.6 Rational Expressions

Concept	Illustration	Study and Review
Definition of a rational expression A quotient of two polynomial expressions is called a **rational expression.** A rational expression is defined whenever the denominator is not equal to zero.	The rational expression $\dfrac{3x}{x+1}$ is defined for all $x \neq -1$.	Example 1 Chapter P Review, Exercises 83, 84
Multiplication and division of rational expressions To find the product of two rational expressions, multiply the numerators, multiply the denominators, and then simplify the answer.	Find the product as follows. $$\frac{x}{x-1} \cdot \frac{x+1}{x} = \frac{x(x+1)}{(x-1)x}$$ $$= \frac{x+1}{x-1}$$	Examples 2–4 Chapter P Review, Exercises 85–88
Addition and subtraction of rational expressions To add and subtract rational expressions, first write them in terms of the same denominator, known as the **least common denominator.**	$$\frac{1}{x-1} + \frac{1}{x+1} = \frac{x+1}{(x-1)(x+1)}$$ $$+ \frac{x-1}{(x-1)(x+1)}$$ $$= \frac{x+1+(x-1)}{(x-1)(x+1)}$$ $$= \frac{2x}{(x+1)(x-1)}$$	Example 5 Chapter P Review, Exercises 89–92
Complex fractions A **complex fraction** is one in which the numerator and/or denominator of the fraction contains a rational expression. To simplify a complex fraction, multiply the numerator and denominator by the LCD of all the denominators.	$$\frac{\dfrac{1}{x} + \dfrac{2}{y}}{\dfrac{1}{y}} = \frac{\dfrac{1}{x} + \dfrac{2}{y}}{\dfrac{1}{y}} \cdot \frac{xy}{xy}$$ $$= \frac{y + 2x}{x}$$	Example 6, 7 Chapter P Review, Exercises 93, 94

Section P.7 **Geometry Review**

Concept	Illustration	Study and Review
Perimeter and area formulas for two-dimensional figures Rectangle with length l and width w: $\qquad$ Perimeter $\quad P = 2l + 2w$ $\qquad\quad$ Area $\quad A = lw$ Square with side of length s: $\qquad$ Perimeter $\quad P = 4s$ $\qquad\quad$ Area $\quad A = s^2$ Circle with radius r: $\qquad$ Circumference $\quad C = 2\pi r$ $\qquad\quad$ Area $\quad A = \pi r^2$ Triangle with base b, sides a, b, c, and height h: $\qquad$ Perimeter $\quad P = a + b + c$ $\qquad\quad$ Area $\quad A = \dfrac{1}{2}bh$ Parallelogram with base b, sides b and s, and height h: $\qquad$ Perimeter $\quad P = 2b + 2s$ $\qquad\quad$ Area $\quad A = bh$	A circle with a radius of 6 inches has an area of $A = \pi(6)^2 = 36\pi$ square inches. Its circumference is $2\pi r = 2\pi(6) = 12\pi$ inches. A triangle with a base of 2 feet and a height of 4 feet has an area of $A = \frac{1}{2}(2)(4) = 4$ square feet.	Example 1 Chapter P Review, Exercises 95, 96
Surface area and volume formulas for three-dimensional figures Rectangular solid with length l, width w, and height h: $\qquad$ Surface area $\quad S = 2(wh + lw + lh)$ $\qquad\quad$ Volume $\quad V = lwh$ Right circular cylinder with radius r and height h: $\qquad$ Surface area $\quad S = 2\pi rh + 2\pi r^2$ $\qquad\quad$ Volume $\quad V = \pi r^2 h$ Sphere with radius r: $\qquad$ Surface area $\quad S = 4\pi r^2$ $\qquad\quad$ Volume $\quad V = \dfrac{4}{3}\pi r^3$ Right circular cone with radius r and height h: $\qquad$ Surface area $\quad S = \pi r(r^2 + h^2)^{1/2} + \pi r^2$ $\qquad\quad$ Volume $\quad V = \dfrac{1}{3}\pi r^2 h$	The volume of a right circular cylinder with radius 3 inches and height 4 inches is $\quad V = \pi r^2 h = \pi(3)^2(4) = 36\pi$ cubic inches. The surface area of a sphere with radius 5 inches is $\quad S = 4\pi r^2 = 4\pi(5)^2 = 100\pi$ square inches.	Example 2 Chapter P Review, Exercises 97–100

Continued

Section P.7 **Geometry Review**

Concept	Illustration	Study and Review
Pythagorean Theorem For a right triangle with legs a and b and hypotenuse c, $\quad a^2 + b^2 = c^2.$ Hypotenuse c / b Leg / $90°$ / a Leg	To find c for a right triangle with legs $a = 3$ and $b = 4$, use the Pythagorean Theorem. $\quad c^2 = a^2 + b^2 = 3^2 + 4^2 = 25 \Longrightarrow c = 5$	Examples 3, 4 Chapter P Review, Exercises 101, 102

Section P.8 **Solving Basic Equations**

Concept	Illustration	Study and Review
Solving equations • To solve an equation, simplify it using basic operations until you arrive at the form $x = c$ for some number c. • If an equation contains fractions, multiply the equation by the LCD to clear the fractions. This makes the equation easier to work with. • If an equation contains decimals, multiply the equation by the smallest power of 10 that will eliminate the decimals. This makes the equation easier to work with. • If an equation contains two variables, solve for one variable in terms of the other.	To solve $2(x + 1) = 5$, remove the parentheses and isolate the x term. $$2(x + 1) = 5$$ $$2x + 2 = 5$$ $$2x = 3$$ $$x = \frac{3}{2}$$ To solve $\frac{x}{4} + 3 = 5$, multiply *both* sides of the equation by 4. $$\frac{x}{4} + 3 = 5$$ $$4\left(\frac{x}{4} + 3\right) = 4(5)$$ $$x + 12 = 20$$ $$x = 8$$	Examples 1–4 Chapter P Review, Exercises 103–110

Chapter P   Review Exercises

Section P.1

In Exercises 1–4, consider the following numbers.

$$\sqrt{3},\ 1.2,\ 3,\ -1.006,\ \frac{3}{2},\ -5,\ 8$$

1. Which are integers?

2. Which are irrational numbers?

3. Which are integers that are not negative?

4. Which are rational numbers that are not integers?

In Exercises 5 and 6, name the property illustrated by each equality.

5. $4 + (5 + 7) = (4 + 5) + 7$

6. $2(x + 5) = 2x + 10$

In Exercises 7–10, graph each interval on the real number line.

7. $[-4, 1)$

8. $\left(-3, \frac{3}{2}\right)$

9. $(-1, \infty)$

10. $(-\infty, -3]$

In Exercises 11 and 12, find the distance between the numbers on the real number line.

11. $-6, 4$

12. $-3.5, 4.7$

In Exercises 13–16, evaluate the expression without using a calculator.

13. $-|-3.7|$

14. $2^3 - 5(4) + 1$

15. $\dfrac{6 + 3^2}{-2^2 + 1}$

16. $-7 + 4^2 \div 8$

In Exercises 17 and 18, evaluate the expression using a calculator, and check your solution by hand.

17. $12(4 - 6) + 14 \div 7$

18. $\dfrac{-5^2 + 6(4)}{2 + 3(4)}$

Section P.2

In Exercises 19 and 20, evaluate the expression for the given value of the variable.

19. $-5(x + 2) - 3, \ x = -3$

20. $4a + 3(2a - 1), \ a = 2$

In Exercises 21–26, simplify the expression and write it using positive exponents. Assume that all variables represent nonzero numbers.

21. $6x^{-2}y^4$

22. $-(7x^3y^2)^2$

23. $\dfrac{4x^3y^{-2}}{x^{-1}y}$

24. $\dfrac{xy^4}{3^{-1}x^3y^{-2}}$

25. $\left(\dfrac{16x^4y^{-2}}{4x^{-2}y}\right)^2$

26. $\left(\dfrac{5x^{-2}y^3}{15x^3y}\right)^{-1}$

In Exercises 27 and 28, express the number in scientific notation.

27. $4{,}670{,}000$

28. 0.000317

In Exercises 29 and 30, express the number in decimal form.

29. 3.001×10^4

30. 5.617×10^{-3}

In Exercises 31 and 32, simplify and write the answer in scientific notation.

31. $(3.2 \times 10^5) \times (2.0 \times 10^{-3})$

32. $\dfrac{4.8 \times 10^{-2}}{1.6 \times 10^{-1}}$

In Exercises 33 and 34, perform the indicated calculation. Round your answer to the correct number of significant figures.

33. 4.01×0.50

34. $\dfrac{4.125}{2.0}$

Section P.3

In Exercises 35–42, simplify the expression. Assume that all variables represent positive real numbers.

35. $\sqrt[3]{375}$

36. $\sqrt{128}$

37. $\sqrt{5x} \cdot \sqrt{10x^2}$

38. $\sqrt{3x^2} \cdot \sqrt{15x}$

39. $\sqrt{\dfrac{50}{36}}$

40. $\sqrt[3]{\dfrac{-96}{125}}$

41. $\sqrt{25x} - \sqrt{36x} + \sqrt{16}$

42. $\sqrt[3]{24} - \sqrt[3]{81} + \sqrt[3]{-64}$

In Exercises 43 and 44, rationalize the denominator.

43. $\dfrac{5}{3 - \sqrt{2}}$

44. $-\dfrac{2}{1 + \sqrt{3}}$

In Exercises 45–48, evaluate the expression.

45. $-16^{1/2}$

46. $(-125)^{1/3}$

47. $64^{3/2}$

48. $(-27)^{2/3}$

In Exercises 49–52, simplify and write your answer using positive exponents.

49. $3x^{1/3} \cdot 12x^{1/4}$

50. $5x^{1/2}y^{1/2} \cdot 4x^{2/3}y$

51. $\dfrac{12x^{2/3}}{4x^{1/2}}$

52. $\dfrac{16x^{1/3}y^{1/2}}{8x^{2/3}y^{3/2}}$

Section P.4

In Exercises 53–62, perform the indicated operations and write your answer as a polynomial in descending order.

53. $(13y^2 + 19y - 9) + (6y^3 + 5y - 3)$

54. $(-11z^2 - 4z - 8) - (6z^3 + 25z + 10)$

55. $(3t^4 - 8) - (-9t^5 + 2)$

56. $(17u^5 + 8u) + (-16u^4 - 21u + 6)$

57. $(4u + 1)(-3u - 10)$

58. $(-2y + 7)(2 - y)$

59. $(8z - 9)(-3z + 8)$

60. $(-3y + 5)(9 + y)$

61. $(-3z + 5)(2z^2 - z + 8)$

62. $(4t + 1)(7t^2 - 6t - 5)$

In Exercises 63–66, find the special product.

63. $(3x + 2)(3x - 2)$

64. $(2x + 5)^2$

65. $(5 - x)^2$

66. $(x + \sqrt{3})(x - \sqrt{3})$

Section P.5

In Exercises 67–82, factor each expression completely.

67. $8z^3 + 4z^2$

68. $125u^3 - 5u^2$

69. $y^2 + 11y + 28$

70. $-y^2 + 2y + 15$

71. $3x^2 - 7x - 20$

72. $2x^2 + 3x - 9$

73. $5x^2 - 8x - 4$

74. $-3x^2 - 10x + 8$

75. $9u^2 - 49$

76. $4y^2 - 25$

77. $z^3 + 8z$

78. $4z^2 - 16$

79. $2x^2 + 4x + 2$

80. $3x^3 - 18x^2 + 27x$

81. $4x^3 + 32$

82. $5y^3 - 40$

Section P.6

In Exercises 83 and 84, simplify the rational expression and indicate the values of the variable for which the expression is defined.

83. $\dfrac{x^2 - 9}{x - 3}$

84. $\dfrac{x^2 + 2x - 15}{x^2 - 25}$

In Exercises 85–88, multiply or divide. Express your answer in lowest terms.

85. $\dfrac{x^2 + 2x + 1}{x^2 - 1} \cdot \dfrac{x^2 - x - 2}{x + 1}$

86. $\dfrac{y^2 - y - 12}{y^2 - 9} \cdot \dfrac{y + 3}{y^2 - 4y}$

87. $\dfrac{3x + 6}{x^2 - 4} \div \dfrac{3x}{x^2 + 4x + 4}$

88. $\dfrac{4x + 12}{x^2 - 9} \div \dfrac{x^2 + 1}{x + 3}$

In Exercises 89–92, add or subtract. Express your answer in lowest terms.

89. $\dfrac{1}{x + 1} + \dfrac{4}{x - 3}$

90. $\dfrac{3}{x - 4} - \dfrac{2}{x^2 - x - 12}$

91. $\dfrac{2x}{x - 3} + \dfrac{1}{x + 3} - 2x^2 - 9$

92. $\dfrac{2x + 1}{x^2 + 3x + 2} - \dfrac{3x - 1}{2x^2 + 3x - 2}$

In Exercises 93 and 94, simplify the complex fraction.

93. $\dfrac{\dfrac{a^2 - b^2}{ab}}{\dfrac{a - b}{b}}$

94. $\dfrac{\dfrac{3}{x - 2} - \dfrac{1}{x + 1}}{\dfrac{2}{x - 1} + \dfrac{3}{x + 1}}$

Section P.7

In Exercises 95–100, compute the given quantity. Round your answer to three decimal places.

95. Area of a parallelogram with base 3 inches and height 4 inches

96. Circumference of a circle with radius 8 inches

97. Volume of a right circular cylinder with radius 7 centimeters and height 4 centimeters

98. Volume of a rectangular solid with length 5 centimeters, width 4 centimeters, and height 2 centimeters

99. Surface area of a sphere with radius 3 inches

100. Surface area of a right circular cone with radius 5 inches and height 3 inches

In Exercises 101 and 102, find the hypotenuse of the right triangle, given the lengths of its legs.

101. $a = 4$, $b = 6$

102. $a = 5$, $b = 8$

Section P.8

In Exercises 103–108, solve the equation.

103. $3(x + 4) - 2(2x + 1) = 13$

104. $-4(x + 2) + 7 = 3x - 1$

105. $\dfrac{3x - 1}{5} + 1 = \dfrac{1}{2}$

106. $\dfrac{x - 3}{2} - \dfrac{2x - 1}{3} = 1$

107. $0.02(x + 4) - 0.1(x - 2) = 0.2$

108. $-0.4x + (x + 3) = 1$

In Exercises 109 and 110, solve the equation for y in terms of x.

109. $3x + y = 5$

110. $2(x - 1) = y - 7$

Applications

111. **Chemistry** If 1 liter of a chemical solution contains 5×10^{-3} gram of arsenic, how many grams of arsenic are in 3.2 liters of the same solution?

112. **Finance** At the beginning of a stock trading day (day 1), the price of Yahoo! stock was $39.09 per share. The price climbed $1.30 at the end of that day and dropped $4.23 at the end of the following day (day 2). What was the share price of Yahoo! stock at the end of day 2? (*Source:* **finance.yahoo.com**)

113. **Physics** If an object is dropped from a height of h feet, it will take $\sqrt{\dfrac{h}{16}}$ seconds to hit the ground. How long will it take a ball dropped from a height of 50 feet to hit the ground?

114. **Investment** Suppose an investment of $2000 is worth $2000(1 + r)^2$ after 2 years, where r is the interest rate. Assume that no additional deposits or withdrawals are made.
   (a) Write $2000(1 + r)^2$ as a polynomial in descending order.

   (b) [calculator icon] If the interest rate is 4%, use a calculator to determine how much the $2000 investment is worth after 2 years. (In the formula $2000(1 + r)^2$, r is assumed to be in decimal form.)

115. **Geometry** A small sphere of radius 2 inches is embedded inside a larger sphere of radius 5 inches. What is the difference in their volumes?

116. **Manufacturing** A box 3 feet long, 5 feet wide, and 2 feet high is to be wrapped in special paper that costs $2 per square foot. Assuming no waste, how much will it cost to wrap the box?

Chapter P   Test

1. Which of the numbers in the set $\{\sqrt{2}, -1, 1.55, \pi, 41\}$ are rational?

2. Name the property illustrated by the equality $3(x + 4) = 3x + 12$.

3. Graph the interval $[-4, 2)$ on the number line.

4. Find the distance between -5.7 and 4.6 on the number line.

5. Evaluate without using a calculator: $\dfrac{2^3 - 6 \cdot 4 - 2}{-3^2 - 5}$

6. Evaluate the expression $-3x^2 + 6x - 1$ for $x = -2$.

7. Express 8,903,000 in scientific notation.

In Exercises 8–12, simplify the expression and write your answer using positive exponents.

8. $-(6x^2y^5)^2$

9. $\left(\dfrac{36x^5y^{-1}}{9x^{-4}y^3}\right)^3$

10. $\sqrt{6x}\sqrt{8x^2}, \; x \geq 0$

11. $-5x^{1/3} \cdot 6x^{1/5}$

12. $\dfrac{45x^{2/3}y^{-1/2}}{5x^{1/3}y^{5/2}}; \; x, y > 0$

In Exercises 13 and 14, simplify without using a calculator.

13. $\sqrt[3]{54} - \sqrt[3]{16}$

14. $(-125)^{2/3}$

15. Rationalize the denominator: $\dfrac{7}{1 + \sqrt{5}}$

In Exercises 16–21, factor each expression completely.

16. $25 - 49y^2$

17. $4x^2 + 20x + 25$

18. $6x^2 - 7x - 5$

19. $4x^3 - 9x$

20. $3x^2 + 8x - 35$

21. $2x^3 + 16$

In Exercises 22–26, perform the operation and simplify.

22. $\dfrac{2x + 4}{x^2 - 9} \cdot \dfrac{2x^2 - 5x - 3}{x^2 - 4}$

23. $\dfrac{5x + 1}{x^2 + 4x + 4} \div \dfrac{5x^2 - 9x - 2}{x^2 + x - 2}$

24. $\dfrac{5}{x^2 - 4} - \dfrac{7}{x + 2}$

25. $\dfrac{3x - 1}{2x^2 - x - 1} - \dfrac{1}{x^2 + 2x - 3}$

26. $\dfrac{\dfrac{5}{x - 2} + \dfrac{3}{x}}{\dfrac{1}{x} - \dfrac{4}{x - 2}}$

27. Calculate the area of a circle with a diameter of 10 inches.

28. Calculate the volume of a right circular cylinder with a radius of 6 centimeters and a height of 10 centimeters.

29. Solve for x: $\dfrac{2x + 1}{2} - \dfrac{3x - 2}{5} = 2$

30. If 1 liter of a chemical solution contains 5×10^{-6} gram of sodium, how many grams of sodium are in 5.7 liters of the same solution? Express your answer in scientific notation.

31. A Bundt pan is made by inserting a cylinder of radius 1 inch and height 4 inches into a larger cylinder of radius 6 inches and height 4 inches. The centers of both cylinders coincide. What is the volume outside the smaller cylinder and inside the larger cylinder?

Functions, Graphs, and Applications

The number of people attending movies in the United States has been rising steadily, according to the Motion Picture Association. Such a trend can be studied mathematically by using the language of functions. See Exercise 63 in Section 1.1 and Exercise 109 in Section 1.3. This chapter will define what functions are, show you how to work with them, and illustrate how they are used in various applications.

1.1 Functions

Objectives

▶ Define a function

▶ Evaluate a function at a certain value

▶ Interpret tabular and graphical representations of a function

▶ Define the domain and range of a function

Table 1.1.1

Gallons Used	Miles Driven
2.5	50
5	100
10	200

Discover *and* Learn

Give an expression for a function that takes an input value x and produces an output value that is 2 greater than 3 times the input value.

A function describes a relationship between two quantities of interest. Many applications of mathematics involve finding a suitable function that can reasonably represent a set of data that has been gathered. Therefore, it is very important that you understand the notion of function and are able to work with the mathematical notation that is an integral part of the definition of a function.

Describing Relationships Between Quantities

To help understand an abstract idea, it is useful to first consider a concrete example and consider how you can think about it in mathematical terms. A car you just bought has a mileage rating of 20 miles per gallon. You would like to know how many miles you can travel given a certain amount of gasoline. Your first attempt to keep track of the mileage is to make a table. See Table 1.1.1.

Note that the table does not state anything about how many miles you can drive if you use 4 gallons, 6 gallons, or 12.5 gallons. To get the most out of this information, it would be convenient to use a *general formula* to express the number of miles driven in terms of the number of gallons of gasoline used. Just how do we go about getting this formula?

First, let's use some variables to represent the items being dicussed.

x is the amount of gasoline used (in gallons).

d is the distance traveled (in miles).

Now, the distance traveled *depends* on the amount of gasoline used. A mathematical way of stating this relationship is to say that d is a **function** of x. We shall define precisely what a function is later in this section.

Instead of saying "distance is a function of x," we can abbreviate even further by using the notation $d(x)$. This notation can be read as "d evaluated at the point x" or "d of x" or "d at x." The variable x is often called the **input variable,** and $d(x)$ the **output variable.** The notion of function can be represented by a diagram, as shown in Figure 1.1.1.

But what is d? For the example we are discussing, we can simply take the mileage rating, which is 20 miles per gallon, and multiply it by the number of gallons used, x, to get the total distance, d. That is,

$$d(x) = 20x. \quad \textit{Expression for distance, d, in terms of number of gallons used, x}$$

We have now derived a mathematical expression that describes how d is related to x. In words, the function d takes a value x and multiplies it by 20. The resulting output value, $d(x)$, is given by $20x$. See Figure 1.1.2.

Figure 1.1.1

Input x ⟹ Function **d** ⟹ Output $d(x)$

Figure 1.1.2

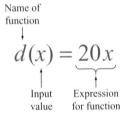

Name of function

$$d(x) = 20x$$

Input value

Expression for function

Example 1 **Working with Functions**

In the previous discussion, we found that the number of miles driven is a function of the number of gallons of gasoline used. This can be written as $d(x) = 20x$. Find how many miles can be driven using

(a) 4 gallons of gasoline.

(b) 13.5 gallons of gasoline.

(c) k gallons of gasoline.

▶Solution In the expression for $d(x)$, which is $d(x) = 20x$, we simply substitute the number of gallons of gasoline used for x. We then have the following.

(a) Miles driven: $d(4) = 20(4) = 80$ miles

(b) Miles driven: $d(13.5) = 20(13.5) = 270$ miles

(c) Miles driven: $d(k) = 20(k) = 20k$ miles

Note the convenience of the function notation. In part (a), 4 takes the place of x, so we write $d(4)$. This means we want the function expression evaluated at 4. A similar remark holds for parts (b) and (c).

✔ *Check It Out 1:* Consider the function in Example 1.

(a) How many miles can be driven if 8.25 gallons of gasoline are used?

(b) Write your answer to part (a) using function notation. ▪

Note The set of parentheses in, say, $d(4)$ does not mean multiplication. It is simply a shorthand way of saying "the function d evaluated at 4." Whether a set of parentheses means multiplication or whether it indicates function notation will usually be clear from the context.

In many applications of mathematics, figuring out the exact relationship between two quantities may not be as obvious as in Example 1. This course is intended to train you to choose appropriate functions for particular situations. This skill will help prepare you to use mathematics in your course of study and at work.

We now give the formal definition of a function.

Definition of a Function

Suppose you have a set of input values and a set of output values.

Definition of a Relation

A **relation** establishes a correspondence between a set of input values and a set of output values in such a way that for each input value, there is *at least* one corresponding output value.

In many situations, both practical and theoretical, it is preferable to have a relationship in which an allowable input value yields *exactly one* output value. Such a relationship is given by a function, defined as follows.

Definition of a Function

A **function** establishes a correspondence between a set of input values and a set of output values in such a way that for each allowable input value, there is *exactly one* corresponding output value. See Figure 1.1.3.

The input variable is called the **independent variable** and the output variable is called the **dependent variable.**

Figure 1.1.3   Function correspondence

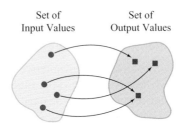

Set of Input Values   Set of Output Values

We see that a function sets up a *correspondence* between the input variable and the output variable. This correspondence set up by the function must produce only one value for the output variable for each value of the input variable. Functions need not always be defined by formulas. A function can be given as a table, pictured as a graph, or simply described in words. As long as the definition of function is satisfied, it does not matter how the function itself is described.

Example 2 **Definition of a Function**

Which of the following correspondences satisfy the definition of a function?

(a) The input value is a letter of the alphabet and the output value is the name of a month beginning with that letter.

(b) The input value is the radius of a circle and the output value is the area of the circle.

▶Solution

(a) If the letter J were input, the output could be January, June, or July. Thus, this correspondence is not a function.

(b) The area A of a circle is related to its radius r by the formula $A = \pi r^2$. For each value of r, only one value of A is output by the formula. Therefore, this correspondence is a function.

✔ *Check It Out 2:* Which of the following are functions?

(a) The input value is the number of days in a month and the output value is the name of the corresponding month(s).

(b) The input value is the diameter of a circle and the output value is the circumference of the circle. ■

Example 3 **Tabular Representation of a Function**

Which of the following tables represents a function? Explain your answer.

Table 1.1.2

Input	Output
−4	0.50
−3	0.50
0	0.50
2.5	0.50
$\frac{13}{2}$	0.50

Table 1.1.3

Input	Output
−4	1
−3	2
−3	−1
2	0
6	−2

▶**Solution** Table 1.1.2 represents a function because each input value has only one corresponding output value. It does *not* matter that the *output* values are *repeated*.

Table 1.1.3 does *not* represent a function because the input value of −3 has *two distinct* output values, 2 and −1.

☑ *Check It Out 3:* Explain why the following table represents a function. ▪

Table 1.1.4

Input	Output
−6	1
−4	4.3
−3	−1
1	6
6	−2

In the process of working out Example 1, we introduced some function notation. The following examples will illustrate the usefulness of function notation.

Example 4 **Evaluating a Function**

Let $f(x) = x^2 - 1$. Evaluate the following.

(a) $f(-2)$ (b) $f\left(\frac{1}{2}\right)$ (c) $f(x + 1)$ (d) $f(\sqrt{5})$ (e) $f(x^3)$

▶**Solution**

(a) $f(-2) = (-2)^2 - 1 = 4 - 1 = 3$

(b) $f\left(\frac{1}{2}\right) = \left(\frac{1}{2}\right)^2 - 1 = \frac{1}{4} - 1 = -\frac{3}{4}$

(c) $f(x + 1) = (x + 1)^2 - 1 = x^2 + 2x + 1 - 1 = x^2 + 2x$

Technology Note

To evaluate a function at a particular numerical value, first store the value of x and then evaluate the function at that value. Figure 1.1.4 shows how to evaluate $x^2 - 1$ at $x = -2$.

Keystroke Appendix:
Section 4

Figure 1.1.4

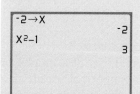

```
-2→X
            -2
X²-1
             3
```

(d) $f(\sqrt{5}) = (\sqrt{5})^2 - 1 = 5 - 1 = 4$

(e) $f(x^3) = (x^3)^2 - 1 = x^6 - 1$

☑ **Check It Out 4:** Let $f(t) = -t^2 + 2$. Evaluate the following.

(a) $f(-1)$

(b) $f(a + 1)$

(c) $f(x^2)$ ■

Observations:

▶ A function can be assigned any arbitrary name. Functions do not always have to be called f.

▶ The variable x in $f(x)$ is a placeholder. It can be replaced by any quantity, as long as the same replacement occurs throughout the expression for the function.

Example 5 gives more examples of evaluating functions.

Example 5 **Evaluating Functions**

Discover *and* **Learn**

Let $g(x) = x^2 - 1$. Show that $g(x + 1)$ is *not* equal to $g(x) + 1$.

Evaluate $g(-3)$ and $g(a^2)$ for the following functions.

(a) $g(x) = \dfrac{\sqrt{1 - x}}{2}$

(b) $g(x) = \dfrac{x + 4}{x - 2}$

▶**Solution**

(a) $g(-3) = \dfrac{\sqrt{1 - (-3)}}{2} = \dfrac{\sqrt{4}}{2} = 1$

$g(a^2) = \dfrac{\sqrt{1 - (a^2)}}{2}$. This expression cannot be simplified further.

(b) $g(-3) = \dfrac{(-3) + 4}{(-3) - 2} = -\dfrac{1}{5}$

$g(a^2) = \dfrac{(a^2) + 4}{(a^2) - 2}$. This expression cannot be simplified further.

☑ **Check It Out 5:** Let $f(x) = \dfrac{x - 3}{x^2 + 1}$. Evaluate $f(-3)$. ■

More Examples of Functions

In Example 1, we examined a function that could be represented by an algebraic expression involving a variable x. Functions also can be represented by tables, graphs, or just a verbal description. Regardless of the representation, the important feature of a function is the correspondence between input and output in such a way that for each valid input value, there is exactly one output value.

In everyday life, information such as postal rates or income tax rates is often given in tables. The following example lists postal rates as a function of weight.

Example 6 Postal Rate Table

Table 1.1.5

Weight, w (ounces)	Rate ($)
$0 < w \leq 1$	0.37
$1 < w \leq 2$	0.60
$2 < w \leq 3$	0.83
$3 < w \leq 4$	1.06
$4 < w \leq 5$	1.29
$5 < w \leq 6$	1.52
$6 < w \leq 7$	1.75
$7 < w \leq 8$	1.98

Table 1.1.5 gives the rates for first-class U.S. mail in 2005 as a function of the weight of a single piece of mail. (*Source:* United States Postal Service)

(a) Identify the input variable and the output variable.

(b) Explain why this table represents a function.

(c) What is the rate for a piece of first-class mail weighing 6.4 ounces?

(d) What are the valid input values for this function?

▶Solution

(a) Reading the problem again, the rate for first-class U.S. mail is *a function of the weight* of the piece of mail. Thus, the input variable is the weight of the piece of first-class mail. The output variable is the rate charged.

(b) This table represents a function because for each input weight, only one rate will be output.

(c) Since 6.4 is between 6 and 7, it will cost $1.75 to mail this piece.

(d) The valid input values for this function are all values of the weight greater than 0 and less than or equal to 8 ounces. The table does not give any information for weights beyond 8 ounces, and weights less than or equal to 0 do not make sense.

✔ *Check It Out 6:* Use the table in Example 6 to calculate the rate for a piece of first-class mail weighing 3.3 ounces. ▪

Functions can also be depicted graphically. In newspapers and magazines, you will see many graphical representations of relationships between quantities. The input variable is on the horizontal axis of the graph, and the output variable is on the vertical axis of the graph. Example 7 shows a graphical representation of a function. We will discuss graphs of functions in more detail in the next section.

Example 7 Graphical Depiction of a Function

Figure 1.1.5 depicts the average high temperature, in degrees Fahrenheit (°F), in Fargo, North Dakota, as a function of the month of the year. (*Source:* weather.yahoo.com)

(a) Let $T(m)$ be the function represented by the graph, where T is the temperature and m is the month of the year. What is $T(\text{May})$?

(b) What are the valid input values for this function?

Figure 1.1.5 Average high temperatures (°F) in Fargo, ND

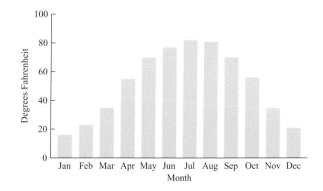

▶Solution

(a) Estimating from the graph, $T(\text{May}) \approx 70°\text{F}$. Thus the average high temperature for May in Fargo is approximately 70°F.

(b) The valid input values are the names of months of the year from January through December. Note that input values and output values do not necessarily have to be numbers.

✔ *Check It Out 7:* In Example 7, estimate $T(\text{August})$. ■

Domain and Range of a Function

It is important to know when a function is defined and when it is not. For example, the function $f(x) = \dfrac{1}{x}$ is not defined for $x = 0$. The set of all input values for which a function is defined is called the **domain.**

> **Definition of Domain**
>
> The **domain** of a function is the set of all input values for which the function will produce a real number.

Similarly, we can consider the set of all output values of a function, called the **range.**

> **Definition of Range**
>
> The **range** of a function is the set of all output values that are possible for the given domain of the function.

Example 8 Finding the Domain of a Function

Find the domain of each of the following functions. Write your answers using interval notation.

(a) $g(t) = t^2 + 4$ (b) $f(x) = \sqrt{4 - x}$ (c) $h(s) = \dfrac{1}{s - 1}$ (d) $h(t) = \dfrac{1}{\sqrt{4 - t}}$

Just In Time

Review square roots in Section P.3.

▶Solution

(a) For the function $g(t) = t^2 + 4$, any value can be substituted for t and a real number will be output by the function. Thus, the domain for g is *all real numbers,* or $(-\infty, \infty)$ in interval notation.

(b) For $f(x) = \sqrt{4 - x}$, recall that the square root of a number is a real number only when the number under the square root sign is greater than or equal to zero. Therefore, we have

$$4 - x \geq 0 \qquad \text{Expression under square root sign must be greater than or equal to zero}$$
$$4 \geq x \qquad \text{Solve for } x$$
$$x \leq 4. \qquad \text{Rewrite inequality}$$

Thus, the domain is the set of all real numbers less than or equal to 4, or $(-\infty, 4]$ in interval notation.

(c) For $h(s) = \dfrac{1}{s-1}$, we see that this expression is defined only when the denominator is *not equal to zero*. In this case, we have $s - 1 \neq 0$, which implies that $s \neq 1$. Thus, the domain consists of the set of all real numbers *not equal to* 1. In interval notation, the domain is $(-\infty, 1) \cup (1, \infty)$.

(d) For $h(t) = \dfrac{1}{\sqrt{4-t}}$, we see that this expression is not defined when the denominator is equal to zero, or when $t = 4$. However, since the square root is not defined for negative numbers, we must have

$$4 - t > 0 \quad \text{Expression under radical is greater than zero}$$
$$4 > t \quad \text{Solve for } t$$
$$t < 4. \quad \text{Rewrite inequality}$$

So, the domain consists of the set of all real numbers *less than* 4, $(-\infty, 4)$.

✔ *Check It Out 8:* Find the domain of each of the following functions. Write your answers using interval notation.
(a) $H(t) = -t^2 - 1$
(b) $g(x) = \sqrt{x-4}$
(c) $h(x) = \dfrac{1}{2x+1}$
(d) $f(t) = \dfrac{2}{\sqrt{t-4}}$ ▪

Note Finding the range of a function algebraically involves techniques discussed in later chapters. However, the next section will show how you can determine the range graphically.

1.1 Key Points

▸ A **function** establishes a correspondence between a set of input values and a set of output values in such a way that for each input value, there is exactly one corresponding output value.

▸ Functions can be represented by mathematical expressions, tables, graphs, and verbal expressions.

▸ The **domain** of a function is the set of all allowable input values for which the function is defined.

▸ The **range** of a function is the set of all output values that are possible for the given domain of the function.

1.1 Exercises

▶**Just in Time Exercises** These exercises correspond to the Just in Time reference in this section. Complete them to review topics relevant to the remaining exercises.

1. True or False: $\sqrt{-4} = 2$

2. $\sqrt{x}$ is a real number when
   (a) $x < 0$ (b) $x \geq 0$ (c) $x \leq 0$

▶**Skills** This set of exercises will reinforce the skills illustrated in this section.

In Exercises 3–14, evaluate $f(3), f(-1)$, and $f(0)$.

3. $f(x) = 5x + 3$ 4. $f(x) = -4x - 1$

5. $f(x) = -\dfrac{7}{2}x + 2$ 6. $f(x) = \dfrac{2}{3}x - 1$

7. $f(x) = x^2 + 2$ 8. $f(x) = -x^2 - 4$

9. $f(x) = -2(x + 1)^2 - 4$ 10. $f(x) = 3(x - 3)^2 + 5$

11. $f(t) = \sqrt{3t + 4}$ 12. $f(t) = \sqrt{2t + 5}$

13. $f(t) = \dfrac{t^2 - 1}{t + 3}$ 14. $f(t) = \dfrac{t^2 + 1}{t - 2}$

In Exercises 15–22, evaluate $f(a), f(a + 1)$, and $f\left(\dfrac{1}{2}\right)$.

15. $f(x) = 4x + 3$ 16. $f(x) = -2x - 1$

17. $f(x) = -x^2 + 4$ 18. $f(x) = -2x^2 + 1$

19. $f(x) = \sqrt{3x - 1}$ 20. $f(x) = \sqrt{x + 1}$

21. $f(x) = \dfrac{1}{x + 1}$ 22. $f(x) = \dfrac{1}{2x + 1}$

In Exercises 23–32, evaluate $g(-x), g(2x)$, and $g(a + h)$.

23. $g(x) = \sqrt{6}$ 24. $g(x) = \sqrt{5}$

25. $g(x) = 2x - 3$ 26. $g(x) = -\dfrac{1}{2}x + 1$

27. $g(x) = 3x^2$ 28. $g(x) = -x^2$

29. $g(x) = \dfrac{1}{x}$ 30. $g(x) = -\dfrac{3}{x}$

31. $g(x) = -x^2 - 3x + 5$ 32. $g(x) = x^2 + 6x - 1$

33. Let $g(t)$ be defined by the following table.
    (a) Evaluate $g(5)$.
    (b) Evaluate $g(0)$.
    (c) Is $g(3)$ defined? Explain.

t	$g(t)$
-2	$\dfrac{4}{3}$
-1	4.5
0	2
2	-1
5	-1

34. Let $h(t)$ be defined by the following table.
    (a) Evaluate $h(-2)$.
    (b) Evaluate $h(4)$.
    (c) Is $h(5)$ defined? Explain.

t	$g(t)$
-2	4
$-\dfrac{1}{2}$	4.5
0	2
3	$\dfrac{4}{3}$
4	-1

In Exercises 35–42, determine whether a function is being described.

35. The length of a side of a square is the input variable and the perimeter of the square is the output variable.

36. A person's height is the input variable and his/her weight is the output variable at a specific point in time.

37. The price of a store product is the input and the name of the product is the output.

38. The perimeter of a rectangle is the input and its length is output.

39. The input variable is the denomination of a U.S. paper bill (1-dollar bill, 5-dollar bill, etc.) and the output variable is the length of the bill.

40. The input variable is the bar code on a product at a store and the output variable is the name of the product.

41. The following input–output table:

Input	Output
-2	$\frac{1}{3}$
-1	-2
0	2
2	-1
5	-1

42. The following input–output table:

Input	Output
-2	$-\frac{1}{3}$
-1	0
-1	2
2	5
5	4

In Exercises 43–56, find the domain of each function. Write your answer in interval notation.

43. $f(x) = x^2 - 4$

44. $g(x) = -x^3 - 2$

45. $f(s) = \dfrac{1}{s + 1}$

46. $h(y) = \dfrac{1}{y + 2}$

47. $f(w) = \dfrac{5}{w - 3}$

48. $H(t) = \dfrac{3}{1 - t}$

49. $h(x) = \dfrac{1}{x^2 - 4}$

50. $f(x) = \dfrac{2}{x^2 - 9}$

51. $g(x) = \sqrt{2 - x}$

52. $F(w) = \sqrt{-4 - w}$

53. $f(x) = \dfrac{1}{x^2 + 1}$

54. $h(s) = \dfrac{3}{s^2 + 3}$

55. $f(x) = \dfrac{2}{\sqrt{x + 7}}$

56. $g(x) = \dfrac{3}{\sqrt{8 - x}}$

▶**Applications** In this set of exercises you will use functions to study real-world problems.

57. **Geometry** The volume of a sphere is given by

$$V(r) = \frac{4}{3}\pi r^3,$$

where r is the radius. Find the volume when the radius is 3 inches.

58. **Geometry** The hypotenuse of a right triangle having sides of lengths 4 and x is given by

$$H(x) = \sqrt{16 + x^2}.$$

Find and interpret the quantity $H(2)$.

59. **Sales** A commissioned salesperson's earnings can be determined by the function

$$S(x) = 1000 + 20x$$

where x is the number of items sold. Find and interpret $S(30)$.

60. **Engineering** The distance between the n supports of a horizontal beam 10 feet long is given by

$$d(n) = \frac{10}{\sqrt{n^2 - 1}}$$

if bending is to be kept to a minimum.
(a) Find $d(2)$, $d(3)$, and $d(5)$.
(b) What is the domain of this function?

61. **Motion and Distance** A car travels 45 miles per hour.
(a) Write the distance traveled by the car (in miles) as a function of the time (in hours).
(b) How far does the car travel in 2 hours? Write this information using function notation.
(c) Find the domain and range of this function.

62. **Business** The following graph gives the number of DVD players sold, in millions of units, in the United States for various years. (*Source:* Consumer Electronics Association)

Number of DVD players sold in the U.S.

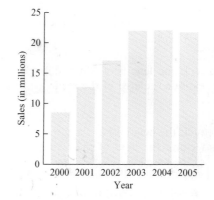

(a) If the number of players sold, S, is a function of the year, estimate $S(2004)$ and interpret it.
(b) What is the domain of this function?
(c) What general trend do you notice in the sale of DVD players?

63. **Film Industry** The following graph gives the amount of money grossed (in millions of dollars) by certain hit movies. (*Source:* moves.yahoo.com)

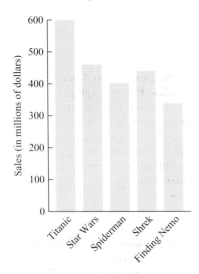

Cumulative gross movie sales

(a) If the dollar amount grossed, *D*, is a function of the movie, estimate *D*(*Finding Nemo*) and interpret it.

(b) What is the domain of this function?

(c) Must the domain of a function always consist of numbers? Explain.

64. **Manufacturing** For each watch manufactured, it costs a watchmaker $20 over and above the watchmaker's fixed cost of $5000.

(a) Write the total manufacturing cost (in dollars) as a function of the number of watches produced.

(b) How much does it cost for 35 watches to be manufactured? Write this information in function notation.

(c) Find the domain of this function that makes sense in the real world.

65. **Geometry** If the length of a rectangle is three times its width, express the area of the rectangle as a function of its width.

66. **Geometry** If the height of a triangle is twice the length of the base, find the area of the triangle in terms of the length of the base.

67. **Physics** A ball is dropped from a height of 100 feet. The height of the ball *t* seconds after it is dropped is given by the function $h(t) = -16t^2 + 100$.

(a) Find $h(0)$ and interpret it.

(b) Find the height of the ball after 2 seconds.

68. **Marine Biology** The amount of coral, in kilograms, harvested in North American waters for selected years is shown on the following graph. (*Source:* United Nations, FAOSTAT data)

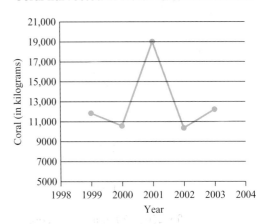

Coral harvested in North American waters

(a) From the graph, in what year(s) was the amount of coral harvested approximately 12,000 kilograms?

(b) From the data in the graph, in what year was the maximum amount of coral harvested?

69. **Consumer Behavior** The following table lists the per capita consumption of high-fructose corn syrup, a sweetener found in many foods and beverages, for selected years between 1970 and 2002. (*Source: Statistical Abstracts of the United States*)

Year	Per Capita Consumption (in pounds)
1970	0.5
1980	19.0
1990	49.6
2000	62.7
2001	62.6
2002	62.8

(a) If the input value is the year and the output value is the per capita consumption of high-fructose corn syrup, explain why this table represents a function. Denote this function by *S*.

(b) Find *S*(2001) and interpret it.

(c) High-fructose corn syrup was developed in the early 1970s and gained popularity as a cheaper alternative to sugar in processed foods. How is this reflected in the given table?

70. Population The following table shows the number of five-year-olds in the United States as of July 1 of the year indicated. Data such as this is often used to determine staffing and funding for schools. (*Source:* U.S. Census Bureau)

Year	Number of Five-Year-Olds (in thousands)
1999	3996
2000	3951
2001	3933
2002	3837
2003	3868
2004	3859
2005	3914
2006	4048

(Figures for 2004–2006 are projections.)

(a) If the input is the year and the output is the number of five-year-olds, explain why this table represents a function. Denote this function by P.

(b) Find $P(2003)$ and interpret it.

(c) Find $P(2006)$ and interpret it.

(d) What trend do you observe in the number of five-year-olds over the period 1999–2006?

71. Automobile Mileage The 2005 Mitsubishi Eclipse has a combined city and highway mileage rating of 26 miles per gallon. Write a formula for the distance an Eclipse can travel (in miles) as a function of the amount of gasoline used (in gallons). (*Source:* www.fueleconomy.gov)

72. Online Shopping On the online auction site eBay, the minimum amount that one may bid on an item is based on the current bid, as shown in the table. (*Source:* www.ebay.com)

Current Bid	Minimum Bid Increment
$1.00–$4.99	$0.25
$5.00–$24.99	$0.50
$25.00–$99.99	$1.00
$100.00–$249.99	$2.50

For example, if the current bid on an item is $7.50, then the next bid must be *at least* $0.50 higher.

(a) Explain why the *minimum* bid increment, I, is a function of the current bid, b.

(b) Find $I(2.50)$ and interpret it.

(c) Find $I(175)$ and interpret it.

(d) Can you find $I(400)$ using this table? Why or why not?

73. Environmental Science Greenhouse gas emissions from motor vehicles, such as carbon dioxide and methane, contribute significantly to global warming. The Environmental Protection Agency (EPA) lists the number of tons of greenhouse gases emitted per year for various models of automobiles. For example, the sports utility vehicle (SUV) Land Rover Freelander (2005 model) emits 10 tons of greenhouse gases per year. The greenhouse gas estimates presented here are full-fuel-cycle estimates; they include all steps in the use of a fuel, from production and refining to distribution and final use. (*Source:* www.fueleconomy.gov)

(a) Write an expression for the total amount of greenhouse gases released by the Freelander as a function of time (in years).

(b) The SUV Mitsubishi Outlander (2005 model) releases 8.2 tons of greenhouse gases per year. It has a smaller engine than the Freelander, which accounts for the decrease in emissions. Write an expression for the total amount of greenhouse gases released by the Outlander as a function of time (in years).

(c) How many more tons of greenhouse gases does the Freelander release over an 8-year period than the Outlander?

▶**Concepts** This set of exercises will draw on the ideas presented in this section and your general math background.

74. Let $f(x) = ax^2 + 5$. Find a if $f(1) = 2$.

75. Let $f(x) = -2x + c$. Find c if $f(3) = 1$.

Find the domain and range of each of the following functions.

76. $f(t) = k$, k is a fixed real number

77. $g(s) = 1 + \sqrt{s + 1}$

78. $H(x) = \dfrac{1}{x}$

1.2 Graphs of Functions

Objectives

▶ Draw graphs of functions defined by a single expression

▶ Determine the domain and range of a function given its graph

▶ Use the vertical line test to determine whether a given graph is the graph of a function

▶ Determine x- and y-intercepts given the graph of a function

A graph is one of the most common ways to represent a function. Many newspaper articles and magazines summarize pertinent data in a graph. In this section, you will learn how to sketch the graphs of various functions that were discussed in Section 1.1. As you learn more about the various types of functions in later chapters, you will build on the ideas presented in this section.

Our first example illustrates the use of graphical representation of data.

Example 1 Graphical Representation of Data

The annual attendance at Yellowstone National Park for selected years between 1960 and 2004 is given in Figure 1.2.1. (*Source:* National Park Service)

(a) Assume that the input values are on the horizontal axis and the output values are on the vertical axis. Identify the input and output variables for the given data.

(b) Explain why the correspondence between year and attendance shown in the graph represents a function.

(c) What trend do you observe in the given data?

▶ Solution

(a) For the given set of data, the input variable is a particular year and the output variable is the number of people who visited Yellowstone National Park that year.

(b) The correspondence between the input and output represents a function because, for any particular year, there is only one number for the attendance.

In this example, the function is not described by an algebraic expression. It is simply represented by a graph.

(c) From the graph, we see that there was an increase in attendance at Yellowstone National Park from 1960 to 2002, and then a slight decrease from 2002 to 2004.

☑ *Check It Out 1:* From the graph, in what year(s) was the annual attendance approximately 2 million? ▪

By examining graphs of data, we can easily observe trends in the behavior of the data. Analysis of graphs, therefore, plays an important role in using mathematics in a variety of settings.

Figure 1.2.1 Attendance at Yellowstone National Park

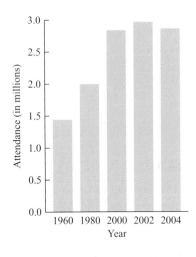

Graphs of Functions Defined by a Single Expression

In order to visualize relationships between input and output variables, we first set up a reference system. On a plane, we draw horizontal and vertical lines, known as **axes.** The intersection of these lines is denoted by the **origin,** $(0, 0)$. The horizontal axis is called the **x-axis** and the vertical axis is called the **y-axis,** although other variable names can be used.

The axes divide the plane into four regions called **quadrants.** The convention for labeling quadrants is to begin with Roman numeral I for the quadrant in the upper right, and continue numbering counterclockwise. See Figure 1.2.2.

To locate a point in the plane, we use an ordered pair of numbers (x, y). The x value gives the horizontal location of the point and the y value gives the vertical location. The first number x is called the **x-coordinate, first coordinate,** or **abscissa.** The second number y is called the **y-coordinate, second coordinate,** or **ordinate.** If an ordered pair is given by another pair of variables, such as (u, v), it is assumed that the first coordinate represents the horizontal location and the second coordinate represents the vertical location. The xy-coordinate system is also called the **Cartesian coordinate system** or the **rectangular coordinate system.**

Figure 1.2.2

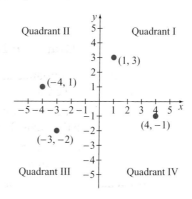

One of the main features of a function is its graph. Recall that if we are given a value for x, then $f(x)$ is the value that is output by the function f. In the xy-plane, the input value, x, is the x-coordinate and the output value, $f(x)$, is the y-coordinate. For all x in the domain of f, the set of points $(x, f(x))$ is called the **graph** of f.

The following example shows the connection between a set of (x, y) values and the definition of a function.

Example 2 Satisfying the Function Definition

Does the following set of points define a function?
$$S = \{(-1, -1), (0, 0), (1, 2), (2, 4)\}$$

▶**Solution** When a point is written in the form (x, y), x is the input variable, or the independent variable; y is the output variable, or the dependent variable. Table 1.2.1 shows the correspondence between the x and y values.

The definition of a function states that for each value of x, there must be exactly one value of y. This definition is satisfied, and thus the set of points S defines a function.

✔ *Check It Out 2:* Does the following set of points define a function?
$$S = \{(-3, 1), (1, 0), (1, 2), (3, 4)\} \quad ▪$$

Table 1.2.1

x	y
-1	-1
0	0
1	2
2	4

To graph a function by hand, we first make a table of x and $f(x)$ values, choosing various values for x. The set of coordinates given by $(x, f(x))$ is then plotted in the xy-plane. This process is illustrated in Example 3.

Example 3 Graphing a Function

Graph the function $f(x) = \frac{2}{3}x - 2$. Use the graph to find the domain and range of f.

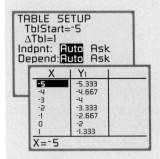
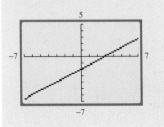
▶**Solution** We first make a table of values of x and $f(x)$ (Table 1.2.2) and then plot the points on the xy-plane. We have chosen multiples of 3 for the x values to make the arithmetic easier. Recall that the y-coordinate corresponding to x is given by $f(x)$. Since the points lie along a line, we draw the line passing through the points to get the graph shown in Figure 1.2.5.

Table 1.2.2

x	$f(x) = \frac{2}{3}x - 2$
−6	−6
−3	−4
0	−2
3	0
6	2

Figure 1.2.5

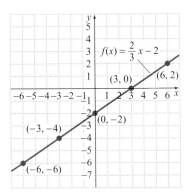

Note that $f(x) = \frac{2}{3}x - 2$ is defined for all values of x. To see this from the graph, move your pencil to any point on the x-axis. Then move your pencil vertically from the x-coordinate of the point you picked to the corresponding y-coordinate. Since this works for *any* x-coordinate you choose, the domain is the set of *all real numbers,* $(-\infty, \infty)$.

The y values are zero, positive numbers, and negative numbers. By examining the graph of f, we see that the range of f is the set of *all real numbers,* $(-\infty, \infty)$.

✔ *Check It Out 3:* Graph the function $g(x) = -3x + 1$ and use the graph to find the domain and range of g. ■

Example 4 Finding Domain and Range from a Graph

Graph the function $g(t) = -2t^2$. Use the graph to find the domain and range of g.

▶**Solution** We first make a table of values of t and $g(t)$ (Table 1.2.3) and then plot the points on the ty-plane, as shown in Figure 1.2.6. Recall that the y-coordinate corresponding to t is given by $g(t)$.

Table 1.2.3

t	$g(t)$
−2	−8
−1	−2
0	0
1	−2
2	−8

Figure 1.2.6

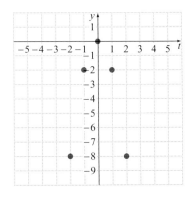

Figure 1.2.7

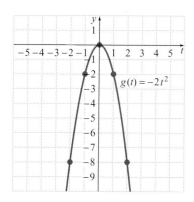

Examine the plotted dots in Figure 1.2.6, which represent certain (t, y) coordinates for the graph of $g(t) = -2t^2$. This set of points seems to give a bowl-shaped picture. We will explore functions such as this in more detail later, but for now, if you draw a curve pointing downward, you will get the graph shown in Figure 1.2.7.

To find the domain of this function, note that $g(t) = -2t^2$ is defined for all values of t. Graphically, there is a y-coordinate on the curve corresponding to *any* t that you choose. Thus the domain is the set of *all real numbers*, $(-\infty, \infty)$.

To determine the range of g, note that the curve lies only in the bottom half of the ty-plane and touches the t-axis at the origin. This means that the y-coordinates take on values that are *less than or equal to zero*. Hence the range is the set of *all real numbers less than or equal to zero*, $(-\infty, 0]$.

By looking at the expression for $g(t)$, $g(t) = -2t^2$, we can come to the same conclusion about the range: t^2 will always be greater than or equal to zero, and when it is multiplied by -2, the end result will always be less than or equal to zero.

✔ *Check It Out 4:* Graph the function $f(x) = x^2 - 2$ and use the graph to find the domain and range of f. ■

Example 5 **Graphing More Functions**

Graph each of the following functions. Use the graph to find the domain and range of f.

(a) $f(x) = \sqrt{4 - x}$

(b) $f(x) = |x|$

▶Solution

(a) We first make a table of values of x and $f(x)$ (Table 1.2.4) and then plot the points on the xy-plane. Connect the dots to get a smooth curve, as shown in Figure 1.2.10.

From Figure 1.2.10, we see the x values for which the function is defined are $x \le 4$. For these values of x, $f(x) = \sqrt{4 - x}$ will be a real number. Thus the domain is $(-\infty, 4]$.

To determine the range, we see that the y-coordinates of the points on the graph of $f(x) = \sqrt{4 - x}$ take on all values greater than or equal to zero. Thus the range is $[0, \infty)$.

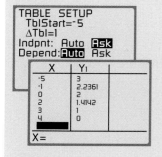

Technology Note

To manually generate a table of values, use the ASK option in the TABLE feature (Figure 1.2.8). The graph of $Y_1 = \sqrt{4 - x}$ is shown in Figure 1.2.9.

Keystroke Appendix: Section 6

Figure 1.2.8

```
TABLE SETUP
 TblStart=-5
 ΔTbl=1
Indpnt: Auto Ask
Depend:Auto Ask
```

X	Y₁
-5	3
-1	2.2361
0	2
2	1.4142
3	1
4	0

X=

Figure 1.2.9

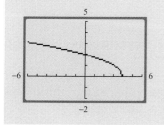

Table 1.2.4

x	$f(x) = \sqrt{4 - x}$
-5	3
-1	$\sqrt{5} \approx 2.236$
0	2
2	$\sqrt{2} \approx 1.414$
3	1
4	0

Figure 1.2.10

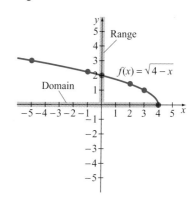

Just In Time

Review absolute value in Section P.1.

(b) The function $f(x) = |x|$ is read as "absolute value of x." It measures the distance of x from zero. Therefore, $|x|$ is always zero or positive. Make a table of values of x and $f(x)$ (see Table 1.2.5) and plot the points. The points form a "V" shape—the graph comes in sharply at the origin, as shown in Figure 1.2.11. This can be confirmed by choosing additional points near the origin. From the graph, we see that the domain is the set of all real numbers, $(-\infty, \infty)$, and the range is the set of all real numbers greater than or equal to zero, $[0, \infty)$.

Table 1.2.5

x	$f(x) = |x|$
-4	4
-2	2
-1	1
0	0
1	1
2	2
4	4

Figure 1.2.11

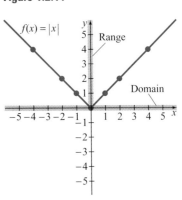

☑ *Check It Out 5:* Graph the function $g(x) = \sqrt{x - 4}$ and use the graph to find the domain and range of g. ■

When a function with some x values excluded is to be graphed, then the corresponding points are denoted by an open circle on the graph. For example, Figure 1.2.12 shows the graph of $f(x) = x^2 + 1$, $x > 0$. To show that $(0, 1)$ is *not* part of the graph, the point is indicated by an open circle.

Figure 1.2.12

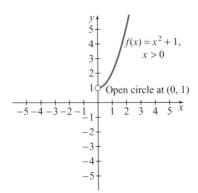

The Vertical Line Test for Functions

We can examine the graph of a set of ordered pairs to determine whether the graph describes a function.

Recall that the definition of a function states that for each value in the domain of the correspondence, there can be only one value in the range. Graphically, this means that any vertical line can intersect the graph of a function at most once.

Figure 1.2.13 is the graph of a function because any vertical line crosses the graph in at most one point. Some sample vertical lines are drawn for reference.

Figure 1.2.14 does *not* represent the graph of a function because a vertical line crosses the graph at more than one point.

Figure 1.2.13   **Figure 1.2.14**

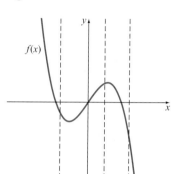   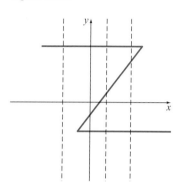

 Vertical Line Test

Use the vertical line test to determine which of the graphs in Figure 1.2.15 are graphs of functions.

Figure 1.2.15

(a)   (b)   (c)

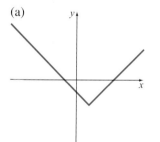

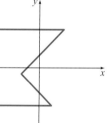   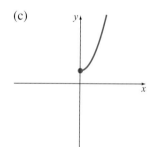

▶Solution Graphs (a) and (c) both represent functions because any vertical line intersects the graphs at most once. Graph (b) does not represent a function because a vertical line intersects the graph at more than one point.

☑ *Check It Out 6:* Does the graph in Figure 1.2.16 represent a function? Explain.

Figure 1.2.16

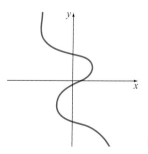

Intercepts and Zeros of Functions

When graphing a function, it is important to know where the graph crosses the x- and y-axes. An **x-intercept** is a point at which the graph of a function crosses the x-axis. In terms of function terminology, the first coordinate of an x-intercept is a value of x such that $f(x) = 0$. Values of x satisfying $f(x) = 0$ are called **zeros** of the function f. The **y-intercept** is the point at which the graph of a function crosses the y-axis. Thus the first coordinate of the y-intercept is 0, and its second coordinate is simply $f(0)$.

Examining the graph of a function can help us to understand the various features of the function that may not be evident from its algebraic expression alone.

Figure 1.2.17

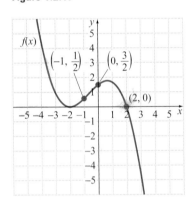

Example 7   Function Values and Intercepts from a Graph

Consider the graph of the function f shown in Figure 1.2.17.

(a) Find $f(-1)$, $f(0)$, and $f(2)$.

(b) Find the domain of f.

(c) What are the x- and y-intercepts of the graph of f?

▶**Solution**

(a) Since $\left(-1, \frac{1}{2}\right)$ lies on the graph of f, the y-coordinate $\frac{1}{2}$ corresponds to $f(-1)$. Thus $f(-1) = \frac{1}{2}$. Similarly, $f(0) = \frac{3}{2}$ and $f(2) = 0$.

(b) From the graph, the domain of f seems to be all real numbers. Unless explicitly stated, graphs are assumed to extend beyond the actual region shown.

(c) The graph of f intersects the x-axis at $x = 2$ and $x = -2$. Thus the x-intercepts are $(-2, 0)$ and $(2, 0)$. The graph of f intersects the y-axis at $y = \frac{3}{2}$, and so the y-intercept is $\left(0, \frac{3}{2}\right)$.

✔ *Check It Out 7:* In Example 7, estimate $f(1)$ and $f(-3)$. ■

1.2 Key Points

▶ For all x in the domain of f, the set of points $(x, f(x))$ is called the **graph** of f.

▶ To sketch the graph of a function, make a table of values of x and $f(x)$ and use it to plot points. Pay attention to the domain of the function.

▶ **Vertical line test:** Any vertical line can intersect the graph of a function at most once.

▶ An **x-intercept** is a point at which the graph of a function crosses the x-axis. The first coordinate of an x-intercept is a value of x such that $f(x) = 0$.

▶ Values of x satisfying $f(x) = 0$ are called **zeros** of the function f.

▶ The **y-intercept** is the point at which the graph of a function crosses the y-axis. The coordinates of the y-intercept are $(0, f(0))$.

1.2 Exercises

▶**Just in Time Exercises** These exercises correspond to the Just in Time references in this section. Complete them to review topics relevant to the remaining exercises.

In Exercises 1–6, evaluate.

1. $|2|$

2. $|-3|$

3. $|(-4)(-1)|$

4. $|(-2)(5)|$

5. $|-3| + 1$

6. $|6| - 8$

▶**Skills** This set of exercises will reinforce the skills illustrated in this section.

In Exercises 7–12, determine whether each set of points determines a function.

7. $S = \{(-2, 1), (1, 5), (1, 2), (6, 1)\}$

8. $S = \{(-4, -1), (1, -1), (2, 0), (3, -1)\}$

9. $S = \left\{\left(-\dfrac{3}{2}, 1\right), (0, 4), (1.4, -2), (0, 1.3)\right\}$

10. $S = \left\{\left(\dfrac{2}{3}, 3\right), (6.7, 1.2), (3.1, 1.4), (4.2, 3.5)\right\}$

11. $S = \{(-5, 2.3), (-4, 3.1), (3, 2.5), (-5, 1.3)\}$

12. $S = \{(-3, -3), (-2, 2), (0, 0), (1, 1)\}$

In Exercises 13–16, fill in the table with function values for the given function, and sketch its graph.

13.

x	-4	-2	0	2	4
$f(x) = -\dfrac{1}{2}x - 4$					

14.

x	-6	-3	0	3	6
$f(x) = \dfrac{1}{3}x + 2$					

15.

x	0	2	$\dfrac{9}{2}$	8	18
$f(x) = \sqrt{2x}$					

16.

x	-16	-9	-4	-1	0
$f(x) = \sqrt{-x} + 3$					

In Exercises 17–46, graph the function without using a graphing utility, and determine the domain and range. Write your answer in interval notation.

17. $f(x) = 2x - 1$

18. $g(x) = -3x + 4$

19. $f(x) = 4x + 5$

20. $g(x) = -5x - 2$

21. $f(x) = -\dfrac{1}{3}x - 4$

22. $f(x) = \dfrac{3}{2}x + 3$

23. $f(x) = -2x + 1.5$

24. $f(x) = 3x - 4.5$

25. $f(x) = 4$

26. $H(x) = 7$

27. $G(x) = 4x^2$

28. $h(x) = -x^2 + 1$

29. $h(s) = -3s^2 + 4$

30. $g(s) = s^2 - 2$

31. $h(x) = \sqrt{x + 4}$

32. $g(t) = \sqrt{t - 3}$

33. $f(x) = \sqrt{3x}$

34. $f(x) = \sqrt{4x}$

35. $f(x) = 2\sqrt{x}$

36. $f(x) = \sqrt{x} + 1$

37. $f(x) = 2|x|$

38. $f(x) = -3|x|$

39. $f(x) = -|x|$

40. $f(x) = |x| + 4$

41. $f(x) = |-2x|$

42. $f(x) = |-4x|$

43. $f(x) = \sqrt{-2x}$

44. $f(x) = \sqrt{-3x}$

45. $H(s) = -s^3 + 1$

46. $f(x) = x^3 - 3$

In Exercises 47–50, determine whether the graph depicts the graph of a function. Explain your answer.

47.

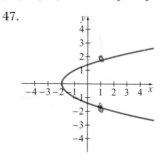

48.

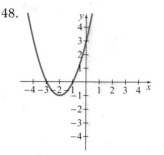

49.
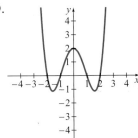

50.
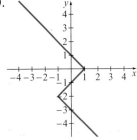

In Exercises 51–54, find the domain and range for each function whose graph is given. Write your answer in interval notation.

51.
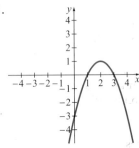

52.
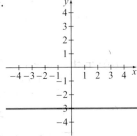

53.
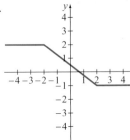

54.
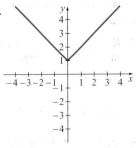

 In Exercises 55–60, use a graphing utility to graph each function. Be sure to adjust your window size to see a complete graph.

55. $f(x) = 2.5|x| + 10$

56. $f(x) = -|1.4\,x| - 15.2$

57. $f(x) = 0.4\sqrt{0.4x - 4.5}$

58. $f(x) = 1.6\sqrt{2.6 - 0.3x}$

59. $f(x) = -2.36x^2 - 9$

60. $f(x) = -2.4x^2 + 8.5$

In Exercises 61–68, for each function f given by the graph, find an approximate value of (a) f(−1), f(0), and f(2); (b) the domain of f; and (c) the x- and y-intercepts of the graph of f.

61.
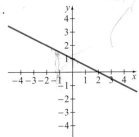

62.
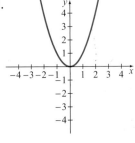

63.
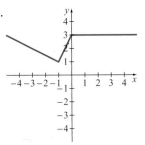

64.
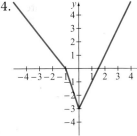

65.
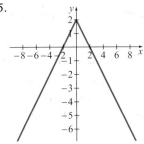

66.
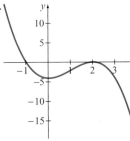

67.
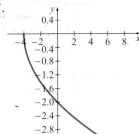

68.
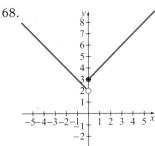

▶**Applications** In this set of exercises you will use graphs to study real-world problems.

69. **NASA Budget** The following graph gives the budget for the National Aeronautics and Space Administration (NASA) for the years 2004–2008. The figures for 2006–2008 are projections. (*Source:* NASA)

Budget for NASA

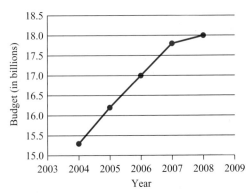

(a) For what year is the NASA budget 17 billion dollars?

(b) By approximately how much did the budget increase from 2004 to 2005?

70. **Geometry** The area of a circle of radius r is given by the function $A(r) = \pi r^2$. Sketch a graph of the function A using values of $r > 0$. Why are negative values of r not used?

71. **Geometry** The volume of a sphere of radius r is given by the function $V(r) = \frac{4}{3}\pi r^3$. Sketch a graph of the function V using values of $r > 0$. Why are negative values of r not used?

72. **Rate and Distance** The time it takes for a person to row a boat 50 miles is given by $T(s) = \dfrac{50}{s}$, where s is the speed at which the boat is rowed. For what values of s does this function make sense? Sketch a graph of the function T using these values of s.

73. **Rate and Distance** A car travels 55 miles per hour. Find and graph the distance traveled by the car (in miles) as a function of the time (in hours). For what values of the input variable does your function make sense?

74. **Construction Costs** It costs $85 per square foot of area to build a house. Find and graph the total cost of building a house as a function of the area (in square feet).

75. **Measuring Distance** In the accompanying figure, a hot air balloon is at point C and an observer is at point A. The balloon is x feet directly above point B, and A is 10 feet to the right of point B. Find and graph the distance $d(x)$ from A to C for $x > 0$.

76. **Ecology** A coastal region with an area of 250 square miles in 2003 has been losing 2.5 square miles of land per year due to erosion. Thus, the area A of the region t years after 2003 is $v(t) = 250 - 2.5t$.
(a) Sketch a graph of v for $0 \le t \le 100$.
(b) What does the x-intercept represent in this problem?

77. **Environmental Science** The sports utility vehicle (SUV) Land Rover Freelander (2005 model) emits 10.1 tons of greenhouse gases per year, while the SUV Mitsubishi Outlander (2005 model) releases 8.1 tons of greenhouse gases per year. (*Source:* www.fueleconomy.gov)
(a) Express the amount of greenhouse gases released by the Freelander as a function of time, and graph the function. What are the units of the input and output variables?
(b) On the same set of coordinate axes as in part (a), graph the amount of greenhouse gases released by the Outlander as a function of time.
(c) Compare the two graphs. What do you observe?

▶**Concepts** This set of exercises will draw on the ideas presented in this section and your general math background.

In Exercises 78–81, graph the pair of functions on the same set of coordinate axes and explain the differences between the two graphs.

78. $f(x) = 2$ and $g(x) = 2x$

79. $h(x) = -2x$ and $g(x) = 2x$

80. $f(x) = -3x^2$ and $g(x) = 3x^2$

81. $f(x) = 3x^2$ and $g(x) = 3x^2 + 1$

82. Explain why the graph of $x = |y|$ is not a function.

83. Is the graph of $f(x) = \sqrt{x + 4}$ the same as the graph of $g(x) = \sqrt{x} + 4$? Explain by sketching their graphs.

In Exercises 84–87, graph the pair of functions on the same set of coordinate axes and find the functions' respective ranges.

84. $f(x) = |x|, g(x) = |x| - 3$

85. $f(x) = x^2 + 4, g(x) = x^2 - 4$

86. $f(x) = \sqrt{2x}, g(x) = \sqrt{x}$

87. $f(x) = 3x + 4, g(x) = 3x + 7$

1.3 Linear Functions

Objectives

▶ Define a linear function

▶ Graph a linear function

▶ Find the slope of a line

▶ Find the equation of a line in both slope-intercept and point-slope forms

▶ Find equations of parallel and perpendicular lines

In the previous sections, you learned some general information about a variety of functions. In this section and those that follow, you will study specific types of functions in greater detail.

One of the most important functions in mathematical applications is a linear function. The following example shows how such a function arises in the calculation of a person's pay.

Example 1 Modeling Weekly Pay

Eduardo is a part-time salesperson at Digitex Audio, a sound equipment store. Each week, he is paid a salary of $200 plus a commission of 10% of the amount of sales he generates that week (in dollars).

(a) What are the input and output variables for this problem?

(b) Express Eduardo's pay for one week as a function of the sales he generates that week.

(c) Make a table of function values for various amounts of sales generated, and use this table to graph the function.

▶**Solution**

(a) Let the variables be defined as follows:

Input variable: x (amount of sales generated in one week, in dollars)

Output variable: $P(x)$ (pay for that week, in dollars)

(b) Eduardo's pay for a given week consists of a fixed portion, $200, plus a commission based on the amount of sales generated that week. Since he receives 10% of the sales generated, the commission portion of his pay is given by $0.10x$. Hence his total pay for the week is given by

$$\text{Pay} = \text{fixed portion} + \text{commission portion}$$
$$P(x) = 200 + 0.10x.$$

(c) Eduardo's pay for a given week depends on the sales he generates. Typically, the total sales generated in one week will be in the hundreds of dollars or more. We make a table of values (Table 1.3.1) and plot the points (Figure 1.3.1).

Table 1.3.1

Sales, x	Pay, $P(x)$
0	200
1000	300
2000	400
3000	500
4000	600

Figure 1.3.1

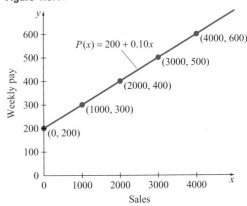

Note that $200 is his salary, which is the fixed portion of his pay. He receives this amount even when no audio equipment is sold, that is, when $x = 0$. The 10% commission is based on his sales—the more he sells, the more he earns.

✔ *Check It Out 1:* Repeat Example 1, but now assume that Eduardo's weekly salary is increased to $500 and his commission is increased to 15% of total sales generated. ■

Definition of a Linear Function

In Example 1, we saw that a salary plus commission can be given by a function of the form $P(x) = 200 + 0.10x$. This is a specific case of a **linear function,** which we now define.

> **Definition of a Linear Function**
>
> A **linear function** $f(x)$ is defined as $f(x) = mx + b$, where m and b are constants.

Linear functions are found in many applications of mathematics. We will explore a number of such applications in the next section. The following examples will show you how to determine whether a function is linear.

Example 2 Determining a Linear Function

Determine which of the following functions are linear.

(a) $f(t) = \dfrac{3}{2}t + 1$ (b) $f(x) = 2x^2$ (c) $f(x) = \pi$

▶Solution

(a) $f(t) = \dfrac{3}{2}t + 1$. This function is linear because it is of the form $mt + b$, with $m = \dfrac{3}{2}$ and $b = 1$.

(b) $f(x) = 2x^2$. This function is *not* linear because x is raised to the second power.

(c) $f(x) = \pi$. This function is linear because it is of the form $mx + b$, with $m = 0$ and $b = \pi$. Recall that π is just a real number (a constant).

✔ *Check It Out 2:* Which of the following are linear functions? For those functions that are linear, identify m and b.

(a) $g(x) = -2x + \dfrac{1}{3}$

(b) $f(x) = x^2 + 4$

(c) $H(x) = 3x$ ■

Slope of a Line

In the definition of a linear function, you might wonder what the significance of the constants m and b are. In Example 1, you saw that Eduardo's pay, P, was given by $P(x) = 200 + 0.10x$. Here b equals 200, which represents the fixed portion of his pay. The quantity m is equal to 0.10, which represents the rate of his commission (the amount he receives *per dollar of sales* he generates in a week). Put another way, for every one dollar increase in sales, Eduardo's pay increases by $0.10.

The graph of a linear function is always a straight line. The quantity m in the definition of a linear function $f(x) = mx + b$ is called the **slope** of the line $y = mx + b$. The slope is the ratio of the change in the output variable to the corresponding change in the input variable. For a *linear function*, this ratio is *constant*.

Let's examine the table given in Example 1 and calculate this ratio.

Table 1.3.2

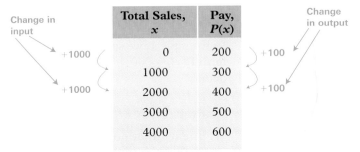

Total Sales, x	Pay, $P(x)$
0	200
1000	300
2000	400
3000	500
4000	600

From Table 1.3.2, we have the following:

$$\text{Slope} = \frac{\text{change in output value}}{\text{change in input value}} = \frac{+100}{+1000} = 0.10$$

We can extend this idea of slope to any nonvertical line. Figure 1.3.2 shows a line $y = mx + b$ and the coordinates of two points (x_1, y_1) and (x_2, y_2) on the line. Here, y_1 and y_2 are the output values and x_1 and x_2 are the input values. The arrows indicate changes in x and y between the points (x_1, y_1) and (y_1, y_2). We define the slope of a line to be the ratio of the change in y to the change in x.

Figure 1.3.2

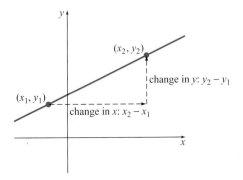

Definition of Slope

The **slope** of a line containing the points (x_1, y_1) and (x_2, y_2) is given by $m = \dfrac{y_2 - y_1}{x_2 - x_1}$, where $x_1 \neq x_2$.

Note Remember the following when calculating the slope of a line:

▶ It does *not* matter which point is called (x_1, y_1) and which is called (x_2, y_2).

▶ As long as two points lie on the same line, the slope will be the same regardless of which two points are used.

Discover *and* **Learn**

Graph the functions
$f(x) = 0.1x$, $g(x) = \frac{1}{2}x$, and
$h(x) = 2x$ on the same set of axes.
What do you notice about the
graphs as m increases?

Example 3 Finding the Slope of a Line

Find the slope of the line passing through the points $(-1, 2)$ and $(-3, 4)$. Plot the points and indicate the slope on your plot.

▶**Solution** The points and the line passing through them are shown in Figure 1.3.3.

Figure 1.3.3

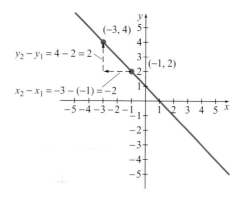

Letting $(x_1, y_1) = (-1, 2)$ and $(x_2, y_2) = (-3, 4)$, we have

$$m = \frac{y_2 - y_1}{x_2 - x_1} \qquad \text{Formula for slope}$$

$$= \frac{4 - 2}{-3 - (-1)} \qquad \text{Substitute values}$$

$$= \frac{2}{-2} = -1. \qquad \text{Simplify}$$

☑ *Check It Out 3:* Find the slope of the line passing through the points $(-5, 3)$ and $(9, 4)$. Plot the points and indicate the slope on your plot. ■

Graphically, the sign of m shows you how the line slants: if $m > 0$, the line slopes upward as the value of x increases. If $m < 0$, the line slopes downward as the value of x increases. This is illustrated in Figure 1.3.4.

Figure 1.3.4

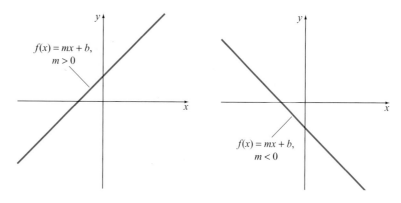

Equation of a Line: Slope-Intercept Form

You might think it would be sufficient to use just m to describe a line, but that is not enough information.

Graph the lines $y_1 = 2x$, $y_2 = 2x + 3$, and $y_3 = 2x - 3$. Note that the value of m is the same for all three lines, but the value of b changes. What do you notice about the differences in the graphs? For each line, find the coordinates of the point where its graph crosses the y-axis.

If two lines have the same slope, then the slant of both lines is the same. However, they may cross the y-axis at different points. Thus, it is important to know the steepness of the line *and* the point where the line crosses the y-axis. For a line $y = mx + b$, the amount of steepness is given by m and the point where the line crosses the y-axis is determined by b; the coordinates of this point are $(0, b)$.

Definition of x- and y-intercepts

▶ The point where the graph of the line $y = mx + b$ crosses the y-axis, $(0, b)$, is called the **y-intercept.** Notice that the x-coordinate of the y-intercept is 0.

▶ The point where the graph of a nonhorizontal line $y = mx + b$ crosses the x-axis is called the **x-intercept.** Since the y-coordinate of the x-intercept is 0, the x-intercept is found by setting the expression $mx + b$ equal to 0 and solving for x.

Note   Even when the variables have labels other than x or y, it is common to still refer to the points where the graph crosses the horizontal or vertical axis as x- or y-intercepts, respectively.

Our discussion leads us to the following form of the equation of a line, which is known as the *slope-intercept* form.

Slope-Intercept Form of the Equation of a Line

The **slope-intercept form** of the equation of a line with slope m and y-intercept $(0, b)$ is given by

$$y = mx + b.$$

Note   We have defined a linear function to be of the form f(x) = mx + b, where m and b can be any real numbers. Since y is often used to represent the output value, you will also see a line represented by the equation y = mx + b.

Point on a Line

A **point** (x_1, y_1) is said to **lie on the line** $y = mx + b$ if $y_1 = mx_1 + b$—that is, if (x_1, y_1) satisfies the equation $y = mx + b$.

Example 4 Equation of a Line in Slope-Intercept Form

Write the equation of the line with a slope of -4 and a y-intercept of $(0, \sqrt{2})$ in slope-intercept form. Does $(0, 1)$ lie on this line? Find the x-intercept of the graph of this line.

▶**Solution** Since the slope, m, is -4 and b is $\sqrt{2}$, the equation of the line in slope-intercept form is

$$y = -4x + \sqrt{2}.$$

Substituting $(x, y) = (0, 1)$, we see that $1 \neq -4(0) + \sqrt{2}$. Thus, $(0, 1)$ does not lie on the line $y = -4x + \sqrt{2}$.

To find the x-intercept of this line, we set $y = f(x) = -4x + \sqrt{2} = 0$. Solving for x, we have $x = \frac{\sqrt{2}}{4}$. Thus $\left(\frac{\sqrt{2}}{4}, 0\right)$ is the x-intercept.

☑ *Check It Out 4:* Write the equation of the line with a slope of 2 and a y-intercept of $\left(0, -\frac{1}{3}\right)$ in slope-intercept form. Does $\left(\frac{1}{6}, 0\right)$ lie on this line? ▪

Equation of a Line: Point-Slope Form

It is not necessary to know the y-intercept to write down the equation of a line. It is sufficient to know the slope and *any* point on the line to write down its equation. To see how we can do this, we first sketch the graph of a line containing a point (x_1, y_1). On this graph, we indicate (x, y) to be any other point on the line, as shown in Figure 1.3.5.

Figure 1.3.5

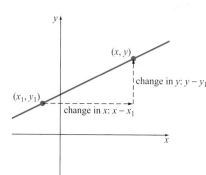

We can derive the equation of the line in Figure 1.3.5 by following these steps:

▶ Write down the formula for the slope of the line passing through the points (x, y) and (x_1, y_1).

$$m = \frac{y - y_1}{x - x_1}$$

▶ Multiply both sides by $x - x_1$.

$$m(x - x_1) = y - y_1$$

Since the slope m and a point on the line (x_1, y_1) are known, we see from the result above that an equation of the line can easily be found.

Equation of a Line in Point-Slope Form

The equation of a line with slope m and containing the point (x_1, y_1) is given by

$$y - y_1 = m(x - x_1)$$

or, equivalently,

$$y = m(x - x_1) + y_1.$$

Using function notation, we can write the linear function as

$$f(x) = m(x - x_1) + y_1.$$

The point-slope form of a line is very useful in many applications, since you may not always be provided the value of b with which to determine the y-intercept. The following examples show how to write the equation of a line in point-slope form.

Example 5 **Equation of a Line in Point-Slope Form**

Find the equation of the line passing through $(2, 6)$ and $(-1, -2)$ in point-slope form. Also, write the equation using function notation.

▶Solution The points and the line passing through them are plotted in Figure 1.3.6. First, we must find the slope. Letting $(x_1, y_1) = (-1, -2)$ and $(x_2, y_2) = (2, 6)$, we have

$$m = \frac{y_2 - y_1}{x_2 - x_1} \qquad \text{Formula for slope}$$

$$= \frac{6 - (-2)}{2 - (-1)} \qquad \text{Substitute values}$$

$$= \frac{8}{3}.$$

Figure 1.3.6

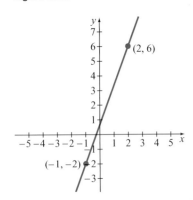

The point-slope formula gives

$$y - y_1 = m(x - x_1)$$

$$y - (-2) = \frac{8}{3}(x - (-1)) \qquad \text{Substitute } m = \frac{8}{3}, \, x_1 = -1, \, y_1 = -2$$

$$y + 2 = \frac{8}{3}(x + 1).$$

You can solve for y to get

$$y = \frac{8}{3}(x + 1) - 2.$$

Note that we could have used the point $(2, 6)$ instead of $(-1, -2)$. The equation of the line would still be the same.

Using function notation, we have

$$f(x) = \frac{8}{3}(x + 1) - 2.$$

> **Just In Time**
>
> *Review equation solving in Section P.8.*

✔ *Check It Out 5:* Find the equation of the line passing through $\left(-\frac{3}{2}, 1\right)$ and $\left(\frac{1}{2}, 2\right)$ in point-slope form. ■

A line given in point-slope form can be written in $y = mx + b$ form (i.e., slope-intercept form) by just rearranging terms.

Example 6 **Rewriting an Equation in Slope-Intercept Form**

Write the equation of the line in Example 5 in slope-intercept form.

▶Solution The equation of the line can be written in slope-intercept form as follows:

$$y = \frac{8}{3}(x + 1) - 2 \qquad \text{Equation in point-slope form}$$

$$= \frac{8}{3}x + \frac{8}{3} - 2 \qquad \text{Distribute } \frac{8}{3}$$

$$= \frac{8}{3}x + \frac{2}{3} \qquad \text{Simplify: } \frac{8}{3} - 2 = \frac{8}{3} - \frac{6}{3} = \frac{2}{3}$$

Thus, $y = \frac{8}{3}x + \frac{2}{3}$ is the desired equation of the line in slope-intercept form.

✔ *Check It Out 6:* Find the equation of the line passing through $\left(-\frac{3}{2}, 1\right)$ and $\left(\frac{1}{2}, 2\right)$ in slope-intercept form. You can use the result of Check It Out 5. ■

The following example illustrates how to find an equation of a line given its graph.

Example **7** **Finding an Equation Given a Graph**

Figure 1.3.7

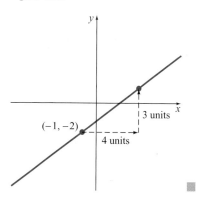

Find the equation of the line whose graph is given in Figure 1.3.7.

▶**Solution** Examining the graph, we note that the coordinates of one point on the line and directions to another point on the line are given. The slope can be calculated from the information that gives the directions from one point to the other. We move 4 units to the right and then 2 units down. Therefore,

$$\text{Slope} = m = \frac{\text{change in } y}{\text{change in } x} = \frac{-2}{+4} = -\frac{1}{2}.$$

Note that the change in y is *negative,* since we move down 2 units. We can now easily use the point-slope form for the equation of the line to get

$y - y_1 = m(x - x_1)$	Point-slope form of equation of a line
$y - 3 = -\frac{1}{2}(x - (-2))$	Substitute $m = -\frac{1}{2}, x_1 = -2, y_1 = 3$
$y = -\frac{1}{2}(x + 2) + 3$	Add 3 to both sides
$y = -\frac{1}{2}x + 2.$	Simplify

Just In Time

Review equation solving in Section P.8.

✔ *Check It Out 7:* Find the equation of the line whose graph is given in Figure 1.3.8.

Figure 1.3.8

Horizontal and Vertical Lines

Two special cases of equations of lines are worth noting: equations of horizontal lines and equations of vertical lines. In both cases, it is useful to see their respective graphs before proceeding to find their equations.

Figure 1.3.9

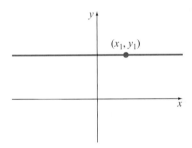

Horizontal Line To find the equation of a horizontal line passing through a point (x_1, y_1), we can first sketch its graph, as shown in Figure 1.3.9.

Since the line is horizontal, a second point on the line, (x_2, y_1), will have the *same* y-coordinate. This is illustrated in Figure 1.3.10.

Therefore, the slope of this line will be

$$m = \frac{y_2 - y_1}{x_2 - x_1} = \frac{y_1 - y_1}{x_2 - x_1} = 0.$$

Another way to see this is to note that since there is no movement in the y direction as we go from the first point to the second, the change in y is 0. Substituting 0 for m in the point-slope formula, we get

$$y - y_1 = 0(x - x_1) \qquad \text{Solve for } y$$
$$y = y_1.$$

Figure 1.3.10

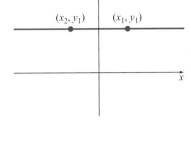

Equation of a Horizontal Line

The **equation of the horizontal line** passing through (x_1, y_1) is

$$y = y_1.$$

Example **8** **Equation of a Horizontal Line**

Find the equation of the line passing through the points $(1, 3)$ and $(-2, 3)$.

▶Solution Plotting the two points given and drawing the line that passes through them, we observe that the line is horizontal (Figure 1.3.11). We can come to the same conclusion by noting that the y-coordinates of the two points are the same.

Figure 1.3.11

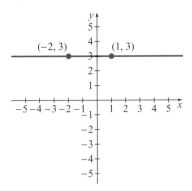

Figure 1.3.12

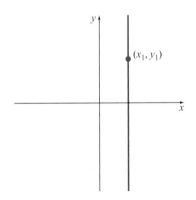

Thus, the equation of the horizontal line is

$$y = 3.$$

✔ *Check It Out 8:* Find the equation of the line passing through the points $(2, -1)$ and $(-6, -1)$. ▪

Vertical Line To find the equation of a vertical line passing through a point (x_1, y_1), we can first sketch its graph, as shown in Figure 1.3.12.

Figure 1.3.13

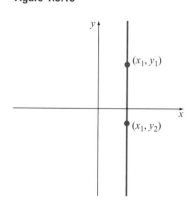

Since the line is vertical, a second point on the line, (x_1, y_2), will have the *same* x-coordinate. This is illustrated in Figure 1.3.13.

If we try to compute the slope of this line, we get

$$m = \frac{y_2 - y_1}{x_2 - x_1} = \frac{y_2 - y_1}{x_1 - x_1} = \frac{y_2 - y_1}{0} \longleftarrow \text{Undefined}$$

which is *undefined*, since there is a zero in the denominator. One way to see this is to note that there is no movement in the x direction as we go from the first point to the second. Thus, we cannot determine the ratio between the change in y and the change in x.

> **Equation of a Vertical Line**
>
> The **equation of the vertical line** passing through (x_1, y_1) is
>
> $$x = x_1.$$

The correspondence between the x- and y-coordinates of the points on a vertical line is *not* a function because the equation of such a line is of the form $x = x_1$ (for some real number x_1) and there are multiple outputs for an input value of $x = x_1$. Strictly speaking, there is no function (linear or otherwise) that represents a vertical line. However, the equation of a vertical line comes up in various contexts, and so it is useful to know and understand it.

Example 9 **Finding the Equation of a Vertical Line**

Find the equation of the line passing through the points $(1, 1)$ and $(1, 3)$.

▶ **Solution** Plotting the two points given and drawing the line that passes through them, we observe that the line is vertical (Figure 1.3.14). We can come to the same conclusion by noting that the x-coordinates of the two points are the same.

Figure 1.3.14

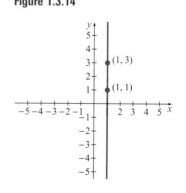

Thus, the equation of the vertical line is

$$x = 1.$$

Since the vertical line does *not* represent a function, we *cannot* write this equation using function notation.

✔ *Check It Out 9:* Find the equation of the line passing through the points $(4, -1)$ and $(4, 2)$. ■

Parallel and Perpendicular Lines

If two nonvertical lines are parallel, they have the same slope. Two vertical lines are always parallel. Also, if you know that two different lines have the same slope, then they are parallel. This leads to the following statement about parallel lines.

Slopes of Parallel Lines

Two different nonvertical lines are parallel if and only if they have the same slope. Vertical lines are always parallel.

If two lines are perpendicular, it can be shown that the following result holds.

Slopes of Perpendicular Lines

Two lines with slopes m_1 and m_2 are perpendicular if and only if

$$m_1 m_2 = -1 \quad \text{or, equivalently,} \quad m_2 = -\frac{1}{m_1}.$$

Thus, perpendicular lines have slopes that are *negative reciprocals* of each other. Vertical and horizontal lines are always perpendicular to each other. The following examples illustrate these properties.

Example 10 Finding the Equation of a Parallel Line

Find an equation of the line parallel to $2x - y = 1$ and passing through $(2, -3)$. Write the equation in slope-intercept form.

▶ Solution First, find the slope of the original line by writing it in slope-intercept form.

$$2x - y = 1$$
$$-y = 1 - 2x \qquad \text{Subtract } 2x$$
$$y = 2x - 1 \qquad \text{Multiply by } -1 \text{ and rearrange terms}$$

Thus, the slope of the original line is

$$m = 2.$$

Since the new line is parallel to the given line, it has the same slope. Using the point-slope form for the equation of a line,

$$y - y_1 = m(x - x_1)$$
$$y - (-3) = 2(x - 2) \qquad \text{Use } m = 2, (x_1, y_1) = (2, -3)$$
$$y + 3 = 2x - 4 \qquad \text{Remove parentheses}$$
$$y = 2x - 7. \qquad \text{Subtract 3 from both sides}$$

Thus, the equation of the parallel line in slope-intercept form is $y = 2x - 7$.

☑ *Check It Out 10:* Find an equation of the line parallel to $-4x + y = 8$ and passing through $(-1, 2)$. Write the equation in slope-intercept form. ■

Technology Note

Graph the lines $Y_1 = 2x - 1$ and $Y_2 = 2x - 7$ to see that they are parallel (Figure 1.3.15).

Keystroke Appendix: Section 7

Figure 1.3.15

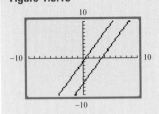

Just In Time

Review equation solving in Section P.8.

Technology Note

Graph the lines
$Y_1 = -\frac{2}{3}x + \frac{1}{3}$ and
$Y_2 = \frac{3}{2}x - 7$ using a
SQUARE window to see
that the lines are perpendicular (Figure 1.3.16). That
is, the units on the x- and
y-axes must be the same
size. In a standard window,
these lines will not appear
to be perpendicular because
the unit for y is smaller than
the unit for x.

Keystroke Appendix:
Section 7

Figure 1.3.16

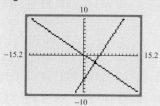

Example 11 Finding the Equation of a Perpendicular Line

Find an equation of the line perpendicular to $2x + 3y - 1 = 0$ and passing through $(4, -1)$. Write the equation in slope-intercept form.

▶**Solution** First, find the slope of the original line by writing the equation in slope-intercept form.

$$2x + 3y - 1 = 0$$
$$3y = -2x + 1 \qquad \text{Isolate 3y term}$$
$$y = -\frac{2}{3}x + \frac{1}{3} \qquad \text{Divide by 3}$$

The slope of the original line is $-\frac{2}{3}$. The line perpendicular to it has slope

$$m = -\frac{1}{-\frac{2}{3}} = \frac{3}{2}. \qquad \text{Negative reciprocal}$$

Using the point-slope form for the equation of a line,

$$y - y_1 = m(x - x_1)$$
$$y - (-1) = \frac{3}{2}(x - 4) \qquad \text{Use } m = \frac{3}{2}, (x_1, y_1) = (4, -1)$$
$$y + 1 = \frac{3}{2}x - 6 \qquad \text{Remove parentheses}$$
$$y = \frac{3}{2}x - 7. \qquad \text{Subtract 1 from each side}$$

Thus, the equation of the perpendicular line in slope-intercept form is $y = \frac{3}{2}x - 7$.

✔ *Check It Out 11:* Find an equation of the line perpendicular to $4x + 2y - 3 = 0$ and passing through $(2, -1)$. Write the equation in slope-intercept form. ▪

1.3 Key Points

▶ A **linear function** $f(x)$ is defined as $f(x) = mx + b$, where m and b are constants.

▶ The **slope** of the line containing the points (x_1, y_1) and (x_2, y_2) is given by
$m = \frac{y_2 - y_1}{x_2 - x_1}$, where $x_1 \neq x_2$.

▶ The **slope-intercept form** of the equation of a line with slope m and y-intercept $(0, b)$ is $y = mx + b$.

▶ The **point-slope form** of the equation of a line with slope m and passing through (x_1, y_1) is $y - y_1 = m(x - x_1)$.

▶ The equation of a **horizontal line** though (x_1, y_1) is given by $y = y_1$.

▶ The equation of a **vertical line** through (x_1, y_1) is given by $x = x_1$.

▶ Nonvertical **parallel lines** have the *same slope*. All vertical lines are parallel to each other.

▶ **Perpendicular lines** have slopes that are *negative reciprocals* of each other. Vertical and horizontal lines are always perpendicular to each other.

1.3 Exercises

▶**Just in Time Exercises** These exercises correspond to the Just in Time references in this section. Complete them to review topics relevant to the remaining exercises.

In Exercises 1–6, solve for y.

1. $y - 4 = 2(y + 1)$

2. $y + 6 = 3(y - 2)$

3. $\frac{1}{2}y + 3 = 4$

4. $\frac{1}{4}y - 1 = -2$

5. $\frac{3}{5}(y - 5) = 9$

6. $\frac{3}{4}(y + 4) = 12$

▶**Skills** This set of exercises will reinforce the skills illustrated in this section.

7. Which of the following are linear functions? Explain your answers.
   (a) $f(t) = 1 + 3t$
   (b) $H(t) = \frac{1}{t} - 1$
   (c) $g(x) = -5x$
   (d) $h(s) = \sqrt{s + 1}$

8. Which of the following are linear functions? Explain your answers.
   (a) $h(s) = \frac{1}{3}s + 1$
   (b) $H(x) = \frac{2}{x^2} + 1$
   (c) $g(x) = 3$
   (d) $f(t) = -3\sqrt{t}$

In Exercises 9–22, find the slope of the line passing through each pair of points (if the slope is defined).

9. $(1, -3)$ and $(0, 4)$

10. $(-1, 2)$ and $(0, -2)$

11. $(1, 3)$ and $(2, 3)$

12. $(4, -1)$ and $(4, 2)$

13. $(0, 1)$ and $(-2, 0)$

14. $(3, 0)$ and $(0, -4)$

15. $(-5, -2)$ and $(-5, 1)$

16. $(4, 1)$ and $(2, 4)$

17. $(0, -2)$ and $\left(0, \frac{1}{2}\right)$

18. $\left(-\frac{1}{2}, 3\right)$ and $(-4, 3)$

19. $\left(\frac{2}{3}, -1\right)$ and $\left(-\frac{1}{3}, -2\right)$

20. $\left(1, \frac{4}{3}\right)$ and $\left(\frac{1}{2}, \frac{2}{3}\right)$

21. $\left(\pi, \frac{\pi}{2}\right)$ and $\left(\frac{\pi}{4}, \frac{\pi}{3}\right)$

22. $\left(\frac{\pi}{2}, \pi\right)$ and $\left(\frac{\pi}{3}, \frac{\pi}{4}\right)$

In Exercises 23 and 24, use the graph of the line to find the slope of the line.

23.   24.

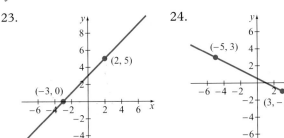

25. Plot the points in Exercise 11, and indicate the direction of the changes in x and y on your plot.

26. Plot the points given in Exercise 12, and indicate the direction of the changes in x and y on your plot.

In Exercises 27–30, check whether each point lies on the line having the equation $y = -2x + 5$.

27. $(2, 1)$ 28. $(0, 3)$

29. $(-1, 0)$ 30. $(1, 3)$

In Exercises 31–42, write the equation of the line in the form $y = mx + b$. Then write the equation using function notation. Find the slope of the line and the x- and y-intercepts.

31. $6x - y = 1$ 32. $-5x - y = -2$

33. $-10x + 5y = 3$ 34. $2x + 3y = -6$

35. $y - 4 = 3(x - 5)$ 36. $y + 3 = -4(x - 1)$

37. $y + 2 = \frac{1}{2}(x + 10)$ 38. $y - 1 = \frac{3}{5}(x - 10)$

39. $y = 5 + 2(x + 4)$ 40. $y = -6 - (x - 1)$

41. $2x - 5y - 10 = 0$ 42. $4x + 3y + 8 = 0$

In Exercises 43–50, write the equation of the line in the form
$y = mx + b$. *Then write the equation using function notation.*
Find the slope and the x- and y-intercepts. Graph the line.

43. $2x + y = 6$ 44. $-3x + y = 4$

45. $4x - 3y = -2$ 46. $3x - 4y = 1$

47. $y - 3 = -2(x - 6)$ 48. $y + 5 = -1(x + 1)$

49. $-5x + 3y - 9 = 0$ 50. $-2x + 4y - 8 = 0$

In Exercises 51–60, find the equation of the line, in point-slope
form, passing through the pair of points.

51. $(2, -1)$ and $(1, 4)$ 52. $(-1, 3)$ and $(0, 1)$

53. $(-1, -4)$ and $(2, -3)$ 54. $(-3, 2)$ and $(5, 0)$

55. $(4, 5)$ and $(7, 5)$ 56. $(10, 8)$ and $(5, 8)$

57. $(1.5, 3.6)$ and $(2.5, 4.5)$ 58. $(4.2, 1.2)$ and $(6.2, 5.7)$

59. $\left(\frac{1}{2}, -1\right)$ and $\left(3, \frac{1}{2}\right)$ 60. $\left(\frac{1}{3}, 2\right)$ and $\left(4, \frac{1}{4}\right)$

In Exercises 61–66, find the equation, in slope-intercept form or in
the form x = c, for some real number c, of each line pictured. Then
write the equation using function notation, if possible.

61.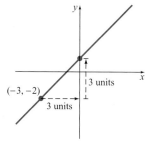

62.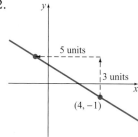

63.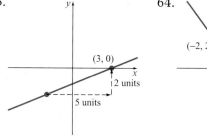

64.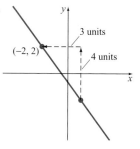

65.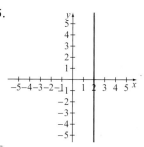

66.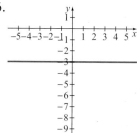

In Exercises 67–72, use a graphing utility to graph each line.
Choose an appropriate window to display the graph clearly.

67. $y = 5.7x + 13.8$

68. $y = -3.2x - 12.6$

69. $3x - y = 15$

70. $0.1x + 0.2y = 2$

71. $y - 0.5 = 2.8(x + 1.7)$

72. $y + 0.9 = -1.4(x - 1.2)$

In Exercises 73–106, find an equation of the line, in slope-
intercept form, having the given properties.

73. Slope: -1; y-intercept: $(0, 2)$

74. Slope: 3; y-intercept: $(0, 5)$

75. x-intercept: $(-3, 0)$; y-intercept: $(0, 1)$

76. x-intercept: $(1, 0)$; y-intercept: $(0, 4)$

77. Slope: $-\frac{1}{2}$; passes through $(1, 2)$

78. Slope: $\frac{2}{3}$; passes through $(2, -3)$

79. Slope: -3; passes through $(3, -1)$

80. Slope: -2; passes through $(1, -1)$

81. Slope: $\frac{3}{4}$; y-intercept: $\left(0, -\frac{3}{2}\right)$

82. Slope: $-\frac{1}{3}$; y-intercept: $(0, 3)$

83. x-intercept: $\left(\frac{1}{2}, 0\right)$; y-intercept: $(0, 3)$

84. x-intercept: $(-3, 0)$; y-intercept: $\left(0, -\frac{3}{2}\right)$

85. Vertical line through $(4, 6)$

86. Horizontal line through $(-1, -4)$

87. Horizontal line through $(2, -1)$

88. Vertical line through $(0, 3)$

89. Horizontal line through $(4, 0.5)$

90. Vertical line through $(-2, 0)$

91. Parallel to the line $y = -3x$ and passing through $(0, -1)$

92. Parallel to the line $y = 2x + 5$ and passing through $(0, 3)$

93. Parallel to the line $y = -\frac{1}{2}x + 2$ and passing through $(4, -1)$

94. Parallel to the line $y = -\frac{1}{3}x - 1$ and passing through $(3, 2)$

95. Parallel to the line $2x + 3y = 6$ and passing through $(-3, 1)$

96. Parallel to the line $-3x + 4y = 8$ and passing through $(8, -2)$

97. Perpendicular to the line $y = \frac{2}{3}x$ and passing through $(0, -1)$

98. Perpendicular to the line $y = -\frac{1}{4}x$ and passing through $(0, -2)$

99. Perpendicular to the line $y = 2x - 1$ and passing through $(-2, 1)$

100. Perpendicular to the line $y = -3x + 2$ and passing through $(1, 4)$

101. Perpendicular to the line $x + 2y = 1$ and passing through $(2, 1)$

102. Perpendicular to the line $2x - y = 1$ and passing through $(-1, 0)$

103. Perpendicular to the line $y = 4$ and passing through $(0, 2)$

104. Perpendicular to the line $x = 3$ and passing through $(2, 1)$

105. Parallel to the line $x = 1$ and passing through $(-2, -5)$

106. Parallel to the line $y = 3$ and passing through $(1, -2)$

▶**Applications** In this set of exercises you will use linear functions to study real-world problems.

107. **Salary** A computer salesperson earns $650 per week plus $50 for each computer sold.
(a) Express the salesperson's earnings for one week as a linear function of the number of computers sold.
(b) Find the values of m and b and interpret them.

108. **Salary** An appliance salesperson earns $800 per week plus $75 for each appliance sold.
(a) Express the salesperson's earnings for one week as a linear function of the number of appliances sold.
(b) Find the values of m and b and interpret them.

109. **Film Industry** The total number of moviegoers, in millions, in the United States has been rising according to the linear function
$$m(t) = 1.6t + 167$$
where t is the number of years since 2003. (*Source:* Motion Picture Association)
(a) According to this function, how many moviegoers will there be in the year 2007?
(b) What does the y-intercept represent for this problem?

110. **Travel** Amy is driving her car to a conference. The function $d(t) = 65t + 100$ represents her total distance from home in miles, and t is the number of hours since 8:00 A.M.
(a) How many miles from home will Amy be at 12:00 P.M.?
(b) What does the y-intercept represent in this problem?

111. **Sales** The number of handbags sold per year since 2003 by Tres Chic Boutique is given by the linear function $h(t) = 160t + 500$. Here, t is the number of years since 2003.
(a) How many handbags were sold in 2006?
(b) What is the y-intercept of this function and what does it represent?
(c) According to the function, in what year will 1200 handbags be sold?

112. **Sales** The number of computers sold per year since 2001 by T.J.'s Computers is given by the linear function $n(t) = 25t + 350$. Here, t is the number of years since 2001.
(a) How many computers were sold in 2005?
(b) What is the y-intercept of this function, and what does it represent?
(c) According to the function, in what year will 600 computers be sold?

113. **Leisure** The admission price to Wonderland Amusement Park has been increasing by $1.50 each year. If the price of admission was $25.50 in 2004, find a linear function that gives the price of admission in terms of t, where t is the number of years since 2004.

114. **Leisure** The admission price to Blue's Water Park has been increasing by $1.75 each year. If the price of admission in 2003 was $28, find a linear function that gives the price of admission in terms of t, where t is the number of years since 2003.

115. **Pricing** The total cost of a car is the sum of the sales price of the car, a sales tax of 6% of the sales price, and $500 for title and tags. Express the total cost of the car as a linear function of the sales price of the car.

116. **Commerce** The total cost of a washing machine is the sum of the selling price of the washing machine, a sales tax of 7.5% of the selling price, and $20 for item disposal. Express the total cost of the washing machine as a linear function of the selling price of the washing machine.

117. **Manufacturing** It costs $5 per watch to manufacture watches, over and above the fixed cost of $5000.
    (a) Express the cost of manufacturing watches as a linear function of the number of watches made.
    (b) What are the domain and range of this function? Keep in mind that this function models a real-world problem.
    (c) Identify the slope and y-intercept of the graph of this linear function and interpret them.
    (d) Using the expression for the linear function you found in part (a), find the total cost of manufacturing 1250 watches.
    (e) Graph the function.

118. **Commerce** It costs $7 per PDA to manufacture PDAs, over and above the fixed cost of $3500.
    (a) Express the cost of manufacturing PDAs as a linear function of the number of PDAs produced.
    (b) What are the domain and range of this function? Keep in mind that this function models a real-world problem.
    (c) Identify the slope and y-intercept of the graph of this linear function. Interpret their meanings in the context of the problem.
    (d) Using the expression for the linear function you found in part (a), find the total cost of manufacturing 2150 PDAs.
    (e) Graph the function.

▶**Concepts** This set of exercises will draw on the ideas presented in this section and your general math background.

119. What happens when you graph $y = x + 100$ in the standard viewing window of your graphing utility? How can you change the window so that you can see a clearer graph?

120. What happens when you graph $y = 100x$ in the standard viewing window of your graphing utility? How can you change the window so that you can see a clearer graph?

121. Sketch by hand the graph of the line with slope $\frac{-4}{5}$ and y-intercept $(0, -1)$. Find the equation of this line.

122. Sketch by hand the graph of the line with slope $\frac{-3}{2}$ and y-intercept $(0, -2)$. Find the equation of this line.

123. Sketch by hand the graph of the line with slope $\frac{5}{3}$ that passes through the point $(-2, 6)$. Find the equation of this line.

124. Sketch by hand the graph of the line with slope $\frac{2}{5}$ that passes through the point $(1, -3)$. Find the equation of this line.

125. Explain why the following table of function values *cannot* be that of a linear function.

t	$g(t)$
-2	-3
-1	-5
0	-8
1	-15

126. Let $f(s) = ms + b$. Find values of m and b such that $f(0) = 2$ and $f(2) = -4$. Write an expression for the linear function $f(s)$. (*Hint:* Start by using the given information to write down the coordinates of two points that satisfy $f(s) = ms + b$.)

127. Let $g(t) = mt + b$. Find m and b such that $g(1) = 4$ and $g(3) = 4$. Write an expression for $g(t)$. (*Hint:* Start by using the given information to write down the coordinates of two points that satisfy $g(t) = mt + b$.)

1.4 Modeling with Linear Functions; Variation

Many problems that arise in business, marketing, and the social and physical sciences are *modeled* using linear functions. This means we *assume* that the quantities we are observing have a linear relationship. In this section, we explore means to analyze such problems. That is, you will learn to take the mathematical ideas in the previous section and *apply* them to a problem. In particular, you will learn to:

▶ Identify input and output variables for a given situation

▶ Find a function that describes the relationship between the input and output variables

▶ Use the function you found to further analyze the problem

In addition to linear models, we will briefly examine models in which the input and output variables have a relationship that is not linear.

Linear Models

A **linear model** is an application whose mathematical analysis results in a linear function. The following example illustrates how such a model may arise.

Example 1 **Modeling Yearly Bonus**

At the end of each year, Jocelyn's employer gives her an annual bonus of $1000 plus $200 for each year she has been employed by the company. Answer the following questions.

(a) Find a linear function that relates Jocelyn's bonus to the number of years of her employment with the company.

(b) Use the function you found in part (a) to calculate the annual bonus Jocelyn will receive after she has worked for the company for 8 years.

(c) Use the function you found in part (a) to calculate how long Jocelyn would have to work at the company for her annual bonus to amount to $3200.

(d) Interpret the slope and y-intercept for this problem, both verbally and graphically.

▶ Solution

(a) We first identify the variables for this problem.

  ▶ t is the input, or independent, variable, denoting the number of years worked.

  ▶ B is the output, or dependent, variable, denoting Jocelyn's bonus.

Next, we proceed to write the equation.

Bonus $= 1000 + 200 \cdot$ number of years employed   *From problem statement*

$\qquad B = 1000 + 200t$ *Substitute the variable letters*

Since B depends on t, we can express the bonus using function notation as follows.

$$B(t) = 1000 + 200t$$

Note that $B(t)$ could also be written as $B(t) = 200t + 1000$. These two forms are equivalent.

(b) Jocelyn's bonus after 8 years of employment is given by $B(8)$:

$$B(8) = 1000 + 200(8) = 1000 + 1600 = 2600$$

Thus, she will receive a bonus of $2600 after 8 years of employment.

(c) To find out when Jocelyn's bonus would amount to $3200, we have to find the value of t such that $B(t) = 3200$.

$3200 = 1000 + 200t$	Substitute 3200 for the bonus
$2200 = 200t$	Isolate the t variable (subtract 1000 from both sides)
$11 = t$	Solve for t (divide both sides by 200)

We see that Jocelyn will need to work for the company for 11 years to receive a bonus of $3200.

(d) Since the function describing Jocelyn's bonus is $B(t) = 1000 + 200t$, the slope of the line $B = 1000 + 200t$ is 200. Recall that slope is the ratio of the change in B to the change in t:

$$\text{Slope} = 200 = \frac{200}{1} = \frac{\text{change in bonus}}{\text{change in years employed}}$$

The slope of 200 signifies that Jocelyn's bonus will increase by $200 each year she works for the company.

The y-intercept for this problem is $(0, 1000)$. It means that Jocelyn will receive a bonus of $1000 at the start of her employment with the company, which corresponds to $t = 0$.

The graphical interpretations of the slope and y-intercept are indicated in Figure 1.4.1.

Figure 1.4.1

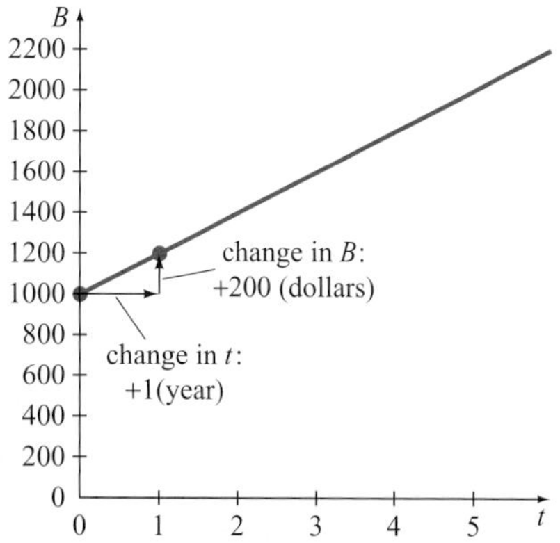

✔ *Check It Out 1:* Rework Example 1 for the case in which Jocelyn's bonus is $1400 plus $300 for every year she has been with the company. ▪

In many models using linear functions, you will be given two data points and asked to find the equation of the line passing through them. We review the procedure in the following example.

Table 1.4.1

x	$f(x)$
1.3	4.5
2.6	1.9

Table 1.4.2

x	$f(x)$
1	4.5
2.5	7.5

Discover *and* Learn

How would you check that the expression for the linear function in Example 2 is correct?

Example 2 Finding a Linear Function Given Two Points

Find the linear function f whose input and output values are given in Table 1.4.1. Evaluate $f(5)$.

▶Solution From the information in the table, we see that the following two points must lie on a line:

$$(x_1, y_1) = (1.3, 4.5) \qquad (x_2, y_2) = (2.6, 1.9)$$

Recall that a y-coordinate is the function value corresponding to a given x-coordinate.

We next calculate the slope:

$$m = \frac{y_2 - y_1}{x_2 - x_1} = \frac{1.9 - 4.5}{2.6 - 1.3} = \frac{-2.6}{1.3} = -2$$

Using the point-slope form for a linear function with $m = -2$ and $(x_1, y_1) = (1.3, 4.5)$,

$$f(x) = m(x - x_1) + y_1 = -2(x - 1.3) + 4.5 = -2x + 2.6 + 4.5 = -2x + 7.1.$$

Evaluating $f(5)$, we have

$$f(5) = -2(5) + 7.1 = -10 + 7.1 = -2.9.$$

✔ *Check It Out 2:* Find the linear function f whose input and output values are given in Table 1.4.2. Evaluate $f(-1)$. ■

Our next example approaches a real-world application in a completely different manner. It is important to keep in mind that these types of problems never present themselves in uniform fashion in the real world. One of the challenges of solving them is to take the various facts that are given and put them into the context of the mathematics that you have learned.

Example 3 Depreciation Model

Example 4 in Section 1.4 builds upon this example. ⋯▶

Table 1.4.3 gives the value of a 2002 Honda Civic Sedan at two different times after its purchase. (*Source:* Kelley Blue Book)

Table 1.4.3

Time After Purchase (years)	Value (dollars)
2	10,000
3	9000

(a) Identify the input and output variables.

(b) Express the value of the Civic as a linear function of the number of years after its purchase.

(c) Using the function found in part (c), find the original purchase price.

(d) Assuming that the value of the car is a linear function of the number of years after its purchase, when will the car's value reach $0?

▶Solution

(a) The input variable, t, is the number of years after purchase of the car. The output variable, v, is the value of the car after t years.

(b) We first compute the slope using the two data points (2, 10,000) and (3, 9000).

$$\text{Slope} = m = \frac{9000 - 10{,}000}{3 - 2} = \frac{-1000}{1} = -1000$$

Then, using the point-slope form for a linear function with (2, 10,000) as the given point, we have

$$v(t) = -1000(t - 2) + 10{,}000 \quad \text{Substitute into point-slope form of equation in function form}$$
$$= -1000t + 2000 + 10{,}000 \quad \text{Remove parentheses}$$
$$= -1000t + 12{,}000. \quad \text{Simplify}$$

(c) The original purchase price would correspond to the value of the car 0 years after purchase. Substituting $t = 0$ into the equation above, we get

$$v(0) = -1000(0) + 12{,}000 = 12{,}000.$$

Thus, the car originally cost \$12,000. Note that this is also the y-intercept of the linear function $v(t) = -1000t + 12{,}000$.

(d) To find out when the car's value will reach \$0, we must set $v(t)$ equal to zero and solve for t.

$$0 = -1000t + 12{,}000 \quad \text{Set } v(t) \text{ equal to zero}$$
$$-12{,}000 = -1000t \quad \text{Isolate the } t \text{ term}$$
$$t = 12 \quad \text{Solve for } t$$

Table 1.4.4

t	$v(t)$
0	12,000
2	10,000
4	8000
6	6000
8	4000
10	2000
12	0

Thus, it will take 12 years for the Honda Civic to reach a value of \$0. To illustrate the decrease in value, we can make a table of values for t and $v(t)$ (Table 1.4.4).

Note that for each 2-year increase in its age, the value of the car goes down by the same amount, \$2000.

☑ *Check It Out 3:* Rework Example 3 for the case in which the value of the car after 5 years is \$4000. Keep its value after 2 years the same as before. ■

The next example graphically explores the car depreciation problem discussed above.

Example 4 Graph of a Depreciation Model

❖••• *This example builds on Example 3 in Section 1.4.*

From Example 3, a 2002 Honda Civic's value over time is given by the linear function $v(t) = -1000t + 12{,}000$, where t denotes the number of years after purchase.

(a) Graph the given function for values of t from 0 through 12. Why must the values of t be greater than or equal to zero?

(b) What is the y-intercept of the graph of this function? Explain its significance for this problem.

(c) Where does the graph cross the t-axis? Find this value of t.

(d) What is the value of the car that corresponds to the point at which the graph crosses the t-axis?

▶Solution

(a) The graph of the function is shown in Figure 1.4.2. The values of t are greater than or equal to zero because the function represents the number of years since 2002.

Figure 1.4.2

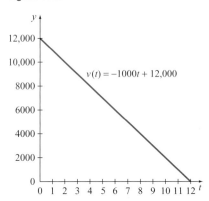

(b) The y-intercept is $(0, 12{,}000)$. It signifies the value of the car at time $t = 0$. In other words, it means that the original purchase price of the car is \$12,000.

(c) From Figure 1.4.2, the graph crosses the t-axis at $t = 12$.

(d) At $t = 12$, the value of the car is \$0. That is, the car will have no value 12 years after purchase. Another way of saying this is $v(12) = 0$. The point $(12, 0)$ is the x-intercept of the graph.

☑ *Check It Out 4:* Repeat Example 4 using the data from Check It Out 3. ▪

Linear Models Using Curve-Fitting

Using mathematics to model real-world situations involves a variety of techniques. We have already used one such technique: taking a problem statement given in English and translating it into a mathematical expression. We used this approach in Example 1 of Section 1.3. The linear function that arises from this problem statement is an exact representation of the problem at hand. However, many real-world problems can only be *approximately* represented by a linear function.

In this section, we will learn how to analyze trends in a set of actual data, and we will gain some experience in generating a model that approximates a set of data. One common technique used in modeling a real-word situation entails finding a suitable mathematical function whose graph closely resembles the plot of a set of data points. This technique is known as **fitting data points to a curve,** or just **curve-fitting.** It is also referred to as **regression.**

Guidelines for Curve-Fitting

▶ Examine the given set of data and decide which variable would be most appropriate as the **input variable** and which as the **output variable.**

▶ Set up a coordinate system and plot the given points, either by hand or by using a graphing utility.

▶ Observe the trend in the data points—does it look like the graph of a function you are familiar with?

Continued

▶ Once you decide on the type of function that your set of data most closely resembles, you must find suitable values of the parameters for the function. For example, if you find that your data represents a linear function $y = mx + b$, then you must find the values of m and b that best approximate the data. The mathematics behind finding the "best" values for m and b is beyond the scope of this text. However, you can use your graphing utility to find these values.

Technology Note

Curve-fitting features can be accessed under the Statistics options in your graphing utility.

Keystroke Appendix:
Section 12

Example 5 Modeling the Relation of Body Weight to Organ Weight

Table 1.4.5 gives the body weights of laboratory rats and the weights of their hearts, in grams. All data points are given to five significant digits. (*Source:* NASA life sciences data archive)

Table 1.4.5

Body Weight (g)	Heart Weight (g)
281.58	1.0353
285.03	1.0534
290.03	1.0726
295.16	1.1034
300.63	1.1842
313.46	1.2673

(a) Let x denote the body weight. Make a scatter plot of the heart weight h versus x.

(b) State, in words, any general observations you can make about the data.

(c) Find an expression for the *linear* function that best fits the given data points.

(d) Compare the actual heart weights for the given body weights with the heart weights predicted by your function. What do you observe?

(e) If a rat weighs 308 grams, use your model to predict its heart weight.

▶ **Solution**

(a) The scatter plot is given in Figure 1.4.3.

Just In Time

Review significant digits in Section P.2.

Figure 1.4.3

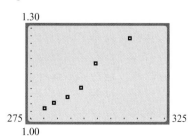

(b) From the scatter plot, it appears that the weight of the heart increases with the weight of the rat.

(c) Using the curve-fitting features of a graphing utility, we obtain the following equation for the "best-fit" line through the data points:

$$h(x) = 0.0075854x - 1.1131$$

All numbers are rounded to five significant digits, since the original data values were given to five significant digits. See Figures 1.4.4 and 1.4.5.

Figure 1.4.4

```
LinReg
y = ax+b
a = .0075853596
b = -1.113118448
```

Figure 1.4.5

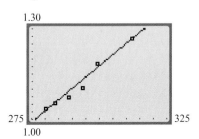

(d) Table 1.4.6 gives the actual heart weights and the heart weights predicted by the model. We see that the values are close, though not exact.

Table 1.4.6

Body Weight (g)	Actual Heart Weight (g)	Predicted Heart Weight (g)
281.58	1.0353	1.0228
285.03	1.0534	1.0490
290.03	1.0726	1.0869
295.16	1.1034	1.1258
300.63	1.1842	1.1673
313.46	1.2673	1.2646

When generating a model from a set of data, you should always compare the given values to those predicted by the model to see if the model is valid.

(e) Substituting the weight of 308 grams into $h(x)$, we have

$$h(308) = 0.0075854(308) - 1.1131 = 1.2232.$$

Thus the rat's predicted heart weight is 1.2232 grams, to five significant digits.

✔ *Check It Out 5:* Use the model in Example 5 to predict the heart weight of a rat weighing 275 grams. ■

> **Note** Linear models should be used only as an approximation for predicting values close to those for which actual data is available. Generally, linear models will be less accurate as we move further away from the given data points.

Direct Variation

Linear models of the form $f(x) = kx$, $k > 0$, are often referred to as models involving **direct variation.** That is, the ratio of the output value to the input value is a positive constant. Models of this type occur frequently in business and scientific applications. For example, if a car travels at a constant speed of 45 miles per hour, then the distance traveled in x hours is given by $f(x) = 45x$. This is a model involving direct variation.

Direct Variation

If a model gives rise to a linear function of the form $f(x) = kx$ or $y = kx$, $k > 0$, we say that the model involves **direct variation.** Other ways of stating this are:

▶ y varies directly as x

▶ y is directly proportional to x

The number k is called the **variation constant** or the **constant of proportionality.**

Example 6 Temperature-Volume Relation

The volume of a fixed mass of gas is directly proportional to its temperature. If the volume of a gas is 40 cc (cubic centimeters) at 25°C (degrees Celsius), find the following.

(a) The variation constant k in the equation

$$V = kT$$

(b) The volume of the same gas at 50°C

(c) The temperature of the same gas if the volume is 30 cc

▶ Solution

(a) To find k, substitute the values for volume and temperature to get

$$40 = k(25) \qquad \text{Use } V = 40 \text{ and } T = 25$$

$$\frac{40}{25} = k \qquad \text{Divide by 25}$$

$$\frac{8}{5} = k. \qquad \text{Simplify}$$

Thus $k = \frac{8}{5}$ and the equation is

$$V = \frac{8}{5}T.$$

(b) Substituting $k = \frac{8}{5}$ in the equation $V = \frac{8}{5}T$,

$$V = \frac{8}{5}(50) = 80.$$

Thus, at 50°C, the volume is 80 cc.

(c) Substitute $V = 30$ into the equation $V = \frac{8}{5}T$.

$$30 = \frac{8}{5}T \Longrightarrow T = \frac{5}{8}(30) = 18.75$$

Thus, the volume is 30 cc at 18.75°C.

✔ *Check It Out 6:* Find the constant k in the equation $V = kT$ if $V = 35$ cc and $T = 45$°C. ■

Inverse Variation

Our focus so far in this section has been on linear models. Many situations, however, are not modeled by linear functions. These applications can be modeled by **nonlinear models.** We briefly discuss one such nonlinear model that occurs frequently in science and business.

Suppose a bus travels a distance of 200 miles. The time t it takes for the bus to reach its destination depends on its speed, r, and the distance traveled. That is, the relationship between r and t is determined by the equation $rt = 200$, or $t = \dfrac{200}{r}$. The greater the speed, the less time it will take to travel 200 miles. This is an example of **inverse variation.**

Inverse Variation

If a model gives rise to a function of the form $f(x) = \dfrac{k}{x}$ or $xy = k$, $k > 0$, we say that the model involves **inverse variation.** Other ways of stating this are:

▶ y varies inversely as x.

▶ y is inversely proportional to x.

The number k is called the **variation constant** or the **constant of proportionality.**

A complete discussion of functions of the form $f(x) = \dfrac{k}{x}$ will be presented in Section 4.6. We next examine a model involving inverse variation.

Example 7 Price-Demand Relation

The price of a product is inversely proportional to its demand. That is, $P = \dfrac{k}{q}$, where P is the price per unit and q is the number of products demanded. If 3000 units are demanded at \$10 per unit, how many units will be demanded at \$6 per unit?

▶**Solution** Since k is a constant, we have

$$k = Pq = (10)(3000) = 30{,}000.$$

If $P = 6$, we have

$$6 = \frac{30{,}000}{q} \qquad \text{Substitute } P = 6,\ k = 30{,}000$$

$$6q = 30{,}000 \qquad \text{Multiply by } q$$

$$q = \frac{30{,}000}{6} = 5000. \qquad \text{Solve for } q$$

Thus, 5000 units will be demanded at \$6 per unit.

✔ *Check It Out 7:* If $P = \dfrac{k}{q}$, where P is the price per unit and q is the number of products demanded, find k if $q = 2500$ and $P = 5$. ■

1.4 Key Points

▶ Begin by reading the problem a couple of times to get an idea of what is going on.

▶ Identify the *input* and *output* variables.

▶ Sometimes, you will be able to write down the linear function for a problem by just reading the problem and "translating" the words into mathematical symbols. This is how we obtained the function in Example 1.

▶ At other times, you will have to look for two data points within the problem to find the slope of your line. Only after you perform this step can you find the linear function. It is critical that you know which will be your input, or independent, variable and which will be your output, or dependent, variable. Then you can write down the data points and plot them.

▶ Sketch a graph of the linear function that you found. Interpret the slope and *y*-intercept both verbally and graphically.

▶ If a model involves *direct variation*, it is a special case of a linear model, with a positive slope and a *y*-intercept of (0, 0).

▶ A model involves *inverse variation* if it gives rise to a function of the form $f(x) = \dfrac{k}{x}$, $k > 0$. Thus it is not linear.

1.4 Exercises

▶**Just in Time Exercises** These exercises correspond to the Just in Time references in this section. Complete them to review topics relevant to the remaining exercises.

In Exercises 1 and 2, find the number of significant figures in each number.

1. 250.03

2. 100.810

In Exercises 3–6, simplify and round your answer to the correct number of significant figures.

3. $256.28 - 251.13$

4. $586.1 - 580.12$

5. $\dfrac{3.09}{0.035}$

6. $\dfrac{4.50}{2.11}$

▶**Skills** This set of exercises will reinforce the skills illustrated in this section.

In Exercises 7–12, for each table of values, find the linear function f having the given input and output values.

7.
x	$f(x)$
3	5
4	9

8.
x	$f(x)$
2	10
6	7

9.
x	$f(x)$
3.1	−2.5
5.6	3.5

10.
x	$f(x)$
1.7	15
3.2	10

11.
x	$f(x)$
30	600
50	900

12.
x	$f(x)$
60	1000
80	1500

In Exercises 13–28, find the variation constant and the corresponding equation for each situation.

13. Let y vary directly as x, and $y = 40$ when $x = 10$.

14. Let y vary directly as x, and $y = 100$ when $x = 20$.

15. Let y vary inversely as x, and $y = 5$ when $x = 2$.

16. Let y vary inversely as x, and $y = 6$ when $x = 8$.

17. Let y vary directly as x, and $y = 35$ when $x = 10$.

18. Let y vary directly as x, and $y = 80$ when $x = 15$.

19. Let y vary inversely as x, and $y = 2.5$ when $x = 6$.

20. Let y vary inversely as x, and $y = 4.2$ when $x = 10$.

21. Let y vary directly as x, and $y = \frac{1}{3}$ when $x = 2$.

22. Let y vary directly as x, and $y = \frac{1}{5}$ when $x = 3$.

23. Let y vary inversely as x, and $y = \frac{3}{4}$ when $x = 8$.

24. Let y vary inversely as x, and $y = \frac{5}{2}$ when $x = 6$.

25. The variable y is directly proportional to x, and $y = 35$ when $x = 7$.

26. The variable y is directly proportional to x, and $y = 48$ when $x = 8$.

27. The variable y is inversely proportional to x, and $y = 14$ when $x = 7$.

28. The variable y is inversely proportional to x, and $y = 4$ when $x = 12$.

▶**Applications** In this set of exercises you will use linear functions and variation to study real-world problems.

29. **Utility Bill** A monthly long-distance bill is $4.50 plus $0.07 for each minute of telephone use. Express the amount of the long-distance bill as a linear function of the number of minutes of use.

30. **Printing** The total cost of printing small booklets is $500 (the fixed cost) plus $2 for each booklet printed. Express the total cost of producing booklets as a function of the number of booklets produced.

31. **Temperature Scales** If 0° Celsius corresponds to 32° Fahrenheit and 100° Celsius corresponds to 212° Fahrenheit, find a linear function that converts a Celsius temperature to a Fahrenheit temperature.

32. **Demand Function** At $5 each, 300 hats will be sold. But at $3 each, 800 hats will be sold. Express the number of hats sold as a linear function of the price per hat.

33. **Physics** The volume of a gas varies directly with the temperature. Find k if the volume is 50 cc at 40°C. What is the volume if the temperature is 60°C?

34. **Tax Rules** According to Internal Revenue Service rules, the depreciation amount of an item is directly proportional to its purchase price. If the depreciation amount of a $2500 piece of equipment is $500, what is the depreciation amount of a piece of equipment purchased at $4000? (*Source:* Internal Revenue Service)

35. **Construction** The rise of a roof (y) is directly proportional to its run (x). If the rise is 5 feet when the run is 8 feet, find the rise when the run is 20 feet.

36. **Electronics Demand** It is known that 10,000 units of a computer chip are demanded at $50 per chip. How many units are demanded at $60 per chip if price varies inversely as the number of chips?

37. **Work and Rate** The time required to do a job, t, varies inversely as the number of people, p, who work on the job. Assume all people work on the job at the same rate. If it takes 10 people to paint the inside of an office building in 5 days, how long will it take 15 people to finish the same job?

38. **Travel** The time t required to drive a fixed distance varies inversely as the speed of the car. If it takes 2 hours to drive from New York to Philadelphia at a speed of 55 miles per hour, how long will it take to drive the same route at a speed of 50 miles per hour?

39. **Rental Cost** A 10-foot U-Haul truck for in-town use rents for $19.95 per day plus $0.99 per mile. You are planning to rent the truck for just one day. (*Source:* www.uhaul.com)

    (a) Write the total cost of rental as a linear function of the number of miles driven.

    (b) Give the slope and y-intercept of the graph of this function and explain their significance.

    (c) How much will it cost to rent the truck if you drive a total of 56 miles?

40. Depreciation The following table gives the value of a personal computer purchased in 2002 at two different times after its purchase.

(a) Express the value of the computer as a linear function of the number of years after its purchase.

Time After Purchase (years)	Value (dollars)
2	1000
3	500

(b) Using the function found in part (a), find the original purchase price.

(c) Assuming that the value of the computer is a linear function of the number of years after its purchase, when will the computer's value reach $0?

(d) Sketch a graph of this function, indicating the x- and y-intercepts.

41. Website Traffic The number of visitors to a popular website grew from 40 million in September 2003 to 58 million in March 2004.

(a) Let t be the number of *months* since September 2003. Plot the given data points with N, the number of visitors, on the vertical axis and t on the horizontal axis. You may find it easier to represent N in millions; that is, represent 40 million visitors as 40 and set the scale for the vertical axis accordingly.

(b) From your plot, find the slope of the line between the two points you plotted.

(c) What does the slope represent?

(d) Find the expression that gives the number of visitors as a linear function of t.

(e) How many visitors are expected in July 2004?

42. Communications A certain piece of communications equipment cost $123 to manufacture in 2004. Since then, its manufacturing cost has been decreasing by $4.50 each year.

(a) If the input variable, t, is the number of years since 2004, find a linear function that gives the manufacturing cost as a function of t.

(b) If the trend continues, what will be the cost of manufacturing the equipment in 2007?

(c) When will the manufacturing cost of the equipment reach $78?

43. Beverage Sales The number of 192-ounce cases of the bottled water Aquafina sold in the United States has been increasing by 50 million cases each year. Suppose 203 million cases of bottled water were sold in 2002. (*Source:* Beverage Marketing Association)

(a) Find a linear function that describes the number (in millions) of 192-ounce cases of Aquafina sold as a function of t. Let t denote the number of years since 2002.

(b) What is the slope of the corresponding line, and what does it signify?

(c) What is the y-intercept of the corresponding line, and what does it signify?

44. Automobile Costs A 2003 Subaru Outback wagon costs $23,500 and gets 22 miles per gallon. Assume that gasoline costs $4 per gallon.

(a) What is the cost of gasoline per mile for the Outback wagon?

(b) Assume that the total cost of owning the car consists of the price of the car and the cost of gasoline. (In reality, the total cost is much more than this.) For the Subaru Outback, find a linear function describing the total cost, with the input variable being the number of miles driven.

(c) What is the slope of the graph of the function in part (b), and what does it signify?

(d) What is the y-intercept of the graph of the function in part (b), and what does it signify?

45. Consumer Behavior Linear models can be used to predict buying habits of consumers. Suppose a survey found that in 2000, 20% of the surveyed group bought designer frames for their eyeglasses. In 2003, the percentage climbed to 29%.

(a) Assuming that the percentage of people buying designer frames is a linear function of time, find an equation for the percentage of people buying designer frames. Let t correspond to the number of years since 2000.

(b) Use your equation to predict the percentage of people buying designer frames in 2006.

(c) Use your equation to predict when the percentage of people who buy designer frames will reach 50%.

(d) Do you think you can use this model to predict the percentage of people buying designer frames in the year 2030? Why or why not?

(e) From your answer to the previous question, what do you think are some limitations of this model?

46. **Economy** In the year 2000, the average hourly earnings of production workers nationwide rose steadily from $13.50 per hour in January to $14.03 per hour in December (*Source:* Bureau of Labor Statistics).

    (a) Create a linear model that expresses average hourly earnings as a function of time, *t*. How would you define *t*?

    (b) Using your function, how much was the average production worker earning in March? in October?

    (c) How fast is the average hourly wage increasing per month?

47. **Traffic Flow** In 2000, the average weekday volume of traffic on a particular stretch of the Princess Parkway was 175,000 vehicles. By 2004, the volume had increased to 200,000 vehicles per weekday.

    (a) By how much did the traffic increase per year? Mathematically, what does this quantity represent?

    (b) Create a linear model for the volume of traffic as a function of time, and use it to determine the average weekday traffic flow for 2006.

48. **Sports Revenue** The revenues for the hockey teams Dallas Stars and New York Rangers for the years 2001 and 2004 are given in the following table. (*Source: Forbes* magazine)

Year	Stars' Revenue ($ million)	Rangers' Revenue ($ million)
2001	85	125
2004	103	118

    (a) Express the revenue, *R*, for the Stars as a linear function of time, *t*. Let *t* correspond to the number of years since 2001.

    (b) Express the revenue, *R*, for the Rangers as a linear function of time, *t*. Let *t* correspond to the number of years since 2001.

    (c) Project the Stars' revenue for the year 2006.

    (d) Project the Rangers' revenue for the year 2006.

    (e) Plot the functions from parts (a) and (b) in the same window of your calculator.

    (f) From your graph, when will both teams generate the same amount of revenue, assuming the same trend continues?

49. **Population Mobility** The following table lists historical mobility rates (the percentage of people who had a change of residence) for selected years from 1960 through 2000. (*Source:* U.S. Census Bureau)

Year	Percentage
1960	20.6
1970	18.7
1980	18.6
1990	16.3
2000	16.1

    (a) What general trend do you notice in these figures?

    (b) Fit a linear function to this set of points, using the number of years since 1960 as the independent variable.

50. **Aging** The following table lists the population of U.S. residents who are 65 years of age or older, in millions. (*Source: Statistical Abstract of the United States*)

Year	Population 65 or Older (in millions)
1990	29.6
1995	31.7
2000	32.6
2003	34.2

    (a) What general trend do you notice in these figures?

    (b) Fit a linear function to this set of points, using the number of years since 1990 as the independent variable.

    (c) Use your function to predict the number of people over 65 in the year 2008.

51. **Higher Education** The following table lists data on the number of college students, both undergraduate and graduate, in the United States for selected years from 1999 through 2005. (*Source:* U.S. Census Bureau)

Year	1999	2002	2003	2005
Population (in millions)	15.2	16.5	16.7	17.5

    (a) What general trend do you notice in these figures?

    (b) Fit a linear function to this set of data, using the number of years since 1999 as the independent variable.

    (c) Use your function to predict the number of college students in the United States in the year 2009.

52. 📊 **Income** The following table lists data on the median household income in the United States for selected years from 1989 through 2003. (*Source:* U.S. Census Bureau)

Year	1989	2000	2002	2003
Median Income (in dollars)	30,056	41,994	43,349	44,368

(a) What general trend do you notice?

(b) Fit a linear function to this set of data, using the number of years since 1989 as the independent variable.

(c) Use your function to predict the year in which the median salary will be $46,000.

53. 📊 **Pricing** Market research of college class ring sales at Salem University revealed the following sales figures for differently priced rings over the past year:

Price (in dollars)	300	425	550	675	800	925
Number of Rings Sold	1252	1036	908	880	432	265

(a) What general trend do you notice in these figures?

(b) Fit a linear function to this set of data, using the price as the independent variable.

(c) Use your function to predict the number of $1000-rings that were sold over the past year.

54. 📊 **Contest** Billy Bob's Hot Dog Emporium has a contest each year. Whoever eats the most hot dogs and buns in 12 minutes is the winner. The following table shows the number of hot dogs eaten by the winner for each year since 2000.

Year	2000	2001	2002	2003	2004	2005	2006
Number of Hot Dogs	14	15	18	17.5	22	23	23.5

(a) What general trend do you notice in these figures?

(b) Fit a linear function to this set of data, using the number of years since 2000 as the independent variable.

(c) Use your function to predict, to the nearest hot dog, the number of hot dogs the winner will consume in the year 2007.

▶**Concepts** This set of exercises will draw on the ideas presented in this section and your general math background.

55. Can $y = 2x + 5$ represent an equation for direct variation?

56. Does the following table of values represent a linear model? Explain.

x	y
−1	4
0	7
1	16
2	30

57. Does the following table of values represent a situation involving direct variation? Explain.

x	y
0	0
1	6
2	12

58. Does the following table of values represent a situation involving inverse variation? Explain.

x	y
1	6
2	9

1.5 Intersections of Lines and Linear Inequalities

Objectives

▶ Algebraically and graphically find the point of intersection of two lines

▶ Understand the advantages of the different approaches for finding points of intersection

▶ Algebraically and graphically solve a linear inequality

▶ Solve a compound inequality

▶ Use intersection of lines and inequalities to model and solve real-world problems

When you use mathematics in a real-world situation, you will often need to find the point where two lines intersect. We begin with one such application, which discusses a comparison between two types of telephone calling plans.

Example 1 Comparing Rate Plans

The Verizon phone company in New Jersey has two plans for local toll calls:

▶ Plan A charges $4.00 per month plus 8 cents per minute for every local toll call.

▶ Plan B charges a flat rate of $20 per month for local toll calls, regardless of the number of minutes of use.

(a) Express the monthly cost for Plan A as a function of the number of minutes used.

(b) Express the monthly cost for Plan B as a function of the number of minutes used.

(c) How many minutes would you have to use per month for the costs of the two plans to be equal?

▶Solution

(a) Let the cost function for Plan A be represented by $A(t)$, where t is the number of minutes used. From the wording of the problem, we have

$$A(t) = 4 + 0.08t.\quad \text{Total monthly cost for Plan A}$$

(b) Let the cost function for Plan B be represented by $B(t)$, where t is the number of minutes used. Since the monthly cost is the same regardless of the number of minutes, we have

$$B(t) = 20.\quad \text{Total monthly cost for Plan B}$$

(c) To find out when the two plans would cost the same, we set $A(t)$ equal to $B(t)$ and solve for t.

$$4 + 0.08t = 20 \quad \text{Set } A(t) = B(t)$$
$$0.08t = 16 \quad \text{Isolate } t \text{ variable}$$
$$t = 200 \quad \text{Solve for } t$$

You would have to use 200 minutes' worth of local toll calls per month for the costs of the two plans to be identical.

✔ *Check It Out 1:* In Example 1, suppose Plan B cost $25 per month and Plan A remained the same. How many minutes would you have to use per month for the costs of the two plans to be equal? ▪

In this section, we will examine problems such as this using both graphical and algebraic approaches.

Just In Time

Review graphs of lines in Sections 1.2 and 1.3.

Finding Points of Intersection

In Example 1 we were asked to find the input value at which two linear functions have the same output value. Graphically, this is the point where the graphs of the two linear functions intersect.

Figure 1.5.1

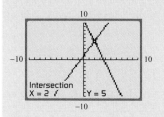

To find the point of intersection of two nonvertical lines algebraically, proceed as
follows.

▶ Solve for y in each equation.

▶ Set the expressions for y equal to each other and solve the resulting equation for x.

The following example shows how to determine the point of intersection of two
lines both algebraically and graphically.

Example 2 Finding a Point of Intersection

Find the point of intersection of the lines given by the equations $y = 2x + 1$ and
$y = -3x + 11$.

▶Solution

Algebraic approach: The point of intersection lies on *both* lines. That is, the same x- and
y-values must satisfy *both* equations $y = 2x + 1$ and $y = -3x + 11$. Since the y values
must be same, we can equate the expressions for y in the two equations:

$$2x + 1 = -3x + 11$$
$$5x = 10 \qquad \text{Rearrange terms}$$
$$x = 2 \qquad \text{Solve for } x$$

Now that we have found $x = 2$, we can substitute this value into either of the original equations to find the y value. We will substitute $x = 2$ into the first equation.

$$y = 2x + 1 = 2(2) + 1 = 5$$

Thus, the point of intersection is $(2, 5)$. Another way of saying this is that the point
$(2, 5)$ is a solution of the pair of equations $y = 2x + 1$ and $y = -3x + 11$.

Graphical approach: Graph the two lines on the same grid, as shown in Figure
1.5.2. From the figure, we see that the intersection occurs at $(2, 5)$. The point $(2, 5)$
lies on both of the lines $y = 2x + 1$ and $y = -3x + 11$.

Figure 1.5.2

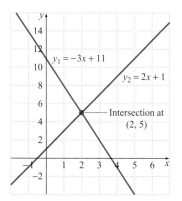

Check To check our solution, we substitute $(2, 5)$ into the two equations and determine if they are *both* satisfied.

$$5 \overset{?}{=} 2(2) + 1 \qquad \text{Substitute x and y into the first equation}$$
$$5 = 5 \qquad \text{The values check in the first equation}$$
$$5 \overset{?}{=} -3(2) + 11 \qquad \text{Substitute x and y into the second equation}$$
$$5 = 5 \qquad \text{The values check in the second equation}$$

✔ *Check It Out 2:* Find the point of intersection of the two lines given by the
equations $y = -2x + 1$ and $y = 4x - 5$ both algebraically and graphically. ▪

Discover *and* Learn

Graphically find the input value at
which the monthly costs for the
two calling plans described in
Example 1 are the same.

Note It is not necessarily the case that any given pair of lines will
intersect at some point—or at only one point. For some illustrations,
see the problems in the *Concepts* section of the exercises for this
section. We will study these types of problems in great detail in
Chapter 6.

Linear Inequalities

In the preceding discussion, we examined methods of finding the points of intersection of lines. It is also useful to determine values of x for which $f(x)$ is *greater than* $g(x)$ or values of x for which $f(x)$ is *less than* $g(x)$.

Reconsider Example 1 about the comparison of two different phone rate plans. The functions for the monthly costs of each plan were determined in Example 1 to be as follows.

$$A(t) = 4 + 0.08t \qquad \text{Cost function for Plan A}$$
$$B(t) = 20 \qquad \text{Cost function for Plan B}$$

The table of values for the costs of both plans, as well as the graphs of the two cost functions, are given in Table 1.5.1 and Figure 1.5.3, respectively.

Table 1.5.1

t	$A(t) = 4 + 0.08t$	$B(t) = 20$
0	4	20
50	8	20
100	12	20
150	16	20
200	20	20
250	24	20
300	28	20

Figure 1.5.3

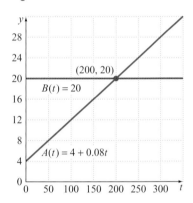

From the table and the graph, we see that Plan A is cheaper if the number of monthly minutes used, t, is between 0 and 200. For t greater than 200 minutes, Plan B is cheaper. For $t = 200$, the two plans will cost the same. For each value of t (other than 200) given in the table, the cost of the cheaper plan is shown in red.

In this discussion, you were able to see that for some values of t, the value of $A(t)$ is less than the value of $B(t)$. Mathematically, this is written as $A(t) < B(t)$. This is an example of an **inequality.** When you find all the values of t at which this inequality is satisfied, you are said to have **solved the inequality.**

Algebraic and Graphical Solutions of Inequalities

In this section, we illustrate how to solve an inequality algebraically. Just as we did with the algebraic method for finding the intersection of lines, we will give a step-by-step procedure for solving inequalities that is guaranteed to work. But first, let's review some properties of inequalities.

Properties of Inequalities

Let a, b, and c be any real numbers.

Addition principle: If $a < b$, then $a + c < b + c$.

Multiplication principle for $c > 0$: If $a < b$, then $ac < bc$ if $c > 0$.

Multiplication principle for $c < 0$: If $a < b$, then $ac > bc$ if $c < 0$. Note that the *direction* of inequality is *reversed* when both sides are multiplied by a negative number.

Similar statements hold true for $a \le b$, $a > b$, and $a \ge b$.

Note The properties of inequalities are somewhat similar to the properties of equations. The main exception occurs when multiplying both sides of an inequality by a negative number: in this case, the direction of the inequality is reversed.

Example 3 shows how to use the properties of inequalities to solve an inequality. The set of values satisfying an inequality is called a **solution set.**

Example 3 Solving a Linear Inequality

Let $y_1(x) = x - 4$ and $y_2(x) = -2x + 2$. Find the values of x at which $y_1(x) > y_2(x)$. Use both an algebraic and a graphical approach.

▶Solution

Algebraic approach:

$$x - 4 > -2x + 2 \qquad \text{Given statement}$$
$$-6 > -3x \qquad \text{Collect like terms}$$
$$\left(-\frac{1}{3}\right)(-6) < \left(-\frac{1}{3}\right)(-3x) \qquad \text{Multiply by } -\frac{1}{3}\text{; switch direction of inequality}$$
$$2 < x \qquad \text{Solve for } x$$
$$x > 2 \qquad \text{Rewrite solution}$$

The set of values of x such that $x > 2$ is the solution set of the inequality. In interval notation, this set is written as $(2, \infty)$.

Figure 1.5.4

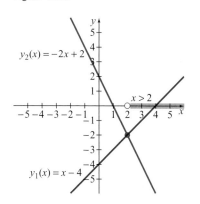

Graphical approach: We want to solve the inequality $y_1(x) > y_2(x)$. Substituting the expressions for $y_1(x)$ and $y_2(x)$, we get

$$x - 4 > -2x + 2.$$

The graphs of $y_1(x) = x - 4$ and $y_2(x) = -2x + 2$ are shown on the same set of axes in Figure 1.5.4. We see that $y_1(x) > y_2(x)$ for $x > 2$.

✔ *Check It Out 3:* Let $y_1(x) = -x + 1$ and $y_2(x) = -3x + 5$. Find the values of x at which $y_1(x) > y_2(x)$. Use both an algebraic and a graphical approach. ■

Note Unlike solving an equation, solving an inequality gives an infinite number of solutions. You cannot really check your solution in the same way that you do for an equation, but you can get an idea of whether your solution is correct by substituting some values from your solution set into the inequality.

You must be careful to choose suitable scales for the x- and y-axes when solving inequalities graphically. Example 3 did not involve any special scaling. However, this will not be the case for every problem that entails solving an inequality. Example 4 illustrates this point.

Example 4 Solving Inequalities: Graphing Considerations

Solve the inequality $40x \le 20x + 100$.

Figure 1.5.5

X	Y1	Y2
0	0	100
1	40	120
2	80	140
3	120	160
4	160	180
5	200	200
6	240	240

X=0

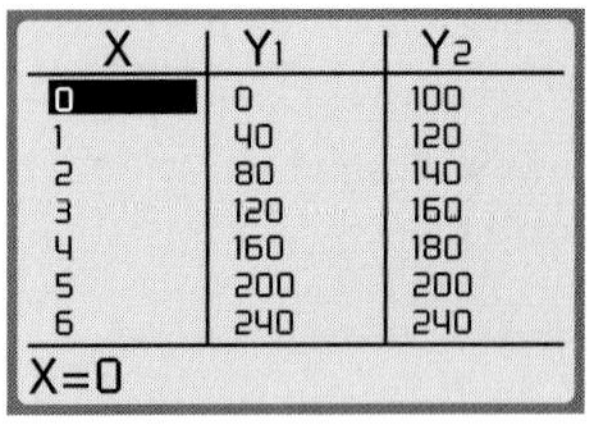

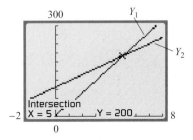

Figure 1.5.6

▶Solution

Algebraic approach:

$$40x \le 20x + 100 \qquad \textit{Given statement}$$

$$20x \le 100 \qquad \textit{Collect like terms}$$

$$\left(\frac{1}{20}\right)20x \le \left(\frac{1}{20}\right)100 \qquad \textit{Multiply by } \frac{1}{20}$$

$$x \le 5 \qquad \textit{Solve for x}$$

Thus, $x \le 5$ is the solution of the inequality $40x \le 20x + 100$. In interval notation, the solution set is $(-\infty, 5]$.

Graphical approach: Let $y_1(x) = 40x$ and $y_2(x) = 20x + 100$. We wish to find the values of x for which $y_1(x)$ is less than or equal to $y_2(x)$. By making a table of x and y values first, we can get a better idea of how to scale the x- and y-axes. See Figure 1.5.5.

The y values are much larger in magnitude than the corresponding x values, so the scale for the x values should be different from the scale for the y values. The window settings, then, must be modified accordingly. From the graph in Figure 1.5.6, the solution is $(-\infty, 5]$.

✔ *Check It Out 4:* Solve the inequality $-30x < 40x + 140$. ■

Compound Inequalities

If two inequalities are joined by the word *and,* then the conditions for both inequalities must be satisfied. Such inequalities are called **compound inequalities.** For example, $-2 \le x + 4$ and $x + 4 < 9$ is a compound inequality that can be abbreviated as $-2 \le x + 4 < 9$.

Example 5 illustrates additional techniques for solving inequalities, including compound inequalities.

Example 5 Solving Additional Types of Inequalities

Solve the following inequalities.

(a) $2x + \frac{5}{2} > 3x - 6$ (b) $-4 \le 3x - 2 < 7$

▶Solution

(a) Solving this inequality involves clearing the fraction. Otherwise, all steps are similar to those used in the previous examples.

$$2x + \frac{5}{2} > 3x - 6 \qquad \textit{Original inequality}$$

$$2\left(2x + \frac{5}{2}\right) > 2(3x - 6) \qquad \textit{Clear fraction: multiply each side by 2}$$

$$4x + 5 > 6x - 12 \qquad \textit{Simplify each side}$$

$$-2x > -17 \qquad \textit{Collect like terms}$$

$$x < \frac{17}{2} \qquad \textit{Divide by } -2\textit{; reverse inequality}$$

Thus, the solution set is the set of all real numbers that are less than $\frac{17}{2}$. In interval notation, this is $\left(-\infty, \frac{17}{2}\right)$.

(b) We solve a compound inequality by working with all parts at once.

$$-4 \le 3x - 2 < 7$$
$$-2 \le 3x \le 9 \qquad \text{Add 2 to each part}$$
$$-\frac{2}{3} \le x \le 3 \qquad \text{Multiply each part by } \frac{1}{3}$$

Thus, the solution set is $\left[-\frac{2}{3}, 3\right]$.

✔ Check It Out 5: Solve the inequality $-\frac{2}{3}x + 4 \le 3x + 5$. ■

To summarize, we see that by using the properties of inequalities, we can algebraically solve any inequality in a manner similar to that used to solve an equation.

Algebraic Approach Versus Graphical Approach: The Advantages and Disadvantages

You should now be able to see some of the advantages and disadvantages of each approach used in solving inequalities, which we summarize here.

The **algebraic approach** has the following **advantages:**

▶ It provides a set of steps to solve *any* problem and so is guaranteed to work.

▶ It gives an exact solution, unless a calculator is used in one or more of the steps.

It also has the following **disadvantages:**

▶ It provides no visual insight into the problem.

▶ It requires you to perform the steps in a mechanical fashion, thereby obscuring an intuitive understanding of the solution.

The **graphical approach** has the following **advantages:**

▶ It allows you to *see* where the inequality is satisfied.

▶ It gives an overall picture of the problem at hand.

It also has the following **disadvantages:**

▶ It does not yield a solution unless the viewing window is chosen properly.

▶ It does not always provide an exact solution—a graphing utility usually gives only an approximate answer.

Note that using *both* approaches simultaneously can give a better idea of the problem at hand. This is particularly true in applications.

Applications

Example 6 illustrates how an inequality can be used in making budget decisions.

Example 6 Budgeting for a Computer

Alicia has a total of $1000 to spend on a new computer system. If the sales tax is 8%, what is the retail price range of computers that she should consider?

Technology Note

The INTERSECT feature can be used to find the intersection of $Y_1 = 1.08x$ and $Y_2 = 1000$. From the graph, you can read the solution to the inequality $Y_1 \leq Y_2$ as [0, 925.93]. See Figure 1.5.7.

Keystroke Appendix:
Section 9

Figure 1.5.7

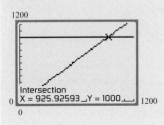

▶**Solution** Let p denote the price of the computer system. The sales tax is then 8% of p, or $0.08p$. The problem can be written and solved as an inequality, as follows.

Price + sales tax ≤ 1000	From problem statement
$p + 0.08p \leq 1000$	Substitute for price and sales tax
$1.08p \leq 1000$	Collect like terms
$p \leq 925.93$	Solve for p

Thus, Alicia can purchase any computer system that has a retail price of less than or equal to $925.93 without having the combination of price and sales tax exceed her budget of $1000.

☑ *Check It Out 6:* Rework Example 6 if Alicia has a total of $1200 to spend on a new computer system. ▪

Example 7 illustrates an application involving weather prediction.

Example 7 Dew Point and Relative Humidity

The dew point is the temperature at which the air can no longer hold the moisture it contains, and so the moisture will condense. The higher the dew point, the more muggy it feels on a hot summer day. The relative humidity measures the moisture in the air at a certain dew point temperature. At a dew point of 70°F, the relative humidity, in percentage points, can be approximated by the linear function

$$RH(x) = -2.58x + 280$$

where x represents the actual temperature. We assume that $x \geq 70$, the dew point temperature. What is the range of temperatures for which the relative humidity is greater than or equal to 40%? (*Source:* National Weather Service)

▶**Solution** We want to solve the inequality $RH(x) \geq 40 \Longrightarrow -2.58x + 280 \geq 40$.

$-2.58x + 280 \geq 40$	
$-2.58x \geq -240$	Subtract 280 from both sides
$x \leq 93.0$	Divide both sides by -2.58

Since we assumed that $x \geq 70$, the solution is

$$70 \leq x \leq 93.0.$$

Table 1.5.2 gives the relative humidity for various values of the temperature above 70°F.

Table 1.5.2

Temperature (°F)	Relative Humidity (%)
72	94.2
75	86.5
78	78.8
80	73.6
85	60.7
95	34.9

Figure 1.5.8

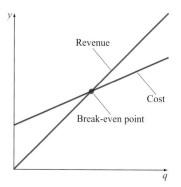

We see that the closer the actual temperature is to the dew point, the higher the relative humidity.

☑ *Check It Out 7:* Rework Example 7 assuming you are interested in the range of temperatures for which the relative humidity is less than 35%. ■

An important application of intersection of lines occurs in business models, when dealing with the production or operating costs of a product and the revenue earned from selling the product. Typical linear cost and production functions are illustrated in Figure 1.5.8. We would like to determine the "break-even point"—that is, the point at which production cost equals revenue. Example 8 explores this topic.

Example 8 Cost and Revenue

To operate a gourmet coffee booth in a shopping mall, it costs $500 (the fixed cost) plus $6 for each pound of coffee bought at wholesale price. The coffee is then sold to customers for $10 per pound.

(a) Find a linear function for the operating cost of selling q pounds of coffee.

(b) Interpret the y-intercept for the cost function.

(c) Find a linear function for the revenue earned by selling q pounds of coffee.

(d) Find the break-even point algebraically.

(e) Graph the two functions on the same set of axes and find the break-even point graphically.

(f) How many pounds of coffee must be sold for the revenue to be greater than the total cost?

▶ Solution

(a) Let $C(q)$ represent the cost of selling q pounds of coffee. From the wording of the problem, we have

$$C(q) = 500 + 6q.$$

(b) The y-intercept of the cost function is $(0, 500)$. This is the amount it costs to operate the booth even if no coffee is bought or sold. This amount is frequently referred to as the *fixed cost*. The variable cost is the cost that depends on the number of pounds of coffee purchased at the wholesale price. The variable cost is added to the fixed cost to get the total cost, $C(q)$.

(c) Since the coffee is sold for $10 per pound, the revenue function $R(q)$ is

$$R(q) = 10q.$$

(d) To find the break-even point algebraically, we set the expressions for the cost and revenue functions equal to each other to get

$$500 + 6q = 10q \qquad \text{Set cost equal to revenue}$$

$$500 = 4q \qquad \text{Collect like terms}$$

$$125 = q. \qquad \text{Solve for } q$$

Thus, the store owner must sell 125 pounds of coffee for the operating cost to equal the revenue. In this case, the production cost is $1250, and so is the revenue.

(e) The two functions are plotted in Figure 1.5.9. Note the scaling of the axes.

Figure 1.5.9

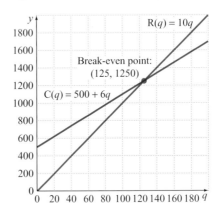

(f) We see from the graph that the revenue is greater than the total cost if more than 125 pounds of coffee is sold. Algebraically, we solve the inequality $R(q) > C(q)$.

$$10q > 500 + 6q \qquad \textit{Substitute expressions for cost and revenue}$$

$$4q > 500 \qquad \textit{Collect like terms}$$

$$q > 125 \qquad \textit{Solve for q}$$

We obtain the same answer: more than 125 pounds of coffee must be sold for the revenue to be greater than the cost.

☑ *Check It Out 8:* Rework Example 8 for the case in which the coffee is sold for $12 per pound. The cost function remains unchanged. Comment on the differences between the new result and the result obtained in Example 8. ■

1.5 Key Points

▶ To find the point of intersection of two lines algebraically, equate the expressions for y in the two equations and solve for x. Then find the corresponding y-value.

▶ To find the point of intersection of two lines graphically, graph both lines and determine where they intersect.

▶ Use the properties of inequalities to solve inequalities algebraically. Remember to reverse the direction of the inequality when multiplying or dividing by a negative number.

▶ Graphically, the solution set of the linear inequality $f(x) > g(x)$ represents the set of all values of x for which the graph of f lies above the graph of g. Similar statements are true for $f(x) \ge g(x)$, $f(x) < g(x)$, and $f(x) \le g(x)$.

1.5 Exercises

▶**Just in Time Exercises** These exercises correspond to the Just in Time references in this section. Complete them to review topics relevant to the remaining exercises.

1. In a function of the form $f(x) = mx + b$, m represents the _____ of the line and b represents the _____.

2. True or False: The slope of the line $y = \frac{1}{2}x + 2$ is 2.

3. True or False: The y-intercept of the line $y = -3x + 1$ is $(0, 1)$.

4. True or False: The graph of the equation $x = -2$ is a horizontal line.

In Exercises 5–12, sketch a graph of the line.

5. $f(x) = 3$

6. $g(x) = -5$

7. $f(x) = x + 3$

8. $g(x) = -2x - 5$

9. $f(x) = -\frac{3}{2}x + 2$

10. $g(x) = \frac{1}{3}x - 1$

11. $y = 0.25x + 10$

12. $y = 0.2x - 1$

▶**Skills** This set of exercises will reinforce the skills illustrated in this section.

In Exercises 13–18, find the point of intersection for each pair of lines both algebraically and graphically.

13. $y = -2x + 4; y = x + 1$

14. $y = x + 2; y = -3x + 2$

15. $y = -x + 4; y = 2x - 5$

16. $y = -2x - 8; y = x - 2$

17. $y = \frac{1}{3}x + 2; y = -\frac{2}{3}x + 5$

18. $y = -\frac{1}{2}x + 5; y = \frac{1}{4}x + 2$

In Exercises 19–36, find the point of intersection for each pair of lines algebraically.

19. $y = -x + 2; y = x + 4$

20. $y = 2x - 5; y = 3x - 6$

21. $y = -2x - 1; y = x + 2$

22. $y = 4x; y = -x + 10$

23. $y = 2x + 6; y = -x - 6$

24. $y = 5x + 1; y = 3x - 1$

25. $y = \frac{1}{3}x - 3; y = -\frac{2}{3}x + 5$

26. $y = -\frac{1}{2}x + 1; y = \frac{1}{4}x + 2$

27. $y = \frac{5}{3}x - 1; y = \frac{3}{2}x - 2$

28. $y = -\frac{2}{5}x + 3; y = \frac{5}{2}x + 4$

29. $y = 0.25x + 6; y = -0.3x - 4$

30. $y = 1.2x - 3; y = x - 2.4$

31. $2x - y = 5; x + y = 16$

32. $-3x + y = -4; 2x - y = 1$

33. $-\frac{3}{2}x - y = 3; x + y = -2$

34. $\frac{1}{5}x - y = -6; 2x - y = 3$

35. $x = -1; y = 4$

36. $x = 3; y = -2$

In Exercises 37 and 38, use the graph to determine the values of x at which $f(x) \geq g(x)$.

37.
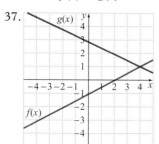

38.
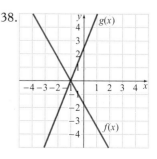

In Exercises 39–42, check whether the indicated value of the independent variable satisfies the given inequality.

39. Value: $x = 1$; Inequality: $x + 1 < 2$

40. Value: $x = \dfrac{1}{2}$; Inequality: $3x + 1 > -1$

41. Value: $s = 3.2$; Inequality: $2s - 1 \le 10$

42. Value: $t = \sqrt{2}$; Inequality: $5 > -t - 1$

In Exercises 43–48, solve the inequality algebraically and graphically. Express your answer in interval notation.

43. $-3x + 2 \le 5x + 10$

44. $2t + 1 > 3t + 4$

45. $8s - 9 \ge 2s + 15$

46. $-x + 6 \le 2x + 9$

47. $2x < 3x - 10$

48. $x \ge 3x - 6$

In Exercises 49–68, solve the inequality. Express your answer in interval notation.

49. $2x + 3 \ge 0$

50. $x - 4 < 0$

51. $4x - 5 > 3$

52. $3x + 1 \le 7$

53. $2 - 2x \ge x - 1$

54. $x + 4 < x - 1$

55. $-4(x + 2) \ge x + 5$

56. $-3(x - 3) < 7x + 1$

57. $\dfrac{x}{3} \le \dfrac{2x}{3} - 1$

58. $-\dfrac{x}{2} > \dfrac{3x}{2} + 3$

59. $\dfrac{1}{3}(x + 1) < x + 3$

60. $-2x - 1 \ge \dfrac{x + 5}{2}$

61. $\dfrac{1}{3}x + 2 \le \dfrac{3}{2}x - 1$

62. $\dfrac{2}{3}x + 3 \le 5$

63. $-2 \le 2x + 1 \le 3$

64. $-4 \le 3x - 2 \le 2$

65. $0 < -x + 5 < 4$

66. $-1 < -2x + 1 < 5$

67. $0 < \dfrac{x + 3}{2} < 3$

68. $1 \le \dfrac{2x - 1}{3} \le 4$

▶**Applications** In this set of exercises you will use the concepts of intersection of lines and linear inequalities to study real-world problems.

Cost and Revenue *In Exercises 69–72, for each set of cost and revenue functions, (a) find the break-even point and (b) calculate the values of q for which revenue exceeds cost.*

69. $C(q) = 2q + 10$;
$R(q) = 4q$

70. $C(q) = 3q + 20$;
$R(q) = 6q$

71. $C(q) = 10q + 200$;
$R(q) = 15q$

72. $C(q) = 8q + 150$;
$R(q) = 10q$

73. **Meteorology** At a dew point of 70°F, the relative humidity, in percentage points, can be approximated by the linear function

$$RH(x) = -2.58x + 280$$

where x represents the actual temperature. We assume that $x \ge 70$, the dew point temperature. What is the range of temperatures for which the relative humidity is greater than or equal to 50%?

74. **Manufacturing** To manufacture boxes, it costs $750 (the fixed cost) plus $2 for each box produced. The boxes are then sold for $4 each.
(a) Find a linear function for the production cost of q boxes.
(b) Interpret the y-intercept of the graph of the cost function.
(c) Find a linear function for the revenue earned by selling q boxes.
(d) Find the break-even point algebraically.
(e) Graph the functions from parts (a) and (c) on the same set of axes and find the break-even point graphically. You will have to adjust the window size and scales appropriately. Compare your result with the result you obtained algebraically.

75. **Film Industry** Films with plenty of special effects are very expensive to produce. For example, *Terminator 3* cost $55 million to make, and another $30 million to market. Suppose an average movie ticket costs $8, and only half of this amount goes to the studio that made the film. How many tickets must be sold for the movie studio to break even for *Terminator 3*? (*Source:* Standford Graduate School of Business)

76. **Pricing Tickets** Sherman is planning to bring in a jazz group of four musicians for a fund-raising concert at Grand State University. The jazz group charges $500 for an appearance, and dinner will be provided to the musicians at a cost of $20 each. In addition, the musicians will be reimbursed for mileage at a rate of $0.30 per

mile. The group will be traveling a total of 160 miles. A ticket for the concert will be priced at $8. How many people must attend the concert for the university to break even?

77. **Special Event Costs** Natasha is the president of the student organization at Grand State University. She is planning a public lecture on free speech by a noted speaker and expects an attendance of 150 people. The speaker charges an appearance fee of $450, and she will be reimbursed for mileage at a rate of $0.30 per mile. She will be traveling a total of 120 miles. The speaker's lunch and dinner will be provided by the organization at a total cost of $45. How much does Natasha need to charge per person for the lecture so that the student organization breaks even?

78. **Communications** A telephone company offers two different long-distance calling plans. Plan A charges a fee of $4.95 per month plus $0.07 for each minute used. Plan B costs $0.10 per minute of use, but has no monthly fee.

(a) Find the total monthly cost of using Plan A as a linear function of the number of minutes used.

(b) Find the total monthly cost of using Plan B as a linear function of the number of minutes used.

(c) Interpret the *y*-intercept of the graph of each cost function.

(d) Calculate algebraically the number of minutes of long-distance calling for which the two plans will cost the same. What will be the monthly charge at that level of usage?

(e) Graph the functions from parts (a) and (b) on the same set of axes and find the number of minutes of long-distance calling for which the two plans will cost the same. You will have to adjust the window size and scales appropriately. What is the monthly cost at that level of usage? Compare your result with the result you found algebraically.

79. **Health and Fitness** A jogger on a pre-set treadmill burns 3.2 calories per minute. How long must she jog to burn at least 200 calories?

80. **Compensation** A salesperson earns $100 a week in salary plus 20% percent commission on total sales. How much must the salesperson generate in sales in one week to earn a total of at least $400 for the week?

81. **Exam Scores** In a math class, a student has scores of 94, 86, 84, and 97 on the first four exams. What must the student score on the fifth exam so that the average of the five tests is greater than or equal to 90? Assume 100 is the maximum number of points on each test.

82. **Sales Tax** The total cost of a certain type of laptop computer ranges from $1200 to $2000. The total cost includes a sales tax of 6%. Set up and solve an inequality to find the range of prices for the laptop before tax.

83. **Cost Comparison** Rental car company A charges a flat rate of $45 per day to rent a car, with unlimited mileage. Company B charges $25 per day plus $0.25 per mile.

(a) Find an expression for the cost of a car rental for one day from Company A as a linear function of the number of miles driven.

(b) Find an expression for the cost of a car rental for one day from Company B as a linear function of the number of miles driven.

(c) Determine algebraically how many miles must be driven so that Company A charges the same amount as Company B. What is the daily charge at this number of miles?

(d) Confirm your algebraic result by checking it graphically.

84. **Car Ownership Costs** In this problem, you will investigate whether it is cost effective to purchase a car that gets better gasoline mileage, even though its purchase price may be higher. A 2003 Subaru Outback wagon costs $23,500 and gets 22 miles per gallon. A 2003 Volkswagen Passat wagon costs $24,110 and gets 25 miles per gallon. Assume that gasoline costs $4 per gallon. (*Sources:* Edmunds.com and U.S. Environmental Protection Agency)

(a) What is the cost of gasoline per mile for the Outback wagon? the Passat wagon?

(b) Assume that the total cost of owning a car consists of the price of the car and the cost of gasoline. For each car, find a linear function describing the total cost, with the input variable being the number of miles driven.

(c) What is the slope of the graph of each function in part (b), and what do the slopes signify?

(d) How many miles would you have to drive for the total cost of the Passat to be the same as that of the Outback?

85. **Education** The overall ratio of students to computers in Maryland public schools declined from 8 to 1 in 2000 to 5 to 1 in 2004. (*Source:* Maryland State Department of Education)

    (a) Write a linear equation that gives the ratio of students to computers in terms of the number of years since 2000. Keep in mind that in your equation, the ratio must be expressed as a single variable and its value must be treated as a single number.

    (b) The state of Maryland would like to achieve a ratio of 3.5 students for every computer. When will the ratio be less than or equal to 3.5?

    (c) 📊 Check your result graphically.

86. **Airplane Manufacturing** The following table shows the market share (percentage of the total market) of airplanes with 100 seats or more for Manufacturer A and Manufacturer B.

	Year	Market Share (%)
Manufacturer A	2000	82
	2005	62
Manufacturer B	2000	10
	2005	35

    (a) Assuming that the market share can be modeled by a linear function, find Manufacturer A's market share as a function of time. Let t denote the number of years since 2000.

    (b) Repeat part (a) for Manufacturer B.

    (c) 📊 Plot the functions from parts (a) and (b) in the same window. What are the trends you observe for the two airline companies?

    (d) If these trends continue, when will the market share for Manufacturer B exceed that for Manufacturer A?

    (e) Set up and solve an inequality to determine when the market share for Manufacturer B will exceed that for Manufacturer A.

    (f) Can you think of events that might change the trend you are observing? Sketch a graph that would reflect a change in trend (this graph may not be linear). Remember that you cannot change the data that already exist for 2000 and 2005!

▶ **Concepts** This set of exercises will draw on the ideas presented in this section and your general math background.

87. What happens when you try to find the intersection of $y = x$ and $y = x + 2$ algebraically? Graph the two lines on the same set of axes. Do they appear to intersect? Why or why not? This is an example of how graphs can help you to see things that are not obvious from algebraic methods. Examples such as this will be discussed in greater detail in a later chapter on systems of linear equations.

88. Find the intersection of the lines $x + y = 2$ and $x - y = 1$. You will have to first solve for y in both equations and then use the methods presented in this section. (This is an example of a *system of linear equations*, a topic that will be explored in greater detail in a later chapter.)

89. What is the intersection of the lines $x + y = 2$ and $2x + 2y = 4$? You will have to first solve for y in both equations. What do you observe when you try to solve the system algebraically? graphically? Examples such as this will be discussed in greater detail in a later chapter on systems of linear equations.

Chapter 1   # Summary

Section 1.1 Functions

Concept	Illustration	Study and Review
Definition of a function A **function** establishes a correspondence between a set of input values and a set of output values in such a way that for each input value, there is exactly one corresponding output value.	The circumference of a circle is given by $2\pi r$, where r is the radius of the circle. This situation describes a function, since there is only one output (circumference) for every input (radius).	Examples 1–7 Chapter 1 Review, Exercises 1–6
Domain and range of a function The **domain** of a function is the set of all allowable input values for which the function is defined. The **range** of a function is the set of all output values that are possible for the given domain of the function.	For $f(x) = 3x^2$, the domain is the set of real numbers and the range is the set of all nonnegative real numbers.	Example 8 Chapter 1 Review, Exercises 7–14, 76

Section 1.2 Graphs of Functions

Concept	Illustration	Study and Review
Graph of a function For all x in the domain of f, the set of points $(x, f(x))$ is called the **graph** of f.	The following is a graph of $\sqrt{4 - x}$. The domain and range are indicated. 	Examples 1–5 Chapter 1 Review, Exercises 15–22, 27–30
The vertical line test for functions Any vertical line can intersect the graph of a function at most once.	The following does *not* represent the graph of a function because the vertical line shown crosses the graph at more than one point. 	Example 6 Chapter 1 Review, Exercises 23–25

Continued

Section 1.2 **Graphs of Functions**

Concept	Illustration	Study and Review
Intercepts and zeros of functions An **x-intercept** is a point at which the graph of a function crosses the x-axis. The first coordinate of an x-intercept is a value of x such that $f(x) = 0$. Values of x satisfying $f(x) = 0$ are called **zeros** of the function f. The **y-intercept** is the point at which the graph of a function crosses the y-axis. The coordinates of the y-intercept are $(0, f(0))$.	Let $f(x) = 3x - 2$. The zero of f is obtained by solving $3x - 2 = 0 \Rightarrow x = \frac{2}{3}$. The x-intercept is $\left(\frac{2}{3}, 0\right)$. The y-intercept is $(0, f(0)) = (0, -2)$.	Example 7 Chapter 1 Review, Exercise 26

Section 1.3 **Linear Functions**

Concept	Illustration	Study and Review
Definition of a linear function A **linear function** $f(x)$ is defined as $f(x) = mx + b$, where m and b are constants.	The functions $f(x) = -2x + 5$, $g(x) = \frac{1}{3}x$, and $h(x) = -4$ are all examples of linear functions.	Examples 1, 2 Chapter 1 Review, Exercises 31–34
Definition of slope The **slope** of a line containing the points (x_1, y_1) and (x_2, y_2) is given by $$m = \frac{y_2 - y_1}{x_2 - x_1}$$ where $x_1 \neq x_2$.	The slope of the line passing through $(-3, 2)$ and $(4, 5)$ is $$m = \frac{y_2 - y_1}{x_2 - x_1} = \frac{5 - 2}{4 - (-3)} = \frac{3}{7}.$$	Example 3 Chapter 1 Review, Exercises 35–40
Equations of lines The **slope-intercept form** of the equation of a line with slope m and y-intercept $(0, b)$ is $y = mx + b$.	In slope-intercept form, the equation of a line with slope -3 and y-intercept $(0, 2)$ is $y = -3x + 2$.	Examples 4–9 Chapter 1 Review, Exercises 41–50, 77, 78
The **point-slope form** of the equation of a line with slope m and passing through (x_1, y_1) is $y - y_1 = m(x - x_1)$.	In point-slope form, the equation of the line with slope -3 and passing through $(1, -2)$ is $y - (-2) = -3(x - 1)$ or, equivalently, $y + 2 = -3(x - 1)$.	
The equation of a **horizontal line** through (x_1, y_1) is given by $y = y_1$. The equation of a **vertical line** through (x_1, y_1) is given by $x = x_1$.	The horizontal line through $(2, -1)$ has the equation $y = -1$, while the vertical line through $(2, -1)$ has the equation $x = 2$.	
Parallel and perpendicular lines Nonvertical **parallel lines** have the *same slope*. All vertical lines are parallel to each other.	The lines $y = -2x + 1$ and $y = -2x - 4$ are parallel to each other because both have a slope of -2.	Examples 10, 11 Chapter 1 Review, Exercises 47–50
Perpendicular lines have slopes that are *negative reciprocals* of each other. Vertical and horizontal lines are always perpendicular to each other.	The lines $y = -2x + 1$ and $y = \frac{1}{2}x + 5$ are perpendicular to each other because -2 and $\frac{1}{2}$ are negative reciprocals of each other.	

Section 1.4 Modeling with Linear Functions; Variation

Concept	Illustration	Study and Review
Guidelines for finding a linear model • Begin by reading the problem a couple of times to get an idea of what is going on. • Identify the *input* and *output* variables. • Sometimes, you will be able to write down the linear function for the problem by just reading the problem and "translating" the words into mathematical symbols. • At other times, you will have to look for two data points within the problem to find the slope of your line. Only after you perform this step can you find the linear function. • Interpret the slope and *y*-intercept both verbally and graphically.	A computer bought for $1500 in 2003 is worth $500 in 2005. Express the value of the computer as a linear function of the number of years after its purchase. The input variable, x, is the number of years after purchase, and the output variable, v, is the value of the computer. The data points for the problem are $(0, 1500)$ and $(2, 500)$. The slope is $$m = \frac{500 - 1500}{2 - 0} = \frac{-1000}{2} = -500.$$ Since the *y*-intercept is given, use the slope-intercept form of the equation to get $v(x) = -500x + 1500$. The slope of -500 states that the value of the computer *decreases* by $500 each year. The *y*-intercept is $(0, 1500)$, and it gives the initial cost of the computer.	Examples 1–5 Chapter 1 Review, Exercises 7, 51, 52, 79, 82, 85
Direct and inverse variation **Direct variation:** A model giving rise to a linear function of the form $f(x) = kx$ or $y = kx$, $k > 0$. **Inverse variation:** A model giving rise to a function of the form $f(x) = \dfrac{k}{x}$ or $xy = k$, $k > 0$. In both models, $k > 0$ is called a **variation constant** or **constant of proportionality.**	The function $f(x) = 5x$, or $y = 5x$, is a direct variation model with constant $k = 5$. The function $f(x) = \dfrac{10}{x}$, or $xy = 10$, is an inverse variation model with $k = 10$.	Examples 6, 7 Chapter 1 Review, Exercises 53–56, 80, 81

Section 1.5 Intersections of Lines and Linear Inequalities

Concept	Illustration	Study and Review
Algebraic method for finding the intersection of two lines To find the point of intersection of two lines algebraically, equate the expressions for *y* in the two equations and solve for *x*. Then find the corresponding *y* value.	To find the point of intersection of the lines $y = 2x$ and $y = -x + 6$ algebraically, equate the expressions for y to get $2x = -x + 6$. Then solve to get $x = 2$. The corresponding y value is $y = 2x = 2(2) = 4$. The point of intersection is $(2, 4)$.	Examples 1, 2 Chapter 1 Review, Exercises 57–64, 83

Continued

Section 1.5 Intersections of Lines and Linear Inequalities

Concept	Illustration	Study and Review
Finding points of intersection by graphing You need to find an input value such that two linear functions have the same output value. Graphically, this is the point at which the graphs of the two lines intersect.	To find the point of intersection of the lines given by the equations $y = 2x$ and $y = -x + 6$, graph both lines on the same grid and locate the intersection point. From the graph, the point of intersection is $(2, 4)$.	Example 2 Chapter 1 Review, Exercises 57–60
Properties of inequalities Let a, b, and c be any real numbers. **Addition principle:** If $a < b$, then $a + c < b + c$. **Multiplication principle for $c > 0$:** If $a < b$, then $ac < bc$ if $c > 0$. **Multiplication principle for $c < 0$:** If $a < b$, then $ac > bc$ if $c < 0$. Note that the *direction* of the inequality is *reversed* when both sides are multiplied by a negative number. Similar statements hold true for $a \le b$, $a > b$, and $a \ge b$.	Using the properties of inequalities, we can solve the inequality $2x \ge -x + 6$. $$2x \ge -x + 6 \Longrightarrow 3x \ge 6 \Longrightarrow x \ge 2$$ The solution set is $[2, \infty)$. This can also be seen from the above graph. Note that the line $y = 2x$ is above the line $y = -x + 6$ for values of x greater than 2. The lines intersect at $x = 2$.	Examples 3–8 Chapter 1 Review, Exercises 65–74, 84

Chapter 1   Review Exercises

Section 1.1 _____

In Exercises 1–6, evaluate (a) $f(4)$, (b) $f(-2)$, (c) $f(a)$, and (d) $f(a + 1)$ for each function.

1. $f(x) = 3x - 1$

2. $f(x) = 2x^2 - 1$

3. $f(x) = \dfrac{1}{x^2 + 1}$

4. $f(x) = \sqrt{x^2 - 4}$

5. $f(x) = |2x + 1|$

6. $f(x) = \dfrac{x + 1}{x - 1}$

In Exercises 7–14, find the domain of the function. Write your answer in interval notation.

7. $h(x) = x^3 + 2$

8. $H(x) = \sqrt{6 - x}$

9. $f(x) = \dfrac{7}{x - 2}$

10. $g(x) = \dfrac{1}{x^2 - 4}$

11. $f(x) = \dfrac{-3}{x^2 + 2}$

12. $f(x) = -\dfrac{1}{(x + 5)^2}$

13. $f(x) = |x| + 1$

14. $g(x) = \dfrac{x}{(x + 2)(x - 3)}$

Section 1.2

In Exercises 15–22, graph the function and determine its domain and range.

15. $f(x) = 3$

16. $h(x) = -2x + 3$

17. $h(x) = 3x + \dfrac{1}{2}$

18. $f(x) = \sqrt{5 - x}$

19. $g(x) = -2x^2 + 3$

20. $G(x) = x^2 - 4$

21. $f(x) = -2|x|$

22. $f(x) = |x| - 2$

In Exercises 23 and 24, determine whether the set of points defines a function.

23. $S = \{(0, 1), (2, 3), (3, 4), (6, 10)\}$

24. $S = \{(-1, 1), (2, 3), (2, -5), (4, 12)\}$

In Exercise 25, determine which of the following are graphs of functions. Explain your answer.

25. (a)

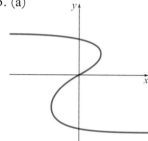

(b)

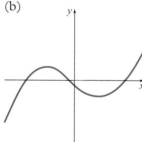

(c)

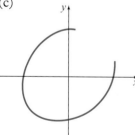

(d)

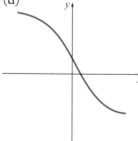

In Exercise 26, evaluate $f(-2)$ and $f(1)$ and find the x- and y-intercepts for f given by the graph.

26.

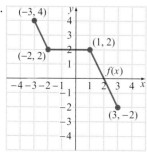

In Exercises 27–30, graph the pair of functions on the same set of coordinate axes, and explain the difference between the two graphs.

27. $f(x) = -3$ and $g(x) = -3x$

28. $g(t) = \dfrac{1}{5}t$ and $h(t) = -\dfrac{1}{5}t$

29. $h(w) = w^2$ and $f(w) = -w^2$

30. $f(t) = 2t^2$ and $g(t) = 2t^2 + 1$

Section 1.3

In Exercises 31–34, determine whether each function is a linear function. Explain your answers.

31. $f(x) = \dfrac{3}{4}x$

32. $H(x) = 4x^3 + 2$

33. $g(t) = \dfrac{3}{t} + 1$

34. $h(x) = \dfrac{1}{\sqrt{x}} - 8$

In Exercises 35–40, for each pair of points, find the slope of the line passing through the points (if the slope is defined).

35. $(-2, 0), (0, 5)$

36. $(1, -6), (-4, 5)$

37. $\left(\dfrac{2}{3}, 1\right), \left(-\dfrac{1}{2}, 3\right)$

38. $(4.1, 5.5), (2.1, -3.5)$

39. $(3, 1), (5, 1)$

40. $(-2, 5), (-2, 7)$

In Exercises 41–50, find an equation of the line with the given properties and express the equation in slope-intercept form. Graph the line.

41. Passing through the point $(4, -1)$ and with slope -2

42. Vertical line through the point $(5, 0)$

43. x-intercept: $(-2, 0)$; y-intercept: $(0, 3)$

44. x-intercept: $(1, 0)$; y-intercept: $(0, -2)$

45. Passing through the points $(-8, -3)$ and $(12, -7)$

46. Passing through the points $(-3, -5)$ and $(0, 5)$

47. Perpendicular to the line $x - y = 1$ and passing through the point $(-1, 2)$

48. Perpendicular to the line $-3x + y = 4$ and passing through the point $(2, 0)$

49. Parallel to the line $x + y = 3$ and passing through the point $(3, -1)$

50. Parallel to the line $-2x + y = -1$ and passing through the point $(0, 3)$

Section 1.4 _____

In Exercises 51 and 52, for each table of values, find the linear function f having the given input and output values.

51.
x	$f(x)$
0	4
3	6

52.
x	$f(x)$
-1	3
4	-2

In Exercises 53–56, find the variation constant and the corresponding equation.

53. Let y vary directly as x, and $y = 25$ when $x = 10$.

54. Let y vary directly as x, and $y = 40$ when $x = 8$.

55. Let y vary inversely as x, and $y = 9$ when $x = 6$.

56. Let y vary inversely as x, and $y = 12$ when $x = 8$.

Section 1.5 _____

In Exercises 57–60, find the point of intersection for each pair of lines both algebraically and graphically.

57. $y = x - 4; y = -x + 2$ 58. $y = -2; y = -3x + 1$

59. $y = 2x - 1; y = -3x - 1$ 60. $2y - x = 9; y = -x$

In Exercises 61–64, find the point of intersection for each pair of lines algebraically.

61. $y = 6x - 4; y = -x + 3$ 62. $y = -2; y = \frac{1}{3}x - 1$

63. $y = x - 1; y = -\frac{5}{2}x - 4$ 64. $y = -\frac{4}{5}x + \frac{1}{2}; y = -\frac{3}{5}x$

In Exercises 65–70, solve the inequality. Express your answer in interval notation.

65. $-7x + 3 > 5x - 2$ 66. $5x - 2 \le 3x + 7$

67. $\frac{1}{3}x - 6 \ge 4x - 1$ 68. $x - 4 \le \frac{2}{5}x$

69. $4 \le \frac{2x + 2}{3} \le 7$ 70. $-1 \le \frac{x - 4}{3} \le 4$

In Exercises 71–74, use a graphing utility to solve the inequality.

71. $3.1x - 0.5 \le -2.2x$

72. $-0.6x + 12 \ge 1.8x$

73. $-3 \le 1.5x + 6 \le 4$

74. $0 \le -3.1x + 6.5 \le 3$

Applications

75. **Elections** The following graph gives the percentage of the voting population who cast their ballots in the U.S. presidential election for the years 1980–2000. (*Source: Statistical Abstract of the United States*)

Percent voting in U.S. presidential elections

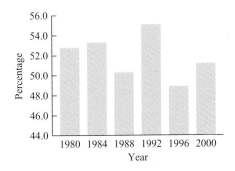

(a) Estimate the percentage of the voting population who cast their ballots in the 2000 election.

(b) In what year during this time period did the maximum percentage of the voting population cast their votes?

76. **Geometry** If the surface area of a sphere is given by $S(r) = 4\pi r^2$, find and interpret $S(3)$. What are the values of r for which this function makes sense, and why?

77. **Salary** A commissioned salesperson's earnings can be determined by the function

$$S(x) = 800 + 0.1x$$

where x is the total amount of sales generated by the salesperson per week. Find and interpret $S(20,000)$.

78. **Rental Costs** Charlie is renting a cargo van for the day. The van costs \$70.00 per day plus \$0.35 for each mile driven.

(a) Write the total cost of the van rental as a linear function of the miles driven.

(b) Find the values of the slope and y-intercept and interpret them.

(c) Find the total cost of the van rental if Charlie drove 300 miles in one day.

79. **Sales** The number of gift-boxed pens sold per year since 2003 by The Pen and Quill Shop is given by the linear function $h(t) = 400 + 80t$. Here, t is the number of years since 2003.

(a) According to the function, how many gift-boxed pens will be sold in 2006?

(b) What is the y-intercept of this function, and what does it represent?

(c) In what year will 1120 gift-boxed pens be sold?

80. **Business** The revenue of a wallet manufacturer varies directly with the quantity of wallets sold. Find the revenue function if the revenue from selling 5000 wallets is $30,000. What would be the revenue if 800 wallets were sold?

81. **Economics** The demand for a product is inversely proportional to its price. If 400 units are demanded at a price of $3 per unit, how many units are demanded at a price of $2 per unit?

82. **Depreciation** The following table gives the value of a computer printer purchased in 2002 at two different times after its purchase.

Time After Purchase (years)	Value (dollars)
2	300
4	200

(a) Express the value of the printer as a linear function of the number of years after its purchase.

(b) Using the function found in part (a), find the original purchase price.

(c) Assuming the value of the printer is a linear function of the number of years after its purchase, when will the printer's value reach $0?

(d) Sketch a graph of this function, indicating the x- and y-intercepts.

83. **Business** If the production cost function for a product is $C(q) = 6q + 240$ and the revenue function is $R(q) = 10q$, find the break-even point.

84. **Business** Refer to Exercise 83. How many products must be sold so that the revenue exceeds the cost?

85. **Music** The following table shows the number of music compact discs (CDs), in millions, sold in the United States for the years 2000 through 2004. (*Source:* Recording Industry Association of America)

Year	Units Sold (in millions)
2000	942
2001	881
2002	803
2003	745
2004	766

(a) What general trend do you notice in these figures?

(b) Fit a linear function to this set of points, using the number of years since 2000 as the independent variable.

(c) Use your function to predict the number of CDs that will be sold in the United States in 2007.

Chapter 1   Test

1. Let $f(x) = -x^2 + 2x$ and $g(x) = \sqrt{x + 6}$. Evaluate each of the following.
   (a) $f(-2)$
   (b) $f(a - 1)$
   (c) $g(3)$
   (d) $g(10)$

2. Find the domain in interval form of $f(x) = -3x$.

3. Find the domain in interval form of $f(x) = \frac{1}{x - 5}$.

4. Sketch the graph of $f(x) = -2x - 3$ and find its domain.

5. Sketch the graph of $f(x) = \sqrt{x + 3}$ and find its domain.

6. Sketch the graph of $f(x) = -x^2 + 4$ and find its domain and range.

7. Determine whether the set of points $S = \{(1, -1), (0, 1), (1, 2), (2, 3)\}$ defines a function.

8. Determine whether the following graph is the graph of a function.

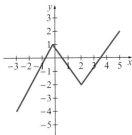

9. Explain why $f(x) = 4x^{-1} + 2$ is not a linear function.

10. Find the slope of the line, if defined, passing through the pair of points.
   (a) $(2, -5)$ and $(4, 2)$
   (b) $(-2, 4)$ and $(-2, 6)$

11. Find the slope-intercept form of the equation of the line passing through the point $(-1, 3)$ and with slope -4.

12. Find the slope-intercept form of the equation of the line passing through the points $(5, -2)$ and $(3, 0)$.

13. Find the equation of the line perpendicular to the line $2y - x = 3$ and passing through $(1, 4)$. Write the equation in slope-intercept form.

14. Find the equation of the line parallel to the line $-4x - y = 6$ and passing through $(-3, 0)$. Write the equation in slope-intercept form.

15. Find the equation of the horizontal line through $(4, -5)$.

16. Find the equation of the vertical line through $(7, -1)$.

17. If y varies directly as x and $y = 36$ when $x = 8$, find the variation constant and the corresponding equation.

18. If y varies inversely as x and $y = 10$ when $x = 7$, find the variation constant and the corresponding equation.

19. Find the point of intersection of the pair of lines $2x + y = 5$ and $x - y = -2$ algebraically and graphically.

20. Solve the inequality $\dfrac{-2x + 3}{4} \le 5$. Express your answer in interval notation.

21. Solve the inequality $-2 \le \dfrac{5x - 1}{2} < 4$. Express your answer in interval notation.

22. A house purchased for \$300,000 in 2006 increases in value by \$15,000 each year.
   (a) Express the value of the house as a linear function of t, the number of years after its purchase.
   (b) According to your function, when will the price of the house reach \$420,000?

23. Julia is comparing two rate plans for cell phones. Plan A charges \$0.18 per minute with no monthly fee. Plan B charges \$8 per month plus \$0.10 per minute. What is the minimum number of minutes per month that Julia must use her cell phone for the cost of Plan B to be less than or equal to that of Plan A?

24. The production cost for manufacturing q units of a product is $C(q) = 3200 + 12q$. The revenue function for selling q units of the same product is $R(q) = 20q$. How many units of the product must be sold to break even?

More About Functions and Equations

Multinational corporations must work with a variety of units and currencies. For example, in order to state their annual profits in euros instead of dollars, they must work with *two* functions—one to determine the profit in dollars, and the other to convert dollars to euros. See Exercise 117 in Section 2.2. This chapter will present topics from coordinate geometry as well as additional properties of functions such as combinations of functions. We will study the graphs of functions in more detail and work with additional types of functions, such as absolute value and piecewise-defined functions.

2.1 Coordinate Geometry: Distance, Midpoints, and Circles

Just in Time

Review the Pythagorean Theorem in Section P.7.

Figure 2.1.1 Right triangle on coordinate plane

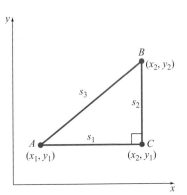

Discover *and* Learn

Find four different points that are a distance of 2 units from the point (0, 0).

Distance Between Two Points

In this section, you will learn how to find the distance between two points and use it to find the equation of a circle. The formula for the distance between two points is based on the Pythagorean Theorem from geometry: In a right triangle, the square of the length of the side opposite the right angle equals the sum of the squares of the lengths of the other two sides.

We place the right triangle on an xy-coordinate system with the two perpendicular sides parallel to the x- and y-axes, and label the sides and vertices as shown in Figure 2.1.1. We then use the Pythagorean Theorem to find s_3, the length of the side opposite the right angle, which is the distance between the points A and B.

$$\text{Length of side } AC = s_1 = x_2 - x_1$$
$$\text{Length of side } BC = s_2 = y_2 - y_1$$
$$\text{Length of side } AB = s_3 = \sqrt{s_1^2 + s_2^2} = \sqrt{(x_2 - x_1)^2 + (y_2 - y_1)^2}$$

Distance Formula

The distance d between the points (x_1, y_1) and (x_2, y_2) is given by

$$d = \sqrt{(x_2 - x_1)^2 + (y_2 - y_1)^2}.$$

Midpoint of a Line Segment

The midpoint of a line segment is the point that is equidistant from the endpoints of the segment.

Midpoint of a Line Segment

The coordinates of the **midpoint** of the line segment joining the points (x_1, y_1) and (x_2, y_2) are

$$\left(\frac{x_1 + x_2}{2}, \frac{y_1 + y_2}{2} \right).$$

Notice that the x-coordinate of the midpoint is the average of the x-coordinates of the endpoints, and the y-coordinate of the midpoint is the average of the y-coordinates of the endpoints.

Example **1** **Calculating Distance and Midpoint**

(a) Find the distance between the points $(3, -5)$ and $(6, 1)$.

(b) Find the midpoint of the line segment joining the points $(3, -5)$ and $(6, 1)$.

▶**Solution**

(a) Using the distance formula with $(x_1, y_1) = (3, -5)$ and $(x_2, y_2) = (6, 1)$,

$$d = \sqrt{(x_2 - x_1)^2 + (y_2 - y_1)^2}$$
$$= \sqrt{(6 - 3)^2 + (1 - (-5))^2}$$
$$= \sqrt{3^2 + 6^2} = \sqrt{9 + 36} = 3\sqrt{5}.$$

(b) Using the midpoint formula with $(x_1, y_1) = (3, -5)$ and $(x_2, y_2) = (6, 1)$, the coordinates of the midpoint are

$$\left(\frac{x_1 + x_2}{2}, \frac{y_1 + y_2}{2}\right) = \left(\frac{3 + 6}{2}, \frac{-5 + 1}{2}\right)$$

$$= \left(\frac{9}{2}, -2\right).$$

☑ *Check It Out 1:*

(a) Find the distance between the points $(1, 2)$ and $(-4, 7)$.

(b) Find the midpoint of the line segment joining the points $(1, 2)$ and $(-4, 7)$. ■

The distance formula is useful in describing the equations of some basic figures. Next we apply the distance formula to find the standard form of the equation of a circle.

Equation of a Circle

Figure 2.1.2

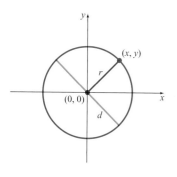

Recall from geometry that a circle is the set of all points in a plane whose distance to a fixed point is a constant. The fixed point is called the **center** of the circle, and the distance from the center to any point on the circle is called the **radius** of the circle. A **diameter** of a circle is a line segment through the center of the circle with endpoints on the circle. The length of the diameter is twice the length of the radius of the circle. See Figure 2.1.2.

Next we find the equation of a circle with center at the origin. The distance from the center of a circle to any point (x, y) on the circle is the radius of the circle, r. This gives us the following.

Distance from center to (x, y) = radius of circle

$$\sqrt{(x - 0)^2 + (y - 0)^2} = r \qquad \text{Apply distance formula}$$

$$x^2 + y^2 = r^2 \qquad \text{Square both sides of equation}$$

Figure 2.1.3

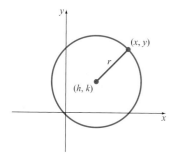

Similarly, we can find the equation of a circle with center at the point (h, k) and radius r, as shown in Figure 2.1.3. To do this, we use the distance formula to represent r in terms of $x, y, h,$ and k.

Distance from center (h, k) to (x, y) = radius of circle

$$\sqrt{(x - h)^2 + (y - k)^2} = r \qquad \text{Apply distance formula}$$

$$(x - h)^2 + (y - k)^2 = r^2 \qquad \text{Square both sides of equation}$$

Equation of a Circle in Standard Form

The circle with center at $(0, 0)$ and radius r is the set of all points (x, y) satisfying the equation

$$x^2 + y^2 = r^2.$$

The circle with center at (h, k) and radius r is the set of all points (x, y) satisfying the equation

$$(x - h)^2 + (y - k)^2 = r^2.$$

Example 2 **Finding the Standard Form of the Equation of a Circle**

Write the standard form of the equation of the circle with center at $(3, -2)$ and radius 5. Sketch the circle.

Figure 2.1.4 Circle with radius 5 and center $(3, -2)$

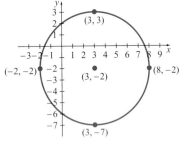

▶**Solution** The center $(3, -2)$ corresponds to the point (h, k) in the equation of the circle. We have

$$(x - h)^2 + (y - k)^2 = r^2 \qquad \textit{Equation of circle}$$
$$(x - 3)^2 + (y - (-2))^2 = 5^2 \qquad \textit{Use h = 3, k = -2, r = 5}$$
$$(x - 3)^2 + (y + 2)^2 = 25. \qquad \textit{Standard form of equation}$$

To sketch the circle, first plot the center $(3, -2)$. Since the radius is 5, we can plot four points on the circle that are 5 units to the left, to the right, up, and down from the center. Using these points as a guide, we can sketch the circle, as shown in Figure 2.1.4.

✔ *Check It Out 2:* Write the standard form of the equation of the circle with center at $(4, -1)$ and radius 3. Sketch the circle. ■

Example 3 **Finding the Equation of a Circle Given a Point on the Circle**

Write the standard form of the equation of the circle with center at $(-1, 2)$ and containing the point $(1, 5)$. Sketch the circle.

Figure 2.1.5

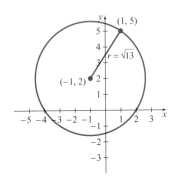

▶**Solution** Since the radius is the distance from $(-1, 2)$ to $(1, 5)$,

$$r = \sqrt{(1 - (-1))^2 + (5 - 2)^2} = \sqrt{4 + 9} = \sqrt{13}.$$

The equation is then

$$(x - h)^2 + (y - k)^2 = r^2 \qquad \textit{Equation of circle}$$
$$(x - (-1))^2 + (y - 2)^2 = (\sqrt{13})^2 \qquad \textit{Use h = -1, k = 2, r = }\sqrt{13}$$
$$(x + 1)^2 + (y - 2)^2 = 13 \qquad \textit{Standard form of equation}$$

The circle is sketched in Figure 2.1.5.

✔ *Check It Out 3:* Write the standard form of the equation of the circle with center at $(3, -1)$ and containing the point $(0, 2)$. ■

The General Form of the Equation of a Circle

The equation of a circle can be written in another form, known as the general form of the equation of a circle. To do so, we start with the standard form of the equation and expand the terms.

$$(x - h)^2 + (y - k)^2 = r^2 \qquad \textit{Equation of circle}$$
$$x^2 - 2hx + h^2 + y^2 - 2ky + k^2 = r^2 \qquad \textit{Expand terms on left side}$$
$$x^2 + y^2 - 2hx - 2ky + h^2 + k^2 - r^2 = 0 \qquad \textit{Rearrange terms in decreasing}$$
$$\textit{powers of x and y}$$

Letting $D = -2h$, $E = -2k$, and $F = h^2 + k^2 - r^2$, we get the **general form of the equation of a circle.**

> **General Form of the Equation of a Circle**
>
> The **general form of the equation of a circle** with center (h, k) and radius r is given by
>
> $$x^2 + y^2 + Dx + Ey + F = 0$$
>
> where $D = -2h$, $E = -2k$, and $F = h^2 + k^2 - r^2$.

Just In Time

Review trinomials that are perfect squares in Section P.5.

If you are given the equation of a circle in general form, you can use the technique of **completing the square** to rewrite the equation in standard form. You can then quickly identify the center and radius of the circle. To complete the square on an expression of the form $x^2 + bx$, add an appropriate number c so that $x^2 + bx + c$ is a perfect square trinomial.

For example, to complete the square on $x^2 + 8x$, you add 16. This gives $x^2 + 8x + 16 = (x + 4)^2$. In general, if the expression is of the form $x^2 + bx$, you add $c = \left(\dfrac{b}{2}\right)^2$ to make $x^2 + bx + c$ a perfect square trinomial. Table 2.1.1 gives more examples.

Table 2.1.1

Begin With	Then Add	To Get
$x^2 + 10x$	$\left(\dfrac{10}{2}\right)^2 = 5^2 = 25$	$x^2 + 10x + 25 = (x + 5)^2$
$y^2 - 8y$	$\left(\dfrac{-8}{2}\right)^2 = (-4)^2 = 16$	$y^2 - 8y + 16 = (y - 4)^2$
$x^2 + 3x$	$\left(\dfrac{3}{2}\right)^2 = \dfrac{9}{4}$	$x^2 + 3x + \dfrac{9}{4} = \left(x + \dfrac{3}{2}\right)^2$
$x^2 + bx$	$\left(\dfrac{b}{2}\right)^2$	$x^2 + bx + \left(\dfrac{b}{2}\right)^2 = \left(x + \dfrac{b}{2}\right)^2$

Example 4 Completing the Square to Write the Equation of a Circle

Write the equation $x^2 + y^2 + 8x - 2y - 8 = 0$ in standard form. Find the coordinates of the center of the circle and find its radius. Sketch the circle.

▶**Solution** To put the equation in standard form, we complete the square on both x and y.

Step 1 Group together the terms containing x and then the terms containing y. Move the constant to the right side of the equation. This gives

$$x^2 + y^2 + 8x - 2y - 8 = 0 \qquad \textit{Original equation}$$
$$(x^2 + 8x) + (y^2 - 2y) = 8. \qquad \textit{Group x, y terms}$$

Step 2 Complete the square for each expression in parentheses by using $\left(\dfrac{8}{2}\right)^2 = 16$ and $\left(\dfrac{-2}{2}\right)^2 = 1$. Remember that any number added to the left side of the equation must also be added to the right side.

$$(x^2 + 8x + 16) + (y^2 - 2y + 1) = 8 + 16 + 1$$
$$(x^2 + 8x + 16) + (y^2 - 2y + 1) = 25$$

Figure 2.1.6 Circle with radius 5 and center $(-4, 1)$

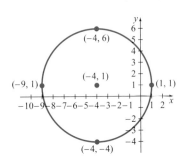

Step 3 Factoring to get $x^2 + 8x + 16 = (x + 4)^2$ and $y^2 - 2y + 1 = (y - 1)^2$, we have

$$(x + 4)^2 + (y - 1)^2 = 25 = 5^2.$$

The equation is now in standard form. The coordinates of the center are $(-4, 1)$ and the radius is 5. Using the center and the radius, we can sketch the circle shown in Figure 2.1.6.

☑ *Check It Out 4:* Write the equation $x^2 + y^2 + 2x - 6y - 6 = 0$ in standard form. Find the coordinates of the center of the circle and find its radius. Sketch the circle. ▦

A circle is not a function because its graph does not pass the vertical line test. However, the equation of a circle can be rewritten to represent two different functions by solving for y. The top half of the circle is given by $y = \sqrt{r^2 - x^2}$, and the bottom half is given by $y = -\sqrt{r^2 - x^2}$. Each equation now represents a function. This fact will be useful when using graphing calculators.

Example **5** **Graphing a Circle with a Graphing Utility**

Use a graphing utility to graph the circle whose equation is given by

$$x^2 + y^2 - 2y - 8 = 0.$$

▶Solution A graphing utility can graph only functions. Therefore, we must rewrite the equation of the circle so that y is given in terms of x.

Step 1 Put the equation in standard form. Following the procedure used in Example 4, the standard form of the equation is

$$x^2 + (y - 1)^2 = 9.$$

The center is at $(0, 1)$ and the radius is 3.

Step 2 Solve for y.

$$
\begin{aligned}
x^2 + (y - 1)^2 &= 9 \\
(y - 1)^2 &= 9 - x^2 \qquad \text{Subtract } x^2 \text{ from both sides} \\
y - 1 &= \pm\sqrt{9 - x^2} \qquad \text{Take square root of both sides} \\
y &= \pm\sqrt{9 - x^2} + 1 \qquad \text{Solve for } y
\end{aligned}
$$

This gives us two functions:

$$Y_1(x) = \sqrt{9 - x^2} + 1 \quad \text{and} \quad Y_2(x) = -\sqrt{9 - x^2} + 1$$

Figure 2.1.7

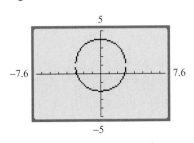

Step 3 Enter the two functions into the graphing calculator. Use a suitable window size that will show the entire circle. Since the center is $(0, 1)$ and the radius is 3, the leftmost and rightmost points of the circle will be $(-3, 1)$ and $(3, 1)$, respectively. The highest and lowest points on the circle will be $(0, 4)$ and $(0, -2)$, respectively. Thus, one possible window size containing these points is $[-4, 4] \times [-5, 5]$. After entering the window size, make sure you use the option for a square screen, or your circle will appear oval. The square option will adjust the original horizontal size of the window. See Figure 2.1.7.

Note that the circle does not close completely. This is because the calculator has only limited resolution. For this graph, there are no highlighted pixels corresponding to $(3, 1)$ and $(-3, 1)$. To see this, trace through the circle. You will notice that the x

values traced are never equal to $x = 3$ or $x = -3$. To get around this problem, you can use a decimal window. This option is discussed in the accompanying Technology Note.

✔ *Check It Out 5:* Use a graphing utility to graph the circle whose equation is given by $x^2 + y^2 + 2x - 8 = 0$. ▪

Technology Note Consider the graph of the circle given by
$$Y_1(x) = \sqrt{9 - x^2} + 1 \quad \text{and} \quad Y_2(x) = -\sqrt{9 - x^2} + 1.$$

In Example 5, we saw that the graphing calculator graphs a circle, but with breaks. One way to avoid this problem is to use a decimal window in which the x values are in increments of 0.1. In this way, $x = -3$ and $x = 3$ will be included in the set of x values generated by the calculator.

Use a decimal window to graph the circle given above. Set Ymin $= -2.1$ and Ymax $= 4.1$ in the WINDOW menu. Otherwise, not all of the circle will show. You should see a picture similar to Figure 2.1.8.

Keystroke Appendix:
Section 7

Figure 2.1.8 Graph of circle using decimal window

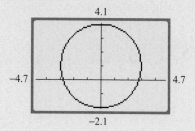

2.1 Key Points

▸ The **distance** d between the points (x_1, y_1) and (x_2, y_2) is given by
$$d = \sqrt{(x_2 - x_1)^2 + (y_2 - y_1)^2}.$$

▸ The coordinates of the **midpoint** of the line segment joining the points (x_1, y_1) and (x_2, y_2) are
$$\left(\frac{x_1 + x_2}{2}, \frac{y_1 + y_2}{2} \right).$$

▸ The **standard form of the equation of a circle** with center at (h, k) and radius r is given by
$$(x - h)^2 + (y - k)^2 = r^2.$$

▸ The **general form of the equation of a circle** is given by
$$x^2 + y^2 + Dx + Ey + F = 0.$$

By completing the square, we can rewrite the general equation in standard form.

2.1 Exercises

▶**Just in Time Exercises** These exercises correspond to the Just in Time references in this section. Complete them to review topics relevant to the remaining exercises.

1. In a right triangle, the side opposite the 90° angle is called the _____.

2. If a right triangle has legs of lengths 3 and 4, what is the length of the hypotenuse?

3. If a right triangle has a leg of length 5 and a hypotenuse of length $\sqrt{34}$, what is the length of the remaining leg?

4. A polynomial of the form $a^2 + 2ab + b^2$ is called a _____.

5. Multiply: $(2x - 5)^2$

6. Factor: $16t^2 - 24t + 9$

▶**Skills** This set of exercises will reinforce the skills illustrated in this section.

In Exercises 7–18, find the distance between each pair of points and the midpoint of the line segment joining them.

7. $(6, 4), (-8, 11)$

8. $(-5, 8), (-10, 14)$

9. $(-4, 20), (-10, 14)$

10. $(4, 3), (-5, 13)$

11. $(1, -1), (5, 5)$

12. $(-5, -2), (6, 10)$

13. $(6, -3), (6, 11)$

14. $(4, 7), (-10, 7)$

15. $\left(-\frac{1}{2}, 1\right), \left(-\frac{1}{4}, 0\right)$

16. $\left(-\frac{3}{4}, 0\right), (-5, 3)$

17. $(a_1, a_2), (b_1, b_2)$

18. $(a_1, 0), (0, b_2)$

In Exercises 19–28, write the standard form of the equation of the circle with the given radius and center. Sketch the circle.

19. $r = 5$; center: $(0, 0)$
20. $r = 3$; center: $(0, 0)$

21. $r = 3$; center: $(-1, 0)$
22. $r = 4$; center: $(0, -2)$

23. $r = 5$; center: $(3, -1)$
24. $r = 3$; center: $(-2, 4)$

25. $r = \frac{3}{2}$; center: $(1, 0)$
26. $r = \frac{5}{3}$; center: $(0, -2)$

27. $r = \sqrt{3}$; center: $(1, 1)$
28. $r = \sqrt{5}$; center: $(-2, -1)$

In Exercises 29–36, write the standard form of the equation of the circle with the given center and containing the given point.

29. Center: $(0, 0)$; point: $(1, 3)$

30. Center: $(0, 0)$; point: $(-2, 1)$

31. Center: $(2, 0)$; point: $(2, 5)$

32. Center: $(0, 3)$; point: $(-2, 3)$

33. Center: $(1, -2)$; point: $(5, 1)$

34. Center: $(-3, 2)$; point: $(3, -2)$

35. Center: $\left(\frac{1}{2}, 0\right)$; point: $(1, 3)$

36. Center: $\left(0, \frac{1}{3}\right)$; point: $(4, 2)$

In Exercises 37–42, what number must be added to complete the square of each expression?

37. $x^2 + 12x$
38. $x^2 - 10x$

39. $y^2 - 5y$
40. $y^2 + 7y$

41. $x^2 - 3x$
42. $y^2 + 5y$

In Exercises 43–56, find the center and radius of the circle having the given equation.

43. $x^2 + y^2 = 36$
44. $x^2 + y^2 = 49$

45. $(x - 1)^2 + (y + 2)^2 = 36$

46. $(x + 3)^2 + (y - 5)^2 = 121$

47. $(x - 8)^2 + y^2 = \dfrac{1}{4}$

48. $x^2 + (y - 12)^2 = \dfrac{1}{9}$

49. $x^2 + y^2 - 6x + 4y - 3 = 0$

50. $x^2 + y^2 + 8x + 2y - 8 = 0$

51. $x^2 + y^2 - 2x + 2y - 7 = 0$

52. $x^2 + y^2 + 8x - 2y + 8 = 0$

53. $x^2 + y^2 - 6x - 4y - 5 = 0$

54. $x^2 + y^2 + 4x - 2y - 7 = 0$

55. $x^2 + y^2 - x = 2$

56. $x^2 + y^2 + 3y = 4$

In Exercises 57–60, find the equation, in standard form, of each circle.

57.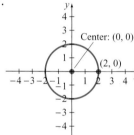

58.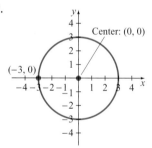

59.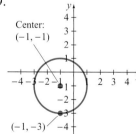

60.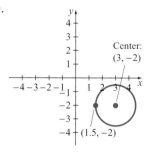

In Exercises 61–64, find the center and radius of the circle having the given equation. Use a graphing utility to graph the circle.

61. $x^2 + y^2 = 6.25$

62. $x^2 + y^2 = 12.25$

63. $(x - 3.5)^2 + y^2 = 10$

64. $(x + 4.2)^2 + (y - 2)^2 = 30$

▶ **Applications** In this set of exercises, you will use the distance formula and the equation of a circle to study real-world problems.

65. **Gardening** A gardener is planning a circular garden with an area of 196π square feet. He wants to plant petunias at the boundary of the circular garden.
   (a) If the center of the garden is at $(0, 0)$, find an equation for the circular boundary.
   (b) If each petunia plant covers 1 foot of the circular boundary, how many petunia plants are needed?

66. **Construction** A circular walkway is to be built around a monument, with the monument as the center. The distance from the monument to any point on the inner boundary of the walkway is 30 feet.

Concentric circles

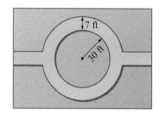

   (a) What is the equation of the inner boundary of the walkway? Use a coordinate system with the monument at $(0, 0)$.
   (b) If the walkway is 7 feet wide, what is the equation of the outer boundary of the walkway?

67. **Concert Seating** A circular stage of diameter 50 feet is to be built for a concert-in-the-round. If the center of the stage is at $(0, 0)$, find an equation for the edge of the stage.

68. **Concert Seating** At the concert in Exercise 67, the first row of seats must be 10 feet from the edge of the stage. Find an equation for the first row of seats.

69. **Distance Measurement** Two people are standing at the same road intersection. One walks directly east at 3 miles per hour. The other walks directly north at 4 miles per hour. How far apart will they be after half an hour?

70. **Distance** Two cars begin at the same road intersection. One drives west at 30 miles per hour and the other drives north at 40 miles per hour. How far apart will they be after 1 hour and 30 minutes?

71. **Treasure Hunt** Mike's fraternity holds a treasure hunt each year to raise money for a charity. Each participant is given a map with the center marked as (0, 0) and the location of each treasure marked with a dot. The map of this year's treasures is given below. The scale of the graph is in miles. If Mike is at treasure A and decides to go to treasure B, how far will he have to travel, to the nearest tenth of a mile, if he takes the shortest route?

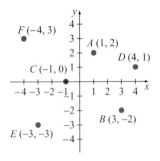

72. **Treasure Hunt** In Exercise 71, another participant, Mary, travels from the starting point at the center to treasure A, then to treasure C, then to treasure E, and finally back to the starting point. Assuming she took the shortest path between each two points, determine the total distance Mary traveled on her treasure hunt.

73. **Engineering** The Howe Truss was developed in about 1840 by the Massachusetts bridge builder William Howe. It is illustrated below and constitutes a section of a bridge. The points labeled A, B, C, D, and E are equally spaced. The points F, G, and H are also equally spaced.

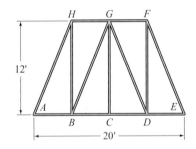

(a) Let the point labeled A be the origin for a coordinate system for this problem. What are the coordinates of points A through H?

(b) Trusses such as the one illustrated here were used to build wooden bridges in the nineteenth century. Use the distance formula to find the total length of all the lumber required to build this truss.

74. **Engineering** The following drawing illustrates a type of roof truss found in many homes.

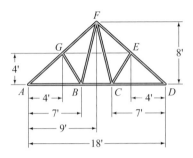

(a) Let the point labeled A be the origin for a coordinate system for this problem. What are the coordinates of points A through G?

(b) Use the distance formula to find the total length of all the lumber required to build this truss.

▶**Concepts** This set of exercises will draw on the ideas presented in this section and your general math background.

75. Write the equation of the circle whose diameter has endpoints at $(-3, 4)$ and $(1, 2)$. Sketch the circle.

76. Find the equation of the circle with center at the origin and a circumference of 7π units.

77. Find the equation of the circle with center at $(-5, -1)$ and an area of 64π.

78. Find the point(s) on the x-axis that is (are) a distance of 7 units from the point $(3, 4)$.

79. Find the point(s) on the y-axis that is (are) a distance of 8 units from the point $(1, -2)$.

2.2 The Algebra of Functions

Objectives

▶ Find the sum or difference of two functions and the corresponding domain

▶ Find the product or quotient of two functions and the corresponding domain

▶ Find the composition of functions

▶ Find the domain of a composite function

▶ Write a function as the composition of two functions

▶ Calculate the difference quotient

Multinational firms, such as the courier service DHL, must deal with conversions between different currencies and/or different systems of weights and measures. Computations involving these conversions use *operations* on functions—operations such as addition and composition. These topics will be studied in detail in this section. They have a wide variety of uses in both practical and theoretical situations.

Arithmetic Operations on Functions

We first examine a cost–revenue problem to see how an arithmetic operation involving functions arises.

Example **1** **Calculating Profit**

The GlobalEx Corporation has revenues modeled by the function $R(t) = 40 + 2t$, where t is the number of years since 2003 and $R(t)$ is in millions of dollars. Its operating costs are modeled by the function $C(t) = 35 + 1.6t$, where t is the number of years since 2003 and $C(t)$ is in millions of dollars. Find the profit function $P(t)$ for GlobalEx Corporation.

▶**Solution** Since profit is equal to revenue minus cost, we can write

$$P(t) = R(t) - C(t).$$

Substituting the expressions for $R(t)$ and $C(t)$ gives

$$P(t) = (40 + 2t) - (35 + 1.6t) = 40 + 2t - 35 - 1.6t = 5 + 0.4t.$$

Thus the profit function is $P(t) = 5 + 0.4t$, where t is the number of years since 2003.

✔ *Check It Out 1:* Find the profit function for GlobalEx Corporation in Example 1 if the revenue and cost functions are given by $R(t) = 42 + 2.2t$ and $C(t) = 34 + 1.5t$. ■

Computing a profit function, as we did in Example 1, is an example of an **arithmetic operation** on functions. We can add, subtract, multiply, or divide two functions. These operations are defined as follows.

Just in Time

Review function notation in Section 1.1.

Arithmetic Operations on Functions

Given two functions f and g, for each x in the domain of *both* f and g, the sum, difference, product, and quotient of f and g are defined as follows.

$$(f + g)(x) = f(x) + g(x)$$
$$(f - g)(x) = f(x) - g(x)$$
$$(fg)(x) = f(x) \cdot g(x)$$
$$\left(\frac{f}{g}\right)(x) = \frac{f(x)}{g(x)}, \qquad \text{where } g(x) \neq 0$$

Discover *and* **Learn**

Show that $(f + g)(x) = (g + f)(x)$ and $(fg)(x) = (gf)(x)$.

> **Note** When combining two functions *f* and *g* using arithmetic operations, the domain of the combined function consists of all real numbers that are common to the domains of both *f* and *g*. When forming the quotient $\frac{f(x)}{g(x)}$, the value of *x* at which $g(x) = 0$ must be excluded from the domain of $\frac{f}{g}$.

Example 2 Arithmetic Operations on Two Functions

Let f and g be two functions defined as follows:

$$f(x) = \frac{2}{x - 4} \quad \text{and} \quad g(x) = 3x - 1$$

Find the following and determine the domain of each.

(a) $(f + g)(x)$

(b) $(f - g)(x)$

(c) $(fg)(x)$

(d) $\left(\dfrac{f}{g}\right)(x)$

▶ Solution

(a) From the definition of the sum of two functions,

$$(f + g)(x) = f(x) + g(x) = \frac{2}{x - 4} + 3x - 1.$$

Just in Time

Review rational expressions in Section P.6.

We can simplify as follows.

$$(f + g)(x) = \frac{2}{x - 4} + (3x - 1)$$

$$= \frac{2 + (3x - 1)(x - 4)}{x - 4} \qquad \text{LCD is } x - 4$$

$$= \frac{2 + (3x^2 - 13x + 4)}{x - 4}$$

$$= \frac{3x^2 - 13x + 6}{x - 4} \qquad \text{Combine like terms}$$

The domain of f is $(-\infty, 4) \cup (4, \infty)$, and the domain of g consists of all real numbers. Thus the domain of $f + g$ consists of $(-\infty, 4) \cup (4, \infty)$, since these values are in the domains of both f and g.

(b) $(f - g)(x) = f(x) - g(x) = \dfrac{2}{x - 4} - (3x - 1)$

$$= \frac{2 - (3x - 1)(x - 4)}{x - 4} \qquad \text{LCD is } x - 4$$

$$= \frac{2 - (3x^2 - 13x + 4)}{x - 4}$$

$$= \frac{-3x^2 + 13x - 2}{x - 4} \qquad \begin{array}{l}\text{Collect like terms; be careful} \\ \text{with the negative sign}\end{array}$$

The domain of $f - g$ consists of $(-\infty, 4) \cup (4, \infty)$, since these values are in the domains of both f and g.

(c) $(fg)(x) = f(x) \cdot g(x) = \left(\dfrac{2}{x - 4}\right)(3x - 1)$

$= \dfrac{6x - 2}{x - 4}$

The domain of fg consists of $(-\infty, 4) \cup (4, \infty)$, since these values are in the domains of both f and g.

(d) $\left(\dfrac{f}{g}\right)(x) = \dfrac{f(x)}{g(x)} = \dfrac{\frac{2}{x - 4}}{3x - 1}$

$= \dfrac{2}{x - 4} \cdot \dfrac{1}{3x - 1}$

$= \dfrac{2}{(x - 4)(3x - 1)}$

The set of x values common to both functions is $(-\infty, 4) \cup (4, \infty)$. In addition, $g(x) = 0$ for $x = \frac{1}{3}$. Thus the domain of $\frac{f}{g}$ is the set of all real numbers such that $x \neq 4$ and $x \neq \frac{1}{3}$. In interval notation, we have

$$\left(-\infty, \frac{1}{3}\right) \cup \left(\frac{1}{3}, 4\right) \cup (4, \infty).$$

✔ *Check It Out 2:* Let f and g be two functions defined as follows:

$$f(x) = \frac{1}{x + 2} \quad \text{and} \quad g(x) = 2x + 1$$

Find (a) $(f + g)(x)$ and (b) $\left(\dfrac{f}{g}\right)(x)$. ▪

Once we find the arithmetic combination of two functions, we can evaluate the new function at any point in its domain, as illustrated in the following example.

Example **3** **Evaluating a Sum, Product, and Quotient of Two Functions**

Let f and g be two functions defined as follows.

$$f(x) = 2x^2 - x \quad \text{and} \quad g(x) = \sqrt{x + 1}$$

Evaluate the following.
(a) $(f + g)(-1)$

(b) $\left(\dfrac{f}{g}\right)(3)$

(c) $(fg)(3)$

▶**Solution**

(a) Compute the sum of the two functions as follows.

$$(f + g)(x) = f(x) + g(x)$$
$$= 2x^2 - x + \sqrt{x + 1}$$

Next, evaluate $(f + g)(x)$ at $x = -1$:

$$(f + g)(-1) = 2(-1)^2 - (-1) + \sqrt{-1 + 1}$$
$$= 2 - (-1) + 0 = 3$$

(b) Compute the quotient of the two functions as follows.

$$\left(\frac{f}{g}\right)(x) = \frac{f(x)}{g(x)}$$
$$= \frac{2x^2 - x}{\sqrt{x + 1}}$$

Next, evaluate $\left(\frac{f}{g}\right)(x)$ at $x = 3$:

$$\left(\frac{f}{g}\right)(3) = \frac{2(3)^2 - 3}{\sqrt{3 + 1}}$$
$$= \frac{2(9) - 3}{2}$$
$$= \frac{15}{2}$$

(c) Compute the product of the two functions as follows.

$$(fg)(x) = f(x) \cdot g(x)$$
$$= (2x^2 - x)(\sqrt{x + 1})$$

Thus, $(fg)(3) = (2(3)^2 - 3)(\sqrt{3 + 1}) = 30$.

☑ *Check It Out 3:* Let f and g be two functions defined as follows.

$$f(x) = 3x - 4 \quad \text{and} \quad g(x) = x^2 + x$$

Evaluate the following.

(a) $(f + g)(2)$

(b) $\left(\frac{f}{g}\right)(2)$ ■

Composition of Functions

When converting between currencies or weights and measures, a function expressed in terms of a given unit must be restated in terms of a new unit. For example, if profit is given in terms of dollars, another function must convert the profit function to a different currency. Successive evaluation of this series of two functions is known as a **composition of functions,** and is of both practical and theoretical importance. We first give a concrete example of such an operation and then give its formal definition.

Example 4 **Evaluating Two Functions Successively Using Tables**

The cost incurred for fuel by GlobalEx Corporation in running a fleet of vehicles is given in Table 2.2.1 in terms of the number of gallons used. However, the European branch of GlobalEx Corporation records its fuel consumption in units of liters rather than gallons. For various quantities of fuel in liters, Table 2.2.2 lists the equivalent quantity of fuel in gallons.

Table 2.2.1

Quantity (gallons)	Cost ($)
30	45
45	67.50
55	82.50
70	105

Table 2.2.2

Quantity (liters)	Quantity (gallons)
113.55	30
170.325	45
208.175	55
264.95	70

Answer the following questions.

(a) Find the cost of 55 gallons of fuel.

(b) Find the cost of 113.55 liters of fuel.

▶Solution

(a) From Table 2.2.1, it is clear that 55 gallons of fuel costs $82.50.

(b) For 113.55 liters of fuel, we must find the equivalent quantity of fuel in gallons before looking up the price.

$$\text{Quantity (liters)} \rightarrow \text{Quantity (gallons)} \rightarrow \text{Cost}$$

From Table 2.2.2, we see that 113.55 liters is equal to 30 gallons. We then refer to Table 2.2.1 to find that 30 gallons of fuel costs the company $45.

To answer the second question, we had to use *two* different tables to look up the value of the cost function. In the previous chapter, we were able to look up function values by using only one table. The following discussion will elaborate on the process of using two tables.

☑ *Check It Out 4:* In Example 4, find the cost of 264.95 liters of fuel. ■

In Example 4, the cost is given in Table 2.2.1 as a function of the number of gallons. We will denote this function by C. The number of gallons can, in turn, be written as a function of the number of liters, as shown in Table 2.2.2. We will denote this function by G. Thus, the cost of x liters of fuel can be written in function notation as

$$C(G(x)).$$

Using this notation, a schematic diagram for Example 4 can be written as follows:

$$\text{Quantity (in liters)} \rightarrow \text{Quantity (in gallons)} \rightarrow \text{Cost}$$
$$x \rightarrow G(x) \qquad\qquad \rightarrow C(G(x))$$

The process of taking a function value, $G(x)$, and using it as an input for another function, $C(G(x))$, is known as the **composition of functions.** Composition of functions comes up often enough to have its own notation, which we now present.

Composition of Functions

The **composition of functions** f and g is a function that is denoted by $f \circ g$ and defined as

$$(f \circ g)(x) = f(g(x)).$$

The domain of $f \circ g$ is the set of all x in the domain of g such that $g(x)$ is in the domain of f.

The function $f \circ g$ is called a **composite function.**

In Example 4, we considered the composition of functions that were given in the form of data tables. In the next example, we will compute the composition of functions that are given by algebraic expressions. While knowing how to algebraically manipulate expressions to obtain a composite function is important, it is perhaps even more important to understand what a composite function actually represents.

Example 5 Finding and Evaluating Composite Functions

Let $f(s) = s^2 + 1$ and $g(s) = -2s$.

(a) Find an expression for $(f \circ g)(s)$ and give the domain of $f \circ g$.

(b) Find an expression for $(g \circ f)(s)$ and give the domain of $g \circ f$.

(c) Evaluate $(f \circ g)(-2)$.

(d) Evaluate $(g \circ f)(-2)$.

▶Solution

(a) The composite function $f \circ g$ is defined as $(f \circ g)(s) = f(g(s))$. Computing the quantity $f(g(s))$ is often the most confusing part. To make things easier, think of $f(s)$ as $f(\Box)$, where the box can contain anything. Then proceed as follows.

$$f(\Box) = (\Box)^2 + 1 \qquad \text{Definition of } f$$
$$f(\boxed{g(s)}) = (\boxed{g(s)})^2 + 1 \qquad \text{Place } g(s) \text{ in the box}$$
$$= (\boxed{-2s})^2 + 1 \qquad \text{Substitute expression for } g(s)$$
$$= 4s^2 + 1 \qquad \text{Simplify: } (-2s)^2 = 4s^2$$

Thus, $(f \circ g)(s) = 4s^2 + 1$. Since the domain of g is all real numbers and the domain of f is also all real numbers, the domain of $f \circ g$ is all real numbers.

(b) The composite function $g \circ f$ is defined as $(g \circ f)(s) = g(f(s))$. This time, we compute $g(f(s))$. Thinking of $g(s)$ as $g(\Box)$, we have

$$g(\Box) = -2(\Box) \qquad \text{Definition of } g$$
$$g(\boxed{f(s)}) = -2(\boxed{f(s)}) \qquad \text{Place } f(s) \text{ in the box}$$
$$= -2(\boxed{s^2 + 1}) \qquad \text{Substitute expression for } f(s)$$
$$= -2s^2 - 2. \qquad \text{Simplify}$$

Thus, $(g \circ f)(s) = -2s^2 - 2$. Note that $(f \circ g)(s)$ is *not* equal to $(g \circ f)(s)$. Since the domain of f is all real numbers and the domain of g is also all real numbers, the domain of $g \circ f$ is all real numbers.

(c) Since $(f \circ g)(s) = 4s^2 + 1$, the value of $(f \circ g)(-2)$ is

$$(f \circ g)(-2) = 4(-2)^2 + 1 = 4(4) + 1 = 16 + 1 = 17.$$

Alternatively, using the expressions for the individual functions,

$$
\begin{aligned}
(f \circ g)(-2) &= f(g(-2)) \\
&= f(4) \qquad\qquad g(-2) = -2(-2) = 4 \\
&= 17. \qquad\qquad f(4) = 4^2 + 1 = 16 + 1 = 17
\end{aligned}
$$

(d) Since $(g \circ f)(s) = -2s^2 - 2$, the value of $(g \circ f)(-2)$ is

$$(g \circ f)(-2) = -2(-2)^2 - 2 = -2(4) - 2 = -8 - 2 = -10.$$

Alternatively, using the expressions for the individual functions,

$$
\begin{aligned}
(g \circ f)(-2) &= g(f(-2)) \\
&= g(5) \qquad\qquad f(-2) = (-2)^2 + 1 = 4 + 1 = 5 \\
&= -10. \qquad\quad g(5) = -2(5) = -10
\end{aligned}
$$

✔ *Check It Out 5:* Let $f(s) = s^2 - 2$ and $g(s) = 3s$. Evaluate $(f \circ g)(-1)$ and $(g \circ f)(-1)$. ■

Just in Time

Review domain of functions in Section 1.1.

The domain of a composite function can differ from the domain of either or both of the two functions from which it is composed, as illustrated in Example 6.

Technology Note

For Example 6(a), enter $Y_1(x) = \frac{1}{x}$ and $Y_2(x) = x^2 - 1$. Then $Y_3(x) = Y_1(Y_2(x))$ defines the composite function $Y_1 \circ Y_2$. Note that the table of values gives "ERROR" as the value of $Y_3(x)$ for $x = -1$ and $x = 1$. (See Figure 2.2.1.) These numbers are not in the domain of $Y_3 = Y_1 \circ Y_2$.

Keystroke Appendix:
Sections 4 and 6

Figure 2.2.1

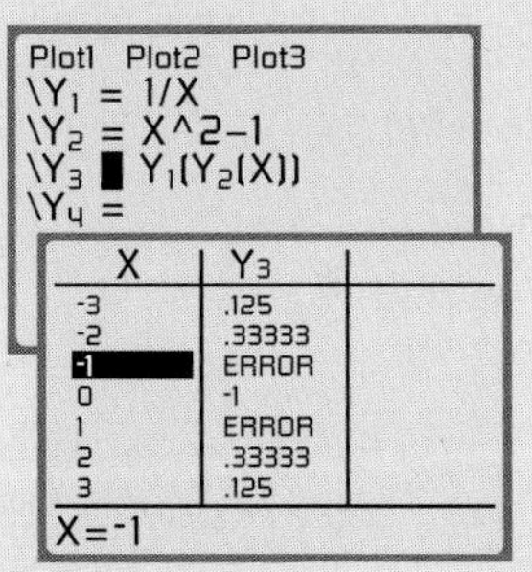

Example	6 **Domains of Composite Functions**

Let $f(x) = \frac{1}{x}$ and $g(x) = x^2 - 1$.

(a) Find $f \circ g$ and its domain.

(b) Find $g \circ f$ and its domain.

▶ **Solution** We first note that the domain of f is $(-\infty, 0) \cup (0, \infty)$ and the domain of g is all real numbers.

(a) To find $f \circ g$, proceed as follows.

$$
\begin{aligned}
(f \circ g)(x) &= f(g(x)) && \text{Definition of } f \circ g \\
&= f(x^2 - 1) && \text{Substitute expression} \\
& && \text{for } g(x) \\[4pt]
&= \frac{1}{x^2 - 1} && \text{Use definition of } f
\end{aligned}
$$

The domain of $f \circ g$ is the set of all x in the domain of g such that $g(x)$ is in the domain of f. The domain of f is $(-\infty, 0) \cup (0, \infty)$. Therefore, every value output by g must be a number other than 0. Thus, we find the numbers for which $x^2 - 1 = 0$ and then *exclude* them. By factoring,

$$x^2 - 1 = (x + 1)(x - 1).$$

Hence

$$x^2 - 1 = 0 \Longrightarrow x + 1 = 0 \quad \text{or} \quad x - 1 = 0 \Longrightarrow x = -1 \quad \text{or} \quad x = 1.$$

The domain of $f \circ g$ is

$$(-\infty, -1) \cup (-1, 1) \cup (1, \infty).$$

(b) To find $g \circ f$, proceed as follows.

$$
\begin{aligned}
(g \circ f)(x) &= g(f(x)) && \text{Definition of } g \circ f \\
&= g\left(\frac{1}{x}\right) && \text{Substitute expression for } f(x) \\
&= \left(\frac{1}{x}\right)^2 - 1 = \frac{1}{x^2} - 1 && \text{Use definition of } g
\end{aligned}
$$

The domain of $g \circ f$ is the set of all x in the domain of f such that $f(x)$ is in the domain of g. Since the domain of g is all real numbers, any value output by f is acceptable. Thus, by definition, the domain of $g \circ f$ is the set of all x in the domain of f, which is $(-\infty, 0) \cup (0, \infty)$.

✔ *Check It Out 6:* Let $f(x) = \sqrt{x}$ and $g(x) = x + 1$. Find $f \circ g$ and its domain. ■

The next example illustrates how a given function can be written as a composition of two other functions.

Example **7** **Writing a Function as a Composition**

If $h(x) = \sqrt{3x^2 - 1}$, find two functions f and g such that $h(x) = (f \circ g)(x) = f(g(x))$.

▶**Solution** The function h takes the square root of the quantity $3x^2 - 1$. Since $3x^2 - 1$ must be calculated before its square root can be taken, let $g(x) = 3x^2 - 1$. Then let $f(x) = \sqrt{x}$. Thus, two functions that can be used for the composition are

$$
f(x) = \sqrt{x} \quad \text{and} \quad g(x) = 3x^2 - 1.
$$

We check this by noting that

$$
\begin{aligned}
h(x) &= (f \circ g)(x) = f(g(x)) \\
&= f(3x^2 - 1) = \sqrt{3x^2 - 1}.
\end{aligned}
$$

✔ *Check It Out 7:* If $h(x) = (x^3 + 9)^5$, find two functions f and g such that $h(x) = (f \circ g)(x) = f(g(x))$. ■

> **Note** In some instances, there can be more than one way to write a function as a composition of two functions.

Difference Quotient

Combining functions is a technique that is often used in calculus. For example, it is used in calculating the **difference quotient** of a function f, which is an expression of the form $\dfrac{f(x + h) - f(x)}{h}$, $h \neq 0$. This is illustrated in the following example.

Example **8** **Computing a Difference Quotient**

Compute $\dfrac{f(x + h) - f(x)}{h}$, $h \neq 0$, for $f(x) = 2x^2 + 1$.

▶Solution First, compute each component by step:

$$f(x + h) = 2(x + h)^2 + 1$$
$$= 2(x^2 + 2xh + h^2) + 1 \qquad \text{Expand } (x + h)^2$$
$$= 2x^2 + 4xh + 2h^2 + 1$$

Next,

$$f(x + h) - f(x) = 2x^2 + 4xh + 2h^2 + 1 - (2x^2 + 1)$$
$$= 4xh + 2h^2.$$

Finally,

$$\frac{f(x + h) - f(x)}{h} = \frac{4xh + 2h^2}{h} = 4x + 2h.$$

☑ *Check It Out 8:* Compute $\dfrac{f(x + h) - f(x)}{h}$, $h \neq 0$, for $f(x) = -x^2 + 4$. ■

2.2 Key Points

▶ Given two functions f and g, for each x in the domain of *both* f and g, the **sum, difference, product,** and **quotient** of f and g are defined as follows.

$$(f + g)(x) = f(x) + g(x)$$
$$(f - g)(x) = f(x) - g(x)$$
$$(fg)(x) = f(x) \cdot g(x)$$
$$\left(\frac{f}{g}\right)(x) = \frac{f(x)}{g(x)}, \qquad \text{where } g(x) \neq 0$$

▶ The **composite function** $f \circ g$ is a function defined as

$$(f \circ g)(x) = f(g(x)).$$

The domain of $f \circ g$ is the set of all x in the domain of g such that $g(x)$ is in the domain of f.

▶ The **difference quotient** of a function f is defined as $\dfrac{f(x + h) - f(x)}{h}$, $h \neq 0$.

2.2 Exercises

▶**Just in Time Exercises** These exercises correspond to the Just in Time references in this section. Complete them to review topics relevant to the remaining exercises.

1. A quotient of two polynomial expressions is called a _____ and is defined whenever the denominator is not equal to _____.

2. True or False: The variable x in $f(x)$ is a placeholder and can be replaced by any quantity as long as the same replacement occurs in the expression for the function.

3. What is the domain of the function $f(x) = x^2 - 3x$?

4. What is the domain of the function $f(x) = \sqrt{x - 1}$?

5. What is the domain of the function $f(x) = \sqrt{x^2 - 9}$?

6. What is the domain of the function

$$f(x) = \frac{x + 2}{x - 1}?$$

▶**Skills** This set of exercises will reinforce the skills illustrated in this section.

In Exercises 7–16, for the given functions f and g, find each composite function and identify its domain.

(a) $(f + g)(x)$

(b) $(f - g)(x)$

(c) $(fg)(x)$

(d) $\left(\dfrac{f}{g}\right)(x)$

7. $f(x) = 3x - 5; g(x) = -x + 3$

8. $f(x) = 2x + 1; g(x) = -5x - 1$

9. $f(x) = x - 3; g(x) = x^2 + 1$

10. $f(x) = x^3; g(x) = 3x^2 + 4$

11. $f(x) = \dfrac{1}{x}; g(x) = \dfrac{1}{2x - 1}$

12. $f(x) = \dfrac{2}{x + 1}; g(x) = \dfrac{-1}{x^2}$

13. $f(x) = \sqrt{x}; g(x) = -x + 1$

14. $f(x) = 2x - 1; g(x) = \sqrt{x}$

15. $f(x) = |x|; g(x) = \dfrac{1}{2x + 5}$

16. $f(x) = \dfrac{2}{x - 4}; g(x) = -|x|$

In Exercises 17–40, let $f(x) = -x^2 + x$, $g(x) = \dfrac{2}{x + 1}$, and $h(x) = -2x + 1$. Evaluate each of the following.

17. $(f + g)(1)$

18. $(f + g)(0)$

19. $(g + h)(0)$

20. $(g + h)(1)$

21. $(f + h)(-2)$

22. $(f + h)(0)$

23. $(f - g)(2)$

24. $(f - g)(-3)$

25. $(g - h)(-2)$

26. $(g - h)(3)$

27. $(h - f)(-1)$

28. $(h - f)(0)$

29. $(fg)(3)$

30. $(fg)(-3)$

31. $(gh)(-3)$

32. $(gh)(0)$

33. $(fh)(-2)$

34. $(fh)(1)$

35. $\left(\dfrac{f}{g}\right)(-2)$

36. $\left(\dfrac{f}{g}\right)(3)$

37. $\left(\dfrac{g}{h}\right)(3)$

38. $\left(\dfrac{g}{h}\right)(-2)$

39. $\left(\dfrac{f}{h}\right)(1)$

40. $\left(\dfrac{h}{f}\right)(2)$

In Exercises 41–48, use f and g given by the following tables of values.

x	-1	0	3	6
$f(x)$	-2	3	4	2

x	-2	1	2	4
$g(x)$	0	6	-2	3

41. Evaluate $f(-1)$.

42. Evaluate $g(4)$.

43. Evaluate $(f \circ g)(-2)$.

44. Evaluate $(f \circ g)(4)$.

45. Evaluate $(g \circ f)(-1)$.

46. Evaluate $(g \circ f)(6)$.

47. Is $(g \circ f)(0)$ defined? Why or why not?

48. Is $(f \circ g)(2)$ defined? Why or why not?

In Exercises 49–66, let $f(x) = x^2 + x$, $g(x) = \sqrt{x}$, and $h(x) = -3x$. Evaluate each of the following.

49. $(f \circ h)(5)$

50. $(f \circ h)(1)$

51. $(f \circ h)(-2)$

52. $(f \circ h)(-1)$

53. $(h \circ g)(4)$

54. $(h \circ g)(0)$

55. $(g \circ h)(-3)$

56. $(g \circ h)(-12)$

57. $(f \circ g)(4)$

58. $(f \circ g)(9)$

59. $(g \circ f)(2)$

60. $(g \circ f)(1)$

61. $(g \circ f)(-3)$

62. $(g \circ f)(-5)$

63. $(h \circ f)(2)$

64. $(h \circ f)(-3)$

65. $(h \circ f)\left(\dfrac{1}{2}\right)$

66. $(h \circ f)\left(\dfrac{3}{2}\right)$

In Exercises 67–86, find expressions for $(f \circ g)(x)$ and $(g \circ f)(x)$. Give the domains of $f \circ g$ and $g \circ f$.

67. $f(x) = -x^2 + 1$; $g(x) = x + 1$

68. $f(x) = 2x + 5$; $g(x) = 3x^2$

69. $f(x) = 4x - 1$; $g(x) = \dfrac{x + 1}{4}$

70. $f(x) = 2x + 3$; $g(x) = \dfrac{x - 3}{2}$

71. $f(x) = 3x^2 + 4x$; $g(x) = x + 2$

72. $f(x) = -2x + 1$; $g(x) = 2x^2 - 5x$

73. $f(x) = \dfrac{1}{x}$; $g(x) = 2x + 5$

74. $f(x) = 3x + 1$; $g(x) = \dfrac{2}{x}$

75. $f(x) = \dfrac{3}{2x + 1}$; $g(x) = 2x^2$

76. $f(x) = 3x^2 + 1$; $g(x) = \dfrac{2}{x + 5}$

77. $f(x) = \sqrt{x + 1}$; $g(x) = -3x - 4$

78. $f(x) = 5x + 1$; $g(x) = \sqrt{x - 3}$

79. $f(x) = |x|$; $g(x) = \dfrac{2x}{x - 1}$

80. $f(x) = |x|$; $g(x) = \dfrac{x}{x - 3}$

81. $f(x) = x^2 - 2x + 1$; $g(x) = x + 1$

82. $f(x) = x - 2$; $g(x) = 2x^2 - x + 3$

83. $f(x) = \dfrac{x^2 + 1}{x^2 - 1}$; $g(x) = |x|$

84. $f(x) = |x|$; $g(x) = \dfrac{x^2 + 3}{x^2 - 4}$

85. $f(x) = \dfrac{1}{x^2 + 1}$; $g(x) = \dfrac{2x + 1}{3x - 1}$

86. $f(x) = \dfrac{-x + 1}{2x + 3}$; $g(x) = \dfrac{1}{x^2 + 1}$

In Exercises 87–96, find two functions f and g such that $h(x) = (f \circ g)(x) = f(g(x))$. Answers may vary.

87. $h(x) = (3x - 1)^2$

88. $h(x) = (-2x + 5)^2$

89. $h(x) = \sqrt[3]{4x^2 - 1}$

90. $h(x) = \sqrt[5]{-x^3 + 8}$

91. $h(x) = \dfrac{1}{2x + 5}$

92. $h(x) = \dfrac{3}{x^2 + 1}$

93. $h(x) = \sqrt{x^2 + 1} + 5$

94. $h(x) = \sqrt[3]{5x + 7} - 2$

95. $h(x) = 4(2x + 9)^5 - (2x + 9)^8$

96. $h(x) = (3x - 7)^{10} + 5(3x - 7)^2$

In Exercises 97–100, let $f(t) = -t^2$ and $g(x) = x^2 - 1$.

97. Evaluate $(f \circ f)(-1)$.

98. Evaluate $(g \circ g)\left(\dfrac{2}{3}\right)$.

99. Find an expression for $(f \circ f)(t)$, and give the domain of $f \circ f$.

100. Find an expression for $(g \circ g)(x)$, and give the domain of $g \circ g$.

In Exercises 101–104, let $f(t) = 3t + 1$ and $g(x) = x^2 + 4$.

101. Evaluate $(f \circ f)(2)$.

102. Evaluate $(g \circ g)\left(\dfrac{1}{2}\right)$.

103. Find an expression for $(f \circ f)(t)$, and give the domain of $f \circ f$.

104. Find an expression for $(g \circ g)(x)$, and give the domain of $g \circ g$.

In Exercises 105–110, find the difference quotient $\dfrac{f(x + h) - f(x)}{h}$, $h \neq 0$, for the given function f.

105. $f(x) = 3x - 1$

106. $f(x) = -2x + 3$

107. $f(x) = -x^2 + x$

108. $f(x) = 3x^2 + 2x$

109. $f(x) = \dfrac{1}{x - 3}, x \neq 3$

110. $f(x) = \dfrac{1}{x + 1}, x \neq -1$

▶**Applications** In this set of exercises, you will use combinations of functions to study real-world problems.

111. **Sports** The Washington Redskins' revenue can be modeled by the function $R(t) = 245 + 40t$, where t is the number of years since 2003 and $R(t)$ is in millions of dollars. The team's operating costs are modeled by the function $C(t) = 170 + 60t$, where t is the number of years since 2003 and $C(t)$ is in millions of dollars. Find the profit function $P(t)$. (*Source:* Associated Press)

112. **Commerce** The following two tables give revenues, in dollars, from two stores for various years. Compute the table for $(f + g)(x)$ and explain what it represents.

Year, x	Revenue, Store 1, f(x)
2002	200,000
2003	210,000
2004	195,000
2005	230,000

Year, x	Revenue, Store 2, g(x)
2002	300,000
2003	320,000
2004	295,000
2005	330,000

113. **Business** The following tables give the numbers of hours billed for various weeks by two lawyers who work at a prestigious law firm. Compute the table for $(f - g)(x)$ and explain what it represents.

Week, x	Hours Billed by Employee 1, f(x)
1	70
2	65
3	73
4	71

Week, x	Hours Billed by Employee 2, g(x)
1	69
2	72
3	70
4	68

114. **Commerce** The number of copies of a popular mystery writer's newest release sold at a local bookstore during each month after its release is given by $n(x) = -5x + 100$. The price of the book during each month after its release is given by $p(x) = -1.5x + 30$. Find $(np)(3)$. Interpret your results.

115. **Education** Let $n(t)$ represent the number of students attending a review session each week, starting with the first week of school. Let $p(t)$ represent the number of tutors scheduled to work during the review session each week. Interpret the amount $\dfrac{n(t)}{p(t)}$.

116. **Real Estate** A salesperson generates \$400,000 in sales for each new home that is sold in a housing development. Her commission is 6% of the total amount of dollar sales.

(a) What is the total amount of sales, $S(x)$, if x is the number of homes sold?

(b) What is the commission, $C(x)$, if x is the number of homes sold?

(c) Interpret the amount $S(x) - C(x)$.

117. **Currency Exchange** The exchange rate from U.S. dollars to euros on a particular day is given by the function $f(x) = 0.82x$, where x is in U.S. dollars. If GlobalEx Corporation has revenue given by the function $R(t) = 40 + 2t$, where t is the number of years since 2003 and $R(t)$ is in millions of dollars, find $(f \circ R)(t)$ and explain what it represents. (*Source:* www.xe.com)

118. **Unit Conversion** The conversion of temperature units from degrees Fahrenheit to degrees Celsius is given by the equation $C(x) = \frac{5}{9}(x - 32)$, where x is given in degrees Fahrenheit. Let $T(x) = 70 + 4x$ denote the temperature, in degrees Fahrenheit, in Phoenix, Arizona, on a typical July day, where x is the number of hours after 6 A.M. Assume the temperature model holds until 4 P.M. of the same day. Find $(C \circ T)(x)$ and explain what it represents.

119. **Geometry** The surface area of a sphere is given by $A(r) = 4\pi r^2$, where r is in inches and $A(r)$ is in square inches. The function $C(x) = 6.4516x$ takes x square inches as input and outputs the equivalent result in square centimeters. Find $(C \circ A)(r)$ and explain what it represents.

120. **Geometry** The perimeter of a square is $P(s) = 4s$, where s is the length of a side in inches. The function $C(x) = 2.54x$ takes x inches as input and outputs the equivalent result in centimeters. Find $(C \circ P)(s)$ and explain what it represents.

▶**Concepts** This set of exercises will draw on the ideas presented in this section and your general math background.

121. Is it true that $(fg)(x)$ is the same as $(f \circ g)(x)$ for any functions f and g? Explain.

122. Give an example to show that $(f \circ g)(x) \neq (g \circ f)(x)$.

123. Let $f(x) = ax + b$ and $g(x) = cx + d$, where a, b, c, and d are constants. Show that $(f + g)(x)$ and $(f - g)(x)$ also represent linear functions.

124. Find $\dfrac{f(x + h) - f(x)}{h}$, $h \neq 0$, for $f(x) = ax + b$, where a and b are constants.

2.3 Transformations of the Graph of a Function

Objectives

▶ Graph vertical and horizontal shifts of the graph of a function

▶ Graph a vertical compression or stretch of the graph of a function

▶ Graph reflections across the x-axis of the graph of a function

▶ Graph a horizontal compression or stretch of the graph of a function

▶ Graph reflections across the y-axis of the graph of a function

▶ Graph combinations of transformations of the graph of a function

▶ Identify an appropriate transformation of the graph of a function from a given expression for the function

In this section, you will see how to create graphs of new functions from the graph of an existing function by simple geometric **transformations.** The general properties of these transformations are very useful for sketching the graphs of various functions.

Throughout this section, we will investigate transformations of the graphs of the functions $|x|$, x^2, and $\sqrt{x}$ (see Figure 2.3.1) and state the general rules of transformations. Since these rules apply to the graph of any function, they will be applied to transformations of the graphs of other functions in later chapters.

Figure 2.3.1 Graphs of some basic functions

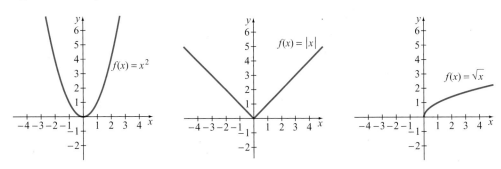

Vertical and Horizontal Shifts of the Graph of a Function

The simplest transformations we can consider are those where we take the graph of a function and simply shift it vertically or horizontally. Such shifts are also called *translations.*

Discover *and* Learn

Table 2.3.1 gives values of
$f(x) = |x|$ for several values of x.

Table 2.3.1

x	f(x)
−3	3
−1	1
0	0
1	1
3	3

(a) Extend the given table to include the values of $g(x) = f(x) + 2$.

(b) Use the values in the table to plot the points $(x, f(x))$ and $(x, g(x))$ on the same set of coordinate axes. What do you observe?

Example 1 Comparing $f(x) = |x|$ and $g(x) = |x| - 2$

Make a table of values for the functions $f(x) = |x|$ and $g(x) = |x| - 2$, for $x = -3, -2, -1, 0, 1, 2, 3$. Use your table to sketch the graphs of the two functions. What are the domain and range of f and g?

▶**Solution** We make a table of function values as shown in Table 2.3.2, and sketch the graphs as shown in Figure 2.3.2.

Table 2.3.2

x	$f(x) = |x|$	$g(x) = |x| - 2$
−3	3	1
−2	2	0
−1	1	−1
0	0	−2
1	1	−1
2	2	0
3	3	1

Figure 2.3.2

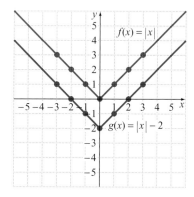

Observations:

▶ Examining the table of function values, we see that the values of $g(x) = |x| - 2$ are 2 units less than the corresponding values of $f(x) = |x|$.

▶ Note that the graph of $g(x) = |x| - 2$ has the same shape as the graph of $f(x) = |x|$, but it is shifted *down by 2 units.*

The domain of both f and g is the set of all real numbers. From the graph, we see that the range of f is $[0, \infty)$ and range of g is $[-2, \infty)$.

✔ *Check It Out 1:* Make a table of values for the functions $f(x) = |x|$ and $g(x) = |x| + 3$, for $x = -3, -2, -1, 0, 1, 2, 3$. Use your table to sketch the graphs of the two functions. What are the domain and range of f and g? ■

The results of Example 1 lead to a general statement about **vertical shifts** of the graph of a function.

Technology Note

Vertical shifts can be seen easily with a graphing calculator. Figure 2.3.4 shows the graphs of $f(x) = |x|$ and $g(x) = |x| - 2$ on the same set of axes, using a decimal window.

Keystroke Appendix:
Section 7

Figure 2.3.4

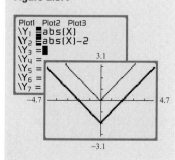

Vertical Shifts of the Graph of $f(x)$

Let f be a function and c be a positive constant.

▶ The graph of $g(x) = f(x) + c$ is the graph of $f(x)$ shifted c units **upward.**

▶ The graph of $g(x) = f(x) - c$ is the graph of $f(x)$ shifted c units **downward.**

See Figure 2.3.3.

Figure 2.3.3 Vertical shifts of f, $c > 0$

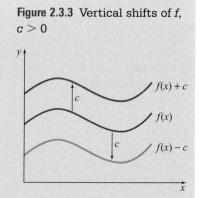

Example 2 **Comparing** $f(x) = |x|$ **and** $g(x) = |x - 2|$

Make a table of values for the functions $f(x) = |x|$ and $g(x) = |x - 2|$, for $x = -3, -2, -1, 0, 1, 2, 3$. Use your table to sketch the graphs of the two functions. What are the domain and range of f and g?

▶Solution We make a table of function values as shown in Table 2.3.3, and sketch the graphs as shown in Figure 2.3.5.

Table 2.3.3

x	-3	-2	-1	0	1	2	3
$f(x) = \|x\|$	3	2	1	0	1	2	3
$g(x) = \|x - 2\|$	5	4	3	2	1	0	1

Figure 2.3.5

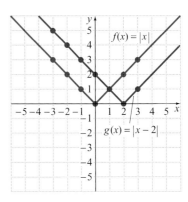

Observations:

▶ Examining the table of function values, we see that the values of $g(x) = |x - 2|$ are the values of $f(x) = |x|$ *shifted to the right by 2 units*. This is illustrated by the numbers in red in the rows labeled $f(x) = |x|$ and $g(x) = |x - 2|$.

▶ The graph of $g(x) = |x - 2|$ is the same as the graph of $f(x) = |x|$, but it is *shifted to the right by 2 units*.

The domain of both f and g is the set of all real numbers. From the graphs, we see that the range of both f and g is $[0, \infty)$.

☑ *Check It Out 2:* Make a table of values for the functions $f(x) = |x|$ and $g(x) = |x - 1|$, for $x = -3, -2, -1, 0, 1, 2, 3$. Use your table to sketch the graphs of the two functions. What are the domain and range of f and g? ▪

The results of Example 2 lead to a general statement about **horizontal shifts** of the graph of a function.

Technology Note

Horizontal shifts can be seen easily with a graphing calculator. Figure 2.3.7 shows the graphs of $f(x) = |x|$ and $g(x) = |x - 2|$ on the same set of axes, using a decimal window.

Keystroke Appendix: Section 7

Figure 2.3.7

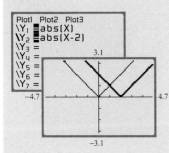

Horizontal Shifts of the Graph of $f(x)$

Let f be a function and c be a positive constant.

▶ The graph of $g(x) = f(x - c)$ is the graph of $f(x)$ shifted c units *to the right*.

▶ The graph of $g(x) = f(x + c)$ is the graph of $f(x)$ shifted c units *to the left*.

See Figure 2.3.6.

Figure 2.3.6 Horizontal shifts of f, $c > 0$

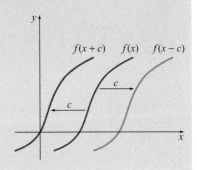

Vertical and horizontal shifts can also be combined to create the graph of a new function.

Example 3 Combining Vertical and Horizontal Shifts of a Function

Discover *and* **Learn**

In Example 3b, verify that the order of the vertical and horizontal translations does not matter by first shifting the graph of $f(x) = |x|$ down by 2 units and then shifting the resulting graph horizontally to the left by 3 units.

Use vertical and/or horizontal shifts, along with a table of values, to graph the following functions.

(a) $g(x) = \sqrt{x + 2}$ (b) $g(x) = |x + 3| - 2$

▶Solution

(a) The graph of $g(x) = \sqrt{x + 2}$ is a horizontal shift of the graph of $f(x) = \sqrt{x}$ by 2 units to the *left*, since $g(x) = \sqrt{x + 2} = f(x + 2) = f(x - (-2))$. Make a table of function values as shown in Table 2.3.4, and use it to sketch the graphs of both functions as shown in Figure 2.3.8.

Table 2.3.4

x	$f(x) = \sqrt{x}$	$g(x) = \sqrt{x + 2} = f(x + 2)$
-3	error	error
-2	error	0
-1	error	1
0	0	$\sqrt{2} \approx 1.414$
1	1	$\sqrt{3} \approx 1.732$
2	$\sqrt{2} \approx 1.414$	2
3	$\sqrt{3} \approx 1.732$	$\sqrt{5} \approx 2.236$

Figure 2.3.8

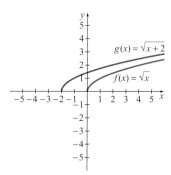

(b) The graph of the function $g(x) = |x + 3| - 2$ consists of both a horizontal shift of the graph of $f(x) = |x|$ by 3 units *to the left* and a vertical shift by 2 units *down*. We make a table of function values as shown in Table 2.3.5. Examining the table of function values, we see that the values of $y = |x + 3|$ are the values of $f(x) = |x|$ shifted *to the left by 3 units*.

To find the values of $g(x) = |x + 3| - 2$, we shift the values of $y = |x + 3|$ *down by 2 units*. We can graph the function $g(x) = |x + 3| - 2$ in two stages—first the horizontal shift and then the vertical shift. See Figure 2.3.9.

The order of the horizontal and vertical translations does not matter. You can verify this in the Discover and Learn on this page.

Table 2.3.5

x	-4	-3	-2	-1	0	1	2
$f(x) = |x|$	4	3	2	1	0	1	2
$y = |x + 3|$	1	0	1	2	3	4	5
$g(x) = |x + 3| - 2$	-1	-2	-1	0	1	2	3

Figure 2.3.9

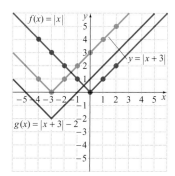

☑ *Check It Out 3:* Graph the function $f(x) = \sqrt{x - 2} + 1$ using transformations. ■

Vertical Scalings and Reflections Across the Horizontal Axis

In this subsection, we will examine what happens when an expression for a function is multiplied by a nonzero constant. First, we make a table of values for the functions $f(x) = x^2$, $g(x) = 2x^2$, and $h(x) = \frac{1}{2}x^2$ for $x = -3, -2, -1, 0, 1, 2, 3$ (Table 2.3.6) and then we sketch their graphs (Figure 2.3.10).

Table 2.3.6

x	$f(x) = x^2$	$g(x) = 2x^2$	$h(x) = \frac{1}{2}x^2$
-3	9	18	4.5
-2	4	8	2
-1	1	2	0.5
0	0	0	0
1	1	2	0.5
2	4	8	2
3	9	18	4.5

Figure 2.3.10

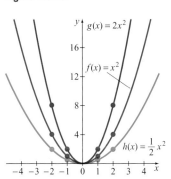

Observations:

▶ The graph of $g(x) = 2x^2 = 2f(x)$ has y-coordinates that are twice those of $f(x) = x^2$, and so $g(x) \geq f(x)$ for all x. The graph of $g(x)$ is thus *vertically stretched* away from the x-axis.

▶ The graph of $h(x) = \frac{1}{2}x^2 = \frac{1}{2}f(x)$ has y-coordinates that are half those of $f(x) = x^2$, and so $h(x) \leq f(x)$ for all x. The graph of $h(x)$ is thus *vertically compressed* toward the x-axis.

We see that multiplying $f(x)$ by a nonzero constant has the effect of scaling the function values. This results in a vertical stretch or compression of the graph of $f(x)$.

The above discussion leads to a general statement about **vertical scalings** of the graph of a function.

Vertical Scalings of the Graph of f(x)

Let f be a function and c be a positive constant.

▶ If $c > 1$, the graph of $g(x) = cf(x)$ is the graph of $f(x)$ **stretched vertically** away from the x-axis, with the y-coordinates of $g(x)$ multiplied by c.

▶ If $0 < c < 1$, the graph of $g(x) = cf(x)$ is the graph of $f(x)$ **compressed vertically** toward the x-axis, with the y-coordinates of $g(x)$ multiplied by c.

See Figure 2.3.11.

Figure 2.3.11 Vertical scalings of f, $c > 0$

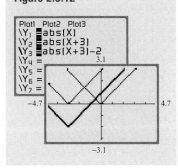

We can also consider the graph of $g(x) = -f(x)$. In this case, the y-coordinate of $(x, g(x))$ will be the negative of the corresponding y-coordinate of $(x, f(x))$. Graphically, this results in a reflection of the graph of $f(x)$ across the x-axis.

Reflection of the Graph of $f(x)$ Across the x-Axis

Let f be a function.

▶ The graph of $g(x) = -f(x)$ is the graph of $f(x)$ **reflected across the x-axis.**

See Figure 2.3.13.

Figure 2.3.13 Reflection of f across x-axis

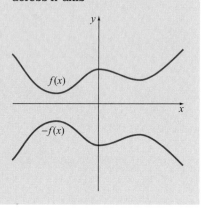

Many graphing problems involve some combination of vertical and horizontal translations, as well as vertical scalings and reflections. These actions on the graph of a function are collectively known as **transformations** of the graph of the function, and are explored in the following examples.

Example 4 Sketching Graphs Using Transformations

Identify the basic function $f(x)$ that is transformed to obtain $g(x)$. Then use transformations to sketch the graphs of both $f(x)$ and $g(x)$.

(a) $g(x) = 3\sqrt{x}$ (b) $g(x) = -2|x|$ (c) $g(x) = -2|x + 1| + 3$

▶**Solution**

(a) The graph of the function $g(x) = 3\sqrt{x}$ is a vertical stretch of the graph of $f(x) = \sqrt{x}$, since the function values of $f(x)$ are multiplied by a factor of 3. Both functions have domain $(0, \infty)$. A table of function values for the two functions is given in Table 2.3.7, and their corresponding graphs are shown in Figure 2.3.14.

Table 2.3.7

x	$f(x) = \sqrt{x}$	$g(x) = 3\sqrt{x}$
0	0	0
1	1	3
2	1.414	4.243
4	2	6
9	3	9

Figure 2.3.14

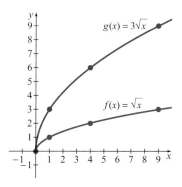

(b) The graph of the function $g(x) = -2|x|$ consists of a vertical stretch and then a reflection across the x-axis of the graph of $f(x) = |x|$, since the values of $f(x)$ are not only doubled but also negated. A table of function values for the two functions is given in Table 2.3.8, and their corresponding graphs are shown in Figure 2.3.15.

Table 2.3.8

x	$f(x) = \|x\|$	$g(x) = -2\|x\|$
-3	3	-6
-2	2	-4
-1	1	-2
0	0	0
1	1	-2
2	2	-4
3	3	-6

Figure 2.3.15

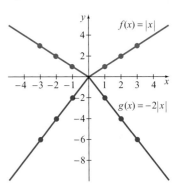

(c) We can view the graph of $g(x)$ as resulting from transformations of the graph of $f(x) = |x|$ in the following manner:

$$f(x) = |x| \quad \rightarrow \quad y_1 = -2|x| \quad \rightarrow \quad y_2 = -2|x + 1| \quad \rightarrow \quad g(x) = -2|x + 1| + 3$$

| Vertical scaling by 2 and reflection across the x-axis | Horizontal shift to the left by 1 unit | Vertical shift upward by 3 units |

The transformation from $f(x) = |x|$ to $y_1 = -2|x|$ was already discussed and graphed in part (b). We now take the graph of $y_1 = -2|x|$ and shift it to the left by 1 unit and then upward by 3 units, as shown in Figure 2.3.16.

Figure 2.3.16

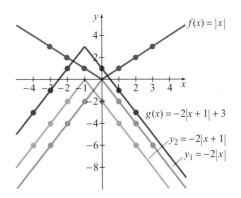

✔ *Check It Out 4:* Use transformations to sketch the graphs of the following functions.

(a) $g(x) = -3x^2$

(b) $h(x) = -|x - 1| + 2$ ■

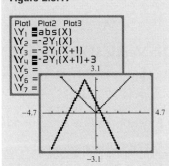

| Example | 5 Using Transformations to Sketch a Graph |

Suppose the graph of a function $g(x)$ is produced from the graph of $f(x) = x^2$ by vertically compressing the graph of f by a factor of $\frac{1}{3}$, then shifting it to the left by 1 unit, and finally shifting it downward by 2 units. Give an expression for $g(x)$, and sketch the graphs of both f and g.

▶Solution   The series of transformations can be summarized as follows.

$$f(x) = x^2 \quad \rightarrow \quad y_1 = \frac{1}{3}x^2 \quad \rightarrow \quad y_2 = \frac{1}{3}(x + 1)^2 \quad \rightarrow \quad g(x) = \frac{1}{3}(x + 1)^2 - 2$$

| Original function | Vertical scaling by $\frac{1}{3}$ | Shift left by 1 unit | Shift down by 2 units |

The function g is then given by

$$g(x) = \frac{1}{3}(x + 1)^2 - 2.$$

The graphs of f and g are given in Figure 2.3.18. Figure 2.3.18 also shows the graph of $y_1 = \frac{1}{3}x^2$. The horizontal and vertical shifts are indicated by arrows.

Figure 2.3.18

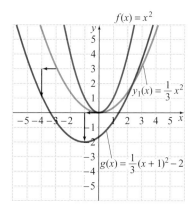

✔ *Check It Out 5:* Let the graph of $g(x)$ be produced from the graph of $f(x) = \sqrt{x}$ by vertically stretching the graph of f by a factor of 3, then shifting it to the left by 2 units, and finally shifting it upward by 1 unit. Give an expression for $g(x)$, and sketch the graphs of both f and g. ■

Horizontal Scalings and Reflections Across the Vertical Axis

The final set of transformations involves stretching and compressing the graph of a function along the horizontal axis. In function notation, we examine the relationship between the graph of $f(x)$ and the graph of $f(cx)$, $c > 0$. A good way to study these types of transformations is to first look at the graph of a function and its corresponding table of function values.

Consider the following function, $f(x)$, defined by the graph shown in Figure 2.3.19 and the corresponding representative values given in Table 2.3.9.

Table 2.3.9

x	-4	-2	0	2	4
$f(x)$	0	2	4	2	0

Figure 2.3.19

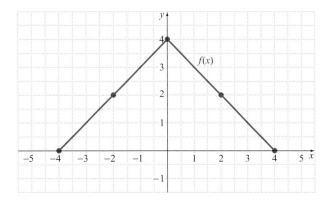

We can make a table of values for $f(2x)$ and $f\left(\frac{1}{2}x\right)$ and sketch their graphs. For $f(2x)$, if $x = 1$, we evaluate $f(2(1)) = f(2) = 2$. We obtained the value for $f(2)$ from Table 2.3.9. The rest of the table for $f(2x)$ is filled in similarly. See Table 2.3.10.

For $f\left(\frac{1}{2}x\right)$, first let $x = 4$. Then evaluate $f\left(\frac{1}{2}(4)\right) = f(2) = 2$ by using the values for $f(x)$ given in Table 2.3.9. The rest of the table for $f\left(\frac{1}{2}x\right)$ is filled in similarly. See Table 2.3.11.

The graph of $f(2x)$ is a **horizontal compression** of the graph of $f(x)$. See Figure 2.3.20. This is also evident from the table of input values for $f(2x)$. The graph of $f\left(\frac{1}{2}x\right)$ is a **horizontal stretching** of the graph of $f(x)$. See Figure 2.3.21. This is also evident from the table of input values for $f\left(\frac{1}{2}x\right)$.

Table 2.3.10

x	-2	-1	0	1	2
$f(2x)$	0	2	4	2	0

Figure 2.3.20 Graph of $f(2x)$

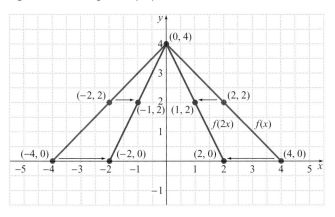

Table 2.3.11

x	-8	-4	0	4	8
$f\left(\frac{1}{2}x\right)$	0	2	4	2	0

Figure 2.3.21 Graph of $f\left(\frac{1}{2}x\right)$

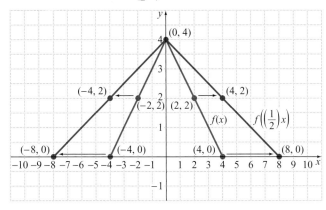

> **Note** It is important to see that horizontal scalings work in an opposite manner to vertical scalings. For example, if x = 1, f(2x) = f(2). The value of f(2) is reached "earlier" (at x = 1) by the function f(2x) than by the function f(x). This accounts for the shrinking effect.

The preceding discussion leads to a general statement about **horizontal scalings** of the graph of a function.

Horizontal Scaling of the Graph of f(x)

Let f be a function and c be a positive constant.

▶ If $c > 1$, the graph of $g(x) = f(cx)$ is the graph of $f(x)$ **compressed horizontally** toward the y-axis, scaled by a factor of $\frac{1}{c}$.

▶ If $0 < c < 1$, the graph of $g(x) = f(cx)$ is the graph of $f(x)$ **stretched horizontally** away from the y-axis, scaled by a factor of $\frac{1}{c}$.

See Figure 2.3.22.

Figure 2.3.22 Horizontal scalings of f, c > 0

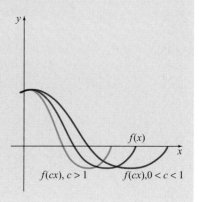

We can also consider the graph of $g(x) = f(-x)$, which is a reflection of the graph of $f(x)$ across the y-axis.

Reflection of the Graph of f(x) Across the y-Axis

Let f be a function.

▶ The graph of $f(x) = f(-x)$ is the graph of $f(x)$ **reflected across the y-axis.**

See Figure 2.3.23.

Figure 2.3.23 Reflection of f across y-axis

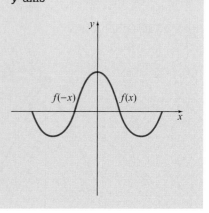

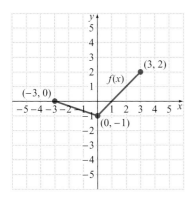

Figure 2.3.24

| Example | 6 Using Transformations to Sketch a Graph |

The graph of $f(x)$ is shown in Figure 2.3.24. Use it to sketch the graphs of (a) $f(3x)$ and (b) $f(-x) + 1$.

▶Solution

(a) The graph of $f(3x)$ is a *horizontal compression* of the graph of $f(x)$. The x-coordinates of $f(x)$ are scaled by a factor of $\frac{1}{3}$. Table 2.3.12 summarizes how each of the key points on the graph of $f(x)$ is transformed to the corresponding point on the graph of $f(3x)$. The points are then used to sketch the graph of $f(3x)$. See Figure 2.3.25.

Table 2.3.12

Point on Graph of $f(x)$		Point on Graph of $f(3x)$
$(-3, 0)$	$\rightarrow$	$\left(\frac{1}{3}(-3), 0\right) = (-1, 0)$
$(0, -1)$	$\rightarrow$	$\left(\frac{1}{3}(0), -1\right) = (0, -1)$
$(3, 2)$	$\rightarrow$	$\left(\frac{1}{3}(3), 2\right) = (1, 2)$

Figure 2.3.25

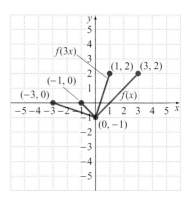

(b) The graph of $f(-x) + 1$ is a *reflection across the y-axis* of the graph of $f(x)$, followed by a *vertical shift upward* of 1 unit. To obtain the graph of $f(-x)$, the x-coordinates of $f(x)$ are negated. To obtain the vertical shift, 1 is then added to the y-coordinates. Table 2.3.13 summarizes how each of the key points on the graph of $f(x)$ is transformed first to the corresponding point on the graph of $f(-x)$, and then to the corresponding point on the graph of $f(-x) + 1$. The points are then used to sketch the graph of $f(-x) + 1$. See Figure 2.3.26.

Table 2.3.13

Point on Graph of $f(x)$		Point on Graph of $f(-x)$		Point on Graph of $f(-x) + 1$
$(-3, 0)$	$\rightarrow$	$(-(-3), 0) = (3, 0)$	$\rightarrow$	$(3, 0 + 1) = (3, 1)$
$(0, -1)$	$\rightarrow$	$(-(0), -1) = (0, -1)$	$\rightarrow$	$(0, -1 + 1) = (0, 0)$
$(3, 2)$	$\rightarrow$	$(-(3), 2) = (-3, 2)$	$\rightarrow$	$(-3, 2 + 1) = (-3, 3)$

Figure 2.3.26

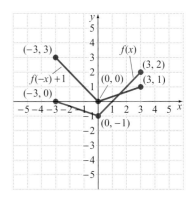

☑ *Check It Out 6:* For $f(x)$ given in Example 6, graph $f\left(\frac{1}{2}x\right)$. ▪

2.3 Key Points

▶ **Vertical shifts:** The graph of $g(x) = f(x) + c, c > 0$, is the graph of $f(x)$ shifted c units *upward;* the graph of $g(x) = f(x) - c, c > 0$, is the graph of $f(x)$ shifted c units *downward.*

▶ **Horizontal shifts:** The graph of $g(x) = f(x - c), c > 0$, is the graph of $f(x)$ shifted c units *to the right;* the graph of $g(x) = f(x + c), c > 0$, is the graph of $f(x)$ shifted c units *to the left.*

▶ **Vertical stretching and compression:** If $c > 1$, the graph of $g(x) = cf(x)$ is the graph of $f(x)$ *stretched vertically* away from the x-axis, with the y-coordinates of $g(x) = cf(x)$ multiplied by c. If $0 < c < 1$, the graph of $g(x) = cf(x)$ is the graph of $f(x)$ *compressed vertically* toward the x-axis, with the y-coordinates of $g(x) = cf(x)$ multiplied by c.

▶ **Reflections:** The graph of $g(x) = -f(x)$ is the graph of $f(x)$ *reflected across the x-axis.* The graph of $g(x) = f(-x)$ is the graph of $f(x)$ *reflected across the y-axis.*

▶ **Horizontal stretching and compression:** If $c > 1$, the graph of $g(x) = f(cx)$ is the graph of $f(x)$ *compressed horizontally* toward the y-axis; if $0 < c < 1$, the graph of $g(x) = f(cx)$ is the graph of $f(x)$ *stretched horizontally* away from the y-axis.

2.3 Exercises

▶**Skills** This set of exercises will reinforce the skills illustrated in this section.

In Exercises 1–30, identify the underlying basic function, and use transformations of the basic function to sketch the graph of the given function.

1. $g(t) = t^2 + 1$

2. $g(t) = t^2 - 3$

3. $f(x) = \sqrt{x} - 2$

4. $g(x) = \sqrt{x} + 1$

5. $h(x) = |x - 2|$

6. $h(x) = |x + 4|$

7. $F(s) = (s + 5)^2$

8. $G(s) = (s - 3)^2$

9. $f(x) = \sqrt{x - 4}$

10. $f(x) = \sqrt{x + 3}$

11. $H(x) = |x - 2| + 1$

12. $G(x) = \sqrt{x + 1} - 2$

13. $S(x) = (x + 3)^2 - 1$

14. $g(x) = (x - 2)^2 + 5$

15. $H(t) = 3t^2$

16. $g(x) = 2\sqrt{x}$

17. $S(x) = -4|x|$

18. $H(x) = -2x^2$

19. $H(s) = -|s| - 3$

20. $F(x) = -\sqrt{x + 4}$

21. $h(x) = -\dfrac{1}{2}|x + 1| - 3$

22. $h(x) = -2|x - 4| + 1$

23. $g(x) = -3(x + 2)^2 - 4$

24. $h(x) = -\dfrac{1}{3}(x - 2)^2 - \dfrac{3}{2}$

25. $f(x) = |2x|$

26. $f(x) = \left|\dfrac{x}{2}\right|$

27. $f(x) = (2x)^2$

28. $f(x) = \left(\dfrac{1}{2}x\right)^2$

29. $g(x) = \sqrt{3x}$

30. $f(x) = \sqrt{2x}$

In Exercises 31–34, explain how each graph is a transformation of the graph of $f(x) = |x|$, and find a suitable expression for the function represented by the graph.

31.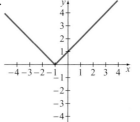

32.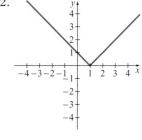

33. 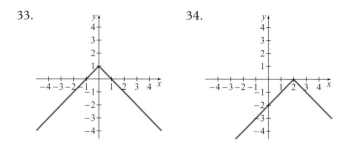 34.

In Exercises 35–38, explain how each graph is a transformation of the graph of $f(x) = x^2$, and find a suitable expression for the function represented by the graph.

35. 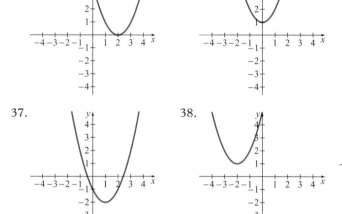 36.

37. 38.

In Exercises 39–46, use the verbal description to find an algebraic expression for the function.

39. The graph of the function $g(t)$ is formed by translating the graph of $f(t) = |t|$ 4 units to the left and 3 units down.

40. The graph of the function $f(t)$ is formed by translating the graph of $h(t) = t^2$ 2 units to the right and 6 units upward.

41. The graph of the function $g(t)$ is formed by vertically scaling the graph of $f(t) = t^2$ by a factor of -3 and moving it to the right by 1 unit.

42. The graph of the function $g(t)$ is formed by vertically scaling the graph of $f(t) = |t|$ by a factor of -2 and moving it to the left by 5 units.

43. The graph of the function $k(t)$ is formed by scaling the graph of $f(t) = \sqrt{t}$ horizontally by a factor of -1 and moving it up 3 units.

44. The graph of the function $h(x)$ is formed by scaling the graph of $g(x) = x^2$ horizontally by a factor of $\frac{1}{2}$ and moving it down 4 units.

45. The graph of the function $h(t)$ is formed by scaling the graph of $f(t) = |t|$ vertically by a factor of $\frac{1}{2}$ and shifting it up 4 units.

46. The graph of the function $g(x)$ is formed by scaling the graph of $f(x) = \sqrt{x}$ vertically by a factor of -1 and horizontally by a factor of -1.

In Exercises 47–54, use the given function f to sketch a graph of the indicated transformation of f. First copy the graph of f onto a sheet of graph paper.

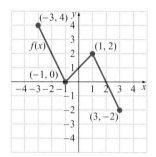

47. $2f(x)$

48. $\dfrac{1}{2}f(x)$

49. $f(x) + 2$

50. $-f(x) - 3$

51. $f(2x)$

52. $f\left(\dfrac{1}{2}x\right)$

53. $f(x - 1) + 2$

54. $-f(x + 2) - 1$

In Exercises 55–62, use the given function f to sketch a graph of the indicated transformation of f. First copy the graph of f onto a sheet of graph paper.

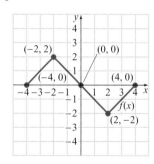

55. $f(x + 2)$

56. $f(x - 1)$

57. $2f(x) + 1$

58. $-f(x) - 1$

59. $f(2x)$

60. $f\left(\dfrac{1}{3}x\right)$

61. $f(x + 1) - 3$

62. $-2f(x + 3)$

In Exercises 63 and 64, use the table giving values for f(x) and g(x) = f(x) + k to find the appropriate value of k.

63.

x	f(x)	g(x) = f(x) + k
−2	−4	−6
−1	−6	−8
0	−9	−11
1	−6	−8
2	−4	−6

64.

x	f(x)	g(x) = f(x) + k
10	3	4.5
9	2	3.5
8	1	2.5
7	2	3.5
6	3	4.5

In Exercises 65 and 66, use the table giving values for f(x) and g(x) = f(x − k) to find the appropriate value of k.

65.

x	−2	−1	0	1	2
f(x)	4	6	8	10	12
g(x) = f(x − k)	2	4	6	8	10

66.

x	−2	−1	0	1	2
f(x)	−4	−5	−6	−5	−4
g(x) = f(x − k)	−5	−6	−5	−4	−3

In Exercises 67 and 68, fill in the missing values of the function f in the table.

67.

x	f(x)	g(x) = f(x) − 3
−2		33
−1		22
0		13
1		6
2		1

68.

x	f(x)	g(x) = f(x) + 2
−2		20
−1		18.5
0		16
1		17.2
2		13

In Exercises 69–76, use a graphing utility to solve the problem.

69. Graph $f(x) = |x + 3.5|$ and $g(x) = |x| + 3.5$. Describe each graph in terms of transformations of the graph of $h(x) = |x|$.

70. Graph $f(x) = (x − 4.5)^2$ and $g(x) = x^2 + 4.5$. Describe each graph in terms of transformations of the graph of $h(x) = x^2$.

71. If $f(x) = \sqrt{x}$, graph $f(x)$ and $f(x − 4.5)$ in the same viewing window. What is the relationship between the two graphs?

72. If $f(x) = |x|$, graph $f(x)$ and $f(0.3x)$ in the same viewing window. What is the relationship between the two graphs?

73. If $f(x) = |x|$, graph $−2f(x)$ and $f(−2x)$ in the same viewing window. Are the graphs the same? Explain.

74. If $f(x) = \sqrt{x}$, graph $3f(x)$ and $f(3x)$ in the same viewing window. Are the graphs the same? Explain.

75. Graph $f(x) = x^3$ and $g(x) = (x − 7)^3$. How can the graph of g be described in terms of the graph of f?

76. Graph the functions $f(x) = |x − 4|$ and $g(x) = f(−x) = |(−x) − 4|$. What relationship do you observe between the graphs of the two functions? Do the same with $f(x) = (x − 2)^2$ and $g(x) = f(−x) = ((−x) − 2)^2$. What type of reflection of the graph of $f(x)$ gives the graph of $g(x) = f(−x)$?

▶**Applications** In this set of exercises, you will use transformations to study real-world problems.

77. **Coffee Sales** Let $P(x)$ represent the price of x pounds of coffee. Assuming the entire amount of coffee is taxed at 6%, find an expression, in terms of $P(x)$, for just the sales tax on x pounds of coffee.

78. **Salary** Let $S(x)$ represent the weekly salary of a salesperson, where x is the weekly dollar amount of sales generated. If the salesperson pays 15% of her salary in federal taxes, express her after-tax salary in terms of $S(x)$. Assume there are no other deductions to her salary.

79. **Printing** The production cost, in dollars, for x color brochures is $C(x) = 500 + 3x$. The fixed cost is $500, since that is the amount of money needed to start production even if no brochures are printed.
    (a) If the fixed cost is decreased by $50, find the new cost function.
    (b) [calculator icon] Graph both cost functions and interpret the effect of the decreased fixed cost.

80. **Geometry** The area of a square is given by $A(s) = s^2$, where s is the length of a side in inches. Compute the expression for $A(2s)$ and explain what it represents.

81. **Physics** The height of a ball thrown upward with a initial velocity of 30 meters per second from an initial height of h meters is given by

    $$s(t) = -16t^2 + 30t + h$$

    where t is the time in seconds.
    (a) If $h = 0$, how high is the ball at time $t = 1$?
    (b) If $h = 20$, how high is the ball at time $t = 1$?
    (c) In terms of shifts, what is the effect of h on the function $s(t)$?

82. **Unit Conversion** Let $T(x)$ be the temperature, in degrees Celsius, of a point on a long rod located x centimeters from one end of the rod (where that end of the rod corresponds to $x = 0$). Temperature can be measured in kelvin (the unit of temperature for the absolute temperature scale) by adding 273 to the temperature in degrees Celsius. Let $t(x)$ be the temperature function in kelvin, and write an expression for $t(x)$ in terms of the function $T(x)$.

▶**Concepts** This set of exercises will draw on the ideas presented in this section and your general math background.

83. The point $(2, 4)$ on the graph of $f(x) = x^2$ has been shifted horizontally to the point $(-3, 4)$. Identify the shift and write a new function $g(x)$ in terms of $f(x)$.

84. The point $(-2, 2)$ on the graph of $f(x) = |x|$ has been shifted horizontally *and* vertically to the point $(3, 4)$. Identify the shifts and write a new function $g(x)$ in terms of $f(x)$.

85. When using transformations with both vertical scaling and vertical shifts, the order in which you perform the transformations matters. Let $f(x) = |x|$.
    (a) Find the function $g(x)$ whose graph is obtained by first vertically stretching $f(x)$ by a factor of 2 and then shifting the result upward by 3 units. A table of values and/or a sketch of the graph will be helpful.
    (b) Find the function $g(x)$ whose graph is obtained by first shifting $f(x)$ upward by 3 units and then multiplying the result by a factor of 2. A table of values and/or a sketch of the graph will be helpful.
    (c) Compare your answers to parts (a) and (b). Explain why they are different.

86. Let $f(x) = 2x + 5$ and $g(x) = f(x + 2) - 4$. Graph both functions on the same set of coordinate axes. Describe the transformation from $f(x)$ to $g(x)$. What do you observe?

2.4 Symmetry and Other Properties of Functions

Objectives

▶ Determine if a function is even, odd, or neither

▶ Given a graph, determine intervals on which a function is increasing, decreasing, or constant

▶ Determine the average rate of change of a function over an interval

In this section, we will study further properties of functions that will be useful in later chapters. These properties will provide additional tools for understanding functions and their graphs.

Even and Odd Functions

You may have observed that the graph of $f(x) = x^2$ is a mirror image of itself when reflected across the y-axis. This is referred to as **symmetry with respect to the y-axis.** How can this symmetry be described using function notation? Let's take a closer look at the table of values (Table 2.4.1) and graph (Figure 2.4.1) for $f(x) = x^2$.

Table 2.4.1

x	$f(x) = x^2$
-3	9
-1.5	2.25
-1	1
0	0
1	1
1.5	2.25
3	9

Figure 2.4.1

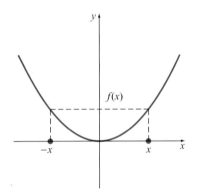

From the figure and the table of selected function values, we see that

▶ the points $(x, f(x))$ and $(-x, f(-x))$ are symmetric with respect to the y-axis.

▶ $f(x) = f(-x)$.

In fact, these statements are true for *every* x in the domain of this function. We summarize our findings as follows.

Discover *and* Learn

Verify that $f(x) = x^4 + 2$ is an even function by checking the definition.

> **Definition of an Even Function**
>
> A function is **symmetric with respect to the y-axis** if
>
> $$f(x) = f(-x) \text{ for each } x \text{ in the domain of } f.$$
>
> Functions having this property are called **even functions.**

Another type of symmetry that occurs is defined as **symmetry with respect to the origin.** Once again, let's see how this new type of symmetry can be described using function notation. Let's use   as an example and examine its graph (Figure 2.4.2) along with a selected set of function values (Table 2.4.2).

Table 2.4.2

x	$f(x) = x^3$
-3	-27
-1.5	-3.375
-1	-1
0	0
1	1
1.5	3.375
3	27

Figure 2.4.2

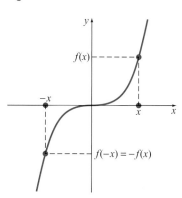

Discover *and* Learn

Verify that $f(s) = s^3 - s$ is an odd function by checking the definition.

Note that when a function is symmetric with respect to the origin, $f(-x) = -f(x)$. For example, for $x = 1.5$, this relationship is highlighted in color in Table 2.4.2. Our findings are summarized as follows.

Definition of an Odd Function

A function is **symmetric with respect to the origin** if

$$f(-x) = -f(x) \text{ for each } x \text{ in the domain of } f.$$

Functions having this property are called **odd functions**.

Technology Note

You can graph a function to see if it is odd, even, or neither and then check your conjecture algebraically. The graph of $h(x) = -x^3 + 3x$ looks as though it is symmetric with respect to the origin (Figure 2.4.3). This is checked algebraically in Example 1(c).

Keystroke Appendix: Section 7

Figure 2.4.3

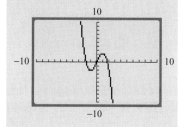

Observations:

▶ A function *cannot* be *both* odd and even at the same time unless it is the function $f(x) = 0$.

▶ There are various other symmetries in addition to those we have discussed here. For example, a function can be symmetric with respect to a vertical line other than the y-axis, or symmetric with respect to a point other than the origin. However, these types of symmetries are beyond the scope of our current discussion.

Example 1 Determining Odd or Even Functions

Using the definitions of odd and even functions, classify the following functions as odd, even, or neither.

(a) $f(x) = |x| + 2$

(b) $g(x) = (x - 4)^2$

(c) $h(x) = x^3 + 3x$

▶Solution

(a) First check to see if f is an even function.

$$f(-x) = |-x| + 2 = |-1||x| + 2 = |x| + 2 = f(x)$$

Since $f(x) = f(-x)$, f is an even function. The graph of f is symmetric with respect to the y-axis, as shown in Figure 2.4.4. This graph verifies what we found by use of algebra alone.

Figure 2.4.4

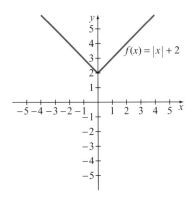

$f(x) = |x| + 2$

(b) For $g(x) = (x - 4)^2$, it helps to first expand the expression for the function, which gives

$$g(x) = (x - 4)^2 = x^2 - 8x + 16.$$

We can then see that

$$g(-x) = (-x)^2 - 8(-x) + 16 = x^2 + 8x + 16.$$

Since $g(x) \neq g(-x)$, g is not even. Using the expression for $g(-x)$ that we have already found, we can see that

$$g(-x) = x^2 + 8x + 16 \quad \text{and} \quad -g(x) = -(x^2 - 8x + 16) = -x^2 + 8x - 16.$$

Since $g(-x) \neq -g(x)$, g is not odd. The fact that this function is neither even nor odd can also be seen from its graph. Figure 2.4.5 shows no symmetry, either with respect to the y-axis or with respect to the origin.

Figure 2.4.5

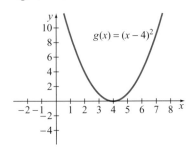

$g(x) = (x - 4)^2$

(c) Since $h(x)$ has an odd-powered term, we will first check to see if it is an odd function.

$$h(-x) = (-x)^3 + 3(-x) = -x^3 - 3x = -(x^3 + 3x) = -h(x)$$

Since $h(-x) = -h(x)$, h is an odd function. Thus it is symmetric with respect to the origin, as verified by the graph in Figure 2.4.6.

Figure 2.4.6

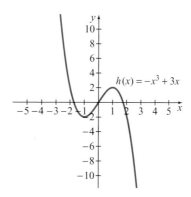

✔ *Check It Out 1:* Decide whether the following functions are even, odd, or neither.
(a) $h(x) = 2|x|$
(b) $f(x) = (x + 1)^2$ ▪

Increasing and Decreasing Functions

An important idea in studying functions is figuring out how the function value, y, changes as x changes. You should already have some intuitive ideas about this quality of a function. The following definition about increasing and decreasing functions makes these ideas precise.

> **Increasing, Decreasing, and Constant Functions**
> ▶ A function f is **increasing** on an open interval I if, for any a, b in the interval, $f(a) < f(b)$ for $a < b$. See Figure 2.4.7.
> ▶ A function f is **decreasing** on an open interval I if, for any a, b in the interval, $f(a) > f(b)$ for $a < b$. See Figure 2.4.8.
> ▶ A function f is **constant** on an open interval I if, for any a, b in the interval, $f(a) = f(b)$. See Figure 2.4.9.

Figure 2.4.7

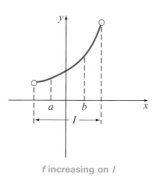

f increasing on I

Figure 2.4.8

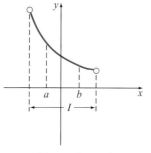

f decreasing on I

Figure 2.4.9

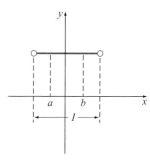

f constant on I

Example 2 **Increasing and Decreasing Functions**

For the function f given in Figure 2.4.10, find the interval(s) on which

(a) f is increasing.

(b) f is decreasing.

(c) f is constant.

Figure 2.4.10

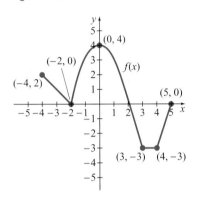

►Solution

(a) From the graph, the function is increasing on the intervals $(-2, 0)$ and $(4, 5)$.

(b) From the graph, the function is decreasing on the intervals $(-4, -2)$ and $(0, 3)$.

(c) From the graph, the function is constant on the interval $(3, 4)$.

✔ *Check It Out 2:* For the function f given in Figure 2.4.11, find the interval(s) on which f is decreasing.

Figure 2.4.11

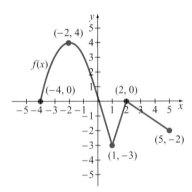

In Chapters 3, 4, and 5, we will discuss increasing and decreasing functions in more detail. In those chapters, we will also examine points at which the graph of a function "turns" from increasing to decreasing or vice versa.

Average Rate of Change

While determining whether a function is increasing or decreasing is of some value, it is of greater interest to figure out how *quickly* a function increases or decreases.

Put another way, we would like to figure out how fast a function *changes*. One quantity that tells us how quickly a function changes is called the **average rate of change.**

Average Rate of Change

The **average rate of change** of a function f on an interval $[x_1, x_2]$ is given by

$$\text{Average rate of change} = \frac{f(x_2) - f(x_1)}{x_2 - x_1}.$$

Example **3** **Determining Average Rate of Change**

Find the average rate of change of $f(x) = 2x^2 + 1$ on the following intervals.
(a) $[-3, -2]$
(b) $[0, 2]$

▶**Solution**

(a) Using $x_1 = -3$ and $x_2 = -2$ in the definition of average rate of change, we have

$$\text{Average rate of change} = \frac{f(x_2) - f(x_1)}{x_2 - x_1}$$

$$= \frac{f(-2) - f(-3)}{-2 - (-3)}$$

$$= \frac{9 - 19}{1}$$

$$= \frac{-10}{1}$$

$$= -10.$$

(b) Using $x_1 = 0$ and $x_2 = 2$ in the definition of average rate of change, we have

$$\text{Average rate of change} = \frac{f(x_2) - f(x_1)}{x_2 - x_1}$$

$$= \frac{f(2) - f(0)}{2 - 0}$$

$$= \frac{9 - 1}{2}$$

$$= \frac{8}{2}$$

$$= 4.$$

✔ *Check It Out 3:* Find the average rate of change of $f(x) = 2x^2 + 1$ on the interval $[3, 4]$. ▪

We will further examine the idea of average rate of change in Example 4.

Table 2.4.3

Gallons Used	Miles Driven
2	40
5	100
10	200

Example **4** **Rate of Change of a Linear Function**

Table 2.4.3 gives the distance traveled (in miles) by a car as a function of the amount of gasoline used (in gallons). Assuming that the distance traveled is a *linear* function of the amount of gasoline used, how many extra miles are traveled for each extra gallon of gasoline used?

▶Solution We are asked a question about how an output changes as its corresponding input changes. This is exactly the information given by the slope. Because the distance traveled is a linear function of the amount of gasoline used, we can use any two sets of values in the table and calculate the slope. We will use (2, 40) and (5, 100).

Increase in amount of gasoline used is $5 - 2 = 3$ gallons.

Increase in distance traveled is $100 - 40 = 60$ miles.

Since there is an increase of 60 miles for 3 gallons, we have the ratio

$$\frac{60 \text{ miles}}{3 \text{ gallons}} = \frac{20 \text{ miles}}{\text{gallon}}.$$

We can summarize our findings by saying that "the distance driven increases by 20 miles for each gallon of gasoline used."

✔ *Check It Out 4:* Show that you reach the same conclusion in Example 4 if the points (5, 100) and (10, 200) are used instead to calculate the slope. ■

Note Example 4 may be familiar to you from Section 1.1. We are examining the same problem, but from a different perspective.

We now show that the rate of change of a linear function, $f(x) = mx + b$, is constant.

$$\text{Rate of change} = \frac{f(x_2) - f(x_1)}{x_2 - x_1} \qquad \text{Definition of rate of change}$$

$$= \frac{mx_2 + b - (mx_1 + b)}{x_2 - x_1} \qquad \text{Substitute } x_2 \text{ and } x_1 \text{ into } f(x)$$

$$= \frac{mx_2 + b - mx_1 - b}{x_2 - x_1} \qquad \text{Remove parentheses}$$

$$= \frac{mx_2 - mx_1}{x_2 - x_1} \qquad \text{Simplify}$$

$$= \frac{m(x_2 - x_1)}{x_2 - x_1} \qquad \text{Factor out } m$$

$$= m \qquad \text{Cancel the term } (x_2 - x_1), \text{ since } x_2 \neq x_1$$

From our discussion, you can see that for a linear function, the average rate of change is exactly the slope. Furthermore, *the average rate of change of a linear function does not depend on the choices of x_1 and x_2*. This is the same as saying that *a linear function has constant slope.*

2.4 Key Points

▶ A function is **symmetric with respect to the y-axis** if

$$f(x) = f(-x) \text{ for each } x \text{ in the domain of } f.$$

Functions having this property are called **even functions**.

▶ A function is **symmetric with respect to the origin** if

$$f(-x) = -f(x) \text{ for each } x \text{ in the domain of } f.$$

Functions having this property are called **odd functions**.

▶ A function f is **increasing** on an open interval I if, for any a, b in the interval, $f(a) < f(b)$ for $a < b$.

▶ A function f is **decreasing** on an open interval I if, for any a, b in the interval, $f(a) > f(b)$ for $a < b$.

▶ A function f is **constant** on an open interval I if, for any a, b in the interval, $f(a) = f(b)$.

▶ The **average rate of change** of a function f on an interval $[x_1, x_2]$ is given by

$$\text{Average rate of change} = \frac{f(x_2) - f(x_1)}{x_2 - x_1}.$$

2.4 Exercises

▶**Skills** This set of exercises will reinforce the skills illustrated in this section.

In Exercises 1–6, classify each function given by its graph as odd, even, or neither.

1.

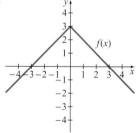

2.

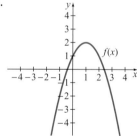

3.

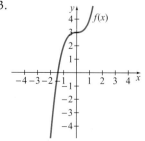

4.

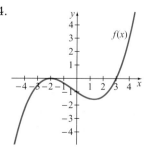

5.

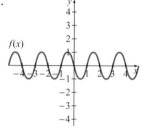

6.

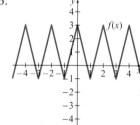

In Exercises 7–10, decide if each function is odd, even, or neither by using the appropriate definitions.

7.

x	-4	-2	0	2	4
$f(x)$	17	5	1	5	17

8.

x	-3	-1	0	1	3
$f(x)$	10	3	-2	4	10

9.

x	-4	-2	0	2	4
$f(x)$	-3	-1	0	1	3

10.

x	-3	-1	0	1	3
$f(x)$	-5	-7	-10	-7	-5

Exercises 11–17 pertain to the function f given by the following graph.

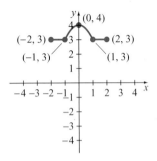

11. Find the domain of f.

12. Find the range of f.

13. Find the y-intercept.

14. Find the interval(s) on which f is increasing.

15. Find the interval(s) on which f is decreasing.

16. Find the interval(s) on which f is constant.

17. Is the function f even, odd, or neither?

Exercises 18–25 pertain to the function f given by the following graph.

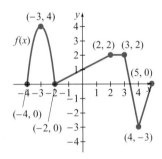

18. Find the interval(s) on which f is increasing.

19. Find the interval(s) on which f is decreasing.

20. Find the interval(s) on which f is constant.

21. Find the y-intercept.

22. Find the average rate of change of f on the interval $[-3, -2]$.

23. Find the average rate of change of f on the interval $[-2, 2]$.

24. Find the average rate of change of f on the interval $[2, 3]$.

25. Find the average rate of change of f on the interval $[4, 5]$.

Exercises 26–31 pertain to the function f given by the following graph.

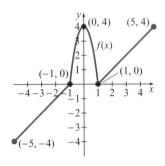

26. Find the interval(s) on which f is increasing.

27. Find the interval(s) on which f is decreasing.

28. Find the average rate of change of f on the interval $[-1, 0]$.

29. Find the average rate of change of f on the interval $[0, 1]$.

30. Find the y-intercept.

31. Is this function odd, even, or neither?

In Exercises 32–43, decide if each function is odd, even, or neither by using the definitions.

32. $f(x) = x + 3$

33. $f(x) = -2x$

34. $f(x) = |3x| - 2$

35. $f(x) = (x + 1)^2$

36. $f(x) = -3x^2 + 1$

37. $f(x) = -x^3 + 1$

38. $f(x) = -|x| + 1$

39. $f(x) = 2x$

40. $f(x) = |x - 1|$

41. $f(x) = x^5 - 2x$

42. $f(x) = (x^2 + 1)(x - 1)$

43. $f(x) = (x^2 - 3)(x^2 - 4)$

In Exercises 44–55, find the average rate of change of each function on the given interval.

44. $f(x) = -2x^2 + 5$; interval: $[-2, -1]$

45. $f(x) = 3x^2 - 1$; interval: $[2, 3]$

46. $f(x) = x^3 + 1$; interval: $[0, 2]$

47. $f(x) = -2x^3$; interval: $[-2, 0]$

48. $f(x) = 2x^2 + 3x - 1$; interval: $[-2, -1]$

49. $f(x) = 3x^3 + x^2 + 4$; interval: $[-2, 0]$

50. $f(x) = -x^4 + 6x^2 - 1$; interval: $[1, 2]$

51. $f(x) = -4x^3 - 3x^2 - 1$; interval: $[0, 2]$

52. $f(x) = 2|x| + 4$; interval: $[3, 5]$

53. $f(x) = |x| - 5$; interval: $[-4, -2]$

54. $f(x) = \sqrt{-x}$; interval: $[-4, -3]$

55. $f(x) = \sqrt{x} + 3$; interval: $[2, 4]$

In Exercises 56–61, use a graphing utility to decide if the function is odd, even, or neither.

56. $f(x) = x^2 - 4x + 1$

57. $f(x) = -2x^2 + 2x + 3$

58. $f(x) = 2x^3 - x$

59. $f(x) = (x + 1)(x - 2)(x + 3)$

60. $f(x) = x^4 - 5x^2 + 4$

61. $f(x) = -x^4 + 4x^2$

▶ **Applications** In this set of exercises, you will use properties of functions to study real-world problems.

62. **Demand Function** The demand for a product, in thousands of units, is given by $d(x) = \dfrac{100}{x}$, where x is the price of the product, $(x > 0)$. Is this an increasing or a decreasing function? Explain.

63. **Revenue** The revenue for a company is given by $R(x) = 30x$, where x is the number of units sold in thousands. Is this an increasing or a decreasing function? Explain.

64. **Depreciation** The value of a computer t years after purchase is given by $v(t) = 2000 - 300t$, where $v(t)$ is in dollars. Find the average rate of change of the value of the computer on the interval $[0, 3]$, and interpret it.

65. **Stamp Collecting** The value of a commemorative stamp t years after purchase appreciates according to the function

$v(t) = 0.37 + 0.05t$, where $v(t)$ is in dollars. Find the average rate of change of the value of the stamp on the interval $[0, 4]$, and interpret it.

66. **Commerce** The following table lists the annual sales of CDs by a small music store for selected years.

Year	Number of Units Sold
2002	10,000
2005	30,000
2006	33,000

Find the average rate of change in sales from 2002 to 2005. Also find the average rate of change in sales from 2005 to 2006. Does the average rate of change stay the same for both intervals? Why would a linear function *not* be useful for modeling these sales figures?

▶ **Concepts** This set of exercises will draw on the ideas presented in this section and your general math background.

67. Fill in the following table for $f(x) = 3x^2 - 2$.

Interval	Average Rate of Change
[1, 2]	
[1, 1.1]	
[1, 1.05]	
[1, 1.01]	
[1, 1.001]	

What do you notice about the average rate of change as the right endpoint of the interval gets closer to the left endpoint of the interval?

68. Fill in the following table for $f(x) = -x^2 + 1$.

Interval	Average Rate of Change
[1, 2]	
[1.9, 2]	
[1.95, 2]	
[1.99, 2]	
[1.999, 2]	

What do you notice about the average rate of change as the left endpoint of the interval gets closer to the right endpoint of the interval?

69. Suppose f is constant on an interval $[a, b]$. Show that the average rate of change of f on $[a, b]$ is zero.

70. If the average rate of change of a function on an interval is zero, does that mean the function is constant on that interval?

71. Let f be decreasing on an interval (a, b). Show that the average rate of change of f on $[c, d]$ is negative, where $a < c < d < b$.

2.5 Equations and Inequalities Involving Absolute Value

Objectives

▶ Express the absolute value of a number in terms of distance on the number line

▶ Solve equations involving absolute value

▶ Solve inequalities involving absolute value

▶ Solve an applied problem involving absolute value

Just in Time

Review absolute value in Section P.1.

The absolute value function can be defined as follows.

$$f(x) = |x| = \begin{cases} x & \text{if } x \geq 0 \\ -x & \text{if } x < 0 \end{cases}$$

Note that $f(x)$ has two different expressions: x if $x \geq 0$ and $-x$ if $x < 0$. You will use *only one* of the two expressions, depending on the value of x. To solve equations and inequalities involving absolute value, it is useful to think of the absolute value function in terms of distance on the number line. Figure 2.5.1 illustrates this concept.

Figure 2.5.1

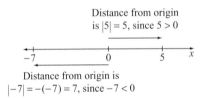

Distance from origin is $|5| = 5$, since $5 > 0$

Distance from origin is $|-7| = -(-7) = 7$, since $-7 < 0$

In this section, we will discuss general methods for finding solutions to equations and inequalities involving absolute value.

Equations Involving Absolute Value

From the definition of the absolute value function, we have the following statement.

Absolute Value Equations

Let $a > 0$. Then the expression

$$|X| = a \text{ is equivalent to } X = a \quad \text{or} \quad X = -a.$$

In the above statement, X can be *any* quantity, not just a single variable. The set of all numbers that satisfy the equation $|X| = a$ is called its **solution set.** We can use this statement to solve equations involving absolute value, as shown in the next example.

Technology Note

In Example 1(a), let
$Y_1(x) = |2x - 3|$ and
$Y_2(x) = 7$, and graph both
functions. Figure 2.5.2
shows one of the solutions,
$x = -2$, which was found
by using the INTERSECT
feature. The second
solution, $x = 5$, can be
found similarly.

Keystroke Appendix:
Section 9

Figure 2.5.2

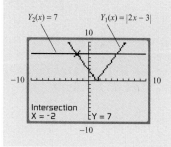

Just in Time

*Review linear inequalities in
Section 1.5.*

Example 1 Equations Involving Absolute Value

Solve the following equations.

(a) $|2x - 3| = 7$

(b) $|x| = -3$

(c) $-|3x + 1| - 3 = -8$

▶Solution

(a) We have the following two equations that, taken together, correspond to the single equation $|2x - 3| = 7$:

$$2x - 3 = 7 \quad \text{or} \quad 2x - 3 = -7$$

The word *or* means that a number x is a solution of the equation $|2x - 3| = 7$ if and only if x is a solution of *at least one* of the two equations $2x - 3 = 7$ or $2x - 3 = -7$. Each of these two equations must be solved separately.

$$2x - 3 = 7 \quad \text{or} \quad 2x - 3 = -7 \quad \text{Write down both equations}$$
$$2x = 10 \qquad\qquad 2x = -4 \quad \text{Add 3 to both sides}$$
$$x = 5 \qquad\qquad x = -2 \quad \text{Divide by 2}$$

Thus, the solution set is $\{-2, 5\}$.

(b) Since the absolute value of any number must be greater than or equal to zero, the equation $|x| = -3$ has no solution.

(c) In order to solve the equation $-|3x + 1| - 3 = -8$, we must first isolate the absolute value term.

$$-|3x + 1| - 3 = -8 \quad \text{Original equation}$$
$$-|3x + 1| = -5 \quad \text{Add 3 to both sides}$$
$$|3x + 1| = 5 \quad \text{Isolate absolute value term}$$

We next apply the definition of absolute value to get the following two equations that, taken together, correspond to the single equation $|3x + 1| = 5$.

$$3x + 1 = 5 \quad \text{or} \quad 3x + 1 = -5 \quad \text{Write down both equations}$$
$$3x = 4 \qquad\qquad 3x = -6 \quad \text{Subtract 1 from both sides}$$
$$x = \frac{4}{3} \qquad\qquad x = -2 \quad \text{Divide by 3}$$

Thus, the solution set is $\left\{\frac{4}{3}, -2\right\}$.

✔ *Check It Out 1:* Solve the equation $|-5x + 2| = 12$. ◼

Inequalities Involving Absolute Value

Solving inequalities involving absolute value is straightforward if you keep in mind the definition of absolute value. Thinking of the absolute value of a number as its distance from the origin (on the number line) leads us to the following statements about inequalities.

Absolute Value Inequalities

Let $a > 0$. Then the inequality

$$|X| < a \text{ is equivalent to } -a < X < a.$$

See Figure 2.5.3.

Figure 2.5.3

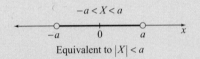

Equivalent to $|X| < a$

Similarly, $|X| \leq a$ is equivalent to $-a \leq X \leq a$.

Let $a > 0$. Then the inequality

$$|X| > a \text{ is equivalent to } X < -a \quad \text{or} \quad X > a.$$

See Figure 2.5.4.

Figure 2.5.4

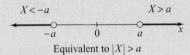

Equivalent to $|X| > a$

Similarly, $|X| \geq a$ is equivalent to $X \leq -a$ or $X \geq a$.

Observations:

▶ In the above statement, X can be *any* expression, not just a single variable.

▶ The pair of equivalent inequalities for $|X| > a$ *must* be written as two *separate* inequalities, and similarly for $|X| \geq a$. ▪

We next show how to solve inequalities involving absolute value.

Example **2** **Inequalities Involving Absolute Value**

Solve the following inequalities and indicate the solution set on a number line.

(a) $|2x - 3| > 7$ (b) $\left| -\dfrac{2}{3}x + 4 \right| \leq 5$ (c) $-4 + |3 - x| > 5$

▶**Solution**

(a) To solve the inequality $|2x - 3| > 7$, we proceed as follows. Since this is a "greater than" absolute value inequality, we must rewrite it as two separate inequalities without an absolute value:

$$2x - 3 < -7 \quad \text{or} \quad 2x - 3 > 7 \quad \text{Rewrite as two separate inequalities}$$
$$2x < -4 \qquad\qquad 2x > 10 \quad \text{Add 3 to both sides}$$
$$x < -2 \qquad\qquad\quad x > 5 \quad \text{Divide by 2}$$

Thus, the solution of $|2x - 3| > 7$ is the set of all x such that $x < -2$ or $x > 5$. In interval notation, the solution set is $(-\infty, -2) \cup (5, \infty)$. The solution is graphed on the number line in Figure 2.5.5.

Figure 2.5.5

Just in Time

Review compound inequalities in Section 1.5.

Technology Note

Let $Y_1(x) = |2x - 3|$ and $Y_2(x) = 7$, and graph both functions. Use the INTERSECT feature to find both points of intersection. From Figure 2.5.6, $Y_1(x) > Y_2(x)$, or, equivalently, $|2x - 3| > 7$, when $x > 5$ or $x < -2$.

Keystroke Appendix:
Section 9

Figure 2.5.6

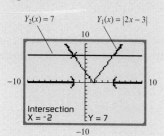

(b) To solve $\left| -\frac{2}{3}x + 4 \right| \le 5$, we first write the inequality as an equivalent expression without an absolute value:

$$-5 \le -\frac{2}{3}x + 4 \le 5.$$

We now solve the inequality for x.

$-5 \le -\dfrac{2}{3}x + 4 \le 5$	Write equivalent expression
$-15 \le -2x + 12 \le 15$	Multiply by 3 to clear fraction
$-27 \le -2x \le 3$	Subtract 12 from each part
$\dfrac{27}{2} \ge x \ge -\dfrac{3}{2}$	Divide by -2; inequalities are reversed
$-\dfrac{3}{2} \le x \le \dfrac{27}{2}$	Rewrite inequality

The solution of $\left| -\frac{2}{3}x + 4 \right| \le 5$ is the set of all x such that $-\frac{3}{2} \le x \le \frac{27}{2}$. In the last step, we turned the solution around and rewrote it so that $-\frac{3}{2}$, the smaller of the two numbers $-\frac{3}{2}$ and $\frac{27}{2}$, comes first. This makes it easier to see how to write the solution in interval notation, which is $\left[-\frac{3}{2}, \frac{27}{2} \right]$. The solution set is graphed on the number line in Figure 2.5.7.

Figure 2.5.7

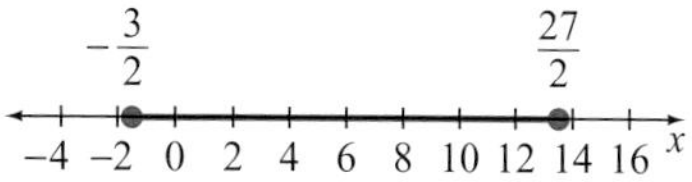

$$-\frac{3}{2} \qquad\qquad \frac{27}{2}$$

$-4\ -2\ \ 0\ \ 2\ \ 4\ \ 6\ \ 8\ \ 10\ 12\ 14\ 16\quad x$

(c) To solve the inequality $-4 + |3 - x| > 5$, first isolate the term containing the absolute value.

$-4 + |3 - x| > 5$	Original inequality
$|3 - x| > 9$	Add 4 to each side

Since this is a "greater than" inequality, we must rewrite it as two separate inequalities without the absolute value.

$3 - x < -9$	or $\quad 3 - x > 9$	Rewrite as two separate inequalities
$-x < -12$	$-x > 6$	Subtract 3 from both sides
$x > 12$	$x < -6$	Multiply by -1; inequalities are reversed

Therefore, the solution is $(-\infty, -6) \cup (12, \infty)$. The solution set is graphed on the number line in Figure 2.5.8.

Figure 2.5.8

$-6 \qquad 0 \qquad 12 \quad x$

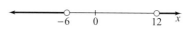 *Check It Out 2:* Solve the inequality $|-2x - 1| \le 6$. Express your answer in interval notation, and graph the solution set on the number line. ◼

Understanding absolute value as distance plays an important role in applications, and in more advanced math courses such as calculus.

Example 3 Distance and Absolute Value

Graph the following on a number line, and write each set using an absolute value inequality.

(a) The set of all x whose distance from 4 is less than 5

(b) The set of all x whose distance from 4 is greater than 5

▶Solution

(a) The set of all x whose distance from 4 is less than 5 is indicated on the number line in Figure 2.5.9. If $x > 4$, then $x - 4$ gives the distance from 4. If $x < 4$, then $4 - x = -(x - 4)$ gives the distance from 4. By the definition of absolute value, $|x - 4|$ gives the distance of x from 4. Thus, the set of all such x that are within 5 units of 4 is given by the inequality

$$|x - 4| < 5.$$

Figure 2.5.9

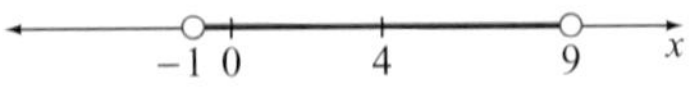

(b) To write an inequality that represents the set of all x whose distance from 4 is greater than 5, we simply set the distance expression, $|x - 4|$, to be greater than 5.

$$|x - 4| > 5$$

The corresponding points on the number line are shown in Figure 2.5.10.

Figure 2.5.10

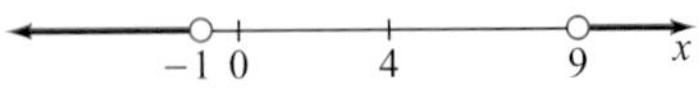

✔ *Check It Out 3:* Graph the following on a number line, and write the set using an absolute value inequality: the set of all x whose distance from 6 is greater than or equal to 3. ■

Example 4 Street Numbers and Absolute Value

In New York City, the east-west streets are numbered consecutively, beginning with the number 1 for the southernmost east-west street and increasing by 1 for every block as you proceed north along an avenue. See Figure 2.5.11. Which east-west streets are within five blocks of 49th Street? Use an absolute value inequality to solve this problem.

▶Solution We want all the east-west streets that are within five blocks of 49th Street. The answer is easy to figure out without using absolute value. The point is, however, to relate something familiar to something abstract—in this case, the definition of absolute value. Recalling that absolute value measures distance, and using the given map, we can write

$$|x - 49| \leq 5$$

since we are interested in the streets that are within five blocks of 49th Street. Solving the inequality, we have

$$-5 \leq x - 49 \leq 5 \Longrightarrow 44 \leq x \leq 54.$$

Thus, the east-west streets that are within five blocks of 49th Street are all the east-west streets from 44th Street to 54th Street, inclusive. Note that the values of x are limited to the positive integers in the interval [44, 54], since we are considering the names of numbered streets.

Figure 2.5.11

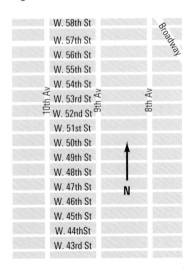

✔ *Check It Out 4:* Referring to Example 4, write an absolute value inequality that indicates the east-west streets that are more than five blocks from 49th Street. ▪

2.5 Key Points

▶ Let $a > 0$. Then the expression

$$|X| = a \text{ is equivalent to } X = a \quad \text{or} \quad X = -a.$$

▶ The expression $|X| < a$ is equivalent to $-a < X < a$. Similarly, the expression $|X| \leq a$ is equivalent to $-a \leq |X| \leq a$.

▶ The expression $|X| > a$ is equivalent to

$$X < -a \quad \text{or} \quad X > a.$$

Similarly, the expression $|X| \geq a$ is equivalent to $X \leq -a$ or $X \geq a$.

2.5 Exercises

▶**Just in Time Exercises** These exercises correspond to the Just in Time references in this section. Complete them to review topics relevant to the remaining exercises.

1. Evaluate $|3|$.

2. Evaluate $|-8|$.

3. Evaluate $|x + 2|$ for $x = -6$.

4. Solve for x: $x - 3 < 5$

5. Solve for x: $3x + 4 > -8$

6. Solve for x: $-2x + 5 > 9$

7. Solve for x: $-3 \leq 2x + 7 \leq 15$

8. Solve for x: $-2 \leq -\frac{1}{3}x \leq 4$

▶**Skills** This set of exercises will reinforce the skills illustrated in this section.

In Exercises 9–28, solve the equation.

9. $|x + 4| = 6$

10. $|x - 2| = 7$

11. $|2x - 4| = 8$

12. $|5 - x| = 1$

13. $|3x + 6| = 9$

14. $|3 - 2x| = 5$

15. $\left|2s - \dfrac{3}{2}\right| = 10$

16. $\left|3s + \dfrac{4}{3}\right| = 9$

17. $4|t + 5| = 16$

18. $-|3t + 2| = -5$

19. $-|t - 3| = 7$

20. $2|2x + 1| = 10$

21. $4|x - 5| = 12$

22. $-2|x + 1| = 6$

23. $|x - 1| + 5 = 9$

24. $|x - 2| + 3 = 8$

25. $1 + |-2x + 5| = 3$

26. $-2 + |4x - 3| = 7$

27. $|x^2 - 8| = 1$

28. $|x^2 - 1| = 3$

In Exercises 29–32, determine whether the given value of x satisfies the inequality.

29. $|x - 2| > 4$; $x = 3$

30. $|x + 2| < 4$; $x = -2$

31. $|3x - 2| \le 4$; $x = \dfrac{3}{2}$

32. $|3x - 2| \ge 2$; $x = 4.1$

In Exercises 33–40, graph the solution set of each inequality on the real number line.

33. $x > -3$

34. $t < 4$

35. $-1 \le s \le 2$

36. $-4 \le x \le -1$

37. $|x| \le \dfrac{4}{3}$

38. $|x| \le 3$

39. $|x| > 7$

40. $|x| > 5$

In Exercises 41–62, solve the inequality. Express your answer in interval notation, and graph the solution set on the number line.

41. $|2x| > 8$

42. $|3x| > 9$

43. $|x + 3| \le 4$

44. $|x - 4| \le 11$

45. $|x - 10| > 6$

46. $|x - 4| < 7$

47. $|2s - 7| > 3$

48. $|3s + 2| \ge 6$

49. $|2 - 3x| \le 10$

50. $|-1 + 7x| \le 13$

51. $\left|\dfrac{1}{2}x + 6\right| \le 5$

52. $\left|\dfrac{2}{3}x - 2\right| < 9$

53. $\left|\dfrac{x + 7}{6}\right| < 5$

54. $\left|\dfrac{x + 5}{8}\right| > 3$

55. $|x - 4| - 2 \ge 6$

56. $|x + 3| - 1 \le 4$

57. $|3x + 7| - 2 < 8$

58. $|4x + 2| + 4 \ge 9$

59. $|t - 6| < 0$

60. $|t - 6| > 0$

61. $|x - 4| < 0.001$

62. $|x - 3| < 0.01$

In Exercises 63–68, use absolute value notation to write an appropriate equation or inequality for each set of numbers.

63. All numbers whose distance from -7 is equal to 3

64. All numbers whose distance from 8 is equal to $\dfrac{5}{4}$

65. All numbers whose distance from 8 is less than 5

66. All numbers whose distance from -4 is less than 7

67. All numbers whose distance from -6.5 is greater than 8

68. All numbers whose distance from 5 is greater than 12.3

69. Use a graphing utility to find the solution(s), if any, of the equation $|x| + 2 = kx$ for the following values of k.
 (a) $k = 3$
 (b) $k = 1$
 (c) $k = -\dfrac{1}{2}$
 (d) $k = 0$

70. Solve $|x + 4| = x$.

71. Solve $|2x| = -x + 4$.

72. Solve $|x + 4| = |2x|$.

73. Solve $|x - 2| = -|x - 3| + 4$.

74. Solve the inequality $|x - 2| \ge |x + 1|$.

75. Solve the inequality $|2x| - 3 \le |x| + 1$.

76. Solve the inequality $|x - 1| \le -\dfrac{1}{2}x + 3$.

▶**Applications** In this set of exercises, you will use absolute value to study real-world problems.

77. **Weather** The average temperature, in degrees Fahrenheit, in Frostbite Falls over the course of a year is given by $|T + 10| < 20$. Solve this inequality and interpret it.

78. **Weather** Over the course of a year, the average daily temperature in Honolulu, Hawaii, varies from 65°F to 80°F. Express this range of temperatures using an absolute value inequality.

79. **Geography** You are located at the center of Omaha, Nebraska. Write an absolute value inequality that gives all points within 30 miles north or south of the center of Omaha. Indicate what point you would use as the origin.

80. **Geography** You are located at the center of Hartford, Connecticut. Write an absolute value inequality that gives all points more than 65 miles east or west of the center of Hartford. Indicate what point you would use as the origin.

81. **Temperature Measurement** A room thermostat is set at 68°F and measures the temperature of the room with an

uncertainty of $\pm 1.5°$F. Assuming the temperature is uniform throughout the room, use absolute value notation to write an inequality for the range of possible temperatures in the room.

82. **Length Measurement** A ruler measures an object with an uncertainty of $\frac{1}{16}$ inch. If a pencil is measured to be 8 inches, use absolute value notation to write an inequality for the range of possible lengths of the pencil.

▶**Concepts** This set of exercises will draw on the ideas presented in this section and your general math background.

83. Explain why $|-3(x + 2)|$ is *not* the same as $-3|x + 2|$.

84. Explain why the expression "$x > 3$ or $x < -2$" *cannot* be written as $3 < x < -2$.

85. Show that $|x - k| = |k - x|$, where k is any real number.

86. Sketch the graph of $f(x) = -3x + 2$ by hand. Use it to graph $g(x) = |f(x)|$. What is the x-intercept of the graph of $g(x)$?

87. Can you think of an absolute value equation with no solution?

88. Explain why $|x| < 0$ has no solution.

2.6 Piecewise-Defined Functions

Objectives

▶ Evaluate a piecewise-defined function

▶ Graph a piecewise-defined function

▶ Solve an applied problem involving a piecewise-defined function

▶ Evaluate and graph the greatest integer function

In many applications of mathematics, the algebraic expression for the output value of a function may be different for different conditions of the input. For example, the formula for commuter train fares may vary according to the time of day traveled. The next example shows how such a situation can be described using function notation.

Example 1 **A Function Describing Train Fares**

A one-way ticket on a weekday from Newark, New Jersey, to New York, New York, costs \$3.30 for a train departing during peak hours and \$2.50 for a train departing during off-peak hours. Peak evening hours are from 4 P.M. to 7 P.M. The rest of the evening is considered to be off-peak. (*Source:* New Jersey Transit)

(a) Describe the fare as a function of the time of day from 4 P.M. to 11 P.M.

(b) How much does a one-way ticket from Newark to New York cost for a train departing Newark at 5 P.M.? Describe this fare in function notation and evaluate.

▶**Solution**

(a) The input variable is the time of day, written as a single number, and the output variable is the fare (in dollars). We consider only the part of the day from 4 P.M. to 11 P.M. We have the following situation:

Fare from 4 P.M. to 7 P.M. (inclusive)   \$3.30

Fare after 7 P.M., up to and including 11 P.M.   \$2.50

Let $F(t)$ represent the fare at time t, where t is between 4 and 11. We see that the function cannot be defined by just one expression for these values of t, because there is a reduction in the fare immediately after 7 P.M. Functions such as this can only be defined *piecewise*. The expressions for these types of functions vary according to the conditions that the input variable must satisfy.

Mathematically, the function is written as follows.

$$F(t) = \begin{cases} 3.30, & \text{if } 4 \le t \le 7 \\ 2.50, & \text{if } 7 < t \le 11 \end{cases}$$

The expression for the function depends on the departure time of the train.

(b) If the train departs at 5 P.M., the fare is given by $F(5)$, which, by the definition of F, is 3.30 (meaning $3.30). This result agrees with the verbal description of the problem.

✔ *Check It Out 1:* In Example 1, evaluate $F(9)$ and interpret it. ■

Functions that are defined using different expressions corresponding to different conditions satisfied by the independent variable are called **piecewise-defined functions**. Example 1 on train fares is an example of a piecewise-defined function. Example 2 gives another example of a piecewise-defined function.

Example 2 Evaluting a Piecewise-Defined Function

Define $H(x)$ as follows:

$$H(x) = \begin{cases} 1, & \text{if } x < 0 \\ -x + 1, & \text{if } 0 < x \le 2 \\ -3, & \text{if } x > 2 \end{cases}$$

Evaluate the following, if defined.

(a) $H(2)$ (b) $H(-6)$ (c) $H(0)$

▶ **Solution** Note that the function H is not given by just one formula. The expression for $H(x)$ will depend on whether $x < 0$, $0 < x \le 2$, or $x > 2$.

(a) To evaluate $H(2)$, first note that $x = 2$, and thus you must use the expression for $H(x)$ corresponding to $0 < x \le 2$, which is $-x + 1$. Therefore, $H(2) = -(2) + 1 = -1$.

(b) To evaluate $H(-6)$, note that $x = -6$. Since x is less than zero, we use the value for $H(x)$ corresponding to $x < 0$. Therefore, $H(-6) = 1$.

(c) Now we are asked to evaluate $H(0)$. We see that $H(x)$ is defined only for $x < 0$, $0 < x \le 2$, or $x > 2$, and that $x = 0$ satisfies none of these three conditions. Therefore, $H(0)$ is not defined.

✔ *Check It Out 2:* For $H(x)$ defined in Example 2, find

(a) $H(4)$.   (b) $H(-3)$. ■

Graphing Piecewise-Defined Functions

We can graph piecewise-defined functions by essentially following the same procedures given thus far. However, you have to be careful about "jumps" in the function that may occur at points at which the function expression changes. Example 3 illustrates this situation.

Example 3 Graphing a Piecewise-Defined Function

Graph the function $H(x)$, defined as follows:

$$H(x) = \begin{cases} x^2, & \text{if } x < 0 \\ x + 1, & \text{if } 0 \le x < 3 \\ -1, & \text{if } x \ge 3 \end{cases}$$

Technology Note

When using a graphing utility, you must give each piece of the piecewise-defined function a different name in the Y= editor, and include the conditions that *x* must satisfy. See Figure 2.6.2. The calculator should be in DOT mode so that the pieces of the graph are not joined. See Figure 2.6.3.

Keystroke Appendix:
Section 7

Figure 2.6.2

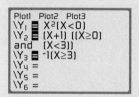

Figure 2.6.3

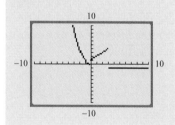

▶**Solution** This function is given by three different expressions, depending on the value of *x*. The graph will thus be constructed in three steps, as follows.

Step 1 If $x < 0$, then $f(x) = x^2$. So we first graph $f(x) = x^2$ on the interval $(-\infty, 0)$. The value $x = 0$ is not included because $f(x) = x^2$ holds true only for $x < 0$. Since $x = 0$ is not part of the graph, we indicate this point by an open circle. See Part I of the graph in Figure 2.6.1.

Step 2 If $0 \le x < 3$, then $f(x) = x + 1$. Thus we graph the line $f(x) = x + 1$ on $[0, 3)$, indicated by Part II of the graph in Figure 2.6.1. Note that $x = 3$ is not part of the graph.

Step 3 If $x \ge 3$, then $f(x) = -1$. Thus we graph the horizontal line $f(x) = -1$ on the interval $[3, \infty)$, indicated by Part III of the graph in Figure 2.6.1.

Figure 2.6.1

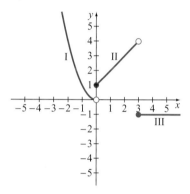

✔ *Check It Out 3:* Graph the function $H(x)$, defined as follows.

$$H(x) = \begin{cases} -4, & \text{if } x \le 0 \\ 4, & \text{if } x > 0 \end{cases}$$

An Application of a Piecewise-Defined Function

In many applications, there may not be a single linear function that holds true for all input values. Such applications include rate structures for telephone services and other utilities. We will now examine such an application.

Example 4 **Describing a Telephone Rate Plan**

The IM-Mobile cell phone company has a rate plan that costs $30 per month and includes up to 500 minutes of usage per month. For each minute over and above 500, the company charges $0.18. Assume the minutes are nonnegative integers.

(a) If you wish to calculate the monthly cost for this plan, what are the input and output variables?

(b) Calculate the cost for 580 minutes of monthly use.

(c) Write down the expression for the function that represents the monthly cost for using this plan.

(d) How much will the plan cost if you use the phone for 400 minutes per month? 720 minutes per month?

▶Solution

(a) The input variable, t, is the number of minutes of use during 1 month. The output variable, C, is the total cost (in dollars) for that month.

(b) Since 580 minutes is 80 minutes over the 500-minute limit, the total cost is

$$\text{Total cost} = 30 + (0.18)(580 - 500) = 30 + (0.18)(80) = 44.40.$$

Thus the monthly bill for 580 minutes of use is $44.40.

(c) The expression for the monthly cost depends on the value of t, as follows:

▸ Flat cost of $30 if $0 \le t \le 500$

▸ Variable cost of $30 + (t - 500)(0.18)$ if $t > 500$. See part (b) for an example of this case.

Using function notation, we have:

$$C(t) = \begin{cases} 30, & \text{if } 0 \le t \le 500 \\ 30 + (0.18)(t - 500), & \text{if } t > 500 \end{cases}$$

(d) If $t = 400$, the cost is $30, since $t < 500$. If $t = 720$, then the total cost is

$$C(720) = 30 + (0.18)(720 - 500) = 30 + (0.18)(220) = \$69.60.$$

✔ *Check It Out 4:* Suppose the IM-Mobile cell phone company has another rate plan that costs $40 per month and includes up to 700 minutes of usage per month. For each minute over and above 700, the company charges $0.10. Write the piecewise-defined function corresponding to this plan. Assume the minutes are nonnegative integers. ▪

The Greatest Integer Function

One type of piecewise-defined function that occurs frequently in mathematics gives a correspondence between a real number x and the largest integer less than or equal to x. It is called the **greatest integer function** and is defined as follows.

Greatest Integer Function

The greatest integer function, denoted by $f(x) = [\![x]\!]$, is defined as the largest integer less than or equal to x. The domain of f is the set of all real numbers. The range of f is the set of integers.

Example 5 Evaluating the Greatest Integer Function

Let $f(x) = [\![x]\!]$.
(a) Find $f(2)$.
(b) Find $f\left(\dfrac{1}{3}\right)$
(c) Find $f\left(-\dfrac{1}{2}\right)$.

▶Solution

(a) $f(2) = [\![2]\!] = 2$ because 2 is the largest integer less than or equal to 2.

(b) Since the largest integer less than or equal to $\frac{1}{3}$ is 0,

$$f\left(\frac{1}{3}\right) = \left[\!\left[\frac{1}{3}\right]\!\right] = 0.$$

(c) Since the largest integer less than or equal to $-\frac{1}{2}$ is -1,

$$f\left(-\frac{1}{2}\right) = \left[\!\left[-\frac{1}{2}\right]\!\right] = -1.$$

☑ *Check It Out 5:* Let $f(x) = [\![x]\!]$. Find $f(-2.5)$. ■

We can sketch a graph of the greatest integer function by using the values given in Table 2.6.1.

Table 2.6.1

x	$f(x) = [\![x]\!]$
$-2 \le x < -1$	-2
$-1 \le x < 0$	-1
$0 \le x < 1$	0
$1 \le x < 2$	1
$2 \le x < 3$	2

For the first entry, if $-2 \le x < -1$, then $[\![x]\!] = -2$ because that is the largest integer less than x. Next, if $-1 \le x < 0$, then $[\![x]\!] = -1$, using the definition of the greatest integer function. Continuing in this way, we obtain the graph given in Figure 2.6.4.

Figure 2.6.4

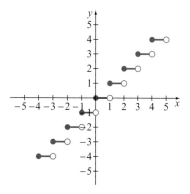

2.6 Key Points

▶ A **piecewise-defined function** is a function whose expression differs according to the value of the independent variable.

▶ To graph a piecewise-defined function, graph the portion corresponding to each expression separately.

▶ The **greatest integer function,** denoted by $f(x) = [\![x]\!]$, is defined as the largest integer less than or equal to x. The domain of f is the set of all real numbers. The range of f is the set of integers.

2.6 Exercises

▶**Skills** This set of exercises will reinforce the skills illustrated in this section.

In Exercises 1–10, evaluate $f(-2), f(0)$, and $f(1)$, if possible, for each function. If a function value is undefined, so state.

1. $f(x) = \begin{cases} 1, & \text{if } x \le 3 \\ -2, & \text{if } x > 3 \end{cases}$ 2. $f(x) = \begin{cases} -1, & \text{if } x < 2 \\ 4, & \text{if } x \ge 2 \end{cases}$

3. $f(x) = \begin{cases} 1, & \text{if } x < -2 \\ \frac{1}{2}, & \text{if } x > -2 \end{cases}$ 4. $f(x) = \begin{cases} \frac{2}{3}, & \text{if } x < 1 \\ -2, & \text{if } x > 1 \end{cases}$

5. $f(x) = \begin{cases} x, & \text{if } x < 0 \\ 1, & \text{if } x \ge 0 \end{cases}$ 6. $f(x) = \begin{cases} -2, & \text{if } x < 1 \\ x^2, & \text{if } x \ge 1 \end{cases}$

7. $f(x) = \begin{cases} -1, & \text{if } x \le -2 \\ 2, & \text{if } -2 < x \le 1 \\ 4, & \text{if } x > 1 \end{cases}$

8. $f(x) = \begin{cases} 0, & \text{if } x < 0 \\ 2, & \text{if } 0 \le x < 2 \\ 4, & \text{if } x \ge 2 \end{cases}$

9. $f(x) = \begin{cases} \sqrt{x}, & \text{if } 0 \le x \le 1 \\ -x + 4, & \text{if } x > 1 \end{cases}$

10. $f(x) = \begin{cases} 2, & \text{if } x < -2 \\ |x|, & \text{if } x \ge 2 \end{cases}$

In Exercises 11–16, evaluate f at the indicated value for $f(x) = [\![x]\!]$.

11. $f(3.5)$ 12. $f(4.2)$

13. $f(\pi)$ 14. $f(\sqrt{2})$

15. $f(-3.4)$ 16. $f\left(-\dfrac{4}{3}\right)$

In Exercises 17–30, graph the function by hand.

17. $F(x) = \begin{cases} 0, & x \le 1 \\ 2, & x > 1 \end{cases}$ 18. $h(x) = \begin{cases} -1, & x < 0 \\ 4, & x \ge 0 \end{cases}$

19. $f(x) = \begin{cases} 1, & x < 0 \\ 0, & 0 \le x < 1 \\ -1, & x \ge 1 \end{cases}$

20. $f(x) = \begin{cases} 0, & x \le -1 \\ 4, & -1 < x \le 2 \\ 1, & x > 2 \end{cases}$

21. $f(x) = \begin{cases} x + 2, & x < 2 \\ 4, & x \ge 2 \end{cases}$

22. $f(x) = \begin{cases} 3, & x \le -1 \\ -x + 2, & x > -1 \end{cases}$

23. $g(x) = \begin{cases} x + 1, & x \le 0 \\ x, & x > 0 \end{cases}$

24. $g(x) = \begin{cases} x + 1, & x < 0 \\ -x + 1, & x \ge 0 \end{cases}$

25. $f(x) = \begin{cases} 1, & x \le 2 \\ x, & 2 < x \le 5 \end{cases}$

26. $f(x) = \begin{cases} -x, & -3 \le x < 3 \\ 2x, & x \ge 3 \end{cases}$

27. $f(x) = \begin{cases} x^2, & -1 \le x \le 2 \\ -2, & 2 < x \le 3 \\ x + 1, & x > 3 \end{cases}$

28. $f(x) = \begin{cases} 3, & -2 \le x \le 1 \\ -x^2, & 1 < x \le 2 \\ 5, & x > 2 \end{cases}$

29. $f(x) = \begin{cases} \sqrt{x}, & 0 \le x \le 4 \\ -x + 4, & 4 < x \le 5 \\ x - 6, & 5 < x \le 10 \end{cases}$

30. $f(x) = \begin{cases} -2, & x < -1 \\ |x|, & -1 \le x \le 2 \\ 2, & 2 < x \le 4 \end{cases}$

In Exercises 31 and 32, evaluate $f(-1), f(0)$, and $f(2)$, if possible, for f given by the graph. If not possible, explain why you cannot do so.

31. 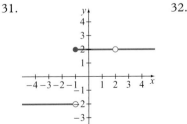   32.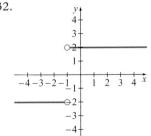

In Exercises 33–36, state the domain, range, x-intercept(s), and y-intercept, and find the expression for f(x).

33.

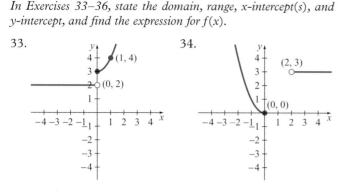

34.

35.

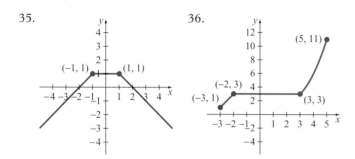

36.

In Exercises 37–40, graph the piecewise-defined function using a graphing utility. The display should be in DOT mode.

37. $f(x) = \begin{cases} 0.5x^2, & \text{if } x \leq 0 \\ -x^2, & \text{if } x > 0 \end{cases}$

38. $f(x) = \begin{cases} x^2, & \text{if } -2 \leq x < 0 \\ -x + 1, & \text{if } 0 \leq x < 2.5 \\ x - 3.5, & \text{if } x \geq 2.5 \end{cases}$

39. $f(x) = \begin{cases} 0, & \text{if } x < 0 \\ [\![x]\!], & \text{if } x \geq 0 \end{cases}$

(*Hint:* The greatest integer function is found under the option INT in a graphing calculator.)

40. $f(x) = 5[\![x]\!] - 2$ (*Hint:* The greatest integer function is found under the option INT in a graphing calculator.)

▶ **Applications** In this set of exercises, you will use piecewise-defined functions to study real-world problems.

41. **Leisure** Wonderland Amusement Park charges the following daily admission prices, $f(x)$ (in dollars), according to the age of the customer, x.

$$f(x) = \begin{cases} 0, & \text{if } 0 < x < 3 \\ 20, & \text{if } 3 \leq x < 12 \\ 25, & \text{if } 12 \leq x < 55 \\ 18, & \text{if } x \geq 55 \end{cases}$$

(a) What is the admission price for a five-year-old?

(b) Find the total admission price for a family of four consisting of two adults, each 32 years old, and two children—one a seven-year-old and the other a two-year-old.

42. **Sports** The attendance at professional basketball games for the years 1995–2003 can be approximated by the following piecewise-defined function. (*Source: Statistical Abstract of the United States*)

$$f(t) = \begin{cases} 19 + 0.4(t - 1995), & \text{if } 1995 \leq t \leq 2000 \\ 21, & \text{if } 2000 < t \leq 2003 \end{cases}$$

Here, $f(t)$ is the attendance in millions, and t is the year.

(a) What was the attendance in 2002?

(b) Evaluate $f(1998)$ and explain what it represents.

43. **Music** The sales of portable CD players in the United States held steady in the years 2000 and 2001, but then slowly declined in the years 2002–2005 (*Source:* Consumer Electronics Association). The sales are given by the function

$$S(t) = \begin{cases} 32, & \text{if } 0 \leq t \leq 1 \\ 32 - 3.75(t - 1), & \text{if } 1 < t \leq 5 \end{cases}$$

where $S(t)$ denotes the number of portable CD players sold in millions, and t denotes the number of years since 2000.

(a) Evaluate $S(4)$ and explain what it represents.

(b) How many portable CD players were sold in 2001?

44. **Printing Costs** The following function gives the cost for printing x copies of a 120-page book. The minimum number of copies printed is 200, and the maximum number printed is 1000.

$$f(x) = \begin{cases} 5x, & \text{if } 200 \leq x < 500 \\ 4.5x, & \text{if } 500 \leq x < 750 \\ 4x, & \text{if } 750 \leq x \leq 1000 \end{cases}$$

(a) Find the cost for printing 400 copies of the book.

(b) Find the cost for printing 620 copies of the book.

(c) Which is more expensive—printing 700 copies of the book or printing 750 copies of the book?

45. **Postage** As of June 30, 2002, the postage rate for first-class mail in the United States was $0.37 for up to 1 ounce and $0.23 for each additional ounce or part thereof. For this class of mail, the maximum weight was 13 ounces. The following function can be used to find the cost of sending a letter, postcard, or small package via first-class mail.

$$P(x) = \begin{cases} 0.37, & \text{if } 0 < x \le 1 \\ 0.23[\![x]\!] + 0.37, & \text{if } 1 < x \le 13 \end{cases}$$

Find the cost of the sending each of the following pieces of first-class mail.

(a) A letter weighing 3.6 ounces

(b) A small package weighing 6.9 ounces

(c) An envelope weighing 5.3 ounces

46. **Commuter Travel** This problem is an extension of Example 1. A one-way ticket on a weekday from Newark, New Jersey, to New York, New York, costs $3.30 for a train departing during peak hours and $2.50 for a train departing during off-peak hours. Peak morning hours are from 6 A.M. to 10 A.M. and peak evening hours are from 4 P.M. to 7 P.M. The rest of the day is considered to be off-peak. (*Source:* New Jersey Transit)

(a) Construct a table that takes the time of day as its input and gives the fare as its output.

(b) Write the fare as a function of the time of day using piecewise function notation.

(c) Graph the function.

47. **Rate Plan** A long-distance telephone company advertises that it charges $1.00 for the first 20 minutes of phone use and 7 cents a minute for every minute beyond the first 20 minutes. Let $C(t)$ denote the total cost of a telephone call lasting t minutes. Assume that the minutes are non-negative integers.

(a) Many people will assume that it will cost only $0.50 to talk for 10 minutes. Why is this incorrect?

(b) Write an expression for the function $C(t)$.

(c) How much will it cost to talk for 5 minutes? 20 minutes? 30 minutes?

48. **Online Shopping** On the online auction site Ebay, the next highest amount that one may bid on an item is based on the current bid, as shown in the table. (*Source:* www.ebay.com)

Current Bid	Bid Increment
$1.00–$4.99	$0.25
$5.00–$24.99	$0.50
$25.00–$99.99	$1.00

For example, if the current bid on an item is $7.50, then the next bid must be at least $0.50 higher.

(a) Explain why the bid increment, I, is a piecewise-defined function of the current bid, b.

(b) Graph this function.

▶**Concepts** This set of exercises will draw on the ideas presented in this section and your general math background.

In Exercises 49 and 50, let f be defined as follows.

$$f(x) = \begin{cases} 0, & \text{if } x \le 1 \\ 2, & \text{if } x > 1 \end{cases}$$

49. Graph $f(x - 1)$.

50. Graph $3f(x)$.

In Exercises 51 and 52, let F be defined as follows.

$$F(x) = \begin{cases} x, & \text{if } 0 \le x \le 4 \\ 4, & \text{if } x > 4 \end{cases}$$

51. Graph $F(x) + 3$.

52. Graph $F(2x)$.

The piecewise-defined function given below is known as the characteristic function, $C(x)$. *It plays an important role in advanced mathematics.*

$$C(x) = \begin{cases} 0, & \text{if } x \le 0 \\ 1, & \text{if } 0 < x < 1 \\ 0, & \text{if } x \ge 1 \end{cases}$$

Use this function for Exercises 53–56.

53. Evaluate $C\left(\dfrac{1}{2}\right)$.

54. Evaluate $3C(0.4)$.

55. Graph $C(x)$.

56. Graph the function $f(x) = xC(x)$.

Chapter 2   Summary

Section 2.1 Coordinate Geometry: Distance, Midpoints, and Circles

Concept	Illustration	Study and Review
Distance between two points The **distance** d between the points (x_1, y_1) and (x_2, y_2) is given by $$d = \sqrt{(x_2 - x_1)^2 + (y_2 - y_1)^2}.$$	The distance between $(-4, 5)$ and $(2, -3)$ is $\sqrt{(2 - (-4))^2 + (-3 - 5)^2} = \sqrt{36 + 64}$ $= 10.$	Example 1 Chapter 2 Review, Exercises 1–4, 110
Midpoint of a line segment The coordinates of the **midpoint** of the line segment joining the points (x_1, y_1) and (x_2, y_2) are $$\left(\frac{x_1 + x_2}{2}, \frac{y_1 + y_2}{2}\right).$$	The midpoint of the line segment joining $(-4, 5)$ and $(2, -3)$ is $$\left(\frac{-4 + 2}{2}, \frac{5 + (-3)}{2}\right) = (-1, 1).$$	Example 1 Chapter 2 Review, Exercises 1–4, 110
Equation of a circle in standard form The circle with center at (h, k) and radius r is the set of all points (x, y) satisfying the equation $$(x - h)^2 + (y - k)^2 = r^2.$$	The equation of the circle with center $(-2, 5)$ and radius 4 is $$(x + 2)^2 + (y - 5)^2 = 16.$$	Examples 2, 3 Chapter 2 Review, Exercises 5–10, 109
General form of the equation of a circle The **general form of the equation of a circle** is given by $$x^2 + y^2 + Dx + Ey + F = 0.$$ By completing the square, we can rewrite the general equation in standard form.	The equation $x^2 + y^2 + 2x + 4y + 1 = 0$ is the equation of a circle in general form. Rewrite as $x^2 + 2x + y^2 + 4y + 1 = 0$ and complete the square to get $(x + 1)^2 + (y + 2)^2 = 4.$	Example 4 Chapter 2 Review, Exercises 11–16, 109

Section 2.2 The Algebra of Functions

Concept	Illustration	Study and Review
Given two functions f and g, then for each x in the domain of *both* f and g, the **sum, difference, product,** and **quotient** of f and g are defined as follows. $$(f + g)(x) = f(x) + g(x)$$ $$(f - g)(x) = f(x) - g(x)$$ $$(fg)(x) = f(x) \cdot g(x)$$ $$\left(\frac{f}{g}\right)(x) = \frac{f(x)}{g(x)}, \quad \text{where } g(x) \neq 0$$	Let $f(x) = x^2$ and $g(x) = 2x + 1$. $(f + g)(x) = x^2 + 2x + 1$ $(f - g)(x) = x^2 - (2x + 1) = x^2 - 2x - 1$ $(fg)(x) = x^2(2x + 1) = 2x^3 + x^2$ $\left(\frac{f}{g}\right)(x) = \dfrac{x^2}{2x + 1}$	Examples 1–3 Chapter 2 Review, Exercises 17–22, 27–29, 111, 113

Continued

Section 2.2 **The Algebra of Functions**

Concept	Illustration	Study and Review
Definition of a composite function Let f and g be functions. The **composite function** $f \circ g$ is defined as $$(f \circ g)(x) = f(g(x)).$$ The domain of $f \circ g$ is the set of all x in the domain of g such that $g(x)$ is in the domain of f.	Let $f(x) = x^2$ and $g(x) = 2x + 1$. Then $f \circ g = f(g(x)) = (2x + 1)^2$.	Examples 4–7 Chapter 2 Review, Exercises 23–26, 30–42, 112
Difference quotient The **difference quotient** of a function f is an expression of the form $$\frac{f(x + h) - f(x)}{h}, \quad h \neq 0.$$	If $f(x) = x^2$, then $$\frac{f(x+h) - f(x)}{h} = \frac{(x+h)^2 - x^2}{h} = 2x + h.$$	Example 8 Chapter 2 Review, Exercises 43–46, 116

Section 2.3 **Transformations of the Graph of a Function**

Concept	Illustration	Study and Review
Vertical and horizontal shifts of the graph of a function Let f be a function and c be a positive constant.		Examples 1–3 Chapter 2 Review, Exercises 47–68, 113
Vertical shifts of the graph of $f(x)$ • The graph of $g(x) = f(x) + c$ is the graph of $f(x)$ shifted c units **upward.** • The graph of $g(x) = f(x) - c$ is the graph of $f(x)$ shifted c units **downward.**		
Horizontal shifts of the graph of $f(x)$ • The graph of $g(x) = f(x - c)$ is the graph of $f(x)$ shifted c units **to the right.** • The graph of $g(x) = f(x + c)$ is the graph of $f(x)$ shifted c units **to the left.**		

Continued

Section 2.3 **Transformations of the Graph of a Function**

Concept	Illustration	Study and Review
Vertical scalings and reflections across the horizontal axis Let f be a function and c be a positive constant. **Vertical scalings of the graph of $f(x)$** • If $c > 1$, the graph of $g(x) = cf(x)$ is the graph of $f(x)$ **stretched vertically** away from the x-axis, with the y-coordinates of $f(x)$ multiplied by c. • If $0 < c < 1$, the graph of $g(x) = cf(x)$ is the graph of $f(x)$ **compressed vertically** toward the x-axis, with the y-coordinates of $f(x)$ multiplied by c. **Reflection of the graph of $f(x)$ across the x-axis** • The graph of $g(x) = -f(x)$ is the graph of $f(x)$ **reflected across the x-axis.**	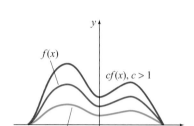 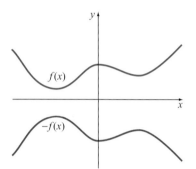	Examples 4, 5 Chapter 2 Review, Exercises 51–68, 113
Horizontal scalings and reflections across the vertical axis Let f be a function and c be a positive constant. **Horizontal scalings of the graph of $f(x)$** • If $c > 1$, the graph of $g(x) = f(cx)$ is the graph of $f(x)$ **compressed horizontally** toward the y-axis, scaled by a factor of $\dfrac{1}{c}$. • If $0 < c < 1$, the graph of $g(x) = f(cx)$ is the graph of $f(x)$ **stretched horizontally** away from the y-axis, scaled by a factor of $\dfrac{1}{c}$. **Reflection of the graph of $f(x)$ across the y-axis** • The graph of $g(x) = f(-x)$ is the graph of $f(x)$ **reflected across the y-axis.**	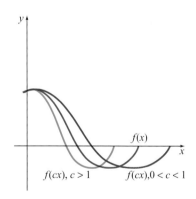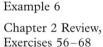 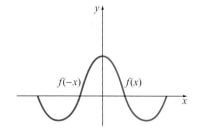	Example 6 Chapter 2 Review, Exercises 56–68

Section 2.4 Symmetry and Other Properties of Functions

Concept	Illustration	Study and Review
Odd and even functions A function is **symmetric with respect to the y-axis** if $f(x) = f(-x)$ for each x in the domain of f. Functions with this property are called **even functions**.	The function $f(x) = x^2 + 2$ is even because $$f(-x) = (-x)^2 + 2 = x^2 + 2.$$ Thus $f(-x)$ is equal to $f(x)$.	Example 1 Chapter 2 Review, Exercises 75–82
A function is **symmetric with respect to the origin** if $f(-x) = -f(x)$ for each x in the domain of f. Functions with this property are called **odd functions**.	The function $f(x) = x^3$ is odd because $$f(-x) = (-x)^3 = -x^3.$$ Thus $f(-x)$ is equal to $-f(x) = -(x^3)$.	
Increasing, decreasing, and constant functions • A function f is **increasing** on an open interval I if, for any a, b in the interval, $f(a) < f(b)$ for $a < b$. • A function f is **decreasing** on an open interval I if, for any a, b in the interval, $f(a) > f(b)$ for $a < b$. • A function f is **constant** on an open interval I if, for any a, b in the interval, $f(a) = f(b)$.	The function f given by the following graph is increasing on $(0, 1)$, decreasing on $(1, 2)$, and constant on $(2, 4)$. 	Example 2 Chapter 2 Review, Exercises 69–74
Average rate of change The **average rate of change** of a function f on an interval $[x_1, x_2]$ is given by $$\text{Average rate of change} = \frac{f(x_2) - f(x_1)}{x_2 - x_1}.$$	The average rate of change of $f(x) = -x^2$ over the interval $[1, 2]$ is given by $$\frac{f(x_2) - f(x_1)}{x_2 - x_1} = \frac{-4 - (-1)}{2 - 1} = -3.$$	Examples 3, 4 Chapter 2 Review, Exercises 83–86

Section 2.5 Equations and Inequalities Involving Absolute Value

Concept	Illustration	Study and Review
Equations involving absolute value Let $a > 0$. Then the expression $|X| = a$ is equivalent to $X = a$ or $X = -a$.	The equation $|-2x + 5| = 7$ is equivalent to $-2x + 5 = 7$ or $-2x + 5 = -7$. Solving $-2x + 5 = 7$ gives $x = -1$. Solving $-2x + 5 = -7$ gives $x = 6$. The solution is $x = -1$ or $x = 6$.	Example 1 Chapter 2 Review, Exercises 87–92

Continued

Section 2.5 Equations and Inequalities Involving Absolute Value

Concept	Illustration	Study and Review
Inequalities involving absolute value Let $a > 0$. Then the following hold. • $\lvert X \rvert < a$ is equivalent to $-a < X < a$. Similarly, $\lvert X \rvert \leq a$ is equivalent to $-a \leq X \leq a$. • $\lvert X \rvert > a$ is equivalent to $\qquad X < -a \text{ or } X > a.$ Similarly, $\lvert X \rvert \geq a$ is equivalent to $X \leq -a$ or $X \geq a$.	The inequality $\lvert -2x + 5 \rvert \leq 7$ is equivalent to $-7 \leq -2x + 5 \leq 7$. Solving the second inequality for x, $-1 \leq x \leq 6$. The solution set is $[-1, 6]$. The inequality $\lvert -2x + 5 \rvert \geq 7$ is equivalent to $-2x + 5 \leq -7$ or $-2x + 5 \geq 7$. Solving the two inequalities *separately* gives the solutions $x \geq 6$ or $x \leq -1$, respectively. The solution set is $(-\infty, -1] \cup [6, \infty)$.	Examples 2–4 Chapter 2 Review, Exercises 93–100, 114

Section 2.6 Piecewise-Defined Functions

Concept	Illustration	Study and Review
Piecewise-defined functions A **piecewise-defined function** is a function whose expression differs according to the value of the independent variable.	The following function has two different definitions, depending on the value of x. $$f(x) = \begin{cases} -5, & \text{if } x < 1 \\ x + 4, & \text{if } x \geq 1 \end{cases}$$	Examples 1–4 Chapter 2 Review, Exercises 105–107, 115
Greatest integer function The **greatest integer function,** denoted by $f(x) = [\![x]\!]$, is defined as the largest integer less than or equal to x. The domain of f is the set of all real numbers. The range of f is the set of integers.	$f(3.5) = [\![3.5]\!] = 3$ because 3 is the largest integer less than 3.5.	Example 5 Chapter 2 Review, Exercise 108

Chapter 2   Review Exercises

Section 2.1 _____

In Exercises 1–4, find the distance between the points and the midpoint of the line segment joining them.

1. $(0, -2)$, $(-3, 3)$

2. $(-4, 6)$, $(8, -10)$

3. $(6, 9)$, $(11, 13)$

4. $(-7, -6)$, $(13, 9)$

In Exercises 5–8, write the standard form of the equation of the circle having the given radius and center. Sketch the circle.

5. $r = 6$; center: $(-1, 2)$

6. $r = 3$; center: $(4, 1)$

7. $r = \dfrac{1}{2}$; center: $(0, -1)$

8. $r = \sqrt{3}$; center: $(-2, 1)$

In Exercises 9–14, find the center and radius of the circle having the given equation. Sketch the circle.

9. $(x - 3)^2 + (y - 2)^2 = \dfrac{1}{4}$

10. $x^2 + (y - 12)^2 = \dfrac{1}{9}$

11. $x^2 + y^2 - 8x + 2y - 5 = 0$

12. $x^2 + y^2 + 2x + 4y - 11 = 0$

13. $x^2 + y^2 - 2x + 2y - 7 = 0$

14. $x^2 + y^2 - 2x + 6y + 1 = 0$

In Exercises 15 and 16, find the center and radius of the circle having the given equation. Use a graphing utility to graph the circle.

15. $x^2 + (y - 2.3)^2 = 17$

16. $(x + 3.5)^2 + (y - 5.1)^2 = 27$

Section 2.2

In Exercises 17–22, for the given functions f and g, find each combination of functions and identify its domain.

(a) $(f + g)(x)$

(b) $(f - g)(x)$

(c) $(fg)(x)$

(d) $\left(\dfrac{f}{g}\right)(x)$

17. $f(x) = 4x^2 + 1;\ g(x) = x + 1$

18. $f(x) = 3x - 1;\ g(x) = x^2 - 4$

19. $f(x) = \left(\dfrac{1}{2x}\right);\ g(x) = \dfrac{1}{x^2 + 1}$

20. $f(x) = \sqrt{x};\ g(x) = \dfrac{1}{\sqrt{x}}$

21. $f(x) = \dfrac{2}{x - 4};\ g(x) = 3x^2$

22. $f(x) = \dfrac{2}{x + 3};\ g(x) = \dfrac{x + 1}{x - 4}$

In Exercises 23–26, use f and g given by the following tables of values.

x	-4	-2	0	3
$f(x)$	-1	0	3	-2

x	-1	0	3
$g(x)$	-2	0	5

23. Evaluate $(f \circ g)(-1)$.

24. Evaluate $(f \circ g)(0)$.

25. Evaluate $(g \circ f)(0)$.

26. Evaluate $(g \circ f)(-4)$.

In Exercises 27–34, let $f(x) = 3x - 1$, $g(x) = -2\sqrt{x}$, and $h(x) = 4x$. Evaluate each of the following.

27. $(f + g)(4)$ 28. $(g - h)(9)$

29. $\dfrac{f}{h}(2)$ 30. $(f \cdot h)(3)$

31. $(f \circ h)(-1)$ 32. $(h \circ f)(2)$

33. $(f \circ g)(9)$ 34. $(g \circ f)(3)$

In Exercises 35–42, find expressions for $(f \circ g)(x)$ and $(g \circ f)(x)$, and give the domains of $f \circ g$ and $g \circ f$.

35. $f(x) = -x^2 + 4;\ g(x) = x - 2$

36. $f(x) = 2x + 5;\ g(x) = \dfrac{x - 5}{2}$

37. $f(x) = -x^2 + 3x;\ g(x) = x - 3$

38. $f(x) = -\dfrac{2}{x};\ g(x) = x + 5$

39. $f(x) = \dfrac{1}{x - 2};\ g(x) = x^2 + x$

40. $f(x) = \sqrt{x + 2};\ g(x) = 2x + 1$

41. $f(x) = \dfrac{x}{x + 3};\ g(x) = |x|$

42. $f(x) = x^2 - 4x + 4;\ g(x) = \dfrac{1}{x}$

In Exercises 43–46, find the difference quotient $\dfrac{f(x+h)-f(x)}{h}$, $h \neq 0$, for each function.

43. $f(x) = 4x - 3$

44. $f(x) = -3x^2$

45. $f(x) = 2x^2 - 3x + 1$

46. $f(x) = \dfrac{1}{x}, \quad x \neq 0$

Section 2.3

In Exercises 47–58, use transformations to sketch the graph of each function.

47. $g(x) = |x| - 6$

48. $F(s) = (s - 5)^2$

49. $H(x) = |x - 1| + 2$

50. $G(x) = (x + 4)^2 - 3$

51. $f(x) = 2\sqrt{x}$

52. $H(s) = -|s|$

53. $F(s) = -(s + 4)^2$

54. $P(x) = -\sqrt{x} + 1$

55. $f(x) = -3(x + 2)^2 + 1$

56. $h(x) = \sqrt{3x}$

57. $h(x) = |2x| - 3$

58. $h(x) = \left| \dfrac{1}{3}x \right|$

In Exercises 59–62, use the verbal description to find an algebraic expression for each function.

59. The graph of the function $g(x)$ is formed by translating the graph of $f(x) = |x|$ 3 units to the right and 1 unit up.

60. The graph of the function $f(x)$ is formed by translating the graph of $h(x) = \sqrt{x}$ 2 units to the right and 3 units up.

61. The graph of the function $g(x)$ is formed by vertically scaling the graph of $f(x) = x^2$ by a factor of 2 and moving the result to the left by 1 unit.

62. The graph of the function $g(x)$ is formed by horizontally compressing the graph of $f(x) = |x|$ by a factor of $\dfrac{1}{2}$ and moving the result up by 2 units.

In Exercises 63–66, use the given graph of f to graph each expression.

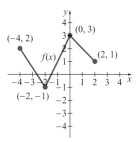

63. $2f(x) - 3$

64. $f(x - 1)$

65. $f(2x)$

66. $-f(x) + 1$

 In Exercises 67 and 68, use a graphing utility.

67. Graph the functions $y_1(x) = (x - 1.5)^2$ and $y_2(x) = x^2 - 1.5$ in the same viewing window and comment on the difference between the two functions in terms of transformations of $f(x) = x^2$.

68. Graph the function $y_1(x) = 4|x + 1.5| - 2.5$ and describe the graph in terms of transformations of $f(x) = |x|$.

Section 2.4

Exercises 69–74 pertain to the function f given by the following graph.

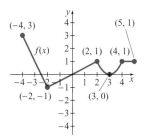

69. Find the interval(s) on which f is increasing.

70. Find the interval(s) on which f is decreasing.

71. Find the interval(s) on which f is constant.

72. Find the average rate of change of f on the interval $[-1, 1]$.

73. Find the average rate of change of f on the interval $[3, 4]$.

74. Find the y-intercept.

In Exercises 75–80, decide if the function is odd, even, or neither, using the definitions.

75. $f(x) = -x$

76. $f(x) = 3x + 4$

77. $f(x) = -x^2 + 5$

78. $f(x) = x^5 + x^2$

79. $f(x) - -3x^5 \quad x$

80. $f(x) = 2x^4 + x^2 + 1$

In Exercises 81 and 82, use a graphing utility to decide if each function is odd, even, or neither. Confirm your result using the definitions.

81. $f(x) = 2x^3 + 3x - 5$

82. $f(x) = -x^4 + x^2 + 3$

In Exercises 83–86, for each function, find the average rate of change on the given interval.

83. $f(x) = -2x + 4$; interval: $[2, 3]$

84. $f(x) = -2x^3 + 3x$; interval: $[-3, -1]$

85. $f(x) = -x^3 + 3$; interval: $[-1, 0]$

86. $f(x) = -x^2 + 2x$; interval: $[-2, 1]$

Section 2.5

In Exercises 87–92, solve the equation.

87. $|x - 5| = 6$

88. $|x + 6| = 7$

89. $\left| 2s - \dfrac{1}{2} \right| = 8$

90. $\left| 4s + \dfrac{3}{2} \right| = 10$

91. $1 + 3|2x - 5| = 4$

92. $-3 + 2|x - 4| = 7$

In Exercises 93–98, solve the inequality. Write your answer in interval notation, and graph the solution set on the number line.

93. $|3x + 10| > 5$

94. $|-x - 8| \le 7$

95. $\left| \dfrac{1}{2}x + 6 \right| \le 5$

96. $\left| \dfrac{2x + 1}{5} \right| \le 1$

97. $3\left| \dfrac{3}{2}x + 1 \right| < 9$

98. $|-2x - 7| + 4 \le 8$

In Exercises 99 and 100, solve the inequality using a graphing utility.

99. $|2x - 3| \le -|x| + 6$

100. $|x| > -|2x| + 4$

Section 2.6

In Exercises 101–104, let $f(x) = \begin{cases} -2x + 1, & \text{if } x < 0 \\ x, & \text{if } x \ge 0 \end{cases}$.

Evaluate the following.

101. $f(0)$

102. $f(-1)$

103. $f(3)$

104. $f\left(\dfrac{1}{2} \right)$

In Exercises 105–108, sketch a graph of the piecewise-defined function.

105. $g(x) = \begin{cases} x, & \text{if } x \le -2 \\ 2.5, & \text{if } x > -2 \end{cases}$

106. $F(x) = \begin{cases} 4, & \text{if } x < 0 \\ x^2, & \text{if } 0 \le x < 2 \\ -3, & \text{if } 2 \le x \le 3 \end{cases}$

107. $f(t) = \begin{cases} 1, & \text{if } t < 4 \\ \sqrt{t - 4}, & \text{if } t \ge 4 \end{cases}$

108. $f(x) = [\![x]\!] - 1$

Applications

109. **Gardening** A circular flower border is to be planted around a statue, with the statue at the center. The distance from the statue to any point on the inner boundary of the flower border is 20 feet. What is the equation of the outer boundary of the border if the flower border is 2 feet wide? Use a coordinate system with the statue at $(0, 0)$.

110. **Distance** Two boats start from the same point. One travels directly west at 30 miles per hour. The other travels directly north at 44 miles per hour. How far apart will they be after an hour and a half?

111. **Revenue** The revenue for a corporation is modeled by the function $R(t) = 150 + 5t$, where t is the number of years since 2002 and $R(t)$ is in millions of dollars. The company's operating costs are modeled by the function $C(t) = 135 + 2.4t$, where t is the number of years since 2002 and $C(t)$ is in millions of dollars. Find the profit function, $P(t)$.

112. **Pediatrics** The length of an infant 21 inches long at birth can be modeled by the linear function $h(t) = 21 + 1.5t$, where t is the age of the infant in months and $h(t)$ is in inches. (*Source:* Growth Charts, Centers for Disease Control)
    (a) What is the slope of this function, and what does it represent?
    (b) What is the length of the infant at 6 months?
    (c) If $f(x) = 2.54x$ is a function that converts inches to centimeters, express the length of the infant in centimeters by using composition of functions.

113. **Commerce** The production cost, in dollars, for producing x booklets is given by $C(x) = 450 + 0.35x$.
    (a) If the fixed cost is decreased by \$40, find the new cost function.
    (b) The shipping and handling cost is 4% of the total production cost $C(x)$. Find the function describing the shipping and handling costs.
    (c) Graph both cost functions and interpret the effect of the decreased fixed cost.

114. **Geography** You are located in the center of Columbus, Ohio. Write an absolute value inequality that gives all points less than or equal to 43 miles east or west of the center of Columbus. Indicate what point you would use as the origin.

115. **Cellular Phone Rates** The Virgin Mobile prepaid cellular phone plan charges \$0.25 per minute of airtime for the first 10 minutes of the day, and \$0.10 per minute of airtime for the rest of the day. (*Source:* Virgin Mobile, 2005)
    (a) How much is the charge if a person used 15 minutes of airtime on a particular day?
    (b) If x is the number of minutes used per day, what is the piecewise-defined function that describes the total daily charge? Assume that x is a whole number.

116. **Profit Growth** The following table lists the gross profits (i.e., profits before taxes) for IBM Corporation for the years 2002–2004. (*Source:* IBM Annual Report)

Year	Gross Profit (in billions of dollars)
2002	30
2003	33
2004	36

Find and interpret the average rate of change in gross profit from 2002 to 2004.

Chapter 2 — Test

1. Given the points $(1, 2)$ and $(-4, 3)$, find the distance between them and the midpoint of the line segment joining them.

2. Write the standard form of the equation of the circle with center $(2, 5)$ and radius 6.

3. Find the center and radius of the circle with equation $x^2 + 2x + y^2 - 4y = 4$. Sketch the circle.

In Exercises 4–11, let $f(x) = x^2 + 2x$ and $g(x) = 2x - 1$.

4. Evaluate $(f + g)(2)$.

5. Evaluate $(g - f)(-1)$.

6. Evaluate $\left(\dfrac{f}{g}\right)(3)$.

7. Evaluate $(f \cdot g)(-2)$.

8. Evaluate $(f \circ g)(0)$.

9. Evaluate $(g \circ f)(-1)$.

10. Find $\left(\dfrac{f}{g}\right)(x)$ and identify its domain.

11. Find $\dfrac{f(x + h) - f(x)}{h}$, $h \neq 0$.

12. Let $f(x) = \dfrac{1}{2x}$ and $g(x) = x^2 - 1$. Find $(f \circ g)(x)$ and identify its domain.

In Exercises 13–16, use transformations to sketch the graph of each function.

13. $f(x) = \left|\dfrac{1}{2}x\right|$

14. $f(x) = |3x| - 1$

15. $f(x) = -2\sqrt{x} + 3$

16. $f(x) = -(x + 2)^2 - 2$

In Exercises 17 and 18, use the verbal description of the function to find its corresponding algebraic expression.

17. The graph of the function $g(x)$ is formed by translating the graph of $f(x) = |x|$ 2 units to the left and 1 unit up.

18. The graph of the function $g(x)$ is formed by compressing the graph of $f(x) = x^2$ horizontally by a factor of $\frac{1}{2}$ and moving it down by 1 unit.

In Exercises 19–21, decide if the function is odd, even, or neither by using the appropriate definitions.

19. $f(x) = 2x + 1$ 20. $f(x) = -3x^2 - 5$

21. $f(x) = 2x^5 + x^3$

In Exercises 22 and 23, use the following graph.

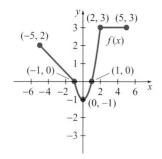

22. Find the interval(s) on which
 (a) f is increasing.
 (b) f is decreasing.
 (c) f is constant.

23. Find the average rate of change of f on the interval $[-2, 1]$.

24. Find the average rate of change of $f(x) = 4x^2 - 5$ on the interval $[2, 3]$.

In Exercises 25 and 26, solve the equation.

25. $\left| 6x + \dfrac{4}{3} \right| = 5$ 26. $|4x - 7| - 3 = 6$

In Exercises 27–29, solve the inequality. Write your answer in interval notation and graph the solution set on the real number line.

27. $\left| \dfrac{5x + 4}{3} \right| \le 8$

28. $2|2x - 5| \ge 6$

29. $|5x + 2| + 6 < 13$

30. Let f be defined as follows.
$$f(x) = \begin{cases} -x + 1, & \text{if } x \le 1 \\ x - 1, & \text{if } x > 1 \end{cases}$$
 (a) Sketch a graph of f.
 (b) Find a single expression for f in terms of absolute value.

31. **Travel** You are located in the center of St. Louis, Missouri. Write an absolute value inequality that gives all the points greater than or equal to 53 miles north or south of the center of St. Louis. Indicate what point you would use as the origin.

32. **Shopping** The revenue for an online shopping website is modeled by the function $R(t) = 200 + 15t$, where t is the number of years since 2004 and $R(t)$ is in millions of dollars.
 (a) Find the revenue for the year 2008.
 (b) The company's operating cost is modeled by $C(t) = 215 + 8.5t$, where t is the number of years since 2004 and $C(t)$ is in millions of dollars. Find the profit function $P(t)$.
 (c) According to the model, when will the profit be equal to 10 million dollars?

33. **Sales** A wholesale nut producer charges $3.50 per pound of cashews for the first 50 pounds. The price then drops to $3 per pound for each pound (or portion thereof) over 50 pounds. Express the cost of x pounds of cashews as a piecewise-defined function.

Quadratic Functions

The percentage of tufted puffin eggs that hatch during a breeding season depends very much on the sea surface temperature of the surrounding area. A slight increase or decrease from the optimal temperature results in a decrease in the number of eggs hatched. This phenomenon can be modeled by a **quadratic function.** See Exercises 77 and 78 in Section 3.1. This chapter explores how quadratic functions and equations arise, how to solve them, and how they are used in various applications.

3.1 Graphs of Quadratic Functions

Objectives

▶ Define a quadratic function

▶ Use transformations to graph a quadratic function in vertex form

▶ Write a quadratic function in vertex form by completing the square

▶ Find the vertex and axis of symmetry of a quadratic function

▶ Identify the maximum or minimum of a quadratic function

▶ Graph a quadratic function in standard form

▶ Solve applied problems using maximum and minimum function values

In Chapter 1, we saw how linear functions can be used to model certain problems we encounter in the real world. However, there exist many problems for which a linear function will not provide an accurate model. To help us analyze a wider set of problems, we will explore another class of functions, known as *quadratic functions*.

Area: An Example of a Quadratic Function

Before giving a formal definition of a quadratic function, we will examine the problem of finding the area of a region with a fixed perimeter. In this way, we will see how this new type of function arises.

Example 1 **Deriving an Expression for Area**

Example 6 in Section 3.1 builds upon this example. ┄┄┊

Traffic authorities have 100 feet of rope to cordon off a rectangular region to form a ticket arena for concert goers who are waiting to purchase tickets. Express the area of this rectangular region as a function of the length of just one of the four sides of the region.

▶**Solution** Since this is a geometric problem, we first draw a diagram of the cordoned-off region, as shown in Figure 3.1.1. Here, l denotes the length of the region and w denotes its width. From geometry, we know that

$$A = lw.$$

We must express the area as a function of just *one* of the two variables l, w. To do so, we must find a relationship between the length and the width. Since 100 feet of rope is available, the perimeter of the enclosed region must equal 100:

$$\text{Perimeter} = 2l + 2w = 100$$

We can solve for w in terms of l as follows.

$$2w = 100 - 2l \qquad \text{Isolate the } w \text{ term}$$
$$w = \frac{1}{2}(100 - 2l) \qquad \text{Solve for } w$$
$$w = 50 - l \qquad \text{Simplify}$$

Now we can write the area as

$$A = lw = l(50 - l) = 50l - l^2.$$

Thus the area, A, is a function of the length l, and its expression is given by $A(l) = -l^2 + 50l$. We will revisit this function later to explore other aspects of this problem.

✔ *Check It Out 1:* Find the area expression for the cordoned region in Example 1 if 120 feet of rope is available. ■

We saw that the function for the area of the rectangular region in Example 1 involves the length l raised to the second power. This is an example of a quadratic function, which we now define.

Figure 3.1.1

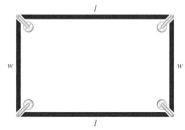

Discover *and* Learn

Use the area function from Example 1 to compute the area for various values of the length l by filling in the following table.

Table 3.1.1

Length	10	20	25	30	40
Area					

What do you observe about the area as the length is varied?

Definition of a Quadratic Function

A function f is a **quadratic function** if it can be expressed in the form

$$f(x) = ax^2 + bx + c$$

where a, b, and c are real numbers and $a \neq 0$. The **domain** of a quadratic function is the set of all real numbers.

Throughout our discussion, a is the coefficient of the x^2 term; b is the coefficient of the x term; and c is the constant term. To help us better understand quadratic functions, it is useful to look at their graphs.

Some General Features of a Quadratic Function

We study the graphs of general quadratic functions by first examining the graph of the function $f(x) = ax^2$, $a \neq 0$. We will later see that the graph of *any* quadratic function can be produced by a suitable combination of transformations of this graph.

Example 2 Sketching the Graph of $f(x) = ax^2$

Consider the quadratic functions $f(x) = x^2$, $g(x) = -x^2$, and $h(x) = 2x^2$. Make a table of values and graph the three functions on the same set of coordinate axes. Find the domain and range of each function. What observations can you make?

▶**Solution** We make a table of values of $f(x)$, $g(x)$, and $h(x)$ for $x = -2, -1, 0, 1, 2$ (see Table 3.1.2) and then use it to graph the functions, as shown in Figure 3.1.2. Note that for a given value of x, the value of $g(x)$ is just the negative of the value of $f(x)$ and the value of $h(x)$ is exactly twice the value of $f(x)$.

Table 3.1.2

x	$f(x) = x^2$	$g(x) = -x^2$	$h(x) = 2x^2$
-2	4	-4	8
-1	1	-1	2
0	0	0	0
1	1	-1	2
2	4	-4	8

Figure 3.1.2

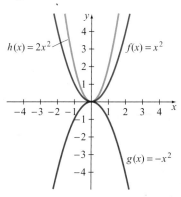

The domain of all three functions f, g, and h is the set of all real numbers, $(-\infty, \infty)$, because x^2 is defined for all real numbers. Since $x^2 \geq 0$, the range of $f(x) = x^2$ is the set of all numbers greater than or equal to zero, $[0, \infty)$. The range of $g(x) = -x^2$ is the set of all numbers less than or equal to zero, $(-\infty, 0]$, and the range of $h(x) = 2x^2$ is the set of all numbers greater than or equal to zero, $[0, \infty)$.

Observations:

▶ The graph of $f(x) = x^2$ opens upward, whereas the graph of $g(x) = -x^2$ opens downward.

▶ The graph of $h(x) = 2x^2$ is vertically scaled by a factor of 2 compared to the graph of $f(x) = x^2$, and the graphs of both f and h open upward.

☑ *Check It Out 2:* Rework Example 2 using the functions $f(x) = x^2$ and $g(x) = \frac{1}{2}x^2$.

Discover *and* Learn

Graph the function $f(x) = (x - 2)^2 + 1$ using a graphing utility with a decimal window. Trace to find the lowest point on the graph. How are its coordinates related to the given expression for $f(x)$?

Quadratic Functions Written in Vertex Form, $f(x) = a(x - h)^2 + k$

In many instances, it is easier to analyze a quadratic function if it is written in the form $f(x) = a(x - h)^2 + k$. This is known as the **vertex form** of the quadratic function.[1]

Quadratic Function in Vertex Form

A quadratic function $f(x) = ax^2 + bx + c$ can be rewritten in the form $f(x) = a(x - h)^2 + k$, known as the **vertex form.** The graph of $f(x) = a(x - h)^2 + k$ is called a **parabola.** Its lowest or highest point is given by (h, k). This point is known as the **vertex** of the parabola.

▶ If $a > 0$, the parabola opens upward, and the vertex is the lowest point on the graph. This is the point where f has its minimum value, called the **minimum point.** The range of f is $[k, \infty)$. See Figure 3.1.3.

▶ If $a < 0$, the parabola opens downward, and the vertex is the highest point on the graph. This is the point where f has its maximum value, called the **maximum point.** See Figure 3.1.4.

Figure 3.1.3   Parabola: $a > 0$

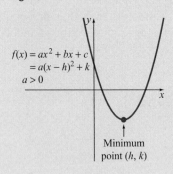

Figure 3.1.4   Parabola: $a < 0$

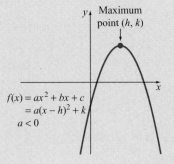

When a quadratic function is expressed in the form $a(x - h)^2 + k$, we can use transformations to help sketch its graph, as shown in the following example.

Example 3 **Sketching a Quadratic Function Using Transformations**

Use transformations to graph $f(x) = -2(x - 1)^2 + 3$. What is the highest point on the graph and how are its coordinates related to the expression for $f(x)$?

---

[1]Some textbooks refer to this form as the *standard form* of the quadratic function.

▶**Solution** Note that the graph of $f(x) = -2(x-1)^2 + 3$ can be written as a series of transformations of the graph of $y = x^2$ as follows:

$$x^2 \;\rightarrow\; -2x^2 \qquad\qquad \rightarrow \qquad\qquad -2(x-1)^2 \qquad \rightarrow \qquad -2(x-1)^2 + 3$$

Vertical stretch by a factor   Horizontal shift   Vertical shift
of 2 and reflection across   1 unit to the right   3 units up
the *x*-axis

Just In Time

Review transformations in Section 2.3.

The transformation of the graph of $y = x^2$ to the graph of $y = -2x^2$ is sketched in Figure 3.1.5.

Figure 3.1.5

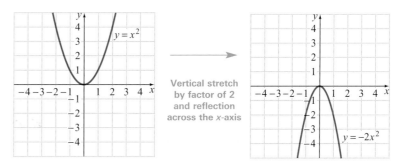

The shifts of the graph of $y = -2x^2$ to the right 1 unit and up 3 units are shown in the series of graphs in Figure 3.1.6.

Figure 3.1.6

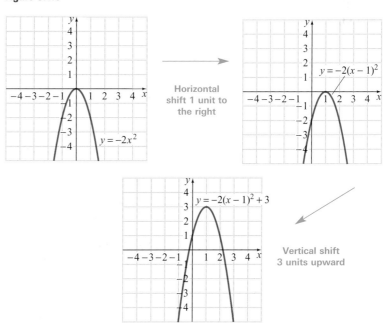

We see that the graph of $f(x) = -2(x-1)^2 + 3$ opens downward, and the highest point on the graph is $(1, 3)$. The *x*- and *y*-coordinates of the highest point correspond to the horizontal and vertical shifts, respectively.

✔ *Check It Out 3:* Rework Example 3 using the function $g(x) = 2(x-2)^2 - 1$. Here, you will find the lowest point on the graph of g. ■

Just In Time

*Review factoring of trinomi-
als that are perfect squares
in Section P.5.*

In Example 3, we observed some major features of a quadratic function by examining it as a transformation of the basic function $f(x) = x^2$. To write *any* quadratic
function $f(x) = ax^2 + bx + c$ in the form $f(x) = a(x - h)^2 + k$, where a, h, and k are
real numbers, you must use a technique known as **completing the square.**

For example, to complete the square on $x^2 + 8x$, you add 16. This gives
$x^2 + 8x + 16 = (x + 4)^2$. In general, if the expression is of the form $x^2 + bx$, you add
$c = \left(\frac{b}{2}\right)^2$ to make $x^2 + bx + c$ a perfect square. If the expression is of the form
$ax^2 + bx$, $a \neq 0$, you factor out a to get $a\left(x^2 + \frac{b}{a}x\right)$ and then complete the square on
the expression *inside* the parentheses. The following example illustrates this method.

Example 4 **Writing a Quadratic Function in Vertex Form**

Let $f(x) = 2x^2 - 4x + 5$.

(a) Use the technique of completing the square to write $f(x) = 2x^2 - 4x + 5$ in the
form $f(x) = a(x - h)^2 + k$.

(b) What is the vertex of the associated parabola? Is it the maximum point or the minimum point?

▶**Solution**

(a) To complete the square for $2x^2 - 4x + 5$, we examine the first two terms:
$2x^2 - 4x$. Factor out the 2 to get $2(x^2 - 2x)$. Then see what number must be
added to $x^2 - 2x$ so that it can be written in the form $(x - h)^2$. The number to
be added is calculated by taking half of the coefficient of x and squaring it.

Number to be added: The square of $\frac{1}{2}(-2)$ is $(-1)^2 = 1$. Thus, the number 1 will
complete the square on $x^2 - 2x$.

Just In Time

*See more examples of
completing the square in
Section 2.1.*

Putting all this together, we have

$2x^2 - 4x + 5 = 2(x^2 - 2x) + 5$ Factor 2 out of x^2- and x-terms

$= 2(x^2 - 2x + 1 - 1) + 5$ Add 1 within parentheses to complete
the square. Subtract 1 within
parentheses to retain the value
of the original expression.

$= 2(x^2 - 2x + 1) - 2(1) + 5$ Regroup terms

$= 2(x - 1)^2 - 2 + 5$ Rewrite $x^2 - 2x + 1$ as $(x - 1)^2$

$= 2(x - 1)^2 + 3.$ Combine -2 and 5

This expression is now in the form $a(x - h)^2 + k$, with $a = 2$, $h = 1$, and $k = 3$.

(b) The vertex of the parabola is $(h, k) = (1, 3)$. Since $a = 1 > 0$, the parabola opens
upward and $(1, 3)$ is the minimum point, as shown in Figure 3.1.7.

Figure 3.1.7

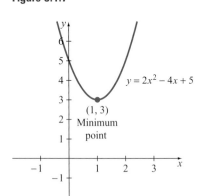

☑ *Check It Out 4:* Use the technique of completing the square to write $f(x) = 3x^2 -
12x + 7$ in the form $f(x) = a(x - h)^2 + k$. Find the vertex of the associated parabola
and determine whether it is the maximum point or the minimum point. ■

Formula for the Vertex of a Quadratic Function

It can be tedious to rewrite a quadratic function in the form $f(x) = a(x - h)^2 - k$ in
order to find the vertex. In this section, we will complete the square on the general

quadratic function $f(x) = ax^2 + bx + c$ so that we can write it in the form $f(x) = a(x - h)^2 + k$. When we do this, we will find a general formula for the values of h and k. We can then use this formula to find the vertex of the parabola associated with the quadratic function $f(x) = ax^2 + bx + c$.

To find the formula for the vertex, we begin by factoring an a out of the first two terms.

$$ax^2 + bx + c = a\left(x^2 + \frac{b}{a}x\right) + c \qquad \text{Factor } a \text{ out of first two terms}$$

$$= a\left(x^2 + \frac{b}{a}x + \left(\frac{b}{2a}\right)^2\right) - a\left(\frac{b}{2a}\right)^2 + c \qquad \text{Complete the square on } x^2 + \frac{b}{a}x$$

Since $x^2 + \frac{b}{a}x + \left(\frac{b}{2a}\right)^2 = \left(x + \frac{b}{2a}\right)^2$ and $a\left(\frac{b}{2a}\right)^2 = \frac{b^2}{4a}$, the above equation becomes

$$a\left(x + \frac{b}{2a}\right)^2 + \left[c - \frac{b^2}{4a}\right] = a(x - h)^2 + k$$

where $h = -\frac{b}{2a}$ and $k = c - \frac{b^2}{4a}$. From the previous discussion, we know that (h, k) is the vertex of the parabola. Thus we have the following result.

Formula for the Vertex of a Quadratic Function

The **vertex** of the parabola associated with the quadratic function $f(x) = ax^2 + bx + c$ is given by (h, k), where $h = -\frac{b}{2a}$ and $k = c - \frac{b^2}{4a}$.

Since (h, k) is a point on the graph of f, we must have $f(h) = k$. It is easier to first calculate the value of h and then calculate $k = f(h) = f\left(-\frac{b}{2a}\right)$. Thus the vertex is given by

$$(h, k) = \left(-\frac{b}{2a}, c - \frac{b^2}{4a}\right)$$

or

$$(h, k) = \left(-\frac{b}{2a}, f\left(-\frac{b}{2a}\right)\right).$$

See Figure 3.1.8.

Figure 3.1.8

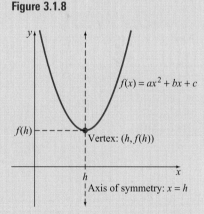

Note that a parabola is symmetric about the vertical line passing through the vertex. That is, the part of the parabola that lies to the left of the line is the mirror image of the part that lies to the right of it. This line is known as the **axis of symmetry.**

Axis of Symmetry

The equation of the **axis of symmetry** of the parabola associated with the quadratic function $f(x) = ax^2 + bx + c$ is given by

$$x = \frac{-b}{2a}.$$

Although you can readily find the vertex of a parabola using the given formula, the technique of completing the square will be quite useful in later sections. The next example shows you how to put together everything you have learned in this section to create the graph of a quadratic function.

Example 5 Sketching a Quadratic Function

Let $f(x) = 2x^2 - 3x - 2$.

(a) Find the vertex of the parabola associated with f.

(b) Find the axis of symmetry of the parabola.

(c) Find two additional points on the graph, and then sketch the graph of f by hand.

(d) Use the graph to find the intervals on which f is increasing and decreasing, and find the domain and range of f.

▶ **Solution**

(a) The vertex is given by $(h, k) = \left(\frac{-b}{2a}, f\left(\frac{-b}{2a} \right) \right)$. Using $a = 2$, $b = -3$,

$$h = \frac{-b}{2a} = \frac{-(-3)}{2(2)} = \frac{3}{4}$$

$$k = f\left(\frac{3}{4} \right) = 2\left(\frac{3}{4} \right)^2 - 3\left(\frac{3}{4} \right) - 2 = \frac{9}{8} - \frac{9}{4} - 2 = \frac{9 - 18 - 16}{8} = \frac{-25}{8}$$

The vertex is therefore $\left(\frac{3}{4}, \frac{-25}{8} \right)$. Since $a = 2 > 0$, the parabola opens upward and the vertex is the minimum point.

(b) The axis of symmetry is given by $x = \frac{3}{4}$.

(c) One point that is easy to calculate is the y-intercept, obtained by substituting $x = 0$ into the expression for $f(x)$. This gives $(0, -2)$ as the y-intercept. The point $(0, -2)$ is $\frac{3}{4}$ unit to the left of the axis of symmetry, so its mirror image will be $\frac{3}{4}$ unit to the right of that line. Thus the x-coordinate of the mirror image must be $\frac{3}{4} + \frac{3}{4} = \frac{3}{2}$. Also, its y-coordinate must be equal to the y-coordinate of the point $(0, -2)$. In this way, reflecting $(0, -2)$ across the axis of symmetry $x = \frac{3}{4}$ gives $\left(\frac{3}{2}, -2 \right)$ as a third point on the parabola. Putting all this together, we get the graph shown in Figure 3.1.10.

Note that the axis of symmetry is *not* part of the graph of the function. It is sketched merely to indicate the symmetry of the graph.

(d) From the graph, we see that the function is decreasing on $\left(-\infty, \frac{3}{4} \right)$ and increasing on $\left(\frac{3}{4}, \infty \right)$. The domain is $(-\infty, \infty)$ and the range is $\left[\frac{-25}{8}, \infty \right)$.

✔ *Check It Out 5:* Rework Example 5 using the function $g(t) = -3t^2 + 6t - 2$. ▪

Note   When finding points on a parabola, if the vertex also happens to be the y-intercept, then you must find a different second point on the parabola and use symmetry to find the third point.

Technology Note

The minimum point (0.7500, −3.125) can be found by using the MINIMUM feature of your graphing utility. See Figure 3.1.9.

Keystroke Appendix: Section 10

Figure 3.1.9

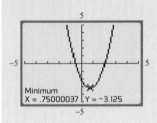

Figure 3.1.10

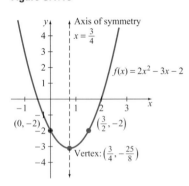

Applications

This section illustrates a number of applications involving quadratic functions. First, we revisit the ticket arena problem from Example 1.

Figure 3.1.11

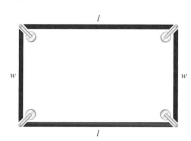

| Example | 6 | **Maximizing Area** |

✦••• *This example builds on Example 1 in Section 3.1.*

Find the dimensions of the enclosed region in Example 1 that will give the maximum area, and determine the maximum area. Recall that the area function was given by $A(l) = -l^2 + 50l$, where l is the length of the region. The region is shown again in Figure 3.1.11.

▶Solution The quadratic function for the area is given by

$$A(l) = -l^2 + 50l.$$

Since $a = -1 < 0$, the graph of the area function is a parabola opening downward. Therefore, the vertex will be the maximum point. The first coordinate of the vertex is

$$h = \frac{-b}{2a} = \frac{-50}{2(-1)} = 25.$$

Thus the maximum area occurs when the length is 25 feet. Since $2l + 2w = 100$, we find that $w = 25$, and so the width is 25 feet as well. Substituting into the area function, we have

$$A(25) = -(25)^2 + 50(25) = 625 \text{ square feet}.$$

Thus the maximum area is 625 square feet. This corresponds to roping off a 25-foot square region to maximize the area, given 100 feet of rope. The graph of the area function and a diagram of the cordoned-off region are given in Figures 3.1.13 and 3.1.14, respectively.

Technology Note

You can use a graphing utility to generate a table of values for the area function and see how the values of the area vary as the length is changed. See Figure 3.1.12.

Keystroke Appendix:
Section 6

Figure 3.1.12

X	Y₁	
5	225	
10	400	
15	525	
20	600	
25	625	
30	600	
35	525	
X=25		

Figure 3.1.13

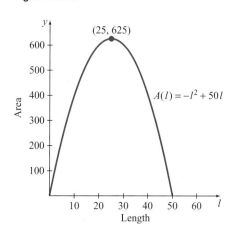

Figure 3.1.14

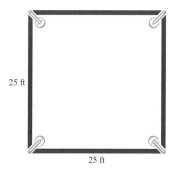

☑ *Check It Out 6:* If the area function of a rectangular region cordoned off with 120 feet of rope is given by $A(l) = -l^2 + 60l$, find the dimensions of the region that will give the maximum area, and determine the maximum area. ▪

Example **7** **Quadratic Model of Health Data**

The chart in Figure 3.1.15 shows the cesarean delivery rates among American women who have not had a previous cesarean delivery. (*Source:* National Center for Health Statistics)

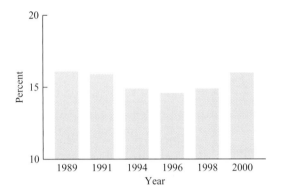

Figure 3.1.15 First-time cesarean delivery rates among American women (in percentage)

(a) Examine the data in the bar graph. Why would a quadratic function be suitable for describing the first-time cesarean delivery rate over the given time period?

(b) The given data can be modeled by the quadratic function $c(t) = 0.0428t^2 - 0.5229t + 16.364$, where $c(t)$ is the first-time cesarean delivery rate given as a percent, and t is the number of years since 1989. Find $c(5)$ and explain what it represents.

(c) According to this model, when did the first-time cesarean delivery rate reach its lowest point?

▶**Solution**

(a) Since the first-time cesarean delivery rate first decreased and then increased, a quadratic function would be suitable for describing the behavior of the data.

(b) Evaluate $c(5)$ to get

$$c(5) = 0.0428(5)^2 - 0.5229(5) + 16.364$$
$$= 14.820.$$

Thus, in 1994, five years after 1989, 14.82% of deliveries among American women who had not had a previous cesarean delivery were cesarean.

(c) The lowest point can be found by computing the coordinates of the vertex of the function $c(t)$:

$$h = \frac{-b}{2a} = \frac{-(-0.5229)}{2(0.0428)} \approx 6.109;$$

$$k = c(h) = c(6.109) \approx 14.77$$

The first-time cesarean delivery rate reached its lowest point of 14.77% at 6.109 years after 1989, or just after the beginning of 1995.

✔ *Check It Out 7:* Use the model in Example 7 to predict the first-time cesarean delivery rate for the year 2001. ▪

Example 8 Modeling Car Mileage

Table 3.1.3 lists the speed at which a car is driven, in miles per hour, and the corresponding gas mileage obtained, in miles per gallon. (*Source:* Environmental Protection Agency)

Table 3.1.3

Speed, x (miles per hour)	Gas Mileage, m (miles per gallon)
5	12
10	17
25	27
45	30
65	25
75	22

(a) Make a scatter plot of m, the gas mileage, versus x, the speed at which the car is driven. What type of trend do you observe—linear or quadratic? Find an expression for the function that best fits the given data points.

(b) Use your function to find the gas mileage obtained when the car is driven at 35 miles per hour.

(c) Find the speed at which the car's gas mileage is at its maximum, using a graphing utility.

▶Solution

(a) Enter the data into your graphing utility and display the scatter plot, as shown in Figure 3.1.16.

Figure 3.1.16   **Figure 3.1.17**

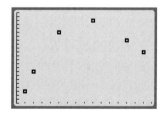   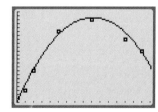

Note that the gas mileage first increases with speed and then decreases. This suggests that a quadratic model would be appropriate. Using the Regression feature of your graphing utility, the expression for the function that best fits this data is

$$m(x) = -0.0108x^2 + 0.981x + 8.05.$$

The coefficients have been rounded to three significant digits. The graph of the function along with the data points (Figure 3.1.17) shows that the function approximates the data reasonably well. Store this function in your calculator as Y_1, for subsequent analysis.

(b) Evaluating the function at $x = 35$ gives a gas mileage of approximately 29.2 miles per gallon. See Figure 3.1.18.

(c) Using the MAXIMUM feature of your graphing utility, the maximum point is approximately (45.5, 30.4), as seen in Figure 3.1.19. Thus, at a speed of 45.5 miles per hour, the gas mileage is at a maximum of 30.4 miles per gallon. Your answer may vary slightly if you use the rounded function instead.

Figure 3.1.18 Mileage at 35 mph

Figure 3.1.19 Maximum mileage

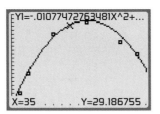

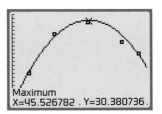

☑ *Check It Out 8:* Use the TABLE feature in ASK mode to evaluate the gas mileage function in Example 8 at the speeds given in the table. Compare your results with the actual gas mileages listed for those speeds, and comment on the validity of the model. ■

Note You can fit a linear or quadratic function to any set of points. It is up to you to decide whether the function is an appropriate model or not. Choosing a reasonable function to model a set of data requires that you be familiar with the properties of different types of functions. As you work with the data sets for the problems in this and other sections, make sure you understand why you are choosing one particular type of function over another.

3.1 Key Points

▶ A **quadratic function** is given by $f(x) = ax^2 + bx + c$, $a \neq 0$.
▶ The graph of the function $f(x) = ax^2 + bx + c$ is called a **parabola.**
▶ If $a > 0$, the parabola opens upward.
▶ If $a < 0$, the parabola opens downward.
▶ The **vertex form** of a quadratic function is given by $f(x) = a(x - h)^2 + k$.
▶ The **vertex** (h, k) of a parabola is given by:

$$h = \frac{-b}{2a}$$

$$k = f(h) = f\left(\frac{-b}{2a}\right) = c - \frac{b^2}{4a}$$

If the parabola opens upward, then $(h, f(h))$ is the lowest point of the parabola (the minimum point). If the parabola opens downward, then $(h, f(h))$ is the highest point of the parabola (the maximum point).

▶ The **axis of symmetry** of a parabola is given by $x = \frac{-b}{2a}$.

3.1 Exercises

▶**Just in Time Exercises** These exercises correspond to the Just in Time references in this section. Complete them to review topics relevant to the remaining exercises.

1. The graph of $g(x) = f(x) + 2$ is the graph of $f(x)$ shifted _____ 2 units.

2. The graph of $g(x) = f(x) - 3$ is the graph of $f(x)$ shifted _____ 3 units.

3. The graph of $g(x) = f(x + 2)$ is the graph of $f(x)$ shifted _____ 2 units.

4. The graph of $g(x) = -f(x)$ is the graph of $f(x)$ reflected about the _____-axis.

5. Factor: $x^2 - 16x + 64$

6. Factor: $25y^2 + 10y + 1$

In Exercises 7 and 8, what number must be added to write the expression in the form $(x + b)^2$?

7. $x^2 + 8x$ 8. $x^2 - 14x$

▶**Skills** This set of exercises will reinforce the skills illustrated in this section.

In Exercises 9–18, graph each pair of functions on the same set of coordinate axes, and find the domain and range of each function.

9. $f(x) = -2x^2$, $g(x) = -x^2$

10. $f(x) = 3x^2$, $g(x) = x^2$

11. $f(x) = \frac{1}{2}x^2$, $g(x) = 2x^2$

12. $f(x) = -x^2$, $g(x) = -\frac{1}{2}x^2$

13. $f(x) = x^2 + 1$, $g(x) = x^2 - 1$

14. $f(x) = -x^2 + 2$, $g(x) = -x^2 - 2$

15. $f(x) = (x + 1)^2$, $g(x) = x^2$

16. $f(x) = -(x - 3)^2$, $g(x) = -x^2$

17. $f(x) = (x + 2)^2$, $g(x) = (x - 2)^2$

18. $f(x) = -(x - 4)^2$, $g(x) = -(x + 4)^2$

In Exercises 19–26, each of the graphs represents a quadratic function. Match the graph with its corresponding expression or description.

a.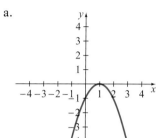

b.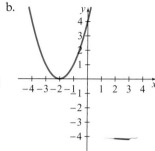

c.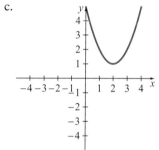

d.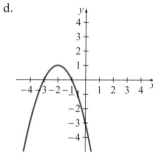

e.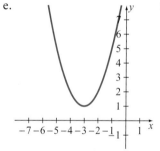

f.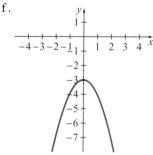

g.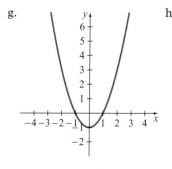

h.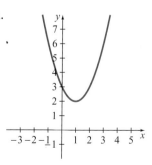

19. $f(x) = (x + 2)^2$ 20. $f(x) = -(x + 2)^2 + 1$

21. The graph of $f(x) = x^2$ shifted 1 unit to the right and reflected across the x-axis

22. The graph of $f(x) = x^2$ shifted 2 units to the right and 1 unit up

23. $f(x) = (x + 3)^2 + 1$

24. $f(x) = (x - 1)^2 + 2$

25. The graph of $f(x) = x^2$ reflected about the y-axis and shifted 3 units down

26. The graph of $f(x) = x^2$ shifted 1 unit up

In Exercises 27–32, use transformations to graph the quadratic function and find the vertex of the associated parabola.

27. $f(x) = (x + 2)^2 - 1$

28. $h(x) = (x - 3)^2 + 2$

29. $f(x) = -(x + 1)^2 - 1$

30. $g(s) = -(s - 2)^2 + 2$

31. $h(x) = -3(x + 4)^2 - 2$

32. $g(t) = 2(t - 3)^2 + 3$

In Exercises 33–40, write each quadratic function in the form $f(x) = a(x - h)^2 + k$ by completing the square. Also find the vertex of the associated parabola and determine whether it is a maximum or minimum point.

33. $g(x) = x^2 + 2x + 5$

34. $h(x) = x^2 - 4x + 6$

35. $w(x) = -x^2 + 6x + 4$

36. $f(x) = -x^2 - 4x + 5$

37. $h(x) = x^2 + x - 3$

38. $g(x) = -x^2 + x - 7$

39. $f(x) = 3x^2 + 6x - 4$

40. $f(x) = -2x^2 + 8x + 3$

In Exercises 41–50, find the vertex and axis of symmetry of the associated parabola for each quadratic function. Then find at least two additional points on the parabola and sketch the parabola by hand.

41. $f(x) = -2x^2 + 4x - 1$

42. $f(x) = x^2 - 6x + 1$

43. $g(x) = -x^2 + 4x - 3$

44. $f(x) = 3x^2 - 12x + 4$

45. $h(x) = x^2 - 3x + 5$

46. $h(x) = -x^2 + x - 2$

47. $f(t) = -16t^2 + 100$

48. $f(x) = 10x^2 - 65$

49. $f(t) = \dfrac{1}{3} - 3t + t^2$

50. $h(t) = 1 - \dfrac{1}{2}t - t^2$

In Exercises 51–58, find the vertex and axis of symmetry of the associated parabola for each quadratic function. Sketch the parabola. Find the intervals on which the function is increasing and decreasing, and find the range.

51. $f(x) = -x^2 + 10x - 8$

52. $h(x) = x^2 + 6x - 7$

53. $f(x) = 2x^2 + 4x - 3$

54. $g(x) = -4x^2 - 8x + 5$

55. $f(x) = -0.2x^2 + 0.4x - 2.2$

56. $f(x) = 0.3x^2 + 0.6x + 1.3$

57. $h(x) = \dfrac{1}{4}x^2 + \dfrac{1}{2}x - 2$

58. $g(x) = -\dfrac{1}{6}x^2 + \dfrac{1}{3}x + 1$

In Exercise 59, refer to graph (a) on page 225.

59. (a) Find the domain of f.
    (b) Find the range of f.
    (c) Does f have a maximum value or a minimum value? What is it?
    (d) Find the equation of the axis of symmetry.
    (e) Find the interval(s) over which f is increasing.
    (f) Find the interval(s) over which f is decreasing.

In Exercise 60, refer to graph (c) on page 225.

60. (a) Find the domain of f.
    (b) Find the range of f.

(c) Does f have a maximum value or a minimum value? What is it?

(d) Find the equation of the axis of symmetry.

(e) Find the interval(s) over which f is increasing.

(f) Find the interval(s) over which f is decreasing.

In Exercise 61, refer to graph (e) *on page 225.*

61. (a) Find the domain of f.

(b) Find the range of f.

(c) Does f have a maximum value or a minimum value? What is it?

(d) Find the equation of the axis of symmetry.

(e) Find the interval(s) over which f is increasing.

(f) Find the interval(s) over which f is decreasing.

In Exercise 62, refer to graph (h) *on page 225.*

62. (a) Find the domain of f.

(b) Find the range of f.

(c) Does f have a maximum value or a minimum value? What is it?

(d) Find the equation of the axis of symmetry.

(e) Find the interval(s) over which f is increasing.

(f) Find the interval(s) over which f is decreasing.

In Exercises 63–67, graph each quadratic function by finding a suitable viewing window with the help of the TABLE feature of a graphing utility. Also find the vertex of the associated parabola using the graphing utility.

63. $y_1(x) = 0.4x^2 + 20$

64. $g(s) = -s^2 - 15$

65. $h(x) = (\sqrt{2})x^2 + x + 1$

66. $h(x) = x^2 + 5x - 20$

67. $s(t) = -16t^2 + 40t + 120$

68. Graph the function $f(t) = t^2 - 4$ in a decimal window. Using your graph, determine the values of t for which $f(t) \geq 0$.

69. Suppose the vertex of the parabola associated with a certain quadratic function is $(2, 1)$, and another point on this parabola is $(3, -1)$.

(a) Find the equation of the axis of symmetry of the parabola.

(b) Use symmetry to find a third point on the parabola.

(c) Sketch the parabola.

70. Examine the following table of values for a quadratic function f.

x	$f(x)$
-2	3
-1	0
0	-1
1	0
2	3

(a) What is the equation of the axis of symmetry of the associated parabola? Justify your answer.

(b) Find the minimum or maximum value of the function and the value of x at which it occurs.

(c) Sketch a graph of the function from the values given in the table, and find an expression for the function.

71. Let $g(s) = -2s^2 + bs$. Find the value of b such that the vertex of the parabola associated with this function is $(1, 2)$.

▶ **Applications** In this set of exercises you will use quadratic functions to study real-world problems.

72. **Landscaping** A rectangular garden plot is to be enclosed with a fence on three of its sides and a brick wall on the fourth side. If 100 feet of fencing material is available, what dimensions will yield the maximum area?

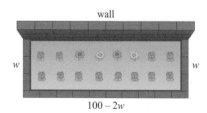

73. **Construction** A rectangular fence is being constructed around a new play area at a local elementary school. If the school has 2000 feet of fencing available for the project, what is the maximum area that can be enclosed for the new play area?

74. **Physics: Ball Height** The height of a ball that is thrown directly upward from a point 200 feet above the ground with an initial velocity of 40 feet per second is given by $h(t) = -16t^2 + 40t + 200$, where t is the amount of time elapsed since the ball was thrown. Here, t is in seconds and $h(t)$ is in feet.

(a) Sketch a graph of h.

(b) When will the ball reach its maximum height, and what is the maximum height?

75. **Physics: Ball Height** A cannonball is fired at an angle of inclination of 45° to the horizontal with a velocity of 50 feet per second. The height h of the cannonball is given by

$$h(x) = \frac{-32x^2}{(50)^2} + x$$

where x is the horizontal distance of the cannonball from the end of the cannon.

(a) How far away from the cannon should a person stand if the person wants to be directly below the cannonball when its height is maximum?

(b) What is the maximum height of the cannonball?

76. **Manufacturing** A carpenter wishes to make a rain gutter with a rectangular cross-section by bending up a flat piece of metal that is 18 feet long and 20 inches wide. The top of the gutter is open.

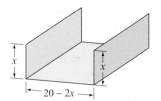

(a) Write an expression for the cross-sectional area in terms of x, the length of metal that is bent upward.

(b) How much metal has to be bent upward to maximize the cross-sectional area? What is the maximum cross-sectional area?

The quadratic function

$$p(x) = -0.387(x - 45)^2 + 2.73(x - 45) - 3.89$$

gives the percentage (in decimal form) of puffin eggs that hatch during a breeding season in terms of x, the sea surface temperature of the surrounding area, in degrees Fahrenheit. Use this information to solve Problems 77 and 78. (Source: Gjerdum et al., *"Tufted puffin reproduction reveals ocean climate variability,"* Proceedings of the National Academy of Sciences, August 2003.)

77. **Ecology** What is the percentage of puffin eggs that will hatch at 49°F? at 47°F?

78. **Ecology** For what temperature is the percentage of hatched puffin eggs a maximum? Find the percentage of hatched eggs at this temperature.

79. **Performing Arts** Attendance at Broadway shows in New York can be modeled by the quadratic function $p(t) = 0.0489t^2 - 0.7815t + 10.31$, where t is the number of years since 1981 and $p(t)$ is the attendance

in millions. The model is based on data for the years 1981–2000. (*Source:* The League of American Theaters and Producers, Inc.)

(a) Use this model to estimate the attendance in the year 1995. Compare it to the actual value of 9 million.

(b) Use this model to predict the attendance for the year 2006.

(c) What is the vertex of the parabola associated with the function p, and what does it signify in relation to this problem?

(d) Would this model be suitable for predicting the attendance at Broadway shows for the year 2025? Why or why not?

(e) Use a graphing utility to graph the function p. What is an appropriate range of values for t?

80. **Maximizing Revenue** A chartered bus company has the following price structure. A single bus ticket costs $30. For each *additional* ticket sold to a group of travelers, the price per ticket is reduced by $0.50. The reduced price applies to all the tickets sold to the group.

(a) Calculate the total cost for one, two, and five tickets.

(b) Using your calculations in part (a) as a guide, find a quadratic function that gives the total cost of the tickets.

(c) How many tickets must be sold to maximize the revenue for the bus company?

81. **Maximizing Revenue** A security firm currently has 5000 customers and charges $20 per month to monitor each customer's home for intruders. A marketing survey indicates that for each dollar the monthly fee is decreased, the firm will pick up an additional 500 customers. Let $R(x)$ represent the revenue generated by the security firm when the monthly charge is x dollars. Find the value of x that results in the maximum monthly revenue.

82. **Construction Analysis** A farmer has 400 feet of fencing material available to make two identical, adjacent rectangular corrals for the farm animals, as pictured. If the farmer wants to maximize the total enclosed area, what will be the dimensions of each corral?

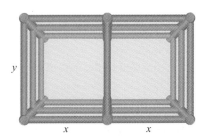

83. **Architecture** There is 24 feet of material available to trim the outside perimeter of a window shaped as shown in the diagram.

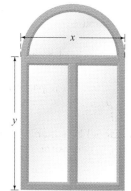

(a) Write an expression for the length of the semi-circular arc (just the curved portion) in terms of x.

(b) Set up an equation relating the amount of available trim material to the lengths of the parts of the window that are to be covered with the trim. Solve this equation for y.

(c) Write an expression for the area of the window in terms of x.

(d) What are the dimensions that will maximize the area of the window? What is the maximum area of the window?

84. **Engineering** When designing buildings, engineers must pay careful attention to how different factors affect the load a structure can bear. The following table gives the load in terms of the weight of concrete that can be borne when threaded rod anchors of various diameters are used to form joints. (*Source:* Simpson Anchor Systems)

Diameter (in.)	Load (lb)
0.3750	2105
0.5000	3750
0.6250	5875
0.7500	8460
0.8750	11,500

(a) Examine the table and explain why the relationship between the diameter and the load is *not* linear.

(b) The function

$$f(x) = 14{,}926x^2 + 148x - 51$$

gives the load (in pounds of concrete) that can be borne when rod anchors of diameter x (in inches) are employed. Use this function to determine the load for an anchor with a diameter of 0.8 inch.

(c) Since the rods are drilled into the concrete, the manufacturer's specifications sheet gives the load in terms of the diameter of the drill bit. This diameter is always 0.125 inch larger than the diameter of the anchor. Write the function in part (b) in terms of the diameter of the drill bit. The loads for the drill bits will be the same as the loads for the corresponding anchors. (*Hint:* Examine the table of values and see if you can present the table in terms of the diameter of the drill bit.)

85. **Physics** A ball is thrown directly upward from ground level at time $t = 0$ (t is in seconds). At $t = 3$, the ball reaches its maximum distance from the ground, which is 144 feet. Assume that the distance of the ball from the ground (in feet) at time t is given by a quadratic function $d(t)$. Find an expression for $d(t)$ in the form $d(t) = a(t - h)^2 + k$ by performing the following steps.

(a) From the given information, find the values of h and k and substitute them into the expression $d(t) = a(t - h)^2 + k$.

(b) Now find a. To do this, use the fact that at time $t = 0$ the ball is at ground level. This will give you an equation having just a as a variable. Solve for a.

(c) Now, substitute the value you found for a into the expression you found in part (a).

(d) Check your answer. Is (3, 144) the vertex of the associated parabola? Does the parabola pass through (0, 0)?

86. **Physics** A child kicks a ball a distance of 9 feet. The maximum height of the ball above the ground is 3 feet. If the point at which the child kicks the ball is the origin and the flight of the ball can be approximated by a parabola, find an expression for the quadratic function that models the ball's path. Check your answer by graphing the function.

87. **Design** One of the parabolic arcs of the fountain pictured has a maximum height of 5 feet. The horizontal range of the arc is 3 feet. The diagram on the left shows one of the arcs placed in an x-y coordinate system. Find an expression for the quadratic function associated with the parabola shown in the diagram. Graph your function as a check.

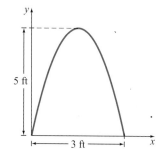

88. **Sociology** The percentage of employed mothers of preschoolers who used a nursery or preschool as the primary source of child care is listed below for selected years between 1990 and 2000. (*Source:* U.S. Census Bureau)

Year	Percentage Using Nursery/ Preschool for Child Care
1990	6.9
1991	7.3
1995	5.9
1997	4.2
2000	3.8

(a) Make a scatter plot of the data, using the number of years since 1990 as the independent variable. Explain why a quadratic function would be a better fit than a linear function for this set of data.

(b) Find the quadratic function that best fits the given data points.

(c) Use the function to predict the percentage of employed mothers using a preschool or nursery for child care in the year 2001.

(d) Compare the actual percentages for the given years with the values predicted by your function. Comment on the reliability of your function as a model.

89. **Accounting** The table below gives the number of electronically filed tax returns for the calendar years listed. (*Source:* Internal Revenue Service)

Year	1995	1998	2000	2003	2005
Number of Returns (in millions)	11.8	24.6	35.4	52.9	68.2

(a) Make a scatter plot of the data, using the number of years since 1995 as the independent variable.

(b) Find the quadratic function that best fits the given data points.

(c) Compare the actual numbers of electronically filed tax returns for the given years with the numbers predicted by your function.

(d) Use your function to determine the number of tax returns filed electronically in the year 2004.

(e) The IRS reported a total of 61.2 million tax returns filed electronically in 2004. How does this number compare with your answer in part (d)?

▶**Concepts** This set of exercises will draw on the ideas presented in this section and your general math background.

90. Why must we have $a \neq 0$ in the definition of a quadratic function?

91. Name at least two features of a quadratic function that differ from those of a linear function.

92. Which of the following points lie(s) on the parabola associated with the function $f(s) = -s^2 + 6$? Justify your answer.
(a) $(3, -1)$ (b) $(0, 6)$ (c) $(2, 1)$

93. Suppose that the vertex and an x-intercept of the parabola associated with a certain quadratic function are given by $(-1, 2)$ and $(4, 0)$, respectively.
(a) Find the other x-intercept.
(b) Find the equation of the parabola.
(c) Check your answer by graphing the function.

94. The range of a quadratic function $g(x) = ax^2 + bx + c$ is given by $(-\infty, 2]$. Is a positive or negative? Justify your answer.

95. A parabola associated with a certain quadratic function f has the point $(2, 8)$ as its vertex and passes through the point $(4, 0)$. Find an expression for $f(x)$ in the form $f(x) = a(x - h)^2 + k$.
(a) From the given information, find the values of h and k.
(b) Substitute the values you found for h and k into the expression $f(x) = a(x - h)^2 + k$.
(c) Now find a. To do this, use the fact that the parabola passes through the point $(4, 0)$. That is, $f(4) = 0$. You should get an equation having just a as a variable. Solve for a.
(d) Substitute the value you found for a into the expression you found in part (b).
(e) Graph the function using a graphing utility and check your answer. Is $(2, 8)$ the vertex of the parabola? Does the parabola pass through $(4, 0)$?

96. Is it possible for a quadratic function to have the set of all real numbers as its range? Explain. (*Hint:* Examine the graph of a general quadratic function.)

3.2 Quadratic Equations

Objectives

► Solve a quadratic equation by factoring

► Find the real zeros of a quadratic function and the *x*-intercepts of its graph

► Solve a quadratic equation by completing the square

► Solve a quadratic equation by using the quadratic formula

► Set up and solve applied problems involving quadratic equations

---

Just In Time

Review factoring in Section P.5.

In the previous section, we discussed the general form of a quadratic function, its graph, and some of its key features. We continue our discussion of quadratic functions by examining solutions of equations containing quadratic expressions.

Definition of a Quadratic Equation

A **quadratic equation** is an equation that can be written in the **standard form**

$$ax^2 + bx + c = 0$$

where a, b, and c are real numbers, with

$$a \neq 0.$$

Solving a Quadratic Equation by Factoring

One of the simplest ways to solve a quadratic equation is by factoring. We need the following rule to justify our procedure for solving equations by factoring.

Zero Product Rule

If a product of real numbers is zero, then at least one of the factors is zero. That is,

$$\text{if } cd = 0, \text{ then } c = 0 \quad \text{or} \quad d = 0.$$

Technology Note

You can use a graphing utility to check that $x = \frac{1}{2}$ and $x = 3$ satisfy $-2x^2 + 7x - 3 = 0$ by successively storing the *x* values and then evaluating the expression $-2x^2 + 7x - 3$ at these values. See Figure 3.2.1.

Keystroke Appendix: Section 4

Figure 3.2.1

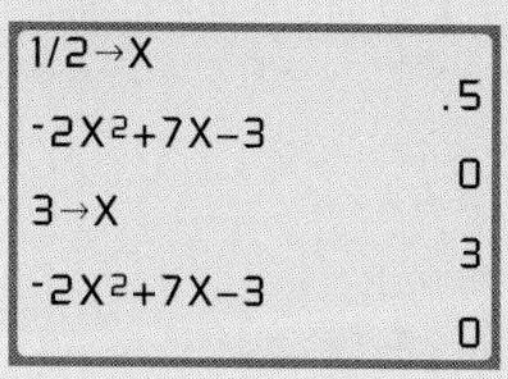

```
1/2→X
              .5
-2X²+7X-3
              0
3→X
              3
-2X²+7X-3
              0
```

Example **1** **Solving an Equation by Factoring**

Solve $-2x^2 + 7x - 3 = 0$ by factoring.

► **Solution** Factoring the left-hand side gives

$$(-2x + 1)(x - 3) = 0.$$

According to the Zero Product Rule, if the product of two factors equals zero, then at least one of the factors is equal to zero. Thus, we set each factor equal to zero and solve for x.

$$-2x + 1 = 0 \implies x = \frac{1}{2}$$

or

$$x - 3 = 0 \implies x = 3$$

The solutions of the equation are $x = \frac{1}{2}$ and $x = 3$. You should check that these values satisfy the original equation.

✔ *Check It Out 1:* Solve $5x^2 - 3x - 2 = 0$ by factoring. ▪

Finding the Zeros of a Quadratic Function and the *x*-Intercepts of Its Graph

The solution of a quadratic equation is connected to certain properties of quadratic functions. We next explore these relationships. A **real zero** of a function is defined as follows.

> ### Definition of Real Zeros
>
> The real number values of x at which $f(x) = 0$ are called the **real zeros** of the function f.

Recall that the points at which the graph of a function crosses the *x*-axis are called **x-intercepts.** The *x*-coordinate of an *x*-intercept is a value of x such that $f(x) = 0$. Thus, a real zero of the function f is the first coordinate (the *x*-coordinate) of an *x*-intercept. The graph of a quadratic function may have one, two, or no *x*-intercepts, as illustrated in Figure 3.2.2.

Figure 3.2.2 Number of *x*-intercepts

Graphs of $f(x) = ax^2 + bx + c$

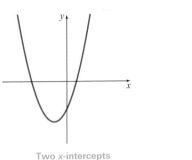

Two *x*-intercepts

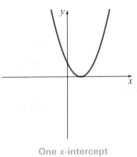

One *x*-intercept

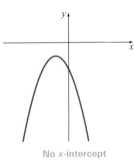

No *x*-intercept

To calculate the *x*-intercepts, we proceed as follows.

Procedure for finding the *x*-intercepts of the graph of f:

▶ Set $f(x)$ equal to zero.

▶ Solve the resulting equation for x.

▶ Each solution is the *x*-coordinate of an *x*-intercept of the graph of f. The *y*-coordinate of an *x*-intercept is always 0.

The second step above—solving the equation—is the one that requires most of the work. If a quadratic expression is easy to factor, then we can find the *x*-intercepts and zeros very quickly. The following example illustrates the use of factoring to find the *x*-intercepts of the graph of a quadratic function.

Just In Time

Review factoring in Section P.5.

Example **2** **Relating *x*-Intercepts to Zeros**

Find the *x*-intercepts of the parabola associated with the function $f(x) = 3x^2 - 5x - 2$. Also find the zeros of f.

▶**Solution** We set $f(x) = 0$ and solve the resulting equation.

$$3x^2 - 5x - 2 = 0 \qquad \text{Set the function equal to zero}$$

$$(3x + 1)(x - 2) = 0 \qquad \text{Factor the left-hand side}$$

Figure 3.2.3

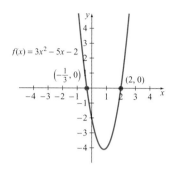

$f(x) = 3x^2 - 5x - 2$

$\left(-\frac{1}{3}, 0\right)$ $(2, 0)$

The product of two factors will equal zero if at least one of the factors is equal to zero. Thus we set each factor equal to zero and solve for x.

$$3x + 1 = 0 \implies x = -\frac{1}{3} \quad \text{or} \quad x - 2 = 0 \implies x = 2$$

Hence, $3x^2 - 5x - 2 = 0$ if $x = -\frac{1}{3}$ or $x = 2$. The x-intercepts are therefore $\left(-\frac{1}{3}, 0\right)$ and $(2, 0)$. The graph of the function crosses the x-axis at these points and no others, as seen in Figure 3.2.3. The zeros of f are the values of x such that $f(x) = 0$. We see that these values are $x = -\frac{1}{3}$ and $x = 2$.

✔ *Check It Out 2:* Find the x-intercepts of the parabola associated with the function $f(x) = 2x^2 - 3x - 2$, and find the zeros of f. ▪

Observations:

▶ The zeros of a function are just values of x, whereas x-intercepts are coordinate pairs.

▶ Finding the solutions of the quadratic equation $ax^2 + bx + c = 0$ involves exactly the same procedure as finding the zeros of the quadratic function $f(x) = ax^2 + bx + c$. It is important that you see this connection between the zeros of a function and the solutions of a related equation.

In Example 2, each of the zeros of the function f is also the first coordinate of an x-intercept of the graph of f. However, as we shall see in a later section, there are quadratic functions whose graphs have no x-intercepts, even though the functions themselves have zeros. The zeros of such functions are not *real* zeros (real numbers x at which $f(x) = 0$). In order to treat quadratic functions of this type, in Section 3.3 we will introduce a set of numbers known as *complex numbers*.

Figure 3.2.4 Two different functions with the same zeros

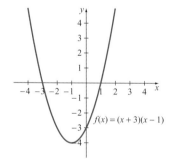

$f(x) = (x + 3)(x - 1)$

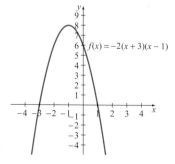

$f(x) = -2(x + 3)(x - 1)$

Example 3 **Zeros and Factors of a Quadratic Function**

Find a possible expression for a quadratic function $f(x)$ with zeros at $x = -3$ and $x = 1$.

▶Solution Since the zeros are obtained by factoring the expression for $f(x)$ and setting it equal to zero, we can work backward by noting that

$$x = -3 \implies x + 3 = 0$$

and

$$x = 1 \implies x - 1 = 0.$$

Thus the factors for $f(x)$ are $x + 3$ and $x - 1$. So,

$$f(x) = (x + 3)(x - 1)$$
$$= x^2 + 2x - 3.$$

This is not the only possible expression for f. Any function of the form

$$a(x + 3)(x - 1), \quad a \neq 0$$

would have the same zeros. Two possible functions are pictured in Figure 3.2.4.

✔ *Check It Out 3:* Find a possible expression for a quadratic function $f(x)$ with zeros at $x = 2$ and $x = -4$. ▪

The previous example illustrates an important connection between the zeros and factors of a quadratic function and the x-intercepts of its graph. This connection will resurface later in the study of polynomial functions.

Solving a Quadratic Equation by Completing the Square

In Example 2, we found the solutions of $3x^2 - 5x - 2 = 0$ by factoring. Unfortunately, not all equations of the form $ax^2 + bx + c = 0$ can be solved by factoring. In this section, we will discuss a general method that can be used to solve all quadratic equations. We first need the following rule regarding square roots.

Principle of Square Roots

If $x^2 = c$, where $c \geq 0$, then $x = \sqrt{c}$ or $x = -\sqrt{c}$.

Example 4 Using the Principle of Square Roots

Solve $-3x^2 + 9 = 0$.

▶**Solution** We use the principle of square roots to solve this equation.

$$-3x^2 + 9 = 0$$
$$-3x^2 = -9 \qquad \text{Subtract 9 from both sides}$$
$$x^2 = 3 \qquad \text{Divide by } -3 \text{ on both sides}$$
$$x = \sqrt{3} \quad \text{or} \quad x = -\sqrt{3} \qquad \text{Apply the principle of square roots}$$

✔ *Check It Out 4:* Solve $4x^2 - 20 = 0$. ▪

Note The principle of square roots can be used to solve a quadratic equation only when the quadratic equation can be rewritten in the form $x^2 = c$. One side of the equation must be a constant, and the other side must be a perfect square.

Just In Time

Review perfect square trinomials in Section P.5.

We now introduce the method of *completing the square* as a tool for solving *any* type of quadratic equation. Recall that completing the square was discussed in Section 3.1, when we discussed writing a quadratic function in vertex form. Since the expression $3x^2 - 6x - 1$ cannot be factored easily, we will use the method of completing the square to solve the equation $3x^2 - 6x - 1 = 0$, as seen in the following example.

Just In Time

See more examples on completing the square in Section 2.1.

Example 5 Solving by Completing the Square

Solve the equation $3x^2 - 6x - 1 = 0$ by using the technique of completing the square. What are the zeros of $f(x) = 3x^2 - 6x - 1$, and what are the x-intercepts of the graph of the function f?

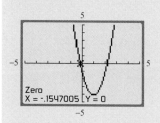

Technology Note

You can solve the equation in Example 5 by using the ZERO feature of a graphing utility. The function $Y_1(x) = 3x^2 - 6x - 1$ is graphed in Figure 3.2.5. The solutions to $Y_1(x) = 0$ are $x \approx 2.1547$ and $x \approx -0.1547$.

Keystroke Appendix:
Section 9

Figure 3.2.5

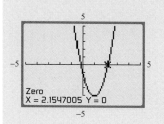

▶ Solution

$3x^2 - 6x - 1 = 0$	Start with the equation
$3x^2 - 6x = 1$	Move the constant to the right side
$x^2 - 2x = \dfrac{1}{3}$	Divide by 3 to get the coefficient of x^2 equal to 1
$x^2 - 2x + 1 = \dfrac{1}{3} + 1$	Complete the square by taking half of -2 and squaring it: $\left(\dfrac{1}{2}(-2)\right)^2 = 1$. Add 1 to both sides.
$(x - 1)^2 = \dfrac{4}{3}$	Write the left side as a perfect square
$x - 1 = \pm \sqrt{\dfrac{4}{3}}$	Use the principle of square roots
$x = 1 \pm \sqrt{\dfrac{4}{3}}$	Solve for x
$x = 1 \pm \dfrac{2\sqrt{3}}{3}$	Simplify the radical

Thus, $x = 1 + \dfrac{2\sqrt{3}}{3} \approx 2.155$ and $x = 1 - \dfrac{2\sqrt{3}}{3} \approx -0.1547$ are the two solutions of the equation. They are also the two zeros of the function $f(x) = 3x^2 - 6x - 1$. The x-intercepts of the graph of this function are $\left(1 + \dfrac{2\sqrt{3}}{3}, 0\right)$ and $\left(1 - \dfrac{2\sqrt{3}}{3}, 0\right)$.

☑ **Check It Out 5:** Solve the equation $2x^2 - 4x - 1 = 0$ by using the technique of completing the square. What are the zeros of $f(x) = 2x^2 - 4x - 1$, and what are the x-intercepts of the graph of the function f? ■

Solving a Quadratic Equation by Using the Quadratic Formula

We now derive a general formula for solving quadratic equations. In the following derivation, we assume $a > 0$. If $a < 0$, we can multiply the equation by -1 and obtain a positive coefficient on x^2.

$ax^2 + bx + c = 0$	Quadratic equation
$a\left(x^2 + \dfrac{b}{a}x\right) + c = 0$	Factor a out of the first two terms on the left side
$a\left[x^2 + \dfrac{b}{a}x + \left(\dfrac{b}{2a}\right)^2\right] - a\left(\dfrac{b}{2a}\right)^2 + c = 0$	Complete the square on $x^2 + \dfrac{b}{a}x$

Since $x^2 + \dfrac{b}{a}x + \left(\dfrac{b}{2a}\right)^2 = \left(x + \dfrac{b}{2a}\right)^2$ and $a\left(\dfrac{b}{2a}\right)^2 = \dfrac{b^2}{4a}$, we have

$$a\left(x + \dfrac{b}{2a}\right)^2 + \left(c - \dfrac{b^2}{4a}\right) = 0$$

$$a\left(x + \dfrac{b}{2a}\right)^2 = -\left(c - \dfrac{b^2}{4a}\right)$$

$$a\left(x + \dfrac{b}{2a}\right)^2 = \dfrac{b^2 - 4ac}{4a} \qquad \text{Simplify the right-hand side}$$

$$\left(x + \dfrac{b}{2a}\right)^2 = \dfrac{b^2 - 4ac}{4a^2} \qquad \text{Divide by } a$$

$$x + \dfrac{b}{2a} = \pm\sqrt{\dfrac{b^2 - 4ac}{4a^2}} \qquad \begin{array}{l}\text{Take the square roots}\\\text{of both sides}\end{array}$$

$$x = -\dfrac{b}{2a} \pm \sqrt{\dfrac{b^2 - 4ac}{4a^2}} \qquad \text{Subtract } \dfrac{b}{2a}$$

$$x = -\dfrac{b}{2a} \pm \dfrac{\sqrt{b^2 - 4ac}}{2a}. \qquad \begin{array}{l}\text{Simplify under the radical:}\\\sqrt{4a^2} = 2a, \text{ since } a > 0\end{array}$$

The Quadratic Formula

The solutions of $ax^2 + bx + c = 0$, with $a \neq 0$, are given by the **quadratic formula**

$$x = \dfrac{-b \pm \sqrt{b^2 - 4ac}}{2a}.$$

These solutions are the **zeros** of the quadratic function

$$f(x) = ax^2 + bx + c.$$

Note Since the quadratic formula gives all the solutions of a quadratic equation, we see that a quadratic equation can have at most two solutions.

To solve a quadratic equation, perform the following steps.

▸ Write the equation in the form $ax^2 + bx + c = 0$.

▸ If possible, factor the left-hand side of the equation to find the solution(s).

▸ If it is not possible to factor, use the quadratic formula to solve the equation. Alternatively, you can complete the square.

Example 6 **Solving an Equation by Using the Quadratic Formula**

Solve the equation $-4x^2 + 3x + \frac{1}{2} = 0$ by using the quadratic formula. What are the zeros of $f(x) = -4x^2 + 3x + \frac{1}{2}$ and the x-intercepts of the graph of f?

▶**Solution** Since the expression $-4x^2 + 3x + \frac{1}{2}$ cannot be readily factored, we use the quadratic formula.

$$x = \frac{-(3) \pm \sqrt{(3)^2 - 4(-4)\left(\frac{1}{2}\right)}}{2(-4)}$$ Substitute $a = -4$, $b = 3$, and $c = \frac{1}{2}$ in the formula

$$= \frac{-3 \pm \sqrt{9 + 8}}{-8}$$ $-(4)(-4)\left(\frac{1}{2}\right) = 8$

$$= \frac{-3 \pm \sqrt{17}}{-8}$$ Simplify under the radical

$$= \frac{3}{8} \pm \frac{\sqrt{17}}{8}$$

Figure 3.2.6

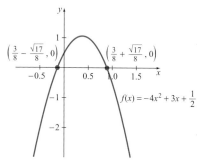

$\left(\frac{3}{8} - \frac{\sqrt{17}}{8}, 0\right)$ $\left(\frac{3}{8} + \frac{\sqrt{17}}{8}, 0\right)$

$f(x) = -4x^2 + 3x + \frac{1}{2}$

The solutions of the equation are $x = \frac{3}{8} + \frac{\sqrt{17}}{8}$ and $x = \frac{3}{8} - \frac{\sqrt{17}}{8}$. Also, the two real zeros of $f(x) = -4x^2 + 3x + \frac{1}{2}$ are $x = \frac{3}{8} + \frac{\sqrt{17}}{8}$ and $x = \frac{3}{8} - \frac{\sqrt{17}}{8}$.

Since the first coordinate of each of the x-intercepts of the graph of f is a real zero of f, we see that the x-intercepts are $\left(\frac{3}{8} + \frac{\sqrt{17}}{8}, 0\right)$ and $\left(\frac{3}{8} - \frac{\sqrt{17}}{8}, 0\right)$. These are indicated in the graph of f in Figure 3.2.6.

☑ *Check It Out 6:* Solve $2x^2 - 4x - 1 = 0$ by using the quadratic formula. What are the zeros of $g(x) = 2x^2 - 4x - 1$ and the x-intercepts of the graph of g? ■

In the quadratic formula, the quantity under the radical, $b^2 - 4ac$, can be positive, negative, or zero. The characteristics of the solutions in each of these three cases will be different, as summarized below.

Types of Solutions of a Quadratic Equation and the Discriminant

The quadratic formula contains the quantity $b^2 - 4ac$ under the radical; this quantity is known as the **discriminant.** Solutions of quadratic equations that are real numbers are called **real solutions.** The number of real solutions of a particular quadratic equation depends on the sign of the discriminant:

▶ If $b^2 - 4ac > 0$, there will be *two* distinct real solutions.

▶ If $b^2 - 4ac = 0$, there will be *one* real solution.

▶ If $b^2 - 4ac < 0$, there will be *no* real solutions.

Solutions that are not real numbers are called *nonreal* solutions; these are part of the complex number system discussed in Section 3.3.

The connection between the sign of the discriminant and the number of x-intercepts is illustrated in Figure 3.2.7. Recall that the first coordinate of each of the x-intercepts of the graph of the quadratic function $f(x) = ax^2 + bx + c$ corresponds to a real zero of f.

Figure 3.2.7 Relationship between discriminant and x-intercepts

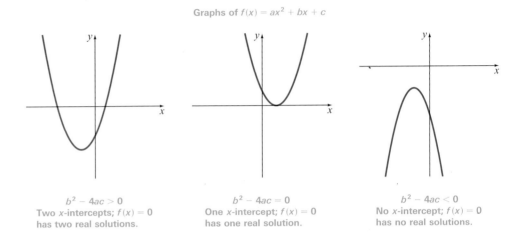

Graphs of $f(x) = ax^2 + bx + c$

$b^2 - 4ac > 0$
Two x-intercepts; $f(x) = 0$
has two real solutions.

$b^2 - 4ac = 0$
One x-intercept; $f(x) = 0$
has one real solution.

$b^2 - 4ac < 0$
No x-intercept; $f(x) = 0$
has no real solutions.

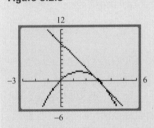

Technology Note

If you use the INTERSECT feature of a graphing utility to find the solution to Example 7, you may get an error message on some calculator models. (See Figure 3.2.8.) This is because of the limitations of the calculator algorithm. You can rewrite the equation as $Y_1(x) = 0$ and use the ZERO feature instead.

Keystroke Appendix: Section 9

Figure 3.2.8

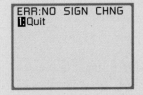

Example 7 The Quadratic Formula and the Discriminant

Use the quadratic formula to solve $-x(x - 3) = -3x + 9$. Find the value of the discriminant.

▶**Solution** We first rewrite the equation in standard form so that we can apply the quadratic formula.

$$-x(x - 3) = -3x + 9 \qquad \text{Given equation}$$
$$-x^2 + 3x = -3x + 9 \qquad \text{Distributive property}$$
$$-x^2 + 6x - 9 = 0 \qquad \text{Add 3x and subtract 9 on both sides}$$

Since the original equation has now been rewritten in standard form, we apply the quadratic formula with $a = -1$, $b = 6$, and $c = -9$.

$$x = \frac{-b \pm \sqrt{b^2 - 4ac}}{2a}$$

$$= \frac{-(6) \pm \sqrt{(6)^2 - 4(-1)(-9)}}{2(-1)} \qquad \text{Substitute } a = -1, b = 6, \text{ and } c = -9$$

$$= \frac{-6 \pm \sqrt{36 - 36}}{-2} \qquad \text{Simplify, using care with the signs}$$

$$= 3$$

Thus, we have only one solution of the quadratic equation, $x = 3$. The value of the discriminant is $b^2 - 4ac = (-6)^2 - 4(-1)(-9) = 0$, as we would expect.

☑ *Check It Out 7:* Use the quadratic formula to solve $x(x - 2) = 2x - 4$. ▪

Technology Note

Finding a suitable window size can be challenging when using a graphing utility for applications. A table of values is helpful, as is an understanding of the application itself. For Example 9, set

$Y_1(x) = -16x^2 + 40x + 80$.

Scroll through the table of values for $X \geq 0$. An acceptable window size is [0, 5] by [0, 120](10). Graph the function and then use the ZERO feature to find the time when the ball reaches the ground. See Figure 3.2.9.

Keystroke Appendix:
Sections 6, 9

Figure 3.2.9

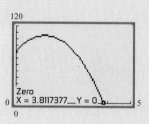

| Example | 8 | **The Quadratic Formula and Nonreal Solutions** |

Use the quadratic formula to find the real solutions of the equation $-2t^2 + 3t = 5$. Find the value of the discriminant.

▶Solution Writing the given equation as $-2t^2 + 3t - 5 = 0$ and applying the quadratic formula gives

$$t = \frac{-(3) \pm \sqrt{(3)^2 - 4(-2)(-5)}}{2(-2)} \quad \text{Substitute } a = -2, b = 3, \text{ and } c = -5$$

$$= \frac{-3 \pm \sqrt{-31}}{-4}. \quad \text{Simplify, using care with the signs}$$

The discriminant, $b^2 - 4ac = -31$, is negative. Thus, there are no real number solutions of this equation. Solutions of quadratic equations in which the discriminant is negative can be found only by using complex numbers, discussed in Section 3.3.

✔ *Check It Out 8:* Use the quadratic formula to solve $x^2 - x + 3 = 0$. Find the value of the discriminant. ▪

Applications

Quadratic equations arise frequently in applications, some of which we now discuss.

| Example | 9 | **Quadratic Model for the Height of a Baseball in Flight** |

The height of a ball thrown vertically upward from a point 80 feet above the ground with a velocity of 40 feet per second is given by $h(t) = -16t^2 + 40t + 80$, where t is the time in seconds since the ball was thrown and $h(t)$ is in feet.

(a) When will the ball be 50 feet above the ground?

(b) When will the ball reach the ground?

(c) For what values of t does this problem make sense (from a physical standpoint)?

▶Solution

(a) Setting $h(t) = 50$ and solving for t, we have

$$-16t^2 + 40t + 80 = 50$$

$$-16t^2 + 40t + 30 = 0 \quad \text{Write in standard form}$$

$$t = \frac{-(40) \pm \sqrt{(40)^2 - 4(-16)(30)}}{2(-16)} \approx 3.10 \text{ or } -0.604. \quad \text{Use the quadratic formula}$$

Since a negative number makes no sense for a value of time, the ball will be 50 feet above the ground in $t = 3.10$ seconds.

(b) When the ball reaches the ground, the height $h(t)$ will be zero. Thus, setting $h(t) = 0$ and solving for t gives

$$-16t^2 + 40t + 80 = 0$$

$$t = \frac{-(40) \pm \sqrt{(40)^2 - 4(-16)(80)}}{2(-16)} \approx 3.81 \text{ or } -1.31.$$

Since a negative number makes no sense for a value of time, the ball will reach the ground in $t \approx 3.81$ seconds.

(c) The problem makes sense for values of t in the interval $[0, 3.81]$. The time t must be greater than or equal to zero. Once the ball hits the ground, the motion of the ball is no longer governed by the given expression.

✔ *Check It Out 9:* In Example 9, when will the ball be 40 feet above the ground?
◼

Example 10 Quadratic Model for Population Mobility

The mobility rate (the percentage of people who changed residence) for the years 1980–2000 can be modeled by the function $m(t) = 0.0109t^2 - 0.34t + 18.54$, where t is the number of years since 1980. (*Source:* U.S. Census Bureau)

(a) What is the y-intercept of this function, and what does it represent?

(b) In what year between 1980 and 2000 was the mobility rate 17%?

▶Solution

(a) Substituting $t = 0$ into the expression for $m(t)$, we see that the y-intercept is $(0, 18.54)$. Since t is the number of years since 1980, the y-intercept tells us that 18.54% of the population changed residence in 1980.

(b) Setting $m(t) = 17$ and solving for t, we have

$$0.0109t^2 - 0.34t + 18.54 = 17$$

$$0.0109t^2 - 0.34t + 1.54 = 0 \qquad \text{Write in standard form}$$

$$t = \frac{-(-0.34) \pm \sqrt{(-0.34)^2 - 4(0.0109)(1.54)}}{2(0.0109)} \approx 5.50 \quad \text{or} \quad 25.7.$$

Since we are interested only in the years between 1980 and 2000, we take the first solution of $t \approx 5.5$. Thus, the mobility rate was 17% a little after 1985.

✔ *Check It Out 10:* Consider Example 10. In what year(s) between 1980 and 2000 was the mobility rate 16%? ◼

3.2 Key Points

▶ The points at which the graph of a quadratic function crosses the x-axis are known as **x-intercepts.**

▶ The x-coordinates of the x-intercepts are found by setting $f(x) = 0$ and solving for x.

▶ Values of x such that $f(x) = 0$ are called **zeros** of f.

▶ **Relationship between the real zeros of f and the x-intercepts of the graph of f:** Every real zero of f is the x-coordinate of an x-intercept of the graph of f, and the x-coordinate of every x-intercept of the graph of f is a real zero of f. See Figure 3.2.2.

▶ You can find the zeros by factoring, by completing the square, or by using the quadratic formula.

▶ A **quadratic equation** can be solved by writing it in the form $ax^2 + bx + c = 0$ and then applying the technique of factoring, completing the square, or using the quadratic formula.

▶ The solutions of the quadratic equation $ax^2 + bx + c = 0$ are the same as the zeros of the function $f(x) = ax^2 + bx + c$.

3.2 Exercises

▶**Just in Time Exercises** These exercises correspond to the Just in Time references in this section. Complete them to review topics relevant to the remaining exercises.

1. Factor: $x^2 - 13x + 40$

2. Factor: $2x^2 - 9x - 35$

3. Factor: $x^2 - 2x + 1$

4. Factor: $4x^2 + 12x + 9$

5. Find the constant term needed to make $x^2 - 6x$ a perfect square trinomial.

6. Find the constant term needed to make $x^2 + 7x$ a perfect square trinomial.

7. Write the expression in the form $(ax + b)^2$: $x^2 - 8x + 16$

8. Write the expression in the form $(ax + b)^2$: $9x^2 - 30x + 25$

▶**Skills** This set of exercises will reinforce the skills illustrated in this section.

In Exercises 9–18, solve the quadratic equation by factoring.

9. $x^2 - 25 = 0$

10. $x^2 - 16 = 0$

11. $x^2 - 7x + 12 = 0$

12. $x^2 - 4x - 21 = 0$

13. $-3x^2 + 12 = 0$

14. $-5x^2 + 45 = 0$

15. $6x^2 - x - 2 = 0$

16. $5x^2 - 7x - 6 = 0$

17. $4x^2 - 4x + 1 = 0$

18. $9x^2 + 6x + 1 = 0$

In Exercises 19–26, factor to find the x-intercepts of the parabola described by the quadratic function. Also find the real zeros of the function.

19. $g(x) = x^2 - 9$

20. $f(t) = -t^2 + 4t$

21. $h(s) = -s^2 + 2s - 1$

22. $f(x) = x^2 + 4x + 4$

23. $g(x) = 2x^2 + 5x - 3$

24. $f(x) = 6x^2 - x - 2$

25. $G(t) = 2t^2 - t - 3$

26. $h(t) = -3t^2 + 10t - 8$

In Exercises 27–32, find a possible expression for a quadratic function $f(x)$ having the given zeros. There can be more than one correct answer.

27. $x = 1$ and $x = -3$

28. $x = -2$ and $x = 4$

29. $x = -3$ and $x = 0$

30. $x = -5$ is the only zero

31. $x = \dfrac{1}{2}$ and $x = 3$

32. $x = 0.4$ and $x = 0.8$

In Exercises 33–40, solve the quadratic equation by completing the square.

33. $x^2 + 4x = -3$

34. $x^2 - 6x = 7$

35. $x^2 - 2x = 4$

36. $x^2 + 8x = 6$

37. $x^2 + x = 2$

38. $x^2 - x = 3$

39. $2x^2 + 8x - 1 = 0$

40. $3x^2 - 6x + 2 = 0$

In Exercises 41–54, solve the quadratic equation by using the quadratic formula. Find only real solutions.

41. $x^2 + 2x - 1 = 0$

42. $x^2 + x - 5 = 0$

43. $-2x^2 + 2x + 1 = 0$

44. $2t^2 + 4t - 5 = 0$

45. $3 - x - x^2 = 0$

46. $-2 + t^2 + t = 0$

47. $2x^2 + x + 2 = 0$

48. $-3x^2 + 2x - 1 = 0$

49. $-l^2 + 40l = 100$

50. $-x^2 + 50x = 300$

51. $\dfrac{1}{2}t^2 - 4t - 3 = 0$

52. $-\dfrac{1}{3}x^2 - 3x + 9 = 0$

53. $-0.75x^2 + 2 = 2x$

54. $0.25x^2 - 0.5x = 1$

In Exercises 55–64, solve the quadratic equation using any method. Find only real solutions.

55. $x^2 - 4 = 0$

56. $x^2 - 9 = 0$

57. $-x^2 + 2x = 1$

58. $x^2 - 4x = -4$

59. $-2x^2 - 1 = 3x$

60. $-3x^2 - 2 = 7x$

61. $x^2 - 2x = 9$

62. $-x^2 - 3x = 1$

63. $(x - 1)(x + 2) = 1$

64. $(x + 1)(x - 2) = 2$

In Exercises 65–70, for each function of the form $f(x) = ax^2 + bx + c$, find the discriminant, $b^2 - 4ac$, and use it to determine the number of x-intercepts of the graph of f. Also determine the number of real solutions of the equation $f(x) = 0$.

65. $f(x) = x^2 - 2x - 1$

66. $f(x) = -x^2 + x + 3$

67. $f(x) = 2x^2 + x + 1$

68. $f(x) = 3x^2 - 4x + 4$

69. $f(x) = x^2 + 2x + 1$

70. $f(x) = -x^2 + 4x - 4$

In Exercises 71–76, sketch a graph of the quadratic function, indicating the vertex, the axis of symmetry, and any x-intercepts.

71. $G(x) = -6x + x^2 + 5$

72. $h(t) = -5t + 3 - t^2$

73. $F(s) = -2s^2 + 3s + 1$

74. $g(t) = 3t^2 - 6t - \dfrac{3}{4}$

75. $g(t) = t^2 + t + 1$

76. $f(t) = -t^2 - 1$

77. Let $f(x) = x^2 + c$.
 (a) Use a graphing utility to sketch the graph of f by choosing different values of c: some positive, some negative, and 0.
 (b) For what value(s) of c will the graph of f have two x-intercepts? Justify your answer.
 (c) For what value(s) of c will f have two real zeros? How is your answer related to part (a)?
 (d) Repeat parts (b) and (c) for the cases of *one* x-intercept and *one* real zero, respectively.
 (e) Repeat parts (b) and (c) for the cases of *no* x-intercepts and *no* real zeros, respectively.

78. Use the *intersect* feature of your graphing calculator to explore the real solution(s), if any, of $x^2 = x + k$ for $k = 0$, $k = -\dfrac{1}{4}$, and $k = -3$. Also use the *zero* feature to explore the solution(s). Relate your observations to the quadratic formula.

▶**Applications** In this set of exercises you will use quadratic equations to study real-world problems.

79. **Physics: Ball Height** The height of a ball after being dropped from a point 100 feet above the ground is given by $h(t) = -16t^2 + 100$, where t is the time in seconds since the ball was dropped, and $h(t)$ is in feet.
 (a) When will the ball be 60 feet above the ground?
 (b) When will the ball reach the ground?
 (c) For what values of t does this problem make sense (from a physical standpoint)?

80. **Physics** The height of a ball after being dropped from the roof of a 200-foot-tall building is given by $h(t) = -16t^2 + 200$, where t is the time in seconds since the ball was dropped, and $h(t)$ is in feet.
 (a) When will the ball be 100 feet above the ground?
 (b) When will the ball reach the ground?
 (c) For what values of t does this problem make sense (from a physical standpoint)?

81. **Manufacturing** A carpenter wishes to make a rain gutter with an open top and a rectangular cross-section by bending up a flat piece of metal that is 18 feet long and

20 inches wide. The top of the gutter is open. How much metal has to be bent upward to obtain a cross-sectional area of 30 square inches?

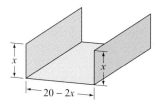

82. **Landscaping** A rectangular garden plot is to be enclosed with a fence on three of its sides and an existing wall on the fourth side. There is 45 feet of fencing material available.
(a) Write an equation relating the amount of available fencing material to the lengths of the three sides that are to be fenced.
(b) Use the equation in part (a) to write an expression for the width of the enclosed region in terms of its length.
(c) For each value of the length given in the following table of possible dimensions for the garden plot, fill in the value of the corresponding width. Use your expression from part (b) and compute the resulting area. What do you observe about the area of the enclosed region as the dimensions of the garden plot are varied?

Length (feet)	Width (feet)	Total Amount of Fencing Material (feet)	Area (square feet)
5		45	
10		45	
15		45	
20		45	
30		45	
k		45	

(d) Write an expression for the area of the garden plot in terms of its length.
(e) Find the dimensions that will yield a garden plot with an area of 145 square feet.

83. **Construction** A rectangular sandbox is to be enclosed with a fence on three of its sides and a brick wall on the fourth side. If 24 feet of fencing material is available, what dimensions will yield an enclosed region with an area of 70 square feet?

84. **Construction** A rectangular plot situated along a river is to be fenced in. The side of the plot bordering the river

will not need fencing. The builder has 100 feet of fencing available.

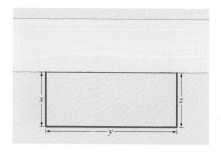

(a) Write an equation relating the amount of fencing material available to the lengths of the three sides of the plot that are to be fenced.
(b) Use the equation in part (a) to write an expression for the width of the enclosed region in terms of its length.
(c) Write an expression for the area of the plot in terms of its length.
(d) Find the dimensions that will yield the maximum area.

85. **Performing Arts** Attendance at Broadway shows in New York can be modeled by the quadratic function $p(t) = 0.0489t^2 - 0.7815t + 10.31$, where t is the number of years since 1981 and $p(t)$ is the attendance in millions of dollars. The model is based on data for the years 1981–2000. When did the attendance reach $12 million? (*Source:* The League of American Theaters and Producers, Inc.)

86. **Leisure** The average amount of money spent on books and magazines per household in the United States can be modeled by the function $r(t) = -0.2837t^2 + 5.547t + 136.7$. Here, $r(t)$ is in dollars and t is the number of years since 1985. The model is based on data for the years 1985–2000. According to this model, in what year(s) was the average expenditure per household for books and magazines equal to $160? (*Source:* U.S. Bureau of Labor Statistics)

87. **Architecture** The diagram of a window pictured below consists of a semicircle mounted on top of a square. What value of x will give a total area of 20 square feet?

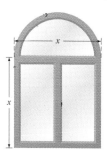

88. **Construction** A farmer has 300 feet of fencing material available to make two adjacent rectangular corrals for the farm animals, as pictured.

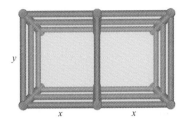

(a) Write an equation relating the amount of available fencing material to the lengths and widths of the two sections that are to be fenced.

(b) Use the equation in part (a) to write an expression for the width, y, in terms of the length, x.

(c) Fill in the following table of possible dimensions for the corrals and the total area of *both* corrals.

Length, x (feet)	Width, y (feet)	Total Amount of Fencing Material (feet)	Area (square feet)
10		300	
15		300	
25		300	
40		300	
45		300	
k		300	

(d) From the table, what do you observe about the area of the enclosed region as the dimensions of the corrals are varied?

(e) Write an expression for the total area of the enclosed corrals in terms of the length x.

(f) Find the dimensions of the corrals that will enclose a total area of 3600 square feet.

89. **Construction** A farmer has 500 feet of fencing material available to construct three adjacent rectangular corrals of equal size for the farm animals, as pictured.

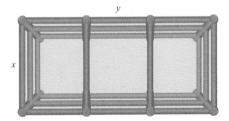

(a) Write an equation relating the amount of fencing material available to the lengths and widths of the three sections that are to be fenced.

(b) Use the equation in part (a) to write an expression for the width, y, in terms of the length, x.

(c) Write an expression for the total area of the enclosed corrals in terms of the length x.

(d) Find the dimensions of the corrals that will enclose the maximum area.

90. **Architecture** The perimeter of the window below is 24 feet. Find the values of x and y if the area of the window is to be 40 square feet.

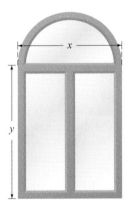

91. **Design** The shape of the Gateway Arch in St. Louis, Missouri, can be *approximated* by a parabola. (The actual shape will be discussed in an exercise in the chapter on exponential functions.) The highest point of the arch is approximately 625 feet above the ground, and the arch has a (horizontal) span of approximately 600 feet.

(a) Set up a coordinate system with the origin at the midpoint of the base of the arch. What are the x-intercepts of the parabola, and what do they represent?

(b) What do the x- and y-coordinates of the vertex of the parabola represent?

(c) Write an expression for the quadratic function associated with the parabola. (*Hint:* Use the vertex form of a quadratic function and find the coefficient a.)

(d) What is the y-coordinate of a point on the arch whose (horizontal) distance from the axis of symmetry of the parabola is 100 feet?

92. **Engineering** One section of a suspension bridge has its weight evenly distributed between two beams that are 600 feet apart and rise 90 feet above the horizontal (see figure). When a cable is strung from the tops of the beams, it takes on the shape of a parabola with its center touching the roadway, as shown in the figure. Let the origin be placed at the point where the cable meets the bridge floor.

(a) Find an equation for the parabola.

(b) Find the height of the cable 100 feet from the center.

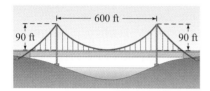

93. **Business** The following table gives the average hotel room rate for selected years from 1990 to 1999. (*Source:* American Hotel and Motel Association)

Year	Rate (in dollars)
1990	57.96
1992	58.91
1994	62.86
1996	70.93
1998	78.62
1999	81.33

(a) What general trend do you notice in these figures?

(b) Fit both a linear and a quadratic function to this set of points, using the number of years since 1990 as the independent variable.

(c) Based on your answer to part (b), which function would you use to model this set of data, and why?

(d) Using the quadratic model, find the year in which the average hotel room rate will be $85.

94. **Communications** The following table lists the total revenue, in billions of dollars, realized by long-distance carriers in the United States from 1996 to 2001. These figures are for calls placed from land-line phones; they do not include revenue realized by calls placed through wireless carriers.

Year	Revenue (billions of dollars)
1996	99.7
1997	100.8
1998	105.1
1999	108.2
2000	109.6
2001	99.3

(*Source:* Federal Communications Commission)

(a) Let t denote the number of years since 1996. Make a scatter plot of the revenue r versus time t. From your plot, what type of trend do you observe—linear or quadratic? Explain.

(b) Find a quadratic function that best fits the given data points.

(c) Find the year after 2001 in which the revenue realized from long-distance calls placed from land-line telephones will be 98 billion dollars.

(d) To what would you attribute the sharp decline in 2001 revenue from calls placed from land-line telephones?

▶**Concepts** This set of exercises will draw on the ideas presented in this section and your general math background.

The following problems can have more than one correct answer.

95. Solving the quadratic equation in Exercise 33 involves finding the zeros of what quadratic function?

96. Solving the quadratic equation in Exercise 34 involves finding the zeros of what quadratic function?

97. Solving the quadratic equation in Exercise 49 involves finding the point(s) of intersection of the graphs of what two functions?

98. Solving the quadratic equation in Exercise 50 involves finding the point(s) of intersection of the graphs of what two functions?

99. Can you write down an expression for a quadratic function whose x-intercepts are given by $(2, 0)$ and $(3, 0)$? Is there more than one possible answer? Explain.

3.3 Complex Numbers and Quadratic Equations

Objectives

▶ Define a complex number

▶ Perform arithmetic with complex numbers

▶ Find the complex zeros of a quadratic function

▶ Find the complex solutions of a quadratic equation

In order to complete our analysis of the zeros of quadratic functions, we must introduce a set of numbers known as the *complex numbers*. Since finding the zeros of a quadratic function is equivalent to solving a related quadratic equation, the introduction of complex numbers will also complete our analysis of quadratic equations.

Note that the graph of the function $f(x) = x^2 + 4$ has no x-intercepts, as seen in Figure 3.3.1.

If we tried to solve the equation

$$x^2 + 4 = 0$$

we would get

$$x^2 = -4$$

which has no real number solutions. The problem is that we cannot take the square root of a negative number and get a real number as an answer.

We need a special number that will produce -1 when it is squared. One such special number is denoted by the letter i, which is an example of what we call an **imaginary number.**

Figure 3.3.1

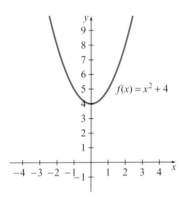

Definition of the Imaginary Number i

The **imaginary number i** is defined as the number such that

$$i = \sqrt{-1} \quad \text{and} \quad i^2 = -1.$$

In the following example, we see how this definition helps us to solve the equation we discussed at the beginning of this section.

Example 1 Imaginary Solutions of a Quadratic Equation

Use the definition of i to solve the equation $x^2 = -4$.

▶Solution Since $i^2 = -1$, we can write

$$x^2 = (-1)(4) = i^2 2^2 \quad \text{Use the definition of } i$$
$$x^2 = (2i)^2 \quad \text{Use the properties of exponents}$$
$$x = \pm 2i. \quad \text{Solve for x}$$

Checking the solutions, we see that

$$x^2 = (2i)^2 = 4i^2 = -4 \quad \text{and} \quad x^2 = (-2i)^2 = 4i^2 = -4.$$

We can therefore conclude that

$$x^2 = -4 \quad \text{for} \quad x = \pm 2i.$$

✔ *Check It Out 1:* Use the definition of i to solve the equation $x^2 = -9$. ■

Definition of Pure Imaginary Numbers

Numbers of the form bi, where b is a real number, are called **pure imaginary numbers.**

Example 2 Pure Imaginary Numbers

Write the following as pure imaginary numbers.

(a) $\sqrt{-36}$ (b) $\sqrt{-8}$ (c) $\sqrt{-\dfrac{1}{4}}$

▶Solution

(a) $\sqrt{-36} = i\sqrt{36} = 6i$, using the fact that $i = \sqrt{-1}$.

(b) $\sqrt{-8} = i\sqrt{8} = 2i\sqrt{2}$ (c) $\sqrt{-\dfrac{1}{4}} = i\sqrt{\dfrac{1}{4}} = \dfrac{1}{2}i$

Note that when i is multiplied by a radical, we place the i in front of the radical so that it is clear that it is not under the radical.

✔ *Check It Out 2:* Write $\sqrt{-25}$, $\sqrt{-108}$, and $\sqrt{-\dfrac{4}{9}}$ as pure imaginary numbers. ▪

Now that we know how to define the square root of a negative number, we can proceed to find the zeros of quadratic functions that we were unable to find in the previous section.

Example 3 Finding the Zeros of a Quadratic Function

Find the zeros of the function $g(t) = t^2 + t + 1$.

▶Solution We note that the graph of g does not cross the horizontal axis. Therefore, g has no real zeros, as seen in Figure 3.3.2.

We apply the quadratic formula to find the solutions of the equation $t^2 + t + 1 = 0$. Here, $a = 1$, $b = 1$, and $c = 1$. Substituting these values into the quadratic formula, we get

$$t = \frac{-1 \pm \sqrt{(1)^2 - 4(1)(1)}}{2(1)} \qquad \text{Quadratic formula}$$

$$= \frac{-1 \pm \sqrt{-3}}{2} \qquad \text{Simplify}$$

$$= -\frac{1}{2} \pm \frac{\sqrt{3}}{2}i. \qquad \text{Definition of imaginary number}$$

Each solution of the equation $t^2 + t + 1 = 0$, namely $t = -\dfrac{1}{2} + \dfrac{\sqrt{3}}{2}i$ and $t = -\dfrac{1}{2} - \dfrac{\sqrt{3}}{2}i$, consists of the sum of a real number and a pure imaginary number. The numbers $-\dfrac{1}{2} \pm \dfrac{\sqrt{3}}{2}i$ are examples of **complex numbers.**

✔ *Check It Out 3:* Find the zeros of the function $h(s) = -s^2 - s - 4$. ▪

Figure 3.3.2

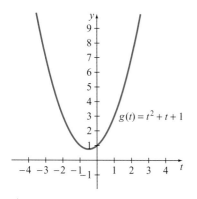

$g(t) = t^2 + t + 1$

Definition of a Complex Number

A **complex number** is a number of the form $a + bi$, where a and b are real numbers.

Note that the complex numbers include all the real numbers and all the pure imaginary numbers.

▶ If $b = 0$ in the above definition, then $a + bi = a$, which is a real number.

▶ If $a = 0$ in the above definition, then $a + bi = bi$, which is a pure imaginary number.

Example 4 Writing Numbers in the Form $a + bi$

Write the following numbers in the form $a + bi$, and identify a and b.

(a) $\sqrt{2}$

(b) $\dfrac{1}{3}i$

(c) $1 + \sqrt{3}$

▶Solution

(a) $\sqrt{2} = \sqrt{2} + 0i$. Note that $a = \sqrt{2}$, $b = 0$.

(b) $\dfrac{1}{3}i = 0 + \dfrac{1}{3}i$. Note that $a = 0$, $b = \dfrac{1}{3}$.

(c) $1 + \sqrt{3} = (1 + \sqrt{3}) + 0i$. Note that $a = 1 + \sqrt{3}$, $b = 0$.

✔ *Check It Out 4:* Write the following numbers in the form $a + bi$:

$\sqrt[3]{5}$, $-1 + \sqrt{2}$, $\dfrac{i}{4}$ ■

Parts of a Complex Number

For a complex number $a + bi$, a is called the **real part** and b is called the **imaginary part**.

Example 5 Real and Imaginary Parts of a Complex Number

What are the real and imaginary parts of the following complex numbers?

(a) $-2 + 3i$

(b) $-i + 4$

(c) $-\sqrt{3}$

▶Solution

(a) Note that $-2 + 3i$ is in the form $a + bi$. Thus the real part is $a = -2$ and the imaginary part is $b = 3$.

(b) First rewrite $-i + 4$ as $4 - i$. This gives 4 as the real part and -1 as the imaginary part.

(c) Note that $-\sqrt{3} = -\sqrt{3} + 0i$. This gives $-\sqrt{3}$ as the real part and 0 as the imaginary part.

✔ *Check It Out 5:* What are the real and imaginary parts of the following complex numbers? $\sqrt[3]{5}$, $-1 + \sqrt{2}$, $\dfrac{i}{4}$ ■

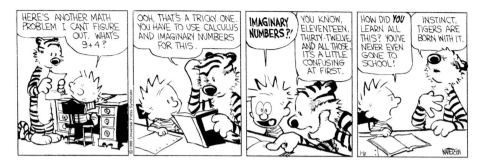

Addition and Subtraction of Complex Numbers

To add two complex numbers, we simply add their real parts to get the real part of their sum, and we add their imaginary parts to get the imaginary part of their sum. To subtract two complex numbers, we subtract their real and imaginary parts instead of adding them.

Example 6 Adding and Subtracting Complex Numbers

Perform the following operations.

(a) $(1 + 2i) + (3 - 5i)$　　(b) $(\sqrt{2} + i) + (-\sqrt{2} - i)$

(c) $i + (-1)$　　　　　　　(d) $(1 + i) - (2 - i)$

▶**Solution**

(a) Grouping the real terms together and the imaginary terms together, we have
$$(1 + 2i) + (3 - 5i) = (1 + 3) + (2i - 5i) = 4 - 3i.$$

(b) Once again, we group the real terms together and the imaginary terms together to get
$$(\sqrt{2} + i) + (-\sqrt{2} - i) = (\sqrt{2} - \sqrt{2}) + (i - i) = 0 + 0i = 0.$$

(c) Note that $i + (-1)$ cannot be simplified any further, and so the answer is $-1 + i$, rewritten in the form $a + bi$.

(d) To calculate $(1 + i) - (2 - i)$, we first distribute the negative sign over the real and imaginary parts of the second complex number and then add:
$$(1 + i) - (2 - i) = 1 + i - 2 + i = -1 + 2i$$

✔ *Check It Out 6:* Perform the following operations:

(a) $3 + 2i - 4 + i$　　(b) $-2 + i$　　(c) $\sqrt{3} + i - \sqrt{3}$ ■

Multiplication of Complex Numbers

To multiply two complex numbers, we apply the rules of multiplication of binomials. This is illustrated in the following example.

Example 7 Multiplying Complex Numbers

Multiply:

(a) $(1 + 3i)(2 - 4i)$　　(b) $\sqrt{-4}\,\sqrt{-9}$

▶**Solution**

(a) $(1 + 3i)(2 - 4i) = 2 - 4i + 6i - 12i^2$ Multiply (use FOIL)

$\qquad\qquad\qquad\quad = 2 + 2i - 12i^2$ Add the real and imaginary parts

$\qquad\qquad\qquad\quad = 2 + 2i - 12(-1)$ Note that $i^2 = -1$

$\qquad\qquad\qquad\quad = 14 + 2i$ Simplify

(b) $\sqrt{-4}\,\sqrt{-9} = (2i)(3i)$ Write as imaginary numbers

$\qquad\qquad\quad = 6i^2 = -6$ Use $i^2 = -1$

Note that we wrote $\sqrt{-4}$ and $\sqrt{-9}$ using imaginary numbers *before* simplifying. The reason for this is explained in the note below.

☑ *Check It Out 7:* Multiply $(-3 + 4i)(5 - 2i)$. ▪

> **Note** As illustrated in Example 7(b), you must be careful when multiplying if there are negative numbers under the radical. In this example, we cannot simply multiply the numbers under the radicals first. The rule $\sqrt{x}\,\sqrt{y} = \sqrt{xy}$ is valid only when x or y is positive. It does not hold when *both* x and y are negative. To avoid this potential source of error, always write square roots of negative numbers in terms of i before simplifying.

Division of Complex Numbers

Before we can define division of complex numbers, we must define the **complex conjugate** of a complex number.

> **Definition of Complex Conjugate**
>
> The **complex conjugate** of a complex number $a + bi$ is given by $a - bi$. The complex conjugate of a complex number has the *same real part* as the original number, but the *negative of the imaginary part.*

The next example will illustrate this definition.

Example 8 Conjugate of a Complex Number

Find the complex conjugates of the following numbers.

(a) $1 + 2i$ (b) $-3i$ (c) 2

▶**Solution**

(a) The complex conjugate of $1 + 2i$ is $1 - 2i$. We simply take the negative of the imaginary part and keep the real part of the original number.

(b) The complex conjugate of $-3i$ is $3i$. The real part here is zero, and so we just negate the imaginary part. Every pure imaginary number is equal to the negative of its complex conjugate.

(c) The complex conjugate of 2 is 2. The real part is 2, and it remains the same. The imaginary part is zero, and it will remain zero when negated. Every real number is equal to its complex conjugate.

✔ *Check It Out 8:* Find the complex conjugate of $-3 - 7i$. ■

Complex conjugates are often abbreviated as simply *conjugates*. The next example will illustrate why conjugates are useful.

Example 9 **Multiplying a Number by Its Conjugate**

Multiply $-3 + 2i$ by its conjugate. What type of number results from this operation?

▶**Solution** The conjugate of $-3 + 2i$ is $-3 - 2i$. Following the rules of multiplication, we have

$$
\begin{aligned}
(-3 + 2i)(-3 - 2i) &= 9 + 6i - 6i - 4i^2 && \text{Use FOIL to multiply} \\
&= 9 + 0i - 4i^2 && \text{Add the real and imaginary parts} \\
&= 9 - 4(-1) && \text{Note that } i^2 = -1 \text{ and } 0i = 0 \\
&= 13. && \text{Simplify}
\end{aligned}
$$

Thus we see that the product of $-3 + 2i$ and its conjugate is a positive real number.

✔ *Check It Out 9:* Multiply $-3 - 7i$ by its conjugate. What type of number results from this operation? ■

> **Note** It can be shown that the product of any nonzero complex number and its conjugate is a positive real number. This fact is extremely useful in the division of complex numbers, as illustrated in the next example.

Example 10 **Dividing Complex Numbers**

Find $\dfrac{2}{-3 + 2i}$.

▶**Solution** The idea is to multiply both the numerator and the denominator of the expression $\dfrac{2}{-3 + 2i}$ by the complex conjugate of the denominator. This will give a real number in the denominator. Thus we have

$$
\begin{aligned}
\frac{2}{-3 + 2i} &= \frac{2}{-3 + 2i} \cdot \frac{-3 - 2i}{-3 - 2i} && \text{Multiply the numerator and denominator by the conjugate of } -3 + 2i \\[2mm]
&= \frac{2(-3 - 2i)}{13} && (-3 + 2i)(-3 - 2i) = 13 \text{ from the previous example} \\[2mm]
&= \frac{2}{13}(-3 - 2i) = \frac{-6}{13} - \frac{4}{13}i. && \text{Simplify}
\end{aligned}
$$

Discover *and* Learn

How would you write the expression $\dfrac{-2 + \sqrt{4 - 4(2)(1)}}{2(2)}$ in the form $a + bi$?

✔ *Check It Out 10:* Find $\dfrac{4}{-3 - 7i}$. ■

Just In Time

Review the quadratic formula in Section 3.2.

Zeros of Quadratic Functions and Solutions of Quadratic Equations

By expanding from the real number system to the complex number system, we see that the zeros of any quadratic function can be computed. This is because we are now able to compute the square root of a negative number; thus we can find nonreal zeros by using the quadratic formula.

We conclude this section by computing the zeros of some quadratic functions.

Example 11 Complex Zeros of Quadratic Functions

Compute the zeros of the quadratic function $f(x) = 3x^2 + x + 1$. Use the zeros to find the x-intercepts, if any, of the graph of the function. Verify your results by graphing the function.

▶ **Solution** We solve $f(x) = 3x^2 + x + 1 = 0$ for x. Noting that the expression cannot be factored easily, we use the quadratic formula to solve for x. Thus

$$x = \frac{-b \pm \sqrt{b^2 - 4ac}}{2a}$$ The quadratic formula

$$= \frac{-(1) \pm \sqrt{(1)^2 - 4(3)(1)}}{2(3)}$$ $a = 3, b = 1,$ and $c = 1$

$$= \frac{-1 \pm \sqrt{-11}}{6} = -\frac{1}{6} \pm \frac{\sqrt{11}}{6}i.$$ Simplify

Figure 3.3.4

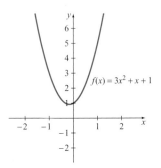

We see that this function has two nonreal zeros: $x = -\frac{1}{6} + \frac{\sqrt{11}}{6}i$ and $x = -\frac{1}{6} - \frac{\sqrt{11}}{6}i$.
Since the zeros are not real numbers, the graph of the function has no x-intercepts. These results are confirmed by examining the graph of $f(x) = 3x^2 + x + 1$ in Figure 3.3.4.

✔ *Check It Out 11:* Rework Example 11 using the quadratic function $g(s) = -3s^2 + 2s - 1.$ ▪

Technology Note

The ZERO and INTERSECT features of a graphing utility will not find complex zeros. However, you can use the quadratic formula to find complex solutions. First set the calculator to $a + bi$ mode. Then store the values of a, b, and c in the home screen. Type in the quadratic formula to find the solutions. See Figure 3.3.5. On some models, you may use a downloadable application that will find complex zeros.

Keystroke Appendix:
Section 11

Figure 3.3.5

```
2→A:-2→B:(3/2)→C
                    1.5
(-B+√(B^2-4AC))/
(2A)
.5+.7071067812i
(-B-√(B^2-4AC))/
(2A)
.5-.7071067812i
```

Example 12 Complex Solutions of a Quadratic Equation

Find all solutions of the quadratic equation $2t^2 - 2t = -\frac{3}{2}$. Relate the solutions of this equation to the zeros of an appropriate quadratic function.

▶ **Solution** First write the equation in standard form.

$$2t^2 - 2t + \frac{3}{2} = 0$$

Apply the quadratic formula to solve for t.

$$x = \frac{-b \pm \sqrt{b^2 - 4ac}}{2a}$$ The quadratic formula

$$= \frac{-(-2) \pm \sqrt{(-2)^2 - 4(2)(3/2)}}{2(2)}$$ $a = 2, b = -2,$ and $c = \frac{3}{2}$

$$= \frac{2 \pm \sqrt{4 - 12}}{4} = \frac{2 \pm \sqrt{-8}}{4} = \frac{2 \pm i\sqrt{8}}{4}$$ Use $\sqrt{-8} = i\sqrt{8}$

$$= \frac{2 \pm 2i\sqrt{2}}{4} = \frac{1}{2} \pm i\frac{\sqrt{2}}{2}$$ Simplify

Figure 3.3.6

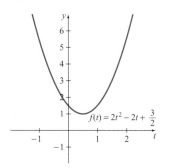

We see that this quadratic equation has two solutions, $t = \frac{1}{2} + i\frac{\sqrt{2}}{2}$ and $t = \frac{1}{2} - i\frac{\sqrt{2}}{2}$, both of which are nonreal. An associated quadratic function is $f(t) = 2t^2 - 2t + \frac{3}{2}$, which thus has two nonreal zeros: $t = \frac{1}{2} + i\frac{\sqrt{2}}{2}$ and $t = \frac{1}{2} - i\frac{\sqrt{2}}{2}$. The graph of f will have no x-intercepts, since f has no real zeros. This is confirmed by the graph of $f(t) = 2t^2 - 2t + \frac{3}{2}$ shown in Figure 3.3.6.

✔ *Check It Out 12:* Find all solutions of the quadratic equation

$$-2t^2 + 3t - 2 = 0.\ ■$$

Throughout this book, we have discussed relevant applications for the topics under study. Complex numbers have applications in engineering, physics, and advanced mathematics. However, these applications require a certain amount of technical background that is beyond the scope of this book, and so we will not discuss them here.

3.3 Key Points

▶ To add two complex numbers, add their corresponding real and imaginary parts.

▶ To subtract two complex numbers, subtract their corresponding real and imaginary parts.

▶ To multiply two complex numbers, apply the rules of multiplication of binomials (FOIL).

▶ To divide two complex numbers, multiply the numerator and denominator by the complex conjugate of the denominator. The complex conjugate of $a + bi$ is $a - bi$.

▶ Complex numbers can be used to find the nonreal zeros of a quadratic function and the nonreal solutions of a quadratic equation by using the quadratic formula.

3.3 Exercises

▶**Just in Time Exercises** These exercises correspond to the Just in Time references in this section. Complete them to review topics relevant to the remaining exercises.

1. Multiply: $(x + 3)(x - 2)$

2. Multiply: $(x - 5)(x + 1)$

3. Multiply: $(x + 4)(x - 4)$

4. Multiply: $(-x + 1)(x - 5)$

In Exercises 5–8, use the quadratic formula to solve the equation.

5. $x^2 - 5x + 3 = 0$

6. $2x^2 + x - 5 = 0$

7. $x^2 - 3x - 2 = 0$

8. $x^2 + x - 4 = 0$

▶**Skills** This set of exercises will reinforce the skills illustrated in this section.

In Exercises 9–14, write the number as a pure imaginary number.

9. $\sqrt{-16}$

10. $\sqrt{-64}$

11. $\sqrt{-12}$

12. $\sqrt{-24}$

13. $\sqrt{-\dfrac{4}{25}}$

14. $\sqrt{-\dfrac{9}{4}}$

In Exercises 15–20, use the definition of i to solve the equation.

15. $x^2 = -16$

16. $x^2 = -25$

17. $-x^2 = 8$

18. $-x^2 = 12$

19. $3x^2 = -30$

20. $5x^2 = -60$

In Exercises 21–28, find the real and imaginary parts of the complex number.

21. 2

22. -3

23. $-\pi i$

24. $i\sqrt{3}$

25. $1 + \sqrt{5}$

26. $\sqrt{7} - 1$

27. $1 + \sqrt{-5}$

28. $\sqrt{-7} - 1$

In Exercises 29–36, find the complex conjugate of each number.

29. -2

30. -5

31. $i - 1$

32. $-2i + 4$

33. $3 + \sqrt{2}$

34. $9 - \sqrt{3}$

35. i^2

36. i^3

In Exercises 37–48, find $x + y$, $x - y$, xy, and x/y.

37. $x = 3i; y = 2 - i$

38. $x = -2i; y = 5 + i$

39. $x = -3 + 5i; y = 2 - 3i$

40. $x = 2 - 9i; y = -4 + 6i$

41. $x = 4 - 5i; y = 3 + 2i$

42. $x = 2 - 7i; y = 11 + 2i$

43. $x = \dfrac{1}{2} - 3i; y = \dfrac{1}{5} + \dfrac{4}{3}i$

44. $x = \dfrac{1}{3} - 2i; y = \dfrac{1}{3} - \dfrac{2}{5}i$

45. $x = -\dfrac{1}{3} + i\sqrt{5}; y = -\dfrac{1}{2} - 2i\sqrt{5}$

46. $x = \dfrac{1}{2} - i\sqrt{3}; y = \dfrac{1}{5} + 3i\sqrt{3}$

47. $x = -3 + i; y = i + \dfrac{1}{2}$

48. $x = -2 - i; y = i + 2$

In Exercises 49–60, compute the zeros of the quadratic function.

49. $f(x) = 2x^2 + 9$

50. $f(x) = 3x^2 + 5$

51. $h(x) = -3x^2 - 10$

52. $f(x) = -3x^2 - 18$

53. $f(x) = -x^2 - x - 1$

54. $g(x) = x^2 - x + 1$

55. $h(t) = 3t^2 - 2t - 9$

56. $f(x) = 2x^2 - x + 8$

57. $f(x) = -2x^2 - 2x + 11$

58. $g(t) = -5t^2 + 2t - 3$

59. $h(x) = 3x^2 + 8x - 16$

60. $f(t) = 2t^2 + 11t + 9$

In Exercises 61–76, find all solutions of the quadratic equation. Relate the solutions of the equation to the zeros of an appropriate quadratic function.

61. $x^2 + 2x + 3 = 0$

62. $-x^2 + x - 5 = 0$

63. $-3x^2 + 2x - 4 = 0$

64. $-2x^2 + 3x - 1 = 0$

65. $5x^2 - 2x + 3 = 0$

66. $-7x^2 + 2x - 1 = 0$

67. $5x^2 = -2x - 3$

68. $7x^2 = -x - 1$

69. $-3x^2 + 8x = 16$

70. $2t^2 + 8t = -9$

71. $-4t^2 + t - \dfrac{1}{2} = 0$

72. $-6t^2 + 2t - \dfrac{1}{3} = 0$

73. $\dfrac{2}{3}x^2 + x = -1$

74. $-\dfrac{3}{4}x^2 - x = 2$

75. $(x + 1)^2 = -25$

76. $(x - 2)^2 = -16$

In Exercises 77–80, use the following definition. A complex number $a + bi$ is often denoted by the letter z. Its conjugate, $a - bi$, is denoted by $\bar{z}$.

77. Show that $z + \bar{z} = 2a$ and $z - \bar{z} = 2bi$.

78. Show that $z\bar{z} = a^2 + b^2$.

79. Show that the real part of z is equal to $\dfrac{z + \bar{z}}{2}$.

80. Show that the imaginary part of z is equal to $\dfrac{z - \bar{z}}{2i}$.

In Exercises 81–84, use the function $f(x) = ax^2 + 2x + 1$, where a is a real number.

81. Find the discriminant $b^2 - 4ac$.

82. For what value(s) of a will f have two real zeros?

83. For what value(s) of a will f have one real zero?

84. For what value(s) of a will f have no real zeros?

In Exercises 85–88, solve the quadratic equation by entering the quadratic formula in the home screen of your graphing utility. (See Technology Note on page 252.)

85. $-0.25x^2 + 1.14x - 2.5 = 0$

86. $0.62t^2 - 1.29t + 1.5 = 0$

87. $3t^2 + \sqrt{19} = 2t$

88. $2x^2 + \sqrt{11} = x$

▶**Concepts** This set of exercises will draw on the ideas presented in this section and your general math background.

89. Consider a parabola that opens upward and has vertex $(0, 4)$.
    (a) Why does the quadratic function associated with such a parabola have *no* real zeros?
    (b) Show that $f(x) = 2x^2 + 4$ is a possible quadratic function associated with such a parabola. Is this the only possible quadratic function associated with such a parabola? Explain.
    (c) Find the zeros of the function f given in part (b).

90. For the functions y_1 and y_2 graphed below, explain why the equation $y_1(x) = y_2(x)$ has no real-valued solutions. Assuming that y_1 is a quadratic function with real coefficients and y_2 is a linear function, explain why the equation $y_1(x) = y_2(x)$ has at least one complex-valued solution.

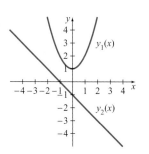

91. In this problem, you will explore the relationship between factoring a quadratic expression over the complex numbers and finding the zeros of the associated quadratic function. This topic will be explained in greater detail in Chapter 4.
    (a) Multiply $(x + i)(x - i)$.
    (b) What are the zeros of $f(x) = x^2 + 1$?
    (c) What is the relationship between your answers to parts (a) and (b)?
    (d) Using your answers to parts (a)–(c) as a guide, how would you factor $x^2 + 9$?
    (e) Using you answers to parts (a)–(d) as a guide, how would you factor $x^2 + c^2$, where c is a positive real number?

92. We know that $i^2 = -1$, but is there a complex number z such that $z^2 = i$? We answer that question in this exercise.
    (a) Calculate $\left(\dfrac{\sqrt{2}}{2}(1 + i)\right)\left(\dfrac{\sqrt{2}}{2}(1 + i)\right)$.
    (b) Use your answer in part (a) to find a complex number z such that $z^2 = i$.

93. Examine the following table of values of a quadratic function.

x	$f(x)$
-2	9
-1	3
0	1
1	3
2	9

    (a) What is the equation of the axis of symmetry of the associated parabola? Explain how you got your answer.
    (b) Find the minimum or maximum value of the function and the value of x at which it occurs.
    (c) Sketch a graph of the function from the values given in the table.
    (d) Does this function have real or nonreal zeros? Explain.

94. Is it possible for a quadratic function with real coefficients to have one real zero and one nonreal zero? Explain. (*Hint:* Examine the quadratic formula.)

3.4 Quadratic Inequalities

Objectives

▶ Interpret quadratic inequalities graphically

▶ Solve quadratic inequalities algebraically

▶ Use quadratic inequalities in applications

Figure 3.4.1

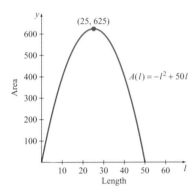

Just In Time

Review Example 1 in Section 3.1

In many applications of quadratic functions, it is useful to know when the values of a function are positive or when they are negative so that we can solve an inequality of the form $f(x) > 0$ or $f(x) < 0$.

Let us explore this idea using the area function $A(l) = -l^2 + 50l$ for the concert ticket venue, given in Example 1 of Section 3.1. Its graph is shown in Figure 3.4.1.

Areas are always positive, so it makes sense to ask the following question: *What values of the length l will give a positive value for the area?* From the graph, we see that the length of the enclosed rectangular region must be greater than 0 and less than 50 feet. The former is obvious, since lengths are always positive. Moreover, recall from Example 1 of Section 3.1 that the total amount of rope available was only 100 feet. Therefore, from a physical standpoint, it makes sense that the length of the enclosed rectangle must be less than 50 feet. We can also look at this from a strictly mathematical standpoint. We found the expression for the width of the cordoned-off region to be $50 - l$. Since widths are always positive, we see that $50 - l > 0$, which gives $50 > l$.

If we wanted to express the statements about l in the previous paragraph in mathematical terms, we would say that the **solution of the inequality** $A(l) > 0$ is the **set of all values** l such that $0 < l < 50$. This is similar to the terminology we used in solving linear inequalities in Section 1.5 and absolute value inequalities in Section 2.5.

> **Definition of a Quadratic Inequality**
>
> A **quadratic inequality** is one that can be written as $ax^2 + bx + c > 0$, where $>$ may be replaced by $\geq$, $<$, or $\leq$.

Graphical Approach to Solving Inequalities

The following example will examine inequalities of the forms $f(x) > 0$, $f(x) < 0$, $f(x) \geq 0$, and $f(x) \leq 0$.

Example 1 **Graphical Solution of a Quadratic Inequality**

For the quadratic function $f(x) = 2x^2 - x - 3$, sketched in Figure 3.4.2, find the values of x for which:

(a) $f(x) \geq 0$ (b) $f(x) > 0$ (c) $f(x) \leq 0$ (d) $f(x) < 0$

Figure 3.4.2

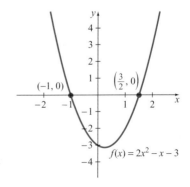

▶ **Solution**

(a) From the graph, we see that the value of $f(x)$ is **positive** if $x > \frac{3}{2}$ or $x < -1$, and the value of $f(x)$ is **zero** if $x = \frac{3}{2}$ or $x = -1$. Thus the solution set for the inequality $f(x) \geq 0$ consists of all x such that $x \geq \frac{3}{2}$ or $x \leq -1$. Using interval notation, the solution set is

$$\left(-\infty, -1\right] \cup \left[\frac{3}{2}, \infty\right).$$

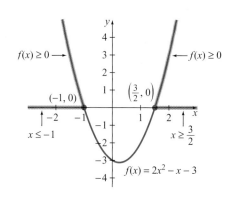

(b) To find the values of x for which $f(x) > 0$, we take the solution set found in part (a) and *exclude* the values of x at which $f(x) = 0$—namely $x = \frac{3}{2}$ and $x = -1$. Thus the solution set consists of all x such that $x > \frac{3}{2}$ or $x < -1$. Using interval notation, the solution set is $\left(-\infty, -1\right) \cup \left(\frac{3}{2}, \infty\right)$.

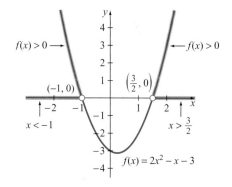

(c) From the graph, we see that the value of $f(x)$ is *negative* if $-1 < x < \frac{3}{2}$. As found in part (a), the value of $f(x)$ is zero if $x = \frac{3}{2}$ or $x = -1$. Thus the solution set for the inequality $f(x) \leq 0$ consists of all x such that $-1 \leq x \leq \frac{3}{2}$. In interval notation, the solution set is $\left[-1, \frac{3}{2}\right]$.

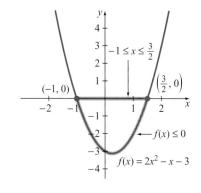

Figure 3.4.3

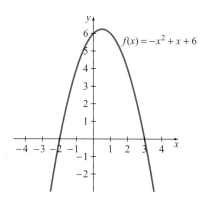

(d) To find the values of x for which $f(x) < 0$, we take the solution set found in part (c) and *exclude* the values of x at which $f(x) = 0$—namely $x = \frac{3}{2}$ and $x = -1$. Thus the solution set consists of all x such that $-1 < x < \frac{3}{2}$. In interval notation, the solution set is $\left(-1, \frac{3}{2}\right)$.

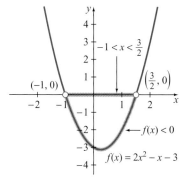

✔ **Check It Out 1:** For the quadratic function $f(x) = -x^2 + x + 6$, sketched in Figure 3.4.3, find the values of x for which (a) $f(x) \geq 0$ and (b) $f(x) < 0$. ■

Algebraic Approach to Solving Inequalities

The graphical approach to solving inequalities enables you to visualize a solution set almost instantly. However, graphing by hand can sometimes be tedious. When using a graphing utility, appropriate choices of scales and window sizes may not always be immediately obvious. To overcome such limitations, we introduce an algebraic method for solving inequalities. The steps taken in using the algebraic approach are illustrated in the following example.

Example 2 Algebraic Solution of a Quadratic Inequality

Solve the inequality $-x^2 + 5x - 4 \leq 0$ algebraically.

▶ Solution

STEPS	EXAMPLE
1. The inequality should be written so that one side consists only of zero.	$-x^2 + 5x - 4 \leq 0$
2. Factor the expression on the nonzero side of the inequality; this will transform it into a product of two linear factors.	$(-x + 4)(x - 1) \leq 0$
3. Find the zeros of the expression on the nonzero side of the inequality—that is, the zeros of $(-x + 4)(x - 1)$. These are the only values of x at which the expression on the nonzero side can change sign. To find the zeros, set each of the factors found in the previous step equal to zero, and solve for x.	$-x + 4 = 0 \implies x = 4$ $x - 1 = 0 \implies x = 1$
4. If the zeros found in the previous step are distinct, use them to break up the number line into three disjoint intervals. Otherwise, break it up into just two disjoint intervals. Indicate these intervals on the number line.	
5. Use a test point in each interval to calculate the sign of the expression on the nonzero side of the inequality for that interval. Indicate these signs on the number line.	
6. Select the interval(s) on which the inequality is satisfied—in this case, $(-\infty, 1] \cup [4, \infty)$.	Solution set: $(-\infty, 1] \cup [4, \infty)$. The endpoints are included because the original inequality reads "less than or equal to zero."

Technology Note

You can use the ZERO feature of a graphing utility to find the zeros of

$$Y_1(x) = -x^2 + 5x - 4.$$

The solution of the inequality

$$-x^2 + 5x - 4 \le 0$$

is then seen to be

$$(-\infty, 1] \cup [4, \infty).$$

See Figure 3.4.4.

Keystroke Appendix:
Sections 6, 9

Figure 3.4.4

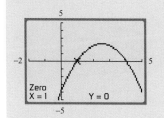

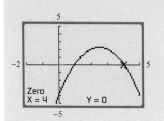

☑ *Check It Out 2:* Solve the inequality $x^2 + 6x - 7 \ge 0$ algebraically. ▪

> **Note**   Unlike solving an equation, solving an inequality gives an infinite number of solutions. You cannot really check your solution in the same way that you can for an equation, but you can get an idea of whether your solution is correct by substituting some values from your solution set (and even some values not in your solution set!) into the inequality.

In many situations, it will be necessary to solve inequalities in which neither side of the inequality consists only of zero. In the next example, we discuss this type of inequality for quadratic functions.

Example 3 **Algebraic Solution of a Quadratic Inequality**

Algebraically solve the inequality $2x^2 - x - 1 > 2x + 1$.

▶**Solution**

Step 1 Manipulate the inequality algebraically so that one side consists only of zero. In this case, subtract $2x + 1$ from both sides. Simplify the expression on the nonzero side if possible.

$$2x^2 - x - 1 > 2x + 1$$
$$2x^2 - x - 1 - 2x - 1 > 0$$
$$2x^2 - 3x - 2 > 0$$

Step 2 Factor the expression $2x^2 - 3x - 2$ to get

$$(2x + 1)(x - 2) > 0.$$

Step 3 Find the zeros of $(2x + 1)(x - 2)$. Take each of the factors found in the previous step, set it equal to zero, and solve for x.

$$2x + 1 = 0 \implies x = -\frac{1}{2}$$
$$x - 2 = 0 \implies x = 2$$

Step 4 Use the zeros found in the previous step to break up the number line into three disjoint intervals. Indicate these intervals on the number line. Use a test point in each interval to calculate the sign of the expression $(2x + 1)(x - 2)$ in that interval. Indicate these signs on the number line, as shown in Figure 3.4.5.

Figure 3.4.5

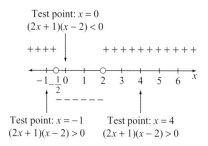

Technology Note

You can use the INTERSECT feature of a graphing utility to see that the graphs of $Y_1(x) = 2x^2 - x - 1$ and $Y_2(x) = 2x + 1$ intersect at (2, 5) and (−0.5, 0). See Figure 3.4.7. Note that the graph of Y_1 lies above the graph of Y_2 if $x < -0.5$ or $x > 2$. You can verify this result numerically by using the TABLE feature in ASK mode with selected values of x. See Figure 3.4.8.

Keystroke Appendix: Sections 6, 9

Figure 3.4.7

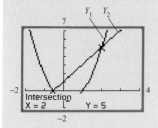

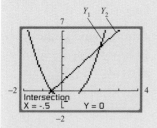

Figure 3.4.8

X	Y₁	Y₂
-2	9	-3
-1	2	-1
-.5	0	0
0	-1	1
1	0	3
2	5	5
3	14	7
X=-2		

Step 5 Since the inequality is satisfied on the interval $\left(-\infty, -\frac{1}{2}\right)$ or the interval $(2, \infty)$, the solution set is $\left(-\infty, -\frac{1}{2}\right) \cup (2, \infty)$. The endpoints are *not* included because the inequality is strictly "greater than." Thus the solution set of the inequality $2x^2 - x - 1 > 2x + 1$ consists of all x such that $x > 2$ or $x < -\frac{1}{2}$. The solution of the inequality is shown graphically in Figure 3.4.6.

Figure 3.4.6

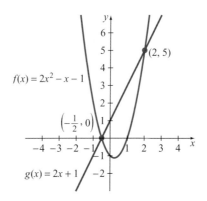

☑ *Check It Out 3:* Algebraically solve the inequality $-2x^2 + 5x < 3$. ▪

Note When using algebraic methods to obtain the solution of an inequality, we must have one side of the inequality consisting only of 0 so that we can apply the Zero Product Rule to find the solution set.

The technique just presented for solving quadratic inequalities algebraically involves factoring the expression on the nonzero side of the inequality. As we know from Section 3.2, not all quadratic expressions are readily factorable. When an expression is unfactorable, we can do one of two things:

▸ Use the quadratic formula to find the zeros of the quadratic expression and then proceed with the final steps of the algebraic approach given in Example 3.

▸ Use a graphical approach to solve the inequality.

The next example illustrates this situation.

Example 4 An Inequality That Is Not Readily Factorable

Solve the inequality $3x^2 - 2x \geq 4$.

▸**Solution** Let us see if we can solve this inequality by using the factoring method.

Step 1 Manipulate the inequality so that the right-hand side consists only of zero by subtracting 4 from both sides.

$$3x^2 - 2x - 4 \geq 0$$

Step 2 Factor $3x^2 - 2x - 4$, if possible, to find its zeros. We find through trial and error that this expression is not easily factorable.

Just In Time

Review a detailed solution using the quadratic formula in Example 6 of Section 3.2.

Step 3 We now have two possibilities: either find the zeros via the quadratic formula or use a graphing utility to find the solution graphically.

Step 4 Algebraic approach: Using the quadratic formula, we find that the two zeros of the expression $3x^2 - 2x - 4$ are

$$x = \frac{1}{3} + \frac{\sqrt{13}}{3} \approx 1.535 \quad \text{and} \quad x = \frac{1}{3} - \frac{\sqrt{13}}{3} \approx -0.869.$$

Next, we proceed as in steps 4–6 of the algebraic approach presented in Example 2. Use the zeros of the expression $3x^2 - 2x - 4$ to break up the number line into three disjoint intervals. Use a test point in each interval to calculate the sign of the expression $3x^2 - 2x - 4$ in that interval. Indicate these signs on the number line. See Figure 3.4.9.

Figure 3.4.9

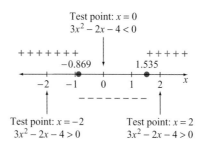

Since the inequality is satisfied on the interval $\left(-\infty, \frac{1}{3} - \frac{\sqrt{13}}{3}\right]$ or the interval $\left[\frac{1}{3} + \frac{\sqrt{13}}{3}, \infty\right)$, the solution set is $\left(-\infty, \frac{1}{3} - \frac{\sqrt{13}}{3}\right] \cup \left[\frac{1}{3} + \frac{\sqrt{13}}{3}, \infty\right)$.

Graphical approach: Using the original inequality $3x^2 - 2x \geq 4$, we graph two functions, $y_1(x) = 3x^2 - 2x$ and $y_2(x) = 4$, as shown in Figure 3.4.10.

Figure 3.4.10

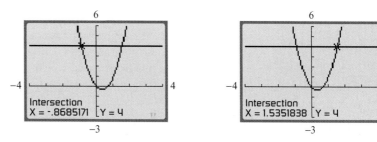

The points of intersection are at $x \approx -0.869$ and $x \approx 1.535$. Examining the graph, we see that $y_1(x) \geq y_2(x)$ if $x \leq -0.869$ or $x \geq 1.535$.

✔ *Check It Out 4:* Solve the inequality $2x^2 + 3x \leq 1$ either algebraically or graphically.

Figure 3.4.12

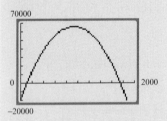

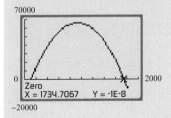

Figure 3.4.13

Example 5 **A Quadratic Inequality with No Real Solution**

Solve the inequality $x^2 + x + 1 < 0$.

▶**Solution** The expression $x^2 + x + 1$ is not readily factorable. Using the quadratic formula to find the zeros, we have

$$x = \frac{-(1) \pm \sqrt{(1)^2 - 4(1)(1)}}{2(1)} = \frac{-1 \pm i\sqrt{3}}{2}.$$

Since there are no real zeros, the graph of $f(x) = x^2 + x + 1$ has no x-intercepts, as seen in Figure 3.4.11. Thus the value of the expression $x^2 + x + 1$ never changes sign. At the test point $x = 0$, the nonzero side of the inequality is positive. Since the expression $x^2 + x + 1$ never changes sign, $x^2 + x + 1 > 0$ for *all* real numbers x, and so the inequality $x^2 + x + 1 < 0$ has no real solution.

Figure 3.4.11

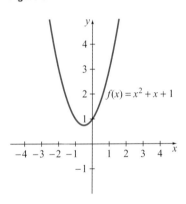

✔ *Check It Out 5:* Solve the inequality $-x^2 - 2 < 0$. ■

An Application of a Quadratic Inequality

In practice, inequalities are frequently used to find a feasible set of values of a variable that satisfy a given condition in the real world, whether it be economic, biological, chemical, physical, or otherwise. Often, graphical and algebraic approaches for solving such inequalities are used in conjunction to find a solution set. The next example shows how we can integrate graphical and algebraic approaches to solve a quadratic inequality that stems from a real-world application.

Example 6 **Earning Profit**

The price s (in dollars) of a portable CD player is given by

$$s(q) = 200 - 0.1q, \qquad 0 \le q \le 2000$$

where q is the number of portable CD players sold per day. It costs $20,000 per day to operate the factory to produce the product and an additional $15 for each portable CD player produced.

(a) Find the daily revenue function, $R(q)$ (the revenue from sales of q units of portable CD players per day).

(b) Find the daily cost function, $C(q)$ (the cost of producing q units of portable CD players per day).

(c) The profit function is given by $P(q) = R(q) - C(q)$. For what values of q will the profit be greater than zero?

▶**Solution**

(a) The daily revenue is given by

$$R(q) = \text{(number of units sold)(price per player)}$$
$$= q(200 - 0.1q)$$
$$= 200q - 0.1q^2$$
$$= -0.1q^2 + 200q.$$

(b) Since it costs $20,000 per day for factory operating costs and an additional $15 for each portable CD player produced, the cost function is

$$C(q) = 20,000 + 15q.$$

(c) The profit function is given by

$$P(q) = R(q) - C(q) = (-0.1q^2 + 200q) - (20,000 + 15q)$$
$$= -0.1q^2 + 185q - 20,000, \quad 0 \le q \le 2000.$$

The zeros of P can be found by using the quadratic formula or a graphing utility. The approximate zeros are

$$q \approx 115.29 \quad \text{and} \quad q \approx 1734.71.$$

Using the techniques outlined in this section, you can check that the value of the profit function $P(q)$ will be greater than zero for

$$115.29 < q < 1734.71.$$

Since the number of portable CD players must be in whole numbers, the manufacturer can produce between 116 and 1734 portable CD players per day to yield a profit.

✔ *Check It Out 6:* Repeat Example 6 for the case in which it costs $25,000 per day to operate the factory, and all other data remain the same. ▪

3.4 Key Points

▶ To solve a quadratic inequality graphically, graph f and examine the regions where $f(x) > 0$, $f(x) = 0$, and $f(x) < 0$.

▶ To solve a quadratic inequality algebraically, solve $ax^2 + bx + c = 0$. This can be done either by factoring or by using the quadratic formula. Use the solutions to divide the number line into disjoint intervals. Use test values in each interval to decide which interval(s) satisfy the inequality.

3.4 Exercises

▶**Skills** This set of exercises will reinforce the skills illustrated in this section.

In Exercises 1–4, use the graph of f to solve the inequality.

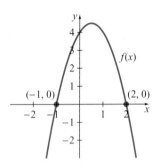

1. $f(x) \geq 0$ 2. $f(x) \leq 0$

3. $f(x) > 0$ 4. $f(x) < 0$

In Exercises 5–8, use the graph of g to solve the inequality.

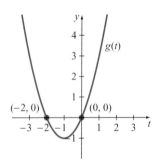

5. $g(t) \geq 0$ 6. $g(t) \leq 0$

7. $g(t) > 0$ 8. $g(t) < 0$

In Exercises 9 and 10, use the graphs of f and g to solve the inequality.

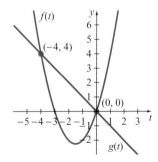

9. $f(t) \geq g(t)$ 10. $f(t) \leq g(t)$

In Exercises 11 and 12, use the graphs of f and g to solve the inequality.

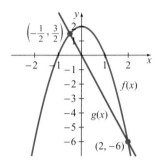

11. $f(x) \geq g(x)$ 12. $f(x) \leq g(x)$

In Exercises 13 and 14, use the following table of test values of the quadratic functions f and g defined on $(-\infty, \infty)$.

t	$f(t)$	$g(t)$
-1	0	2
-0.5	0.5	1.25
0	1	1
0.5	1.5	1.25
1	2	2
1.5	2.5	3.25
2	3	5
2.5	3.5	7.25
3	4	10

13. Find the region(s) where $f(t) \geq g(t)$.

14. Find the region(s) where $f(t) \leq g(t)$.

In Exercises 15–28, solve the inequality by factoring.

15. $x^2 - 1 \leq 0$ 16. $x^2 - 9 < 0$

17. $2x^2 + 3x \geq 5$ 18. $-2x^2 - 3x > 2$

19. $-3x^2 + x \leq -2$ 20. $6x^2 - 5x < 6$

21. $2x^2 < x + 1$ 22. $6x^2 \geq 13x - 5$

23. $5x^2 - 8x \geq 4$ 24. $-3x^2 \leq -7x - 6$

25. $10x^2 \leq -13x + 3$ 26. $12x^2 + 5x - 2 \geq 0$

27. $-x^2 + 2x - 1 < 0$ 28. $x^2 + 4x + 4 > 0$

In Exercises 29–38, solve the inequality algebraically or graphically.

29. $2x^2 - 3x < 1$

30. $-x^2 - 3x > -1$

31. $x^2 - 4 \geq x$

32. $x^2 - 9 \geq 2x$

33. $3x^2 - x - 1 \geq 0$

34. $-2x^2 + 2x + 3 \geq 0$

35. $x^2 + 1 < 0$

36. $-x^2 - 4 > 0$

37. $x^2 + 2x + 1 \geq 0$

38. $x^2 - x + 1 \geq 0$

▶**Applications** In this set of exercises you will use quadratic inequalities to study real-world problems.

39. **Landscaping** A rectangular garden plot is to be enclosed with a fence on three of its sides and a brick wall on the fourth side. There is 100 feet of fencing available. Let w denote the width of the fenced plot, as illustrated. For what range of values of w will the area of the enclosed region be less than or equal to 1200 square feet?

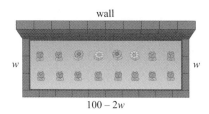
wall
w w
$100 - 2w$

40. **Business** The price s (in dollars) of a product is given by $s(q) = 100 - 0.1q$, $0 \leq q \leq 1000$, where q is the number of units sold per day. It costs \$10,000 per day to operate the factory and an additional \$12 for each unit produced.
  (a) Find the daily revenue function, $R(q)$.
  (b) Find the daily cost function, $C(q)$.
  (c) The profit function is given by $P(q) = R(q) - C(q)$. For what values of q will the profit be greater than or equal to zero?

41. **Manufacturing** A carpenter wishes to make a rain gutter with a rectangular cross-section by bending up a flat piece of metal that is 18 feet long and 20 inches wide. The top of the gutter is open. What values of x, the length of metal bent up, will give a cross-sectional area of at most 30 square inches?

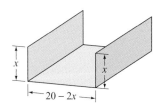
x
x
$20 - 2x$

42. **Physics** The height of a ball that is thrown directly upward from a point 200 feet above the ground with an initial velocity of 40 feet per second is given by $h(t) = -16t^2 + 40t + 200$, where t is the amount of time elapsed since the ball was thrown; t is in seconds and $h(t)$ is in feet. For what values of t will the height of the ball be below 100 feet?

43. **Performing Arts** Attendance at Broadway shows in New York can be modeled by the quadratic function $p(t) = 0.0489t^2 - 0.7815t + 10.31$, where t is the number of years since 1981 and $p(t)$ is the attendance in millions. The model is based on data for the years 1981–2000. For which years was the attendance above 8 million? (*Source:* The League of American Theaters and Producers, Inc.)

44. **Leisure** The average amount of money spent on books and magazines per household in the United States can be modeled by the function $r(t) = -0.2837t^2 + 5.5468t + 136.68$. Here, $r(t)$ is in dollars and t is the number of years since 1985. The model is based on data for the years 1985–2000. In what year(s) was the average expenditure per household for books and magazines greater than \$160? (*Source:* U.S. Bureau of Labor Statistics)

45. **Summation** If n is a positive integer, the sum $1 + 2 + \cdots + n$ is equal to $\frac{n(n + 1)}{2}$. For what values of n will the sum $1 + 2 + \cdots + n$ be greater than or equal to 45?

▶**Concepts** This set of exercises will draw on the ideas presented in this section and your general math background.

46. For what value(s) of c will the inequality $x^2 + c > 0$ have all real numbers as its solution? Explain.

47. For what value(s) of a will the inequality $ax^2 \leq 0$ have all real numbers as its solution? Explain.

48. For what value(s) of a will the inequality $ax^2 < 0$ have no real-valued solution? Explain.

49. Explain why $(x + 1)^2 \leq 0$ has a solution, whereas $(x + 1)^2 < 0$ has no real-valued solution.

50. Give graphical and algebraic explanations of why $x^2 + 1 < -x$ has no real-valued solution.

3.5 Equations That Are Reducible to Quadratic Form; Rational and Radical Equations

Objectives

▶ Solve polynomial equations by reducing them to quadratic form

▶ Solve equations containing rational expressions

▶ Solve equations containing radical expressions

▶ Solve applied problems

In this section, we will solve equations containing polynomial, rational, and radical expressions by algebraically manipulating them to resemble quadratic equations. Note that the techniques covered in this section apply only to a select class of equations. Chapter 4 on polynomial, rational, and radical functions discusses general graphical solutions of such equations in more detail.

Solving Equations by Reducing Them to Quadratic Form

Recall that to solve a quadratic equation, you can either factor a quadratic polynomial or use the quadratic formula. Some types of polynomial equations can be solved by using a substitution technique to reduce them to quadratic form. Then, either the quadratic formula or factoring is used to solve the equation. We illustrate this technique in the following examples.

Technology Note

To solve the equation $3x^4 + 5x^2 - 2 = 0$ with a graphing utility, graph the function $f(x) = 3x^4 + 5x^2 - 2$ and use the ZERO feature. Note that the approximate locations of the *real* solutions, $x \approx \pm 0.5774$, can be found by graphical methods, but that the locations of the imaginary zeros cannot be found in this way, since imaginary solutions are not visible on a graph. See Figure 3.5.1.

Keystroke Appendix:
Section 9

Figure 3.5.1

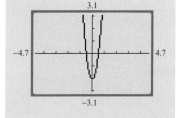

Example 1 Solving Using Substitution

Solve the equation $3x^4 + 5x^2 - 2 = 0$.

▶**Solution** Since the powers of x in this equation are all even, we can use the substitution

$$u = x^2.$$

This gives

$$3x^4 + 5x^2 - 2 = 0$$
$$3(x^2)^2 + 5(x^2) - 2 = 0$$
$$3u^2 + 5u - 2 = 0$$

The equation involving u is quadratic and is easily factored:

$$3u^2 + 5u - 2 = 0$$
$$(3u - 1)(u + 2) = 0$$

This implies that

$$u = -2 \quad \text{or} \quad u = \frac{1}{3}.$$

We now go back and find the values of x that correspond to these solutions for u.

$$x^2 = u = -2 \implies x = \pm\sqrt{-2} = \pm i\sqrt{2}$$

$$x^2 = u = \frac{1}{3} \implies x = \pm\sqrt{\frac{1}{3}} = \pm\frac{\sqrt{3}}{3}$$

Therefore, the original equation has two real and two imaginary solutions.

☑ *Check It Out 1:* Solve the equation $2x^4 + x^2 - 3 = 0$. ▪

Technology Note

To solve the equation $t^6 - t^3 - 2 = 0$ with a graphing utility, graph the function $Y_1(x) = x^6 - x^3 - 2$ and use the ZERO feature to find the real zeros. Note that the independent variable must be x when using a graphing utility. Figure 3.5.2 shows the approximate solution $x \approx 1.2599$. The other solution, $x = -1$, can also be readily found.

You can also use the INTERSECT feature to find the real solutions of $t^6 - t^3 = 2$ by graphing $Y_1(x) = x^6 - x^3$ and $Y_2(x) = 2$.

Keystroke Appendix:
Section 9

Figure 3.5.2

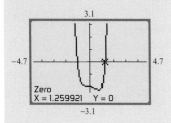

Just In Time

Review rational expressions in Section P.6.

Example 2 **Solving Using Substitution**

Solve the equation $t^6 - t^3 = 2$ for real values of t.

▶**Solution** Here, too, we will try to find a substitution that will reduce the equation to a quadratic. We note that only cubic powers of t appear in the equation, so we can use the substitution

$$u = t^3.$$

Thus we have

$$(t^3)^2 - (t^3) = 2$$
$$u^2 - u = 2$$
$$u^2 - u - 2 = 0$$
$$(u - 2)(u + 1) = 0$$
$$u = -1 \quad \text{or} \quad u = 2.$$

Converting back to t, we have

$$t^3 = u = -1 \implies t = -1$$
$$t^3 = u = 2 \implies t = \sqrt[3]{2}$$

Thus the real-valued solutions are $t = -1$ and $t = \sqrt[3]{2}$. Both of the equations $t^3 = -1$ and $t^3 = 2$ have complex-valued solutions as well, but we will not discuss those here.

☑ *Check It Out 2:* Solve the equation $3t^6 + 5t^3 = 2$ for real values of t. ■

Solving Equations Containing Rational Expressions

Many equations involving rational expressions can be reduced to linear or quadratic equations. To do so, we multiply both sides of the equation by the least common denominator (LCD) of each term in the equation. The process is illustrated in the following example.

Example 3 **Solving an Equation Containing Rational Expressions**

Solve $\dfrac{1}{x} = \dfrac{2}{x - 2} + 3$.

▶**Solution** The LCD of the three terms in this equation is $x(x - 2)$. Note that the term 3 has a denominator of 1. We proceed as follows.

$$\frac{1}{x} = \frac{2}{x - 2} + 3 \qquad \text{Original equation}$$

$$x(x - 2)\frac{1}{x} = x(x - 2)\frac{2}{x - 2} + x(x - 2)(3) \qquad \begin{array}{l}\text{Multiply } both \text{ sides of the}\\\text{equation by the LCD}\end{array}$$

$$(x - 2)(1) = 2x + 3x(x - 2) \qquad \text{Cancel like factors}$$

$$x - 2 = 2x + 3x^2 - 6x \qquad \text{Expand products}$$

$$-3x^2 + 5x - 2 = 0 \qquad \begin{array}{l}\text{Standard form of a quadratic}\\\text{equation}\end{array}$$

$$(-3x + 2)(x - 1) = 0 \qquad \text{Factor}$$

Technology Note

You can check solutions with a graphing calculator by storing each value of x and then evaluating the expression on each side of the equation for that value of x. Figure 3.5.3 shows the check for $x = \frac{2}{3}$. You can similarly check that $x = 1$ is also a solution.

In order to use a graphing utility to solve rational equations, you must interpret graphs of *rational functions*, discussed in Chapter 4. Thus, we will not discuss graphical solutions of rational equations at this point.

Keystroke Appendix:
Section 4

Figure 3.5.3

```
2/3→X
           .6666666667
1/X
                    1.5
2/(X−2)+3
                    1.5
```

Apply the Zero Product Rule to get

$$-3x + 2 = 0 \implies x = \frac{2}{3} \quad \text{or} \quad x - 1 = 0 \implies x = 1.$$

Thus the possible solutions are $x = \frac{2}{3}$ and $x = 1$. We must check both possibilities in the *original* equation.

Check $x = \frac{2}{3}$:

$$\frac{1}{x} = \frac{2}{x - 2} + 3 \qquad \text{Original equation}$$

$$\frac{1}{\frac{2}{3}} \overset{?}{=} \frac{2}{\frac{2}{3} - 2} + 3 \qquad \text{Let } x = \frac{2}{3}$$

$$\frac{3}{2} \overset{?}{=} \frac{2}{-\frac{4}{3}} + 3$$

$$\frac{3}{2} \overset{?}{=} -\frac{3}{2} + 3$$

$$\frac{3}{2} = \frac{3}{2} \qquad x = \frac{2}{3} \text{ checks}$$

Check $x = 1$:

$$\frac{1}{x} = \frac{2}{x - 2} + 3 \qquad \text{Original equation}$$

$$\frac{1}{1} \overset{?}{=} \frac{2}{1 - 2} + 3 \qquad \text{Let } x = 1$$

$$1 = 1 \qquad x = 1 \text{ checks}$$

Thus, $x = \frac{2}{3}$ and $x = 1$ are solutions.

☑ *Check It Out 3:* Solve $-\dfrac{2}{x + 3} = \dfrac{1}{x} - \dfrac{3}{2}$. ■

Solving Equations Containing Radical Expressions

The next example shows how to solve an equation involving variables under a square root symbol. The main idea is to isolate the term containing the radical and then square both sides of the equation to eliminate the radical.

Example **4** **Solving an Equation Containing One Radical**

Solve $\sqrt{3x + 1} + 2 = x - 1$.

▶Solution

$$\sqrt{3x + 1} + 2 = x - 1 \qquad \text{Original equation}$$

$$\sqrt{3x + 1} = x - 3 \qquad \text{Isolate the radical term by subtracting 2}$$

$$3x + 1 = x^2 - 6x + 9 \qquad \text{Square both sides}$$

$$0 = x^2 - 9x + 8 \qquad \text{Quadratic equation in standard form}$$

$$0 = (x - 8)(x - 1) \qquad \text{Factor}$$

The only possible solutions are $x = 1$ and $x = 8$. To determine whether they actually are solutions, we substitute them for x—one at a time—in the *original* equation:

Check $x = 8$: $\sqrt{3x + 1} + 2 = \sqrt{3(8) + 1} + 2 = 7$ and $x - 1 = 8 - 1 = 7$

Therefore, $x = 8$ is a solution. Now check $x = 1$:

Check $x = 1$: $\sqrt{3x + 1} + 2 = \sqrt{3(1) + 1} + 2 = 4$ but $x - 1 = 1 - 1 = 0$

Since $4 \neq 0$, $x = 1$ is not a solution of the original equation. Therefore, the only solution is $x = 8$.

✔ *Check It Out 4:* Solve $\sqrt{4x + 5} - 1 = x + 1$. ▪

> **Note** *When solving equations containing radicals, we often obtain extraneous solutions. This occurs as a result of modifying the original equation (such as by raising both sides of the equation to some power) in the course of the solution process. Therefore, it is very important to check all possible solutions by substituting each of them into the original equation.*

In the case of Example 4, the extraneous solution crept in when we squared both sides of the equation $\sqrt{3x + 1} = x - 3$, which gave $3x + 1 = x^2 - 6x + 9$. Note that $x = 1$ is a solution of the latter but not of the former.

If a radical equation contains two terms with variables under the radicals, isolate one of the radicals and raise both sides to an appropriate power. If a radical term containing a variable still remains, repeat the process. The next example illustrates the technique.

Example 5 Solving an Equation Containing Two Radicals

Solve $\sqrt{3x + 1} - \sqrt{x + 4} = 1$.

▶**Solution**

$$\sqrt{3x + 1} - \sqrt{x + 4} = 1 \qquad \text{Original equation}$$
$$\sqrt{3x + 1} = 1 + \sqrt{x + 4} \qquad \text{Isolate a radical}$$
$$3x + 1 = 1 + 2\sqrt{x + 4} + (x + 4) \qquad \text{Square both sides}$$
$$3x + 1 = x + 5 + 2\sqrt{x + 4} \qquad \text{Combine like terms}$$
$$2x - 4 = 2\sqrt{x + 4} \qquad \text{Isolate the radical}$$
$$4x^2 - 16x + 16 = 4x + 16 \qquad \text{Square both sides}$$
$$4x^2 - 20x = 0 \qquad \text{Quadratic equation in standard form}$$
$$4x(x - 5) = 0 \qquad \text{Factor the left-hand side}$$

Apply the Zero Product Rule to get

$$4x = 0 \implies x = 0$$
$$x - 5 = 0 \implies x = 5.$$

The only possible solutions are $x = 0$ and $x = 5$. To determine whether they actually are solutions, we substitute them for x—one at a time—in the *original* equation:

Check $x = 0$: $\sqrt{3x + 1} - \sqrt{x + 4} = \sqrt{3(0) + 1} - \sqrt{0 + 4} = -1 \neq 1$.

Technology Note

When using a graphing utility to solve the equation $\sqrt{3x + 1} + 2 = x - 1$, you can use the INTERSECT feature with $Y_1(x) = \sqrt{3x + 1} + 2$ and $Y_2(x) = x - 1$. Since you are solving the *original equation* with the graphing utility, you will not get any extraneous solutions. See Figure 3.5.4.

Keystroke Appendix: Section 9

Figure 3.5.4

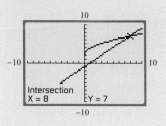

Intersection
X = 8   Y = 7

Therefore, $x = 0$ is not a solution. Now check $x = 5$:

Check $x = 5$: $\sqrt{3x + 1} - \sqrt{x + 4} = \sqrt{3(5) + 1} - \sqrt{(5) + 4} = 4 - 3 = 1$

Thus $x = 5$ is a solution of the original equation. Therefore, the only solution is $x = 5$.

☑ *Check It Out 5:* Solve $\sqrt{2x - 1} - \sqrt{x + 3} = -1$. ▪

Applications

Rational and radical equations occur in a variety of applications. In the following examples we examine two such applications.

Example 6 Average Cost

A theater club arranged a chartered bus trip to a play at a cost of $350. To lower costs, 10 nonmembers were invited to join the trip. The bus fare per person then decreased by $4. How many theater club members are going on the trip?

▶Solution First, identify the variable and the relationships among the many quantities mentioned in the problem.

Variable: The number of club members going on the trip, denoted by x
Total cost: $350
Number of people on the trip: $x + 10$
Original cost per club member: $\dfrac{350}{x}$
New cost per person: $\dfrac{350}{x} - 4$

Equation: (New cost per person)(number of people on trip) = total cost

We thus have the equation

$$\left(\frac{350}{x} - 4\right)(x + 10) = 350 \qquad \text{(Cost per person)(number of people) = total cost}$$

$$\left(\frac{350 - 4x}{x}\right)(x + 10) = 350 \qquad \text{Write the first factor as a single fraction}$$

$$(350 - 4x)(x + 10) = 350x \qquad \text{Multiply by } x$$

$$350x - 4x^2 - 40x + 3500 = 350x \qquad \text{Expand the left side}$$

$$-4x^2 - 40x + 3500 = 0 \qquad \text{Make the right side of the equation zero}$$

$$-4(x^2 + 10x - 875) = 0 \qquad \text{Factor out } -4$$

$$x^2 + 10x - 875 = 0 \qquad \text{Divide by } -4 \text{ on both sides}$$

$$(x + 35)(x - 25) = 0. \qquad \text{Factor}$$

Setting each factor equal to zero,

$$x + 35 = 0 \implies x = -35$$
$$x - 25 = 0 \implies x = 25.$$

Only the positive value of x makes sense, and so there are 25 members of the theater club going on the trip. You should check this solution in the original equation.

☑ *Check It Out 6:* Check the solution to Example 6. ▪

The next problem will illustrate an application of a radical equation.

Example 7 Distance and Rate

Jennifer is standing on one side of a river that is 3 kilometers wide. Her bus is located on the opposite side of the river. Jennifer plans to cross the river by rowboat and then jog the rest of the way to reach the bus, which is 10 kilometers down the river from a point B directly across the river from her current location (point A). If she can row 5 kilometers per hour and jog 7 kilometers per hour, at which point on the other side of the river should she dock her boat so that it will take her a total of exactly 2 hours to reach her bus? Assume that Jennifer's path on each leg of the trip is a straight line and that there is no river current or wind speed.

▶ **Solution** First we draw a figure illustrating the problem. See Figure 3.5.5.

Figure 3.5.5

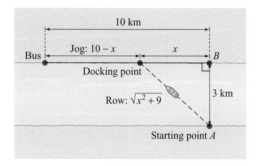

Recall that distance = speed × time, so time = $\frac{\text{distance}}{\text{speed}}$. From the diagram, the distance rowed is $\sqrt{x^2 + 9}$ and the distance jogged is $10 - x$. Thus,

$$\text{Time to row} = \frac{\text{distance rowed}}{\text{rowing speed}} = \frac{\sqrt{x^2 + 9}}{5}$$

$$\text{Time to jog} = \frac{\text{distance jogged}}{\text{jogging speed}} = \frac{10 - x}{7}.$$

Since the total time must equal 2 hours, we have

$$\frac{\sqrt{x^2 + 9}}{5} + \frac{10 - x}{7} = 2 \qquad \text{Time rowed + time jogged = 2 hours}$$

$$7(\sqrt{x^2 + 9}) + 5(10 - x) = 70 \qquad \text{Multiply by 35 to clear fractions}$$

$$7(\sqrt{x^2 + 9}) + 50 - 5x = 70 \qquad \text{Distribute the 5}$$

$$7(\sqrt{x^2 + 9}) = 20 + 5x \qquad \text{Isolate the radical expression}$$

$$49(x^2 + 9) = 400 + 200x + 25x^2 \qquad \text{Square both sides}$$

$$49x^2 + 441 = 400 + 200x + 25x^2 \qquad \text{Distribute the 49}$$

$$24x^2 - 200x + 41 = 0. \qquad \text{Write the quadratic equation in standard form}$$

Using the quadratic formula to solve the equation, we have

$$x = \frac{-(-200) \pm \sqrt{(-200)^2 - 4(24)(41)}}{2(24)} \approx 0.2103 \quad \text{or} \quad 8.123.$$

In order to reach the bus in 2 hours, Jennifer should dock the boat either 0.2103 kilometers along the river from point B or 8.123 kilometers along the river from point B.

Graphical approach: Note that Jennifer's total travel time is a function of x, the distance of the docking point from point B:

$$t(x) = \frac{\sqrt{x^2 + 9}}{5} + \frac{10 - x}{7}$$

It is useful first to generate a table of function values to see how the value of x affects the value of $t(x)$:

Table 3.5.1

x (kilometers)	0	1.5	3	4.5	6	7.5	9	10
$t(x)$ (hours)	2.029	1.885	1.849	1.867	1.913	1.973	2.040	2.088

From Table 3.5.1, we see that as x increases, the amount of time required to reach the bus first decreases and then increases. This is because there is a trade-off between the total distance traveled and the two different speeds at which Jennifer travels, one for rowing and the other for jogging. It will take her exactly 2 hours to reach her bus for x somewhere between 0 and 1.5 kilometers, or for x somewhere between 7.5 and 9 kilometers.

To get a precise solution of the equation $t(x) = \dfrac{\sqrt{x^2 + 9}}{5} + \dfrac{10 - x}{7} = 2$, use the INTERSECT feature of your graphing utility and graph $y_1(x) = \dfrac{\sqrt{x^2 + 9}}{5} + \dfrac{10 - x}{7}$ and $y_2(x) = 2$. The two graphical solutions, pictured in Figure 3.5.6, agree with the algebraic solutions. The window size, the choice of which was guided by the table of function values, is $[0, 10]$ by $[1.75, 2.25](0.25)$. The graphical solution enables you to observe how the total time changes as a function of distance. This additional information is not available when solving an equation algebraically.

Figure 3.5.6

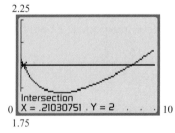

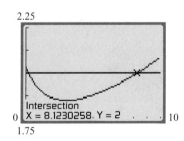

✔ **Check It Out 7:** Rework Example 7 for the case in which Jennifer can jog at a speed of 8 kilometers per hour, with all other information remaining the same. ■

Discover *and* Learn

In Example 7, use a graphing utility to find the minimum time it takes Jennifer to get to the bus.

In practice, problems such as the one in Example 7 would ask for the minimum time it takes to reach a destination. You can solve such problems by hand only by using calculus. However, if you are using a graphing utility, the MINIMUM feature can be used to find the minimum amount of time it takes to make the trip.

3.5 Key Points

▸ If an equation contains a polynomial, use a substitution such as $u = x^2$ or $u = x^3$ to reduce the given equation to quadratic form. Then solve by using factoring or the quadratic formula. You should *always* check your solutions.

▸ If an equation involves rational expressions, multiply both sides of the equation by the LCD of the terms in the equation. Solve the resulting quadratic or linear equation and check your solutions.

▸ If an equation involves square roots, isolate the radical term on one side and square both sides. If, after squaring, there is still another radical, repeat the process. Solve the resulting quadratic equation and check your solutions.

▸ If using a graphing utility, you must input the expressions in the original equation. Solutions found by a graphing utility using the original expressions will never be extraneous solutions.

3.5 Exercises

▸**Just in Time Exercises**   These exercises correspond to the Just in Time references in this section. Complete them to review topics relevant to the remaining exercises.

1. True or False: The function $f(x) = \dfrac{2x}{x - 3}$ is undefined at $x = 3$.

2. True or False: The function $f(x) = \dfrac{2x}{x - 3}$ is undefined at $x = 0$.

In Exercises 3–6, multiply.

3. $(x - 2)\left(\dfrac{3}{x - 2}\right)$

4. $x^2\left(\dfrac{5}{x}\right)$

5. $x(x - 5)\left(\dfrac{2}{x}\right)$

6. $(2x + 1)(x - 3)\left(\dfrac{x + 7}{x - 3}\right)$

▸**Skills** This set of exercises will reinforce the skills illustrated in this section.

In Exercises 7–18, solve the polynomial equation. In Exercises 7–14, find all solutions. In Exercises 15–18, find only real solutions. Check your solutions.

7. $x^4 - 49 = 0$

8. $x^4 - 25 = 0$

9. $x^4 - 10x^2 = -21$

10. $x^4 - 5x^2 = 24$

11. $6s^4 - s^2 - 2 = 0$

12. $4s^4 + 11s^2 - 3 = 0$

13. $4x^4 - 7x^2 = 2$

14. $-x^4 + 2x^2 - 1 = 0$

15. $x^6 - 4x^3 = 5$

16. $x^6 = x^3 + 6$

17. $3t^6 + 14t^3 = -8$

18. $2x^6 - 7x^3 = 7$

In Exercises 19–34, solve the rational equation. Check your solutions.

19. $\dfrac{2}{3} + \dfrac{3}{5} = \dfrac{2}{x}$

20. $\dfrac{1}{4} - \dfrac{3}{2} = \dfrac{3}{x}$

21. $-\dfrac{2}{3x} + \dfrac{1}{x} = \dfrac{1}{4}$

22. $\dfrac{1}{2x} + \dfrac{4}{5} = \dfrac{3}{x}$

23. $\dfrac{1}{x^2} - \dfrac{3}{x} = 10$

24. $\dfrac{1}{x^2} - \dfrac{7}{x} = 18$

25. $\dfrac{2x}{x - 1} - \dfrac{3}{x} = 2$

26. $-\dfrac{3x}{x + 2} + \dfrac{1}{x} = 2$

27. $\dfrac{3}{x+1} + \dfrac{2}{x-3} = 4$

28. $\dfrac{1}{2x-3} - \dfrac{x}{x-1} = 2$

29. $\dfrac{1}{x^2-x-6} + \dfrac{3}{x+2} = \dfrac{-4}{x-3}$

30. $\dfrac{1}{x^2+4x-5} + \dfrac{6}{x+5} = \dfrac{1}{x-1}$

31. $\dfrac{x}{2x^2+x-3} + \dfrac{1}{x-1} = \dfrac{3}{2x+3}$

32. $\dfrac{x}{3x^2+5x-2} - \dfrac{5}{x+2} = \dfrac{-1}{3x-1}$

33. $\dfrac{x-3}{2x-4} + \dfrac{1}{x^2-4} = \dfrac{1}{x+2}$

34. $\dfrac{x-1}{3x+3} - \dfrac{9}{x^2-1} = \dfrac{2}{x+1}$

In Exercises 35–50, solve the radical equation to find all real solutions. Check your solutions.

35. $\sqrt{x+3} = 5$

36. $\sqrt{x+2} = 6$

37. $\sqrt{x^2+1} = \sqrt{17}$

38. $\sqrt{x^2+3} = \sqrt{28}$

39. $\sqrt{x^2+6x-1} = 3$

40. $\sqrt{x^2-5x+4} = 10$

41. $\sqrt{x+1} + 2 = x$

42. $\sqrt{2x-1} + 2 = x$

43. $\sqrt[3]{x+3} = 5$

44. $\sqrt[3]{5x-3} = \sqrt[3]{4}$

45. $\sqrt[4]{x-1} = 2$

46. $\sqrt[4]{2x+1} = 3$

47. $\sqrt{2x+3} - \sqrt{x-2} = 2$

48. $\sqrt{x+3} - \sqrt{x+2} = 4$

49. $\sqrt{x+3} + \sqrt{x-5} = 4$

50. $\sqrt{x+10} - \sqrt{x-1} = 3$

In Exercises 51–54, solve the equation to find all real solutions. Check your solutions.

51. $x - 4\sqrt{x} = -3$ (*Hint:* Use $u = \sqrt{x}$.)

52. $x - 6\sqrt{x} = -5$ (*Hint:* Use $u = \sqrt{x}$.)

53. $3x^{2/3} + 2x^{1/3} - 1 = 0$ (*Hint:* Use $u = x^{1/3}$.)

54. $2x^{2/3} - 5x^{1/3} - 3 = 0$ (*Hint:* Use $u = x^{1/3}$.)

In Exercises 55–60, use a graphing utility to find all real solutions. You may need to adjust the window size manually or use the ZOOMFIT feature to get a clear graph.

55. Solve $\sqrt{2.35-x} + 1.8 = 2.75$.

56. Solve $\sqrt{x-1.95} - 3.6 = -2.5$.

57. Solve $\sqrt{x-0.8} + \sqrt{0.25x+0.9} = 1.6$.

58. Solve $\sqrt{0.3x+0.95} - \sqrt{0.75x-0.5} = -0.3$.

59. Graphically solve $\sqrt{x+1} = x + k$ for $k = \frac{1}{2}$, 1, and 2. How many solutions does the equation have for each value of k?

60. Graphically solve $\sqrt{x-k} = x$ for $k = -2, 0$, and 2. How many solutions does the equation have for each value of k?

▶**Applications** In this set of exercises you will use radical and rational equations to study real-world problems.

61. **Average Cost** Four students plan to rent a minivan for a weekend trip and share equally in the rental cost of the van. By adding two more people, each person can save $10 on his or her share of the cost. How much is the total rental cost of the van?

62. **Work Rate** Two painters are available to paint a room. Working alone, the first painter can paint the room in 5 hours. The second painter can paint the room in 4 hours working by herself. If they work together, they can paint

the room in t hours. To find t, we note that in 1 hour, the first painter paints $\frac{1}{5}$ of the room and the second painter paints $\frac{1}{4}$ of the room. If they work together, they paint $\frac{1}{t}$ portion of the room. The equation is thus

$$\frac{1}{5} + \frac{1}{4} = \frac{1}{t}.$$

Find t, the time it takes both painters to paint the room working together.

63. **Work Rate** Two water pumps work together to fill a storage tank. If the first pump can fill the tank in 6 hours and the two pumps working together can fill the tank in 4 hours, how long would it take to fill the storage tank using just the second pump? (*Hint:* To set up an equation, refer to the preceding problem.)

64. **Engineering** In electrical circuit theory, the formula

$$\frac{1}{R} = \frac{1}{R_1} + \frac{1}{R_2}$$

is used to find the total resistance R of a circuit when two resistors with resistances R_1 and R_2 are connected in parallel. In such a parallel circuit, if the total resistance R is $\frac{8}{3}$ ohms and R_2 is twice R_1, find the resistances R_1 and R_2.

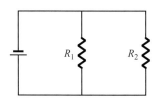

Use the following figure for Exercises 65–68.

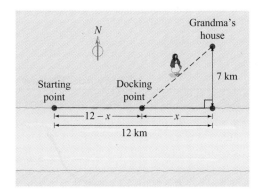

65. **Distance and Rate** To get to her grandmother's house, Little Red Riding Hood first rows a boat along a river at a speed of 10 kilometers per hour and then walks through the woods at 4 kilometers per hour. If she starts from a point that is 7 kilometers south and 12 kilometers west of Grandma's house, at what point along the river should she dock her boat so that she can reach the house in a total of exactly 3 hours? Assume that all paths are straight paths.

66. **Distance and Rate** The Big Bad Wolf would like to beat Little Red Riding Hood to her grandmother's house. The Big Bad Wolf can row a boat at a speed of 12 kilometers per hour and can walk through the woods at 5 kilometers per hour. If the wolf starts from the same point as Little Red Riding Hood (see Exercise 65), at what point along the river should the wolf dock his boat so that he will reach the house in a total of exactly 2.5 hours? Assume that all paths are straight paths.

67. **Minimizing Travel Time** For the conditions given in Exercise 65, at what point along the river should Little Red Riding Hood dock her boat so that she can reach the house in the *least possible time*?

68. **Minimizing Travel Time** For the conditions given in Exercise 66, at what point along the river should the wolf dock his boat so that he can reach the house in the *least possible time*?

▶ **Concepts** This set of exercises will draw on the ideas presented in this section and your general math background.

69. Explain what is wrong with the following steps for solving a radical equation.

$$\sqrt{x + 1} - 2 = 0$$
$$(x + 1) + 4 = 0$$
$$x = -5$$

70. Without doing any calculations, explain why

$$\sqrt{x + 1} = -2$$

does *not* have a solution.

71. How many zeros, real and nonreal, does the function $f(x) = x^4 - 1$ have? How many x-intercepts does the graph of f have?

Chapter 3   # Summary

Section 3.1 Graphs of Quadratic Functions

Concept	Illustration	Study and Review
Definition of a quadratic function A function f is a **quadratic function** if it can be expressed in the form $f(x) = ax^2 + bx + c$, where a, b, and c are real numbers and $a \neq 0$.	$f(x) = 2x^2 + 3x - 2$ and $g(x) = 3 - x + 6x^2$ are examples of quadratic functions.	Example 1 Chapter 3 Review, Exercises 1, 2
Graph of a quadratic function The graph of the function $f(x) = ax^2 + bx + c$ is called a **parabola.** If $a > 0$, the parabola opens upward. If $a < 0$, the parabola opens downward.	The graph of $f(x) = -3x^2 + 1$ opens downward, since $a = -3$ is negative. The graph of $g(x) = 2x^2 - 3x$ opens upward, since $a = 2$ is positive.	Example 2 Chapter 3 Review, Exercises 1, 2
Vertex form of a quadratic function The **vertex form** of a quadratic function is given by $f(x) = a(x - h)^2 + k$. The graph of any quadratic function can be represented as a series of transformations of the graph of the basic function $y = x^2$.	The quadratic function $f(x) = 3(x + 2)^2 - 1$ is written in vertex form. The graph of f can be obtained from the graph of $y = x^2$ by a vertical stretch by a factor of 3, then a horizontal shift of 2 units to the left, and finally a vertical shift of 1 unit down.	Examples 3, 4 Chapter 3 Review, Exercises 3–6
Vertex of a parabola The **vertex** (h, k) of a parabola is given by $h = \dfrac{-b}{2a}$ and $k = f(h) = f\left(\dfrac{-b}{2a}\right) = c - \dfrac{b^2}{4a}$. The **axis of symmetry** of the parabola is given by $x = \dfrac{-b}{2a}$.		Examples 5–8 Chapter 3 Review, Exercises 7–20

Section 3.2 Quadratic Equations

Concept	Illustration	Study and Review
Definition of a quadratic equation A **quadratic equation** is an equation that can be written in the **standard form** $$ax^2 + bx + c = 0$$ where a, b, and c are real numbers with $a \neq 0$.	$x - 2x^2 = 1$ is a quadratic equation, since it can be rewritten as $-2x^2 + x - 1 = 0$.	Definition on p. 231
Zero Product Rule If a product of real numbers is zero, then at least one of the factors is zero.	If $(2x + 1)(x - 2) = 0$, then $2x + 1 = 0$ or $x - 2 = 0$.	Definition on p. 231

Continued

Section 3.2 **Quadratic Equations**

Concept	Illustration	Study and Review
Solving a quadratic equation by factoring First write the quadratic equation in standard form. Then factor the nonzero side of the equation, if possible. Use the Zero Product Rule to find the solution(s).	To solve $2x^2 - 3x - 2 = (2x + 1)(x - 2) = 0$, set $2x + 1 = 0$ to get $x = -\frac{1}{2}$ and set $x - 2 = 0$ to get $x = 2$. The solutions are $x = 2$ and $x = -\frac{1}{2}$.	Example 1 Chapter 3 Review, Exercises 21–26
Finding the zeros of a quadratic function and the *x*-intercepts of its graph The real number values of x at which $f(x) = 0$ are called the **real zeros** of the function f. The x-coordinate of an x-intercept is a value of x such that $f(x) = 0$.	The real zeros of $f(x) = (x - 1)(x + 2)$ are $x = 1$ and $x = -2$. The x-intercepts of the graph of f are $(1, 0)$ and $(-2, 0)$.	Examples 2, 3 Chapter 3 Review, Exercises 27–30
Principle of square roots If $x^2 = c$, where $c \geq 0$, then $x = \pm\sqrt{c}$.	The solution of $x^2 = 12$ is $x = \pm\sqrt{12} = \pm 2\sqrt{3}$.	Example 4 Chapter 3 Review, Exercises 31–34
Solving quadratic equations by completing the square To solve $x^2 + bx = k$ by completing the square, add $\left(\frac{b}{2}\right)^2$ to both sides of the equation. Then apply the principle of square roots to solve.	To solve $x^2 + 6x = 4$, add $\left(\frac{6}{2}\right)^2 = 9$ to both sides of the equation to get $$x^2 + 6x + 9 = (x + 3)^2 = 13.$$ Take the square root of both sides to get $x + 3 = \pm\sqrt{13}$. The solutions are $x = -3 + \sqrt{13}$ and $x = -3 - \sqrt{13}$.	Example 5 Chapter 3 Review, Exercises 31–34
Solving quadratic equations by using the quadratic formula The solutions of $ax^2 + bx + c = 0$, with $a \neq 0$, are given by the **quadratic formula** $$x = \frac{-b \pm \sqrt{b^2 - 4ac}}{2a}.$$	Using the quadratic formula to solve $2x^2 - 2x - 1 = 0$, we have $$x = \frac{-(-2) \pm \sqrt{(-2)^2 - 4(2)(-1)}}{2(2)}.$$ Simplifying, the solutions are $$x = \frac{1}{2} + \frac{\sqrt{3}}{2}, \frac{1}{2} - \frac{\sqrt{3}}{2}.$$	Examples 6–10 Chapter 3 Review, Exercises 35–42
The discriminant The quantity $b^2 - 4ac$ under the radical in the quadratic formula is known as the **discriminant.** The number of solutions of $ax^2 + bx + c = 0$ can be determined as follows.	The equation $-3x^2 + 4x + 1 = 0$ has *two* distinct, real solutions because $b^2 - 4ac = 16 - (4)(-3)(1) = 28$ is positive. The equation $x^2 + x + 1 = 0$ has *no* real solutions because $b^2 - 4ac = 1 - (4)(1)(1) = -3$ is negative.	Examples 7, 8 Chapter 3 Review, Exercises 43–46

$b^2 - 4ac$	Number of solutions
Positive	Two distinct, real solutions
Zero	One real solution
Negative	No real solutions

Section 3.3 **Complex Numbers and Quadratic Equations**

Concept	Illustration	Study and Review
Definition of a complex number The number i is defined as $\sqrt{-1}$. A **complex number** is a number of the form $a + bi$, where a and b are real numbers. If $a = 0$, then the number is a **pure imaginary number.**	The number $3 - 4i$ is a complex number with $a = 3$ and $b = -4$. The number $i\sqrt{2}$ is a complex number with $a = 0$ and $b = \sqrt{2}$. It is also a pure imaginary number.	Examples 1–5 Chapter 3 Review, Exercises 47–50
Addition and subtraction of complex numbers To add two complex numbers, add their corresponding real and imaginary parts. To subtract two complex numbers, subtract their corresponding real and imaginary parts.	Addition: $$(6 + 2i) + (-4 - 3i)$$ $$= (6 + (-4)) + (2i + (-3i))$$ $$= 2 - i$$ Subtraction: $$(-3 + i) - (7 + 5i)$$ $$= -3 + i - 7 - 5i$$ $$= -10 - 4i$$	Example 6 Chapter 3 Review, Exercises 55–60
Multiplication of complex numbers To multiply two complex numbers, apply the rules of multiplication of binomials.	$$(1 + 2i)(3 - i) = 3 - i + 6i - 2i^2$$ $$= 5 + 5i$$	Example 7 Chapter 3 Review, Exercises 55–60
Conjugates and division of complex numbers The **complex conjugate** of a complex number $a + bi$ is given by $a - bi$. The quotient of two complex numbers, written as a fraction, can be found by multiplying the numerator and denominator by the complex conjugate of the denominator.	$$\frac{3 - 2i}{1 + 2i} = \frac{(3 - 2i)(1 - 2i)}{(1 + 2i)(1 - 2i)}$$ $$= -\frac{1}{5} - \frac{8}{5}i$$	Examples 8–10 Chapter 3 Review, Exercises 51–60
Zeros of quadratic functions and solutions of quadratic equations By using complex numbers, one can find the nonreal zeros of a quadratic function and the nonreal solutions of a quadratic equation by using the quadratic formula.	The nonreal solutions of the equation $2x^2 + x + 1 = 0$ are $$x = \frac{-(1) \pm \sqrt{(1)^2 - 4(2)(1)}}{2(2)}$$ $$= -\frac{1}{4} + i\frac{\sqrt{7}}{4}, -\frac{1}{4} - i\frac{\sqrt{7}}{4}.$$	Examples 11, 12 Chapter 3 Review, Exercises 61–64

Section 3.4 **Quadratic Inequalities**

Concept	Illustration	Study and Review
Quadratic inequality A **quadratic inequality** is of the form $ax^2 + bx + c > 0$, where $>$ may be replaced by $\geq$, $<$, or $\leq$.	$-3x^2 - 2x + 1 \geq 0$ and $2x^2 - 1 < 0$ are both examples of quadratic inequalities.	Definition on p. 256

Continued

Section 3.4 Quadratic Inequalities

Concept	Illustration	Study and Review
Graphical approach to solving inequalities By examining the graph of a quadratic function f, it is possible to see where $f(x) > 0$, $f(x) = 0$, and $f(x) < 0$.	To solve $x^2 - 1 > 0$, observe that the graph of $f(x) = x^2 - 1$ is above the x-axis for $x > 1$ and $x < -1$. 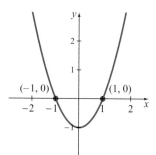	Example 1 Chapter 3 Review, Exercises 65–68
Algebraic approach to solving inequalities Solve $ax^2 + bx + c = 0$. Use the solutions to divide the number line into disjoint intervals. Use test values in each interval to decide which intervals satisfy the inequality.	To solve $x^2 - 1 > 0$ algebraically, solve $x^2 - 1 = 0$ to get $x = \pm 1$. By using a test value for x in each of the intervals $(-\infty, -1)$, $(-1, 1)$, and $(1, \infty)$, you can observe that $x^2 - 1 > 0$ in $(-\infty, -1)$ or $(1, \infty)$.	Examples 2–6 Chapter 3 Review, Exercises 69–78

Section 3.5 Equations That Are Reducible to Quadratic Form; Rational and Radical Equations

Concept	Illustration	Study and Review
Equations that are reducible to quadratic form Use a substitution such as $u = x^2$ or $u = x^3$ to reduce the given equation to quadratic form. Then solve by using factoring or the quadratic formula. You should *always* check your solutions.	To solve $t^4 - 2t^2 - 3 = 0$, let $u = t^2$ to get $$u^2 - 2u - 3 = (u - 3)(u + 1) = 0.$$ Thus $u = t^2 = 3$ or $u = t^2 = -1$. Solving for t, $$t = \sqrt{3}, \ -\sqrt{3}, \ i, \ -i.$$	Examples 1, 2 Chapter 3 Review, Exercises 79–82
Equations containing rational expressions When an equation involves rational expressions, multiply both sides of the equation by the least common denominator (LCD) of all terms in the equation. This results in a linear or quadratic equation.	To solve $\frac{1}{x} - \frac{2}{x^2} = -6$, multiply both sides of the equation by the LCD, x^2. Rearranging terms, $$6x^2 + x - 2 = 0$$ $$(3x + 2)(2x - 1) = 0.$$ Thus $x = \frac{1}{2}$ or $-\frac{2}{3}$. Both values of x check in the *original* equation.	Examples 3, 6 Chapter 3 Review, Exercises 83–86
Equations containing radical expressions When an equation contains a square root symbol, isolate the radical term and square both sides. Solve the resulting quadratic equation and check your solution(s).	To solve $\sqrt{x + 1} - x = -1$, rewrite as $\sqrt{x + 1} = x - 1$ and square both sides to get $$x + 1 = x^2 - 2x + 1$$ $$-x^2 + 3x = -x(x - 3) = 0.$$ Possible values for x are $x = 0$ and $x = 3$. Only $x = 3$ checks in the *original* equation, and it is therefore the only solution.	Examples 4, 5, 7 Chapter 3 Review, Exercises 87–90

Chapter 3

Review Exercises

Section 3.1

In Exercises 1 and 2, graph each pair of functions on the same set of coordinate axes, and find the domain and range of each function.

1. $f(x) = x^2$, $g(x) = 3x^2$

2. $f(x) = -x^2$, $g(x) = -\frac{1}{3}x^2$

In Exercises 3–6, use transformations to graph the quadratic function and find the vertex of the associated parabola.

3. $f(x) = (x + 3)^2 - 1$ 4. $g(x) = (x - 1)^2 + 4$

5. $f(x) = -2x^2 - 1$ 6. $g(x) = 3(x + 1)^2 + 3$

In Exercises 7–10, write the quadratic function in the form $f(x) = a(x - h)^2 + k$ by completing the square. Also find the vertex of the associated parabola and determine whether it is a maximum or minimum point.

7. $f(x) = x^2 - 4x + 3$ 8. $g(x) = 3 - 6x + x^2$

9. $f(x) = 4x^2 + 8x - 1$ 10. $g(x) = -3x^2 + 12x + 5$

In Exercises 11–16, find the vertex and axis of symmetry of the associated parabola for each quadratic function. Then find at least two additional points on the parabola and sketch the parabola by hand.

11. $f(x) = 3(x + 1)^2 - 6$ 12. $f(t) = -2(t + 3)^2$

13. $f(s) = -s^2 - 3s + 1$ 14. $f(x) = 1 - 4x + 3x^2$

15. $f(x) = \frac{2}{3}x^2 + x - 3$ 16. $f(t) = \frac{1}{4}t^2 - 2t + 1$

In Exercises 17–20, find the vertex and axis of symmetry of the associated parabola for each quadratic function. Sketch the parabola. Find the intervals on which the function is increasing and decreasing, and find the range.

17. $f(x) = x^2 - 2x + 1$ 18. $g(x) = -2x^2 - 3x$

19. $g(x) = \frac{1}{2}x^2 - 2x + 5$ 20. $f(x) = -\frac{2}{3}x^2 + x - 1$

Section 3.2

In Exercises 21–26, solve the quadratic equation by factoring.

21. $x^2 - 9 = 0$ 22. $2x^2 - 8 = 0$

23. $x^2 - 9x + 20 = 0$ 24. $x^2 + x - 12 = 0$

25. $6x^2 - x - 12 = 0$

26. $-6x^2 - 5x + 4 = 0$

In Exercises 27–30, factor to find the x-intercepts of the parabola described by the quadratic function. Also find the zeros of the function.

27. $h(x) = -2x^2 - 3x + 5$ 28. $f(x) = x^2 - 4x + 4$

29. $f(x) = -3x^2 - 5x + 2$ 30. $g(x) = 2x^2 - 7x + 3$

In Exercises 31–34, solve the quadratic equation by completing the square. Find only real solutions.

31. $x^2 - 4x - 2 = 0$ 32. $-x^2 - 2x = -5$

33. $x^2 + 3x - 7 = 0$ 34. $-2x^2 + 8x = 1$

In Exercises 35–42, solve the quadratic equation by using the quadratic formula. Find only real solutions.

35. $-3x^2 - x + 3 = 0$ 36. $-x^2 + 2x + 2 = 0$

37. $t^2 - t + 5 = 0$ 38. $-2x^2 + x + 4 = 0$

39. $-3t^2 - 2t + 4 = 0$ 40. $\frac{4}{3}x^2 + x = 2$

41. $-s^2 - \sqrt{2}s = -\frac{1}{2}$ 42. $-(x + 1)(x - 4) = -6$

In Exercises 43–46, for each function of the form $f(x) = ax^2 + bx + c$, find the discriminant, $b^2 - 4ac$, and use it to determine the number of x-intercepts of the graph of f. Also determine the number of real solutions of the equation $f(x) = 0$.

43. $f(x) = x^2 - 6x + 4$ 44. $f(x) = -2x^2 - 7x$

45. $f(x) = x^2 - 6x + 9$ 46. $f(x) = -x^2 - x - 2$

Section 3.3

In Exercises 47–50, find the real and imaginary parts of the complex number.

47. $\sqrt{3}$ 48. $-\frac{3}{2}i$

49. $7 - 2i$ 50. $-1 - \sqrt{-5}$

In Exercises 51–54, find the complex conjugate of each number.

51. $\frac{1}{2}$ 52. $3 - i$

53. $i + 4$ 54. $1 - \sqrt{2} + 3i$

In Exercises 55–60, find $x + y$, $x - y$, xy, and x/y.

55. $x = 1 + 4i$,
   $y = 2 - 3i$

56. $x = 3 + 2i$,
   $y = -4 + 3i$

57. $x = 1.5 - 3i$,
   $y = 2i - 1$

58. $x = -\sqrt{2} + i$,
   $y = -3$

59. $x = \dfrac{1}{2}i - 1$, $y = i + \dfrac{3}{2}$

60. $x = -i$,
   $y = -\sqrt{-3} + 1$

In Exercises 61–64, find all solutions of the quadratic equation. Relate the solutions of the equation to the zeros of an appropriate quadratic function.

61. $-x^2 + x - 3 = 0$

62. $-2x^2 = -x + 1$

63. $-\dfrac{4}{5}t - 1 = t^2$

64. $2t^2 - \sqrt{13} = t$

Section 3.4

In Exercises 65–68, use the graph of f to solve the inequality.

65. $f(x) \le 0$

66. $f(x) \ge 0$

67. $f(x) < 0$

68. $f(x) > 0$

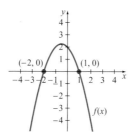

In Exercises 69–74, solve the inequality by factoring.

69. $3x^2 - 12 \ge 0$

70. $-4x^2 + 36 < 0$

71. $4x^2 + 21x + 5 \le 0$

72. $-6x^2 - 5x > -4$

73. $2x^2 - 11x \le -12$

74. $2x^2 + 4x + 2 < 0$

In Exercises 75–78, solve the inequality algebraically or graphically.

75. $-2x^2 + 7x > 3$

76. $x^2 - 3x - 1 \le 0$

77. $(x + 1)(x - 1) \ge 3x$

78. $4x^2 + 3 < 0$ (*Hint:* Examine the corresponding graph.)

Section 3.5

In Exercises 79–82, solve the polynomial equation.

79. $x^4 - 11x^2 + 24 = 0$

80. $x^6 - 4x^3 = 21$

81. $-3x^4 + 11x^2 = -4$

82. $6x^2 - 7x = 3$

In Exercises 83–86, solve the rational equation.

83. $\dfrac{3x}{x + 1} + \dfrac{1}{x} = \dfrac{5}{2}$

84. $\dfrac{5}{x + 2} + x = -8$

85. $\dfrac{1}{x^2 + 2x - 3} + \dfrac{4}{x + 3} = 1$

86. $\dfrac{2}{x^2 + 3x - 4} + \dfrac{1}{x - 1} = -\dfrac{6}{x + 4}$

In Exercises 87–90, solve the radical equation.

87. $\sqrt{2x + 1} - x = -1$

88. $\sqrt{3x + 4} + 2 = x$

89. $\sqrt{3x - 5} - \sqrt{x - 3} = 2$

90. $\sqrt{2x + 3} + \sqrt{x + 6} = 6$

Applications

91. Construction A rectangular play yard is to be enclosed with a fence on three of its sides and a brick wall on the fourth side. If 120 feet of fencing material is available, what dimensions will yield the maximum area?

92. Economics The dollar value of toys, games, and sporting goods imported into the United States can be modeled by the quadratic function

$$s(t) = 0.1525t^2 + 0.3055t + 18.66$$

where t is the number of years since 1998 and $s(t)$ is the dollar amount of the imports in billions of dollars. The model is based on data for the years 1998–2002, inclusive. (*Source: Statistical Abstract of the United States*)

(a) Use this model to estimate the dollar amount of imported toys, games, and sporting goods for the year 2001. Compare your estimate to the actual value of $20.9 billion.

(b) Use this model to predict the dollar amount of imported toys, games, and sporting goods for the year 2006.

(c) Use a graphing utility to graph the function s. What is an appropriate range of values for t?

93. Business Expenditures The percentage of total operating expenses incurred by airlines for airline food can be modeled by the function

$$f(t) = -0.0055t^2 + 0.116t + 2.90$$

where t is the number of years since 1980. The model is based on data for selected years from 1980 to 2000. (*Source: Statistical Abstract of the United States*)

(a) What is the y-intercept of the graph of this function, and what does it signify in relation to this problem?

(b) In what year between 1980 and 2000 was the expenditure for airline food 2% of the total operating expenses?

(c) Is this model reliable as a long-term indicator of airline expenditures for airline food as a percentage of total operating expenses? Justify your answer.

94. **Revenue** The following table lists the total revenue, in millions of dollars, realized by payphone providers from 1996 to 2001. (*Source:* Federal Communications Commission)

(a) Let t denote the number of years since 1996. Make a scatter plot of the revenue r versus time t.

(b) From your plot, what type of trend do you observe—linear or quadratic? Explain.

(c) Find the best-fit function for the given data points.

(d) Find the year between 1996 and 2001 during which the revenue realized by payphone providers was highest.

(e) To what would you attribute the sharp decline in payphone revenue in recent years?

Year	Revenue (millions of dollars)
1996	357
1997	933
1998	1101
1999	1213
2000	972
2001	836

Chapter 3   Test

1. Write $f(x) = 2x^2 - 4x + 1$ in the form $f(x) = a(x - h)^2 + k$. Find the vertex of the associated parabola and determine if it is a maximum or a minimum point.

In Exercises 2–4, find the vertex and axis of symmetry of the parabola represented by $f(x)$. Sketch the graph of f and find its range.

2. $f(x) = -(x - 1)^2 + 2$ 3. $f(x) = x^2 + 4x + 2$

4. $f(x) = -2x^2 + 8x - 4$

5. Sketch a graph of $f(x) = 3x^2 + 6x$. Find the vertex, axis of symmetry, and intervals on which the function is increasing or decreasing.

6. Find the x-intercepts of the graph of $f(x) = x^2 - 5x - 6$ by factoring. Also, find the zeros of f.

7. Solve $2x^2 - 4x - 3 = 0$ by completing the square.

In Exercises 8 and 9, solve the equation using the quadratic formula.

8. $3x^2 + x - 1 = 0$ 9. $-2x^2 + 2x + 3 = 0$

In Exercises 10–12, find all real solutions using any method.

10. $3x^2 - x - 4 = 0$ 11. $-x^2 + x = 5$

12. $2x^2 + 2x - 5 = 0$

13. Find the real and imaginary parts of the complex number $4 - \sqrt{-2}$.

In Exercises 14–16, perform the indicated operations and write in the form $a + bi$.

14. $5 + 4i - (6 + 2i)$ 15. $(3 - 4i)(-2 + i)$

16. $\dfrac{2 + i}{3 - 2i}$

In Exercises 17 and 18, find all solutions, real or complex, of the equation.

17. $x^2 + 2x + 3 = 0$ 18. $-2x^2 + x - 1 = 0$

19. Solve the inequality $3x^2 - 4x - 15 < 0$.

20. If $f(x) = x^2 + bx + 1$, determine the values of b for which the graph of f would have no x-intercepts.

In Exercises 21–23, find all solutions, real or complex, of the equation.

21. $6x^4 - 5x^2 - 4 = 0$

22. $\dfrac{1}{2x + 1} + \dfrac{3}{x - 2} = \dfrac{5}{2x^2 - 3x - 2}$

23. $\sqrt{2x - 1} + \sqrt{x + 4} = 6$

24. A rectangular garden plot is to be enclosed with a short fence on three of its sides and a brick wall on the fourth side. If 40 feet of fencing material is available, what dimensions will yield an enclosed region of 198 square feet?

25. The height of a ball after being dropped from a point 256 feet above the ground is given by $h(t) = -16t^2 + 256$, where t is the time in seconds since the ball was dropped and $h(t)$ is in feet.

(a) Find and interpret $h(0)$.

(b) When will the ball reach the ground?

(c) For what values of t will the height of the ball be at least 192 feet?

Polynomial and Rational Functions

Boxes can be manufactured in many shapes and sizes. A polynomial function can be used in constructing a box to meet a volume specification. See Example 1 in Section 4.1 for an example of such an application. In this section, we extend our study of functions by exploring polynomial and rational functions. These functions are used in applications when linear or quadratic models will not suffice. They also play an important role in advanced mathematics.

4.1 Graphs of Polynomial Functions

Objectives

▶ Define a polynomial function

▶ Determine end behavior

▶ Find *x*-intercepts and zeros by factoring

▶ Sketch a graph of a polynomial function

▶ Solve applied problems using polynomials

In Chapters 1 and 3, we discussed linear and quadratic functions in detail. Recalling that linear functions are of the form $f(x) = mx + b$ and quadratic functions are of the form $f(x) = ax^2 + bx + c$, we might ask whether new functions can be defined with x raised to the *third* power, the *fourth* power, or even higher powers. The answer is *yes*!

In fact, we can define functions with x raised to any power. When the powers are nonnegative integers, such as 0, 1, 2, 3, ..., the resulting functions are known as **polynomial functions.** Linear and quadratic functions are special types of polynomial functions. Before discussing polynomial functions in more detail, we will investigate a problem in which a function arises that is neither linear nor quadratic.

Example 1   Volume: An Example of a Polynomial Function

Example 7 in Section 4.1 builds upon this example. ⋯▸

Gift Horse, Inc., manufactures various types of decorative gift boxes. The bottom portion of one such box is made by cutting a small square of length x inches from each corner of a 10-inch by 10-inch piece of cardboard and folding up the sides. See Figure 4.1.1. Find an expression for the volume of the resulting box.

Figure 4.1.1

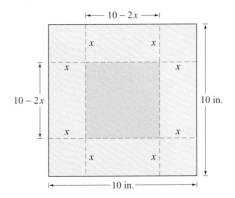

Just In Time

Review operations on polynomials in Section P.4, and the formula for volume in Section P.7.

▶**Solution** Recall that the volume of a rectangular box is given by the formula

$$V = \text{length} \times \text{width} \times \text{height}.$$

From Figure 4.1.1, the expressions for the length, width, and height of the box are given by the following.

$$l = 10 - 2x$$
$$w = 10 - 2x$$
$$h = x$$

Substituting into the expression for the volume, we get

$$V(x) = \text{length} \times \text{width} \times \text{height} = (10 - 2x)(10 - 2x)x.$$

Multiplying the terms in the parentheses and simplifying gives

$$V(x) = 4x^3 - 40x^2 + 100x.$$

Note that the volume function contains the variable x raised to the *third* power as well as to the first and second powers. This volume function is called a **cubic function** because

Discover *and* Learn

For what values of x is the volume expression in Example 1 defined? To answer this question, it may be helpful to look at Figure 4.1.1.

the highest power of x that occurs is 3. We will see in this chapter that a cubic function has properties that are quite different from those of linear and quadratic functions.

☑ *Check It Out 1:* Find an expression for the volume of the box in Example 1 if the piece of cardboard measures 8 inches by 8 inches. ■

We next give a precise definition of a polynomial function.

Definition of a Polynomial Function

A function f is said to be a **polynomial function** if it can be written in the form

$$f(x) = a_n x^n + a_{n-1}x^{n-1} + \cdots + a_1 x + a_0$$

where $a_n \neq 0$, n is a nonnegative integer, and $a_0, a_1, \ldots, a_n$ are real-valued constants. The domain of f is the set of all real numbers.

Just In Time

Review polynomials in Section P.4.

Some of the constants that appear in the definition of a polynomial function have specific names associated with them:

▶ The nonnegative integer n is called the **degree** of the polynomial. Polynomials are usually written in **descending order,** with the exponents decreasing from left to right.

▶ The constants $a_0, a_1, \ldots, a_n$ are called **coefficients.**

▶ The term $a_n x^n$ is called the **leading term,** and the coefficient a_n is called the **leading coefficient.**

▶ A function of the form $f(x) = a_0$ is called a **constant polynomial** or a **constant function.**

Example 2 Identifying Polynomial Functions

Which of the following functions are polynomial functions? For those that are, find the degree and the coefficients, and identify the leading coefficient.

(a) $g(x) = 3 + 5x$ 　　(b) $h(s) = 2s(s^2 - 1)$ 　　(c) $f(x) = \sqrt{x^2 + 1}$

▶**Solution**

(a) The function $g(x) = 3 + 5x = 5x + 3$ is a polynomial function of degree 1 with coefficients $a_0 = 3$ and $a_1 = 5$. The leading coefficient is $a_1 = 5$. This is a linear function.

(b) Simplifying gives $h(s) = 2s(s^2 - 1) = 2s^3 - 2s$. This is a polynomial of degree 3. The coefficients are $a_3 = 2$, $a_2 = 0$, $a_1 = -2$, and $a_0 = 0$. The leading coefficient is $a_3 = 2$.

(c) The function $f(x) = \sqrt{x^2 + 1} = (x^2 + 1)^{1/2}$ is not a polynomial function because the expression $x^2 + 1$ is raised to a fractional exponent, and $(x^2 + 1)^{1/2}$ cannot be written as a sum of terms in which x is raised to nonnegative-integer powers.

☑ *Check It Out 2:* Rework Example 2 for the following functions.

(a) $f(x) = 6$

(b) $g(x) = (x + 1)(x - 1)$

(c) $h(t) = \sqrt{t} + 3$ ■

The rest of this section will be devoted to exploring the graphs of polynomial functions. These graphs have no breaks or holes, and no sharp corners. Figure 4.1.2 illustrates the graphs of several functions and indicates which are the graphs of polynomial functions.

Figure 4.1.2

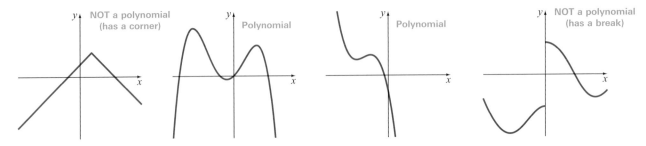

Polynomials of the Form $f(x) = x^n$ and Their End Behavior

The graphs of polynomial functions can be quite varied. We begin by examining polynomial functions with just one term, x^n, since they are the simplest. Consider the functions $f(x) = x^3$ and $g(x) = x^4$. Table 4.1.1 gives some values of these functions. Their corresponding graphs are given in Figure 4.1.3.

Table 4.1.1

x	$f(x) = x^3$	$g(x) = x^4$
-100	-10^6	10^8
-10	-10^3	10^4
-2	-8	16
-1	-1	1
0	0	0
1	1	1
2	8	16
10	10^3	10^4
100	10^6	10^8

Figure 4.1.3

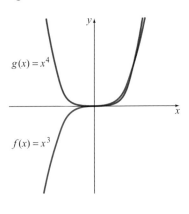

Observations:

▶ The function $f(x) = x^3$ is a polynomial function of degree 3; its *domain* is the set of all real numbers, $(-\infty, \infty)$, because every real number can be cubed. From the graph and the table, the *range* of this function seems to be the set of all real numbers, $(-\infty, \infty)$.

▶ The function $g(x) = x^4$ is a polynomial function of degree 4; its *domain* is $(-\infty, \infty)$, because every real number can be raised to the fourth power. From the graph and the table, the *range* of this function seems to be $[0, \infty)$. This is to be expected, since $x^4 \geq 0$.

To investigate these functions further, it is useful to examine the trend in the function value as the value of x gets larger and larger in magnitude. This is known as determining the **end behavior** of a function. Basically, we ask the following question: How does $f(x)$ behave as the value of x increases to positive infinity ($x \to \infty$) or decreases to negative infinity ($x \to -\infty$)?

Using the information from Table 4.1.1, we can summarize the end behavior of the functions $f(x) = x^3$ and $g(x) = x^4$ in Table 4.1.2.

Table 4.1.2

Function	Behavior of Function as $x \to -\infty$	Behavior of Function as $x \to +\infty$
$f(x) = x^3$	$f(x) \to -\infty$	$f(x) \to +\infty$
$g(x) = x^4$	$g(x) \to +\infty$	$g(x) \to +\infty$

The graphs and their end behaviors are shown in Figure 4.1.4.

Figure 4.1.4

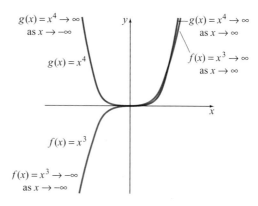

Discover *and* Learn

Confirm the end behavior of $f(x) = x^3$ and $g(x) = x^4$ using a graphing utility. Examine the table of values as well as the graph.

We now summarize the properties of the general function $f(x) = x^n$, n a positive integer.

Properties of $f(x) = x^n$

For $f(x) = x^n$, n an odd, positive integer, the following properties hold.

▶ Domain: $(-\infty, \infty)$
▶ Range: $(-\infty, \infty)$
▶ End behavior: As $x \to \infty$, $f(x) = x^n \to \infty$. As $x \to -\infty$, $f(x) = x^n \to -\infty$.

See Figure 4.1.5.

For $f(x) = x^n$, n an even, positive integer, the following properties hold.

▶ Domain: $(-\infty, \infty)$
▶ Range: $[0, \infty)$
▶ End behavior: As $x \to \infty$, $f(x) = x^n \to \infty$. As $x \to -\infty$, $f(x) = x^n \to \infty$.

See Figure 4.1.6.

Figure 4.1.5

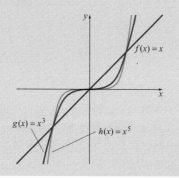

Figure 4.1.6

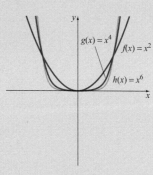

Just In Time

Review transformations in Section 2.3.

In Example 3 we graph some transformations of the basic function $f(x) = x^n$.

Transformations of Polynomial Functions

Example **3** **Transformations of $f(x) = x^n$**

Graph the following functions using transformations.

(a) $g(x) = (x - 1)^4$

(b) $h(x) = -x^3 + 1$

▶Solution

(a) The graph of $g(x) = (x - 1)^4$ is obtained by horizontally shifting the graph of $f(x) = x^4$ to the right by 1 unit, as shown in Figure 4.1.7.

Figure 4.1.7

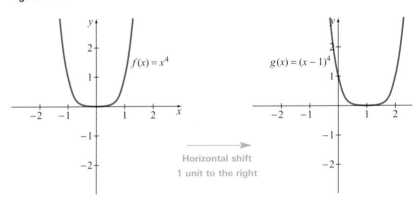

(b) The graph of $h(x) = -x^3 + 1$ can be thought as a series of transformations of the graph of $f(x) = x^3$, as shown in Figure 4.1.8.

Figure 4.1.8

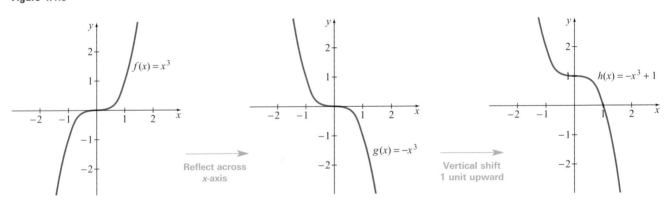

☑ *Check It Out 3:* Graph $f(x) = x^4 + 2$ using transformations. ■

The Leading Term Test for End Behavior

So far we have discussed the end behavior of polynomials with just one term. What about polynomial functions with more than one term? We will examine the end behavior of two functions, $f(x) = -2x^3$ and $g(x) = -2x^3 + 8x$. Table 4.1.3 gives some values for these two functions.

Table 4.1.3

x	$f(x) = -2x^3$	$g(x) = -2x^3 + 8x$
-1000	2×10^9	$1{,}999{,}992{,}000$
-100	$2{,}000{,}000$	$1{,}999{,}200$
-10	2000	1920
-5	250	210
0	0	0
5	-250	-210
10	-2000	-1920
100	$-2{,}000{,}000$	$-1{,}999{,}200$
1000	-2×10^9	$1{,}999{,}992{,}000$

The graphs of the two functions are given in Figure 4.1.9. Notice that they are nearly indistinguishable. For very large values of $|x|$, the magnitude of $-2x^3$ is *much* larger than that of $8x$; therefore, the $8x$ term makes a very small contribution to the value of $g(x)$. Thus the values of $f(x)$ and $g(x)$ are almost the same. For small values of $|x|$, the values of $f(x)$ and $g(x)$ are indistinguishable on the graph because of the choice of vertical scale.

Figure 4.1.9

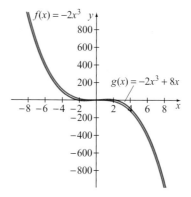

We now generalize our observations about the end behavior of polynomial functions.

Leading Term Test for End Behavior

Given a polynomial function of the form

$$f(x) = a_n x^n + a_{n-1} x^{n-1} + \cdots + a_1 x + a_0, \; a_n \neq 0$$

the *end behavior* of f is determined by the *leading term* of the polynomial, $a_n x^n$.

The shape of the graph of $f(x) = a_n x^n$ will resemble the shape of the graph of $y = x^n$ if $a_n > 0$, and it will resemble the shape of the graph of $y = -x^n$ if $a_n < 0$. (Refer back to Example 3 for a graph of $y = -x^3$.)

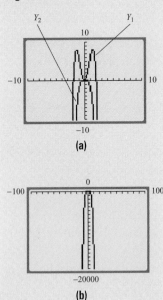
Figure 4.1.10 summarizes our discussion of the end behavior of nonconstant polynomials.

Figure 4.1.10

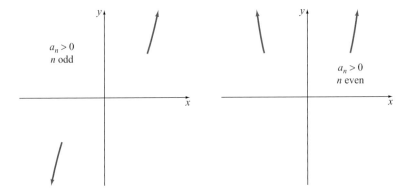

The shape of the graph in the middle region cannot be determined using the leading term test.

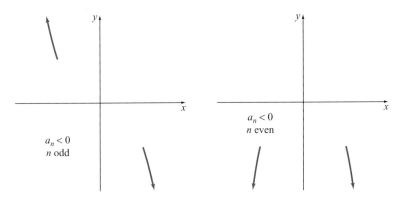

Note *Only the end behavior of a function can be sketched using the leading term test. The shape of the graph for small and moderate values of $|x|$ cannot be discerned from this test.*

Example 4 **Determining End Behavior**

Determine the end behavior of the following functions by examining the leading term.

(a) $f(x) = x^3 + 3x^2 + x$

(b) $g(t) = -2t^4 + 8t^2$

▶**Solution**

(a) For $f(x) = x^3 + 3x^2 + x$, the leading term is x^3. Thus, for large values of $|x|$, we expect $f(x)$ to behave like $y = x^3$: $f(x) \to +\infty$ as $x \to +\infty$ and $f(x) \to -\infty$ as $x \to -\infty$.

(b) For $g(t) = -2t^4 + 8t^2$, the leading term is $-2t^4$. Thus, for large values of $|t|$, we expect $g(t)$ to behave like $y = -2t^4$: $g(t) \to -\infty$ as $t \to +\infty$ and $g(t) \to -\infty$ as $t \to -\infty$.

☑ *Check It Out 4:* Determine the end behavior of the following functions by examining the leading term.

(a) $h(x) = -3x^3 + x$

(b) $s(x) = 2x^2 + 1$ ▪

Finding Zeros and *x*-Intercepts by Factoring

Just In Time

Review factoring in Section P.5 and x-intercepts and zeros in Section 3.2.

In addition to their end behavior, an important feature of the graphs of polynomial functions is the location of their *x*-intercepts. In this section, we graph polynomial functions whose *x*-intercepts can be found easily by factoring.

Recall from Section 3.2 the following connection between the real zeros of a function and the *x*-intercepts of its graph: the real number values of *x* satisfying $f(x) = 0$ are called the *real zeros* of the function *f*. Each of these values of *x* is the first coordinate of an *x*-intercept of the graph of the function.

Example 5 **Finding Zeros and *x*-Intercepts**

Find the zeros of $f(x) = 2x^3 - 18x$ and the corresponding *x*-intercepts of the graph of *f*.

▶Solution To find the zeros of *f*, solve the equation $f(x) = 0$:

$$2x^3 - 18x = 0 \qquad \text{Set expression for f equal to zero}$$
$$2x(x^2 - 9) = 0 \qquad \text{Factor out 2x}$$
$$2x(x + 3)(x - 3) = 0 \qquad \text{Factor } x^2 - 9 = (x + 3)(x - 3)$$
$$2x = 0 \Longrightarrow x = 0$$
$$x + 3 = 0 \Longrightarrow x = -3 \qquad \text{Set each factor equal to zero and solve for x}$$
$$x - 3 = 0 \Longrightarrow x = 3$$

The zeros of *f* are $x = 0$, $x = -3$, and $x = 3$. The *x*-intercepts of the graph of *f* are $(0, 0)$, $(-3, 0)$, and $(3, 0)$.

☑ *Check It Out 5:* Find the zeros of $f(x) = -3x^3 + 12x$ and the corresponding *x*-intercepts of the graph of *f*. ▪

Hand-Sketching the Graph of a Polynomial Function

If a polynomial function can be easily factored, then we can find the *x*-intercepts and sketch the function by hand using the following procedure.

Hand-Sketching the Graph of a Polynomial Function

Step 1 Determine the end behavior of the function.

Step 2 Find the *y*-intercept and plot it.

Step 3 Find and plot the *x*-intercepts of the graph of the function. These points will divide the *x*-axis into smaller intervals.

Step 4 Find the sign and value of $f(x)$ for a test value *x* in each of these intervals. Plot these test values.

Step 5 Use the plotted points and the end behavior to sketch a smooth graph of the function. Plot additional points if needed.

A sketch of the graph of a polynomial is **complete** if it shows all the x-intercepts and the y-intercept and illustrates the correct end behavior of the function. Finer details of the graph of a polynomial function will be discussed in the next section.

Example 6 Sketching a Polynomial Function

Find the zeros of the function $f(x) = x^3 - x$ and the x-intercepts of its graph. Use the x-intercepts and the end behavior of the function to sketch the graph of the function by hand.

▶**Solution**

Step 1 Determine the end behavior. For $|x|$ large, $f(x)$ behaves like $y = x^3$; $f(x) \to +\infty$ as $x \to +\infty$ and $f(x) \to -\infty$ as $x \to -\infty$. See Figure 4.1.12.

Figure 4.1.12

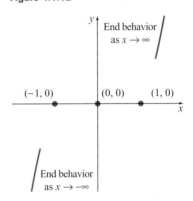

Step 2 Find the y-intercept. Since $f(0) = 0$, the y-intercept is $(0, 0)$.

Step 3 Find the x-intercepts. The zeros of this function are found by setting $f(x)$ equal to 0 and solving for x.

$$x^3 - x = 0 \qquad \text{\textit{Set expression equal to zero}}$$
$$x(x^2 - 1) = x(x + 1)(x - 1) = 0 \qquad \text{\textit{Factor left side completely}}$$
$$x = 0$$
$$x + 1 = 0 \Longrightarrow x = -1 \qquad \text{\textit{Use the Zero Product Rule}}$$
$$x - 1 = 0 \Longrightarrow x = 1$$

Thus the zeros are $x = 0$, $x = -1$, and $x = 1$ and the x-intercepts are $(0, 0)$, $(-1, 0)$, and $(1, 0)$. See Figure 4.1.12.

Step 4 Determine the signs of function values. We still have to figure out what the graph looks like in between the x-intercepts. The three x-intercepts break the x-axis into four intervals:

$$(-\infty, -1), (-1, 0), (0, 1), \text{ and } (1, \infty)$$

Table 4.1.4 lists the value of $f(x)$ for at least one value of x, called the *test value*, in each of these intervals. It suffices to choose just one test value in each interval, since the sign of the function value is unchanged within an interval.

Table 4.1.4

Interval	Test Value, x	Function Value, $f(x) = x^3 - x$	Sign of $f(x)$
$(-\infty, -1)$	-2	-6	$-$
$(-1, 0)$	-0.5	0.375	$+$
$(0, 1)$	0.5	-0.375	$-$
$(1, \infty)$	2	6	$+$

The signs given in Table 4.1.4 are summarized on the number line in Figure 4.1.13.

Figure 4.1.13

Plotting the x-intercepts and the test values, we have the partial sketch shown in Figure 4.1.14(a).

Step 5 Sketch the entire graph. By plotting the points given in Table 4.1.4 and using the sign of the function value in each of the intervals, we can sketch the graph of the function, as shown in Figure 4.1.14(b).

Figure 4.1.14

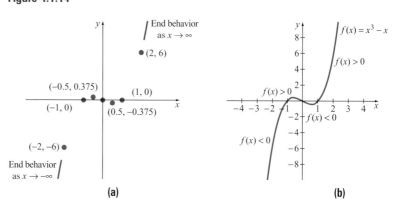

(a)　　　　　(b)

☑ *Check It Out 6:* Determine the end behavior and the x- and y-intercepts of the graph of the function $f(x) = -3x^3 + 27x$. Use this information to sketch a graph of the function by hand. ■

Example 7 **Graphing a Volume Function**

⋯ *This example builds on Example 1 in Section 4.1.*

Graph the volume function of the open box obtained by cutting a square of length x from each corner of a 10-inch by 10-inch piece of cardboard and then folding up the sides. Use your graph to determine the values of x for which the expression makes sense in the context of the problem.

▶ **Solution** From Example 1, the expression for the volume of the open box is $V(x) = 4x^3 - 40x^2 + 100x = x(10 - 2x)(10 - 2x)$, where x is the length of the square cut from each corner.

The x-intercepts of the graph of this function are found by setting $V(x)$ equal to zero.

$$x(10 - 2x)(10 - 2x) = 0 \qquad \text{Set } V(x) \text{ equal to zero}$$

$$x = 0 \qquad \text{Use the Zero Product Rule}$$

$$10 - 2x = 0 \Longrightarrow x = 5$$

Thus the x-intercepts are $(0, 0)$ and $(5, 0)$. The x-intercepts divide the x-axis into three intervals. Table 4.1.5 gives the value of the function at one point in each of these intervals.

Table 4.1.5

Interval	Test Value, x	Function Value, $V(x)$	Sign of $V(x)$
$(-\infty, 0)$	-1	-144	$-$
$(0, 5)$	2	72	$+$
$(5, \infty)$	6	24	$+$

Figure 4.1.15 Dashed sections indicate that the function values are beyond realistic limits.

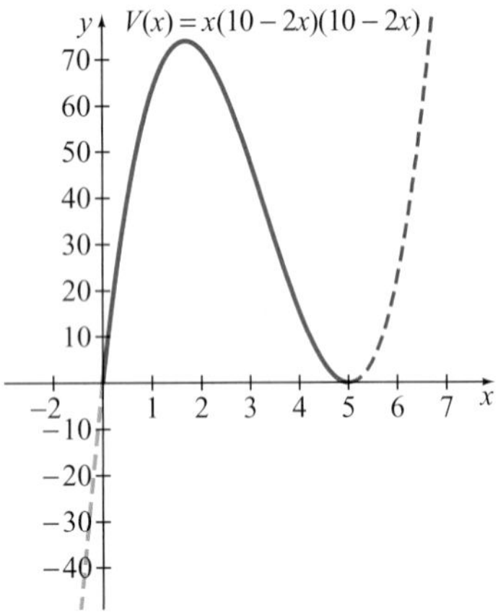

Using the end behavior along with the x-intercepts and the data given in the table, we obtain the graph shown in Figure 4.1.15.

We see that x cannot be negative because it represents a length. From the graph, we see that $V(x)$ is positive for values of x such that $0 < x < 5$ or $x > 5$. Because the piece of cardboard is only 10 inches by 10 inches, we cannot cut out a square more than 5 inches on a side from each corner. Thus the only allowable values for x are $x \in (0, 5)$. We exclude $x = 0$ and $x = 5$ because a solid object with zero volume is meaningless.

Mathematically, the function $V(x)$ is defined for all values of x, whether or not they make sense in the context of the problem.

☑ *Check It Out 7:* Graph the volume function of the open box obtained by cutting a square of length x from each corner of an 8-inch by 8-inch piece of cardboard and then folding up the sides. ■

Application of Polynomials

We conclude this section with an example of an application of a polynomial function.

Example 8 Model of College Attendance

Using data for the years 1990–2004, college attendance by recent high school graduates can be modeled by the cubic polynomial $f(x) = -0.644x^3 + 14.1x^2 - 58.4x + 1570$, where x is the number of years since 1990 and $f(x)$ is the number of students in thousands. (*Source:* National Center for Education Statistics)

(a) Evaluate and interpret $f(0)$.

(b) Determine the number of recent high school graduates attending college in 2002.

(c) Determine the end behavior of this function, and use it to explain why this model is not valid for long-term predictions.

▶**Solution**

(a) Substituting $x = 0$ into the given polynomial, we get $f(0) = 1570$. This value is in *thousands*. Thus there are 1,570,000 recent high school graduates who attended college in 1990 ($x = 0$ corresponds to the year 1990).

(b) For the year 2002, the corresponding x-value is $x = 12$. Substituting $x = 12$ into the expression for $f(x)$ gives

$$f(12) = -0.644(12)^3 + 14.1(12)^2 - 58.4(12) + 1570 \approx 1790.$$

Thus there were approximately 1,790,000 recent high school graduates who attended college in 2002.

(c) Using the leading term test, the end behavior of $f(x)$ resembles that of $y = -0.644x^3$, which would imply *negative* numbers of students in the long term. This is unrealistic and so this function is not valid for making predictions for years that are much beyond 2004. In general, polynomial functions should not be used for predictions too far beyond the interval of time they model.

✔ *Check It Out 8:* Use the model in Example 8 to determine the number of recent high school graduates who attended college in the year 2000. ■

4.1 Key Points

▶ A function f is a **polynomial function** if it can be expressed in the form $f(x) = a_n x^n + a_{n-1}x^{n-1} + \cdots + a_1 x + a_0$, where $a_n \neq 0$, n is a nonnegative integer, and $a_0, a_1, \ldots, a_n$ are real numbers.

▶ The **degree** of the polynomial is n. The constants $a_0, a_1, \ldots, a_n$ are called **coefficients.** Polynomials are usually written in **descending order,** with the exponents decreasing from left to right.

▶ A function of the form $f(x) = a_0$ is called a **constant polynomial** or a **constant function.**

▶ The **Leading Term Test for End Behavior:** The **end behavior** of a polynomial function $f(x)$ is determined by its **leading term,** $a_n x^n$. The coefficient a_n is called the **leading coefficient.**

▶ Polynomials of the form $f(x) = (x - k)^n + k$ can be graphed by translations of the graph of $f(x) = x^n$.

▶ To sketch a polynomial function $f(x)$ by hand, use the following procedure.

1. Determine the end behavior of the function.

2. Find the y-intercept and plot it.

3. Find and plot the x-intercepts of the graph of the function; these points divide the x-axis into smaller intervals.

4. Find the sign and value of $f(x)$ for a test value x in each of these intervals. Plot these test values.

5. Use the plotted points and the end behavior to sketch a smooth graph of the function. Plot additional points, if needed.

4.1 Exercises

▶**Just in Time Exercises** These exercises correspond to the Just in Time references in this section. Complete them to review topics relevant to the remaining exercises.

1. The _____ of a polynomial is the highest power to which a variable is raised.

2. What is the degree of the polynomial $5x^4 - 2x - 7$?

3. The _____ of $f(x)$ are the values of x such that $f(x) = 0$.

4. Find the x-intercept of $f(x) = 3x + 9$.

5. Find the y-intercept of $f(x) = 3x + 9$.

6. Multiply: $x^3(x^2 - 3)(x + 1)$

7. Factor: $x^3 - 3x^2 - 4x$

8. Factor: $2x^3 - 50x$

9. The graph of $f(x) = x^2 + 3$ is the graph of $y = x^2$ shifted _____ 3 units.

10. The graph of $f(x) = (x - 4)^2$ is the graph of $y = x^2$ shifted _____ 4 units.

▶**Skills** This set of exercises will reinforce the skills illustrated in this section.

In Exercises 11–14, determine whether the graph represents the graph of a polynomial function. Explain your reasoning.

11.

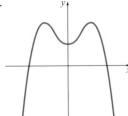

12.

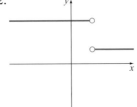

13.

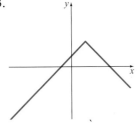

14.

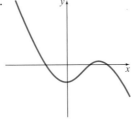

In Exercises 15–22, determine whether the function is a polynomial function. If so, find the degree. If not, state the reason.

15. $f(x) = -x^3 + 3x^3 + 1$

16. $f(s) = 4s^5 - 5s^3 + 6s - 1$

17. $f(t) = \sqrt{t}$

18. $g(t) = \dfrac{1}{t}$

19. $f(x) = 5$

20. $g(x) = -2$

21. $f(x) = -(x + 1)^3$

22. $g(x) = (x - 1)^2$

In Exercises 23–32, determine the end behavior of the function.

23. $f(t) = 7t$

24. $g(x) = -2x$

25. $f(x) = -2x^3 + 4x - 1$

26. $g(x) = 3x^4 + 2x^2 - 1$

27. $H(x) = -5x^4 + 3x^2 + x - 1$

28. $h(x) = 5x^6 - 3x^3$

29. $g(x) = -10x^3 + 3x^2 + 5x - 2$

30. $f(x) = 3x^3 - 4x^2 + 5$

31. $f(s) = \dfrac{7}{2}s^5 - 14s^3 + 10s$

32. $f(s) = -\dfrac{3}{4}s^4 + 8s^2 - 3s - 16$

In Exercises 33–44, sketch the polynomial function using transformations.

33. $f(x) = x^3 - 2$

34. $f(x) = x^4 - 1$

35. $f(x) = \dfrac{1}{2}x^3$

36. $g(x) = -\dfrac{1}{2}x^4$

37. $g(x) = (x - 2)^3$

38. $h(x) = (x + 1)^4$

39. $h(x) = -2x^5 - 1$

40. $f(x) = 3x^4 + 2$

41. $f(x) = -(x + 1)^3 - 2$

42. $f(x) = (x - 2)^4 + 1$

43. $h(x) = -\dfrac{1}{2}(x + 1)^3 - 2$ 44. $h(x) = \dfrac{1}{2}(x - 2)^4 - 1$

 In Exercises 45–48, find a function of the form $y = cx^k$ that has the same end behavior as the given function. Confirm your results with a graphing utility.

45. $g(x) = -5x^3 - 4x^2 + 4$ 46. $h(x) = 6x^3 - 4x^2 + 7x$

47. $f(x) = 1.5x^5 - 10x^2 + 14x$

48. $g(x) = -3.6x^4 + 4x^2 + x - 20$

In Exercises 49–64, for each polynomial function,

(a) *find a function of the form $y = cx^k$ that has the same end behavior.*

(b) *find the x- and y-intercept(s) of the graph.*

(c) *find the interval(s) on which the value of the function is positive.*

(d) *find the interval(s) on which the value of the function is negative.*

(e) *use the information in parts (a)–(d) to sketch a graph of the function.*

49. $f(x) = -2x^3 + 8x$

50. $f(x) = 3x^3 - 27x$

51. $g(x) = (x - 3)(x + 4)(x - 1)$

52. $f(x) = (x + 1)(x - 2)(x + 3)$

53. $f(x) = -\dfrac{1}{2}(x^2 - 4)(x^2 - 1)$

54. $f(x) = (x^2 - 4)(x + 1)(x - 3)$

55. $f(x) = x^3 - 2x^2 - 3x$

56. $g(x) = x^3 + x^2 - 3x$

57. $f(x) = -x(2x + 1)(x - 3)$

58. $g(x) = 2x(x - 2)(2x - 1)$

59. $f(x) = -(x^2 - 1)(x - 2)(x + 3)$

60. $f(x) = x(x^2 - 4)(x + 1)$

61. $g(x) = 2x^2(x + 3)$

62. $f(x) = -3x^2(x - 1)$

63. $f(x) = (2x + 1)(x - 3)(x^2 + 1)$

64. $g(x) = -(x - 2)(3x - 1)(x^2 + 1)$

In Exercises 65–68, for each polynomial function graphed below, find (a) the x-intercepts of the graph of the function, if any; (b) the y-intercept of the graph of the function; (c) whether the power of the leading term is odd or even; and (d) the sign of the leading coefficient.

65.

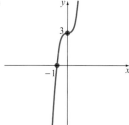

66.

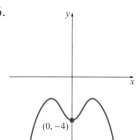

67.

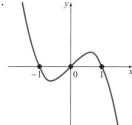

68.

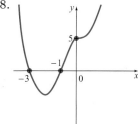

69. Consider the function $f(x) = 0.001x^3 + 2x^2$. Answer the following questions.

(a) Graph the function in a standard window of a graphing utility. Explain why this window setting does not give a complete graph of the function.

(b) Using the x-intercepts and the end behavior of the function, sketch an approximate graph of the function by hand.

(c) Find a graphing window that shows a correct graph for this function.

▶ **Applications** In this set of exercises, you will use polynomial functions to study real-world problems.

70. **Wine Industry** The following model gives the supply of wine from France, based on data for the years 1994–2001:

$$w(x) = 0.0437x^4 - 0.661x^3 + 3.00x^2 - 4.83x + 62.6$$

where $w(x)$ is in kilograms per capita and x is the number of years since 1994. (*Source:* Food and Agriculture Organization of the United Nations)

(a) According to this model, what was the per capita wine supply in 1994? How close is this value to the actual value of 62.5 kilograms per capita?

(b) Use this model to compute the wine supply from France for the years 1996 and 2000.

(c) The actual wine supplies for the years 1996 and 2000 were 60.1 and 54.6 kilograms per capita, respectively. How do your calculated values compare with the actual values?

(d) Use end behavior to determine if this model will be accurate for long-term predictions.

71. **Criminology** The numbers of burglaries (in thousands) in the United States can be modeled by the following cubic function, where x is the number of years since 1985.

$$b(x) = 0.6733x^3 - 22.18x^2 + 113.9x + 3073$$

(a) What is the y-intercept of the graph of $b(x)$, and what does it signify?

(b) Find $b(6)$ and interpret it.

(c) Use this model to predict the number of burglaries that occurred in the year 2004.

(d) Why would this cubic model be inaccurate for predicting the number of burglaries in the year 2040?

72. **Wildlife Conservation** The number of species on the U.S. endangered species list during the years 1998–2005 can be modeled by the function

$$f(t) = 0.308t^3 - 5.20t^2 + 32.2t + 921$$

where t is the number of years since 1998. (*Source:* U.S. Fish and Wildlife Service)

(a) Find and interpret $f(0)$.

(b) How many species were on the list in 2004?

(c) Use a graphing utility to graph this function for $0 \le t \le 7$. Judging by the trend seen in the graph, is this model reliable for long-term predictions? Why or why not?

73. **Manufacturing** An open box is to be made by cutting four squares of equal size from a 10-inch by 15-inch rectangular piece of cardboard (one at each corner) and then folding up the sides.

(a) Let x be the length of a side of the square cut from each corner. Find an expression for the volume of the box in terms of x. Leave the expression in factored form.

(b) What is a realistic range of values for x? Explain.

74. **Construction** A cylindrical container is to be constructed so that the *sum* of its height and its diameter is 10 feet.

(a) Write an equation relating the height of the cylinder, h, to its radius, r. Solve the equation for h in terms of r.

(b) The volume of a cylinder is given by $V = \pi r^2 h$. Use your answer from part (a) to express the volume of the cylindrical container in terms of r alone. Leave your expression in factored form so that it will be easier to analyze.

(c) What are the values of r for which this problem makes sense? Explain.

75. **Manufacturing** A rectangular container with a square base is constructed so that the *sum* of the height and the perimeter of the base is 20 feet.

(a) Write an equation relating the height, h, to the length of a side of the base, s. Solve the equation for h in terms of s.

(b) Use your answer from part (a) to express the volume of the container in terms of s alone. Leave your expression in factored form so that it will be easier to analyze.

(c) What are the values of s for which this problem makes sense? Explain.

▶**Concepts** This set of exercises will draw on the ideas presented in this section and your general math background.

76. Explain why the following graph is *not* a complete graph of the function $p(x) = 0.01x^3 + x^2$.

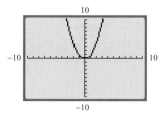

77. Show that all polynomial functions have a y-intercept. Can the same be said of x-intercepts?

78. Can the graph of a function with range $[4, \infty)$ cross the x-axis?

79. Explain why all polynomial functions of odd degree must have range $(-\infty, \infty)$.

80. Explain why all polynomial functions of odd degree must have at least one real zero.

4.2 More on Graphs of Polynomial Functions and Models

Objectives

▶ Define the multiplicity of a zero of a polynomial

▶ Check for symmetry of polynomial functions

▶ Know about the existence of local extrema

▶ Sketch a complete graph of a polynomial

▶ Relate zeros, x-intercepts, and factors of a polynomial

▶ Model with polynomial functions

In the previous section, we learned to sketch the graph of a polynomial function by determining the end behavior, the x-intercepts, and the sign of the function between the x-intercepts. However, we did not discuss how the graph might *look* between the x-intercepts—we simply found the sign of the y-coordinates of all points in each of those intervals. In this section we will examine the finer properties of the graphs of polynomial functions. These include:

▶ Examining the behavior of the polynomial function at its x-intercepts.

▶ Observing any types of symmetry in the graph of the polynomial function.

▶ Locating the peaks and valleys of the graph of a polynomial function, known as maxima and minima, by using a graphing utility.

A complete analysis of the graph of a polynomial function involves calculus, which is beyond the scope of this book.

Multiplicities of Zeros

The number of times a linear factor $x - a$ occurs in the completely factored form of a polynomial expression is known as the **multiplicity** of the real zero a associated with that factor. For example, $f(x) = (x + 1)(x - 3)^2 = (x + 1)(x - 3)(x - 3)$ has two real zeros: $x = -1$ and $x = 3$. The zero $x = -1$ has multiplicity 1 and the zero $x = 3$ has multiplicity 2, since their corresponding factors are raised to the powers 1 and 2, respectively. A formal definition of the multiplicity of a zero of a polynomial is given in Section 4.5.

Use your graphing utility to sketch a graph of $f(x) = (x + 1)^3(x - 2)^2$. What are the zeros of this function, and what are their multiplicities? Does the graph of the function cross the x-axis at the corresponding x-intercepts, or just touch it?

The multiplicity of a real-valued zero of a polynomial and the graph of the polynomial at the corresponding x-intercept have a close connection, as we will see next.

Multiplicities of Zeros and Behavior at x-Intercepts

▶ If the multiplicity of a real zero of a polynomial function is **odd,** the graph of the function **crosses** the x-axis at the corresponding x-intercept.

▶ If the multiplicity of a real zero of a polynomial function is **even,** the graph of the function **touches,** but does not cross, the x-axis at the corresponding x-intercept.

See Figure 4.2.1.

Figure 4.2.1

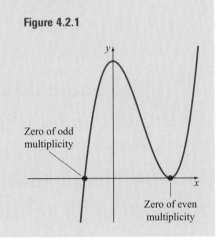

Zero of odd multiplicity

Zero of even multiplicity

Figure 4.2.2

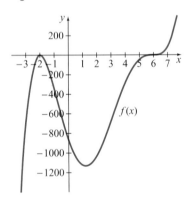

Figure 4.2.3

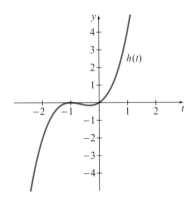

Example 1 Multiplicities of Zeros

Determine the multiplicities of the real zeros of the following functions. Does the graph cross the x-axis or just touch it at the x-intercepts?

(a) $f(x) = (x - 6)^3(x + 2)^2$ (b) $h(t) = t^3 + 2t^2 + t$

▶**Solution**

(a) Recall that the zeros of f are found by setting $f(x)$ equal to zero and solving for x.

$$(x - 6)^3(x + 2)^2 = 0$$
$$(x - 6)^3 = 0 \quad \text{or} \quad (x + 2)^2 = 0$$

Thus, $x = 6$ or $x = -2$.

The zero $x = 6$ has multiplicity 3, since the corresponding factor, $x - 6$, is raised to the third power. Because the multiplicity is odd, the graph will cross the x-axis at $(6, 0)$.

The zero $x = -2$ has multiplicity 2. Because this zero has even multiplicity, the graph of f only touches the x-axis at $(-2, 0)$. This is verified by the graph of f shown in Figure 4.2.2.

(b) To find the zeros of h, we first need to factor the expression and then set it equal to zero. This gives

$$t^3 + 2t^2 + t = t(t^2 + 2t + 1) = t(t + 1)^2 = 0$$
$$t = 0 \quad \text{or} \quad (t + 1)^2 = 0$$

Thus, $t = 0$ or $t = -1$.

We see that $t = 0$ has multiplicity 1, and so the graph will cross the t-axis at $(0, 0)$. The zero $t = -1$ has multiplicity 2, and the graph will simply touch the t-axis at $(-1, 0)$. This is verified by the graph of h given in Figure 4.2.3.

✔ *Check It Out 1:* Rework Example 1 for $g(x) = x^2(x - 5)^2$. ■

Just In Time

Review symmetry in Section 2.4.

Symmetry of Polynomial Functions

Recall that a function is *even* if $f(x) = f(-x)$ for all x in the domain of f. An even function is symmetric with respect to the y-axis. A function is *odd* if $f(x) = -f(-x)$ for all x in the domain of f. An odd function is symmetric with respect to the origin. If a polynomial function happens to be odd or even, we can use that fact to help sketch its graph.

Example **2** **Checking for Symmetry**

Check whether $f(x) = x^4 + x^2 + x$ is odd, even, or neither.

▶**Solution** Because $f(x)$ has an even-powered term, we will check to see if it is an even function.

$$f(-x) = (-x)^4 + (-x)^2 + (-x) = x^4 + x^2 - x$$

So, $f(x) \neq f(-x)$.
    Next we check whether it is an odd function.

$$-f(-x) = -(-x)^4 - (-x)^2 - (-x) = -x^4 - x^2 + x$$

So, $f(x) \neq -f(-x)$. Thus, f is neither even nor odd. It is not symmetric with respect to the y-axis or the origin.

☑ *Check It Out 2:* Decide whether the following functions are even, odd, or neither.
(a) $h(t) = -t^4 + t$
(b) $g(s) = s^3 + 8s$ ■

Finding Local Extrema and Sketching a Complete Graph

You may have noticed that the graphs of most of the polynomials we have examined so far have peaks and valleys. The peaks and valleys are known as **local maxima** and **local minima,** respectively. Together they are known as **local extrema.** They are also referred to as **turning points.** The term *local* is used because the values are not necessarily the maximum and minimum values of the function over its entire domain.

    Finding the precise locations of local extrema requires the use of calculus. If you sketch a graph by hand, you can at best get a rough idea of the locations of the local extrema. One way to get a rough idea of the locations of local extrema is by plotting additional points in the intervals in which such extrema may exist. If you are using a graphing utility, you can find the local extrema rather accurately.

    In addition to the techniques presented in the previous section, we can now use multiplicity of zeros and symmetry to help us sketch the graphs of polynomials.

Example **3** **Sketching a Complete Graph**

Sketch a complete graph of $f(x) = -2x^4 + 8x^2$.

▶**Solution**

Step 1 This function's end behavior is similar to that of $y = -2x^4$: as
        $x \to \pm\infty$, $f(x) \to -\infty$.

Step 2 The y-intercept is $(0, 0)$.

Step 3 Find the x-intercepts of the graph of the function.

$$-2x^4 + 8x^2 = 0 \qquad \text{Set expression equal to zero}$$

$$-2x^2(x^2 - 4) = -2x^2(x + 2)(x - 2) = 0 \qquad \text{Factor left side completely}$$

$$x^2 = 0 \Longrightarrow x = 0 \qquad x = 0 \text{ is a zero of multiplicity 2}$$

$$x + 2 = 0 \Longrightarrow x = -2$$

$$x - 2 = 0 \Longrightarrow x = 2$$

Thus the x-intercepts of the graph of this function are $(0, 0)$, $(-2, 0)$, and $(2, 0)$. Because $x = 0$ is a zero of multiplicity 2, the graph *touches*, but *does not cross*, the x-axis at $(0, 0)$. The zeros at $x = -2$ and $x = 2$ are of multiplicity 1, so the graph *crosses* the x-axis at $(-2, 0)$ and $(2, 0)$.

Step 4 Check for symmetry. Because $f(x)$ has an even-powered term, we will check to see if it is an even function $f(-x) = -2(-x)^4 + 8(-x)^2 = -2x^4 + 8x^2 = f(x)$. Since $f(x) = f(-x)$, f is an even function. The graph is symmetric with respect to the y-axis.

Figure 4.2.4

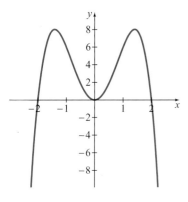

Step 5 Make a table of test values for x in the intervals $(0, 2)$ and $(2, \infty)$ to determine the sign of f. See Table 4.2.1. The positive values of x are sufficient because the graph is symmetric with respect to the y-axis. We simply reflect the graph across the y-axis to complete the graph.

Table 4.2.1

Interval	Test Value, x	Function Value, $f(x) = -2x^4 + 8x^2$	Sign of $f(x)$
$(0, 2)$	1	6	$+$
$(2, \infty)$	3	-90	$-$

Putting the information from all of these steps together, we obtain the graph in Figure 4.2.4.

✔ *Check It Out 3:* Sketch a complete graph of $f(x) = x^5 - x^3$. ■

Discover *and* **Learn**

Use a graphing utility to find the local extrema of $f(x) = x^5 - x^3$.

Technology Note   Examining Figure 4.2.4, we see that there must be one local maximum between $x = -2$ and $x = 0$, and another local maximum between $x = 0$ and $x = 2$. The value $x = 0$ gives rise to a local minimum. Using the MAXIMUM and MINIMUM features of your graphing utility, you can determine the local extrema. One of the maximum values is shown in Figure 4.2.5. The local extrema are summarized in Table 4.2.2.

Figure 4.2.5

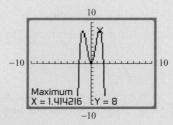

Maximum
X = 1.414216 Y = 8

Keystroke Appendix:
Sections 7, 8, 10

Table 4.2.2

x	$f(x)$	Type of Extremum
-1.4142	8	Local maximum
0	0	Local minimum
1.4142	8	Local maximum

Relationship Between x-Intercepts and Factors

Thus far we have used the factored form of a polynomial to find the x-intercepts of its graph. Next we try to find a possible expression for a polynomial function *given* the x-intercepts of the graph. To do so, we need the following important fact, which will be discussed in detail in Section 4.5.

Number of Real Zeros

The **number of real zeros** of a polynomial function f of degree n is less than or equal to n, counting multiplicity. The graph of f can cross the x-axis no more than n times.

We also note the following relationship between the real zeros of a polynomial, its x-intercepts, and its factors.

Zeros, x-Intercepts, and Factors

If c is a real zero of a polynomial function $f(x)$—that is, $f(c) = 0$—and c is a zero of multiplicity k, then

(a) $(c, 0)$ is an x-intercept of the graph of f, and

(b) $(x - c)^k$ is a factor of $f(x)$.

Example 4 Finding a Polynomial Given Its Zeros

Find a polynomial $f(x)$ of degree 4 that has zeros at -1 and 2, each of multiplicity 1, and a zero at -2 of multiplicity 2.

▶Solution Use the fact that if c is a zero of f, then $x - c$ is a factor of f. Because -1 and 2 are zeros of multiplicity 1, the corresponding factors of f are $x - (-1)$ and $x - 2$. Because -2 is a zero of multiplicity 2, the corresponding factor is $(x - (-2))^2$. A possible fourth-degree polynomial satisfying all of the given conditions is

$$f(x) = (x + 1)(x - 2)(x + 2)^2.$$

This is not the only possible expression. Any nonzero multiple of $f(x)$ will satisfy the same conditions. Thus any polynomial of the form

$$f(x) = a(x + 1)(x - 2)(x + 2)^2, \quad a \neq 0$$

will suffice.

✔ *Check It Out 4:* Find a polynomial $f(x)$ of degree 3 that has zeros 1, 2, and -1, each of multiplicity 1. ■

Modeling with Polynomial Functions

Just as we found linear and quadratic functions of best fit, we can also find polynomials of best fit. In practice, polynomials of degree greater than 4 are rarely used, since they have many turning points and so are not suited for modeling purposes. Even cubic and quartic (degree 4) polynomials provide a good model only for values of the independent variable that are close to the given data.

Just In Time

Review curve fitting in Section 1.4.

| Example | **5 Modeling HMO Enrollment** |

The data in Table 4.2.3 represents the total number of people in the United States enrolled in a health maintenance organization (HMO) for selected years from 1990 to 2004. (*Source:* National Center for Health Statistics)

Table 4.2.3

Year	Years Since 1990, x	HMO Enrollees (in millions)
1990	0	33.0
1992	2	36.1
1994	4	45.1
1996	6	59.1
1998	8	76.6
2000	10	80.9
2002	12	76.1
2004	14	68.8

(a) Use a graphing utility to draw a scatter plot of the data using x, the number of years since 1990, as the independent variable. What degree polynomial best models this data?

(b) Find $h(x)$, the cubic function of best fit, for the data, and graph the function.

(c) Use the function found in part (b) to predict the number of people enrolled in an HMO for the year 2006.

▶**Solution**

(a) The scatter plot is given in Figure 4.2.6. From the plot, we see that a cubic polynomial may be a good choice to model the data since the plot first curves upward and then curves downward.

(b) Using the CUBIC REGRESSION feature of your graphing utility (Figure 4.2.7 (a)), the cubic function of best fit is

$$h(x) = -0.0797x^3 + 1.33x^2 - 0.515x + 32.5.$$

Because the given data values have three significant digits, we have retained three significant digits in each of the coefficients. The data points, along with the graph, are shown in Figure 4.2.7(b). Observe that the model fits the data fairly closely.

Figure 4.2.6

Just In Time

Review significant digits in Section P.3.

Figure 4.2.7

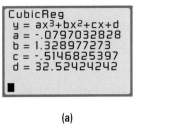

(a)

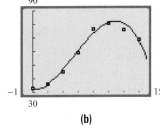

(b)

(c) To predict the enrollment in 2006, use $x = 2006 - 1990 = 16$.

$$h(x) = -0.0797(16)^3 + 1.33(16)^2 - 0.515(16) + 32.5 \approx 38.3$$

According to this model, there will be approximately 38.3 million people enrolled in an HMO in the year 2006.

✔ *Check It Out 5:* Use the model found in Example 5 to estimate the number of people enrolled in an HMO in the year 2003. Round to the nearest tenth of a million. ■

4.2 Key Points

▶ The number of times a linear factor $x - a$ occurs in the completely factored form of a polynomial expression is known as the **multiplicity** of the real zero a associated with that factor.

1. If the multiplicity of a real zero of a polynomial function is **odd,** the graph of the function **crosses** the x-axis at the corresponding x-intercept.

2. If the multiplicity of a real zero of a polynomial function is **even,** the graph of the function **touches,** but does not cross, the x-axis at the corresponding x-intercept.

▶ The **number of real zeros** of a polynomial function f of degree n is less than or equal to n, counting multiplicity.

▶ A function is **even** if $f(x) = f(-x)$. Its graph is symmetric with respect to the y-axis.

▶ A function is **odd** if $f(x) = -f(-x)$. Its graph is symmetric with respect to the origin.

▶ The peaks and valleys present in the graphs of polynomial functions are known as **local maxima** and **local minima,** respectively. Together they are known as **local extrema.**

▶ If c is a real zero of a polynomial function $f(x)$—i.e., $f(c) = 0$—and c is a zero of multiplicity k, then

1. $(c, 0)$ is an x-intercept of the graph of f, and
2. $(x - c)^k$ is a factor of $f(x)$.

4.2 Exercises

▶**Just in Time Exercises** These exercises correspond to the Just in Time references in this section. Complete them to review topics relevant to the remaining exercises.

1. A function is symmetric with respect to the _____ if $f(x) = f(-x)$ for each x in the domain of f. Functions having this property are called _____ functions.

2. A function is symmetric with respect to the _____ if $f(x) = -f(-x)$ for each x in the domain of f. Functions having this property are called _____ functions.

In Exercises 3–6, classify each function as odd, even, or neither.

3. $f(x) = x^2 + 2$

4. $h(x) = 3|x|$

5. $g(x) = -x$

6. $f(x) = x^3$

▶**Skills** This set of exercises will reinforce the skills illustrated in this section.

In Exercises 7–14, determine the multiplicities of the real zeros of the function. Comment on the behavior of the graph at the x-intercepts. Does the graph cross or just touch the x-axis? You may check your results with a graphing utility.

7. $f(x) = (x - 2)^2(x + 5)^5$

8. $g(s) = (s + 6)^4(s - 3)^3$

9. $h(t) = t^2(t - 1)(t + 2)$

10. $g(x) = x^3(x + 2)(x - 3)$

11. $f(x) = x^2 + 2x + 1$

12. $h(s) = s^2 - 2s + 1$

13. $g(s) = 2s^3 + 4s^2 + 2s$

14. $h(x) = 2x^3 - 4x^2 + 2x$

In Exercises 15–22, determine what type of symmetry, if any, the function illustrates. Classify the function as odd, even, or neither.

15. $g(x) = x^4 + 2x^2 - 1$

16. $h(x) = 2x^4 - x^2 + 2$

17. $f(x) = -3x^3 + 1$

18. $g(x) = x^3 - 2$

19. $f(x) = -x^3 + 2x$

20. $g(x) = x^3 - 3x$

21. $h(x) = -2x^4 + 3x^2 - 1$

22. $g(x) = 3x^4 - 2x^2 + 1$

In Exercises 23–26, use the graph of the polynomial function to find the real zeros of the corresponding polynomial and to determine whether their multiplicities are even or odd.

23.

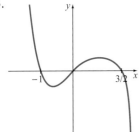

24.

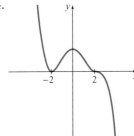

25.

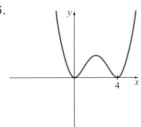

26.

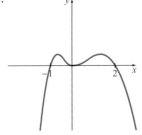

In Exercises 27–38, for each polynomial function, find (a) the end behavior; (b) the y-intercept; (c) the x-intercept(s) of the graph of the function and the multiplicities of the real zeros; (d) the symmetries of the graph of the function, if any; and (e) the intervals on which the function is positive or negative. Use this information to sketch a graph of the function. Factor first if the expression is not in factored form.

27. $f(x) = x^2(x - 1)$

28. $h(x) = x(x - 2)^2$

29. $f(x) = (x - 2)^2(x + 2)$

30. $g(x) = (x + 1)(x - 2)^2$

31. $g(x) = (x + 1)^2(x - 2)(x + 3)$

32. $f(x) = (x - 1)(x + 2)^2(x + 1)$

33. $g(x) = -2(x + 1)^2(x - 3)^2$

34. $f(x) = -3(x - 2)^2(x + 1)^2$

35. $f(x) = x^3 + 4x^2 + 4x$

36. $f(x) = -x^3 - 2x^2 - x$

37. $h(x) = -2x^4 + 4x^3 + 2x^2$

38. $f(x) = 3x^4 - 6x^3 + 3x^2$

In Exercises 39–46, find an expression for a polynomial function $f(x)$ having the given properties. There can be more than one correct answer.

39. Degree 3; zeros -2, 5, and 6, each of multiplicity 1

40. Degree 3; zeros -6, 0, and 3, each of multiplicity 1

41. Degree 4; zeros 2 and 4, each of multiplicity 2

42. Degree 4; zeros 2 and -3, each of multiplicity 1; zero at 5 of multiplicity 2

43. Degree 3; zero at 2 of multiplicity 1; zero at -3 of multiplicity 2

44. Degree 3; zero at 5 of multiplicity 3

45. Degree 5; zeros at -2 and -1, each of multiplicity 1; zero at 5 of multiplicity 3

46. Degree 5; zeros at -3 and 1, each of multiplicity 2; zero at 4 of multiplicity 1

 In Exercises 47–50, graph the polynomial function using a graphing utility. Then (a) approximate the x-intercept(s) of the graph of the function; (b) find the intervals on which the function is positive or negative; (c) approximate the values of x at which a local maximum or local minimum occurs; and (d) discuss any symmetries.

47. $f(x) = -x^3 + 3x + 1$

48. $f(x) = x^3 + x^2 + \dfrac{1}{2}$

49. $f(x) = x^4 + 2x^3 - 1$

50. $f(x) = -x^4 + 3x - 1$

▶ **Applications** In this set of exercises, you will use polynomials to study real-world problems.

51. **Geometry** A rectangular solid has height h and a square base. One side of the square base is 3 inches greater than the height.
   (a) Find an expression for the volume of the solid in terms of h.
   (b) Sketch a graph of the volume function.
   (c) For what values of h does the volume function make sense?

52. **Manufacturing** An open box is to be made by cutting four squares of equal size from a 12-inch by 12-inch square piece of cardboard (one at each corner) and then folding up the sides.

(a) Let x be the length of a side of the square cut from each corner. Find an expression for the volume of the box in terms of x.

(b) Sketch a graph of the volume function.

(c) Find the value of x that gives the maximum volume for the box.

53. **Economics** Gross Domestic Product (GDP) is the market value of all final goods and services produced within a country during a given time period. The following fifth-degree polynomial approximates the per capita GDP (the average GDP per person) for the United States for the years 1933 to 1950.

$$g(x) = 0.294x^5 - 12.2x^4 + 169x^3 - 912x^2 + 2025x + 4508$$

where $g(x)$ is in 1996 dollars and x is the number of years since 1933. Note that when dollar amounts are measured over time, they are converted to the dollar value for a specific base year. In this case, the base year is 1996. (*Source: Economic History Services*)

(a) Use this model to calculate the per capita GDP (in 1996 dollars) for the years 1934, 1942, and 1949. What do you observe?

(b) Explain why this model may not be suitable for predicting the per capita GDP for the year 2002.

(c) Use your graphing utility to find the year(s), during the period 1933–1950, when the GDP reached a local maximum.

54. **Foreign Economies** *Coir* is a fiber obtained from the husk of a coconut. It is used chiefly in making rope and floor mats. The amount of coir exported from India during the years 1995 to 2001 is summarized in the following table. (*Source:* Food and Agriculture Organization of the United Nations)

Year	Quantity Exported (metric tons)
1995	1,577
1996	963
1997	1,691
1998	3,268
1999	4,323
2000	5,768
2001	11,538

(a) Make a scatter plot of the data, and find the cubic function of best fit for this data set. Let x be the number of years since 1995.

(b) Use the cubic function to estimate the quantity of Indian coir exported in 2003.

(c) Use the cubic function to estimate the quantity of coir exported in 2001. How close is this value to the actual data value?

(d) Explain why the cubic function is not adequate for describing the long-term trend in exports of Indian coir.

55. **Environment** Sulfur dioxide (SO_2) is emitted by power-generating plants and is one of the primary sources of acid rain. The following table gives the total annual SO_2 emissions from the 263 highest-emitting sources for selected years. (*Source:* Environmental Protection Agency)

Year	Annual SO_2 Emissions (millions of tons)
1980	9.4
1985	9.3
1990	8.7
1994	7.4
1996	4.8
1998	4.7
2000	4

(a) Let t denote the number of years since 1980. Make a scatter plot of sulfur dioxide emissions versus t.

(b) Find an expression for the cubic curve of best fit for this data.

(c) Plot the cubic model for the years 1980–2005. Remember that for the years 2001–2005, the curve gives only a *projection*.

(d) Forecast the amount of SO_2 emissions for the year 2005 using the cubic function from part (b).

(e) Do you think the projection found in part (d) is attainable? Why or why not?

(f) The Clean Air Act was passed in 1990, in part to implement measures to reduce the amount of sulfur dioxide emissions. According to the model presented here, have these measures been successful? Explain.

▶ **Concepts** This set of exercises will draw on the ideas presented in this section and your general math background.

56. Sketch the graph of a cubic polynomial function with *exactly* two real zeros. There can be more than one correct answer.

57. Find a polynomial function whose zeros are $x = 0$, 1, and -1. Is your answer the only correct answer? Why or why not? You may confirm your answer with a graphing utility.

58. Find a polynomial function whose graph crosses the x-axis at $(2, 0)$ and $(1, 0)$. Is your answer the only correct answer? Why or why not? You may confirm your answer with a graphing utility.

In Exercises 59–62, use the given information to (a) sketch a possible graph of the polynomial function; (b) indicate on your graph roughly where the local maxima and minima, if any, might occur; (c) find a possible expression for the polynomial; and (d) use a graphing utility to check your answers to parts (a)–(c).

59. The polynomial $p(x)$ has real zeros at $x = -1$ and $x = 3$, and the graph crosses the x-axis at both of these zeros. As $x \to \pm\infty$, $p(x) \to \infty$.

60. The only points at which the graph of the polynomial $f(s)$ crosses the s-axis are $(-1, 0)$ and $(2, 0)$, and the only point at which it just touches the s-axis is $(0, 0)$. The function is positive on the intervals $(-\infty, -1)$ and $(2, \infty)$.

61. The real zeros of the polynomial $h(x)$ are $x = 3$ and $x = 0.5$, each of multiplicity 1, and $x = \sqrt{2}$, of multiplicity 2. As $|x|$ gets large, $h(x) \to +\infty$.

62. The polynomial $q(x)$ has exactly one real zero and no local maxima or minima.

4.3 Division of Polynomials; the Remainder and Factor Theorems

Objectives

▶ Perform long division of polynomials

▶ Perform synthetic division of polynomials

▶ Apply the Remainder and Factor Theorems

In previous courses, you may have learned how to factor polynomials using various techniques. Many of these techniques apply only to special kinds of polynomial expressions. For example, in the previous two sections of this chapter, we dealt only with polynomials that could easily be factored to find the zeros and x-intercepts.

A process called **long division of polynomials** can be used to find the zeros and x-intercepts of polynomials that cannot readily be factored. After learning this process, we will use the long division algorithm to make a general statement about the factors of a polynomial.

Long Division of Polynomials

Long division of polynomials is similar to long division of numbers. When dividing polynomials, we obtain a quotient and a remainder. Just as with numbers, if the remainder is 0, then the divisor is a factor of the dividend.

Example 1 Determining Factors by Division

Divide to determine whether $x - 1$ is a factor of $x^2 - 3x + 2$.

▶ **Solution** Here, $x^2 - 3x + 2$ is the *dividend* and $x - 1$ is the *divisor*.

Step 1 Set up the division as follows.

$$\text{Divisor} \rightarrow x - 1 \overline{)x^2 - 3x + 2} \leftarrow \text{Dividend}$$

Step 2 Divide the leading term of the dividend (x^2) by the leading term of the divisor (x). The result (x) is the first term of the quotient, as illustrated below.

$$\begin{array}{r} x \leftarrow \text{First term of quotient} \\ x - 1 \overline{)x^2 - 3x + 2} \end{array}$$

Step 3 Take the first term of the quotient (x) and multiply it by the divisor, which gives $x^2 - x$. Put this result in the second row.

$$
\begin{array}{r}
x \phantom{{}- 3x + 2} \\
x - 1 \overline{) x^2 - 3x + 2} \\
\underline{x^2 - x} \quad \leftarrow \text{Multiply } x \text{ by divisor}
\end{array}
$$

Step 4 Subtract the second row from the first row, which gives $-2x$, and bring down the 2 from the dividend. Treat the resulting expression ($-2x + 2$) as though it were a new dividend, just as in long division of numbers.

$$
\begin{array}{r}
x \phantom{{}- 3x + 2} \\
x - 1 \overline{) x^2 - 3x + 2} \\
\underline{x^2 - x} \\
-2x + 2 \quad \leftarrow (x^2 - 3x + 2) - (x^2 + x) \text{ (Watch your signs!)}
\end{array}
$$

Step 5 Continue as in Steps 1–4, but in Step 2, divide the leading term of the expression in the bottom row (i.e., the leading term of $-2x + 2$) by the leading term of the divisor. This result, -2, is the second term of the quotient.

$$
\begin{array}{r}
x - 2 \\
x - 1 \overline{) x^2 - 3x + 2} \\
\underline{x^2 - x} \\
\text{Leading term is } -2x \rightarrow \quad -2x + 2 \\
\text{Multiply } -2 \text{ by divisor} \rightarrow \quad \underline{-2x + 2} \\
(-2x + 2) - (-2x + 2) \rightarrow \quad 0
\end{array}
$$

Thus, dividing $x^2 - 3x + 2$ by $x - 1$ gives $x - 2$ as the *quotient* and 0 as the *remainder*. This tells us that $x - 1$ is a factor of $x^2 - 3x + 2$.

Check Check your answer: $(x - 2)(x - 1) = x^2 - 3x + 2$.

☑ *Check It Out 1:* Find the quotient and remainder when the polynomial $x^2 + x - 6$ is divided by $x - 2$. ■

We can make the following statement about the relationships among the dividend, the divisor, the quotient, and the remainder:

$$(\text{Divisor} \times \text{quotient}) + \text{remainder} = \text{dividend}$$

This very important result is stated formally as follows.

The Division Algorithm

Let $p(x)$ be a polynomial divided by a nonzero polynomial $d(x)$. Then there exist a quotient polynomial $q(x)$ and a remainder polynomial $r(x)$ such that

$$p(x) = d(x)q(x) + r(x) \quad \text{or, equivalently,} \quad \frac{p(x)}{d(x)} = q(x) + \frac{r(x)}{d(x)}$$

where either $r(x) = 0$ or the degree of $r(x)$ is less than the degree of $d(x)$.

The following result illustrates the relationship between factors and remainders.

Factors and Remainders

Let a polynomial $p(x)$ be divided by a nonzero polynomial $d(x)$, with a **quotient polynomial** $q(x)$ and a **remainder polynomial** $r(x)$. If $r(x) = 0$, then $d(x)$ and $q(x)$ are both *factors* of $p(x)$.

Example 2 shows how polynomial division can be used to factor a polynomial that cannot be factored using the methods that you are familiar with.

Example 2 Long Division of Polynomials

Find the quotient and remainder when $2x^4 + 7x^3 + 4x^2 - 7x - 6$ is divided by $2x + 3$.

▶ **Solution** We follow the same steps as before, but condense them in this example.

Step 1 ▶ Divide the leading term of the dividend $(2x^4)$ by the leading term of the divisor $(2x)$. The result, x^3, is the first term of the quotient.

▶ Multiply the first term of the quotient, x^3, by the divisor and put the result, $2x^4 + 3x^3$, in the second row.

▶ Subtract the second row from the first row, just as in division of numbers.

$$
\begin{array}{r}
x^3 \\
2x+3\overline{)2x^4 + 7x^3 + 4x^2 - 7x - 6} \\
\underline{2x^4 + 3x^3} \quad \leftarrow \text{Multiply } x^3 \text{ by divisor} \\
4x^3 + 4x^2 - 7x - 6 \quad \leftarrow \text{Subtract}
\end{array}
$$

Step 2 Divide the leading term of the expression in the bottom row, $4x^3$, by the leading term of the divisor. Multiply the result, $2x^2$, by the divisor and subtract.

$$
\begin{array}{r}
x^3 + 2x^2 \\
2x+3\overline{)2x^4 + 7x^3 + 4x^2 - 7x - 6} \\
\underline{2x^4 + 3x^3} \\
4x^3 + 4x^2 - 7x - 6 \\
\underline{4x^3 + 6x^2} \quad \leftarrow \text{Multiply } 2x^2 \text{ by divisor} \\
-2x^2 - 7x - 6 \quad \leftarrow \text{Subtract}
\end{array}
$$

Step 3 Divide the leading term of the expression in the bottom row, $-2x^2$, by the leading term of the divisor. Multiply the result, $-x$, by the divisor and subtract.

$$
\begin{array}{r}
x^3 + 2x^2 - x \\
2x+3\overline{)2x^4 + 7x^3 + 4x^2 - 7x - 6} \\
\underline{2x^4 + 3x^3} \\
4x^3 + 4x^2 - 7x - 6 \\
\underline{4x^3 + 6x^2} \\
-2x^2 - 7x - 6 \\
\underline{-2x^2 - 3x} \quad \leftarrow \text{Multiply } -x \text{ by divisor} \\
-4x - 6 \quad \leftarrow \text{Subtract (Be careful with signs!)}
\end{array}
$$

Step 4 Divide the leading term of the expression in the bottom row, $-4x$, by the leading term of the divisor. Multiply the result, -2, by the divisor and subtract.

$$
\begin{array}{r}
x^3 + 2x^2 - x - 2 \\
2x + 3\overline{)2x^4 + 7x^3 + 4x^2 - 7x - 6} \\
\underline{2x^4 + 3x^3} \\
4x^3 + 4x^2 - 7x - 6 \\
\underline{4x^3 + 6x^2} \\
-2x^2 - 7x - 6 \\
\underline{-2x^2 - 3x} \\
-4x - 6 \\
\underline{-4x - 6} \quad \leftarrow \text{Multiply } -2 \text{ by divisor} \\
0 \quad \leftarrow \text{Subtract (Be careful with signs!)}
\end{array}
$$

Thus, dividing $2x^4 + 7x^3 + 4x^2 - 7x - 6$ by $2x + 3$ gives a quotient $q(x)$ of $x^3 + 2x^2 - x - 2$ with a remainder $r(x)$ of 0. Equivalently,

$$\frac{2x^4 + 7x^3 + 4x^2 - 7x - 6}{2x + 3} = x^3 + 2x^2 - x - 2.$$

Check You can check that $(2x + 3)(x^3 + 2x^2 - x - 2) = 2x^4 + 7x^3 + 4x^2 - 7x - 6$.

☑ *Check It Out 2:* Find the quotient and remainder when $6x^3 - x^2 - 3x + 1$ is divided by $2x - 1$. ■

Thus far, none of the long division problems illustrated have produced a remainder. Example 3 illustrates a long division that results in a remainder.

Example 3 Long Division with Remainder

Find the quotient and remainder when $p(x) = 6x^3 + x - 1$ is divided by $d(x) = x + 2$. Write your answer in the form $\frac{p(x)}{d(x)} = q(x) + \frac{r(x)}{d(x)}$.

▶Solution The steps for long division should by now be fairly clear to you.

Step 1 Note that the expression $6x^3 + x - 1$ does not have an x^2 term; i.e., the coefficient of the x^2 term is 0. Thus, when we set up the division, we write the x^2 term as $0x^2$. We then divide $6x^3$ by the leading term of the divisor.

$$
\begin{array}{r}
6x^2 \\
x + 2\overline{)6x^3 + 0x^2 + x - 1} \\
\underline{6x^3 + 12x^2} \quad \leftarrow \text{Multiply } 6x^2 \text{ by divisor} \\
-12x^2 + x - 1 \quad \leftarrow \text{Subtract}
\end{array}
$$

Step 2 Next we divide $-12x^2$ by the leading term of the divisor.

$$
\begin{array}{r}
6x^2 - 12x \\
x + 2\overline{)6x^3 + 0x^2 + x - 1} \\
\underline{6x^3 + 12x^2} \\
-12x^2 + x - 1 \\
\underline{-12x^2 - 24x} \quad \leftarrow \text{Multiply } -12x \text{ by divisor} \\
25x - 1 \quad \leftarrow \text{Subtract}
\end{array}
$$

Step 3 Finally, we divide $25x$ by the leading term of the divisor.

$$
\begin{array}{r}
6x^2 - 12x + 25 \\
x + 2 \overline{\smash{\big)}\, 6x^3 + 0x^2 + x - 1} \\
\underline{6x^3 + 12x^2} \\
-12x^2 + x - 1 \\
\underline{-12x^2 - 24x} \\
25x - 1 \\
\underline{25x + 50} \quad \leftarrow \text{Multiply 25 by divisor} \\
-51 \quad \leftarrow \text{Subtract}
\end{array}
$$

Thus, the quotient is $q(x) = 6x^2 - 12x + 25$ and the remainder is $r(x) = -51$. Equivalently,

$$
\frac{6x^3 + x - 1}{x + 2} = 6x^2 - 12x + 25 - \frac{51}{x + 2}.
$$

In this example, long division yields a nonzero remainder. Thus $x + 2$ is *not* a factor of $6x^3 + x - 1$.

☑ *Check It Out 3:* Find the quotient and remainder when $3x^3 + x^2 - 1$ is divided by $x + 1$. Write your answer in the form $\dfrac{p(x)}{d(x)} = q(x) + \dfrac{r(x)}{d(x)}$. ■

Synthetic Division

Synthetic division is a compact way of dividing polynomials when the divisor is of the form $x - c$. Instead of writing out all the terms of the polynomial, we work only with the coefficients. We illustrate this shorthand form of polynomial division using the problem from Example 3.

Example **4** **Synthetic Division**

Use synthetic division to divide $6x^3 + x - 1$ by $x + 2$.

▶ **Solution** Because the divisor is of the form $x - c$, we can use synthetic division. Note that $c = -2$.

Step 1 Write down the coefficients of the dividend in a row, from left to right, and then place the value of c (which is -2) in that same row, to the left of the leading coefficient of the dividend.

$$
\text{Value of } c \rightarrow \quad \underline{-2} \begin{vmatrix} \; \end{vmatrix} \; 6 \quad 0 \quad 1 \quad -1 \quad \leftarrow \text{Coefficients of the dividend}
$$

Step 2 Bring down the leading coefficient of the dividend, 6, and multiply it by c, which is -2.

$$
\begin{array}{r}
\underline{-2} \begin{vmatrix} \; \end{vmatrix} \; 6 \quad 0 \quad 1 \quad -1 \\
\downarrow -12 \quad \leftarrow \text{(b) Then multiply 6 by } -2 \\
6 \nearrow
\end{array}
$$

(a) First bring down 6 →

Step 3 Place the result, -12, below the coefficient of the next term of the dividend, 0, and add.

$$
\begin{array}{r}
\underline{-2} \begin{vmatrix} \; \end{vmatrix} \; 6 \quad 0 \quad 1 \quad -1 \\
\downarrow -12 \\
\hline
6 \; -12 \quad \leftarrow \text{Add 0 and } -12
\end{array}
$$

Step 4 Apply Steps 2(b) and 3 to the result, which is −12.

$$\begin{array}{r|rrrr} -2 & 6 & 0 & 1 & -1 \\ & \downarrow & -12 & 24 & \\ \hline & 6 & -12 \nearrow 25 & & \end{array}$$ ← Multiply −12 by −2
← Add 1 and 24

Step 5 Apply Steps 2(b) and 3 to the result, which is 25.

$$\begin{array}{r|rrrr} -2 & 6 & 0 & 1 & -1 \\ & \downarrow & -12 & 24 & -50 \\ \hline & 6 & -12 & 25 \nearrow & -51 \end{array}$$ ← Multiply 25 by −2
← Add −1 and −50

Step 6 The last row consists of the coefficients of the quotient polynomial. The remainder is the last number in the row. The degree of the first term of the quotient is one less than the degree of the dividend. So, the 6 in the last row represents $6x^2$; the −12 represents −12x; and the 25 represents the constant term. Thus we have

$$q(x) = 6x^2 - 12x + 25.$$

The remainder $r(x)$ is −51, the last number in the bottom row.

✔ *Check It Out 4:* Use synthetic division to divide $-2x^3 + 3x^2 - 1$ by $x - 1$. ▪

The Remainder and Factor Theorems

We now examine an important connection between a polynomial $p(x)$ and the remainder we get when $p(x)$ is divided by $x - c$. In Example 3, division of $p(x) = 6x^3 + x - 1$ by $x + 2$ yielded a remainder of −51. Also, $p(-2) = -51$. This is a consequence of the Remainder Theorem, formally stated as follows.

The Remainder Theorem

When a polynomial $p(x)$ is divided by $x - c$, the remainder is equal to the value of $p(c)$.

Because the remainder is equal to $p(c)$, synthetic division provides a quick way to evaluate $p(c)$. This is illustrated in the next example.

Example 5 Applying the Remainder Theorem

Let $p(x) = -2x^4 + 6x^3 + 3x - 1$. Use synthetic division to evaluate $p(2)$.

▶Solution From the Remainder Theorem, $p(2)$ is the remainder obtained when $p(x)$ is divided by $x - 2$. Following the steps outlined in Example 4, we have the following.

$$\begin{array}{r|rrrrr} 2 & -2 & 6 & 0 & 3 & -1 \\ & \downarrow & -4 & 4 & 8 & 22 \\ \hline & -2 & 2 & 4 & 11 & 21 \end{array}$$

Because the remainder is 21, we know from the Remainder Theorem that $p(2) = 21$.

✔ *Check It Out 5:* Let $p(x) = 3x^4 - x^2 + 3x - 1$. Use synthetic division to evaluate $p(-2)$. ▪

The **Factor Theorem** is a direct result of the Remainder Theorem.

The Factor Theorem

The term $x - c$ is a *factor* of a polynomial $p(x)$ if and only if $p(c) = 0$.

The Factor Theorem makes an important connection between zeros and factors. It states that if we have a *linear factor*—i.e., a factor of the form $x - c$—of a polynomial $p(x)$, then $p(c) = 0$. That is, *c is a zero of the polynomial $p(x)$*. It also works the other way around: if *c* is a zero of the polynomial $p(x)$, then *$x - c$ is a factor of $p(x)$*.

Example 6 **Applying the Factor Theorem**

Determine whether $x + 3$ is a factor of $2x^3 + 3x - 2$.

▶**Solution** Let $p(x) = 2x^3 + 3x - 2$. Because $x + 3$ is in the form $x - c$, we can apply the Factor Theorem with $c = -3$.

Evaluating, $p(-3) = -65$. By the Factor Theorem, $x + 3$ is *not* a factor of $2x^3 + 3x - 2$ because $p(-3) \neq 0$.

✔ *Check It Out 6:* Determine whether $x - 1$ is a factor of $2x^3 - 4x + 2$. ▪

4.3 Key Points

▶ Let $p(x)$ be a polynomial divided by a nonzero polynomial $d(x)$. Then there exist a quotient polynomial $q(x)$ and a remainder polynomial $r(x)$ such that

$$p(x) = d(x)q(x) + r(x).$$

▶ **Synthetic division** is a shorter way of dividing polynomials when the divisor is of the form $x - c$.

▶ The **Remainder Theorem** states that when a polynomial $p(x)$ is divided by $x - c$, the remainder is equal to the value of $p(c)$.

▶ The **Factor Theorem** states that the term $x - c$ is a *factor* of a polynomial $p(x)$ if and only if $p(c) = 0$.

4.3 Exercises

▶**Skills** This set of exercises will reinforce the skills illustrated in this section.

In Exercises 1–14, find the quotient and remainder when the first polynomial is divided by the second. You may use synthetic division wherever applicable.

1. $2x^2 + 13x + 15$; $x + 5$
2. $2x^2 - 7x + 3$; $x - 3$
3. $2x^3 - x^2 - 8x + 4$; $2x - 1$
4. $3x^3 + 2x^2 - 3x - 2$; $3x + 2$
5. $x^3 - 3x^2 + 2x - 4$; $x + 2$
6. $x^3 + 2x^2 - x - 3$; $x - 3$
7. $-3x^4 + x^2 - 2$; $3x - 1$
8. $2x^4 - x^3 + x^2 - x$; $2x + 1$
9. $x^6 + 1$; $x + 1$
10. $-x^3 + x$; $x - 5$
11. $x^3 + 2x^2 - 5$; $x^2 - 2$
12. $-x^3 - 3x^2 + 6$; $x^2 + 1$
13. $x^5 - x^4 + 2x^3 + x^2 - x + 1$; $x^3 + x - 1$
14. $-2x^5 + x^4 - x^3 + 2x^2 - 1$; $x^3 + x^2 + 1$

In Exercises 15–20, write each polynomial in the form $p(x) = d(x)q(x) + r(x)$, where $p(x)$ is the given polynomial and $d(x)$ is the given factor. You may use synthetic division wherever applicable.

15. $x^2 + x + 1$; $x + 1$

16. $x^2 + x + 1$; $x - 1$

17. $3x^3 + 2x - 8$; $x - 4$

18. $4x^3 - x + 4$; $x - 2$

19. $x^6 - 3x^5 + x^4 - 2x^2 - 5x + 6$; $x^2 + 2$

20. $-x^6 + 4x^5 - x^3 + x^2 + x - 8$; $x^2 + 4$

In Exercises 21–28, use synthetic division to find the function values.

21. $f(x) = x^3 - 7x + 5$; find $f(3)$ and $f(5)$.

22. $f(x) = -2x^3 + 4x^2 - 7$; find $f(4)$ and $f(-3)$.

23. $f(x) = -2x^4 - 10x^3 - 3x + 10$; find $f(-1)$ and $f(2)$.

24. $f(x) = -x^4 + 3x^3 - 2x - 4$; find $f(-2)$ and $f(3)$.

25. $f(x) = x^5 - 2x^3 + 12$; find $f(3)$ and $f(-2)$.

26. $f(x) = -2x^5 + x^4 + x^2 - 2$; find $f(-3)$ and $f(4)$.

27. $f(x) = x^4 - 2x^2 + 1$; find $f\left(\dfrac{1}{2}\right)$.

28. $f(x) = -x^4 + 3x^2 - 2x$; find $f\left(\dfrac{3}{2}\right)$.

In Exercises 29–38, determine whether $q(x)$ is a factor of $p(x)$. Here, $p(x)$ is the first polynomial and $q(x)$ is the second polynomial. Justify your answer.

29. $x^3 - 7x + 6$; $x - 3$

30. $x^3 - 5x^2 + 8x - 4$; $x + 2$

31. $x^3 - 7x + 6$; $x + 3$

32. $x^3 - 5x^2 + 8x - 4$; $x - 2$

33. $x^5 - 3x^3 + 2x - 8$; $x - 4$

34. $-2x^4 - 7x^3 + 5$; $x + 2$

35. $x^4 - 50$; $x - 5$

36. $2x^5 - 1$; $x - 2$

37. $3x^3 - 48x - 4x^2 + 64$; $x + 4$

38. $x^3 + 9x + x^2 + 9$; $x + 1$

▶**Concepts** This set of exercises will draw on the ideas presented in this section and your general math background.

39. Given the following graph of a polynomial function $p(x)$, find a linear factor of $p(x)$.

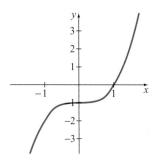

40. Consider the following graph of a polynomial function $p(x)$.

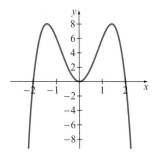

   (a) Evaluate $p(2)$.
   (b) Is $x - 2$ a factor of $p(x)$? Explain.
   (c) Find the remainder when $p(x)$ is divided by $x - 2$.

41. Find the remainder when $x^7 + 7$ is divided by $x - 1$

42. Find the remainder when $x^8 - 3$ is divided by $x + 1$.

43. Let $x - \dfrac{1}{2}$ be a factor of a polynomial function $p(x)$. Find $p\left(\dfrac{1}{2}\right)$.

44. For what value(s) of k do you get a remainder of 15 when you divide $kx^3 + 2x^2 - 10x + 3$ by $x + 2$?

45. For what value(s) of k do you get a remainder of -2 when you divide $x^3 - x^2 + kx + 3$ by $x + 1$?

46. Why is the Factor Theorem a direct result of the Remainder Theorem?

4.4 Real Zeros of Polynomials; Solutions of Equations

Objectives

▶ Find the rational zeros of a polynomial

▶ Solve polynomial equations by finding zeros

▶ Know and apply Descartes' Rule of Signs

Recall the techniques we have used so far to find zeros of polynomials and solve polynomial equations algebraically: factoring, the quadratic formula, and the methods of long division and synthetic division. In this section and the next, we will learn properties of polynomial functions that, taken together with factoring and methods of division, will enable us to find the zeros of a broader range of polynomials. This will in turn assist us in solving a broader range of polynomial equations.

To find the zeros of polynomials of degree 2, you can use the quadratic formula. However, for polynomials of degree greater than 2, it can be challenging to find the zeros as there are no easy-to-use formulas like the quadratic formula. For polynomials of degree 5 or greater, formulas for finding zeros do not even exist! However, it is possible to find zeros of polynomials of degree 3 or greater if we can factor the polynomials or if they are of a special type. We will study these special types of polynomials and their corresponding equations in this section.

A computer or graphing calculator is useful for finding the zeros of general polynomial functions. If you are using a graphing utility, you will be able to find the zeros of a wider range of polynomials and solve a wider range of polynomial equations.

Example **1** **Using a Known Zero to Factor a Polynomial**

Show that $x = 4$ is a zero of $p(x) = 3x^3 - 4x^2 - 48x + 64$. Use this fact to completely factor $p(x)$.

▶**Solution** Evaluate $p(4)$ to get

$$p(4) = 3(4)^3 - 4(4)^2 - 48(4) + 64 = 0.$$

We see that $x = 4$ is a zero of $p(x)$. By the Factor Theorem, $x - 4$ is therefore a factor of $p(x)$. Using synthetic division, we obtain $p(x) = (x - 4)(3x^2 + 8x - 16)$. The second factor, $3x^2 + 8x - 16$, is a quadratic expression that can be factored further:

$$3x^2 + 8x - 16 = (3x - 4)(x + 4)$$

Therefore,

$$p(x) = 3x^3 - 4x^2 - 48x + 64 = (x - 4)(3x - 4)(x + 4).$$

☑ *Check It Out 1:* Show that $x = -2$ is a zero of $p(x) = 2x^3 + x^2 - 5x + 2$. Use this fact to completely factor $p(x)$. ■

The following statements summarize the key connections among the factors and real zeros of a polynomial and the x-intercepts of its graph.

> **Zeros, Factors, and *x*-Intercepts of a Polynomial**
>
> Let $p(x)$ be a polynomial function and let c be a real number. Then the following are equivalent statements. That is, if one of the following statements is true, then the other two statements are also true. Similarly, if one of the following statements is false, then the other two statements are also false.
>
> ▶ $p(c) = 0$
>
> ▶ $x - c$ is a factor of $p(x)$.
>
> ▶ $(c, 0)$ is an x-intercept of the graph of $p(x)$.

| Example | 2 | **Relating Zeros, Factors, and *x*-Intercepts** |

Fill in Table 4.4.1, where p, h, and g are polynomial functions.

Table 4.4.1

	Function	Zero	x-Intercept	Factor
(a)	$p(x)$		(5, 0)	
(b)	$h(x)$			$x - 3$
(c)	$g(x)$	-1		

▶**Solution**

(a) Because $(5, 0)$ is an *x*-intercept of the graph of $p(x)$, the corresponding zero is 5 and the corresponding factor is $x - 5$.

(b) Because $x - 3$ is a factor of $h(x)$, the corresponding zero is 3 and the corresponding *x*-intercept is $(3, 0)$.

(c) Because -1 is a zero of $g(x)$, the corresponding *x*-intercept is $(-1, 0)$ and the corresponding factor is $x - (-1)$, or $x + 1$.

The results are given in Table 4.4.2.

Table 4.4.2

	Function	Zero	x-Intercept	Factor
(a)	$p(x)$	5	(5, 0)	$x - 5$
(b)	$h(x)$	3	(3, 0)	$x - 3$
(c)	$g(x)$	-1	$(-1, 0)$	$x + 1$

✔ *Check It Out 2:* Fill in Table 4.4.3, where p, h, and g are polynomial functions.

Table 4.4.3

Function	Zero	x-Intercept	Factor
$p(x)$			$x + 6$
$h(x)$	4		
$g(x)$		(2, 0)	

The Rational Zero Test

We have seen that a relationship exists between the factors of a polynomial and its zeros. But we still do not know how to *find* the zeros of a given polynomial. Remember that once we find the zeros, we can readily factor the polynomial. The following fact is useful when discussing zeros of a polynomial.

> **Number of Real Zeros of a Polynomial**
>
> A nonconstant polynomial function $p(x)$ of degree n has at most n real zeros, where each zero of multiplicity k is counted k times.

In general, finding *all* the zeros of any given polynomial by hand is not possible. However, we can use a theorem called the Rational Zero Theorem to find out whether a polynomial with *integer coefficients* has any *rational* zeros—that is, rational numbers that are zeros of the polynomial.

Consider $p(x) = 10x^2 - 29x - 21 = (5x + 3)(2x - 7)$. The zeros of $p(x)$ are $\frac{7}{2}$ and $-\frac{3}{5}$. Notice that the numerator of each of the zeros is a factor of the constant term of the polynomial, -21. Also notice that the denominator of each zero is a factor of the leading coefficient of the polynomial, 10. This observation can be generalized to the rational zeros of *any* polynomial of degree n with integer coefficients. This is summarized by the **Rational Zero Theorem.**

> **The Rational Zero Theorem**
>
> If $f(x) = a_n x^n + a_{n-1} x^{n-1} + \cdots + a_1 x + a_0$ is a polynomial with integer coefficients, and $\frac{p}{q}$ is a rational zero of f with p and q having no common factor other than 1, then p is a factor of a_0 and q is a factor of a_n.

Just In Time

Review the real number system in Section P.I.

The Rational Zero Theorem can be used in conjunction with long division to find *all* the real zeros of a polynomial. This technique is discussed in the next example.

Example 3 Applying the Rational Zero Theorem

Find all the real zeros of

$$p(x) = 3x^3 - 6x^2 - x + 2.$$

▶**Solution**

Step 1 First, list all the possible rational zeros.

We consider all possible factors of 2 for the numerator of a rational zero, and all possible factors of 3 for the denominator. The factors of 2 are ± 1 and ± 2, and the factors of 3 are ± 1 and ± 3. So the possible rational zeros are

$$\pm 1,\ \pm 2,\ \pm \frac{1}{3},\ \pm \frac{2}{3}.$$

Step 2 The value of $p(x)$ at each of the possible rational zeros is summarized in Table 4.4.4.

Table 4.4.4

x	-2	-1	$-\frac{2}{3}$	$-\frac{1}{3}$	$\frac{1}{3}$	$\frac{2}{3}$	1	2
$p(x)$	-44	-6	-0.88889	1.55556	1.11111	-0.44444	-2	0

Only $x = 2$ is an actual zero. Thus, $x - 2$ is a factor of $p(x)$.

Figure 4.4.1

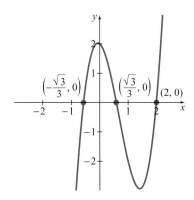

Step 3 Divide $p(x)$ by $x - 2$ using synthetic division:

$$\begin{array}{r|rrrr} 2 & 3 & -6 & -1 & 2 \\ & & 6 & 0 & 2 \\ \hline & 3 & 0 & -1 & 0 \end{array}$$

Thus, $p(x) = 3x^3 - 6x^2 - x + 2 = (x - 2)(3x^2 - 1)$.

Step 4 To find the other zeros, solve the quadratic equation $3x^2 - 1 = 0$ which gives $x = \pm\dfrac{\sqrt{3}}{3}$. Note that the Rational Zero Theorem does *not* give these two zeros, since they are *irrational*.

Thus the three zeros of p are 2, $\dfrac{\sqrt{3}}{3}$, and $-\dfrac{\sqrt{3}}{3}$. These are the only zeros, because a cubic polynomial function can have at most three zeros. The graph of $p(x)$ given in Figure 4.4.1 indicates the locations of the zeros. The graph can be sketched using the techniques presented in Sections 4.1 and 4.2.

✔ *Check It Out 3:* Find all the real zeros of $p(x) = 4x^3 + 4x^2 - x - 1$. ■

 Example 4 Using a Graphing Utility to Locate Zeros

Use a graphing utility to find all the real zeros of $p(x) = 3x^3 - 6x^2 - x + 2$.

Figure 4.4.2

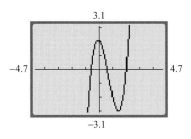

▶**Solution** We can quickly locate all the *rational* zeros of the polynomial by using a graphing utility in conjunction with the Rational Zero Theorem. First, list all the possible rational zeros.

$$\pm 1, \ \pm 2, \ \pm\frac{1}{3}, \ \pm\frac{2}{3}$$

The zeros range from -2 to 2. When we graph the function $p(x) = 3x^3 - 6x^2 - x + 2$ in a decimal window, we immediately see that 2 is a probable zero. The possibilities $x = \pm 1$ can be excluded right away. See Figure 4.4.2. Consult Sections 5 and 9 of the Keystroke Appendix for details.

Using the graphing calculator to evaluate the function, we see that $p(2) = 0$. Note that $\pm\dfrac{1}{3}$ are *not* zeros. It may happen that the polynomial has an irrational zero very close to the suspected rational zero.

Using the ZERO feature of your graphing utility, you will find that the other zeros are approximately ± 0.5774. See Figure 4.4.3 for one of the zeros. Thus we have the following zeros: $x = 2$, $x \approx -0.5774$, and $x \approx 0.5774$. The numbers ± 0.5774 are approximations to the exact values $\pm\dfrac{\sqrt{3}}{3}$ found algebraically in Example 3. Because this is a cubic polynomial, it cannot have more than three zeros. So we have found all the zeros of $p(x) = 3x^3 - 6x^2 - x + 2$.

Figure 4.4.3

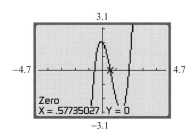

✔ *Check It Out 4:* Use a graphing utility to find all the real zeros of

$$p(x) = x^3 - x^2 - 7x - 2. \quad ■$$

Finding Solutions of Polynomial Equations by Finding Zeros

Because any equation can be rewritten so that the right-hand side is zero, solving an equation is identical to finding the zeros of a suitable function.

Example 5 **Solving a Polynomial Equation**

Solve the equation $3x^4 - 8x^3 - 9x^2 + 22x = -8$.

▶**Solution**

Algebraic Approach: We first write the equation in the form $p(x) = 0$.

$$3x^4 - 8x^3 - 9x^2 + 22x + 8 = 0$$

The task now is to find all zeros of $p(x) = 3x^4 - 8x^3 - 9x^2 + 22x + 8$. We can use the Rational Zero Theorem to list all the possible rational zeros.

$$\frac{\text{Factors of 8}}{\text{Factors of 3}} = \frac{\pm 1, \pm 2, \pm 4, \pm 8}{\pm 1, \pm 3} = \pm 1, \pm 2, \pm 4, \pm 8, \pm\frac{1}{3}, \pm\frac{2}{3}, \pm\frac{4}{3}, \pm\frac{8}{3}$$

Next we evaluate $p(x)$ at each of these possibilities, either by direct evaluation or by synthetic division. We try the integer possibilities first, from the smallest in magnitude to the largest. We find that $p(2) = 0$, and so $x - 2$ is a factor of $p(x)$. Using synthetic division, we factor out the term $x - 2$.

$$
\begin{array}{r|rrrrr}
2 & 3 & -8 & -9 & 22 & 8 \\
  & \downarrow & 6 & -4 & -26 & -8 \\
\hline
  & 3 & -2 & -13 & -4 & 0
\end{array}
$$

Thus,

$$3x^4 - 8x^3 - 9x^2 + 22x + 8 = (x - 2)(3x^3 - 2x^2 - 13x - 4).$$

Next we try to factor $q(x) = 3x^3 - 2x^2 - 13x - 4$. Checking the possible rational roots listed earlier, we find that none of the other integers on the list is a zero of $q(x)$. We try the fractions and find that $q\left(-\frac{1}{3}\right) = 0$. Now we can use synthetic division with $q(x)$ as the dividend to factor out the term $\left(x + \frac{1}{3}\right)$.

$$
\begin{array}{r|rrrr}
-\dfrac{1}{3} & 3 & -2 & -13 & -4 \\
  & \downarrow & -1 & 1 & 4 \\
\hline
  & 3 & -3 & -12 & 0
\end{array}
$$

Thus,

$$3x^3 - 2x^2 - 13x - 4 = \left(x + \frac{1}{3}\right)(3x^2 - 3x - 12) = 3\left(x + \frac{1}{3}\right)(x^2 - x - 4).$$

Note that we factored the quadratic expression $3x^2 - 3x - 12$ as $3(x^2 - x - 4)$. To find the remaining zeros, solve the equation $x^2 - x - 4 = 0$. Using the quadratic formula, we find that

$$x^2 - x - 4 = 0 \Longrightarrow x = \frac{1}{2} \pm \frac{1}{2}\sqrt{17}.$$

Because these two zeros are *irrational*, they did not appear in the list of possible rational zeros.

Thus the solutions to the equation $3x^4 - 8x^3 - 9x^2 + 22x = -8$ are

$$x = 2, \quad x = -\frac{1}{3}, \quad x = \frac{1}{2} + \frac{\sqrt{17}}{2}, \quad x = \frac{1}{2} - \frac{\sqrt{17}}{2}.$$

Graphical Approach: 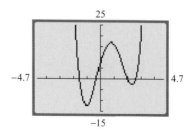 To solve the equation $3x^4 - 8x^3 - 9x^2 + 22x = -8$, graph the function $p(x) = 3x^4 - 8x^3 - 9x^2 + 22x + 8$ and find its zero(s). Using a window size of $[-4.7, 4.7] \times [-15, 25](5)$, we get the graph shown in Figure 4.4.4.

Figure 4.4.4

It looks as though there is a zero at $x = 2$. Using the graphing calculator, we can verify that, indeed, $p(2) = 0$. Consult Section 9 of the Keystroke Appendix for the ZERO Feature.

Using the ZERO feature, we find that there is another zero, close to 2, at $x \approx 2.5616$, as shown in Figure 4.4.5. The two negative zeros, which can be found by using the graphical solver twice, are $x \approx -1.5616$ and $x \approx -0.3333$. Thus the four zeros are

$$x = 2, \ x \approx -0.3333, \ x \approx 2.5616, \text{ and } x \approx -1.5616.$$

Figure 4.4.5

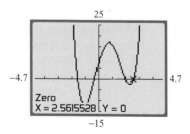

✔ *Check It Out 5:* Solve the equation $2x^4 + 3x^3 - 6x^2 = 5x - 6.$ ▪

Descartes' Rule of Signs

An nth-degree polynomial can have *at most* n real zeros. But many nth-degree polynomials have fewer real zeros. For example, $p(x) = x(x^2 + 1)$ has only one real zero, and $p(x) = -x^4 - 16$ has no real zeros. To get a better idea of the number of real zeros of a polynomial, a rule that uses the signs of the coefficients was developed by the French mathematician Rene Descartes around 1637. For a polynomial written in descending order, the number of **variations in sign** is the number of times that successive coefficients are of different signs. This concept plays a key role in Descartes' rule.

For instance, the polynomial $p(x) = -3x^4 + 6x^3 + x^2 - x + 1$ has *three* variations in sign, illustrated as follows.

$$p(x) = -3x^4 + 6x^3 + x^2 - x + 1$$
$$\underbrace{\qquad}_{1} \quad \underbrace{\qquad}_{2} \underbrace{\ }_{3}$$

We now state Descartes' Rule of Signs, without proof.

> **Descartes' Rule of Signs**
>
> Let $p(x)$ be a polynomial function with real coefficients and a nonzero constant term. Let k be the number of variations in sign of $p(x)$, and let m be the number of variations in sign of $p(-x)$.
>
> ► The number of **positive zeros** of p is either equal to k or less than k by an even integer.
>
> ► The number of **negative zeros** of p is either equal to m or less than m by an even integer.

Note In Descartes' Rule, the number of positive and negative zeros includes multiplicity. For example, if a zero has multiplicity 2, it counts as two zeros.

Example 6 Applying Descartes' Rule of Signs

Use Descartes' Rule of Signs to determine the number of positive and negative zeros of $p(x) = -3x^4 + 4x^2 - 3x + 2$.

► **Solution** First we determine the variations in sign of $p(x)$.

$$p(x) = -3x^4 + 4x^2 - 3x + 2$$

Because $p(x)$ has three variations in sign, the number of *positive zeros* of p is equal to either 3 or less than 3 by an even integer. Therefore, the number of positive zeros is 3 or 1, since a negative number of zeros does not make sense.

Next we determine the variations in sign of $p(-x)$.

$$p(-x) = -3(-x)^4 + 4(-x)^2 - 3(-x) + 2 = -3x^4 + 4x^2 + 3x + 2$$

Because $p(-x)$ has one variation in sign, the number of negative zeros of p is equal to 1.

✔ *Check It Out 6:* Use Descartes' Rule of Signs to determine the number of positive and negative zeros of $p(x) = 4x^4 - 3x^3 + 2x - 1$. ■

4.4 Key Points

► Let $p(x)$ be a polynomial function and let c be a real number. Then the following statements are equivalent.

1. $p(c) = 0$
2. $x - c$ is a factor of $p(x)$.
3. $(c, 0)$ is an x-intercept of the graph of $p(x)$.

► The **Rational Zero Theorem** gives the possible rational zeros of a polynomial with integer coefficients.

► **Descartes' Rule of Signs** gives the number of positive and negative zeros of a polynomial $p(x)$ by examining the variations in sign of the coefficients of $p(x)$ and $p(-x)$, respectively.

4.4 Exercises

▶**Just in Time Exercises** These exercises correspond to the Just in Time references in this section. Complete them to review topics relevant to the remaining exercises.

1. A rational number is a number that can be expressed as the division of two _____.

2. True or False: $\sqrt{2}$ is a rational number.

3. True or False: 0.33333 . . . is a rational number.

4. True or False: $-\dfrac{2}{3}$ is a rational number.

▶**Skills** This set of exercises will reinforce the skills illustrated in this section.

In Exercises 5–10, for each polynomial, determine which of the numbers listed next to it are zeros of the polynomial.

5. $p(x) = (x - 10)^8$, $x = 6, -10, 10$

6. $p(x) = (x + 6)^{10}$, $x = 6, -6, 0$

7. $g(s) = s^2 + 4$, $s = -2, 2$

8. $f(x) = x^2 + 9$, $x = -3, 3$

9. $f(x) = x^3 + 2x^2 - 3x - 6$; $x = \sqrt{3}, -\sqrt{2}$

10. $f(x) = x^3 + 2x^2 - 2x - 4$; $x = \sqrt{2}, -\sqrt{3}$

In Exercises 11–18, show that the given value of x is a zero of the polynomial. Use the zero to completely factor the polynomial.

11. $p(x) = x^3 - 5x^2 + 8x - 4$; $x = 2$

12. $p(x) = x^3 - 7x + 6$; $x = 2$

13. $p(x) = -x^4 - x^3 + 18x^2 + 16x - 32$; $x = 1$

14. $p(x) = 2x^3 - 11x^2 + 17x - 6$; $x = \dfrac{1}{2}$

15. $p(x) = 3x^3 - 2x^2 + 3x - 2$; $x = \dfrac{2}{3}$

16. $p(x) = 2x^3 - x^2 + 6x - 3$; $x = \dfrac{1}{2}$

17. $p(x) = 3x^3 + x^2 + 24x + 8$; $x = -\dfrac{1}{3}$

18. $p(x) = 2x^5 + x^4 - 2x - 1$; $x = -\dfrac{1}{2}$

In Exercises 19–22, fill in the following table, where f, p, h, and g are polynomial functions.

	Function	Zero	x-Intercept	Factor
19.	$f(x)$		$(-2, 0)$	
20.	$p(x)$			$x + 5$
21.	$h(x)$	-4		
22.	$g(x)$	6		

In Exercises 23–34, find all the real zeros of the polynomial.

23. $P(x) = x^3 + 2x^2 - 5x - 6$

24. $P(x) = 2x^3 + 3x^2 - 8x + 3$

25. $P(x) = x^4 - 13x^2 - 12x$

26. $Q(s) = s^4 - s^3 + s^2 - 3s - 6$

27. $P(s) = 4s^4 - 25s^2 + 36$

28. $P(t) = 6t^3 - 4t^2 + 3t - 2$

29. $f(x) = 4x^4 + 11x^3 + x^2 + 11x - 3$

30. $G(x) = 2x^3 + x^2 - 16x - 15$

31. $P(x) = 7x^3 + 2x^2 - 28x - 8$

32. $Q(x) = x^4 - 8x^2 - 9$

33. $h(x) = x^4 + 3x^3 - 8x^2 - 22x - 24$

34. $f(x) = x^5 - 7x^4 + 10x^3 + 14x^2 - 24x$

In Exercises 35–42, find all real solutions of the polynomial equation.

35. $x^3 + 2x^2 + 2x = -1$ 36. $3x^3 - 7x^2 = -5x + 1$

37. $x^3 - 6x^2 + 5x = -12$ 38. $4x^3 - 16x^2 + 19x = -6$

39. $2x^3 - 3x^2 = 11x - 6$ 40. $2x^3 - x^2 - 18x = -9$

41. $x^4 + x^3 - x = 1$ 42. $6x^4 + 11x^3 - 3x^2 = 2x$

In Exercises 43–52, use Descartes' Rule of Signs to determine the number of positive and negative zeros of p. You need not find the zeros.

43. $p(x) = 4x^4 - 5x^3 + 6x - 3$

44. $p(x) = x^4 + 6x^3 - 7x^2 + 2x - 1$

45. $p(x) = -2x^3 + x^2 - x + 1$

46. $p(x) = -3x^3 + 2x^2 - x - 1$

47. $p(x) = 2x^4 - x^3 - x^2 + 2x + 5$

48. $p(x) = 3x^4 - 2x^3 + 3x^2 - 4x + 1$

49. $p(x) = x^5 + 3x^4 - 4x^2 + 10$

50. $p(x) = 2x^5 - 6x^3 + 7x^2 - 8$

51. $p(x) = x^6 + 4x^3 - 3x + 7$

52. $p(x) = 5x^6 - 7x^5 + 4x^3 - 6$

In Exercises 53–59, graph the function using a graphing utility, and find its zeros.

53. $f(x) = x^3 - 3x^2 - 3x - 4$

54. $g(x) = 2x^5 + x^4 - 2x - 1$

55. $h(x) = 4x^3 - 12x^2 + 5x + 6$

56. $p(x) = -x^4 - x^3 + 18x^2 + 16x - 32$

57. $p(x) = -2x^4 + 13x^3 - 23x^2 + 3x + 9$

58. $f(x) = x^3 + x^2 + x - 3.1x^2 - 2.5x - 4$

59. $p(x) = x^3 + (3 + \sqrt{2})x^2 + 4x + 6.7$

▶**Applications** In this set of exercises, you will use polynomials to study real-world problems.

60. **Geometry** A rectangle has length $x^2 - x + 6$ units and width $x + 1$ units. Find x such that the area of the rectangle is 24 square units.

61. **Geometry** The length of a rectangular box is 10 inches more than the height, and its width is 5 inches more than the height. Find the dimensions of the box if the volume is 168 cubic inches.

62. **Manufacturing** An open rectangular box is constructed by cutting a square of length x from each corner of a 12-inch by 15-inch rectangular piece of cardboard and then folding up the sides. For this box, x must be greater than or equal to 1 inch.
   (a) What is the length of the square that must be cut from each corner if the volume of the box is to be 112 cubic inches?
   (b) What is the length of the square that must be cut from each corner if the volume of the box is to be 150 cubic inches?

63. **Manufacturing** The height of a right circular cylinder is 5 inches more than its radius. Find the dimensions of the cylinder if its volume is 1000 cubic inches.

▶**Concepts** This set of exercises will draw on the ideas presented in this section and your general math background.

64. The following is the graph of a cubic polynomial function. Find an expression for the polynomial function with leading coefficient 1 that corresponds to this graph. You may check your answer by using a graphing utility.

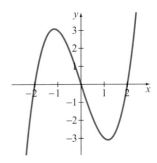

65. Find at least two different cubic polynomials whose only real zero is -1. Graph your answers to check them.

66. Let $p(x) = x^5 + x^3 - 2x$.
   (a) Show that p is symmetric with respect to the origin.
   (b) Find a zero of p by inspection of the polynomial expression.
   (c) Use a graphing utility to find the other zeros.
   (d) How do you know that you have found *all* the zeros of p?

4.5 The Fundamental Theorem of Algebra; Complex Zeros

Objectives

▶ Understand the statement and consequences of the Fundamental Theorem of Algebra

▶ Factor polynomials with real coefficients over the complex numbers

▶ Understand the connections among the real zeros, x-intercepts, and factors of a polynomial

In previous sections of this chapter, we examined ways to find the zeros of a polynomial function. We also saw that there is a close connection between the zeros of a polynomial and its factors. In this section, we will make these observations precise by presenting some known facts about the zeros of a polynomial. We will also expand our search for zeros to include complex zeros of polynomials.

The Fundamental Theorem of Algebra

Recall that the solutions of the equation $P(x) = 0$ are known as *zeros* of the polynomial function P. Another name for a solution of a polynomial equation is *root*. A famous theorem about the existence of a solution to a polynomial equation was proved by the mathematician Karl Friedrich Gauss in 1799. It is stated as follows.

> ### The Fundamental Theorem of Algebra
>
> Every nonconstant polynomial function with real or complex coefficients has at least one complex zero.

Many proofs of this theorem are known, but they are beyond the scope of this text.

Note   The Fundamental Theorem of Algebra states only that a solution exists. It does not tell you how to find the solution.

In order to find the exact number of zeros of a polynomial, the following precise definition of the multiplicity of a zero of a polynomial function is needed. Recall that multiplicity was discussed briefly in Section 4.2.

> ### Definition of Multiplicity of a Zero
>
> A zero c of a polynomial P of degree $n > 0$ has **multiplicity** k if $P(x) = (x - c)^k Q(x)$, where $Q(x)$ is a polynomial of degree $n - k$ and c is not a zero of $Q(x)$.

The following example will help you unravel the notation used in the definition of multiplicity.

Example 1 Determining the Multiplicity of a Zero

Let $h(x) = x^3 + 2x^2 + x$.

(a) What is the value of the multiplicity k of the zero at $x = -1$?

(b) Write $h(x)$ in the form $h(x) = (x + 1)^k Q(x)$. What is $Q(x)$?

Chapter 4 ■ Polynomial and Rational Functions

▶**Solution**

(a) Factoring $h(x)$, we obtain

$$x^3 + 2x^2 + x = x(x^2 + 2x + 1) = x(x + 1)^2.$$

Thus the zero at $x = -1$ is of multiplicity 2; that is, $k = 2$.

(b) We have

$$h(x) = (x + 1)^2 Q(x) = (x + 1)^2 x$$

where $Q(x) = x$. Note that the degree of $Q(x)$ is $n - k = 3 - 2 = 1$. Hence we see that the various aspects of the definition of multiplicity are verified.

✔ **Check It Out 1:** For the function $h(x) = x^4 - 2x^3 + x^2$, what is the value of the multiplicity at $x = 0$? ■

Factorization and Zeros of Polynomials with Real Coefficients

The Factorization Theorem gives information on factoring a polynomial with real coefficients.

> **The Factorization Theorem**
>
> Any polynomial P with *real* coefficients can be factored uniquely into linear factors and/or **irreducible quadratic factors,** where an irreducible quadratic factor is one that cannot be factored any further using real numbers. The two zeros of each irreducible quadratic factor are **complex conjugates** of each other.

Example 2 illustrates the Factorization Theorem.

Example **2** **Factorization of a Polynomial**

Using the fact that $x = -2$ is a zero of f, factor $f(x) = x^3 + 2x^2 + 7x + 14$ into linear and irreducible quadratic factors.

▶**Solution** Because $x = -2$ is a zero of f, we know that $x + 2$ is a factor of $f(x)$. Dividing $f(x) = x^3 + 2x^2 + 7x + 14$ by $x + 2$, we have

$$f(x) = (x + 2)(x^2 + 7).$$

Since $x^2 + 7$ cannot be factored any further using real numbers, the factorization is complete as far as real numbers are concerned. The factor $x^2 + 7$ is an example of an irreducible quadratic factor.

✔ **Check It Out 2:** Using the fact that $t = 6$ is a zero of h, factor $h(t) = t^3 - 6t^2 + 5t - 30$ into linear and irreducible quadratic factors. ■

Just In Time

Review complex numbers in Section 3.3.

If we allow factorization over the complex numbers, then we can use the Fundamental Theorem of Algebra to write a polynomial $p(x) = a_n x^n + a_{n-1} x^{n-1} + a_{n-2} x^{n-2} + \cdots + a_1 x + a_0$ in terms of factors of the form $x - c$, where c is a complex zero of $p(x)$. To do so, let c_1 be a complex zero of the polynomial $p(x)$. The existence of c_1 is guaranteed by the Fundamental Theorem of Algebra. Since $p(c_1) = 0$, $x - c_1$ is a factor of $p(x)$ by the Factorization Theorem, and

$$p(x) = (x - c_1)q_1(x)$$

where $q_1(x)$ is a polynomial of degree less than n.

Assuming the degree of $q_1(x)$ is greater than or equal to 1, $q_1(x)$ has a complex zero c_2. Then,

$$q_1(x) = (x - c_2)q_2(x).$$

Thus

$$p(x) = (x - c_1)q_1(x)$$
$$= (x - c_1)(x - c_2)q_2(x) \qquad \text{Substitute } q_1(x) = (x - c_2)q_2(x)$$

This process can be continued until we get a complete factored form:

$$p(x) = a_n(x - c_1)(x - c_2) \cdots (x - c_n)$$

In general, the c_i's may not be distinct.

We have thus established the following result.

The Linear Factorization Theorem

Let $p(x) = a_n x^n + a_{n-1}x^{n-1} + a_{n-2}x^{n-2} + \cdots + a_1 x + a_0$, where $n \geq 1$ and $a_n \neq 0$. Then

$$p(x) = a_n(x - c_1)(x - c_2) \cdots (x - c_n).$$

The numbers $c_1, c_2, \ldots, c_n$ are complex, possibly real, and not necessarily distinct.

Thus every polynomial $p(x)$ of degree $n \geq 1$ has exactly n zeros, if multiplicities and complex zeros are counted.

Example 3 illustrates factoring over the complex numbers and finding complex zeros.

Example 3 Factorization Over the Complex Numbers and Complex Zeros

Factor $f(x) = x^3 + 2x^2 + 7x + 14$ over the complex numbers, and find all complex zeros.

▶**Solution** From Example 2, we have

$$f(x) = (x + 2)(x^2 + 7).$$

But $x^2 + 7 = (x + i\sqrt{7})(x - i\sqrt{7})$. Thus the factorization over the complex numbers is given by

$$f(x) = (x + 2)(x + i\sqrt{7})(x - i\sqrt{7}).$$

Setting $f(x) = 0$, the zeros are $x = -2$, $x = i\sqrt{7}$, and $x = -i\sqrt{7}$. The zeros $x = i\sqrt{7}$ and $x = -i\sqrt{7}$ are complex conjugates of each other.

✔ *Check It Out 3:* Factor $h(t) = t^3 - 6t^2 + 5t - 30$ over the complex numbers. Use the fact that $t = 6$ is a zero of h. ▪

> **Note** The statements discussed thus far regarding the zeros and factors of a polynomial do *not* tell us how to find the factors or zeros.

Finding a Polynomial Given Its Zeros

So far, we have been given a polynomial and have been asked to factor it and find its zeros. If the zeros of a polynomial are given, we can reverse the process and find a factored form of the polynomial using the Linear Factorization Theorem.

Example 4 Finding an Expression for a Polynomial

Find a polynomial $p(x)$ of degree 4 with $p(0) = -9$ and zeros $x = -3$, $x = 1$, and $x = 3$, with $x = 3$ a zero of multiplicity 2. For this polynomial, is it possible for the zeros other than 3 to have a multiplicity greater than 1?

▶**Solution** By the Factorization Theorem, $p(x)$ is of the form

$$p(x) = a(x - (-3))(x - 1)(x - 3)^2 = a(x + 3)(x - 1)(x - 3)^2$$

where a is the leading coefficient, which is still to be determined. Since we are given that $p(0) = -9$, we write down this equation first.

$$p(0) = -9$$
$$a(0 + 3)(0 - 1)(0 - 3)^2 = -9 \quad \text{Substitute 0 in expression for } p$$
$$a(3)(-1)(-3)^2 = -9 \quad \text{Simplify}$$
$$-27a = -9$$
$$a = \frac{1}{3} \quad \text{Solve for } a$$

Thus the desired polynomial is $p(x) = \frac{1}{3}(x + 3)(x - 1)(x - 3)^2$. It is not possible for the zeros other than 3 to have multiplicities greater than 1 because the number of zeros already adds up to 4, counting the multiplicity of the zero at $x = 3$, and p is a polynomial of degree 4.

✔ *Check It Out 4:* Rework Example 4 for a polynomial of degree 5 with $p(0) = 32$ and zeros -2, 4, and 1, where -2 is a zero of multiplicity 2 and 1 is a zero of multiplicity 2. ▪

We have already discussed how to find any possible *rational* zeros of a polynomial. If rational zeros exist, we can use synthetic division to help factor the polynomial. We can use a graphing utility when a polynomial of degree greater than 2 has only irrational or complex zeros.

4.5 Key Points

▶ The **Fundamental Theorem of Algebra** states that every nonconstant polynomial function with real or complex coefficients has at least one complex zero.
▶ A zero c of a polynomial P of degree $n > 0$ has **multiplicity** k if $P(x) = (x - c)^k Q(x)$, where $Q(x)$ is a polynomial of degree $n - k$ and c is not a zero of $Q(x)$.

▶ Every polynomial $p(x)$ of degree $n > 0$ has exactly n zeros, if multiplicities and complex zeros are counted. $p(x)$ can be written as $p(x) = a_n(x - c_1)(x - c_2) \cdots (x - c_n)$, where $c_1, c_2, \ldots, c_n$ are complex numbers.

▶ **Factorization Theorem** Any polynomial P with *real* coefficients can be factored uniquely into linear factors and/or **irreducible quadratic factors**. The two zeros of each irreducible quadratic factor are **complex conjugates** of each other.

4.5 Exercises

▶**Just in Time Exercises** These exercises correspond to the Just in Time references in this section. Complete them to review topics relevant to the remaining exercises.

1. A _____ number is a number of the form $a + bi$, where a and b are real numbers.

2. The complex conjugate of the number $a + bi$ is _____.

3. Find the conjugate of the complex number $2 + 3i$.

4. Find the conjugate of the complex number $4 - i$.

5. Find the conjugate of the complex number $3i$.

6. Find the conjugate of the complex number $i\sqrt{7}$.

▶**Skills** This set of exercises will reinforce the skills illustrated in this section.

In Exercises 7–10, for each polynomial function, list the zeros of the polynomial and state the multiplicity of each zero.

7. $g(x) = (x - 1)^3(x - 4)^5$

8. $f(t) = t^5(t - 3)^2$

9. $f(s) = (s - \pi)^{10}(s + \pi)^3$

10. $h(x) = (x - \sqrt{2})^{13}(x + \sqrt{2})^7$

In Exercises 11–22, find all the zeros, real and nonreal, of the polynomial. Then express $p(x)$ as a product of linear factors.

11. $p(x) = 2x^2 - 5x + 3$ 12. $p(x) = 2x^2 - x - 6$

13. $p(x) = x^3 + 5x$ 14. $p(x) = x^3 + 7x$

15. $p(x) = x^2 - \pi^2$ 16. $p(x) = x^2 - 2$

17. $p(x) = x^2 - \sqrt{3}$ 18. $p(x) = x^2 - 5$

19. $p(x) = x^2 + 9$ 20. $p(x) = x^2 + 4$

21. $p(x) = x^4 - 9$ (*Hint:* Factor first as a difference of squares.)

22. $p(x) = x^4 - 16$ (*Hint:* Factor first as a difference of squares.)

In Exercises 23–28, one zero of each polynomial is given. Use it to express the polynomial as a product of linear and irreducible quadratic factors.

23. $x^3 - 2x^2 + x - 2$; zero: $x = 2$

24. $x^3 - x^2 + 4x - 4$; zero: $x = 1$

25. $2x^3 - 9x^2 - 11x + 30$; zero: $x = 5$

26. $2x^3 - 9x^2 + 7x + 6$; zero: $x = 2$

27. $x^4 - 5x^3 + 7x^2 - 5x + 6$; zero: $x = 3$

28. $x^4 + 2x^3 - 2x^2 + 2x - 3$; zero: $x = -3$

In Exercises 29–38, one zero of each polynomial is given. Use it to express the polynomial as a product of linear factors over the complex numbers. You may have already factored some of these polynomials into linear and irreducible quadratic factors in the previous group of exercises.

29. $x^4 + 4x^3 - x^2 + 16x - 20$; zero: $x = -5$

30. $x^4 - 6x^3 + 9x^2 - 24x + 20$; zero: $x = 5$

31. $x^3 - 2x^2 + x - 2$; zero: $x = 2$

32. $x^3 - x^2 + 4x - 4$; zero: $x = 1$

33. $2x^3 - 9x^2 - 11x + 30$; zero: $x = 5$

34. $2x^3 - 9x^2 + 7x + 6$; zero: $x = 2$

35. $x^4 - 5x^3 + 7x^2 - 5x + 6$; zero: $x = 2$

36. $x^4 + 2x^3 - 2x^2 + 2x - 3$; zero: $x = -3$

37. $x^4 + 4x^3 - x^2 + 16x - 20$; zero: $x = -5$

38. $x^4 - 6x^3 + 9x^2 - 24x + 20$; zero: $x = 5$

In Exercises 39–44, find an expression for a polynomial $p(x)$ with real coefficients that satisfies the given conditions. There may be more than one possible answer.

39. Degree 2; $x = 2$ and $x = -1$ are zeros

40. Degree 2; $x = \frac{1}{2}$ and $x = \frac{3}{4}$ are zeros

41. Degree 3; $x = 1$ is a zero of multiplicity 2; the origin is the y-intercept

42. Degree 3; $x = -2$ is a zero of multiplicity 2; the origin is an x-intercept

43. Degree 4; $x = 1$ and $x = \frac{1}{3}$ are both zeros of multiplicity 2

44. Degree 4; $x = -1$ and $x = -3$ are zeros of multiplicity 1 and $x = \frac{1}{3}$ is a zero of multiplicity 2

▶**Concepts** This set of exercises will draw on the ideas presented in this section and your general math background.

45. One of the zeros of a certain quadratic polynomial with real coefficients is $1 + i$. What is its other zero?

46. The graph of a certain cubic polynomial function f has one x-intercept at $(1, 0)$ that crosses the x-axis, and another x-intercept at $(-3, 0)$ that touches the x-axis but does not cross it. What are the zeros of f and their multiplicities?

47. Explain why there cannot be two different points at which the graph of a cubic polynomial touches the x-axis without crossing it.

48. Why can't the numbers i, $2i$, 1, and 2 be the set of zeros for some fourth-degree polynomial with real coefficients?

49. The graph of a polynomial function is given below. What is the lowest possible degree of this polynomial? Explain. Find a possible expression for the function.

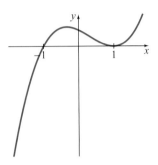

50. The graph of a polynomial function is given below. What is the lowest possible degree of this polynomial? Explain. Find a possible expression for the function.

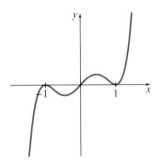

4.6 Rational Functions

Just In Time

Review rational expressions in Section P.6.

Thus far, we have seen various situations in the real world that give rise to linear, quadratic, and other polynomial functions. Our study of these types of functions helped us to explore mathematical models in greater detail.

In this section, we extend our study of functions to include a type of model that arises when a function is defined by a rational expression. You may want to review rational expressions in Section P.6 before you begin this section.

We now explore a model involving a rational expression.

Example 1 Average Cost

Suppose it costs $45 a day to rent a car with unlimited mileage.

(a) What is the average cost per mile per day?

(b) What happens to the average cost per mile per day as the number of miles driven per day increases?

▶**Solution**

(a) Let x be the number of miles driven per day. The average cost per mile per day will depend on the number of miles driven per day, as follows.

$$A(x) = \text{Average cost per mile per day} = \frac{\text{total cost per day}}{\text{miles driven per day}} = \frac{45}{x}$$

(b) To see what happens to the average cost per mile per day as the number of miles driven per day increases, we create a table of values (Table 4.6.1) and the corresponding graph.

Table 4.6.1

Miles Driven	Average Cost
0	undefined
$\frac{1}{2}$	90
1	45
5	9
10	4.5
100	0.45
1000	0.045

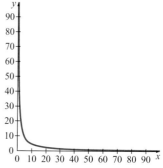

We can see that as the number of miles driven per day increases, the average cost per mile per day goes down. Note that this function is defined only when $x > 0$, since you cannot drive a negative number of miles, and the average cost of driving 0 miles is not defined.

Although the average cost keeps getting smaller as x gets larger, it will never equal zero. None of the functions we have studied so far exhibits this type of behavior.

✔ *Check It Out 1:* What is the average cost per mile per day if the daily cost of renting a car with unlimited mileage is $50? ▪

The type of behavior exhibited by the function in Example 1 is typical of functions known as **rational functions,** which we now define.

Definition of a Rational Function

A **rational function** $r(x)$ is defined as a quotient of two polynomials $p(x)$ and $h(x)$,

$$r(x) = \frac{p(x)}{h(x)}$$

where $h(x)$ is not the constant zero function.

The **domain** of a rational function consists of all real numbers for which the denominator is not equal to zero. We will be especially interested in the behavior of the rational function very close to the value(s) of x at which the denominator is zero.

Before examining rational functions in general, we will look at some specific examples.

Example 2 Analyzing a Simple Rational Function

Let $f(x) = \dfrac{1}{x - 1}$.

(a) What is the domain of f?

(b) Make a table of values of x and $f(x)$. Include values of x that are near 1 as well as larger values of x.

(c) Graph f by hand.

(d) Comment on the behavior of the graph.

▶Solution

(a) The domain of f is the set of all values of x such that the denominator, $x - 1$, is not equal to zero. This is true for all $x \neq 1$. In interval notation, the domain is $(-\infty, 1) \cup (1, \infty)$.

(b) Table 4.6.2 is a table of values of x and $f(x)$. Note that it contains some values of x that are close to 1 as well as some larger values of x.

(c) Graphing the data in Table 4.6.2 gives us Figure 4.6.1. Note that there is no value for $f(x)$ at $x = 1$.

Table 4.6.2

x	$f(x) = \dfrac{1}{x - 1}$
-100	-0.009901
-10	-0.090909
0	-1
0.5	-2
0.9	-10
1	undefined
1.1	10
1.5	2
2	1
10	0.111111
100	0.010101

Figure 4.6.1

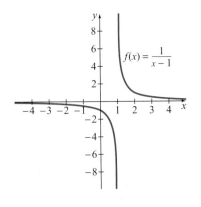

(d) Examining the table or graph of $f(x) = \dfrac{1}{x-1}$, we see that as the absolute value of x gets large, the value of the function approaches zero, though it never actually reaches zero. Also, as the value of x approaches 1, the absolute value of the *function* gets large.

☑ *Check It Out 2:* What is the domain of $f(x) = \dfrac{1}{x}$? Sketch a graph of f. ■

When considering rational functions, we will often use a pair of facts about the relationship between large numbers (numbers that are large in absolute value) and their reciprocals; these facts are informally stated as follows.

LARGE-Small Principle

$$\frac{1}{\text{LARGE}} = \text{small}; \qquad \frac{1}{\text{small}} = \text{LARGE}$$

We will refer to this pair of facts as the **LARGE-small principle.**

Vertical Asymptotes

We have already noted that sometimes a rational function is not defined for certain real values of the input variable. We also noted that the value of the function gets very large near the values of x at which the function is undefined. (See Example 1.) In this section, we examine closely what the graph of a rational function looks like near these values.

As we saw in Example 2, the absolute value of $f(x) = \dfrac{1}{x-1}$ *near* $x = 1$ is very large; at $x = 1$, $f(x)$ *is not defined.* We say that the line with equation $x = 1$ is a **vertical asymptote** of the graph of f. The line $x = 1$ is indicated by a dashed line. It is *not* part of the graph of the function $f(x) = \dfrac{1}{x-1}$. See Figure 4.6.2.

Figure 4.6.2

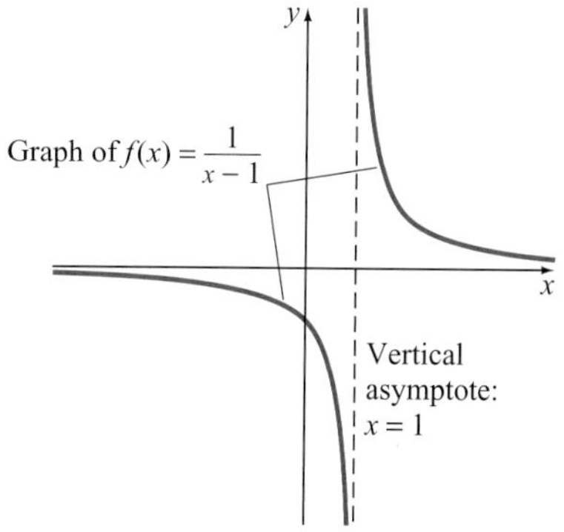

Graph of $f(x) = \dfrac{1}{x-1}$

Vertical asymptote: $x = 1$

Vertical Asymptote

To find the **vertical asymptote(s)** of a rational function $r(x) = \dfrac{p(x)}{q(x)}$, first make sure that $p(x)$ and $q(x)$ have no common factors. Then the vertical asymptotes occur at the values of x at which $q(x) = 0$. At these values of x, the graph of the function will approach positive or negative infinity.

Note *A case in which p(x) and q(x) have common factors is given in Example 8.*

Technology Note

Graphing calculators often do a poor job of graphing rational functions. Figure 4.6.3 shows the graph of $f(x) = \dfrac{1}{x - 1}$ using three different settings.

Figure 4.6.3

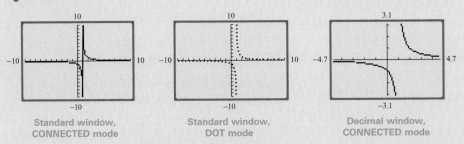

| Standard window, CONNECTED mode | Standard window, DOT mode | Decimal window, CONNECTED mode |

The vertical line that appears in the leftmost display when the standard window setting is used is *not* the vertical asymptote. Because the default setting of the calculator is CONNECTED mode, the calculator display connects the negative and positive values. This is *not* an acceptable graph.

The problem of the connecting vertical line can be avoided by setting the calculator to DOT mode, as in the middle display, or by using a decimal window in CONNECTED mode, as in the third display.

Keystroke Appendix:

Sections 7 and 8

For now, we will use algebraic methods to find the vertical asymptotes of the graphs of several rational functions. Later, we will use them to sketch graphs.

Example 3 **Finding Vertical Asymptotes**

Find all vertical asymptotes of the following functions.

(a) $f(x) = \dfrac{2x}{x + 1}$ 　　(b) $f(x) = \dfrac{x + 2}{x^2 - 1}$ 　　(c) $f(x) = \dfrac{x^2 - 3}{2x + 1}$

▶Solution

(a) For the function $f(x) = \dfrac{2x}{x + 1}$, we see that the numerator and denominator have no common factors, so we can set the denominator equal to zero and find the vertical asymptote(s):

$$x + 1 = 0 \Longrightarrow x = -1$$

Thus the line $x = -1$ is the only vertical asymptote of the function.

(b) The numerator and denominator of $f(x) = \frac{x+2}{x^2-1}$ have no common factors, so we can set the denominator equal to zero and find the vertical asymptote(s). In this case, the denominator can be factored, so we will apply the Zero Product Rule:

$$x^2 - 1 = 0 \Longrightarrow (x+1)(x-1) = 0 \Longrightarrow x = 1, -1$$

This function has *two* vertical asymptotes: the line $x = 1$ and the line $x = -1$.

(c) For the final example, we see again that the numerator and denominator have no common factors. The vertical asymptote is the line $x = -\frac{1}{2}$, since $x = -\frac{1}{2}$ is the solution of the equation $2x + 1 = 0$.

✔ *Check It Out 3:* Find all vertical asymptotes of $f(x) = \frac{3x}{x^2 - 9}$. ■

End Behavior of Rational Functions and Horizontal Asymptotes

Just as we did with polynomial functions, we can examine the end behavior of rational functions. We will use this information later to help us sketch complete graphs of rational functions.

We can examine what happens to the values of a rational function $r(x)$ as $|x|$ gets large. This is the same as determining the end behavior of the rational function. For example, as $x \to \infty$, $f(x) = \frac{1}{x-1} \to 0$, because the denominator becomes large in magnitude but the numerator stays constant at 1. Similarly, as $x \to -\infty$, $f(x) = \frac{1}{x-1} \to 0$. These are instances of the LARGE-small principle. When such behavior occurs, we say that $y = 0$ is a **horizontal asymptote** of the function $f(x) = \frac{1}{x-1}$. Not all rational functions have a horizontal asymptote, and if they do, it need not be $y = 0$. The following gives the necessary conditions for a rational function to have a horizontal asymptote.

Horizontal Asymptotes of Rational Functions

Let $r(x)$ be a rational function given by

$$r(x) = \frac{p(x)}{q(x)} = \frac{a_n x^n + a_{n-1} x^{n-1} + \cdots + a_1 x + a_0}{b_m x^m + b_{m-1} x^{m-1} + \cdots + b_1 x + b_0}.$$

Here, $p(x)$ is a polynomial of degree n and $q(x)$ is a polynomial of degree m. Assume that $p(x)$ and $q(x)$ have no common factors.

▶ If $n < m$, $r(x)$ approaches zero for large values of $|x|$. The line $y = 0$ is the horizontal asymptote of the graph of $r(x)$.

▶ If $n = m$, $r(x)$ approaches a nonzero constant $\frac{a_n}{b_m}$ for large values of $|x|$. The line $y = \frac{a_n}{b_m}$ is the horizontal asymptote of the graph of $r(x)$.

▶ If $n > m$, $r(x)$ has no horizontal asymptote.

Example 4 **Finding Horizontal Asymptotes**

Find the horizontal asymptote, if it exists, for each the following rational functions. Use a table and a graph to discuss the end behavior of each function.

(a) $f(x) = \dfrac{x + 2}{x^2 - 1}$ (b) $f(x) = \dfrac{2x}{x + 1}$ (c) $f(x) = \dfrac{x^2 - 3}{2x + 1}$

▶Solution

(a) The degree of the numerator is 1 and the degree of the denominator is 2. Because $1 < 2$, the line $y = 0$ is a horizontal asymptote of the graph of f as shown in Figure 4.6.4. We can generate a table of values for $f(x) = \dfrac{x + 2}{x^2 - 1}$ for $|x|$ large, as shown in Table 4.6.3.

Table 4.6.3

x	$f(x)$
−1000	−0.00099800
−100	−0.0098009
−50	−0.019208
50	0.020808
100	0.010201
1000	0.0010020

Figure 4.6.4

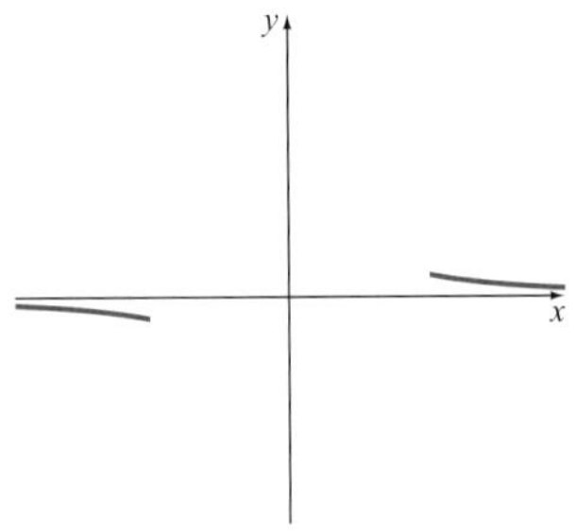

Note that as $x \to +\infty$, $f(x)$ gets close to zero. Once again, it will never reach zero; it will be slightly above zero. As $x \to -\infty$, $f(x)$ gets close to zero again, but this time it will be slightly below zero.

Recalling the end behavior of polynomials, for large values of $|x|$, the value of the numerator $x + 2$ is about the same as x, and the value of the denominator $x^2 - 1$ is approximately the same as x^2. We then have

$$f(x) = \frac{x + 2}{x^2 - 1} \approx \frac{x}{x^2} = \frac{1}{x} \to 0 \text{ for large values of } |x|$$

by the LARGE-small principle. This is what we observed in Table 4.6.3 and the corresponding graph.

(b) The degree of the numerator is 1 and the degree of the denominator is 1. Because $1 = 1$, the line $y = \dfrac{a_1}{b_1} = \dfrac{2}{1} = 2$ is a horizontal asymptote of the graph of f as shown in Figure 4.6.5. We can generate a table of values for $f(x) = \dfrac{2x}{x + 1}$ for large values of $|x|$, as shown in Table 4.6.4. We include both positive and negative values of x.

Table 4.6.4

x	$f(x)$
−1000	2.0020
−100	2.0202
−50	2.0408
50	1.9608
100	1.9802
1000	1.9980

Figure 4.6.5

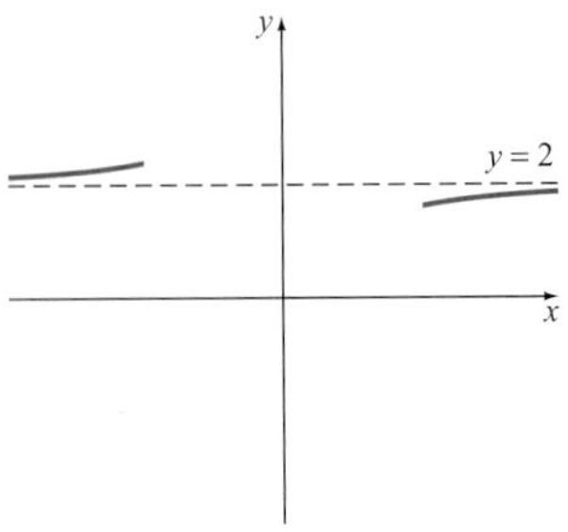

From Table 4.6.4 and Figure 4.6.5, we observe that as $x \to +\infty$, $f(x)$ gets close to 2. It will never reach 2, but it will be slightly below 2. As $x \to -\infty$, $f(x)$ gets close to 2 again, but this time it will be slightly above 2.

To justify the end behavior, for large values of $|x|$, the numerator $2x$ is just $2x$ and the value of the denominator $x + 1$ is approximately the same as x. We then have $f(x) = \dfrac{2x}{x + 1} \approx \dfrac{2x}{x} = 2$ for large values of $|x|$. This is what we observed in Table 4.6.4 and the corresponding graph.

(c) The degree of the numerator is 2 and the degree of the denominator is 1. Because $2 > 1$, the graph of $f(x)$ has no horizontal asymptote. A table of values for $f(x) = \dfrac{x^2 - 3}{2x + 1}$ for $|x|$ large is given in Table 4.6.5. Unlike in parts (a) and (b), the values in the table do not seem to tend to any *one* particular number for $|x|$ large. However, you will notice that the values of $f(x)$ are very close to the values of $\dfrac{1}{2}x$ as $|x|$ gets large. We will discuss the behavior of functions such as this later in this section.

Table 4.6.5

x	$f(x)$
-1000	-500.25
-500	-250.25
-100	-50.236
-50	-25.222
50	24.723
100	49.736
500	249.75
1000	499.75

✔ *Check It Out 4:* Find the horizontal asymptote of the rational function
$$f(x) = \frac{3x}{x^2 - 9}. \quad ■$$

Graphs of Rational Functions

The features of rational functions that we have discussed so far, combined with some additional information, can be used to sketch the graphs of rational functions. The procedure for doing so is summarized next, followed by examples.

Sketching the Graph of a Rational Function

Step 1 Find the vertical asymptotes, if any, and indicate them on the graph.

Step 2 Find the horizontal asymptotes, if any, and indicate them on the graph.

Step 3 Find the x- and y-intercepts and plot these points on the graph. For a rational function, the x-intercepts occur at those points in the domain of f at which

$$f(x) = \frac{p(x)}{q(x)} = 0.$$

This means that $p(x) = 0$ at the x-intercepts.
To calculate the y-intercept, evaluate $f(0)$, if $f(0)$ is defined.

Step 4 Use the information in Steps 1–3 to sketch a partial graph. That is, find function values for points near the vertical asymptote(s) and sketch the behavior near the vertical asymptote(s). Also sketch the end behavior.

Step 5 Determine whether the function has any symmetries.

Step 6 Plot some additional points to help you complete the graph.

Example 5 **Graphing Rational Functions**

Sketch a graph of $f(x) = \dfrac{2x}{x+1}$.

▶**Solution**

Steps 1 and 2 The vertical and horizontal asymptotes of this function were computed in Examples 3 and 4.

Vertical Asymptote	$x = -1$
Horizontal Asymptote	$y = 2$

Step 3 To find the y-intercept, evaluate $f(0) = \dfrac{2(0)}{(0)+1} = 0$. The y-intercept is $(0, 0)$.

To find the x-intercept, we find the points at which the numerator, $2x$, is equal to zero. This happens at $x = 0$. Thus the x-intercept is $(0, 0)$.

x-Intercept	$(0, 0)$
y-Intercept	$(0, 0)$

Step 4 We now find values of $f(x)$ near the vertical asymptote, $x = -1$. Note that we have chosen some values of x that are slightly to the right of $x = -1$ and some that are slightly to the left of $x = -1$. See Table 4.6.6.

Table 4.6.6

x	-1.5	-1.1	-1.01	-1.001	-1	-0.999	-0.99	-0.9	-0.5
$f(x)$	6	22	202	2002	undefined	-1998	-198	-18	-2

From Table 4.6.6, we see that the value of $f(x)$ increases to $+\infty$ as x approaches -1 from the left, and decreases to $-\infty$ as x approaches -1 from the right. The end behavior of f is as follows: $f(x) \to 2$ as $x \to \pm\infty$. Next we sketch the information collected so far. See Figure 4.6.6.

Figure 4.6.6

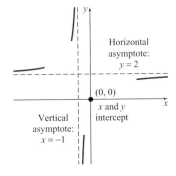

Step 5 Check for symmetry. Because

$$f(-x) = \frac{2(-x)}{(-x)+1} = -\frac{2x}{-x+1}$$

$f(-x) \neq f(x)$ and $f(x) \neq -f(-x)$. Thus the graph of this function has no symmetries.

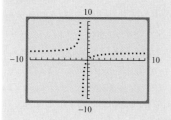
Step 6 There is an x-intercept at $(0, 0)$ and a vertical asymptote at $x = -1$, so we choose values in the intervals $(-\infty, -1)$, $(-1, 0)$, and $(0, \infty)$ to fill out the graph. See Table 4.6.7.

Table 4.6.7

x	-5	$-\dfrac{1}{2}$	1	2	5
$f(x)$	2.5	-2	1	1.3333	1.6667

Plotting these points and connecting them with a smooth curve gives the graph shown in Figure 4.6.8. The horizontal and vertical asymptotes are *not* part of the graph of f. They are shown on the plot to indicate the behavior of the graph of f.

Figure 4.6.8

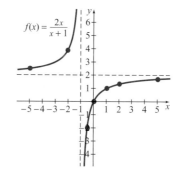

$$f(x) = \frac{2x}{x + 1}$$

✔ *Check It Out 5:* Sketch the graph $f(x) = \dfrac{x - 1}{3x + 1}$. ■

Example 6 Graphing Rational Functions

Sketch the graph of $f(x) = \dfrac{x + 2}{x^2 - 1}$.

▶Solution

Steps 1 and 2 The vertical and horizontal asymptotes of this function were computed in Examples 3 and 4.

Vertical Asymptotes	$x = 1, x = -1$
Horizontal Asymptote	$y = 0$

Step 3 The y-intercept is at $(0, -2)$ because

$$f(0) = \frac{0 + 2}{0^2 - 1} = -2.$$

The x-intercept is at $(-2, 0)$ because $x + 2 = 0$ at $x = -2$.

x-Intercept	$(-2, 0)$
y-Intercept	$(0, -2)$

Step 4 Find some values of $f(x)$ near the vertical asymptotes $x = -1$ and $x = 1$. See Table 4.6.8.

Table 4.6.8

x	$f(x)$	x	$f(x)$
-1.1	4.2857	0.9	-15.2632
-1.01	49.2537	0.99	-150.2513
-1.001	499.2504	0.999	-1500.2501
-1	undefined	1	undefined
-0.999	-500.7504	1.001	1499.7501
-0.99	-50.7538	1.01	149.7512
-0.9	-5.7895	1.1	14.7619

Observations:

▶ The value of $f(x)$ increases to $+\infty$ as x approaches -1 from the left.

▶ The value of $f(x)$ decreases to $-\infty$ as x approaches -1 from the right.

▶ The value of $f(x)$ decreases to $-\infty$ as x approaches 1 from the left.

▶ The value of $f(x)$ increases to $+\infty$ as x approaches 1 from the right.

We can use the information collected so far to sketch the graph shown in Figure 4.6.9.

Figure 4.6.9

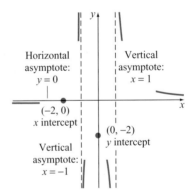

Step 5 Check for symmetry. Because

$$f(-x) = \frac{(-x) + 2}{(-x)^2 - 1} = \frac{-x + 2}{x^2 - 1},$$

$f(x) \neq f(-x)$ and $f(x) \neq -f(-x)$. So the graph of this function has no symmetries.

Step 6 Choose some additional values to fill out the graph. There is an x-intercept at $(-2, 0)$ and vertical asymptotes at $x = \pm 1$, so we choose at least one value in each of the intervals $(-\infty, -2)$, $(-2, -1)$, $(-1, 1)$, and $(1, \infty)$. See Table 4.6.9.

Table 4.6.9

x	-3	-1.5	-0.5	0.5	2
$f(x)$	-0.1250	0.4000	-2.0000	-3.3333	1.3333

Plotting these points and connecting them with a smooth curve gives the graph in Figure 4.6.10.

Figure 4.6.10

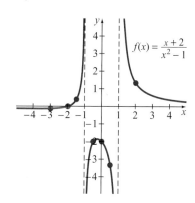

$f(x) = \dfrac{x+2}{x^2-1}$

✔ *Check It Out 6:* Sketch the graph of $f(x) = \dfrac{3x}{x^2 - 9}$. ■

Rational Functions with Slant Asymptotes

Thus far we have sketched the graphs of a few rational functions that have had horizontal asymptotes, and examined the end behavior of some that have not. If the degree of the numerator of a rational function is 1 greater than the degree of the denominator, the graph of the function will have what is known as a **slant asymptote.** Using long division, we can write

$$r(x) = \frac{p(x)}{q(x)} = ax + b + \frac{s(x)}{q(x)}$$

where ax and b are the first two terms of the quotient and $s(x)$ is the remainder. Because the degree of $s(x)$ is less than the degree of $q(x)$, the value of the function $\frac{s(x)}{q(x)}$ approaches 0 as $|x|$ goes to infinity. Thus $r(x)$ will resemble the line $y = ax + b$ as $x \to \infty$ or $x \to -\infty$. The end-behavior analysis that we performed earlier, without using long division, gives only the ax expression for the line. Next we show how to find the equation of the asymptotic line $y = ax + b$.

Example 7 Rational Functions with Slant Asymptotes

Find the slant asymptote of the graph of $r(x) = \dfrac{6x^2 - x - 1}{2x + 1}$, and sketch the complete graph of $r(x)$.

▶Solution Performing the long division, we can write

$$r(x) = \frac{6x^2 - x - 1}{2x + 1} = 3x - 2 + \frac{1}{2x + 1}.$$

For large values of $|x|$, $\frac{1}{2x + 1} \to 0$, and so the graph of $r(x)$ resembles the graph of the line $y = 3x - 2$. The equation of the slant asymptote is thus $y = 3x - 2$.

We can use the information summarized in Table 4.6.10 to sketch the complete graph shown in Figure 4.6.11.

Table 4.6.10

x-Intercept(s)	$\left(\frac{1}{2}, 0\right)$ and $\left(-\frac{1}{3}, 0\right)$ (obtained by setting the numerator $6x^2 - x - 1 = (3x + 1)(2x - 1)$ equal to zero and solving for x)
y-Intercept	$(-1, 0)$, since $r(0) = -1$.
Vertical Asymptote	$x = -\frac{1}{2}$ (obtained by setting the denominator $2x + 1$ equal to zero and solving for x)
Slant Asymptote	$y = 3x - 2$
Additional Points	$(-2, -8.3333), (-0.4, 1.8),$ $(0.4, -0.2444), (2, 4.2)$

Figure 4.6.11

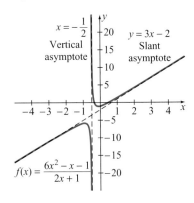

✔ **Check It Out 7:** Find the slant asymptote of the graph of $r(x) = \dfrac{x^2 - 3x - 4}{x - 5}$, and sketch the complete graph of $r(x)$. ■

Rational Functions with Common Factors

We now examine the graph of a rational function in which the numerator and denominator have a common factor.

Example 8 Rational Function with Common Factors

Sketch the complete graph of $r(x) = \dfrac{x - 1}{x^2 - 3x + 2}$.

▶Solution Factor the denominator of $r(x)$ to obtain

$$r(x) = \frac{x - 1}{x^2 - 3x + 2}$$
$$= \frac{x - 1}{(x - 2)(x - 1)}.$$

The function is undefined at $x = 1$ and $x = 2$ because those values of x give rise to a zero denominator. If $x \neq 1$, then the factor $x - 1$ can be divided out to obtain

$$r(x) = \frac{1}{x - 2}, \quad x \neq 1, 2.$$

The features of the graph of $r(x)$ are summarized in Table 4.6.11.

Table 4.6.11

x-Intercept	None
y-Intercept	$\left(0, -\dfrac{1}{2}\right)$
Vertical Asymptote	$x = 2$
Horizontal Asymptote	$y = 0$
Undefined at	$x = 2, x = 1$

Figure 4.6.12

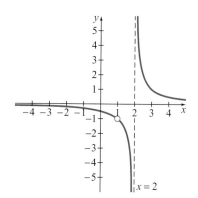

Even though $r(x)$ is undefined at $x = 1$, it does *not* have a vertical asymptote there. To see why, examine Table 4.6.12, which gives the values of $r(x)$ at some values of x close to 1.

Table 4.6.12

x	0.9	0.99	0.999	1.001	1.01	1.1
$r(x)$	-0.9091	-0.9901	-0.999	-1.001	-1.01	-1.111

The values of $r(x)$ near $x = 1$ do not tend to infinity. In fact, the table of values suggests that $r(x)$ gets close to -1 as x gets close to 1. Thus the graph of $r(x) = \dfrac{1}{x - 2}$, $x \neq 1, 2$, will have a *hole* at $(1, -1)$, as shown in Figure 4.6.12. You can obtain the value -1 by substituting the value $x = 1$ into $\dfrac{1}{x - 2}$. The justification for this substitution involves theorems of calculus and is beyond the scope of this discussion.

☑ *Check It Out 8:* Sketch the complete graph of $r(x) = \dfrac{x - 3}{x^2 - 5x + 6}$. ▪

4.6 Key Points

▶ A **rational function** $r(x)$ is defined as a quotient of two polynomials $p(x)$ and $h(x)$, $r(x) = \frac{p(x)}{h(x)}$, where $h(x)$ is not the constant zero function.

▶ The **domain** of a rational function consists of all real numbers for which the denominator is not equal to zero.

▶ The **vertical asymptotes** of $r(x) = \frac{p(x)}{h(x)}$ occur at the values of x at which $h(x) = 0$, assuming $p(x)$ and $h(x)$ have no common factors.

▶ A **horizontal asymptote** of a rational function $r(x) = \frac{p(x)}{h(x)}$ exists when the degree of $p(x)$ is less than or equal to the degree of $h(x)$.

▶ A **slant asymptote** of a rational function $r(x) = \frac{p(x)}{h(x)}$ exists when the degree of $p(x)$ is 1 higher than the degree of $h(x)$.

4.6 Exercises

▶**Just in Time Exercises** These exercises correspond to the Just in Time references in this section. Complete them to review topics relevant to the remaining exercises.

1. A rational expression is a quotient of two _____.

2. For which values of x is the following rational expression defined?

$$\frac{x + 2}{(x - 1)(x + 5)}$$

In Exercises 3–6, simplify each rational expression.

3. $\dfrac{x^2 - 2x + 1}{2 - x - x^2}$

4. $\dfrac{x^2 + 2x - 15}{x^2 - 9x + 18}$

5. $\dfrac{x^2 - 1}{x^2 - 2x - 3}$

6. $\dfrac{x + 2}{x^2 + 3x + 2}$

▶**Skills** This set of exercises will reinforce the skills illustrated in this section.

In Exercises 7–20, for each function, find the domain and the vertical and horizontal asymptotes (if any).

7. $h(x) = \dfrac{-2}{x + 6}$

8. $F(x) = \dfrac{4}{x - 3}$

9. $g(x) = \dfrac{3}{x^2 - 4}$

10. $f(x) = \dfrac{2}{x^2 - 9}$

11. $f(x) = \dfrac{-x^2 + 9}{-2x^2 + 8}$

12. $h(x) = \dfrac{-3x^2 + 12}{x^2 - 9}$

13. $h(x) = \dfrac{1}{(x - 2)^2}$

14. $G(x) = \dfrac{-2}{(x + 4)^2}$

15. $h(x) = \dfrac{3x^2}{x + 1}$

16. $f(x) = \dfrac{-2x^2}{x - 1}$

17. $f(x) = \dfrac{2x + 7}{2x^2 + 5x - 3}$

18. $f(x) = \dfrac{3x + 5}{x^2 - x - 2}$

19. $f(x) = \dfrac{x + 1}{x^2 + 1}$

20. $h(x) = \dfrac{x + 2}{4 + x^2}$

In Exercises 21–26, for the graph of the function, find the domain, the vertical and horizontal asymptotes (if any), and the x- and y-intercepts (if any).

21.
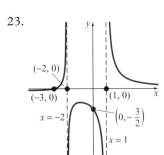

22.
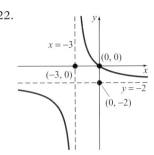

23.
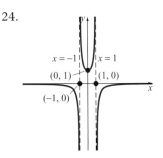

24.

25.

26.

In Exercises 27–30, for each function, fill in the given tables.

27. Let $f(x) = \dfrac{2}{x + 1}$.

(a) Fill in the following table for values of x near -1. What do you observe about the value of $f(x)$ as x approaches -1 from the right? from the left?

x	-1.5	-1.1	-1.01	-0.99	-0.9	-0.5
$f(x)$						

(b) Complete the following table. What happens to the value of $f(x)$ as x gets very large and positive?

x	10	50	100	1000
$f(x)$				

(c) Complete the following table. What happens to the value of $f(x)$ as x gets very large and negative?

x	-1000	-100	-50	-10
$f(x)$				

28. Let $f(x) = \dfrac{1}{3 - x}$.

(a) Fill in the following table for values of x near 3. What do you observe about the value of $f(x)$ as x approaches 3 from the right? from the left?

x	2.5	2.9	2.99	3.01	3.1	3.5
$f(x)$						

(b) Complete the following table. What happens to the value of $f(x)$ as x gets very large and positive?

x	10	50	100	1000
$f(x)$				

(c) Complete the following table. What happens to the value of $f(x)$ as x gets very large and negative?

x	-1000	-100	-50	-10
$f(x)$				

29. Let $f(x) = \dfrac{2x^2 - 1}{x^2}$.

(a) Fill in the following table for values of x near zero. What do you observe about the value of $f(x)$ as x approaches zero from the right? from the left?

x	-0.5	-0.1	-0.01	0.01	0.1	0.5
$f(x)$						

(b) Complete the following table. What happens to the value of $f(x)$ as x gets very large and positive?

x	10	50	100	1000
$f(x)$				

(c) Complete the following table. What happens to the value of $f(x)$ as x gets very large and negative?

x	-1000	-100	-50	-10
$f(x)$				

30. Let $f(x) = \dfrac{-3x^2 + 4}{x^2}$.

(a) Fill in the following table for values of x near zero. What do you observe about the value of $f(x)$ as x approaches zero from the right? from the left?

x	-0.5	-0.1	-0.01	0.01	0.1	0.05
$f(x)$						

(b) Complete the following table. What happens to the value of $f(x)$ as x gets very large and positive?

x	10	50	100	1000
$f(x)$				

(c) Complete the following table. What happens to the value of $f(x)$ as x gets very large and negative?

x	-1000	-100	-50	-10
$f(x)$				

In Exercises 31–52, sketch a graph of the rational function. Indicate any vertical and horizontal asymptote(s) and all intercepts.

31. $f(x) = \dfrac{1}{x - 2}$

32. $f(x) = \dfrac{1}{x + 3}$

33. $f(x) = \dfrac{-12}{x + 6}$

34. $f(x) = \dfrac{-10}{x + 2}$

35. $f(x) = \dfrac{12}{3 - x}$

36. $f(x) = \dfrac{8}{4 - x}$

37. $f(x) = \dfrac{3}{(x + 1)^2}$

38. $h(x) = \dfrac{-9}{(x - 3)^2}$

39. $g(x) = \dfrac{3 - x}{x + 4}$

40. $g(x) = \dfrac{2 - x}{x + 3}$

41. $g(x) = \dfrac{x + 4}{x - 1}$

42. $g(x) = \dfrac{x + 5}{x - 2}$

43. $h(x) = \dfrac{-2x}{(x - 1)(x + 4)}$

44. $f(x) = \dfrac{x}{(x - 3)(x - 1)}$

45. $f(x) = \dfrac{3x^2}{x^2 - x - 2}$

46. $f(x) = \dfrac{-4x^2}{x^2 - x - 6}$

47. $f(x) = \dfrac{x - 1}{2x^2 - 5x - 3}$

48. $f(x) = \dfrac{x - 2}{2x^2 + x - 3}$

49. $f(x) = \dfrac{x^2 + x - 6}{x^2 - 1}$

50. $f(x) = \dfrac{x^2 + 3x + 2}{x^2 - 9}$

51. $h(x) = \dfrac{1}{x^2 + 1}$

52. $h(x) = \dfrac{2}{x^2 + 4}$

In Exercises 53–64, sketch a graph of the rational function and find all intercepts and slant asymptotes. Indicate all asymptotes on the graph.

53. $g(x) = \dfrac{x^2}{x + 4}$

54. $g(x) = \dfrac{4x^2}{x + 3}$

55. $g(x) = \dfrac{-3x^2}{x - 5}$

56. $h(x) = \dfrac{-x^2}{x - 3}$

57. $h(x) = \dfrac{4 - x^2}{x}$

58. $h(x) = \dfrac{x^2 - 9}{x}$

59. $h(x) = \dfrac{x^2 + x + 1}{x - 1}$

60. $h(x) = \dfrac{x^2 + 2x + 1}{x + 3}$

61. $f(x) = \dfrac{3x^2 + 5x - 2}{x + 1}$

62. $g(x) = \dfrac{2x^2 + 11x + 5}{x - 3}$

63. $h(x) = \dfrac{x^3 + 1}{x^2 + 3x}$

64. $h(x) = \dfrac{x^3 - 1}{x^2 - 2x}$

In Exercises 65–70, sketch a graph of the rational function involving common factors and find all intercepts and asymptotes. Indicate all asymptotes on the graph.

65. $f(x) = \dfrac{3x + 9}{x^2 - 9}$

66. $f(x) = \dfrac{2x - 4}{x^2 - 4}$

67. $f(x) = \dfrac{x^2 + x - 2}{x^2 + 2x - 3}$

68. $f(x) = \dfrac{2x^2 - 5x + 2}{x^2 - 5x + 6}$

69. $f(x) = \dfrac{x^2 + 3x - 10}{x - 2}$

70. $f(x) = \dfrac{x^2 + 2x + 1}{x + 1}$

▶**Applications** In this set of exercises, you will use rational functions to study real-world problems.

71. **Drug Concentration** The concentration $C(t)$ of a drug in a patient's bloodstream t hours after administration is given by

$$C(t) = \frac{10t}{1 + t^2}$$

where $C(t)$ is in milligrams per liter.
(a) What is the drug concentration in the patient's bloodstream 8 hours after administration?
(b) Find the horizontal asymptote of $C(t)$ and explain its significance.

72. **Environmental Costs** The annual cost, in millions of dollars, of removing arsenic from drinking water in the United States can be modeled by the function

$$C(x) = \frac{1900}{x}$$

where x is the concentration of arsenic remaining in the water, in micrograms per liter. A microgram is 10^{-6} gram. (*Source:* Environmental Protection Agency)
(a) Evaluate $C(10)$ and explain its significance.
(b) Evaluate $C(5)$ and explain its significance.
(c) What happens to the cost function as x gets closer to zero?

73. **Rental Costs** A truck rental company charges a daily rate of $15 plus $0.25 per mile driven. What is the average cost per mile of driving x miles per day? Use this expression to find the average cost per mile of driving 50 miles per day.

74. **Printing** To print booklets, it costs $300 plus an additional $0.50 per booklet. What is the average cost per booklet of printing x booklets? Use this expression to find the average cost per booklet of printing 1000 booklets.

75. **Phone Plans** A wireless phone company has a pricing scheme that includes 250 minutes worth of phone usage in the basic monthly fee of $30. For each minute over and above the first 250 minutes of usage, the user is charged an additional $0.60 per minute.
(a) Let x be the number of minutes of phone usage per month. What is the expression for the average cost per minute if the value of x is in the interval $(0, 250)$?
(b) What is the expression for the average cost per minute if the value of x is above 250?
(c) If phone usage in a certain month is 600 minutes, what is the average cost per minute?

76. **Health** Body-mass index (BMI) is a measure of body fat based on height and weight that applies to both adult

males and adult females. It is calculated using the following formula:

$$\text{BMI} = \frac{703w}{h^2}$$

where w is the person's weight in pounds and h is the person's height in inches. A BMI in the range 18.5–24.9 is considered normal. (*Source:* National Institutes of Health)

(a) Calculate the BMI for a person who is 5 feet 5 inches tall and weighs 140 pounds. Is this person's BMI within the normal range?

(b) Calculate the weight of a person who is 6 feet tall and has a BMI of 24.

(c) Calculate the height of a person who weighs 170 pounds and has a BMI of 24.3.

77. **Metallurgy** How much pure gold should be added to a 2-ounce alloy that is presently 25% gold to make it 60% gold?

78. **Manufacturing** A packaging company wants to design an open box with a square base and a volume of exactly 30 cubic feet.

(a) Let x denote the length of a side of the base of the box, and let y denote the height of the box. Express the total surface area of the box in terms of x and y.

(b) Write an equation relating x and y to the total volume of 30 cubic feet.

(c) Solve the equation in part (b) for y in terms of x.

(d) Now write an expression for the surface area in terms of just x. Call this function $S(x)$.

(e) Fill in the following table giving the value of the surface area for the given values of x.

x	1	2	3	4	5	6
$S(x)$						

(f) What do you observe about the total surface area as x increases? From your table, approximate the value of x that would give the minimum surface area.

(g) Use a graphing utility to find the value of x that would give the minimum surface area.

79. **Manufacturing** A gift box company wishes to make a small open box by cutting four equal squares from a 3-inch by 5-inch card, one from each corner.

(a) Let x denote the length of the square cut from each corner. Write an expression for the volume of the box in terms of x. Call this function $V(x)$. What is the realistic domain of this function?

(b) Write an expression for the surface area of the box in terms of x. Call this function $S(x)$.

(c) Write an expression in terms of x for the ratio of the volume of the box to its surface area. Call this function $r(x)$.

(d) Fill in the following table giving the values of $r(x)$ for the given values of x.

x	0.2	0.4	0.6	0.8	1.0	1.2	1.4
$r(x)$							

(e) What do you observe about the ratio of the volume to the surface area as x increases? From your table, approximate the value of x that would give the maximum ratio of volume to surface area.

(f) Use a graphing utility to find the value of x that would give the maximum ratio of volume to surface area.

▶ **Concepts** This set of exercises will draw on the ideas presented in this section and your general math background.

80. Sketch a possible graph of a rational function $r(x)$ of the following description: the graph of r has a horizontal asymptote $y = -2$ and a vertical asymptote $x = 1$, with y-intercept at $(0, 0)$.

81. Sketch a possible graph of a rational function $r(x)$ of the following description: the graph of r has a horizontal asymptote $y = -2$ and a vertical asymptote $x = 1$, with y-intercept at $(0, 0)$ and x-intercept at $(2, 0)$.

82. Give a possible expression for a rational function $r(x)$ of the following description: the graph of r has a horizontal asymptote $y = 2$ and a vertical asymptote $x = 1$, with y-intercept at $(0, 0)$. It may be helpful to sketch the graph of r first. You may check your answer with a graphing utility.

83. Give a possible expression for a rational function $r(x)$ of the following description: the graph of r has a horizontal asymptote $y = 0$ and a vertical asymptote $x = 0$, with no x- or y-intercepts. It may be helpful to sketch the graph of r first. You may check your answer with a graphing utility.

84. Give a possible expression for a rational function $r(x)$ of the following description: the graph of r is symmetric with respect to the y-axis; it has a horizontal asymptote $y = 0$ and a vertical asymptote $x = 0$, with no x- or y-intercepts. It may be helpful to sketch the graph of r first. You may check your answer with a graphing utility.

85. Explain why the following output from a graphing utility is not a complete graph of the function

$$f(x) = \frac{1}{(x - 10)(x + 3)}.$$

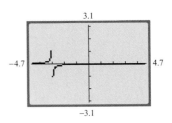

4.7 Polynomial and Rational Inequalities

Objectives

▶ Solve a polynomial inequality

▶ Solve a rational inequality

In this section, we will solve inequalities involving polynomial and rational expressions. The technique we will use is similar to the one we used to solve quadratic inequalities in Section 3.4.

Polynomial Inequalities

A **polynomial inequality** can be written in the form $a_n x^n + a_{n-1} x^{n-1} + \cdots + a_0 \,\Box\, 0$, where the symbol inside the box can be $>$, $\geq$, $<$, or $\leq$. Using factoring, we can solve certain polynomial inequalities, as shown in Example 1.

Just In Time

Review quadratic inequalities in Section 3.4.

Example 1 Solving a Polynomial Inequality

Solve the following inequality.

$$x^3 - 2x^2 - 3x \leq 0$$

▶**Solution**

Step 1 One side of the inequality is already zero. Therefore, we factor the nonzero side.

$$x(x^2 - 2x - 3) \leq 0 \qquad \text{Factor out } x$$
$$x(x - 3)(x + 1) \leq 0 \qquad \text{Factor inside parentheses}$$

Step 2 Determine the values at which $x(x - 3)(x + 1)$ equals zero. These values are $x = 0, x = 3$, and $x = -1$. Because the expression $x(x - 3)(x + 1)$ can change sign only at these three values, we form the following intervals.

$$(-\infty, -1), (-1, 0), (0, 3), (3, \infty)$$

Step 3 Make a table with these intervals in the first column. Choose a test value in each interval and determine the sign of each factor of the polynomial expression in that interval. See Table 4.7.1.

Table 4.7.1

Interval	Test Value	Sign of x	Sign of $x - 3$	Sign of $x + 1$	Sign of $x(x - 3)(x + 1)$
$(-\infty, -1)$	-2	$-$	$-$	$-$	$-$
$(-1, 0)$	$-\dfrac{1}{2}$	$-$	$-$	$+$	$+$
$(0, 3)$	1	$+$	$-$	$+$	$-$
$(3, \infty)$	4	$+$	$+$	$+$	$+$

Figure 4.7.1

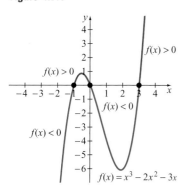

$f(x) = x^3 - 2x^2 - 3x$

Step 4 From Table 4.7.1, we observe that $x(x - 3)(x + 1) \leq 0$ for all x in the intervals

$$(-\infty, -1] \cup [0, 3].$$

So the solution of the inequality is $(-\infty, -1] \cup [0, 3]$. We have included the endpoints of the interval because we want to find the values of x that make the expression $x(x - 3)(x + 1)$ less than *or equal to* zero.

We can confirm our results by graphing the function $f(x) = x^3 - 2x^2 - 3x$ and observing where the graph lies above the x-axis and where it intersects the x-axis. See Figure 4.7.1.

✔ *Check It Out 1:* Solve the inequality $x(x^2 - 4) \geq 0$. ■

Example **2** **An Application of a Polynomial Inequality**

A box with a square base and a height 3 inches less than the length of one side of the base is to be built. What lengths of the base will produce a volume greater than or equal to 16 cubic inches?

▶**Solution** Let x be the length of the square base. Then the volume is given by

$$V(x) = x \cdot x \cdot (x - 3) = x^2(x - 3).$$

Since we want the volume to be greater than or equal to 16 cubic inches, we solve the inequality $x^2(x - 3) \geq 16$.

$$x^2(x - 3) \geq 16$$
$$x^3 - 3x^2 \geq 16 \qquad \text{Remove parentheses}$$
$$x^3 - 3x^2 - 16 \geq 0 \qquad \text{Set right-hand side equal to zero}$$

Next we factor the nonzero side of the inequality. The expression does not seem to be factorable by any of the elementary techniques, so we use the Rational Zero Theorem to factor, if possible.

The possible rational zeros are

$$x = \pm 16, \pm 8, \pm 4, \pm 2, \pm 1.$$

We see that $x = 4$ is a zero. Using synthetic division, we can write

$$p(x) = x^3 - 3x^2 - 16 = (x - 4)(x^2 + x + 4).$$

The expression $x^2 + x + 4$ cannot be factored further over the real numbers. Thus we have only the intervals $(-\infty, 4)$ and $(4, \infty)$ to check. For x in $(-\infty, 4)$, a test value of 0 yields $p(0) < 0$. For x in $(4, \infty)$, a test value of 5 yields $p(0) > 0$. We find that $p(x) = (x - 4)(x^2 + x + 4) \geq 0$ for all x in the interval $[4, \infty)$.

Thus, for our box, a square base whose side is greater than or equal to 4 inches will produce a volume greater than or equal to 16 cubic inches. The corresponding height will be 3 inches less than the length of the side.

✔ *Check It Out 2:* Rework Example 2 if the volume of the box is to be greater than or equal to 50 cubic inches. ■

Rational Inequalities

In some applications, and in more advanced mathematics courses, it is important to know how to solve an inequality involving a rational function, referred to as a **rational inequality.** Let $p(x)$ and $q(x)$ be polynomial functions with $q(x)$ not equal to zero.

A rational inequality can be written in the form $\frac{p(x)}{q(x)} \,\square\, 0$, where the symbol inside the box can be $>, \geq, <,$ or $\leq$. Example 3 shows how we solve such an inequality.

Example **3** **Solving a Rational Inequality**

Solve the inequality $\dfrac{x^2 - 9}{x + 5} > 0$.

▶Solution

Step 1 The right-hand side is already zero. We factor the numerator on the left side.

$$\frac{(x + 3)(x - 3)}{x + 5} > 0$$

Step 2 Determine the values at which the numerator $(x + 3)(x - 3)$ equals zero. These values are $x = 3$ and $x = -3$. Also, the denominator is equal to zero at $x = -5$. Because the expression $\dfrac{(x + 3)(x - 3)}{x + 5}$ can change sign only at these three values, we form the following intervals.

$$(-\infty, -5), (-5, -3), (-3, 3), (3, \infty)$$

Step 3 Make a table of these intervals. Choose a test value in each interval and determine the sign of each factor of the inequality in that interval. See Table 4.7.2.

Technology Note

A graphing utility can be used to confirm the results of Example 3. The graph of $f(x) = \dfrac{x^2 - 9}{x + 5}$ lies above the x-axis for x in $(-5, -3)$ or $(3, \infty)$. See Figure 4.7.2.

Keystroke Appendix:
Sections 7 and 8

Figure 4.7.2

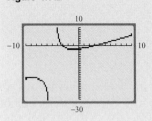

Table 4.7.2

Interval	Test Value	Sign of $x + 3$	Sign of $x - 3$	Sign of $x + 5$	Sign of $\dfrac{(x + 3)(x - 3)}{x + 5}$
$(-\infty, -5)$	-6	$-$	$-$	$-$	$-$
$(-5, -3)$	-4	$-$	$-$	$+$	$+$
$(-3, 3)$	0	$+$	$-$	$+$	$-$
$(3, \infty)$	4	$+$	$+$	$+$	$+$

Step 4 The inequality $\dfrac{(x + 3)(x - 3)}{x + 5} > 0$ is satisfied for all x in the intervals

$$(-5, -3) \cup (3, \infty).$$

The numbers -3 and 3 are not included because the inequality symbol is *less than*, not *less than or equal to*.

☑ *Check It Out 3:* Solve the inequality $\dfrac{x - 1}{x + 2} < 0$. ■

Example **4** **Solving a Rational Inequality**

Solve the following inequality.

$$\frac{2x + 5}{x - 2} \geq x$$

Technology Note

Using the INTERSECTION feature of a graphing utility, find the intersection points of $Y_1 = \frac{2x + 5}{x - 2}$ and $Y_2 = x$. The values of x for which $Y_1 \geq Y_2$ will be the solution set. One of the intersection points is given in Figure 4.7.3.

Keystroke Appendix:
Sections 7, 8, 9

Figure 4.7.3

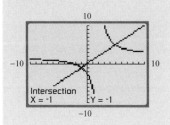

▶Solution

Step 1   Set the right-hand side equal to zero by subtracting x from both sides. Then simplify.

$$\frac{2x + 5}{x - 2} - x \geq 0$$

$$\frac{2x + 5 - x(x - 2)}{x - 2} \geq 0 \qquad \textit{Combine terms using x − 2 as the LCD}$$

$$\frac{2x + 5 - x^2 + 2x}{x - 2} \geq 0 \qquad \textit{Simplify}$$

$$\frac{-x^2 + 4x + 5}{x - 2} \geq 0 \qquad \textit{Collect like terms}$$

$$\frac{x^2 - 4x - 5}{x - 2} \leq 0 \qquad \textit{Multiply by −1. Note reversal of inequality.}$$

Step 2   Factor the numerator on the left side.

$$\frac{(x + 1)(x - 5)}{x - 2} \leq 0$$

Step 3   Determine the values at which the numerator $(x + 1)(x - 5)$ equals zero. These values are $x = -1$ and $x = 5$. Also, the denominator is equal to zero at $x = 2$. Because the expression $\frac{(x + 1)(x - 5)}{x - 2}$ can change sign only at these three values, we form the following intervals.

$$(-\infty, -1), (-1, 2), (2, 5), (5, \infty)$$

Step 4   Make a table of these intervals. Choose a test value in each interval and determine the sign of each factor of the inequality in that interval. See Table 4.7.3.

Table 4.7.3

Interval	Test Value	Sign of $x + 1$	Sign of $x - 2$	Sign of $x - 5$	Sign of $\dfrac{(x + 1)(x - 5)}{(x - 2)}$
$(-\infty, -1)$	-2	$-$	$-$	$-$	$-$
$(-1, 2)$	0	$+$	$-$	$-$	$+$
$(2, 5)$	3	$+$	$+$	$-$	$-$
$(5, \infty)$	6	$+$	$+$	$+$	$+$

Step 5   The inequality $\frac{(x + 1)(x - 5)}{x - 2} \leq 0$ is satisfied for all x in the intervals

$$(-\infty, -1] \cup (2, 5].$$

We have included the endpoints -1 and 5 because the inequality states *less than or equal to*. The endpoint 2 is *not* included because division by zero is undefined.

✔ *Check It Out 4:* Solve the following inequality.

$$\frac{2x - 1}{x + 3} \leq 0 \ ■$$

4.7 Key Points

▶ A **polynomial inequality** can be written in the form

$$a_n x^n + a_{n-1} x^{n-1} + \cdots + a_0 \,\square\, 0$$

where the symbol inside the box can be $>, \geq, <,$ or $\leq$.

▶ Let $p(x)$ and $q(x)$ be polynomial functions with $q(x)$ not equal to zero. A **rational inequality** can be written in the form $\dfrac{p(x)}{q(x)} \,\square\, 0$, where the symbol inside the box can be $>, \geq, <,$ or $\leq$.

▶ To solve polynomial and rational inequalities, follow these steps:

1. Rewrite the inequality with zero on the right-hand side.
2. Factor the nonzero side and set each factor equal to zero.
3. The resulting zeros divide the x-axis into test intervals. Test the sign of the inequality in each of these intervals, and choose those intervals that satisfy the inequality.

4.7 Exercises

▶**Just in Time Exercises** These exercises correspond to the Just in Time references in this section. Complete them to review topics relevant to the remaining exercises.

In Exercises 1–6, solve the quadratic inequality.

1. $x^2 \leq 4$
2. $y^2 \geq 9$
3. $x^2 + x - 6 \leq 0$
4. $x^2 + x - 20 \geq 0$
5. $3x^2 \geq 2x + 5$
6. $2x^2 \leq 3 - x$

▶**Skills** This set of exercises will reinforce the skills illustrated in this section.

In Exercises 7–26, solve the polynomial inequality.

7. $2x(x + 5)(x - 3) \geq 0$
8. $(x + 1)^2(x - 2) \leq 0$
9. $x^3 - 16x < 0$
10. $x^3 - 9x > 0$
11. $x^3 - 4x^2 \geq 0$
12. $x^3 + 2x^2 + x < 0$
13. $x^3 + 5x^2 + 4x < 0$
14. $x^3 + 4x^2 + 4x < 0$
15. $(x - 2)(x^2 - 4) < 0$
16. $(x - 3)(x^2 - 25) < 0$
17. $(x + 2)(x^2 - 4x + 5) \geq 0$
18. $(x + 3)(x^2 - 3x + 2) \geq 0$

19. $x^4 - x^2 > 3$
20. $x^4 - 3x^2 < 10$
21. $x^3 - 4x \leq -x^2 + 4$
22. $x^3 - 7x \leq -6$
23. $x^3 \leq 4x$
24. $x^3 \geq x$
25. $x^3 > 2x^2 + 3x$
26. $x^3 < 4x^2 - 4x$

In Exercises 27–46, solve the rational inequality.

27. $\dfrac{x + 2}{x - 1} \leq 0$
28. $\dfrac{x - 4}{2x + 1} > 0$
29. $\dfrac{x^2 - 4}{x - 3} \leq 0$
30. $\dfrac{4x^2 - 9}{x + 2} < 0$
31. $\dfrac{x(x + 1)}{1 + x^2} \geq 0$
32. $\dfrac{x^2 - 2x - 3}{x^2 + 2x + 1} \leq 0$
33. $\dfrac{4 - x}{x - 1} > x$
34. $\dfrac{-8}{x + 3} < -2x$
35. $\dfrac{1}{x} \leq \dfrac{1}{2x - 1}$
36. $\dfrac{2}{x + 1} > \dfrac{1}{x - 2}$
37. $\dfrac{3}{x - 1} \leq 2$
38. $\dfrac{-1}{2x + 1} \geq 1$

39. $\dfrac{x-1}{x+2} \ge 0$

40. $\dfrac{3x+6}{x-3} < 0$

41. $\dfrac{1}{2x+1} \le 0$

42. $\dfrac{-1}{3x-1} > 0$

43. $\dfrac{x+1}{x^2-9} < 0$

44. $\dfrac{x^2-4}{x+5} \ge 0$

45. $\dfrac{x+1}{x-3} \le \dfrac{x-2}{x+4}$

46. $\dfrac{x-2}{x+2} > \dfrac{x+5}{x-1}$

▶**Applications** In this set of exercises, you will use polynomial and rational inequalities to study real-world problems.

47. **Geometry** A rectangular solid has a square base and a height that is 2 inches less than the length of one side of the base. What lengths of the base will produce a volume greater than or equal to 32 inches?

48. **Manufacturing** A rectangular box with a rectangular base is to be built. The length of one side of the rectangular base is 3 inches more than the height of the box, while the length of the other side of the rectangular base is 1 inch more than the height. For what values of the height will the volume of the box be greater than or equal to 40 cubic inches?

49. **Drug Concentration** The concentration $C(t)$ of a drug in a patient's bloodstream t hours after administration is given by

$$C(t) = \frac{4t}{3+t^2}$$

where $C(t)$ is in milligrams per liter. During what time interval will the concentration be greater than 1 milligram per liter?

50. **Printing Costs** To print booklets, it costs $400 plus an additional $0.50 per booklet. What is the minimum number of booklets that must be printed so that the average cost per booklet is less than $0.55?

▶**Concepts** This set of exercises will draw on the ideas presented in this section and your general math background.

51. To solve the inequality $x(x+1)(x-1) < 2$, a student starts by setting up the following inequalities.

$$x < 2; \quad x+1 < 2; \quad x-1 < 2$$

Why is this the *wrong* way to start the problem? What is the correct way to start this problem?

52. To solve the inequality $\dfrac{x}{x+1} \ge 2$, a student first "simplifies" the problem by multiplying both sides by $x+1$ to get

$$x \ge 2(x+1).$$

Why is this an incorrect way to start the problem?

53. Find a polynomial $p(x)$ such that $p(x) > 0$ has the solution set $(0, 1) \cup (3, \infty)$. There may be more than one correct answer.

54. Find polynomials $p(x)$ and $q(x)$, with $q(x)$ not a constant function, such that $\dfrac{p(x)}{q(x)} \ge 0$ has the solution set $[3, \infty)$. There may be more than one correct answer.

Chapter 4   # Summary

Section 4.1 **Graphs of Polynomial Functions**

Concept	Illustration	Study and Review
Definition of a polynomial function A function f is a **polynomial function** if it can be expressed in the form $f(x) = a_n x^n + a_{n-1} x^{n-1} + \cdots + a_1 x + a_0$, where $a_n \neq 0$, n is a nonnegative integer, and $a_0, a_1, \ldots, a_n$ are real numbers.	$f(x) = -5x^5 + 2x^4 - \dfrac{1}{3}x^2 - 2$ and $g(x) = 6x^2 + 4x$ are examples of polynomial functions.	Examples 1, 2 Chapter 4 Review, Exercises 1–4
Terminology involving polynomials In the definition of a polynomial function, • the nonnegative integer n is called the **degree** of the polynomial. • the constants $a_0, a_1, \ldots, a_n$ are called **coefficients.** • the term $a_n x^n$ is called the **leading term,** and the coefficient a_n is called the **leading coefficient.**	For $p(x) = -5x^5 + 2x^4 - \dfrac{1}{3}x^2 - 2$, the degree of $p(x)$ is 5, the leading term is $-5x^2$, and the leading coefficient is -5.	Example 2 Chapter 4 Review, Exercises 1–4
The leading term test for end behavior For sufficiently large values of $|x|$, the leading term of a polynomial function $f(x)$ will be much larger in magnitude than any of the subsequent terms of $f(x)$.	For large values of $|x|$, the graph of $f(x) = 3x^4 + 4x^2 - 5$ resembles the graph of $g(x) = 3x^4$.	Example 4 Chapter 4 Review, Exercises 5–10
Connection between zeros and x-intercepts The real number values of x satisfying $f(x) = 0$ are called **real zeros** of the function f. Each real zero x is the first coordinate of an x-intercept of the graph of the function.	The zeros of $f(x) = x^2 - 4$ are found by solving the equation $x^2 - 4 = 0$. Factoring and applying the Zero Product Rule gives $(x + 2)(x - 2) = 0 \Longrightarrow x = 2, -2$. The x-intercepts are $(-2, 0)$ and $(2, 0)$.	Example 5 Chapter 4 Review, Exercises 11–16
Hand-sketching a polynomial function For a polynomial written in factored form, use the following procedure to sketch the function. **Step 1** Determine the end behavior of the function. **Step 2** Find and plot the y-intercept. **Step 3** Find and plot the x-intercepts of the graph of the function; these points divide the x-axis into smaller intervals. **Step 4** Find the sign and value of $f(x)$ for a test value x in each of these intervals. Plot these test values. **Step 5** Use the plotted points and the end behavior to sketch a smooth graph of the function. Plot additional points, if needed.	Let $f(x) = -x^3 + 4x$. To determine the end behavior, note that for large $|x|$, $f(x) \approx -x^3$. The y-intercept is $(0, 0)$. Find the x-intercepts by solving $-x^3 + 4x = -x(x^2 - 4) = 0$ to get $x = 0, -2, 2$. Tabulate the sign and value of $f(x)$ in each subinterval and sketch the graph. 	Examples 6, 7 Chapter 4 Review, Exercises 11–16

Section 4.2 More on Graphs of Polynomial Functions and Models

Concept	Illustration	Study and Review
Multiplicities of zeros The number of times a linear factor $x - a$ occurs in the completely factored form of a polynomial expression is known as the **multiplicity** of the real zero a associated with that factor. The number of real zeros of a polynomial $f(x)$ of degree n is less than or equal to n, counting multiplicity.	$f(x) = (x + 5)^2(x - 2)$ has two real zeros: $x = -5$ and $x = 2$. The zero $x = -5$ has multiplicity 2 and the zero $x = 2$ has multiplicity 1.	Example 1 Chapter 4 Review, Exercises 17–22
Multiplicities of zeros and behavior at the **x-intercept** • If the multiplicity of a real zero of a polynomial function is **odd,** the graph of the function **crosses** the x-axis at the corresponding x-intercept. • If the multiplicity of a real zero of a polynomial function is **even,** the graph of the function **touches,** but does not cross, the x-axis at the corresponding x-intercept.	Because $x = -5$ is a zero of multiplicity 2, the graph of $f(x) = (x + 5)^2(x - 2)$ touches the x-axis at $(-5, 0)$. Because $x = 2$ is a zero of multiplicity 1, the graph cross the x-axis at $(2, 0)$. $f(x) = (x + 5)^2(x - 2)$	Example 1 Chapter 4 Review, Exercises 17–22
Finding local extrema and sketching a complete graph The peaks and valleys of the graphs of most polynomial functions are known as **local maxima** and **local minima,** respectively. Together, they are known as **local extrema** and can be located by using a graphing utility.	For $f(x) = (x + 5)^2(x - 2)$, we can use a graphing utility to find that the local minimum occurs at $x \approx -0.3333$ and the local maximum occurs at $x = -5$.	Example 3 Chapter 4 Review, Exercises 23–30

Section 4.3 Division of Polynomials; the Remainder and Factor Theorems

Concept	Illustration	Study and Review
The division algorithm Let $p(x)$ be a polynomial divided by a nonzero polynomial $d(x)$. Then there exist a quotient polynomial $q(x)$ and a remainder polynomial $r(x)$ such that $\quad p(x) = d(x)q(x) + r(x).$ The remainder $r(x)$ is either equal to zero or its degree is less than the degree of $d(x)$.	Using long division, when $p(x) = x^3 - 1$ is divided by $d(x) = x - 1$, the quotient is $q(x) = x^2 + x + 1$. The remainder is $r(x) = 0$.	Examples 1–4 Chapter 4 Review, Exercises 31–34

Continued

Section 4.3 Division of Polynomials; the Remainder and Factor Theorems

Concept	Illustration	Study and Review
Synthetic division Synthetic division is a compact way of dividing polynomials when the divisor is of the form $x - c$.	To divide $x^3 - 8$ by $x - 2$, write out the coefficients of $x^3 - 8$ and place the 2 on the left. Proceed as indicated below. $$\begin{array}{r\|rrrr} 2 & 1 & 0 & 0 & -8 \\ & & 2 & 4 & 8 \\ \hline & 1 & 2 & 4 & 0 \end{array}$$ The quotient is $x^2 + 2x + 4$, with a remainder of 0.	Example 4 Chapter 4 Review, Exercises 31–34
The Remainder Theorem When a polynomial $p(x)$ is divided by $x - c$, the remainder is equal to the value of $p(c)$.	When $p(x) = x^3 - 9x$ is divided by $x - 3$, the remainder is $p(3) = 0$.	Example 5 Chapter 4 Review, Exercises 35–38
The Factor Theorem The term $x - c$ is a **factor** of a polynomial $p(x)$ if and only if $p(c) = 0$.	The term $x - 3$ is a factor of $p(x) = x^3 - 9x$ because $p(3) = 3^3 - 9(3) = 27 - 27 = 0$.	Example 6 Chapter 4 Review, Exercises 35–38

Section 4.4 Real Zeros of Polynomials; Solutions of Equations

Concept	Illustration	Study and Review
Zeros, factors, and x-intercepts of a polynomial Let $p(x)$ be a polynomial function and let c be a real number. Then the following are equivalent statements: • $p(c) = 0$ • $x - c$ is a factor of $p(x)$. • $(c, 0)$ is an x-intercept of the graph of $p(x)$.	For the polynomial $p(x) = x^3 - 9x$, we have | Zero, c	$p(c)$	x-Intercept, $(c, 0)$	Factor, $x - c$	 |---|---|---|---| | 0	0	(0, 0)	x	 | −3	0	(−3, 0)	$x + 3$	 | 3	0	(3, 0)	$x - 3$	 	Examples 1, 2 Chapter 4 Review, Exercises 39–42
The Rational Zero Theorem If $f(x) = a_n x^n + a_{n-1}x^{n-1} + \cdots + a_1 x + a_0$ is a polynomial with integer coefficients and $\frac{p}{q}$ is a rational zero of f with p and q having no common factor other than 1, then p is a factor of a_0 and q is a factor of a_n.	The possible rational zeros of $f(x) = 2x^3 - 6x + x^2 - 3$ are $$\pm\frac{3}{2}, \pm\frac{1}{2}, \pm 3, \pm 1.$$ Of these, only $x = -\frac{1}{2}$ is an actual zero of $f(x)$.	Examples 3, 4 Chapter 4 Review, Exercises 43–46

Continued

Section 4.4 Real Zeros of Polynomials; Solutions of Equations

Concept	Illustration	Study and Review
Finding solutions of polynomial equations by finding zeros Because any equation can be rewritten so that the right-hand side is zero, solving an equation is identical to finding the zeros of a suitable function.	To solve $x^3 = -2x^2 - x$, rewrite the equation as $x^3 + 2x^2 + x = 0$. Factor to get $x(x^2 + 2x + 1) = x(x + 1)^2 = 0$. The solutions are $x = 0$ and $x = -1$.	Example 5 Chapter 4 Review, Exercises 47–50
Descartes' Rule of Signs The number of *positive* real zeros of a polynomial $p(x)$, counting multiplicity, is either equal to the number of variations in sign of $p(x)$ or less than that number by an even integer. The number of *negative* real zeros of a polynomial $p(x)$, counting multiplicity, is either equal to the number of variations in sign of $p(-x)$ or less than that number by an even integer.	For $p(x) = x^3 + 3x^2 - x + 1$, the number of positive real zeros is either two or zero. Because $p(-x) = -x^3 + 3x^2 + x + 1$, the number of negative real zeros is only one.	Example 6 Chapter 4 Review, Exercises 51, 52

Section 4.5 The Fundamental Theorem of Algebra; Complex Zeros

Concept	Illustration	Study and Review
The Fundamental Theorem of Algebra Every nonconstant polynomial function with real or complex coefficients has at least one complex zero.	According to this theorem, $f(x) = x^2 + 1$ has at least one complex zero. This theorem does not tell you what the zero is—only that a complex zero *exists*.	Definition on page 325
Definition of multiplicity of zeros A zero c of a polynomial P of degree $n > 0$ has **multiplicity** k if $P(x) = (x - c)^k Q(x)$, where $Q(x)$ is a polynomial of degree $n - k$ and c is not a zero of $Q(x)$.	If $p(x) = (x + 4)^2(x - 2)$, $c = -4$ is a zero of multiplicity $k = 2$, with $Q(x) = x - 2$. The degree of $Q(x)$ is $n - k = 3 - 2 = 1$.	Examples 1, 2 Chapter 4 Review, Exercises 53–56
Factorization over the complex numbers Every polynomial $p(x)$ of degree $n > 0$ has exactly n zeros, if multiplicities and complex zeros are counted.	The polynomial $p(x) = (x - 1)^2(x^2 + 1) = x^2(x + i)(x - i)$ has four zeros, since the zero $x = 1$ is counted twice and there are two complex zeros, $x = i$ and $x = -i$.	Example 2 Chapter 4 Review, Exercises 53–56
Polynomials with real coefficients **The Factorization Theorem** Any polynomial P with *real* coefficients can be factored uniquely into linear factors and/or **irreducible quadratic factors.**	The polynomial $p(x) = x^3 + x = x(x^2 + 1)$ has a linear factor, x. It also has an irreducible quadratic factor, $x^2 + 1$, that cannot be factored further using real numbers.	Examples 3, 4 Chapter 4 Review, Exercises 53–56

Section 4.6 **Rational Functions**

Concept	Illustration	Study and Review
Definition of a rational function A **rational function** $r(x)$ is defined as a quotient of two polynomials $p(x)$ and $h(x)$, $r(x) = \frac{p(x)}{h(x)}$, where $h(x)$ is not the constant zero function. The **domain** of a rational function consists of all real numbers for which $h(x)$ is not equal to zero.	The functions $f(x) = \frac{1}{x^2}$, $g(x) = \frac{2x}{3 + x^3}$, and $h(x) = \frac{x^2 - 1}{x^2 - 9}$ are all examples of rational functions.	Examples 1, 2
Vertical asymptotes The vertical asymptotes of $r(x) = \frac{p(x)}{h(x)}$ occur at the values of x at which $h(x) = 0$, assuming $p(x)$ and $h(x)$ have no common factors.	The function $f(x) = \frac{1}{(x - 4)^2}$ has a vertical asymptote at $x = 4$. The function $g(x) = \frac{x}{x^2 - 1}$ has vertical asymptotes at $x = 1$ and $x = -1$.	Examples 3, 5–7 Chapter 4 Review, Exercises 57–64
Horizontal asymptotes and end behavior Let $r(x) = \frac{p(x)}{h(x)}$, where $p(x)$ and $h(x)$ are polynomials of degrees n and m, respectively. • If $n < m$, $r(x)$ approaches zero for large values of $|x|$. The line $y = 0$ is the **horizontal asymptote** of the graph of $r(x)$. • If $n = m$, $r(x)$ approaches a nonzero constant $\frac{a_n}{b_n}$ for large values of $|x|$. The line $y = \frac{a_n}{b_n}$ is the **horizontal asymptote** of the graph of $r(x)$. • If $n > m$, then $r(x)$ does not have a horizontal asymptote.	The function $r(x) = \frac{x}{x + 1}$ has $x = -1$ as a vertical asymptote. The line $y = 1$ is a horizontal asymptote because the degrees of the polynomials in the numerator and the denominator are equal. 	Example 4 Chapter 4 Review, Exercises 57–64
Slant asymptotes If the degree of the numerator of a rational function is 1 greater than the degree of the denominator, the graph of the function has a **slant asymptote.**	The function $r(x) = \frac{x^2 - 1}{x + 1}$ has the line $y = x - 1$ as its slant asymptote.	Example 7 Chapter 4 Review, Exercises 65, 66

Section 4.7 **Polynomial and Rational Inequalities**

Concept	Illustration	Study and Review
Polynomial inequalities A **polynomial inequality** can be written in the form $a_n x^n + a_{n-1} x^{n-1} + \cdots + a_0 \, \square \, 0$, where the symbol inside the box can be $>$, $\geq$, $<$, or $\leq$.	To solve $$x(x + 1)(x - 1) > 0$$ set each factor on the left equal to zero, giving $x = -1, 0, 1$. Then choose a test value in each of the intervals $(-\infty, -1)$, $(-1, 0)$, $(0, 1)$, and $(1, \infty)$. The value of the polynomial is positive in the intervals $(-1, 0) \cup (1, \infty)$.	Examples 1, 2 Chapter 4 Review, Exercises 67–70

Continued

Section 4.7 **Polynomial and Rational Inequalities**

Concept	Illustration	Study and Review
Rational inequalities Let $p(x)$ and $q(x)$ be polynomial functions with $q(x)$ not equal to zero. A **rational inequality** can be written in the form $\frac{p(x)}{q(x)} \,\Box\, 0$, where the symbol inside the box can be $>$, $\geq$, $<$, or $\leq$.	To solve $$\frac{x}{x+1} > 0$$ set each factor in the numerator and denominator equal to zero, giving $x = -1, 0$. Then choose a test value in each of the intervals $(-\infty, -1)$, $(-1, 0)$, and $(0, \infty)$. The value of $\frac{x}{x+1}$ is positive in the intervals $(-\infty, -1) \cup (0, \infty)$.	Examples 3, 4 Chapter 4 Review, Exercises 71–74

Chapter 4

Review Exercises

Section 4.1 _____

In Exercises 1–4, determine whether the function is a polynomial function. If so, find its degree, its coefficients, and its leading coefficient.

1. $f(x) = -x^3 - 6x^2 + 5$ 2. $f(s) = s^5 + 6s - 1$

3. $f(t) = \sqrt{t+1}$ 4. $g(t) = \dfrac{1}{t^2}$

In Exercises 5–10, determine the end behavior of the function.

5. $f(x) = -3x^3 + 5x + 9$ 6. $g(t) = 5t^4 - 6t^2 + 1$

7. $H(s) = -6s^4 - 3s$ 8. $g(x) = -x^3 + 2x - 1$

9. $h(s) = 10s^5 - 2s^2$ 10. $f(t) = 7t^2 - 4$

In Exercises 11–16, for each polynomial function, find (a) the end behavior; (b) the x- and y-intercepts of the graph of the function; (c) the interval(s) on which the value of the function is positive; and (d) the interval(s) on which the value of the function is negative. Use this information to sketch a graph of the function. Factor first if the expression is not in factored form.

11. $f(x) = -(x-1)(x+2)(x+4)$

12. $g(x) = (x-3)(x-4)(x-1)$

13. $f(t) = t(3t-1)(t+4)$

14. $g(t) = 2t(t+4)\left(t + \dfrac{3}{2}\right)$

15. $f(x) = 2x^3 + x^2 - x$

16. $g(x) = -x^3 - 6x^2 + 7x$

Section 4.2 _____

In Exercises 17–22, determine the multiplicities of the real zeros of the function. Comment on the behavior of the graph at the x-intercepts. Does the graph cross or just touch the x-axis?

17. $f(x) = (x+2)^3(x+7)^2$ 18. $g(s) = (s+8)^5(s-1)^2$

19. $h(t) = -t^2(t+1)(t-2)$ 20. $g(x) = -x^3(x^2-16)$

21. $f(x) = x^3 + 2x^2 + x$ 22. $h(s) = s^7 - 16s^3$

In Exercises 23–30, for each polynomial function, find (a) the x- and y-intercepts of the graph of the function; (b) the multiplicities of each of the real zeros; (c) the end behavior; and (d) the intervals on which the function is positive or negative. Use this information to sketch a graph of the function.

23. $f(x) = x^2(2x+1)$

24. $h(t) = -t(t+4)^2$

25. $f(x) = \left(x - \dfrac{1}{2}\right)^2(x-4)$

26. $f(x) = (x-7)^2(x+2)(x-3)$

27. $f(t) = (t+2)(t-1)(t^2+1)$

28. $g(s) = \left(s - \dfrac{1}{2}\right)(s+3)(s^2+4)$

29. $g(x) = x^4 - 3x^3 - 18x^2$

30. $h(t) = -2t^5 + 4t^4 + 2t^3$

Section 4.3

In Exercises 31–34, write each polynomial in the form $p(x) = d(x)q(x) + r(x)$, *where* $p(x)$ *is the given polynomial and* $d(x)$ *is the given factor. You may use synthetic division wherever applicable.*

31. $-4x^2 + x - 7; \; x - 4$

32. $5x^3 + 2x + 4; \; x + 2$

33. $x^5 - x^4 + x^2 - 3x + 1; \; x^2 + 3$

34. $-4x^3 - x^2 + 2x + 1; \; 2x + 1$

In Exercises 35–38, find the remainder when the first polynomial, $p(x)$, is divided by the second polynomial, $d(x)$. Determine whether $d(x)$ is a factor of $p(x)$ and justify your answer.

35. $-x^3 + 7x + 6; \; x + 1$ 36. $2x^3 + x^2 + 8x; \; x + 2$

37. $x^3 - 7x + 6; \; x + 3$ 38. $x^{10} - 1; \; x - 1$

Section 4.4

In Exercises 39–42, show that the given value of x is a zero of the polynomial. Use the zero to completely factor the polynomial over the real numbers.

39. $p(x) = x^3 - 6x^2 + 3x + 10; \; x = 2$

40. $p(x) = -x^3 - 7x - 8; \; x = -1$

41. $p(x) = -x^4 + x^3 + 4x^2 + 5x + 3; \; x = 3$

42. $p(x) = x^4 - x^2 + 6x; \; x = -2$

In Exercises 43–46, find all real zeros of the polynomial using the Rational Zero Theorem.

43. $P(x) = 2x^3 - 3x^2 - 2x + 3$

44. $P(t) = -t^3 - 5t^2 + 4t + 20$

45. $h(x) = x^3 - 3x^2 + x - 3$

46. $f(x) = x^3 + 3x^2 - 9x + 5$

In Exercises 47–50, solve the equation for real values of x.

47. $2x^3 + 9x^2 - 6x = 5$

48. $x^3 = 21x - 20$

49. $x^3 - 7x^2 = -14x + 8$

50. $x^4 - 9x^2 - 2x^3 + 2x = -8$

In Exercises 51 and 52, use Descartes' Rule of Signs to determine the number of positive and negative zeros of p. You need not find the zeros.

51. $p(x) = -x^4 + 2x^3 - 7x - 4$

52. $p(x) = x^5 + 3x^4 - 8x^2 - x - 3$

Section 4.5

In Exercises 53–56, find all the zeros, real and nonreal, of the polynomial. Then express $p(x)$ as a product of linear factors.

53. $p(x) = x^3 - 25x$ 54. $p(x) = x^3 - 4x^2 - x + 4$

55. $p(x) = x^4 - 8x^2 - 9$

56. $p(x) = x^3 + 2x^2 + 4x + 8$

Section 4.6

57. Let $f(x) = \dfrac{1}{(x + 1)^2}$.

(a) Fill in the following table for values of x near -1. What do you observe about the value of $f(x)$ as x approaches -1 from the right? from the left?

x	-1.5	-1.1	-1.01	-0.99	-0.9	-0.5
$f(x)$						

(b) Complete the following table. What happens to the value of $f(x)$ as x gets very large and positive?

x	10	50	100	1000
$f(x)$				

(c) Complete the following table. What happens to the value of $f(x)$ as x gets very large and negative?

x	-1000	-100	-50	-10
$f(x)$				

58. Let $f(x) = \dfrac{1}{(3 - x)^2}$.

(a) Fill in the following table for values of x near 3. What do you observe about the value of $f(x)$ as x approaches 3 from the right? from the left?

x	2.5	2.9	2.99	3.01	3.1	3.5
$f(x)$						

(b) Complete the following table. What happens to the value of $f(x)$ as x gets very large and positive?

x	10	50	100	1000
$f(x)$				

(c) Complete the following table. What happens to the value of $f(x)$ as x gets very large and negative?

x	-1000	-100	-50	-10
$f(x)$				

In Exercises 59–66, for each rational function, find all asymptotes and intercepts, and sketch a graph.

59. $f(x) = \dfrac{2}{x-1}$

60. $f(x) = \dfrac{3x}{x+5}$

61. $h(x) = \dfrac{1}{x^2-4}$

62. $g(x) = \dfrac{2x^2}{x^2-1}$

63. $g(x) = \dfrac{x-2}{x^2-2x-3}$

64. $h(x) = \dfrac{x^2-2}{x^2-4}$

65. $r(x) = \dfrac{x^2-1}{x+2}$

66. $p(x) = \dfrac{x^2-4}{x-1}$

Section 4.7

In Exercises 67–74, solve the inequality.

67. $-x(x+1)(x^2-9) > 0$

68. $x^2(x+2)(x-3) \le 0$

69. $x^3 + 4x^2 \le -x + 6$

70. $9x^3 - x \ge -9x^2 + 1$

71. $\dfrac{x^2-1}{x+1} \le 0$

72. $\dfrac{x^2+2x-3}{x-3} \le 0$

73. $\dfrac{4x-2}{3x-1} \ge 2$

74. $-\dfrac{2}{x+3} \le x$

Applications

75. **Manufacturing** An open box is to be made by cutting four squares of equal size from an 8-inch by 11-inch rectangular piece of cardboard (one from each corner) and then folding up the sides.

(a) Let x be the length of a side of the square cut from each corner. Find an expression for the volume of the box in terms of x. Leave the expression in factored form.

(b) What is a realistic range of values for x? Explain.

(c) Use a graphing utility to find an approximate value of x that will yield the maximum volume.

76. **Design** A pencil holder in the shape of a right circular cylinder is to be designed with the specification that the sum of its radius and height must equal 8 inches.

(a) Let r denote the radius. Find an expression for the volume of the cylinder in terms of r. Leave the expression in factored form.

(b) What is a realistic range of values for r? Explain.

(c) Use a graphing utility to find an approximate value of r that will yield the maximum volume.

77. **Geometry** The length of a rectangular solid is 3 inches more than its height, and its width is 4 inches more than its height. Write and solve a polynomial equation to determine the height of the solid such that the volume is 60 cubic inches.

78. **Average Cost** A truck rental company charges a daily rate of $20 plus $0.25 per mile driven.

(a) What is the average cost per mile of driving x miles per day?

(b) Use the expression in part (a) to find the average cost per mile of driving 100 miles per day.

(c) Find the horizontal asymptote of this function and explain its significance.

(d) How many miles must be driven per day if the average cost per mile is to be less than $0.30 per day?

79. **Geometry** A rectangular solid has a square base and a height that is 1 inch less than the length of one side of the base. Set up and solve a polynomial inequality to determine the lengths of the base that will produce a volume greater than or equal to 48 cubic inches.

80. **Consumer Complaints** The following table gives the number of consumer complaints against U.S. airlines for the years 1997–2003. (*Source: Statistical Abstract of the United States*)

Year	Number of Complaints
1997	6394
1998	7980
1999	17,345
2000	20,564
2001	14,076
2002	7697
2003	4600

(a) Let t be the number of years since 1997. Make a scatter plot of the given data, with t as the input variable and the number of complaints as the output.

(b) Find the fourth-degree polynomial function $p(t)$ of best fit for the set of points plotted in part (a). Graph it along with the data.

(c) Use the function $p(t)$ from part (b) to predict the number of complaints in 2006.

(d) For what value of t, $0 \le t \le 6$, will $p(t) = 5000$?

Chapter 4 Test

1. Determine the degree, coefficients, and leading coefficient of the polynomial $p(x) = 3x^5 + 4x^2 - x + 7$.

2. Determine the end behavior of $p(x) = -8x^4 + 3x - 1$.

3. Determine the real zeros of $p(x) = -2x^2(x^2 - 9)$ and their multiplicities. Also find the x-intercepts and determine whether the graph of p crosses or touches the x-axis.

In Exercises 4–7, find
(a) the x- and y-intercepts of the graph of the polynomial;
(b) the multiplicities of each of the real zeros;
(c) the end behavior; and
(d) the intervals on which the function is positive and the intervals on which the function is negative.
Use this information to sketch a graph of the function.

4. $f(x) = -2x(x - 2)(x + 1)$

5. $f(x) = (x + 1)(x - 2)^2$

6. $f(x) = -3x^3 - 6x^2 - 3x$

7. $f(x) = 2x^4 + 5x^3 + 2x^2$

In Exercises 8 and 9, write each polynomial in the form $p(x) = d(x)q(x) + r(x)$, where $p(x)$ is the given polynomial and $d(x)$ is the given factor.

8. $3x^4 - 6x^2 + x - 1; x^2 + 1$

9. $-2x^5 + x^4 - 4x^2 + 3; x - 1$

10. Find the remainder when $p(x) = x^4 + x - 2x^3 - 2$ is divided by $x - 2$. Is $x - 2$ a factor of $p(x)$? Explain.

11. Use the fact that $x = 3$ is a zero of $p(x) = x^4 - x - 3x^3 + 3$ to completely factor $p(x)$.

In Exercises 12 and 13, find all real zeros of the given polynomial.

12. $p(x) = x^3 - 3x - 2x^2 + 6$

13. $q(x) = 2x^4 + 9x^3 + 14x^2 + 9x + 2$

In Exercises 14 and 15, find all real solutions of the given equation.

14. $2x^3 + 5x^2 = 2 - x$

15. $x^4 - 4x^3 + 2x^2 + 4x = 3$

16. Use Descartes' Rule of Signs to determine the number of positive and negative zeros of $p(x) = -x^5 + 4x^4 - 3x^2 + x + 8$. You need not find the zeros.

In Exercises 17 and 18, find all zeros, both real and nonreal, of the polynomial. Then express $p(x)$ as a product of linear factors.

17. $p(x) = x^5 - 16x$

18. $p(x) = x^4 - x^3 - 2x^2 - 4x - 24$

In Exercises 19–21, find all asymptotes and intercepts, and sketch a graph of the rational function.

19. $f(x) = \dfrac{-6}{2(x + 3)}$

20. $f(x) = \dfrac{-2x}{x - 2}$

21. $f(x) = \dfrac{2}{2x^2 - 3x - 2}$

In Exercises 22–24, solve the inequality.

22. $(x^2 - 4)(x + 3) \le 0$

23. $\dfrac{x^2 - 4x - 5}{x + 2} > 0$

24. $\dfrac{4}{3x + 1} \ge 2$

25. The radius of a right circular cone is 2 inches more than its height. Write and solve a polynomial equation to determine the height of the cone such that the volume is 144π cubic inches.

26. A couple rents a moving van at a daily rate of $50 plus $0.25 per mile driven.
    (a) What is the average cost per mile of driving x miles per day?
    (b) If the couple drives 250 miles per day, find the average cost per mile.

Exponential and Logarithmic Functions

Population growth can be modeled in the initial stages by an *exponential function,* a type of function in which the independent variable appears in the *exponent.* A simple illustration of this type of model is given in Example 1 of Section 5.2, as well as Exercises 30 and 31 of Section 5.6. This chapter will explore exponential functions. These functions are useful for studying applications in a variety of fields, including business, the life sciences, physics, and computer science. The exponential functions are also invaluable in the study of more advanced mathematics.

5.1 Inverse Functions

Just in Time

Review composition of functions in Section 2.2.

Table 5.1.1

Quantity in Gallons	Equivalent Quantity in Liters
30	113.55
45	170.325
55	208.175
70	264.95

Table 5.1.2

Liters, x	Gallons, $G(x)$
113.55	30
170.325	45
208.175	55
264.95	70

Inverse Functions

Section 2.2 treated the composition of functions, which entails using the output of one function as the input for another. Using this idea, we can sometimes find a function that will undo the action of another function—a function that will use the output of the original function as input, and will in turn output the number that was input to the original function. A function that undoes the action of a function f is called the **inverse** of f.

As a concrete example of undoing the action of a function, Example 1 presents a function that converts a quantity of fuel in gallons to an equivalent quantity of that same fuel in liters.

Example 1 Unit Conversion

Table 5.1.1 lists certain quantities of fuel in gallons and the corresponding quantities in liters.

(a) There are 3.785 liters in 1 gallon. Find an expression for a function $L(x)$ that will take the number of gallons of fuel as its input and give the number of liters of fuel as its output.

(b) Rewrite Table 5.1.1 so that the number of liters is the input and the number of gallons is the output.

(c) Find an expression for a function $G(x)$ that will take the number of liters of fuel as its input and give the number of gallons of fuel as its output.

(d) Find an expression for $(L \circ G)(x)$.

► Solution

(a) To convert from gallons to liters, we *multiply* the number of gallons by 3.785. The function $L(x)$ is thus given by

$$L(x) = 3.785x$$

where x is the number of *gallons* of fuel.

(b) Since we already have the amount of fuel in gallons and the corresponding amount in liters, we simply interchange the columns of the table so that the *input* is the quantity of fuel in *liters* and the *output* is the equivalent quantity of fuel in *gallons*. See Table 5.1.2.

(c) To convert from liters to gallons, we *divide* the number of liters by 3.785. The function $G(x)$ is thus given by

$$G(x) = \frac{x}{3.785}$$

where x is the number of *liters* of fuel.

(d) $(L \circ G)(x) = L(G(x)) = 3.785\left(\frac{x}{3.785}\right) = x$

When L and G are composed, the output (the number of liters of fuel) is simply the same as the input.

☑ *Check It Out 1:* In Example 1, find an expression for $(G \circ L)(x)$ and comment on your result. What quantity does x represent in this case? ▪

Example 1 showed the action of a function (conversion of liters to gallons) and the "undoing" of that action (conversion of gallons to liters). This "undoing" process is called *finding the inverse* of a function. We next give the formal definition of the inverse of a function.

Inverse of a Function

Let *f* be a function. A function *g* is said to be the **inverse function** of *f* if the domain of *g* is equal to the range of *f* and, for every *x* in the domain of *f* and every *y* in the domain of *g*,

$$g(y) = x \quad \text{if and only if} \quad f(x) = y.$$

The notation for the inverse function of *f* is f^{-1}. Equivalently,

$$f^{-1}(y) = x \quad \text{if and only if} \quad f(x) = y.$$

The notation f^{-1} does NOT mean $\frac{1}{f}$.

Note   *The above definition does not tell you how to find the inverse of a function f, or if such an inverse function even exists. All it does is assert that if f has an inverse function, then the inverse function must have the stated properties.*

Discover *and* Learn

Let *f* be a function whose inverse function exists. Use the definition of an inverse function to find an expression for $(f^{-1} \circ f)(x)$.

The idea of an inverse can be illustrated graphically as follows. Consider evaluating $f^{-1}(4)$ using the graph of the function *f* given in Figure 5.1.1. From the definition of an inverse function, $f^{-1}(y) = x$ if and only if $f(x) = y$. Thus, to evaluate $f^{-1}(4)$, we have to determine the value of *x* that produces $f(x) = 4$. From the graph of *f*, we see that $f(2) = 4$, so $f^{-1}(4) = 2$.

Figure 5.1.1

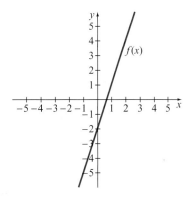

We saw in Example 1 that $(L \circ G)(x) = x$ for the functions *L* and *G* given there. From the definition of an inverse function, we can conclude that *L* and *G* must be inverse functions of each other. The following definition generalizes this property.

Composition of a Function and Its Inverse

If f is a function with an inverse function f^{-1}, then

▶ for every x in the domain of f, $f^{-1}(f(x))$ is defined and $f^{-1}(f(x)) = x$.

▶ for every x in the domain of f^{-1}, $f(f^{-1}(x))$ is defined and $f(f^{-1}(x)) = x$.

See Figure 5.1.2. If g is any function having the same properties with respect to f as those stated here for f^{-1}, then f and g are inverse functions of one another (i.e., $f^{-1} = g$ and $g^{-1} = f$).

Figure 5.1.2 Relationship between f and f^{-1}

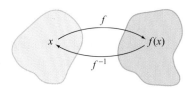

Example 2 Verifying Inverse Functions

Verify that the following functions are inverses of each other.

$$f(x) = 3x - 4; \quad g(x) = \frac{x}{3} + \frac{4}{3}$$

▶**Solution** To check that f and g are inverses of each other, we verify that $(f \circ g)(x) = x$ and $(g \circ f)(x) = x$. We begin with $(f \circ g)(x)$.

$$(f \circ g)(x) = f(g(x)) = 3\left(\frac{x}{3} + \frac{4}{3}\right) - 4 = x + 4 - 4 = x$$

Next, we calculate $(g \circ f)(x)$.

$$(g \circ f)(x) = g(f(x)) = \frac{3x - 4}{3} + \frac{4}{3} = x - \frac{4}{3} + \frac{4}{3} = x$$

Thus, by definition, f and g are inverses of each other.

☑ *Check It Out 2:* Verify that $f(x) = 2x - 9$ and $g(x) = \frac{x}{2} + \frac{9}{2}$ are inverses of each other. ■

Example 1 showed how to find the inverse of a function defined by a table. We next show how to find the inverse of a function defined by an algebraic expression.

Example 3 Finding the Inverse of a Function

Find the inverse of the function $f(x) = -2x + 3$, and check that your result is valid.

▶**Solution** Before we illustrate an algebraic method for determining the inverse of f, let's examine what f does: it takes a number x, multiplies it by -2, and adds 3 to the result. The inverse function will *undo* this sequence (in order to get back to x): it will start with the value output by f, then *subtract* 3, and then *divide the result* by -2.

This inverse can be found by using a set of algebraic steps.

STEPS	EXAMPLE
1. Start with the expression for the given function f.	$f(x) = -2x + 3$
2. Replace $f(x)$ with y.	$y = -2x + 3$
3. Interchange the variables x and y so that the input variable for the inverse function f^{-1} is x and the output variable is y.	$x = -2y + 3$
4. Solve for y. Note that this gives us the same result that was described in words before.	$x = -2y + 3$ $x - 3 = -2y$ $\dfrac{x-3}{-2} = y$
5. The inverse function f^{-1} is now given by y, so replace y with $f^{-1}(x)$ and simplify the expression for $f^{-1}(x)$.	$f^{-1}(x) = \dfrac{x-3}{-2} = -\dfrac{1}{2}x + \dfrac{3}{2}$

To check that the function $f^{-1}(x) = -\frac{1}{2}x + \frac{3}{2}$ is the inverse of f, find the expressions for $(f \circ f^{-1})(x)$ and $(f^{-1} \circ f)(x)$.

$$(f \circ f^{-1})(x) = f(f^{-1}(x)) = -2\left(-\frac{1}{2}x + \frac{3}{2}\right) + 3 = x - 3 + 3 = x$$

$$(f^{-1} \circ f)(x) = f^{-1}(f(x)) = -\frac{1}{2}(-2x + 3) + \frac{3}{2} = x - \frac{3}{2} + \frac{3}{2} = x$$

Since $(f \circ f^{-1})(x) = x = (f^{-1} \circ f)(x)$, the functions f and f^{-1} are inverses of each other.

✔ **Check It Out 3:** Find the inverse of the function $f(x) = 4x - 5$, and check that your result is valid. ■

One-to-One Functions

Thus far in this chapter, we have dealt only with functions that have an inverse, and so you may have the impression that every function has an inverse, or that the task of determining the inverse of a function is always easy and straightforward. Neither of these statements is true. We will now study the conditions under which a function can have an inverse.

Table 5.1.3 lists the values of the function $f(x) = x^2$ for selected values of x. The inverse function would output the value of x such that $f(x) = y$. However, in this particular case, the value of 4 that is output by f corresponds to *two* input values: 2 and -2. Therefore, the inverse of f would have to output two different values (*both 2 and* -2) for the same input value of 4. Since *no* function is permitted to do that, we see that the function $f(x) = x^2$ does *not* have an inverse function.

It seems reasonable to assert that the only functions that have inverses are those for which different input values always produce different output values, since each value output by f would correspond to only *one* value input by f.

We now make the above discussion mathematically precise.

Table 5.1.3

x	$f(x) = x^2$
-2	4
-1	1
0	0
1	1
2	4

Definition of a One-to-One Function

For a function f to have an inverse function, f must be **one-to-one.** That is,

$$\text{if } f(a) = f(b), \text{ then } a = b.$$

Graphically, any horizontal line can cross the graph of a one-to-one function at most once. The reasoning is as follows: if there exists a horizontal line that crosses the graph of f more than once, that means a single output value of f corresponds to two different input values of f. Thus f is no longer one-to-one.

The above comment gives rise to the *horizontal line test.*

The Horizontal Line Test

A function f is one-to-one if every horizontal line intersects the graph of f at most once.

In Figure 5.1.3, the function f does *not* have an inverse because there are horizontal lines that intersect the graph of f more than once. The function h *has* an inverse because *every* horizontal line intersects the graph of h exactly once.

Figure 5.1.3 Horizontal line test

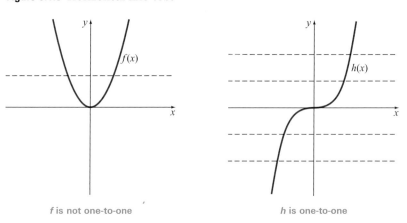

f is not one-to-one h is one-to-one

Example 4 **Checking Whether a Function Is One-to-One**

Which of the following functions are one-to-one and therefore have an inverse?

(a) $f(t) = -3t + 1$

(b) The function f given graphically in Figure 5.1.4

(c) The function f given by Table 5.1.4

Figure 5.1.4

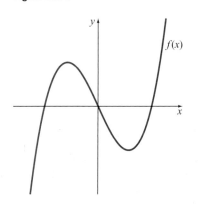

Table 5.1.4

t	$f(t)$
-4	17
-3	19
-1	0
1	21
5	0

▶ **Solution**

(a) To show that f is one-to-one, we show that if $f(a) = f(b)$, then $a = b$.

$$f(a) = f(b) \qquad \text{Assumption}$$
$$-3a + 1 = -3b + 1 \qquad \text{Evaluate } f(a) \text{ and } f(b)$$
$$-3a = -3b \qquad \text{Subtract 1 from each side}$$
$$a = b \qquad \text{Divide each side by } -3$$

We have shown that if $f(a) = f(b)$, then $a = b$. Thus f is one-to-one and does have an inverse.

(b) In Figure 5.1.5, we have drawn a horizontal line that intersects the graph of f more than once, and it is easy to see that there are other such horizontal lines. Thus, according to the horizontal line test, f is not one-to-one and does not have an inverse.

Figure 5.1.5 Horizontal line test

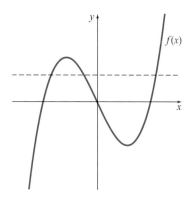

(c) Note that $f(-1) = 0 = f(5)$. Since two different inputs yield the same output, f is not one-to-one and does not have an inverse.

✔ *Check It Out 4:* Show that the function $f(x) = 4x - 6$ is one-to-one. ▪

Graph of a Function and Its Inverse

We have seen how to obtain some simple inverse functions algebraically. Visually, the graph of a function f and its inverse are *mirror images* of each other, with the line $y = x$ being the mirror. This property is stated formally as follows.

> **Graph of a Function and Its Inverse**
>
> The graphs of a function f and its inverse function f^{-1} are symmetric with respect to the line $y = x$.

This symmetry is illustrated in Example 5.

Example **5** **Graphing a Function and Its Inverse**

Graph the function $f(x) = -2x + 3$ and its inverse, $f^{-1}(x) = -\frac{1}{2}x + \frac{3}{2}$, on the same set of axes, using the same scale for both axes. What do you observe?

Technology Note

The functions f and f^{-1} from Example 5 and the line $y = x$ are graphed in Figure 5.1.7 using a standard window setting followed by the SQUARE option.

Keystroke Appendix:
Section 7

Figure 5.1.7

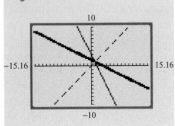

▶**Solution** The graphs of f and f^{-1} are shown in Figure 5.1.6. The inverse of f was computed in Example 3. Recall that the inverse of a function is found by using the output values of the original function as input, and using the input values of the original function as output. For instance, the points $(3, -3)$ and $(-1, 5)$ on the graph of $f(x) = -2x + 3$ are reflected to the points $(-3, 3)$ and $(5, -1)$, respectively, on the graph of $f^{-1}(x) = -\frac{1}{2}x + \frac{3}{2}$. It is this interchange of inputs and outputs that causes the graphs of f and f^{-1} to be symmetric with respect to the line $y = x$.

Figure 5.1.6

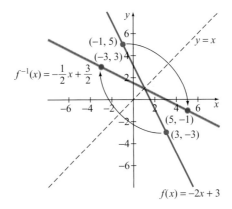

☑ *Check It Out 5:* Graph the function $f(x) = 3x - 2$ and its inverse, $f^{-1}(x) = \frac{1}{3}x + \frac{2}{3}$, on the same set of axes, using the same scale for both axes. What do you observe? ▪

Example **6** **Finding an Inverse Function and Its Graph**

Let $f(x) = x^3 + 1$.
(a) Show that f is one-to-one.   (b) Find the inverse of f.
(c) Graph f and its inverse on the same set of axes.

▶**Solution**
(a) The graph of $f(x) = x^3 + 1$, which is shown in Figure 5.1.8, passes the horizontal line test. Thus f is one-to-one.

Figure 5.1.8

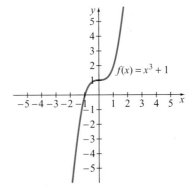

Just in Time

Review cube roots in Section P.3.

(b) To find the inverse, we proceed as follows.

Step 1 Start with the definition of the given function.

$$y = x^3 + 1$$

Step 2 Interchange the variables x and y.

$$x = y^3 + 1$$

Step 3 Solve for y.

$$x - 1 = y^3$$
$$\sqrt[3]{x - 1} = y$$

Step 4 The expression for the inverse function $f^{-1}(x)$ is now given by y.

$$f^{-1}(x) = \sqrt[3]{x - 1}$$

Thus the inverse of $f(x) = x^3 + 1$ is $f^{-1}(x) = \sqrt[3]{x - 1}$.

(c) Points on the graph of f^{-1} can be found by interchanging the x and y coordinates of points on the graph of f. Table 5.1.5 lists several points on the graph of f. The graphs of f and its inverse are shown in Figure 5.1.9.

Table 5.1.5

x	$f(x) = x^3 + 1$
-1.5	-2.375
-1	0
0	1
1.5	4.375

Figure 5.1.9

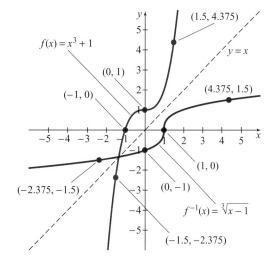

✔ *Check It Out 6:* Rework Example 6 for the function $f(x) = -x^3 + 2$. ▨

Restriction of Domain to Find an Inverse Function

We saw earlier that the function $f(x) = x^2$ does not have an inverse. We can, however, define a new function g from f by restricting the domain of g to only nonnegative numbers x, so that g will have an inverse. This technique is shown in the next example.

Example **7** **Restriction of Domain to Find an Inverse Function**

Show that the function $g(x) = x^2$, $x \geq 0$, has an inverse, and find the inverse.

▶**Solution** Recall that the function $f(x) = x^2$, whose domain consists of the set of all real numbers, has *no* inverse because f is not one-to-one. However, here we are examining the function g, which has the same function expression as f but is defined *only for $x \geq 0$*. From the graph of g in Figure 5.1.10, we see by the horizontal line test that g is one-to-one and thus has an inverse.

Figure 5.1.10

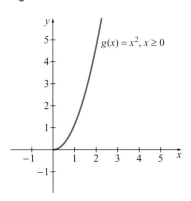

$g(x) = x^2, x \geq 0$

To find the inverse, we proceed as follows.

$$y = x^2, \quad x \geq 0 \qquad \textit{Definition of the given function}$$
$$x = y^2, \quad x \geq 0 \qquad \textit{Interchange the variables x and y}$$
$$\sqrt{x} = y, \quad x \geq 0 \qquad \textit{Take the square root of both sides}$$

Since $x \geq 0$, $\sqrt{x}$ is a real number. The inverse function g^{-1} is given by y.

$$g^{-1}(x) = \sqrt{x} = x^{1/2}, \quad x \geq 0$$

☑ *Check It Out 7:* Show that the function $f(x) = x^4$, $x \geq 0$, has an inverse, and find the inverse. ■

5.1 Key Points

▶ A function g is said to be the **inverse function** of f if the domain of g is equal to the range of f and, for every x in the domain of f and every y in the domain of g,

$$g(y) = x \quad \text{if and only if} \quad f(x) = y.$$

The notation for the inverse function of f is f^{-1}.

▶ To verify that two functions f and g are inverses of each other, check whether $(f \circ g)(x) = x$ and $(g \circ f)(x) = x$.

▶ A function f is **one-to-one** if $f(a) = f(b)$ implies $a = b$. For a function to have an inverse, it must be one-to-one.

▶ The graphs of a function f and its inverse function f^{-1} are symmetric with respect to the line $y = x$.

▶ Many functions that are not one-to-one can be restricted to an interval on which they are one-to-one. Their inverses are then defined on this restricted interval.

5.1 Exercises

▶**Just in Time Exercises** These exercises correspond to the Just in Time references in this section. Complete them to review topics relevant to the remaining exercises.

1. The composite function $f \circ g$ is defined as $(f \circ g)(x) = $ _____.

   (a) $f(g(x))$
   (b) $g(f(x))$

In Exercises 2–4, find $(f \circ g)(x)$.

2. $f(x) = x + 2,\ g(x) = x^2 - 2$

3. $f(x) = \dfrac{1}{x},\ g(x) = \dfrac{1}{x + 3}$

4. $f(x) = \sqrt{x},\ g(x) = (x + 4)^2$

In Exercises 5–8, simplify the expression.

5. $\sqrt[3]{x^3}$

6. $\sqrt[3]{7x^3}$

7. $\sqrt[3]{4\left(\dfrac{1}{4}x^3\right)}$

8. $\sqrt[3]{2\left(\dfrac{1}{2}x^3 + \dfrac{5}{2}\right) - 5}$

▶**Skills** This set of exercises will reinforce the skills illustrated in this section.

In Exercises 9–18, verify that the given functions are inverses of each other.

9. $f(x) = -x - 3;\ g(x) = -x - 3$

10. $f(x) = x + 7;\ g(x) = x - 7$

11. $f(x) = 6x;\ g(x) = \dfrac{1}{6}x$

12. $f(x) = -8x;\ g(x) = -\dfrac{1}{8}x$

13. $f(x) = -3x + 8;\ g(x) = -\dfrac{1}{3}x + \dfrac{8}{3}$

14. $f(x) = \dfrac{1}{2}x + 1;\ g(x) = 2x - 2$

15. $f(x) = x^3 + 2;\ g(x) = \sqrt[3]{x - 2}$

16. $f(x) = x^3 - 4;\ g(x) = \sqrt[3]{x + 4}$

17. $f(x) = x^2 + 3,\ x \geq 0;\ g(x) = \sqrt{x - 3}$

18. $f(x) = x^2 - 7,\ x \leq 0;\ g(x) = -\sqrt{x + 7}$

In Exercises 19–22, state whether each function given by a table is one-to-one. Explain your reasoning.

19.

x	$f(x)$
-3	6
-2	-8
0	0
1	8
3	-6

20.

x	$f(x)$
-3	4
-1	7
0	4
1	5
3	12

21.

x	$f(x)$
-2	-6
-1	5
0	9
1	4
2	9

22.

x	$f(x)$
-2	-9
-1	-8
0	-7
1	-6
2	-5

In Exercises 23–28, state whether each function given graphically is one-to-one.

23.

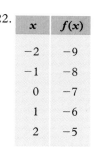

24.

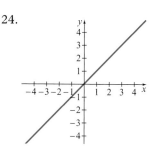

25.

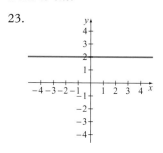

26.

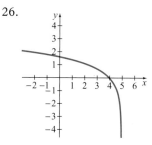

27.

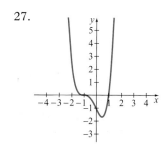

28.

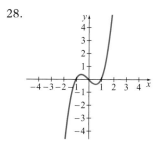

In Exercises 29–34, state whether each function is one-to-one.

29. $f(x) = -3x + 2$

30. $f(x) = \frac{4}{3}x + 1$

31. $f(x) = 2x^2 - 3$

32. $f(x) = -3x^2 + 1$

33. $f(x) = -2x^3 + 4$

34. $f(x) = -\frac{1}{3}x^3 - 5$

In Exercises 35–58, find the inverse of the given function. Then graph the given function and its inverse on the same set of axes.

35. $f(x) = -\frac{2}{3}x$

36. $g(x) = \frac{4}{3}x$

37. $f(x) = -4x + \frac{1}{5}$

38. $f(s) = 2s - \frac{9}{5}$

39. $f(x) = x^3 - 6$

40. $f(x) = -x^3 + 4$

41. $f(x) = \frac{1}{2}x - 4$

42. $f(x) = -\frac{3}{4}x + 2$

43. $g(x) = -x^2 + 8, x \geq 0$

44. $g(x) = -x^2 + 3, x \leq 0$

45. $g(x) = x^2 - 5, x \leq 0$

46. $g(x) = x^2 - 6, x \geq 0$

47. $f(x) = -2x^3 + 7$

48. $g(x) = 3x^3 - 5$

49. $f(x) = -4x^5 + 9$

50. $g(x) = 2x^5 - 6$

51. $f(x) = \frac{1}{x}$

52. $g(x) = \frac{-1}{2x}$

53. $g(x) = (x - 1)^2, x \geq 1$

54. $g(x) = (x + 2)^2, x \geq -2$

55. $f(x) = \sqrt{x + 3}, x \geq -3$

56. $f(x) = \sqrt{x - 4}, x \geq 4$

57. $f(x) = \frac{2x}{x - 1}$

58. $f(x) = \frac{x + 3}{x}$

In Exercises 59–62, the graph of a one-to-one function f is given. Draw the graph of the inverse function f^{-1}. Copy the given graph onto a piece of graph paper and use the line $y = x$ to help you sketch the inverse. Then give the domain and range of f and f^{-1}.

59.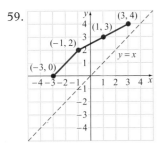

60.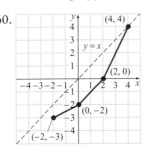

61.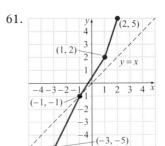

62.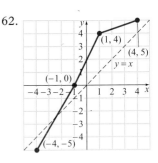

In Exercises 63–66, evaluate the given quantity by referring to the function f given in the following table.

x	f(x)
-2	1
-1	2
0	0
1	-1
2	-2

63. $f^{-1}(1)$

64. $f^{-1}(2)$

65. $f^{-1}(f^{-1}(-2))$

66. $f^{-1}(f^{-1}(1))$

In Exercises 67–70, evaluate the given quantity by referring to the function f given by the following graph.

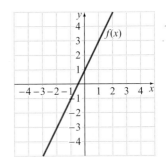

67. $f^{-1}(1)$

68. $f^{-1}(3)$

69. $f^{-1}(f^{-1}(-3))$

70. $f^{-1}(f^{-1}(3))$

▶**Applications** In this set of exercises, you will use inverse functions to study real-world problems.

71. **Converting Liquid Measures** Find a function that converts x gallons into quarts. Find its inverse and explain what it does.

72. **Shopping** When you buy products at a store, the Universal Product Code (UPC) is scanned and the price is output by a computer. The price is a function of the UPC. Why? Does this function have an inverse? Why or why not?

73. **Economics** In economics, the demand function gives the price p as a function of the quantity q. One example of a demand function is $p = 100 - 0.1q$. However, mathematicians tend to think of the price as the input variable and the quantity as the output variable. How can you take this example of a demand function and express q as a function of p?

74. **Physics** After t seconds, the height of an object dropped from an initial height of 100 feet is given by $h(t) = -16t^2 + 100$, $t \geq 0$.
   (a) Why does h have an inverse?
   (b) Write t as a function of h and explain what it represents.

75. **Fashion** A woman's dress size in the United States can be converted to a woman's dress size in France by using the function $f(s) = s + 30$, where s takes on all even values from 2 to 24, inclusive. (*Source:* www.onlineconversion.com)
   (a) What is the range of f?

   (b) Find the inverse of f and interpret it.

76. **Temperature** When measuring temperature, 100° Celsius (C) is equivalent to 212° Fahrenheit (F). Also, 0°C is equivalent to 32°F.
   (a) Find a linear function that converts Celsius temperatures to Fahrenheit temperatures.

   (b) Find the inverse of the function you found in part (a). What does this inverse function accomplish?

▶**Concepts** This set of exercises will draw on the ideas presented in this section and your general math background.

77. The following is the graph of a function f.

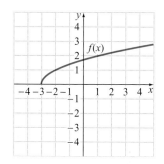

   (a) Find x such that $f(x) = 1$. You will have to approximate the value of x from the graph.
   (b) Let g be the inverse of f. Approximate $g(2)$ from the graph.
   (c) Sketch the graph of g.

78. Two students have an argument. One says that the inverse of the function f given by the expression $f(x) = 6$ is the function g given by the expression $g(x) = \frac{1}{6}$; the other claims that f has no inverse. Who is correct and why?

79. Do all linear functions have inverses? Explain.

80. If the graph of a function f is symmetric with respect to the y-axis, can f be one-to-one? Explain.

81. If a function f has an inverse and the graph of f lies in Quadrant IV, in which quadrant does the graph of f^{-1} lie?

82. If a function f has an inverse and the graph of f lies in Quadrant III, in which quadrant does the graph of f^{-1} lie?

83. Give an example of an odd function that is not one-to-one.

84. Give an example of a function that is its own inverse.

85. The function $f(x) = x^6$ is not one-to-one. How can the domain of f be restricted to produce a one-to-one function?

86. The function $f(x) = |x + 2|$ is not one-to-one. How can the domain of f be restricted to produce a one-to-one function?

5.2 Exponential Functions

Objectives

▶ Define an exponential function

▶ Sketch the graph of an exponential function

▶ Identify the main properties of an exponential function

▶ Define the natural exponential function

▶ Find an exponential function suitable to a given application

Bacteria such as *E. coli* reproduce by splitting into two identical pieces. This process is known as *binary fission*. If there are no other constraints to its growth, the bacteria population over time can be modeled by a function known as an **exponential function**. We will study the features of this type of function in this section. You will see that its properties are quite different from those of a polynomial or a rational function.

Example **1** **Modeling Bacterial Growth**

Example 1 in Section 5.3 builds upon this example. ⋯⟫

Suppose a bacterium splits into two bacteria every hour.

(a) Fill in Table 5.2.1, which shows the number of bacteria present, $P(t)$, after t hours.

Table 5.2.1

t (hours)	0	1	2	3	4	5	6	7	8
$P(t)$ (number of bacteria)									

(b) Find an expression for $P(t)$.

▶Solution

(a) Since we start with one bacterium and each bacterium splits into two bacteria every hour, the population is *doubled* every hour. This gives us Table 5.2.2.

Table 5.2.2

t (hours)	0	1	2	3	4	5	6	7	8
$P(t)$ (number of bacteria)	1	2	4	8	16	32	64	128	256

(b) To find an expression for $P(t)$, note that the number of bacteria present after t hours will be double the number of bacteria present an hour earlier. This gives

$$P(1) = 2(1) = 2^1; \quad P(2) = 2(P(1)) = 2(2) = 4 = 2^2;$$
$$P(3) = 2(P(2)) = 2(2^2) = 2^3; \quad P(4) = 2(P(3)) = 2(2^3) = 2^4; \ldots$$

Following this pattern, we find that $P(t) = 2^t$. Here, the *independent variable*, t, is in the *exponent*. This is quite different from the functions we examined in the previous chapters, where the independent variable was raised to a *fixed power*. The function $P(t)$ is an example of an exponential function.

✔ *Check It Out 1:* In Example 1, evaluate $P(9)$ and interpret your result. ■

Next we give the formal definition of an exponential function, and examine its properties.

Just in Time

Review properties of exponents in Sections P.2 and P.3.

Definition of an Exponential Function

We now briefly recall some properties of exponents. You already know how to calculate quantities such as 2^3 or $1.5^{1/2}$ or $3^{2/3}$. In each of these expressions, the exponent is either an integer or a rational number. Actually, *any real number* can be used as an exponent in an expression of the form Ca^b (where a, b, and C are real numbers), provided certain conditions are satisfied: a must be a nonnegative number whenever the exponent b is (1) an irrational number or (2) a rational number of the form $\frac{p}{q}$, where p and q are integers and q is even. Also, b must be nonzero if $a = 0$. For example, the expression $3^{\sqrt{2}}$ represents a real number, but the expressions $(-2)^{\sqrt{3}}$ and 0^0 do not.

We will take these general properties of exponents for granted, since their verification is beyond the scope of this discussion. All the properties of integer and rational exponents apply to real-valued exponents as well.

Motivated by the above discussion and Example 1, we now present a definition of an exponential function.

Definition of an Exponential Function

An **exponential function** is a function of the form

$$f(t) = Ca^t$$

where a and C are constants such that $a > 0$, $a \neq 1$, and $C \neq 0$. The domain of the exponential function is the set of all real numbers. The range will vary depending on the values of C and a.

The number a is known as the **base** of the exponential function.

In the following two examples, we graph some exponential functions.

Example 2 Graphing an Exponential Function

Make a table of values for the exponential function $f(x) = 2^x$. Use the table to sketch the graph of the function. What happens to the value of the function as $x \to \pm\infty$? What is the range of the function?

▶**Solution** We first make a table of values for $f(x)$. See Table 5.2.3. Note that the domain of f is the set of all real numbers, and so there are no specific values of the independent variable that must be excluded. (In Example 1, we excluded negative values of the independent variable, since a negative number of hours makes no sense.)

We then plot the points and connect them with a smooth curve. See Figure 5.2.1.

Technology Note

When graphing an exponential function, you will need to adjust the window size so that you can see how rapidly the y-value increases. In Figure 5.2.2, the graph of $f(x) = 2^x$ uses a window size of $[-5, 5]$ by $[0, 35](5)$.

Keystroke Appendix:
Section 7

Figure 5.2.2

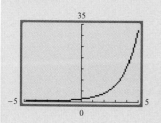

Table 5.2.3

x	$f(x) = 2^x$
-10	$2^{-10} = \dfrac{1}{2^{10}} \approx 0.000977$
-5	$2^{-5} = \dfrac{1}{2^5} = 0.03125$
-2	$2^{-2} = 0.25$
-1	$2^{-1} = 0.5$
0	$2^0 = 1$
1	$2^1 = 2$
2	$2^2 = 4$
5	$2^5 = 32$
10	$2^{10} = 1024$

Figure 5.2.1

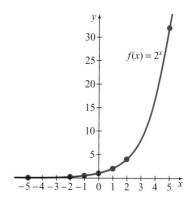

Observations:

▶ This function represents an example of exponential *growth*, since the value of the function increases as x increases.

▶ As $x \to +\infty$, the value of $f(x)$ gets very large.

▶ As $x \to -\infty$, the value of $f(x)$ gets extremely small but never reaches zero. For example, $2^{-1000} = \frac{1}{2^{1000}}$, which is quite small but still positive.

Note that when you use a calculator, you may sometimes get 0 instead of an extremely small value. This is because of the limited precision of the calculator. It does *not* mean that the actual value is zero!

▶ The graph of $f(x) = 2^x$ has a *horizontal asymptote* at $y = 0$. This means that the graph of f gets very close to the line $y = 0$, but never touches it.

Range of Function To determine the range of f, we note the following: 2 raised to any power is positive, and every positive number can be expressed as 2 raised to some power. Thus the range of f is the set of all positive numbers, or $(0, \infty)$ in interval notation.

Discover *and* **Learn**

For $g(x) = 2x$, make a table of function values for x ranging from -10 to 10. Sketch the graph of g. How does it differ from the graph of $f(x) = 2^x$?

✔ *Check It Out 2:* Rework Example 2 for the function $g(x) = 3^x$. ■

We now make some general observations about the graphs of exponential functions.

Properties of Exponential Functions

Given an *exponential function* $f(x) = Ca^x$ with $C > 0$, the function will exhibit one of the following two types of behavior, depending on the value of the base a:

If $a > 1$ and $C > 0$:

$f(x) = Ca^x \to +\infty$ as $x \to +\infty$

The *domain* is the set of *all real numbers*.

The *range* is the set of *all positive numbers*.

The x-axis is a *horizontal asymptote;* the graph of f approaches the x-axis as $x \to -\infty$, but does not touch or cross it.

The function is *increasing* on $(-\infty, \infty)$ and illustrates *exponential growth.*

See Figure 5.2.3.

Figure 5.2.3

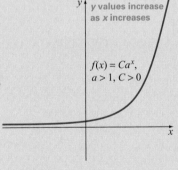

If $0 < a < 1$ and $C > 0$:

$f(x) = Ca^x \to 0$ as $x \to +\infty$

The *domain* is the set of *all real numbers*.

The *range* is the set of *all positive numbers*.

The x-axis is a *horizontal asymptote;* the graph of f approaches the x-axis as $x \to +\infty$, but does not touch or cross it.

The function is *decreasing* on $(-\infty, \infty)$ and illustrates *exponential decay.*

See Figure 5.2.4.

Figure 5.2.4

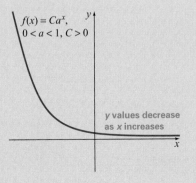

Example 3 Graphing an Exponential Function

Make a table of values for the exponential function $f(x) = 5\left(\frac{1}{3}\right)^x = 5(3^{-x})$. Use the table to sketch a graph of the function. What happens to the value of the function as $x \to \pm\infty$? What is the range of the function?

▶ **Solution** With the help of a calculator, we can make a table of values for $f(x)$, as shown in Table 5.2.4. Note that the domain of f is the set of all real numbers, and so there are no specific values of the independent variable that must be excluded.

We then plot the points and connect them with a smooth curve. See Figure 5.2.5.

Figure 5.2.5

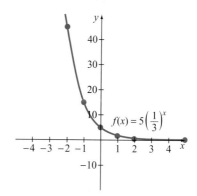

Table 5.2.4

x	$f(x) = 5\left(\dfrac{1}{3}\right)^x$
-10	$5\left(\dfrac{1}{3}\right)^{-10} = 295{,}245$
-5	$5\left(\dfrac{1}{3}\right)^{-5} = 1215$
-2	$5\left(\dfrac{1}{3}\right)^{-2} = 45$
-1	$5\left(\dfrac{1}{3}\right)^{-1} = 15$
0	$5\left(\dfrac{1}{3}\right)^{0} = 5$
1	$5\left(\dfrac{1}{3}\right)^{1} \approx 1.6667$
2	$5\left(\dfrac{1}{3}\right)^{2} \approx 0.5556$
5	$5\left(\dfrac{1}{3}\right)^{5} \approx 0.0206$
10	$5\left(\dfrac{1}{3}\right)^{10} \approx 0.0000847$

Observations:

▶ This function represents an example of exponential *decay*, since the value of the function decreases as x increases.

▶ Note that as $x \to -\infty$, the value of $f(x)$ gets very large. For example, if $x = -1000$, then $5\left(\frac{1}{3}\right)^{-1000} = 5(3^{1000})$, which is quite large.

▶ As $x \to +\infty$, the value of $f(x)$ gets extremely small but never reaches zero. For example, $5\left(\frac{1}{3}\right)^{1000} = \frac{5}{3^{1000}}$, which is quite small but still positive.

▶ The graph of $f(x) = 5\left(\frac{1}{3}\right)^x$ has a *horizontal asymptote* at $y = 0$, since the graph of f gets very close to the line $y = 0$ but never touches it.

Range of Function To determine the range of f, we note the following: $\frac{1}{3}$ raised to any power is positive, and every positive number can be expressed as $\frac{1}{3}$ raised to some power. Since $\left(\frac{1}{3}\right)^x$ is multiplied by the positive number 5, the range of f is the set of all positive numbers, or $(0, \infty)$ in interval notation. You can also see this graphically.

☑ *Check It Out 3:* Rework Example 3 for the function $g(x) = 6^{-x}$. ■

Example 4 Graphing an Exponential Function

Make a table of values for the function $h(x) = -(2)^{2x}$ and sketch a graph of the function. Find the domain and range. Describe the behavior of the function as x approaches $\pm\infty$.

▶Solution Note that

$$h(x) = -2^{2x}$$
$$= -(2^2)^x$$
$$= -4^x.$$

We can make a table of values of $h(x)$, as shown in Table 5.2.5. We then plot the points and connect them with a smooth curve. See Figure 5.2.6.

Be careful when calculating the values of the function. For example,

$$h(-2) = -[(4)^{-2}]$$
$$= -\left(\frac{1}{4^2}\right)$$
$$= -0.0625.$$

The negative sign in front of the 4 is applied only after the exponentiation is performed.

Table 5.2.5

x	$h(x) = -2^{2x} = -4^x$
-10	-9.536×10^{-7}
-5	-0.000977
-2	-0.0625
0	-1
1	-4
2	-16
5	-1024
10	$-1,048,576$

Figure 5.2.6

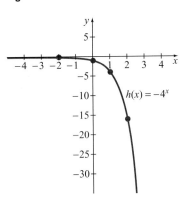

Observations:

▶ The y-intercept is $(0, -1)$.

▶ The domain of h is the set of all real numbers.

▶ From the sketch of the graph, we see that the *range* of h is the set of all *negative real numbers*, or $(-\infty, 0)$ in interval notation.

▶ As $x \to +\infty$, $h(x) \to -\infty$.

▶ As $x \to -\infty$, $h(x) \to 0$. Thus, the *horizontal asymptote* is the line $y = 0$.

✔ *Check It Out 4:* Make a table of values for the function $h(s) = -(3)^s$ and sketch a graph of the function. Find the domain and range. Describe the behavior of the function as the independent variable approaches $\pm\infty$. ▓

The Number *e* and the Natural Exponential Function

There are some special numbers that occur frequently in the study of mathematics. For instance, in geometry, you would have encountered π, an irrational number. Recall that an irrational number is one that cannot be written in the form of a terminating decimal or a repeating decimal.

Another irrational number that occurs frequently is the number e, which is defined as the number that the quantity $\left(1 + \frac{1}{x}\right)^x$ approaches as x approaches infinity. The nonterminating, nonrepeating decimal representation of the number e is

$$e = 2.7182818284\ldots.$$

The fact that the quantity $\left(1 + \frac{1}{x}\right)^x$ tends to level off as x increases can be seen by examining the graph of $A(x) = \left(1 + \frac{1}{x}\right)^x$, shown in Figure 5.2.7.

Figure 5.2.7

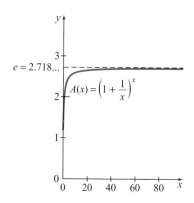

We can define an exponential function with e as the base. The graph of the exponential function $f(x) = e^x$ has the same general shape as that of $f(x) = 2^x$ or $f(x) = 3^x$.

Example **5** **Exponential Function with Base e**

Make a table of values for the function $g(x) = e^x + 3$ and sketch a graph of the function. Find the y-intercept, domain, and range. Describe the behavior of the function as x approaches $\pm\infty$.

▶Solution Make a table of values for $g(x)$ by choosing various values of x, as shown in Table 5.2.6. Use the e^x button on your calculator to help find the function values. Then plot the points and connect them with a smooth curve. See Figure 5.2.8.

Table 5.2.6

x	$g(x) = e^x + 3$
-10	3.000
-5	3.007
-2	3.135
-1	3.368
0	4.000
1	5.718
2	10.389
5	151.413
10	22,029.466

Figure 5.2.8

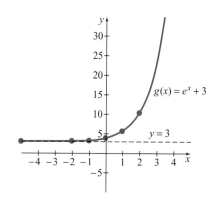

Discover *and* Learn

Use a graphing utility to graph the functions $f(x) = e^x$, $g(x) = 3^x$, and $h(x) = (1.05)^x$, with x ranging from -5 to 5. Which function rises most steeply? Which function rises least steeply? Where does the graph of $f(x) = e^x$ lie relative to the graphs of $g(x) = 3^x$ and $h(x) = (1.05)^x$? Why do all three graphs have the same y-intercept?

Observations:

▶ The y-intercept is $(0, 4)$.

▶ The *domain* of g is the set of *all real numbers*. Since the graph of g is always above the line $y = 3$ but comes closer and closer to it as the value of x decreases, the *range* of g is the set of *all real numbers strictly greater than 3*, or $(3, \infty)$.

▶ The graph of $g(x) = e^x + 3$ is the graph of e^x shifted upward by 3 units.

▶ As $x \to +\infty$, $g(x) \to \infty$.

▶ As $x \to -\infty$, $g(x) \to 3$. Thus, the *horizontal asymptote* is $y = 3$.

✔ *Check It Out 5:* Rework Example 5 for the function $h(x) = e^{-x}$. ■

Applications of Exponential Functions

Exponential functions are extremely useful in a variety of fields, including finance, biology, and the physical sciences. This section will illustrate the usefulness of exponential functions in analyzing some real-world problems. We will study additional applications in later sections on exponential equations and exponential and logarithmic models.

Example 6 Depreciation of an Automobile

Depreciation is a process by which something loses value. For example, the depreciation rate of a Honda Civic (two-door coupe) is about 8% per year. This means that each year the Civic will *lose* 8% of the value it had the previous year. If the Honda Civic was purchased for $10,000, make a table of its value over the first 5 years after purchase. Find a function that gives its value t years after purchase, and sketch a graph of the function. (*Source:* Kelley Blue Book)

▶**Solution** Note that if the car loses 8% of its value each year, then it retains 92% of its value from the previous year. Using this fact, we can generate Table 5.2.7.

Technology Note

Use a table of values to find a suitable window to graph $Y_1(x) = 10000(0.92)^x$. One possible window size is $[0, 30](5)$ by $[0, 11000](1000)$. See Figure 5.2.9.

Keystroke Appendix: Sections 6 and 7

Figure 5.2.9

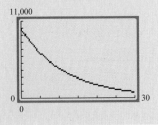

Table 5.2.7

Years Since Purchase	Expression for Value	Value
0	10,000	$10,000.00
1	$0.92 \cdot 10,000$	$9200.00
2	$0.92(0.92 \cdot 10,000) = 0.92^2 (10,000)$	$8464.00
3	$0.92(0.92^2 \cdot 10,000) = 0.92^3 (10,000)$	$7786.88
4	$0.92(0.92^3 \cdot 10,000) = 0.92^4 (10,000)$	$7163.93
5	$0.92(0.92^4 \cdot 10,000) = 0.92^5 (10,000)$	$6590.82

We would like to find a function of the form $v(t) = Ca^t$, where $v(t)$ is the value of the Honda Civic t years after purchase. Note that $v(0) = C = 10,000$. From Table 5.2.7, we see that $a = 0.92$ because this is the factor that relates the value of the car at the end of a given year to its value at the end of the previous year. Thus

$$v(t) = Ca^t = 10,000(0.92)^t.$$

Since $0 < 0.92 < 1$, we can expect this function to decrease over time. This function represents an example of **exponential decay,** as is confirmed by sketching the graph of $v(t)$. See Figure 5.2.10.

Figure 5.2.10

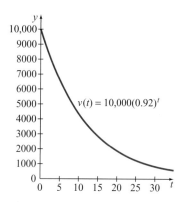

$$v(t) = 10,000(0.92)^t$$

✔ *Check It Out 6:* If a Ford Focus was purchased for $12,000 and depreciates at a rate of 10% per year, find an exponential function that gives the value of the Focus t years after purchase, and sketch a graph of the function. ■

Observations:

Note the following differences between a linear model of depreciation (see Section 1.4) and an exponential model of depreciation.

▶ In a linear model of depreciation, a fixed dollar amount is *subtracted* each year from the previous year's value. In an exponential model, a positive constant less than 1 is *multiplied* each year by the previous year's value.

▶ For an exponential model of depreciation, the value *never* reaches zero; for a linear model, the value eventually *does* reach zero. In this sense, an exponential model of depreciation is more realistic than a linear model.

We will now consider an example of an exponential function that occurs in banking.

Example 7 An Application of Exponential Functions

A bank has advertised an interest rate of 5% compounded annually. That is, at the end of each year, all the interest earned during that year is added to the principal, and so the interest for that year will be included in the balance on which the following year's interest is computed. If $100 is invested at the beginning of the year, how much money will be in the account after (a) 1 year? (b) 2 years? (c) t years? Assume that no withdrawals or additional deposits are made.

▶Solution

(a) After the first year, the total amount of money in the account will be

$$\text{Total} = \text{amount invested} + \text{interest earned in first year}$$

$= 100 + (0.05)(100)$ Interest calculated on $100

$= 100(1 + 0.05)$ Factor out 100

$= 100(1.05) = 105.$ Simplify

(b) In the second year, the interest will be calculated on the total amount of money in the account at the end of the first year. This includes the interest earned during the first year. Thus, the *total amount of money in the account after 2 years* will be

Total = balance at end of first year + interest earned in second year

$$= 105 + (0.05)(105)$$ Interest calculated on $105 (total balance at end of first year)

$$= 105(1 + 0.05)$$ Factor out 105

$$= 105(1.05) = 110.25.$$ Simplify

Note that the amount of money in the account at the end of the second year can be written as $100(1.05)^2$. Writing it in this form will be helpful in the next part of the discussion.

(c) To calculate the total amount of money in the account after t years, it is useful to make a table, since the process above will be repeated many times. We write the expressions in exponential form so that we can observe a pattern. See Table 5.2.8.

Table 5.2.8

Year	Amount at Start of Year	Interest Earned During Year	Amount at End of Year
1	100	$(0.05)(100)$	$100(1.05)$
2	$100(1.05)$	$(0.05)(100(1.05))$	$100(1.05)(1 + 0.05) = 100(1.05)^2$
3	$100(1.05)^2$	$(0.05)(100(1.05)^2)$	$100(1.05)^2(1 + 0.05) = 100(1.05)^3$
4	$100(1.05)^3$	$(0.05)(100(1.05)^3)$	$100(1.05)^3(1 + 0.05) = 100(1.05)^4$
⋮	⋮	⋮	⋮

The dots indicate that the table can be continued indefinitely. From the numbers we have listed so far, observe that the amount in the account after 2 years is $100(1.05)^2$; after three years, it is $100(1.05)^3$; and after 4 years, it is $100(1.05)^4$. The amount present at the end of each year is 1.05 times the amount present at the beginning of the year.

We can conclude from this pattern that the *amount of money in the account after t years* will be $A(t) = 100(1.05)^t$ dollars. This function represents an instance of **exponential growth.**

☑ *Check It Out 7:* In Example 7, how much money will be in the account after 10 years? ■

Discover *and* Learn

Evaluate the function $f(t) = 100(1.05)^t$ for $t = 0, 10, 20,$ and 30. Do the same for $g(t) = 100 + 1.05t$. Make a table to summarize the values. What observations can you make?

Note   Example 7 shows how an amount of money invested can grow over a period of time. Note that the amount in the account at the beginning of each year is multiplied by a constant of 1.05 to obtain the amount in the account at the end of that year.

In the preceding discussion, interest was compounded annually. It can also be compounded in many other ways—quarterly, monthly, daily, and so on. The following is a general formula that applies to interest compounded n times a year.

Discover *and* **Learn**

Use a graphing utility to sketch graphs of the functions
$f(t) = 100(1.05)^t$ and
$g(t) = 100 + 1.05t$ on the same set of axes, with t ranging from 0 to 30. You have to be careful with the scaling of the y-axis so that the graph of g is not "squashed." What do you observe?

Compounded Interest

Suppose an amount P is invested in an account that pays interest at rate r, and the interest is compounded n times a year. Then, after t years, the amount in the account will be

$$A(t) = P\left(1 + \frac{r}{n}\right)^{nt}.$$

When interest is compounded *continuously* (i.e., the interest is compounded as soon as it is earned rather than being compounded at discrete intervals of time, such as once a month or once a year), a different formula (using e as the base) is needed to calculate the total amount of money in the account.

Continuously Compounded Interest

Suppose an amount P is invested in an account that pays interest at rate r, and the interest is compounded *continuously*. Then, after t years, the amount in the account will be

$$A(t) = Pe^{rt}.$$

Example 8 Computing the Value of a Savings Account

Suppose $2500 is invested in a savings account. Find the following quantities.

(a) The amount in the account after 4 years if the interest rate is 5.5% compounded monthly

(b) The amount in the account after 4 years if the interest rate is 5.5% compounded continuously

▶ Solution

(a) Here, $P = 2500$, $r = 0.055$, $t = 4$, and $n = 12$. Substituting this data, we obtain

$$A(t) = P\left(1 + \frac{r}{n}\right)^{nt} = 2500\left(1 + \frac{0.055}{12}\right)^{(12)(4)} \approx 3113.63.$$

There will be $3113.63 in the account after 4 years if the interest is compounded monthly.

(b) Here, $P = 2500$, $r = 0.055$, and $t = 4$. Since the interest is compounded continuously, we have
$$A(t) = Pe^{rt} = 2500e^{0.055(4)} \approx 3115.19.$$

There will be $3115.19 in the account after 4 years if the interest is compounded continuously. Note that this amount is just slightly more than the amount obtained in part (a).

✔ *Check It Out 8:* Suppose $3000 is invested in a savings account. Find the following quantities.

(a) The amount in the account after 3 years if the interest rate is 6.5% compounded monthly

(b) The amount in the account after 3 years if the interest rate is 6.5% compounded continuously ▪

Example 9 **Finding Doubling Time for an Investment**

Use a graphing utility to find out how long it will take an investment of $2500 to double if the interest rate is 5.5% compounded monthly.

▶Solution Since $r = 0.055$ and $n = 12$, the expression for the amount in the account after t years is given by

$$A(t) = P\left(1 + \frac{r}{n}\right)^{nt} = 2500\left(1 + \frac{0.055}{12}\right)^{12t}.$$

We are interested in the value of t for which the total amount in the account will be equal to twice the initial investment. The initial investment is $2500, so twice this amount is $5000. We must therefore solve the equation

$$5000 = 2500\left(1 + \frac{0.055}{12}\right)^{12t}.$$

This equation cannot be solved by any of the algebraic means studied so far. However, you can solve it by using the INTERSECT feature of your graphing utility with $y_1(t) = 5000$ and $y_2(t) = 2500\left(1 + \frac{0.055}{12}\right)^{12t}$. You will need to choose your horizontal and vertical scales appropriately by using a table of values. As shown in Figure 5.2.11, a window size of [0, 15] by [2500, 5500](250) works well.

Figure 5.2.11

X	Y₁	Y₂
7	5000	3670.8
8	5000	3877.9
9	5000	4096.6
10	5000	4327.7
11	5000	4571.8
12	5000	4829.7
13	5000	5102.1

X=13

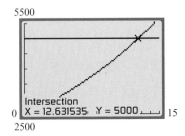

The solution is $t \approx 12.63$. Thus it will take approximately 12.63 years for the initial investment of $2500 to double, at a rate of 5.5% compounded monthly.

☑ *Check It Out 9:* Use a graphing utility to find out how long it will take an investment of $3500 to double if the interest rate is 6% compounded monthly. ■

Note The powerful features of your graphing utility can enable you to solve equations for which you may not yet know an algebraic technique. However, you need good analytical skills to approximate an accurate solution. These skills include setting up a good viewing window so that you can see the point(s) of intersection of a pair of graphs. Knowledge of the behavior of the functions involved is crucial to finding a suitable window.

5.2 Key Points

▸ An **exponential function** is a function of the form

$$f(x) = Ca^x$$

where a and C are constants such that $a > 0$, $a \neq 1$, and $C \neq 0$.

▸ If $a > 1$ and $C > 0$, $f(x) = Ca^x \to +\infty$ as $x \to +\infty$. The function is **increasing** on $(-\infty, \infty)$ and illustrates **exponential growth.**

▸ If $0 < a < 1$ and $C > 0$, $f(x) = Ca^x \to 0$ as $x \to +\infty$. The function is **decreasing** on $(-\infty, \infty)$ and represents **exponential decay.**

▸ Suppose an amount P is invested in an account that pays interest at rate r, and the interest is compounded n times a year. Then, after t years, the amount in the account will be

$$A(t) = P\left(1 + \frac{r}{n}\right)^{nt}.$$

▸ Suppose an amount P is invested in an account that pays interest at rate r, and the interest is compounded *continuously*. Then, after t years, the amount in the account will be

$$A(t) = Pe^{rt}.$$

5.2 Exercises

▸**Just in Time Exercises** These exercises correspond to the Just in Time references in this section. Complete them to review topics relevant to the remaining exercises.

In Exercises 1–6, evaluate the expression.

1. 5^3

2. $8^{1/3}$

3. 2^{-2}

4. $3^{1/2}$

5. $2(3^2)$

6. $2^3 2^5$

▸**Skills** This set of exercises will reinforce the skills illustrated in this section.

In Exercises 7–16, evaluate each expression to four decimal places using a calculator.

7. $2.1^{1/3}$

8. $3.2^{1/2}$

9. $4^{1.6}$

10. $6^{2.5}$

11. $3^{\sqrt{2}}$

12. $2^{\sqrt{3}}$

13. e^3

14. e^6

15. $e^{-2.5}$

16. $e^{-3.2}$

In Exercises 17–36, sketch the graph of each function.

17. $f(x) = 4^x$

18. $f(x) = 5^x$

19. $g(x) = \left(\frac{1}{4}\right)^x$

20. $g(x) = \left(\frac{1}{5}\right)^x$

21. $f(x) = 2(3)^{-x}$

22. $f(x) = 4(2)^{-x}$

23. $f(x) = 2e^x$

24. $g(x) = 5e^x$

25. $f(x) = 2 + 3e^x$

26. $f(x) = 5 + 2e^x$

27. $g(x) = 10(2)^x$

28. $h(x) = -5(3)^x$

29. $f(x) = -2\left(\frac{1}{3}\right)^x$

30. $h(x) = 4\left(\frac{2}{3}\right)^x$

31. $f(x) = 3^{2x}$

32. $g(x) = 2^{3x}$

33. $f(x) = -4(3)^x + 1$

34. $f(x) = -2(3)^x + 1$

35. $f(x) = 2^{-x} - 1$

36. $f(x) = 3^{-x} + 1$

In Exercises 37–44, sketch the graph of each function and find (a) the y-intercept; (b) the domain and range; (c) the horizontal asymptote; and (d) the behavior of the function as x approaches ±∞.

37. $f(x) = -5^x$

38. $f(x) = 2^{-x}$

39. $f(x) = 3 - 2^x$

40. $g(x) = 6 - 5^x$

41. $f(x) = 7e^x$

42. $g(x) = -4e^{2x}$

43. $g(x) = 3e^{-x} - 4$

44. $h(x) = 10e^{-x} + 2$

In Exercises 45–48, state whether the graph represents an exponential function of the form $f(x) = Ca^x$, $a > 0$, $a \neq 1$, $C \neq 0$. Explain your reasoning.

45.
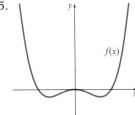

46.
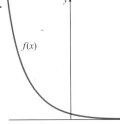

47.
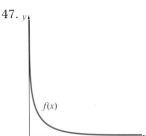

48.
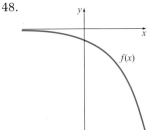

In Exercises 49–52, match the graph with one of the exponential functions. (a) $f(x) = -2^x$; (b) $g(x) = 2^x$; (c) $h(x) = 2^x - 3$; (d) $k(x) = -2^x + 3$

49.
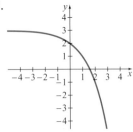

50.
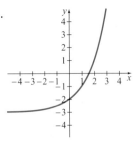

51.
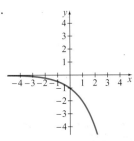

52.
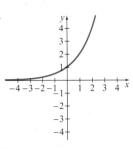

 In Exercises 53–58, use a graphing utility to solve each equation for x.

53. $5 = 3^x$

54. $7 = 4^x$

55. $10 = 2^{-x}$

56. $20 = 100(5)^{-x}$

57. $100 = 50e^{0.06x}$

58. $25 = 50e^{-0.05x}$

59. Consider the function $f(x) = xe^{-x}$.
   (a) Use a graphing utility to graph this function, with x ranging from −5 to 5. You may need to scroll through the table of values to set an appropriate scale for the vertical axis.
   (b) What are the domain and range of f?
   (c) What are the x- and y-intercepts, if any, of the graph of this function?
   (d) Describe the behavior of the function as x approaches ±∞.

60. Consider the function $f(x) = e^{-x^2}$.
   (a) Use a graphing utility to graph this function, with x ranging from −5 to 5. You may need to scroll through the table of values to set an appropriate scale for the vertical axis.
   (b) What are the domain and range of f?.
   (c) Does f have any symmetries?
   (d) What are the x- and y-intercepts, if any, of the graph of this function?
   (e) Describe the behavior of the function as x approaches ±∞.

▶**Applications** In this set of exercises, you will use exponential functions to study real-world problems.

Compound Interest *In Exercises 61–64, for an initial deposit of $1500, find the total amount in a bank account after 5 years for the interest rates and compounding frequencies given.*

61. 6% compounded annually

62. 3% compounded semiannually

63. 6% compounded monthly

64. 3% compounded quarterly

Banking *In Exercises 65–68, for an initial deposit of $1500, find the total amount in a bank account after t years for the interest rates and values of t given, assuming continuous compounding of interest.*

65. 6% interest; $t = 3$

66. 7% interest; $t = 4$

67. 3.25% interest; $t = 5.5$

68. 4.75% interest; $t = 6.5$

In Exercises 69–72, fill in the table according to the given rule and find an expression for the function represented by the rule.

69. **Salary** The annual salary of an employee at a certain company starts at $10,000 and is increased by 5% at the end of every year.

Years at Work	Annual Salary
0	
1	
2	
3	
4	

70. **Population Growth** A population of cockroaches starts out at 100 and doubles every month.

Month	Population
0	
1	
2	
3	
4	

71. **Depreciation** An automobile purchased for $20,000 depreciates at a rate of 10% per year.

Years Since Purchase	Value
0	
1	
2	
3	
4	

72. **Ecology** A rainforest with a current area of 10,000 square kilometers loses 5% of its area every year.

Years in the Future	Area of Rainforest (km²)
0	
1	
2	
3	
4	

73. **Depreciation** The depreciation rate of a Mercury Sable is about 30% per year. If the Sable was purchased for $18,000, make a table of its values over the first 5 years after purchase. Find a function that gives its value t years after purchase, and sketch a graph of the function. (*Source:* Kelley Blue Book)

74. **Depreciation** The depreciation rate of a Toyota Camry is about 8% per year. If the Camry was purchased for $25,000, make a table of its values over the first 4 years after purchase. Find a function that gives its value t years after purchase, and sketch a graph of the function. (*Source:* Kelley Blue Book)

75. **Savings Bonds** U.S. savings bonds, Series EE, pay interest at a rate of 3% compounded quarterly. How much would a bond purchased for $1000 be worth after 10 years? These bonds stop paying interest after 30 years. Why do you think this is so? (*Hint:* Think about how much this bond would be worth after 80 years.)

76. **Savings Bonds** U.S. savings bonds, Series EE, are purchased at half their face value (for example, a $50 savings bond would be purchased for $25) and are guaranteed to reach their face value after 17 years. If a Series EE savings bond with a face value of $500 is held until it reaches its face value, what is the interest rate? The interest on savings bonds is compounded quarterly. You will need to use a graphing utility to solve this problem.

77. **Salary** The average hourly wage for construction workers was $17.48 in 2000 and has risen at a rate of 2.7% annually. (*Source:* Bureau of Labor Statistics)

(a) Find an expression for the average hourly wage as a function of time t. Measure t in years since 2000.

(b) Using your answer to part (a), make a table of predicted values for the average hourly wage for the years 2000–2007.

(c) The actual average hourly wage for 2003 was $18.95. How does this value compare with the predicted value found in part (b)?

78. **Pharmacology** When a drug is administered orally, the amount of the drug present in the bloodstream of the patient can be modeled by a function of the form

$$C(t) = ate^{-bt}$$

where $C(t)$ is the concentration of the drug in milligrams per liter (mg/L), t is the number of hours since the drug was administered, and a and b are positive constants. For a 300-milligram dose of the asthma drug aminophylline, this function is

$$C(t) = 4.5te^{-0.275t}.$$

(*Source: Merck Manual of Diagnosis and Therapy*)

(a) How much of this drug is present in the bloodstream at time $t = 0$? Why does this answer make sense in the context of the problem?

(b) How much of this drug is present in the bloodstream after 1 hour?

(c) Sketch a graph of this function, either by hand or using a graphing utility, with t ranging from 0 to 20.

(d) What happens to the value of the function as $t \to \infty$? Does this make sense in the context of the problem? Why?

(e) Use a graphing utility to find the time when the concentration of this drug reaches its maximum.

(f) Use a graphing utility to determine when the concentration of this drug reaches 3 mg/L for the second time. (This will occur after the concentration peaks.)

79. **Design** The height (in feet) of the point on the Gateway Arch in Saint Louis that is directly above a given point along the base of the arch can be written as a function of the distance x (also in feet) of the latter point from the midpoint of the base:

$$h(x) = -34.38(e^{-0.01x} + e^{0.01x}) + 693.76$$

(*Source:* National Park Service)

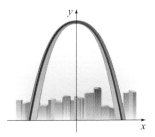

(a) What is the maximum value of this function?

(b) Evaluate $h(100)$.

(c) Graph the function $h(x)$ using a graphing utility. Choose a suitable window size so that you can see the entire arch. For what value(s) of x is $h(x)$ equal to 300 feet?

▶ **Concepts** This set of exercises will draw on the ideas presented in this section and your general math background.

80. In the definition of the exponential function, why is $a = 1$ excluded?

81. Consider the function $f(x) = 2 + e^{-x}$.
(a) What number does $f(x)$ approach as $x \to +\infty$?
(b) How could you use the graph of this function to confirm the answer to part (a)?

82. The graph of the function $f(x) = Ca^x$ passes through the points $(0, 12)$ and $(2, 3)$.
(a) Use $f(0)$ to find C.
(b) Is this function increasing or decreasing? Explain.
(c) Now that you know C, use $f(2)$ to find a. Does your value of a confirm your answer to part (b)?

83. Consider the two functions $f(x) = 2x$ and $g(x) = 2^x$.
(a) Make a table of values for $f(x)$ and $g(x)$, with x ranging from -1 to 4 in steps of 0.5.
(b) Find the interval(s) on which $2x < 2^x$.
(c) Find the interval(s) on which $2x > 2^x$.
(d) Using your table from part (a) as an aid, state what happens to the value of $f(x)$ if x is increased by 1 unit.
(e) Using your table from part (a) as an aid, state what happens to the value of $g(x)$ if x is increased by 1 unit.
(f) Using your answers from parts (c) and (d) as an aid, explain why the value of $g(x)$ is increasing much faster than the value of $f(x)$.

84. Explain why the function $f(x) = 2^x$ has no vertical asymptotes (review Section 4.6).

5.3 Logarithmic Functions

Bacterial Growth Revisited

We begin this section by examining the bacterial growth problem given in Example 1 of Section 5.2 in a different light. This will motivate us to define a new type of function known as a **logarithmic function.**

Example 1 Bacterial Growth

❖⋯ *This example builds on Example 1 of Section 5.2.*

A bacterium splits into two bacteria every hour. How many hours will it take for the bacterial population to reach 128?

▶Solution Note that in this example, we are given the ending population and must figure out how long it takes to reach that population. Table 5.3.1 gives the population for various values of the time t, in hours. (See Example 1 from Section 5.2 for details.)

Table 5.3.1

t (hours)	0	1	2	3	4	5	6	7	8
$P(t) = 2^t$ (number of bacteria)	1	2	4	8	16	32	64	128	256

From the table, we see that the bacterial population reaches 128 after 7 hours. Put another way, we are asked to find the *exponent* t such that $2^t = 128$. The answer is $t = 7$.

✔ *Check It Out 1:* Use the table in Example 1 to determine when the bacterial population will reach 64. ■

When you are given the output of an exponential function and asked to find the exponent, or the corresponding input, you are taking the inverse of the exponential function. This inverse function is called the **logarithmic function.** We next present the definition of a logarithm, and then study its properties.

Definition of Logarithm

The formal definition of a logarithm follows.

Definition of Logarithm

Let $a > 0$, $a \neq 1$. If $x > 0$, then the **logarithm of x with respect to base a** is denoted by $y = \log_a x$ and defined by

$$y = \log_a x \quad \text{if and only if} \quad x = a^y.$$

The number a is known as the **base.** Thus the functions $f(x) = a^x$ and $g(x) = \log_a x$ are inverses of each other. That is,

$$a^{\log_a x} = x \quad \text{and} \quad \log_a a^x = x.$$

This formal definition of a logarithm does *not* tell us how to calculate the value of $\log_a x$; it simply gives a definition for such a number.

Observations:

▶ The number denoted by $\log_a x$ is defined to be the unique exponent y that satisfies the equation $a^y = x$.

▶ Substituting for y, the definition of logarithm gives

$$a^y = a^{\log_a x} = x.$$

Thus, a logarithm is an exponent. To understand the definition of logarithm, it is helpful to go back and forth between a logarithmic statement (such as $\log_2 8 = 3$) and its corresponding exponential statement (in this case, $2^3 = 8$).

Example 2 Equivalent Exponential Statements

Complete the following table by filling in the exponential statements that are equivalent to the given logarithmic statements.

Logarithmic Statement	Exponential Statement
$\log_3 9 = 2$	
$\log_5 \sqrt{5} = \dfrac{1}{2}$	
$\log_2 \dfrac{1}{4} = -2$	
$\log_a b = k, a > 0$	

▶**Solution** To find the exponential statement, we use the fact that the logarithmic equation $y = \log_a x$ is equivalent to the exponential equation $a^y = x$.

Just in Time

Review rational exponents in Section P.3.

Logarithmic Statement	Question to Ask Yourself	Exponential Statement
$\log_3 9 = 2$	To what power must 3 be raised to produce 9? The answer is 2.	$3^2 = 9$
$\log_5 \sqrt{5} = \dfrac{1}{2}$	Note that the square root of a number is the same as the number raised to the $\frac{1}{2}$ power. To what power must 5 be raised to produce $\sqrt{5}$? The answer is $\frac{1}{2}$.	$5^{1/2} = \sqrt{5}$
$\log_2 \dfrac{1}{4} = -2$	To what power must 2 be raised to produce $\frac{1}{4}$? The answer is -2.	$2^{-2} = \dfrac{1}{4}$
$\log_a b = k, a > 0$	To what power must a be raised to produce b? The answer is k.	$a^k = b$

✔ *Check It Out 2:* Rework Example 2 for the following logarithmic statements.

Logarithmic Statement	Exponential Statement
$\log_3 27 = 3$	
$\log_4 \dfrac{1}{4} = -1$	

■

Example 3 **Equivalent Logarithmic Statements**

Complete the following table by filling in the logarithmic statements that are equivalent to the given exponential statements.

Exponential Statement	Logarithmic Statement
$4^0 = 1$	
$10^{-1} = 0.1$	
$6^{1/3} = \sqrt[3]{6}$	
$a^k = v, a > 0$	

▶**Solution** To find the logarithmic statements, we use the fact that the logarithmic equation $y = \log_a x$ is equivalent to the exponential equation $a^y = x$.

Exponential Statement	Question to Ask Yourself	Logarithmic Statement
$4^0 = 1$	What is the logarithm of 1 with respect to base 4? The answer is 0.	$\log_4 1 = 0$
$10^{-1} = 0.1$	What is the logarithm of 0.1 with respect to base 10? The answer is -1.	$\log_{10} 0.1 = -1$
$6^{1/3} = \sqrt[3]{6}$	What is the logarithm of $\sqrt[3]{6}$ with respect to base 6? The answer is $\frac{1}{3}$.	$\log_6 \sqrt[3]{6} = \dfrac{1}{3}$
$a^k = v, a > 0$	What is the logarithm of v with respect to base a? The answer is k.	$\log_a v = k$

✔ *Check It Out 3:* Rework Example 3 for the following exponential statements.

Exponential Statement	Logarithmic Statement
$4^3 = 64$	
$10^{1/2} = \sqrt{10}$	

■

You must understand the definition of a logarithm and its relationship to an exponential expression in order to evaluate logarithms without using a calculator. The next example will show you how to evaluate logarithms using the definition.

Example 4 Evaluating Logarithms Without Using a Calculator

Evaluate the following without using a calculator. If there is no solution, so state.

(a) $\log_5 125$ (b) $\log_{10} \dfrac{1}{100}$ (c) $\log_a a^4, a > 0$

(d) $3^{\log_3 5}$ (e) $\log_{10}(-1)$

▶ **Solution**

(a) Let $y = \log_5 125$. The equivalent exponential equation is $5^y = 125$. To find y, note that $125 = 5^3$, so

$$5^y = 125 \implies y = 3.$$

Thus $\log_5 125 = 3$.

(b) Let $y = \log_{10} \dfrac{1}{100}$; equivalently, $10^y = \dfrac{1}{100}$. Since $\dfrac{1}{100} = 10^{-2}$, we find that $y = -2$. Thus $\log_{10} \dfrac{1}{100} = -2$.

(c) Let $y = \log_a a^4, a > 0$; equivalently, $a^y = a^4$. Thus $y = 4$ and $\log_a a^4 = 4$.

(d) To evaluate $3^{\log_3 5}$, we note from the definition of a logarithm that $a^{\log_a x} = x$. Using $a = 3$ and $x = 5$, we see that $3^{\log_3 5} = 5$. This is an illustration of the fact that the exponential and logarithmic functions are inverses of each other.

(e) Let $y = \log_{10}(-1)$; equivalently, $10^y = -1$. However, 10 raised to any real number is always positive. Thus the equation $10^y = -1$ has no solution and $\log_{10}(-1)$ does not exist. It is not possible to take the logarithm of a negative number.

✔ *Check It Out 4:* Evaluate the following without using a calculator: (a) $\log_6 36$, (b) $\log_b b^{1/3}$ $(b > 0)$, and (c) $10^{\log_{10} 9}$. ■

Note *Examples 1–4 reiterate the fact that the exponential and logarithmic functions are inverses of each other.*

Example 5 Solving an Equation Involving a Logarithm

Use the definition of a logarithm to find the value of x.

(a) $\log_4 16 = x$

(b) $\log_3 x = -2$

▶ **Solution**

(a) Using the definition of a logarithm, the equation $\log_4 16 = x$ can be written as

$$4^x = 16.$$

Since $16 = 4^2$, we see that $x = 2$.

(b) Using the definition of a logarithm, the equation $\log_3 x = -2$ can be written as

$$3^{-2} = x.$$

Since $3^{-2} = \dfrac{1}{9}$, we obtain $x = \dfrac{1}{9}$.

✔ *Check It Out 5:* Solve the equation $\log_5 x = 3$. ■

Common Logarithms and Natural Logarithms

Certain bases for logarithms occur so often that they have special names. The logarithm with respect to base 10 is known as the **common logarithm;** it is abbreviated as **log** (without the subscript 10).

Discover *and* **Learn**

Graph the function $y = 10^x$ using a window size of [0, 1] (0.25) by [0, 10]. Use the TRACE feature until you reach a y value of approximately 4. What is the value of x at this point? Explain why this value should be very close to the value of log 4 that you can find using the LOG key on your calculator.

> **Definition of a Common Logarithm**
>
> $$y = \log x \quad \text{if and only if} \quad x = 10^y.$$

The LOG key on your calculator evaluates the common logarithm of a number. For example, using a calculator, log $4 \approx 0.6021$, rounded to four decimal places.

Just as the exponential function $y = 10^x$ is the inverse of the logarithmic function with respect to base 10, the exponential function $y = e^x$ is the inverse of the logarithmic function with respect to base e. This base occurs so often in applications that the logarithm with respect to base e is called the **natural logarithm;** it is abbreviated as **ln.**

> **Definition of a Natural Logarithm**
>
> $$y = \ln x \quad \text{if and only if} \quad x = e^y.$$

The natural logarithm of a number can be found by pressing the LN key on your calculator. For example, using a calculator, ln $3 \approx 1.0986$, rounded to four decimal places.

Example 6 Evaluating Common and Natural Logarithms

Without using a calculator, evaluate the following expressions.

(a) log 10,000   (b) $\ln e^{1/2}$ (c) $e^{\ln a}$, $a > 0$

▶ Solution

(a) To find log 10,000, we find the power to which 10 must be raised to get 10,000. Since $10,000 = 10^4$, we get log $10,000 = 4$.

(b) Once again, we ask the question, "To what power must e be raised to get $e^{1/2}$?" The answer is $\frac{1}{2}$. Thus

$$\ln e^{1/2} = \frac{1}{2}.$$

(c) By the definition of the natural logarithm,

$$e^{\ln a} = a, \quad a > 0.$$

✔ *Check It Out 6:* Without using a calculator, find log $10^{2/3}$ and ln $e^{4/3}$. ▪

The inverse relationship between the exponential and logarithmic functions can be seen clearly with the help of a graphing calculator, as illustrated in Example 7.

Example **7** **Evaluating a Logarithm Graphically**

Use the definition of the natural logarithm and the graph of the exponential function $f(x) = e^x$ to find an approximate value for $\ln 10$. Use a window size of $[0, 3]$ by $[0, 12]$. Compare your solution with the answer you obtain using the LN key on your calculator.

▶Solution To evaluate $\ln 10$, we must find an exponent x such that

$$e^x = 10.$$

We are given the output value of 10 and asked to find the input value of x. This is exactly the process of finding an inverse. To solve for x, use the INTERSECT feature of your graphing calculator with $Y_1(x) = e^x$ and $Y_2(x) = 10$. The solution is $x \approx 2.3026$. To reiterate, $e^{2.3026} \approx 10$, implying that $\ln 10 \approx 2.3026$. The LN key gives the same answer. See Figure 5.3.1.

Figure 5.3.1

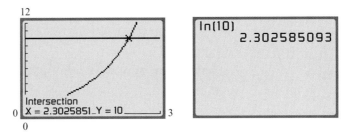

✔ *Check It Out 7:* Use the definition of the natural logarithm and the graph of the exponential function $f(x) = e^x$ to find an approximate value for $\ln 8$. Compare your answer with the answer you obtain using the LN key on your calculator. ▪

Change-of-Base Formula

Since your scientific or graphing calculator has special keys only for common logarithms and natural logarithms, you must use a change-of-base formula to calculate logarithms with respect to other bases.

> **Change-of-Base Formula**
>
> To write a logarithm with base a in terms of a logarithm with base 10 or base e, we use the formulas
>
> $$\log_a x = \frac{\log_{10} x}{\log_{10} a}$$
>
> $$\log_a x = \frac{\ln x}{\ln a}$$
>
> where $x > 0$, $a > 0$, and $a \neq 1$.

It makes no difference whether you choose the change-of-base formula with the common logarithm or the natural logarithm. The two formulas will give the same result for the value of $\log_a x$.

Example 8 Using the Change-of-Base Formula

Use the change-of-base formula with the indicated logarithm to calculate the following.

(a) $\log_6 15$, using common logarithm

(b) $\log_7 0.3$, using natural logarithm

▶**Solution**

(a) Using the change-of-base formula with the common logarithm, with $x = 15$ and $a = 6$,

$$\log_6 15 = \frac{\log_{10} 15}{\log_{10} 6}$$

$$\approx \frac{1.176}{0.7782} \approx 1.511.$$

(b) Using the change-of-base formula with the natural logarithm, with $x = 0.3$ and $a = 7$,

$$\log_7 0.3 = \frac{\ln 0.3}{\ln 7}$$

$$\approx \frac{-1.2040}{1.9459} \approx -0.6187.$$

✔ *Check It Out 8:* Compute $\log_6 15$ using the change-of-base formula with the natural logarithm and show that you get the same result as in Example 8(a). ▪

Graphs of Logarithmic Functions

Consider the function $f(x) = \log x$. We will graph this function after making a table of function values. To fill in the table, ask yourself the question, "10 raised to what power equals x?" First, let $x = 0$. Since 10 raised to any power does not equal zero, the answer to the above question is *none*. Thus, $\log 0$ is undefined. Next, suppose $x = 0.001$. Since $0.001 = 10^{-3}$, the answer to the question is -3. This procedure can be used to fill in the rest of Table 5.3.2. By plotting the points in Table 5.3.2, we obtain the graph shown in Figure 5.3.2.

Table 5.3.2

x	$f(x) = \log x$
0	Undefined
10^{-10}	-10
0.001	-3
$\dfrac{1}{10}$	-1
1	0
10	1
100	2
1.00×10^7	7

Figure 5.3.2

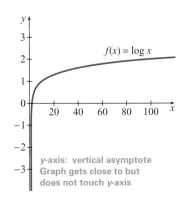

$f(x) = \log x$

y-axis: vertical asymptote
Graph gets close to but
does not touch *y*-axis

Observations:

▶ Although the x values in the table range in magnitude from 10^{-10} to 10^7, the y values range only from -10 to 7. Thus the logarithmic function can take inputs of widely varying magnitude and yield output values that are much closer together in magnitude. This is reflected in the horizontal and vertical scales of the graph.

▶ As x gets very close to 0, the graph of the function gets very close to the line $x = 0$ but does not touch it. The function is *not* defined at $x = 0$.

▶ Since we cannot solve the equation $10^y = x$ for y when x is negative or zero, $y = \log_{10} x$ is defined only for *positive real numbers.*

▶ From the graph, we see that $f(x) \to -\infty$ as $x \to 0$. Thus the y-axis is a vertical asymptote of the graph of f.

▶ Is there a horizontal asymptote? It can be shown that $\log x \to +\infty$ as $x \to +\infty$, although the value of $\log x$ grows fairly slowly. Therefore, the graph has no horizontal asymptote.

▶ If the two columns of Table 5.3.2 were interchanged, we would have a table of values of the function $g(x) = 10^x$ because the functions $g(x) = 10^x$ and $f(x) = \log x$ are inverses of each other.

We next summarize the properties of the logarithmic function with respect to any base a, $a > 0$, $a \neq 1$.

Properties of Logarithmic Functions

Figure 5.3.3

$f(x) = \log_a x$, $a > 1$:

Domain: all *positive* real numbers; $(0, \infty)$

Range: all real numbers; $(-\infty, \infty)$

Vertical asymptote: $x = 0$ (the y-axis)

Increasing on $(0, \infty)$

Inverse function of $f(x) = a^x$

See Figure 5.3.3.

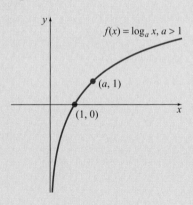

Figure 5.3.4

$f(x) = \log_a x$, $0 < a < 1$:

Domain: all *positive* real numbers; $(0, \infty)$

Range: all real numbers; $(-\infty, \infty)$

Vertical asymptote: $x = 0$ (the y-axis)

Decreasing on $(0, \infty)$

Inverse function of $f(x) = a^x$

See Figure 5.3.4.

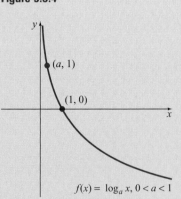

Just in Time

See Section 5.1 to review graphs of inverses.

Logarithmic functions with bases between 0 and 1 are rarely used in practice.

Since $y = \log_a x$ and $y = a^x$ are inverses of each other, the graph of $y = \log_a x$ can be obtained by reflecting the graph of $y = a^x$ across the line $y = x$. See Figure 5.3.5.

Figure 5.3.5

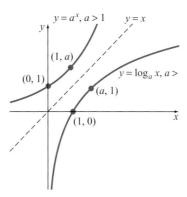

Example 9 Graphing Logarithmic Functions

Find the domain of each of the following logarithmic functions. Sketch a graph of each function and find its range. Indicate the vertical asymptote.

(a) $g(t) = \ln(-t)$

(b) $f(x) = 3 \log_2(x - 1)$

▶Solution

(a) The function $g(t) = \ln(-t)$ is defined only when $-t > 0$, which is equivalent to $t < 0$. Thus its domain is the set of all negative real numbers, or $(-\infty, 0)$. Using this information, we can generate a table (Table 5.3.3) using a suitable set of t values, and then use a calculator to find the corresponding function values. Once we have done this, we can sketch the graph, as shown in Figure 5.3.6.

Table 5.3.3

t	$g(t) = \ln(-t)$
0	Undefined
-0.025	-3.689
-0.5	-0.693
-1	0.000
$-e$	1.000
-5	1.609

Figure 5.3.6

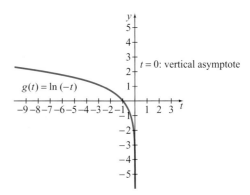

The graph of g is the same as the graph of $h(t) = \ln t$ reflected across the y-axis. The range of g is the set of all real numbers, or $(-\infty, \infty)$ in interval notation. We see from the graph that as t gets close to 0, the value of $\ln(-t)$ approaches $-\infty$. Therefore, the graph has a vertical asymptote at $t = 0$.

Technology Note

If you graph $y = \log x$ on a calculator, you will find that the graph will stop at a certain point near the vertical asymptote ($x = 0$). See Figure 5.3.7. This is because the calculator can plot only a finite number of points and cannot go beyond a certain limit. However, from the foregoing discussion, you know that the value of $y = \log x$ approaches $-\infty$ as x gets close to zero.

Keystroke Appendix:
Section 7

Figure 5.3.7

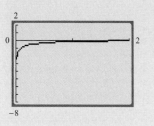

Technology Note

In order to graph $f(x) = 3 \log_2 (x - 1)$ on a calculator, you must first use the change-of-base formula to rewrite the logarithm with respect to base 10. In the Y= editor, you must press 3 LOG X, T, Θ, n − 1) ÷ LOG 2) ENTER . See Figure 5.3.9.

Keystroke Appendix:
Sections 4, 7

Figure 5.3.9

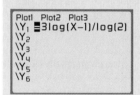

(b) The function $f(x) = 3 \log_2(x - 1)$ is defined only when $x - 1 > 0$, which is equivalent to $x > 1$. Thus its domain is the set of all real numbers greater than 1, or $(1, \infty)$. We can use transformations to graph this function.

Note that the graph of $f(x) = 3 \log_2(x - 1)$ is the same as the graph of $g(x) = \log_2 x$ shifted to the right by 1 unit and then vertically stretched by a factor of 3. See Figure 5.3.8. The range of f is the set of all real numbers, or $(-\infty, \infty)$ in interval notation. As x gets close to 1, the value of $3 \log_2(x - 1)$ approaches $-\infty$. Therefore, the graph has a vertical asymptote at $x = 1$.

Figure 5.3.8

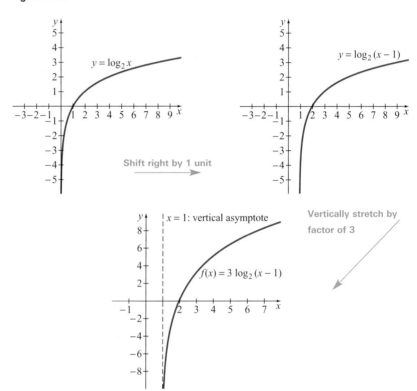

☑ *Check It Out 9:* Rework Example 9 for the function $g(x) = -\log x$. ■

Applications of Logarithmic Functions

The fact that logarithmic functions grow very slowly is an attractive feature for modeling certain applications. We examine some of these applications in the following examples. Additional applications of logarithmic functions will be studied in the last two sections of this chapter.

Example 10 Earthquakes and the Richter Scale

Since the intensities of earthquakes vary widely, they are measured on a logarithmic scale known as the Richter scale using the formula

$$R(I) = \log\left(\frac{I}{I_0}\right)$$

where I represents the actual intensity of the earthquake, and I_0 is a baseline intensity used for comparison. The Richter scale gives the *magnitude* of an earthquake.

Because of the logarithmic nature of this function, an *increase of a single unit* in the value of $R(I)$ represents a *tenfold increase* in the intensity of the earthquake. A recording of 7, for example, corresponds to an intensity that is 10 times as large as the intensity of an earthquake with a recording of 6. A quake that registers 2 on the Richter scale is the smallest quake normally felt by human beings. Earthquakes with a Richter value of 6 or more are commonly considered major, while those that have a magnitude of 8 or more on the Richter scale are classified as "great." (*Source:* U.S. Geological Survey)

(a) If the intensity of an earthquake is 100 times the baseline intensity I_0, what is its magnitude on the Richter scale?

(b) A 2003 earthquake in San Simeon, California, registered 6.5 on the Richter scale. Express its intensity in terms of I_0.

(c) A 2004 earthquake in central Japan registered 5.4 on the Richter scale. Express its intensity in terms of I_0. What is the ratio of the intensity of the 2003 San Simeon quake to the intensity of this quake?

▶**Solution**

(a) If the intensity of an earthquake is 100 times the baseline intensity I_0, then $I = 100I_0$. Substituting this expression for I in the formula for $R(I)$, we have

$$R(I) = R(100I_0) = \log\left(\frac{100I_0}{I_0}\right) = \log 100 = 2.$$

Thus the earthquake has a magnitude of 2 on the Richter scale.

(b) Substituting 6.5 for $R(I)$ in the formula $R(I) = \log\left(\frac{I}{I_0}\right)$ gives $6.5 = \log\left(\frac{I}{I_0}\right)$. Rewriting this equation in exponential form gives

$$10^{6.5} = \frac{I}{I_0}$$

from which it follows that

$$I = 10^{6.5}I_0 \approx 3{,}162{,}278I_0.$$

Therefore, the San Simeon earthquake had an intensity nearly 3.2 million times that of the baseline intensity I_0.

(c) Substituting 5.4 for $R(I)$ in the formula $R(I) = \log\left(\frac{I}{I_0}\right)$ gives $5.4 = \log\left(\frac{I}{I_0}\right)$. Rewriting this equation in exponential form, we find that

$$10^{5.4} = \frac{I}{I_0}.$$

Thus $I = 10^{5.4}I_0 \approx 251{,}189I_0$. Therefore, the Japan earthquake had an intensity about 250,000 times that of the baseline intensity I_0. Comparing the intensity of this earthquake with that of the San Simeon earthquake, we find that the ratio is

$$\frac{\text{Intensity of San Simeon quake}}{\text{Intensity of Japan quake}} = \frac{3{,}162{,}278I_0}{251{,}189I_0} \approx 12.6.$$

The San Simeon quake was 12.6 times as intense as the Japan quake.

☑ *Check It Out 10:* Find the intensity in terms of I_0 of a quake that measured 7.2 on the Richter scale. ■

Just in Time

Review scientific notation in Section P.2.

Example 11 Distances of Planets

Table 5.3.4 lists the distances from the sun of various planets in our solar system, as well as the nearest star, Alpha Centauri. Find the common logarithm of each distance.

Table 5.3.4

Planet or Star	Distance from Sun (miles)
Earth	9.350×10^7
Jupiter	4.862×10^8
Pluto	3.670×10^9
Alpha Centauri	2.543×10^{13}

▶**Solution** We first compute the common logarithms of the distances given in the table. For example,

$$\log(9.350 \times 10^7) \approx 7.971.$$

Table 5.3.5 summarizes the results.

Table 5.3.5

Planet or Star	Distance from Sun (miles)	Logarithm of Distance
Earth	9.350×10^7	7.971
Jupiter	4.862×10^8	8.687
Pluto	3.670×10^9	9.565
Alpha Centauri	2.543×10^{13}	13.41

Note that the distances of these celestial bodies from the sun vary widely—the longest distance given in the table exceeds the shortest distance by several powers of 10, and the ratio of these two distances is about 270,000. However, the common logarithms of the distances vary by much less than the distances themselves—the logarithms range from 7.971 to 13.41.

✔ *Check It Out 11:* Saturn is 9.3×10^8 miles from the sun. Find the common logarithm of this distance. ■

Discover *and* Learn

Keep in mind that each unit increase in the common logarithm of a distance represents a tenfold increase in the actual distance. Saturn is 10 times farther from the sun than Earth is. How is this fact reflected in the common logarithms of their respective distances from the sun?

When numerical values of a dependent variable have widely varying magnitudes and must be plotted on a single graph, we often take the logarithms of those values and plot them on a scale known as a **logarithmic scale.** On such a scale, each unit increase in the common logarithm of a numerical value represents a *tenfold increase* in the value itself. Logarithmic scales play an important role in the graphing of scientific, engineering, and financial data.

5.3 Key Points

▶ Let $a > 0$, $a \neq 1$. If $x > 0$, then the **logarithm of x with respect to base a** is denoted by $y = \log_a x$ and defined by

$$y = \log_a x \quad \text{if and only if} \quad x = a^y.$$

▶ The exponential and logarithmic functions are inverses of each other. That is,

$$a^{\log_a x} = x \quad \text{and} \quad \log_a a^x = x.$$

▶ If the base of a logarithm is 10, the logarithm is a **common logarithm:**

$$y = \log x \quad \text{if and only if} \quad x = 10^y$$

▶ If the base of a logarithm is e, the logarithm is a **natural logarithm:**

$$y = \ln x \quad \text{if and only if} \quad x = e^y$$

▶ To write a logarithm with base a in terms of a logarithm with base 10 or base e, use one of the following.

$$\log_a x = \frac{\log_{10} x}{\log_{10} a} \qquad \log_a x = \frac{\ln x}{\ln a}$$

▶ For $a > 0$, $a \neq 1$, a **logarithmic function** is defined as $f(x) = \log_a x$. The domain of f is all positive real numbers. The range of f is all real numbers.

5.3 Exercises

▶ **Just in Time Exercises** These exercises correspond to the Just in Time references in this section. Complete them to review topics relevant to the remaining exercises.

In Exercises 1–4, rewrite using rational exponents.

1. $\sqrt{3}$

2. $\sqrt{5}$

3. $\sqrt[3]{10}$

4. $\sqrt[3]{12}$

In Exercises 5–8, f and g are inverses of each other.

5. True or False? $(f \circ g)(x) = x$

6. True or False? The domain of f equals the domain of g.

7. True or False? The domain of f equals the range of g.

8. True or False? f is a one-to-one function.

9. Write 8,450,000 in scientific notation.

10. Write 1,360,000,000,000 in scientific notation.

▶ **Skills** This set of exercises will reinforce the skills illustrated in this section.

In Exercises 11 and 12, complete the table by filling in the exponential statements that are equivalent to the given logarithmic statements.

11.

Logarithmic Statement	Exponential Statement
$\log_3 1 = 0$	
$\log 10 = 1$	
$\log_5 \dfrac{1}{5} = -1$	
$\log_a x = b, a > 0$	

12.

Logarithmic Statement	Exponential Statement
$\log_2 4 = 2$	
$\log 100 = 2$	
$\log_7 \dfrac{1}{49} = -2$	
$\log \alpha = \beta, \alpha > 0$	

In Exercises 13 and 14, complete the table by filling in the logarithmic statements that are equivalent to the given exponential statements.

13.

Exponential Statement	Logarithmic Statement
$3^4 = 81$	
$5^{1/3} = \sqrt[3]{5}$	
$6^{-1} = \dfrac{1}{6}$	
$a^v = u,$ $a > 0$	

14.

Exponential Statement	Logarithmic Statement
$3^5 = 243$	
$7^{1/2} = \sqrt{7}$	
$6^{-2} = \dfrac{1}{36}$	
$10^\alpha = \beta$	

In Exercises 15–34, evaluate each expression without using a calculator.

15. $\log 10{,}000$
16. $\log 0.001$

17. $\log \sqrt[3]{10}$
18. $\log \sqrt{10}$

19. $\ln e^2$
20. $\ln \sqrt{e}$

21. $\ln e^{1/3}$
22. $\ln \dfrac{1}{e}$

23. $\log 10^{x+y}$
24. $\ln e^{x-z}$

25. $\log 10^k$
26. $\ln e^w$

27. $\log_2 \sqrt{2}$
28. $\log_7 49$

29. $\log_3 \dfrac{1}{81}$
30. $\log_7 \dfrac{1}{49}$

31. $\log_{1/2} 4$
32. $\log_{1/3} 9$

33. $\log_4 4^{x^2+1}$
34. $\log_6 6^{6x}$

In Exercises 35–42, evaluate the expression to four decimal places using a calculator.

35. $2 \log 4$
36. $-3 \log 6$

37. $\ln \sqrt{2}$
38. $\ln \pi$

39. $\log 1400$
40. $\log 2500$

41. $2 \log \dfrac{1}{5}$
42. $-\ln \dfrac{2}{3}$

In Exercises 43–50, use the change-of-base formula to evaluate each logarithm using a calculator. Round answers to four decimal places.

43. $\log_3 1.25$
44. $\log_3 2.75$

45. $\log_5 0.5$
46. $\log_5 0.65$

47. $\log_2 12$
48. $\log_2 20$

49. $\log_7 150$
50. $\log_7 230$

In Exercises 51–56, use the definition of a logarithm to solve for x.

51. $\log_2 x = 3$
52. $\log_5 \sqrt{5} = x$

53. $\log_3 x = \dfrac{1}{3}$
54. $\log_6 x = -2$

55. $\log_x 216 = 3$
56. $\log_x 9 = \dfrac{1}{2}$

In Exercises 57–72, find the domain of each function. Use your answer to help you graph the function, and label all asymptotes and intercepts.

57. $f(x) = 2 \log x$
58. $f(x) = 4 \ln x$

59. $f(x) = 4 \log_3 x$
60. $f(x) = 3 \log_5 x$

61. $g(x) = \log x - 3$
62. $h(x) = \ln x + 2$

63. $f(x) = \log_4(x + 1)$
64. $f(x) = \log_5(x - 2)$

65. $f(x) = \ln(x + 4)$
66. $f(x) = \log(x - 3)$

67. $g(x) = 2 \log_3(x - 1)$
68. $f(x) = -\log_2(x + 3)$

69. $f(t) = \log_{1/3} t$
70. $g(s) = \log_{1/2} s$

71. $f(x) = \log |x|$
72. $g(x) = \ln(x^2)$

73. Use the following graph of $f(x) = 10^x$ to estimate $\log 7$. Explain how you obtained your answer.

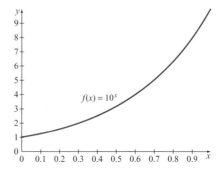

74. Use the following graph of $f(x) = e^x$ to estimate ln 10. Explain how you obtained your answer.

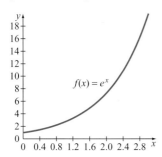

$f(x) = e^x$

In Exercises 75–78, match the description with the correct graph. Each description applies to exactly one of the four graphs.

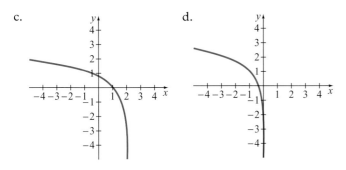

a.

b.

c.

d.

75. Graph of a logarithmic function with vertical asymptote at $x = 2$ and domain $x < 2$

76. Graph of a logarithmic function with vertical asymptote at $x = 1$ and domain $x > 1$

77. Graph of $f(x) = \ln x$ reflected across the x-axis and shifted down 1 unit

78. Graph of $f(x) = \ln x$ reflected across the y-axis and shifted up 1 unit

In Exercises 79–82, solve each equation graphically and express the solution as an appropriate logarithm to four decimal places. If a solution does not exist, explain why.

79. $10^t = 7$

80. $e^t = 6$

81. $4(10^x) = 20$

82. $e^t = -3$

▶**Applications** In this set of exercises, you will use logarithms to study real-world problems.

Exercises 83–87 refer to the following. The magnitude of an earthquake is measured on the Richter scale using the formula

$$R(I) = \log\left(\frac{I}{I_0}\right)$$

where I represents the actual intensity of the earthquake and I_0 is a baseline intensity used for comparison.

83. **Richter Scale** If the intensity of an earthquake is 10,000 times the baseline intensity I_0, what is its magnitude on the Richter scale?

84. **Richter Scale** If the intensity of an earthquake is a million times the baseline intensity I_0, what is its magnitude on the Richter scale?

85. **Great Earthquakes** The great San Francisco earthquake of 1906, the most powerful earthquake in Northern California's recorded history, is estimated to have registered 7.8 on the Richter scale. (*Source:* U.S. Geological Survey) Express its intensity in terms of I_0.

86. **Great Earthquakes** In 1984, another significant earthquake in San Francisco registered 6.1 on the Richter scale. Express its intensity in terms of I_0.

87. **Earthquake Intensity** What is the ratio of the intensity of a quake that measures 7.1 on the Richter scale to the intensity of one that measures 4.2?

Exercises 88 and 89 refer to the following. The pH of a chemical solution is given by pH $= -\log[H^+]$, where $[H^+]$ is the concentration of hydrogen ions in the solution, in units of moles per liter. (One mole is 6.02×10^{23} molecules.)

88. **Chemistry** Find the pH of a solution for which $[H^+] = 0.001$ mole per liter.

89. **Chemistry** Find the pH of a solution for which $[H^+] = 10^{-4}$ mole per liter.

90. **Astronomy** The brightness of a star is designated on a numerical scale called *magnitude*, which is defined by the formula

$$M(I) = -\log_{2.5} \frac{I}{I_0}$$

where I is the energy intensity of the star and I_0 is the baseline intensity used for comparison. A decrease of 1 unit in magnitude represents an increase in energy intensity of a factor of 2.5. (*Source:* National Aeronautics and Space Agency)

(a) If the star Spica has magnitude 1, find its intensity in terms of I_0.

(b) The star Sirius, the brightest star other than the sun, has magnitude -1.46. Find its intensity in terms of I_0. What is the ratio of the intensity of Sirius to that of Spica?

91. **Computer Science** Computer programs perform many kinds of sorting. It is preferable to use the least amount of computer time to do the sorting, where the measure of computer time is the number of operations the computer needs to perform. Two methods of sorting are the **bubble sort** and the **heap sort**. It is known that the bubble sort algorithm requires approximately n^2 operations to sort a list of n items, while the heap sort algorithm requires approximately $n \log_{10} n$ operations to sort n items.

(a) To sort 100 items, how many operations are required by the bubble sort? by the heap sort?

(b) Make a table listing the number of operations required for the bubble sort to sort a list of n items, with n ranging from 5 to 20, in steps of 5. If the number of items sorted is doubled from 10 to 20, what is the corresponding increase in the number of operations?

(c) Rework part (b) for the heap sort.

(d) Which algorithm, the bubble sort or the heap sort, is more efficient? Why?

(e) In the same viewing window, graph the functions that give the number of operations for the bubble sort and for the heap sort. Let n range from 1 to 20. Which function is growing faster, and why? Note that you will have to choose the vertical scale carefully so that the $n \log n$ function does not get "squashed."

92. **Ecology** The pH scale measures the level of acidity of a solution on a logarithmic scale. A pH of 7.0 is considered neutral. If the pH is less than 7.0, then the solution is acidic. The lower the pH, the more acidic the solution. Since the pH scale is logarithmic, a single unit *decrease* in pH represents a *tenfold increase* in the acidity level.

(a) The average pH of rainfall in the northeastern part of the United States is 4.5. Normal rainfall has a pH of 5.5. Compared to normal rainfall, how many times more acidic is the rainfall in the northeastern United States, on average? Explain. (*Source:* U.S. Environmental Protection Agency)

(b) Because of increases in the acidity of rain, many lakes in the northeastern United States have become more acidic. The degree to which acidity can be tolerated by fish in these lakes depends on the species. The yellow perch can easily tolerate a pH of 4.0, while the common shiner cannot easily tolerate pH levels below 6.0. Which species is more likely to survive in a more acidic environment, and why? What is the ratio of the acidity levels that are easily tolerated by the yellow perch and the common shiner? Explain. (*Source:* U.S. Environmental Protection Agency)

▶**Concepts** This set of exercises will draw on the ideas presented in this section and your general math background.

93. Explain why log 400 is between 2 and 3, without using a calculator.

94. Explain why ln 4 is between 1 and 2, without using a calculator.

In Exercises 95–98, explain how you would use the following table of values for the function $f(x) = 10^x$ to find the given quantity.

x	$f(x) = 10^x$
0.4771	3
0.5	$\sqrt{10}$
3	1000
-0.3010	0.5
$\sqrt{10}$	1452

95. log 1000   96. log 3

97. log 0.5   98. log $\sqrt{10}$

99. The graph of $f(x) = a \log x$ passes through the point $(10, 3)$. Find a and thus the complete expression for f. Check your answer by graphing f.

100. The graph of $f(x) = A \ln x + B$ passes through the points $(1, 2)$ and $(e, 4)$.

(a) Find A and B using the given points.

(b) Check your answer by graphing f.

101. Sketch graphs of the two functions to show that $\log_{1/2} x = -\log_2 x$. (The equality can be established algebraically by techniques in the following section.)

102. Find the domains of $f(x) = 2 \ln x$ and $g(x) = \ln x^2$. Graph these functions in separate viewing windows. Where are the graphs identical? Explain in terms of the domain you found for each function.

5.4 Properties of Logarithms

Objectives

▶ Define the various properties of logarithms

▶ Combine logarithmic expressions

▶ Use properties of logarithms in an application

In the previous section you were introduced to logarithms and logarithmic functions. We continue our study of logarithms by examining some of their special properties.

Product Property of Logarithms

If you compute log 3.6 and log 36 using a calculator, you will note that the value of log 36 exceeds the value of log 3.6 by only 1 unit, even though 36 is 10 times as large as 3.6. This curious fact is actually the result of a more general property of logarithms, which we now present.

Discover *and* Learn

Make a table of values for the functions $f(x) = \log(10x)$ and $g(x) = \log x + 1$ for $x = 0.5$, 1, 5, 10, and 100. What do you observe?

Product Property of Logarithms

Let $x, y > 0$ and $a > 0, a \neq 1$. Then

$$\log_a(xy) = \log_a x + \log_a y.$$

Because logarithms are exponents, and multiplication of a pair of exponential expressions with the same base can be carried out by adding the exponents, we see that the logarithm of a product "translates" into a sum of logarithms. We derive the product property of common logarithms as follows.

$$xy = a^{\log_a(xy)} \qquad \text{Logarithmic and exponential functions are inverses}$$

Again using the inverse relationship of logarithmic and exponential functions, we have $x = a^{\log_a x}$ and $y = a^{\log_a y}$. So an alternative expression for xy is $a^{\log_a x} \times a^{\log_a y}$. Carrying out the multiplication in this expression, we obtain

$$a^{\log_a x} \times a^{\log_a y} = a^{\log_a x + \log_a y}. \qquad \text{Add exponents, since the bases are the same}$$

Finally, we equate the two expressions for xy.

$$a^{\log_a x + \log_a y} = a^{\log_a(xy)} \qquad \text{Equate expressions for xy}$$

$$\log_a x + \log_a y = \log_a(xy) \qquad \text{Equate exponents, since the bases are the same}$$

Example 1 Using the Product Property to Calculate Logarithms

Given that $\log 2.5 \approx 0.3979$ and $\log 3 \approx 0.4771$, calculate the following logarithms *without* the use of a calculator. Then check your answers using a calculator.

(a) log 25

(b) log 75

▶ Solution

(a) Because we can write 25 as 2.5×10, and we are given an approximate value for log 2.5, we have

$$\log 25 = \log(2.5 \times 10) \qquad \text{Write 25 as a product}$$

$$= \log 2.5 + \log 10 \qquad \text{Use product property}$$

$$\approx 0.3979 + 1 = 1.3979. \qquad \text{Substitute and simplify}$$

(b) We can write 75 as 7.5×10. This can, in turn, be written as $(3 \times 2.5) \times 10$, since $7.5 = 3 \times 2.5$. Using the approximate values of log 2.5 and log 3, we have

$$\log 75 = \log((3 \times 2.5) \times 10) \qquad \textit{Write 75 as a product}$$
$$= \log(3 \times 2.5) + \log 10 \qquad \textit{Use product property twice}$$
$$= \log 3 + \log 2.5 + \log 10$$
$$\approx 0.4771 + 0.3979 + 1 = 1.8750. \qquad \textit{Substitute and simplify}$$

✔ *Check It Out 1:* Using the approximate value of log 2.5 from Example 1, calculate log 2500 without using your calculator. Check your answer using a calculator. ▪

Discover *and* **Learn**

Tables of common logarithms contain the logarithms of numbers from 1 to 9.9999. How would you use such tables to calculate the common logarithms of numbers not in this range? (*Hint:* Look at Example 1.)

Before the widespread use of calculators, the product property of logarithms was used to calculate products of large numbers. Textbooks that covered the topic of logarithms contained tables of logarithms in the appendix to aid in the calculation. Nowadays, the main purpose of presenting properties of logarithms is to impart an understanding of the nature of logarithms and their applications.

Power Property of Logarithms

We next present the power property of logarithms.

Power Property of Logarithms

Let $x > 0$, $a > 0$, $a \neq 1$, and let k be any real number. Then

$$\log_a x^k = k \log_a x.$$

It is important to note that the power property holds true only when $x > 0$. We can illustrate this using the case in which $a = e$ and $k = 2$, so that the power property gives $\ln x^2 = 2 \ln x$.

Consider the functions $f(x) = \ln x^2$ and $g(x) = 2 \ln x$, which are graphed in Figure 5.4.1. The domain of f is $(-\infty, 0) \cup (0, \infty)$, whereas the domain of g is $(0, \infty)$, since $2 \ln x$ is undefined if x is negative. From the graphs, we observe that these functions are equal only on their common domain, which is the set of all positive real numbers. Thus the power property, illustrated by $\ln x^2 = 2 \ln x$, holds true only when $x > 0$.

Figure 5.4.1

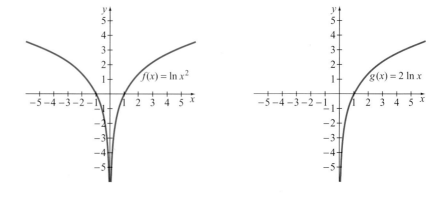

Just in Time

Review rational exponents in Section P.3.

Example **2** **Simplifying Logarithmic Expressions**

Simplify the following expressions, if possible, by eliminating exponents and radicals. Assume $x, y > 0$.

(a) $\log(xy^{-3})$

(b) $\ln(3x^{1/2}\sqrt[3]{y})$

(c) $(\ln x)^{1/3}$

▶ Solution

(a) We use the product property, followed by the power property.

$$\log(xy^{-3}) = \log x + \log y^{-3} \qquad \textit{Product property}$$
$$= \log x - 3\log y \qquad \textit{Power property}$$

(b) Again we begin by applying the product property.

$$\ln(3x^{1/2}\sqrt[3]{y}) = \ln 3 + \ln x^{1/2} + \ln \sqrt[3]{y}$$

Using the fact that $\sqrt[3]{y} = y^{1/3}$ and applying the power property, we find that

$$\ln(3x^{1/2}\sqrt[3]{y}) = \ln 3 + \frac{1}{2}\ln x + \frac{1}{3}\ln y.$$

(c) We are asked to simplify $(\ln x)^{1/3}$. Although it may look as if we could use the power property to do so, the power property applies *only* to logarithms of the form $\ln x^a$, and the given expression is of the form $(\ln x)^a$. Therefore, we cannot simplify $(\ln x)^{1/3}$. Applying the power property to expressions such as this is a common mistake.

✔ *Check It Out 2:* Simplify $\log(4x^{-1/3}\sqrt{y})$ by eliminating exponents and radicals. Assume $x, y > 0$. ■

Quotient Property of Logarithms

We now derive another property of logarithms, known as the quotient property. Let $x, y > 0$. Then

$$\log_a \frac{x}{y} = \log_a(xy^{-1}) \qquad \textit{Write } \frac{x}{y} \textit{ as } xy^{-1}$$
$$= \log_a x + \log_a y^{-1} \qquad \textit{Product property of logarithms}$$
$$= \log_a x - \log_a y. \qquad \textit{Power property of logarithms}$$

Thus we see that $\log_a \frac{x}{y} = \log_a x - \log_a y$. This is known as the **quotient property of logarithms.**

> **Quotient Property of Logarithms**
>
> Let $x, y > 0$, $a > 0$, and $a \neq 1$. Then
>
> $$\log_a \frac{x}{y} = \log_a x - \log_a y.$$

| Example | **3** Writing an Expression as a Sum or Difference of Logarithms |

Write each expression as a sum and/or difference of logarithmic expressions. Eliminate exponents and radicals wherever possible.

(a) $\log\left(\dfrac{x^3 y^2}{100}\right)$, $x, y > 0$ (b) $\log_a\left(\dfrac{\sqrt{x^2 + 2}}{(x + 1)^3}\right)$, $x > -1$

▶**Solution**

(a) Use the quotient property, the product property, and the power property, in that order.

$$\log\left(\dfrac{x^3 y^2}{100}\right) = \log(x^3 y^2) - \log 100 \qquad \text{Quotient property}$$

$$= \log x^3 + \log y^2 - \log 100 \qquad \text{Product property}$$

$$= 3 \log x + 2 \log y - 2 \qquad \text{Power property; also log 100 = 2}$$

(b) Use the quotient property, followed by the power property.

$$\log_a\left(\dfrac{\sqrt{x^2 + 2}}{(x + 1)^3}\right) = \log_a \sqrt{x^2 + 2} - \log_a(x + 1)^3 \qquad \text{Quotient property}$$

$$= \dfrac{1}{2} \log_a(x^2 + 2) - 3 \log_a(x + 1) \qquad \text{Power property; also}$$
$$\sqrt{x^2 + 2} = (x^2 + 2)^{1/2}$$

Because the logarithm of a sum *cannot* be simplified further, we cannot eliminate the exponent in the expression $(x^2 + 2)$.

✔ *Check It Out 3:* Write $\log\left(\dfrac{\sqrt{x - 1}}{x^2 + 4}\right)$ as a sum and/or difference of logarithmic expressions. Eliminate exponents and radicals wherever possible. Assume $x > 1$. ▪

Combining Logarithmic Expressions

The properties of logarithms can also be used to combine sums and differences of logarithms into a single expression. This will be useful in the next section, where we solve exponential and logarithmic equations.

| Example | **4** Writing an Expression as a Single Logarithm |

Write each expression as the logarithm of a single quantity.

(a) $\log_a 3 + \log_a 6$, $a > 0$ (b) $\dfrac{1}{3} \ln 64 + \dfrac{1}{2} \ln x$, $x > 0$

(c) $3 \log 5 - 1$ (d) $\log_a x + \dfrac{1}{2} \log_a(x^2 + 1) - \log_a 3$, $a > 0$, $x > 0$

▶**Solution**

(a) Using the product property, write the sum of logarithms as the logarithm of a product.

$$\log_a 3 + \log_a 6 = \log_a(3 \times 6) = \log_a 18$$

Discover *and* **Learn**

Common errors involving logs
Give an example to verify each statement.
(a) $\log(x + y)$ *does not equal* $\log x + \log y$
(b) $\log(xy)$ *does not equal* $(\log x)(\log y)$
(c) $(\log x)^k$ *does not equal* $k \log x$

(b) Use the power property first, and then the product property.

$$\frac{1}{3} \ln 64 + \frac{1}{2} \ln x = \ln 64^{1/3} + \ln x^{1/2} \quad \text{Power property}$$

$$= \ln 4 + \ln x^{1/2} \quad \text{Write } 64^{1/3} \text{ as } 4$$

$$= \ln(4x^{1/2}) \quad \text{Product property}$$

(c) Write 1 as log 10 (so that every term is expressed as a logarithm) and then apply the power property, followed by the quotient property.

$$3 \log 5 - 1 = 3 \log 5 - \log 10 \quad \text{Write 1 as log 10}$$

$$= \log 5^3 - \log 10 \quad \text{Power property}$$

$$= \log\left(\frac{5^3}{10}\right) \quad \text{Quotient property}$$

$$= \log \frac{125}{10} = \log 12.5 \quad \text{Simplify}$$

(d) Use the power property, the product property, and the quotient property, in that order.

$$\log_a x + \frac{1}{2} \log_a(x^2 + 1) - \log_a 3 = \log_a x + \log_a(x^2 + 1)^{1/2} - \log_a 3 \quad \text{Power property}$$

$$= \log_a(x(x^2 + 1)^{1/2}) - \log_a 3 \quad \text{Product property}$$

$$= \log_a\left(\frac{x(x^2 + 1)^{1/2}}{3}\right) \quad \text{Quotient property}$$

✔ *Check It Out 4:* Write $3 \log_a x - \log_a(x^2 + 1)$ as the logarithm of a single expression. Assume $x > 0$. ■

Applications of Logarithms

Logarithms occur in a variety of applications. Example 5 explores an application of logarithms that occurs frequently in chemistry and biology.

Example 5 Measuring the pH of a Solution

The pH of a solution is a measure of the concentration of hydrogen ions in the solution. This concentration, which is denoted by $[H^+]$, is given in units of moles per liter, where one mole is 6.02×10^{23} molecules.

Since the concentration of hydrogen ions can vary widely (often by several powers of 10) from one solution to another, the pH scale was introduced to express the concentration in more accessible terms. The pH of a solution is defined as

$$pH = -\log [H^+].$$

(a) Find the pH of solution A, whose hydrogen ion concentration is 10^{-4} mole/liter.

(b) Find the pH of solution B, whose hydrogen ion concentration is 4.1×10^{-8} mole/liter.

(c) If a solution has a pH of 9.2, what is its concentration of hydrogen ions?

▶Solution

(a) Using the definition of pH, we have

$$pH = -\log 10^{-4} = -(\log 10^{-4}) = -(-4 \log 10) = 4 \log 10 = 4(1) = 4.$$

(b) Again using the definition of pH, we have

$$pH = -\log(4.1 \times 10^{-8}) = -(\log 4.1 + \log 10^{-8}) = -(\log 4.1 - 8 \log 10)$$
$$= -(\log 4.1 - 8(1)) = -(\log 4.1 - 8) \approx -(0.613 - 8) = -(-7.387) = 7.387.$$

Note that solution B has a higher pH, but a smaller concentration of hydrogen ions, than solution A. As the concentration of hydrogen ions *decreases* in a solution, the solution is said to become more *basic*. Likewise, if the concentration of hydrogen ions *increases* in a solution, the solution is said to become more *acidic*.

(c) Here we are given the pH of a solution and must find $[H^+]$. We proceed as follows.

$$9.2 = -\log[H^+] \qquad \text{Set pH to 9.2 in definition of pH}$$
$$-9.2 = \log[H^+] \qquad \text{Isolate log expression}$$
$$10^{-9.2} = [H^+] \qquad \text{Use definition of logarithm}$$

Thus, the concentration of hydrogen is $10^{-9.2} \approx 6.310 \times 10^{-10}$ mole/liter. Note how we used the definition of logarithm to solve the logarithmic equation $-9.2 = \log[H^+]$ in a single step.

☑ *Check It Out 5:* Find the pH of a solution whose hydrogen ion concentration is 3.2×10^{-8} mole/liter. ■

5.4 Key Points

In the following statements, let $x, y > 0$, let k be any real number, and let $a > 0$, $a \neq 1$.

▶ From the definition of logarithm,

$$k = \log_a x \quad \text{implies} \quad a^k = x.$$

Also,

$$a^k = x \quad \text{implies} \quad k = \log_a x.$$

▶ Using the definition of logarithm,

$$a^{\log_a x} = x.$$

▶ $\log_a a = 1, \quad \log_a 1 = 0$

▶ **The product, power, and quotient properties**

$$\log_a(xy) = \log_a x + \log_a y \qquad \text{Product property of logarithms}$$
$$\log_a x^k = k \log_a x \qquad \text{Power property of logarithms}$$
$$\log_a \frac{x}{y} = \log_a x - \log_a y \qquad \text{Quotient property of logarithms}$$

5.4 Exercises

▶**Just in Time Exercises** These exercises correspond to the Just in Time references in this section. Complete them to review topics relevant to the remaining exercises.

In Exercises 1–4, rewrite using rational exponents.

1. $\sqrt[5]{x}$

2. $\sqrt[3]{z}$

3. $\sqrt[5]{x^3}$

4. $\sqrt[3]{y^2}$

5. True or False? $x^{-1} = \dfrac{1}{x}$

6. True or False? $\dfrac{y}{x^3} = x^3 y^{-1}$

▶**Skills** This set of exercises will reinforce the skills illustrated in this section.

In Exercises 7–14, use log 2 ≈ 0.3010, log 5 ≈ 0.6990, and log 7 ≈ 0.8451 to evaluate each logarithm without using a calculator. Then check your answer using a calculator.

7. log 35

8. log 14

9. $\log \dfrac{2}{5}$

10. $\log \dfrac{5}{7}$

11. $\log \sqrt{2}$

12. $\log \sqrt{5}$

13. log 125

14. log 8

In Exercises 15–20, use the properties of logarithms to simplify each expression by eliminating all exponents and radicals. Assume that x, y > 0.

15. $\log(xy^3)$

16. $\log(x^3 y^2)$

17. $\log \sqrt[3]{x} \sqrt[4]{y}$

18. $\log \sqrt[5]{x^2} \sqrt{y^5}$

19. $\log \dfrac{\sqrt[4]{x}}{y^{-1}}$

20. $\log \dfrac{\sqrt[3]{x}}{y^2}$

In Exercises 21–30, write each logarithm as a sum and/or difference of logarithmic expressions. Eliminate exponents and radicals and evaluate logarithms wherever possible. Assume that a, x, y, z > 0 and a ≠ 1.

21. $\log \dfrac{x^2 y^5}{10}$

22. $\log \dfrac{x^5 y^4}{1000}$

23. $\ln \dfrac{\sqrt[3]{x^2}}{e^2}$

24. $\ln \dfrac{\sqrt[4]{y^3}}{e^5}$

25. $\log_a \dfrac{\sqrt{x^2 + y}}{a^3}$

26. $\log_a \dfrac{\sqrt{x^3 y + 1}}{a^4}$

27. $\log_a \sqrt{\dfrac{x^6}{y^3 z^5}}$

28. $\log_a \sqrt{\dfrac{z^5}{xy^4}}$

29. $\log \sqrt[3]{\dfrac{xy^3}{z^5}}$

30. $\log \sqrt[3]{\dfrac{x^3 z^5}{10 y^2}}$

In Exercises 31–46, write each expression as a logarithm of a single quantity and then simplify if possible. Assume that each variable expression is defined for appropriate values of the variable(s). Do not use a calculator.

31. log 6.3 − log 3

32. log 4.1 + log 3

33. $\log 3 + \log x + \log \sqrt{y}$

34. $\ln y - \ln 2 + \ln \sqrt{x}$

35. $3 \log x + \dfrac{1}{2} \log y - \log z$

36. $\ln 4 - 1$

37. log 8 + 1

38. 3 log x + 2

39. 2 ln y + 3

40. $\dfrac{1}{3} \log_4 8x^9 - \log_4 x^2$

41. $\dfrac{1}{4} \log_3 81 y^8 + \log_3 y^3$

42. $\ln(x^2 - 9) - \ln(x + 3)$

43. $\ln(x^2 - 1) - \ln(x - 1)$

44. $\dfrac{1}{2} [\log(x^2 - 1) - \log(x + 1)] + \log x$

45. $\dfrac{1}{3} [\log(x^2 - 9) - \log(x - 3)] - \log x$

46. $\dfrac{3}{2} \log 16x^4 - \dfrac{1}{2} \log y^8$

In Exercises 47–52, let b = log k. Write each expression in terms of b. Assume k > 0.

47. log 10k

48. log 100k

49. $\log k^3$

50. $\log k^4$

51. $\log \dfrac{1}{k}$

52. $\log \dfrac{1}{k^3}$

In Exercises 53–64, simplify each expression. Assume that each variable expression is defined for appropriate values of x. Do not use a calculator.

53. $\log 10^{\sqrt{2}}$

54. $\log 10^{2x}$

55. $\ln e^{\sqrt{3}}$

56. $\ln e^{(x+1)}$

57. $10^{\log(5x)}$

58. $e^{\ln(5x^2-1)}$

59. $10^{\log(3x+1)}$

60. $e^{\ln(2x+1)}$

61. $\log_2 8$

62. $\log_5 625$

63. $\log_a \sqrt[5]{a^2}$, $a > 0$, $a \neq 1$

64. $\log_b \sqrt[3]{b}$, $b > 0$, $b \neq 1$

In Exercises 65–68, use a graphing utility with a decimal window.

65. Graph $f(x) = \log 10x$ and $g(x) = \log x$ on the same set of axes. Explain the relationship between the two graphs in terms of the properties of logarithms.

66. Graph $f(x) = \log 0.1x$ and $g(x) = \log x$ on the same set of axes. Explain the relationship between the two graphs in terms of the properties of logarithms.

67. Graph $f(x) = \ln e^2 x$ and $g(x) = \ln x$ on the same set of coordinate axes. Explain the relationship between the two graphs in terms of the properties of logarithms.

68. Graph $f(x) = \log x - \log(x-1)$ and $g(x) = \log \frac{x}{x-1}$ on the same set of axes.
    (a) What are the domains of the two functions?
    (b) For what values of x do these two functions agree?
    (c) To what extent does this pair of functions exhibit the quotient property of logarithms?

▶**Applications** In this set of exercises, you will use properties of logarithms to study real-world problems.

Chemistry *Refer to the definition of pH in Example 5 to solve Exercises 69–73.*

69. Suppose solution A has a pH of 5 and solution B has a pH of 9. What is the ratio of the concentration of hydrogen ions in solution A to the concentration of hydrogen ions in solution B?

70. Find the pH of a solution with $[H^+] = 4 \times 10^{-5}$.

71. Find the pH of a solution with $[H^+] = 6 \times 10^{-8}$.

72. Find the hydrogen ion concentration of a solution with a pH of 7.2.

73. Find the hydrogen ion concentration of a solution with a pH of 3.4.

Noise Levels *Use the following information for Exercises 74–76. The decibel (dB) is a unit that is used to express the relative loudness of two sounds. One application of this is the relative value of the output power of an amplifier with respect to the input power. Since power levels can vary greatly in magnitude, the relative value D of power level P_1 with respect to power level P_2 is given (in units of dB) in terms of the logarithm of their ratio, as follows.*

$$D = 10 \log \frac{P_1}{P_2}$$

The values P_1 and P_2 are expressed in the same units, such as watts (W).

74. If $P_1 = 20\,W$ and $P_2 = 0.3\,W$, find the relative value of P_1 with respect to P_2, in units of dB.

75. If an amplifier's output power is 10 W and the input power is 0.5 W, what is the relative value of the output with respect to the input, in units of dB?

76. Use the properties of logarithms to show that the relative value of one power level with respect to another, expressed in units of dB, is actually a *difference* of two quantities.

▶**Concepts** This set of exercises will draw on the ideas presented in this section and your general math background.

77. Consider the function $f(x) = 2^x$.
    (a) Sketch the graph of f.
    (b) What are the domain and range of f?
    (c) Graph the inverse function.
    (d) What are the domain and range of the inverse function?

78. Consider the function $f(x) = x^3$.
    (a) Sketch the graph of f.
    (b) What are the domain and range of f?
    (c) Graph the inverse function.
    (d) What are the domain and range of the inverse function?

79. Graph $f(x) = e^{\ln x}$ and $g(x) = x$ on the same set of axes.
    (a) What are the domains of the two functions?
    (b) For what values of x do these two functions agree?

80. Graph $f(x) = \ln e^x$ and $g(x) = x$ on the same set of axes.
    (a) What are the domains of the two functions?
    (b) For what values of x do these two functions agree?

81. Let $a > 1$. Can $(-3, 1)$ lie on the graph of $\log_a x$? Why or why not?

5.5 Exponential and Logarithmic Equations

Objectives

▶ Solve exponential equations

▶ Solve applied problems using exponential equations

▶ Solve logarithmic equations

▶ Solve applied problems using logarithmic equations

Just in Time

Review one-to-one functions in Section 5.1.

Exponential Equations

In Section 5.3, Example 1, we introduced logarithms by seeking a solution to an equation of the form

$$2^t = 128.$$

Equations with variables in the exponents occur quite frequently and are called **exponential equations.** In this section, we will illustrate some algebraic techniques for solving these types of equations by using logarithms.

Since the exponential and logarithmic functions are inverses of each other, they are one-to-one functions. We will use the following one-to-one property to solve exponential and logarithmic equations.

> **One-to-One Property**
>
> For any $a > 0$, $a \neq 1$,
>
> $$a^x = a^y \quad \text{implies} \quad x = y.$$

Example 1 Solving an Exponential Equation

Solve the equation

$$2^t = 128.$$

▶**Solution** Since 128 can be written as a power of 2, we have

$2^t = 128$	Original equation
$2^t = 2^7$	Write 128 as power of 2
$t = 7$	Equate exponents using the one-to-one property

The solution of the equation is $t = 7$, which is the same solution we found in Example 1, Section 5.3, by using a slightly different approach.

✔ *Check It Out 1:* Solve the equation $2^t = 512$. ■

In some cases, the two sides of an exponential equation cannot be written easily in the form of exponential expressions with the same base. In such cases, we take the logarithm of both sides with respect to a suitable base, and then use the one-to-one property to solve the equation.

Technology Note

To solve the equation in Example 2 with a calculator, graph $Y_1(x) = 10^{2x-1}$ and $Y_2(x) = 3^x$ and use the INTERSECT feature. See Figure 5.5.1.

Keystroke Appendix: Section 9

Figure 5.5.1

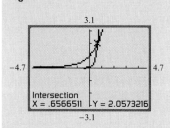

Example 2 Solving an Exponential Equation

Solve the equation $10^{2x-1} = 3^x$.

▶**Solution** We begin by taking the logarithm of both sides. Since 10 is a base of one of the expressions, we will take the logarithm, base 10, on each side.

$$10^{2x-1} = 3^x \qquad \text{Original equation}$$
$$\log 10^{2x-1} = \log 3^x \qquad \text{Take common logarithm of both sides}$$
$$(2x - 1)\log 10 = x \log 3 \qquad \text{Power property of logarithms}$$
$$2x - 1 = x \log 3 \qquad \log 10 = 1$$
$$2x - x \log 3 = 1 \qquad \text{Collect like terms}$$
$$x(2 - \log 3) = 1 \qquad \text{Factor out x}$$
$$x = \frac{1}{2 - \log 3} \approx 0.6567 \qquad \text{Divide both sides by } (2 - \log 3)$$

You can use your calculator to verify that this is indeed the solution of the equation.

✔ **Check It Out 2:** Solve the equation in Example 2 by taking $\log_3$ of each side in the second step and making other modifications as appropriate. If you solve the equation correctly, your answer will match that found in Example 2. ■

The next example shows how to manipulate an equation before taking the logarithm of both sides of the equation.

Example 3 Solving an Exponential Equation

Solve the equation $3e^{2t} + 6 = 24$.

▶**Solution** To solve this equation, we first isolate the exponential term.

$$3e^{2t} + 6 = 24 \qquad \text{Original equation}$$
$$3e^{2t} = 18 \qquad \text{Subtract 6 from both sides to isolate the exponential expression}$$
$$e^{2t} = 6 \qquad \text{Divide both sides by 3}$$

Because e is the base in the exponential expression that appears on the left-hand side of the equation, we will use natural logarithms.

$$\ln e^{2t} = \ln 6 \qquad \text{Take natural logarithm of both sides}$$
$$2t \ln e = \ln 6 \qquad \text{Power property of logarithms}$$
$$2t = \ln 6 \qquad \ln e = 1$$
$$t = \frac{\ln 6}{2} \approx 0.8959 \qquad \text{Divide both sides by 2}$$

✔ **Check It Out 3:** Solve the equation $4e^{3t} - 10 = 26$. ■

Applications of Exponential Equations

Exponential equations occur frequently in applications. We'll explore some of these applications in the examples that follow.

Example 4 Continuous Compound Interest

Suppose a bank pays interest at a rate of 5%, compounded continuously, on an initial deposit of $1000. How long does it take for an investment of $1000 to grow to a total of $1200, assuming that no withdrawals or additional deposits are made?

Technology Note

To check the answer to
Example 4 with a calculator,
first create a table of values
listing the amount in the
account at various times.
The amount of $1200 will
be reached somewhere
between 3 and 4 years.
Graph the functions
$Y_1(x) = 1000e^{0.05x}$ and
$Y_2(x) = 1200$ and find the
intersection point. Use the
table to choose a suitable
window size, such as [0, 10]
by [1000, 1500](100). See
Figure 5.5.2.

Keystroke Appendix:
Section 9

Figure 5.5.2

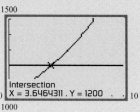

▶**Solution** As stated in Section 5.2, the amount A of money in the account after t years with continuous compounding of interest is given by

$$A = Pe^{rt}$$

where P is the initial deposit and r is the interest rate. Thus we start by substituting the given data.

$1200 = 1000\,e^{0.05t}$	$A = 1200, P = 1000, r = 0.05$
$1.2 = e^{0.05t}$	Divide both sides by 1000 to isolate the exponential expression
$\ln 1.2 = 0.05t \ln e$	Take natural log of both sides
$\ln 1.2 = 0.05t$	$\ln e = 1$
$t = \dfrac{\ln 1.2}{0.05} \approx 3.646$	Solve for t

Thus, it will take about 3.65 years for the amount of money in the account to reach $1200.

✔ *Check It Out 4:* In Example 4, how long will it take for the amount of money in the account to reach $1400? ■

It is fairly easy to make algebraic errors when solving exponential equations, but there are ways to avoid coming up with a solution that is unreasonable. For an application problem involving compound interest, for example, common sense can come in handy: you know that the number of years cannot be negative or very large. In Example 4, the total interest earned is $200. This is 20% of $1000, and even at 5% simple interest (not compounded), this amount can be reached in 4 years. Compounding continuously will lessen the time somewhat. Thus you can see that 3.65 years is a reasonable solution just by making estimations such as this. Using a table of values is another way to see if your answer makes sense.

The following example examines a model in which the exponential function decreases over time.

Example 5 Cost of Computer Disk Storage

Computer storage and memory are calculated using a *byte* as a unit. A kilobyte (KB) is 1000 bytes and a megabyte (MB) is 1,000,000 bytes. The costs of computer storage and memory have decreased exponentially since the 1990s. For example, the cost of computer storage over time can be modeled by the exponential function

$$C(t) = 10.7(0.48)^t$$

where t is the number of years since 1997 and $C(t)$ is the cost at time t, given in cents per megabyte. (*Source:* www.microsoft.com)

(a) How much did a megabyte of computer storage cost in 2002?

(b) When did the cost of computer storage decrease to 1 cent per megabyte?

▶**Solution** Note that the cost function is of the form $y = Ka^t$, with $a = 0.48 < 1$, making it a decreasing function.

(a) Because 2002 is 5 years after 1997, substitute 5 for t in the cost function.

$$C(5) = 10.7(0.48)^5 \approx 0.273$$

Thus, in 2002, one megabyte of storage cost approximately 0.273 cent.

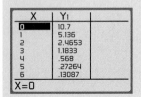

(b) To compute when the cost of one megabyte of computer storage reached 1 cent, set the cost function equal to 1 and solve for t.

$$1 = 10.7(0.48)^t \qquad \textit{Set cost function equal to 1}$$

$$\frac{1}{10.7} = (0.48)^t \qquad \textit{Isolate exponential expression}$$

$$\log \frac{1}{10.7} = t \log 0.48 \qquad \textit{Take common logarithm of both sides}$$

$$t = \frac{\log \dfrac{1}{10.7}}{\log 0.48} \approx 3.23 \quad \textit{Solve for t}$$

Therefore, computer storage cost only 1 cent per megabyte approximately 3.23 years after 1997, or sometime in the beginning of the year 2000.

☑ **Check It Out 5:** Using the model in Example 5, how much will a megabyte of storage cost in 2006? ■

The next example uses logarithms to find the parameters associated with a function that models bacterial population growth.

Example 6 Bacterial Growth

Suppose a colony of bacteria doubles its initial population of 10,000 in 10 hours. Assume the function that models this growth is given by $P(t) = P_0 e^{kt}$, where t is given in hours and P_0 is the initial population.

(a) Find the population at time $t = 0$.   (b) Find the value of k.

(c) What is the population at time $t = 20$?

▶**Solution**

(a) The population at $t = 0$ is simply the initial population of 10,000. Thus, the function modeling this bacteria colony's growth is $P(t) = 10{,}000e^{kt}$. We still have to find k, which is done in the next step.

(b) To find k, we need to write an equation with k as the only variable. Using the fact that the population doubles in 10 years, we have

$$10{,}000e^{k(10)} = 20{,}000 \qquad \textit{Substitute t = 10 and P(10) = 20,000}$$

$$e^{10k} = 2 \qquad \textit{Divide both sides by 10,000}$$

$$10k \ln e = \ln 2 \qquad \textit{Take natural logarithm of both sides}$$

$$10k = \ln 2 \qquad \textit{ln e = 1}$$

$$k = \frac{\ln 2}{10} \approx 0.0693 \qquad \textit{Solve for k}$$

(c) Using the expression for $P(t)$ from part (a) and the value of k from part (b), we have

$$P(t) = 10{,}000e^{0.0693t}.$$

Evaluating the function at $t = 20$ gives

$$P(20) = 10{,}000e^{0.0693(20)} \approx 40{,}000.$$

Note that this value is twice 20,000, the population at time $t = 10$. So, the population doubles every 10 hours.

Check It Out 6: When will the population of the bacteria colony in Example 6 reach 50,000? ■

Equations Involving Logarithms

When an equation involves logarithms, we can use the inverse relationship between exponents and logarithms to solve it, although this approach sometimes yields extraneous solutions. Recall the following property:

$$a^{\log_a x} = x, \ a, x > 0, \ a \neq 1$$

Example 7 illustrates the use of this inverse relationship between exponents and logarithms, together with an operation known as exponentiation, to solve an equation involving logarithms. When we exponentiate both sides of an equation, we choose a suitable base and then raise that base to the expression on each side of the equation.

Example 7 **Solving a Logarithmic Equation**

Solve the equation $4 + \log_3 x = 6$.

▶**Solution**

$4 + \log_3 x = 6$	Original equation
$\log_3 x = 2$	Isolate logarithmic expression
$3^{\log_3 x} = 3^2$	Exponentiate both sides (base 3)
$x = 9$	Inverse property: $3^{\log_3 x} = x$

The solution is $x = 9$. You can check this solution by substituting 9 for x in the original equation.

 Check It Out 7: Solve the equation $-2 + \log_2 x = 3$. ■

Example 8 **Solving an Equation Containing Two Logarithmic Expressions**

Solve the equation $\log 2x + \log(x + 4) = 1$.

▶**Solution**

$\log 2x + \log(x + 4) = 1$	Original equation
$\log 2x(x + 4) = 1$	Combine logarithms using the product property
$10^{\log 2x(x+4)} = 10^1$	Exponentiate both sides (base 10)
$2x(x + 4) = 10$	Inverse property: $10^{\log a} = a$
$2x^2 + 8x - 10 = 0$	Write as a quadratic equation in standard form
$2(x^2 + 4x - 5) = 0$	Factor out a 2
$2(x + 5)(x - 1) = 0$	Factor completely

Setting each factor equal to 0, we find that the only possible solutions are $x = -5$ and $x = 1$. We check each of these possible solutions by substituting them into the original equation.

Check $x = -5$: Since $\log(2(-5)) + \log(-5 + 4) = \log(-10) + \log(-1)$, and logarithms of negative numbers are not defined, $x = -5$ is *not* a solution.

Technology Note

To solve the equation in Example 8 with a calculator, graph $Y_1(x) = \log 2x + \log(x + 4)$ and $Y_2(x) = 1$ and use the INTERSECT feature. See Figure 5.5.4. There are no extraneous solutions because we are directly solving the original equation rather than the quadratic equation obtained through algebra in Example 8.

Keystroke Appendix:
Section 9

Figure 5.5.4

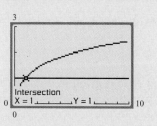

Check $x = 1$: $\log(2(1)) + \log(1 + 4) = \log 2 + \log 5$
$$= \log(2 \times 5)$$
$$= \log 10 = 1$$

Thus $x = 1$ satisfies the equation and is the only solution.

☑ *Check It Out 8:* Solve the equation $\log x + \log(x - 3) = 1$ and check your solution(s). ■

Example 9 **Solving an Equation Involving a Natural Logarithm**

Solve the equation $\ln x = 2 + \ln(x - 1)$.

▶**Solution**

$$\ln x = 2 + \ln(x - 1) \qquad \text{Original equation}$$
$$\ln x - \ln(x - 1) = 2 \qquad \text{Gather logarithmic expressions on one side}$$
$$\ln\left(\frac{x}{x - 1}\right) = 2 \qquad \text{Quotient property of logarithms}$$
$$e^{\ln (x/(x-1))} = e^2 \qquad \text{Exponentiate both sides (base e)}$$
$$\frac{x}{x - 1} = e^2 \qquad \text{Inverse property}$$

We now need to solve for x.

$$x = e^2(x - 1) \qquad \text{Clear fraction: multiply both sides by $x - 1$}$$
$$x - e^2 x = -e^2 \qquad \text{Gather x terms on one side}$$
$$x(1 - e^2) = -e^2 \qquad \text{Factor out x}$$
$$x = -\frac{e^2}{(1 - e^2)} \approx 1.1565 \qquad \text{Solve for x}$$

You can check this answer by substituting it into the *original* equation.

☑ *Check It Out 9:* Solve the equation $\ln x = 1 + \ln(x - 2)$ and check your solution. ■

Application of Logarithmic Equations

Logarithmic functions can be used to model phenomena for which the growth is rapid at first and then slows down. For instance, the total revenue from ticket sales for a movie will grow rapidly at first and then continue to grow, but at a slower rate. This is illustrated in Example 10.

Example 10 **Box Office Revenue**

The cumulative box office revenue from the movie *Finding Nemo* can be modeled by the logarithmic function

$$R(x) = 78.05 \ln(x + 1) + 114.3$$

where x is the number of weeks since the movie opened and $R(x)$ is given in millions of dollars. How many weeks after the opening of the movie was the cumulative revenue equal to \$300 million? (*Source:* movies.yahoo.com)

Technology Note

Let $Y_1(x) = 78.05 \ln(x + 1) + 114.3$ and $Y_2(x) = 300$. Use the INTERSECT feature of a calculator to solve the problem in Example 10. A window size of [0, 20](2) by [100, 400](25) was used in the graph in Figure 5.5.5. The graphical solution ($X = 9.798$) differs in the third decimal place from the answer obtained algebraically because we rounded off in the third step of the algebraic solution.

Keystroke Appendix:
Section 9

Figure 5.5.5

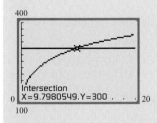

▶**Solution** We set $R(x)$ equal to 300 and solve the resulting logarithmic equation.

$$300 = 78.05 \ln(x + 1) + 114.3 \quad \text{Original equation}$$
$$185.7 = 78.05 \ln(x + 1) \quad \text{Subtract 114.3 from both sides}$$
$$2.379 \approx \ln(x + 1) \quad \text{Divide both sides by 78.05 and round the result}$$
$$e^{2.379} \approx e^{\ln(x+1)} \quad \text{Exponentiate both sides}$$
$$e^{2.379} \approx x + 1 \quad \text{Inverse property}$$
$$x \approx e^{2.379} - 1 \approx 9.794 \quad \text{Solve for } x$$

Thus, by around 9.794 weeks after the opening of the movie, $300 million in total revenue had been generated.

☑ *Check It Out 10:* In Example 10, when did the cumulative revenue reach $200 million?

5.5 Key Points

▶ **One-to-one property:** For any $a > 0$, $a \neq 1$,

$$a^x = a^y \quad \text{implies} \quad x = y.$$

▶ To **solve an exponential equation,** take the logarithm of both sides of the equation and use the power property of logarithms to solve the resulting equation.

▶ To **solve a logarithmic equation,** isolate the logarithmic term, exponentiate both sides of the equation, and use the inverse relationship between logarithms and exponents to solve the resulting equation.

5.5 Exercises

▶**Just in Time Exercises** These exercises correspond to the Just in Time references in this section. Complete them to review topics relevant to the remaining exercises.

1. True or False? Suppose f is a one-to-one function with domain all real numbers. Then there is only one solution to the equation $f(x) = 4$.

2. True or False? $f(x) = 2x + 3$ is not a one-to-one function.

3. True or False? $f(x) = e^x$ is not a one-to-one function.

4. True or False? $f(x) = \ln x$ is a one-to-one function.

▶**Skills** This set of exercises will reinforce the skills illustrated in this section.

In Exercises 5–34, solve the exponential equation. Round to three decimal places, when needed.

5. $5^x = 125$

6. $7^{2x} = 49$

7. $10^x = 1000$

8. $10^x = 0.0001$

9. $4^x = \dfrac{1}{16}$

10. $6^x = \dfrac{1}{216}$

11. $4e^x = 36$

12. $5e^x = 60$

13. $2^x = 5$

14. $3^x = 7$

15. $3(1.3^x) = 5$

16. $6(0.9^x) = 7$

17. $10^x = 2^{-x+4}$

18. $3^{-x} = 10^{-4x+1}$

19. $3^{-2x-1} = 2^x$

20. $5^{x+5} = 3^{-2x+1}$

21. $1000e^{0.04x} = 2000$

22. $250e^{0.05x} = 400$

23. $5e^x + 7 = 32$

24. $4e^x + 6 = 22$

25. $2(0.8^x) - 3 = 8$

26. $4(1.2^x) - 4 = 9$

27. $e^{x^2+1} - 2 = 3$

28. $5 + e^{x^2+1} = 8$

29. $9 - e^{x^2-1} = 2$

30. $10^{2x^2+1} - 8 = 4$

31. $1.7e^{0.5x} = 3.26$

32. $4e^x = -x + 3$

33. $xe^{-x} + e^x = 2$

34. $e^x + e^{-x} = -x + 4$

In Exercises 35–60, solve the logarithmic equation and eliminate any extraneous solutions. If there are no solutions, so state.

35. $\log x = 0$

36. $\ln x = 1$

37. $\ln(x - 1) = 2$

38. $\ln(x + 1) = 3$

39. $\log(x + 2) = 1$

40. $\log(x - 2) = 3$

41. $\log_3(x + 4) = 2$

42. $\log_5(x + 3) = 1$

43. $\log(x + 1) + \log(x - 1) = 0$

44. $\log(x + 3) + \log(x - 3) = 0$

45. $\log x + \log(x + 3) = 1$

46. $\log x + \log(2x - 1) = 1$

47. $\log_2 x = 2 - \log_2(x - 3)$

48. $\log_5 x = 1 - \log_5(x - 4)$

49. $\ln(2x) = 1 + \ln(x + 3)$

50. $\log_3 x = 2 + \log_3(x - 2)$

51. $\log(3x + 1) - \log(x^2 + 1) = 0$

52. $\log(x + 5) - \log(4x^2 + 5) = 0$

53. $\log(2x + 5) + \log(x + 1) = 1$

54. $\log(3x + 1) + \log(x + 1) = 1$

55. $\log_2(x + 5) = \log_2(x) + \log_2(x - 3)$

56. $\ln 2x - \ln(x^2 + 1) = \ln 1$

57. $2 \ln x + \ln(x - 1) = 3.1$

58. $-\ln x - \ln(x + 2) = 2.5$

59. $\log |x - 2| + \log |x| = 1.2$

60. $\ln x = (x - 2)^2$

▶**Applications** In this set of exercises, you will use exponential and logarithmic equations to study real-world problems.

Banking *In Exercises 61–66, determine how long it takes for the given investment to double if r is the interest rate and the interest is compounded continuously. Assume that no withdrawals or further deposits are made.*

61. Initial amount: \$1500; $r = 6\%$

62. Initial amount: \$3000; $r = 4\%$

63. Initial amount: \$4000; $r = 5.75\%$

64. Initial amount: \$6000; $r = 6.25\%$

65. Initial amount: \$2700; $r = 7.5\%$

66. Initial amount: \$3800; $r = 5.8\%$

Banking *In Exercises 67–72, find the interest rate r if the interest on the initial deposit is compounded continuously and no withdrawals or further deposits are made.*

67. Initial amount: \$1500; Amount in 5 years: \$2000

68. Initial amount: \$3000; Amount in 3 years: \$3600

69. Initial amount: \$4000; Amount in 8 years: \$6000

70. Initial amount: \$6000; Amount in 10 years: \$12,000

71. Initial amount: $8500; Amount in 5 years: $10,000

72. Initial amount: $12,000; Amount in 20 years: $25,000

73. **Bacterial Growth** Suppose the population of a colony of bacteria doubles in 12 hours from an initial population of 1 million. Find the growth constant k if the population is modeled by the function $P(t) = P_0 e^{kt}$. When will the population reach 4 million? 8 million?

74. **Bacterial Growth** Suppose the population of a colony of bacteria doubles in 20 hours from an initial population of 1 million. Find the growth constant k if the population is modeled by the function $P(t) = P_0 e^{kt}$. When will the population reach 4 million? 8 million?

75. **Computer Science** In 1965, Gordon Moore, then director of Intel research, conjectured that the number of transistors that fit on a computer chip doubles every few years. This has come to be known as *Moore's Law*. Analysis of data from Intel Corporation yields the following model of the number of transistors per chip over time:

$$s(t) = 2297.1 e^{0.3316t}$$

where $s(t)$ is the number of transistors per chip and t is the number of years since 1971. (*Source:* Intel Corporation)

(a) According to this model, what was the number of transistors per chip in 1971?

(b) How long did it take for the number of transistors to double?

76. **Depreciation** The value of a 2003 Toyota Corolla is given by the function

$$v(t) = 14,000(0.93)^t$$

where t is the number of years since its purchase and $v(t)$ is its value in dollars. (*Source:* Kelley Blue Book)

(a) What was the Corolla's initial purchase price?

(b) What percent of its value does the Toyota Corolla lose each year?

(c) How long will it take for the value of the Toyota Corolla to reach $12,000?

77. **Depreciation** The value of a 2006 S-type Jaguar is given by the function

$$v(t) = 43,173(0.8)^t$$

where t is the number of years since its purchase and $v(t)$ is its value in dollars. (*Source:* Kelley Blue Book)

(a) What was the Jaguar's initial purchase price?

(b) What percentage of its value does the Jaguar S-type lose each year?

(c) How many years will it take for the Jaguar S-type to reach a value of $22,227?

78. **Film Industry** The cumulative box office revenue from the movie *Terminator 3* can be modeled by the logarithmic function

$$R(x) = 26.203 \ln x + 90.798$$

where x is the number of weeks since the movie opened and $R(x)$ is given in millions of dollars. How many weeks after the opening of the movie did the cumulative revenue reach $140 million? (*Source:* movies.yahoo.com)

79. **Physics** Plutonium is a radioactive element that has a *half-life* of 24,360 years. The **half-life** of a radioactive substance is the time it takes for *half* of the substance to decay (which means the other half will still exist after that length of time). Find an exponential function of the form $f(t) = Ae^{kt}$ that gives the amount of plutonium left after t years if the initial amount of plutonium is 10 pounds. How long will it take for the plutonium to decay to 2 pounds?

Chemistry *Exercises 80 and 81 refer to the following. The pH of a solution is defined as* $\mathrm{pH} = -\log [\mathrm{H}^+]$. *The concentration of hydrogen ions,* $[\mathrm{H}^+]$, *is given in moles per liter, where one mole is equal to* 6.02×10^{23} *molecules.*

80. What is the concentration of hydrogen ions in a solution that has a pH of 6.2?

81. What is the concentration of hydrogen ions in a solution that has a pH of 1.5?

82. **Geology** The 1960 earthquake in Chile registered 9.5 on the Richter scale. Find the energy E (in Ergs) released by using the following model, which relates the energy in Ergs to the magnitude R of an earthquake. (*Source:* National Earthquake Information Center, U.S. Geological Survey)

$$\log E = 11.4 + (1.5)R$$

83. **Acoustics** The decibel (dB) is a unit that is used to express the relative loudness of two sounds. One application of decibels is the relative value of the output power of an amplifier with respect to the input power. Since power levels can vary greatly in magnitude, the *relative value D* of power level P_1 with respect to power level P_2 is given (in units of dB) in terms of the logarithm of their ratio as follows:

$$D = 10 \log \frac{P_1}{P_2}$$

where the values of P_1 and P_2 are expressed in the same units, such as watts (W). If $P_2 = 75$ W, find the value of P_1 at which $D = 0.7$.

84. **Depreciation** A new car that costs $25,000 depreciates to 80% of its value in 3 years.

(a) Assume the depreciation is linear. What is the linear function that models the value of this car t years after purchase?

(b) Assume the value of the car is given by an exponential function $y = Ae^{kt}$, where A is the initial price of the car. Find the value of the constant k and the exponential function.

(c) Using the linear model found in part (a), find the value of the car 5 years after purchase. Do the same using the exponential model found in part (b).

(d) ▨ Graph both models using a graphing utility. Which model do you think is more realistic, and why?

85. **Horticulture** Pesticides decay at different rates depending on the pH level of the water contained in the pesticide solution. The pH scale measures the acidity of a solution. The lower the pH value, the more acidic the solution. When produced with water that has a pH of 6.0, the pesticide chemical known as malathion has a *half-life* of 8 days; that is, *half* the initial amount of malathion will remain after 8 days. However, if it is produced with water that has a pH of 7.0, the half-life of malathion decreases to 3 days. (*Source:* Cooperative Extension Program, University of Missouri)

(a) Assume the initial amount of malathion is 5 milligrams. Find an exponential function of the form $A(t) = A_0 e^{kt}$ that gives the amount of malathion that remains after t days if it is produced with water that has a pH of 6.0.

(b) Assume the initial amount of malathion is 5 milligrams. Find an exponential function of the form $B(t) = B_0 e^{kt}$ that gives the amount of malathion that remains after t days if it is produced with water that has a pH of 7.0.

(c) How long will it take for the amount of malathion in each of the solutions in parts (a) and (b) to decay to 3 milligrams?

(d) If the malathion is to be stored for a few days before use, which of the two solutions would be more effective, and why?

(e) ▨ Graph the two exponential functions in the same viewing window and describe how the graphs illustrate the differing decay rates.

Investment *In Exercises 86–89, use the following table, which illustrates the growth over time of an amount of money deposited in a bank account.*

t (years)	Amount ($)
0	5000
1	5309.18
2	5637.48
4	6356.25
6	7166.65
10	9110.59
12	10,272.17

86. What is the amount of the initial deposit?

87. From the table, approximately how long does it take for the initial investment to earn a total of $600 in interest?

88. From the table, approximately how long does it take for the amount of money in the account to double?

89. Assume that the amount of money in the account at time t (in years) is given by $V(t) = P_0 e^{rt}$, where P_0 is the initial deposit and r is the interest rate. Find the exponential function and the value of r.

▶ **Concepts** This set of exercises will draw on the ideas presented in this section and your general math background.

90. Do the equations $\ln x^2 = 1$ and $2 \ln x = 1$ have the same solutions? Explain.

91. Explain why the equation $2e^x = -1$ has no solution.

92. What is wrong with the following step?

$$\log x + \log(x + 1) = 0 \Rightarrow x(x + 1) = 0$$

93. What is wrong with the following step?

$$2^{x+5} = 3^{4x} \Rightarrow x + 5 = 4x$$

In Exercises 94–97, solve using any method, and eliminate extraneous solutions.

94. $\ln(\log x) = 1$

95. $e^{\log x} = e$

96. $\log_5 |x - 2| = 2$

97. $\ln |2x - 3| = 1$

5.6 Exponential, Logistic, and Logarithmic Models

Objectives

▶ Construct an exponential decay model

▶ Use curve-fitting for an exponential model

▶ Use curve-fitting for a logarithmic model

▶ Define a logistic model

▶ Use curve-fitting for a logistic model

We have examined some applications of exponential and logarithmic functions in the previous sections. In these applications, we were usually given the expression for the function. In this section, we will study how an appropriate model can be selected when we are given some *data* about a certain problem. This is how real-world models often arise.

In simple cases, it is possible to come up with an appropriate model through paper-and-pencil work. For more complicated data sets, we will need to use technology to find a suitable model. We will examine both types of problems in this section. In addition, we will investigate other functions, closely related to the exponential function, that are useful in solving real-world applications.

Exponential Growth and Decay

Recall the following facts from Section 5.2 about exponential functions.

Just in Time

Review properties of exponential functions in Section 5.2.

Properties of Exponential Functions

▶ An exponential function of the form $f(x) = Ca^x$, where $C > 0$ and $a > 1$, models **exponential growth.** See Figure 5.6.1.

Figure 5.6.1

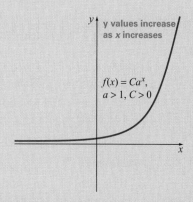

▶ An exponential function of the form $f(x) = Ca^x$, where $C > 0$ and $0 < a < 1$, models **exponential decay.** See Figure 5.6.2.

Figure 5.6.2

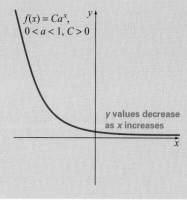

When modeling exponential growth and decay without the use of technology, it is more convenient to state the appropriate functions using the base e. If $a > 1$, we can write $a = e^k$, where k is some constant such that $k > 0$. If $a < 1$, we can write $a = e^k$, where k is some constant such that $k < 0$. We then have the following.

Modeling Growth and Decay with Base e

▶ An exponential function of the form

$$f(x) = Ce^{kx}, \text{ where } C > 0 \text{ and } k > 0$$

models **exponential growth.**

▶ An exponential function of the form

$$f(x) = Ce^{kx}, \text{ where } C > 0 \text{ and } k < 0$$

models **exponential decay.**

The exponential function is well suited to many models in the social, life, and physical sciences. The following example discusses the decay rate of the radioactive metal strontium-90. This metal has a variety of commercial and research uses. For example, it is used in fireworks displays to produce the red flame color. (*Source: Argonne National Laboratories*)

Example 1 Modeling Radioactive Decay

It takes 29 years for an initial amount A_0 of strontium-90 to break down into half the initial amount, $\frac{A_0}{2}$. That is, the *half-life* of strontium-90 is 29 years.

(a) Given an initial amount of A_0 grams of strontium-90 at time $t = 0$, find an exponential decay model, $A(t) = A_0 e^{kt}$, that gives the amount of strontium-90 at time t, $t \geq 0$.

(b) Calculate the time required for the initial amount of strontium-90 to decay to $\frac{1}{10}A_0$.

▶**Solution**

(a) The exponential model is given by $A(t) = A_0 e^{kt}$. After 29 years, the amount of strontium-90 is $\frac{1}{2}A_0$. Putting this information together, we have

$$A(t) = A_0 e^{kt} \qquad \text{\textit{Given model}}$$

$$A(29) = \frac{1}{2}A_0 = A_0 e^{k(29)} \qquad \text{\textit{At } t = 29, A(29) = \frac{1}{2}A_0}$$

$$\frac{1}{2} = e^{k(29)} \qquad \text{\textit{Divide both sides of equation by } A_0}$$

$$\ln\left(\frac{1}{2}\right) = 29k \qquad \text{\textit{Take natural logarithm of both sides}}$$

$$\frac{\ln\frac{1}{2}}{29} = k \qquad \text{\textit{Solve for } k}$$

$$k \approx -0.02390 \qquad \text{\textit{Approximate } k \text{ to four significant digits}}$$

Because this is a decay model, we know that $k < 0$. Thus the decay model is

$$A(t) = A_0 e^{-0.02390t}.$$

(b) We must calculate t such that $A(t) = \frac{1}{10}A_0$. We proceed as follows.

$$A(t) = A_0 e^{-0.02390t} \qquad \text{Given model}$$

$$\frac{1}{10}A_0 = A_0 e^{-0.02390t} \qquad \text{Substitute } A(t) = \frac{1}{10}A_0$$

$$\frac{1}{10} = e^{-0.02390t} \qquad \text{Divide both sides of equation by } A_0$$

$$\ln\left(\frac{1}{10}\right) = -0.02390t \qquad \text{Take natural logarithm of both sides}$$

$$\frac{\ln\frac{1}{10}}{-0.02390} = t \qquad \text{Solve for } t$$

$$t \approx 96.34 \qquad \text{Approximate } t \text{ to four significant digits}$$

Thus, in approximately 96.34 years, one-tenth of the original amount of strontium-90 will remain.

✔ **Check It Out 1:** Radium-228 is a radioactive metal with a half-life of 6 years. Find an exponential decay model, $A(t) = A_0 e^{kt}$, that gives the amount of radium-228 at time t, $t \geq 0$. ▪

Example 2 discusses population growth assuming an exponential model.

Example 2 Modeling Population Growth

The population of the United States is expected to grow from 282 million in 2000 to 335 million in 2020. (*Source:* U.S. Census Bureau)

(a) Find a function of the form $P(t) = Ce^{kt}$ that models the population growth. Here, t is the number of years after 2000, and $P(t)$ is the population in millions.

(b) Use your model to predict the population of the United States in 2010.

▶**Solution**

(a) If t is the number of years after 2000, we have the following two data points: $(0, 282)$ and $(20, 335)$ First, we find C.

$$P(t) = Ce^{kt} \qquad \text{Given equation}$$
$$P(0) = Ce^{k(0)} = 282 \qquad \text{Population at } t = 0 \text{ is 282 million}$$
$$C = 282 \qquad \text{Because } Ce^{k(0)} = C$$

Thus, the model is $P(t) = 282e^{kt}$. Next, we find k.

$$P(t) = 282e^{kt} \qquad \text{Given equation}$$
$$P(20) = 282e^{k(20)} = 335 \qquad \text{Population at } t = 20 \text{ is 335 million}$$
$$282e^{20k} = 335$$

$$e^{20k} = \frac{335}{282} \qquad \text{Isolate the exponential term}$$

$$20k = \ln\frac{335}{282} \qquad \text{Take natural logarithm of both sides}$$

$$k = \frac{\ln\frac{335}{282}}{20} \approx 0.00861 \qquad \text{Solve for } k \text{ and approximate to three significant digits}$$

Thus the function is $P(t) = 282e^{0.00861t}$.

(b) We substitute $t = 10$ into the function from part (a), since 2010 is 10 years after 2000.

$$P(10) = 282e^{0.00861(10)} \approx 307$$

Thus, in 2010, the population of the United States will be about 307 million.

✔ *Check It Out 2:* Use the function in Example 2 to estimate the population of the United States in 2015. ■

Models Using Curve-Fitting

In this section we will explore real-world problems that can be analyzed using only the curve-fitting, or regression, capabilities of your graphing utility. We will discuss examples of exponential, logistic, and logarithmic models.

Example 3 Growth of the National Debt

Table 5.6.1 shows the United States national debt (in billions of dollars) for selected years from 1975 to 2005. (*Source:* U.S. Department of the Treasury)

Technology Note

Curve-fitting features are available under the Statistics option on most graphing calculators.

Keystroke Appendix: Section 12

Table 5.6.1

Years Since 1975	National Debt (in billions of dollars)
0	576.6
5	930.2
10	1946
15	3233
20	4974
25	5674
30	7933

(a) Make a scatter plot of the data and find the exponential function of the form $f(x) = Ca^x$ that best fits this data.

(b) From the model in part (a), what is the projected national debt for the year 2010?

▶Solution

(a) We can use a graphing utility to graph the data points and find the best-fitting exponential model. Figure 5.6.3 shows a scatter plot of the data, along with the exponential curve.

Figure 5.6.3

The exponential function that best fits this data is given by

$$d(x) = 681.2(1.093)^x.$$

(b) To find the projected debt in 2010, we calculate $d(35)$:

$$d(35) = 681.2(1.093)^{35} \approx 15,310$$

Thus the projected national debt in 2010 will be approximately $15,310 *billion* dollars, or $15.31 *trillion* dollars.

✔ *Check It Out 3:* Use the model in Example 3 to project the national debt in the year 2012. ▪

Example 4 Modeling Loads in a Structure

When designing buildings, engineers must pay careful attention to how different factors affect the load a structure can carry. Table 5.6.2 gives the load in pounds of concrete when a 1-inch-diameter anchor is used as a joint. The table summarizes the relation between the load and how deep the anchor is drilled into the concrete. (*Source:* Simpson Anchor Systems)

(a) From examining the table, what is the general relationship between the depth of the anchor and the load?

(b) Make a scatter plot of the data and find the natural logarithmic function that best fits the data.

(c) If an anchor were drilled 10 inches deep, what is the resulting load that could be carried?

(d) What is the minimum depth an anchor should be drilled in order to sustain a load of 9000 pounds?

Table 5.6.2

Depth (in.)	Load (lb)
4.5	5020
6.75	10,020
9	15,015
12	17,810
15	20,600

▶Solution

(a) From examining the table, we see that the deeper the anchor is drilled, the heavier the load that can be sustained. However, the sustainable load increases rapidly at first, and then increases slowly. Thus a logarithmic model seems appropriate.

(b) We can use a graphing utility to plot the data points and find the best-fitting logarithmic model. Figure 5.6.4 shows a scatter plot of the data, along with the logarithmic curve.

Figure 5.6.4

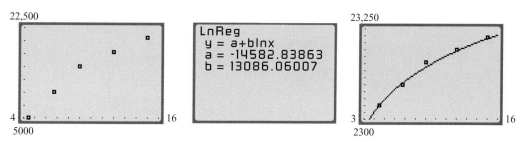

The logarithmic function that best fits this data is given by

$$L(x) = 13,086 \ln(x) - 14,583$$

where x is the depth the anchor is drilled.

(c) To find the sustainable load when an anchor is drilled 10 inches deep, we evaluate $L(10)$.

$$L(10) = 13{,}086 \ln(10) - 14{,}583 \approx 15{,}548$$

The resulting load is approximately 15,548 pounds. We can check that this value is reasonable by comparing it with the data in the table.

(d) To find the minimum depth, we set the expression for the load to 9000 and solve for x.

$$13{,}086 \ln(x) - 14{,}583 = 9000 \qquad \textit{Set load expression equal to 9000}$$
$$13{,}086 \ln x = 23{,}583 \qquad \textit{Add 14,583 to both sides}$$
$$\ln x = \frac{23{,}583}{13{,}086} \approx 1.8021 \qquad \textit{Divide both sides by 13,086}$$
$$x = e^{1.8021} \approx 6.0623$$

Thus, the anchor must be drilled to a depth of at least 6.0623 inches to sustain a load of 9000 pounds. Drilling to a greater depth simply means the anchor will sustain more than 9000 pounds.

We also could have found the solution with a graphing utility by storing the logarithmic function as y_1 and finding its intersection with the equation $y_2 = 9000$.

✔ **Check It Out 4:** Find the solution to part (d) of Example 4 using a graphing utility. ■

The Logistic Model

In the previous section, we examined population growth models by using an exponential function. However, it seems unrealistic that any population would simply tend to infinity over a long period of time. Other factors, such as the ability of the environment to support the population, would eventually come into play and level off the population. Thus we need a more refined model of population growth that takes such issues into account.

One function that models this behavior is known as a **logistic function.** It is defined as

$$f(x) = \frac{c}{1 + ae^{-bx}}$$

where a, b, and c are constants determined from a given set of data. Finding these constants involves using the *logistic regression* feature of your graphing utility. The graph of the logistic function is shown in Figure 5.6.5. We can examine some properties of this function by letting $f(x) = \frac{3}{1 + 2e^{-0.5x}}$. Table 5.6.3 lists some values for $f(x)$.

Figure 5.6.5 Graph of logistic function

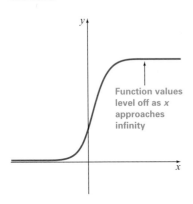

Function values level off as x approaches infinity

Table 5.6.3

x	$f(x) = \dfrac{3}{1 + 2e^{-0.5x}}$
−10	0.0101
−5	0.1183
−1	0.6981
0	1.0000
1	1.3556
2	1.7284
5	2.5769
10	2.9601
25	2.9999

Observations:

▶ As x increases, the function values approach 3. This is because $2e^{-0.5x}$ will get very small in magnitude as x increases to $+\infty$, making the denominator of $f(x)$, $1 + 2e^{-0.5x}$, very close to 1. Because $f(x) = \frac{3}{1 + 2e^{-0.5x}}$, the values of $f(x)$ will approach 3 as x increases to $+\infty$.

▶ As x decreases, the function values approach 0. This is because $2e^{-0.5x}$ will get very large in magnitude as x decreases to $-\infty$, making the denominator of $f(x)$, $1 + 2e^{-0.5x}$, very large in magnitude. Because $f(x) = \frac{3}{1 + 2e^{-0.5x}}$, the values of $f(x)$ will approach 0 as x decreases to $-\infty$.

In Example 5, a logistic function is used to analyze a set of data.

Example 5 Logistic Population Growth

Table 5.6.4

Year	Population (millions)
1970	191
1980	242
1990	296
2000	347

Table 5.6.4 gives the population of South America for selected years from 1970 to 2000. (*Source:* U.S. Census Bureau)

(a) Use a graphing utility to make a scatter plot of the data and find the logistic function of the form $f(x) = \dfrac{c}{1 + ae^{-bx}}$ that best fits the data. Let x be the number of years after 1970.

(b) Using this model, what is the projected population in 2020? How does it compare with the projection of 421 million given by the U.S. Census Bureau?

▶**Solution**

(a) Figure 5.6.6 shows a scatter plot of the data, along with the logistic curve. The logistic function that best fits this data is given by

$$p(x) = \frac{537}{1 + 1.813e^{-0.04x}}.$$

Figure 5.6.6

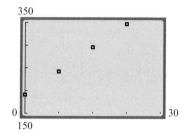

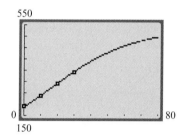

(b) To find the projected population in 2020, we calculate $p(50)$:

$$p(50) = \frac{537}{1 + 1.813e^{-0.04(50)}} \approx 431$$

This model predicts that there will be approximately 431 million people in South America in 2020. The statisticians who study these types of data use a more sophisticated type of analysis, of which curve-fitting is only a part. The Census Bureau's projection of 421 million is relatively close to the population predicted by our model.

✔ *Check It Out 5:* Using the model found in Example 5, what is the projected population of South America in 2040? How does it compare with the projection of 468 million given by the U.S. Census Bureau? ■

5.6 Key Points

▶ An exponential function of the form $f(x) = Ce^{kx}$, where $C > 0$ and $k > 0$, models **exponential growth.**

▶ An exponential function of the form $f(x) = Ce^{kx}$, where $C > 0$ and $k < 0$, models **exponential decay.**

▶ A **logarithmic function** can be used to model the growth rate of a phenomenon that grows rapidly at first and then more slowly.

▶ A **logistic function** of the form $f(x) = \dfrac{c}{1 + ae^{-bx}}$ is used to model the growth rate of a phenomenon whose growth must eventually level off.

5.6 Exercises

▶**Just in Time Exercises** These exercises correspond to the Just in Time references in this section. Complete them to review topics relevant to the remaining exercises.

1. An exponential function of the form $f(x) = Ca^x$, where $C > 0$ and $a > 1$, models exponential _____.

2. An exponential function of the form $f(x) = Ca^x$, where $C > 0$ and $a < 1$, models exponential _____.

3. Let $f(x) = 5e^x$. As $x \to \infty$, $f(x) \to$ _____.

4. Let $f(x) = 5e^x$. As $x \to -\infty$, $f(x) \to$ _____.

5. Let $f(x) = \left(\frac{1}{3}\right)^x$. As $x \to \infty$, $f(x) \to$ _____.

6. Let $f(x) = \left(\frac{1}{3}\right)^x$. As $x \to -\infty$, $f(x) \to$ _____.

▶**Skills** This set of exercises will reinforce the skills illustrated in this section.

In Exercises 7–10, use $f(t) = 10e^{-t}$.

7. Evaluate $f(0)$.

8. Evaluate $f(2)$.

9. For what value of t will $f(t) = 5$?

10. For what value of t will $f(t) = 2$?

In Exercises 11–14, use $f(t) = 4e^t$.

11. Evaluate $f(1)$.

12. Evaluate $f(3)$.

13. For what value of t will $f(t) = 8$?

14. For what value of t will $f(t) = 10$?

In Exercises 15–18, use $f(x) = \dfrac{10}{1 + 2e^{-0.3x}}$.

15. Evaluate $f(0)$.   16. Evaluate $f(1)$.

17. Evaluate $f(10)$.   18. Evaluate $f(12)$.

In Exercises 19–22, use $f(x) = 3 \ln x - 4$.

19. Evaluate $f(e)$.

20. Evaluate $f(1)$.

21. For what value of x will $f(x) = 2$?

22. For what value of x will $f(x) = 3$?

In Exercises 23–26, match the description to one of the graphs (a)–(d).

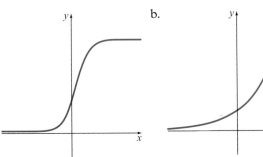

a.   b.

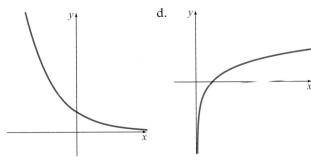

c.   d.

23. Exponential decay model

24. Logarithmic growth model

25. Logistic growth model

26. Exponential growth model

▶**Applications** In this set of exercises, you will use exponential, logistic, and logarithmic models to study real-world problems.

27. **Chemistry** It takes 5700 years for an initial amount A_0 of carbon-14 to break down into half the amount, $\frac{A_0}{2}$.
    (a) Given an initial amount of A_0 grams of carbon-14 at time $t = 0$, find an exponential decay model, $A(t) = A_0 e^{kt}$, that gives the amount of carbon-14 at time t, $t \geq 0$.
    (b) Calculate the time required for A_0 grams of carbon-14 to decay to $\frac{1}{3} A_0$.

28. **Chemistry** The half-life of plutonium-238 is 88 years.
    (a) Given an initial amount of A_0 grams of plutonium-238 at time $t = 0$, find an exponential decay model, $A(t) = A_0 e^{kt}$, that gives the amount of plutonium-238 at time t, $t \geq 0$.

(b) Calculate the time required for A_0 grams of plutonium-238 to decay to $\frac{1}{3} A_0$.

29. Archaeology At an excavation site, an archaeologist discovers a piece of wood that contains 70% of its initial amount of carbon-14. Approximate the age of the wood. (Refer to Exercise 27.)

30. Population Growth The population of the United States is expected to grow from 282 million in 2000 to 335 million in 2020. (*Source:* U.S. Census Bureau)

(a) Find a function of the form $P(t) = Ce^{kt}$ that models the population growth of the United States. Here, t is the number of years since 2000 and $P(t)$ is in millions.

(b) Assuming the trend in part (a) continues, in what year will the population of the United States be 300 million?

31. Population Growth The population of Florida grew from 16.0 million in 2000 to 17.4 million in 2004. (*Source:* U.S. Census Bureau)

(a) Find a function of the form $P(t) = Ce^{kt}$ that models the population growth. Here, t is the number of years since 2000 and $P(t)$ is in millions.

(b) Use your model to predict the population of Florida in 2010.

32. Temperature Change A frozen pizza with a temperature of 30°F is placed in a room with a steady temperature of 75°F. In 15 minutes, the temperature of the pizza has risen to 40°F.

(a) Find an exponential model, $F(t) = Ce^{kt}$, that models the temperature of the pizza in degrees Fahrenheit, where t is the time in minutes after the pizza is removed from the freezer.

(b) How long will it take for the temperature of the pizza to reach 55°F?

33. Housing Prices The median price of a new home in the United States rose from $123,000 in 1990 to $220,000 in 2004. Find an exponential function $P(t) = Ce^{kt}$ that models the growth of housing prices, where t is the number of years since 1990. (*Source:* National Association of Home Builders)

34. Economics Due to inflation, a dollar in the year 1994 is worth $1.28 in 2005 dollars. Find an exponential function $v(t) = Ce^{kt}$ that models the value of a 1994 dollar t years after 1994. (*Source:* Inflationdata.com)

35. Depreciation The purchase price of a 2006 Ford F150 longbed pickup truck is $23,024. After 1 year, the price of the Ford F150 is $17,160. (*Source:* Kelley Blue Book)

(a) Find an exponential function, $P(t) = Ce^{kt}$, that models the price of the truck, where t is the number of years since 2006.

(b) What will be the value of the Ford F150 in the year 2009?

36. Health Sciences The spread of the flu in an elementary school can be modeled by a logistic function. The number of children infected with the flu virus t days after the first infection is given by

$$N(t) = \frac{150}{1 + 4e^{-0.5t}}.$$

(a) How many children were initially infected with the flu?

(b) How many children were infected with the flu virus after 5 days? after 10 days?

37. Wildlife Conservation The population of white-tailed deer in a wildlife refuge t months after their introduction into the refuge can be modeled by the logistic function

$$N(t) = \frac{300}{1 + 14e^{-0.05t}}.$$

(a) How many deer were initially introduced into the refuge?

(b) How many deer will be in the wildlife refuge 10 months after introduction?

38. Commerce The following table gives the sales, in billions of current dollars, for restaurants in the United States for selected years from 1970 to 2005. (*Source:* National Restaurant Association Fact Sheet, 2005)

Year	1970	1985	1995	2005
Sales	42.8	173.7	295.7	475.8

(a) Make a scatter plot of the data and find the exponential function of the form $f(x) = Ca^x$ that best fits the data. Let x be the number of years since 1970.

(b) Why must a be greater than 1 in your model?

(c) Using your model, what are the projected sales for restaurants in the year 2008?

(d) Do you think this model will be accurate over the long term? Explain.

39. **Tourism** The following table shows the tourism revenue for China, in billions of dollars, for selected years since 1990. (*Source:* World Tourism Organization)

Year	Revenue (billions of dollars)
1990	2.218
1995	8.733
1996	10.200
1998	12.602
2000	16.231
2002	20.385

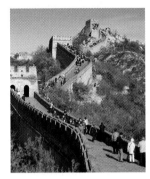

(a) Make a scatter plot of the data and find the exponential function of the form $f(x) = Ca^x$ that best fits the data. Let x be the number of years since 1990.

(b) Using this model, what is the projected revenue from tourism in the year 2008?

(c) Do you think this model will be accurate over the long term? Explain.

40. **Health Sciences** The spread of a disease can be modeled by a logistic function. For example, in early 2003, there was an outbreak of an illness called SARS (Severe Acute Respiratory Syndrome) in many parts of the world. The following table gives the *total number of cases* in Canada for the weeks following March 20, 2003. (*Source:* World Health Organization) (*Note:* The total number of cases dropped from 149 to 140 between weeks 3 and 4 because some of the cases thought to be SARS were reclassified as other diseases.)

Weeks Since March 20, 2003	Total Cases
0	9
1	62
2	132
3	149
4	140
5	216
6	245
7	252
8	250

(a) Explain why a logistic function would suit this data well.

(b) Make a scatter plot of the data and find the logistic function of the form $f(x) = \dfrac{c}{1 + ae^{-bx}}$ that best fits the data.

(c) What does c signify in your model?

(d) The World Health Organization declared in July 2003 that SARS no longer posed a threat in Canada. By analyzing this data, explain why that would be so.

41. **Car Racing** The following table lists the qualifying speeds, in miles per hour, of the Indianapolis 500 car race winners for selected years from 1931 to 2005. (*Source:* www.indy500.com)

Year	Qualifying Speed (mph)
1931	107
1941	121
1951	135
1961	145
1971	174
1981	200
1991	224
2005	228

(a) Explain why a logistic function would fit this data well.

(b) Make a scatter plot of the data and find the logistic function of the form $f(x) = \dfrac{c}{1 + ae^{-bx}}$ that best fits the data. Let x be the number of years since 1931.

(c) What does c signify in your model?

(d) Using your model, what is the projected qualifying speed for the winner in 2008?

42. **Heat Loss** The following table gives the temperature, in degrees Celsius, of a cup of hot water sitting in a room with constant temperature. The data was collected over a period of 30 minutes. (*Source:* www.phys.unt.edu, Dr. James A. Roberts)

Time (min)	Temperature (degrees Celsius)
0	95
1	90.4
5	84.6
10	73
15	64.7
20	59
25	54.5
29	51.4

(a) Make a scatter plot of the data and find the exponential function of the form $f(t) = Ca^t$ that best fits the data. Let t be the number of minutes the water has been cooling.

(b) Using your model, what is the projected temperature of the water after 1 hour?

43. **Oil Prices** The following table gives the price per barrel of crude oil for selected years from 1992 to 2006. (*Source:* www.ioga.com/special/crudeoil-Hist.htm)

Year	Price (dollars)
1992	19.25
1996	20.46
2000	27.40
2004	37.41
2006	58.30

(a) Make a scatter plot of the data and find the exponential function of the form $P(t) = Ca^t$ that best fits the data. Let t be the number of years since 1992.

(b) Using your model, what is the projected price per barrel of crude oil in 2009?

44. **Campaign Spending** The following table gives the total amount spent by all candidates in each presidential election, beginning in 1988. Each amount listed is in millions. (*Source:* Federal Election Commission)

Year	Price (millions of dollars)
1988	495
1992	550
1996	560
2000	649.5
2004	1,016.5

(a) Make a scatter plot of the data and find the exponential function of the form $P(t) = Ca^t$ that best fits the data. Let t be the number of years since 1988.

(b) Using your model, what is the projected total amount all candidates will spend during the 2012 presidential election?

45. **Environmental Science** The cost of removing chemicals from drinking water depends on how much of the chemical can safely be left behind in the water. The following table lists the annual removal costs for arsenic in

terms of the concentration of arsenic in the drinking water. (*Source:* Environmental Protection Agency)

Arsenic Concentration (micrograms per liter)	Annual Cost (millions of dollars)
3	645
5	379
10	166
20	65

(a) Interpret the data in the table. What is the relation between the amount of arsenic left behind in the removal process and the annual cost? (One microgram is equal to 10^{-6} gram.)

(b) Make a scatter plot of the data and find the exponential function of the form $C(x) = Ca^x$ that best fits the data. Here, x is the arsenic concentration.

(c) Why must a be less than 1 in your model?

(d) Using your model, what is the annual cost to obtain an arsenic concentration of 12 micrograms per liter?

(e) It would be best to have the smallest possible amount of arsenic in the drinking water, but the cost may be prohibitive. Use your model to calculate the annual cost of processing such that the concentration of arsenic is only 2 micrograms per liter of water. Interpret your result.

46. **Prenatal Care** The following data gives the percentage of women who smoked during pregnancy for selected years from 1994 to 2002. (*Source:* National Center for Health Statistics)

Year	Percent Smoking During Pregnancy
1994	14.6
1996	13.6
1998	12.9
2000	12.2
2001	12.0
2002	11.4

(a) From examining the table, what is the general relationship between the year and the percentage of women smoking during pregnancy?

(b) Let t be the number of years after 1993. Here, t starts at 1 because ln 0 is undefined. Make a scatter plot of the data and find the natural logarithmic function of the form $p(t) = a \ln t + b$ that best fits the data. Why must a be negative?

(c) Project the percentage of women who will smoke during pregnancy in the year 2007.

▶**Concepts** This set of exercises will draw on the ideas presented in this section and your general math background.

47. Refer to Example 1. Without solving an equation, how would you figure out when the amount of strontium-90 would reach one-fourth the initial amount A_0?

48. The value c in the logistic function $f(x) = \dfrac{c}{1 + ae^{-bx}}$ is sometimes called the *carrying capacity*. Can you give a reason why this term is used?

49. Explain why the function $f(t) = e^{(1/2)t}$ cannot model exponential decay.

50. For the logistic function $f(x) = \dfrac{c}{1 + ae^{-bx}}$, show that $f(x) > 0$ for all x if a and c are positive.

Chapter 5   Summary

Section 5.1 Inverse Functions

Concept	Illustration	Study and Review
Definition of an inverse function Let f be a function. A function g is said to be the **inverse function** of f if the domain of g is equal to the range of f and, for every x in the domain of f and every y in the domain of g, $\quad g(y) = x$ if and only if $f(x) = y$. The notation for the inverse function of f is f^{-1}.	The inverse of $f(x) = 4x + 1$ is $f^{-1}(x) = \dfrac{x-1}{4}$.	Examples 1, 2 Chapter 5 Review, Exercises 1–12
Composition of a function and its inverse If f is a function with an inverse function f^{-1}, then • for every x in the domain of f, $f^{-1}(f(x))$ is defined and $f^{-1}(f(x)) = x$. • for every x in the domain of f^{-1}, $f(f^{-1}(x))$ is defined and $f(f^{-1}(x)) = x$.	Let $f(x) = 4x + 1$ and $f^{-1}(x) = \dfrac{x-1}{4}$. Note that $f^{-1}(f(x)) = \dfrac{(4x+1)-1}{4} = x$. Similarly, $f(f^{-1}(x)) = 4\left(\dfrac{x-1}{4}\right) + 1 = x$.	Examples 2, 3 Chapter 5 Review, Exercises 1–4
One-to-one function A function f is **one-to-one** if $f(a) = f(b)$ implies $a = b$. For a function to have an inverse, it must be one-to-one.	The function $f(x) = x^3$ is one-to-one, whereas the function $g(x) = x^2$ is not.	Example 4 Chapter 5 Review, Exercises 5–12
Graph of a function and its inverse The graphs of a function f and its inverse function f^{-1} are symmetric with respect to the line $y = x$.	The graphs of $f(x) = 4x + 1$ and $f^{-1}(x) = \dfrac{x-1}{4}$ are pictured. Note the symmetry about the line $y = x$.	Examples 5, 6 Chapter 5 Review, Exercises 13–16

Continued

Section 5.1 **Inverse Functions**

Concept	Illustration	Study and Review
Restriction of domain to find an inverse Many functions that are not one-to-one can be restricted to an interval on which they *are* one-to-one. Their inverses are then defined on this restricted interval.	The function $f(x) = x^2$, $x \geq 0$, is one-to-one because the domain is restricted to $[0, \infty)$.	Example 7 Chapter 5 Review, Exercise 16

Section 5.2 **Exponential Functions**

Concept	Illustration	Study and Review
Definition of an exponential function An **exponential function** is a function of the form $$f(x) = Ca^x$$ where a and C are constants such that $a > 0$, $a \neq 1$, and $C \neq 0$. The domain of the exponential function is the set of all real numbers.	The functions $f(x) = -15(3)^x$ and $g(x) = \left(\frac{1}{2}\right)^x$ are both examples of exponential functions.	Examples 1, 2 Chapter 5 Review, Exercises 17–24
Properties of exponential functions • If $a > 1$ and $C > 0$, $f(x) = Ca^x \to +\infty$ as $x \to +\infty$. The function is **increasing** on $(-\infty, \infty)$ and illustrates **exponential growth.** • If $0 < a < 1$ and $C > 0$, $f(x) = Ca^x \to 0$ as $x \to +\infty$. The function is **decreasing** on $(-\infty, \infty)$ and represents **exponential decay.**	$f(x) = Ca^x$, $a > 1$, $C > 0$ y values increase as x increases $f(x) = Ca^x$, $0 < a < 1$, $C > 0$ y values decrease as x increases	Examples 3–6 Chapter 5 Review, Exercises 17–24, 95, 96
Application: Periodic compounded interest Suppose an amount P is invested in an account that pays interest at rate r, and the interest is compounded n times a year. Then, after t years, the amount in the account will be $$A(t) = P\left(1 + \frac{r}{n}\right)^{nt}.$$	An amount of $1000 invested at 7% compounded quarterly will yield, after 5 years, $$A = 1000\left(1 + \frac{0.07}{4}\right)^{4(5)} \approx 1414.78.$$	Examples 7, 8 Chapter 5 Review, Exercises 91, 92

Continued

Section 5.2 **Exponential Functions**

Concept	Illustration	Study and Review
Application: Continuous compounded interest Suppose an amount P is invested in an account that pays interest at rate r, and the interest is compounded *continuously*. Then, after t years, the amount in the account will be $$A(t) = Pe^{rt}.$$ The number e is defined as the number that the quantity $\left(1 + \dfrac{1}{n}\right)^n$ approaches as n approaches infinity. The nonterminating, nonrepeating decimal representation of the number e is $$e = 2.7182818284\ldots.$$	An amount of \$1000 invested at 7% compounded continuously will yield, after 5 years, $$A = 1000e^{(0.07)(5)} \approx 1419.07.$$	Examples 7, 8 Chapter 5 Review, Exercises 93, 94

Section 5.3 **Logarithmic Functions**

Concept	Illustration	Study and Review
Definition of logarithm Let $a > 0$, $a \neq 1$. If $x > 0$, then the **logarithm of x with respect to base a** is denoted by $y = \log_a x$ and defined by $$y = \log_a x \quad \text{if and only if} \quad x = a^y.$$	The statement $3 = \log_5 125$ is equivalent to the statement $5^3 = 125$. Here, 5 is the base.	Examples 1–5 Chapter 5 Review, Exercises 25–36
Common logarithms and natural logarithms If the base of a logarithm is 10, the logarithm is a common logarithm: $$y = \log x \quad \text{if and only if} \quad x = 10^y$$ If the base of a logarithm is e, the logarithm is a natural logarithm: $$y = \ln x \quad \text{if and only if} \quad x = e^y$$	• $\log 0.001 = -3$ because $10^{-3} = 0.001$. • $\ln e = 1$ because $e^1 = e$.	Examples 6, 7 Chapter 5 Review, Exercises 37–40
Change-of-base formula To write a logarithm with base a in terms of a logarithm with base 10 or base e, use $$\log_a x = \frac{\log_{10} x}{\log_{10} a}$$ $$\log_a x = \frac{\ln x}{\ln a}$$ where $x > 0$, $a > 0$, and $a \neq 1$.	The logarithm $\log_3 15$ can be written as $$\frac{\log_{10} 15}{\log_{10} 3} \quad \text{or} \quad \frac{\ln 15}{\ln 3}.$$	Example 8 Chapter 5 Review, Exercises 41–44

Continued

Section 5.3 **Logarithmic Functions**

Concept	Illustration	Study and Review
Graphs of logarithmic functions $\quad f(x) = \log_a x,\ a > 1$ Domain: all *positive* real numbers, $(0, \infty)$ Range: all real numbers, $(-\infty, \infty)$ Vertical asymptote: $x = 0$ (the y-axis) Increasing on $(0, \infty)$ Inverse function of $y = a^x$		Examples 9–11 Chapter 5 Review, Exercises 45–48
$\quad f(x) = \log_a x,\ 0 < a < 1$ Domain: all *positive* real numbers, $(0, \infty)$ Range: all real numbers, $(-\infty, \infty)$ Vertical asymptote: $x = 0$ (the y-axis) Decreasing on $(0, \infty)$ Inverse function of $y = a^x$		

Section 5.4 **Properties of Logarithms**

Concept	Illustration	Study and Review
Product property of logarithms Let $x, y > 0$ and $a > 0$, $a \neq 1$. Then $\quad \log_a(xy) = \log_a x + \log_a y.$	$\log_5(35) = \log_5 7 + \log_5 5$	Examples 1, 4, 5 Chapter 5 Review, Exercises 49–64
Power property of logarithms Let $x > 0$, $a > 0$, $a \neq 1$, and let k be any real number. Then $\quad \log_a x^k = k \log_a x.$	$\log 7^{1/2} = \dfrac{1}{2} \log 7$	Examples 2, 4, 5 Chapter 5 Review, Exercises 49–64
Quotient property of logarithms Let $x, y > 0$ and $a > 0$, $a \neq 1$. Then $\quad \log_a \dfrac{x}{y} = \log_a x - \log_a y.$	$\log_2\left(\dfrac{5}{7}\right) = \log_2 5 - \log_2 7$	Examples 3, 4, 5 Chapter 5 Review, Exercises 49–64

Section 5.5 **Exponential and Logarithmic Equations**

Concept	Illustration	Study and Review
One-to-one property For any $a > 0$, $a \neq 1$, $\quad a^x = a^y$ implies $x = y$.	Use the one-to-one property to solve $3^t = 81$. $$3^t = 81$$ $$3^t = 3^4$$ $$t = 4$$	Example 1 Chapter 5 Review, Exercises 65–67
Solving exponential equations To solve an exponential equation, take the logarithm of both sides of the equation and use the power property of logarithms to solve the resulting equation.	Solve $3^{2x} = 13$. $$3^{2x} = 13$$ $$2x \log 3 = \log 13$$ $$2x = \frac{\log 13}{\log 3}$$ $$x = \frac{\log 13}{2 \log 3} \approx 1.167$$	Examples 2–6 Chapter 5 Review, Exercises 68–74
Solving logarithmic equations To solve a logarithmic equation, isolate the logarithmic term, exponentiate both sides of the equation, and use the inverse relationship between logarithms and exponents to solve the resulting equation.	Solve $1 + \log x = 5$. $$1 + \log x = 5$$ $$\log x = 4$$ $$x = 10^4$$	Examples 7–10 Chapter 5 Review, Exercises 75–84

Section 5.6 **Exponential, Logistic, and Logarithmic Models**

Concept	Illustration	Study and Review
Exponential models • An exponential function of the form $f(x) = Ce^{kx}$, where $C > 0$ and $k > 0$, models **exponential growth.** • An exponential function of the form $f(x) = Ce^{kx}$, where $C > 0$ and $k < 0$, models **exponential decay.**	The function $f(x) = 100e^{0.1x}$ models exponential growth, whereas the function $f(x) = 100e^{-0.1x}$ models exponential decay.	Examples 1–3 Chapter 5 Review, Exercises 85, 86, 100, 101
Logarithmic models A **logarithmic function** can be used to model a phenomenon that grows rapidly at first and then grows more slowly.	The function $f(x) = 100 \ln x + 20$ models logarithmic growth.	Example 4 Chapter 5 Review, Exercises 87, 88, 97, 98
Logistic models A **logistic function** $f(x) = \frac{c}{1 + ae^{-bx}}$ is used to model a growth phenomenon that must eventually level off.	The function $f(x) = \frac{200}{1 + 4e^{-0.3x}}$ models logistic growth. For large positive values of x, $f(x)$ levels off at 200.	Example 5 Chapter 5 Review, Exercises 89, 90, 99

Chapter 5

Review Exercises

Section 5.1

In Exercises 1–4, verify that the functions are inverses of each other.

1. $f(x) = 2x + 7$; $g(x) = \dfrac{x - 7}{2}$

2. $f(x) = -x + 3$; $g(x) = -x + 3$

3. $f(x) = 8x^3$; $g(x) = \dfrac{\sqrt[3]{x}}{2}$

4. $f(x) = -x^2 + 1$, $x \geq 0$; $g(x) = \sqrt{1 - x}$

In Exercises 5–12, find the inverse of each one-to-one function.

5. $f(x) = -\dfrac{4}{5}x$

6. $g(x) = \dfrac{2}{3}x$

7. $f(x) = -3x + 6$

8. $f(x) = -2x - \dfrac{5}{3}$

9. $f(x) = x^3 + 8$

10. $f(x) = -2x^3 + 4$

11. $g(x) = -x^2 + 8$, $x \geq 0$

12. $g(x) = 3x^2 - 5$, $x \geq 0$

In Exercises 13–16, find the inverse of each one-to-one function. Graph the function and its inverse on the same set of axes, making sure the scales on both axes are the same.

13. $f(x) = -x - 7$

14. $f(x) = 2x + 1$

15. $f(x) = -x^3 + 1$

16. $f(x) = x^2 - 3$, $x \geq 0$

Section 5.2

In Exercises 17–24, sketch the graph of each function. Label the y-intercept and a few other points (by giving their coordinates). Determine the domain and range and describe the behavior of the function as $x \to \pm\infty$.

17. $f(x) = -4^x$

18. $f(x) = -3^x$

19. $g(x) = \left(\dfrac{2}{3}\right)^x$

20. $g(x) = \left(\dfrac{3}{5}\right)^x$

21. $f(x) = 4e^x$

22. $g(x) = -3e^x + 2$

23. $g(x) = 2e^{-x} + 1$

24. $h(x) = 5e^{-x} - 3$

Section 5.3

25. Complete the table by filling in the exponential statements that are equivalent to the given logarithmic statements.

Logarithmic Statement	Exponential Statement
$\log_3 9 = 2$	
$\log 0.1 = -1$	
$\log_5 \dfrac{1}{25} = -2$	

26. Complete the table by filling in the logarithmic statements that are equivalent to the given exponential statements.

Exponential Statement	Logarithmic Statement
$3^5 = 243$	
$4^{1/5} = \sqrt[5]{4}$	
$8^{-1} = \dfrac{1}{8}$	

In Exercises 27–36, evaluate each expression without using a calculator.

27. $\log_5 625$

28. $\log_6 \dfrac{1}{36}$

29. $\log_9 81$

30. $\log_7 \dfrac{1}{7}$

31. $\log \sqrt{10}$

32. $\ln e^{1/2}$

33. $\ln \sqrt[3]{e}$

34. $\ln e^{-1}$

35. $\log 10^{x+2}$

36. $\ln e^{5x}$

In Exercises 37–40, evaluate each expression to four decimal places using a calculator.

37. $4 \log 2$

38. $-6 \log 7.3$

39. $\ln \sqrt{8}$

40. $\ln\left(\dfrac{\pi}{2}\right)$

In Exercises 41–44, use the change-of-base formula to evaluate each expression using a calculator. Round your answers to four decimal places.

41. $\log_3 4.3$

42. $\log_4 6.52$

43. $\log_6 0.75$

44. $\log_5 0.85$

In Exercises 45–48, find the domain of each function. Use your answer to help you graph the function. Find all asymptotes and intercepts.

45. $f(x) = \log x - 6$

46. $f(x) = \ln(x - 4)$

47. $f(x) = 3 \log_4 x$

48. $f(x) = \log_5 x + 4$

Section 5.4 _____

In Exercises 49–52, use the following. Given that $\log 3 \approx 0.4771$, $\log 5 \approx 0.6990$, and $\log 7 \approx 0.8451$, evaluate the following logarithms without the use of a calculator. Then check your answer using a calculator.

49. $\log 21$

50. $\log 15$

51. $\log\left(\dfrac{5}{3}\right)$

52. $\log \sqrt{3}$

In Exercises 53–58, write each expression as a sum and/or difference of logarithmic expressions. Eliminate exponents and radicals and evaluate logarithms wherever possible. Assume that $a, x, y, z > 0$ and $a \neq 1$.

53. $\log \sqrt[4]{x} \sqrt[3]{y}$

54. $\ln \sqrt[3]{x^5} \sqrt{y^3}$

55. $\log_a \sqrt{\dfrac{x^6}{y^3 z^5}}$

56. $\log_a \sqrt{\dfrac{z^5}{xy^4}}$

57. $\ln \sqrt[3]{\dfrac{xy^3}{z^5}}$

58. $\log \sqrt[3]{\dfrac{x^3 z^5}{10y^2}}$

In Exercises 59–64, write each expression as a logarithm of a single quantity and then simplify if possible. Assume that each variable expression is defined for appropriate values of the variable(s).

59. $\ln(x^2 - 3x) - \ln(x - 3)$

60. $\log_a(x^2 - 4) - \log_a(x + 2), a > 0, a \neq 1$

61. $\dfrac{1}{4}[\log(x^2 - 1) - \log(x + 1)] + 3 \log x$

62. $\dfrac{2}{3} \log 9x^3 - \dfrac{1}{4} \log 16y^8$

63. $2 \log_3 x^2 - \dfrac{1}{3} \log_3 \sqrt{x}$

64. $2 \ln(x^2 + 1) + \dfrac{1}{2} \ln x^4 - 3 \ln x$

Section 5.5 _____

In Exercises 65–74, solve each exponential equation.

65. $5^x = 625$

66. $6^{2x} = 1296$

67. $7^x = \dfrac{1}{49}$

68. $4e^x + 6 = 38$

69. $25e^{0.04x} = 100$

70. $3(1.5^x) - 2 = 9$

71. $4^{2x+3} = 16$

72. $5^{3x+2} = \dfrac{1}{5}$

73. $e^{2x+1} = 4$

74. $2^{x-1} = 10$

In Exercises 75–84, solve each logarithmic equation and eliminate any extraneous solutions.

75. $\ln(2x - 1) = 0$

76. $\log(x + 3) - \log(2x - 4) = 0$

77. $\ln x^{1/2} = 2$

78. $2 \log_3 x = -4$

79. $\log x + \log(2x - 1) = 1$

80. $\log(x + 6) = \log x^2$

81. $\log(3x + 1) - \log(x^2 + 1) = 0$

82. $\log_3 x + \log_3(x + 8) = 2$

83. $\log_4 x + \log_4(x + 3) = 1$

84. $\log(3x + 10) = 2 \log x$

Section 5.6 _____

In Exercises 85–90, evaluate $f(0)$ and $f(3)$ for each function. Round your answer to four decimal places.

85. $f(x) = 4e^{-2.5x}$

86. $f(x) = 30e^{1.2x}$

87. $f(x) = 20 \ln(x + 2) + 1$

88. $f(x) = 10 \ln(2x + 1) - 2$

89. $f(x) = \dfrac{100}{1 + 4e^{-0.2x}}$

90. $f(x) = \dfrac{200}{1 + 5e^{-0.5x}}$

Applications

Banking *In Exercises 91 and 92, for an initial deposit of $1500, find the total amount in a bank account after 6 years for the interest rates and compounding frequencies given.*

91. 5% compounded quarterly

92. 8% compounded semiannually

Banking *In Exercises 93 and 94, for an initial deposit of $1500, find the total amount in a bank account after t years for the interest rates and values of t given. Assume continuous compounding of interest.*

93. 8% interest; $t = 4$

94. 3.5% interest; $t = 5$

95. **Depreciation** The depreciation rate of a Ford Focus is about 25% per year. If the Focus was purchased for $17,000, make a table of its values over the first 5 years after purchase. Find a function that gives its value t years after purchase, and sketch a graph of the function. (*Source:* www.edmunds.com)

96. **Tuition Savings** To save for her newborn daughter's education, Jennifer invests $4000 at an interest rate of 4% compounded monthly. What is the value of this investment after 18 years, assuming no additional deposits or withdrawals are made?

Geology *The magnitude of an earthquake is measured on the Richter scale using the formula*

$$R(I) = \log\left(\frac{I}{I_0}\right)$$

where I represents the actual intensity of the earthquake and I_0 is a baseline intensity used for comparison.

97. If the intensity of an earthquake is 10 times the baseline intensity I_0, what is its magnitude on the Richter scale?

98. On October 8, 2005, a devastasting earthquake affected areas of northern Pakistan. It registered a magnitude of 7.6 on the Richter scale. Express its intensity in terms of I_0. (*Source:* U.S. Geological Survey)

99. **Ecology** The number of trout in a pond t months after their introduction into the pond can be modeled by the logistic function

$$N(t) = \frac{450}{1 + 9e^{-0.3t}}.$$

(a) How many trout were initially introduced into the pond?

(b) How many trout will be in the pond 15 months after introduction?

(c) Graph this function for $0 \le t \le 30$. What do you observe as t increases?

(d) How many months after introduction will the number of trout in the pond be equal to 400?

100. **Global Economy** One measure of the strength of a country's economy is the country's collective purchasing power. In 2000, China had a purchasing power of 2.5 trillion dollars. If the purchasing power was forecasted to grow at a rate of 7% per year, find a function of the form $P(t) = Ca^t$, $a > 1$, that models China's purchasing power at time t. Here, t is the number of years since 2000. (*Source:* Proceedings of the National Academy of Sciences)

101. **Music** In the first few years following the introduction of a popular product, the number of units sold per year can increase exponentially. Consider the following table, which shows the sales of portable MP3 players for the years 2000–2005. (*Source:* Consumer Electronics Association)

Year	Number of Units Sold (in millions)
2000	0.510
2001	0.724
2002	1.737
2003	3.031
2004	6.972
2005	10.052

(a) Make a scatter plot of the data and find the exponential function of the form $f(x) = Ca^x$ that best fits the data. Let x denote the number of years since 2000.

(b) Use the function from part (a) to predict the number of MP3 players sold in 2007.

(c) Use the function from part (a) to determine the year when the number of MP3 players sold equals 21 million units.

Chapter 5   Test

1. Verify that the functions $f(x) = 3x - 1$ and $g(x) = \dfrac{x + 1}{3}$ are inverses of each other.

2. Find the inverse of the one-to-one function
$$f(x) = 4x^3 - 1.$$

3. Find $f^{-1}(x)$ given $f(x) = x^2 - 2$, $x \geq 0$. Graph f and f^{-1} on the same set of axes.

In Exercises 4–6, sketch a graph of the function and describe its behavior as $x \to \pm\infty$.

4. $f(x) = -3^x + 1$

5. $f(x) = 2^{-x} - 3$

6. $f(x) = e^{-2x}$

7. Write in exponential form: $\log_6 \dfrac{1}{216} = -3$.

8. Write in logarithmic form: $2^5 = 32$.

In Exercises 9 and 10, evaluate the expression without using a calculator.

9. $\log_8 \dfrac{1}{64}$

10. $\ln e^{3.2}$

11. Use a calculator to evaluate $\log_7 4.91$ to four decimal places.

12. Sketch the graph of $f(x) = \ln(x + 2)$. Find all asymptotes and intercepts.

In Exercises 13 and 14, write the expression as a sum or difference of logarithmic expressions. Eliminate exponents and radicals when possible.

13. $\log \sqrt[3]{x^2 y^4}$

14. $\ln(e^2 x^2 y)$

In Exercises 15 and 16, write the expression as a logarithm of a single quantity, and simplify if possible.

15. $\ln(x^2 - 4) - \ln(x - 2) + \ln x$

16. $4 \log_2 x^{1/3} + 2 \log_2 x^{1/3}$

In Exercises 17–22, solve.

17. $6^{2x} = 36^{3x-1}$

18. $4^x = 7.1$

19. $4e^{x+2} - 6 = 10$

20. $200e^{0.2t} = 800$

21. $\ln(4x + 1) = 0$

22. $\log x + \log(x + 3) = 1$

23. For an initial deposit of $3000, find the total amount in a bank account after 6 years if the interest rate is 5%, compounded quarterly.

24. Find the value in 3 years of an initial investment of $4000 at an interest rate of 7%, compounded continuously.

25. The depreciation rate of a laptop computer is about 40% per year. If a new laptop computer was purchased for $900, find a function that gives its value t years after purchase.

26. The magnitude of an earthquake is measured on the Richter scale using the formula $R(I) = \log\left(\dfrac{I}{I_0}\right)$, where I represents the actual intensity of the earthquake and I_0 is a baseline intensity used for comparison. If an earthquake registers 6.2 on the Richter scale, express its intensity in terms of I_0.

27. The number of college students infected with a cold virus in a dormitory can be modeled by the logistic function $N(t) = \dfrac{120}{1 + 3e^{-0.4t}}$, where t is the number of days after the breakout of the infection.
    (a) How many students were initially infected?
    (b) Approximately how many students will be infected after 10 days?

28. The population of a small town grew from 28,000 in 2004 to 32,000 in 2006. Find a function of the form $P(t) = Ce^{kt}$ that models this growth, where t is the number of years since 2004.

Systems of Equations and Inequalities

Maintaining good health and nutrition translates into targeting values for calories, cholesterol, and fat, and can be modeled by a system of linear equations. Exercise 44 in Section 6.2 uses a system of linear equations to determine the nutritional value of slices of different types of pizza. In this chapter you will study how systems of equations and inequalities arise, how to solve them, and how they are used in various applications.

6.1 Systems of Linear Equations and Inequalities in Two Variables

Objectives

▶ Solve systems of linear equations in two variables by elimination

▶ Understand what is meant by an inconsistent system of linear equations in two variables

▶ Understand what is meant by a dependent system of linear equations in two variables

▶ Relate the nature of the solution(s) of a system of linear equations in two variables to the graphs of the equations of the system

▶ Graph and solve systems of linear inequalities in two variables

▶ Use systems of equations and systems of inequalities in two variables to solve applied problems

Suppose you want to set your physical fitness goals by incorporating various types of activities. You can actually formulate the problem mathematically by using different variables, one for each activity. We study a problem of this type in Example 1, using the comic strip character Cathy as our subject.

Example **1** **Application of Physical Fitness**

Example 2 in Section 6.1 builds upon this example. ⋯⫶

Cathy would like to burn off all the calories from a granola bar she just ate. She wants to spend 30 minutes on the treadmill, in some combination of walking and running. At the walking speed, she burns off 3 calories per minute. At the running speed, she burns off 8 calories per minute. Set up the problem that must be solved to answer the following question: If the granola bar has 180 calories, write the equations to determine how many minutes Cathy should spend on each activity to burn off all the calories from the granola bar.

▶**Solution** We want to determine the amount of time spent on walking and the amount of time spent on running. Define the following variables to represent the times spent on these two activities.

$$x\text{: number of minutes spent walking}$$

$$y\text{: number of minutes spent running}$$

Using these variables, we formulate equations from the given information. First, we know that the total amount of time spent on the treadmill is 30 minutes. This gives

$$x + y = 30.$$

In the statement of the problem, we see that another piece of information is given—the total number of calories burned. So we next find an expression for this quantity.

$$\text{Total calories burned} = \text{calories burned from walking} + \text{calories burned from running}$$

$$= \left(3\,\frac{\text{cal}}{\text{min}} \times \text{minutes walking} \right) + \left(8\,\frac{\text{cal}}{\text{min}} \times \text{minutes running} \right)$$

$$= 3x + 8y$$

We are given that the total number of calories burned is 180. Our two equations are thus:

$$\begin{cases} x + y = 30 & \textit{Total time equals 30 minutes} \\ 3x + 8y = 180 & \textit{Total calories burned equals 180} \end{cases}$$

We have placed a large brace to the left of these equations to indicate that they form a set of two conditions that must be satisfied by the same ordered pair of numbers (x, y). We will find a solution to this system of equations later in this section.

☑ *Check It Out 1:* If Cathy spends 15 minutes walking and 15 minutes running, will these numbers solve *both* equations from Example 1? Explain. ▪

The general form of a linear equation in the variables x and y is $Ax + By = C$, where A, B, and C are constants such that A and B are not both zero. A **system of linear equations** in x and y is a collection of equations that can be put into this form. For example, a system of two linear equations in the variables x and y can be written as

$$\begin{cases} Ax + By = E \\ Cx + Dy = F \end{cases}$$

where A, B, C, and D are not all zero. A **solution** of such a system of linear equations is an ordered pair of numbers (x, y) that satisfies all the equations in the system. Graphically, it is the point of intersection of two lines.

Recall that in Section 1.5, we found the intersection of two lines by solving each of the equations for y in terms of x and then equating them to solve for x. This is known as the **substitution method.** This method is limited in scope, since it can be readily applied only to systems of two linear equations in two variables. We next introduce a method for solving a system of two linear equations in two variables that can be extended to systems of more than two linear equations and/or more than two variables.

Just in Time

Review intersection of lines in Section 1.5.

Solving Systems of Linear Equations in Two Variables by Elimination

The process of **elimination** can be used to solve systems of two linear equations in two variables. The idea is to eliminate one of the variables by combining the equations in a suitable manner. Example 2 illustrates this procedure.

Example 2 Using the Elimination Method to Solve a System

⋮• *This example builds on Example 1 in Section 6.1.*

Use the method of elimination to solve the following system of equations.

$$\begin{cases} x + y = 30 \\ 3x + 8y = 180 \end{cases}$$

This is the system of equations we formulated for the problem in Example 1.

▶**Solution**

STEPS	EXAMPLE
1. Write the system of equations in the form $$\begin{cases} Ax + By = E \\ Cx + Dy = F \end{cases}$$	$$\begin{cases} x + y = 30 \\ 3x + 8y = 180 \end{cases}$$
2. Eliminate x from the second equation. To do so, multiply the first equation by -3, so that the coefficients of x are negatives of one another. Then add the two equations.	$-3(x + y = 30)$ $3x + 8y = 180$ Adding the two equations gives $$\begin{aligned} -3x - 3y &= -90 \\ +\quad 3x + 8y &= 180 \\ \hline 5y &= 90 \end{aligned}$$
3. Replace the second equation in the original system with the result of Step 2, and retain the *original* first equation.	$$\begin{cases} x + y = 30 \\ 5y = 90 \end{cases}$$
4. The second equation has only one variable, y. Solve this equation for y.	$$\begin{cases} 5y = 90 \\ y = 18 \end{cases}$$
5. Substitute this value of y into the first equation and solve for x.	$x + y = 30$ $x + (18) = 30$ $x = 12$
6. Write out and interpret the solution.	We obtain $x = 12$ and $y = 18$. Cathy must spend 12 minutes walking and 18 minutes running to burn off all the calories from the granola bar.

☑ *Check It Out 2:* Rework Example 2 for the case in which the total number of calories burned is 215. ▪

Discover *and* **Learn**

Graph the lines $y = -x + 5$ and $y = -x + 8$. Do they intersect? Why or why not?

Inconsistent Systems of Linear Equations in Two Variables

Sometimes a system of linear equations in two variables has no solution. The following example illustrates this case.

Example 3 A System of Linear Equations with No Solution

The sum of two numbers is 20, and twice the sum of the numbers is 50. Find the two numbers, if they exist, and examine the situation graphically.

▶Solution It is easy to see that if the sum of two numbers is 20, then twice their sum is 40, not 50. Thus the two conditions are mutually exclusive. Let's examine this example algebraically.

$$\begin{cases} x + y = 20 & \text{Sum of numbers is 20} \\ 2(x + y) = 50 & \text{Twice their sum is 50} \end{cases}$$

This can be written as

$$\begin{cases} x + y = 20 \\ 2x + 2y = 50 \end{cases}.$$

Multiplying the first equation by -2 and then adding the result to the second equation gives the following.

$$\begin{array}{r} -2x - 2y = -40 \\ + \quad 2x + 2y = 50 \\ \hline 0 = 10 \end{array}$$

The equation $0 = 10$ is false; thus the original system of equations has no solution. We had already deduced this by examining the problem.

To solve this problem graphically, we first solve for y in terms of x in both equations. This gives

$$\begin{cases} y = -x + 20 \\ y = -x + 25 \end{cases}.$$

The graphs of these two functions are given in Figure 6.1.1.

Figure 6.1.1

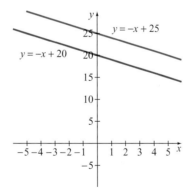

Note that the slope of each of these lines is -1, but they have different y-intercepts. Thus the lines are parallel and will never intersect. This confirms our algebraic approach, which indicated that there is no solution.

☑ *Check It Out 3:* Show both algebraically and graphically that the following system of equations has no solution.

$$\begin{cases} y = -2x + 5 \\ y = -2x + 2 \end{cases} ■$$

When a system of equations has no solution, the system is said to be **inconsistent.** Graphically, this means that the lines that represent the equations are *parallel* and will never intersect.

Dependent Systems of Linear Equations in Two Variables

So far we have seen that it is possible for a system of linear equations to have one solution or no solution. The final case is a system of linear equations in two variables that has an *infinite* number of solutions.

Example 4 **A System of Linear Equations with an Infinite Number of Solutions**

The sum of two numbers is 20, and twice the sum of the two numbers is 40. Find the two numbers, if they exist, and examine the situation graphically.

▶Solution Because the sum of the two numbers is 20, twice their sum must equal 40. Thus, the second condition tells us nothing new. The solution is the set of *all* pairs of numbers whose sum is 20—and there are infinitely many such pairs!

This idea of there being infinitely many solutions may be difficult to grasp at first, so we will examine the situation algebraically.

$$\begin{cases} x + y = 20 & \text{Sum of numbers is 20} \\ 2(x + y) = 40 & \text{Twice their sum is 40} \end{cases}$$

This system can be rewritten as follows.

$$\begin{cases} x + y = 20 \\ 2x + 2y = 40 \end{cases}$$

We will solve this system by elimination. Multiplying the first equation by -2 and then adding the result to the second equation gives

$$0 = 0.$$

Table 6.1.1

x	y	$x + y$
10	10	20
2	18	20
-30	50	20
$-\dfrac{5}{2}$	$\dfrac{45}{2}$	20

While it is certainly true that $0 = 0$, this result gives us no useful information. Thus, we will resort to using only one of the original equations, $x + y = 20$, to find *both* x and y. There are infinitely many ordered pairs (x, y) that satisfy this equation. A few of them are given in Table 6.1.1.

Because the equation $x + y = 20$ does not have a unique solution, we settle for solving this equation for one of the variables in terms of the other. For example, we can write y in terms of x as follows:

$$y = -x + 20$$

Here, x can be any real number, and the system of equations has infinitely many solutions.

To solve this problem graphically, we first solve each of the original equations for y in terms of x.

$$\begin{cases} y = -x + 20 \\ y = -x + 20 \end{cases}$$

Figure 6.1.2

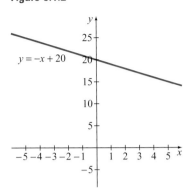

The equations are identical, and so are their graphs, as shown in Figure 6.1.2. Every point on the entire line $y = -x + 20$ is a solution of both of the original equations. This confirms our algebraic approach, which indicated that there are infinitely many pairs (x, y) that satisfy the equation $y = -x + 20$.

☑ *Check It Out 4:* Find some additional points, other than those given in Table 6.1.1 from Example 4, that solve the following system.

$$\begin{cases} x + y = 20 \\ 2x + 2y = 40 \end{cases} ■$$

When a system of linear equations has infinitely many solutions, the system is said to be **dependent**. In the case of a system of two linear equations in two variables, this happens when the equations can be derived from each other. Graphically, this means that the lines that represent the equations are *identical*.

Nature of Solutions of Systems of Linear Equations in Two Variables

The discussion up to this point can be summarized as follows.

Nature of Solutions of Systems of Linear Equations in Two Variables

Let a system of linear equations in two variables be given as follows:

$$\begin{cases} Ax + By = E \\ Cx + Dy = F \end{cases}$$

where A, B, C, and D are not all zero. The solution of this system will have one of the following properties:

▶ Exactly one solution (independent system)
▶ No solution (inconsistent system)
▶ Infinitely many solutions (dependent system)

Figure 6.1.3 summarizes the ways in which two lines can intersect, and indicates the nature of the solutions of the corresponding system of two linear equations in two variables.

Figure 6.1.3

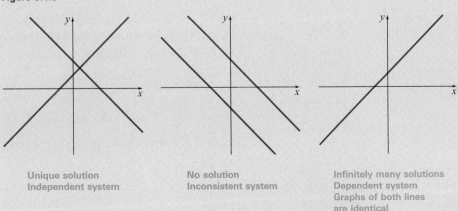

Unique solution
Independent system

No solution
Inconsistent system

Infinitely many solutions
Dependent system
Graphs of both lines
are identical

Example 5 **Solving Systems of Linear Equations**

Use the method of elimination to solve the following systems of linear equations, if a solution exists.

(a) $\begin{cases} -3x + 2y = -5 \\ 5x + 4y = 12 \end{cases}$ (b) $\begin{cases} -2x + y = 4 \\ 4x - 2y = -8 \end{cases}$ (c) $\begin{cases} -2x + y = 4 \\ -6x + 3y = 10 \end{cases}$

▶**Solution**

(a) To solve the system

$$\begin{cases} -3x + 2y = -5 \\ 5x + 4y = 12 \end{cases}$$

note that the coefficients of x in these equations (namely, -3 and 5) are of opposite sign, but neither of them is a multiple of the other. In order to eliminate x, we will find positive numbers a and b such that $a \cdot |-3| = b \cdot 5$, and then multiply the first equation by a and the second equation by b. The least common multiple of $|-3|$ and 5 is 15, so we can let $a = 5$ (since $5 \cdot -3 = -15$) and $b = 3$ (since $3 \cdot 5 = 15$). Performing the multiplication, we obtain the following.

$$\begin{aligned} -15x + 10y &= -25 && \text{5 times the first equation} \\ \underline{15x + 12y} &= \underline{36} && \text{3 times the second equation} \\ 22y &= 11 && \text{5 times the first equation plus 3 times the second equation} \end{aligned}$$

The equation $22y = 11$ is equivalent to the equation $y = \frac{1}{2}$. Replacing the second equation with $y = \frac{1}{2}$ and retaining the original first equation, we can write

$$\begin{cases} -3x + 2y = -5 \\ \quad\quad\; y = \dfrac{1}{2} \end{cases}.$$

Substituting $\frac{1}{2}$ for y in the first equation and solving for x gives

$$-3x + 2\left(\frac{1}{2}\right) = -5 \Longrightarrow -3x + 1 = -5 \Longrightarrow -3x = -6 \Longrightarrow x = 2.$$

Thus the solution of this system of equations is $x = 2, y = \frac{1}{2}$. You should check that this is indeed the solution.

(b) To solve the system

$$\begin{cases} -2x + y = 4 \\ 4x - 2y = -8 \end{cases}$$

we can eliminate the variable x by multiplying the first equation by 2 and then adding the result to the second equation.

$$\begin{aligned} -4x + 2y &= 8 && \text{2 times the first equation} \\ \underline{4x - 2y} &= \underline{-8} && \text{The second equation} \\ 0 &= 0 && \text{2 times the first equation plus the second equation} \end{aligned}$$

Replacing the second equation with $0 = 0$ and retaining the original first equation, we can write

$$\begin{cases} -2x + y = 4 \\ \quad\quad\; 0 = 0 \end{cases}.$$

The second equation ($0 = 0$) is true but gives us no useful information. Thus we have a dependent system with infinitely many solutions. Solving the first equation for y in terms of x, we get

$$-2x + y = 4 \Longrightarrow y = 2x + 4.$$

Thus the solution set is the set of all ordered pairs (x, y) such that $y = 2x + 4$, where x is any real number. That is, the solution set consists of all points (x, y) on the line $y = 2x + 4$.

(c) To solve the system

$$\begin{cases} -2x + y = 4 \\ -6x + 3y = 10 \end{cases}$$

we can eliminate the variable x by multiplying the first equation by -3 and then adding the result to the second equation.

$$\begin{array}{ll} 6x - 3y = -12 & \text{-3 times the first equation} \\ \underline{-6x + 3y = 10} & \text{The second equation} \\ 0 = -2 & \text{-3 times the first equation} \\ & \text{plus the second equation} \end{array}$$

Replacing the second equation with $0 = -2$ and retaining the original first equation, we can write

$$\begin{cases} -2x + y = 4 \\ 0 = -2 \end{cases}.$$

The second equation, $0 = -2$, is false. Thus this system of equations is inconsistent and has no solution.

✔ *Check It Out 5:* Solve by elimination.

$$\begin{cases} -2x + y = 6 \\ 5x - 2y = 10 \end{cases}$$

Solving a Linear Inequality in Two Variables

An example of a linear inequality in two variables is $2x + y > 0$. A solution of this inequality is any point (x, y) that satisfies the inequality. For example, $(1, 10)$ and $(-1, 4)$ are points that satisfy the inequality. However, $(0, -1)$ does not satisfy it.

Definition of a Linear Inequality

An inequality in the variables x and y is a **linear inequality** if it can be written in the form $Ax + By \leq C$ or $Ax + By < C$ (or $Ax + By \geq C$, $Ax + By > C$), where A, B, and C are constants such that A and B are not both zero. The **solution set** of such a linear inequality is the set of all points (x, y) that satisfy the inequality.

To find the solution set of a linear inequality, we first graph the corresponding equality (i.e., the equation we get by replacing the inequality symbol with an equals sign). This graph divides the xy-plane into two regions. One of these regions will satisfy the inequality. Example 6 illustrates how an inequality in two variables is solved.

Example **6** **Solving a Linear Inequality**

Graph the solution set of each of the following linear inequalities.

(a) $x \geq 3$ 　　(b) $y < 2x$

▶Solution

(a) First graph the line $x = 3$. This line separates the xy-plane into two half-planes. Choose a point *not* on the line and see if it satisfies the inequality $x \geq 3$. We choose $(0, 0)$ because it is easy to check.

$$x \overset{?}{\geq} 3 \Longrightarrow 0 \overset{?}{\geq} 3$$

Because the inequality $0 \geq 3$ is false, the half-plane that contains the point $(0, 0)$ does *not* satisfy the inequality. Thus all the points in the *other* half-plane do satisfy the inequality, and so this region is shaded. The points *on* the line $x = 3$ also satisfy the inequality $x \geq 3$, so this line is graphed as a solid line. See Figure 6.1.4.

Figure 6.1.4

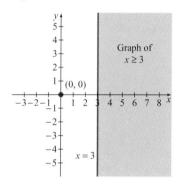

(b) Graph the line $y = 2x$. Choose a point *not* on the line and see if it satisfies the inequality $y < 2x$. We choose $(2, 0)$ because it is easy to check. Note that we cannot use $(0, 0)$ because it lies on the line $y = 2x$.

$$y \overset{?}{<} 2x \Longrightarrow 0 \overset{?}{<} 2(2)$$

Because the inequality $0 < 4$ is true, the half-plane that contains the point $(2, 0)$ is shaded. The line $y = 2x$ is graphed as a dashed line because it is not included in the inequality, as seen in Figure 6.1.5.

Figure 6.1.5

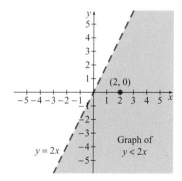

✔ *Check It Out 6:* Graph the inequality $y \geq x + 5$. ■

Solving Systems of Linear Inequalities in Two Variables

The **solution set of a system of linear inequalities** in the variables x and y consists of the set of all points (x, y) in the intersection of the solution sets of the individual inequalities of the system. To find the graph of the solution set of the system, graph the solution set of each inequality and then find the intersection of the shaded regions.

Example **7**   **Graphing Systems of Linear Inequalities**

Graph the following system of linear inequalities.

$$\begin{cases} x + 3y \le 6 \\ -x + y \ge -2 \end{cases}$$

▶**Solution** First graph the inequality $x + 3y \le 6$. Do this by graphing the line $x + 3y = 6$ and then using a test point not on the line to determine which half-plane satisfies the inequality. The test point $(0, 0)$ satisfies the inequality, and so the half-plane containing this point is shaded. All the points on the line $x + 3y = 6$ also satisfy the inequality, so this line is graphed as a solid line. See Figure 6.1.6.

Technology Note

With a graphing utility, you can graph both lines in Example 7 by first solving each equation for y. You then choose the appropriate line marker to shade the portion of the plane above or below the graph of the line. See Figure 6.1.7.

Keystroke Appendix: Section 7

Figure 6.1.7

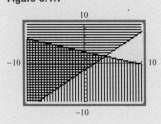

Figure 6.1.6

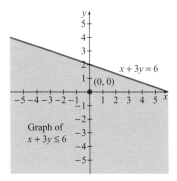

Next graph the inequality $-x + y \ge -2$. Do this by graphing the line $-x + y = -2$ as a solid line and then using a test point not on the line to determine which half-plane satisfies the inequality. The test point $(0, 0)$ satisfies the inequality, and so the half-plane containing this point is shaded. See Figure 6.1.8.

Figure 6.1.8

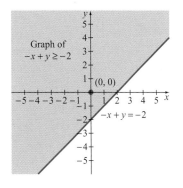

Although we have graphed the two inequalities separately for the sake of clarity, you should graph the second inequality on the same set of coordinate axes as the first inequality, as shown in Figure 6.1.9. The region where the two separate shaded regions overlap is the region that satisfies the system of inequalities. Every point in this overlapping region satisfies *both* of the given inequalities.

Figure 6.1.9

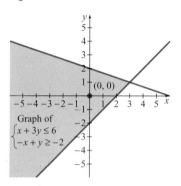

✔ *Check It Out 7:* Graph the following system of linear inequalities.

$$\begin{cases} x - 2y \geq 4 \\ -2x + y \leq 3 \end{cases}$$

A Linear Programming Application

In many business and industrial applications, some quantity such as profit or output may need to be optimized subject to certain conditions. These conditions, which are referred to as **constraints,** can often be expressed in the form of linear inequalities. **Linear programming** is a procedure that yields all the solutions of a given system of linear inequalities that correspond to achieving an important goal, such as maximizing profit or minimizing cost. Actual applications, such as airline scheduling, usually involve thousands of variables and numerous constraints, and are solved using sophisticated computer algorithms. In this section, we focus on problems involving only two variables and relatively few constraints.

Example 8 illustrates how a system of linear inequalities can be set up to solve a simple optimization problem.

Example 8 **Maximizing Calories Burned by Exercising**

Cathy can spend at most 30 minutes on the treadmill, in some combination of running and walking. To warm up and cool down, she must spend at least 8 minutes walking. At the walking speed, she burns off 3 calories per minute. At the running speed, she burns off 8 calories per minute. Set up the problem that must be solved to answer the following question: How many minutes should Cathy spend on each activity (walking and running) to maximize the total number of calories burned?

▶Solution First define the variables.

x: number of minutes spent walking

y: number of minutes spent running

Using these variables, we formulate inequalities that express the stated constraints. The first constraint is that the total amount of time spent on the treadmill can be at most 30 minutes.

$$x + y \leq 30 \qquad \text{First constraint}$$

In addition, we know that Cathy must spend at least 8 minutes walking. This gives us a second constraint.

$$x \geq 8 \qquad \textit{Second constraint}$$

Because x and y represent time, they must be nonnegative. This gives us two more constraints.

$$x \geq 0, \quad y \geq 0 \qquad \textit{Third and fourth constraints}$$

The system of inequalities that corresponds to this set of constraints is given below. The set of points (x, y) for which *all four constraints* are satisfied is depicted by the shaded region of the graph in Figure 6.1.10.

Figure 6.1.10

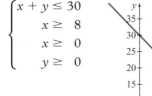

$$\begin{cases} x + y \leq 30 \\ \quad x \geq \quad 8 \\ \quad x \geq \quad 0 \\ \quad y \geq \quad 0 \end{cases}$$

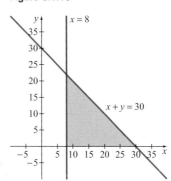

The number of calories burned can be expressed algebraically as

$$C = 3x + 8y.$$

Thus the ultimate goal, or *objective*, is to find the point(s) (x, y) within the shaded region at which the value of C attains its maximum value.

☑ *Check It Out 8:* Rework Example 8 for the case in which the additional constraint that Cathy can spend at most 15 minutes running is imposed. ▪

To solve optimization problems such as the one that was introduced in Example 8, we need the following important theorem.

Fundamental Theorem of Linear Programming

Given a system of linear inequalities in the variables x and y that express the constraints for an optimization problem, we can define the following.

▶ The **feasible set** for the optimization problem is the solution set of the system of linear inequalities.

▶ A **corner point** is a point at which two or more of the boundary lines of the feasible set intersect.

▶ A **linear objective function** is a function of the form $C = Ax + By$, where A and B are constants that are not both zero.

Then the objective function attains its maximum value (if a maximum value exists) at one or more of the corner points; the minimum value (if it exists) is also attained at one or more of the corner points.

Using this theorem, we can now solve the problem that was set up in Example 8.

Example 9 **Maximizing an Objective Function with Constraints**

Maximize the objective function $C = 3x + 8y$ subject to the following constraints.

$$\begin{cases} x + y \leq 30 \\ \quad\; x \geq\; 8 \\ \quad\; x \geq\; 0 \\ \quad\; y \geq\; 0 \end{cases}$$

▶Solution The setup for this problem was given in Example 8. By the fundamental theorem of linear programming, we know that the objective function attains its maximum value at one or more of the corner points of the feasible set. Each of the corner points is a point of intersection of the boundary lines, which are illustrated on the graph in Figure 6.1.11.

Figure 6.1.11

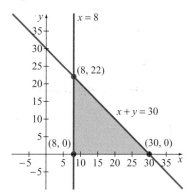

Table 6.1.2

(x, y) (minutes)	C = 3x + 8y (calories)
(8, 0)	24
(8, 22)	200
(30, 0)	90

Table 6.1.2 lists values of the objective function C at the corner points. From the table, we see that the maximum number of calories that will be burned under the given constraints is 200, and that Cathy will have to spend 8 minutes walking and 22 minutes running to burn this many calories. Note that the minimum amount of time needed to satisfy the walking constraint (8 minutes) turns out to be the value of x in the optimal solution. This makes sense from a physical standpoint, because walking burns off fewer calories per minute than running does.

✔ *Check It Out 9:* Rework Example 9 for the case in which Cathy is required to spend at least 6 minutes walking. ■

6.1 Key Points

▶ To solve a system of linear equations by elimination, eliminate one of the variables from the two equations to get one equation in one variable. Solve this equation and substitute the result into the first equation. The second variable is solved for by substitution.

▶ A system of linear equations can have one solution, no solution, or infinitely many solutions.

▶ An inequality is a **linear inequality** if it can be written in the form $Ax + By \leq C$ or $Ax + By < C$, where A, B, are C are real numbers, not all zero. (The symbols $\leq$ and $<$ may be replaced with $\geq$ or $>$.) The **solution set** of a linear inequality is the set of points (x, y) that satisfy the inequality.

▶ A **solution set of a system of linear inequalities** is the set of points (x, y) consisting of the intersection of the solution sets of the individual inequalities.

▶ The **fundamental theorem of linear programming** states that the maximum or minimum of a linear objective function occurs at the corner point of the boundary of the solution set of a system of linear inequalities.

6.1 Exercises

▶**Just in Time Exercises** These exercises correspond to the Just in Time references in this section. Complete them to review topics relevant to the remaining exercises.

In Exercises 1–6, find the intersection of the two lines.

1. $x = -1, y = 2$

2. $x + y = 7, y = 3$

3. $x = 3, y = 2x - 1$

4. $6x + 3y = 9, x - y = 3$

5. $2x - y = 6, x + 2y = 8$

6. $x + y = 4, x - y = 12$

▶**Skills** This set of exercises will reinforce the skills illustrated in this section.

In Exercises 7–12, verify that each system of equations has the indicated solution.

7. $\begin{cases} x + 5y = -6 \\ -x + 2y = -8 \end{cases}$
   Solution: $x = 4, y = -2$

8. $\begin{cases} -x - 2y = 5 \\ -2x + y = -5 \end{cases}$
   Solution: $x = 1, y = -3$

9. $\begin{cases} 2x - 3y = -6 \\ -x + 2y = 4 \end{cases}$
   Solution: $x = 0, y = 2$

10. $\begin{cases} 2x - y = 2 \\ 6x - 5y = 8 \end{cases}$
    Solution: $x = \dfrac{1}{2}, y = -1$

11. $\begin{cases} x - y = 5 \\ -2x + 2y = -10 \end{cases}$
    Solution: $x = a, y = a - 5$ (for every real number a)

12. $\begin{cases} 2x - y = 1 \\ 8x - 4y = 4 \end{cases}$
    Solution: $x = a, y = -1 + 2a$ (for every real number a)

In Exercises 13–30, use elimination to solve each system of equations. Check your solution.

13. $\begin{cases} x + 2y = 7 \\ -2x + y = 1 \end{cases}$

14. $\begin{cases} -x - y = -7 \\ 3x + 4y = 24 \end{cases}$

15. $\begin{cases} -3x + y = 5 \\ 6x - y = -8 \end{cases}$

16. $\begin{cases} 5x + 3y = -1 \\ -10x + 2y = 26 \end{cases}$

17. $\begin{cases} 2x + y = 5 \\ 4x + 2y = 3 \end{cases}$

18. $\begin{cases} -3x + 4y = 9 \\ 6x - 8y = 3 \end{cases}$

19. $\begin{cases} 3x - y = 9 \\ x + y = -1 \end{cases}$

20. $\begin{cases} 5x - 3y = 23 \\ x + y = -13 \end{cases}$

21. $\begin{cases} x + y = 5 \\ -2x - 2y = -10 \end{cases}$

22. $\begin{cases} -3x + 4y = 9 \\ 9x - 12y = -27 \end{cases}$

23. $\begin{cases} 3x - 6y = 2 \\ y = -3 \end{cases}$

24. $\begin{cases} -2x = -4 \\ -4x + 3y = -3 \end{cases}$

25. $\begin{cases} -2x - 3y = 0 \\ 3x + 5y = -2 \end{cases}$

26. $\begin{cases} 5x - 2y = -3 \\ 3x - y = 1 \end{cases}$

27. $\begin{cases} -3x + 2y = \dfrac{1}{2} \\ 4x + y = 3 \end{cases}$

28. $\begin{cases} 4x + 3y = \dfrac{9}{2} \\ 5x + y = 7 \end{cases}$

29. $\begin{cases} \dfrac{1}{2}x + \dfrac{1}{3}y = -2 \\ \dfrac{1}{4}x + \dfrac{2}{3}y = -\dfrac{5}{2} \end{cases}$
    (*Hint:* Clear fractions first to simplify the arithmetic.)

30. $\begin{cases} \dfrac{1}{5}x - \dfrac{3}{2}y = 4 \\ -\dfrac{2}{3}x + \dfrac{1}{2}y = -\dfrac{13}{3} \end{cases}$
    (*Hint:* Clear fractions first to simplify the arithmetic.)

In Exercises 31–36, write each system of equations in the form
$\begin{cases} Ax + By = E \\ Cx + Dy = F \end{cases}$*, and then solve the system.*

31. $\begin{cases} 3(x + y) = 1 \\ \quad -2x = -y + 2 \end{cases}$

32. $\begin{cases} 2x = -3y + 4 \\ \dfrac{x + y}{3} = 1 \end{cases}$

33. $\begin{cases} -4y = x + 5 \\ \dfrac{x}{3} + \dfrac{y}{2} = 1 \end{cases}$

34. $\begin{cases} 3y = 7 - x \\ -\dfrac{2x}{3} + \dfrac{3y}{2} = 5 \end{cases}$

35. $\begin{cases} \dfrac{x + 1}{2} + \dfrac{y - 1}{3} = 1 \\ 3x + \quad y = 7 \end{cases}$

36. $\begin{cases} -2x - \quad y = -7 \\ \dfrac{2x}{3} + \dfrac{y + 1}{2} = 1 \end{cases}$

 In Exercises 37–40, use a graphing utility to approximate the solution set of each system. If there is no solution, state that the system is inconsistent.

37. $\begin{cases} 0.3x = y - 4 \\ 0.5x + y = 1 \end{cases}$

38. $\begin{cases} 1.9x = y + 2.6 \\ -0.5x - y = 1.7 \end{cases}$

39. $\begin{cases} 1.2x - 0.4y = -2 \\ 0.5x + 1.3y = 3.2 \end{cases}$

40. $\begin{cases} -3.2x + 2.5y = -5.3 \\ 1.6x - 2.8y = \quad 4.7 \end{cases}$

In Exercises 41 and 42, find a system of equations whose solution is indicated graphically.

41.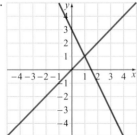

42.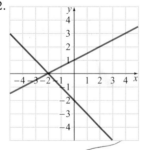

In Exercises 43–52, graph the solution set of each inequality.

43. $x < -2$

44. $y \geq 3$

45. $-2x + y \geq -4$

46. $-3x + y \leq 2$

47. $x > y + 6$

48. $y < 2x + 7$

49. $3x - 4y > 12$

50. $6x - 2y \leq 24$

51. $3x + 5y > 10$

52. $7x \leq 10x + 3$

In Exercises 53–70, graph the solution set of each system of inequalities.

53. $\begin{cases} -2x + y \leq 8 \\ -x + y \geq 2 \end{cases}$

54. $\begin{cases} x + y \leq 10 \\ 3x - y \geq 6 \end{cases}$

55. $\begin{cases} y \geq -2 \\ 5x + 2y \leq 10 \end{cases}$

56. $\begin{cases} x \leq 10 \\ x + y \geq 7 \end{cases}$

57. $\begin{cases} -x + y < 3 \\ x + y > -5 \end{cases}$

58. $\begin{cases} -x + y > 5 \\ x + y < 1 \end{cases}$

59. $\begin{cases} x + y \leq 4 \\ -4x + 2y \geq -1 \end{cases}$

60. $\begin{cases} x \leq 5 \\ x \geq 2 \\ y > 1 \end{cases}$

61. $\begin{cases} y \leq -2 \\ y \geq -5 \\ x \leq -1 \end{cases}$

62. $\begin{cases} -2x + 3y \leq 7 \\ -2x + 9y \geq 1 \\ 2x + 3y \leq 11 \end{cases}$

63. $\begin{cases} x + y \leq 4 \\ -x + y \leq 4 \\ x + 5y \geq 8 \end{cases}$

64. $\begin{cases} 3x - y \geq -1 \\ -2x - y \leq 4 \end{cases}$

65. $\begin{cases} -3x + y \leq -1 \\ 4x + y \leq 6 \end{cases}$

66. $\begin{cases} -x - y \geq 3 \\ 2x - y \leq 1 \end{cases}$

67. $\begin{cases} x \leq 0 \\ -5x + 4y \leq 20 \\ 3x + 4y \geq -12 \end{cases}$

68. $\begin{cases} x \geq 0 \\ 2x + 3y \leq 2 \\ -4x + 3y \geq -4 \end{cases}$

69. $\begin{cases} -\dfrac{4}{3}x + y \geq -5 \\ -4x + 3y \leq 13 \\ x + y \geq 2 \end{cases}$

70. $\begin{cases} -\dfrac{3}{2}x + y \geq -3 \\ 2x + y \leq 4 \\ 2x + y \geq -3 \end{cases}$

In Exercises 71–74, solve the optimization problem.

71. Maximize $P = 10x + 8y$ subject to the following constraints.

$$\begin{cases} x \geq 5 \\ y \geq 2 \\ x \leq 9 \\ y \leq 10 \end{cases}$$

72. Maximize $P = 12x + 15y$ subject to the following constraints.

$$\begin{cases} x \geq 3 \\ y \geq 1 \\ x \leq 10 \\ y \leq 14 \end{cases}$$

73. Minimize $P = 16x + 10y$ subject to the following constraints.

$$\begin{cases} y \geq 2x \\ x \geq 5 \\ x \geq 0 \\ y \geq 0 \end{cases}$$

74. Minimize $P = 20x + 30y$ subject to the following constraints.

$$\begin{cases} 3x + y \leq 9 \\ y \geq x \\ y \geq 2 \\ x \geq 0 \end{cases}$$

▶**Applications** In this set of exercises, you will use systems of linear equations and inequalities to study real-world problems.

75. **Film Industry** The total revenue generated by a film comes from two sources: box-office ticket sales and the sale of merchandise associated with the film. It is estimated that for a very popular film such as *Spiderman* or *Harry Potter*, the revenue from the sale of merchandise is four times the revenue from ticket sales. Assume this is true for the film *Spiderman*, which grossed a total of $3 billion. Find the revenue from ticket sales and the revenue from the sale of merchandise. (*Source: The Economist*)

76. **Nutrition** According to health professionals, the daily intake of fat in a diet that consists of 2000 calories per day should not exceed 50 grams. The total fat content of a meal that consists of a Whopper and a medium order of fries exceeds this limit by 14 grams. Two Whoppers and a medium order of fries have a total fat content of 111 grams. Set up and solve a system of equations to find the fat content of a Whopper and the fat content of a medium order of fries. (*Source:* www.burgerking.com)

77. **Nutrition** The following table lists the caloric content of a typical fast-food meal.

Food (single serving)	Calories
Cheeseburger	330
Medium order of fries	450
Medium cola (21 oz)	220

(*Source:* www.mcdonalds.com)

(a) After a lunch that consists of a cheeseburger, a medium order of fries, and a medium cola, you decide to burn off a quarter of the total calories in the meal by some combination of running and walking. You know that running burns 8 calories per minute and walking burns 3 calories per minute. If you exercise for a total of 40 minutes, how many minutes should you spend on each activity?

(b) Rework part (a) for the case in which you exercise for a total of only 20 minutes. Do you get a realistic solution? Explain your answer.

78. **Criminology** In 2004, there were a total of 3.38 million car thefts and burglaries in the United States. The number of burglaries exceeded the number of car thefts by 906,000. Find the number of burglaries and the number of car thefts. (*Source:* Federal Bureau of Investigation)

79. **Ticket Pricing** An airline charges $380 for a round-trip flight from New York to Los Angeles if the ticket is purchased at least 7 days in advance of travel. Otherwise, the price is $700. If a total of 80 tickets are purchased at a total cost of $39,040, find the number of tickets sold at each price.

80. **Utilities** In a residential area serviced by a utility company, the percentage of single-family homes with central air conditioning was 4 percentage points higher than 5 times the percentage of homes without central air. What percentage of these homes had central air-conditioning, and what percentage did not?

81. **Mixture** A chemist wishes to make 10 gallons of a 15% acid solution by mixing a 10% acid solution with a 25% acid solution.

(a) Let x and y denote the total volumes (in gallons) of the 10% and 25% solutions, respectively. Using the variables x and y, write an equation for the total volume of the 15% solution (the mixture).

(b) Using the variables x and y, write an equation for the total volume of acid in the mixture by noting that

Volume of acid in 15% solution = volume of acid in 10% solution + volume of acid in 25% solution.

(c) Solve the system of equations from parts (a) and (b), and interpret your solution.

(d) Is it possible to obtain a 5% acid solution by mixing a 10% solution with a 25% solution? Explain without solving any equations.

82. **Financial Planning** A couple has $20,000 to invest for their child's future college expenses. Their accountant recommends placing at least $12,000 in a high-yield mutual fund and at most $6000 in a low-yield mutual fund.
    (a) Use x to denote the amount of money placed into the high-yield fund. Use y to denote the amount of money placed into the low-yield fund. Write a system of linear inequalities that describes the possible amounts in each type of account.
    (b) Graph the region that represents all possible amounts the couple could place into each account if they wish to follow the accountant's advice.

83. **Financial Planning** A couple has $10,000 to invest for their child's wedding. Their accountant recommends placing at least $6000 in a high-yield investment and no more than $4000 in a low-yield investment.
    (a) Use x to denote the amount of money placed into the high-yield investment. Use y to denote the amount of money placed into the low-yield investment. Write a system of linear inequalities that describes the possible amounts the couple could invest in each type of venture.
    (b) Graph the region that represents all possible amounts the couple could put into each investment if they wish to follow the accountant's advice.

84. **Ticket Pricing** The gymnasium at a local high school has 1000 seats. The state basketball championship game will be held there next Friday night. Tickets to the game will be priced as follows: $10 if purchased in advance and $15 if purchased at the door. The booster club feels that at least 100 tickets will be sold at the advance ticket price. Total sales of at least $5000 are expected.
    (a) Use x to denote the number of tickets sold in advance. Use y to denote the number of tickets sold at the door. Write a system of linear inequalities that describes the possible numbers of tickets sold at each price.
    (b) Graph the region that represents all possible combinations of ticket sales.

85. **Ticket Pricing** The auditorium at the library has 200 seats. Tickets to a lecture by a local author are priced as follows: $6 if purchased in advance and $8 if purchased at the door. According to past records, at least 40 tickets will be sold in advance, and total sales of at least $960 are expected.
    (a) Use x to denote the number of tickets sold in advance. Use y to denote the number of tickets sold at the door. Write a system of linear inequalities that describes the possible numbers of tickets sold at each price.
    (b) Graph the region that represents all possible combinations of ticket sales.

86. **Inventory Control** A store sells two brands of mobile phones, Brand A and Brand B. Past sales records indicate that the store must have at least twice as many Brand B phones in stock as Brand A phones. The store must have at least two Brand A phones in stock. The maximum inventory the store can carry is 20 phones.
    (a) Use x to denote the number of Brand A phones in stock. Use y to denote the number of Brand B phones in stock. Write a system of linear inequalities that describes the possible numbers of each brand of phone in stock.
    (b) Graph the region that represents all possible numbers of Brand A and Brand B phones the store could have in stock.

87. **Inventory Control** A grocery store sells two types of carrots, organic and non-organic. Past sales records indicate that the store must stock at least three times as many pounds of organic carrots as non-organic carrots. The store needs to have at least 5 pounds of non-organic carrots in stock. The maximum inventory of carrots the store can carry is 30 pounds.
    (a) Use x to denote the number of pounds of organic carrots in stock. Use y to denote the number of pounds of non-organic carrots in stock. Write a system of linear inequalities that describes the possible numbers of pounds of each type of carrot the store could keep in stock.
    (b) Graph the region that represents all possible amounts of organic and non-organic carrots the store could have in stock.

88. **Maximizing Profit** An electronics firm makes a clock radio in two different models: one (model 380) with a battery backup feature and the other (model 360) without. It takes 1 hour and 15 minutes to manufacture each unit of the model 380 radio, and only 1 hour to manufacture each unit of the model 360. At least 500 units of the model 360 radio are to be produced. The manufacturer realizes a profit per radio of $15 for the model 380 and only $10 for the model 360. If at most 2000 hours are to be allocated to the manufacture of the two models combined, how many of each model should be made to maximize the total profit?

89. **Maximizing Profit** A telephone company manufactures two different models of phones: Model 120 is cordless and Model 140 is not cordless. It takes 1 hour to manufacture the cordless model and 1 hour and 30 minutes to manufacture the traditional phone. At least 300 of the cordless models are to be produced. The manufacturer realizes a profit per phone of $12 for Model 120 and $10 for Model 140. If at most 1000 hours are to be allocated to the manufacture of the two models combined, how many of each model should be produced to maximize the total profit?

90. **Minimizing Commuting Time** Bill can't afford to spend more than $90 per month on transportation to and from work. The bus fare is only $1.50 one way, but it takes Bill 1 hour and 15 minutes to get to work by bus. If he drives the 20-mile round trip, his one-way commuting time is reduced to 1 hour, but it costs him $.45 per mile. If he works at least 20 days per month, how often does he need to drive in order to minimize his commuting time and keep within his monthly budget?

91. **Minimizing Commuting Time** Sarah can't afford to spend more than $90 per month on transportation to and from work. The bus fare is only $1.50 one way, but it takes Sarah 1 hour and 15 minutes to get to work by bus. If she drives the 15-mile round trip, her one-way commuting time is reduced to 40 minutes, but it costs her $.40 per mile. If she works at least 20 days a month, how often does she have to drive in order to minimize her commuting time and keep within her monthly budget?

92. **Maximizing Profit** A cosmetics company makes a profit of 15 cents on a tube of lipstick and a profit of 8 cents on a tube of lip gloss. To meet dealer demand, the company needs to produce between 300 and 800 tubes of lipstick and between 100 and 300 tubes of lip gloss. The maximum number of tubes of lipstick and lip gloss the company can produce per day is 800. How many of each type of beauty product should be produced to maximize profit?

93. **Maximizing Profit** A golf club manufacturer makes a profit of $3 on a driver and a profit of $2 on a putter. To meet dealer demand, the company needs to produce between 20 and 50 drivers and between 30 and 50 putters each day. The maximum number of clubs produced each day by the company is 80. How many of each type of club should be produced to maximize profit?

94. **Maximizing Profit** A farmer has 90 acres available for planting corn and soybeans. The cost of seed per acre is $4 for corn and $6 for soybeans. To harvest the crops, the farmer will need to hire some temporary help. It will cost the farmer $20 per acre to harvest the corn and $10 per acre to harvest the soybeans. The farmer has $480 available for seed and $1400 available for labor. His profit is $120 per acre of corn and $150 per acre of soybeans. How many acres of each crop should the farmer plant to maximize the profit?

95. **Maximizing Profit** A farmer has 110 acres available for planting cucumbers and peanuts. The cost of seed per acre is $5 for cucumbers and $6 for peanuts. To harvest the crops, the farmer will need to hire some temporary help. It will cost the farmer $30 per acre to harvest the cucumbers and $20 per acre to harvest the peanuts. The farmer has $300 available for seed and $1200 available for labor. His profit is $100 per acre of cucumbers and $125 per acre of peanuts. How many acres of each crop should the farmer plant to maximize the profit?

96. **Minimizing Cost** Joi and Cheyenne are planning a party for at least 50 people. They are going to serve hot dogs and hamburgers. Each hamburger costs $1 and each hot dog costs $.50. Joi thinks that each person will eat only one item, either a hot dog or a hamburger. She also estimates that they will need at least 15 hot dogs and at least 20 hamburgers. How many hamburgers and how many hot dogs should Joi and Cheyenne buy if they want to minimize their cost?

97. **Minimizing Cost** Tara is planning a party for at least 100 people. She is going to serve two types of appetizers: mini pizzas and mini quiche. Each mini pizza costs $.50 and each mini quiche costs $.60. Tara thinks that each person will eat only one item, either a mini pizza or a mini quiche. She also estimates that she will need at least 60 mini pizzas and at least 20 mini quiche. How many mini pizzas and how many mini quiche should Tara order to minimize her cost?

98. **Mixture** You wish to make a 1-pound blend of two types of coffee, Kona and Java. The Kona costs $8 per pound and the Java costs $5 per pound. The blend will sell for $7 per pound.

    (a) Let k and j denote the amounts (in pounds) of Kona and Java, respectively, that go into making a 1-pound blend. One equation that must be satisfied by k and j is

    $$k + j = 1.$$

    Both k and j must be between 0 and 1. Why?

    (b) Using the variables k and j, write an equation that expresses the fact that the total cost of 1 pound of the blend will be $7.

    (c) Solve the system of equations from parts (a) and (b), and interpret your solution.

    (d) To make a 1-pound blend of Kona and Java that costs $7.50 per pound, which type of coffee would you use more of? Explain without solving any equations.

99. **Economics** A supply function for widgets is modeled by $P(q) = aq + b$, where q is the number of widgets supplied and $P(q)$ is the total price of q widgets, in dollars. It is known that 200 widgets can be supplied for $40 and 100 widgets can be supplied for $25. Use a system of linear equations to find the constants a and b in the expression for the supply function.

▶**Concepts** This set of exercises will draw on the ideas presented in this section and your general math background.

100. The sum of money invested in two savings accounts is $1000. If both accounts pay 4% interest compounded annually, is it possible to earn a total of $50 in interest in the first year?
    (a) Explain your answer in words.
    (b) Explain your answer using a system of equations.

101. The following is a system of three equations in only two variables.
$$\begin{cases} x - y = 1 \\ x + y = 1 \\ 2x - y = 1 \end{cases}$$

(a) Graph the solution of each of these equations.
(b) Is there a single point at which *all three lines* intersect?
(c) Is there one ordered pair (x, y) that satisfies *all three equations*? Why or why not?

102. State the conditions that a system of three equations in two variables must satisfy in order to have just one solution. Give an example of such a system, and illustrate your example graphically.

103. Adult and children's tickets for a certain show sell for $8 each. A total of 1000 tickets are sold, with total sales of $8000. Is it possible to figure out exactly how many of each type of ticket were sold? Why or why not?

6.2 Systems of Linear Equations in Three Variables

Objectives

▶ Solve systems of linear equations in three variables by elimination

▶ Solve systems of linear equations in three variables using Gaussian elimination

▶ Understand what is meant by an inconsistent system of linear equations in three variables

▶ Understand what is meant by a dependent system of linear equations in three variables

▶ Use systems of linear equations in three variables to solve applied problems

Many employers offer their employees the option of participating in a retirement plan, such as a 401(k). It is not uncommon for participants in these plans to be faced with building an investment portfolio from a menu of literally dozens of investment instruments.

Our first example shows how a system of linear equations in three variables arises in developing an investment plan targeted toward a particular level of risk. Later in this section, we will present an elimination method that can be used to solve this and other systems of linear equations in three variables.

Example **1** **Making Investment Decisions Using Linear Equations**

Example 4 in Section 6.2 builds upon this example. ⋯⟩

An investment counselor would like to advise her client about three specific investment instruments: a stock-based mutual fund, a corporate bond, and a savings bond. The counselor wants to distribute the total amount of the investment among the individual instruments in the portfolio according to the client's tolerance for risk. Risk factors for individual instruments can be quantified on a scale of 1 to 5, with 1 being the most risky. The risk factors associated with each investment instrument are summarized in Table 6.2.1.

Table 6.2.1

Investment Instrument	Risk Factor
Stock-based mutual fund	2
Corporate bond	4
Savings bond	5

The client can tolerate an overall risk level of 3.5. In addition, the client wants to invest the same amount of money in the corporate bond as in the savings bond. Set up a system of equations that would determine the percentage of the total investment that should be allocated to each instrument.

▶Solution First we define the variables.

x: percentage (in decimal form) invested in mutual fund

y: percentage (in decimal form) invested in corporate bond

z: percentage (in decimal form) invested in savings bond

Note that we now have *three* variables, corresponding to the three different instruments. Using these variables, we formulate equations from the given information.

Formulating the first equation
We know that the total percentage should add up to 1, because the percentage is assumed to be in decimal form. This gives

$$x + y + z = 1.$$

Formulating the second equation
Next we create an expression to represent the overall risk.

$$\text{Overall risk} = 2x + 4y + 5z$$

To obtain this equation, we multiplied the percentages allocated to the individual instruments in the portfolio by their corresponding risk factors. The client can tolerate an overall risk level of 3.5, so our second equation is

$$2x + 4y + 5z = 3.5.$$

Formulating the third equation
Finally, we use the fact that the percentage invested in the corporate bond must equal the percentage invested in the savings bond. This gives us a third equation.

$$\text{Percentage invested in corporate bond} = \text{percentage invested in savings bond}$$
$$y = z$$

Thus we have the following system of three equations in three variables.

$$\begin{cases} x + y + z = 1 \\ 2x + 4y + 5z = 3.5 \\ y = z \end{cases}$$

We will solve this system of equations later in this section, after discussing a procedure known as Gaussian elimination.

✔ *Check It Out 1:* If the client in Example 1 invests 20% of the total investment in the corporate bond, 20% in the savings bond, and 60% in the mutual fund, will all three equations be satisfied? Why or why not? ■

Solving Systems of Linear Equations in Three Variables

The general form of a linear equation in the variables x, y, and z is $Ax + By + Cz = D$, where A, B, C, and D are constants such that A, B and C are not all zero. A **system of linear equations** in these variables is a collection of equations that can be put into this form. A solution of such a system of linear equations is an **ordered triple** of numbers (x, y, z) that satisfies all the equations in the system.

We now introduce a process known as **Gaussian elimination,** which can be used to find the solution(s) of a system of linear equations in three variables. Gaussian elimination, named after the German mathematician Karl Friedrich Gauss (1777–1855), is an extension of the method of elimination for solving systems of two linear equations in two variables.

We will first apply this method to a simple system of equations whose solution is easy to find.

Example **2** **Solving a System of Linear Equations in Three Variables**

Solve the following system of equations.

$$\begin{cases} x + y - z = 10 \\ \quad\quad y + z = \;\; 0 \\ \quad\quad\quad\quad z = \;\; 2 \end{cases}$$

▶**Solution** From the last equation, we immediately have

$$z = 2.$$

Note that the middle equation contains only the variables y and z. Since we know that $z = 2$, we can substitute this value of z into the second equation to find y.

$y + z = \;\; 0$	*The middle equation*
$y + 2 = \;\; 0$	*Substitute z = 2*
$y = -2$	*Solve for y*

Now that we have found both z and y, we can find x by using the first equation. Substituting $y = -2$ and $z = 2$ into the first equation, we have

$$x + (-2) - 2 = 10 \Longrightarrow x - 2 - 2 = 10 \Longrightarrow x - 4 = 10 \Longrightarrow x = 14.$$

Thus our solution is $x = 14$, $y = -2$, $z = 2$. You may use substitution to verify that these values of x, y, and z do indeed satisfy all three equations.

✔ *Check It Out 2:* Solve the following system of equations.

$$\begin{cases} x - y + z = \;\; 10 \\ \quad\quad y - z = \;\; 0 \\ \quad\quad\quad\quad z = -1 \end{cases} \blacksquare$$

The procedure illustrated in Example 2 is known as **back-substitution.** That is, we start with a known value of one variable and successively work backward to find the values of the other variables.

We were able to solve the system of equations in Example 2 in a systematic fashion because the equations were arranged in a convenient manner. Most systems of equations that we will encounter are not so nicely arranged. Fortunately, in many cases we can obtain a system of equations similar to the one in Example 2 just by using simple algebraic manipulation. We can then easily find the solution.

The following operations will be used in solving systems of linear equations by Gaussian elimination.

Operations Used in Manipulating a System of Linear Equations

1. Interchange two equations in the system.

2. Multiply one equation in the system by a nonzero constant.

3. Multiply one equation in the system by a nonzero constant and add the result to another equation in the system.

Whenever these operations are performed on a given system of equations, the resulting system of equations is equivalent to the original system. That is, the two systems will have the same solution(s).

Note We will not necessarily use all three types of operations in a particular example or exercise. You should study the examples given here to get an understanding of when a certain type of operation should be used. Also, there are many correct ways to apply the sequence of operations to a given system of equations to obtain the solution(s).

Example 3 Solving a System of Linear Equations Using Gaussian Elimination

Solve the system of equations.

$$\begin{cases} -2x + 2y + z = -6 \\ x - 2y + 2z = -1 \\ 3x + 2y - z = 3 \end{cases}$$

▶**Solution** Begin by labeling the equations.

$$\begin{cases} -2x + 2y + z = -6 & (1) \\ x - 2y + 2z = -1 & (2) \\ 3x + 2y - z = 3 & (3) \end{cases}$$

Our first goal is to eliminate the variable x from the last two equations. Check to see if the coefficient of x in either of the last two equations is 1. If so, we will interchange that equation with Equation (1) to make the elimination easier.

Step 1 Interchange Equations (1) and (2), since the coefficient of x in Equation (2) is 1.

$$\begin{cases} x - 2y + 2z = -1 & (4) = (2) \\ -2x + 2y + z = -6 & (5) = (1) \\ 3x + 2y - z = 3 & (3) \end{cases}$$

We have relabeled the first two equations to make the subsequent discussion less confusing.

Step 2 Eliminate x from Equation (5).

▶ Multiply Equation (4) by 2 and add the result to Equation (5):

$$
\begin{array}{ll}
2x - 4y + 4z = -2 & 2 \cdot (4) \\
\underline{-2x + 2y + z = -6} & (5) \\
- 2y + 5z = -8 & 2 \cdot (4) + (5)
\end{array}
$$

▶ Replace Equation (5) with the result (and relabel the equation as Equation (6)), and retain Equation (4):

$$
\begin{cases}
x - 2y + 2z = -1 & (4) \\
- 2y + 5z = -8 & (6) = 2 \cdot (4) + (5) \\
3x + 2y - z = 3 & (3)
\end{cases}
$$

Step 3 Eliminate x from Equation (3).

▶ Multiply Equation (4) by -3 and add the result to Equation (3):

$$
\begin{array}{ll}
-3x + 6y - 6z = 3 & -3 \cdot (4) \\
\underline{3x + 2y - z = 3} & (3) \\
8y - 7z = 6 & -3 \cdot (4) + (3)
\end{array}
$$

▶ Replace Equation (3) with the result (and relabel the equation as Equation (7)), and retain Equation (4):

$$
\begin{cases}
x - 2y + 2z = -1 & (4) \\
- 2y + 5z = -8 & (6) \\
8y - 7z = 6 & (7) = -3 \cdot (4) + (3)
\end{cases}
$$

Our next goal is to eliminate the variable y from the third equation. In doing so, we will make use of the second equation. Note that both the second and third equations contain only the two variables y and z.

Step 4 Eliminate y from Equation (7).

▶ Multiply Equation (6) by 4 and add the result to Equation (7):

$$
\begin{array}{ll}
-8y + 20z = -32 & 4 \cdot (6) \\
\underline{8y - 7z = 6} & (7) \\
13z = -26 & 4 \cdot (6) + (7)
\end{array}
$$

▶ Replace Equation (7) with the result (and relabel the equation as Equation (8)), and retain Equation (6):

$$
\begin{cases}
x - 2y + 2z = -1 & (4) \\
- 2y + 5z = -8 & (6) \\
13z = -26 & (8) = 4 \cdot (6) + (7)
\end{cases}
$$

We now have a system of equations that is very similar in structure to the one given in Example 2: Equation (8) contains only the variable z; Equation (6) contains only the variables y and z; and Equation (4) contains all three variables (x, y, and z). To find the value of each of these variables, we will use back-substitution.

Step 5 Use Equation (8) to solve for z:

$$13z = -26 \Longrightarrow z = -2.$$

Proceeding as in Example 2, substitute this value into Equation (6) to solve for y:

$$-2y + 5(-2) = -8 \Longrightarrow -2y - 10 = -8 \Longrightarrow -2y = 2 \Longrightarrow y = -1.$$

Substitute $z = -2$ and $y = -1$ into Equation (4) to solve for x:

$$x - 2(-1) + 2(-2) = -1 \Longrightarrow x + 2 - 4 = -1 \Longrightarrow x - 2 = -1 \Longrightarrow x = 1.$$

Thus the solution is $x = 1$, $y = -1$, $z = -2$. You should verify that these values of x, y, and z satisfy all three equations.

☑ *Check It Out 3:* Solve the system of equations.

$$\begin{cases} 2x + 2y + z = 1 \\ -3x - 2y + 2z = -5 \\ x + 2y - z = 2 \end{cases} ▪$$

Note In using Gaussian elimination to solve a system of three linear equations in the variables x, y, and z, we can always eliminate the variable x first and then eliminate the variable y. If the system has a unique solution, we will end up with an equation in the variable z alone. This systematic procedure will always work.

Every linear equation in three variables is represented by a plane in three-dimensional space. Therefore, the solution set of a system of linear equations in three variables can be thought of geometrically as the set of points at which all the planes associated with the equations in the system intersect. Figure 6.2.1 summarizes the ways in which three planes can intersect and indicates the nature of the solution(s) of the corresponding system of three linear equations in three variables.

Figure 6.2.1

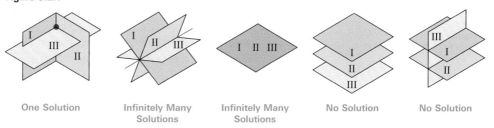

One Solution   Infinitely Many Solutions   Infinitely Many Solutions   No Solution   No Solution

The Gaussian elimination method presented in this section can be generalized to find the solutions of systems of linear equations in more than three variables. However, it is next to impossible to visualize the geometric representation of a linear equation in more than three variables.

Example 4 Solving the Investment Problem

⁝••• *This example builds on Example 1 in Section 6.2.*

Solve the following system of equations, which we formulated in Example 1. Express the value of each of the variables in your solution as a percentage, and use those percentages to interpret the solution in the context of Example 1.

$$\begin{cases} x + y + z = 1 \\ 2x + 4y + 5z = 3.5 \\ y = z \end{cases}$$

▶ Solution

Step 1 Write the system of equations in the general form, with the variables on the left and the constants on the right, and label the equations.

$$\begin{cases} x + y + z = 1 & (1) \\ 2x + 4y + 5z = 3.5 & (2) \\ y - z = 0 & (3) \end{cases}$$

Step 2 Equation (2) contains the variable x, but Equation (3) does not. Thus we eliminate x from Equation (2) only: Multiply Equation (1) by -2 and add the result to Equation (2).

$$\begin{cases} x + y + z = 1 & (1) \\ 2y + 3z = 1.5 & (4) = -2 \cdot (1) + (2) \\ y - z = 0 & (3) \end{cases}$$

Step 3 Since y has a coefficient of 1 in Equation (3), interchange Equations (4) and (3) as a prelude to eliminating y. The sole purpose of switching these equations is to make the arithmetic easier.

$$\begin{cases} x + y + z = 1 & (1) \\ y - z = 0 & (5) = (3) \\ 2y + 3z = 1.5 & (6) = (4) \end{cases}$$

Step 4 Eliminate y from Equation (6): Multiply Equation (5) by -2 and add the result to Equation (6).

$$\begin{cases} x + y + z = 1 & (1) \\ y - z = 0 & (5) \\ 5z = 1.5 & (7) = -2 \cdot (5) + (6) \end{cases}$$

Step 5 Solving the last equation, we get $z = 0.3$. Substituting $z = 0.3$ into the second equation yields $y = 0.3$. Because $x + y + z = 1$, substituting $y = 0.3$ and $z = 0.3$ gives $x = 0.4$. When expressed as percentages, the values of these variables are $x = 40\%$, $y = 30\%$, and $z = 30\%$.

The interpretation of this solution with respect to the application presented in Example 1 is as follows: 40% of the total investment is allocated to the mutual fund (x), and the remaining 60% is split evenly between the corporate bond (y) and the savings bond (z).

✔ *Check It Out 4:* Solve the following system of equations.

$$\begin{cases} x + y + z = 1 \\ 2x + 3y + 5z = 2 \\ z = x \end{cases}$$ ■

Systems of Linear Equations in Three Variables with Either Infinitely Many Solutions or No Solution

When manipulating a system of linear equations, it is possible to end up with an equation of the form $0 = 0$. Such an equation is of no value and you must work with the remaining two equations, as illustrated in the following example.

Example 5 A System with Infinitely Many Solutions

Use Gaussian elimination to solve the following system of linear equations.

$$\begin{cases} -3x + y - 2z = 1 \\ 5x - y + 4z = -1 \end{cases}$$

▶ **Solution** First, label the equations.

$$\begin{cases} -3x + y - 2z = 1 & (1) \\ 5x - y + 4z = -1 & (2) \end{cases}$$

The coefficients of x in this system of equations are of opposite sign, but neither of them is a multiple of the other. The least common multiple of -3 and 5 is 15, so we will multiply Equation (2) by 3 and then replace Equation (2) with the result.

$$\begin{cases} -3x + y - 2z = 1 & (1) \\ 15x - 3y + 12z = -3 & (3) = 3 \cdot (2) \end{cases}$$

The next step is to eliminate x from the second equation: Multiply Equation (1) by 5 and add the result to Equation (3).

$$\begin{cases} -3x + y - 2z = 1 & (1) \\ 2y + 2z = 2 & (4) = 5 \cdot (1) + (3) \end{cases}$$

Because there is no third equation, we cannot eliminate y. Instead, we use the second equation to solve for y in terms of z:

$$2y + 2z = 2 \Longrightarrow 2y = -2z + 2 \Longrightarrow y = -z + 1.$$

Now we can substitute $-z + 1$ for y in Equation (1), which will yield an equation in the variables x and z. After simplifying, we will solve for x in terms of z.

$$-3x + (-z + 1) - 2z = 1$$
$$-3x - 3z + 1 = 1$$
$$-3x = 3z \Longrightarrow x = -z$$

The solution set is thus the set of all x, y, z such that $x = -z$ and $y = -z + 1$, where z is any real number. Since the expression for at least one of the variables x or y depends explicitly on z, this system of equations is dependent and has infinitely many solutions.

Individual solutions can be found by choosing specific values for z. For example, if $z = 0$, then $y = 1$ and $x = 0$. If $z = 2$, then $y = -1$ and $x = -2$.

✔ *Check It Out 5:* Use Gaussian elimination to solve the following system of linear equations.

$$\begin{cases} -x + y - z = 3 \\ 4x - 5y + z = -2 \end{cases} ■$$

Example 6 **A System of Equations with No Solution**

Use Gaussian elimination to solve the following system of linear equations.

$$\begin{cases} x \quad\;\; + \;\; z = 3 \\ 3x + y + 2z = 0 \\ 4x + y + 3z = 5 \end{cases}$$

▶Solution First, label the equations.

$$\begin{cases} x \quad\;\; + \;\; z = 3 & (1) \\ 3x + y + 2z = 0 & (2) \\ 4x + y + 3z = 5 & (3) \end{cases}$$

Next, eliminate x from Equations (2) and (3).

$$\begin{cases} x \quad\;\; + z = \quad\; 3 & (1) \\ \quad\;\; y - z = -9 & (4) = -3 \cdot (1) + (2) \\ \quad\;\; y - z = -7 & (5) = -4 \cdot (1) + (3) \end{cases}$$

Finally, eliminate y from Equation (5).

$$\begin{cases} x \quad\;\; + z = \quad\; 3 & (1) \\ \quad\;\; y - z = -9 & (4) \\ \quad\quad\;\; 0 = \quad\; 2 & (6) = -1 \cdot (4) + (5) \end{cases}$$

The last equation $(0 = 2)$ is false. Thus this system of equations is inconsistent and has no solution.

☑ *Check It Out 6:* Use Gaussian elimination to solve the following system of linear equations.

$$\begin{cases} x + y - \quad\; z = -4 \\ 3x \quad\quad + 2z = \quad\; 1 \\ -2x + y - 3z = \quad\; 6 \end{cases} ▨$$

6.2 Key Points

▶ **Gaussian elimination** is used to solve a system of linear equations in three or more variables.

▶ The following operations are used in the elimination process for solving systems of equations.

   1. Interchange two equations in the system.

   2. Multiply an equation by a nonzero constant.

   3. Multiply an equation by a nonzero constant and add the result to another equation.

▶ A system of three or more linear equations can have **one solution, no solution,** or **infinitely many solutions.**

6.2 Exercises

▶**Skills** This set of exercises will reinforce the skills illustrated in this section.

In Exercises 1–6, verify that each system of equations has the indicated solution.

1. $\begin{cases} 3x \quad\quad - z = 3 \\ 2x + y - 2z = 0 \\ 3x - 2y + z = 7 \end{cases}$
   Solution: $x = 1, y = -2, z = 0$

2. $\begin{cases} x - 2y - 4z = 4 \\ -2x \quad\quad - 3z = -7 \\ 5x + y - 2z = 5 \end{cases}$
   Solution: $x = 2, y = -3, z = 1$

3. $\begin{cases} 5x - y + 3z = -13 \\ x - y + 2z = -9 \\ 4x - y + z = -5 \end{cases}$
   Solution: $x = 0, y = 1, z = -4$

4. $\begin{cases} -3x + 2y - z = -1 \\ 2x \quad\quad + 5z = 18 \\ -4x + 2y + z = 8 \end{cases}$
   Solution: $x = -1, y = 0, z = 4$

5. $\begin{cases} x + 2y - 3z = 1 \\ 2x + 3y + z = -3 \end{cases}$
   Solution: $x = -9 - 11z, y = 5 + 7z$ (for every real number z)

6. $\begin{cases} x - 3y + z = 4 \\ -x + 2y - 2z = -7 \end{cases}$
   Solution: $x = 13 - 4z, y = 3 - z$ (for every real number z)

In Exercises 7–10, use back-substitution to solve the system of linear equations.

7. $\begin{cases} x - y + z = 1 \\ -2y - z = 0 \\ z = -2 \end{cases}$

8. $\begin{cases} x + y - z = -1 \\ -y - 3z = -2 \\ 2z = 4 \end{cases}$

9. $\begin{cases} -2u + v + 3w = -1 \\ v - w = 1 \\ 3w = 9 \end{cases}$

10. $\begin{cases} 2u - 3v + w = -1 \\ v - w = 1 \\ -2w = 8 \end{cases}$

In Exercises 11–38, use Gaussian elimination to solve the system of linear equations. If there is no solution, state that the system is inconsistent.

11. $\begin{cases} x + 2y = 2 \\ -x + y + 3z = 4 \\ 3y - 3z = -6 \end{cases}$

12. $\begin{cases} x - 2y + z = 2 \\ -x + 4y + 3z = -8 \\ -6y + 2z = 4 \end{cases}$

13. $\begin{cases} x + 2y = 0 \\ -2x + 4y + 8z = 8 \\ 3x - 3z = -9 \end{cases}$

14. $\begin{cases} x + 3y - z = 4 \\ -x - 2y = -8 \\ 2x + 4y - z = 10 \end{cases}$

15. $\begin{cases} 5x + y = 0 \\ x - z = 2 \\ 4y + z = -2 \end{cases}$

16. $\begin{cases} 4x - y = 2 \\ y - 2z = 1 \\ 6x - z = 3 \end{cases}$

17. $\begin{cases} x - 4y - 3z = -3 \\ y + z = 2 \\ x + 3y + 3z = 0 \end{cases}$

18. $\begin{cases} 4x + y - 2z = 6 \\ -x - y + z = -2 \\ 3x - z = 5 \end{cases}$

19. $\begin{cases} 3x + 2y + 3z = 1 \\ x - y - z = 1 \\ x + 4y + 5z = -1 \end{cases}$

20. $\begin{cases} -2x + 4z = 4 \\ x - 2y = 2 \\ -3x - 2y + 4z = -2 \end{cases}$

21. $\begin{cases} x \quad\;\;\; + 2z = \;\;\; 0 \\ \quad\; y + \;\; z = -1 \\ x + 8y + 4z = \;\;\; 1 \end{cases}$

22. $\begin{cases} x + y + 4z = -1 \\ 2x + y + 2z = \;\;\; 3 \\ 3x \quad\;\;\; - 6z = 12 \end{cases}$

23. $\begin{cases} 5x \quad\;\;\; + 3z = \quad\; 2 \\ x - 4y + \;\; z = \quad\; 6 \\ x + 8y + \;\; z = -10 \end{cases}$

24. $\begin{cases} 4x \quad\;\;\; + z = \;\;\; 0 \\ 3x - 5y + z = -1 \\ x + 5y \quad\;\;\; = -1 \end{cases}$

25. $\begin{cases} 5x + 6y - 2z = 2 \\ 2x - \;\; y + \;\; z = 2 \\ x + 4y - 2z = 0 \end{cases}$

26. $\begin{cases} -3u - 2v + \;\; w = 1 \\ 2u - \;\; v + 2w = 1 \\ u + \;\; v - \;\; w = 0 \end{cases}$

27. $\begin{cases} 3u - 2v + 2w = -2 \\ -u + 4v + \;\; w = -1 \\ 5u + 3v + 5w = \quad\; 1 \end{cases}$

28. $\begin{cases} 3r + s + 2t = \quad\; 5 \\ -2r - s + \;\; t = -1 \\ 4r \quad\;\;\; + 2t = \quad\; 6 \end{cases}$

29. $\begin{cases} 4r + \;\; s - 2t = \quad\; 7 \\ 3r - \;\; s + \;\; t = \quad\; 6 \\ -6r + 3s - 2t = -1 \end{cases}$

30. $\begin{cases} x + \;\; y + 2z = 4 \\ -3x + 2y - \;\; z = 3 \end{cases}$

31. $\begin{cases} x - 2y - z = \;\; 7 \\ 2x - 3y + z = 10 \end{cases}$

32. $\begin{cases} r - s - t = -3 \\ 3r + s + t = -5 \end{cases}$

33. $\begin{cases} r + 2s \quad\;\;\; = 1 \\ 3r + 5s + 4t = 7 \end{cases}$

34. $\begin{cases} -2x + \;\; y - z = 2 \\ 5x - 2y + z = 3 \end{cases}$

35. $\begin{cases} -3x - 2y - z = \;\; 7 \\ 3x + 3y + z = -3 \end{cases}$

36. $\begin{cases} x - 2y - 3z = 2 \\ 5x - 3y - \;\; z = 3 \\ 6x - 5y - 4z = 5 \end{cases}$

37. $\begin{cases} x - 2y - \;\; 5z = -3 \\ 3x - 6y - \;\; 7z = \quad\; 1 \\ -2x + 4y + 12z = -4 \end{cases}$

38. $\begin{cases} x - 4y + 2z = -2 \\ y - \;\; z = \quad\; 2 \\ 3x - 6y + 2z = \quad\; 3 \end{cases}$

▶**Applications** In this set of exercises, you will use systems of equations to study real-world problems.

39. **Investments** An investor would like to build a portfolio from three specific investment instruments: a stock-based mutual fund, a high-yield bond, and a certificate of deposit (CD). Risk factors for individual instruments can be quantified on a scale of 1 to 5, with 1 being the most risky. The risk factors associated with the particular instruments chosen by this investor are summarized in the following table.

Type of Investment	Risk Factor
Stock-based mutual fund	2
High-yield bond	1
CD	5

The investor can tolerate an overall risk level of 2.7. In addition, the amount of money invested in the mutual fund must equal the sum of the amounts invested in the high-yield bond and the CD. Determine the percentage of the total investment that should be allocated to each instrument.

40. **Clothing** Princess Clothing, Inc., has the following yardage requirements for making a single blouse, dress, or skirt.

	Fabric (yd)	Lining (yd)	Trim (yd)
Blouse	2	1	1
Dress	4	2	1
Skirt	3	1	0

There are 52 yards of fabric, 24 yards of lining, and 10 yards of trim available. How many blouses, dresses, and skirts can be made, assuming that all the fabric, lining, and trim is used up?

41. Electrical Wiring In an electrical circuit in which two resistors are connected in series, the formula for the total resistance R is $R = R_1 + R_2$, where R_1 and R_2 are the resistances of the individual resistors. Consider three resistors A, B, and C. The total resistance when A and B are connected in series is 55 ohms. The total resistance when B and C are connected in series is 80 ohms. The sum of the resistances of B and C is four times the resistance of A. Find the resistances of A, B, and C.

42. Computers The Hi-Tech Computer Company builds three types of computers: basic, upgrade, and high-power. Component requirements for each model are given below.

	Peripheral Cards	Processors	Disk Drives
Basic	3	1	1
Upgrade	4	1	2
High-power	5	2	2

A recent shipment of parts contained 125 peripheral cards, 40 processors, and 50 disk drives. How many of each computer model can be built if all the parts are to be used up?

43. Car Rentals A car rental company structures its rates according to the specific day(s) of the week for which a car is rented. The rate structure is as follows.

Daily rental fee, Level A	Monday through Thursday
Daily rental fee, Level B	Friday
Daily rental fee, Level C	Saturday and Sunday

The total rental fee for each of three different situations is given in the following table.

Days Rented	Total Rental Fee
Thurs., Fri., Sat., Sun.	$140
Wed., Thurs., Fri.	$125
Fri., Sat., Sun.	$95

What is the daily rental fee at each of the three levels (A, B, and C)?

44. Nutrition JoAnna and her friends visited a popular pizza place to inquire about the cholesterol content of some of their favorite kinds of pizza. They were told that two slices of pepperoni pizza and one slice of Veggie Delight contain a total of 65 milligrams of cholesterol. Also, the amount of cholesterol in one slice of Meaty Delight exceeds that in a slice of Veggie Delight by 20 milligrams. Finally, the total amount of cholesterol in three slices of Meaty Delight and one slice of Veggie Delight is 120 milligrams. How many milligrams of cholesterol are there in each slice of pepperoni, Veggie Delight, and Meaty Delight? (*Source:* www.pizzahut.com)

45. Tourism In 2003, the top three U.S. states visited by foreign tourists were Florida, New York, and California. The total tourism market share of these three states was 68.7%. Florida's share was equal to that of New York. California's share was lower than Florida's by 1.2 percentage points. What was the market share of each of these three states? (*Source:* U.S. Department of Commerce)

▶**Concepts** This set of exercises will draw on the ideas presented in this section and your general math background.

46. Perform Gaussian elimination on the following system of linear equations, where a is some unspecified constant.

$$\begin{cases} x - y + z = 1 \\ -x + 2y - 2z = 3 \\ y - z = a \end{cases}$$

(a) For what value(s) of a does this system have infinitely many solutions?

(b) For what value(s) of a does this system have no solution?

47. The graph of the function $f(x) = ax^2 + bx + c$ is a parabola that passes through the points $(-1, -3)$, $(1, 1)$, and $(-2, -8)$, where a, b, and c are constants to be determined.

(a) Because $f(-1) = a(-1)^2 + b(-1) + c = a - b + c$ and the value of $f(-1)$ is given to be -3, one linear equation satisfied by a, b, and c is

$$a - b + c = -3.$$

Give two more linear equations satisfied by a, b, and c.

(b) Solve the system of three linear equations satisfied by a, b, and c (the equation you were given together with the two equations that you found).

(c) Substitute your values of a, b, and c into the expression for $f(x)$ and check that the graph of f passes through the given points.

48. Use the steps outlined in Exercise 47 to find the equation of the parabola that passes through the points $(0, 1)$, $(2, -3)$, and $(-3, -8)$.

49. Use the steps outlined in Exercise 47 to find the equation of the parabola that passes through the point $(2, 6)$ and has $(1, 1)$ as its vertex.

6.3 Solving Systems of Equations Using Matrices

Objectives

▶ Understand what a matrix is

▶ Perform Gaussian elimination using matrices

▶ Perform Gauss-Jordan elimination to solve a system of linear equations

▶ Solve systems of equations arising from applications

In the previous section, the choices of arithmetic operations performed in the course of solving systems of linear equations depended only on the coefficients of the variables in those equations, but not on the variables themselves. Thus it would make sense to refine the elimination process so that we would need to keep track only of the coefficients and the constant terms. In this section, we introduce *matrices* precisely for this purpose.

A **matrix** is just a table that consists of rows and columns of numbers. The numbers in a matrix are called **entries;** there is an entry at the intersection of any given row and column. The following is an example of a matrix with three rows and four columns.

$$\begin{bmatrix} -3 & 6 & 2 & 6 \\ 2 & 0 & -1 & 4 \\ 0 & 1 & 8 & \frac{2}{3} \end{bmatrix}$$

A matrix is always enclosed within square brackets, to distinguish it from other types of tables.

Any linear equation can be represented as a row of numbers, with all but one of the numbers being the coefficients and the last number being the constant term.

Consider the following system of equations.

$$\begin{cases} -x + y - 2z = -7 \\ 3x + z = 5 \\ 3y + 2z = -2 \end{cases}$$

We can represent this system as a matrix with three rows and four columns. In the first equation, the coefficients of x, y, and z are -1, 1, and -2, respectively, and the constant term is -7. Thus the first row of the matrix will be

$$-1 \ 1 \ -2 \ -7.$$

Similarly, we have the following for the second and third rows, respectively.

$$3 \quad 0 \quad 1 \quad 5$$
$$0 \quad 3 \quad 2 \quad -2$$

There is no term in the second equation that explicitly contains the variable y, so the coefficient of y in that equation is 0, corresponding to $0y$. Similarly, the coefficient of x in the third equation is 0.

Thus the matrix for this system of equations is

$$\begin{bmatrix} -1 & 1 & -2 & -7 \\ 3 & 0 & 1 & 5 \\ 0 & 3 & 2 & -2 \end{bmatrix}.$$

Discover *and* Learn

Construct the augmented matrix for the following system of equations.

$$\begin{cases} 2x - y + z = 4 \\ x + y = 2 \\ -y + z = -6 \end{cases}$$

The vertical bar inside the matrix separates the coefficients from the constant terms. Such a matrix is known as the **augmented matrix** for the system of equations from which it was constructed.

Gaussian Elimination Using Matrices

Now that we have introduced a compact notation for handling systems of linear equations, we will introduce operations on matrices that will enable us to solve such systems of equations. The following operations are called **elementary row operations.** These operations correspond to the operations you performed on systems of linear equations in Section 6.2.

> ### Elementary Row Operations On Matrices
>
> 1. Interchange two rows of the matrix.
> 2. Multiply one row of the matrix by a nonzero constant.
> 3. Multiply one row of the matrix by a nonzero constant and add the result to another row of the matrix.

Example 1 Performing Elementary Row Operations

Perform the indicated row operations (independently of one another, not in succession) on the following augmented matrix.

$$\begin{bmatrix} 0 & 3 & 1 & -1 \\ 1 & -2 & -1 & 2 \\ 0 & 6 & -4 & 5 \end{bmatrix}$$

(a) Interchange rows 1 and 2.

(b) Multiple the first row by -2.

(c) Multiply the first row by -2 and add the result to the third row. Retain the original first row.

▶Solution

Before performing the indicated row operations, label the rows of the matrix.

$$\begin{bmatrix} 0 & 3 & 1 & -1 \\ 1 & -2 & -1 & 2 \\ 0 & 6 & -4 & 5 \end{bmatrix} \begin{matrix} (1) \\ (2) \\ (3) \end{matrix}$$

(a) In the original matrix, interchanging rows (1) and (2) gives

$$\begin{bmatrix} 1 & -2 & -1 & 2 \\ 0 & 3 & 1 & -1 \\ 0 & 6 & -4 & 5 \end{bmatrix}.$$

(b) Simply multiply each entry in row (1) of the original matrix by -2.

$$\begin{bmatrix} 0 & -6 & -2 & 2 \\ 1 & -2 & -1 & 2 \\ 0 & 6 & -4 & 5 \end{bmatrix} \qquad \text{New row} = -2 \cdot \text{row (1)}$$

(c) For this computation, there are two steps.

Step 1 Multiply each entry in row (1) by -2, and then add the entries in the resulting row to the corresponding entries in row (3).

$$\begin{array}{rrrr} 0 & -6 & -2 & 2 \\ + \; 0 & 6 & -4 & 5 \\ \hline 0 & 0 & -6 & 7 \end{array}$$

Step 2 Replace row (3) with the result of Step 1 (and relabel the third row), and retain row (1).

$$\begin{bmatrix} 0 & 3 & 1 & | & -1 \\ 1 & -2 & -1 & | & 2 \\ 0 & 0 & -6 & | & 7 \end{bmatrix}$$ New row $= -2 \cdot$ row (1) + row (3)

☑ *Check It Out 1:* Rework Example 1, part (c), for the following augmented matrix.

$$\begin{bmatrix} 0 & 0 & -2 & | & -1 \\ -2 & 4 & 1 & | & 0 \\ 3 & 4 & -2 & | & 3 \end{bmatrix}$$ ▪

By performing elementary row operations, we can convert the augmented matrix for any system of linear equations to a matrix that is in a special form known as **row-echelon form**.

Row-Echelon Form of an Augmented Matrix

1. In each nonzero row (i.e., in each row that has at least one nonzero entry to the left of the vertical bar), the first nonzero entry is 1. This 1 is called the **leading 1.**
2. For any two successive nonzero rows, the leading 1 in the higher of the two rows must be farther to the left than the leading 1 in the lower row.
3. Each zero row (i.e., each row in which every entry to the left of the vertical bar is 0) lies below all the nonzero rows.

The following matrices are in row-echelon form. The leading 1's are given in red, and the stars represent real numbers.

$$\begin{bmatrix} 1 & \star & \star & | & \star \\ 0 & 1 & \star & | & \star \\ 0 & 0 & 1 & | & \star \end{bmatrix} \quad \begin{bmatrix} 1 & \star & \star & | & \star \\ 0 & 1 & \star & | & \star \\ 0 & 0 & 0 & | & 0 \end{bmatrix} \quad \begin{bmatrix} 1 & \star & \star & | & \star \\ 0 & 0 & 0 & | & 1 \\ 0 & 0 & 0 & | & 0 \end{bmatrix}$$

Once the augmented matrix for a system of linear equations is in row-echelon form, back-substitution is used to solve the system.

Note   We will not necessarily use all three types of elementary row operations in any particular example. You should study all the examples given here to see when a certain type of operation would arise.

Example 2 Gaussian Elimination Using Matrices

Apply Gaussian elimination to a matrix to solve the following system of equations.

$$\begin{cases} x + y + z = 1 \\ 2x + 4y + 5z = 3.5 \\ y = z \end{cases}$$

This is the system of equations we formulated in Example 1 of Section 6.2 and solved in Example 4 of that same section.

►Solution

Step 1 Write the system of equations in the general form, with the variables on the left and the constants on the right. For each variable, make sure that all the terms containing that variable are aligned vertically with one another.

$$\begin{cases} x + y + z = 1 \\ 2x + 4y + 5z = 3.5 \\ y - z = 0 \end{cases}$$

In each of the remaining steps, the corresponding equations are displayed to the right of the matrix.

Step 2 Construct the augmented matrix for this system of equations, and label the rows of the matrix.

$$\begin{bmatrix} 1 & 1 & 1 & | & 1 \\ 2 & 4 & 5 & | & 3.5 \\ 0 & 1 & -1 & | & 0 \end{bmatrix} \begin{matrix} (1) \\ (2) \\ (3) \end{matrix} \qquad \begin{matrix} x + y + z = 1 \\ 2x + 4y + 5z = 3.5 \\ y - z = 0 \end{matrix}$$

Step 3 Eliminate the 2 in the second row, first column: Multiply row (1) by -2 and add the result to row (2).

$$\begin{bmatrix} 1 & 1 & 1 & | & 1 \\ 0 & 2 & 3 & | & 1.5 \\ 0 & 1 & -1 & | & 0 \end{bmatrix} \begin{matrix} (1) \\ (2') = -2 \cdot (1) + (2) \\ (3) \end{matrix} \qquad \begin{matrix} x + y + z = 1 \\ 2y + 3z = 1.5 \\ y - z = 0 \end{matrix}$$

Step 4 To obtain a 1 in the second row, second column, swap rows (2') and (3).

$$\begin{bmatrix} 1 & 1 & 1 & | & 1 \\ 0 & 1 & -1 & | & 0 \\ 0 & 2 & 3 & | & 1.5 \end{bmatrix} \begin{matrix} (1) \\ (2'') = (3) \\ (3') = (2') \end{matrix} \qquad \begin{matrix} x + y + z = 1 \\ y - z = 0 \\ 2y + 3z = 1.5 \end{matrix}$$

Step 5 Eliminate the 2 in the third row, second column: Multiply row (2'') by -2 and add the result to row (3').

$$\begin{bmatrix} 1 & 1 & 1 & | & 1 \\ 0 & 1 & -1 & | & 0 \\ 0 & 0 & 5 & | & 1.5 \end{bmatrix} \begin{matrix} (1) \\ (2'') \\ (3'') = -2 \cdot (2'') + (3') \end{matrix} \qquad \begin{matrix} x + y + z = 1 \\ y - z = 0 \\ 5z = 1.5 \end{matrix}$$

Step 6 To get a 1 in the third row, third column, multiply row (3'') by $\frac{1}{5}$.

$$\begin{bmatrix} 1 & 1 & 1 & | & 1 \\ 0 & 1 & -1 & | & 0 \\ 0 & 0 & 1 & | & 0.3 \end{bmatrix} \begin{matrix} (1) \\ (2'') \\ (3''') = \frac{1}{5} \cdot (3'') \end{matrix} \qquad \begin{matrix} x + y + z = 1 \\ y - z = 0 \\ z = 0.3 \end{matrix}$$

Step 7 Now perform the back-substitution. From the third row, $z = 0.3$. From the second row, $y - z = 0$. Substitute $z = 0.3$ into this equation to solve for y:

$$y - 0.3 = 0 \Longrightarrow y = 0.3.$$

From the first row,

$$x + y + z = 1.$$

Substitute $z = 0.3$ and $y = 0.3$ into this equation to solve for x:

$$x + 0.3 + 0.3 = 1 \Longrightarrow x + 0.6 = 1 \Longrightarrow x = 0.4.$$

The solution to the system of equations is $x = 0.4$, $y = 0.3$, $z = 0.3$, which is the same as the solution found in Example 4 of Section 6.2.

Technology Note

Steps 3 and 4 of Example 2 are illustrated in Figure 6.3.1 using a graphing utility. Note that after each row operation is performed on matrix A, the resulting matrix is also called A.

Keystroke Appendix:
Section 13

Figure 6.3.1

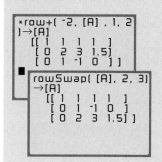

✔ *Check It Out 2:* Apply Gaussian elimination to a matrix to solve the following system of equations.

$$\begin{cases} 2x + 2y + z = 1 \\ -3x - 2y + 2z = -5 \\ x + 2y - z = 2 \end{cases}$$ ■

Figure 6.3.2 gives a schematic illustration of the steps in the Gaussian elimination process in which nonzero entries of an augmented matrix are eliminated. Additional operations, such as swapping two rows of the matrix and/or multiplying a nonzero row by a nonzero constant, may be needed to get the leading 1 in one or more of the nonzero rows of the matrix.

Figure 6.3.2 Gaussian Elimination

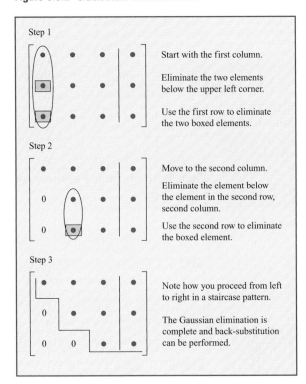

Step 1

Start with the first column.

Eliminate the two elements below the upper left corner.

Use the first row to eliminate the two boxed elements.

Step 2

Move to the second column.

Eliminate the element below the element in the second row, second column.

Use the second row to eliminate the boxed element.

Step 3

Note how you proceed from left to right in a staircase pattern.

The Gaussian elimination is complete and back-substitution can be performed.

Example 3 A System of Equations with Infinitely Many Solutions

Apply Gaussian elimination to a matrix to solve the following system of equations.

$$\begin{cases} 2x - 4y = -8 \\ 3x - 6z = 4 \end{cases}$$

▶Solution

Step 1 Construct the augmented matrix for the given system of equations, and label the rows of the matrix.

$$\left[\begin{array}{ccc|c} 2 & -4 & 0 & -8 \\ 3 & 0 & -6 & 4 \end{array}\right] \quad \begin{array}{c} (1) \\ (2) \end{array}$$

Step 2 To get a 1 in the first row, first column, multiply row (1) by $\frac{1}{2}$.

$$\begin{bmatrix} 1 & -2 & 0 & | & -4 \\ 3 & 0 & -6 & | & 4 \end{bmatrix} \quad \begin{array}{l} (1') = \frac{1}{2} \cdot (1) \\ (2) \end{array}$$

Step 3 Eliminate the 3 in the second row, first column: multiply row (1') by -3 and add the result to row (2).

$$\begin{bmatrix} 1 & -2 & 0 & | & -4 \\ 0 & 6 & -6 & | & 16 \end{bmatrix} \quad \begin{array}{l} (1') \\ (2') = -3 \cdot (1') + (2) \end{array}$$

Step 4 To get a 1 in the second row, second column, multiply row (2') by $\frac{1}{6}$.

$$\begin{bmatrix} 1 & -2 & 0 & | & -4 \\ 0 & 1 & -1 & | & \frac{8}{3} \end{bmatrix} \quad \begin{array}{l} (1') \\ (2'') = \frac{1}{6} \cdot (2') \end{array}$$

Because there are only two rows to work with, stop here and perform back-substitution. From the second row,

$$y - z = \frac{8}{3}.$$

Solve this equation for y in terms of z.

$$y - z = \frac{8}{3} \Longrightarrow y = z + \frac{8}{3}$$

From the first row,

$$x - 2y + 0z = -4.$$

Substitute $z + \frac{8}{3}$ for y in this equation and simplify. Then solve for x in terms of z.

$$x - 2\left(z + \frac{8}{3}\right) = -4$$

$$x - 2z - \frac{16}{3} = -4$$

$$x - 2z = \frac{16}{3} - 4$$

$$x - 2z = \frac{16}{3} - \frac{12}{3}$$

$$x - 2z = \frac{4}{3}$$

$$x = 2z + \frac{4}{3}$$

The solution set is $x = 2z + \frac{4}{3}$ and $y = z + \frac{8}{3}$, where z is any real number. Since the expression for at least one of the variables x or y depends explicitly on z, this system of equations is dependent and has infinitely many solutions.

☑ *Check It Out 3:* Give two specific solutions of the system of equations given in Example 3. ▪

Technology Note

The augmented matrix in Example 3 is shown in row-echelon form in Figure 6.3.3. It differs from the augmented matrix we obtained in Example 3 because the row-echelon form of an augmented matrix is not unique. Fractions are used to display the result so that it is easier to read and easier to check against the result obtained by hand.

Keystroke Appendix: Section 13

Figure 6.3.3

```
ref( [A] )▶Frac
 [[ 1  0  -2  4/3]
  [ 0  1  -1  8/3]]
■
```

Example **4** **A System of Equations with No Solution**

Apply Gaussian elimination to a matrix to solve the following system of equations.

$$\begin{cases} x + y - z = 1 \\ -x \quad\quad + z = 2 \\ x + 2y - z = 0 \end{cases}$$

▶Solution

Step 1 Construct the augmented matrix for the given system of equations, and label the equations.

$$\begin{bmatrix} 1 & 1 & -1 & | & 1 \\ -1 & 0 & 1 & | & 2 \\ 1 & 2 & -1 & | & 0 \end{bmatrix} \quad \begin{matrix} (1) \\ (2) \\ (3) \end{matrix}$$

Step 2 Eliminate the -1 and 1 in the second and third rows of the first column. Multiply the first row by 1 and add it to the second row. Multiply the first row by -1 and add it to the third row.

$$\begin{bmatrix} 1 & 1 & -1 & | & 1 \\ 0 & 1 & 0 & | & 3 \\ 0 & 1 & 0 & | & -1 \end{bmatrix} \quad \begin{matrix} (1) \\ (2') = (1) + (2) \\ (3') = -(1) + (3) \end{matrix}$$

Step 3 Eliminate the 1 in the third row, second column: Multiply the second row by -1 and add it to the third row.

$$\begin{bmatrix} 1 & 1 & -1 & | & 1 \\ 0 & 1 & 0 & | & 3 \\ 0 & 0 & 0 & | & -4 \end{bmatrix} \quad \begin{matrix} (1) \\ (2') \\ (3'') = -(2') + (3') \end{matrix}$$

Step 4 We stop here because the last row is equivalent to the equation

$$0x + 0y + 0z = -4 \Longrightarrow 0 = -4.$$

The equation $0 = -4$ is false. Thus the systems of equations has no solution.

✔ *Check It Out 4:* Use a matrix to solve the following system of equations.

$$\begin{cases} x \quad\quad - z = 1 \\ x - y \quad\quad = 2 \\ 2x - y - z = 4 \end{cases} ▪$$

Gauss-Jordan Elimination

An augmented matrix can be converted to a matrix that is in a special type of row-echelon form known as **reduced row-echelon form.** The distinctive feature of a matrix in reduced row-echelon form is that each leading 1 is the only nonzero entry in its column. The following augmented matrices are in reduced row-echelon form. The leading 1's are given in red.

$$\begin{bmatrix} 1 & 0 & 0 & | & -3 \\ 0 & 1 & 0 & | & \frac{1}{4} \\ 0 & 0 & 1 & | & \frac{13}{3} \end{bmatrix} \quad \begin{bmatrix} 1 & 0 & 0 & 0 & | & 2 \\ 0 & 1 & 0 & 0 & | & \frac{1}{2} \\ 0 & 0 & 1 & 0 & | & 0 \end{bmatrix} \quad \begin{bmatrix} 1 & 0 & 4 & | & -3 \\ 0 & 1 & -2 & | & \frac{1}{4} \\ 0 & 0 & 0 & | & 0 \end{bmatrix}$$

Discover *and* **Learn**

Solve the system of linear equations that has the following augmented matrix.

$$\begin{bmatrix} 1 & 0 & 0 & | & -2 \\ 0 & 1 & 0 & | & 1 \\ 0 & 0 & 1 & | & 0 \end{bmatrix}$$

When the augmented matrix for a system of linear equations is in reduced row-echelon form, you can read off the solution directly. Of course, you have to perform more computations to get the matrix into that form.

The process used to reduce an augmented matrix to reduced row-echelon form is known as **Gauss-Jordan elimination.** The following example illustrates this process, which is an extension of Gaussian elimination.

Example 5 Solving a System of Equations Using Gauss-Jordan Elimination

Use Gauss-Jordan elimination to solve the following system of linear equations.

$$\begin{cases} 3x + y - 10z = -8 \\ x + y - 2z = -4 \\ -2x + 9z = 5 \end{cases}$$

▶**Solution** Construct the augmented matrix for this system of equations, and label the rows of the matrix.

$$\begin{bmatrix} 3 & 1 & -10 & | & -8 \\ 1 & 1 & -2 & | & -4 \\ -2 & 0 & 9 & | & 5 \end{bmatrix} \quad \begin{matrix} (1) \\ (2) \\ (3) \end{matrix}$$

First we will reduce the augmented matrix to row-echelon form via ordinary Gaussian elimination. Then we will complete the process of reducing the matrix to reduced row-echelon form by using additional elementary row operations.

Step 1 To get a 1 in the first row, first column, swap rows (1) and (2).

$$\begin{bmatrix} 1 & 1 & -2 & | & -4 \\ 3 & 1 & -10 & | & -8 \\ -2 & 0 & 9 & | & 5 \end{bmatrix} \quad \begin{matrix} (1') = (2) \\ (2') = (1) \\ (3) \end{matrix}$$

Step 2 Eliminate the 3 in the second row, first column: Multiply row (1′) by −3 and add the result to row (2′). Then eliminate the −2 in the third row, first column: multiply row (1′) by 2 and add the result to row (3).

$$\begin{bmatrix} 1 & 1 & -2 & | & -4 \\ 0 & -2 & -4 & | & 4 \\ 0 & 2 & 5 & | & -3 \end{bmatrix} \quad \begin{matrix} (1') \\ (2'') = -3 \cdot (1') + (2') \\ (3') = 2 \cdot (1') + (3) \end{matrix}$$

Step 3 To get a 1 in the second row, second column, multiply row (2″) by $-\frac{1}{2}$.

$$\begin{bmatrix} 1 & 1 & -2 & | & -4 \\ 0 & 1 & 2 & | & -2 \\ 0 & 2 & 5 & | & -3 \end{bmatrix} \quad \begin{matrix} (1') \\ (2''') = -\frac{1}{2} \cdot (2'') \\ (3') \end{matrix}$$

Step 4 Eliminate the 2 in the third row, second column: Multiply row (2‴) by −2 and add the result to row (3′).

$$\begin{bmatrix} 1 & 1 & -2 & | & -4 \\ 0 & 1 & 2 & | & -2 \\ 0 & 0 & 1 & | & 1 \end{bmatrix} \quad \begin{matrix} (1') \\ (2''') \\ (3'') = -2 \cdot (2''') + (3') \end{matrix}$$

As far as Gaussian elimination is concerned, we are done and can find the solution by back-substitution. For Gauss-Jordan elimination, we need to continue with the row operations. What remains is elimination of the nonzero entries at the locations indicated in red in the following matrix.

$$\begin{bmatrix} 1 & 1 & -2 & -4 \\ 0 & 1 & 2 & -2 \\ 0 & 0 & 1 & 1 \end{bmatrix} \begin{matrix} (1') \\ (2''') \\ (3'') \end{matrix}$$

Step 5 Eliminate the 2 in the second row, third column: Multiply row (3″) by −2 and add the result to row (2‴). Then eliminate the −2 in the first row, third column: Multiply row (3″) by 2 and add the result to row (1′).

$$\begin{bmatrix} 1 & 1 & 0 & -2 \\ 0 & 1 & 0 & -4 \\ 0 & 0 & 1 & 1 \end{bmatrix} \begin{matrix} (1'') = 2 \cdot (3'') + (1') \\ (2'''') = -2 \cdot (3'') + (2''') \\ (3'') \end{matrix}$$

Step 6 Eliminate the 1 in the first row, second column: Multiply row (2‴‴) by −1 and add the result to row (1″).

$$\begin{bmatrix} 1 & 0 & 0 & 2 \\ 0 & 1 & 0 & -4 \\ 0 & 0 & 1 & 1 \end{bmatrix} \begin{matrix} (1''') = -1 \cdot (2'''') + (1'') \\ (2'''') \\ (3'') \end{matrix}$$

The matrix is now in reduced row-echelon form and corresponds to the following system of linear equations.

$$\begin{cases} x & = 2 \\ y & = -4 \\ z & = 1 \end{cases}$$

Thus the solution is

$$x = 2, \ y = -4, \ z = 1.$$

Instead of writing down the system of equations, we could have read the solution from the matrix directly.

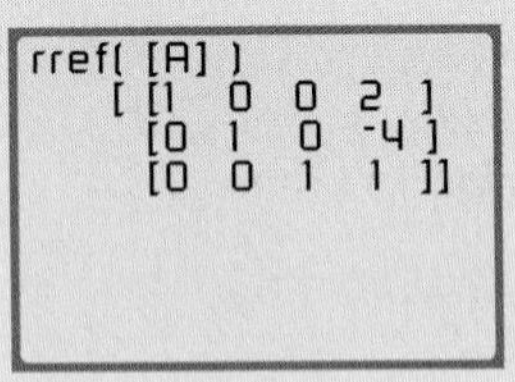

✔ *Check It Out 5:* Use Gauss-Jordan elimination to solve the following system of equations.

$$\begin{cases} x - y - 2z = 0 \\ 2x + y + z = 3 \\ 3y + 2z = 3 \end{cases} ■$$

You may have noticed that performing Gauss-Jordan elimination by hand is quite tedious, even for a system of equations in only three variables. The reduced row-echelon form of a matrix can be obtained easily by using a graphing utility.

Technology Note

In Figure 6.3.4, a graphing utility gives the reduced row-echelon form for the augmented matrix in Example 5.

Keystroke Appendix: Section 13

Figure 6.3.4

```
rref( [A] )
   [ [1  0  0  2 ]
     [0  1  0  -4]
     [0  0  1  1 ]]
```

Example 6 Finding a Solution from a Reduced Matrix

Find the solution set of the system of linear equations in the variables *x*, *y*, *z*, and *u* (in that order) that has the following augmented matrix.

$$\begin{array}{cccc} x & y & z & u \\ \end{array}$$
$$\begin{bmatrix} 1 & 0 & 0 & 3 & 0 \\ 0 & 1 & 0 & 5 & 1 \\ 0 & 0 & 1 & -6 & 4 \end{bmatrix}$$

▶**Solution** The given matrix is in reduced row-echelon form. The corresponding system of linear equations consists of *three* equations in *four* variables.

$$\begin{cases} x & + 3u = 0 & \text{From first row} \\ \quad y & + 5u = 1 & \text{From second row} \\ \quad z - 6u = 4 & \text{From third row} \end{cases}$$

Solving for x, y, and z in terms of u, we have

$$x = -3u, \quad y = -5u + 1,$$

and

$$z = 6u + 4.$$

The solution set is $x = -3u$, $y = -5u + 1$, and $z = 6u + 4$, where u is any real number. Since the expression for at least one of the variables x, y, or z depends explicitly on u, this system of equations is dependent and has infinitely many solutions.

✔ *Check It Out 6:* Find the solution set of the system of linear equations in the variables x, y, z, and u (in that order) that has the following augmented matrix.

$$\begin{array}{cccc} x & y & z & u \end{array}$$
$$\left[\begin{array}{cccc|c} 1 & 0 & 0 & -2 & 3 \\ 0 & 1 & 0 & 1 & -4 \\ 0 & 0 & 1 & -2 & 0 \end{array}\right] ■$$

Applications

Systems of linear equations have a number of applications in a wide variety of fields. Let's solve an application in manufacturing.

Example 7 **Systems of Equations in Manufacturing**

The Quilter's Corner is a company that makes quilted wall hangings, pillows, and bedspreads. The process of quilting involves cutting, sewing, and finishing. For each of the three products, the numbers of hours spent on each task are given in Table 6.3.1.

Table 6.3.1

	Cutting (hr)	Sewing (hr)	Finishing (hr)
Wall hanging	1.5	1.5	1
Pillow	1	1	0.5
Bedspread	4	3	2

Every week, there are 17 hours available for cutting, 15 hours available for sewing, and 9 hours available for finishing. How many of each item can be made per week if all the available time is to be used?

▶Solution First, define the variables.

w: number of wall hangings

p: number of pillows

b: number of bedspreads

The total number of hours spent on cutting is given by

$$\text{Total cutting hours} = \text{cutting for wall hanging} + \text{cutting for pillow}$$
$$+ \text{ cutting for bedspread}$$
$$= 1.5w + p + 4b.$$

The total amount of time available for cutting is 17 hours, so our first equation is

$$1.5w + p + 4b = 17.$$

Since there are 15 hours available for sewing and 9 hours available for finishing, our other two equations are

$$1.5w + \quad p + 3b = 15$$
$$w + 0.5p + 2b = \quad 9.$$

Thus the system of equations to be solved is

$$\begin{cases} 1.5w + \quad p + 4b = 17 \\ 1.5w + \quad p + 3b = 15. \\ \quad w + 0.5p + 2b = \quad 9 \end{cases}$$

We will solve this system by applying Gaussian elimination to the corresponding augmented matrix.

$$\begin{bmatrix} 1.5 & 1 & 4 & | & 17 \\ 1.5 & 1 & 3 & | & 15 \\ 1 & 0.5 & 2 & | & 9 \end{bmatrix} \quad \begin{matrix} (1) \\ (2) \\ (3) \end{matrix}$$

Step 1 To get a 1 in the first row, first column, swap rows (1) and (3).

$$\begin{bmatrix} 1 & 0.5 & 2 & | & 9 \\ 1.5 & 1 & 3 & | & 15 \\ 1.5 & 1 & 4 & | & 17 \end{bmatrix} \quad \begin{matrix} (1') = (3) \\ (2) \\ (3') = (1) \end{matrix}$$

Step 2 Eliminate the 1.5 in the second row, first column: Multiply row $(1')$ by -1.5 and add the result to row (2). Then eliminate the 1.5 in the third row, first column: Multiply row $(1')$ by -1.5 and add the result to row $(3')$.

$$\begin{bmatrix} 1 & 0.5 & 2 & | & 9 \\ 0 & 0.25 & 0 & | & 1.5 \\ 0 & 0.25 & 1 & | & 3.5 \end{bmatrix} \quad \begin{matrix} (1') \\ (2') = -1.5 \cdot (1') + (2) \\ (3'') = -1.5 \cdot (1') + (3') \end{matrix}$$

Step 3 To get a 1 in the second row, second column, multiply row $(2')$ by $\dfrac{1}{0.25}$.

$$\begin{bmatrix} 1 & 0.5 & 2 & | & 9 \\ 0 & 1 & 0 & | & 6 \\ 0 & 0.25 & 1 & | & 3.5 \end{bmatrix} \quad \begin{matrix} (1') \\ (2'') = \dfrac{1}{0.25} \cdot (2') \\ (3'') \end{matrix}$$

Step 4 Eliminate the 0.25 in the third row, second column: Multiply row (2″) by -0.25 and add the result to row (3″).

$$\begin{bmatrix} 1 & 0.5 & 2 & | & 9 \\ 0 & 1 & 0 & | & 6 \\ 0 & 0 & 1 & | & 2 \end{bmatrix} \quad \begin{matrix} (1') \\ (2'') \\ (3''') = -0.25 \cdot (2'') + (3'') \end{matrix}$$

Step 5 Because the matrix is in row echelon form, perform back-substitution. From the third row, $b = 2$; from the second row, $p = 6$. From the first row,

$$w + 0.5p + 2b = 9.$$

Substituting $b = 2$ and $p = 6$ into this equation yields

$$w + 0.5(6) + 2(2) = 9.$$

Simplifying and then solving for w, we get

$$w + 0.5(6) + 2(2) = 9 \Longrightarrow w + 3 + 4 = 9 \Longrightarrow w + 7 = 9 \Longrightarrow w = 2.$$

Thus 2 wall hangings, 6 pillows, and 2 bedspreads can be made per week.

✔ *Check It Out 7:* Rework Example 7 for the case in which 36 hours are available for cutting, 30 hours are available for sewing, and 19 hours are available for finishing. ■

We recommend using Gaussian elimination with back-substitution when working by hand because this method is likely to result in fewer arithmetic errors than Gauss-Jordan elimination. If technology is used, however, Gauss-Jordan elimination, using the **rref** feature of a graphing utility, is much quicker.

6.3 Key Points

▸ In an **augmented matrix** corresponding to a system of equations, the rows of the left-hand side are formed by taking the coefficients of the variables in each equation, and the constants appear on the right-hand side.

▸ The following **elementary row operations** are used when performing Gaussian elimination on matrices.

1. Interchange two rows of the matrix.

2. Multiply a row of the matrix by a nonzero constant.

3. Multiply a row of the matrix by a nonzero constant and add the result to another row.

▸ The following matrices are in **row-echelon form:** there are zeros beneath the first 1 in each row.

$$\begin{bmatrix} 1 & -2 & 2 & | & 2 \\ 0 & 1 & 3 & | & -1 \\ 0 & 0 & 1 & | & 4 \end{bmatrix} \quad \begin{bmatrix} 1 & -1 & 1 & | & 5 \\ 0 & 1 & 2 & | & -3 \\ 0 & 0 & 0 & | & 0 \end{bmatrix}$$

▶ The following matrices are in **reduced row-echelon form:** there are zeros above and beneath the first 1 in each row.

$$\begin{bmatrix} 1 & 0 & 0 & | & 3 \\ 0 & 1 & 0 & | & 4 \\ 0 & 0 & 1 & | & -2 \end{bmatrix} \quad \begin{bmatrix} 1 & 0 & 3 & | & 4 \\ 0 & 1 & 2 & | & -2 \\ 0 & 0 & 0 & | & 0 \end{bmatrix}$$

▶ In **Gauss-Jordan elimination,** a matrix is converted to reduced row-echelon form by using elementary row operations.

6.3 Exercises

▶**Skills** This set of exercises will reinforce the skills illustrated in this section.

In Exercises 1–8, construct the augmented matrix for each system of equations. Do not solve the system.

1. $\begin{cases} 4x + y - 2z = 6 \\ -x - y + z = -2 \\ 3x - z = 4 \end{cases}$

2. $\begin{cases} 3r + s + 2t = -1 \\ -2r - s + t = 3 \\ 4r + 2t = -2 \end{cases}$

3. $\begin{cases} 3x - 2y + z = -1 \\ x + y - 4z = 3 \\ -2x - y + 3z = 0 \end{cases}$

4. $\begin{cases} -x + 5y - z = 6 \\ x - 4y + 2z = 3 \\ 3x - y + 5z = -1 \end{cases}$

5. $\begin{cases} 6x - 2y + z = 0 \\ -5x + y - 3z = -2 \\ 2x - 3y + 5z = 7 \end{cases}$

6. $\begin{cases} -2x - y - z = 5 \\ x + y + z = 0 \\ 3x + 2y + 7z = -8 \end{cases}$

7. $\begin{cases} x + y + 2z = -3 \\ -3x + 2y + z = 1 \end{cases}$

8. $\begin{cases} -2x + 6z = -1 \\ -3x + 2y + z = 0 \end{cases}$

In Exercises 9–14, perform the indicated row operations (independently of one another, not in succession) on the following augmented matrix.

$$\begin{bmatrix} 1 & -2 & 0 & | & -1 \\ 2 & -8 & -2 & | & 1 \\ 3 & 5 & 1 & | & 2 \end{bmatrix}$$

9. Multiply the second row by $\frac{1}{2}$.

10. Multiply the second row by $-\frac{1}{2}$.

11. Switch rows 1 and 2.

12. Switch rows 1 and 3.

13. Multiply the first row by 2 and add the result to the second row.

14. Multiply the first row by -3 and add the result to the third row.

In Exercises 15–22, for each matrix, construct the corresponding system of linear equations. Use the variables listed above the matrix, in the given order. Determine whether the system is consistent or inconsistent. If it is consistent, give the solution(s).

15.
$$\begin{array}{cc} x & y \\ \begin{bmatrix} 1 & 0 & | & -7 \\ 0 & 1 & | & 3 \end{bmatrix} \end{array}$$

16.
$$\begin{array}{cc} x & y \\ \begin{bmatrix} 1 & 0 & | & 3 \\ 0 & 0 & | & 1 \end{bmatrix} \end{array}$$

17.
$$\begin{array}{cc} x & y \\ \begin{bmatrix} 2 & 0 & | & 6 \\ 0 & 1 & | & 5 \end{bmatrix} \end{array}$$

18.
$$\begin{array}{ccc} x & y & z \end{array}$$
$$\left[\begin{array}{ccc|c} 3 & -2 & 5 & 2 \\ 1 & 0 & -1 & -3 \\ 0 & 0 & 0 & 8 \end{array}\right]$$

19.
$$\begin{array}{ccc} x & y & z \end{array}$$
$$\left[\begin{array}{ccc|c} 1 & 0 & 3 & 5 \\ 0 & 1 & -2 & -2 \\ 0 & 0 & 0 & 0 \end{array}\right]$$

20.
$$\begin{array}{ccc} x & y & z \end{array}$$
$$\left[\begin{array}{ccc|c} 1 & 0 & -2 & 7 \\ 0 & 1 & 4 & 3 \\ 0 & 0 & 0 & 0 \end{array}\right]$$

21.
$$\begin{array}{cccc} x & y & z & u \end{array}$$
$$\left[\begin{array}{cccc|c} 1 & 0 & 0 & -5 & 2 \\ 0 & 1 & 0 & -2 & -3 \\ 0 & 0 & 1 & 3 & 5 \end{array}\right]$$

22.
$$\begin{array}{cccc} x & y & z & u \end{array}$$
$$\left[\begin{array}{cccc|c} 1 & 0 & 0 & -4 & -3 \\ 0 & 1 & 0 & -2 & 1 \\ 0 & 0 & 1 & 3 & -10 \end{array}\right]$$

In Exercises 23–52, apply elementary row operations to a matrix to solve the system of equations. If there is no solution, state that the system is inconsistent.

23. $\begin{cases} x + 3y = 10 \\ -2x - 5y = -12 \end{cases}$

24. $\begin{cases} -x - y = -10 \\ 3x + 4y = 24 \end{cases}$

25. $\begin{cases} -2x + y = -5 \\ 4x - y = 6 \end{cases}$

26. $\begin{cases} x + 2y = 5 \\ 2x - 3y = 3 \end{cases}$

27. $\begin{cases} -x + y = 2 \\ 7x - 4y = -2 \end{cases}$

28. $\begin{cases} 5x + 3y = 7 \\ 2x + y = 2 \end{cases}$

29. $\begin{cases} x + 2y = 1 \\ x + 5y = -2 \end{cases}$

30. $\begin{cases} 2x - 3y = -1 \\ -3x + 2y = 6 \end{cases}$

31. $\begin{cases} 5x + 3y = -1 \\ -10x - 6y = 2 \end{cases}$

32. $\begin{cases} 2x + 4y = 5 \\ -4x - 8y = -10 \end{cases}$

33. $\begin{cases} 2x - 3y = 7 \\ 4x - 6y = 9 \end{cases}$

34. $\begin{cases} -x + y = 5 \\ 4x - 4y = 10 \end{cases}$

35. $\begin{cases} x + y + z = 3 \\ x + 2y + z = 5 \\ -2x + y - z = 4 \end{cases}$

36. $\begin{cases} x + y - z = 0 \\ 3x + 2y - z = -1 \\ -2x + y - 2z = -1 \end{cases}$

37. $\begin{cases} x + 3y - 2z = 4 \\ -5x - 3y - 2z = -8 \\ x - y - z = 0 \end{cases}$

38. $\begin{cases} x + 2y - 2z = -7 \\ 2x + 5y - 2z = -10 \\ x - 2y - 3z = -9 \end{cases}$

39. $\begin{cases} 3x - 4y = 14 \\ x - y + 2z = 14 \\ -x + 4z = 18 \end{cases}$

40. $\begin{cases} 2x - 5z = -15 \\ -x - 2y + z = 5 \\ x + y = -1 \end{cases}$

41. $\begin{cases} x + 2y - z = 2 \\ -2x + y - 3z = 6 \\ -x + 3y - 4z = 8 \end{cases}$

42. $\begin{cases} x + 3y = 2 \\ 5x + 12y + 3z = 1 \\ -4x - 9y - 3z = 1 \end{cases}$

43. $\begin{cases} -x + 2y - 3z = 2 \\ 2x + 3y + 2z = 1 \\ 3x + y + 5z = 1 \end{cases}$

44. $\begin{cases} x + 2y + z = -3 \\ 3x + y - 2z = 2 \\ 4x + 3y - z = 0 \end{cases}$

45. $\begin{cases} -x + 4y + 3z = 6 \\ 2x - 8y - 4z = 8 \end{cases}$

46. $\begin{cases} 3u + 5v - 3w = 1 \\ u + 2v - w = -2 \end{cases}$

47. $\begin{cases} 3r + 4s - 8t = 14 \\ 2r - 2s + 4t = 28 \end{cases}$

48. $\begin{cases} -3x + y + 24z = -9 \\ 2x + 2y - 8z = 6 \end{cases}$

49. $\begin{cases} 3x + 4y - 8z = 10 \\ -6x - 8y + 16z = 20 \end{cases}$

50. $\begin{cases} 2u - 3v - 2w = 4 \\ u + 2v + w = -3 \end{cases}$

51. $\begin{cases} z + 2y = 0 \\ z - 5x = -1 \\ 3x + 2y = 3 \end{cases}$

(*Hint:* Be careful with the order of the variables.)

52. $\begin{cases} x + 4z = -3 \\ x - 5y = 0 \\ z + 4y = 2 \end{cases}$

(*Hint:* Be careful with the order of the variables.)

▶**Applications** In this set of exercises, you will use the method of solving linear systems using matrices to study real-world problems.

53. **Mixture** JoAnn's Coffee wants to sell a new blend of three types of coffee: Colombian, Java, and Kona. The company plans to market this coffee in 10-pound bags

and has established the following specifications for the product:

▶ The amount of Colombian coffee in the blend is to be twice the amounts of Java and Kona combined.

▶ The amount of Kona coffee in the blend is to be half the amount of Java.

How many pounds of each type of coffee will go into each bag of the blend?

54. **Sports Equipment** The athletic director of a local high school is ordering equipment for spring sports. He needs to order twice as many baseballs as softballs. The total number of balls he must order is 300. How many of each type should he order?

55. **Take-Out Orders** A boy scout troop orders eight pizzas. Cheese pizzas cost $5 each and pepperoni pizzas cost $6 each. The scout leader paid a total of $43 for the pizzas. How many of each type did he order?

56. **Merchandise Sales** An electronics store carries two brands of video cameras. For a certain week, the number of Brand A video cameras sold was 10 less than twice the number of Brand B cameras sold. Brand A cameras cost $200 and Brand B cameras cost $350. If the total revenue generated that week from the sale of both types of cameras was $16,750, how many of each type were sold?

57. **Merchandise Sales** A grocery store carries two brands of diapers. For a certain week, the number of boxes of Brand A diapers sold was 4 more than the number of boxes of Brand B diapers sold. Brand A diapers cost $10 per box and Brand B diapers cost $12 per box. If the total revenue generated that week from the sale of diapers was $172, how many of each brand did the store sell?

58. **Design** Sarita is designing a mobile in which three objects will be suspended from a lightweight rod, as illustrated below.

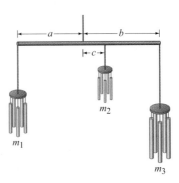

The weights of the objects are given as $m_1 = 2$ ounces, $m_2 = 1$ ounce, and $m_3 = 2.5$ ounces. Use a system of linear equations to determine the values of a, b, and c (in inches) such that the mobile will meet the following three requirements:

▶ The distance between object 1 and object 3 must be 20 inches.

▶ To balance the mobile, Sarita must position the objects in such a way that

$$m_1 a = m_2 b + m_3 c.$$

▶ Object 1 must hang three times further from the support (horizontally speaking) than object 2.

59. **Investments** A financial advisor offers three specific investment instruments: a stock-based mutual fund, a high-yield bond, and a certificate of deposit (CD). Risk factors for individual instruments can be quantified on a scale of 1 to 5, with 1 being the most risky. The risk factors associated with these particular instruments are summarized in the following table.

Type of Investment	Risk Factor
Stock-based mutual fund	3
High-yield bond	1
CD	5

One of the advisor's clients can tolerate an overall risk level of 3.5. In addition, the client stipulates that the amount of money invested in the mutual fund must equal the sum of the amounts invested in the high-yield bond and the CD. To satisfy the client's requirements, what percentage of the total investment should be allocated to each instrument?

60. **Utilities** Privately owned, single-family homes in a small town were heated with gas, electricity, or oil. The percentage of homes heated with electricity was 9 times the percentage heated with oil. The percentage of homes heated with gas was 40 percentage points higher than the percentage heated with oil and the percentage heated with electricity combined. Find the percentage of homes heated with each type of fuel.

61. **Electrical Engineering** An electrical circuit consists of three resistors connected in series. The formula for the total resistance R is given by $R = R_1 + R_2 + R_3$, where R_1, R_2, and R_3 are the resistances of the individual resistors. In a circuit with two resistors A and B connected in series, the total resistance is 60 ohms. The total resistance

when B and C are connected in series is 100 ohms. The sum of the resistances of B and C is 2.5 times the resistance of A. Find the resistances of A, B, and C.

62. **Gardening** Mr. Greene, a gardener, is mixing organic fertilizers consisting of bone meal, cottonseed meal, and poultry manure. The percentages of nitrogen (N), phosphorus (P), and potassium (K) in each fertilizer are given in the table below.

	Nitrogen (%)	Phosphorus (%)	Potassium (%)
Bone meal	4	12	0
Cottonseed meal	6	2	1
Poultry manure	4	4	2

If Mr. Greene wants to produce a 10-pound mix containing 5% nitrogen content and 6% phosphorus content, how many pounds of each fertilizer should he use?

▶**Concepts** This set of exercises will draw on the ideas presented in this section and your general math background.

63. Consider the following augmented matrix. For what value(s) of a does the corresponding system of linear equations have infinitely many solutions? One solution? Explain your answers.

$$\begin{bmatrix} 1 & 0 & 0 & | & -2 \\ 0 & 1 & 0 & | & 5 \\ 0 & 0 & a & | & 0 \end{bmatrix}$$

64. Consider the following system of equations.

$$\begin{cases} x + y = 3 \\ -x + y = 1 \\ 2x + y = 4 \end{cases}$$

Use Gauss-Jordan elimination to find the solution, if it exists. Interpret your answer in terms of the graphs of the given equations.

65. Consider the following system of equations.

$$\begin{cases} x + y = 3 \\ -x + y = 1 \\ 2x + y = 6 \end{cases}$$

Use Gauss-Jordan elimination to show that this system has no solution. Interpret your answer in terms of the graphs of the given equations.

66. Consider the following system of equations.

$$\begin{cases} x + y = 3 \\ -2x - 2y = -6 \\ -x - y = -3 \end{cases}$$

Use Gauss-Jordan elimination to show that this system has infinitely many solutions. Interpret your answer in terms of the graphs of the given equations.

67. Consider the following system of equations.

$$\begin{cases} 6u + 6v - 3w = -3 \\ 2u + 2v - w = -1 \end{cases}$$

(a) Show that each of the equations in this system is a multiple of the other equation.

(b) Explain why this system of equations has infinitely many solutions.

(c) Express w as an equation in u and v.

(d) Give two solutions of this system of equations.

68. Consider the following system of equations.

$$\begin{cases} x + y = 1 \quad (1) \\ x \quad - z = 0 \quad (2) \\ y + z = 1 \quad (3) \end{cases}$$

(a) Find constants b and c such that Equation (1) can be expressed as

Equation (1) = b [Equation (2)] + c [Equation (3)].

(b) Solve the system.

(c) Give three individual solutions such that $0 < z < 1$.

6.4 Operations on Matrices

Objectives

▶ Define a matrix

▶ Perform matrix addition and scalar multiplication

▶ Find the product of two matrices

▶ Use matrices and matrix operations in applications

In Section 6.3, we represented systems of equations by matrices in order to streamline the bookkeeping involved in solving the systems. In this section, we will study the properties of matrices in more detail. We begin with an example of how a matrix can be used to present labor statistics in compact form.

Example **1** **A Matrix Reflecting Unemployment Data**

In 2004, the unemployment rates for people (20 years old and older) in the labor force in the United States were 5.0% for males and 4.9% for females. In 2005, the rates were 4.4% for males and 4.6% for females. Summarize this information in a table and in a corresponding matrix. (*Source:* Bureau of Labor Statistics)

▶**Solution** We can organize the given information in the form of a table, as shown in Table 6.4.1.

The corresponding matrix is

Table 6.4.1

	Male	Female
2004	5.0%	4.9%
2005	4.4%	4.6%

$$\begin{bmatrix} 5.0 & 4.9 \\ 4.4 & 4.6 \end{bmatrix}.$$

In this matrix, each row represents a particular year and each column represents a particular gender. Every entry of the matrix represents a percentage, so we have omitted the percent signs. Note that we could have constructed the matrix in such a way that each row represented a particular gender and each column a particular year.

☑ *Check It Out 1:* Construct the matrix for Example 1 in such a way that each row represents a particular gender and each column represents a particular year. ▩

A **matrix** is a rectangular array of numbers. A matrix that has m rows and n columns has **dimensions** $m \times n$ (pronounced "m by n"); such a matrix is referred to as an **$m \times n$ matrix.**

An uppercase letter of the alphabet is typically used to name a matrix (as in matrix A). The corresponding lowercase letter (in this case, a) is used in conjunction with a pair of positive integers to refer to a particular entry of the matrix by the row and column in which it is located. For example, a_{23} denotes the entry in the second row, third column of matrix A. This is shown schematically for an $m \times n$ matrix in the figure below. The entry in the ith row and jth column of matrix A is denoted by a_{ij} (where i ranges from 1 through m, and j from 1 through n).

$$A = \begin{bmatrix} a_{11} & a_{12} & \cdots & a_{1n} \\ a_{21} & a_{22} & \cdots & a_{2n} \\ \vdots & \vdots & \vdots & \vdots \\ a_{m1} & a_{m2} & \cdots & a_{mn} \end{bmatrix}$$

The next example will familiarize you with this basic terminology and notation.

Example 2 Solving an $m \times n$ Matrix

Let

$$B = \begin{bmatrix} -2 & 4 & 7 & 0 \\ 0 & 5 & 12 & 6 \\ -8 & -7 & 0 & 1 \end{bmatrix}.$$

Find the following.

(a) The dimensions of B

(b) The value of b_{31}

(c) The value of b_{13}

▶Solution

(a) Because matrix B has three rows and four columns, it has dimensions 3×4.

(b) The value of b_{31} (the entry in the third row, first column) is -8.

(c) The value of b_{13} (the entry in the first row, third column) is 7.

✔ *Check It Out 2:* Let

$$A = \begin{bmatrix} -2 & 4 & -5 \\ -9 & -2 & 1 \\ 12 & 3 & 0 \end{bmatrix}.$$

Find the dimensions of A and the values of a_{12} and a_{33}. ■

Matrix Addition and Scalar Multiplication

Just as with numbers, we can define equality, addition, and subtraction for matrices—but only for matrices that have the *same dimensions.*

Technology Note

Matrices can be added using a graphing utility. See Figure 6.4.1. Attempting to add matrices that do not have the same dimensions will yield an error message. See Figure 6.4.2.

Keystroke Appendix: Section 13

Figure 6.4.1

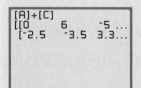

Figure 6.4.2

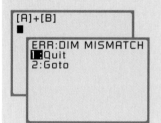

Equality of Matrices

Two $m \times n$ matrices A and B are said to be equal, which is denoted by $A = B$, if their corresponding entries are equal (that is, for all i, j, $a_{ij} = b_{ij}$).

Addition and Subtraction of Matrices

Let A and B be $m \times n$ matrices. Then the sum of A and B, which is denoted by $A + B$, is the matrix that is formed by simply adding corresponding entries. Note that $A + B$ is also an $m \times n$ matrix. The difference of A and B, which is denoted by $A - B$, is defined analogously.

Example 3 Adding and Subtracting Matrices

Let

$$A = \begin{bmatrix} -1 & 2 & 0 \\ 0 & -3.5 & 1 \end{bmatrix}, \quad B = \begin{bmatrix} 3 & -2 \\ -1 & 4.2 \\ 2 & -6 \end{bmatrix}, \quad C = \begin{bmatrix} 1 & 4 & -5 \\ -2.5 & 0 & 2.3 \end{bmatrix}.$$

Perform the following operations, if defined.

(a) $A + B$

(b) $C - A$

(c) $A + C$

▶Solution

(a) The sum $A + B$ is not defined because A has dimensions 2×3 and B has dimensions 3×2.

(b) The difference $C - A$ is defined because A and C have the same dimensions. To find $C - A$, we subtract the entries of A from the corresponding entries of C.

$$C - A = \begin{bmatrix} 1 - (-1) & 4 - 2 & -5 - 0 \\ -2.5 - 0 & 0 - (-3.5) & 2.3 - 1 \end{bmatrix} = \begin{bmatrix} 2 & 2 & -5 \\ -2.5 & 3.5 & 1.3 \end{bmatrix}$$

(c) The sum $A + C$ is defined because A and C have the same dimensions. To find $A + C$, we add corresponding entries.

$$A + C = \begin{bmatrix} -1 + 1 & 2 + 4 & 0 + (-5) \\ 0 + (-2.5) & -3.5 + 0 & 1 + 2.3 \end{bmatrix} = \begin{bmatrix} 0 & 6 & -5 \\ -2.5 & -3.5 & 3.3 \end{bmatrix}$$

✔ *Check It Out 3:* Let

$$A = \begin{bmatrix} 3 & -2 \\ 1 & 2.5 \\ -0.5 & -3 \end{bmatrix}, \quad B = \begin{bmatrix} -3 & 4 \\ 2.1 & 2 \\ 0 & 4 \end{bmatrix}.$$

Perform the following operations.

(a) $A + B$ (b) $A - B$ (c) $B - A$ ■

Another easily defined operation on matrices consists of multiplying an entire matrix by a single number, known as a *scalar*. This operation is known as **scalar multiplication.**

> **Scalar Multiplication**
>
> The product of an $m \times n$ matrix A and a scalar c, which is denoted by cA, is found by multiplying every entry of A by c. The product is also an $m \times n$ matrix.

Example 4 Scalar Multiplication to Calculate Sales Tax

The following matrix gives the pre-tax sales figures (in millions of dollars) for sales of taxable merchandise sold by the Metro department stores in Baltimore and Annapolis for the years 2002–2004. The sales tax rate in Maryland is 5%. Use scalar multiplication to calculate the amount of sales tax generated from sales at each store for the years 2002–2004.

$$
\begin{array}{c}
 \\
2002 \\
2003 \\
2004
\end{array}
\begin{array}{cc}
\text{Baltimore} & \text{Annapolis} \\
\left[\begin{array}{cc}
1.7 & 1.3 \\
1.8 & 1.5 \\
1.9 & 1.6
\end{array}\right]
\end{array}
$$

▶**Solution** Since the sales tax rate in Maryland is 5%, we must multiply each of the sales figures given in the matrix by 0.05 to get the amount of the corresponding sales tax. This amounts to scalar multiplication of the matrix by 0.05. Thus the matrix that gives the amount of sales tax (in millions of dollars) generated from sales at each store for each of the given years is

$$
\begin{array}{c}
 \\
2002 \\
2003 \\
2004
\end{array}
\begin{array}{cc}
\text{Baltimore} & \text{Annapolis} \\
\left[\begin{array}{cc}
0.085 & 0.065 \\
0.09 & 0.075 \\
0.095 & 0.08
\end{array}\right]
\end{array}.
$$

It is quite likely that data of this type has been incorporated into the company's annual report. When displaying figures such as this, it is better to do so in a manner that is understandable to the reader. For example, it would be better to write $0.085 million as $85,000. The following matrix gives the sales tax figures in dollars.

$$
\begin{array}{c}
 \\
2002 \\
2003 \\
2004
\end{array}
\begin{array}{cc}
\text{Baltimore} & \text{Annapolis} \\
\left[\begin{array}{cc}
85{,}000 & 65{,}000 \\
90{,}000 & 75{,}000 \\
95{,}000 & 80{,}000
\end{array}\right]
\end{array}
$$

✔ *Check It Out 4:* Rework Example 4 for the case in which the sales tax rate is 6%. ■

Matrix Multiplication

We have discussed addition and subtraction of matrices as well as multiplication of a matrix by a scalar. But what about multiplication of one matrix by another? Would we just multiply corresponding entries? The answer is: Not really. We have to define matrix multiplication in a way that makes it useful for applications and for further study in mathematics. Example 5 illustrates an application that will motivate our definition of matrix multiplication.

Example 5 Calculating Total Number of Unemployed Adults

The percentages of men and women (20 years old and older) in the United States who were not employed in 2005 have been converted to decimals and are given in the following matrix. (*Source:* Bureau of Labor Statistics)

$$\begin{array}{cc} & \text{Male} \quad \text{Female} \\ 2005 & [0.044 \quad 0.046] = R \end{array}$$

If there were 76 million men and 66 million women in the labor force in 2005, what is the total number of people who were unemployed in 2005? (A person is defined to be *unemployed* if he/she is not employed but is actively looking for a job; the *labor force* includes people who are actually employed as well as those who are defined as being unemployed.)

▶Solution From the given matrix, the unemployment rate in 2005 was 4.4% for men and 4.6% for women. The numbers of men and women in the labor force that year are summarized in the following matrix.

$$\begin{array}{c} \text{Male} \\ \text{Female} \end{array} \begin{bmatrix} 76 \text{ million} \\ 66 \text{ million} \end{bmatrix} = L$$

The total number of people (in millions) who were unemployed in 2005 is given by

$$\text{Total unemployed} = \text{total unemployed men } + \text{ total unemployed women}$$

$$= 4.4\% \text{ of } 76 \text{ (million)} + 4.6\% \text{ of } 66 \text{ (million)}$$

$$= 0.044(76) + 0.046(66)$$

$$= 6.38.$$

Thus 6.38 million people were unemployed in 2005.

Now take a close look at the operations involved in computing this total. We took matrix R, which has just one row and contains the *rates* of unemployment for males and females, and multiplied each of its entries by the appropriate entry of matrix L, which has just one column and contains the *numbers* of males and females in the labor force. We then added the two separate products to get the total number of people who were unemployed.

This process of computing certain products and then finding their sum lies at the very heart of matrix multiplication.

✔ *Check It Out 5:* Find the total number of people unemployed in 2005 if the data are presented as follows.

$$\begin{array}{cc} & \text{Female} \quad \text{Male} \\ 2005 & [0.046 \quad 0.044] = R \end{array}$$

$$\begin{array}{c} \text{Female} \\ \text{Male} \end{array} \begin{bmatrix} 66 \text{ million} \\ 76 \text{ million} \end{bmatrix} = L \ ◼$$

We will begin our discussion of matrix multiplication using the simplest types of matrices: a row matrix and a column matrix. Multiplication of a row matrix by a column matrix will form the basis for matrix multiplication.

A **row matrix** has just one row and any number of columns. The following is an example of a row matrix with four columns.

$$R = \begin{bmatrix} 3 & 2.5 & \frac{1}{3} & -2 \end{bmatrix}$$

Similarly, a **column matrix** has just one column and any number of rows. The following is an example of a column matrix with four rows.

$$C = \begin{bmatrix} 2 \\ -2 \\ 3 \\ 1.7 \end{bmatrix}$$

Multiplication of a Row Matrix by a Column Matrix

Let R be a row matrix with n columns:

$$R = \begin{bmatrix} r_{11} & r_{12} & \cdots & r_{1n} \end{bmatrix}$$

and let C be a column matrix with n rows:

$$C = \begin{bmatrix} c_{11} \\ c_{21} \\ \vdots \\ c_{n1} \end{bmatrix}.$$

Then the product RC is defined as

$$RC = r_{11}c_{11} + r_{12}c_{21} + \cdots + r_{1n}c_{n1}. \tag{6.1}$$

The number of columns in the row matrix R *must be the same as* the number of rows in the column matrix C for the product RC to be defined. The product is just a number (namely, the sum of the products $r_{11}c_{11}, r_{12}c_{21}, \ldots, r_{1n}c_{n1}$).

Note   The product CR, which has not yet been defined, is not the same as the product RC. In matrix multiplication, the order of the matrices matters.

Example 6 Finding the Product of a Row Matrix and a Column Matrix

Find the product RC for the following matrices.

$$R = \begin{bmatrix} 3 & 2.5 & \frac{1}{3} & -2 \end{bmatrix} \quad \text{and} \quad C = \begin{bmatrix} 2 \\ -2 \\ 3 \\ 1.7 \end{bmatrix}$$

▶Solution The product RC is calculated as follows.

$$RC = \begin{bmatrix} 3 & 2.5 & \frac{1}{3} & -2 \end{bmatrix} \begin{bmatrix} 2 \\ -2 \\ 3 \\ 1.7 \end{bmatrix}$$

$$= \begin{bmatrix} r_{11} & r_{12} & r_{13} & r_{14} \end{bmatrix} \begin{bmatrix} c_{11} \\ c_{21} \\ c_{31} \\ c_{41} \end{bmatrix}$$

$$= r_{11}c_{11} + r_{12}c_{21} + r_{13}c_{31} + r_{14}c_{41}$$

$$= 3(2) + 2.5(-2) + \left(\frac{1}{3}\right)(3) + (-2)(1.7)$$

$$= 6 - 5 + 1 - 3.4 = -1.4$$

✔ *Check It Out 6:* Find the product RC for the following matrices.

$$R = \begin{bmatrix} 4 & -1.5 & \frac{1}{2} & 2 \end{bmatrix} \quad \text{and} \quad C = \begin{bmatrix} 0 \\ -3 \\ 4 \\ 1.5 \end{bmatrix} \blacksquare$$

Now that we have defined multiplication of a row matrix by a column matrix, we proceed to define matrix multiplication in general. The idea is to break up the multiplication of a pair of matrices into steps, where each step consists of multiplying some row of the first matrix by some column of the second matrix, using the rule given earlier. Example 7 illustrates the procedure.

Example 7 The Product of Two Matrices

Find the product AB for the following matrices.

$$A = \begin{bmatrix} 3 & 4 & 0 & -2 \\ -1 & 0 & -3 & 2 \end{bmatrix}, \quad B = \begin{bmatrix} 5 & 0 \\ -4 & 1 \\ 2 & 6 \\ 0 & -2 \end{bmatrix}$$

▶Solution

Step 1 Multiply the *first row of A* by the *first column of B*, using the rule for multiplying a row matrix by a column matrix. The result, which is a single number, will be the entry in the *first row, first column* of the *product matrix*.

$$\begin{bmatrix} 3 & 4 & 0 & -2 \\ \star & \star & \star & \star \end{bmatrix} \begin{bmatrix} 5 & \star \\ -4 & \star \\ 2 & \star \\ 0 & \star \end{bmatrix} = [3(5) + 4(-4) + 0(2) + (-2)(0)] = \begin{bmatrix} -1 & \\ & \end{bmatrix}$$

Step 2 Multiply the *first row of A* by the *second column of B*. The result will be the entry in the *first row, second column* of the product matrix.

$$\begin{bmatrix} 3 & 4 & 0 & -2 \\ \star & \star & \star & \star \end{bmatrix} \begin{bmatrix} \star & 0 \\ \star & 1 \\ \star & 6 \\ \star & -2 \end{bmatrix} = [-1 \quad 3(0) + 4(1) + 0(6) + (-2)(-2)] = \begin{bmatrix} -1 & 8 \end{bmatrix}$$

Step 3 Since we have exhausted all the columns of *B* (for multiplication by the first row of *A*), we next multiply the *second row of A* by the *first column of B*. The result will be the entry in the *second row, first column* of the product matrix.

$$\begin{bmatrix} \star & \star & \star & \star \\ -1 & 0 & -3 & 2 \end{bmatrix} \begin{bmatrix} 5 & \star \\ -4 & \star \\ 2 & \star \\ 0 & \star \end{bmatrix} = \begin{bmatrix} -1 & 8 \\ (-1)(5) + 0(-4) + (-3)(2) + 2(0) & \end{bmatrix}$$

$$= \begin{bmatrix} -1 & 8 \\ -11 & \end{bmatrix}$$

Step 4 Multiply the *second row of A* by the *second column of B*. The result will be the entry in the *second row, second column* of the product matrix.

$$\begin{bmatrix} \star & \star & \star & \star \\ -1 & 0 & -3 & 2 \end{bmatrix} \begin{bmatrix} \star & 0 \\ \star & 1 \\ \star & 6 \\ \star & -2 \end{bmatrix} = \begin{bmatrix} -1 & 8 \\ -11 & (-1)(0) + 0(1) + (-3)(6) + 2(-2) \end{bmatrix}$$

$$= \begin{bmatrix} -1 & 8 \\ -11 & -22 \end{bmatrix}$$

Thus the product matrix is

$$AB = \begin{bmatrix} -1 & 8 \\ -11 & -22 \end{bmatrix}.$$

Recall that *A* has dimensions 2 × 4 and *B* has dimensions 4 × 2. The product *AB* has dimensions 2 × 2, where the first 2 arises from the number of rows in *A* and the second 2 arises from the number of columns in *B*.

☑ *Check It Out 7:* Find the product *AB* for the following matrices.

$$A = \begin{bmatrix} -2 & 0 & 3 \\ 0 & 4 & -1 \end{bmatrix}, \quad B = \begin{bmatrix} -3 & 2 \\ 1 & 0 \\ -2 & 5 \end{bmatrix} \blacksquare$$

The method that was used to multiply the two matrices in Example 7 can be extended to find the product of any two matrices, provided the number of columns in the first matrix equals the number of rows in the second matrix.

Multiplication of Matrices

Multiplication of an $m \times p$ matrix A by a $p \times n$ matrix B is a product matrix AB with m rows and n columns.

Furthermore, the entry in the ith row, jth column of the product matrix AB is given by the product of the ith row of A and the jth column of B. (Multiplication of a row matrix by a column matrix is defined in Equation (6.1).)

In matrix multiplication, the order of the matrices matters. In general, the product AB *does not equal* the product BA. Furthermore, it is possible that AB is defined but BA is not. For instance, if A is a 2×3 matrix and B is a 3×4 matrix, then AB is defined (since A has three columns and B has three rows) but BA is undefined (since B has four columns and A has only two rows).

Example 8 Matrix Products

Let

$$A = \begin{bmatrix} 1 & -1 & 0 \\ 2 & -5 & 1 \end{bmatrix}, \quad B = \begin{bmatrix} 6 & -2 \\ 2 & -5 \\ -3 & -1 \\ 1 & -4 \end{bmatrix}, \quad C = \begin{bmatrix} 0 & -4 \\ 2 & 6 \\ 0 & 1 \end{bmatrix}.$$

Calculate the following, if defined.

(a) AC

(b) BA

(c) AB

▶ Solution

(a) For the product AC, we have a 2×3 matrix (A) multiplied by a 3×2 matrix (C). This multiplication is defined because the number of columns of A is the same as the number of rows of C. The result is a 2×2 matrix.

$$AC = \begin{bmatrix} 1 & -1 & 0 \\ 2 & -5 & 1 \end{bmatrix} \begin{bmatrix} 0 & -4 \\ 2 & 6 \\ 0 & 1 \end{bmatrix} = \begin{bmatrix} -2 & -10 \\ -10 & -37 \end{bmatrix}$$

(b) For the product BA, we have a 4×2 matrix multiplied by a 2×3 matrix. This multiplication is defined because the number of columns of B is the same as the number of rows of A. The result is a 4×3 matrix.

$$BA = \begin{bmatrix} 6 & -2 \\ 2 & -5 \\ -3 & -1 \\ 1 & -4 \end{bmatrix} \begin{bmatrix} 1 & -1 & 0 \\ 2 & -5 & 1 \end{bmatrix} = \begin{bmatrix} 2 & 4 & -2 \\ -8 & 23 & -5 \\ -5 & 8 & -1 \\ -7 & 19 & -4 \end{bmatrix}$$

(c) The product AB is not defined, since A has three columns and B has four rows.

✔ *Check It Out 8:* In Example 8, find the product CA, if it is defined. ■

Technology Note

Matrices can be multiplied using a graphing utility. Attempting to multiply matrices whose dimensions are incompatible will yield an error message. See Figure 6.4.3.

Keystroke Appendix:
Section 13

Figure 6.4.3

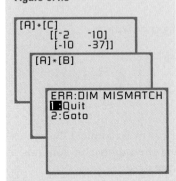

```
[A]*[C]
    [[-2   -10]
     [-10  -37]]
  [A]*[B]

    ERR:DIM MISMATCH
    1:Quit
    2:Goto
```

Example 9 illustrates an application of matrix multiplication to manufacturing.

Example 9 Application of Matrix Multiplication to Manufacturing

A manufacturer of women's clothing makes four different outfits, each of which utilizes some combination of fabrics A, B, and C. The yardage of each fabric required for each outfit is given in matrix F.

$$
\begin{array}{c}
 \\
 \\
\text{Outfit 1} \\
\text{Outfit 2} \\
\text{Outfit 3} \\
\text{Outfit 4}
\end{array}
\begin{array}{ccc}
\text{Fabric A} & \text{Fabric B} & \text{Fabric C} \\
(\text{yd}) & (\text{yd}) & (\text{yd})
\end{array}
$$

$$
\begin{array}{c}
\text{Outfit 1} \\
\text{Outfit 2} \\
\text{Outfit 3} \\
\text{Outfit 4}
\end{array}
\begin{bmatrix}
1.2 & 0.4 & 1.5 \\
0.8 & 0.6 & 2.1 \\
1.5 & 0.3 & 1.8 \\
2.2 & 0.7 & 1.5
\end{bmatrix} = F
$$

The cost of each fabric (in dollars per yard) is given in matrix C.

$$
\begin{array}{c}
\text{Fabric A} \\
\text{Fabric B} \\
\text{Fabric C}
\end{array}
\begin{bmatrix}
8 \\
4 \\
10
\end{bmatrix} = C
$$

Find the total cost of fabric for each outfit.

▶Solution Each row of matrix F gives the fabric requirements for a certain outfit. To determine the total cost of fabric for each outfit, we multiply matrix F by matrix C.

$$
FC = \begin{bmatrix}
1.2 & 0.4 & 1.5 \\
0.8 & 0.6 & 2.1 \\
1.5 & 0.3 & 1.8 \\
2.2 & 0.7 & 1.5
\end{bmatrix}
\begin{bmatrix}
8 \\
4 \\
10
\end{bmatrix} =
\begin{bmatrix}
26.2 \\
29.8 \\
31.2 \\
35.4
\end{bmatrix}
$$

The total cost of fabric for each outfit is given in Table 6.4.2.

Table 6.4.2

Outfit	Total Cost of Fabric
1	$26.20
2	$29.80
3	$31.20
4	$35.40

✔ *Check It Out 9:* Rework Example 9 for the case in which the cost of each fabric (in dollars per yard) is given by the following matrix.

$$
\begin{array}{c}
\text{Fabric A} \\
\text{Fabric B} \\
\text{Fabtic C}
\end{array}
\begin{bmatrix}
6 \\
5 \\
8
\end{bmatrix}
$$ ■

A Cryptography Application

One application of matrix multiplication is in *cryptography*, the science of encoding and decoding messages. We will consider a simple coding scheme: The letters of the alphabet will be assigned the numbers 1 through 26 (in numerical order), and a space will be assigned the number 0. See Table 6.4.3.

Table 6.4.3

Space	A	B	C	D	E	F	G	H	I	J	K	L	M
0	1	2	3	4	5	6	7	8	9	10	11	12	13

	N	O	P	Q	R	S	T	U	V	W	X	Y	Z
	14	15	16	17	18	19	20	21	22	23	24	25	26

Example 10 shows how a message can be encoded with the help of an encoding matrix.

Example 10 Cryptography Application

Use the matrix below to encode the message **HELP ME**.

$$\begin{bmatrix} -1 & 2 & 0 & -2 \\ 3 & -7 & 2 & 6 \\ 2 & -4 & 1 & 7 \\ 1 & -2 & 0 & 1 \end{bmatrix}$$

▶Solution Because the encoding matrix has only four columns, the message must first be parsed into "words" of four characters apiece. We get two such words (**HELP** and _**ME**_). An underscore indicates a space. The underscore that follows the letters of **ME** has been inserted as a filler since the total number of characters in the message has to be a multiple of 4. Replacing the spaces and the letters of the alphabet with their assigned numbers, we get a column matrix for each word.

$$\textbf{HELP} \to \begin{bmatrix} 8 \\ 5 \\ 12 \\ 16 \end{bmatrix}$$

$$\_\textbf{ME}\_ \to \begin{bmatrix} 0 \\ 13 \\ 5 \\ 0 \end{bmatrix}$$

We then encode the word **HELP** by multiplying the encoding matrix by the column matrix for **HELP**.

$$\begin{bmatrix} -1 & 2 & 0 & -2 \\ 3 & -7 & 2 & 6 \\ 2 & -4 & 1 & 7 \\ 1 & -2 & 0 & 1 \end{bmatrix}\begin{bmatrix} 8 \\ 5 \\ 12 \\ 16 \end{bmatrix} = \begin{bmatrix} -30 \\ 109 \\ 120 \\ 14 \end{bmatrix}$$

Similarly, the word **_ME_** is encoded as follows.

$$\begin{bmatrix} -1 & 2 & 0 & -2 \\ 3 & -7 & 2 & 6 \\ 2 & -4 & 1 & 7 \\ 1 & -2 & 0 & 1 \end{bmatrix} \begin{bmatrix} 0 \\ 13 \\ 5 \\ 0 \end{bmatrix} = \begin{bmatrix} 26 \\ -81 \\ -47 \\ -26 \end{bmatrix}$$

The receiver of the information would get these two 4×1 matrices of encoded information. How would the receiver then decode the message? This question will be answered in the next section.

☑ *Check It Out 10:* Rework Example 10 for the message **GOT SPY**. ■

6.4 Key Points

▸ A **matrix** is a rectangular array of real numbers with m rows and n columns.

▸ The **dimension** of the given matrix is $m \times n$, since the matrix has m rows and n columns.

▸ Two matrices can be added or subtracted only if their dimensions are the same. The result is found by adding or subtracting the corresponding entries in each matrix.

▸ The **scalar product** of an $m \times n$ matrix A and a real number c is found by multiplying all the elements of the matrix A by c.

▸ The **product** of an $m \times p$ matrix A and a $p \times n$ matrix B is a product matrix AB with m rows and n columns.

6.4 Exercises

▸**Skills** This set of exercises will reinforce the skills illustrated in this section.

In Exercises 1–6, use the following matrix.

$$A = \begin{bmatrix} -1 & 2 & 0 & 4 \\ 2.1 & -7 & 9 & 0 \\ 1 & 0 & -\frac{2}{3} & \pi \end{bmatrix}$$

1. Find a_{11}.

2. Find a_{22}.

3. Determine the dimensions of A.

4. Find a_{31}.

5. Find a_{34}.

6. Why is a_{43} not defined?

In Exercises 7–10, indicate whether each statement is True or False. Explain your answers.

7. *Some* matrices that do not have the same dimensions can be added.

8. *Some* matrices that do not have the same dimensions can be multiplied.

9. If A is a 2×4 matrix and B is a 4×3 matrix, then the product AB is a 2×3 matrix.

10. If A is a 2×4 matrix and B is a 4×3 matrix, then the product BA is a 4×4 matrix.

In Exercises 11–28, perform the given operations (if defined) on the matrices.

$$A = \begin{bmatrix} 1 & -3 & \frac{1}{3} \\ 5 & 0 & -2 \end{bmatrix}, \quad B = \begin{bmatrix} 8 & 0 \\ 3 & -2 \\ 2 & -6 \end{bmatrix}, \quad C = \begin{bmatrix} -4 & 5 \\ 0 & 1 \\ -2 & 7 \end{bmatrix}$$

If an operation is not defined, state the reason.

11. $B + C$

12. $C - B$

13. $2B + C$

14. $B + 2C$

15. $-3C + B$

16. $C - 2B$

17. $A + 2B$

18. $3B - C$

19. AB

20. AC

21. BC

22. CA

23. $\dfrac{1}{2}A$

24. $\dfrac{2}{3}C$

25. $A(B + C)$

26. $(B + C)A$

27. $C(AB)$

28. $B(AC)$

In Exercises 29–34, for the given matrices A and B, evaluate (if defined) the expressions (a) AB, (b) 3B − 2A, and (c) BA. For any expression that is not defined, state the reason.

29. $A = \begin{bmatrix} -4 & 2 \\ -1 & 0 \end{bmatrix}$; $B = \begin{bmatrix} 1 \\ -3 \end{bmatrix}$

30. $A = \begin{bmatrix} -5 \\ 4 \end{bmatrix}$; $B = \begin{bmatrix} -1 & \dfrac{1}{2} \\ 0 & -6 \end{bmatrix}$

31. $A = \begin{bmatrix} 0 & 4 \\ -6 & 7 \end{bmatrix}$; $B = \begin{bmatrix} 3 & -7 \\ 2 & -1 \end{bmatrix}$

32. $A = \begin{bmatrix} 9 & -4 \\ 7 & -3 \end{bmatrix}$; $B = \begin{bmatrix} 4 & 0 \\ -7 & 5 \end{bmatrix}$

33. $A = \begin{bmatrix} 3 & 0 & -2 \\ 7 & -6 & -1 \\ 5 & 2 & -1 \end{bmatrix}$; $B = \begin{bmatrix} 4 & -2 \\ 1 & 0 \\ 9 & 3 \end{bmatrix}$

34. $A = \begin{bmatrix} 6 & -7 & 0 \\ 2 & 3 & -4 \\ 1 & 0 & -2 \end{bmatrix}$; $B = \begin{bmatrix} 7 & -2 \\ 3 & 0 \\ 2 & 1 \end{bmatrix}$

In Exercises 35–42, for the given matrices A, B, and C, evaluate the indicated expression.

35. $A = \begin{bmatrix} 6 & -1 \\ 5 & 1 \end{bmatrix}$; $B = \begin{bmatrix} 2 \\ 4 \end{bmatrix}$; $C = \begin{bmatrix} 3 \\ -2 \end{bmatrix}$; $AB + AC$

36. $A = \begin{bmatrix} 8 & 3 \\ -4 & 7 \end{bmatrix}$; $B = \begin{bmatrix} -1 \\ 1 \end{bmatrix}$; $C = \begin{bmatrix} -4 \\ 3 \end{bmatrix}$;
$A(C - B)$

37. $A = \begin{bmatrix} 1 & 2 \\ 0 & -5 \end{bmatrix}$; $B = \begin{bmatrix} 7 & -4 \\ -4 & -7 \end{bmatrix}$; $C = \begin{bmatrix} 2 & 1 \\ 0 & -9 \end{bmatrix}$;
$CA - B$

38. $A = \begin{bmatrix} 3 & -8 \\ 2 & 4 \end{bmatrix}$; $B = \begin{bmatrix} -6 & 0 \\ 0 & -6 \end{bmatrix}$; $C = \begin{bmatrix} 3 & 5 \\ -2 & 6 \end{bmatrix}$;
$(A + 2B)C$

39. $A = \begin{bmatrix} 1 & 3 \\ -4 & 2 \\ 6 & 0 \end{bmatrix}$; $B = \begin{bmatrix} -2 & 1 & 4 \\ 3 & -1 & 1 \end{bmatrix}$;
$C = \begin{bmatrix} -4 & 1 \\ 5 & -7 \end{bmatrix}$; $2C + BA$

40. $A = \begin{bmatrix} 5 & -3 \\ -1 & 4 \\ 6 & 0 \end{bmatrix}$; $B = \begin{bmatrix} 0 & 5 & 8 \\ 1 & 0 & -4 \end{bmatrix}$;
$C = \begin{bmatrix} 6 & 8 \\ -5 & 3 \end{bmatrix}$; $BA - CC$

41. $A = \begin{bmatrix} 4 & 1 \\ 0 & 2 \\ 5 & 1 \end{bmatrix}$; $B = \begin{bmatrix} 4 & 3 \\ -6 & 2 \\ 3 & -1 \end{bmatrix}$;
$C = \begin{bmatrix} 1 & 2 & 3 \\ -2 & -3 & -1 \\ 3 & 1 & 2 \end{bmatrix}$; $C(B - A)$

42. $A = \begin{bmatrix} 3 & 1 \\ 2 & 5 \\ -2 & 1 \end{bmatrix}$; $B = \begin{bmatrix} -5 & -3 \\ 1 & 6 \\ 8 & 3 \end{bmatrix}$;
$C = \begin{bmatrix} 2 & 1 & 1 \\ 0 & -1 & 7 \\ 3 & 0 & -3 \end{bmatrix}$; $CB + 2A$

In Exercises 43–46, answer the question pertaining to the matrices

$$A = \begin{bmatrix} a & b \\ c & d \\ e & f \end{bmatrix} \quad and \quad B = \begin{bmatrix} g & h & i \\ j & k & l \end{bmatrix}.$$

43. Let $P = AB$, and find p_{11} and p_{33} without performing the entire multiplication of matrix A by matrix B.

44. Let $Q = BA$, and find q_{11} and q_{22} without performing the entire multiplication of matrix B by matrix A.

45. Let $P = AB$, and find p_{32} and p_{23} without performing the entire multiplication of matrix A by matrix B.

46. Let $Q = BA$, and find q_{12} and q_{21} without performing the entire multiplication of matrix B by matrix A.

In Exercises 47–52, find A^2 (the product AA) and A^3 (the product $(A^2)A$).

47. $A = \begin{bmatrix} 2 & -1 \\ 1 & 0 \end{bmatrix}$

48. $A = \begin{bmatrix} 3 & 1 \\ 0 & -1 \end{bmatrix}$

49. $A = \begin{bmatrix} -4 & 0 \\ 0 & 3 \end{bmatrix}$

50. $A = \begin{bmatrix} 1 & 1 \\ -1 & 2 \end{bmatrix}$

51. $A = \begin{bmatrix} 3 & 0 & 0 \\ 0 & 1 & 1 \\ -4 & 1 & 0 \end{bmatrix}$

52. $A = \begin{bmatrix} 2 & -1 & 1 \\ 0 & 1 & 0 \\ 0 & 3 & 2 \end{bmatrix}$

53. If $A = \begin{bmatrix} 0 & 1 \\ a & 0 \end{bmatrix}$ and $B = \begin{bmatrix} 0 & a \\ 1 & 0 \end{bmatrix}$, for what value(s) of

   a does $AB = \begin{bmatrix} 1 & 0 \\ 0 & 1 \end{bmatrix}$?

54. If $A = \begin{bmatrix} 4a + 5 & -1 \\ -4 & -7 \end{bmatrix}$ and $B = \begin{bmatrix} 7 & 0 \\ -4 & -8 \end{bmatrix}$, for what

   value(s) of a does $2B - 3A = \begin{bmatrix} 2 & 3 \\ 4 & 5 \end{bmatrix}$?

55. If $A = \begin{bmatrix} 2 & 1 \\ 1 & 3 \end{bmatrix}$ and $B = \begin{bmatrix} 2 & 2a + b \\ b - a & 6 \end{bmatrix}$, for what

   values of a and b does $AB = BA$?

56. If $A = \begin{bmatrix} 1 & 0 & 1 \\ 0 & 0 & 1 \\ 2 & -1 & 0 \end{bmatrix}$ and $B = \begin{bmatrix} 0 & 3 & -1 \\ -1 & 2 & 0 \\ 0 & 0 & 1 \end{bmatrix}$, for what

   values of a and b does $AB = \begin{bmatrix} 0 & 2a + 2b + 1 & 0 \\ 3a + 4b & 0 & 1 \\ 1 & 4 & -2 \end{bmatrix}$?

57. If $A = \begin{bmatrix} a^2 - 3a + 3 & 1 \\ 0 & 2b + 5 \end{bmatrix}$ and $B = \begin{bmatrix} 0 & 1 \\ 1 & 0 \end{bmatrix}$, for

   what values of a and b does $AB = \begin{bmatrix} 1 & 1 \\ 1 & 0 \end{bmatrix}$?

58. If $A = \begin{bmatrix} 3 & 16 & 5 \\ 4 & 3 & 6 \end{bmatrix}$ and

   $B = \begin{bmatrix} 1 & a^2 - 2a - 7 & 2 \\ b^2 - 5b - 4 & 1 & 3 \end{bmatrix}$, for what values of

   a and b does $A - 2B = \begin{bmatrix} 1 & 0 & 1 \\ 0 & 1 & 0 \end{bmatrix}$?

▶**Applications** In this set of exercises, you will use matrices to study real-world problems.

59. **Gas Prices** At a certain gas station, the prices of regular and high-octane gasoline are $2.40 per gallon and $2.65 per gallon, respectively. Use matrix scalar multiplication to compute the cost of 12 gallons of each type of fuel.

60. **Taxi Fares** A cab company charges $4.50 for the first mile of a passenger's fare and $1.50 for every mile thereafter. If it is snowing, the fare is increased to $5.50 for the first mile and $1.75 for every mile thereafter. All distances are rounded up to the nearest full mile. Use matrix addition and scalar multiplication to compute the fare for a 6.8-mile trip on both a fair-weather day and a day on which it is snowing.

61. **Economics** Matrix G gives the U.S. gross domestic product for the years 1999–2001.

   GDP
   (billions of $)

   $\begin{array}{c} 1999 \\ 2000 \\ 2001 \end{array} \begin{bmatrix} 9274.3 \\ 9824.6 \\ 10{,}082.2 \end{bmatrix} = G$

   The finance, retail, and agricultural sectors contributed 20%, 9%, and 1.4%, respectively, to the gross domestic product in those years. These percentages have been converted to decimals and are given in matrix P. (*Source:* U.S. Bureau of Economic Analysis)

   Finance Retail Agriculture
   $[0.2 \quad 0.09 \quad 0.014] = P$

   (a) Compute the product GP.
   (b) What does GP represent?
   (c) Is the product PG defined? If so, does it represent anything meaningful? Explain.

62. **Tuition** Three students take courses at two different colleges, Woosamotta University (WU) and Frostbite Falls Community College (FFCC). WU charges $200 per credit hour and FFCC charges $120 per credit hour. The number of credits taken by each student at each college is given in the following table.

Student	Credits WU	FFCC
1	12	6
2	3	9
3	8	8

Use matrix multiplication to find the total tuition paid by each student.

63. Furniture A furniture manufacturer makes three different pieces of furniture, each of which utilizes some combination of fabrics A, B, and C. The yardage of each fabric required for each piece of furniture is given in matrix F.

$$\begin{array}{c} \\ \text{Sofa} \\ \text{Loveseat} \\ \text{Chair} \end{array} \begin{array}{ccc} \text{Fabric A} & \text{Fabric B} & \text{Fabric C} \\ (\text{yd}) & (\text{yd}) & (\text{yd}) \end{array} \\ \begin{bmatrix} 10.5 & 2 & 1 \\ 8 & 1.5 & 1 \\ 4 & 1 & 0.5 \end{bmatrix} = F$$

The cost of each fabric (in dollars per yard) is given in matrix C.

$$\begin{array}{c} \text{Fabric A} \\ \text{Fabric B} \\ \text{Fabric C} \end{array} \begin{bmatrix} 10 \\ 6 \\ 5 \end{bmatrix} = C$$

Find the total cost of fabric for each piece of furniture.

64. Business A family owns and operates three businesses. On their income-tax return, they have to report the depreciation deductions for the three businesses separately. In 2004, their depreciation deductions consisted of use of a car, plus depreciation on 5-year equipment (on which one-fifth of the original value is deductible per year) and 10-year equipment (on which one-tenth of the original value is deductible per year). The car use (in miles) for each business in 2004 is given in the following table, along with the original value of the depreciable 5- and 10-year equipment used in each business that year.

Business	Car Use (miles)	Original Value, 5-Year Equipment ($)	Original Value, 10-Year Equipment ($)
1	3200	9850	435
2	8800	12,730	980
3	6880	2240	615

The depreciation deduction for car use in 2004 was 37.5 cents per mile. Use matrix multiplication to determine the total depreciation deduction for each business in 2004.

65. Shopping Keith and two of his friends, Sam and Cody, take advantage of a sidewalk sale at a shopping mall. Their purchases are summarized in the following table.

Name	Quantity Shirt	Quantity Sweater	Quantity Jacket
Keith	3	2	1
Sam	1	2	2
Cody	2	1	2

The sale prices are \$14.95 per shirt, \$18.95 per sweater, and \$24.95 per jacket. In their state, there is no sales tax on purchases of clothing. Use matrix multiplication to determine the total expenditure of each of the three shoppers.

Cryptography *In Exercises 66–68, use the encoding matrix from Example 10 to encode the following messages.*

66. SPY SENT

67. TOM IS SPY

68. PLEASE HURRY

▶**Concepts** This set of exercises will draw on the ideas presented in this section and your general math background.

69. Let $A = \begin{bmatrix} 1 & -1 \\ 1 & -1 \end{bmatrix}$ and $B = \begin{bmatrix} 1 & 1 \\ 1 & 1 \end{bmatrix}$. What is the product AB? Is it true that if A and B are matrices such that AB is defined and all the entries of AB are zero, then either all the entries of A must be zero or all the entries of B must be zero? Explain.

70. Show that $A + B = B + A$ for any two matrices A and B for which addition is defined.

71. Let $I = \begin{bmatrix} 1 & 0 \\ 0 & 1 \end{bmatrix}$ and $A = \begin{bmatrix} 2 & -1 \\ 1 & 0 \end{bmatrix}$. Calculate AI and IA. What do you observe?

72. Let $I = \begin{bmatrix} 1 & 0 \\ 0 & 1 \end{bmatrix}$. Show that $IA = AI$, where A is any 2×2 matrix.

73. Let $A = \begin{bmatrix} 1 & 2 \\ 3 & 4 \end{bmatrix}$ and $B = \begin{bmatrix} 0 & 1 \\ 1 & 1 \end{bmatrix}$. What are the products AB and BA? Is it true that if A and B are 2×2 matrices, then $AB = BA$? Explain.

6.5 Matrices and Inverses

Objectives

▶ Define a square matrix

▶ Define an identity matrix

▶ Define the inverse of a square matrix

▶ Find the inverse of a 2 × 2 matrix

▶ Find the inverse of a 3 × 3 matrix

▶ Use inverses to solve systems of linear equations

▶ Use inverses in applications

During World War II, Sir Alan Turing, a now-famous mathematician, was instrumental in breaking the code used by the Germans to communicate military secrets. Such a decoding process entails *working backward* from the cryptic information in order to reveal the text of the original message. We have already seen how matrices can be used to encode a message. In Example 6, we will see how an encoded message can be decoded by use of an operation known as *matrix inversion,* which we will study in detail in this section. Before introducing the topic of matrix inversion, we will address two important concepts related to it: **square matrices** and **identity matrices.**

Identity Matrices

Recall that $a \cdot 1 = a = 1 \cdot a$ for any nonzero number a. The number 1 is called a *multiplicative identity* because multiplication of any nonzero number a by 1 gives that same nonzero number, a. We can extend this idea to **square matrices**—matrices that have the same number of rows and columns. As you might guess, an **identity matrix** will contain some 1's. In fact, the identity matrix for 2 × 2 matrices is

$$I = \begin{bmatrix} 1 & 0 \\ 0 & 1 \end{bmatrix}.$$

For any 2 × 2 matrix A, it can be shown that $AI = IA = A$. That is, A multiplied by the identity matrix I results in the same matrix A. In Example 1, the property that $IA = A$ is demonstrated for a particular 2 × 2 matrix A.

Example 1 Checking a Matrix Inverse

Let

$$A = \begin{bmatrix} -2 & 3 \\ 4 & 7 \end{bmatrix}.$$

Show that $IA = A$.

▶**Solution** We perform the multiplication of matrix I by matrix A.

$$IA = \begin{bmatrix} 1 & 0 \\ 0 & 1 \end{bmatrix} \begin{bmatrix} -2 & 3 \\ 4 & 7 \end{bmatrix} = \begin{bmatrix} -2 & 3 \\ 4 & 7 \end{bmatrix}$$

Clearly, the product matrix IA is equal to matrix A.

✔ *Check It Out 1:* For matrix A from Example 1, show that $AI = A$. ■

For every positive integer $n \geq 2$, the $n \times n$ identity matrix is

$$n \text{ columns}$$

$$I = \begin{bmatrix} 1 & 0 & \cdots & 0 \\ 0 & 1 & \cdots & 0 \\ \vdots & \vdots & \ddots & \vdots \\ 0 & 0 & \cdots & 1 \end{bmatrix} \; n \text{ rows}$$

Discover *and* Learn

You might have been tempted to define the 2 × 2 identity matrix as $I = \begin{bmatrix} 1 & 1 \\ 1 & 1 \end{bmatrix}$. Why does this *not* work?

Every diagonal entry of the $n \times n$ identity matrix is 1, and every off-diagonal entry is 0.

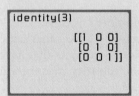

Inverse of a Matrix

Recall that every nonzero number a has a (multiplicative) inverse—namely, the reciprocal of a.

$$a \times \frac{1}{a} = 1 = \frac{1}{a} \times a$$

It would seem only natural to ask whether every nonzero matrix (i.e., every matrix with at least one nonzero entry) has an inverse: Given any nonzero matrix A, is there some matrix B such that $AB = I = BA$, where I is an identity matrix of appropriate size? The answer is that only square matrices can have an inverse, although not *every* square matrix has one!

Before finding the inverse of a specific matrix, we will consider some general aspects of the problem of finding the inverse of a 2 × 2 matrix A. We will denote the entries of A by a, b, c, and d: $A = \begin{bmatrix} a & b \\ c & d \end{bmatrix}$. The goal is to find a 2 × 2 matrix B that satisfies the matrix equation $AB = I$, where I is the 2 × 2 identity matrix. We will denote the entries of B by x, y, u, and v: $B = \begin{bmatrix} x & u \\ y & v \end{bmatrix}$. Thus we want to solve the matrix equation

$$\begin{bmatrix} a & b \\ c & d \end{bmatrix} \begin{bmatrix} x & u \\ y & v \end{bmatrix} = \begin{bmatrix} 1 & 0 \\ 0 & 1 \end{bmatrix}.$$

Using the rules of matrix multiplication, we can break up this matrix equation into two separate matrix equations.

$$\begin{bmatrix} a & b \\ c & d \end{bmatrix} \begin{bmatrix} x \\ y \end{bmatrix} = \begin{bmatrix} 1 \\ 0 \end{bmatrix}$$

$$\begin{bmatrix} a & b \\ c & d \end{bmatrix} \begin{bmatrix} u \\ v \end{bmatrix} = \begin{bmatrix} 0 \\ 1 \end{bmatrix}$$

Each of these two matrix equations corresponds to a system of two linear equations in two variables, the first system being

$$\begin{cases} ax + by = 1 \\ cx + dy = 0 \end{cases}$$

and the second system being

$$\begin{cases} au + bv = 0 \\ cu + dv = 1 \end{cases}.$$

Thus x and y can be found by solving the first system, and u and v can be found by solving the second system. The coefficients (a, b, c, d) are just the entries of matrix A, so they are the same for both systems. The two systems can be solved simultaneously by applying Gauss-Jordan elimination to the following augmented matrix.

$$\left[\begin{array}{cc|cc} a & b & 1 & 0 \\ c & d & 0 & 1 \end{array} \right]$$

This is a modified version of the augmented matrices that were introduced in Section 6.3. Note that the part of this augmented matrix that lies to the left of the vertical bar is matrix A, and the part that lies to the right of the bar is the 2 × 2 identity matrix.

Example 2 illustrates the use of Gauss-Jordan elimination to calculate the inverse of a 2×2 matrix.

Example 2 Finding the Inverse of a 2 × 2 Matrix

Find the inverse of $A = \begin{bmatrix} 1 & -3 \\ -2 & 5 \end{bmatrix}$.

▶**Solution** We need to find a matrix $B = \begin{bmatrix} x & u \\ y & v \end{bmatrix}$ such that $AB = I$, where I is the 2×2 identity matrix. That is, we need to solve the matrix equation

$$\begin{bmatrix} 1 & -3 \\ -2 & 5 \end{bmatrix} \begin{bmatrix} x & u \\ y & v \end{bmatrix} = \begin{bmatrix} 1 & 0 \\ 0 & 1 \end{bmatrix}.$$

Thus we have to solve the following two matrix equations.

$$\begin{bmatrix} 1 & -3 \\ -2 & 5 \end{bmatrix} \begin{bmatrix} x \\ y \end{bmatrix} = \begin{bmatrix} 1 \\ 0 \end{bmatrix}$$

$$\begin{bmatrix} 1 & -3 \\ -2 & 5 \end{bmatrix} \begin{bmatrix} u \\ v \end{bmatrix} = \begin{bmatrix} 0 \\ 1 \end{bmatrix}$$

To solve for u and v at the same time that we solve for x and y, we will apply Gauss-Jordan elimination to the following augmented matrix:

$$\left[\begin{array}{cc|cc} 1 & -3 & 1 & 0 \\ -2 & 5 & 0 & 1 \end{array} \right]$$

The goal is to reduce the part of the augmented matrix that lies to the left of the vertical bar to the 2×2 identity matrix. If we succeed, the part that lies to the right of the vertical bar will be the inverse of A.

Since we are using Gauss-Jordan elimination, we will begin by labeling the rows of the augmented matrix.

$$\left[\begin{array}{cc|cc} 1 & -3 & 1 & 0 \\ -2 & 5 & 0 & 1 \end{array} \right] \quad \begin{array}{l} (1) \\ (2) \end{array}$$

There is already a 1 in the first row, first column, so the first step consists of eliminating the -2 in the second row, first column: Multiply row (1) by 2 and add the result to row (2).

$$\left[\begin{array}{cc|cc} 1 & -3 & 1 & 0 \\ 0 & -1 & 2 & 1 \end{array} \right] \quad \begin{array}{l} (1) \\ (2') = 2 \cdot (1) + (2) \end{array}$$

The next step consists of getting a 1 in the second row, second column: Multiply row (2') by -1.

$$\left[\begin{array}{cc|cc} 1 & -3 & 1 & 0 \\ 0 & 1 & -2 & -1 \end{array} \right] \quad \begin{array}{l} (1) \\ (2'') = -1 \cdot (2') \end{array}$$

The final step is to eliminate the -3 in the first row, second column: Multiply row (2'') by 3 and add the result to row (1).

$$\left[\begin{array}{cc|cc} 1 & 0 & -5 & -3 \\ 0 & 1 & -2 & -1 \end{array} \right] \quad \begin{array}{l} (1') = 3 \cdot (2'') + (1) \\ (2'') \end{array}$$

The part of the final augmented matrix that lies to the right of the vertical bar is

$$B = \begin{bmatrix} -5 & -3 \\ -2 & -1 \end{bmatrix}.$$

This is the inverse of matrix A. You should check that it is indeed the inverse (i.e., that $AB = I = BA$, where I is the 2×2 identity matrix).

✔ *Check It Out 2:* Use Gauss-Jordan elimination to find the inverse of $A = \begin{bmatrix} -3 & 5 \\ -1 & 2 \end{bmatrix}$.

Example **3** **Checking Whether Two Matrices Are Inverses**

If

$$A = \begin{bmatrix} 2 & 0 & -1 \\ 1 & 4 & 0 \\ 0 & -2 & 0 \end{bmatrix} \quad \text{and} \quad B = \begin{bmatrix} 0 & 1 & 2 \\ 0 & 0 & -\frac{1}{2} \\ -1 & 2 & 4 \end{bmatrix}$$

show that $AB = I$, where I is the 3×3 identity matrix.

▶ **Solution** We have to show that

$$AB = \begin{bmatrix} 1 & 0 & 0 \\ 0 & 1 & 0 \\ 0 & 0 & 1 \end{bmatrix}.$$

Multiplying A by B, we get

$$AB = \begin{bmatrix} 2 & 0 & -1 \\ 1 & 4 & 0 \\ 0 & -2 & 0 \end{bmatrix} \begin{bmatrix} 0 & 1 & 2 \\ 0 & 0 & -\frac{1}{2} \\ -1 & 2 & 4 \end{bmatrix} = \begin{bmatrix} 0+0+1 & 2+0-2 & 4+0-4 \\ 0+0+0 & 1+0+0 & 2-2+0 \\ 0+0+0 & 0+0+0 & 0+1+0 \end{bmatrix}$$

$$= \begin{bmatrix} 1 & 0 & 0 \\ 0 & 1 & 0 \\ 0 & 0 & 1 \end{bmatrix}.$$

✔ *Check It Out 3:* Rework Example 3 for the case in which

$$A = \begin{bmatrix} -1 & 0 & -2 \\ 1 & 1 & 1 \\ 1 & 1 & 0 \end{bmatrix} \quad \text{and} \quad B = \begin{bmatrix} -1 & -2 & 2 \\ 1 & 2 & -1 \\ 0 & 1 & -1 \end{bmatrix}.$$

Now that we have seen an example of the inverse of a 2×2 matrix and an example of the inverse of a 3×3 matrix, we will formally define the inverse of a square matrix.

Inverse of an $n \times n$ Matrix

For $n \geq 2$, the inverse of an $n \times n$ matrix A is an $n \times n$ matrix B such that

$$AB = I = BA$$

where I is the $n \times n$ identity matrix. If such an inverse exists, it is denoted by A^{-1}.

To determine whether a given $n \times n$ matrix A has an inverse, we construct the (modified) augmented matrix in which the part that lies to the left of the vertical bar is matrix A, and the part that lies to the right of the bar is the $n \times n$ identity matrix. Then we perform Gauss-Jordan elimination to reduce the augmented matrix to reduced row-echelon form. Matrix A has an inverse if and only if the final augmented matrix has no zero row (i.e., it has no row in which all the entries to the left of the vertical bar are zeros).

Example 4 illustrates the use of Gauss-Jordan elimination to find the inverse of a 3×3 matrix.

Example 4 Finding the Inverse of a 3 × 3 Matrix

Find the inverse of

$$A = \begin{bmatrix} 3 & 3 & 9 \\ 1 & 0 & 2 \\ -2 & 3 & 0 \end{bmatrix}.$$

▶ **Solution** First, construct the (modified) augmented matrix with A to the left of the vertical bar and the 3×3 identity matrix to the right of the bar. Label the rows.

$$\left[\begin{array}{ccc|ccc} 3 & 3 & 9 & 1 & 0 & 0 \\ 1 & 0 & 2 & 0 & 1 & 0 \\ -2 & 3 & 0 & 0 & 0 & 1 \end{array}\right] \quad \begin{array}{l} (1) \\ (2) \\ (3) \end{array}$$

Step 1 To get a 1 in the first row, first column, swap rows (1) and (2).

$$\left[\begin{array}{ccc|ccc} 1 & 0 & 2 & 0 & 1 & 0 \\ 3 & 3 & 9 & 1 & 0 & 0 \\ -2 & 3 & 0 & 0 & 0 & 1 \end{array}\right] \quad \begin{array}{l} (1') = (2) \\ (2') = (1) \\ (3) \end{array}$$

Step 2 Eliminate the 3 in the second row, first column: Multiply row (1′) by −3 and add the result to row (2′). Then eliminate the −2 in the third row, first column: Multiply row (1′) by 2 and add the result to row (3).

$$\left[\begin{array}{ccc|ccc} 1 & 0 & 2 & 0 & 1 & 0 \\ 0 & 3 & 3 & 1 & -3 & 0 \\ 0 & 3 & 4 & 0 & 2 & 1 \end{array}\right] \quad \begin{array}{l} (1') \\ (2'') = -3 \cdot (1') + (2') \\ (3') = 2 \cdot (1') + (3) \end{array}$$

Step 3 To get a 1 in the second row, second column, multiply row (2″) by $\frac{1}{3}$.

$$\left[\begin{array}{ccc|ccc} 1 & 0 & 2 & 0 & 1 & 0 \\ 0 & 1 & 1 & \frac{1}{3} & -1 & 0 \\ 0 & 3 & 4 & 0 & 2 & 1 \end{array}\right] \quad \begin{array}{l} (1') \\ (2''') = \frac{1}{3} \cdot (2'') \\ (3') \end{array}$$

Step 4 Eliminate the 3 in the third row, second column: Multiply row (2‴) by −3 and add the result to row (3′).

$$\left[\begin{array}{ccc|ccc} 1 & 0 & 2 & 0 & 1 & 0 \\ 0 & 1 & 1 & \frac{1}{3} & -1 & 0 \\ 0 & 0 & 1 & -1 & 5 & 1 \end{array}\right] \quad \begin{array}{l} (1') \\ (2''') \\ (3'') = -3 \cdot (2''') + (3') \end{array}$$

Step 5 Eliminate the 2 in the first row, third column: Multiply row $(3'')$ by -2 and add the result to row $(1')$. Then eliminate the 1 in the second row, third column: Multiply row $(3'')$ by -1 and add the result to row $(2''')$.

$$\begin{bmatrix} 1 & 0 & 0 & \bigm| & 2 & -9 & -2 \\ 0 & 1 & 0 & \bigm| & \frac{4}{3} & -6 & -1 \\ 0 & 0 & 1 & \bigm| & -1 & 5 & 1 \end{bmatrix} \quad \begin{array}{l} (1'') = -2 \cdot (3'') + (1') \\ (2'''') = -1 \cdot (3'') + (2''') \\ (3'') \end{array}$$

Thus the inverse of A is

$$A^{-1} = \begin{bmatrix} 2 & -9 & -2 \\ \frac{4}{3} & -6 & -1 \\ -1 & 5 & 1 \end{bmatrix}.$$

✔ *Check It Out 4:* Use Gauss-Jordan elimination to find the inverse of

$$A = \begin{bmatrix} -1 & 0 & -2 \\ 3 & 1 & 5 \\ 1 & 1 & 0 \end{bmatrix}. \ ■$$

As you can see, finding inverses by hand can get quite tedious. For larger matrices, it is better to use technology. We will illustrate this in Example 6 at the end of this section.

Using Inverses to Solve Systems of Equations

Recall that if a is any nonzero number, we solve an algebraic equation of the form $ax = b$ for x by multiplying both sides of the equation by the reciprocal of a.

$$ax = b \Longrightarrow (1/a)ax = (1/a)b \Longrightarrow x = (1/a)b$$

Can we extend this idea to matrices? That is, if we have matrices A, B, and X that satisfy the matrix equation $AX = B$, can we solve for X? The answer is that if A has an inverse, we can.

$$AX = B \Longrightarrow A^{-1}AX = A^{-1}B \Longrightarrow X = A^{-1}B$$

Solution of a System of Linear Equations

If a system of linear equations is written in the form $AX = B$, with A an $n \times n$ matrix and A^{-1} its inverse, then

$$X = A^{-1}B$$

For example, suppose we have a system of three linear equations in three variables. We can let A be the matrix of coefficients (a 3×3 matrix), and we can place the constant terms into a 3×1 matrix B. Then we can place the variables into a 3×1 matrix X. We illustrate this in the following example.

Example 5 Solving a System of Equations Using Inverses

Use the inverse of a matrix to solve the following system of equations.

$$\begin{cases} 3x + 3y + 9z = 6 \\ x \quad\quad\ + 2z = 0 \\ -2x + 3y \quad\quad = 1 \end{cases}$$

▶**Solution** We first write the system of equations in the form of a matrix equation $AX = B$.

$$\begin{bmatrix} 3 & 3 & 9 \\ 1 & 0 & 2 \\ -2 & 3 & 0 \end{bmatrix} \begin{bmatrix} x \\ y \\ z \end{bmatrix} = \begin{bmatrix} 6 \\ 0 \\ 1 \end{bmatrix}$$

Here,

$$A = \begin{bmatrix} 3 & 3 & 9 \\ 1 & 0 & 2 \\ -2 & 3 & 0 \end{bmatrix}, \quad X = \begin{bmatrix} x \\ y \\ z \end{bmatrix}, \quad B = \begin{bmatrix} 6 \\ 0 \\ 1 \end{bmatrix}.$$

Recall that we found the inverse of A in Example 4. Thus, the solution of the matrix equation $AX = B$ is

$$X = A^{-1}B = \begin{bmatrix} 2 & -9 & -2 \\ \frac{4}{3} & -6 & -1 \\ -1 & 5 & 1 \end{bmatrix} \begin{bmatrix} 6 \\ 0 \\ 1 \end{bmatrix} = \begin{bmatrix} 10 \\ 7 \\ -5 \end{bmatrix}.$$

Hence the solution of the given system of equations is $x = 10, y = 7, z = -5$.

✔ *Check It Out 5:* Use the inverse of a matrix to solve the following system of equations.

$$\begin{cases} -3x \quad\quad - 2z = -2 \\ x + y + 5z = \quad 3 \ ■ \\ x + y \quad\quad = -1 \end{cases}$$

Applications

Matrix inverses have a variety of applications. In Example 6 we use the inverse of a matrix to decode a message.

Example 6 Decoding a Message Using a Matrix Inverse

Suppose you receive an encoded message in the form of two 4×1 matrices C and D.

$$C = \begin{bmatrix} -30 \\ 109 \\ 120 \\ 14 \end{bmatrix}, \quad D = \begin{bmatrix} 26 \\ -81 \\ -47 \\ -26 \end{bmatrix}$$

Find the original message if the encoding was done using the matrix

$$A = \begin{bmatrix} -1 & 2 & 0 & -2 \\ 3 & -7 & 2 & 6 \\ 2 & -4 & 1 & 7 \\ 1 & -2 & 0 & 1 \end{bmatrix}.$$

▶**Solution** To decode the message, we need to find two 4×1 matrices X and Y such that

$$C = AX \quad \text{and} \quad D = AY.$$

To do this, we can use the inverse of A:

$$X = A^{-1}C \quad \text{and} \quad Y = A^{-1}D.$$

Using a graphing utility, we obtain

$$A^{-1} = \begin{bmatrix} 15 & -2 & 4 & 14 \\ 7 & -1 & 2 & 6 \\ 5 & 0 & 1 & 3 \\ -1 & 0 & 0 & -1 \end{bmatrix}.$$

Thus we have the following.

$$X = A^{-1}C = \begin{bmatrix} 15 & -2 & 4 & 14 \\ 7 & -1 & 2 & 6 \\ 5 & 0 & 1 & 3 \\ -1 & 0 & 0 & -1 \end{bmatrix} \begin{bmatrix} -30 \\ 109 \\ 120 \\ 14 \end{bmatrix} = \begin{bmatrix} 8 \\ 5 \\ 12 \\ 16 \end{bmatrix}$$

$$Y = A^{-1}D = \begin{bmatrix} 15 & -2 & 4 & 14 \\ 7 & -1 & 2 & 6 \\ 5 & 0 & 1 & 3 \\ -1 & 0 & 0 & -1 \end{bmatrix} \begin{bmatrix} 26 \\ -81 \\ -47 \\ -26 \end{bmatrix} = \begin{bmatrix} 0 \\ 13 \\ 5 \\ 0 \end{bmatrix}$$

Each positive integer from 1 through 26 corresponds to a letter of the alphabet ($A \leftrightarrow 1$, $B \leftrightarrow 2$, and so on), and 0 corresponds to a space. Applying this correspondence to the matrices X and Y, we find that the decoded message is **HELP ME**. Recall that this is the message that was encoded in Example 10 of Section 6.4.

☑ *Check It Out 6:* Decode the following message, which was encoded using matrix A from Example 6.

$$\begin{bmatrix} 2 \\ 21 \\ 29 \\ -7 \end{bmatrix}$$

6.5 Key Points

▸ The $n \times n$ identity matrix is:

n columns

$$I = \begin{bmatrix} 1 & 0 & \cdots & 0 \\ 0 & 1 & \cdots & 0 \\ \vdots & \vdots & \ddots & \vdots \\ 0 & 0 & \cdots & 1 \end{bmatrix} \quad n \text{ rows}$$

▸ The inverse of a $n \times n$ matrix A, denoted by A^{-1}, is another $n \times n$ matrix such that

$$AA^{-1} = A^{-1}A = I$$

where I is the $n \times n$ identity matrix.

▸ If we have a system of equations that can be expressed in the form $AX = B$, with A an $n \times n$ matrix and A^{-1} its inverse, then we have the following.

$$AX = B \Longrightarrow A^{-1}AX = A^{-1}B \Longrightarrow X = A^{-1}B$$

6.5 Exercises

▶**Skills** This set of exercises will reinforce the skills illustrated in this section.

In Exercises 1–6, verify that the matrices are inverses of each other.

1. $\begin{bmatrix} 5 & 2 \\ -3 & -1 \end{bmatrix}$, $\begin{bmatrix} -1 & -2 \\ 3 & 5 \end{bmatrix}$

2. $\begin{bmatrix} 3 & 4 \\ 5 & 7 \end{bmatrix}$, $\begin{bmatrix} 7 & -4 \\ -5 & 3 \end{bmatrix}$

3. $\begin{bmatrix} -6 & 5 \\ 4 & -3 \end{bmatrix}$, $\begin{bmatrix} \frac{3}{2} & \frac{5}{2} \\ 2 & 3 \end{bmatrix}$

4. $\begin{bmatrix} -3 & 2 \\ -4 & 2 \end{bmatrix}$, $\begin{bmatrix} 1 & -1 \\ 2 & -\frac{3}{2} \end{bmatrix}$

5. $\begin{bmatrix} -1 & 3 & -1 \\ 0 & -5 & 2 \\ 1 & 0 & 0 \end{bmatrix}$, $\begin{bmatrix} 0 & 0 & 1 \\ 2 & 1 & 2 \\ 5 & 3 & 5 \end{bmatrix}$

6. $\begin{bmatrix} 1 & 1 & 0 \\ -1 & 1 & 0 \\ 1 & 0 & 1 \end{bmatrix}$, $\begin{bmatrix} \frac{1}{2} & -\frac{1}{2} & 0 \\ \frac{1}{2} & \frac{1}{2} & 0 \\ -\frac{1}{2} & \frac{1}{2} & 1 \end{bmatrix}$

In Exercises 7–22, find the inverse of each matrix.

7. $\begin{bmatrix} 2 & 3 \\ 1 & 1 \end{bmatrix}$

8. $\begin{bmatrix} 4 & 5 \\ 1 & 1 \end{bmatrix}$

9. $\begin{bmatrix} -1 & 3 \\ -1 & 4 \end{bmatrix}$

10. $\begin{bmatrix} 3 & 4 \\ 1 & 2 \end{bmatrix}$

11. $\begin{bmatrix} 5 & 3 \\ 3 & 2 \end{bmatrix}$

12. $\begin{bmatrix} 5 & 3 \\ 4 & 2 \end{bmatrix}$

13. $\begin{bmatrix} 4 & 0 & 5 \\ 0 & 1 & -6 \\ 3 & 0 & 4 \end{bmatrix}$

14. $\begin{bmatrix} 0 & 1 & 0 \\ 3 & 5 & 2 \\ 1 & 2 & 1 \end{bmatrix}$

15. $\begin{bmatrix} -1 & -1 & -1 \\ 3 & 3 & 4 \\ 0 & 1 & 0 \end{bmatrix}$

16. $\begin{bmatrix} 1 & -1 & 0 \\ -2 & 0 & 1 \\ -2 & 5 & -1 \end{bmatrix}$

17. $\begin{bmatrix} 4 & -2 & 1 \\ -2 & 1 & 2 \\ 1 & 2 & 4 \end{bmatrix}$

18. $\begin{bmatrix} 1 & 0 & 0 \\ 0 & 1 & 2 \\ 1 & 0 & 1 \end{bmatrix}$

19. $\begin{bmatrix} 0 & \frac{1}{2} & \frac{1}{2} \\ \frac{1}{2} & 0 & \frac{1}{2} \\ \frac{1}{2} & \frac{1}{2} & 0 \end{bmatrix}$

20. $\begin{bmatrix} 1 & 3 & 0 \\ -1 & -1 & -1 \\ 0 & 3 & -2 \end{bmatrix}$

21. $\begin{bmatrix} 1 & -1 & 0 & 3 \\ 0 & 1 & -2 & 0 \\ -3 & 3 & 1 & -10 \\ 0 & -1 & 2 & 1 \end{bmatrix}$

22. $\begin{bmatrix} 1 & -1 & -2 & 0 \\ 0 & 1 & 0 & 3 \\ -3 & 3 & 7 & -1 \\ 0 & -1 & 0 & -2 \end{bmatrix}$

In Exercises 23 and 24, use one of the matrices given in Exercise 5 to solve the system of equations.

23. $\begin{cases} -x + 3y - z = 6 \\ \quad\;\; -5y + 2z = -2 \\ x \qquad\qquad = 4 \end{cases}$

24. $\begin{cases} -x + 3y - z = 0 \\ \quad\;\; -5y + 2z = -3 \\ x \qquad\qquad = 5 \end{cases}$

In Exercises 25 and 26, use one of the matrices given in Exercise 6 to solve the system of equations.

25. $\begin{cases} x + y = -2 \\ -x + y = 1 \\ x + z = -1 \end{cases}$

26. $\begin{cases} x + y = -4 \\ -x + y = 2 \\ x + z = 6 \end{cases}$

In Exercises 27–46, use matrix inversion to solve the system of equations.

27. $\begin{cases} x - y = -2 \\ -3x + 4y = 5 \end{cases}$

28. $\begin{cases} -x - y = -2 \\ 7x + 6y = 1 \end{cases}$

29. $\begin{cases} 2x + 4y = 1 \\ x + y = -2 \end{cases}$

30. $\begin{cases} x + 2y = -4 \\ -x - y = 5 \end{cases}$

31. $\begin{cases} 3x + 7y = -11 \\ x + 2y = -3 \end{cases}$

32. $\begin{cases} 4x - 3y = 1 \\ 2x - y = -1 \end{cases}$

33. $\begin{cases} 2x - 5y = -7 \\ -3x + 2y = -6 \end{cases}$

34. $\begin{cases} 3x + 2y = -4 \\ 4x + y = 3 \end{cases}$

35. $\begin{cases} x + 2y = 3 \\ 3x + 4y = 3 \end{cases}$

36. $\begin{cases} 7x + 5y = 9 \\ -2x + 3y = -7 \end{cases}$

37. $\begin{cases} x - 3y + 2z = -1 \\ y + z = 4 \\ 2x - 6y + 3z = 3 \end{cases}$

38. $\begin{cases} x - 4y + z = 7 \\ 2x + 9y = -1 \\ y - z = 0 \end{cases}$

39. $\begin{cases} x - y + z = 5 \\ y + 2z = -1 \\ -2x + 3y + z = 6 \end{cases}$

40. $\begin{cases} x + 2z = -3 \\ -2x + y - 7z = 2 \\ x + 3z = 4 \end{cases}$

41. $\begin{cases} 3x - 6y + 2z = -6 \\ x + 2y + 3z = -1 \\ y - z = 5 \end{cases}$

42. $\begin{cases} 2x - y - z = 1 \\ 4x - y + z = -5 \\ x - 3y - 4z = 2 \end{cases}$

43. $\begin{cases} x - 2y - z = \frac{3}{2} \\ 2x - 3y + 2z = -3 \\ -3x + 6y + 4z = 1 \end{cases}$

44. $\begin{cases} x + 5y - 3z = -2 \\ -3x - 16y + 7z = -\frac{1}{2} \\ -x - 5y + 4z = 0 \end{cases}$

45. $\begin{cases} x - y + w = -3 \\ y - 2w = 0 \\ -2x + 2y + z - 3w = 1 \\ -y + 3w = 0 \end{cases}$

46. $\begin{cases} x - y + z + 2w = -3 \\ y - 2z = 0 \\ -2x + 2y - z - w = 1 \\ y - 2z + w = 0 \end{cases}$

In Exercises 47–52, find the inverse of A^2 and the inverse of A^3 (where A^2 is the product AA and A^3 is the product $(A^2)A$).

47. $A = \begin{bmatrix} 1 & 1 \\ 0 & 1 \end{bmatrix}$

48. $A = \begin{bmatrix} 1 & 0 \\ 2 & 1 \end{bmatrix}$

49. $A = \begin{bmatrix} 2 & 1 \\ 0 & -1 \end{bmatrix}$

50. $A = \begin{bmatrix} 1 & 1 \\ 2 & -1 \end{bmatrix}$

51. $A = \begin{bmatrix} 2 & 0 & 0 \\ 0 & 1 & 2 \\ 0 & 0 & 1 \end{bmatrix}$

52. $A = \begin{bmatrix} 1 & 1 & 0 \\ -1 & 1 & 0 \\ 0 & 0 & 1 \end{bmatrix}$

▶**Applications** In this set of exercises, you will use inverses of matrices to study real-world problems.

53. **Theater** There is a two-tier pricing system for tickets to a certain play: one price for adults, and another for children. One customer purchases 12 tickets for adults and 6 for children, for a total of $174. Another customer purchases 8 tickets for adults and 3 for children, for a total of $111. Use the inverse of an appropriate matrix to compute the price of each type of ticket.

54. **Hourly Wage** A firm manufactures metal boxes for electrical outlets. The hourly wage for cutting the metal for the boxes is different from the wage for forming the boxes from the cut metal. In a recent week, one worker spent 16 hours cutting metal and 24 hours forming it, and another worker spent 20 hours on each task. The first worker's gross pay for that week was $784, and the second worker grossed $770. Use the inverse of an appropriate matrix to determine the hourly wage for each task.

55. **Nutrition** Liza, Megan, and Blanca went to a popular pizza place. Liza ate two slices of cheese pizza and one slice of Veggie Delite, for a total of 550 calories. Megan ate one slice each of cheese pizza, Meaty Delite, and Veggie Delite, for a total of 620 calories. Blanca ate one slice of Meaty Delite and two slices of Veggie Delite, for a total of 570 calories. (*Source:* www.pizzahut.com) Use the inverse of an appropriate matrix to determine the number of calories in each slice of cheese pizza, Meaty Delite, and Veggie Delite.

56. **Nutrition** For a diet of 2000 calories per day, the total fat content should not exceed 60 grams per day. An order of two beef burrito supremes and one plate of nachos supreme exceeds this limit by 2 grams. An order of one beef burrito supreme and one bean tostada contains a total of 28 grams of fat. Also, an order of one beef burrito supreme, one plate of nachos supreme, and one bean tostada contains a total of 54 grams of fat. (*Source:* www.tacobell.com) Use the inverse of an appropriate matrix to determine the fat content (in grams) of each beef burrito supreme, each bean tostada, and each plate of nachos supreme.

57. **Quilting** A firm manufactures "patriotic" patchwork quilts in three different patterns. The patches are all of the same dimensions and come in three different solid colors: red, white, and blue. For the top layer of each quilt, the firm uses 9 yards of material. For each pattern, the fractions of the total number of squares used for the three colors are given in the table, along with the total cost of the fabric for the top layer of the quilt.

Pattern	Fraction of Squares			Total Cost ($)
	Red	White	Blue	
1	$\frac{1}{4}$	$\frac{5}{12}$	$\frac{1}{3}$	67.50
2	$\frac{1}{3}$	$\frac{1}{3}$	$\frac{1}{3}$	69.00
3	$\frac{1}{4}$	$\frac{1}{2}$	$\frac{1}{4}$	65.25

Use the inverse of an appropriate matrix to determine the cost of the fabric (per yard) for each color.

Cryptography *In Exercises 58–61, find the decoding matrix for each encoding matrix.*

58. $\begin{bmatrix} 1 & -3 \\ 1 & -2 \end{bmatrix}$ 59. $\begin{bmatrix} 5 & 7 \\ 2 & 3 \end{bmatrix}$

60. $\begin{bmatrix} 1 & 1 & 4 & 1 \\ 2 & -3 & 4 & 1 \\ 3 & -4 & 6 & 2 \\ -1 & 0 & -2 & -1 \end{bmatrix}$

61. $\begin{bmatrix} 1 & 0 & -1 & 1 \\ -2 & 3 & 3 & 7 \\ 2 & 0 & -6 & 0 \\ -1 & 1 & 2 & 2 \end{bmatrix}$

Cryptography *In Exercises 62–65, decode the message, which was encoded using the matrix*

$$\begin{bmatrix} 1 & -2 & 3 \\ -2 & 3 & -4 \\ 2 & -4 & 5 \end{bmatrix}.$$

62. $\begin{bmatrix} 29 \\ -47 \\ 45 \end{bmatrix}, \begin{bmatrix} 62 \\ -90 \\ 99 \end{bmatrix}$

63. $\begin{bmatrix} 52 \\ -77 \\ 86 \end{bmatrix}, \begin{bmatrix} -24 \\ 38 \\ -53 \end{bmatrix}, \begin{bmatrix} 19 \\ -38 \\ 38 \end{bmatrix}$

64. $\begin{bmatrix} -5 \\ 0 \\ -11 \end{bmatrix}, \begin{bmatrix} 20 \\ -36 \\ 38 \end{bmatrix}$

65. $\begin{bmatrix} 6 \\ -16 \\ 7 \end{bmatrix}, \begin{bmatrix} 28 \\ -32 \\ 31 \end{bmatrix}$

▶**Concepts** This set of exercises will draw on the ideas presented in this section and your general math background.

66. Find the inverse of

$$\begin{bmatrix} a & 0 & 0 \\ 0 & b & 0 \\ 0 & 0 & c \end{bmatrix}$$

where a, b, and c are *all* nonzero. Would this matrix have an inverse if $a = 0$? Explain.

67. Find the inverse of

$$\begin{bmatrix} a & a & a \\ 0 & 1 & 0 \\ 0 & 0 & 1 \end{bmatrix}$$

where a is nonzero. Evaluate this inverse for the case in which $a = 1$.

68. Compute $A(BC)$ and $(AB)C$, where

$$A = \begin{bmatrix} 3 & -1 \\ 0 & 2 \end{bmatrix}, \quad B = \begin{bmatrix} 1 & 4 \\ 0 & 1 \end{bmatrix}, \quad \text{and} \quad C = \begin{bmatrix} -1 & 0 \\ 3 & 1 \end{bmatrix}.$$

What do you observe?

Exercises 69–74 involve positive-integer powers of a square matrix A. A^2 is defined as the product AA; for $n \geq 3$, A^n is defined as the product $(A^{n-1})A$.

69. Find $(A^2)^{-1}$ and $(A^{-1})^2$, where $A = \begin{bmatrix} 1 & -2 \\ -1 & 3 \end{bmatrix}$. What do you observe?

70. Use the definition of the inverse of a matrix, together with the fact that $(AB)^{-1} = A^{-1}B^{-1}$, to show that $(A^2)^{-1} = (A^{-1})^2$ for every square matrix A.

71. Find $(A^3)^{-1}$ and $(A^{-1})^3$, where $A = \begin{bmatrix} -5 & -1 \\ 4 & 1 \end{bmatrix}$. What do you observe?

72. For $n \geq 3$ and a square matrix A, express the inverse of A^n in terms of A^{-1}. (*Hint:* See Exercise 70.)

73. Let $A = \begin{bmatrix} 4 & 1 \\ 3 & 1 \end{bmatrix}$. Find the inverses of A^2 and A^3 without computing the matrices A^2 and A^3. (*Hint:* See Exercises 70 and 72.)

74. Let $A = \begin{bmatrix} 0 & 1 \\ 1 & 0 \end{bmatrix}$.
    (a) Find A^2, A^3, and A^4.
    (b) Find the inverse of A without applying Gauss-Jordan elimination. (*Hint:* Use the answer to part (a).)
    (c) For this particular matrix A, what do you observe about A^n for $n = 3, 5, 7, \ldots$?
    (d) For this particular matrix A, what do you observe about A^n for $n = 2, 4, 6, \ldots$?

Exercises 75 and 76 involve the use of matrix multiplication to transform one or more points. This technique, which can be applied to any set of points, is used extensively in computer graphics.

75. Let $A = \begin{bmatrix} 0 & 1 \\ 1 & 0 \end{bmatrix}$ and $B = \begin{bmatrix} 2 \\ -1 \end{bmatrix}$.
    (a) Calculate the product matrix AB.
    (b) On a single coordinate system, plot the point $(2, -1)$ and the point whose coordinates (x, y) are the entries of the product matrix found in part (a). Explain geometrically what the matrix multiplication did to the point $(2, -1)$.
    (c) How would you undo the multiplication in part (a)?

76. Consider a series of points (x_0, y_0), (x_1, y_1), (x_2, y_2), ... such that, for every nonnegative integer i, the point (x_{i+1}, y_{i+1}) is found by applying the matrix $\begin{bmatrix} 1 & -2 \\ 1 & -3 \end{bmatrix}$ to the point (x_i, y_i).

$$\begin{bmatrix} x_{i+1} \\ y_{i+1} \end{bmatrix} = \begin{bmatrix} 1 & -2 \\ 1 & -3 \end{bmatrix}\begin{bmatrix} x_i \\ y_i \end{bmatrix}$$

    (a) Find (x_1, y_1) if $(x_0, y_0) = (2, -1)$.
    (b) Find (x_2, y_2) if $(x_0, y_0) = (4, 6)$. (*Hint:* Find (x_1, y_1) first.)
    (c) Use the inverse of an appropriate matrix to find (x_0, y_0) if $(x_3, y_3) = (2, 3)$.

6.6 Determinants and Cramer's Rule

Objectives

▶ Calculate the determinant of a matrix

▶ Use Cramer's Rule to solve systems of linear equations in two and three variables

In this section, we introduce **Cramer's Rule,** which gives a formula for the solution of a system of linear equations. We will also discuss the limitations of Cramer's Rule and show that, from a practical standpoint, it is suitable only for solving systems of equations in a few variables. Before presenting Cramer's Rule, we will address an important concept related to it: the determinant of a square matrix.

Determinant of a Square Matrix

We first define the **determinant** of a 2×2 matrix.

Determinant of a 2 × 2 Matrix

Let A be a 2×2 matrix:

$$A = \begin{bmatrix} a & b \\ c & d \end{bmatrix}$$

where a, b, c, and d represent numbers. Then the **determinant** of A, which is denoted by $|A|$, is defined as

$$|A| = \begin{vmatrix} a & b \\ c & d \end{vmatrix} = ad - bc.$$

Note that $|A|$ is just a number, whereas A is a matrix.

Example 1 Evaluating a Determinant

Evaluate $\begin{vmatrix} -3 & -2 \\ 4 & 6 \end{vmatrix}$.

▶**Solution** We see that $a = -3$, $b = -2$, $c = 4$, and $d = 6$. Thus

$$\begin{vmatrix} -3 & -2 \\ 4 & 6 \end{vmatrix} = (-3)(6)-(-2)(4) = -18 - (-8) = -18 + 8 = -10.$$

✔ *Check It Out 1:* Evaluate $\begin{vmatrix} 2 & -3 \\ 5 & -7 \end{vmatrix}$. ■

For $n > 2$, the determinant of an $n \times n$ matrix is obtained by breaking down the computation into a sequence of steps, each of which entails finding determinants of smaller matrices than those involved in the preceding step. Before explaining how to do this, we define two new terms: **minor** and **cofactor.**

Definition of Minor and Cofactor

Let A be an $n \times n$ matrix, and let a_{ij} be the entry in the ith row, jth column of A.

The **minor** of a_{ij} (denoted by M_{ij}) is the determinant of the matrix obtained by deleting the ith row and the jth column of A.

The **cofactor** of a_{ij} (denoted by C_{ij}) is defined as

$$C_{ij} = (-1)^{i+j}M_{ij}.$$

Note that M_{ij} and C_{ij} are a pair of numbers associated with a particular entry of matrix A (namely, a_{ij}). Example 2 illustrates how to find minors and cofactors.

Example 2 Calculating Minors and Cofactors

Let

$$A = \begin{bmatrix} 0 & 1 & 3 \\ -2 & 5 & 7 \\ 4 & 0 & -1 \end{bmatrix}.$$

(a) Find M_{12} and C_{12}.

(b) Find M_{13} and C_{13}.

▶**Solution**

(a) To find M_{12}, let $i = 1$ and $j = 2$.

STEPS	EXAMPLE
1. Delete the first row of A (since $i = 1$) and the second column of A (since $j = 2$). The remaining entries of A are given in red.	$A = \begin{bmatrix} 0 & 1 & 3 \\ -2 & 5 & 7 \\ 4 & 0 & -1 \end{bmatrix}$
2. Find the determinant of the 2×2 matrix that is formed by the remaining entries of A.	$M_{12} = \begin{vmatrix} -2 & 7 \\ 4 & -1 \end{vmatrix}$ $= (-2)(-1) - (7)(4)$ $= 2 - 28 = -26$

Because $C_{ij} = (-1)^{i+j}M_{ij}$, the cofactor C_{12} is

$$C_{12} = (-1)^{1+2}M_{12} = (-1)^3(-26) = (-1)(-26) = 26.$$

(b) To find M_{13}, let $i = 1$ and $j = 3$. Then delete the first row of A (since $i = 1$) and the third column of A (since $j = 3$). The remaining entries of A are shown in red.

$$A = \begin{bmatrix} 0 & 1 & 3 \\ -2 & 5 & 7 \\ 4 & 0 & -1 \end{bmatrix}$$

The determinant of the 2×2 matrix that is formed by the remaining entries of A is

$$M_{13} = \begin{vmatrix} -2 & 5 \\ 4 & 0 \end{vmatrix} = (-2)(0) - (5)(4) = 0 - 20 = -20.$$

The cofactor C_{13} is

$$C_{13} = (-1)^{1+3}M_{13} = (-1)^4(-20) = 1(-20) = -20.$$

✔ *Check It Out 2:* For matrix A from Example 2, find M_{23} and C_{23}. ■

Now that we have defined minors and cofactors, we can find the determinant of an $n \times n$ matrix.

Finding the Determinant of an $n \times n$ Matrix

Step 1 Choose any row or column of the matrix, preferably a row or column in which at least one entry is zero. This is the row or column by which the determinant will be *expanded*.

Step 2 Multiply each entry in the chosen row or column by its cofactor.

Step 3 The determinant is the *sum* of all the products found in Step 2. It can be shown that the determinant is independent of the choice of row or column made in Step 1.

The determinant of an $n \times n$ matrix A is denoted by $|A|$. Example 3 illustrates the procedure for evaluating the determinant of a 3×3 matrix.

Technology Note

Access the MATRIX menu of your graphing utility to compute the determinant of a matrix. Figure 6.6.1 shows the determinant of matrix A from Example 3.

Keystroke Appendix:
Section 13

Figure 6.6.1

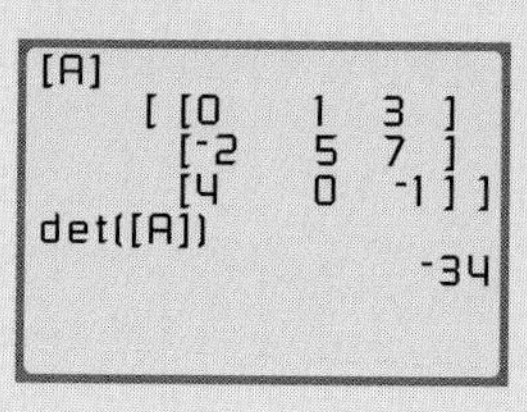

```
[A]
    [ [0    1    3 ]
      [-2   5    7 ]
      [4    0   -1 ] ]
det([A])
                 -34
```

Example 3 Determinant of a 3×3 Matrix

Evaluate the determinant of A.

$$A = \begin{bmatrix} 0 & 1 & 3 \\ -2 & 5 & 7 \\ 4 & 0 & -1 \end{bmatrix}$$

▶**Solution** We will expand the determinant of A by the first row of A.

$$|A| = \begin{vmatrix} 0 & 1 & 3 \\ -2 & 5 & 7 \\ 4 & 0 & -1 \end{vmatrix} = a_{11}C_{11} + a_{12}C_{12} + a_{13}C_{13}$$

Multiply each element in the first row by its cofactor

$$= (0)C_{11} + 1(26) + 3(-20)$$

$C_{12} = 26$ and $C_{13} = -20$, from Example 2

$$= 0 + 26 - 60 = -34$$

Because $a_{11} = 0$, the product $a_{11}C_{11}$ is zero regardless of the value of C_{11}. So we need not compute C_{11}.

☑ *Check It Out 3:* Evaluate the determinant of A.

$$A = \begin{bmatrix} -1 & 0 & 3 \\ 0 & 2 & 5 \\ -2 & 1 & 0 \end{bmatrix} ■$$

Cramer's Rule for Systems of Two Linear Equations in Two Variables

We will now consider some general aspects of solving a system of two linear equations in two variables, called a 2×2 system. We will also see how determinants arise in that process. Consider the following system of equations.

$$\begin{cases} ax + by = e \\ cx + dy = f \end{cases}$$

We first solve for the variable x by eliminating the variable y. For purposes of illustration, we assume that a, b, c and d are all nonzero. We multiply the first equation by d and the second equation by $-b$ and then add the resulting equations.

$$\begin{array}{rcl} adx + bdy & = & ed \\ -bcx - bdy & = & -bf \\ \hline adx - bcx & = & ed - bf \end{array}$$

Solving for x in the last equation,

$$adx - bcx = ed - bf \Longrightarrow (ad - bc)x = ed - bf \Longrightarrow x = \frac{ed - bf}{ad - bc}$$

provided $ad - bc \neq 0$. The expression for x can be rewritten in terms of determinants.

$$x = \frac{ed - bf}{ad - bc} = \frac{\begin{vmatrix} e & b \\ f & d \end{vmatrix}}{\begin{vmatrix} a & b \\ c & d \end{vmatrix}}$$

We can solve for y in a similar manner. We now state Cramer's Rule for a 2×2 system of equations.

Cramer's Rule for a 2 × 2 System of Linear Equations

The solution of the system of equations

$$\begin{cases} ax + by = e \\ cx + dy = f \end{cases}$$

is

$$x = \frac{\begin{vmatrix} e & b \\ f & d \end{vmatrix}}{\begin{vmatrix} a & b \\ c & d \end{vmatrix}}, \quad y = \frac{\begin{vmatrix} a & e \\ c & f \end{vmatrix}}{\begin{vmatrix} a & b \\ c & d \end{vmatrix}}$$

provided $\begin{vmatrix} a & b \\ c & d \end{vmatrix}$ (which is $ad - bc$) is nonzero.

The solution can be written more compactly as follows.

$$x = \frac{D_x}{D}, \quad y = \frac{D_y}{D}$$

where $D_x = \begin{vmatrix} e & b \\ f & d \end{vmatrix}$, $D_y = \begin{vmatrix} a & e \\ c & f \end{vmatrix}$, $D = \begin{vmatrix} a & b \\ c & d \end{vmatrix}$, and $D \neq 0$.

Note If the quantity $ad - bc$ is nonzero, the system of equations has the unique solution given by Cramer's Rule. If $ad - bc = 0$, Cramer's Rule cannot be applied. If such a system can be solved, it is dependent and has infinitely many solutions.

Example 4 **Solving a System of Two Linear Equations Using Cramer's Rule**

Use Cramer's Rule, if applicable, to solve the following system of equations.

$$\begin{cases} 2x - 3y = 6 \\ 5x + y = 7 \end{cases}$$

▶Solution We first calculate D to determine whether it is nonzero.

$$D = \begin{vmatrix} 2 & -3 \\ 5 & 1 \end{vmatrix} = (2)(1) - (-3)(5) = 2 - (-15) = 2 + 15 = 17$$

Since $D = 17 \neq 0$, we can apply Cramer's Rule to solve this system of equations. Calculating the other determinants that we need, we obtain the following.

$$D_x = \begin{vmatrix} 6 & -3 \\ 7 & 1 \end{vmatrix} = (6)(1) - (-3)(7) = 6 - (-21) = 6 + 21 = 27$$

$$D_y = \begin{vmatrix} 2 & 6 \\ 5 & 7 \end{vmatrix} = (2)(7) - (6)(5) = 14 - 30 = -16$$

Thus

$$x = \frac{D_x}{D} = \frac{27}{17}, \quad y = \frac{D_y}{D} = \frac{-16}{17}.$$

✔ *Check It Out 4:* Use Cramer's Rule, if applicable, to solve the following system of equations.

$$\begin{cases} 3x + y = -3 \\ -4x - 2y = 5 \end{cases} \blacksquare$$

Example 5 **Applicability of Cramer's Rule**

Explain why the following system of equations *cannot* be solved by Cramer's Rule.

$$\begin{cases} -x + 2y = -1 \\ 5x - 10y = 5 \end{cases}$$

▶Solution We first calculate D:

$$D = \begin{vmatrix} -1 & 2 \\ 5 & -10 \end{vmatrix} = (-1)(-10) - (2)(5) = 10 - 10 = 0.$$

Because $D = 0$, Cramer's Rule cannot be applied. However, this system of equations can be solved by elimination. It is actually a dependent system, so it has infinitely many solutions.

✔ *Check It Out 5:* Can the following system of equations be solved by Cramer's Rule? Why or why not?

$$\begin{cases} 3x + y = 0 \\ 4x - y = 3 \end{cases} \blacksquare$$

Cramer's Rule for Systems of Three Linear Equations in Three Variables

Cramer's Rule for a system of three linear equations in three variables, called a 3×3 system, is very similar in form to Cramer's Rule for a 2×2 system. The main difference is the size of the square matrices whose determinants have to be evaluated.

> **Cramer's Rule for a 3×3 System of Linear Equations**
>
> The solution of the system of equations
>
> $$\begin{cases} a_{11}x + a_{12}y + a_{13}z = b_1 \\ a_{21}x + a_{22}y + a_{23}z = b_2 \\ a_{31}x + a_{32}y + a_{33}z = b_3 \end{cases}$$
>
> is
>
> $$x = \frac{D_x}{D}, \quad y = \frac{D_y}{D}, \quad z = \frac{D_z}{D}$$
>
> where
>
> $$D_x = \begin{vmatrix} b_1 & a_{12} & a_{13} \\ b_2 & a_{22} & a_{23} \\ b_3 & a_{32} & a_{33} \end{vmatrix}, \quad D_y = \begin{vmatrix} a_{11} & b_1 & a_{13} \\ a_{21} & b_2 & a_{23} \\ a_{31} & b_3 & a_{33} \end{vmatrix}, \quad D_z = \begin{vmatrix} a_{11} & a_{12} & b_1 \\ a_{21} & a_{22} & b_2 \\ a_{31} & a_{32} & b_3 \end{vmatrix}, \quad \text{and}$$
>
> $$D = \begin{vmatrix} a_{11} & a_{12} & a_{13} \\ a_{21} & a_{22} & a_{23} \\ a_{31} & a_{32} & a_{33} \end{vmatrix}$$
>
> provided $D \neq 0$.

Example 6 **Cramer's Rule for a System of Three Linear Equations**

Use Cramer's Rule to solve the following system of equations.

$$\begin{cases} 3x + y - z = -1 \\ x - y + 2z = 7 \\ -2x + y + z = -2 \end{cases}$$

▶ **Solution** We first compute D to determine whether it is nonzero.

$$D = \begin{vmatrix} 3 & 1 & -1 \\ 1 & -1 & 2 \\ -2 & 1 & 1 \end{vmatrix}$$

Expanding the determinant by the first row of the associated matrix, we find that

$$D = \begin{vmatrix} 3 & 1 & -1 \\ 1 & -1 & 2 \\ -2 & 1 & 1 \end{vmatrix} = (-1)^{1+1}(3)\begin{vmatrix} -1 & 2 \\ 1 & 1 \end{vmatrix}$$

$$+ (-1)^{1+2}(1)\begin{vmatrix} 1 & 2 \\ -2 & 1 \end{vmatrix} + (-1)^{1+3}(-1)\begin{vmatrix} 1 & -1 \\ -2 & 1 \end{vmatrix}$$

$$= (1)(3)(-1 - 2) + (-1)(1)[1 - (-4)] + (1)(-1)(1 - 2)$$

$$= 3(-3) - (1)(5) + (-1)(-1) = -9 - 5 + 1 = -13.$$

Technology Note

To check the answer to Example 6, take the matrices whose determinants are D, D_x, D_y, and D_z and enter them into your calculator as matrices A, B, C, and D, respectively. Then compute the quotients of their determinants. See Figure 6.6.2.

Keystroke Appendix:
Section 13

Figure 6.6.2

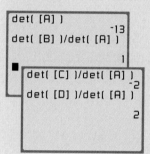

```
det( [A] )
                   -13
det( [B] )/det( [A] )
                    1
■ det( [C] )/det( [A] )
                   -2
  det( [D] )/det( [A] )
                    2
```

Because $D \neq 0$, we can proceed to compute the other determinants and then use them to compute the solution. We have computed all the determinants by expanding them along the first row of the pertinent matrix.

$$D_x = \begin{vmatrix} -1 & 1 & -1 \\ 7 & -1 & 2 \\ -2 & 1 & 1 \end{vmatrix} = (-1)^{1+1}(-1)\begin{vmatrix} -1 & 2 \\ 1 & 1 \end{vmatrix}$$

$$+ (-1)^{1+2}(1)\begin{vmatrix} 7 & 2 \\ -2 & 1 \end{vmatrix} + (-1)^{1+3}(-1)\begin{vmatrix} 7 & -1 \\ -2 & 1 \end{vmatrix}$$

$$= (1)(-1)(-1-2) + (-1)(1)[7-(-4)] + (1)(-1)(7-2)$$

$$= (-1)(-3) - (1)(11) + (-1)(5)$$

$$= 3 - 11 - 5 = -13$$

$$D_y = \begin{vmatrix} 3 & -1 & -1 \\ 1 & 7 & 2 \\ -2 & -2 & 1 \end{vmatrix} = (-1)^{1+1}(3)\begin{vmatrix} 7 & 2 \\ -2 & 1 \end{vmatrix}$$

$$+ (-1)^{1+2}(-1)\begin{vmatrix} 1 & 2 \\ -2 & 1 \end{vmatrix} + (-1)^{1+3}(-1)\begin{vmatrix} 1 & 7 \\ -2 & -2 \end{vmatrix}$$

$$= (1)(3)[7-(-4)] + (-1)(-1)[1-(-4)]$$

$$+ (1)(-1)[-2-(-14)]$$

$$= (3)(11) - (-1)(5) + (-1)(12)$$

$$= 33 + 5 - 12 = 26$$

$$D_z = \begin{vmatrix} 3 & 1 & -1 \\ 1 & -1 & 7 \\ -2 & 1 & -2 \end{vmatrix} = (-1)^{1+1}(3)\begin{vmatrix} -1 & 7 \\ 1 & -2 \end{vmatrix}$$

$$+ (-1)^{1+2}(1)\begin{vmatrix} 1 & 7 \\ -2 & -2 \end{vmatrix} + (-1)^{1+3}(-1)\begin{vmatrix} 1 & -1 \\ -2 & 1 \end{vmatrix}$$

$$= (1)(3)(2-7) + (-1)(1)[-2-(-14)] + (1)(-1)(1-2)$$

$$= (3)(-5) - (1)(12) + (-1)(-1)$$

$$= -15 - 12 + 1 = -26$$

We then have the following result.

$$x = \frac{D_x}{D} = \frac{-13}{-13} = 1$$

$$y = \frac{D_y}{D} = \frac{26}{-13} = -2$$

$$z = \frac{D_z}{D} = \frac{-26}{-13} = 2$$

You should check that these values of x, y, and z satisfy the given system of equations.

☑ *Check It Out 6:* Use Cramer's Rule to solve the following system of equations.

$$\begin{cases} -2x + y & = 0 \\ x - y + 2z = -2 \\ y + z = 1 \end{cases}$$ ▪

Limitations of Cramer's Rule

Cramer's Rule can be generalized to systems of n linear equations in n variables (called $n \times n$ systems) for all $n \geq 2$; however, use of Cramer's Rule to solve systems in more than two or three variables is computationally impractical, because the number of determinants to be evaluated grows much faster than the number of variables. For example, to compute the solution of a 10×10 system with Cramer's Rule, we must evaluate the determinants of eleven 10×10 matrices. Each 10×10 determinant will in turn have ten 9×9 determinants to be evaluated. Each 9×9 determinant will in turn have nine 8×8 determinants to be evaluated. This sequence of computations continues all the way through the stage at which determinants of 2×2 matrices are evaluated—and there will be a huge number of these.

In real-world applications, a 10×10 system of equations is considered very small. But even for such a "small" system, the use of Cramer's Rule entails performing many more arithmetic operations than does Gaussian elimination. Thus Cramer's Rule is used only for systems with two, three, or four variables, and only in a very limited set of applications, such as computer graphics. For larger systems of equations, Cramer's Rule quickly becomes inefficient, even with extremely fast and powerful supercomputers.

6.6 Key Points

▶ The **determinant** of a 2×2 matrix $A = \begin{bmatrix} a & b \\ c & d \end{bmatrix}$ is defined as

$$|A| = \begin{vmatrix} a & b \\ c & d \end{vmatrix} = ad - bc.$$

▶ Minors and cofactors are defined as follows: Let A be an $n \times n$ matrix, and let a_{ij} be the entry in the ith row, jth column of A. The **minor** of a_{ij} (denoted by M_{ij}) is the determinant of the matrix that is obtained by deleting the ith row and jth column of A. The **cofactor** of a_{ij} (denoted by C_{ij}) is defined as

$$C_{ij} = (-1)^{i+j} M_{ij}.$$

▶ The determinants of larger matrices can be found by cofactor expansion.

▶ **Cramer's Rule** gives a formula for obtaining the unique solution of a system of linear equations.

6.6 Exercises

▶ **Skills** This set of exercises will reinforce the skills illustrated in this section.

In Exercises 1–8, evaluate the determinant of A.

1. $A = \begin{bmatrix} -3 & 1 \\ 2 & 4 \end{bmatrix}$

2. $A = \begin{bmatrix} 5 & 2 \\ -2 & 4 \end{bmatrix}$

3. $A = \begin{bmatrix} \frac{1}{2} & 3 \\ 2 & -6 \end{bmatrix}$

4. $A = \begin{bmatrix} -3 & -\frac{1}{4} \\ 8 & 2 \end{bmatrix}$

5. $A = \begin{bmatrix} 4 & 1 \\ -3 & 8 \end{bmatrix}$

6. $A = \begin{bmatrix} 6 & -2 \\ 5 & 3 \end{bmatrix}$

7. $A = \begin{bmatrix} \frac{1}{3} & -2 \\ 4 & 9 \end{bmatrix}$

8. $A = \begin{bmatrix} 5 & -\frac{2}{5} \\ 10 & 2 \end{bmatrix}$

In Exercises 9–12, find the given minor and cofactor pertaining to the matrix

$$\begin{bmatrix} -3 & 0 & 2 \\ 1 & 5 & -4 \\ 0 & 6 & 5 \end{bmatrix}.$$

9. M_{11} and C_{11}

10. M_{23} and C_{23}

11. M_{32} and C_{32}

12. M_{21} and C_{21}

In Exercises 13–22, evaluate the determinant of the matrix.

13. $\begin{bmatrix} 0 & 1 & -2 \\ 5 & -2 & 3 \\ 0 & 6 & 5 \end{bmatrix}$

14. $\begin{bmatrix} -7 & 5 & 0 \\ 0 & 3 & 0 \\ -3 & -2 & 2 \end{bmatrix}$

15. $\begin{bmatrix} -2 & 3 & 5 \\ 6 & -1 & 0 \\ 0 & 1 & -2 \end{bmatrix}$

16. $\begin{bmatrix} -5 & 4 & 9 \\ 1 & 0 & -2 \\ 0 & 7 & 3 \end{bmatrix}$

17. $\begin{bmatrix} 0 & 0 & 0 \\ -7 & 3 & 4 \\ 6 & 3 & 4 \end{bmatrix}$

18. $\begin{bmatrix} -5 & 4 & 9 \\ 1 & 0 & -2 \\ 0 & -5 & 7 \end{bmatrix}$

19. $\begin{bmatrix} 1 & 1 & 1 \\ 2 & 2 & 2 \\ 3 & 3 & 3 \end{bmatrix}$

20. $\begin{bmatrix} 1 & 1 & 1 \\ 1 & 2 & 4 \\ 1 & 3 & 9 \end{bmatrix}$

21. $\begin{bmatrix} -2 & 2 & 0 \\ 0 & -1 & 1 \\ -4 & 5 & 2 \end{bmatrix}$

22. $\begin{bmatrix} 0 & -1 & -3 \\ -2 & 0 & 4 \\ -1 & 0 & 5 \end{bmatrix}$

In Exercises 23–28, solve for x.

23. $\begin{vmatrix} -1 & x \\ 3 & -4 \end{vmatrix} = -2$

24. $\begin{vmatrix} 5 & -1 \\ x & 2 \end{vmatrix} = 13$

25. $\begin{vmatrix} -1 & 0 & 2 \\ 0 & 5 & 3 \\ 0 & x & -2 \end{vmatrix} = -2$

26. $\begin{vmatrix} 5 & 0 & 0 \\ -3 & x & 1 \\ 2 & 8 & -3 \end{vmatrix} = -70$

27. $\begin{vmatrix} 2 & -3 & 5 \\ x & 0 & -4 \\ 3 & 2 & 1 \end{vmatrix} = 39$

28. $\begin{vmatrix} 4 & 3 & x \\ -2 & 8 & 1 \\ 5 & 2 & 1 \end{vmatrix} = 353$

In Exercises 29–46, use Cramer's Rule to solve the system of equations.

29. $\begin{cases} -3x - y = 5 \\ 4x + y = 2 \end{cases}$

30. $\begin{cases} -x + y = -3 \\ -2x + y = -2 \end{cases}$

31. $\begin{cases} 4x - 2y = 7 \\ 3x - y = 1 \end{cases}$

32. $\begin{cases} x - y = -3 \\ 4x + y = 0 \end{cases}$

33. $\begin{cases} x - 2y = 4 \\ -3x + 4y = -8 \end{cases}$

34. $\begin{cases} 7x - y = -8 \\ -x + 3y = 4 \end{cases}$

35. $\begin{cases} 4x + y = -7 \\ 5x + 4y = -6 \end{cases}$

36. $\begin{cases} x + 2y = -3 \\ -x - y = -2 \end{cases}$

37. $\begin{cases} 1.4x + 2y = 0 \\ 3.5x + 3y = -9.7 \end{cases}$

38. $\begin{cases} 2.5x - 0.5y = -7.2 \\ -x + y = 5.6 \end{cases}$

39. $\begin{cases} x + y = 1 \\ x - z = 0 \\ -y + z = 0 \end{cases}$

40. $\begin{cases} x - z = 0 \\ -y - z = -1 \\ x + y + z = 0 \end{cases}$

41. $\begin{cases} 5x + 3z = 3 \\ -2x + y + z = -1 \\ -3y + z = 7 \end{cases}$

42. $\begin{cases} -2x + 3y - z = 7 \\ x - 2z = -7 \\ -3y + z = -1 \end{cases}$

43. $\begin{cases} 3x - 5y + z = -14 \\ -3x + 7y - 4z = 9 \\ 2x + z = 6 \end{cases}$

44. $\begin{cases} x + 4z = -3 \\ -2x + y = -10 \\ x - 2y - z = 7 \end{cases}$

45. $\begin{cases} 3x + y + z = 1 \\ 2x + y - z = -\dfrac{3}{2} \\ x + 3y - z = -5 \end{cases}$

46. $\begin{cases} x + 2y + z = 0 \\ -x + y + 3z = \dfrac{5}{2} \\ 4x + y - z = -\dfrac{3}{2} \end{cases}$

▶**Concepts** This set of exercises will draw on the ideas presented in this section and your general math background.

47. Without computing the following determinant, explain why its value must be zero.

$$\begin{vmatrix} 1 & 2 & -1 \\ 0 & 0 & 0 \\ 3 & -2 & 1 \end{vmatrix}$$

48. For what value(s) of k can Cramer's Rule be used to solve the following system of equations?

$$\begin{cases} 2x - y = 2 \\ kx + 3y = 4 \end{cases}$$

49. Verify that $x = 1$, $y = 2$, $z = 0$ is a solution of the following system of equations.

$$\begin{cases} x + 2y = 5 \\ 4x + y - z = 6 \\ -2x - 4y = -10 \end{cases}$$

Even though there is a solution, explain why Cramer's Rule cannot be used to solve this system.

6.7 Partial Fractions

Objective

▶ Compute the partial fraction decomposition of a rational expression

You already know how to take individual rational expressions and write their sum as a rational expression. In some situations, however, there is a need to apply this process in reverse; that is, a given rational expression must be *decomposed* (rewritten) as a sum of simpler rational expressions. For example, $\frac{5x + 1}{x^2 - 1}$ can be decomposed as $\frac{2}{x + 1} + \frac{3}{x - 1}$. The technique used to decompose a rational expression into simpler expressions is known as **partial fraction decomposition.** The decomposition procedure depends on the form of the denominator. In this section, we cover four different cases; in each case, it is assumed that the degree of the numerator is less than the degree of the denominator.

Case 1: Denominator Can Be Factored Over the Real Numbers as a Product of Distinct Linear Factors

Let $r(x) = \frac{P(x)}{Q(x)}$, where $P(x)$ and $Q(x)$ are polynomials and the degree of $P(x)$ is less than the degree of $Q(x)$, and let $a_1, \ldots, a_n$ and $b_1, \ldots, b_n$ be real numbers such that

$$Q(x) = (a_1 x + b_1)(a_2 x + b_2) \cdots (a_n x + b_n).$$

Then there exist real numbers $A_1, \ldots, A_n$ such that the partial fraction decomposition of $r(x)$ is

$$\frac{A_1}{a_1 x + b_1} + \frac{A_2}{a_2 x + b_2} + \cdots + \frac{A_n}{a_n x + b_n}.$$

For every i, the portion of the partial fraction decomposition of $r(x)$ that pertains to $(a_i x + b_i)$ is the fraction $\frac{A_i}{a_i x + b_i}$.

In the above definition, the constants $A_1, A_2, \ldots, A_n$ are determined by solving a system of linear equations, as illustrated in the following example.

Example 1 Nonrepeated Linear Factors

Compute the partial fraction decomposition of $\frac{12}{x^2 - 4}$.

▶**Solution** Factoring the denominator, we get $x^2 - 4 = (x + 2)(x - 2)$. Since both factors are linear and neither of them is a repeated factor,

$$\frac{12}{(x + 2)(x - 2)} = \frac{A}{x + 2} + \frac{B}{x - 2}.$$

Multiplying both sides of this equation by $(x + 2)(x - 2)$, we obtain the following.

$$12 = A(x - 2) + B(x + 2)$$
$$12 = Ax - 2A + Bx + 2B \qquad \text{Expand}$$
$$12 = (A + B)x + (-2A + 2B) \qquad \text{Combine like terms}$$

The expression $(A + B)x + (-2A + 2B)$ is a polynomial that consists of two terms: a term in x and a constant term. The coefficient of x is $A + B$, and the constant term is $-2A + 2B$. Note that 12 can be considered as a polynomial that has just a constant term, 12. Since these two polynomials are equal, the coefficients of like powers must be equal. Equating coefficients of like powers of x, we obtain

$$0 = A + B \qquad \text{Equate coefficients of } x$$
$$12 = -2A + 2B. \qquad \text{Equate constant terms}$$

From the first equation, $A = -B$. Substituting $-B$ for A in the second equation, we have

$$12 = -2(-B) + 2B$$
$$12 = 2B + 2B$$
$$12 = 4B$$
$$3 = B$$

Because $B = 3$ and $A = -B$, we see that $A = -3$. Thus the partial fraction decomposition of $\dfrac{12}{x^2 - 4}$ is

$$\frac{-3}{x + 2} + \frac{3}{x - 2}.$$

✔ *Check It Out 1:* Compute the partial fraction decomposition of $\dfrac{-x - 1}{x^2 - x}$. ■

Case 2: Denominator Has At Least One Repeated Linear Factor

Let $r(x) = \dfrac{P(x)}{Q(x)}$, where $P(x)$ and $Q(x)$ are polynomials and the degree of $P(x)$ is less than the degree of $Q(x)$. For every repeated linear factor of $Q(x)$, there exist real numbers a and b and a positive integer $m \geq 2$ such that

▶ $(ax + b)^m$ is a factor of $Q(x)$, and
▶ $(ax + b)^{m+1}$ is not a factor of $Q(x)$.

The portion of the partial fraction decomposition of $r(x)$ that pertains to $(ax + b)^m$ is a sum of m fractions of the following form:

$$\frac{A_1}{ax + b} + \frac{A_2}{(ax + b)^2} + \cdots + \frac{A_m}{(ax + b)^m}$$

where $A_1, \ldots, A_m$ are real numbers. The nonrepeated linear factors of $Q(x)$ are treated as specified in Case 1.

Example 2 Repeated Linear Factors

Write the partial fraction decomposition of $\dfrac{4x^2 - 7x + 1}{x^3 - 2x^2 + x}$.

▶**Solution** Factoring the denominator, we have

$$x^3 - 2x^2 + x = x(x^2 - 2x + 1) = x(x-1)^2.$$

Because every factor is linear and $(x-1)^2$ is the only repeated linear factor,

$$\frac{4x^2 - 7x + 1}{x(x-1)^2} = \frac{A}{x} + \frac{B}{x-1} + \frac{C}{(x-1)^2}.$$

Multiplying both sides of this equation by the LCD, $x(x-1)^2$, we obtain the following.

$$
\begin{aligned}
4x^2 - 7x + 1 &= A(x-1)^2 + Bx(x-1) + Cx \\
&= A(x^2 - 2x + 1) + Bx^2 - Bx + Cx \\
&= Ax^2 - 2Ax + A + Bx^2 - Bx + Cx \\
&= (A+B)x^2 + (-2A - B + C)x + A
\end{aligned}
$$

Equating coefficients, we have

$$
\begin{aligned}
4 &= A + B && \text{Equate coefficients of } x^2 \\
-7 &= -2A - B + C && \text{Equate coefficients of } x \\
1 &= A. && \text{Equate constant terms}
\end{aligned}
$$

Because $A = 1$ and $A + B = 4$, we have $B = 3$. Substituting the values of A and B into the second equation leads to the following result.

$$
\begin{aligned}
-7 &= -2(1) - 3 + C \\
-7 &= -2 - 3 + C \\
-7 &= -5 + C \\
-2 &= C
\end{aligned}
$$

Thus the partial fraction decomposition of $\frac{4x^2 - 7x + 1}{x^3 - 2x^2 + x}$ is

$$\frac{1}{x} + \frac{3}{x-1} - \frac{2}{(x-1)^2}.$$

✔ *Check It Out 2:* Compute the partial fraction decomposition of $\frac{x^2 - 2x - 6}{x^2(x+3)}$. ■

Case 3: Denominator Has At Least One Non-Repeated Irreducible Quadratic Factor

Let $r(x) = \frac{P(x)}{Q(x)}$, where $P(x)$ and $Q(x)$ are polynomials and the degree of $P(x)$ is less than the degree of $Q(x)$. For every quadratic factor of $Q(x)$, there exist real numbers a, b, and c such that

▶ $ax^2 + bx + c$ is a factor of $Q(x)$, and

▶ $ax^2 + bx + c$ cannot be factored over the real numbers, and is called an **irreducible quadratic factor.**

The portion of the partial fraction decomposition of $r(x)$ that pertains to $ax^2 + bx + c$ is a fraction of the form

$$\frac{Ax + B}{ax^2 + bx + c}$$

where A and B are real numbers. Each of the linear factors of $Q(x)$ is treated as specified in Case 1 or Case 2, as appropriate.

Example 3 Checking for Irreducible Quadratic Factors

Check whether the quadratic polynomial $2x^2 + x + 3$ is irreducible.

▶**Solution** The quadratic polynomial $2x^2 + x + 3$ can be factored over the real numbers if and only if the quadratic equation $2x^2 + x + 3 = 0$ has real solutions. To determine whether a quadratic polynomial has real solutions, we evaluate the discriminant, $b^2 - 4ac$, where a and b are the coefficients of the x^2 and x terms, respectively, and c is the constant term. For the polynomial $2x^2 + x + 3$,

$$a = 2, b = 1, c = 3.$$

The discriminant is

$$b^2 - 4ac = (1)^2 - 4(2)(3)$$
$$= 1 - 24$$
$$= -23.$$

Because $b^2 - 4ac < 0$, the equation

$$2x^2 + x + 3 = 0$$

has no real solutions. Thus $2x^2 + x + 3$ cannot be factored over the real numbers and is therefore irreducible.

✔ *Check It Out 3:* Check whether the quadratic polynomial $2x^2 + x - 3$ is irreducible. ■

Example 4 An Irreducible Quadratic Factor

Write the partial fraction decomposition of $\dfrac{3x^2 - 2x + 13}{(x + 1)(x^2 + 5)}$.

▶**Solution** Because $x + 1$ is a linear factor and $x^2 + 5$ is an irreducible quadratic factor, we have

$$\frac{3x^2 - 2x + 13}{(x + 1)(x^2 + 5)} = \frac{A}{x + 1} + \frac{Bx + C}{x^2 + 5}.$$

Multiplying both sides of this equation by the LCD, $(x + 1)(x^2 + 5)$, we obtain

$$3x^2 - 2x + 13 = A(x^2 + 5) + (Bx + C)(x + 1). \qquad \text{Equation (1)}$$

This equation holds true for any value of x. Because the term $(Bx + C)(x + 1)$ has a linear factor $(x + 1)$, we can easily solve for at least one of the coefficients by substituting the zero of the linear factor for x. Letting $x = -1$ (because -1 is the zero of $x + 1$), we have

$$3(-1)^2 - 2(-1) + 13 = A((-1)^2 + 5) + (B(-1) + C)(-1 + 1)$$
$$3 + 2 + 13 = A(1 + 5) + (-B + C)(0)$$
$$18 = 6A + 0$$
$$3 = A.$$

Substituting $A = 3$ into Equation (1), we get

$$3x^2 - 2x + 13 = 3(x^2 + 5) + (Bx + C)(x + 1)$$

$$3x^2 - 2x + 13 = 3x^2 + 15 + Bx^2 + Cx + Bx + C$$

$$-2x + 13 = Bx^2 + (B + C)x + 15 + C.$$

Equating coefficients,

$$0 = B \qquad \text{Equate coefficients of } x^2$$

$$-2 = B + C \qquad \text{Equate coefficients of } x$$

$$13 = 15 + C. \qquad \text{Equate constant terms}$$

Because $B = 0$ and $B + C = -2$, we see that $C = -2$. Thus the partial fraction decomposition of $\dfrac{3x^2 - 2x + 13}{(x + 1)(x^2 + 5)}$ is

$$\frac{3}{x + 1} - \frac{2}{x^2 + 5}.$$

✔️ *Check It Out 4:* Compute the partial fraction decomposition of $\dfrac{x^2 + 3x - 8}{(x^2 + 1)(x - 3)}$. ▪

Case 4: Denominator Has At Least One Irreducible, Repeated Quadratic Factor

Let $r(x) = \dfrac{P(x)}{Q(x)}$, where $P(x)$ and $Q(x)$ are polynomials and the degree of $P(x)$ is less than the degree of $Q(x)$. For every repeated irreducible quadratic factor of $Q(x)$, there exist real numbers a, b, and c and a positive integer $m \geq 2$ such that

▶ $(ax^2 + bx + c)^m$ is a factor of $Q(x)$,

▶ $(ax^2 + bx + c)^{m+1}$ is not a factor of $Q(x)$, and

▶ $ax^2 + bx + c$ cannot be factored over the real numbers.

The portion of the partial fraction decomposition of $r(x)$ that pertains to $(ax^2 + bx + c)^m$ is a sum of m fractions of the following form:

$$\frac{A_1 x + B_1}{ax^2 + bx + c} + \frac{A_2 x + B_2}{(ax^2 + bx + c)^2} + \cdots + \frac{A_m x + B_m}{(ax^2 + bx + c)^m}$$

where $A_1, \ldots, A_m$ and $B_1, \ldots, B_m$ are real numbers. Each of the linear factors of $Q(x)$ is treated as specified in Case 1 or Case 2, as appropriate. The nonrepeated irreducible quadratic factors of $Q(x)$ are treated as specified in Case 3.

Example 5 Repeated Irreducible Quadratic Factors

Find the partial fraction decomposition of $\dfrac{-2x^3 + x^2 - 6x + 7}{(x^2 + 3)^2}$.

▶**Solution** Because $x^2 + 3$ is a repeated irreducible quadratic factor, we have

$$\frac{-2x^3 + x^2 - 6x + 7}{(x^2 + 3)^2} = \frac{Ax + B}{x^2 + 3} + \frac{Cx + D}{(x^2 + 3)^2}.$$

Multiplying both sides of this equation by the LCD, $(x^2 + 3)^2$, we obtain the following.

$$-2x^3 + x^2 - 6x + 7 = (Ax + B)(x^2 + 3) + Cx + D$$

$$-2x^3 + x^2 - 6x + 7 = Ax^3 + Bx^2 + 3Ax + 3B + Cx + D$$

$$-2x^3 + x^2 - 6x + 7 = Ax^3 + Bx^2 + (3A + C)x + 3B + D$$

Equating coefficients,

$-2 = A$	Equate coefficients of x^3
$1 = B$	Equate coefficients of x^2
$-6 = 3A + C$	Equate coefficients of x
$7 = 3B + D$	Equate constant terms

Substituting $A = -2$ into the third equation, we get

$$-6 = 3(-2) + C \Longrightarrow -6 = -6 + C \Longrightarrow 0 = C.$$

Substituting $B = 1$ into the fourth equation, we get

$$7 = 3(1) + D \Longrightarrow 7 = 3 + D \Longrightarrow 4 = D.$$

Thus the partial fraction decomposition is

$$\frac{-2x^3 + x^2 - 6x + 7}{(x^2 + 3)^2} = \frac{-2x + 1}{x^2 + 3} + \frac{4}{(x^2 + 3)^2}.$$

✔ *Check It Out 5:* Compute the partial fraction decomposition of $\dfrac{-2x^2 + 1}{(x^2 + 1)^2}$. ■

6.7 Key Points

▶ The technique used to decompose a rational expression into simpler expression is known as **partial fraction decomposition.**

▶ There are four cases of partial fraction decomposition, each of which has a distinct type of decomposition.

1. The denominator can be factored over the real numbers as a product of distinct linear factors.

2. The denominator can be factored over the real numbers as a product of linear factors, at least one of which is a repeated linear factor.

3. The denominator can be factored over the real numbers as a product of distinct irreducible quadratic factors and possibly one or more linear factors.

4. The denominator can be factored over the real numbers as a product of irreducible quadratic factors, at least one of which is a repeated quadratic factor, and possibly one or more linear factors.

6.7 Exercises

▶**Skills** This set of exercises will reinforce the skills illustrated in this section.

In Exercises 1–10, write just the form of the partial fraction decomposition. Do not solve for the constants.

1. $\dfrac{3}{x^2 - x - 3}$

2. $\dfrac{-1}{x^2 - 3x}$

3. $\dfrac{4x}{(x + 5)^2}$

4. $\dfrac{-2}{x^2(x - 1)}$

5. $\dfrac{x + 3}{(x^2 + 2)(2x + 1)}$

6. $\dfrac{-6x + 7}{(x^2 + x + 1)(x + 5)}$

7. $\dfrac{3x - 1}{x^4 - 16}$

8. $\dfrac{2x^2 - 5}{x^4 - 1}$

9. $\dfrac{x + 6}{3x^3 + 6x^2 + 3x}$

10. $\dfrac{3x - 2}{2x^4 + 4x^3 + 2x^2}$

In Exercises 11–16, determine whether the quadratic expression is reducible.

11. $x^2 + 5$

12. $x^2 - 9$

13. $x^2 + x + 1$

14. $2x^2 + 2x + 3$

15. $x^2 + 4x + 4$

16. $x^2 + 6x + 9$

In Exercises 17–40, write the partial fraction decomposition of each rational expression.

17. $\dfrac{8}{x^2 - 16}$

18. $\dfrac{-4}{x^2 - 4}$

19. $\dfrac{2}{2x^2 - x}$

20. $\dfrac{3}{x^2 + 3x + 2}$

21. $\dfrac{x}{x^2 + 5x + 6}$

22. $\dfrac{5x - 7}{x^2 - 4x - 5}$

23. $\dfrac{-3x^2 + 2 - 3x}{x^3 - x}$

24. $\dfrac{x^2 + 4}{x^3 - 4x}$

25. $\dfrac{-2x + 6}{x^2 - 2x + 1}$

26. $\dfrac{x - 1}{x^2 + 4x + 4}$

27. $\dfrac{-x^2 + 2x + 4}{x^3 + 2x^2}$

28. $\dfrac{-x^2 + 3x - 9}{x^3 - 3x^2}$

29. $\dfrac{-2x^2 - 3x - 4}{(x - 1)(x + 2)^2}$

30. $\dfrac{4x^2 - 5x - 5}{(x - 3)(x + 1)^2}$

31. $\dfrac{4x + 1}{(x + 2)(x^2 + 3)}$

32. $\dfrac{x^2 + 3}{(x - 1)(x^2 + 1)}$

33. $\dfrac{-3x + 3}{(x + 2)(x^2 + x + 1)}$

34. $\dfrac{4x + 4}{(x - 1)(x^2 + x + 1)}$

35. $\dfrac{x^2 + 3}{x^4 - 1}$

36. $\dfrac{6x^2 - 8x + 24}{x^4 - 16}$

37. $\dfrac{-x^2 - 2x - 2}{(x^2 + 2)^2}$

38. $\dfrac{2x^2 - x + 2}{(x^2 + 1)^2}$

39. $\dfrac{x^3 - 3x^2 - x - 3}{x^4 - 1}$

40. $\dfrac{-2x^3 + x^2 + 8x + 4}{x^4 - 16}$

▶**Applications** In this set of exercises, you will use partial fractions to study real-world problems.

41. **Environment** The concentration of a pollutant in a lake t hours after it has been dumped there is given by

$$C(t) = \frac{t^2}{t^3 + 125}, \ t \geq 0.$$

Chemists originally constructed the formula for $C(t)$ by modeling the concentration of the pollutant as a sum of at least two rational functions, where each term in the sum represents a different chemical process. Determine the individual terms in that sum.

Engineering *In engineering applications, partial fraction decomposition is used to compute the Laplace transform. The independent variable is usually s. Compute the partial fraction decomposition of each of the following expressions.*

42. $\dfrac{1}{s(s + 1)}$

43. $\dfrac{2}{s(s^2 + 1)}$

▶**Concepts** This set of exercises will draw on the ideas presented in this section and your general math background.

44. What is wrong with the following decomposition?

$$\frac{x}{x^2(x - 1)^2} = \frac{A}{x^2} + \frac{B}{(x - 1)^2}$$

45. Explain why the following decomposition is incorrect.

$$\frac{1}{x(x^2 + 2x - 3)} = \frac{A}{x} + \frac{Bx + C}{x^2 + 2x - 3}$$

46. Find the partial fraction decomposition of $\dfrac{1}{x(x^2 + a^2)}$.

47. Find the partial fraction decomposition of $\dfrac{1}{(x - c)^2}$.

6.8 Systems of Nonlinear Equations

Objectives

▶ Solve systems of nonlinear equations by substitution

▶ Solve systems of nonlinear equations by elimination

▶ Solve systems of nonlinear equations using technology

▶ Use systems of nonlinear equations to solve applied problems

A system of equations in which one or more of the equations is not linear is called a **system of nonlinear equations.** Unlike systems of linear equations, there are no systematic procedures for solving *all* nonlinear systems. In this section, we give strategies for solving some simple nonlinear systems in just two variables. However, many nonlinear systems cannot be solved algebraically and must be solved using technology, as shown in Example 3.

We first discuss the substitution method, which works well when one of the equations in the system is quadratic and the other equation is linear.

Solving Nonlinear Systems by Substitution

We now outline how to solve a system of nonlinear equations using substitution.

> **Substitution Strategy for Solving a Nonlinear System**
>
> **Step 1** In one of the equations, solve for one variable in terms of the other.
> **Step 2** Substitute the resulting expression into the other equation to get one equation in one variable.
> **Step 3** Solve the resulting equation for that variable.
> **Step 4** Substitute each solution into either of the original equations and solve for the other variable.
> **Step 5** Check your solution.

Example **1** **Solving a Nonlinear System Using Substitution**

Solve the following system.

$$x^2 + 4y^2 = 25$$
$$x - 2y + 1 = 0$$

▶**Solution** Take the simpler second equation and solve it for x in terms of y. Then substitute the resulting expression for x into the first equation.

$x = 2y - 1$	Solve second equation for x in terms of y
$(2y - 1)^2 + 4y^2 = 25$	Substitute for x in first equation
$4y^2 - 4y + 1 + 4y^2 = 25$	Expand left-hand side
$8y^2 - 4y - 24 = 0$	Collect like terms
$4(2y^2 - y - 6) = 0$	Factor out 4
$2y^2 - y - 6 = 0$	Divide by 4
$(2y + 3)(y - 2) = 0$	Factor

Setting each factor equal to 0 and solving, we have

$$2y + 3 = 0 \implies y = -\frac{3}{2}; \quad y - 2 = 0 \implies y = 2.$$

Next, substitute each of these values of y into the second equation in the original system. You will obtain a value of x for each value of y.

$$x - 2\left(-\frac{3}{2}\right) + 1 = 0 \qquad \text{Substitute } y = -\frac{3}{2}$$

$$x = -4 \qquad \text{Solve for } x$$

Thus the first solution is $(x, y) = \left(-4, -\frac{3}{2}\right)$. Next, substitute $y = 2$ into the second equation; this gives $x = 2(2) - 1 = 3$. So the second solution is $(3, 2)$. The two solutions are $(3, 2)$ and $\left(-4, -\frac{3}{2}\right)$. You should check that both solutions satisfy the system of equations.

✔ *Check It Out 1:* Solve the following system.

$$x^2 + y^2 = 2$$

$$y = -x \ ▪$$

We next introduce an elimination method that is useful when both of the equations in a system are quadratic.

Solving Nonlinear Systems by Elimination

Next we outline how to solve a system of nonlinear equations using elimination.

Elimination Strategy for Solving a Nonlinear System

Step 1 Eliminate one variable from both equations by suitably combining the equations.

Step 2 Solve one of the resulting equations for the variable that was retained.

Step 3 Substitute each solution into either of the original equations and solve for the variable that was eliminated.

Step 4 Check your solution.

Example *2* **Solving a Nonlinear System by Elimination**

Solve the following system of equations.

$$x^2 + y^2 = 4 \qquad (1)$$

$$y = -2x^2 + 2 \qquad (2)$$

▶**Solution** First note that substituting the expression for y into the first equation will give a polynomial equation that has an x^4 term, which is difficult to solve. Instead, we use the elimination technique.

Rewrite the system with the variables on one side and the constants on the other.

$$x^2 + y^2 = 4 \qquad (1)$$

$$2x^2 + y = 2 \qquad (2)$$

Eliminate the x^2 term from Equation (2) by multiplying Equation (1) by -2 and adding the result to Equation (2). This gives the following.

$$-2x^2 - 2y^2 = -8 \qquad -2 \cdot (1)$$

$$\underline{\quad 2x^2 + y = \quad 2 \qquad (2)}$$

$$-2y^2 + y = -6$$

The last equation is a quadratic equation in y, which can be solved as follows.

$$-2y^2 + y = -6$$

$$-2y^2 + y + 6 = 0 \qquad \text{Add 6 to both sides}$$

$$2y^2 - y - 6 = 0 \qquad \text{Multiply by } -1 \text{ to write in standard form}$$

$$(2y + 3)(y - 2) = 0 \qquad \text{Factor}$$

Setting each factor equal to 0 and solving, we have

$$2y + 3 = 0 \Longrightarrow y = -\frac{3}{2}; \quad y - 2 = 0 \Longrightarrow y = 2.$$

Substitute $y = -\frac{3}{2}$ into Equation (2) and solve for x.

$$y = -2x^2 + 2$$

$$-\frac{3}{2} = -2x^2 + 2 \qquad \text{Substitute } y = -\frac{3}{2}$$

$$-\frac{7}{2} = -2x^2 \qquad \text{Subtract 2 from each side}$$

$$\frac{7}{4} = x^2 \qquad \text{Divide by } -2$$

$$\pm\frac{\sqrt{7}}{2} = x \qquad \text{Take square root}$$

Thus we have two solutions corresponding to $y = -\frac{3}{2}$: $\left(\frac{\sqrt{7}}{2}, -\frac{3}{2}\right)$ and $\left(-\frac{\sqrt{7}}{2}, -\frac{3}{2}\right)$.

Substituting $y = 2$ into Equation (2) and solving for x gives

$$y = -2x^2 + 2 \Longrightarrow 2 = -2x^2 + 2 \Longrightarrow x = 0.$$

Thus we have $(0, 2)$ as another solution. You should check that these solutions satisfy the original system of equations.

The solutions are $\left(\frac{\sqrt{7}}{2}, -\frac{3}{2}\right)$, $\left(-\frac{\sqrt{7}}{2}, -\frac{3}{2}\right)$, and $(0, 2)$.

 Check It Out 2: Solve the following system.

$$x^2 + y^2 = 4 \qquad (1)$$

$$y = x^2 - 1 \qquad (2) \ ■$$

Solving Nonlinear Systems Using Technology

The next example shows how graphing technology can be used to find the solution of a system of equations that is not easy to solve algebraically.

Example **3** **Using a Graphing Utility to Solve a Nonlinear System of Equations**

Solve the following system of equations.

$$xy = 1 \qquad (1)$$

$$y = -x^2 + 3 \qquad (2)$$

▶**Solution** In Equation (1), solve for y in terms of x, which gives $y = \frac{1}{x}$. This is a rational function with a vertical asymptote at $x = 0$. The graph of the function from Equation (2) is a parabola that opens downward. Enter these functions as Y_1 and Y_2 in a graphing utility, and use a decimal window to display the graphs. The graphs are shown in Figure 6.8.1.

Figure 6.8.1

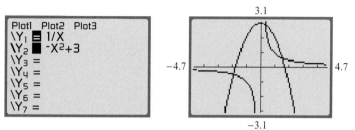

The solution of the system of equations is given by the intersection of the two graphs. From Figure 6.8.1, there are three points of intersection. By zooming out, we can make sure that there are no others. Using the INTERSECT feature, we find the intersection points, one by one. The results are shown in Figure 6.8.2. The approximate solutions are $(0.3473, 2.879)$, $(1.532, 0.6527)$, and $(-1.879, -0.5321)$.

Figure 6.8.2

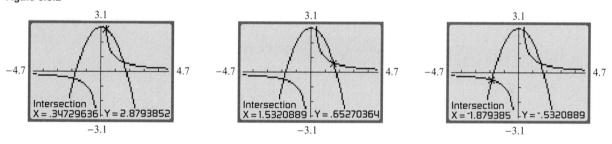

Note that if we had tried to solve this simple system by substitution, we would have obtained

$$\frac{1}{x} = -x^2 + 3.$$

Multiplying by x on both sides and rearranging terms, we obtain the cubic equation $-x^3 + 3x - 1 = 0$. This equation cannot be solved by any of the algebraic techniques discussed in this text. Hence, even some simple-looking nonlinear systems of equations can be difficult or even impossible to solve algebraically.

✔ *Check It Out 3:* Solve the following system of equations.

$$xy = 1 \qquad (1)$$
$$y = x^2 - 5 \qquad (2) \; ■$$

Application of Nonlinear Systems

We next examine a geometry application that involves a nonlinear system of equations.

| Example | 4 **An Application of Geometry** |

A right triangle with a hypotenuse of $2\sqrt{15}$ inches has an area of 15 square inches. Find the lengths of the other two sides of the triangle. See Figure 6.8.3.

Figure 6.8.3

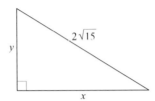

▶Solution Referring to Figure 6.8.3 and using the given information, we can write the following system of equations.

$$\frac{1}{2}xy = 15 \qquad\qquad \text{Area of triangle} \qquad (1)$$

$$x^2 + y^2 = \left(2\sqrt{15}\right)^2 = 60 \qquad \text{Pythagorean Theorem} \qquad (2)$$

We can solve this nonlinear system of equations by substitution. In Equation (1), solve for y.

$$\frac{1}{2}xy = 15 \Longrightarrow xy = 30 \Longrightarrow y = \frac{30}{x}$$

Substituting $y = \frac{30}{x}$ into Equation (2), we have

$$x^2 + y^2 = 60 \qquad \text{Equation (2)}$$

$$x^2 + \left(\frac{30}{x}\right)^2 = 60 \qquad \text{Substitute } y = \frac{30}{x}$$

$$x^2 + \frac{900}{x^2} = 60 \qquad \text{Simplify}$$

$$x^4 + 900 = 60x^2 \qquad \text{Multiply each side by } x^2$$

$$x^4 - 60x^2 + 900 = 0 \qquad \text{Move } x^2 \text{ term to left-hand side}$$

$$(x^2 - 30)(x^2 - 30) = 0. \qquad \text{Factor}$$

Setting the factor $x^2 - 30$ equal to 0 and solving for x gives

$$x^2 - 30 = 0 \Longrightarrow x = \pm\sqrt{30}.$$

We ignore the negative solution for x, since lengths must be positive. Thus $x = \sqrt{30}$. Using $x^2 + y^2 = 60$, we find that

$$y^2 = 60 - x^2 = 60 - 30 = 30.$$

Therefore, $y = \sqrt{30}$, and so the other two sides of the triangle are both of length $\sqrt{30}$ inches.

☑ *Check It Out 4:* A right triangle with a hypotenuse of $2\sqrt{13}$ inches has an area of 12 square inches. Find the lengths of the other two sides of the triangle. ■

6.8 Key Points

▶ A system of equations in which one or more of the equations is not linear is called a **system of nonlinear equations.**

▶ Systems of nonlinear equations can be solved using **substitution** or **elimination.**

▶ Systems of nonlinear equations can also be solved using graphing utilities.

6.8 Exercises

▶**Skills** This set of exercises will reinforce the skills illustrated in this section.

In Exercises 1–34, find all real solutions of the system of equations. If no real solution exists, so state.

1. $\begin{cases} x^2 + y^2 = 13 \\ y = x + 1 \end{cases}$

2. $\begin{cases} x^2 + y^2 = 10 \\ y = x + 2 \end{cases}$

3. $\begin{cases} 5x^2 + y^2 = 9 \\ y = 2x \end{cases}$

4. $\begin{cases} x^2 + 4y^2 = 16 \\ y = \dfrac{1}{2}x \end{cases}$

5. $\begin{cases} 2x^2 - y = 1 \\ y = 5x + 2 \end{cases}$

6. $\begin{cases} x^2 - y = 3 \\ y = 3x + 7 \end{cases}$

7. $\begin{cases} 9x^2 + 4y = 4 \\ 3x + 4y = -2 \end{cases}$

8. $\begin{cases} x^2 + y = 4 \\ 2x + y = 1 \end{cases}$

9. $\begin{cases} 3x^2 - 10y = 5 \\ x - y = -2 \end{cases}$

10. $\begin{cases} 2x^2 - 3y = 2 \\ x - 2y = -2 \end{cases}$

11. $\begin{cases} x^2 + 2y = -2 \\ -2x + y = 1 \end{cases}$

12. $\begin{cases} 4x^2 + y = 2 \\ -4x + y = 3 \end{cases}$

13. $\begin{cases} x^2 + y^2 = 8 \\ xy = -4 \end{cases}$

14. $\begin{cases} x^2 + y^2 = 61 \\ xy = 30 \end{cases}$

15. $\begin{cases} x^2 + y^2 = 9 \\ 2x^2 + y = 15 \end{cases}$

16. $\begin{cases} x^2 + y^2 = 16 \\ -3x + y^2 = 6 \end{cases}$

17. $\begin{cases} 5x^2 - 2y^2 = 10 \\ 3x^2 + 4y^2 = 6 \end{cases}$

18. $\begin{cases} 2x^2 + 3y^2 = 4 \\ 6x^2 + 5y^2 = -8 \end{cases}$

19. $\begin{cases} x^2 - y^2 + 2y = 1 \\ 5x^2 - 3y^2 = 17 \end{cases}$

20. $\begin{cases} x^2 - y^2 + 2x = 4 \\ 7x^2 - 5y^2 = 8 \end{cases}$

21. $\begin{cases} x^2 + y^2 = 4 \\ x^2 + 4y^2 = 1 \end{cases}$

22. $\begin{cases} x^2 + y^2 = 6 \\ y = 4 \end{cases}$

23. $\begin{cases} 2x^2 - 5x - 4y^2 = -4 \\ x^2 - 3y^2 = 4 \end{cases}$

24. $\begin{cases} x^2 + y^2 + 4y = 1 \\ x^2 - y^2 = 3 \end{cases}$

25. $\begin{cases} x^2 + (y-1)^2 = 9 \\ x^2 + \quad\; y^2 = 4 \end{cases}$

26. $\begin{cases} (x+2)^2 + y^2 = 6 \\ \quad\; x^2 + y^2 = -2 \end{cases}$

27. $\begin{cases} x^2 - (y-3)^2 = 7 \\ x^2 + \quad\; y^2 = 16 \end{cases}$

28. $\begin{cases} (x+1)^2 - y^2 = -9 \\ \quad\; x^2 - y^2 = -16 \end{cases}$

29. $\begin{cases} x^2 + 3xy - 2x = -10 \\ \quad\; 2xy + x = -14 \end{cases}$
    (*Hint:* Eliminate the xy term.)

30. $\begin{cases} 3x^2 + 2xy + x = 8 \\ \quad\quad\; xy + x = 3 \end{cases}$
    (*Hint:* Eliminate the xy term.)

31. $\begin{cases} 2^{3x^2 - y^2} = 4 \\ x^2 + y^2 = 14 \end{cases}$

32. $\begin{cases} x + y^2 = 25 \\ e^{3xy} = 1 \end{cases}$

33. $\begin{cases} \log_{10}(2y^2) + \log_{10}(x^3) = 3 \\ \quad\quad\quad\quad\quad\quad\; xy = 5 \end{cases}$

34. $\begin{cases} \log_3(y^2) - \log_3 x = 2 \\ \quad\; 3x - 2y \quad\;\; = 3 \end{cases}$

In Exercises 35–40, graph both equations by hand and find their point(s) of intersection, if any.

35. $\begin{cases} x^2 - 9y = -18 \\ -2x + 3y = 3 \end{cases}$

36. $\begin{cases} (x+2)^2 + 4y = 17 \\ \quad\; 3x + y = 1 \end{cases}$

37. $\begin{cases} (x-5)^2 + (y+3)^2 = 8 \\ \quad\quad\; x + y \quad\quad = 2 \end{cases}$

38. $\begin{cases} x^2 + y^2 - 8y = -15 \\ 2x - y \quad\quad\;\; = -6 \end{cases}$

39. $\begin{cases} (x-3)^2 \quad\quad + y^2 + 2y = 7 \\ \quad\; x^2 - 6x + y^2 + 10y = -26 \end{cases}$

40. $\begin{cases} x^2 + 2x + \quad\quad y^2 + 12y = -33 \\ x^2 + 8x + (y+6)^2 \quad\quad = 9 \end{cases}$

In Exercises 41–48, solve the system using a graphing utility. Round all values to three decimal places.

41. $\begin{cases} y = x^3 - 2x - 1 \\ y = 3x^2 - 2 \end{cases}$

42. $\begin{cases} y = x^3 - 3x + 2 \\ y = 4x^2 - 1 \end{cases}$

43. $\begin{cases} y = -x^2 + 3 \\ y = 3^x \end{cases}$

44. $\begin{cases} y = -x^2 + 2 \\ y = 2^x \end{cases}$

45. $\begin{cases} \quad\;\; 2xy = 8 \\ x^2 + y^2 = 3 \end{cases}$

46. $\begin{cases} xy = 1 \\ \; y = x^2 - 7 \end{cases}$

47. $\begin{cases} 5x^2 - y = 10 \\ 9x^2 + y^2 = 25 \end{cases}$

48. $\begin{cases} 2x^2 - y = 2 \\ 4x^2 + y^2 = 16 \end{cases}$

▶ **Applications** In this set of exercises, you will use non-linear systems of equations to study real-world problems.

49. **Design** The perimeter of a rectangular garden is 80 feet and the area it encloses is 336 square feet. Find the length and width of the garden.

50. **Design** The perimeter of a rectangular garden is 54 feet, and its area is 180 square feet. Find the length and width of the garden.

51. **Geometry** A right triangle with a hypotenuse of $\sqrt{89}$ has an area of 20 square inches. Find the lengths of the other two sides of the triangle.

52. **Geometry** A right triangle with a hypotenuse of $2\sqrt{10}$ inches has an area of 6 square inches. Find the lengths of the other two sides of the triangle.

53. **Manufacturing** A manufacturer wants to make a can in the shape of a right circular cylinder with a volume of 6.75π cubic inches and a lateral surface area of 9π square inches. The lateral surface area includes only the area of the curved surface of the can, not the areas of the flat (top and bottom) surfaces. Find the radius and height of the can.

54. **Manufacturing** A manufacturer wants to make a can in the shape of a right circular cylinder with a volume of 45π cubic inches and a lateral surface area of 30π square inches. The lateral surface area includes only the area of the curved surface of the can, not the area of the flat (top and bottom) surfaces. Find the radius and height of the can.

55. **Geometry** The volume of a paper party hat, shaped in the form of a right circular cone, is 36π cubic inches. If the radius of the cone is one-fourth the height of the cone, find the radius and the height.

56. **Geometry** The volume of a super-size ice cream cone, shaped in the form of a right circular cone, is 8π cubic inches. If the radius of the cone is one-third the height of the cone, find the radius and the height of the cone.

57. **Number Theory** The sum of the squares of two positive integers is 85. If the squares of the integers differ by 13, find the integers.

58. **Number Theory** The sum of the squares of two positive integers is 74. If the squares of the integers differ by 24, find the integers.

59. **Landscaping** The area of a rectangular property is 300 square feet. Its length is three times its width. There is a rectangular swimming pool centered within the property. The dimensions of the property are twice the corresponding dimensions of the pool. The portion of the property that lies outside the pool is paved with concrete. What are the dimensions of the property and of the pool? What is the area of the paved portion?

60. **Landscaping** The area of a rectangular property is 1800 square feet; its length is twice its width. There is a rectangular swimming pool centered within the property. The dimensions of the property are one and one-third times the corresponding dimensions of the pool. The portion of the property that lies outside the pool is paved with concrete. What are the dimensions of the property and of the pool? What is the area of the paved portion?

▶ **Concepts** This set of exercises will draw on the ideas presented in this section and your general math background.

61. For what value(s) of b does the following system of equations have two distinct, real solutions?
$$\begin{cases} y = -x^2 + 2 \\ y = x + b \end{cases}$$

62. Consider the following system of equations.
$$\begin{cases} y = x^2 + 6x - 4 \\ y = b \end{cases}$$
For what value(s) of b do the graphs of the equations in this system have
(a) exactly one point of intersection?
(b) exactly two points of intersection?
(c) no point of intersection?

63. Explain why the following system of equations has no solution.
$$\begin{cases} (x + y)^2 = 36 \\ xy = 18 \end{cases}$$
(*Hint:* Expand the expression $(x + y)^2$.)

64. Give a graphical explanation of why the following system of equations has no real solutions.
$$\begin{cases} x^2 + y^2 = 1 \\ y = x^2 + 3 \end{cases}$$

65. Consider the following system of equations.
$$\begin{cases} x^2 + y^2 = r^2 \\ (x - h)^2 + y^2 = r^2 \end{cases}$$
Let r be a (fixed) positive number. For what value(s) of h does this system have
(a) exactly one real solution?
(b) exactly two real solutions?
(c) infinitely many real solutions?
(d) no real solution?
(*Hint:* Visualize the graphs of the two equations.)

Chapter 6   # Summary

Section 6.1 **Systems of Linear Equations and Inequalities in Two Variables**

Concept	Illustration	Study and Review
Solving a system by elimination Eliminate one of the variables from the two equations to get one equation in one variable. Solve this equation and substitute the result into one of the original equations. The second variable is solved for by substitution.	Given the following system of equations $$\begin{cases} x + y = 0 \\ x - 2y = 3 \end{cases}$$ eliminate y from both equations by multiplying the first equation by 2 and adding it to the second equation. This gives $3x = 3 \Longrightarrow x = 1$. Substituting $x = 1$ into the first equation, we obtain $y = -1$. The solution is $(1, -1)$.	Examples 1–5 Chapter 6 Review, Exercises 1–4
Solutions of a linear inequality An inequality is a **linear inequality** if it can be written in the form $Ax + By \leq C$ or $Ax + By < C$, where A, B, and C are real numbers, not all zero. (The symbols $\leq$ and $<$ may be replaced with $\geq$ or $>$.) The **solution set** of a linear inequality is the set of points (x, y) that satisfy the inequality.	The following is the graph of the inequality $y < 2x$. 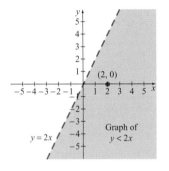	Example 6 Chapter 6 Review, Exercises 5–8
Solutions of a system of linear inequalities The **solution set of a system of linear inequalities** is the set of points (x, y) consisting of the intersection of the solution sets of the individual inequalities. To find the solution set of the system, graph the solution set of each inequality and find the intersection of the shaded regions.	The following is the graph of the system of inequalities $$\begin{cases} x + 3y \leq 6 \\ -x + y \geq -2 \end{cases}.$$ 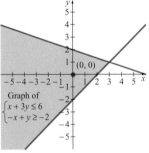	Example 7 Chapter 6 Review, Exercises 9, 10
Fundamental theorem of linear programming The **fundamental theorem of linear programming** states that the maximum or minimum of a linear objective function occurs at a corner point of the boundary of the solution set of a system of linear inequalities (the feasible set).	The maximum of the function $C = 3x + 5y$ over a feasible set with corner points $(1, 2)$, $(3, 4)$, and $(0, 0)$ occurs at $(3, 4)$. The value of C is 29 at $(3, 4)$.	Examples 8, 9 Chapter 6 Review, Exercises 11, 12

Section 6.2 Systems of Linear Equations in Three Variables

Concept	Illustration	Study and Review
Gaussian elimination Gaussian elimination is used to solve a system of linear equations in three or more variables. The following operations are used in the elimination process for solving systems of equations. **Operations for manipulating systems of equations** • Interchange two equations in the system. • Multiply one equation in the system by a nonzero constant. • Multiply one equation in the system by a nonzero constant and add the result to another equation in the system.	To solve $$\begin{cases} x \quad - z = 2 \\ \quad y + 2z = 1 \\ -2x \quad - z = -4 \end{cases}$$ multiply the first equation by 2 and add the result to the third equation to get $$\begin{cases} x \quad - z = 2 \\ \quad y + 2z = 1. \\ \quad - 3z = 0 \end{cases}$$ Solve the third equation for z to get $z = 0$. Substitute $z = 0$ into the second equation to get $y = 1$. Using the first equation, substitute $z = 0$ to get $x = 2$. The solution is $x = 2, y = 1, z = 0$.	Examples 1–6 Chapter 6 Review, Exercises 13–16

Section 6.3 Solving Systems of Equations Using Matrices

Concept	Illustration	Study and Review
Augmented matrix for a linear system In an **augmented matrix** corresponding to a system of equations, the rows of the left-hand side are formed by taking the coefficients of the variables in each equation, and the constants appear on the right-hand side.	The augmented matrix for the linear system $$\begin{cases} x \quad - z = 2 \\ \quad y + 2z = 1 \\ -2x \quad - z = -4 \end{cases}$$ is $$\begin{bmatrix} 1 & 0 & -1 & 2 \\ 0 & 1 & 2 & 1 \\ -2 & 0 & -1 & -4 \end{bmatrix}.$$	Discussion in Section 6.3 Chapter 6 Review, Exercises 17, 18
Gaussian elimination using matrices The following operations are used when performing Gaussian elimination on matrices. **Elementary row operations on matrices** • Interchange two rows of the matrix. • Multiply a row of the matrix by a nonzero constant. • Multiply a row of the matrix by a nonzero constant and add the result to another row. These operations are used to solve a system of linear equations in matrix form.	In the matrix above, multiplying the first row by 2 and adding the result to the third row gives $$\begin{bmatrix} 1 & 0 & -1 & 2 \\ 0 & 1 & 2 & 1 \\ 0 & 0 & -3 & 0 \end{bmatrix}.$$ The bottom row gives $z = 0$. Substituting into the second and first equations gives $y = 1$ and $x = 2$. The solution to the corresponding system of equations is $x = 2, y = 1, z = 0$.	Examples 1–4, 7 Chapter 6 Review, Exercises 19–22

Continued

Section 6.3 **Solving Systems of Equations Using Matrices**

Concept	Illustration	Study and Review
Gauss-Jordan elimination We can continue the elementary row operations in the previous example so that the first 1 in each row has a 0 above and below it.	In the preceding example, multiply the last row by $-\frac{1}{3}$. Then use row operations to eliminate the 2 and -1 directly above it. This gives $$\begin{bmatrix} 1 & 0 & 0 & 2 \\ 0 & 1 & 0 & 1 \\ 0 & 0 & 1 & 0 \end{bmatrix}$$ The solution can be read off the matrix as $x = 2, y = 1,$ and $z = 0.$	Examples 5, 6 Chapter 6 Review, Exercises 19–22

Section 6.4 **Operations on Matrices**

Concept	Illustration	Study and Review
Definition of a matrix A **matrix** is a rectangular array of real numbers with m rows and n columns. The **dimension** of such a matrix is $m \times n$. A matrix is usually referred to by a capital letter.	The following matrix A has dimensions 3×4. $$A = \begin{bmatrix} 4 & -2 & -3 & 1 \\ 0 & -1.5 & 9 & -10 \\ -3.7 & 15 & 0 & -1 \end{bmatrix}$$	Examples 1, 2 Chapter 6 Review, Exercises 25–28
Addition and subtraction of matrices Two matrices can be added or subtracted only if their dimensions are the same. The resulting matrix, which will be of the same size, is obtained by adding (subtracting) the corresponding elements of the two matrices.	If $B = \begin{bmatrix} -2 & 1 \\ 3 & -2 \\ 2 & 4 \end{bmatrix}$ and $C = \begin{bmatrix} 6 & 0 \\ 0 & 1 \\ 7 & 1 \end{bmatrix}$, $$B + C = \begin{bmatrix} 4 & 1 \\ 3 & -1 \\ 9 & 5 \end{bmatrix}.$$	Example 3 Chapter 6 Review, Exercises 29, 30
Scalar multiplication of matrices The product of an $m \times n$ matrix A and a scalar c is found by multiplying all the elements of A by c.	For the matrix B above, $2B = \begin{bmatrix} -4 & 2 \\ 6 & -4 \\ 4 & 8 \end{bmatrix}.$	Example 4 Chapter 6 Review, Exercises 31, 32
Matrix multiplication The product of an $m \times p$ matrix A and a $p \times n$ matrix B is a product matrix AB with m rows and n columns. Furthermore, the element in the ith row, jth column of the product matrix AB is given by the product of the ith row of A and the jth column of B.	The product of $A = \begin{bmatrix} -1 & 2 \\ 5 & 0 \\ 3 & 2 \end{bmatrix}$ and $B = \begin{bmatrix} -2 & 1 \\ 3 & -2 \end{bmatrix}$ is $$AB = \begin{bmatrix} 8 & -5 \\ -10 & 5 \\ 0 & -1 \end{bmatrix}.$$	Examples 5–10 Chapter 6 Review, Exercises 33, 34

Section 6.5 Matrices and Inverses

Concept	Illustration	Study and Review
Inverse of an $n \times n$ matrix The inverse of an $n \times n$ matrix A, denoted by A^{-1}, is another $n \times n$ matrix such that $$AA^{-1} = A^{-1}A = I$$ where I is the $n \times n$ identity matrix.	The inverse of $A = \begin{bmatrix} 2 & 1 \\ 5 & 3 \end{bmatrix}$ is $A^{-1} = \begin{bmatrix} 3 & -1 \\ -5 & 2 \end{bmatrix}$ because $$AA^{-1} = \begin{bmatrix} 1 & 0 \\ 0 & 1 \end{bmatrix}$$ the 2×2 identity matrix I. You can check that $A^{-1}A = I$.	Examples 1–4, 6 Chapter 6 Review, Exercises 35–40
Using inverses to solve systems If a system of equations can be expressed in the form $AX = B$, with A an $n \times n$ matrix and A^{-1} its inverse, then we have the following. $$AX = B \Longrightarrow A^{-1}AX = A^{-1}B \Longrightarrow$$ $$X = A^{-1}B$$	If $\begin{bmatrix} 2 & 1 \\ 5 & 3 \end{bmatrix}\begin{bmatrix} x \\ y \end{bmatrix} = \begin{bmatrix} 2 \\ 1 \end{bmatrix}$, then $\begin{bmatrix} x \\ y \end{bmatrix} = \begin{bmatrix} 3 & -1 \\ -5 & 2 \end{bmatrix}\begin{bmatrix} 2 \\ 1 \end{bmatrix} = \begin{bmatrix} 5 \\ -8 \end{bmatrix}$.	Example 5 Chapter 6 Review, Exercises 41, 42

Section 6.6 Determinants and Cramer's Rule

Concept	Illustration	Study and Review
Determinants The **determinant** of a 2×2 matrix $A = \begin{bmatrix} a & b \\ c & d \end{bmatrix}$ is defined as $$\lvert A \rvert = \begin{vmatrix} a & b \\ c & d \end{vmatrix} = ad - bc.$$ The determinants of larger matrices can be found by cofactor expansion. **Cramer's Rule** is a set of formulas for solving systems of linear equations with exactly one solution.	$\begin{vmatrix} 2 & -3 \\ 1 & 4 \end{vmatrix} = (2)(4) - (-3)(1) = 11$	Examples 1–5 Chapter 6 Review, Exercises 43–50
Minors and cofactors Let A be an $n \times n$ matrix, and let a_{ij} be the entry in the ith row, jth column of A. The **minor** of a_{ij} (denoted by M_{ij}) is the determinant of the matrix that is obtained by deleting the ith row and jth column of A. The **cofactor** of a_{ij} (denoted by C_{ij}) is defined as $C_{ij} = (-1)^{i+j}M_{ij}$.	Let $A = \begin{bmatrix} 1 & -2 & 0 \\ 3 & 2 & 1 \\ 0 & 1 & -1 \end{bmatrix}$. The minor of a_{12} is $M_{12} = \begin{vmatrix} 3 & 1 \\ 0 & -1 \end{vmatrix} = -3$. The cofactor of a_{12} is $C_{12} = (-1)^{1+2}M_{12} = (-1)(-3) = 3$.	Examples 2, 3, 6 Chapter 6 Review, Exercises 45, 46, 49, 50

Section 6.7 **Partial Fractions**

Concept	Illustration	Study and Review
Partial fraction decomposition Partial fraction decomposition is a technique used to decompose a rational expression into simpler expressions. There are four cases: 1. The denominator can be factored over the real numbers as a product of distinct linear factors. 2. The denominator can be factored over the real numbers as a product of linear factors, at least one of which is a repeated linear factor. 3. The denominator can be factored over the real numbers as a product of distinct irreducible quadratic factors and possibly one or more linear factors. 4. The denominator can be factored over the real numbers as a product of irreducible quadratic factors, at least one of which is a repeated quadratic factor, and possibly one or more linear factors.	The forms of the decompositions are as follows. 1. $\dfrac{2}{(x+1)(x-1)} = \dfrac{A}{x+1} + \dfrac{B}{x-1}$ 2. $\dfrac{2}{(x+1)^2(x-1)}$ $\quad = \dfrac{A}{x+1} + \dfrac{B}{(x+1)^2} + \dfrac{C}{x-1}$ 3. $\dfrac{2}{(x^2+1)(x+1)} = \dfrac{Ax+B}{x^2+1} + \dfrac{C}{x+1}$ 4. $\dfrac{2}{(x^2+1)^2} = \dfrac{Ax+B}{x^2+1} + \dfrac{Cx+D}{(x^2+1)^2}$	Examples 1–5 Chapter 6 Review, Exercises 51–54

Section 6.8 **Systems of Nonlinear Equations**

Concept	Illustration	Study and Review
Systems of nonlinear equations A system of equations in which one or more of the equations is not linear is called a **system of nonlinear equations.** Systems of nonlinear equations can be solved using **substitution** or **elimination.** Systems of nonlinear equations can also be solved using graphing utilities.	To solve $$\begin{aligned} x^2 - y &= 6 \\ y &= x \end{aligned}$$ substitute $y = x$ into the first equation and solve $x^2 - x = 6$. Use factoring to solve for x: $x = 3$ or $x = -2$. The solutions are $(x, y) = (3, 3)$ and $(x, y) = (-2, -2)$. Both solutions check in the original system of equations.	Examples 1–4 Chapter 6 Review, Exercises 55–58

Chapter 6   Review Exercises

Section 6.1 _____

In Exercises 1–4, solve each system of equations by elimination and check your solution.

1. $\begin{cases} -x - y = -7 \\ 3x + 4y = 24 \end{cases}$

2. $\begin{cases} -3x + 4y = 9 \\ 6x - 8y = 3 \end{cases}$

3. $\begin{cases} x + y = 5 \\ -2x - 2y = -10 \end{cases}$

4. $\begin{cases} 3x - 6y = 2 \\ y = -3 \end{cases}$

In Exercises 5–8, graph the solution set of each inequality.

5. $y > 3x - 7$

6. $x < -y + 4$

7. $2x - y \geq 1$

8. $x - 3y \leq 6$

In Exercises 9 and 10, graph the solution set of each system of inequalities.

9. $\begin{cases} x \leq 8 \\ 2x - y \geq 6 \end{cases}$

10. $\begin{cases} x \leq 5 \\ x \geq 2 \\ y > 1 \end{cases}$

In Exercises 11 and 12, solve the optimization problem.

11. Maximize $P = 5x + 7y$ subject to the following constraints.

$$\begin{cases} x \geq 1 \\ y \leq 2 \\ x \leq 7 \\ y \leq 10 \end{cases}$$

12. Maximize $P = 10x + 20y$ subject to the following constraints.

$$x + 3y \leq 12$$
$$y \leq x$$
$$y \geq 1$$
$$x \geq 0$$

Section 6.2

In Exercises 13–16, solve the system of linear equations using Gaussian elimination.

13. $\begin{cases} x - 2y + z = 2 \\ -x + 4y + 3z = -8 \\ -6y + 2z = 4 \end{cases}$

14. $\begin{cases} 4x - y = 2 \\ y - 2z = 1 \\ 6x - z = 3 \end{cases}$

15. $\begin{cases} x + y + 2z = 4 \\ -3x + 2y - z = 3 \end{cases}$

16. $\begin{cases} 2x + y - z = 2 \\ 5x - 2y + z = 3 \end{cases}$

Section 6.3

In Exercises 17 and 18, represent the system of equations in augmented matrix form. Do not solve the system.

17. $\begin{cases} 4x + y + z = 0 \\ -y + 2z = -1 \\ x + z = 3 \end{cases}$

18. $\begin{cases} x + 2y - 5z = 3 \\ 3x + z = -1 \end{cases}$

In Exercises 19–22, solve the system of equations using matrices and row operations. If there is no solution, so state.

19. $\begin{cases} -x - y = -10 \\ 3x + 4y = 24 \end{cases}$

20. $\begin{cases} x + 2y + z = -3 \\ 3x + y - 2z = 2 \\ 4x + 3y - z = 0 \end{cases}$

21. $\begin{cases} x + y - z = 0 \\ 3x + 2y - z = -1 \\ -2x + y - 2z = -1 \end{cases}$

22. $\begin{cases} -x + 4y + 3z = 8 \\ 2x - 8y - 4z = 3 \end{cases}$

In Exercises 23 and 24, for each augmented matrix, construct the corresponding system of linear equations. Use the variables listed above the matrix, in the given order. Determine whether the system is consistent or inconsistent. If it is consistent, give the solution(s).

23. $\begin{array}{ccc} x & y & z \end{array}$
$\left[\begin{array}{ccc|c} 1 & 0 & -2 & 3 \\ 0 & 1 & 1 & 5 \\ 0 & 0 & 0 & 0 \end{array}\right]$

24. $\begin{array}{ccc} x & y & z \end{array}$
$\left[\begin{array}{ccc|c} 1 & 0 & 0 & 7 \\ 0 & 1 & 0 & 3 \\ 0 & 0 & 1 & 2 \end{array}\right]$

Section 6.4

In Exercises 25–28, use the following matrix A.

$$A = \left[\begin{array}{cccc} 0 & -3 & 1 & 8 \\ -4.2 & -5 & 8 & 4 \\ 5 & 1.9 & \frac{4}{5} & \sqrt{2} \end{array}\right]$$

25. Find a_{11}.

26. Find a_{22}.

27. Determine the dimensions of A.

28. Find a_{31}.

In Exercises 29–34, let A, B, and C be as given. Perform the indicated operation(s), if defined. If not defined, state the reason.

$$A = \begin{bmatrix} 4 & -1 & \frac{1}{2} \\ 0 & 3 & -1 \end{bmatrix}, \quad B = \begin{bmatrix} -4 & 1 \\ 2 & -3 \\ 2 & -6 \end{bmatrix}, \quad C = \begin{bmatrix} 3 & 5 \\ 1 & -1 \\ 2 & -3 \end{bmatrix}$$

29. $B + C$

30. $C - B$

31. $2B + C$

32. $B - 3C$

33. AB

34. AC

Section 6.5

In Exercises 35–40, find the inverse of each matrix.

35. $\begin{bmatrix} 4 & 5 \\ 1 & 1 \end{bmatrix}$

36. $\begin{bmatrix} 5 & 3 \\ 3 & 2 \end{bmatrix}$

37. $\begin{bmatrix} -4 & 1 \\ -3 & 1 \end{bmatrix}$

38. $\begin{bmatrix} 4 & 3 \\ 5 & 4 \end{bmatrix}$

39. $\begin{bmatrix} 1 & -1 & 0 \\ -2 & 0 & 1 \\ -2 & 5 & -1 \end{bmatrix}$

40. $\begin{bmatrix} 4 & -2 & 1 \\ -2 & 1 & 2 \\ 1 & 2 & 4 \end{bmatrix}$

In Exercises 41 and 42, use inverses to solve the system of equations.

41. $\begin{cases} x + 2y = -4 \\ -x - y = 5 \end{cases}$

42. $\begin{cases} x - 3y + 2z = -1 \\ y + z = 4 \\ 2x - 6y + 3z = 3 \end{cases}$

Section 6.6

In Exercises 43–46, find the determinant of the matrix.

43. $A = \begin{bmatrix} 4 & 2 \\ -3 & 1 \end{bmatrix}$

44. $A = \begin{bmatrix} -2 & -3 \\ 1 & 5 \end{bmatrix}$

45. $A = \begin{bmatrix} -7 & 5 & 0 \\ 0 & 3 & 0 \\ -3 & -2 & 2 \end{bmatrix}$

46. $A = \begin{bmatrix} -2 & 3 & 5 \\ 6 & -1 & 0 \\ 0 & 1 & -2 \end{bmatrix}$

In Exercises 47–50, use Cramer's Rule to solve the system of equations.

47. $\begin{cases} -x - y = -2 \\ 2x + y = 0 \end{cases}$

48. $\begin{cases} -3x + 2y = 1 \\ -2x - y = 2 \end{cases}$

49. $\begin{cases} -2x + y + z = 0 \\ y + z = 4 \\ -3y + z = 1 \end{cases}$

50. $\begin{cases} x + 3y - z = 3 \\ x - z = 0 \\ x - y = 2 \end{cases}$

Section 6.7

In Exercises 51–54, write the partial fraction decomposition of each rational expression.

51. $\dfrac{-3x + 8}{x^2 + 5x + 6}$

52. $\dfrac{-2x^2 - x + 1}{x^3 - x^2}$

53. $-\dfrac{9}{(x + 2)(x^2 + 5)}$

54. $\dfrac{-x^3 - x + x^2 + 2}{x^4 + 4x^2 + 4}$

Section 6.8

In Exercises 55–58, find all real solutions of the system of equations. If no real solution exists, so state.

55. $\begin{cases} x^2 + y^2 = 9 \\ y = x + 1 \end{cases}$

56. $\begin{cases} x^2 - 2y^2 = 4 \\ 2x^2 + 5y^2 = 12 \end{cases}$

57. $\begin{cases} (x + 3)^2 + (y - 4)^2 = 25 \\ x^2 - 8x + (y + 3)^2 = 9 \end{cases}$

58. $\begin{cases} (x - 1)^2 + (y + 1)^2 = 0 \\ x^2 + y^2 + 2y = 0 \end{cases}$

Applications

59. Networks Telephone and computer networks transmit messages from one relay point to another. Suppose 450 messages are relayed to a single point on a network, and at that point the messages are distributed between two separate lines. The capacity of one line is 3.5 times the capacity of the other. Find the number of messages carried by each line.

60. Maximizing Revenue The cows on a dairy farm yield a total of 2400 gallons of milk per week. The farmer receives $1 per gallon for the portion that is sold as milk. The rest is used in the production of cheese, which the farmer sells for $5 per pound. It takes 1 gallon of milk to make a pound of cheese. If at least 25% of the milk is to be sold as milk, how many gallons of milk should go into the production of cheese in order to maximize the farmer's total revenue?

61. Nutrition The members of the Nutrition Club at Grand State University researched the lunch menu at their college cafeteria. They found that a meal consisting of a soft taco, a tostada, and a side dish of rice totaled 600 calories. Also, a meal consisting of two soft tacos and a tostada totaled 580 calories. Finally, a meal consisting of just a soft taco and rice totaled 400 calories. (*Source: www.tacobell.com*) Determine the numbers of calories in a soft taco, a tostada, and a side dish of rice.

62. Manufacturing Princess Clothing, Inc., has the following yardage requirements for making a single blazer, pair of trousers, or skirt.

	Fabric (yd)	Lining (yd)	Trim (yd)
Blazer	2.5	2	0.5
Trousers	3	2	1
Skirt	3	1	0

There are 42 yards of fabric, 25 yards of lining, and 7 yards of trim available. How many blazers, pairs of trousers, and skirts can be made, assuming that all the fabric, lining, and trim is used up?

63. Geometry The sum of the areas of two squares is 549 square inches. The length of a side of the larger square is 3 inches more than the length of a side of the smaller square. Find the length of a side of each square.

Chapter 6   Test

In Exercises 1 and 2, solve the system of equations by elimination and check your solution.

1. $\begin{cases} x + 2y = 4 \\ 4x - y = -11 \end{cases}$ 2. $\begin{cases} -2x + 3y = -8 \\ 5x - 2y = 9 \end{cases}$

3. Graph the solution set of the following system of inequalities.
$$\begin{cases} x \ge 1 \\ y \le 4 \\ 3x + 2y \ge 6 \end{cases}$$

4. Maximize $P = 15x + 10y$ subject to the following constraints.
$$\begin{aligned} 2x + y &\le 8 \\ y &\le 2x \\ y &\le 1 \\ y &\ge 0 \\ x &\ge 0 \end{aligned}$$

In Exercises 5 and 6, solve the system of linear equations by using Gaussian elimination.

5. $\begin{cases} 2x - y + z = -3 \\ -3x + z = 11 \\ x + 2y = -5 \end{cases}$

6. $\begin{cases} -x + 2y - z = 0 \\ 2x - y = 2 \end{cases}$

In Exercises 7 and 8, use matrices A, B, and C given below. Perform the indicated operations, if defined. If an operation is not defined, state the reason.

$$A = \begin{bmatrix} 5 & 0 & 3 \\ 2 & -4 & 3 \end{bmatrix}, \quad B = \begin{bmatrix} -3 & 4 \\ -1 & 2 \\ 7 & -3 \end{bmatrix}, \quad C = \begin{bmatrix} 8 & 3 \\ -6 & 5 \\ 5 & -2 \end{bmatrix}$$

7. $B - 2C$ 8. CA

In Exercises 9 and 10, find the inverse of the matrix.

9. $\begin{bmatrix} 3 & 5 \\ 1 & 2 \end{bmatrix}$

10. $\begin{bmatrix} 1 & 4 & 2 \\ -2 & 1 & 0 \\ 0 & 2 & 1 \end{bmatrix}$

11. Solve the following system of equations using an inverse.

$$\begin{cases} x + 3y - z = 0 \\ x + 4y + z = -2 \\ 2x + 6y - z = 1 \end{cases}$$

12. Find the determinant of $A = \begin{bmatrix} 2 & -3 & 1 \\ 0 & 5 & 2 \\ -4 & 2 & 0 \end{bmatrix}$.

In Exercises 13 and 14, use Cramer's Rule to solve the system of equations.

13. $\begin{cases} x + 3y = 6 \\ x + y = 2 \end{cases}$

14. $\begin{cases} x - 2y - z = 0 \\ -x - y = 3 \\ x + z = -1 \end{cases}$

15. Write the partial fraction decomposition of
$$\frac{-x^2 + 2x + 2}{x^3 + 2x^2 + x}.$$

16. Write the partial fraction decomposition of
$$\frac{5x^2 - 7x + 6}{(x^2 + 1)(x - 3)}.$$

In Exercises 17 and 18, find all real solutions of the system of equations. If no real solution exists, so state.

17. $\begin{cases} x^2 + y^2 = 4 \\ y - 3x = 0 \end{cases}$

18. $\begin{cases} (x + 1)^2 + (y - 1)^2 = 9 \\ x^2 + 2x + y^2 = 9 \end{cases}$

19. A computer manufacturer produces two types of laptops, the CX100 and the FX100. It takes 2 hours to manufacture each unit of the FX100 and only 1 hour to manufacture each unit of the CX100. At least 100 of the FX100 models are to be produced. At most 800 hours are to be allocated to the manufacture of the two models combined. If the company makes a profit of $100 on each FX100 model and a profit of $150 on each CX100 model, how many of each model should the company produce to maximize its profit?

20. Ten airline tickets were purchased for a total of $14,200. Each first-class ticket cost $2500. Business class and coach tickets cost $1500 and $300, respectively. Twice as many coach tickets were purchased as business class tickets. How many of each type of ticket were purchased?

21. A cell phone company charges $0.10 per minute during prime time and $0.05 per minute for non-prime time. The numbers of minutes used by three different customers are as follows.

	Prime-Time Minutes	Non-Prime-Time Minutes
Customer 1	200	400
Customer 2	300	200
Customer 3	150	300

Perform an appropriate matrix multiplication to find the total amount paid by each customer.

Conic Sections

A stronomical objects emit radiation, which is not detectable by optical telescopes. These objects are usually studied using a radio telescope, which consists of a dish with a parabolic cross-section. Because a parabola exhibits a reflective property, the incoming radio waves are reflected off the surface of the dish and focused onto a single point. Exercise 70 in Section 7.1 asks you to find the equation of a parabola for the cross-section of a reflecting telescope. This chapter will examine a class of equations known as the conic sections. The conic sections are widely used in engineering, astronomy, physics, architecture, and navigation because they possess reflective and other unique properties.

7.1 The Parabola

Objectives

▶ Define a parabola

▶ Find the focus, directrix, and axis of symmetry of a parabola

▶ Determine the equation of a parabola and write it in standard form

▶ Translate a parabola in the *xy*-plane

▶ Sketch a parabola

▶ Understand the reflective property of a parabola

This chapter covers a group of curves known as the *conic sections*. These curves are formed by intersecting a right circular cone with a plane. The circular cone consists of two halves, called **nappes,** which intersect in only one point, called the **vertex.** The conic sections are often abbreviated as simply the *conics.*

Figures 7.1.1 and 7.1.2 show how the conic sections are formed.

Figure 7.1.1

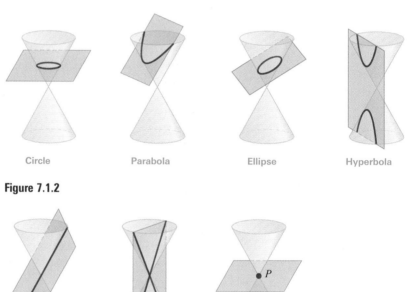

Circle   Parabola   Ellipse   Hyperbola

Figure 7.1.2

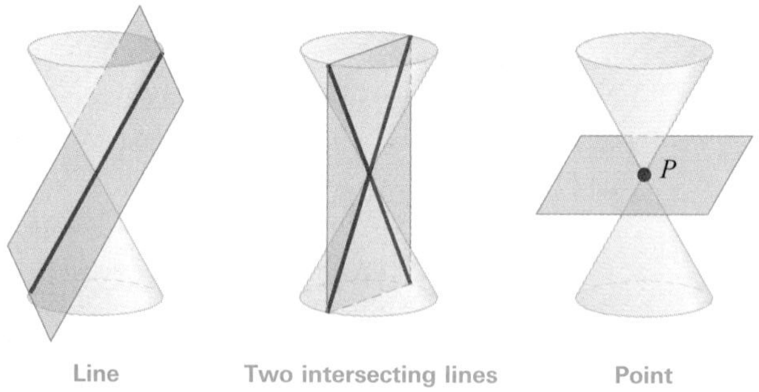

Line   Two intersecting lines   Point

Observations:

▶ If the plane is perpendicular to the axis of the cone but does not pass through the vertex, the conic is a **circle.**

▶ If the plane is tilted and intersects only one nappe of the cone, the conic is a single curve. If the curve is closed, the conic is an **ellipse;** otherwise, it is a **parabola.**

▶ If the plane is parallel to the axis of the cone and does not pass through the vertex, it intersects the cone in a pair of non-intersecting open curves. Each curve intersects just one nappe of the cone. Such a conic (the two curves taken together) is called a **hyperbola,** and each curve is called a **branch** of the hyperbola.

Conic sections are found in many applications. For example, planets travel in elliptic orbits, automobile headlights use parabolic reflectors, and some atomic particles follow hyperbolic paths. You have already studied equations of circles—and, to some extent, equations of parabolas—in earlier parts of this textbook. This chapter will treat the parabola in greater detail, and it will cover the ellipse and the hyperbola as well. The conics in Figure 7.1.2—a point, a line, and a pair of intersecting lines—are called **degenerate forms;** they are studied in plane geometry.

As you read through the mathematical descriptions of the conics, be sure to refer to the corresponding figures so that you can make connections between the descriptions and the figures.

In this section, we will define parabolas differently than we did in Chapter 3. Later on, we will make connections between the two definitions.

> **Definition of a Parabola**
>
> A **parabola** is the set of all points (x, y) in a plane such that the distance of (x, y) from a fixed line is equal to the distance of (x, y) from a fixed point that is not on the fixed line. The fixed line and the fixed point, which lie in the plane of the parabola, are called the **directrix** and the **focus,** respectively.

Figure 7.1.3

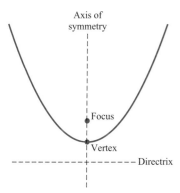

The **axis of symmetry** of a parabola is the line that is perpendicular to the directrix and passes through the focus of the parabola. The **vertex** of a parabola is the midpoint of the segment of the axis of symmetry that has one endpoint at the focus and the other endpoint on the directrix. Figure 7.1.3 illustrates these key features of a parabola.

The only parabolas discussed in this section are those whose axis of symmetry is either vertical (parallel to the y-axis) or horizontal (parallel to the x-axis). We will use the definition of a parabola to derive the equations of such parabolas, beginning with those whose vertex is at the origin.

Parabolas with Vertex at the Origin

Suppose a parabola has its vertex at the origin and the y-axis as its axis of symmetry. Suppose also that the equation of its directrix is $y = -p$, for some $p > 0$. Because the vertex of the parabola is a point *on* the parabola, the distance from the vertex to the directrix is the same as the distance from the vertex to the focus. Thus the coordinates of the focus are $(0, p)$.

Let P be any point on the parabola with coordinates (x, y), and let dist(PF) denote the distance from P to the focus F. To find the distance from P to the directrix, we draw a perpendicular line segment from P to the line $y = -p$ (the directrix). Let M be the point of intersection of that line segment and the directrix, and let dist(PM) denote the distance from P to M. Then the coordinates of M are $(x, -p)$, as shown in Figure 7.1.4.

Figure 7.1.4

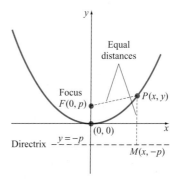

Just In Time

Review the distance formula in Section 2.1.

To derive the equation of the parabola, we apply the definition of a parabola along with the distance formula. From the definition of a parabola, we have

$$\text{Distance from } P \text{ to focus } F = \text{distance from } P \text{ to directrix.}$$

That is,

$$\text{dist}(PF) = \text{dist}(PM).$$

Applying the distance formula,

$$\text{dist}(PF) = \sqrt{(x-0)^2 + (y-p)^2} = \sqrt{x^2 + (y-p)^2} \qquad \text{Distance from } P \text{ to focus } F$$

$$\text{dist}(PM) = \sqrt{(x-x)^2 + (y-(-p))^2} = \sqrt{0^2 + (y+p)^2} \qquad \text{Distance from } P \text{ to directrix}$$

$$\sqrt{x^2 + (y-p)^2} = \sqrt{(y+p)^2} \qquad \text{Equate dist}(PF) \text{ and dist}(PM)$$

$$x^2 + y^2 - 2py + p^2 = y^2 + 2py + p^2 \qquad \text{Square both sides and expand}$$

$$x^2 = 4py \qquad \text{Combine like terms}$$

Thus the standard form of the equation of an upward-opening parabola with vertex at the origin is $x^2 = 4py$, where $p > 0$. The derivations of the standard forms of the equations of other classes of parabolas with vertex at the origin (those that open downward, to the right, or to the left) are similar.

The main features and properties of parabolas with vertex at the origin and a vertical or horizontal axis of symmetry are summarized as follows. Their graphs are shown in Figures 7.1.5 and 7.1.6.

Parabola with Vertex at the Origin

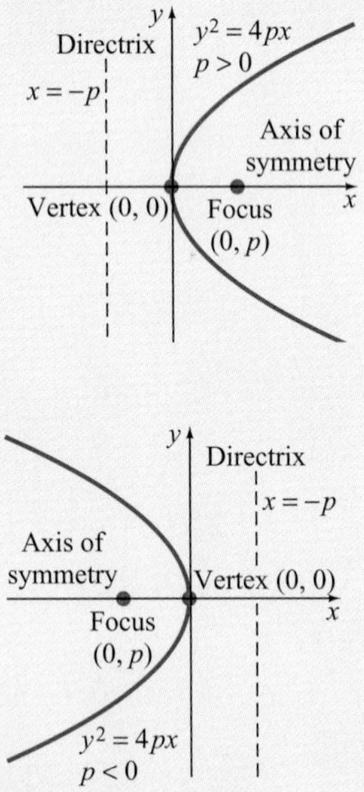

	Vertical Axis of Symmetry	**Horizontal Axis of Symmetry**
Graph	**Figure 7.1.5**	**Figure 7.1.6**
Equation	$x^2 = 4py$	$y^2 = 4px$
Direction of opening	Upward if $p > 0$, downward if $p < 0$	To the right if $p > 0$, to the left if $p < 0$
Vertex	$(0, 0)$	$(0, 0)$
Focus	$(0, p)$	$(p, 0)$
Directrix	The line $y = -p$	The line $x = -p$
Axis of symmetry	y-axis	x-axis

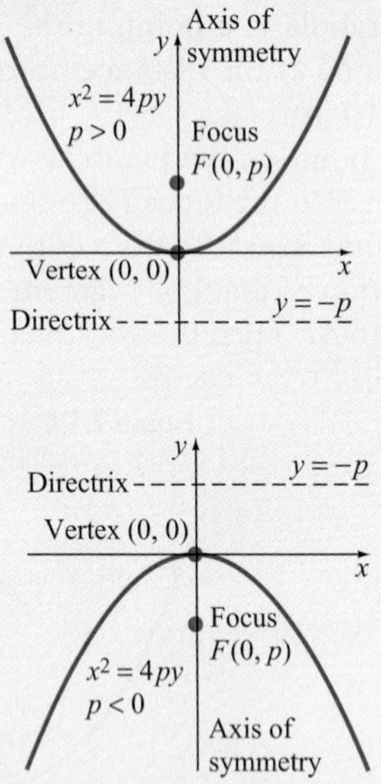

We will use the following observations in solving problems involving parabolas.

Observations:

▶ A parabola opens toward the focus and away from the directrix.

▶ The axis of symmetry of a parabola is perpendicular to the directrix.

▶ The axis of symmetry intersects the parabola in only one point, namely, the vertex.

▶ The directrix does *not* intersect the parabola, and it is *not* part of the parabola.

▶ The focus of a parabola lies on the axis of symmetry, but it is *not* part of the parabola.

▶ If a parabola opens to the right or to the left, the variable y in the equation of the parabola is not a function of x, since the graph of such an equation does not pass the vertical line test.

Example 1 Parabola with Vertex at the Origin

Consider the parabola with vertex at the origin defined by the equation $y = \frac{1}{6}x^2$.

(a) Find the coordinates of the focus.

(b) Find the equations of the directrix and the axis of symmetry.

(c) Find the value(s) of a for which the point $(a, 4)$ is on the parabola.

(d) Sketch the parabola, and indicate the focus and the directrix.

▶ Solution

(a) First write the equation in standard form by multiplying both sides by 6:

$$y = \frac{1}{6}x^2 \Longrightarrow x^2 = 6y$$

This equation is in the form $x^2 = 4py$. To find the focus, we must first calculate p.

$$x^2 = 6y = 4py \Longrightarrow 6 = 4p \Longrightarrow p = \frac{3}{2}$$

Because the parabola is of the form $x^2 = 4py$ and $p > 0$, the parabola opens upward. The focus is $(0, p) = \left(0, \frac{3}{2}\right)$.

(b) The equation of the directrix is $y = -p$; substituting the value of p, we obtain $y = -\frac{3}{2}$. The axis of symmetry of this parabola is the y-axis, which is given by the equation $x = 0$.

(c) Since the point $(a, 4)$ is on the parabola, we will substitute its coordinates into the equation $y = \frac{1}{6}x^2$ (we could just as well substitute the coordinates into the equation $x^2 = 6y$).

$$4 = \frac{1}{6}a^2 \qquad \text{Point } (a, 4) \text{ satisfies equation of parabola}$$

$$24 = a^2 \qquad \text{Multiply both sides by 6}$$

$$a = \pm\sqrt{24} = \pm 2\sqrt{6} \qquad \text{Take square root of both sides}$$

Thus the points $\left(2\sqrt{6}, 4\right)$ and $\left(-2\sqrt{6}, 4\right)$ lie on the parabola.

(d) The vertex of the parabola is at $(0, 0)$. Plot the vertex, along with the points $\left(2\sqrt{6}, 4\right)$ and $\left(-2\sqrt{6}, 4\right)$ from part (c). Draw a parabola through these three points. To find more points, construct a table (see Table 7.1.1). Then indicate the focus and the directrix, as shown in Figure 7.1.7.

Table 7.1.1

x	y
-6	6
-2	$\dfrac{2}{3}$
0	0
2	$\dfrac{2}{3}$
6	6

Figure 7.1.7

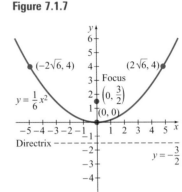

✔ *Check It Out 1:* Find the coordinates of the focus of the parabola $y = -\dfrac{1}{8}x^2$, as well as the equations of the directrix and the axis of symmetry. Sketch the parabola. ■

Example 2 illustrates how to obtain the equation of a parabola given a description of its features.

Example 2 Finding the Equation of a Parabola

Determine the equation in standard form of the parabola with vertex at the origin and directrix $x = 3$. Sketch the parabola, and indicate the focus and the directrix.

▶Solution The directrix, $x = 3$, is a vertical line. The vertex of this parabola is at the origin, which is to the left of the directrix. Thus the parabola must open to the left, since a parabola always opens *away* from the directrix. The focus is at $(-3, 0)$, since the vertex is midway between the focus and the directrix. With this information, we can make a preliminary sketch, as shown in Figure 7.1.8. Because the focus is at

Figure 7.1.8

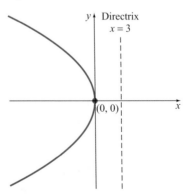

Technology Note

Use a graphing utility to graph the equation $y^2 = -12x$, which defines a parabola that opens to the left. In the equation $y^2 = -12x$, the variable y is not a function of x because the graph of the equation does not satisfy the vertical line test. Graphing utilities accept only functions, so in order to graph this equation, we must first solve for y:

$$y^2 = -12x \implies y = \pm\sqrt{-12x}.$$

This gives us the following *two* functions:

$$Y_1(x) = \sqrt{-12x} \quad \text{and}$$
$$Y_2(x) = -\sqrt{-12x}.$$

The graph of the function Y_1 produces the top half of the parabola, and the graph of Y_2 produces the bottom half. These two functions are defined *only* for $x \le 0$. Graphing the two functions in the same window, using a window size of $[-6, 2] \times [-8, 8]$, gives the graph shown in Figure 7.1.10.

Figure 7.1.10

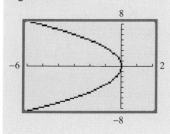

Just In Time

Review translations of graphs in Section 2.3.

$(-3, 0) = (p, 0)$, we have $p = -3$. As a check, we know that p must be negative if the parabola opens to the left. Next we find the equation of the parabola.

$y^2 = 4px$	Standard form of equation of parabola opening to the left
$y^2 = 4(-3)x = -12x$	Substitute $p = -3$

The equation of the parabola is $y^2 = -12x$. To get a more accurate sketch, find points (x, y) that are on the parabola.

$$x = -3 \implies y^2 = -12(-3) = 36 \implies y = \pm 6$$
$$x = -2 \implies y^2 = -12(-2) = 24 \implies y = \pm\sqrt{24} \approx \pm 4.899$$

The graph is shown in Figure 7.1.9.

Figure 7.1.9

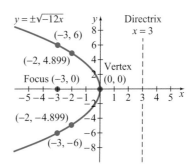

☑ *Check It Out 2:* Determine the equation in standard form of the parabola with vertex at the origin and directrix $x = -2$. ▪

Parabolas with Vertex at the Point (h, k)

Recall the following facts about translations, or shifts, of graphs.

▶ If the variable x in an equation of the form $x^2 = 4py$ or $y^2 = 4px$ is replaced by $(x - h)$, the graph of the equation is translated horizontally by $|h|$ units. The translation is to the right if $h > 0$, and to the left if $h < 0$.

▶ If the variable y in an equation of the form $x^2 = 4py$ or $y^2 = 4px$ is replaced by $(y - k)$, the graph of the equation is translated vertically by $|k|$ units. The translation is upward if $k > 0$, and downward if $k < 0$.

The standard form of the equation of a parabola with vertex at the origin and a vertical axis of symmetry is $x^2 = 4py$, where $p \neq 0$. Using the given facts about translations, we see that the standard form of the equation of a parabola with vertex at the point (h, k) and a vertical axis of symmetry is $(x - h)^2 = 4p(y - k)$, where $p \neq 0$. A similar equation can be written for a parabola with vertex at the point (h, k) and a horizontal axis of symmetry.

The main features and properties of parabolas with vertex at the point (h, k) and a vertical or horizontal axis of symmetry are summarized as follows. Their graphs are shown in Figures 7.1.11 and 7.1.12.

Parabolas with Vertex at (h, k)

	Vertical Axis of Symmetry	**Horizontal Axis of Symmetry**
Graph	**Figure 7.1.11** 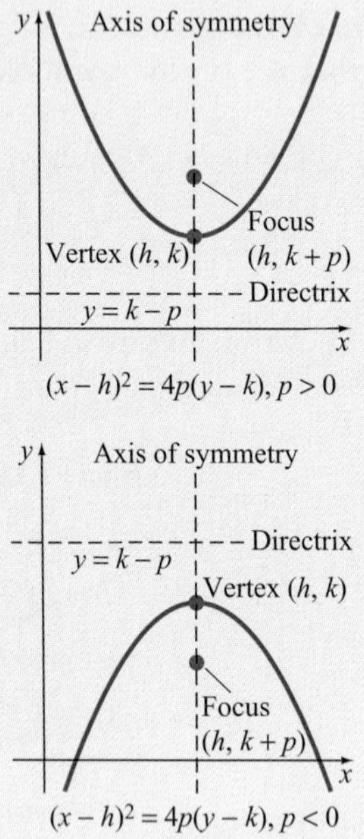 $(x - h)^2 = 4p(y - k), p > 0$ $(x - h)^2 = 4p(y - k), p < 0$	**Figure 7.1.12** 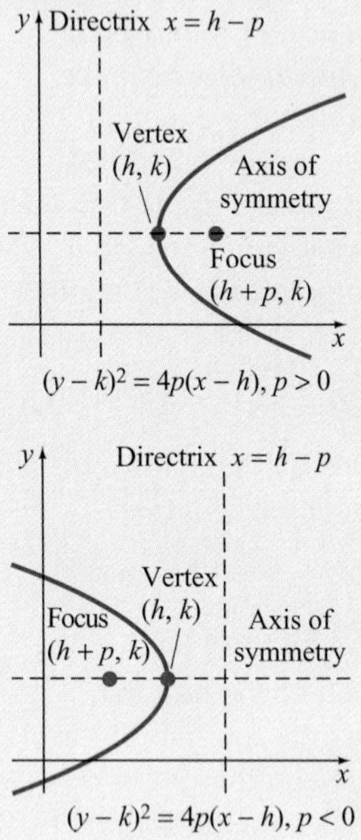 $(y - k)^2 = 4p(x - h), p > 0$ $(y - k)^2 = 4p(x - h), p < 0$
Equation	$(x - h)^2 = 4p(y - k)$	$(y - k)^2 = 4p(x - h)$
Direction of opening	Upward if $p > 0$, downward if $p < 0$	To the right if $p > 0$, to the left if $p < 0$
Vertex	(h, k)	(h, k)
Focus	$(h, k + p)$	$(h + p, k)$
Directrix	The line $y = k - p$	The line $x = h - p$
Axis of symmetry	The line $x = h$	The line $y = k$

Discover *and* **Learn**

Derive the standard form of the equation of a parabola with vertex at the point (h, k) and a horizontal axis of symmetry.

Just In Time

Review equations of horizontal and vertical lines in Section 1.3.

Example 3 **Finding the Equation of a Parabola with Vertex at (h, k)**

Determine the equation in standard form of the parabola with directrix $y = 7$ and focus at $(-3, 3)$. Sketch the parabola.

▶ **Solution** Use the given information to determine the orientation of the parabola and the location of the vertex. Plotting the information as you proceed can be very helpful.

Step 1 First find the axis of symmetry of the parabola, which is perpendicular to its directrix. Here, the directrix is the horizontal line $y = 7$, so the axis of symmetry is vertical. Because the focus lies on the axis of symmetry, the value of h must be equal to the x-coordinate of the focus, which is -3. Thus the equation of the axis of symmetry is $x = -3$.

Step 2 Next, find the vertex. The vertex (h, k) is midway between the focus and the directrix and lies on the axis of symmetry. Because $(-3, 7)$ is the point on the directrix that lies on the axis of the symmetry of the parabola, we obtain the coordinates of the vertex by finding the midpoint of the segment that joins the focus to the point $(-3, 7)$.

$$\text{Vertex: } (h, k) = \left(-3, \frac{3 + 7}{2}\right) = (-3, 5)$$

Step 3 Find the orientation of the parabola to find the correct form of its equation. Since the vertex lies below the directrix, the parabola opens downward, away from the directrix.

A preliminary sketch of the parabola is given in Figure 7.1.13. The standard form of the equation of the directrix of a parabola with a vertical axis of symmetry and vertex (h, k) is $y = k - p$. We can use this equation to find the value of p.

Figure 7.1.13 Preliminary sketch

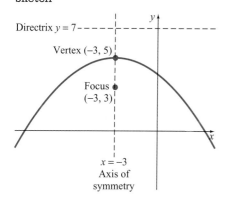

$y = k - p$	*General equation of directrix in terms of k and p*
$y = 7$	*Given equation of directrix*
$k - p = 7$	*Equate expressions for y*
$5 - p = 7$	*Substitute k = 5*
$p = -2$	*Solve for p*

Substituting $(h, k) = (-3, 5)$ and $p = -2$ into the equation $(x - h)^2 = 4p(y - k)$ gives us the equation of the parabola in standard form.

$$(x + 3)^2 = -8(y - 5)$$

To generate other points on the parabola, we solve for y to obtain

$$y = \frac{(x + 3)^2 - 40}{-8}$$

$$= -\frac{1}{8}(x + 3)^2 + 5.$$

Substituting $x = -7$ yields $y = 3$, for example. The fact that the parabola is symmetric with respect to the line $x = -3$ tells us that $(1, 3)$ is also a point on the parabola. Table 7.1.2 shows the coordinates of additional points on the parabola. The graph of the parabola is shown in Figure 7.1.14.

Table 7.1.2

x	y
-7	3
-5	4.5
-3	5
-1	4.5
1	3

Figure 7.1.14

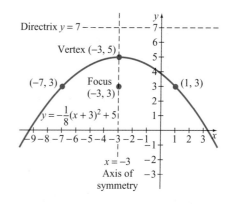

☑ *Check It Out 3:* Determine the equation in standard form of the parabola with directrix $y = 9$ and focus at $(-2, 5)$. Sketch the parabola. ■

Just In Time

Review completing the square in Section 2.1.

Writing the Equation of a Parabola in Standard Form

Sometimes, the equation of a parabola is not given in standard form. In such a case, we complete the square on the given equation to put it in standard form. The procedure is outlined below.

Transforming the Equation of a Parabola into Standard Form

Step 1 Gather the y terms on one side of the equation and the x terms on the other side.

Step 2 Put the constant(s) on the same side as the variable that is raised to only the first power.

Step 3 Complete the square on the variable that is raised to the second power.

The above procedure is illustrated in Example 4.

Example 4 **Completing the Square to Write the Equation of a Parabola**

Find the vertex and focus of the parabola defined by the equation $-4x + 3y^2 + 12y - 8 = 0$. Determine the equation of the directrix and sketch the parabola.

▶**Solution** The equation of the parabola first must be rewritten in standard form. We can proceed as follows.

$$-4x + 3y^2 + 12y - 8 = 0 \qquad \textit{Given equation}$$
$$3y^2 + 12y = 4x + 8 \qquad \textit{Keep y terms on left side}$$
$$3(y^2 + 4y) = 4x + 8 \qquad \textit{Factor out 3 on left side}$$

Find the number a that will make $y^2 + 4y + a$ a perfect square.

$$3(y^2 + 4y + 4) = 4x + 8 + 12 \qquad \textit{Complete the square}$$

The square was completed on the expression *inside* the parentheses by using $\left(\frac{4}{2}\right)^2 = 4$. In order to keep the equation balanced, $12 = 3 \cdot 4$ was added to the *right* side as well.

$$3(y + 2)^2 = 4x + 20 = 4(x + 5) \qquad \textit{Simplify and factor}$$
$$(y + 2)^2 = \frac{4}{3}(x + 5) \qquad \textit{Write equation in standard form}$$

By comparing the standard form of the equation of this parabola with the general form $(y - k)^2 = 4p(x - h)$, we can see that $(h, k) = (-5, -2)$ and $4p = \frac{4}{3} \Longrightarrow p = \frac{1}{3}$. Because $p > 0$, the parabola opens to the right. The graph is shown in Figure 7.1.15.

Figure 7.1.15

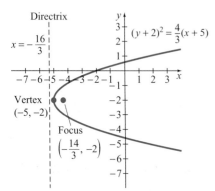

Vertex (h, k):	$(-5, -2)$
Focus $(h + p, k)$:	$\left(-\dfrac{14}{3}, -2\right)$
Equation of directrix, $x = h - p$:	$x = -\dfrac{16}{3}$

✔ *Check It Out 4:* Find the vertex and focus of the parabola defined by the equation $x^2 - 4x - 4y = 0$. Also give the equation of the directrix. ■

Applications

Parabolas are found in many applications, such as the design of flashlights, fluorescent lamps, suspension bridge cables, and the study of the paths of projectiles. Also, parabolas exhibit a remarkable reflective property. Rays of light that emanate from the focus of a mirror with a parabolic cross-section will bounce off the mirror and travel parallel to its axis of symmetry, and vice versa (rays that travel parallel to the axis of symmetry and reach the surface of the mirror will bounce off and pass through the focus). See Figure 7.1.16.

Figure 7.1.16

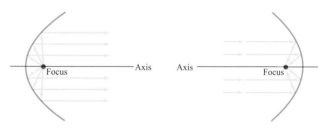

Example 5 examines the design of a headlight reflector.

Example 5 Headlight Design

Figure 7.1.17

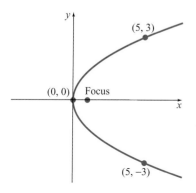

The cross-section of a headlight reflector is in the shape of a parabola. The reflector is 6 inches in diameter and 5 inches deep, as illustrated in Figure 7.1.17.

(a) Find an equation of the parabola, using the position of the vertex of the parabola as the origin of your coordinate system.

(b) The bulb for the headlight is positioned at the focus. Find the position of the bulb.

▶Solution

(a) Setting up the problem with respect to a suitable coordinate system is the first task. Using the vertex of the parabola as the origin, and orienting the coordinate axes in such a way that the focus of the parabola lies on the positive x-axis, we obtain the coordinate system shown in Figure 7.1.18. The points $(5, 3)$ and $(5, -3)$ are on the parabola, since the reflector is 5 inches deep and the cross-sectional diameter is 6 inches. Because the axis of symmetry is horizontal, the standard form of the equation of the parabola is $y^2 = 4px$. Because the point $(5, 3)$ lies on the parabola, we can solve for p by substituting $(x, y) = (5, 3)$ into this equation.

$$4(p)(5) = (3)^2 \Longrightarrow 20p = 9 \Longrightarrow p = \frac{9}{20}$$

Figure 7.1.18

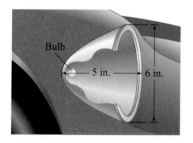

Thus the equation of the parabola is $y^2 = \frac{9}{5}x$.

(b) The coordinates of the focus are $(p, 0) = \left(\frac{9}{20}, 0\right)$, so the bulb should be placed $\frac{9}{20}$ of an inch away from the vertex.

✔ *Check It Out 5:* Rework Example 5 for the case in which the reflector is 8 inches in diameter and 6 inches deep. ■

7.1 Key Points

▶ Parabola with vertical axis of symmetry:

Equation	$(x - h)^2 = 4p(y - k)$
Opening	Upward if $p > 0$, downward if $p < 0$
Vertex	(h, k)
Focus	$(h, k + p)$
Directrix	$y = k - p$
Axis of symmetry	$x = h$

If the vertex is at the origin, then $(h, k) = (0, 0)$.

▶ Parabola with horizontal axis of symmetry:

Equation	$(y - k)^2 = 4p(x - h)$
Opening	To the right if $p > 0$, to the left if $p < 0$
Vertex	(h, k)
Focus	$(h + p, k)$
Directrix	$x = h - p$
Axis of symmetry	$y = k$

If the vertex is at the origin, then $(h, k) = (0, 0)$.

7.1 Exercises

▶**Just in Time Exercises** These exercises correspond to the Just in Time references in this section. Complete them to review topics relevant to the remaining exercises.

1. True or False: The distance between two points (a, b) and (c, d) is given by the formula
$$d = \sqrt{(a - c)^2 + (b - d)^2}.$$

2. Find the distance between the points $(1, 3)$ and $(-5, 3)$.

3. Find the distance between the points $(2, 6)$ and $(-8, 6)$.

4. Write the equation of the function $f(x)$ that is obtained by shifting the graph of $g(x) = x^2$ to the left 3 units.

5. Write the equation of the function $f(x)$ that is obtained by shifting the graph of $g(x) = x^2$ to the right 1 unit.

6. True or False: The graph of the equation $x = 3$ is a horizontal line.

7. True or False: The graph of the equation $y = -2$ is a horizontal line.

8. Write the equation of the vertical line that passes through the point $(2, -5)$.

In Exercises 9 and 10, complete the square.

9. $x^2 + 6x$

10. $x^2 - 10x$

▶**Skills** This set of exercises will reinforce the skills illustrated in this section.

In Exercises 11–14, match each graph (labeled a, b, c, and d) with the appropriate equation.

a.

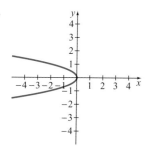

b.

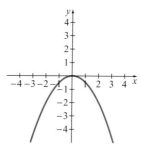

c.

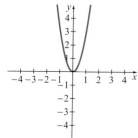

d.

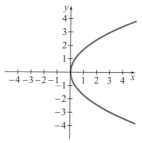

11. $y = 3x^2$

12. $y^2 = 3x$

13. $x^2 = -2y$

14. $y^2 = -\dfrac{x}{2}$

In Exercises 15–38, find the vertex and focus of the parabola that satisfies the given equation. Write the equation of the directrix, and sketch the parabola.

15. $y^2 = 12x$

16. $x^2 = -16y$

17. $y^2 = 8x$

18. $x^2 = 8y$

19. $y = -\dfrac{1}{4}x^2$

20. $y^2 = -4x$

21. $8x = 3y^2$

22. $-4y = 3x^2$

23. $5y^2 = 4x$

24. $-3x^2 = 16y$

25. $y^2 = -12(x - 2)$

26. $x^2 = 8(y + 3)$

27. $(x - 5)^2 = -4(y + 1)$

28. $(y + 3)^2 = 8(x + 2)^2$

29. $(y - 4)^2 = -(x - 1)$

30. $(x + 7)^2 = -(y + 2)$

31. $y^2 - 4y + 4x = 0$

32. $x^2 + 2x - 4y = 0$

33. $x^2 - 2x = -y$

34. $y^2 + 6y = x$

35. $x^2 + 6x + 4y + 25 = 0$

36. $y^2 - 4y + 2x - 6 = 0$

37. $x^2 - 5y + 1 = 8x$

38. $y^2 + 4x + 22 = -10y$

In Exercises 39–62, determine the equation in standard form of the parabola that satisfies the given conditions.

39. Directrix $x = -1$; vertex at $(0, 0)$

40. Directrix $x = -3$; vertex at $(0, 0)$

41. Directrix $y = -3$; vertex at $(0, 0)$

42. Directrix $y = 2$; vertex at $(0, 0)$

43. Focus at $(0, 2)$; vertex at $(0, 0)$

44. Focus at $(0, -3)$; vertex at $(0, 0)$

45. Focus at $(-2, 0)$; vertex at $(0, 0)$

46. Focus at $(4, 0)$; vertex at $(0, 0)$

47. Opens upward; distance of focus from x-axis is 4; vertex at $(0, 0)$

48. Opens downward; distance of directrix from x-axis is 6; vertex at $(0, 0)$

49. Opens to the left; distance of directrix from y-axis is 5; vertex at $(0, 0)$

50. Opens to the right; distance of focus from y-axis is 3; vertex at $(0, 0)$

51. Vertex at $(-2, 1)$; directrix $y = -2$

52. Vertex at $(4, -1)$; directrix $x = 6$

53. Focus at $(0, 4)$; directrix $y = -4$

54. Focus at $(0, -5)$; directrix $y = 5$

55. Horizontal axis of symmetry; vertex at $(0, 0)$; passes through the point $(1, 3)$

56. Vertical axis of symmetry; vertex at $(0, 0)$; passes through the point $(-2, 6)$

57. Vertical axis of symmetry; vertex at $(4, 3)$; passes through the point $(5, 2)$

58. Horizontal axis of symmetry; vertex at $(-7, -5)$; passes through the point $(2, -1)$

59. Opens upward; vertex at (4, 1); passes through the point (8, 3)

60. Opens to the right; vertex at (−3, 2); passes through the point (5, 10)

61. Opens downward; vertex at (5, 3); passes through the point (2, 0)

62. Opens to the left; vertex at (2, −4); passes through the point (−1, 3)

►**Applications** In this set of exercises, you will use parabolas to study real-world problems.

63. **Satellite Technology** Suppose that a satellite receiver with a parabolic cross-section is 36 inches across and 16 inches deep. How far from the vertex must the receptor unit be located to ensure that it is at the focus of the parabola?

64. **Landscaping** A sprinkler set at ground level shoots water upward in a parabolic arc. If the highest point of the arc is 9 feet above the ground, and the water hits the ground at a maximum of 6 feet from the sprinkler, what is the equation of the parabola in standard form?

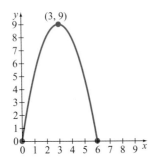

65. **Design** A monument to a famous orator is in the shape of a parabola. The base of the monument is 96 inches wide. The vertex of the parabola is at the top of the monument, which is 144 inches above the base.

(a) How high is the parabola above the locations on the base of the monument that are 32 inches from the ends?

(b) What locations on the base of the monument are 63 inches below the parabola?

66. **Audio Technology** A microphone with a parabolic cross-section is formed by revolving the portion of the parabola $10y = x^2$ between the lines $x = 7$ and $x = −7$ about its axis of symmetry. The sound receiver should be placed at the focus for best reception. Find the location of the sound receiver.

67. **Road Paving** The surface of a footpath over a hill is in the shape of a parabola. The straight-line distance between the ends of the path, which are at the base of the hill, is 60 feet. The path is 5 feet higher at the middle than at the ends. How high above the base of the hill is a point on the path that is located at a horizontal distance of 12 feet from an end of the path?

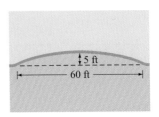

68. **Physics** Neglecting air resistance, the path of an object thrown upward at an angle of 45 degrees is given by the equation

$$y = y_0 + x - g\frac{x^2}{v_0^2}.$$

Here, x denotes the horizontal position of the object relative to the point from which the object leaves the thrower's hand (hence $x = 0$ at that point), y denotes the corresponding height of the object above the ground, and y_0, v_0, and g are constants. The constants y_0 and v_0 are the height above the ground and the speed, respectively, at which the object leaves the thrower's hand, and g is Earth's gravitational constant. Assume the object is thrown at a speed of 16 feet per second, at an angle of 45 degrees, from a point 4 feet above the ground. The value of g is 32 feet per second squared.

(a) Give the equation of the parabolic path of the object.

(b) What are the coordinates of the vertex of the parabola? (*Hint:* Write the equation from part (a) in standard form.)

(c) What is the physical significance of the vertex of the parabola?

(d) When the object comes back down to a height of 4 feet above the ground, how far is it from the point from which it was thrown?

69. **Physics** Suppose an object is thrown at a speed of 16 feet per second (and at an angle of 45 degrees) from a point 4 feet above the surface of a hypothetical planet whose gravitational constant is only half as large as that of planet Earth (see Exercise 68).

(a) Rework Exercise 68 using the appropriate value of g.

(b) Compare your answers to parts (b) and (d) of Exercise 68 to the answers you found here. Are they the same or different? Explain.

70. Parabolic Mirror A reflecting telescope has a mirror with a parabolic cross-section (see the accompanying schematic, which is not drawn to scale). The border of the wide end of the mirror is a circle of radius 50 inches, and the *focal length* of the mirror, which is defined as the distance of the focus from the vertex, is 300 inches.

Schematic of Parabolic Mirror

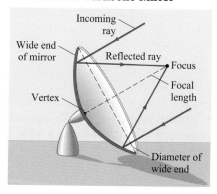

(a) How far is the vertex from the center of the wide end of the mirror?

(b) Astronomers characterize reflecting telescopes by their *focal ratio*, which is the ratio of the focal length to the aperture (the diameter of the wide end of the mirror). What is the focal ratio of this telescope?

▶ **Concepts** This set of exercises will draw on the ideas presented in this section and your general math background.

71. A parabola has directrix $y = -2$ and passes through the point $(4, -7)$. Find the distance of the point $(4, -7)$ from the focus of the parabola.

72. Find the distance of the point $(1, -4)$ from the axis of symmetry of the parabola with vertex $(3, 2)$, if the axis of symmetry is horizontal.

73. Find the distance of the point $(0, 5)$ from the axis of symmetry of the parabola with focus $(-2, 1)$ and directrix $x = 4$.

74. Find the distance of the point $(11, 7)$ from the focus of the parabola with vertex $(5, 3)$ and directrix $x = 2$.

75. Write an equation that is satisfied by the set of points whose distances from the point $(-2, 0)$ and the line $x = 2$ are equal.

76. A parabola has axis of symmetry parallel to the y-axis and passes through the points $(-7, 4)$, $(-5, 5)$, and $(3, 29)$. Determine the equation of this parabola. (*Hint:* Use the general equation of a parabola from Chapter 3 together with the information you learned in Chapter 6 about solving a system of equations.)

77. What point in the xy-plane is the mirror image of the point $(2, 8)$ with respect to the axis of symmetry of a parabola that has vertex $(-5, 3)$ and opens to the right? (*Hint:* Use what you learned in Chapter 3 about the mirror image of a point with respect to a line.)

78. What point in the xy-plane is the mirror image of the point $(4, -6)$ with respect to the axis of symmetry of a parabola that has focus $(-1, 1)$ and directrix $x = -5$? (*Hint:* Use what you learned in Chapter 3 about the mirror image of a point with respect to a line.)

7.2 The Ellipse

Objectives

▶ Define an ellipse

▶ Find the foci, vertices, and major and minor axes of an ellipse

▶ Determine the equation of an ellipse and write it in standard form

▶ Translate an ellipse in the xy-plane

▶ Sketch an ellipse

When a planet revolves around the sun, it traces out an oval-shaped orbit known as an *ellipse*. Once we derive the basic equation of an ellipse, we can use it to study a variety of applications, from planetary motion to elliptical domes.

Definition of an Ellipse

An **ellipse** is the set of all points (x, y) in a plane such that the sum of the distances of (x, y) from two fixed points in the plane is equal to a fixed positive number d, where d is greater than the distance between the fixed points. Each of the fixed points is called a **focus** of the ellipse. Together, they are called the **foci.**

The **major axis** of an ellipse is the line segment that passes through the two foci and has both of its endpoints on the ellipse. The midpoint and the endpoints of the major axis are called the **center** and the **vertices,** respectively, of the ellipse. The **minor axis**

To draw an ellipse, take a piece of string and tack down its ends at any two points F_1 and F_2 such that the string is longer than the distance between F_1 and F_2. With your pencil point, pull the string taut. Move the pencil around the thumbtacks, all the while keeping the string taut. This action will trace out an ellipse with foci at F_1 and F_2. See Figure 7.2.2.

Figure 7.2.2   Drawing an ellipse

of an ellipse is the line segment that is perpendicular to the major axis, passes through the center of the ellipse, and has both of its endpoints on the ellipse. See Figure 7.2.1.

Figure 7.2.1

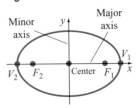

Foci: F_1 and F_2

The only ellipses discussed in this section are those whose major axes are either horizontal (parallel to the x-axis) or vertical (parallel to the y-axis). We will use the definition of an ellipse to derive the equations of such ellipses, beginning with those whose centers are at the origin.

Equation of an Ellipse Centered at the Origin

Suppose an ellipse is centered at the origin, with its foci F_1 and F_2 on the x-axis. Assume that F_1 lies to the right of F_2. If c is the distance from the center to either focus, then F_1 and F_2 are located at $(c, 0)$ and $(-c, 0)$, respectively. Now let (x, y) be any point on the ellipse, and let d_1 and d_2 be the distances from (x, y) to F_1 and F_2, respectively. See Figure 7.2.3.

Figure 7.2.3

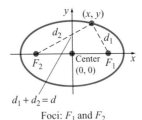

$d_1 + d_2 = d$

Foci: F_1 and F_2

By the definition of an ellipse, there is some fixed positive number d such that $d_1 + d_2 = d$. The individual distances d_1 and d_2 will change from one point on the ellipse to another, but their sum d will not.

Now, because d is the distance between the vertices and the vertices are symmetric with respect to the center of the ellipse, the distance from the center of the ellipse to either vertex is $\frac{d}{2}$. Thus the coordinates of the vertices are $(a, 0)$ and $(-a, 0)$, where $a = \frac{d}{2}$. The major axis is a segment of the x-axis because the endpoints of the major axis are the vertices. Notice that $a > c$, since the vertices are _on_ the ellipse but the foci are _interior_ to it.

Because $a = \frac{d}{2}$, we can write $d = 2a$. Thus the equation $d_1 + d_2 = 2a$ is satisfied by all the points on the ellipse. Writing this equation and then applying the distance formula, we obtain

$$d_1 + d_2 = 2a$$
$$\sqrt{(x - c)^2 + (y - 0)^2} + \sqrt{(x + c)^2 + (y - 0)^2} = 2a.$$

Isolating the second radical and squaring both sides gives

$$\sqrt{(x+c)^2+y^2}=2a-\sqrt{(x-c)^2+y^2}$$

$$(x+c)^2+y^2=4a^2-4a\sqrt{(x-c)^2+y^2}+(x-c)^2+y^2.$$

Isolating the radical, we get

$$4a\sqrt{(x-c)^2+y^2}=(x-c)^2-(x+c)^2+4a^2.$$

Next we expand the right-hand side and then combine like terms.

$$4a\sqrt{(x-c)^2+y^2}=x^2-2cx+c^2-(x^2+2cx+c^2)+4a^2$$

$$4a\sqrt{(x-c)^2+y^2}=4a^2-4cx$$

Dividing by 4 and then squaring both sides gives

$$a\sqrt{(x-c)^2+y^2}=a^2-cx$$

$$a^2[(x-c)^2+y^2]=a^4-2a^2cx+c^2x^2.$$

Next, expand the left-hand side and combine like terms.

$$a^2[(x^2-2cx+c^2)+y^2]=a^4-2a^2cx+c^2x^2$$

$$a^2x^2-2a^2cx+a^2c^2+a^2y^2=a^4-2a^2cx+c^2x^2$$

$$a^2x^2-c^2x^2+a^2y^2=a^4-a^2c^2$$

Factoring both sides, we get

$$(a^2-c^2)x^2+a^2y^2=a^2(a^2-c^2).$$

Because $a>c$ and a and c are both positive, we see that $a^2>c^2$. Thus there is some positive number b such that $a^2-c^2=b^2$. Substituting b^2 for a^2-c^2 in the last equation and then dividing by a^2b^2 gives us the following result.

$$b^2x^2+a^2y^2=a^2b^2$$

$$\frac{x^2}{a^2}+\frac{y^2}{b^2}=1$$

This is the standard form of the equation of an ellipse with center at $(0,0)$ and foci on the x-axis at $(c,0)$ and $(-c,0)$, where c is related to a and b via the equation $b^2=a^2-c^2$. Thus $c^2=a^2-b^2$. Moreover, since c^2 is positive, we see that $a^2-b^2>0$ and so $a^2>b^2$. Since $a,b>0$, $a>b$.

The minor axis of the ellipse is a segment of the y-axis, and so the x-coordinate of every point on the minor axis is zero. Moreover, the endpoints of the minor axis lie on the ellipse, so their y-coordinates can be found by substituting $x=0$ into the equation of the ellipse.

$$\frac{0^2}{a^2}+\frac{y^2}{b^2}=1\Longrightarrow\frac{y^2}{b^2}=1\Longrightarrow y^2=b^2\Longrightarrow y=\pm b$$

Thus the endpoints of the minor axis are $(0,b)$ and $(0,-b)$.

The derivation of the standard form of the equation of an ellipse with center at $(0,0)$ and foci on the y-axis is similar. The main features and properties of ellipses with center at the origin and a horizontal or vertical major axis are summarized next. Their graphs are shown in Figures 7.2.4 and 7.2.5.

Ellipses with Center at the Origin

	Horizontal Major Axis	**Vertical Major Axis**
Graph	**Figure 7.2.4**	**Figure 7.2.5**
Equation	$\dfrac{x^2}{a^2} + \dfrac{y^2}{b^2} = 1$, $a > b > 0$	$\dfrac{x^2}{b^2} + \dfrac{y^2}{a^2} = 1$, $a > b > 0$
Center	$(0, 0)$	$(0, 0)$
Foci	$(-c, 0)$, $(c, 0)$, $c = \sqrt{a^2 - b^2}$	$(0, -c)$, $(0, c)$, $c = \sqrt{a^2 - b^2}$
Vertices	$(-a, 0)$, $(a, 0)$	$(0, -a)$, $(0, a)$
Major axis	Segment of x-axis from $(-a, 0)$ to $(a, 0)$	Segment of y-axis from $(0, -a)$ to $(0, a)$
Minor axis	Segment of y-axis from $(0, -b)$ to $(0, b)$	Segment of x-axis from $(-b, 0)$ to $(b, 0)$

We will use the following observations to solve problems involving ellipses centered at the origin.

Observations:

Assume that $a > b > 0$, where $\dfrac{1}{a^2}$ and $\dfrac{1}{b^2}$ are the coefficients of the quadratic terms in the standard form of the equation of an ellipse. Then the following statements hold.

▶ The major axis of the ellipse is horizontal if the x^2 term in the equation is the term with a coefficient equal to $\dfrac{1}{a^2}$.

▶ The major axis of the ellipse is vertical if the y^2 term in the equation is the term with a coefficient equal to $\dfrac{1}{a^2}$.

▶ The vertices of the ellipse are the endpoints of the major axis, and are a distance of a units from the center.

▶ The distance of either endpoint of the minor axis from the center of the ellipse is b, and the length of the minor axis is $2b$.

▶ The foci of the ellipse lie on the major axis, but they are *not* part of the ellipse. The distance of either focus from the center is $c = \sqrt{a^2 - b^2}$.

▶ The center of the ellipse lies on the major axis, midway between the foci, but the center is *not* part of the ellipse.

▶ In the equation of an ellipse, the variable y is not a function of x, because the graph of such an equation does not pass the vertical line test.

Example 1 **Ellipse Centered at the Origin**

Consider the ellipse that is centered at the origin and defined by the equation

$$16x^2 + 25y^2 = 400.$$

(a) Write the equation of the ellipse in standard form, and determine the orientation of the major axis.

(b) Find the coordinates of the vertices and the foci.

(c) Sketch the ellipse, and indicate the vertices and foci.

▶Solution

(a) Divide the given equation by 400 to get a 1 on one side.

$$\frac{x^2}{25} + \frac{y^2}{16} = 1$$

Since $25 > 16$ and $a^2 > b^2$, we have

$$a^2 = 25, \quad b^2 = 16, \quad c^2 = a^2 - b^2 = 25 - 16 = 9.$$

Using the fact that a, b, and c are all positive, we obtain

$$a = 5, \quad b = 4, \quad c = 3.$$

The x^2 term is the term with a coefficient equal to $\frac{1}{a^2}$, so the major axis is horizontal.

(b) Because the major axis is horizontal and the ellipse is centered at the origin, the vertices are

$$(a, 0) = (5, 0) \quad \text{and} \quad (-a, 0) = (-5, 0).$$

The foci also lie on the major axis, so they are

$$(c, 0) = (3, 0) \quad \text{and} \quad (-c, 0) = (-3, 0).$$

(c) First plot the endpoints of the major and minor axes and then sketch an ellipse through these points. The endpoints of the major axis are the vertices, $(\pm 5, 0)$, and the endpoints of the minor axis are $(0, \pm b) = (0, \pm 4)$. See Figure 7.2.6. To obtain additional points on the ellipse, solve for y in the original equation to get $y = \frac{\pm\sqrt{400 - 16x^2}}{5}$. Then make a table of values. See Table 7.2.1.

Table 7.2.1

x	y
-5	0
-3	± 3.2
0	± 4
3	± 3.2
5	0

Figure 7.2.6

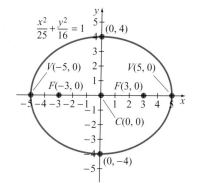

✔ *Check It Out 1:* Determine the orientation of the major axis of the ellipse given by the equation $\frac{x^2}{4} + \frac{y^2}{9} = 1$. Find the vertices and foci and sketch the ellipse. ■

Example **2** **Finding the Equation of an Ellipse**

Determine the equation in standard form of the ellipse with center at the origin, one focus at $\left(0, -\frac{3}{2}\right)$, and one vertex at $(0, 2)$. Sketch the ellipse.

▶**Solution**

Step 1 Determine the orientation of the ellipse. Because the focus and the vertex lie on the y-axis, the major axis is a segment of the y-axis, and so the ellipse is oriented vertically.

Step 2 Next find the values of a and b. Since one of the vertices is at $(0, 2)$, the distance from the center to either vertex is 2. Thus $a = 2$.

Recall that a and b are related by the equation $b^2 = a^2 - c^2$, where c is the distance of either focus from the center of the ellipse. One of the foci is at $\left(0, -\frac{3}{2}\right)$, so $c = \frac{3}{2}$. Thus

$$b^2 = a^2 - c^2 = (2)^2 - \left(\frac{3}{2}\right)^2 = 4 - \frac{9}{4} = \frac{7}{4} \implies b = \frac{\sqrt{7}}{2}.$$

We take b to be the positive square root of $\frac{7}{4}$ because b was defined to be positive.

Step 3 Find the equation. Because the major axis is vertical, the equation of the ellipse is $\frac{x^2}{b^2} + \frac{y^2}{a^2} = 1$, where $a^2 = 4$ and $b^2 = \frac{7}{4}$. We thus obtain

$$\frac{x^2}{\frac{7}{4}} + \frac{y^2}{4} = 1.$$

Step 4 Sketch the ellipse by plotting the vertices and the endpoints of the minor axis, as shown in Figure 7.2.7.

Figure 7.2.7

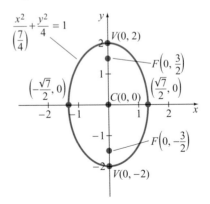

✔ *Check It Out 2:* Determine the equation in standard form of the ellipse with center at the origin, one focus at $(4, 0)$, and one vertex at $(5, 0)$. ■

Example 3 Graphing an Ellipse with a Graphing Utility

Use a graphing utility to graph the equation $16x^2 + 25y^2 = 400$.

▶Solution This is the equation of the ellipse we sketched in Example 1. Because the variable y in the equation of an ellipse is not a function of x, we must first solve the equation for y and then enter two separate functions into the calculator, one for the top half of the ellipse and the other for the bottom half.

$$16x^2 + 25y^2 = 400 \qquad \text{Given equation}$$
$$25y^2 = 400 - 16x^2 \qquad \text{Isolate } y^2 \text{ term}$$
$$y^2 = \frac{400 - 16x^2}{25} \qquad \text{Divide by 25}$$
$$y = \pm\frac{\sqrt{400 - 16x^2}}{5} \qquad \text{Take square root of both sides}$$

We now have the *two* functions

$$Y_1(x) = \frac{\sqrt{400 - 16x^2}}{5} \quad \text{and} \quad Y_2(x) = -\frac{\sqrt{400 - 16x^2}}{5}.$$

Graphing these two functions using a window size of $[-9, 9] \times [-6, 6]$ and the SQUARE window option, we get the graph shown in Figure 7.2.8. The gaps in the graph of the ellipse are due to the limited resolution of the calculator.

Figure 7.2.8

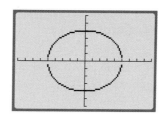

✔ *Check It Out 3:* Use a graphing utility to graph the equation $\frac{x^2}{4} + \frac{y^2}{9} = 1$. ▪

Equation of an Ellipse with Center at the Point (h, k)

Recall the following facts about translations.

▶ If the variable x in an equation of the form $\frac{x^2}{a^2} + \frac{y^2}{b^2} = 1$ or $\frac{x^2}{b^2} + \frac{y^2}{a^2} = 1$ is replaced by $(x - h)$, the graph of the equation is translated horizontally by $|h|$ units. The translation is to the right if $h > 0$, and to the left if $h < 0$.

▶ If the variable y in an equation of the form $\frac{x^2}{a^2} + \frac{y^2}{b^2} = 1$ or $\frac{x^2}{b^2} + \frac{y^2}{a^2} = 1$ is replaced by $(y - k)$, the graph of the equation is translated vertically by $|k|$ units. The translation is upward if $k > 0$, and downward if $k < 0$.

The standard form of the equation of an ellipse with center at the origin and a horizontal major axis is $\frac{x^2}{a^2} + \frac{y^2}{b^2} = 1$, where $a > b > 0$. Using the given facts about translations, we can see that the standard form of the equation of an ellipse with center at the point (h, k) and a horizontal major axis is $\frac{(x - h)^2}{a^2} + \frac{(y - k)^2}{b^2} = 1$, where $a > b > 0$. A similar equation can be written for an ellipse with center at the point (h, k) and a

Technology Note

Some calculator models have built-in features to graph ellipses without having to first solve for y. Consult your manual for details.

vertical major axis. The main features and properties of ellipses with center at the point (h, k) and a horizontal or vertical major axis are summarized as follows. Their graphs are shown in Figures 7.2.9 and 7.2.10.

Ellipses with Center at (h, k)

	Horizontal Major Axis	**Vertical Major Axis**
Graph	Figure 7.2.9	Figure 7.2.10
Equation	$\dfrac{(x-h)^2}{a^2} + \dfrac{(y-k)^2}{b^2} = 1, a > b > 0$	$\dfrac{(x-h)^2}{b^2} + \dfrac{(y-k)^2}{a^2} = 1, a > b > 0$
Center	(h, k)	(h, k)
Foci	$(h-c, k), (h+c, k), c = \sqrt{a^2 - b^2}$	$(h, k-c), (h, k+c), c = \sqrt{a^2 - b^2}$
Vertices	$(h-a, k), (h+a, k)$	$(h, k-a), (h, k+a)$
Major axis	Segment of the line $y = k$ from $(h-a, k)$ to $(h+a, k)$	Segment of the line $x = h$ from $(h, k-a)$ to $(h, k+a)$
Minor axis	Segment of the line $x = h$ from $(h, k-b)$ to $(h, k+b)$	Segment of the line $y = k$ from $(h-b, k)$ to $(h+b, k)$

Discover *and* **Learn**

Derive the standard form of the equation of an ellipse with vertex at the point (h, k) and a vertical major axis.

Example 4 Finding the Equation of an Ellipse with Vertex at (h, k)

Determine the equation in standard form of the ellipse with foci at $(4, 1)$ and $(4, -5)$ and one vertex at $(4, 3)$. Sketch the ellipse.

▶Solution

Step 1 Determine the orientation. The x-coordinate of each focus is 4, so the foci lie on the vertical line $x = 4$. Thus the major axis of the ellipse is vertical, and the standard form of its equation is $\dfrac{(x-h)^2}{b^2} + \dfrac{(y-k)^2}{a^2} = 1$.

Step 2 Next determine the center (h, k) and the values of a and b. Because the center is the midpoint of the line segment that has its endpoints at the foci, we obtain

$$\text{Center} = (h, k) = \left(\frac{4+4}{2}, \frac{1 + (-5)}{2} \right) = (4, -2).$$

The quantity a is the distance from either vertex to the center. Using the vertex at $(4, 3)$ and the distance formula, we find that the distance from $(4, 3)$ to $(4, -2)$ is

$$a = \sqrt{(4 - 4)^2 + (3 - (-2))^2} = 5.$$

Recall that a and b are related by the equation $b^2 = a^2 - c^2$, where c is the distance from the center to either focus. Using the distance formula to compute the distance from the center to the focus at $(4, 1)$ gives

$$c = \sqrt{(4 - 4)^2 + (-2 - 1)^2} = 3.$$

Thus

$$b^2 = a^2 - c^2 = (5)^2 - (3)^2 = 25 - 9 = 16 \Longrightarrow b = 4.$$

We take b to be the positive square root of 16 because b was defined to be positive.

Step 3 To find the equation, substitute the values of h, k, a, and b into the standard form of the equation.

$$\frac{(x - h)^2}{b^2} + \frac{(y - k)^2}{a^2} = 1 \qquad \textit{Standard form of equation}$$

$$\frac{(x - 4)^2}{16} + \frac{(y + 2)^2}{25} = 1 \qquad \textit{Substitute } h = 4, k = -2, a = 5, b = 4$$

The vertices are

$$(h, k + a) = (4, -2 + 5) = (4, 3) \text{ and } (h, k - a) = (4, -2 - 5) = (4, -7).$$

The first vertex checks with the given information.

The endpoints of the minor axis are

$$(h + b, k) = (4 + 4, -2) = (8, -2) \text{ and } (h - b, k) = (4 - 4, -2) = (0, -2).$$

Step 4 To sketch the ellipse, we plot the vertices and the endpoints of the minor axis. These four points can be used to sketch the ellipse, as shown in Figure 7.2.11.

Figure 7.2.11

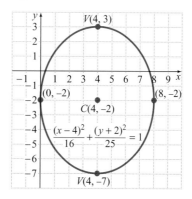

✔ *Check It Out 4:* Determine the equation in standard form of the ellipse with foci at $(3, 2)$ and $(3, -2)$ and one vertex at $(3, 3)$. ■

Example **5** **Completing the Square to Write the Equation of an Ellipse**

Write the equation in standard form of the ellipse defined by the equation $5x^2 - 30x + 25y^2 + 50y + 20 = 0$. Sketch the ellipse.

▶Solution To obtain the standard form of the equation, first complete the squares on both x and y. As the first step in doing this, move the constant to the right side of the equation.

$$5x^2 - 30x + 25y^2 + 50y = -20$$

Next factor out the common factor in the x terms, and then do the same for the y terms. Then complete the square on each expression in parentheses.

$$5(x^2 - 6x + 9) + 25(y^2 + 2y + 1) = -20 + 45 + 25$$

The squares were completed using $\left(\frac{-6}{2}\right)^2 = 9$ and $\left(\frac{2}{2}\right)^2 = 1$, respectively. On the right side, we added 45 ($5 \cdot 9$) and 25 ($25 \cdot 1$). Factoring the left side of the equation and simplifying the right side gives

$$5(x - 3)^2 + 25(y + 1)^2 = 50.$$

Dividing by 50 on both sides, we obtain

$$\frac{(x - 3)^2}{10} + \frac{(y + 1)^2}{2} = 1.$$

This is the standard form of the equation of the ellipse, which implies that the center of the ellipse is $(3, -1)$. Because $10 > 2$ and $a^2 > b^2$, we see that $a^2 = 10$ and $b^2 = 2$. Thus $a = \sqrt{10}$ and $b = \sqrt{2}$. The major axis is horizontal, since the $(x - 3)^2$ term is the term with a coefficient equal to $\frac{1}{a^2}$.

The vertices are

$$(h + a, k) = \left(3 + \sqrt{10}, -1\right) \approx (6.162, -1) \quad \text{and}$$
$$(h - a, k) = \left(3 - \sqrt{10}, -1\right) \approx (-0.162, -1).$$

The endpoints of the minor axis are

$$(h, k + b) = \left(3, -1 + \sqrt{2}\right) \approx (3, 0.414) \quad \text{and}$$
$$(h, k - b) = \left(3, -1 - \sqrt{2}\right) \approx (3, -2.414).$$

Finally, we plot the vertices and the endpoints of the minor axis and use these four points to sketch the ellipse. See Figure 7.2.12.

Figure 7.2.12

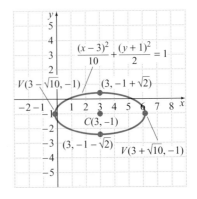

Just in Time

Review completing the square in Section 2.1.

☑ *Check It Out 5:* Write the equation in standard form of the ellipse defined by the equation $4x^2 - 8x + y^2 = 8$. ■

Applications

Ellipses occur in many science and engineering applications. The orbit of each planet in our solar system is an ellipse with the sun at one of the foci. The orbit of our moon is an ellipse with Earth at one focus. Some comets (for example, Halley's comet) travel in elliptical orbits.

Example 6 Elliptical Orbit of Halley's Comet

The orbit of Halley's comet is elliptical, with the sun at one of the foci. The length of the major axis of the orbit is approximately 36 astronomical units (AU), and the length of the minor axis is approximately 9 AU (1 AU ≈ 92,600,000 miles). Find the equation in standard form of the path of Halley's comet, using the origin as the center of the ellipse and a segment of the x-axis as the major axis.

▶Solution The major axis is a segment of the x-axis and its length is 36. This length can be expressed as $2a$, where a is the distance from either vertex of the ellipse to the center. Equating $2a$ and 36, we find that $a = 18$. Thus $a^2 = 324$.

The minor axis is a segment of the y-axis and its length is 9. This length can be expressed as $2b$, where b is the distance from either endpoint of the minor axis to the center of the ellipse. Equating $2b$ and 9, we obtain $b = 4.5$. Thus $b^2 = 20.25$.

Because the ellipse is centered at the origin, the equation in standard form of the orbit is

$$\frac{x^2}{324} + \frac{y^2}{20.25} = 1.$$

☑ *Check It Out 6:* In Example 6, find the distance (in AU) from the sun to the center of the ellipse. (*Hint:* The sun is one of the foci of the elliptical orbit of Halley's comet.) ■

7.2 Key Points

▶ Ellipse with a horizontal major axis:

Equation	$\dfrac{(x-h)^2}{a^2} + \dfrac{(y-k)^2}{b^2} = 1$
Center	(h, k)
Foci	$(h - c, k), (h + c, k), c = \sqrt{a^2 - b^2}$
Vertices	$(h - a, k)$ and $(h + a, k)$
Major axis	Parallel to x-axis between $(h - a, k)$ and $(h + a, k)$
Minor axis	Parallel to y-axis between $(h, k - b)$ and $(h, k + b)$

If the center is at the origin, then $(h, k) = (0, 0)$.

▶ Ellipse with a vertical major axis:

Equation	$\dfrac{(x-h)^2}{b^2} + \dfrac{(y-k)^2}{a^2} = 1$
Center	(h, k)
Foci	$(h, k-c), (h, k+c), c = \sqrt{a^2 - b^2}$
Vertices	$(h, k-a)$ and $(h, k+a)$
Major axis	Parallel to y-axis between $(h, k-a)$ and $(h, k+a)$
Minor axis	Parallel to x-axis between $(h-b, k)$ and $(h+b, k)$

If the center is at the origin, then $(h, k) = (0, 0)$.

7.2 Exercises

▶**Just in Time Exercises** These exercises correspond to the Just in Time references in this section. Complete them to review topics relevant to the remaining exercises.

1. True or False: The distance between the points (a, b) and $(0, 4)$ is given by $d = \sqrt{a^2 + (b-4)^2}$.

2. True or False: The distance between the points (a, b) and $(2, -7)$ is given by $d = \sqrt{(a+7)^2 + (b-2)^2}$.

3. Find the distance between the points $(3, 4)$ and $(1, -2)$.

4. Find the distance between the points $(5, -1)$ and $(5, -8)$.

In Exercises 5 and 6, what number must be added to complete the square?

5. $x^2 - 12x$

6. $y^2 + 24y$

▶**Skills** This set of exercises will reinforce the skills illustrated in this section.

In Exercises 7–10, match each graph (labeled a, b, c, and d) with the appropriate equation.

a.
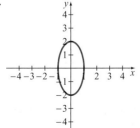

b.
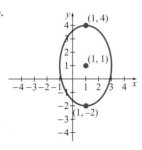

c.
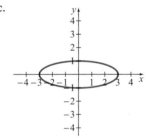

d.
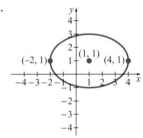

7. $\dfrac{x^2}{9} + y^2 = 1$

8. $x^2 + \dfrac{y^2}{4} = 1$

9. $\dfrac{(x-1)^2}{9} + \dfrac{(y-1)^2}{4} = 1$

10. $\dfrac{(x-1)^2}{4} + \dfrac{(y-1)^2}{9} = 1$

In Exercises 11–34, find the center, vertices, and foci of the ellipse that satisfies the given equation, and sketch the ellipse.

11. $\dfrac{x^2}{25} + \dfrac{y^2}{16} = 1$

12. $\dfrac{x^2}{9} + \dfrac{y^2}{25} = 1$

13. $\dfrac{x^2}{16} + \dfrac{y^2}{25} = 1$

14. $\dfrac{x^2}{25} + \dfrac{y^2}{9} = 1$

15. $\dfrac{x^2}{9} + \dfrac{y^2}{4} = 1$

16. $\dfrac{x^2}{9} + \dfrac{y^2}{16} = 1$

17. $\dfrac{4x^2}{9} + \dfrac{9y^2}{16} = 1$

18. $\dfrac{16x^2}{49} + \dfrac{25y^2}{81} = 1$

19. $\dfrac{(x+3)^2}{9} + \dfrac{(y+1)^2}{16} = 1$

20. $\dfrac{(x+5)^2}{25} + \dfrac{(y-3)^2}{16} = 1$

21. $\dfrac{(x-1)^2}{100} + \dfrac{(y+1)^2}{36} = 1$

22. $\dfrac{(x+6)^2}{64} + \dfrac{y^2}{100} = 1$

23. $\dfrac{(x-1)^2}{16} + \dfrac{(y+2)^2}{9} = 1$

24. $\dfrac{(x-6)^2}{36} + \dfrac{(y-3)^2}{25} = 1$

25. $3x^2 + 4y^2 = 12$

26. $25x^2 + 12y^2 = 300$

27. $5x^2 + 2y^2 = 10$

28. $4x^2 + 7y^2 = 28$

29. $4x^2 + y^2 - 24x - 8y + 48 = 0$

30. $x^2 + 9y^2 + 6x - 36y + 36 = 0$

31. $5x^2 + 9y^2 - 20x + 54y + 56 = 0$

32. $9x^2 + 16y^2 + 36x - 16y - 104 = 0$

33. $25x^2 + 16y^2 - 200x + 96y + 495 = 0$

34. $9x^2 + 4y^2 + 90x - 16y + 216 = 0$

In Exercises 35–40, use a graphing utility to graph the given equation.

35. $3x^2 + 7y^2 = 20$

36. $8x^2 + 3y^2 = 15$

37. $\dfrac{x^2}{6} + \dfrac{y^2}{11} = 1$

38. $\dfrac{x^2}{4} + \dfrac{y^2}{13} = 1$

39. $\dfrac{(x-1)^2}{5} + \dfrac{(y+3)^2}{6} = 1$

40. $\dfrac{(x+4)^2}{7} + \dfrac{(y-1)^2}{3} = 1$

In Exercises 41–48, determine the equation in standard form of the ellipse centered at the origin that satisfies the given conditions.

41. One vertex at $(6, 0)$; one focus at $(3, 0)$

42. One vertex at $(7, 0)$; one focus at $(2, 0)$

43. Minor axis of length $\sqrt{39}$; foci at $(-5, 0)$, $(5, 0)$

44. Minor axis of length 8; foci at $(0, -5)$, $(0, 5)$

45. Minor axis of length 7; major axis of length 9; major axis vertical

46. Minor axis of length 6; major axis of length 14; major axis horizontal

47. One endpoint of minor axis at $(2, 0)$; major axis of length 18

48. One endpoint of minor axis at $(0, -4)$: major axis of length 12

In Exercises 49–56, determine the equation in standard form of the ellipse that satisfies the given conditions.

49. Center at $(-2, 4)$; one vertex at $(-6, 4)$; one focus at $(1, 4)$

50. Center at $(2, 1)$; one vertex at $(7, 1)$; one focus at $(-2, 1)$

51. Vertices at $(-3, 4)$, $(-3, -2)$; foci at $(-3, 3)$, $(-3, 1)$

52. Vertices at $(5, 6)$, $(5, -4)$; foci at $(5, 4)$, $(5, -2)$

53. Major axis of length 8; foci at $(4, 1)$, $(4, -3)$

54. Center at $(-9, 3)$; one focus at $(-5, 3)$; one vertex at $(-3, 3)$

55. One endpoint of minor axis at $(7, -4)$; center at $(7, -8)$; major axis of length 12

56. One endpoint of minor axis at $(-6, 1)$; one vertex at $(-3, -4)$; major axis of length 10

In Exercises 57–60, determine the equations in standard form of two different ellipses that satisfy the given conditions.

57. Center at $(0, 0)$; major axis of length 9; minor axis of length 5

58. Center at $(3, 2)$; major axis of length 11; minor axis of length 6

59. One endpoint of minor axis at $(5, 0)$; minor axis of length 8; major axis of length 12

60. Major axis of length 10; minor axis of length 4; one vertex at $(-5, -3)$; major axis horizontal

▶**Applications** In this set of exercises, you will use ellipses to study real-world problems.

61. **Medical Technology** A lithotripter is a device that breaks up kidney stones by propagating shock waves through water in a chamber that has an elliptical cross-section. High-frequency shock waves are produced at one focus, and the patient is positioned in such a way that the kidney stones are at the other focus. On striking a point on the boundary of the chamber, the shock waves are reflected to the other focus and break up the kidney stones. Find the coordinates of the foci if the center of the ellipse is at the origin, one vertex is at $(6, 0)$, and one endpoint of the minor axis is at $(0, -2.5)$.

62. **Design** An arched bridge over a 20-foot stream is in the shape of the top half of an ellipse. The highest point of the bridge is 5 feet above the base. How high is a point on the bridge that is 5 feet (horizontally) from one end of the base of the bridge?

63. **Astronomy** The orbit of the moon around Earth is an ellipse, with Earth at one focus. If the major axis of the orbit is 477,736 miles and the minor axis is 477,078 miles, find the maximum and minimum distances from Earth to the moon.

64. **Reflective Property** An ellipse has a reflective property. A sound wave that passes through one focus of an ellipse and strikes some point on the ellipse will be reflected through the other focus. This reflective property has been used in architecture to design whispering galleries, such as the one in the Capitol building in Washington, D.C. A weak whisper at one focus can be heard clearly at the other focus, but nowhere else in the room. The dome of the Capitol has an elliptical cross-section. If the length of the major axis is 400 feet and the highest point of the dome is 60 feet above the floor, where should two

senators stand to hear each other whisper? Assume that the ellipse is centered at the origin.

65. **Physics** A laser is located at one focus of an ellipse. A sheet of metal, which is only a fraction of an inch wide and serves as a reflecting surface, lines the entire ellipse and is located at the same height above the ground as the laser. A very narrow beam of light is emitted by the laser. When the beam strikes the metal, it is reflected toward the other focus of the ellipse. If the foci are 20 feet apart and the shorter dimension of the ellipse is 12 feet, how great a distance is traversed by the beam of light from the time it is emitted by the laser to the time it reaches the other focus?

66. **Graphic Design** A graphic artist draws a schematic of an elliptically-shaped logo for an IT firm. The shorter dimension of the logo is 10 inches. If the foci are 8 inches apart, what is the longer dimension of the logo?

67. **Leisure** One of the holes at a miniature golf course is in the shape of an ellipse. The teeing-off point and the cup are located at the foci, which are 12 feet apart. The minor axis of the ellipse is 8 feet long. If on the first shot a player hits the ball off to either side at ground level, and the ball rolls until it reaches some point on the (elliptical) boundary, the ball will bounce off and proceed toward the cup. Under such circumstances, what is the total distance traversed by the ball if it just barely reaches the cup?

68. **Horse Racing** An elliptical track is used for training race horses and their jockeys. Under normal circumstances, two coaches are stationed in the interior of the track, one at each focus, to observe the races and issue commands to the jockeys. One day, when only one coach was on duty, a horse threw its rider just as it reached the vertex closest to the vacant observation post. The coach at the other post called to the horse, which dutifully came running straight toward her. How far did the horse run before reaching the coach if the minor axis of the ellipse is 600 feet long and each observation post is 400 feet from the center of the interior of the track?

69. **Construction** A cylindrically-shaped vent pipe with a diameter of 4 inches is to be attached to the roof of a house. The pipe first has to be cut at an angle that matches the slope of the roof, which is $\frac{3}{4}$. Once that has been done, the lower edge of the pipe will be elliptical. If the shorter dimension of the lower edge of the pipe is to be parallel to the base of the roof, what is the longer dimension?

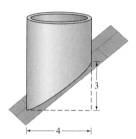

▶**Concepts** This set of exercises will draw on the ideas presented in this section and your general math background.

70. For the ellipse with equation $\frac{x^2}{16} + \frac{y^2}{9} = 1$, find the distance from either endpoint of the major axis to either endpoint of the minor axis.

71. For the ellipse with equation $\frac{(x+2)^2}{25} + \frac{(y-3)^2}{144} = 1$, find the distance from either endpoint of the major axis to either endpoint of the minor axis.

72. Write the equation that is satisfied by the set of points whose distances from the points $(3, 0)$ and $(-3, 0)$ add up to 8.

73. Write the equation that is satisfied by the set of points whose distances from the points $(0, 5)$ and $(0, -5)$ average to 6.

74. What is the set of all points that satisfy an equation of the form $\frac{x^2}{a^2} + \frac{y^2}{b^2} = 1$ if $a = b > 0$?

75. The *eccentricity* of an ellipse is defined as $e = \frac{c}{a}$ $\left(= \frac{\sqrt{a^2 - b^2}}{a} \right)$, where a, b, and c are as defined in this section. Since $0 < c < a$, the value of e lies between 0 and 1. In ellipses that are long and thin, b is small compared to a, so the eccentricity is close to 1. In ellipses that are nearly circular, b is almost as large as a, so the eccentricity is close to 0. What is the eccentricity of the ellipse with equation $\frac{x^2}{9} + \frac{y^2}{25} = 1$? Does this ellipse have a greater or lesser eccentricity than the ellipse with equation $\frac{x^2}{16} + \frac{y^2}{25} = 1$?

7.3 The Hyperbola

Objectives

▶ Define a hyperbola

▶ Find the foci, vertices, transverse axis, and asymptotes of a hyperbola

▶ Determine the equation of a hyperbola and write it in standard form

▶ Translate a hyperbola in the *xy*-plane

▶ Sketch a hyperbola

The last of the conic sections that we will discuss is the hyperbola. We first give its definition and derive its equation. The rest of this section will analyze various equations of hyperbolas, and conclude with an application to sculpture.

Definition of a Hyperbola

A **hyperbola** is the set of all points (x, y) in a plane such that the absolute value of the difference of the distances of (x, y) from two fixed points in the plane is equal to a fixed positive number d, where d is less than the distance between the fixed points. Each of the fixed points is called a **focus** of the hyperbola. Together, they are called the **foci.**

A hyperbola consists of a pair of non-intersecting, open curves, each of which is called a **branch** of the hyperbola. The two points on the hyperbola that lie on the line passing through the foci are known as the **vertices** of the hyperbola. The **transverse axis** of a hyperbola is the line segment that has the vertices as its endpoints. See Figure 7.3.1.

Figure 7.3.1

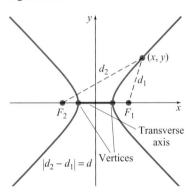

The only hyperbolas discussed in this section are those whose transverse axes are either horizontal or vertical. We will use the definition of a hyperbola to derive the equations of such hyperbolas, beginning with those whose centers are at the origin.

Equation of a Hyperbola Centered at the Origin

Suppose a hyperbola is centered at the origin, with its foci F_1 and F_2 on the x-axis. If c is the distance from the center to either focus, then F_1 and F_2 are located at $(c, 0)$ and $(-c, 0)$, respectively. Let (x, y) be any point on the hyperbola, and let d_1 and d_2 be the distances from (x, y) to F_1 and F_2, respectively. See Figure 7.3.2.

Figure 7.3.2

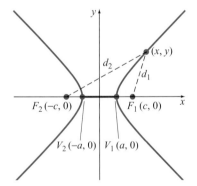

Now d is the distance between the vertices, and the vertices are symmetric with respect to the center of the hyperbola, so the distance from the center of the hyperbola to either vertex is $\frac{d}{2}$. Thus the coordinates of the vertices are $(a, 0)$ and $(-a, 0)$, where $a = \frac{d}{2}$.

Because $a = \frac{d}{2}$, we can write $d = 2a$. Hence the equation $|d_2 - d_1| = 2a$ is satisfied by all points on the hyperbola.

$$|d_2 - d_1| = 2a \Longrightarrow d_2 - d_1 = \pm 2a$$
$$\sqrt{(x + c)^2 + y^2} - \sqrt{(x - c)^2 + y^2} = \pm 2a.$$

Just In Time

Review the distance formula in Section 2.1.

Isolating the first radical and squaring both sides gives

$$\sqrt{(x + c)^2 + y^2} = \pm 2a + \sqrt{(x - c)^2 + y^2}$$
$$(x + c)^2 + y^2 = 4a^2 \pm 4a\sqrt{(x - c)^2 + y^2}$$
$$+ (x - c)^2 + y^2.$$

We then switch sides and isolate the radical.

$$4a^2 \pm 4a\sqrt{(x - c)^2 + y^2} + (x - c)^2 + y^2 = (x + c)^2 + y^2$$
$$\pm 4a\sqrt{(x - c)^2 + y^2} = (x + c)^2 - (x - c)^2 - 4a^2$$

Expanding the right-hand side and combining like terms gives

$$\pm 4a\sqrt{(x - c)^2 + y^2} = x^2 + 2cx + c^2$$
$$- (x^2 - 2cx + c^2) - 4a^2$$
$$\pm 4a\sqrt{(x - c)^2 + y^2} = 4cx - 4a^2.$$

Next, we divide by 4 and square both sides.

$$\pm a\sqrt{(x - c)^2 + y^2} = cx - a^2$$
$$a^2[(x - c)^2 + y^2] = c^2x^2 - 2a^2cx + a^4$$

Expanding the left-hand side and combining like terms, we get

$$a^2[(x^2 - 2cx + c^2) + y^2] = c^2x^2 - 2a^2cx + a^4$$
$$a^2x^2 - 2a^2cx + a^2c^2 + a^2y^2 = c^2x^2 - 2a^2cx + a^4$$
$$a^2x^2 + a^2c^2 + a^2y^2 = c^2x^2 + a^4.$$

Switching sides and isolating the constant terms then gives

$$c^2x^2 + a^4 = a^2x^2 + a^2c^2 + a^2y^2$$
$$c^2x^2 - a^2x^2 - a^2y^2 = a^2c^2 - a^4.$$

Factoring both sides leads to the equation

$$(c^2 - a^2)x^2 - a^2y^2 = a^2(c^2 - a^2).$$

Because $c > a$ and a and c are both positive, we can write $c^2 > a^2$. Thus there is some positive number b such that $c^2 - a^2 = b^2$. Substituting b^2 for $c^2 - a^2$ in the last equation and then dividing by a^2b^2, we obtain

$$b^2x^2 - a^2y^2 = a^2b^2$$
$$\frac{x^2}{a^2} - \frac{y^2}{b^2} = 1.$$

This is the standard form of the equation of a hyperbola with center at $(0, 0)$ and foci on the x-axis at $(c, 0)$ and $(-c, 0)$, where c is related to a and b via the equation $b^2 = c^2 - a^2$. Thus $c^2 = a^2 + b^2$.

Solving the equation $\dfrac{x^2}{a^2} - \dfrac{y^2}{b^2} = 1$ for y, we obtain

$$y = \pm \frac{b}{a}\sqrt{x^2 - a^2}.$$

As x becomes very large, the quantity $x^2 - a^2$ approaches x^2. Therefore, the graph of the equation $y = \dfrac{b}{a}\sqrt{x^2 - a^2}$ approaches the line $y = \dfrac{b}{a}x$, and the graph of the equation $y = -\dfrac{b}{a}\sqrt{x^2 - a^2}$ approaches the line $y = -\dfrac{b}{a}x$. These lines, $y = \dfrac{b}{a}x$ and $y = -\dfrac{b}{a}x$, are called the **asymptotes** of the hyperbola, and pass through the center of the hyperbola.

The standard equation of a hyperbola with center at $(0, 0)$ and foci on the y-axis at $(0, c)$ and $(0, -c)$ is $\frac{y^2}{a^2} - \frac{x^2}{b^2} = 1$, where $a^2 + b^2 = c^2$, using a similar derivation. The main features and properties of hyperbolas with center at the origin and a horizontal or vertical transverse axis are summarized next. Their graphs are shown in Figures 7.3.3 and 7.3.4.

Hyperbolas with Center at the Origin

	Horizontal Transverse Axis	**Vertical Transverse Axis**
Graph	Figure 7.3.3 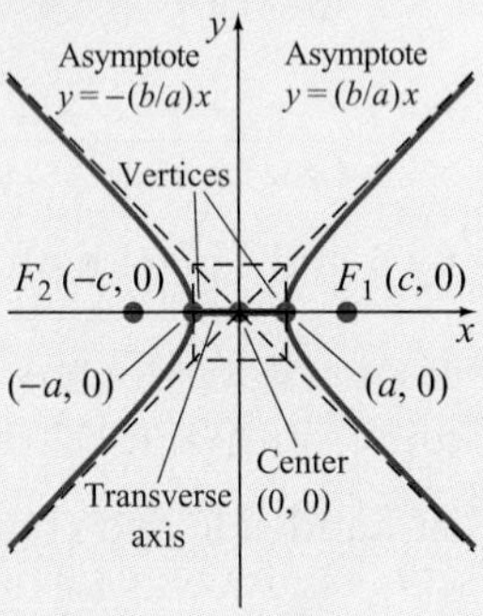	Figure 7.3.4 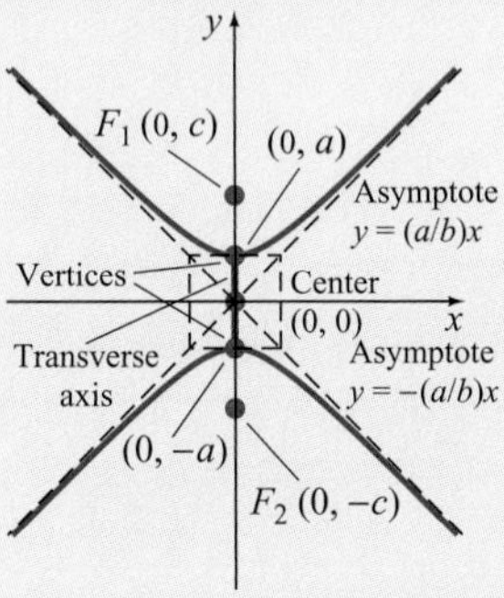
Equation	$\frac{x^2}{a^2} - \frac{y^2}{b^2} = 1$, $a, b > 0$	$\frac{y^2}{a^2} - \frac{x^2}{b^2} = 1$, $a, b > 0$
Center	$(0, 0)$	$(0, 0)$
Foci	$(-c, 0)$, $(c, 0)$, $c = \sqrt{a^2 + b^2}$	$(0, -c)$, $(0, c)$, $c = \sqrt{a^2 + b^2}$
Vertices	$(-a, 0)$, $(a, 0)$	$(0, -a)$, $(0, a)$
Transverse axis	Segment of x-axis from $(-a, 0)$ to $(a, 0)$	Segment of y-axis from $(0, -a)$ to $(0, a)$
Asymptotes	The lines $y = \frac{b}{a}x$ and $y = -\frac{b}{a}x$	The lines $y = \frac{a}{b}x$ and $y = -\frac{a}{b}x$

Observations:
- The transverse axis of a hyperbola is horizontal if the coefficient of the x^2 term is positive in the standard form of the equation.
- The transverse axis of a hyperbola is vertical if the coefficient of the y^2 term is positive in the standard form of the equation.
- The vertices of a hyperbola are a distance of a units from the center.
- Each branch of a hyperbola opens *away* from the center and *toward* one of the foci.
- The center of a hyperbola lies on the transverse axis, midway between the vertices, but the center is *not* part of the hyperbola.
- In the equation of a hyperbola, the variable y is not a function of x, since the graph of such an equation does not pass the vertical line test.

In Example 1, we show how the asymptotes of a hyperbola can be used to help sketch its graph.

Example 1 **Hyperbola Centered at the Origin**

Consider the hyperbola centered at the origin and defined by the equation

$$16x^2 - 9y^2 = 144.$$

(a) Write the equation of the hyperbola in standard form, and determine the orientation of the transverse axis.

(b) Find the coordinates of the vertices and foci.

(c) Find the equations of the asymptotes.

(d) Sketch the hyperbola, and indicate the vertices, foci, and asymptotes.

▶ Solution

(a) Divide the equation by 144 to get a 1 on one side, and then simplify.

$$\frac{16x^2}{144} - \frac{9y^2}{144} = 1 \implies \frac{x^2}{9} - \frac{y^2}{16} = 1$$

Because the y^2 term is subtracted from the x^2 term, the values of a^2 and b^2 are 9 and 16, respectively, and the transverse axis is a segment of the x-axis. Thus the transverse axis is horizontal, so the hyperbola opens to the side.

(b) The vertices are the endpoints of the transverse axis, so the vertices and foci lie on the x-axis. Using $a^2 = 9$ and $b^2 = 16$, we find that

$$c^2 = a^2 + b^2 = 16 + 9 = 25.$$

Using the fact that a, b, and c are all positive, we have

$$a = 3, \quad b = 4, \quad \text{and} \quad c = 5.$$

Vertices: $(\pm a, 0) \implies (3, 0)$ and $(-3, 0)$

Foci: $(\pm c, 0) \implies (5, 0)$ and $(-5, 0)$

(c) The equations of the asymptotes are given by

$$y = \frac{b}{a}x = \frac{4}{3}x \quad \text{and} \quad y = -\frac{b}{a}x = -\frac{4}{3}x.$$

Figure 7.3.5

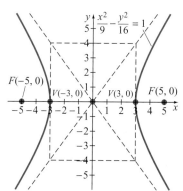

(d) To sketch the hyperbola, proceed as follows.

▶ Plot and label the vertices and the foci.

▶ Draw a rectangular box that is centered at $(0, 0)$ and has horizontal and vertical dimensions of $2a = 6$ and $2b = 8$, respectively.

▶ Graph the asymptotes, each of which is a line that passes through the center of the hyperbola *and* through one pair of opposite corners of the box.

▶ Sketch the hyperbola so that it passes through the vertices and approaches the asymptotes. See Figure 7.3.5.

✔ *Check It Out 1:* For the hyperbola given by the equation $\frac{x^2}{4} - \frac{y^2}{4} = 1$, find the coordinates of the vertices and foci and the equations of the asymptotes. Sketch the hyperbola. ■

In the next example, we derive the equation of a hyperbola given one of its foci and one of its vertices.

Example 2 Finding the Equation of a Hyperbola

Determine the equation in standard form of the hyperbola with center at $(0, 0)$, one focus at $(0, 4)$, and one vertex at $(0, -1)$. Find the other focus and the other vertex. Sketch a graph of the hyperbola by finding and plotting some additional points that lie on the hyperbola.

▶ **Solution** The given focus and vertex are on the y-axis. Because the transverse axis is a segment of the y-axis, the equation of the hyperbola is of the form $\frac{y^2}{a^2} - \frac{x^2}{b^2} = 1$. Because one of the foci is at $(0, 4)$, $c = 4$. Now, a is the distance from $(0, 0)$ to the vertex. Since there is a vertex at $(0, -1)$, $a = 1$. Substituting the values of a and c, we find that

$$b^2 = c^2 - a^2$$
$$= (4)^2 - (1)^2$$
$$= 16 - 1 = 15.$$

Thus the equation of the hyperbola in standard form is

$$\frac{y^2}{a^2} + \frac{x^2}{b^2} = \frac{y^2}{1} - \frac{x^2}{15} = 1.$$

Because the center is at $(0, 0)$, the other focus is at $(0, -c) = (0, -4)$ and the other vertex is at $(0, a) = (0, 1)$.

To find some additional points on the hyperbola, we solve the equation of the hyperbola for y to obtain

$$y = \pm\sqrt{1 + \frac{x^2}{15}}.$$

The expression for y is defined for all real numbers x. Substituting $x = 5$ gives

$$y = \pm\sqrt{1 + \frac{(5)^2}{15}} = \pm\sqrt{1 + \frac{5}{3}} = \pm\sqrt{\frac{8}{3}} \Longrightarrow y \approx \pm 1.633.$$

Similarly, substituting $x = -5$ gives $y \approx \pm 1.633$. We can sketch the hyperbola by plotting the two vertices, $(0, \pm 1)$, along with the four points $(5, \pm 1.633)$ and $(-5, \pm 1.633)$. See Figure 7.3.6.

Figure 7.3.6

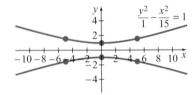

✔ *Check It Out 2:* Determine the equation in standard form of the hyperbola with center at $(0, 0)$, one focus at $(0, 4)$, and one vertex at $(0, 3)$. ■

Example 3 Graphing a Hyperbola Using a Graphing Utility

Use a graphing utility to graph the equation $16x^2 - 9y^2 = 144$.

▶**Solution** This is the equation of the hyperbola that we graphed by hand in Example 1. To use a graphing utility, we must first solve the equation for y.

$$16x^2 - 9y^2 = 144$$

$$-9y^2 = -16x^2 + 144 \qquad \text{Isolate } y^2 \text{ term}$$

$$y^2 = \frac{16x^2 - 144}{9} \qquad \text{Divide by } -9$$

$$y = \pm\sqrt{\frac{16x^2 - 144}{9}} \qquad \text{Take square root of both sides}$$

This gives us two functions,

$$Y_1(x) = \sqrt{\frac{16x^2 - 144}{9}} \quad \text{and} \quad Y_2(x) = -\sqrt{\frac{16x^2 - 144}{9}}.$$

Graphing these two functions using a window size of $[-10, 10]$ by $[-10, 10]$ and the SQUARE window option gives the graph shown in Figure 7.3.7.

Figure 7.3.7

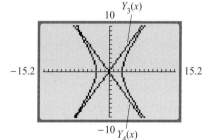

You can graph the asymptotes in the same window by entering the functions

$$Y_3(x) = \frac{4}{3}x \quad \text{and} \quad Y_4(x) = -\frac{4}{3}x.$$

To visualize the concept of an asymptote, use ASK mode to generate a table of values for the functions Y_1 and Y_3, using 100, 1000, and 10,000 as the values of x. For $|x|$ large, we see that the values of Y_1 are very close to those of Y_3. The same can be done for the functions Y_1 and Y_4, using -100, -1000, and $-10,000$ as the values of x. See Figure 7.3.8.

Figure 7.3.8

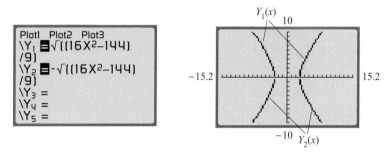

 Check It Out 3: Use a graphing utility to graph the equation $\dfrac{x^2}{4} - \dfrac{y^2}{4} = 1$. ▦

Equation of a Hyperbola with Center at the Point (h, k)

The main features and properties of hyperbolas with center at the point (h, k) and a horizontal or vertical transverse axis are summarized next. Their graphs are shown in Figures 7.3.9 and 7.3.10.

Hyperbolas with Center at (h, k)

	Horizontal Transverse Axis	**Vertical Transverse Axis**
Graph	**Figure 7.3.9** 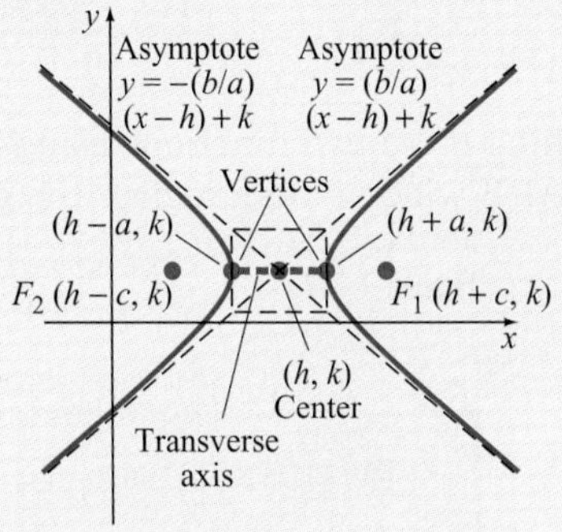	**Figure 7.3.10** 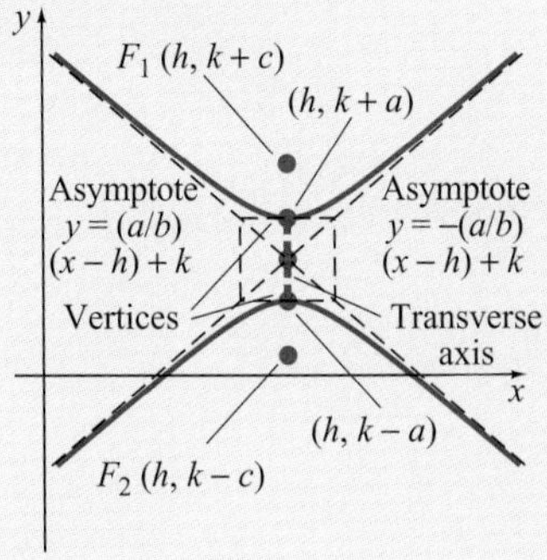
Equation	$\dfrac{(x - h)^2}{a^2} - \dfrac{(y - k)^2}{b^2} = 1, a, b > 0$	$\dfrac{(y - k)^2}{a^2} - \dfrac{(x - h)^2}{b^2} = 1, a, b > 0$
Center	(h, k)	(h, k)
Foci	$(h - c, k), (h + c, k), c = \sqrt{a^2 + b^2}$	$(h, k - c), (h, k + c), c = \sqrt{a^2 + b^2}$
Vertices	$(h - a, k), (h + a, k)$	$(h, k - a), (h, k + a)$
Transverse axis	Segment of the line $y = k$ from $(h - a, k)$ to $(h + a, k)$	Segment of the line $x = h$ from $(h, k - a)$ to $(h, k + a)$
Asymptotes	The lines $y = \dfrac{b}{a}(x - h) + k$ and $y = -\dfrac{b}{a}(x - h) + k$	The lines $y = \dfrac{a}{b}(x - h) + k$ and $y = -\dfrac{a}{b}(x - h) + k$

Example 4 **Finding the Equation of a Hyperbola with Vertex at (h, k)**

Determine the equation in standard form of the hyperbola with foci at $(4, 3)$ and $(4, -7)$ and a transverse axis of length 6. Find the asymptotes and sketch the hyperbola.

▶ **Solution**

Step 1 Determine the orientation of the hyperbola. Because the foci are located on the line $x = 4$, the transverse axis is vertical. You can see this by plotting the foci. Thus the standard form of the equation of the hyperbola is

$$\frac{(y - k)^2}{a^2} - \frac{(x - h)^2}{b^2} = 1.$$

Step 2 Next we must determine the center (h, k) and the values of a and b. Because the center is the midpoint of the line segment that has its endpoints at the foci, the coordinates of the center are

$$(h, k) = \left(\frac{4 + 4}{2}, \frac{3 + (-7)}{2} \right) = (4, -2).$$

To find a, we note that the length of the transverse axis is given as 6. The distance from either vertex to the center is half the length of the transverse axis, so $a = 3$. Because c is the distance from either focus to the center, we obtain

$$c = \text{distance from } (4, 3) \text{ to } (4, -2) = 3 - (-2) = 5.$$

To find b, we substitute the values of a and c.

$$b^2 = c^2 - a^2 = (5)^2 - (3)^2 = 25 - 9 = 16 \Longrightarrow b = 4$$

Substituting the values of h, k, a^2, and b^2, we find that the equation of the hyperbola is

$$\frac{(y - k)^2}{a^2} - \frac{(x - h^2)}{b^2} = \frac{(y + 2)^2}{9} - \frac{(x - 4)^2}{16} = 1.$$

Step 3 Now we can find the asymptotes and vertices. The asymptotes are

$$y = \frac{a}{b}(x - h) + k = \frac{3}{4}(x - 4) - 2 = \frac{3}{4}x - 5 \quad \text{and}$$

$$y = -\frac{a}{b}(x - h) + k = -\frac{3}{4}(x - 4) - 2 = -\frac{3}{4}x + 1.$$

The vertices are located at $(4, -2 + 3) = (4, 1)$ and $(4, -2 - 3) = (4, -5)$.

Step 4 Sketch the hyperbola as follows.

▶ Plot and label the vertices.

▶ Draw a rectangular box that is centered at $(h, k) = (4, -2)$ and has horizontal and vertical dimensions of $2b = 8$ and $2a = 6$, respectively.

▶ Graph the asymptotes, each of which is a line that passes through the center of the hyperbola *and* through one pair of opposite corners of the box.

▶ Sketch the hyperbola so that it passes through the vertices and approaches the asymptotes. See Figure 7.3.11.

Figure 7.3.11

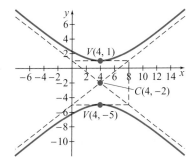

✔ *Check It Out 4:* Determine the equation in standard form of the hyperbola with foci at $(4, 3)$ and $(-4, 3)$ and a transverse axis of length 4. ■

Just In Time

Review completing the square in Section 2.1.

Example **5** **Completing the Square to Write the Equation of a Hyperbola**

Write the equation in standard form of the hyperbola defined by the equation $x^2 - 4y^2 + 2x - 24y = 39$. Find the vertices and the foci, and sketch the hyperbola.

▶**Solution** First, rewrite the given equation by grouping the x terms and the y terms separately and then factoring -4 out of the y terms.

$$(x^2 + 2x) - 4(y^2 + 6y) = 39 \qquad \text{Group x and y terms and factor}$$
$$\text{-4 out of y terms}$$

$$(x^2 + 2x + 1) - 4(y^2 + 6y + 9) = 39 + 1 - 36 \qquad \text{Complete the square on both x}$$
$$\text{and y}$$

$$(x + 1)^2 - 4(y + 3)^2 = 4 \qquad \text{Factor}$$

$$\frac{(x + 1)^2}{4} - \frac{(y + 3)^2}{1} = 1 \qquad \text{Write in standard form}$$

From the standard form of the equation, we see that the center of the hyperbola is at $(h, k) = (-1, -3)$ and that $a^2 = 4$ and $b^2 = 1$. Thus $a = 2$ and $b = 1$.

Since the $(y - k)^2$ term is subtracted from the $(x - h)^2$ term, the transverse axis is horizontal. Thus the vertices lie on the line $y = -3$.

Vertices: $(h + a, k) = (-1 + 2, -3) = (1, -3)$ and $(h - a, k) = (-1 - 2, -3) = (-3, -3)$

To find the foci, we determine c, the distance from the center to either focus.

$$c^2 = a^2 + b^2 = 4 + 1 = 5 \Longrightarrow c = \sqrt{5}$$

Foci: $(h + c, k) = \left(-1 + \sqrt{5}, -3\right)$ and $(h - c, k) = \left(-1 - \sqrt{5}, -3\right)$

The asymptotes are

$$y = \frac{b}{a}(x - h) + k = \frac{1}{2}(x + 1) - 3 \quad \text{and} \quad y = -\frac{b}{a}(x - h) + k = -\frac{1}{2}(x + 1) - 3.$$

To sketch the hyperbola, first plot and label the center of the hyperbola. Then draw a rectangular box that is centered at $(h, k) = (-1, -3)$ and has horizontal and vertical dimensions of $2a = 4$ and $2b = 2$, respectively. Graph the asymptotes, each of which is a line that passes through the center of the hyperbola *and* through one pair of opposite corners of the box. Sketch the hyperbola so that it passes through the vertices and approaches the asymptotes. See Figure 7.3.12.

Figure 7.3.12

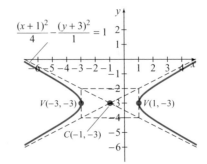

Check It Out 5: Write the equation in standard form of the hyperbola defined by the equation $x^2 - y^2 + 2x - 4y = 12$. ■

Application of Hyperbola

In Example 6, we examine a sculpture that is designed in the shape of a hyperbola.

Example **6** **Application to Sculpture**

The front face of a wire frame sculpture is in the shape of the branches of a hyperbola that opens to the side. The transverse axis of the hyperbola is 40 inches long. If the base of the sculpture is 60 inches below the transverse axis and one of the asymptotes has a slope of $\frac{3}{2}$, how wide is the sculpture at the base?

Figure 7.3.13

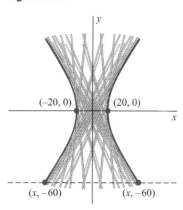

▶**Solution** First sketch a hyperbola with transverse axis on the x-axis, centered at $(0, 0)$. See Figure 7.3.13. We want to find the x-coordinates of two points on the hyperbola such that the y-coordinate of those two points is -60. The distance between the two points will give the width of the sculpture at the base. Because the length of the transverse axis is 40 inches, the vertices are $(20, 0)$ and $(-20, 0)$.

The next step is to find the equation of the hyperbola. Because the transverse axis is horizontal, we have the equation

$$\frac{x^2}{a^2} - \frac{y^2}{b^2} = 1.$$

Now we must find a and b. The point $(20, 0)$ is a vertex, so $a = 20$. Using the fact that the slope of an asymptote of the hyperbola is $\frac{3}{2}$, we obtain

$$\frac{b}{a} = \frac{b}{20} = \frac{3}{2} \Longrightarrow b = 30.$$

Thus the equation of the hyperbola is

$$\frac{x^2}{20^2} - \frac{y^2}{30^2} = 1.$$

Now, if the base of the sculpture is 60 inches below the transverse axis, we need to find x such that $(x, -60)$ is on the hyperbola.

$$\frac{x^2}{20^2} - \frac{y^2}{30^2} = 1 \qquad \text{Equation of hyperbola}$$

$$\frac{x^2}{20^2} - \frac{(-60)^2}{30^2} = 1 \qquad \text{Substitute } y = -60$$

$$\frac{x^2}{400} - \frac{3600}{900} = 1 \qquad \text{Simplify}$$

$$\frac{x^2}{400} - 4 = 1$$

$$x^2 - 1600 = 400 \qquad \text{Multiply both sides of equation by 400}$$

$$x^2 = 2000 \Longrightarrow x = \pm 20\sqrt{5}$$

The distance between $\left(-20\sqrt{5}, -60\right)$ and $\left(20\sqrt{5}, -60\right)$ is $40\sqrt{5}$. Thus the sculpture is $40\sqrt{5} \approx 89.44$ inches wide at the base.

✔ *Check It Out 6:* In Example 6, if the top of the sculpture is 30 inches above the transverse axis, how wide is the sculpture at the top? ■

7.3 Key Points

▶ Hyperbola with a horizontal transverse axis:

Equation	$\dfrac{(x - h)^2}{a^2} - \dfrac{(y - k)^2}{b^2} = 1, a, b > 0$
Center	(h, k)
Foci	$(h - c, k), (h + c, k), c = \sqrt{a^2 + b^2}$
Vertices	$(h - a, k), (h + a, k)$
Transverse axis	Parallel to the x-axis between $(h - a, k)$ and $(h + a, k)$
Asymptotes	$y = \dfrac{b}{a}(x - h) + k$ and $y = -\dfrac{b}{a}(x - h) + k$

▶ Hyperbola with a vertical transverse axis:

Equation	$\dfrac{(y - k)^2}{a^2} - \dfrac{(x - h)^2}{b^2} = 1, a, b > 0$
Center	(h, k)
Foci	$(h, k - c), (h, k + c), c = \sqrt{a^2 + b^2}$
Vertices	$(h, k - a), (h, k + a)$
Transverse axis	Parallel to the y-axis between $(h, k - a)$ and $(h, k + a)$
Asymptotes	$y = \dfrac{a}{b}(x - h) + k$ and $y = -\dfrac{a}{b}(x - h) + k$

7.3 Exercises

▶**Just in Time Exercises** These exercises correspond to the Just in Time references in this section. Complete them to review topics relevant to the remaining exercises.

In Exercises 1–3, find the distance between the given points.

1. (c, d) and (w, v) 2. $(2, 1)$ and $(-5, -1)$

3. $(-3, 5)$ and $(0, 5)$

In Exercises 4–6, complete the square.

4. $y^2 + 8y$

5. $x^2 + 22x$

6. $x^2 - 2x$

▶**Skills** This set of exercises will reinforce the skills illustrated in this section.

In Exercises 7–10, match each graph (labeled a, b, c, and d) with the appropriate equation.

a.
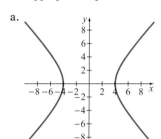

b.
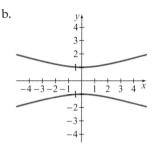

c.
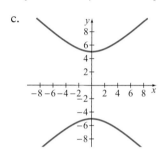

d.
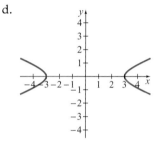

7. $\dfrac{x^2}{9} - y^2 = 1$

8. $y^2 - \dfrac{x^2}{9} = 1$

9. $y^2 - x^2 = 25$

10. $x^2 - y^2 = 16$

In Exercises 11–34, find the center, vertices, foci, and asymptotes of the hyperbola that satisfies the given equation, and sketch the hyperbola.

11. $\dfrac{x^2}{16} - \dfrac{y^2}{9} = 1$

12. $\dfrac{y^2}{16} - \dfrac{x^2}{9} = 1$

13. $\dfrac{y^2}{9} - \dfrac{x^2}{16} = 1$

14. $\dfrac{4x^2}{9} - \dfrac{9y^2}{16} = 1$

15. $\dfrac{y^2}{64} - \dfrac{x^2}{36} = 1$

16. $\dfrac{y^2}{4} - \dfrac{x^2}{4} = 1$

17. $x^2 - y^2 = 9$

18. $4x^2 - 9y^2 = 36$

19. $\dfrac{(x+3)^2}{16} - \dfrac{(y+1)^2}{9} = 1$

20. $\dfrac{(x-6)^2}{9} - \dfrac{(y+4)^2}{16} = 1$

21. $\dfrac{(x+5)^2}{25} - \dfrac{(y+1)^2}{16} = 1$

22. $\dfrac{(y-4)^2}{25} - \dfrac{(x-2)^2}{9} = 1$

23. $y^2 - (x+4)^2 = 1$

24. $\dfrac{x^2}{16} - \dfrac{(y+5)^2}{4} = 1$

25. $\dfrac{(y-3)^2}{9} - \dfrac{(x+1)^2}{25} = 1$

26. $\dfrac{(y-3)^2}{25} - \dfrac{(x+1)^2}{9} = 1$

27. $\dfrac{(x+2)^2}{144} - \dfrac{(y-3)^2}{25} = 1$

28. $\dfrac{(y-3)^2}{25} - \dfrac{(x+2)^2}{144} = 1$

29. $36x^2 - 16y^2 = 225$

30. $9y^2 - 16x^2 = 100$

31. $9x^2 + 54x - y^2 = 0$

32. $y^2 - 9x^2 - 18x = 18$

33. $8x^2 - 32x - y^2 - 6y = 41$

34. $4y^2 - 16y - x^2 + 12x = 29$

In Exercises 35–48, determine the equation in standard form of the hyperbola that satisfies the given conditions.

35. Vertices at $(3, 0)$, $(-3, 0)$; foci at $(4, 0)$, $(-4, 0)$

36. Vertices at $(0, 2)$, $(0, -2)$; foci at $(0, 3)$, $(0, -3)$

37. Foci at $(4, 0)$, $(-4, 0)$; asymptotes $y = \pm 2x$

38. Foci at $(0, 5)$, $(0, -5)$; asymptotes $y = \pm x$

39. Foci at $(2, 0)$, $(-2, 0)$; passes through the point $(2, 3)$

40. Foci at $(5, 0)$, $(-5, 0)$; passes through the point $(3, 0)$

41. Vertices at $(-4, 0)$, $(4, 0)$; passes through the point $(8, 2)$

42. Vertices at $(0, 5)$, $(0, -5)$; passes through the point $\left(12, 5\sqrt{2}\right)$

43. Foci at $(-3, -6)$, $(-3, -2)$; slope of one asymptote is 1

44. Foci at $(4, -2)$, $(-2, -2)$; slope of one asymptote is $\dfrac{\sqrt{5}}{2}$

45. Vertices at $(5, -2)$, $(1, -2)$; slope of one asymptote is $\dfrac{5}{2}$

46. Vertices at $(4, 6)$, $(-4, 6)$; slope of one asymptote is -2

47. Transverse axis of length 10; center at $(1, -4)$; one focus at $(9, -4)$

48. Transverse axis of length 6; one vertex at $(7, 8)$; one focus at $(7, 5)$

In Exercises 49–54, use a graphing utility to graph the given equation.

49. $x^2 - 5y^2 = 10$

50. $3y^2 - 2x^2 = 15$

51. $\dfrac{x^2}{5} - \dfrac{y^2}{7} = 1$

52. $\dfrac{y^2}{10} - \dfrac{x^2}{12} = 1$

53. $\dfrac{(x + 1)^2}{15} - \dfrac{(y - 3)^2}{3} = 1$

54. $\dfrac{(y - 4)^2}{8} - \dfrac{x^2}{13} = 1$

In Exercises 55–58, determine the equations in standard form of two different hyperbolas that satisfy the given conditions.

55. Center at $(0, 0)$; transverse axis of length 12; slope of one asymptote is 4

56. Center at $(-3, -6)$; distance of one vertex from center is 5; distance of one focus from center is 7

57. Transverse axis of length 6; transverse axis vertical; one vertex at $(-1, 1)$; distance of one focus from nearest vertex is 4

58. Transverse axis of length 12; transverse axis horizontal; one vertex at $(6, 5)$; slope of one asymptote is -5

▶**Applications** In this set of exercises, you will use hyperbolas to study real-world problems.

59. **Interior Design** The service counter in the reference area of a college library forms one branch of a hyperbola, and the service counter in the circulation area forms the other branch. The vertices of the hyperbola are 16 feet apart. Using a coordinate system set up such that the transverse axis of the hyperbola is horizontal, the equations of the asymptotes are $y = \pm\dfrac{3}{4}x$. How far apart are the reference librarian's desk and the circulation librarian's desk if the desks are located at the foci of the hyperbola?

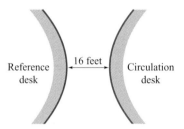

60. **Video Games** A video game designer is creating a video game about a bank robbery. In the game, a person robs a bank and then flees. The robber's path forms one branch of a hyperbola. At the time of the robbery, the bank's security officer is stationed at one focus of the hyperbola and a police cruiser is parked at the opposite vertex (the vertex closer to the opposite focus). If the length of the transverse axis of the hyperbola is 70 millimeters and the foci are 120 millimeters apart, how far from the police cruiser was the bank's security officer when the robbery occurred?

61. **Art** The front face of a small paperweight sold by a modern art store is in the shape of a hyperbola, and includes both of its branches. The transverse axis of the paperweight has a length of 4 inches and is oriented horizontally. If the base of the paperweight is 2 inches below the transverse axis and one of the asymptotes has a slope of 1, how wide is the paperweight at the base?

62. **Physics** Because positively-charged particles repel each other, there is a limit on how close a small, positively-charged particle can get to the nucleus of a heavy atom. (A nucleus is positively charged.) As a result, the smaller particle follows a hyperbolic path in the neighborhood of the nucleus. If the asymptotes of the hyperbola have slopes of ± 1, what is the overall change in the direction of the path of the smaller particle as it first approaches the nucleus of the heavy atom and ultimately recedes from it?

63. **Astronomy** The path of a certain comet is known to be hyperbolic, with the sun at one focus. Assume that a space station is located 13 million miles from the sun and at the center of the hyperbola, and that the comet is 5 million miles from the space station at its point of closest approach. Find the equation of the hyperbola if the coordinate system is set up so that the sun lies on the x-axis and the origin coincides with the center of the hyperbola.

▶ **Concepts** This set of exercises will draw on the ideas presented in this section and your general math background.

64. For the hyperbola defined by the equation $25y^2 - 16x^2 = 400$, solve for x in terms of y and then use your expression for x to determine the equations of the asymptotes.

65. Find the domain of the function $f(x) = \sqrt{\dfrac{16x^2 - 144}{9}}$.

   This is the top half of the hyperbola discussed in Example 1.

66. The hyperbolas $\dfrac{x^2}{a^2} - \dfrac{y^2}{b^2} = 1$ and $\dfrac{y^2}{a^2} - \dfrac{x^2}{b^2} = 1$ are called conjugate hyperbolas. Show that one hyperbola is the reflection of the other in the line $y = x$.

67. Let a, b, and c be as defined in this section. For a rectangle with a length of $2a$ and a width of $2b$, express the length of a diagonal of the rectangle in terms of c alone.

68. What are the slopes of the asymptotes of a hyperbola that satisfies an equation of the form $\dfrac{x^2}{a^2} - \dfrac{y^2}{b^2} = 1$ if $a = b > 0$? At what angle do the asymptotes intersect?

69. Write the equation for the set of points whose distances from the points $(3, 2)$ and $(3, -8)$ differ by 6.

70. The *eccentricity* of a hyperbola is defined as $e = \dfrac{c}{a}\left(= \dfrac{\sqrt{a^2 + b^2}}{a}\right)$, where a, b, and c are as defined in this section. Because $0 < a < c$, the value of e is greater than 1. In hyperbolas that are highly curved at the vertices, b is small compared to a, so the eccentricity is close to 1. In hyperbolas that are nearly flat at the vertices, b is much larger than a, so the eccentricity is large. What is the eccentricity of the hyperbola with equation $\dfrac{x^2}{9} - \dfrac{y^2}{16} = 1$? Does this hyperbola have a greater or lesser eccentricity than the hyperbola with equation $\dfrac{x^2}{9} - \dfrac{y^2}{81} = 1$?

71. There are hyperbolas other than the types studied in this section. For example, some hyperbolas satisfy an equation of the form $xy = c$, where c is a nonzero constant. In which quadrant(s) of the coordinate plane does the hyperbola with equation $xy = 10$ lie? the hyperbola with equation $xy = -10$?

Chapter 7   # Summary

Section 7.1 **The Parabola**

Concept	Illustration	Study and Review
Parabola with vertical axis of symmetry	$(x - h)^2 = 4p(y - k), p > 0$ $(x - h)^2 = 4p(y - k), p < 0$	Examples 1, 4 Chapter 7 Review, Exercises 1–16
Parabola with horizontal axis of symmetry	$(y - k)^2 = 4p(x - h), p > 0$ $(y - k)^2 = 4p(x - h), p < 0$	Examples 2, 3, 5, 6 Chapter 7 Review, Exercises 1–16

Parabola with vertical axis of symmetry

Equation	$(x - h)^2 = 4p(y - k)$
Opening	Upward if $p > 0$, downward if $p < 0$
Vertex	(h, k)
Focus	$(h, k + p)$
Directrix	$y = k - p$
Axis of symmetry	$x = h$

Parabola with horizontal axis of symmetry

Equation	$(y - k)^2 = 4p(x - h)$
Opening	To the right if $p > 0$, to the left if $p < 0$
Vertex	(h, k)
Focus	$(h + p, k)$
Directrix	$x = h - p$
Axis of symmetry	$y = k$

Section 7.2 **The Ellipse**

Concept	Illustration	Study and Review
Ellipse with horizontal major axis	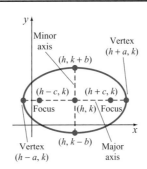	Examples 1, 3 Chapter 7 Review, Exercises 18–34

Ellipse with horizontal major axis

Equation	$\dfrac{(x-h)^2}{a^2} + \dfrac{(y-k)^2}{b^2} = 1$
Center	(h, k)
Foci	$(h-c, k), (h+c, k),$ $c = \sqrt{a^2 - b^2}$
Vertices	$(h-a, k)$ and $(h+a, k)$
Major axis	Parallel to x-axis between $(h-a, k)$ and $(h+a, k)$
Minor axis	Parallel to y-axis between $(h, k-b)$ and $(h, k+b)$

Ellipse with vertical major axis

Equation	$\dfrac{(x-h)^2}{b^2} + \dfrac{(y-k)^2}{a^2} = 1$
Center	(h, k)
Foci	$(h, k-c), (h, k+c),$ $c = \sqrt{a^2 - b^2}$
Vertices	$(h, k-a)$ and $(h, k+a)$
Major axis	Parallel to y-axis between $(h, k-a)$ and $(h, k+a)$
Minor axis	Parallel to x-axis between $(h-b, k)$ and $(h+b, k)$

Examples 2, 4

Chapter 7 Review, Exercises 18–34

Section 7.3 **The Hyperbola**

Concept	Illustration	Study and Review
Hyperbola with horizontal transverse axis	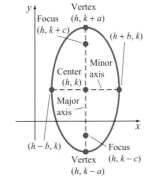	Examples 2, 5, 6 Chapter 7 Review, Exercises 35–50

Hyperbola with horizontal transverse axis

Equation	$\dfrac{(x-h)^2}{a^2} - \dfrac{(y-k)^2}{b^2} = 1,$ $a, b > 0$
Center	(h, k)
Foci	$(h-c, k), (h+c, k),$ $c = \sqrt{a^2 + b^2}$
Vertices	$(h-a, k), (h+a, k)$
Transverse axis	Parallel to x-axis between $(h-a, k)$ and $(h+a, k)$
Asymptotes	$y = \dfrac{b}{a}(x-h) + k$ and $y = -\dfrac{b}{a}(x-h) + k$

Continued

Section 7.3 **The Hyperbola**

Concept	Illustration	Study and Review
Hyperbola with vertical transverse axis	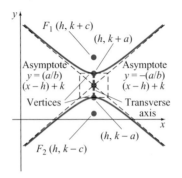	Examples 1, 4
Equation $\dfrac{(y-k)^2}{a^2} - \dfrac{(x-h)^2}{b^2} = 1,$ $a, b > 0$		Chapter 7 Review, Exercises 35–50
Center (h, k)		
Foci $(h, k - c), (h, k + c),$ $c = \sqrt{a^2 + b^2}$		
Vertices $(h, k - a), (h, k + a)$		
Transverse axis Parallel to y-axis between $(h, k - a)$ and $(h, k + a)$		
Asymptotes $y = \dfrac{a}{b}(x - h) + k$ and $y = -\dfrac{a}{b}(x - h) + k$		

Review Exercises

Chapter 7

Section 7.1 _____

In Exercises 1–10, find the vertex, focus, and directrix of the parabola. Sketch the parabola.

1. $x^2 = 12y$ 2. $(x - 7)^2 = y - 1$

3. $y^2 = 4x$ 4. $(y - 4)^2 = -(x + 2)$

5. $x^2 - 6x = y$ 6. $x^2 + 2x + 3y = 5$

7. $y^2 + 10y + 5x = 0$ 8. $y^2 - 8y + 4 = 4x$

9. $7y^2 = x$ 10. $x^2 + 3x - y = 0$

In Exercises 11–16, find the equation of the parabola with vertex at (h, k) satisfying the given conditions. Sketch the parabola.

11. Vertical axis of symmetry; vertex at $(0, 0)$; passes through the point $(-2, -6)$

12. Focus at $(0, 2)$; directrix $y = -2$

13. Focus at $(0, -5)$; directrix $y = 5$

14. Horizontal axis of symmetry; vertex at $(0, 0)$; passes through the point $(1, 6)$

15. Vertex at $(2, 0)$; directrix $x = -2$

16. Vertex at $(1, -4)$; directrix $x = 6$

Section 7.2 _____

In Exercises 17–28, find the vertices and foci of the ellipse given by the equation. Sketch the ellipse.

17. $\dfrac{x^2}{4} + \dfrac{y^2}{1} = 1$ 18. $\dfrac{(x - 1)^2}{64} + \dfrac{y^2}{25} = 1$

19. $\dfrac{(x + 2)^2}{25} + \dfrac{(y - 1)^2}{16} = 1$ 20. $\dfrac{x^2}{4} + \dfrac{y^2}{9} = 1$

21. $\dfrac{(x + 2)^2}{9} + \dfrac{y^2}{16} = 1$ 22. $\dfrac{(x - 1)^2}{4} + \dfrac{(y - 2)^2}{9} = 1$

23. $3x^2 + y^2 = 7$ 24. $2x^2 + 3y^2 = 10$

25. $x^2 - 4x + 6y^2 - 20 = 0$

26. $4y^2 - 32y + 3x^2 + 12x + 40 = 0$

27. $4y^2 - 24y + 7x^2 + 42x + 71 = 0$

28. $3x^2 + 18x + 2y^2 + 8y + 17 = 0$

In Exercises 29–34, find the equation in standard form of the ellipse centered at (0, 0).

29. One vertex at $(0, 5)$; one focus at $(0, 3)$

30. Major axis of length 8; foci at $(0, -2)$, $(0, 2)$

31. One vertex at $(0, 6)$; one focus at $(0, 4)$

32. One vertex at $(5, 0)$; one focus at $(3, 0)$

33. Major axis of length 10; foci at $(-3, 0)$, $(3, 0)$

34. One vertex at $(-6, 0)$; one focus at $(1, 0)$

Section 7.3

In Exercises 35–44, find the center, vertices, foci, and asymptotes for each hyperbola. Sketch the hyperbola.

35. $\dfrac{y^2}{36} - \dfrac{x^2}{25} = 1$

36. $\dfrac{(y-1)^2}{25} - \dfrac{x^2}{36} = 1$

37. $\dfrac{(y+2)^2}{4} - \dfrac{(x-1)^2}{4} = 1$

38. $\dfrac{x^2}{9} - \dfrac{y^2}{9} = 1$

39. $\dfrac{x^2}{4} - \dfrac{(y+2)^2}{9} = 1$

40. $\dfrac{(x-1)^2}{4} - \dfrac{(y-3)^2}{1} = 1$

41. $16y^2 - x^2 - 4x = 20$

42. $y^2 + 10y - 4x^2 - 16x + 5 = 0$

43. $8x^2 - y^2 = 64$

44. $7x^2 + 42x - 5y^2 - 50y = 307$

In Exercises 45–50, find the standard form of the equation of each hyperbola.

45. Foci at $(0, 6)$, $(0, -6)$; transverse axis of length 4

46. Vertices at $(-2, -6)$, $(-2, 2)$; one focus at $(-2, -9)$

47. Transverse axis of length 6; center at $(3, -4)$; one focus at $(3, 1)$

48. Vertices at $(3, 0)$, $(-3, 0)$; slope of one asymptote is -4

49. Transverse axis of length 8; center at $(-4, 2)$; one focus at $(2, 2)$

50. Foci at $(-7, -1)$, $(3, -1)$; slope of one asymptote is -3

Applications

51. **Gardening** The foliage in a planter takes the form of a parabola with a base that is 10 feet across. If the plant at the center is 6 feet tall and the heights of the plants taper off to zero toward each end of the base, how far from the center are the plants that are 4 feet tall?

52. **Water Flow** As a stream of water goes over the edge of a cliff, it forms a waterfall that takes the shape of one-half of a parabola. The equation of the parabola is $y = -64x^2$, where x and y are measured with respect to the edge of the cliff. How high is the cliff if the water hits the ground below at a horizontal distance of 3 feet from the edge of the cliff?

53. **Astronomy** The orbit of the planet Pluto around the sun is an ellipse with a major axis of about 80 astronomical units (AU), where 1 AU is approximately 92,600,000 miles. If the distance between the foci is one-fourth the length of the major axis, how long (in AU) is the minor axis?

54. **Sporting Equipment** The head of a tennis racket is in the shape of an ellipse. The ratio of the length of its major axis to the length of its minor axis is $\dfrac{17}{13}$. If the longer dimension of the head of the racket is 15 inches, how far is each focus from the center?

55. **Landscaping** The borders of the plantings in the two sections of a mathematician's garden form the branches of a hyperbola whose vertices are 16 feet apart. There are only two rosebushes in the garden, one at each focus of the hyperbola. If the transverse axis is horizontal and the slope of one of the asymptotes is $\dfrac{7}{8}$, how far apart are the rosebushes?

Chapter 7   # Test

In Exercises 1 and 2, find the vertex, focus, and directrix of the parabola. Sketch the parabola.

1. $(y + 3)^2 = -12(x + 1)$

2. $y^2 - 4y + 4x = 0$

In Exercises 3 and 4, find the equation, in standard form, of the parabola with vertex at (h, k) satisfying the given conditions.

3. Vertical axis of symmetry; vertex at $(0, -2)$; passes through the point $(4, 0)$

4. Focus at $(3, 0)$; directrix $x = -3$

In Exercises 5 and 6, find the vertices and foci of each ellipse. Sketch the ellipse.

5. $\dfrac{(x + 2)^2}{16} + \dfrac{(y - 3)^2}{25} = 1$

6. $\dfrac{x^2}{9} + \dfrac{(y + 1)^2}{4} = 1$

In Exercises 7 and 8, find the equation of the ellipse in standard form.

7. $4x^2 + 8x + y^2 = 0$

8. Foci at $(0, 3)$, $(0, -3)$; one vertex at $(0, -5)$

In Exercises 9 and 10, find the center, vertices, foci, and asymptotes for each hyperbola. Sketch the hyperbola.

9. $\dfrac{y^2}{4} - x^2 = 1$

10. $\dfrac{(x + 2)^2}{9} - \dfrac{y^2}{16} = 1$

In Exercises 11 and 12, find the standard form of the equation of the hyperbola.

11. Foci at $(3, 0)$, $(-3, 0)$; transverse axis of length 2

12. Vertices at $(0, 1)$, $(0, -1)$; slope of one asymptote is -3

13. A museum plans to design a whispering gallery whose cross-section is the top half of an ellipse. Because of the reflective properties of ellipses, a whisper from someone standing at one focus can be heard by a person standing at the other focus. If the dimensions of the gallery are as shown, how far from the gallery's center should two people stand so that they can hear each other's whispers?

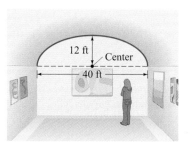

14. A small microphone with a parabolic cross-section measures 2 inches across and 4 inches deep. The sound receiver should be placed at the focus for best reception. How far from the vertex of the parabola should the sound receiver be placed?

More Topics in Algebra

I n a bicycle race with many participants, you might ask how many different possibilities exist for the first-place and second-place finishes. Determining the possibilities involves a counting technique known as the multiplication principle. See Example 1 in Section 8.4. In this chapter you will study various topics in algebra, including sequences, counting methods, probability, and mathematical induction. These concepts are used in a variety of applications, such as finance, sports, pharmacology, and biology, as well as in advanced courses in mathematics.

8.1 Sequences

Objectives

▶ Define and identify an arithmetic sequence

▶ Define and identify a geometric sequence

▶ Compute the terms of an arithmetic or a geometric sequence

▶ Apply arithmetic and geometric sequences to word problems

Many investment firms provide information about their investment products using charts similar to that in Figure 8.1.1. This chart applies to a hypothetical scenario in which a savings bond pays interest at the rate of 7% compounded annually. It illustrates the enormous difference in earnings that can result from investing the same amount of money in two different investment strategies.

In the first strategy, all the income from the investment during any given year is withdrawn from the account at the end of that year. In the second strategy, the income is reinvested each year. From the chart, it is clear that after 20 years the yield on the investment from the second strategy is much higher than the yield from the first strategy. In this section, we discuss the mathematics needed to analyze these types of situations, and we see why the results for the two strategies are so different.

Figure 8.1.1

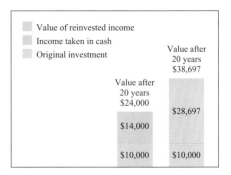

| Example | **1** | **Investment with Interest Not Reinvested** |

Example 5 in Section 8.1 builds upon this example. ⋯▶

According to Figure 8.1.1, the $10,000 bond investment yields interest at an annual rate of 7%. In the first strategy, the interest income for any given year is withdrawn from the account at the end of that year, and so the interest is not reinvested. Make a table in which you list the interest along with the total value of the investment at the end of each of the first 5 years.

▶**Solution** Each year, interest in the amount of $700 is paid on the fixed investment of $10,000. The total value of the investment-related assets at the end of any given year is simply the sum of the interest earned during that year and the total value of the investment-related assets at the end of the previous year. The information is summarized in Table 8.1.1.

Table 8.1.1

Year	Interest ($)	Total Value ($)
0	0	10,000
1	700	10,000 + 700 = 10,700
2	700	10,700 + 700 = 11,400
3	700	11,400 + 700 = 12,100
4	700	12,100 + 700 = 12,800
5	700	12,800 + 700 = 13,500

✔ *Check It Out 1:* Rework Example 1 for the case in which the annual rate of interest is 5%. ■

In Example 1, we listed the total value of the investment-related assets at the ends of only whole numbers of years (specifically, 0, 1, 2, 3, 4, and 5 years) after the initial investment. For the purpose of making lists of this kind, it would be convenient to use a function that is defined only for nonnegative integers. A function of this type is known as a *sequence*. The definition of a sequence follows.

> **Definition of a Sequence**
>
> A **sequence** is a function $f(n)$ whose domain is the set of all nonnegative integers (i.e., $n = 0, 1, 2, 3, \ldots$) and whose range is a subset of the set of all real numbers. The numbers $f(0), f(1), f(2), \ldots$ are called the **terms** of the sequence.
>
> For a nonnegative number n, it is conventional to denote the term that corresponds to n by a_n rather than by $f(n)$. We shall use both notations in our discussion.

In Example 1, suppose we want to find the total value of the investment-related assets after 20 years. Rather than explicitly compute the values of $f(0), f(1), f(2), \ldots, f(20)$ one by one, which is rather tedious, it would be more useful to find a general rule for the terms of the sequence. This is what we did for functions in Chapter 1. Next we will study some special types of sequences for which such a general rule exists.

Arithmetic Sequences

Example 2 The Rule for an Investment with Interest Not Reinvested

Figure 8.1.1 showed a $10,000 investment that yielded interest at an annual rate of 7%. In the first strategy, the interest income was not reinvested. Find a rule to generate the total value of the investment-related assets after n years.

▶**Solution** From Table 8.1.1, you can see that

$$\text{Total value after 1 year} = 10{,}000 + (1)700 = 10{,}700$$

since $700 interest is paid at the end of the first year. Also,

$$\text{Total value after 4 years} = 10{,}000 + (4)700 = 12{,}800.$$

Note that adding $700 to the total value of the investment at the end of each year has the same effect as *adding a suitable multiple* of $700 to the initial investment of $10,000. Now we can find a rule for the total value of the investment after n years.

$$\text{The value after } n \text{ years} = 10{,}000 + (n)700 = 10{,}000 + 700n$$

If we denote the total value of the investment-related assets after n years by $f(n)$, we find that the rule is

$$f(n) = 10{,}000 + 700n.$$

A rule such as this is much more compact than a table.

☑ *Check It Out 2:* Rework Example 2 for the case in which the annual rate of interest is 5%. ■

In Example 2, we took a starting amount of $10,000 and added a fixed amount of $700 to it every year. A sequence in which we set the starting value a_0 to a certain number and *add* a fixed number d to any term of the sequence to get the next term is known as an **arithmetic sequence.**

Discover *and* **Learn**

Use function notation to write the general form of the rule for an arithmetic sequence.

Definition of Arithmetic Sequence

Each term of an **arithmetic sequence** is given by the rule

$$a_n = a_0 + nd, \ n = 0, 1, 2, 3, \ldots$$

where a_0 is the starting value of the sequence and d is the common difference between successive terms. That is,

$$d = a_1 - a_0 = a_2 - a_1 = a_3 - a_2 = \ldots.$$

Example 3 Using the First Few Terms to Find a Rule

Table 8.1.2 gives the first four terms of an arithmetic sequence.

Table 8.1.2

n	0	1	2	3
a_n	−1	−3	−5	−7

(a) Find the rule corresponding to the given sequence.

(b) Use your rule to find a_{10}.

▶**Solution**

(a) We first find the common difference d between successive terms of the sequence. Note that

$$a_1 - a_0 = a_2 - a_1 = a_3 - a_2 = -2.$$

Thus, $d = -2$. Since $a_0 = -1$, the general rule is then given by

$$a_n = a_0 + dn = -1 + (-2)n \implies a_n = -1 - 2n.$$

(b) To find a_{10}, simply substitute $n = 10$ into the rule, which gives

$$a_{10} = -1 - 2(10) = -21.$$

Discover *and* **Learn**

The rule for the arithmetic sequence in Example 3 is given by $a_n = -1 - 2n$. We can write the rule as a function of n: $f(n) = -2n - 1$, where n is any nonnegative integer. What is the domain of f? What is the range of f? Plot the graph of f. How is it different from the graph of the function $g(x) = -2x - 1$, where x is any real number?

☑ *Check It Out 3:* Check that the rule generated in part (a) of Example 3 gives the correct values for the four terms listed in Table 8.1.2. ■

Example 4 Using Any Two Terms to Find a Rule

Suppose two terms of an arithmetic sequence are $a_8 = 10$ and $a_{12} = 26$. What is the rule for this sequence?

Table 8.1.3

n	a_n
8	10
9	$10 + d$
10	$10 + 2d$
11	$10 + 3d$
12	$10 + 4d = 26$

Technology Note

A graphing utility in SEQUENCE mode can be used to graph the terms of a sequence. Figure 8.1.2 shows the graph of the sequence $a_n = -1 - 2n$, $n = 0, 1, 2, 3, \ldots$.

Keystroke Appendix:
Section 14

Figure 8.1.2

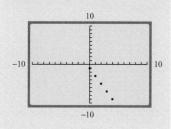

▶**Solution** Because this is an arithmetic sequence, we know that the difference between any two successive terms is a constant, d. We can use the given terms to construct Table 8.1.3.

Thus $10 + 4d = 26$. To find d, we solve this equation.

$$10 + 4d = 26 \Longrightarrow d = 4$$

The rule for the sequence is then $a_n = a_0 + 4n$. We still need to find a_0, the starting value. We could construct another table and work our way backward, but that would be too tedious. Instead, we use the information available to us to find an equation in which a_0 is the only unknown.

$a_n = a_0 + 4n$	Known rule so far
$10 = a_0 + 4(8)$	Substitute $n = 8$ and $a_8 = 10$
$10 = a_0 + 32$	Simplify
$-22 = a_0$	Solve for a_0

Thus the rule for this sequence is given by

$$a_n = -22 + 4n.$$

✔ *Check It Out 4:*

(a) Check that the rule in Example 4 is correct.

(b) Find the rule for the arithmetic sequence with terms $a_9 = 15$ and $a_{12} = 6$. ■

Geometric Sequences

We saw that arithmetic sequences are formed by setting the starting value a_0 equal to a certain number and then *adding* a fixed number d to any term of the sequence to get the next term. Another way to generate a sequence is to set the starting value a_0 equal to a certain number and then *multiply* any term of the sequence by a fixed number r to get the next term. A simple example is when we start with the number 2 and keep multiplying by 2, which gives the sequence 2, 4, 8, 16, ... A sequence generated in this manner is called a **geometric sequence.**

We now examine the second strategy of the investment scenario discussed earlier in this section.

Example 5 **The Rule for an Investment with Interest Reinvested**

⋮▪▪▪ *This example builds on Example 1 in Section 8.1.*

Suppose $10,000 is invested in a savings bond that pays 7% interest compounded annually, and suppose all of the interest is reinvested. This is the second investment strategy for the hypothetical scenario illustrated by the chart in Figure 8.1.1.

(a) Make a table in which you list the interest along with the total value of the investment at the end of each of the first 5 years.

(b) Compare your table with Table 8.1.1 in Example 1.

(c) Find the general rule for the total amount in the account after n years.

(d) Find the value of the investment after 20 years. Compare your result with the amount given in Figure 8.1.1.

Just in Time

*Review compound interest
in Section 5.2.*

▶**Solution**

(a) The total value after 1 year is

$$\text{Initial value} + 7\% \text{ interest}$$

$= 10{,}000 + (0.07)(10{,}000)$	Substitute data
$= (1 + 0.07)(10{,}000)$	Factor out 10,000
$= (1.07)(10{,}000).$	Simplify

Because the interest is compounded annually, the total value after 2 years is the sum of the total value after 1 year plus 7% of that value:

$$\text{Total value after 1 year} + 7\% \text{ of that total}$$

$= (1.07)(10{,}000) + (0.07)(1.07)(10{,}000)$	Substitute data
$= (1 + 0.07)(1.07)(10{,}000)$	Factor out 1.07(10,000)
$= 1.07(1.07)(10{,}000)$	Simplify
$= (1.07)^2(10{,}000).$	

Continuing in this manner, we can generate Table 8.1.4. We have written the total value both in exponential notation and as a numerical value.

Table 8.1.4

Year	Interest ($)	Total Value ($)
0	0.00	10,000.00
1	700.00	$(1.07)(10{,}000) = 10{,}700.00$
2	749.00	$(1.07)^2(10{,}000) = 11{,}449.00$
3	801.43	$(1.07)^3(10{,}000) \approx 12{,}250.43$
4	857.53	$(1.07)^4(10{,}000) \approx 13{,}107.96$
5	917.56	$(1.07)^5(10{,}000) \approx 14{,}025.52$

(b) In Example 1, the total value of the investment-related assets after 5 years was $13,500. When the interest is compounded, the total value after 5 years is $14,025.52. Thus reinvesting the interest income rather than spending it yields a higher return after 5 years. Note that the initial investments and the interest rates for the two investment strategies were identical.

(c) Note from Table 8.1.4 that the total value at the end of n years is 1.07^n times the initial investment of $10,000. If a_n is the amount after n years, then the rule for a_n is given by

$$a_n = (1.07)^n(10{,}000) = (10{,}000)(1.07)^n.$$

(d) After 20 years, the investment is worth

$$10{,}000(1.07)^{20} \approx \$38{,}696.84.$$

This is equal to the amount given in Figure 8.1.1.

✔ *Check It Out 5:* Rework parts (c) and (d) of Example 5 for the case in which the annual rate of interest is 5%. ■

In Example 5, any term of the sequence could have been found by *multiplying* the previous term by a fixed number. A sequence obtained in this manner is called a **geometric sequence,** which we now define.

Definition of a Geometric Sequence

A **geometric sequence** is defined by the rule

$$a_n = a_0 r^n, \quad n = 0, 1, 2, 3, \ldots$$

where a_0 is the starting value of the sequence and r is the *fixed number* by which any term of the sequence is multiplied to obtain the next term. If $r \neq 0$, this is equivalent to saying that r is the *fixed ratio* of successive terms:

$$r = \frac{a_1}{a_0} = \frac{a_2}{a_1} = \cdots = \frac{a_i}{a_{i-1}} = \cdots$$

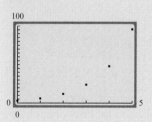

Example 6 Using Two Consecutive Terms to Find a Rule

Find the rule for the geometric sequence with terms $a_2 = 12$ and $a_3 = 24$.

▶**Solution** Because a_2 and a_3 are successive terms of the sequence, $a_3 = ra_2$. To find r, we solve this equation.

$$r = \frac{a_3}{a_2} = \frac{24}{12} = 2$$

The rule for the sequence is then $a_n = a_0(2)^n$.

Next we find a_0, the starting value. We use the information available to us to write an equation in which a_0 is the only unknown.

$a_n = a_0(2)^n$	Known rule so far
$12 = a_0(2)^2$	Substitute $n = 2$ and $a_2 = 12$
$12 = 4a_0$	Simplify
$3 = a_0$	Solve for a_0

Hence, the rule for this sequence is

$$a_n = 3(2)^n.$$

☑ *Check It Out 6:*

(a) Verify that the rule in Example 6 is correct.

(b) Find the rule for the geometric sequence with terms $a_3 = 54$ and $a_4 = 162$. ▪

Example 7 Using Two Nonconsecutive Terms to Find a Rule

Find the rule for the geometric sequence with terms $a_2 = 32$ and $a_5 = 4$.

▶**Solution** Unlike in Example 6, we do not have successive terms here. To find the ratio r between successive terms, we can set up a table to summarize the information we are given. See Table 8.1.5.

From the last row, we have

$$32r^3 = 4 \Longrightarrow r^3 = \frac{1}{8} \Longrightarrow r = \frac{1}{2}.$$

Table 8.1.5

n	a_n
2	32
3	$32r$
4	$r(32r) = 32r^2$
5	$r(32r^2) = 32r^3 = 4$

Thus the rule for this sequence is $a_n = a_0 \left(\frac{1}{2}\right)^n$. We still need to find a_0, the starting value. We can use the information available to us to construct an equation in which a_0 is the only unknown:

$$a_n = a_0 \left(\frac{1}{2}\right)^n \qquad \text{Known rule so far}$$

$$32 = a_0 \left(\frac{1}{2}\right)^2 \qquad \text{Substitute } n = 2 \text{ and } a_2 = 32$$

$$32 = a_0 \left(\frac{1}{4}\right) \qquad \text{Simplify}$$

$$a_0 = 128 \qquad \text{Solve for } a_0$$

Hence, the rule for this geometric sequence is

$$a_n = 128 \left(\frac{1}{2}\right)^n.$$

Note that the ratio $r = \frac{1}{2} < 1$. The terms of this sequence get smaller as n gets larger.

✔ *Check It Out 7:*

(a) Verify that the rule in Example 7 is correct.

(b) Find the rule for the geometric sequence with terms $a_3 = 3$ and $a_6 = 81$. ■

Applications of Arithmetic and Geometric Sequences

Sequences are used in a wide variety of situations. In particular, they are used to model processes in which there are discrete jumps from one stage to the next. The following examples will examine applications from music and biology.

Example **8** A Vibrating String

Table 8.1.6

Harmonic	Frequency (Hz)
Fundamental	55
Second	110
Third	165
Fourth	220

A string of a musical stringed instrument vibrates in many modes, called *harmonics*. Associated with each harmonic is its *frequency*, which is expressed in units of cycles per second, or Hertz (Hz). The harmonic with the lowest frequency is known as the fundamental mode; all the other harmonics are known as overtones. Table 8.1.6 lists the frequencies corresponding to the first four harmonics of a string of a certain instrument.

(a) What type of sequence do the frequencies form: arithmetic, geometric, or neither?

(b) Find the frequency corresponding to the fifth harmonic.

(c) Find the frequency corresponding to the nth harmonic.

▶Solution

(a) Because the *difference* between successive harmonics is constant at 55 Hz, the sequence is arithmetic.

(b) From Table 8.1.6, the fourth harmonic is 220 Hz. Thus the fifth harmonic is

$$220 + d = 220 + 55 = 275 \text{ Hz}.$$

Table 8.1.7

Harmonic	Frequency (Hz)
Fundamental	35
Second	70
Third	140
Fourth	280

(c) The common difference is $d = 55$. Denote the frequency of the nth harmonic by f_n, $n = 1, 2, 3, \ldots$. From the table, $f_1 = 55$, $f_2 = 110 = 55(2)$, $f_3 = 165 = 55(3)$, and so on. Thus we have

$$f_n = 55n, \; n = 1, 2, 3, \ldots.$$

Note that this sequence begins at $n = 1$.

☑ *Check It Out 8:* Rework Example 8 for the case in which the frequencies of the first four harmonics are as shown in Table 8.1.7. ▪

| Example | 9 | DNA Fragments |

When scientists conduct tests using DNA, they often need larger samples of DNA than can be readily obtained. A method of duplicating specific fragments of DNA, known as Polymerase Chain Reaction (PCR), was invented in 1983 by biochemist Kary Mullis, who later won the Nobel Prize in Chemistry for his work. After each cycle of the PCR process, the number of DNA fragments doubles. This procedure has a number of applications, including diagnosis of genetic diseases and investigation of criminal activities.

Suppose a biologist begins with a sample of 1000 DNA fragments.

(a) What type of sequence is generated by the repeated application of the PCR process?

(b) How many DNA fragments will there be after five cycles of the PCR process?

(c) How many DNA fragments will there be after n cycles of the PCR process?

(d) Laboratories typically require millions of DNA fragments to conduct proper tests. How many cycles of the PCR process are needed to produce one million fragments?

(e) Each cycle of the PCR process takes approximately 30 minutes. How long will it take to generate one million DNA fragments?

▶ Solution

(a) Because the number of DNA fragments doubles after each cycle of the PCR process, the resulting sequence of the number of DNA fragments is geometric.

(b) The doubling of DNA fragments after each cycle is summarized in Table 8.1.8.

Table 8.1.8

Number of Cycles	Number of Fragments
0	1000
1	$2(1000) = 2000$
2	$2(2000) = 2^2(1000) = 4000$
3	$2^3(1000) = 8000$
4	$2^4(1000) = 16,000$
5	$2^5(1000) = 32,000$

From the table, the number of DNA fragments after five cycles is

$$1000(2)^5 = 32,000.$$

(c) Let $F(n)$ denote the number of DNA fragments after n cycles. Because the common ratio is 2 and the initial number of fragments is 1000, we can write

$$F(n) = 1000(2)^n.$$

(d) To compute the number of cycles needed to produce one million DNA fragments, solve the following equation for n.

$$1000(2)^n = 1,000,000 \qquad \text{Set } F(n) \text{ equal to 1,000,000}$$
$$2^n = 1000 \qquad \text{Isolate exponential term: divide}$$
$$\text{both sides of equation by 1000}$$

Because n appears as an exponent, use logarithms to solve for n.

$$\log 2^n = \log 1000 \qquad \text{Take common log of both sides}$$
$$n \log 2 = \log 1000$$
$$n = \frac{\log 1000}{\log 2} = \frac{3}{\log 2} \qquad \text{Solve for } n$$
$$n \approx 9.966$$

The number of cycles must be a whole number. Thus 10 cycles of the PCR process are needed to produce at least one million DNA fragments.

(e) Each cycle lasts 30 minutes, so it will take $10(30) = 300$ minutes, or 5 hours, to produce one million DNA fragments.

✔ *Check It Out 9:* How long will it take to produce one million fragments of DNA if the biologist in Example 9 starts with 2000 fragments of DNA? ■

8.1 Key Points

▶ A **sequence** is a function $f(n)$ whose domain is the set of all nonnegative integers and whose range is a subset of the set of all real numbers. The numbers $f(0)$, $f(1), f(2), \ldots$ are called the **terms** of the sequence.

▶ Each term of an **arithmetic sequence** is given by the rule

$$a_n = a_0 + nd, \, n = 0, 1, 2, 3, \ldots$$

where a_0 is the starting value of the sequence and d is the common difference between successive terms.

▶ A **geometric sequence** is defined by the rule

$$a_n = a_0 r^n, \, n = 0, 1, 2, 3, \ldots$$

where a_0 is the initial value of the sequence and $r \neq 0$ is the *fixed ratio* between successive terms.

8.1 Exercises

▶**Just in Time Exercises** These exercises correspond to the Just in Time references in this section. Complete them to review topics relevant to the remaining exercises.

1. True or False: If an amount P is invested in an account that pays interest at rate r and the interest is compounded n times per year, then after t years the amount in the account will be

$$A(t) = P\left(1 + \frac{r}{n}\right)^{nt}.$$

2. If interest on an account is compounded quarterly, what is the value of n in the compound interest formula?

3. A bank is advertising an account with an interest rate of 4%, compounded semiannually. A customer opens an account with an initial deposit of $100 and makes no more deposits or withdrawals. What will the account balance be at the end of 2 years?

4. A bank is advertising an account with an interest rate of 6%, compounded yearly. A customer opens an account with an initial deposit of $100 and makes no more deposits or withdrawals. What will the account balance be at the end of 2 years?

▶**Skills** This set of exercises will reinforce the skills illustrated in this section.

In Exercises 5–20, find the terms a_0, a_1, and a_2 for each sequence.

5. $a_n = 4 + 6n$

6. $a_n = 3 + 5n$

7. $a_n = -5 + 3n$

8. $a_n = -3 + 2n$

9. $a_n = -4 - 4n$

10. $a_n = -3 - 3n$

11. $a_n = 8 - 2n$

12. $a_n = 5 - 3n$

13. $a_n = 7(4^n)$

14. $a_n = 3(2^n)$

15. $a_n = 5(3^n)$

16. $a_n = 6(5^n)$

17. $a_n = -2(3^n)$

18. $a_n = -4(6^n)$

19. $a_n = -3(5^n)$

20. $a_n = -5(2^n)$

In Exercises 21–32, find the rule for the arithmetic *sequence having the given terms.*

21.

n	0	1
a_n	−3	3

22.

n	0	1
$f(n)$	−1	4

23.

n	6	7
$g(n)$	8	12

24.

n	3	4
b_n	−8	−9

25.

n	10	12
b_n	10	16

26.

n	8	11
h_n	9	3

27. The common difference d is 5 and $a_9 = 55$.

28. The common difference d is 2 and $a_5 = 24$.

29. The common difference d is −2 and $a_8 = 5$.

30. The common difference d is −3 and $a_{10} = 7$.

31. The common difference d is $\frac{1}{2}$ and $b_6 = 13$.

32. The common difference d is $\frac{1}{4}$ and $c_8 = 7$.

In Exercises 33–44, find the rule for the geometric *sequence having the given terms.*

33.

n	0	1
a_n	3	6

34.

n	0	1
a_n	4	8

35.

n	2	3
$h(n)$	32	128

36.

n	3	4
$g(n)$	−8	−16

37.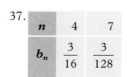

n	4	7
b_n	$\frac{3}{16}$	$\frac{3}{128}$

38.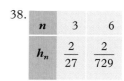

n	3	6
h_n	$\frac{2}{27}$	$\frac{2}{729}$

39. The common ratio r is 2 and $a_5 = 128$.

40. The common ratio r is 3 and $a_4 = -162$.

41. The common ratio r is $\frac{3}{2}$ and $b_4 = \frac{81}{4}$.

42. The common ratio r is $\frac{4}{3}$ and $c_3 = -\frac{64}{27}$.

43. The common ratio r is 5 and $a_4 = 2500$.

44. The common ratio r is 4 and $a_6 = 12{,}288$.

In Exercises 45–50, fill in the missing terms of each arithmetic *sequence.*

45.

n	0	1	2	3
a_n		5	8	

46.

n	0	1	2	3
a_n		5	6	

47.

n	0	1	2	3
a_n		2	6	

48.

n	0	1	2	3
a_n		−2	1	

49.

n	0	1	2	3
a_n		3	1	

50.

n	0	1	2	3
a_n		−9	−15	

In Exercises 51–56, fill in the missing terms of each geometric *sequence.*

51.

n	0	1	2	3
a_n		3	9	

52.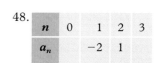

n	0	1	2	3
a_n		5	25	

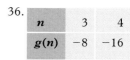
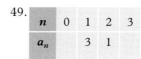

53.

n	0	1	2	3
a_n		10	50	

54.

n	0	1	2	3
a_n		12	36	

55.

n	0	1	2	3
a_n		$\dfrac{1}{4}$	$\dfrac{1}{8}$	

56.

n	0	1	2	3
a_n		$\dfrac{1}{9}$	$\dfrac{1}{27}$	

In Exercises 57–68, state whether the sequence is arithmetic or geometric.

57. $1, 3, 5, 7, \ldots$

58. $4, 10, 16, 22, \ldots$

59. $2, 6, 18, 54, \ldots$

60. $8, 5, 2, -1, \ldots$

61. $-7, -11, -15, -19, \ldots$

62. $3, 15, 75, 375, \ldots$

63. $\dfrac{1}{2}, \dfrac{1}{6}, \dfrac{1}{18}, \ldots$

64. $0.929, 0.939, 0.949, \ldots$

65. $\dfrac{111}{1000}, \dfrac{115}{1000}, \dfrac{119}{1000}, \ldots$

66. $0.9, 0.81, 0.729, \ldots$

67. $0.4, 0.8, 1.6, 3.2, \ldots$

68. $0.4, 0.9, 1.4, 1.9, \ldots$

▶**Applications** In this set of exercises, you will use sequences to study real-world problems.

69. Investment An income-producing investment valued at $2000 pays interest at an annual rate of 6%. Assume that the interest is taken out as income and therefore is not compounded.
(a) Make a table in which you list the initial investment along with the total value of the investment-related assets (initial investment plus total interest earned) at the end of each of the first 4 years.
(b) What is the total value of the investment-related assets after n years?

70. Investment An income-producing investment valued at $3000 pays interest at an annual rate of 4.5%. Assume that the interest is taken out as income and therefore is not compounded.
(a) Make a table in which you list the initial investment along with the total value of the investment-related assets (initial investment plus total interest earned) at the end of each of the first 4 years.
(b) What is the total value of the investment-related assets after n years?

71. Knitting Knitting, whether by hand or by machine, uses a sequence of stitches and proceeds row by row. Suppose you knit 100 stitches for the bottommost row and in-

crease the number of stitches in each row thereafter by 4. This is a standard way to make the sleeve portion of a sweater.
(a) What type of sequence does the number of stitches in each row produce: arithmetic, geometric, or neither?
(b) Find a rule that gives the number of stitches in the nth row.
(c) How many rows must be knitted to end with a row of 168 stitches?

72. Knitting New trends in knitting involve creating vibrant patterns with geometric shapes. Suppose you want to knit a large right triangle. You start with 85 stitches and decrease each row thereafter by 2 stitches.
(a) What type of sequence does the number of stitches in each row produce: arithmetic, geometric, or neither?
(b) Find a rule that gives the number of stitches for the nth row.
(c) How many rows must be knitted to end with a row of just one stitch?

73. Music In music, the frequencies of a certain sequence of tones that are an octave apart are

$$55 \text{ Hz}, \ 110 \text{ Hz}, \ 220 \text{ Hz}, \ldots$$

where Hz (Hertz) is a unit of frequency (1 Hz = 1 cycle per second).
(a) Is this an arithmetic or a geometric sequence? Explain.
(b) Compute the next two terms of the sequence.
(c) Find a rule for the frequency of the nth tone.

74. Sports The men's and women's U.S. Open tennis tournaments are elimination tournaments. Each tournament starts with 128 players in 64 separate matches. After the first round of competition, 64 players are left. The process continues until the final championship match has been played.
(a) What type of sequence gives the number of players left after each round?
(b) How many rounds of competition are there in each tournament?

75. Salary An employee starting with an annual salary of $40,000 will receive a salary increase of 4% at the end of each year. What type of sequence would you use to find her salary after 6 years on the job? What is her salary after 6 years?

76. Salary An employee starting with an annual salary of $40,000 will receive a salary increase of $2000 at the end of each year. What type of sequence would you use to find his salary after 5 years on the job? What is his salary after 5 years?

77. **Biology** A cell divides into two cells every hour.
    (a) How many cells will there be after 4 hours if we start with 10,000 cells?
    (b) Is this a geometric sequence or an arithmetic sequence?
    (c) How long will it take for the number of cells to equal 1,280,000?

78. **Salary** Joan is offered two jobs with differing salary structures. Job A has a starting salary of $30,000 with an increase of 4% per year. Job B has a starting salary of $35,000 with an increase of $500 per year. During what years will Job A pay more? During what years will Job B pay more?

79. **Geometry** A sequence of square boards is made as follows. The first board has dimensions 1 inch by 1 inch, the second has dimensions 2 inches by 2 inches, the third has dimensions 3 inches by 3 inches, and so on.
    (a) What type of sequence is formed by the *perimeters* of the boards? Explain.
    (b) Write a rule for the sequence formed by the *areas* of the boards. Is the sequence arithmetic, geometric, or neither? Explain your answer.

80. **Social Security** The following table gives the average monthly Social Security payment, in dollars, for retired workers for the years 2000 to 2003. (*Source:* Social Security Administration)

Year	2000	2001	2002	2003
Amount	843	881	917	963

    (a) Is this sequence better approximated by an arithmetic sequence or a geometric sequence? Explain.
    (b) Use the regression capabilities of your graphing calculator to find a suitable function that models this data. Make sure that n represents the number of years after 2000.

81. **Recreation** The following table gives the amount of money, in billions of dollars, spent on recreation in the United States from 1999 to 2002. (*Source:* Bureau of Economic Analysis)

Year	1999	2000	2001	2002
Amount ($ billions)	546.1	585.7	603.4	633.9

Assume that this sequence of expenditures approximates an arithmetic sequence.
    (a) If n represents the number of years since 1999, use the linear regression capabilities of your graphing calculator to find a function of the form $f(n) = a_0 + nd$, $n = 0, 1, 2, 3, \ldots$, that models these expenditures.
    (b) Use your model to project the amount spent on recreation in 2007.

▶**Concepts** This set of exercises will draw on the ideas presented in this section and your general math background.

82. Is $4, 4, 4, \ldots$ an arithmetic sequence, a geometric sequence, or both? Explain.

83. What are the terms of the sequence generated by the expression $a_n = a_0 + nd$, $d = 0$?

84. What are the terms of the sequence generated by the expression $a_n = a_0 r^n$, $r = 1$?

85. You are given two terms of a sequence: $a_1 = 1$ and $a_3 = 9$.
    (a) Find the rule for this sequence, assuming it is arithmetic.
    (b) Find the rule for this sequence, assuming it is geometric.
    (c) Find $a_4, \ldots, a_8$ for the sequence in part (a).
    (d) Find $a_4, \ldots, a_8$ for the sequence in part (b).
    (e) Which sequence grows faster, the one in part (a) or the one in part (b)? Explain your answer.

86. Consider the sequence
$$1, 10, 100, 1000, 10,000, \ldots.$$
In this an arithmetic sequence or a geometric sequence? Explain. Now take the common logarithm of each term in this sequence. Is the new sequence arithmetic or geometric? Explain.

87. Find the next two terms in the geometric sequence whose first three terms are $(1 + x)$, $(1 + x)^2$, and $(1 + x)^3$. What is the common ratio r in this case?

88. Suppose a, b, and c are three consecutive terms in an arithmetic sequence. Show that $b = \frac{a + c}{2}$.

89. If $a_0, a_1, a_2, \ldots$ is a geometric sequence, what kind of sequence is $a_0^3, a_1^3, a_2^3, \ldots$? Explain your reasoning.

8.2 Sums of Terms of Sequences

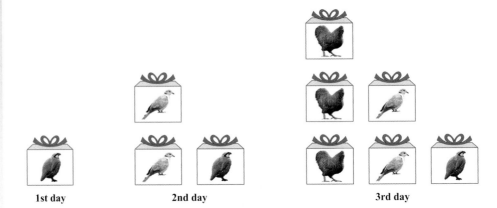

1st day   2nd day   3rd day

On the first day of Christmas, my true love sent to me
a partridge in a pear tree.
On the second day of Christmas, my true love sent to me
two turtle doves and a partridge in a pear tree.
On the third day of Christmas, my true love sent to me
three French hens, two turtle doves, and a partridge in a pear tree.
...

Source: William Henry Husk, Songs of the Nativity, 1868.

In the holiday song *Twelve Days of Christmas*, what is the total number of gifts the lucky person will receive on the twelfth day? This is an example of finding the *sum* of numbers in a sequence. This section will examine sums of numbers generated by special types of sequences.

Sum of Terms of an Arithmetic Sequence

Example 1 **Direct Calculation of the Sum of Terms of a Sequence**

Consider the song *Twelve Days of Christmas*, discussed at the beginning of this section.

(a) How many gifts will be received on the twelfth day?

(b) If the pattern in the song continues, can you find a general formula for the number of gifts received on the *n*th day, where *n* is even?

▶Solution

(a) In this problem, we consider the sequence $1, 2, 3, 4, \ldots, n$. We see that $a_n = n$, $n = 1, 2, 3, \ldots$. This is an arithmetic sequence with a common difference of 1. To find the number of gifts received on the twelfth day, we certainly could obtain the answer by adding the numbers a_1 through a_{12}. However, we will approach the problem in a way that will help us to find a general formula for the total number of gifts received on any particular day.

From Figure 8.2.1, we see that the first and last (twelfth) numbers in the sequence add to 13. The second and eleventh numbers also add to 13.

Figure 8.2.1

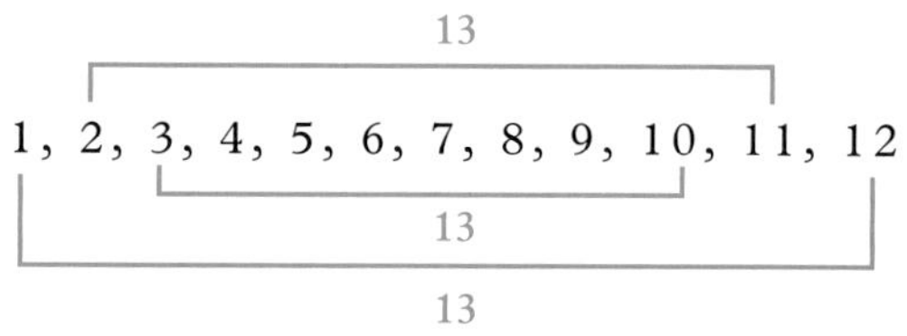

The same is true of four more sums: the sum of the third and tenth numbers, the sum of the fourth and ninth numbers, the sum of the fifth and eighth numbers, and the sum of the sixth and seventh numbers each add to 13. We thus have six pairs of numbers, each of which adds to 13. The sum is

$$1 + 2 + 3 + \cdots + 12 = (1 + 12) + (2 + 11) + (3 + 10) + \cdots + (6 + 7)$$
$$= (6)(13) = 78.$$

Thus the fortunate person receiving the gifts will receive a total of 78 gifts on the twelfth day.

(b) To find the total number of gifts received on the nth day, where n is even, we find the sum $1 + 2 + 3 + \cdots + n$. See Figure 8.2.2.

Figure 8.2.2

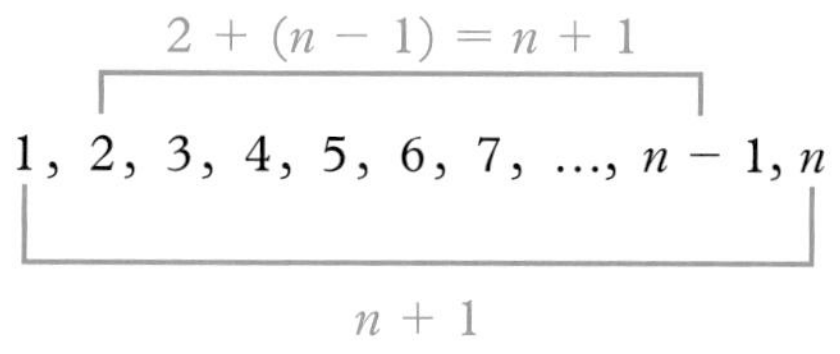

Applying the technique from part (a) gives

$$1 + 2 + 3 + \cdots + (n - 1) + n$$

$$= (1 + n) + (2 + (n - 1)) + \cdots + \left(\frac{n}{2} + \left(\frac{n}{2} + 1\right)\right) \qquad \text{See Figure 8.2.2}$$

$$= \underbrace{(n + 1) + (n + 1) + \cdots + (n + 1)}_{\frac{n}{2} \text{ times}} \qquad \text{Evaluate within parentheses}$$

$$= \frac{n}{2}(n + 1) = \frac{n(n + 1)}{2}. \qquad \text{There are } \frac{n}{2} \text{ pairs of numbers}$$

The number of gifts given on the nth day is $\frac{n(n + 1)}{2}$.

Because we are given that n is even, we have $\frac{n}{2}$ pairs of numbers to sum in this calculation.

☑ *Check It Out 1:* Assuming the gift-giving pattern in Example 1 continues, find the number of gifts that would be received on the fifteenth day. ■

If n is odd, you can use the technique from Example 1 to find the sum of the first n positive integers, but then you will have a leftover term to deal with. However, it can be shown that the same formula holds for the sum regardless of whether n is even or odd. This leads to the following result.

Discover *and* **Learn**

Compute the sum of the first n positive integers for various values of n. Plot your results as a function of n. What do you observe about the graph? What type of function does it represent?

Sum of Numbers from 1 to n

Let n be a positive integer. Then

$$1 + 2 + 3 + \cdots + n = \frac{n(n+1)}{2}.$$

The type of formula discussed in Example 1 arises in many situations. We can use it to find the general formula for the sum of the first n terms of *any* arithmetic sequence. Let $a_j = a_0 + jd, j = 0, 1, 2, \ldots,$ be an arithmetic sequence, and let n be a positive integer. To find the sum of the terms a_0 through a_{n-1}, proceed as follows.

We wish to find

$$S_n = a_0 + a_1 + a_2 + \cdots + a_{n-1}.$$

The quantities that we know are a_0, d, and n. If we can obtain a formula that contains just these quantities, we can compute the sum of any number of terms very easily. From the general form for a_j, we have

$$S_n = a_0 + a_1 + a_2 + \cdots + a_{n-1}$$
$$= a_0 + (a_0 + d) + (a_0 + 2d) + \cdots + (a_0 + (n-1)d)$$
$$= \underbrace{a_0 + a_0 + \cdots + a_0}_{n \text{ times}} + d + 2d + \cdots + (n-1)d \quad \text{Collect like terms}$$
$$= (n)a_0 + d(1 + 2 + \cdots + (n-1)) \quad \text{Factor out } d$$
$$= (n)a_0 + d\left(\frac{(n-1)(n)}{2}\right) \quad \text{Use formula for } 1 + 2 + \cdots + (n-1)$$
$$= \frac{n}{2}(2a_0 + d(n-1)). \quad \text{Factor out } \frac{n}{2}$$

Discover *and* **Learn**

Check your understanding of the formula given at the right by computing the sum of the first 20 terms of the arithmetic sequence

$3, 6, 9, 12, \ldots.$

We can stop at this point, since we have found a formula for the sum that contains only the known quantities, a_0, d, and n. We can also write another formula for S_n by using the fact that $a_{n-1} = a_0 + d(n-1)$. This gives

$$S_n = a_0 + a_1 + a_2 + \cdots + a_{n-1}$$
$$= \frac{n}{2}(2a_0 + d(n-1)) \quad \text{Use formula for sum}$$
$$= \frac{n}{2}(a_0 + (a_0 + d(n-1))) = \frac{n}{2}(a_0 + a_{n-1}). \quad \text{Substitute } a_{n-1} = a_0 + d(n-1)$$

We now have the following formula for the sum of the first n terms of an arithmetic sequence.

Sum of the First n Terms of an Arithmetic Sequence

Let $a_j = a_0 + jd, j = 0, 1, 2, \ldots$, be an arithmetic sequence, and let n be a positive integer. The sum of the first n terms of the sequence (the terms a_0 through a_{n-1}) is given by

$$S_n = a_0 + a_1 + \cdots + a_{n-1} = \frac{n}{2}(2a_0 + d(n - 1)) = \frac{n}{2}(a_0 + a_{n-1}).$$

You can use either of these forms, depending on the type of information you are given.

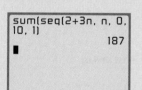

Technology Note

Using the SUM and SEQUENCE features of your graphing utility, you can calculate the sum in Example 2. See Figure 8.2.3.

Keystroke Appendix: Section 14

Figure 8.2.3

```
sum(seq(2+3n, n, 0,
10, 1)
                187
■
```

Example 2 Calculating the Sum from the First Few Terms

Find the sum of the first 11 terms of the arithmetic sequence

$$2, 5, 8, 11, 13, \ldots.$$

▶ **Solution** For the given sequence, the first term, a_0, equals 2 and the common difference, d, equals 3. The sequence can be written as

$$a_j = 2 + 3j, j = 0, 1, 2, 3, \ldots.$$

The sum of the first 11 terms is then found by using the sum formula.

$$S_n = \frac{n}{2}(2a_0 + d(n-1)) \qquad \text{We know } a_0, d, \text{ and } n$$

$$= \frac{11}{2}(2(2) + 3(11-1)) \qquad \text{Substitute } a_0 = 2, n = 11, \text{ and } d = 3$$

$$= \frac{11}{2}(34) = 187 \qquad \text{Simplify}$$

✔ **Check It Out 2:** Find the sum of the first 20 terms of the arithmetic sequence

$$2, 4, 6, 8, 10, \ldots. \ ■$$

The next example uses an alternate formula for finding the sum of an arithmetic sequence.

Example 3 Calculating the Sum from the Initial and Final Terms

Find the sum of all even numbers between 2 and 100, inclusive.

▶ **Solution** Here, we are given the initial term and the final term of the sequence. Thus the formula

$$S_n = \frac{n}{2}(a_0 + a_{n-1})$$

can be used. There are 50 even numbers between 2 and 100, inclusive. Substituting $a_0 = 2$, $a_{49} = 100$, and $n = 50$, we have

$$S_{50} = \frac{n}{2}(a_0 + a_{n-1}) = \frac{n}{2}(a_0 + a_{49}) = \frac{50}{2}(2 + 100) = 2550.$$

✔ *Check It Out 3:* Find the sum of all odd numbers between 3 and 51, inclusive. ■

Summation Notation

It can be very cumbersome to write out the sum of the terms of a sequence (such as the sum $1 + 2 + \cdots + n$) in the form of a string of numbers separated by plus signs. Fortunately, there is a shorthand notation, known as **summation notation,** that can be used to indicate a sum. The capital Greek letter sigma, Σ, is used to denote a sum. In addition, a variable known as an **index,** ranging over some set of consecutive integers, keeps track of the terms in the sum.

Consider the following example of summation notation.

$$\text{The sum } 1 + 2 + \cdots + n \text{ would be written as} \sum_{i=1}^{n} i.$$

The details of this notation are illustrated below.

$$\text{Index } i \text{ ends at } n$$
$$\downarrow$$
$$\text{Summation symbol} \to \sum_{i=1}^{n} i^2 \leftarrow \text{Expression to be evaluated and summed}$$
$$\uparrow$$
$$\text{Index } i \text{ starts at 1}$$

The best way to understand summation notation is to work with it. The next two examples illustrate the use of summation notation.

Example 4 Expanding a Sum Given in Summation Notation

Expand the following sums given in summation notation. Do not evaluate.

(a) $\displaystyle\sum_{i=2}^{5} 3i$ (b) $\displaystyle\sum_{i=3}^{6} 7$

▶Solution

(a) We see that i goes from 2 to 5. Make a table of values corresponding to each term in the sum. From Table 8.2.1,

$$\sum_{i=2}^{5} 3i = 6 + 9 + 12 + 15.$$

Table 8.2.1

i	$3i$
2	6
3	9
4	12
5	15

(b) To expand $\displaystyle\sum_{i=3}^{6} 7$ we first observe that i goes from 3 to 6. Every term in this sum is 7 (i.e., the numbers to be summed do not depend on i, which simply "counts" the terms). Therefore,

$$\sum_{i=3}^{6} 7 = \underbrace{7}_{i=3} + \underbrace{7}_{i=4} + \underbrace{7}_{i=5} + \underbrace{7}_{i=6}.$$

✔ *Check It Out 4:* Expand the sum given by $\displaystyle\sum_{i=3}^{7} 4i$. Do not evaluate. ■

Example 5 Writing an Expanded Sum in Summation Notation

Write the sum $-2 - 4 - 6 - 8 - \cdots - 20$ using summation notation and evaluate.

▶Solution Note that each term in the sum can be written as $-2i$, and that i ranges from 1 to 10. Therefore, this sum can be written as

$$\sum_{i=1}^{10} -2i.$$

When using summation notation, the index i is always incremented in steps of 1 as it goes from its starting value to its ending value.

Using the formula for the sum of terms in an arithmetic sequence,

$$S = \frac{10}{2}[-2 + -20] = -110.$$

✔ *Check It Out 5:* Write the sum $4 + 8 + 12 + \cdots + 36$ using summation notation, and evaluate. ▪

Sum of Terms of a Geometric Sequence

Just as with arithmetic sequences, it is possible to find the sum of a finite number of terms of a geometric sequence.

Recall that a geometric sequence is given by the rule $a_j = a_0 r^j$, $j = 0, 1, 2, \ldots$. We want to examine the sum of the first n terms. That is,

$$\sum_{j=0}^{n-1} a_0 r^j = a_0 + a_0 r + a_0 r^2 + \cdots + a_0 r^{n-1}.$$

Note We have used the letter j here for the index. The actual letter used for the index does not matter, as long as we are consistent—the variable used for the index must be the same as the variable used for the expression that is to be evaluated and summed.

To find a formula for this sum, we first write the sum as follows.

$$S_n = \sum_{j=0}^{n-1} a_0 r^j = a_0(1 + r + r^2 + r^3 + \cdots + r^{n-1})$$

It turns out that

$$1 - r^n = (1 - r)(1 + r + r^2 + r^3 + \cdots + r^{n-1}).$$

You can easily check this equation for $n = 1, 2, 3$. It holds true for all other positive integers n as well. Therefore, if $r \neq 1$, we can write

$$\frac{1 - r^n}{1 - r} = 1 + r + r^2 + r^3 + \cdots + r^{n-1}.$$

Substituting this expression into the expression for S_n, we then have the formula for the sum of terms of a geometric sequence.

Discover *and* **Learn**

Check the formula for S_n for
$n = 1, 2, 3$.

The Sum of Terms of a Geometric Sequence

$$S_n = \sum_{j=0}^{n-1} a_0 r^j = a_0(1 + r + r^2 + r^3 + \cdots + r^{n-1})$$

$$= a_0 \left(\frac{1 - r^n}{1 - r} \right)$$

Example 6 Direct Application of the Summation Formula

Find the sum of the first five terms of the sequence $a_j = \frac{1}{2}(3)^j, j = 0, 1, 2, \ldots$.

▶Solution Here, $a_0 = \frac{1}{2}$, $r = 3$, and $n = 5$. Using the formula for the sum of terms of a geometric sequence, we have

$$S_5 = \frac{1}{2}(1 + 3 + 3^2 + 3^3 + 3^4)$$

$$= \frac{1}{2}\left(\frac{1 - 3^5}{1 - 3}\right)$$

$$= \frac{1}{2}\left(\frac{1 - 243}{-2}\right) = \frac{121}{2}.$$

☑ *Check It Out 6:* Find the sum of the first six terms of the sequence $a_j = 1024\left(\frac{1}{4}\right)^j$, $j = 0, 1, 2, \ldots$. ■

Example 7 Finding the Sum if the Formula Cannot Be Directly Applied

Find the sum $\sum_{j=1}^{5} 3(2)^{j-1}$.

▶Solution Note that this sum is not quite in the form to which our formula for the sum of terms of a geometric sequence can be applied, because

▶ the value of the index begins with 1 instead of 0; and

▶ the exponent is $j - 1$ rather than j.

To see how we can approach this problem, first write out the terms in the sum.

$$\sum_{j=1}^{5} 3(2)^{j-1} = 3(2)^0 + 3(2)^1 + 3(2)^2 + 3(2)^3 + 3(2)^4$$

$$= 3 + 3(2) + 3(2)^2 + 3(2)^3 + 3(2)^4$$

$$= 3\left(\frac{1 - 2^5}{1 - 2}\right) \qquad a_0 = 3, n = 5, r = 2$$

$$= 93$$

☑ *Check It Out 7:* Find the sum $\sum_{i=2}^{5} 5(3)^{i-2}$. ■

When computing a sum of terms of a sequence, it often helps to write out the first few terms and the last few terms. Then it should be clear what should be substituted into the formula for the sum. This is particularly true for applications.

Infinite Geometric Series

The sum of the terms of an infinite geometric sequence is called an **infinite geometric series.** If the terms of the geometric sequence are increasing—for instance, if $r > 1$—then the sum will increase to infinity. However, it can be shown that if $|r| < 1$, then the infinite geometric series has a finite sum.

> **The Sum of an Infinite Geometric Series**
>
> If $|r| < 1$, then the infinite geometric series
> $$a_0 + a_0 r + a_0 r^2 + a_0 r^3 + \cdots + a_0 r^{n-1} + \cdots$$
> has the sum
> $$S = \sum_{i=0}^{\infty} a_0 r^i = \frac{a_0}{1 - r}.$$

Example 8 Sum of an Infinite Geometric Series

Determine whether each of the following infinite geometric series has a sum. If so, find the sum.

(a) $2 + 4 + 8 + 16 + \cdots$ (b) $1 + \dfrac{1}{2} + \dfrac{1}{4} + \dfrac{1}{8} + \dfrac{1}{16} + \cdots$

▶ Solution

(a) For this geometric series, $r = 2$ because each term is twice the previous term. Thus $|r| = 2$. The series does not have a sum because $|r| > 1$.

(b) For this geometric series, each term is one-half the previous term. Thus $r = \dfrac{1}{2}$. Because $|r| = \dfrac{1}{2}$ is less than 1, the series has a sum. We use the formula for the sum of an infinite geometric series to find the sum.
$$S = \frac{a_0}{1 - r} = \frac{1}{1 - \left(\dfrac{1}{2}\right)} = \frac{1}{\dfrac{1}{2}} = 2$$

✔ *Check It Out 8:* Find the following sum: $2 + \dfrac{1}{2} + \dfrac{1}{8} + \dfrac{1}{32} + \cdots$ ■

Applications

The notion of summing terms of a particular sequence occurs in a variety of applications. Two such applications are discussed in the next two examples.

Example 9 Seating Capacity

An auditorium has 30 seats in the front row. Each subsequent row has two seats more than the row directly in front of it. If there are 12 rows in the auditorium, how many seats are there altogether?

▶Solution From the statement of the problem, we see that the second row must have 32 seats, the third row must have 34 seats, and so on. Therefore, this is an arithmetic sequence with $a_0 = 30$ and common difference $d = 2$. Also, $n = 12$ because there are 12 rows of seats. We then have

$$S_{12} = \frac{n}{2}(2a_0 + d(n - 1)) \qquad \text{We know } a_0, d, \text{ and } n$$

$$= \frac{12}{2}(2(30) + 2(12 - 1)) \qquad \text{Substitute } a_0 = 30, n = 12, \text{ and } d = 2$$

$$= \frac{12}{2}(60 + 2(11)) = 492. \qquad \text{Simplify}$$

Thus there are 492 seats in the auditorium.

☑ *Check It Out 9:* Rework Example 9 for the case in which there are 25 seats in the front row, each subsequent row has two seats more than the row directly in front of it, and there are 10 rows. ■

Annuities are investments in which a fixed amount of money is invested each year. The interest earned on the investment is compounded annually. Example 10 discusses a specific case of an annuity.

Just In Time

Review compound interest in Example 5 in Section 8.1.

Example **10** Annuity

Suppose $2000 is deposited initially (and at the *end* of each year) into an annuity that pays 5% interest compounded annually. What is the total amount in the account at the end of 10 years?

▶Solution The interest earned on the $2000 that is deposited in any given year will not begin to be paid until a year later. Thus, by the end of the tenth year, only 9 years' worth of interest will have been paid on the amount deposited at the end of the first year, and only 1 year's worth of interest will have been paid on the amount deposited at the end of the ninth year. No interest will have been paid on the amount deposited at the end of the tenth year, since that deposit will just have been made.

This information is summarized in Table 8.2.2. The compound interest formula has been used to calculate the total value of each $2000 deposit for 10 years after the account was opened.

Table 8.2.2

Years After Opening Account	0	1	· · ·	9	10
Deposit ($)	2000	2000	· · ·	2000	2000
Value at End of Tenth Year ($)	$2000(1.05)^{10}$	$2000(1.05)^9$	· · ·	$2000(1.05)$	2000

To find the total amount, we must add all these amounts:

$$\text{Amount at end of 10 years} = 2000 + 2000(1.05) + \cdots + 2000(1.05)^9 + 2000(1.05)^{10}$$

Using $a_0 = 2000$, $n = 11$, and $r = 1.05$ in the geometric sum formula, we find that this sum is

$$S = 2000\left(\frac{1 - (1.05)^{11}}{1 - 1.05}\right)$$

$$= 28{,}413.57.$$

Thus there will be $28,413.57 in the account at the end of 10 years.

✔ *Check It Out 10:* Rework Example 10 for the case in which the interest rate is 4% compounded annually. ■

8.2 Key Points

▶ Let $a_j = a_0 + jd$, $j = 0, 1, 2, ...,$ be an **arithmetic sequence,** and let n be a positive integer. The **sum of the n terms** from a_0 to a_{n-1} is given by

$$S_n = a_0 + a_1 + \cdots + a_{n-1} = \frac{n}{2}(2a_0 + d(n-1))$$

$$= \frac{n}{2}(a_0 + a_{n-1}).$$

▶ The **summation symbol** is indicated by the Greek letter Σ (sigma).

Index i ends at n
↓
Summation symbol → $\sum_{i=1}^{n} i^2$ ← Expression to be evaluated and summed
↑
Index i starts at 1

▶ Let $a_j = a_0 r^j$, $j = 0, 1, 2, ...,$ be a **geometric sequence.** The **sum of the n terms** from a_0 to a_{n-1} is given by

$$S_n = \sum_{j=0}^{n-1} a_0 r^j = a_0\left(\frac{1 - r^n}{1 - r}\right).$$

▶ If $|r| < 1$, then the **infinite geometric series**

$$a_0 + a_0 r + a_0 r^2 + a_0 r^3 + \cdots + a_0 r^{n-1} + \cdots$$

has the sum $S = \sum_{i=0}^{\infty} a_0 r^i = \dfrac{a_0}{1 - r}.$

8.2 Exercises

▶**Just in Time Exercises** These exercises correspond to the Just in Time references in this section. Complete them to review topics relevant to the remaining exercises.

1. True or False: If a bank is advertising an account that pays 7% interest compounded quarterly, interest will be computed 6 times per year.

2. True or False: A bank is advertising an account that pays 6% interest, compounded monthly. A customer opens an account of this type and leaves money in the account for 5 years. Over the 5 years, interest will be deposited into the account 60 times.

3. A bank is advertising an account with an interest rate of 2%, compounded semiannually. A customer opens an account with an initial deposit of $1000 and makes no more deposits or withdrawals. What will the account balance be at the end of 2 years?

4. A bank is advertising an account with an interest rate of 5%, compounded quarterly. A customer opens an account with an initial deposit of $1000 and makes no more deposits or withdrawals. What will the account balance be at the end of 2 years?

▶**Skills** This set of exercises will reinforce the skills illustrated in this section.

In Exercises 5–10, find the sum of the first 14 terms of each arithmetic sequence.

5. 3, 6, 9, 12, 15, …

6. 4, 8, 12, 16, 20, …

7. −6, −1, 4, 9, …

8. −8, −5, −2, 1, 4, …

9. 2, 7, 12, 17, 22, …

10. 6, 13, 20, 27, …

In Exercises 11–16, find the sum of the first eight terms of each geometric sequence.

11. 3, 6, 12, 24, …

12. 4, 8, 16, 32, …

13. 6, 3, $\dfrac{3}{2}$, $\dfrac{3}{4}$, …

14. 12, 4, $\dfrac{4}{3}$, $\dfrac{4}{9}$, …

15. 2, 3, $\dfrac{9}{2}$, $\dfrac{27}{4}$, …

16. 12, 16, $\dfrac{64}{3}$, $\dfrac{256}{9}$, …

In Exercises 17–42, find the sum.

17. $3 + 6 + 9 + \cdots + 90$

18. $1 + 5 + 9 + \cdots + 53$

19. $10 + 13 + 16 + \cdots + 55$

20. $8 + 12 + 16 + \cdots + 72$

21. $4 + 9 + 14 + \cdots + (5n + 4)$

22. $1 + 3 + 5 + \cdots + (2n + 1)$

23. $1 + 5 + 25 + \cdots + 78{,}125$

24. $1 + 4 + 16 + \cdots + 1024$

25. $2 + 6 + 18 + \cdots + 1458$

26. $7 + 14 + 28 + \cdots + 896$

27. Sum of the odd integers from 5 to 125, inclusive

28. Sum of the odd integers from 35 to 105, inclusive

29. Sum of the odd integers from 27 to 115, inclusive

30. Sum of the odd integers from 51 to 205, inclusive

31. Sum of the even integers from 4 to 130, inclusive

32. Sum of the even integers from 8 to 160, inclusive

33. Sum of the even integers from 10 to 102, inclusive

34. Sum of the even integers from 20 to 200, inclusive

35. $\displaystyle\sum_{i=0}^{6} (2i)$

36. $\displaystyle\sum_{i=0}^{5} (-3i)$

37. $\displaystyle\sum_{i=0}^{10} (5 + 2i)$

38. $\displaystyle\sum_{i=0}^{7} (2 + 4i)$

39. $\displaystyle\sum_{i=0}^{4} \left(\dfrac{1}{2}\right)^{i}$

40. $\displaystyle\sum_{k=0}^{5} \left(\dfrac{2}{3}\right)^{k}$

41. $\displaystyle\sum_{i=0}^{4} 8(3^{i})$

42. $\displaystyle\sum_{i=0}^{5} 5(2^{i})$

In Exercises 43–54, (a) write using summation notation, and (b) find the sum.

43. $2 + 4 + 6 + \cdots + 40$

44. $1 + 4 + 7 + \cdots + 58$

45. $a + 2a + 3a + \cdots + 60a$

46. $2z + 4z + 6z + \cdots + 20z$

47. $2 + 4 + 8 + \cdots + 1024$

48. $3 + 9 + 27 + \cdots + 59{,}049$

49. $a + a^2 + a^3 + \cdots + a^{40}$

50. $2z + 6z^3 + 18z^5 + \cdots + 486z^{11}$

51. The sum of the first 25 terms of the sequence defined by $a_n = 2.5n$, $n = 0, 1, 2, \ldots$

52. The sum of the first 50 terms of the sequence defined by $a_n = 6.5n$, $n = 0, 1, 2, \ldots$

53. The sum of the first 45 terms of the sequence defined by $a_n = (0.5)^n$, $n = 0, 1, 2, \ldots$

54. The sum of the first 60 terms of the sequence defined by $a_n = (0.4)^n$, $n = 0, 1, 2, \ldots$

In Exercises 55–66, evaluate the sum. For each sum, state whether it is arithmetic or geometric. Depending on your answer, state the value of d or r.

55. $\displaystyle\sum_{k=0}^{6} (2k + 1)$

56. $\displaystyle\sum_{k=0}^{8} (3k - 1)$

57. $\displaystyle\sum_{k=0}^{20} (0.5k)$

58. $\displaystyle\sum_{k=0}^{18} (0.25k)$

59. $\displaystyle\sum_{k=0}^{6} 2^{k-1}$

60. $\displaystyle\sum_{k=0}^{7} \left(\frac{3}{4}\right)^{k+1}$

61. $\displaystyle\sum_{k=5}^{10} (0.5)^{k-4}$

62. $\displaystyle\sum_{k=4}^{9} (0.25)^{k-3}$

63. $\displaystyle\sum_{k=1}^{5} 2(0.5)^{k}$

64. $\displaystyle\sum_{k=1}^{9} 3(0.25)^{k}$

65. $\displaystyle\sum_{k=0}^{5} (3(k + 2) + 3(k - 1))$

66. $\displaystyle\sum_{k=0}^{6} (2(2k + 4) - 2(k + 1))$

In Exercises 67–76, determine whether the infinite geometric series has a sum. If so, find the sum.

67. $6 + 3 + \dfrac{3}{2} + \dfrac{3}{4} + \cdots$

68. $8 + 4 + 2 + 1 + \cdots$

69. $9 + 3 + 1 + \dfrac{1}{3} + \dfrac{1}{9} + \cdots$

70. $12 + 3 + \dfrac{3}{4} + \dfrac{3}{12} + \cdots$

71. $4 + 8 + 16 + 32 + \cdots$

72. $\dfrac{1}{2} + \dfrac{3}{2} + \dfrac{9}{2} + \dfrac{27}{2} + \cdots$

73. $\displaystyle\sum_{k=0}^{\infty} 2(0.5)^{k}$

74. $\displaystyle\sum_{k=0}^{\infty} 3(0.25)^{k}$

75. $\displaystyle\sum_{k=0}^{\infty} 3\left(\frac{1}{5}\right)^{k}$

76. $\displaystyle\sum_{k=0}^{\infty} 5\left(\frac{1}{8}\right)^{k}$

▶**Applications** In this set of exercises, you will use sequences and their sums to study real-world problems.

77. **Stacking Displays** A store clerk is told to stack cookie boxes in a pyramid pattern for a store display, as pictured.

Etc.

(a) If the clerk has 55 boxes, how many boxes must be placed in the bottom row if all the boxes are to be displayed at one time?

(b) The store manager gives the clerk 15 more boxes and tells her to use all 70 boxes to build a display in the same pyramid pattern. The clerk replies that that would be impossible. Explain why she is correct.

(c) The clerk offers to start with 70 boxes and make a display in the same pyramid pattern—and to do it in such a way that as few boxes as possible will be left over. How many boxes will be in the bottom row? How many boxes will be left over?

78. **Communication** Many large corporations have in place an emergency telephone chain in which one employee in each division is designated to be the first called in the case of an emergency. That employee then calls three employees within the division, each of whom in turn calls three employees, and so on. The chain stops once all the employees have been notified of the emergency.

  (a) Write the first five terms of the sequence that represents the number of people called at each step of the chain. (The first step consists of just the "designated" employee being called.) Is this an arithmetic sequence or a geometric sequence?

  (b) Use an appropriate formula to answer the following question: How many steps of the chain are needed to notify all the employees of a corporate division with 600 employees?

  (c) Explain why this method of notification is very efficient.

79. **Education Savings** The parents of a newborn child decide to start saving for the child's college education. At the end of each calendar year, they put $1500 into an Educational Savings Account (ESA) that pays 6% interest compounded annually. What will be the total amount in the account 18 years after they make their initial deposit?

80. **Retirement Savings** Maria is a recent college graduate who wants to take advantage of an individual retirement account known as a Roth IRA. In order to build savings for her retirement, she wants to put $2500 at the end of each calendar year into an IRA that pays 5.5% interest compounded annually. If she stays with this plan, what will be the total amount in the account 40 years after she makes her initial deposit?

81. **Physics** A ball dropped to the floor from a height of 10 feet bounces back up to a point that is three-fourths as high. If the ball continues to bounce up and down, and if after each bounce it reaches a point that is three-fourths as high as the point reached on the previous bounce, calculate the total distance the ball travels from the time it is dropped to the time it hits the floor for the third time.

82. **Television Piracy** The loss of revenue to an industry due to piracy can be staggering. For example, a newspaper article reported that the pay television industry in Asia lost nearly $1.3 billion in potential revenue in 2003 because of the use of stolen television signals. The loss was projected to grow at a rate of 10% per year. (*Source: The Financial Times*)

  (a) Assuming the projection was accurate, how much did the pay television industry in Asia lose in the years 2004, 2005, and 2006?

  (b) Assuming the projected trend has prevailed to the present time and will continue into the future, what is the projected loss in revenue for the year that is n years after 2003?

  (c) Find the total loss of revenue for the years 2003 to 2012, inclusive.

83. **Literature** The following poem (*As I Was Going to St. Ives*, circa 1730) refers to the name of a quaint old village in Cornwall, England. (*Source:* www.rhymes.org.uk)

  > As I was going to St. Ives
  > I met a man with seven wives.
  > Every wife had seven sacks,
  > Every sack had seven cats,
  > Every cat had seven kits.
  > Kits, cats, sacks, and wives,
  > How many were going to St. Ives?

  (a) Use the sum of a *sequence* of numbers to express the number of people and objects (combined) that the author of this poem encountered while going to St. Ives. Do not evaluate the sum. Is this the sum of terms of an arithmetic sequence or a geometric sequence? Explain.

  (b) Use an appropriate formula to find the sum from part (a).

84. **Dimensions** A carpet warehouse needs to calculate the diameter of a rolled carpet given its length, width, and thickness. If the diameter of the carpet roll can be predicted ahead of time, the warehouse will know how much to order so as not to exceed warehouse capacity. Assume that the carpet is rolled lengthwise. The cross-section of the carpet roll is then a spiral. To simplify the problem, approximate the spiral cross-section by a set of n concentric circles whose radii differ by the thickness t. Calculate the number of circles n using the fact that the sum of the circumferences of the n circles must equal the given length. How can you find the diameter once you know n?

▶**Concepts** This set of exercises will draw on the ideas presented in this section and your general math background.

85. The first term of an arithmetic sequence is 4. The sum of the first three terms of the sequence is 24. Use summation notation to express the sum of the first eight terms of this sequence, and use an appropriate formula to find the sum.

86. Find the following sum:

$$1 + 2\left(1 + \frac{1}{2}\right) + 3\left(1 + \frac{1}{3}\right) + \cdots + 50\left(1 + \frac{1}{50}\right)$$

(*Hint:* Expand first.)

87. Given two terms of an arithmetic sequence, $a_2 = 14$ and $a_6 = 2$, find $\sum_{k=1}^{8} a_k$. (*Hint:* First find d and a_0.)

88. Given two terms of a geometric sequence, $a_0 = -1$ and $a_3 = 27$, find $\sum_{j=0}^{5} a_j$.

89. For a geometric sequence, find a_0 if $\sum_{j=0}^{4} a_j = 3$ and $r = \frac{1}{2}$.

8.3 General Sequences and Series

Objectives

▶ Generate terms of a general sequence

▶ Find a rule for a sequence given a few terms

▶ Generate terms of a recursively defined sequence

▶ Find a rule for a recursively defined sequence

▶ Calculate partial sums of terms of a sequence

▶ Apply general sequences to word problems

The number of spirals in the head of a sunflower forms a sequence that does not fit the pattern of an arithmetic or a geometric sequence. This is only one example of the many different types of sequences that can be studied. In this section, we will discuss sequences in a more general setting. The Fibonacci sequence illustrated by the sunflower pattern will be discussed in Example 6.

Sequences Given by a Rule

Recall that the terms of arithmetic and geometric sequences are generated by specific types of rules. We can generate other types of sequences simply by using other kinds of rules. The following examples illustrate some of the types of sequences that can be generated in this way.

Example 1 Generating a Sequence from a Rule

Find the first four terms of each of the following sequences.

(a) $a_n = n^2$, $n = 0, 1, 2, 3, \ldots$

(b) $f(n) = \dfrac{1}{n + 1}$, $n = 0, 1, 2, 3, \ldots$

▶ Solution

(a) To find the first four terms, successively substitute $n = 0, 1, 2, 3$ into the formula $a_n = n^2$, which gives

$$a_0 = (0)^2 = 0, \ a_1 = (1)^2 = 1, \ a_2 = (2)^2 = 4, \ a_3 = (3)^2 = 9.$$

(b) Substitute $n = 0, 1, 2, 3$ into $f(n) = \dfrac{1}{n + 1}$, which gives

$$a_0 = \frac{1}{0 + 1} = 1, \ a_1 = \frac{1}{1 + 1} = \frac{1}{2}, \ a_2 = \frac{1}{2 + 1} = \frac{1}{3}, \ a_3 = \frac{1}{3 + 1} = \frac{1}{4}.$$

☑ *Check It Out 1:* Find the first four terms of the sequence defined by $a_n = 1 + 2n^2$, $n = 0, 1, 2, 3, \ldots$. ▪

Example 2 Using the First Few Terms to Find the Rule

Assuming that the pattern continues, find a rule for the sequence whose first four terms are as given.

(a) 1, 8, 27, 64

(b) $1, \dfrac{1}{4}, \dfrac{1}{9}, \dfrac{1}{16}$

▶Solution

(a) The given terms are all perfect cubes. Thus the rule for this sequence is

$$a_n = n^3, \quad n = 1, 2, 3, \dots.$$

Note that this sequence starts with $n = 1$.

(b) Examining the terms, we see that each term is the reciprocal of a perfect square. Thus the rule is

$$a_n = \frac{1}{n^2}, \quad n = 1, 2, 3, \dots.$$

✔ *Check It Out 2:* Find a rule for the sequence whose first four terms are

$$1, \frac{1}{8}, \frac{1}{27}, \frac{1}{64}. \quad ■$$

Alternating Sequences

In an **alternating sequence,** the terms *alternate* between positive and negative numbers. The next example involves finding a rule for a simple sequence of this type.

Example 3 Finding the Rule for an Alternating Sequence

Assuming that the given pattern continues, find a rule for the sequence whose terms are given by

$$1, -1, 1, -1, 1, -1, \dots.$$

▶Solution The terms of this sequence consist only of 1 and −1, with the two numbers alternating. One way to write a rule for this sequence is

$$a_n = (-1)^n, \quad n = 0, 1, 2, 3, \dots.$$

This rule works because −1 raised to an even power will equal 1, while −1 raised to an odd power will equal −1. Since the value of n alternates between even and odd numbers, the rule $a_n = (-1)^n$ produces the sequence

$$1, -1, 1, -1, 1, -1, \dots.$$

✔ *Check It Out 3:* Find a rule for the sequence whose terms are given by

$$-1, 1, -1, 1, -1, 1, \dots. \quad ■$$

Table 8.3.1

n	0	1	2	3	4
$f(n)$					

Example 4 **Generating and Graphing Terms of a Sequence**

Example 4 **Generating and Graphing Terms of a Sequence**

Let $f(n) = (-1)^n(n^2 + 1)$. Fill in Table 8.3.1 and plot the first five terms of the sequence.

▶ Solution To fill in the table, substitute $n = 0, 1, 2, 3, 4$ (in succession) into the expression for $f(n)$.

$$f(0) = (-1)^0((0)^2 + 1) = \quad (1)(1) = \quad 1$$
$$f(1) = (-1)^1((1)^2 + 1) = \quad (-1)(2) = \quad -2$$
$$f(2) = (-1)^2((2)^2 + 1) = \quad (1)(5) = \quad 5$$
$$f(3) = (-1)^3((3)^2 + 1) = (-1)(10) = -10$$
$$f(4) = (-1)^4((4)^2 + 1) = \quad (1)(17) = \quad 17$$

Table 8.3.2

n	0	1	2	3	4
$f(n)$	1	-2	5	-10	17

See Table 8.3.2.

The first five terms of the sequence are plotted in Figure 8.3.1. Note that the dots are *not* connected. Because the function f is a sequence, its domain consists of the set of all nonnegative integers. Thus f is not defined for any number that lies *between* two consecutive nonnegative integers.

Figure 8.3.1

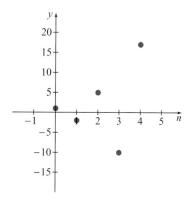

Discover *and* **Learn**

Graph the sequence $u(n) = (-1)^n \dfrac{n}{n^2 + 3}$, with n ranging from 1 to 50. What do you observe about the function values as n gets larger?

☑ *Check It Out 4:* Let $f(n) = (-1)^n\left(\dfrac{1}{n}\right)$, $n = 1, 2, 3, \ldots$. Fill in Table 8.3.3 and plot the first five terms of the sequence.

Table 8.3.3

n	1	2	3	4	5
$f(n)$					

Recursively Defined Sequences

Some sequences are defined by expressing the nth term a_n as a function of one or more of the preceding terms rather than as a function of n alone, and specifying the value of the first term of the sequence, or the values of the first several terms. A sequence generated in this manner is called a **recursively defined sequence.** Example 5 illustrates such a sequence.

Example 5 Generating Terms of a Recursively Defined Sequence

Find the first four terms of the sequence defined recursively by

$$a_0 = 1, \quad a_n = 3 + a_{n-1}, \quad n = 1, 2, 3, \dots .$$

What type of sequence is generated? Find a rule for the nth term of the sequence that depends only on n.

▶ **Solution** The general rule for a_n is defined using a_{n-1}, the term just preceding a_n. The first four terms are

$a_0 = 1$	Given
$a_1 = 3 + a_0 = 3 + 1 = 4$	Rule for a_n with $n = 1$
$a_2 = 3 + a_1 = 3 + 4 = 7$	Rule for a_n with $n = 2$
$a_3 = 3 + a_2 = 3 + 7 = 10$	Rule for a_n with $n = 3$

The terms of this sequence are thus

$$1, 4, 7, 10, \dots .$$

This is an arithmetic sequence with a common difference of 3. Using the formula for the nth term of an arithmetic sequence, we can write

$$a_n = a_0 + nd = 1 + 3n.$$

Discover _and_ Learn

Find the first four terms of the sequence defined recursively by

$a_0 = 1, a_n = 3a_{n-1},$

$n = 1, 2, 3, \dots .$

What type of sequence is generated? Find a rule for the nth term of the sequence that depends only on n.

✔ *Check It Out 5:* Consider the recursively defined sequence

$$b_0 = 1, \quad b_n = 1 - b_{n-1}, \quad n = 1, 2, 3, \dots .$$

(a) Find the first five terms of the sequence.

(b) What is the range of the function f that corresponds to this sequence (i.e., the function $f(n) = b_n$, $n = 1, 2, 3, \dots$)? ▪

The next example deals with the Fibonacci sequence, which was discussed briefly at the beginning of this section.

Example 6 Generating Terms of the Fibonacci Sequence

Find the first five terms of the Fibonacci sequence, defined recursively as follows.

$$f_0 = 1, f_1 = 1, f_n = f_{n-1} + f_{n-2}, \quad n = 2, 3, \dots$$

▶ **Solution** The recursive definition of the Fibonacci sequence states that the term f_n (for $n \geq 2$) is the sum of the *two preceding terms*. We can compute the first five terms as follows.

$f_0 = 1; \quad f_1 = 1$	Given
$f_2 = f_1 + f_0 = 1 + 1 = 2$	Substitute $n = 2$
$f_3 = f_2 + f_1 = 2 + 1 = 3$	Substitute $n = 3$
$f_4 = f_3 + f_2 = 3 + 2 = 5$	Substitute $n = 4$

Hence, the first five terms of the Fibonacci sequence are

$$1, 1, 2, 3, 5.$$

✔ *Check It Out 6:* Generate the terms f_5, f_6, f_7, and f_8 of the Fibonacci sequence. ▪

Partial Sums

It is sometimes necessary to find the sum of two or more consecutive terms of a given sequence of numbers. In Section 8.2, we found sums of terms of arithmetic and geometric sequences. In this section, we extend the discussion to include sums of terms of other types of sequences.

Suppose we want to denote the sum of the first n terms of the sequence $a_0, a_1, a_2, \ldots$. Using the summation notation from Section 8.2, we have the following.

---

nth Partial Sum of a Sequence

Let n be a positive integer. The sum of the first n terms of the sequence $a_0, a_1, a_2, \ldots$ (the terms a_0 through a_{n-1}) is denoted by $\displaystyle\sum_{i=0}^{n-1} a_i$. In other words,

$$\sum_{i=0}^{n-1} a_i \quad \text{means} \quad a_0 + a_1 + \cdots + a_{n-1}.$$

This sum is referred to as the **nth partial sum** of the sequence.

Some partial sums will be computed in the following examples.

Example 7 Direct Calculation of a Partial Sum of a Sequence

Compute each sum.

(a) $\displaystyle\sum_{i=0}^{4} i^2$ (b) $\displaystyle\sum_{i=0}^{3} (-1)^i (2i + 1)$

▶Solution

(a) We consecutively substitute 0, 1, 2, 3, and 4 for i in the expression i^2 and then add the results.

$$\sum_{i=0}^{4} i^2 = 0 + 1^2 + 2^2 + 3^2 + 4^2 = 0 + 1 + 4 + 9 + 16 = 30$$

(b) By consecutively substituting 0, 1, 2, and 3 for i in the expression $(-1)^i (2i + 1)$ and then adding the results, we obtain

$$\sum_{i=0}^{3} (-1)^i (2i + 1) = (-1)^0 (2(0) + 1) + (-1)^1 (2(1) + 1) + (-1)^2 (2(2) + 1)$$
$$+ (-1)^3 (2(3) + 1)$$
$$= 1(1) + (-1)(3) + (1)(5) + (-1)(7) = 1 - 3 + 5 - 7 = -4.$$

✔ *Check It Out 7:* Compute the following sums.

(a) $\displaystyle\sum_{j=0}^{5} (j + 1)$

(b) $\displaystyle\sum_{k=0}^{4} (6k - 5)$ ▪

Applications

Sequences are extremely useful in studying events that happen at regular intervals. This is usually done with a recursive sequence. We illustrate this idea in the next example.

Example **8** **Drug Dose**

An initial dose of 40 milligrams of the pain reliever acetaminophen is given to a patient. Subsequent doses of 20 milligrams each are administered every 5 hours. Just before each 20-milligram dose is given, the amount of acetaminophen in the patient's bloodstream is 25% of the total amount in the bloodstream just after the previous dose was administered.

(a) Let a_0 represent the initial amount of the drug in the bloodstream and, for $n \geq 1$, let a_n represent the amount in the bloodstream immediately after the nth 20-milligram dose is given. Make a table of values for a_0 through a_6.

(b) Plot the values you tabulated in part (a). What do you observe?

(c) With the aid of the values you tabulated in part (a), find a recursive definition of a_n.

▶ Solution

(a) We construct Table 8.3.4 as follows.

Table 8.3.4

n	a_n, Amount in Bloodstream (mg)	Comments
0	40	Initial amount
1	$(0.25)(40) + 20 = 30$	Amount remaining from prior dose + new dose
2	$(0.25)(30) + 20 = 27.5$	Amount remaining from prior dose + new dose
3	$(0.25)(27.5) + 20 = 26.875$	Amount remaining from prior dose + new dose
4	$(0.25)(26.875) + 20 \approx 26.72$	Amount remaining from prior dose + new dose
5	$(0.25)(26.72) + 20 \approx 26.68$	Amount remaining from prior dose + new dose
6	$(0.25)(26.68) + 20 \approx 26.67$	Amount remaining from prior dose + new dose

Figure 8.3.2

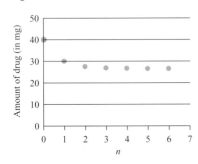

(b) The values of a_n, the amount of acetaminophen in the bloodstream after the nth 20-milligram dose, are plotted in Figure 8.3.2.

Note that as n increases, the amount of the drug in the bloodstream seems to approach a constant amount of approximately 26.67 milligrams. This is known as reaching **steady state.**

(c) Examining the table of values closely, we see that

$$a_n = (0.25)(\text{previous amount}) + 20 = 0.25a_{n-1} + 20.$$

✔ *Check It Out 8:* Rework Example 8 for the case in which the initial dose of acetaminophen is 40 milligrams, subsequent doses of 20 milligrams are given every 5 hours thereafter, and, just before each 20-milligram dose is administered, the amount of the drug in the patient's bloodstream is only 20% of the total amount in the bloodstream just after the previous dose was given. ▪

8.3 Key Points

▶ A **general rule for the terms** a_n of a sequence can be used to generate terms of the sequence.

▶ In an **alternating sequence,** the terms alternate between positive and negative values.

▶ In a **recursively defined sequence,** the nth term is defined by using the preceding terms, along with the first term or first several terms.

▶ The **nth partial sum** of the sequence $a_0, a_1, a_2, \ldots$ is denoted by $\sum_{i=0}^{n-1} a_i$.

8.3 Exercises

▶**Skills** This set of exercises will reinforce the skills illustrated in this section.

In Exercises 1–18, find the first five terms of the sequence.

1. $a_n = -4n + 6, n = 0, 1, 2, 3, \ldots$

2. $a_n = 6n + 2, n = 0, 1, 2, 3, \ldots$

3. $a_n = -4\left(\dfrac{1}{3}\right)^n, n = 0, 1, 2, 3, \ldots$

4. $a_n = -\left(\dfrac{5}{2}\right)^n, n = 0, 1, 2, 3, \ldots$

5. $b_n = n^2 + 4, n = 0, 1, 2, 3, \ldots$

6. $b_n = -6n^3 + 1, n = 0, 1, 2, 3, \ldots$

7. $f(n) = \dfrac{n}{2n^2 + 1}, n = 1, 2, 3, \ldots$

8. $f(n) = \dfrac{n}{-3n^2 - 1}, n = 1, 2, 3, \ldots$

9. $a_n = (-1)^n e^{n+1}, n = 0, 1, 2, 3, \ldots$

10. $a_n = (-1)^n 2^{n+1}, n = 0, 1, 2, 3, \ldots$

11. $g(n) = \dfrac{2^n + 1}{n^2 + 1}, n = 1, 2, 3, \ldots$

12. $g(n) = \dfrac{3^n - 1}{n^2}, n = 1, 2, 3, \ldots$

13. $a_n = \sqrt{2n + 4}, n = 0, 1, 2, 3, \ldots$

14. $a_n = \dfrac{25}{n}, n = 1, 2, 3, \ldots$

15. $a_n = n^2 + 2, n = 0, 1, 2, 3, \ldots$

16. $a_n = 2 + n, n = 0, 1, 2, 3, \ldots$

17. $a_n = 2n^3, n = 0, 1, 2, 3, \ldots$

18. $a_n = (2n^3)^{1/3}, n = 0, 1, 2, 3, \ldots$

In Exercises 19–28, find a rule for each sequence whose first four terms are given. Assume that the given pattern will continue.

19. $-2, -6, -10, -14, \ldots$ 20. $-3, 2, 7, 12, \ldots$

21. $1, \dfrac{1}{2}, \dfrac{1}{4}, \dfrac{1}{8}, \ldots$ 22. $1, \dfrac{1}{3}, \dfrac{1}{9}, \dfrac{1}{27}, \ldots$

23. $1, \sqrt{2}, \sqrt{3}, 2, \ldots$ 24. $\sqrt{2}, 2, \sqrt{6}, \sqrt{8}, \ldots$

25. $1, 0.4, 0.16, 0.064, \ldots$ 26. $1, \dfrac{1}{4}, \dfrac{1}{16}, \dfrac{1}{64}, \ldots$

27. $\sqrt{3}, \sqrt{6}, 3, \sqrt{12}, \ldots$ 28. $2, 2\sqrt{2}, 2\sqrt{3}, 4$

In Exercises 29–36, find the first four terms of the recursively defined sequence.

29. $a_0 = 6$; $a_n = a_{n-1} - 2$, $n = 1, 2, 3, \ldots$

30. $a_0 = -3$; $a_n = a_{n-1} + 2.5$, $n = 1, 2, 3, \ldots$

31. $b_1 = 4$; $b_n = \frac{1}{4}b_{n-1}$, $n = 2, 3, 4, \ldots$

32. $b_1 = 2$; $b_n = \frac{3}{2}b_{n-1}$, $n = 2, 3, 4, \ldots$

33. $a_0 = -1$; $a_n = a_{n-1} + n$, $n = 1, 2, 3, \ldots$

34. $a_0 = 4$; $a_n = a_{n-1} - n$, $n = 1, 2, 3, \ldots$

35. $b_1 = \sqrt{3}$; $b_n = \sqrt{b_{n-1} + 3}$, $n = 2, 3, 4, \ldots$

36. $b_1 = \sqrt{3}$; $b_n = \sqrt{\dfrac{b_{n-1}}{3}}$, $n = 2, 3, 4, \ldots$

In Exercises 37–40, find the first four terms of the recursively defined sequence. Find the rule for a_n in terms of just n.

37. $a_0 = -4$; $a_n = a_{n-1} + 2$, $n = 1, 2, 3, \ldots$

38. $a_0 = 3$; $a_n = a_{n-1} + 1.5$, $n = 1, 2, 3, \ldots$

39. $b_1 = 6$; $b_n = \frac{1}{2}b_{n-1}$, $n = 2, 3, 4, \ldots$

40. $b_1 = 7$; $b_n = \frac{3}{4}b_{n-1}$, $n = 2, 3, 4, \ldots$

▶**Applications** In this set of exercises, you will use general sequences and series to study real-world problems.

41. **Games** A popular electronic game originally sold for $200. The price of the game was adjusted annually by just enough to keep up with inflation. Assume that the rate of inflation was 4% per year.
   (a) For $n \geq 0$, let p_n be the price of the game n years after it was put on the market. Define p_n recursively.
   (b) Find an expression for p_n in terms of just n.
   (c) What was the price of the game 3 years after it was put on the market?
   (d) How many years after the game was put on the market did the price first exceed $250?
   (e) How many years after the game was put on the market did the price first exceed $300?

42. **Compensation** A certain company rewards its employees with annual bonuses that grow with the number of years the employee remains with the company. At the end of the first full year of employment, an employee's bonus is $1000. At the end of each full year beyond the first, the employee receives $1000 plus 50% of the previous year's bonus.
   (a) What bonuses does an employee receive after each of the first four full years of employment?
   (b) For $n \geq 0$, let b_n be the bonus received after n full years of employment. Define b_n recursively.

43. **Manufacturing** One product line offered by a window manufacturer consists of rectangular windows that are 36 inches in height. The widths of the windows range from 18 inches to 54 inches, in increments of half an inch.
   (a) Give an expression for the perimeter P_n of the nth window in this product line, where P_0 is the perimeter of the smallest window, P_1 is the perimeter of the second-smallest window, and so on.
   (b) Give an expression for the area A_n of the nth window in this product line, where A_0 is the area of the smallest window, A_1 is the area of the second-smallest window, and so on.
   (c) What are the dimensions, perimeter, and area of the sixth-smallest window in this product line?
   (d) If the smallest window sells for $250 and the prices of the windows are graduated at the rate of $10 for every half-inch of additional width, what is the total cost (ignoring sales tax) of one 19.5-inch-wide window and five 36-inch-wide windows?

44. **Horticulture** The Morales family bought a Christmas tree. As soon as they got the tree home and set it up, they put 3 quarts of water into the tree holder. Every day thereafter, they awoke to find that half of the water from the previous day was gone, so they added a quart of water.
   (a) For $n \geq 0$, let w_n be the volume of water in the tree holder (just after water was added) n days after the tree was set up in the home of the Morales family. Define w_n recursively.
   (b) How many days after the tree was initially set up did the family awake to find that the water level had dipped below the 1.1-quart mark for the first time?
   (c) When, if ever, did the family awake to find that the water level had dipped below the 1-quart mark for the first time? Explain.

45. **Fundraising** During a recent month, students contributed money at school for the benefit of flood victims in another part of the country. One enterprising student, Matt, asked his aunt to donate money on his behalf. She agreed that on each day that Matt contributed, she would match his

donation plus donate 10 cents more. There were 21 school days during the month in question. From the second school day on, Matt donated 3 cents more than he gave on the previous school day. In total, Matt and his aunt contributed $17.22.

(a) How much money did Matt contribute on the first school day of the month in question?

(b) What was Matt's total contribution for that month?

(c) How much did Matt's aunt donate on his behalf?

46. **Distribution** Kara gives a pencil to every child who comes to her house trick-or-treating on Halloween. The first year she did this, she bought 120 pencils, which turned out to be one-third more pencils than she needed. Kara kept the extras to hand out the next year. The second year, she bought x new pencils (to add to the supply she had left over from the first year). One-fourth of all the pencils she had to give to trick-or-treaters the second year (the new pencils plus the extras from the first year) were left over. The third year, she again bought x new pencils, and one-fifth of the total number available for handout that year were left over.

(a) How many trick-or-treaters went to Kara's house the first year, and how many pencils were left over that year?

(b) Give expressions (in terms of x) for the total number of pencils available for handout the second year and the number of children who came to Kara's house trick-or-treating that year.

(c) Give expressions (in terms of x) for the total number of pencils available for handout the third year and the number of trick-or-treaters who went to Kara's house that year.

(d) If Kara had 14 pencils left over the third year, what is the value of x?

(e) Use the value of x that you found in part (d) to determine the number of children who came to Kara's house trick-or-treating the second year and the number who came the third year.

▶**Concepts** This set of exercises will draw on the ideas presented in this section and your general math background.

47. Find the smallest positive integer n such that $\left|\sum_{i=0}^{n-1} b_i\right| \geq 81$ if $b_n = 3n^2 - 2$, $n = 0, 1, 2, \ldots$.

48. For $n \geq 1$, let V_n be the volume of a cube that is n units on a side. Using summation notation, give an expression for the sum of V_n over the first six positive integers n, and find the sum.

49. A sequence $a_0, a_1, a_2, \ldots$ has the property that $a_n = 3a_{n-1} + 2$ for $n = 1, 2, 3, \ldots$. If $a_3 = 134$, what is the value of a_0?

50. A sequence $b_0, b_1, b_2, \ldots$ has the property that $b_n = c\left(\dfrac{n+3}{n+2}\right)b_{n-1}$ for $n = 1, 2, 3, \ldots$, where c is a positive constant to be determined. Find c if $b_2 = 25$ and $b_4 = 315$.

51. If $a_n = 1 - (a_{n-1})^3$ for $n = 1, 2, 3, \ldots$, for what value(s) of a_0 is the sequence $a_0, a_1, a_2, \ldots$ an alternating sequence?

52. If $a_n = \sqrt{a_{n-1}} + \dfrac{1}{1000}$ for $n = 1, 2, 3, \ldots$, for what value(s) of a_0 are all the terms of the sequence $a_0, a_1, a_2, \ldots$ defined?

8.4 Counting Methods

Objectives

▶ Use the multiplication principle for counting

▶ Use permutations in a counting problem

▶ Use combinations in a counting problem

▶ Distinguish between permutations and combinations

Many applications involve counting the number of ways certain events can occur. For example, consider a bicycle race with many participants. One could ask how many possibilities exist for the first-place and second-place finishes. Answering this question involves the use of *counting strategies*, which we discuss in this section.

Multiplication Principle

Basic to almost all counting problems is a simple rule that relies on multiplication. We can think of this rule in terms of setting up slots and filling each slot with one of the factors in the multiplication. In some cases, different slots can contain the same number; in other cases, each slot must contain a unique number. We illustrate this concept in the next two examples.

Example 1 **Filling Slots with Numbers: No Repetition**

Example 9 in Section 8.4 builds upon this example. ⋯⋙

Suppose 10 members of a cycling club are practicing for a race. From among these members, how many possibilities are there for the first-place and second-place finishes?

▶Solution Let us make two slots, one for first place and one for second place, and decide how many possibilities there are for each slot. From the given information, there are 10 possibilities for the first-place finish, since any of the 10 participants could win the race. For *each* possible first-place winner, there are only 9 possibilities for the second-place finish, since no one can finish in both first place *and* second place. See Table 8.4.1.

Table 8.4.1

Number of Possibilities for First Place	Number of Possibilities for Second Place
10	9

To get the total number of possibilities for the first-place and second-place finishes, we must *multiply* the number of possibilities for the first slot by the number of possibilities for the second slot. Because there are 10 of the former and 9 of the latter, we write

$$\text{Total} = \text{first-place possibilities} \times \text{second-place possibilities}$$
$$= (10)(9) = 90.$$

Thus there are 90 different ways of filling the first- and second-place positions.

☑ *Check It Out 1:* Find the total number of possibilities for the first-place, second-place, and third-place finishes in a horse race in which 12 horses compete. ■

Example 2 **Filling Slots with Numbers That Can Be Repeated**

You wish to form a three-digit number, with each digit ranging from 1 to 9. You may use the same digit more than once. How many three-digit numbers can you make?

▶Solution We use the same strategy as in Example 1, in the sense of filling slots. Now, however, we have *three* slots to fill, one for each of the three digits in the number we are forming. There are nine possibilities for each slot, since

▶ each digit is to be from the set $\{1, 2, \ldots, 9\}$; and

▶ any number in that set may fill any slot.

See Table 8.4.2.

Table 8.4.2

Possibilities for First Digit	Possibilities for Second Digit	Possibilities for Third Digit
9	9	9

By the multiplication technique used in Example 1, there are 81 possibilities for just the first two digits. For each of these 81 two-digit numbers, there are nine possibilities for the third digit.

Total = possibilities for first × possibilities for second × possibilities for third
$$= 9 \times 9 \times 9 = 729$$

Thus there are 729 different three-digit numbers that can be formed from the digits 1 through 9.

Notice how this example differs from Example 1. Here, we can use the same digit more than once, and so we have nine possibilities for each slot. In the first example, we had to reduce the number of possibilities for the second slot by 1.

☑ *Check It Out 2:* How many three-digit numbers can be formed if the first digit can be anything but zero and there is no restriction on the second and third digits? ▪

The preceding two examples illustrate an important rule known as the **multiplication principle**.

Multiplication Principle

Suppose there are n slots to fill. Let a_1 be the number of possibilities for the first slot, a_2 the number of possibilities for the second slot, and so on. Then the total number of ways in which the n slots can be filled is

$$a_1 \times a_2 \times a_3 \times \cdots \times a_n.$$

Permutations

Certain types of counting situations happen so frequently that they have a specific name attached to them. For example, one may be interested in knowing the number of ways in which a particular set of objects can be ordered or arranged. Each way of ordering or arranging a certain set of objects is called a **permutation** of those objects.

Example **3** **Filling Just As Many Slots As There Are Objects**

How many different four-block towers can be built with a given set of four different-colored blocks?

▶Solution Think of the problem as that of arranging four blocks in four slots, from bottom to top.

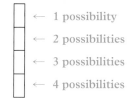

Using the multiplication principle and the fact that any block can be used only once, we see that there are $(4)(3)(2)(1) = 24$ different towers that can be built.

☑ *Check It Out 3:* In how many ways can the letters of the word TRAIN be arranged? ▪

The product of the first n natural numbers occurs so frequently that it has a special name, **n factorial,** which is written as $n!$.

Definition of n factorial ($n!$)

Let n be a positive integer. Then

$$n! = n(n-1)(n-2)\ldots(3)(2)(1).$$

The quantity $0!$ is defined to be 1.

Using this definition, we can say that there are $4!$ ways to build the four-block tower in Example 3.

Example 4 Using Factorial Notation

Use factorial notation to calculate the number of ways in which seven people can be arranged in a row for a photograph.

▶Solution There are seven slots available for the seven people. The first slot has seven possibilities, the second slot has only six possibilities (no one can be in both the first position *and* the second position for the photograph), and so on. Using the multiplication principle, we see that

$$\text{Number of arrangements} = (7)(6)(5)\cdots(1) = 7!.$$

✔ *Check It Out 4:* Use factorial notation to give the number of ways in which 16 people can form a line while waiting to get on a ride at an amusement park. ■

Arrangements also exist in which the number of slots is less than the number of available objects. How do we use factorials to count the arrangements in this case? The next example explores such a situation.

Example 5 Filling Fewer Slots Than There Are Objects

A store manager has six different candy boxes, but spaces for only four of them on the shelf. In how many ways can the boxes be arranged horizontally on the shelf?

▶Solution Since there are four "slots" on the shelf, we can create the following diagram:

| one of 6 boxes | one of 5 boxes | one of 4 boxes | one of 3 boxes |

Therefore, the number of ways in which the boxes can be arranged is

$$(6)(5)(4)(3) = 360.$$

✔ *Check It Out 5:* A total of seven people want to hone up on their archery skills, and there are only three targets available for them to practice on. In how many ways can the targets be assigned if each target is to be used by only one archer? ■

Recall that a permutation is an arrangement of objects. Motivated by Example 5, we now give the formal definition of a permutation.

Discover *and* Learn

Show that the two expressions for $P(n, r)$ given at the right are equivalent.

Technology Note

Using the Probability menu, you can calculate $P(5, 3)$ using your graphing utility. See Figure 8.4.1.

Keystroke Appendix: Section 4

Figure 8.4.1

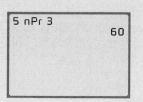

Definition of Permutation

Given n *distinct* objects and r slots to fill, the number of ***r*-permutations of *n* objects** (i.e., the number of permutations of r objects from the collection of n objects) is denoted by $P(n, r)$.

$$P(n, r) = (n)(n - 1) \cdots (n - r + 1)$$

Using the definition of factorial, the expression for $P(n, r)$ can also be written as

$$P(n, r) = \frac{n!}{(n - r)!}.$$

Note that the definition of $P(n, r)$ is really just a new name for an application of the multiplication principle. Other common notations for $P(n, r)$ include $P_{n, r}$ and $_nP_r$.

The following example illustrates the use of permutation notation.

Example 6 Using Permutation Notation

A fast-food restaurant holds a promotion in which each customer scratches off three boxes on a ticket. Each box contains a picture of one of five items: a burger, a bag of fries, a shake, a pie, or a salad. The item in the first box is free. The item in the second box can be purchased at a 50% discount, and the item in the third box can be bought at 25% off. Using permutation notation, determine the number of different tickets that are possible if each picture can be used only once per ticket.

▶ Solution On any given ticket, there are pictures of three of the five items, in some order or other. Thus each ticket represents a permutation of three objects from a collection of five objects. Therefore, the number of different tickets is

$$P(5, 3) = (5)(4)(3) = 60.$$

Note that the second formula for $P(5, 3)$ gives the same answer:

$$P(5, 3) = \frac{5!}{(5 - 3)!} = \frac{5!}{2!} = \frac{120}{2} = 60$$

Use whichever formula you find to be more helpful.

✔ *Check It Out 6:* Use permutation notation to find the number of ways in which eight teachers' aides at an elementary school can be assigned to cafeteria duty during a 5-day week. Only one aide has cafeteria duty each day, and no one takes cafeteria duty more than once during the week. ■

Certain problems can be solved only by combining different counting strategies. Thus it is important to read problems carefully and think them through, and not just memorize formulas. The next example illustrates this point.

Example 7 Filling Two Sets of Slots That Are Independent of Each Other

A group photograph is taken with four children in the front row and five adults in the back row. How many different photographs are possible?

▶Solution Break up the problem into two parts.

1. There are 4! ways to arrange the children in the front row.

2. There are 5! ways to arrange the adults in the back row.

How can we use these two pieces of information?

Note that for each arrangement of the children in the front, there are 5! ways to arrange the adults in the back. Since there are 4! ways to arrange the children in the front, we have the following.

Total number of photographs = (ways to arrange kids)(ways to arrange adults)

$$= 4!\,5! = (24)(120) = 2880$$

☑ *Check It Out 7:* There are two special parking areas in a small company: one with three spaces for SUVs and one with five spaces for cars. There are three employees who drive their SUVs to work and five who drive their cars. How many different arrangements of employee vehicles in the special lots are possible? ■

Combinations

In a permutation, we are counting the number of ways in which a certain set of objects can be *ordered*. However, the way in which objects are ordered is not always relevant. The next example illustrates this concept.

Example 8 Selecting Objects When Order Is Irrelevant

You have three textbooks on your desk: history (H), English (E), and mathematics (M). You choose two of them to put in your book bag. In how many ways can you do this?

▶Solution Let us first list all the different ways in which the books can be chosen. See Table 8.4.3.

Table 8.4.3

Ways to Arrange History, English	Ways to Arrange History, Math	Ways to Arrange English, Math
HE	HM	EM
EH	MH	ME

In the book bag, the order doesn't matter. The only thing that counts is which two books are in the bag. Therefore, HE and EH in Table 8.4.3 count as only one possibility, and similarly for HM and MH and for EM and ME. Therefore,

$$\text{Number of ways} = \frac{\text{ways to arrange any two of three books}}{\text{ways to arrange any set of two books}}$$

$$= \frac{(3)(2)}{2!} = \frac{6}{2} = 3.$$

Thus there are 3 ways to choose two books to go into the book bag.

☑ *Check It Out 8:* Rework Example 8 for the case in which you have four textbooks (history, English, mathematics, and biology) and you put two of them in your book bag. ■

Definition of Combination

Let r and n be positive integers with $r \leq n$. A selection of r objects from a set of n objects, without regard to the order of the selected objects, is called a **combination.** A combination is denoted by

$$C(n, r) \quad \text{or} \quad \binom{n}{r}.$$

To compute $C(n, r)$ or $\binom{n}{r}$, we use the following strategy.

$$\binom{n}{r} = \text{ways to choose } r \text{ objects from } n \text{ objects}$$

$$= \frac{\text{ways to fill } r \text{ slots}}{\text{ways to arrange any set of } r \text{ elements}}$$

$$= \frac{(n)(n-1)\cdots(n-r+1)}{r!}$$

Note   Other common notations for $C(n, r)$ include $C_{n,r}$ and $_nC_r$. Note that $\binom{n}{r}$ is not the fraction $\left(\frac{n}{r}\right)$.

There is another useful formula for computing $C(n, r)$.

Formula for Computing $C(n, r)$

$$C(n, r) = \binom{n}{r} = \text{ways to choose } r \text{ objects from } n \text{ objects}$$

$$= \frac{\text{ways to fill } r \text{ slots}}{\text{ways to arrange any set of } r \text{ elements}}$$

$$= \frac{P(n, r)}{r!} = \frac{n!}{(n-r)!\, r!}$$

We use the idea of combinations in the next two examples.

Technology Note

Using the Probability menu, you can calculate $C(10, 2)$ using your graphing utility. See Figure 8.4.2.

Keystroke Appendix:
Section 4

Figure 8.4.2

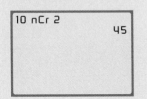

```
10 nCr 2
                45
```

Discover *and* **Learn**

When computing factorials of large numbers, a calculator may give an overflow error. In such a case, it may be easier to simplify the expression for the permutation or combination before evaluating it. To see this, evaluate $\frac{100!}{95!5!}$ on your calculator. What happens? How can you simplify the expression to get the correct value using your calculator?

Example 9 **Using Combination Notation**

⋯ *This example builds on Example 1 in Section 8.4.*

Suppose a cycling club has 10 members. From among the members of the club, how many ways are there to select a two-person committee? How is this situation different from that in Example 1?

▶Solution Within the committee, the order of the members chosen does not matter. All that matters is that there are two committee members. Thus, the number of ways to choose the committee members is

$$C(10, 2) = \frac{(10)(9)}{2!} = 45.$$

We can also use the formula

$$C(10, 2) = \frac{10!}{8!\,2!} = 45.$$

In Example 1, we computed the number of possibilities for first and second place in a race. Those are two *distinct* positions. There, order mattered; in this example it does not.

✔ *Check It Out 9:* A physics lab is equipped with 12 workstations. There are 15 students in one of the physics classes. Assume that all the workstations are in use during the lab period, and that each workstation is being used by a student from the class. If only one student is stationed at each workstation, how many possibilities are there for the particular group of students using the workstations? ■

Example 10 **Selecting Objects from Two Different Sets**

A six-member student board is formed with three male students and three female students. If there are five male candidates and six female candidates, how many different student boards are possible?

▶Solution We need to break this problem into separate parts.

▶ In how many ways can three male board members be chosen from the five male candidates?
  Because order does not matter, the number of ways this can be done is

$$\binom{5}{3} = \frac{(5)(4)(3)}{3!} = 10.$$

▶ In how many ways can three female board members be chosen from the six female candidates?
  Because order does not matter, the number of ways this can be done is

$$\binom{6}{3} = \frac{6!}{(6-3)!\,3!} = \frac{(6)(5)(4)}{3!} = 20.$$

Here we used the second formula to calculate the number of combinations.

▶ For *each* group of three male board members, there are 20 possible groups of three females. Using the multiplication principle, the total number of possible student boards is calculated as follows.

Total number of student boards = (number of three-male groups)
× (number of three-female groups)

$$= \binom{5}{3}\binom{6}{3} = (10)(20) = 200$$

☑ *Check It Out 10:* A person who takes advantage of the combo special at a certain restaurant gets his or her choice of two of the eight main dishes on the menu and three of the six side dishes. How many different combos are there to choose from? ▪

8.4 Key Points

▶ **The multiplication principle:** Suppose there are n slots to fill. Let a_1 be the number of possibilities for the first slot, a_2 the number of possibilities for the second slot, and so on, with a_k representing the number of possibilities for the kth slot. Then the total number of ways that the n slots can be filled is

$$\text{Total possibilities} = a_1 \times a_2 \times a_3 \times \cdots \times a_n.$$

▶ For n a positive integer, $n! = n(n-1)(n-2)\cdots(3)(2)(1)$, with $0! = 1$.

▶ The number of **r-permutations of n objects** is given by

$$P(n, r) = \frac{n!}{(n-r)!}.$$

In a permutation, order matters.

▶ When the order of the objects does not matter, the selection of r objects at a time from a set of n objects is called a **combination:**

$$C(n, r) = \frac{n!}{(n-r)!\, r!}$$

8.4 Exercises

▶**Skills** This set of exercises will reinforce the skills illustrated in this section.

In Exercises 1–22, evaluate.

1. $4!$

2. $6!$

3. $\dfrac{5!}{2!}$

4. $\dfrac{5!}{3!}$

5. $\dfrac{6!}{4!}$

6. $\dfrac{7!}{3!}$

7. $P(4, 3)$

8. $P(5, 3)$

9. $P(7, 5)$

10. $P(8, 4)$

11. $P(8, 5)$

12. $P(9, 4)$

13. $C(4, 3)$

14. $C(5, 3)$

15. $C(8, 5)$

16. $C(9, 4)$

17. $\begin{pmatrix} 8 \\ 6 \end{pmatrix}$

18. $\begin{pmatrix} 9 \\ 7 \end{pmatrix}$

19. $\begin{pmatrix} 8 \\ 8 \end{pmatrix}$

20. $\begin{pmatrix} 8 \\ 0 \end{pmatrix}$

21. $\begin{pmatrix} 100 \\ 99 \end{pmatrix}$

22. $\begin{pmatrix} 100 \\ 1 \end{pmatrix}$

23. Write out all possible three-letter arrangements of the letters B, C, Z.

24. Write out all possible two-letter arrangements of letters selected from B, C, Z.

25. John, Maria, Susan, and Angelo want to form a subcommittee consisting of only three of them. List all the different subcommittees possible.

26. List all the possible ways in which two marbles from a set of three marbles labeled 1, 2, and 3 can be chosen.

27. How many different photographs are possible if four children line up in a row?

28. How many different photographs are possible if six college students line up in a row?

29. How many different photographs are possible if two children sit in the front row and three adults sit in the back row?

30. How many different photographs are possible if four children sit in the front row and two adults sit in the back row?

31. How many different three-person committees can be formed in a club with 12 members?

32. How many different four-person committees can be formed in a club with 12 members?

33. In how many different ways can Sara give her friend Brittany two pieces of candy from a bag containing 10 different pieces of candy?

34. In how many different ways can Jason give his friend Dylan three pieces of candy from a bag containing eight different pieces of candy?

35. In how many different ways can four people be chosen to receive a prize package from a group of 20 people at the grand opening of a local supermarket?

36. In how many different ways can five people be chosen to receive a prize package from a group of 50 people at the grand opening of a local supermarket?

37. There are eight books in a box. There is space on a bookshelf for only three books. How many different three-book arrangements are possible?

▶**Applications** In this set of exercises, you will use counting methods to study real-world problems.

38. **Investments** A stock analyst plans to include in her portfolio stocks from four of the 10 top-performing companies featured in a finance journal. In how many ways can she do this?

39. **Book List** An editor received a short list of 20 books that his company is considering for publication. If he can only choose six of these books to be published this year, in how many different ways can he choose?

40. **Collectibles** A doll collector has a collection of 22 different dolls. She wants to display four of them on her living room shelf. In how many different ways can she display the dolls?

41. **Collectibles** If the collector in Exercise 40 decides to give one of her dolls to each of her four nieces, in how many different ways can she give the dolls to her nieces?

42. **Sports** The manager of a baseball team of 12 players wants to assign infield positions (first base, second base, third base, catcher, pitcher, and shortstop). In how many different ways can the manager make the assignments if each of the players can play any infield position?

43. **Crafts** Mary is making a gift basket for her friend Kate's birthday. She is planning to include two different eye shadow packs, two different lipsticks, and one blush. The store in which she is shopping has 10 different eye shadow packs, five different lipsticks, and three different blushes available for purchase. In how many different ways can Mary make up her gift basket?

44. **Sports** Adnan is purchasing supplies for a weekend fishing trip. He needs to buy three different lures, one spool of line, and one rod-and-reel. The store in which he is shopping has 25 different lures, three different spools of line, and seven different rod-and-reels. In how many different ways can Adnan purchase what he needs for his trip?

45. **Horse Racing** In how many different orders can 15 horses in the Kentucky Derby finish in the top three spots if there are no ties?

46. **Modern Dance** The choreographer Twyla Tharp has 11 male and 11 female dancers in her dance company. Suppose she wants to arrange a dance consisting of a lead pair of a male and a female dancer. In how many ways can she do this, assuming all dancers are qualified for the lead? (*Source:* www.twylatharp.org)

47. **Sports** The Lake Wobegon Little League has to win four of their seven games to earn an "above average" certificate of distinction. In how many ways can this be done?

48. **Hardware** A combination lock can be opened by turning the dial of the lock to three predetermined numbers ranging from 0 to 35. A number can be used more than once. How many different three-number arrangements are possible? Why is a combination lock *not* a good name for this type of lock?

49. **License Plates** How many different license plates can be made by using two letters, followed by three digits, followed by one letter?

50. **Airline Routes** There are five different airline routes from New York to Minneapolis and seven different airline routes from Minneapolis to Los Angeles. How many different trips are possible from New York to Los Angeles with a connection in Minneapolis?

51. **Lottery** The lottery game Powerball is played by choosing six different numbers from 1 through 53, and an extra number from 1 through 44 for the "Powerball." How many different combinations are possible? (*Source:* Iowa State Lottery)

52. **Card Game** A standard card deck has 52 cards. How many five-card hands are possible from a standard deck?

53. **Card Game** A standard card deck has 52 cards. A bridge hand has 13 cards. How many bridge hands are possible from a standard deck?

54. **Card Game** How many five-card hands consisting of all red cards are possible from a standard deck of 52 cards?

55. **Photography** A wedding photographer lines up four people plus the bride and groom for a photograph. If the bride and groom stand side-by-side, how many different photographs are possible?

56. **Computer Security** A password for a computer system consists of six characters. Each character must be a digit or a letter of the alphabet. Assume that passwords are *not* case-sensitive. How many passwords are possible? How many passwords are possible if a password must contain at least one digit? (*Hint for second part:* How many passwords are there containing just letters?)

57. **Computer Security** Rework Exercise 56 for the case in which the passwords *are* case-sensitive.

58. **Retail Store** The Woosamotta University bookstore sells "W. U." T-shirts in four sizes: S, M, L, and XL. Both the blue and yellow shirts are available in all four sizes, but the red shirts come in only small and medium. What is the minimum number of W. U. T-shirts the bookstore should stock if it wishes to have available at least one of each size and color?

59. **Car Options** You are shopping for a new car and have narrowed your search to three models, two colors, and four optional features. These are detailed below.

Car Models	Honda Civic, Hyundai Elantra, Ford Focus
Colors	White, red
Optional Features	Alarm system, CD/MP3 player, sun roof, custom wheels

How many different cars are there to choose from

(a) if you want no optional features?

(b) if you want all four optional features?

(c) if you want *any two* of the optional features?

(d) if you want *at most two* of the optional features?

60. **Board Game** In the board game Mastermind, one of two players chooses at most four pegs to place in a row of four slots, and then hides the colors and positions of the pegs from his opponent. Each peg comes in one of six colors, and the player can use a color more than once. Also, one or more of the slots can be left unfilled.

(a) How many different ways are there to arrange the pegs in the four-slot row? In this game, the order in which the pegs are arranged matters.

(b) The Mastermind website states: "With 2401 combinations possible, it's a mind-bending challenge every time!" Is *combination* the appropriate mathematical term to use here? Explain. This is an instance of how everyday language and mathematical language can be contradictory. (*Source:* www.pressman.com)

▶**Concepts** This set of exercises will draw on the ideas presented in this section and your general math background.

61. Given n points ($n \geq 3$) such that no three of them lie on the same line, how many different line segments can be drawn connecting exactly two of the n points?

62. Write out all the different four-digit numbers possible using the numbers 1, 1, 2, 3. Why is your number of possibilities *not* equal to 4!?

63. A diagonal of a polygon is defined as a line segment with endpoints at a pair of nonadjacent vertices of the polygon. How many diagonals does a pentagon have? an octagon? an n-gon (that is, a polygon with n sides)?

64. How many different six-letter arrangements are there of the letters in the word PIPPIN? This exercise involves a slightly different strategy than the strategies discussed in the Examples.

   ▶ First draw six slots for six letters. In how many ways can you put the three P's in the slots?
   ▶ You have three slots left over. In how many ways can you place the two I's?
   ▶ The last slot, by default, will contain the N.
   ▶ Put together the information outlined above to come up with a solution.

65. Use the strategy outlined in Exercise 64 to find the number of different 11-letter arrangements of the letters in the word MISSISSIPPI.

8.5 Probability

Objectives

▶ Define and identify outcomes and events
▶ Calculate probabilities of equally likely events
▶ Define and identify mutually exclusive events
▶ Calculate probabilities of mutually exclusive events
▶ Calculate probabilities of complements of events
▶ Apply probabilities to problems involving collected data

The notion of chance is something we deal with every day. Weather forecasts, Super Bowl predictions, and auto insurance rates all incorporate elements of chance. The mathematical study of chance behavior is called **probability.** This section will cover some basic ideas in the study of probability.

Basic Notions of Probability

Our next example examines a person's chance of winning a prize on a game show.

Example 1 Calculating the Chance of Winning a Specific Prize

Example 4 in Section 8.5 builds upon this example. ⋯▶

During a certain episode, the television game show *Wheel of Fortune* had a wheel with 24 sectors. One sector was marked "Trip to Hawaii" and two of the other sectors were marked "Bonus." What is a contestant's chance of winning the trip to Hawaii? What is his or her chance of winning a bonus? Assume that the wheel is equally likely to land on any of the 24 sectors.

▶**Solution** Because only one of the 24 sectors is marked "Trip to Hawaii," there is a $\frac{1}{24}$ chance of winning that prize, since we have assumed that the wheel is equally likely to land on any of the 24 sectors.

Using the same argument, the chances of winning a bonus are $\frac{2}{24}$ or $\frac{1}{12}$.

✔ *Check It Out 1:* In Example 1, if three of the 24 sectors are marked "Bankrupt," what is a contestant's chance of landing on a "Bankrupt" sector? ■

In order to study these notions further, we need to introduce two new terms, **outcome** and **event.**

Definition of Outcome and Event

▸ An **outcome** is a possible result of an experiment. Here, an experiment simply denotes an activity that yields random results.

▸ An **event** is a collection of outcomes.

Example 2 **Rolling a Six-Sided Die**

What are the possible outcomes when a six-sided die is rolled and the number on the top face is recorded? What is the event that the number on the top face is even?

▸**Solution** Each face of a die is marked with a unique number from 1 to 6, so the possible outcomes are

$$1, 2, 3, 4, 5, 6.$$

The event that the number on the top face is even is

$$\{2, 4, 6\}.$$

✔ *Check It Out 2:* What is the event that a six-sided die is rolled and the number on the top face is at least 2 but no greater than 5? ■

An event is generally given in set notation. The set consisting of all possible outcomes for a certain situation under consideration is called the **sample space,** which is often denoted by the letter S. For example, the sample space in Example 2 is $S = \{1, 2, 3, 4, 5, 6\}$.

Example 3 **Tossing a Coin Three Times in Succession**

A coin is tossed three times and the sequence of heads and tails that occurs is recorded.

(a) What is the sample space for this experiment?

(b) What is the event that *at least* two heads occur?

(c) What is the event that *exactly* two heads occur?

▸**Solution**

(a) Using the multiplication principle, we know that there are a total of eight possible outcomes. Next we list the possible outcomes. A tree diagram is helpful for doing this. It lists the possibilities for the first toss and then branches out to list the possibilities for the second and third tosses, respectively. See Figure 8.5.1.
  From the tree diagram, we see that the sample space is

$$\{HHH, HHT, HTH, HTT, THH, THT, TTH, TTT\}.$$

(b) The event that *at least* two heads occur is

$$\{HHH, HHT, HTH, THH\}.$$

(c) The event that *exactly* two heads occur is

$$\{HHT, HTH, THH\}.$$

Note that the event in part (c) is different from the event in part (b). This illustrates the importance of paying careful attention to the *wording* of probability problems.

Figure 8.5.1

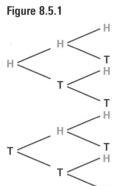

☑ *Check It Out 3:* A coin is tossed four times and the sequence of heads and tails that occurs is recorded.

(a) What is the sample space for this experiment?

(b) What is the event that *exactly three* tails occur?

(c) What is the event that *at most two* heads occur? ▪

We are now in a good position to give a formal definition of probability.

Definition of Probability

Assume that a sample space S consists of equally likely outcomes. Then the probability of an event E, denoted by $P(E)$, is defined as

$$P(E) = \frac{\text{number of outcomes in } E}{\text{number of outcomes in } S}.$$

Note that $P(E)$ must satisfy

$$0 \leq P(E) \leq 1.$$

Note Remember these important points:
- ▸ If $P(E) = 1$, then the event E is certain to happen.
- ▸ If $P(E) = 0$, then the event E will *not* happen.

Throughout this textbook, we have been discussing functions. The probability of an event can also be thought of as a function. Because an event is described as a set, the probability function has a particular set as its domain and the numbers in the interval [0, 1] as its range.

We now calculate the probabilities of some of the events described in the previous examples.

Example 4 Probability of Winning a Specific Prize

⋮⋯ *This example builds on Example 1 in Section 8.5.*

What is the probability of winning a trip to Hawaii in the *Wheel of Fortune* game in Example 1? Use the probability formula to find the answer.

▸Solution We know that the total number of outcomes is 24, since there are 24 different sectors in which the wheel can stop. Let E be the event that the wheel stops in the "trip to Hawaii" sector. Calculate $P(E)$ as follows.

$$P(E) = \frac{\text{number of outcomes in } E}{\text{number of outcomes in } S} = \frac{1}{24}$$

This is exactly what we figured out intuitively in Example 1. Using mathematical terminology helps us to generalize our ideas and make them precise.

☑ *Check It Out 4:* Use the definition of probability to find the probability of winning a bonus in the *Wheel of Fortune* game in Example 1. ■

Example 5 Probability of a Specific Event in a Coin-Tossing Experiment

Suppose a coin is tossed three times. What is the probability of obtaining at least two heads?

▶ Solution Here, E is the event of obtaining at least two heads. To find the probability of E, we use the answers to parts (a) and (b) of Example 3.

$$P(E) = \frac{\text{number of ways to get at least two heads}}{\text{number of outcomes in } S} = \frac{4}{8} = \frac{1}{2}$$

☑ *Check It Out 5:* What is the probability of obtaining at most three heads when a coin is tossed four times? ■

Mutually Exclusive Events

Many applications entail finding a probability that involves two events. If the events have no overlap, they are said to be **mutually exclusive.** Next we examine some pairs of events that may or may not be mutually exclusive.

Example 6 Deciding Whether Events Are Mutually Exclusive

Decide which of the following pairs of events are mutually exclusive.
(a) "Drawing a queen" and "drawing a king" from a deck of 52 cards.
(b) "Drawing a queen" and "drawing a spade" from a deck of 52 cards.

▶ Solution

(a) A card cannot be both a king and a queen. Therefore, the event of drawing a king and the event of drawing a queen have no overlap and are mutually exclusive.

(b) The event of drawing a queen overlaps with the event of drawing a spade because drawing the queen of spades is an element of both events. Therefore, these two events are *not* mutually exclusive.

☑ *Check It Out 6:* Decide which of the following pairs of events are mutually exclusive.
(a) When rolling a die, "rolling a 2" and "rolling the smallest possible even number."
(b) "Drawing a club" and "drawing a red card" from a deck of 52 cards. ■

When two events are mutually exclusive, the probability of one or the other occurring is easy to compute—you simply add up the two respective probabilities.

> **Computing the Probability of Mutually Exclusive Events**
>
> Let F and G be two *mutually exclusive events.* Then
> $$P(F \text{ or } G) = P(F) + P(G).$$

Discover *and* **Learn**

How would you find the probability of drawing a queen or a spade from a deck of 52 cards?

Example 7 Combining the Probabilities of Mutually Exclusive Events

Find the probability of drawing a queen or a king from a deck of 52 cards.

▶Solution Drawing a king and drawing a queen are mutually exclusive events, since they cannot both happen on one draw. See Example 6 for details. Therefore,

$$P(\text{queen or king}) = P(\text{queen}) + P(\text{king})$$
$$= \frac{4}{52} + \frac{4}{52}$$
$$= \frac{8}{52} = \frac{2}{13}.$$

The probability of drawing a queen or a king from a deck of 52 cards is $\frac{2}{13}$.

✔ *Check It Out 7:* Find the probability of tossing three heads or three tails in three tosses of a coin. ■

Example 8 Rolling a Pair of Six-Sided Dice

Example 10 in Section 8.5 builds upon this example. ⋯⋗

Figure 8.5.2 shows all the possibilities for the numbers on the top faces when rolling a pair of dice.

Figure 8.5.2

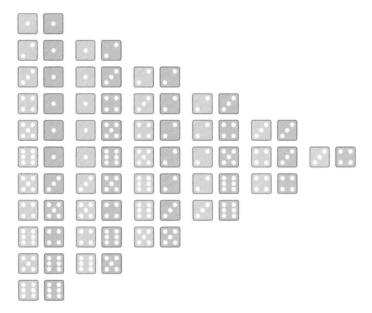

Find the following.
(a) The probability of rolling a sum of 5
(b) The probability of rolling a sum of 6 or 7
(c) The probability of rolling a sum of 13

▶Solution We examine Figure 8.5.2 to help us answer the questions. Note that there are 36 different possible outcomes when rolling the two dice. All 36 outcomes are equally likely.

(a) Probability of rolling a sum of 5: All the possibilities resulting in a sum of 5 are listed in the fourth row of the figure. Written in set notation, the event of rolling a sum of 5 is

$$\{(4, 1), (1, 4), (3, 2), (2, 3)\}.$$

Because all of the possibilities are equally likely, the probability of this event is

$$P(\text{rolling a sum of 5}) = \frac{4}{36} \quad \text{or} \quad \frac{1}{9}.$$

(b) Probability of rolling a sum of 6 or 7: These two events are mutually exclusive, since they cannot happen at the same time. From the figure, there are five ways to roll a sum of 6 and six ways to roll a sum of 7. Therefore,

$$P(\text{rolling a sum of 6 or 7}) = P(\text{rolling a sum of 6}) + P(\text{rolling a sum of 7})$$

$$= \frac{5}{36} + \frac{6}{36} = \frac{11}{36}.$$

(c) Probability of rolling a sum of 13: It is not possible to roll a sum of 13 with two six-sided dice whose faces are uniquely numbered from 1 to 6. Therefore,

$$P(\text{rolling a sum of 13}) = \frac{0}{36} = 0.$$

✔ *Check It Out 8:* When rolling two dice, find the probability of rolling a sum that is an even number less than or equal to 6. ■

Complement of an Event

This section will deal with the probability of an event *not* happening, which is known as the **complement** of the event. Complements of events occur all the time in daily life. Example 9 explores one such everyday application.

Example 9 **Calculating the Probability of the Complement of an Event**

Suppose you are told that the probability of rain today is 0.6. What is the probability that it will *not* rain?

▶Solution Because the total probability must be 1, the probability of *not* raining is

$$P(\text{no rain}) = 1 - P(\text{rain}) = 1 - 0.6 = 0.4.$$

✔ *Check It Out 9:* If the probability of passing an English course is 0.8, what is the probability of not passing the English course? ■

Definition of the Complement of an Event

The set of all outcomes in a sample space that do *not* belong to event E is called the **complement** of E and is denoted by E'.

Discover *and* **Learn**

If you draw a box to represent some event *E*, how would you represent *E*'?

By definition, *E* and *E*' are mutually exclusive. Thus, because either *E* or *E*' is certain to happen, we can write

$$P(E \text{ or } E') = P(E) + P(E') = 1.$$

This gives us a way to calculate $P(E')$:

$$P(E') = 1 - P(E).$$

Complements are very helpful in determining certain types of probabilities.

Example 10 Using the Complement of an Event to Find a Probability

✧··· *This example builds on Example 8 in Section 8.5.*

Refer to Figure 8.5.2 in Example 8, which lists all the possible outcomes of rolling two dice. What is the probability of rolling a sum of *at least* 4?

▶**Solution** It is much easier to figure out the number of ways in which the sum is *not* at least 4. The sum is *not* at least 4 when the sum is equal to 2 or 3. There are only three ways in which this can happen. Thus,

$$P(not \text{ rolling a sum of at least 4}) = \frac{3}{36}.$$

Using complements,

$$P(\text{rolling a sum of at least 4}) = 1 - P(not \text{ rolling a sum of at least 4})$$

$$= \frac{33}{36} = \frac{11}{12}.$$

☑ *Check It Out 10:* In tossing a coin four times, what is the probability of getting at least two tails? ■

Calculating Probabilities from Percentages

In real life, probabilities are often calculated from data that is expressed in terms of percentages. In fact, probabilities are often quoted in terms of percentages. To be consistent with the mathematical definition of probability, we will convert all percentages to their equivalent values between 0 and 1.

Example 11 Converting Percentages to Probabilities

Every spring, the National Basketball Association holds a lottery to determine which team will get first pick of its number 1 draft choice from a pool of college players. The teams with poorer records have a higher chance of winning the lottery than those with better records. Table 8.5.1 lists the percentage chance that each team had of getting first pick of its number 1 draft choice for the year 2002. (*Source:* National Basketball Association)

Table 8.5.1

Team	Chance of Winning First Pick (%)
Golden State Warriors	22.5
Chicago Bulls	22.5
Memphis Grizzlies	15.7
Denver Nuggets	12.0
Houston Rockets	8.9
Cleveland Cavaliers	6.4
New York Knicks	4.4
Atlanta Hawks	2.9
Phoenix Suns	1.5
Miami Heat	1.4
Washington Wizards	0.7
L.A. Clippers	0.6
Milwaukee Bucks	0.5

Find the probability of

(a) the Bulls or the Warriors getting first draft pick.

(b) the Clippers *not* getting first draft pick.

▶Solution First note that these probabilities are not equal—the teams with the poorer records have a higher chance of getting first pick. Also, convert the percentages into their respective decimal equivalents.

(a) The probability of the Bulls or the Warriors getting first draft pick is

$$P(\text{Bulls or Warriors}) = P(\text{Bulls}) + P(\text{Warriors}) = 0.225 + 0.225 = 0.45.$$

We have used the fact that the two events are mutually exclusive, since the two teams cannot both get first draft pick.

(b) Using the formula for computing the complement of an event, the probability of the Clippers *not* getting first draft pick is

$$P(not \text{ Clippers}) = 1 - P(\text{Clippers}) = 1 - 0.006 = 0.994.$$

Thus the probability is very high that the Clippers will *not* get first draft pick.

✔ *Check It Out 11:* Use Table 8.5.1 from Example 11 to find the probability of

(a) the Atlanta Hawks or the L.A. Clippers getting first draft pick.

(b) the New York Knicks *not* getting first draft pick. ▪

8.5 Key Points

▶ An **outcome** is any possibility resulting from an experiment. Here, an experiment simply denotes an activity yielding random results.

▶ An **event** is a collection (set) of outcomes.

▶ A **sample space** is the set consisting of all possible outcomes for a certain situation under consideration.

▶ Assume that a sample space S consists of equally likely outcomes. Then the **probability of an event E,** denoted by $P(E)$, is defined as

$$P(E) = \frac{\text{number of outcomes in } E}{\text{number of outcomes in } S}.$$

Note that $0 \le P(E) \le 1$.

▶ Two events F and G are **mutually exclusive** if they have no overlap. In this case,

$$P(F \text{ or } G) = P(F) + P(G).$$

▶ The set of all outcomes in a sample space that do *not* belong to event E is called the **complement** of E and is denoted by E'. The probability of E' is $P(E') = 1 - P(E)$.

8.5 Exercises

▶**Skills** This set of exercises will reinforce the skills illustrated in this section.

In Exercises 1–4, consider the following experiment: toss a coin twice and record the sequence of heads and tails.

1. What is the sample space (for tossing a coin twice)?

2. What is the event that you get at least one head?

3. What is the complement of the event that you get at least one head?

4. Calculate the probability of the event in Exercise 3.

In Exercises 5–8, consider the following experiment: roll a die and record the number on the top face.

5. What is the event that the number on the top face is odd?

6. What is the complement of the event that the number on the top face is odd?

7. What is the probability that the number on the top face is greater than or equal to 5?

8. What is the probability that the number on the top face is less than 1?

In Exercises 9–12, consider the following experiment: draw a single card from a standard deck of 52 cards.

9. What is the event that the card is a spade?

10. What is the complement of the event that the card is a spade? Describe in words only.

11. What is the probability that the card drawn is the ace of spades?

12. What is the probability that the card drawn is the 2 of clubs?

In Exercises 13–16, consider the following experiment: pick one coin out of a bag that contains one quarter, one dime, one nickel, and one penny.

13. Give the sample space (for picking one coin out of the bag).

14. What is the complement of the event that the coin you pick has a value of 10 cents?

15. What is the probability of picking a nickel?

16. What is the probability of picking a quarter or a penny?

In Exercises 17–22, answer True or False.

17. When rolling a die, "rolling a 2" and "rolling an even number" are mutually exclusive events.

18. When randomly picking a card from a standard deck of 52 cards, "picking a queen" and "picking a jack" are mutually exclusive events.

19. Consider the roll of a die. The complement of the event "rolling an even number" is "rolling a 1, a 3, or a 5."

20. Consider randomly picking a card from a standard deck of 52 cards. The complement of the event "picking a black card" is "picking a heart."

21. When picking one coin at random from a bag that contains one quarter, one dime, one nickel, and one penny, "picking a coin with a value of more than one cent" and "picking a penny" are mutually exclusive events.

22. Consider picking one coin from a bag that contains one quarter, one dime, one nickel, and one penny. The complement of the event "picking a quarter or a nickel" is "picking a dime or a nickel."

▶ **Applications** In this set of exercises, you will use probability to study real-world problems.

23. **Coin Toss** A coin is tossed four times and the number of heads that appear is counted. Fill in the following table listing the probabilities of obtaining various numbers of heads. What do you observe? Are all of these outcomes equally likely?

Number of Heads	Probability
0	
1	
2	
3	
4	

24. **Cards** If a card is drawn from a standard deck of 52 cards, what is the probability that it is a heart?

25. **Cards** If a card is drawn from a standard deck of 52 cards, what is the probability that it is an ace?

26. **Candy Colors** Students in a college math class counted 29 packages (1.5 ounces each) of plain M&M'S® and recorded the following color distribution.

Color	Red	Blue	Green	Yellow	Brown	Orange	Total
Number of M&M'S	278	157	261	265	549	139	1649

If one M&M is drawn at random from the total, find the following probabilities.
(a) The probability of getting a red candy
(b) The probability of getting a red or a green candy
(c) The probability of *not* getting a blue candy

27. **Cards** What is the probability of drawing the 4 of clubs from a standard deck of 52 cards?

28. **Card Game** During the play of a card game, you have seen 20 of the 52 cards in the deck and none of them is the 4 of clubs. You need the 4 of clubs to win the game. What is the probability that you will win the game on the next card drawn?

29. **Cards** What is the probability of drawing a face card (a face card is a jack, queen, or king) from a standard deck of 52 cards?

30. **Cards** What is the probability of drawing a red face card (a face card is a jack, queen, or king) from a standard deck of 52 cards?

In Exercises 31–34, use counting principles from Section 8.4 to calculate the number of outcomes.

31. **Dice Games** A pair of dice, one blue and one green, are rolled and the number showing on the top of each die is recorded. What is the probability that the sum of the numbers on the two dice is 7?

32. **Dice Games** Refer to Exercise 31. What is the probability that the sum is 10?

33. **Movie Theater Seating** A group of friends, five girls and five boys, wants to go to the movies on Friday night. The friends select, at random, two of their group to go to the ticket office to purchase the tickets. What is the probability that the two selected are both boys?

34. **Movie Theater Seating** Refer to Exercise 33. What is the probability that the two selected are a boy and a girl?

Phone Numbers *Exercises 35–38 involve dialing the last four digits of a phone number that has an area code of 907 and an exchange of 316. The exchange consists of the first three digits of the seven-digit phone number.*

35. How many outcomes are there for dialing the last four digits of a phone number?

36. How many possible outcomes are in the event that the first three (of the last four) digits you dial are 726, in that order?

37. What is the probability that the (last four) digits you dial are different from one another?

38. What is the probability that all of the (last four) digits you dial are different from all the digits of the area code *and* different from all the digits of the exchange? Assume each digit can be repeated.

39. Refer to Exercise 33. What is the probability that Ann (who is one of the five girls) is selected?

40. The 10 friends in Exercise 33 all have different last names. The seats they purchased for the movie are numbered 1 through 10. If the tickets are distributed among the friends at random, what is the probability the friends will be seated in alphabetical order from seat 1 to seat 10?

41. **Card Game** During the play of a card game, you see 20 of 52 cards in the deck drawn and discarded and none of them is a black 4. You need a black 4 to win the game. What is the probability that you will win the game on the next card drawn?

42. **Roulette** A roulette wheel has 38 sectors. Two of the sectors are green and are numbered 0 and 00, respectively, and the other 36 sectors are equally divided between red and black. The wheel is spun and a ball lands in one of the 38 sectors.
    (a) What is the probability of the ball landing in a red sector?
    (b) What is the probability of the ball landing in a green sector?
    (c) If you bet $1 on a red sector and the ball lands in a red sector, you will win another $1. Otherwise, you will lose the dollar that you bet. Do you think this is a fair game? That is, do you have the same chance of wining as you do of losing? Why or why not?

43. **Dart Game** Many probabilities are computed by using ratios of areas. This exercise illustrates such a scenario. What is the probability of hitting the shaded inner region of the dart board in the figure if all of the points within the larger circle are equally likely to be hit by the dart? You may assume that the dart will never land anywhere outside the larger circle.

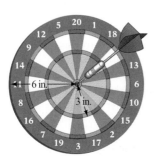

44. **Card Probabilities** Suppose five cards are drawn from a standard deck of 52 cards. Find the probability that all five cards are black. (*Hint:* Use the counting principles from Section 8.4.)

45. **Card Probabilities** If a card is drawn from a standard deck of 52 cards, the probability of drawing a king or a spade is *not* $\frac{17}{52}$. Explain. What is the correct answer?

Slot Machines *A slot machine has four reels, with 10 symbols on each reel. Assume that there is exactly one cherry symbol on each reel. Use this information and the counting principles from Section 8.4 when solving Exercises 46–48.*

46. What is the probability of getting four cherries?

47. What is the probability of getting exactly three cherries?

48. What is the probability of getting at least one cherry?

▶ **Concepts** This set of exercises will draw on the ideas presented in this section and your general math background.

49. Consider rolling a pair of dice. Which, if either, of the following events has a higher probability: "rolling a sum that is odd" or "rolling a sum that is even?"

50. Each card in a standard deck of 52 cards belongs to one of four different suits: hearts, diamonds, spades, or clubs. There are 13 cards in each suit. Consider a scenario in which you draw five cards from the deck, one at a time, and record only the suit to which each card drawn belongs.
    (a) Describe the sample space.
    (b) What is the probability that the set of five cards you draw consists of two spades, one heart, one diamond, and one club (drawn in any order)?
    (c) What is the probability that exactly two of the five cards you draw are from the same suit?

51. In a telephone survey, people are asked whether they have seen each of four different films. Their answers for each film (yes or no) are recorded.
    (a) What is the sample space?
    (b) What is the probability that a respondent has seen exactly two of the four films?
    (c) Assuming that all outcomes are equally likely, what is the probability that a respondent has seen all four films?

52. Assume that the probability of winning $5 in the lottery (on one lottery ticket) for any given week is $\frac{1}{50}$, and consider the following argument. "Henry buys a lottery ticket every week, but he hasn't won $5 in any of the previous 49 weeks, so he is assured of winning $5 this week." Is this a valid argument? Explain.

In Exercises 53 and 54, consider a bag that contains eight coins: three quarters, two dimes, one nickel, and two pennies.

53. Assume that two coins are chosen from the bag.
    (a) How many ways are there to choose two coins from the bag?
    (b) What is the probability of choosing two coins of equal value?

54. Assume that two coins are picked out of the bag, one at a time, and the first coin is put back into the bag before the second coin is chosen.
    (a) How many outcomes are there? (*Hint:* Count the possibilities for the first coin and the possibilities for the second coin.)
    (b) What is the probability of picking two coins of equal value?

8.6 The Binomial Theorem

Objectives

▶ Calculate the variable parts of terms in a binomial expansion

▶ Calculate binomial coefficients

▶ Expand a binomial using the Binomial Theorem

▶ Find the ith term of a binomial expansion

▶ Relate combinations to the Binomial Theorem

When applying algebraic techniques in order to solve a problem, it is sometimes necessary to write a quantity of the form $(a + b)^n$ as the sum of its terms. Because $a + b$ is a binomial, this process is called a **binomial expansion.** You already know the following:

$$(a + b)^1 = a + b$$

$$(a + b)^2 = a^2 + 2ab + b^2$$

Building on these expansions, we can write out $(a + b)^3$ as

$$(a + b)^3 = (a + b)(a + b)^2.$$

We then can expand $(a + b)^2$, multiply $(a + b)$ by the result, and simplify.

$$(a + b)^3 = (a + b)(a + b)^2$$
$$= (a + b)(a^2 + 2ab + b^2)$$
$$= a^3 + 3a^2b + 3ab^2 + b^3$$

As the exponent on the binomial increases to numbers larger than 3, multiplying out the entire expression to find all the terms becomes more tedious. Fortunately, in this section we discuss a way to find all the terms of the expansion without having to multiply repeatedly.

Observations:

Note the following about the expansion of $(a + b)^n$, where n is a positive integer.

▶ The first term of the expansion is a^n, and the last term is b^n.

▶ For each successive term after a^n, the exponent on b increases by 1 while the exponent on a decreases by 1.

▶ For any term in the expansion, the sum of the exponents on a and b is n.

Example 1 **Finding the Variable Parts of the Terms in a Binomial Expansion**

Consider the expansion of $(a + b)^4$.

(a) Write down the variable parts of all the terms that occur in the expansion.

(b) What is the sum of the exponents on a and b for each term of the expansion?

▶Solution

(a) First write

$$(a + b)^4 = (a + b)(a + b)(a + b)(a + b).$$

From the second bulleted item in the preceding list of observations, the variable parts of all the terms that occur in the expansion are

$$a^4, a^3b, a^2b^2, ab^3, b^4.$$

(b) The sum of the exponents on a and b for each term of the expansion is 4.

✔ *Check It Out 1:* Consider the expansion of $(x - y)^5$.

(a) Write down the variable parts of all the terms that occur in the expansion.

(b) What is the sum of the exponents on x and y for each term of the expansion? ■

The Binomial Theorem

Recall the definition of n factorial from Section 8.4.

Definition of n Factorial

Let n be a positive integer. Then

$$n! = n(n - 1)(n - 2) \cdots (3)(2)(1).$$

The quantity 0! is defined to be 1.

Definition of Binomial Coefficient

Let n and r be nonnegative integers with $r \leq n$. In the expansion of $(a + b)^n$, the coefficient of the term $a^{n-r}b^r$ is

$$\binom{n}{r} = \frac{n!}{(n - r)!\, r!}.$$

Example 2 **Calculating Factorials and Binomial Coefficients**

Evaluate the following expressions.

(a) 5!

(b) $\binom{6}{3}$

(c) $\binom{5}{2}$

▶Solution

(a) $5! = 5 \cdot 4 \cdot 3 \cdot 2 \cdot 1 = 120$

(b) Applying the formula for binomial coefficients,

$$\binom{6}{3} = \frac{6!}{(6-3)!\,3!} = \frac{6!}{3!\,3!}$$

$$= \frac{6 \cdot 5 \cdot 4 \cdot 3 \cdot 2 \cdot 1}{(3 \cdot 2 \cdot 1)(3 \cdot 2 \cdot 1)} = 20.$$

(c) $\binom{5}{2} = \frac{5!}{(5-2)!\,2!} = \frac{5!}{3!\,2!}$

$$= \frac{5 \cdot 4 \cdot 3 \cdot 2 \cdot 1}{(3 \cdot 2 \cdot 1)(2 \cdot 1)} = 10$$

✔ *Check It Out 2:* Evaluate $\binom{6}{4}$. ■

Now that we have a method for finding each of the coefficients in the expansion of $(a + b)^n$, we present the Binomial Theorem.

Binomial Theorem

Let n be a positive integer. Then

$$(a + b)^n = \sum_{i=0}^{n} \binom{n}{i} a^{n-i} b^i$$

$$= \binom{n}{0} a^n + \binom{n}{1} a^{n-1}b + \binom{n}{2} a^{n-2}b^2 + \cdots + \binom{n}{n} b^n.$$

Virtually any variable or constant can take the place of a and b in the Binomial Theorem. The next two examples illustrate the use of this theorem.

Example 3 Expanding a Binomial Raised to the Fourth Power

Expand $(3 + 2y)^4$.

▶Solution We apply the Binomial Theorem with $a = 3$, $b = 2y$, and $n = 4$.

$$(3 + 2y)^4 = \binom{4}{0}(3)^4 + \binom{4}{1}(3)^3(2y) \quad \text{Substitute } a = 3, b = 2y, \text{ and } n = 4$$

$$+ \binom{4}{2}(3)^2(2y)^2 + \binom{4}{3}(3)(2y)^3 + \binom{4}{4}(2y)^4$$

$$= (1)(81) + (4)(27)(2y) + (6)(9)(4y^2) + (4)(3)(8y^3)$$

$$+ (1)(16y^4) \qquad \binom{4}{0} = \binom{4}{4} = 1, \binom{4}{1} = \binom{4}{3} = 4, \text{ and } \binom{4}{2} = 6$$

$$= 81 + 216y + 216y^2 + 96y^3 + 16y^4 \quad \text{Simplify}$$

$$= 16y^4 + 96y^3 + 216y^2 + 216y + 81$$

✔ *Check It Out 3:* Expand $(2y - x)^4$. ■

Example 4 **Expanding a Binomial Raised to the Fifth Power**

Expand $(2z - y)^5$.

▶Solution First write $2z - y$ as $2z + (-y)$. We apply the Binomial Theorem with $a = 2z$, $b = -y$, and $n = 5$.

$$(2z - y)^5 = (2z + (-y))^5 = \binom{5}{0}(2z)^5 + \binom{5}{1}(2z)^4(-y)^1 + \binom{5}{2}(2z)^3(-y)^2$$

$$+ \binom{5}{3}(2z)^2(-y)^3 + \binom{5}{4}(2z)^1(-y)^4 + \binom{5}{5}(-y)^5$$

$$= (1)(32z^5) + (5)(16z^4)(-y) + (10)(8z^3)(y^2) + (10)(4z^2)(-y^3)$$

$$+ (5)(2z)(y^4) + (1)(-y^5)$$

$$= 32z^5 - 80z^4y + 80z^3y^2 - 40z^2y^3 + 10zy^4 - y^5 \quad \text{Simplify}$$

✔ *Check It Out 4:* Expand $(3u + 2v)^5$. ■

The *i*th Term of a Binomial Expansion

In many instances, we may be interested only in a particular term or terms of a binomial expansion. In such cases, we can use the following formula, which is a direct result of examining the individual terms of the Binomial Theorem.

> **The *i*th Term of a Binomial Expansion**
>
> Let n and i be positive integers such that $1 \leq i < n + 1$. Then the ith term of $(a + b)^n$ is given by
>
> $$\binom{n}{i - 1}a^{n-i+1}b^{i-1}.$$
>
> Note that the exponent on b is one less than the number of the term.

Example 5 **Finding a Specific Term of a Binomial Expansion**

Find the fourth term in the expansion of $(3x + 5)^6$.

▶Solution Using the formula for the ith term of a binomial expansion with $a = 3x$, $b = 5$, $n = 6$, and $i = 4$, we obtain

$$\binom{6}{4 - 1}(3x)^{6-4+1}(5)^{4-1} = \binom{6}{3}(3x)^3(5)^3$$

$$= (20)(27)x^3(125) = 67,500x^3.$$

✔ *Check It Out 5:* Find the second term in the expansion of $(3x + 5)^6$. ■

Combinations and the Binomial Theorem

If you studied combinations in Section 8.4, you may have noticed that the formula for the binomial expansion uses exactly the same notation as combinations. This is not a coincidence. To illustrate this relationship, we consider the expansion of $(a + b)^3$ using counting methods.

Example 6 **Using Counting Methods to Generate a Binomial Expansion**

Justify the equality $(a + b)^3 = a^3 + 3a^2b + 3ab^2 + b^3$ by using a counting argument.

▶**Solution** We first write

$$(a + b)^3 = (a + b)(a + b)(a + b).$$

When this product is multiplied out, we see that each term of the expansion will be a product of a's and b's. The total number of a's and b's in each term is exactly three. We can make three slots, each corresponding to either an a or a b:

☐ ☐ ☐

There is only one possible way to have a's in all three slots. Thus there is only one term of the form a^3. How many ways are there to have two a's and one b? Multiplying two a's and one b gives a term of the form a^2b. Note that once we choose the two a's, the placement of the b is automatic. There are $C(3, 2) = 3$ ways to choose two a's for the three slots. Thus there are *three* ways to get a term of the form a^2b:

aab   aba   baa

Therefore, a^2b must have a coefficient of 3. Using the same argument, we see that there are three ways to get a term of the form ab^2. Finally, there is only one way to get b's in all three slots. Putting this all together, we have

$$(a + b)^3 = a^3 + 3a^2b + 3ab^2 + b^3.$$

☑ *Check It Out 6:* Use a counting argument to show that in the expansion of $(a + b)^5$, the coefficient of a^2b^3 is $\begin{pmatrix} 5 \\ 2 \end{pmatrix}$. Show that this is equal to $\begin{pmatrix} 5 \\ 3 \end{pmatrix}$. ■

8.6 Key Points

▶ In the expansion of $(a + b)^n$, the **coefficient of the term** $a^{n-r}b^r$ is

$$\begin{pmatrix} n \\ r \end{pmatrix} = \frac{n!}{(n - r)! \, r!}$$

where $r \leq n$ and r and n are integers.

▶ **The Binomial Theorem:** Let n be a positive integer. Then

$$(a + b)^n = \sum_{i=0}^{n} \begin{pmatrix} n \\ i \end{pmatrix} a^{n-i}b^i$$

$$= \begin{pmatrix} n \\ 0 \end{pmatrix} a^n + \begin{pmatrix} n \\ 1 \end{pmatrix} a^{n-1}b + \begin{pmatrix} n \\ 2 \end{pmatrix} a^{n-2}b^2 + \cdots + \begin{pmatrix} n \\ n \end{pmatrix} b^n.$$

▶ The ***i*th term** of $(a + b)^n$ is given by

$$\begin{pmatrix} n \\ i - 1 \end{pmatrix} a^{n-i+1}b^{i-1}$$

where n and i are positive integers, with $1 \leq i < n + 1$.

8.6 Exercises

▶**Skills** This set of exercises will reinforce the skills illustrated in this section.

In Exercises 1–4, write down the variable parts of the terms in the expansion of the binomial.

1. $(a + b)^5$

2. $(a + b)^6$

3. $(x + y)^7$

4. $(x + y)^8$

In Exercises 5–16, evaluate each expression.

5. $4!$

6. $6!$

7. $\dfrac{3!}{2!}$

8. $\dfrac{4!}{3!}$

9. $\dbinom{6}{2}$

10. $\dbinom{5}{3}$

11. $\dbinom{7}{5}$

12. $\dbinom{7}{4}$

13. $\dbinom{10}{10}$

14. $\dbinom{10}{0}$

15. $\dbinom{100}{0}$

16. $\dbinom{100}{100}$

In Exercises 17–28, use the binomial theorem to expand the expression.

17. $(x + 2)^4$

18. $(x - 3)^3$

19. $(2x - 1)^3$

20. $(2x + 3)^4$

21. $(3 + y)^5$

22. $(4 - z)^4$

23. $(x - 3z)^4$

24. $(2z + y)^3$

25. $(x^2 + 1)^3$

26. $(x^2 - 2)^3$

27. $(y - 2x)^4$

28. $(z + 4x)^5$

In Exercises 29–42, use the Binomial Theorem to find the indicated term or coefficient.

29. The coefficient of x^3 when expanding $(x + 4)^5$

30. The coefficient of y^2 when expanding $(y - 3)^5$

31. The coefficient of x^5 when expanding $(3x + 2)^6$

32. The coefficient of y^4 when expanding $(2y + 1)^7$

33. The coefficient of x^6 when expanding $(x + 1)^8$

34. The coefficient of y^7 when expanding $(y - 3)^{10}$

35. The third term in the expansion of $(x - 4)^6$

36. The fourth term in the expansion of $(x + 3)^6$

37. The sixth term in the expansion of $(x + 4y)^5$

38. The seventh term in the expansion of $(a + 2b)^6$

39. The fifth term in the expansion of $(3x - 2)^6$

40. The fifth term in the expansion of $(3x + 1)^8$

41. The fourth term in the expansion of $(4x - 2)^6$

42. The fourth term in the expansion of $(3x - 1)^8$

▶**Concepts** This set of exercises will draw on the ideas presented in this section and your general math background.

43. Show that $\dbinom{n}{r} = \dbinom{n}{n - r}$, where $0 \le r \le n$, with n and r integers.

44. Show that $\dbinom{n}{0} = 1$.

45. Evaluate the following.

$$\dbinom{4}{0}\left(\dfrac{1}{3}\right)^4 + \dbinom{4}{1}\left(\dfrac{1}{3}\right)^3\left(\dfrac{2}{3}\right)$$

$$+ \dbinom{4}{2}\left(\dfrac{1}{3}\right)^2\left(\dfrac{2}{3}\right)^2 + \dbinom{4}{3}\left(\dfrac{1}{3}\right)\left(\dfrac{2}{3}\right)^3$$

$$+ \dbinom{4}{4}\left(\dfrac{2}{3}\right)^4$$

8.7 Mathematical Induction

Many mathematical facts are established by first observing a pattern, then making a conjecture about the general nature of the pattern, and finally *proving* the conjecture. In order to *prove* a conjecture, we use existing facts, combine them in such a way that they are relevant to the conjecture, and proceed in a logical manner until the truth of the conjecture is established.

For example, let us make a conjecture regarding the sum of the first n even integers. First, we look for a pattern:

$$2 = 2$$
$$2 + 4 = 6$$
$$2 + 4 + 6 = 12$$
$$2 + 4 + 6 + 8 = 20$$
$$2 + 4 + 6 + 8 + 10 = 30$$
$$\vdots$$

Table 8.7.1

n	Sum of First n Even Integers
1	2
2	6
3	12
4	20
5	30

From the equations above, we can build Table 8.7.1. The numbers in the "sum" column in the table can be factored as follows: $2 = 1 \cdot 2, 6 = 2 \cdot 3, 12 = 3 \cdot 4, 20 = 4 \cdot 5$, and $30 = 5 \cdot 6$. Noting the values of n to which the factorizations correspond, we make our conjecture:

The sum of the first n even integers is $n(n + 1)$.

According to our calculations, this conjecture holds true for n up to and including 5. But does it hold true for all n? To establish the pattern for all values of n, we must *prove* the conjecture. Simply substituting various values of n is not feasible because we would have to verify the statement for infinitely many n. A more practical proof technique is needed. We next introduce a proof method called **mathematical induction,** which is typically used to prove statements such as this.

Mathematical Induction

Before giving a formal definition of mathematical induction, we take our discussion of the sum of the first n even integers and introduce some new notation that we will need in order to work with this type of proof.

First, the conjecture is given a name: P_n. The subscript n means that the conjecture depends on n. Stating our conjecture, we write

P_n: The sum of the first n even integers is $n(n + 1)$.

For some specific values of n, the conjecture reads as follows:

P_8: The sum of the first 8 even integers is $8 \cdot 9 = 72$.

P_{12}: The sum of the first 12 even integers is $12 \cdot 13 = 156$.

P_k: The sum of the first k even integers is $k(k + 1)$.

P_{k+1}: The sum of the first $k + 1$ even integers is $(k + 1)(k + 2)$.

We now state the principle of mathematical induction, which we will need to complete the proof of our conjecture.

> **The Principle of Mathematical Induction**
>
> Let n be a natural number and let P_n be a statement that depends on n. If
>
> 1. P_1 is true, and
> 2. for all positive integers k, P_{k+1} can be shown to be true if P_k is assumed to be true, then P_n is true for all natural numbers n.

The underlying scheme behind proof by induction consists of two key pieces:

1. Proof of the base case: proving that P_1 is true
2. Use the assumption that P_k is true for a general value of k to show that P_{k+1} is true

Taken together, these two pieces prove that P_n holds true for every natural number n. The assumption that P_k is true is known as the **induction hypothesis.**

In proving statements by induction, we often have to take an expression containing the variable k and replace k with $k + 1$. Example 1 illustrates this process.

Example 1 **Replacing k with $k + 1$ in an Algebraic Expression**

Replace k with $k + 1$ in the following.

(a) $3^k - 1$ (b) $\dfrac{k(k + 1)(2k + 1)}{6}$

▶Solution

(a) Replacing k by $k + 1$, we obtain

$$3^{k+1} - 1.$$

(b) Replacing k by $k + 1$ and simplifying, we obtain

$$\frac{(k + 1)((k + 1) + 1)(2(k + 1) + 1)}{6}$$
$$= \frac{(k + 1)(k + 2)(2k + 3)}{6}.$$

✔ *Check It Out 1:* Replace k by $k + 1$ in $2k(k + 2)$. ■

We now return to the conjecture we made at the beginning of this section, and prove it by induction.

Example 2 **Proving a Formula by Induction**

Prove the following formula by induction:

$$2 + 4 + \cdots + 2n = n(n + 1)$$

▶Solution This is just the statement that we conjectured earlier, but in the form of an equation. Recall that we denoted this statement by P_n, so we denote the proposed equation by P_n as well.

First we must prove that P_n is true for $n = 1$. We do this by replacing every n in P_n with a 1, and then demonstrating that the result is true.

$$P_1: 2(1) = 1(1 + 1)$$

Since $2(1) = 1(1 + 1)$, we see that P_1 is true.

Next we state P_k and assume that P_k is true.

$$P_k: 2 + 4 + \cdots + 2k = k(k + 1)$$

Finally, we state P_{k+1} and use the assumption that P_k is true to prove that P_{k+1} holds true as well.

$$P_{k+1}: 2 + 4 + \cdots + 2k + 2(k + 1) = (k + 1)(k + 2)$$

To prove P_{k+1}, we start with the expression on the left-hand side of P_{k+1} and show that it is equal to the expression on the right-hand side.

$2 + 4 + \cdots + 2k + 2(k + 1)$	Left-hand side of P_{k+1}
$= k(k + 1) + 2(k + 1)$	Induction hypothesis: P_k is true
$= k^2 + k + 2k + 2$	Expand
$= k^2 + 3k + 2$	Combine like terms
$= (k + 1)(k + 2)$	Factor

We see that the result, $(k + 1)(k + 2)$, is the expression on the right-hand side of P_{k+1}. Thus, by mathematical induction, P_n is true for all natural numbers n.

✔ *Check It Out 2:* Prove the following formula by mathematical induction:

$$1 + 3 + 5 + \cdots + (2n - 1) = n^2 \ ■$$

Example 3 Proving a Summation Formula by Induction

Prove the following formula by induction:

$$1 + 2 + 3 + \cdots + n = \frac{n(n + 1)}{2}$$

▶**Solution** First denote the proposed equation by P_n and prove that it holds true for $n = 1$. Replacing every n with a 1, we get

$$P_1: 1 = \frac{1(1 + 1)}{2}.$$

Clearly this is true, so P_1 holds.

Next state P_k and assume that P_k is true.

$$P_k: 1 + 2 + 3 + \cdots + k = \frac{k(k + 1)}{2}$$

Finally, state P_{k+1} and use the assumption that P_k is true to prove that P_{k+1} holds true as well.

$$P_{k+1}: 1 + 2 + 3 + \cdots + k + (k + 1) = \frac{(k + 1)(k + 2)}{2}$$

Show that the expression on the left-hand side of P_{k+1} is equal to the expression on the right-hand side.

$$1 + 2 + 3 + \cdots + k + k + 1 \qquad \text{Left-hand side of } P_k$$

$$= \frac{k(k+1)}{2} + k + 1 \qquad \text{Induction hypothesis: } P_k \text{ is true}$$

$$= \frac{k(k+1) + 2(k+1)}{2} \qquad \text{Use common denominator}$$

$$= \frac{k^2 + k + 2k + 2}{2} \qquad \text{Expand}$$

$$= \frac{k^2 + 3k + 2}{2} \qquad \text{Combine like terms}$$

$$= \frac{(k+1)(k+2)}{2} \qquad \text{Factor}$$

We see that the result, $\frac{(k+1)(k+2)}{2}$, is the expression on the right-hand side of P_{k+1}. Thus, by mathematical induction, P_n is true for all natural numbers n.

✔ *Check It Out 3:* Prove by induction: $2 + 5 + 8 + \cdots + (3n - 1) = \frac{1}{2}n(3n + 1)$. ■

Example 4 Proving a Formula for Partial Sums by Induction

Prove by induction:

$$1 + 2 + 2^2 + 2^3 + \cdots + 2^{n-1} = 2^n - 1$$

▶**Solution** First denote the proposed equation by P_n and prove that it holds true for $n = 1$ by replacing every n with a 1.

$$P_1: 1 = 2^1 - 1$$

It is easy to see that P_1 is true.
   Next state P_k and assume that P_k is true.

$$P_k: 1 + 2 + 2^2 + 2^3 + \cdots + 2^{k-1} = 2^k - 1$$

Finally, state P_{k+1} and use the induction hypothesis (the assumption that P_k is true) to show that P_{k+1} holds true as well.

$$P_{k+1}: 1 + 2 + 2^2 + 2^3 + \cdots + 2^{k-1} + 2^k = 2^{k+1} - 1$$

$$1 + 2 + 2^2 + 2^3 + \cdots + 2^{k-1} + 2^k \qquad \text{Left-hand side of } P_{k+1}$$

$$= 2^k - 1 + 2^k \qquad \text{Induction hypothesis: } P_k \text{ is true}$$

$$= 2(2^k) - 1 \qquad \text{Combine like terms}$$

$$= 2^{k+1} - 1 \qquad \text{Simplify}$$

We see that the result, $2^{k+1} - 1$, is the expression on the right-hand side of P_{k+1}. Thus, by mathematical induction, P_n is true for all natural numbers n.

✔ *Check It Out 4:* Prove by induction: $1 + 4 + 4^2 + \cdots + 4^{n-1} = \frac{4^n - 1}{3}$. ■

You may wonder how we get the formulas to prove by induction in the first place. Many of these formulas are arrived at by first examining patterns and then coming up with a general formula using various mathematical facts. A complete discussion of how to *obtain* these formulas is beyond the scope of this book.

8.7 Key Points

The **Principle of Mathematical Induction** is stated as follows: Let n be a natural number and let P_n be a statement that depends on n. If

1. P_1 is true, and
2. for all positive integers k, P_{k+1} can be shown to be true if P_k is assumed to be true,

then P_n is true for all natural numbers n.

8.7 Exercises

▶**Skills** This set of exercises will reinforce the skills illustrated in this section.

In Exercises 1–4, replace k by k + 1 in each expression.

1. $k(k+1)(k+2)$

2. $3^k - 1$

3. $\dfrac{k}{k+1}$

4. $\dfrac{3}{1+k^2}$

In Exercises 5–25, prove the statement by induction.

5. $3 + 5 + \cdots + (2n+1) = n(n+2)$

6. $2 + 6 + 10 + \cdots + (4n-2) = 2n^2$

7. $1 + 4 + 7 + \cdots + (3n-2) = \dfrac{n(3n-1)}{2}$

8. $5 + 4 + 3 + \cdots + (6-n) = \dfrac{1}{2}n(11-n)$

9. $7 + 5 + 3 + \cdots + (9-2n) = -n^2 + 8n$

10. $3 + 9 + 15 + \cdots + (6n-3) = 3n^2$

11. $2 + 5 + 8 + \cdots + (3n-1) = \dfrac{1}{2}n(3n+1)$

12. $1^2 + 2^2 + 3^2 + \cdots + n^2 = \dfrac{n(n+1)(2n+1)}{6}$

13. $1^3 + 2^3 + \cdots + n^3 = \dfrac{n^2(n+1)^2}{4}$

14. $1 + 2 + 2^2 + \cdots + 2^{n-1} = 2^n - 1$

15. $1^2 + 3^2 + \cdots + (2n-1)^2 = \dfrac{n(2n-1)(2n+1)}{3}$

16. $\dfrac{1}{1\cdot2} + \dfrac{1}{2\cdot3} + \dfrac{1}{3\cdot4} + \cdots + \dfrac{1}{n(n+1)} = \dfrac{n}{n+1}$

17. $1\cdot2 + 2\cdot3 + 3\cdot4 + \cdots + n(n+1) = \dfrac{n(n+1)(n+2)}{3}$

18. $1 + 3 + 3^2 + \cdots + 3^{n-1} = \dfrac{3^n-1}{2}$

19. $1 + 5 + 5^2 + \cdots + 5^{n-1} = \dfrac{5^n-1}{4}$

20. $1 + r + r^2 + \cdots + r^{n-1} = \dfrac{r^n-1}{r-1}$, r a positive integer, $r \neq 1$

21. $3^n - 1$ is divisible by 2.

22. $n^3 - n + 3$ is divisible by 3.

23. $n^2 + 3n$ is divisible by 2.

24. $n^2 + n$ is even.

25. $2^n > n$

▶**Concepts** This set of exercises will draw on the ideas presented in this section and your general math background.

Induction is not the only method of proving that a statement is true. Exercises 26–29 suggest alternate methods for proving statements.

26. By factoring $n^2 + n$, n a natural number, show that $n^2 + n$ is divisible by 2.

27. By factoring $a^3 - b^3$, a and b positive integers, show that $a^3 - b^3$ is divisible by $a - b$.

28. Prove that $1 + 4 + 7 + \cdots + (3n - 2) = \frac{n(3n - 1)}{2}$ by using the formula for the sum of terms of an arithmetic sequence.

29. Prove that $1 + 4 + 4^2 + \cdots + 4^{n-1} = \frac{4^n - 1}{3}$ by using the formula for the sum of terms of a geometric sequence.

Chapter 8   Summary

Section 8.1 **Sequences**

Concept	Illustration	Study and Review
Sequence A **sequence** is a function $f(n)$ whose domain is the set of all nonnegative integers and whose range is a subset of the set of all real numbers. The numbers $f(0), f(1), f(2), \ldots$ are called the **terms** of the sequence.	1, 4, 7, 10, … is a sequence. 1, 4, and 7 are some of the terms.	
Definition of an arithmetic sequence Each term of an **arithmetic sequence** is given by the rule $a_n = a_0 + nd, \ n = 0, 1, 2, 3, \ldots$ where a_0 is the starting value of the sequence and d is the *common difference* between successive terms.	The sequence 4, 7, 10, 13, … is an arithmetic sequence because $a_n = 4 + 3n$, $n = 0, 1, 2, 3, \ldots$, with $d = 3$.	Examples 1–4, 8 Chapter 8 Review, Exercises 1–4
Definition of a geometric sequence A **geometric sequence** is defined by the rule $a_n = a_0 r^n, n = 0, 1, 2, 3, \ldots$ where a_0 is the initial value of the sequence and $r \neq 0$ is the *fixed ratio* between successive terms.	The sequence 4, 8, 16, 32, … is a geometric sequence because $a_n = 4(2)^n$, $n = 0, 1, 2, 3, \ldots$, with $r = 2$.	Examples 5–7, 9 Chapter 8 Review, Exercises 5–8

Section 8.2 Sums of Terms of Sequences

Concept	Illustration	Study and Review
Sum of the first _n_ terms of an arithmetic sequence Let $a_j = a_0 + jd$, $j = 0, 1, 2, \ldots$, be an arithmetic sequence, and let n be a positive integer. The sum of the n terms from a_0 to a_{n-1} is given by $S_n = a_0 + a_1 + \cdots + a_{n-1}$ $\quad = \dfrac{n}{2}(2a_0 + d(n-1))$ $\quad = \dfrac{n}{2}(a_0 + a_{n-1}).$	Using $d = 2$, $a_0 = 3$, and $n = 5$, $3 + 5 + 7 + 9 + 11 = \dfrac{5}{2}(3 + 11)$ $\quad\quad = 35.$	Examples 1–3, 9 Chapter 8 Review, Exercises 9–12
Summation notation The **summation symbol** is indicated by the Greek letter Σ (sigma). Index i ends at n Summation symbol $\rightarrow \displaystyle\sum_{i=1}^{n} i \leftarrow$ Expression to be evaluated and summed Index i starts at 1	The sum $2 + 4 + 6 + 8 + 10$ can be written as $\displaystyle\sum_{i=1}^{5} 2i$.	Examples 4, 5 Chapter 8 Review, Exercises 13, 14
Sum of the first _n_ terms of a geometric sequence Let $a_j = a_0 r^j$, $j = 0, 1, 2, \ldots$, be a geometric sequence. The sum of the n terms from a_0 to a_{n-1} is given by $S_n = \displaystyle\sum_{j=0}^{n-1} a_0 r^j = a_0\left(\dfrac{1 - r^n}{1 - r}\right).$	Using $r = 2$, $n = 4$, and $a_0 = 6$, $6 + 12 + 24 + 48 = 6\left(\dfrac{1 - 2^4}{1 - 2}\right) = 90.$	Examples 6, 7, 10 Chapter 8 Review, Exercises 13–20
Sum of an infinite geometric series If $|r| < 1$, then the infinite geometric series $a_0 + a_0 r + a_0 r^2 + a_0 r^3 + \cdots + a_0 r^{n-1} + \cdots$ has the sum $S = \displaystyle\sum_{i=0}^{\infty} a_0 r^i = \dfrac{a_0}{1 - r}.$	The sum $1 + \dfrac{1}{2} + \dfrac{1}{4} + \dfrac{1}{8} + \cdots = \dfrac{1}{1 - \dfrac{1}{2}} = 2$ because $\lvert r \rvert = \left\lvert \dfrac{1}{2} \right\rvert < 1.$	Example 8 Chapter 8 Review, Exercises 21–24

Section 8.3 General Sequences and Series

Concept	Illustration	Study and Review
Sequences given by a rule A **general rule for the terms** a_n of a sequence can be used to generate terms of the sequence.	The first four terms of the sequence generated by the rule $a_n = n^2 + 1$, $n = 0, 1, 2, 3, \ldots$, are 1, 2, 5, 10.	Examples 1, 2 Chapter 8 Review, Exercises 25–34
Alternating sequences In an **alternating sequence,** the terms alternate between positive and negative values.	The sequence $a_n = (-1)^n 2n$, $n = 0, 1, 2, 3, \ldots$, is an alternating sequence. The first four terms of this sequence are 0, −2, 4, −6.	Examples 3, 4 Chapter 8 Review, Exercises 33, 34
Recursively defined sequences **Recursively defined sequences** define the nth term by using the preceding terms, along with the first term or first several terms.	The sequence $a_n = 2a_{n-1}$, $a_0 = 1$, is a recursively defined sequence. The first five terms of this sequence are 1, 2, 4, 8, 16.	Examples 5, 6, 8 Chapter 8 Review, Exercises 35–38
Partial sums The nth **partial sum** of a sequence $\{a_n\}$ is given by $$\sum_{i=0}^{n-1} a_i.$$	The partial sum $\sum_{i=0}^{3} (i^2 + 1)$ is $1 + 2 + 5 + 10 = 18.$	Example 7 Chapter 8 Review, Exercises 39, 40

Section 8.4 Counting Methods

Concept	Illustration	Study and Review
Multiplication principle Suppose there are n slots to fill. Let a_1 be the number of possibilities for the first slot, a_2 the number of possibilities for the second slot, and so on, with a_k representing the number of possibilities for the kth slot. Then the total number of ways in which the n slots can be filled is Total possibilities $= a_1 \times a_2 \times a_3 \times \cdots \times a_n$.	If there are 10 people and four chairs are available, the number of possible seating arrangements is $10 \cdot 9 \cdot 8 \cdot 7 = 5040$, assuming all four chairs are filled.	Examples 1, 2 Chapter 8 Review, Exercises 43–45
Definition of $n!$ (n factorial) Let n be a positive integer. Then $$n! = n(n-1)(n-2) \cdots (3)(2)(1).$$ The quantity $0! = 1$.	The number 4! is $4! = 4 \cdot 3 \cdot 2 \cdot 1 = 24$.	Examples 3, 4 Chapter 8 Review, Exercises 41, 42
Definition of permutation Given n *distinct* objects and r slots to fill, the number of r-**permutations of n objects** is given by $P(n, r) = \frac{n!}{(n-r)!}$. In a permutation, order matters.	If there are five distinct objects and three slots to fill, and the order matters, the number of ways to permute the five objects is $P(5, 3) = \frac{5!}{2!} = 60$.	Examples 5–7 Chapter 8 Review, Exercises 41, 43–45, 48

Continued

Section 8.4 **Counting Methods**

Concept	Illustration	Study and Review
Definition of combination Let r and n be positive integers with $r \leq n$. When the order of the objects does not matter, the selection of r objects at a time from a set of n objects is called a **combination:** $C(n, r) = \dfrac{n!}{(n - r)!r!}$.	If order does not matter, the number of ways to select three objects from a collection of five objects is $C(5, 3) = \dfrac{5!}{2!\,3!} = 10$. Note the difference between $C(5, 3)$ and $P(5, 3)$.	Examples 8–10 Chapter 8 Review, Exercises 46, 47

Section 8.5 **Probability**

Concept	Illustration	Study and Review
Basic terminology • An **outcome** is any possibility resulting from an experiment. Here, an experiment simply denotes an activity yielding random results. • An **event** is a collection (set) of outcomes. • A **sample space** is the set consisting of all possible outcomes for a certain situation under consideration.	You roll a six-sided die and record the side of the die that lands face up. • The possible **outcomes** are rolling any whole number between 1 and 6. • One possible **event** is rolling an even number. • The **sample space** for the die-rolling experiment is $\{1, 2, 3, 4, 5, 6\}$.	Examples 1–5 Chapter 8 Review, Exercises 49–53
Definition of probability Assume that a sample space S consists of equally likely outcomes. Then the **probability** of an event E, denoted by $P(E)$, is defined as $\qquad P(E) = \dfrac{\text{number of outcomes in } E}{\text{number of outcomes in } S}.$ Note that $0 \leq P(E) \leq 1$.	In the die-rolling experiment, the probability of rolling an even number is $\dfrac{3}{6} = \dfrac{1}{2}$.	Examples 1–5 Chapter 8 Review, Exercises 54, 55
Mutually exclusive events Two events F and G are **mutually exclusive** if they have no overlap. In this case, $\qquad P(F \text{ or } G) = P(F) + P(G).$	In the die-rolling experiment, the event of rolling an even number and the event of rolling an odd number are mutually exclusive.	Examples 6–8 Chapter 8 Review, Exercises 52, 56
Complement of an event The set of all outcomes in a sample space that do *not* belong to event E is called the **complement** of E and is denoted by E'. The probability of E' is $P(E') = 1 - P(E)$.	In the die-rolling experiment, the complement of rolling a 1 is rolling any whole number from 2 through 6.	Examples 9, 10 Chapter 8 Review, Exercise 53

Section 8.6 **The Binomial Theorem**

Concept	Illustration	Study and Review
Binomial coefficients Let n and r be nonnegative integers with $r \le n$. In the expansion of $(a + b)^n$, the **coefficient** of the term $a^{n-r}b^r$ is $$\binom{n}{r} = \frac{n!}{(n-r)!\,r!}.$$	In the expansion of $(3 + y)^3$, the coefficient of $3y^2$ is $$\binom{3}{2} = \frac{3!}{1!\,2!} = 3.$$	Examples 1, 2 Chapter 8 Review, Exercises 57–64
The Binomial Theorem Let n be a positive integer. Then $$(a + b)^n = \sum_{i=0}^{n} \binom{n}{i} a^{n-i} b^i$$ $$= \binom{n}{0} a^n + \binom{n}{1} a^{n-1}b$$ $$+ \binom{n}{2} a^{n-2}b^2 + \cdots + \binom{n}{n} b^n.$$ The ith term of $(a + b)^n$ is given by $\binom{n}{i-1} a^{n-i+1} b^{i-1}$, where n and i are positive integers, with $1 \le i < n + 1$.	$(3 + y)^3 = \binom{3}{0}(3)^3 + \binom{3}{1}(3)^2(y)$ $+ \binom{3}{2}(3)^1(y)^2 + \binom{3}{3}(3)^0(y)^3$ $= 27 + 27y + 9y^2 + y^3$ The second term of $(3 + y)^3$ is $$\binom{3}{1}3^{3-2+1}y^{2-1} = 3(9)y = 27y.$$	Examples 3–5 Chapter 8 Review, Exercises 57–64

Section 8.7 **Mathematical Induction**

Concept	Illustration	Study and Review
Principle of mathematical induction Let n be a natural number and let P_n be a statement that depends on n. If 1. P_1 is true, and 2. for all positive integers k, P_{k+1} can be shown to be true if P_k is assumed to be true then P_n is true for all natural numbers n.	P_1 is the *base case*, which must be proved; P_k is the *induction hypothesis* and is assumed. Then P_{k+1} is proved.	Examples 1–4 Chapter 8 Review, Exercises 65–70

Chapter 8

Review Exercises

Section 8.1

In Exercises 1–4, find a rule for an arithmetic sequence that fits the given information.

1.

n	0	1
a_n	7	9

2.

n	0	1
$f(n)$	2	−1

3. The common difference d is 4 and $a_2 = 9$.

4. The common difference d is −2 and $a_6 = 10$.

In Exercises 5–8, find a rule for a geometric sequence that fits the given information.

5.

n	0	1
a_n	5	10

6.

n	0	1
a_n	8	4

7. The common ratio r is 3 and $a_3 = 54$.

8. The common ratio r is $\frac{1}{3}$ and $b_2 = 9$.

Section 8.2

In Exercises 9–12, find the sum of the terms of the arithmetic sequence using the summation formula.

9. $1 + 5 + 9 + \cdots + 61$

10. $2 + 7 + 12 + \cdots + 102$

11. The sum of the first 20 terms of the sequence defined by $a_n = 4 + 5n$, $n = 0, 1, 2, \ldots$

12. The sum of the first 15 terms of the sequence defined by $a_n = 2 - 3n$, $n = 0, 1, 2, \ldots$

In Exercises 13–16, find the sum of the terms of the geometric sequence using the summation formula.

13. $4 + 8 + 16 + 32 + 64 + 128 + 256$

14. $0.6 + 1.8 + 5.4 + 16.2$

15. $\sum_{i=0}^{5} \left(\frac{1}{2}\right)^n$

16. $\sum_{i=0}^{7} (3)^n$

In Exercises 17–20, (a) write using summation notation and (b) find the sum using the summation formula.

17. The first 25 terms of the series defined by $a_n = 3n$, $n = 0, 1, 2, 3, \ldots$

18. The first 10 terms of the series defined by $a_n = 2.5n$, $n = 0, 1, 2, 3, \ldots$

19. The first eight terms of the series defined by $a_n = 2^n$, $n = 0, 1, 2, 3, \ldots$

20. $a + a^2 + a^3 + \cdots + a^{16}$, $a \neq 0$, $a \neq 1$

In Exercises 21–24, determine whether the infinite geometric series has a sum. If so, find the sum.

21. $1 + \frac{1}{3} + \frac{1}{9} + \cdots$

22. $27 + 9 + 3 + 1 + \frac{1}{3} + \cdots$

23. $1 + 1.1 + 1.21 + 1.331 + \cdots$

24. $3 + 2 + \frac{4}{3} + \frac{8}{9} + \cdots$

Section 8.3

In Exercises 25–34, find the first five terms of the sequence.

25. $a_n = -2n + 5$, $n = 0, 1, 2, 3, \ldots$

26. $a_n = 4n + 1$, $n = 0, 1, 2, 3, \ldots$

27. $a_n = -3\left(\frac{1}{2}\right)^n$, $n = 0, 1, 2, 3, \ldots$

28. $a_n = -\left(\frac{2}{3}\right)^n$, $n = 0, 1, 2, 3, \ldots$

29. $b_n = -n^2 + 1$, $n = 0, 1, 2, 3, \ldots$

30. $b_n = 3n^3 - 1$, $n = 0, 1, 2, 3, \ldots$

31. $f(n) = \frac{n}{n^2 + 1}$, $n = 1, 2, 3, \ldots$

32. $f(n) = \dfrac{1}{2n^2 - 1}, \; n = 1, 2, 3, \ldots$

33. $a_n = (-1)^n \left(\dfrac{1}{3}\right)^n, \; n = 1, 2, 3, \ldots$

34. $a_n = (-1)^n(2n + 3), \; n = 1, 2, 3, \ldots$

In Exercises 35–38, find the first four terms of the recursively defined sequence.

35. $a_0 = 4; \; a_n = a_{n-1} + 3, \; n = 1, 2, 3, \ldots$

36. $a_0 = -2; \; a_n = a_{n-1} + 0.5, \; n = 1, 2, 3, \ldots$

37. $a_0 = -1; \; a_n = a_{n-1} + 2n, \; n = 1, 2, 3, \ldots$

38. $b_1 = \sqrt{2}; \; b_n = \sqrt{b_{n-1} + 2}, \; n = 2, 3, 4, \ldots$

In Exercises 39 and 40, find the partial sum.

39. $\displaystyle\sum_{i=0}^{3} 2i^2$

40. $\displaystyle\sum_{i=0}^{3} (-1)^i(2i + 1)$

Section 8.4

In Exercises 41 and 42, evaluate.

41. (a) $\dfrac{4!}{2!}$ (b) $P(5, 4)$ (c) $P(4, 4)$ (d) $P(6, 3)$

42. (a) $\dfrac{5!}{3!}$ (b) $C(5, 4)$ (c) $\dbinom{4}{4}$ (d) $C(6, 3)$

43. How many different photographs are possible if five adults line up in a row?

44. How many different photographs with exactly four college students are possible if there are six college students to select from? Here, order matters.

45. How many different photographs are possible if four children stand in the front row and three adults stand in the back row?

46. How many three-person committees can be formed in a club with 10 members?

47. How many five-letter words, including nonsense words, can be made from the letters in the word TABLE?

48. You have three English books, two history books, and three physics books.
  (a) How many ways are there to arrange all these books on a shelf?
  (b) How many ways are there to arrange the books if they must be grouped together by subject?

Section 8.5

In Exercises 49–54, use this scenario: A coin is tossed three times and the sequence of heads and tails is recorded.

49. List all the possible outcomes for this problem.

50. What is the event that you get at least one head?

51. What is the sample space for this problem?

52. Are the events of getting all tails and getting all heads mutually exclusive?

53. What is the complement of the event that you get at least one head?

54. Calculate the probability of the event that you get at least one head.

55. A bag consists of three red marbles, two white marbles, and four green marbles. If one marble is chosen randomly, what is the probability that the marble is white?

56. You are dealt one card from an ordinary deck of 52 cards. What is the probability that it is an ace of hearts or an ace of spades?

Section 8.6

In Exercises 57–60, expand each expression using the Binomial Theorem.

57. $(x + 3)^4$

58. $(2x + 1)^3$

59. $(3x + y)^3$

60. $(3z - 2w)^4$

In Exercises 61–64, use the Binomial Theorem to find the indicated term or coefficient.

61. The coefficient of x^4 when expanding $(x + 3)^5$

62. The coefficient of y^3 when expanding $(2y + 1)^5$

63. The second term in the expansion of $(x + 2y)^3$

64. The third term in the expansion of $(y - z)^4$

Section 8.7

In Exercises 65–70, use mathematical induction to prove the statement.

65. $1 + \dfrac{1}{2} + \dfrac{1}{4} + \cdots + \dfrac{1}{2^n} = 2 - \dfrac{1}{2^n}$

66. $1 + 3 + 6 + \cdots + \dfrac{n(n+1)}{2} = \dfrac{n(n+1)(n+2)}{6}$

67. $(n+1)^2 + n$ is odd.

68. $n^3 + 2n$ is divisible by 3.

69. $3 + 6 + 9 + \cdots + 3n = \dfrac{3}{2}n(n+1)$

70. $1 + 5 + 9 + \cdots + (4n-3) = n(2n-1)$

Applications

71. **Compensation** Carolyn has worked as an accountant for the same firm for the past 8 years. Her annual starting salary with her current employer was $36,000. Each year, on the anniversary of her first day on the job with this firm, she has been given a raise of $1500. What was her annual salary just after her fifth anniversary with the company?

72. **Basketball Playoffs** An elimination basketball tournament is held, with 32 teams participating. All the teams play in the first round, with each team playing against just one other team. The losing teams in the first round are eliminated, and the winning teams advance to the second round. This process continues for additional rounds, until all but one team have been eliminated. How many rounds of games were played in the tournament?

73. **Graduation** All 69 graduates of a middle school attended their commencement ceremony, and they all posed for the class photograph. They formed six rows for the photo, with one row in back of another. The number of graduates in each row beyond the first (front) row increased by 1. How many graduates were in the back row of the photo?

74. **Sports** There are 11 swimmers on a swim team. How many different ways could a swim team be formed from a pool of 18 swimmers, assuming that every swimmer is qualified to be on the team?

75. **Clothing** A men's plaid shirt is available in three different color schemes, each of which is manufactured in 10 different sizes. If you wanted to buy this shirt, how many different combinations of color scheme and size would you have to choose from?

76. **Telephone Directory** Julio wants to contact a friend who has long since moved out of his area. Julio doesn't have his friend's mailing address or e-mail address, and all he knows about the friend's current phone number is the area code and the exchange. (The exchange consists of the first three digits of the seven-digit phone number.) Julio has no choice but to dial phone numbers at random. What is the maximum number of phone numbers he would have to dial in order to reach his friend?

77. **Card Game** Kim takes part in a card game in which every player is dealt a hand of five cards from a standard 52-card deck. What is the probability that the hand dealt to her consists of five red cards, if she is dealt the first five cards?

78. **Cryptography** A secret code is made up of a sequence of four letters of the alphabet. What is the probability that all four letters of the code are identical?

Chapter 8   Test

1. Find a rule for an arithmetic sequence with $a_0 = 8$ and $a_1 = 11$.

2. Find a rule for an arithmetic sequence with common difference $d = 5$ and $a_4 = 27$.

3. Find a rule for a geometric sequence with $a_0 = 15$ and $a_1 = 5$.

4. Find a rule for a geometric sequence with common ratio $r = \dfrac{2}{3}$ and $a_2 = 2$.

5. Find the following sum using a formula for the sum of terms: $5 + 8 + 11 + \cdots + 104$.

6. Find the sum of the first 30 terms of the sequence defined by $a_n = -2 + 4n$, $n = 0, 1, 2, \ldots$.

7. Write the following sequence using summation notation, and find its sum: $a_n = 5n$, $n = 0, 1, 2, 3, \ldots, 9$.

8. Find the first four terms of the sequence defined by $a_n = -2n^3 + 2$, $n = 0, 1, 2, 3, \ldots$.

9. Find the first five terms of the sequence defined by $a_n = (-1)^n \left(\frac{1}{4}\right)^n$, $n = 0, 1, 2, 3, \ldots$.

10. Write the first five terms of the following recursively defined sequence: $a_0 = 4$; $a_n = a_{n-1} - n$, $n = 1, 2, 3, \ldots$.

11. Evaluate the following partial sum: $\sum_{i=0}^{3} (3i + 2)$

12. Evaluate the following.

   (a) $\dfrac{5!}{2!}$ (b) $C(6, 4)$

   (c) $P(7, 1)$ (d) $P(6, 4)$

13. How many committees of three people are possible in a club with 12 members?

14. How many six-letter words, including nonsense words, can be made with the letters in the word SAMPLE?

15. If you have four mathematics books, two biology books, and three chemistry books, how many ways are there to arrange the books on one row of a bookshelf if they must be grouped by subject?

In Exercises 16–18, consider a bag containing four red marbles, five blue marbles, and two white marbles. One marble is drawn randomly from the bag and its color is recorded.

16. List all the possible outcomes for this problem.

17. What is the probability that a white marble is drawn?

18. What is the probability that a blue marble is *not* drawn?

19. You are dealt one card from an ordinary deck of 52 cards. What is the probability that it is the queen of hearts or the jack of spades?

20. Expand using the Binomial Theorem: $(3x + 2)^4$

21. Prove by induction: $2 + 6 + 10 + \cdots + (4n - 2) = 2n^2$

22. A clothing store sells a "fashion kit" consisting of three pairs of pants, four shirts, and two jackets. How many different outfits, each consisting of a pair of pants, a shirt, and a jacket, can be put together using this kit?

23. A license plate number consists of two letters followed by three nonzero digits. How many different license plates are possible, assuming no letter or digit can be used more than once?

Keystroke Guide for the TI-83/84 Calculator Series

This appendix is designed to help you effectively use your TI-83 or TI-84 series calculator to explore the mathematical ideas and applications in this textbook. The keystrokes for both the TI-83 and TI-84 families of calculators are the same. Only the colors of the keys differ.

Review Sections A.1–A.4 to learn about the basic operations you can perform on your calculator, especially if you are a new user. These sections are prerequisite to the later material in the appendix, which shows specific keystrokes for corresponding examples in the textbook. Keystrokes are grouped by main topics, such as "graphing functions" or "solving equations."

Today's calculators have many features, and so there is often more than one way to work a problem. Most of the keystrokes in this appendix illustrate only one technique, but you should feel free to explore other ways to accomplish the same task.

A.1 Keys on Your Calculator

The row of buttons just below the screen is used to create graphs and tables. See Figure A.1.1 (taken from the TI-84).

Figure A.1.1

A second set of keys is used for navigation and to access various mathematical functions. See Figure A.1.2. On the TI-84, blue is used for the $\boxed{\text{2ND}}$ function keys. On the TI-83, yellow is used for the $\boxed{\text{2ND}}$ function keys.

Figure A.1.2

The $\boxed{\text{2ND}}$ key accesses the blue function above each key on the TI-84.

The $\boxed{\text{ALPHA}}$ key accesses the green letter or character above each key on the TI-84.

Arrow keys move the cursor on the screen.

The $\boxed{\text{MATH}}$ key brings up a menu with various mathematical functions.

Throughout this appendix, the keystrokes corresponding to the functions above a key will be denoted by $\boxed{\text{2ND}}$ $\boxed{\text{KEY (Name of function above key)}}$. For example, the keystroke for the CALCULATE menu will be given by $\boxed{\text{2ND}}$ $\boxed{\text{TRACE (CALC)}}$.

A.2 Getting Started

Initializing Your Calculator

Calculator On/Off   Turn the calculator *on* with the $\boxed{\text{ON}}$ button. Turn the calculator *off* with the $\boxed{\text{2ND}}$ $\boxed{\text{ON (OFF)}}$ button.

Home Screen   When you turn the calculator on, the *Home Screen* is displayed. This is where you enter expressions and instructions to compute numeric results. You can always get to the Home Screen from another window mode by pressing (2ND) (MODE (QUIT)). Press (CLEAR) to clear the Home Screen.

The Cursor   A blinking box called a *cursor* determines the current position on the screen and is moved around by the arrow keys.

Changing the Screen Contrast   Press (2ND) and the up arrow key to make the display darker, or press (2ND) and the down arrow key to make it lighter.

Initializing the MODE   In the MODE menu, accessed by pressing (MODE), highlight the first entry in each row unless directed to do otherwise. See Figure A.2.1. Press (2ND) (MODE (QUIT)) to exit the MODE menu.

Figure A.2.1

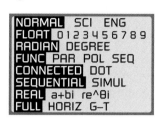

Arithmetic Operations

Calculations   Key in the expression in the Home Screen and then press (ENTER). The standard arithmetic operation symbols are used.

Subtraction Symbol and Negative Sign   These are *different* keys. To enter a *negative number*, use ((-)). This key appears directly beneath the **3** key. To *subtract*, use (-), directly above the (+) key.

Order of Operations   Working outward from the inner parentheses, operations are performed from left to right. Exponentiation and any operations under a radical symbol are evaluated first, followed by multiplications and divisions, and then additions and subtractions.

If you want to change the algebraic order, you must use parentheses. Parentheses also must be used around the numerator and denominator in fractions. See Section P.1 for more information.

Figure A.2.2

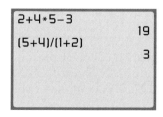

Example 1 Evaluating Simple Expressions

Use a calculator to evaluate the following.

(a) $2 + 4 \cdot 5 - 3$

(b) $\dfrac{5 + 4}{1 + 2}$

▶Solution

(a) Press **2** (+) **4** (×) **5** (-) **3**. The answer is 19. See Figure A.2.2.

(b) Press (( ) **5** (+) **4** ( )) (÷) (( ) **1** (+) **2** ( )). The answer is 3. Note that the numerator and denominator must be entered using parentheses. See Figure A.2.2.   ■

Menus and Submenus

The TI-83 and TI-84 Plus operate using *menus* and *submenus*. When you press a menu key such as (MATH) on the calculator, the submenus are listed in the top row of the screen. The highlighted submenu is displayed. Use the right and left arrow keys to move to the other submenus. To exit a menu, press (2ND) (MODE (QUIT)). The following example shows how to access a menu or submenu item.

Figure A.2.3

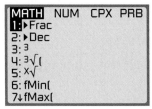

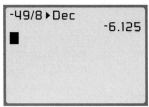

Example 2 **Using the MATH Menu**

Change $\dfrac{-49}{8}$ into decimal form using the MATH menu.

▶Solution There are two ways to access a MATH menu item. Use whichever you prefer. The keystrokes in this appendix use Method A to access menu items.

Method A Use a number to select the menu item. Enter [(−)] **49** [÷] **8** on the Home Screen. Press [MATH] and then **2** for 2:▶Dec. Press [ENTER]. The answer is −6.125. See Figure A.2.3.

Method B Use an arrow key to select the menu item. Enter [(−)] **49** [÷] **8** on the Home Screen. Press [MATH] and use the down arrow key to move to 2:▶Dec. Press [ENTER] [ENTER]. The answer is −6.125. ■

A.3 Editing and Deleting

The following explains how to edit entries in the command line.

Change the Current Entry   Move the blinking cursor to the current entry and type in the new entry, which replaces the old entry.

Delete the Current Entry   Move the cursor to the character and press [DEL].

Insert a New Entry   Move the cursor to the character after the insertion point and press [2ND] [DEL (INS)] to type in new text or symbols.

Edit a Previous Entry   In the Home Screen, press [2ND] [ENTER (ENTRY)] to recall the latest entry, and edit it as explained above. You may continue to press [2ND] [ENTER (ENTRY)] to recall even earlier entries.

Clear Data   To completely delete data from memory, press [2ND] [+ (MEM)] and then **2** for 2:Delete and delete from any of the given menus. Press [2ND] [MODE (QUIT)] to exit the menu.

A.4 Entering and Evaluating Common Expressions

Expressions can be entered on the Home Screen or in the equation editor by pressing the [Y=] editor. Expressions that can be readily evaluated are usually entered on the Home Screen. Order of operations always applies, so you must use parentheses if you wish to change the order. See Section P.1 for details.

Your calculator contains many *built-in functions*, such as LN and e^x. Built-in functions can be accessed via the keyboard, the various menus, or the Catalog, which is a menu containing an alphabetical list of all functions. When using built-in functions, a left parenthesis is often included so that you only have to enter the input value and then type the right parenthesis to complete the expression.

The following table illustrates various examples of entering expressions. You may not yet have studied the expressions for numbers 8–14, but you can come back to them as needed.

EXPRESSION (ENTERED ON HOME SCREEN)	EXAMPLE	KEYSTROKES (PRESS ENTER AFTER EACH ENTRY)
1. Rational expression	$\dfrac{2}{x-1}$	**2** ÷ ((X, T, θ, n − **1**)
2. Change decimal to fraction	0.5	**0.5** MATH ; press **1** for ▸Frac
3. Absolute value	|5|	MATH ▷ ; press **1** for 1:abs(and then press **5** and)
4. Square root	$\sqrt{12}$	2ND $x^2(\sqrt{\ })$ **12**)
5. Root from MATH menu	$\sqrt[4]{16}$	**4** MATH ; press **5** for 5:$\sqrt[x]{\ }$ and then press (**16**)
Root from Home Screen	$\sqrt[4]{16}$	**16** ^ (**1** ÷ **4**)
6. Square	6^2	**6** x^2 or **6** ^ **2**
7. Power	x^7	X, T, θ, n ^ **7**
8. Natural exponential function	e^3	2ND LN (e^x) **3**)
9. Natural logarithm	$\ln(x-2)$	LN X, T, θ, n − **2**)
10. Common logarithm	$\log(x+1)$	LOG X, T, θ, n + **1**)
11. Logarithm to base b (use change-of-base formula)	$\log_2 x$	(LN X, T, θ, n) ÷ (LN **2**)
12. Factorial	7!	**7** MATH ▷ ▷ ▷ ; press **4** for 4:!
13. Combination	$_8C_4$	**8** MATH ▷ ▷ ▷ ; press **3** for 3:nCr and then press **4**
14. Permutation	$_8P_3$	**8** MATH ▷ ▷ ▷ ; press **2** for 2:nPr and then press **3**
15. Scientific notation	3×10^4	**3** 2ND ,(EE) **4**

Evaluating Variable Expressions

You can store the value of a variable and then use it to evaluate expressions. From the Home Screen, you can assign the variable any name from A to Z.

Figure A.4.1

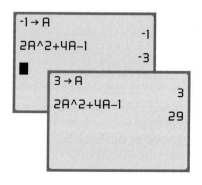

Example **1** **Evaluating an Expression Containing One Variable**

Assign the value of -1 to the variable A and evaluate $2A^2 + 4A - 1$. Then evaluate the expression for $A = 3$.

▶Solution Access the Home Screen.

1. Store the value of -1 by pressing (−) **1** STO ALPHA **A** ENTER.

2. Enter the expression as follows: **2** ALPHA **A** ^ **2** + **4** ALPHA **A** − **1** ENTER. The calculator will display the answer of -3. See Figure A.4.1.

3. If you want to evaluate the expression for A = 3, store the value of 3 in A as follows: 3 [STO] [ALPHA] **A** [ENTER]. Now recall the variable expression by pressing [2ND] [ENTER (ENTRY)] [2ND] [ENTER (ENTRY)] [ENTER]. The answer is 29. See Figure A.4.1. ▨

You can also evaluate expressions containing more than one variable.

Example 2 **Evaluating an Expression Containing Many Variables**

Evaluate the expression $\dfrac{A + 2B^2}{3C}$ for A = −2, B = 1, and C = $\dfrac{1}{2}$.

▶Solution

1. Assign a value to each variable:

[(−)] **2** [STO] [ALPHA] **A** [ALPHA] [:] **1** [STO] [ALPHA] **B** [ALPHA] [:]

1 [÷] **2** [STO] [ALPHA] **C** [ENTER]

Only the value of C is displayed, but the other values will be stored in memory. The colon allows you to enter multiple statements on one line.

2. Enter the expression:

[(] [ALPHA] **A** [+] **2** [ALPHA] **B** [^] **2** [)] [÷] [(] **3** **C** [)] [ENTER]

The answer of 0 is displayed. See Figure A.4.2. The fractional expression must be entered using parentheses to separate the numerator and denominator. You can change the values of the variables and reevaluate the expression as shown in the previous example. ▨

Figure A.4.2

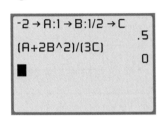

Using the CATALOG Function

The CATALOG function, the function above the **0** key, is an alphabetic list of all the functions and symbols available on the calculator. Most of them are in a menu or on the keyboard, but you can use the CATALOG when you forget which menu you need.

Figure A.4.3

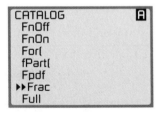

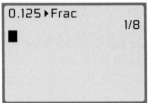

Example 3 **Using the CATALOG**

Use the CATALOG to convert 0.125 to a fraction.

▶Solution

1. Enter 0.125 on the Home Screen: **0.125.**
2. Press [2ND] [0 (CATALOG)]. Since you want to convert to a fraction, your function begins with the letter F. Press [COS (F)]. You do not need to press [ALPHA] or [2ND] because the CATALOG is set up to directly accept alphabetical input. Use [▽] to scroll to the command ▶Frac.
3. Press [ENTER] [ENTER]. The answer 1/8 is displayed. See Figure A.4.3. ▨

A.5 Entering and Evaluating Functions

You must enter functions into the Y= Editor in order to generate tables and graphs. Once the function is entered, it can be evaluated at different values of the input variable. When defining a function in the Y= Editor, only X is allowed as the input variable. If your function uses a different letter for the input variable, you must rewrite the function in terms of X before entering it into the Y= Editor.

Figure A.5.1

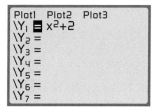

Example **1** **Evaluating a Function**

Enter the function $f(x) = x^2 + 2$ and find $f(3)$ and $f(-1)$.

▶Solution

1. Press $\boxed{Y=}$ to access the editor. In the Y= Editor, enter the definition for Y_1 as follows: $\boxed{X, T, \theta, n}$ $\boxed{\wedge}$ **2** $\boxed{+}$ **2**. Note that the equal sign is highlighted. See Figure A.5.1.

2. To evaluate $Y_1(3)$:

   (a) Press $\boxed{2ND}$ $\boxed{MODE\ (QUIT)}$ to return to the Home Screen.

   (b) To access the function Y_1, press $\boxed{VARS}$ and use the right arrow key to highlight the Y-VARS submenu.

   (c) Press **1** to access the FUNCTION menu and press **1** for the function Y_1. This is the only way to access any named function.

   (d) You will now be on the Home Screen showing Y_1. Complete the expression by typing $\boxed{(}$ **3** $\boxed{)}$ $\boxed{ENTER}$. The answer, 11, is displayed. See Figure A.5.2.

Figure A.5.2

 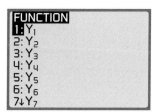 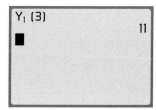

3. To evaluate $Y_1(-1)$, press $\boxed{2ND}$ $\boxed{ENTER\ (ENTRY)}$ to recall the previous line. Move the cursor to highlight **3**, delete it, and replace it with $\boxed{(-)}$ **1**. Press $\boxed{ENTER}$ to get the answer, 3.

Alternate Method   Enter the function as in Step 1. You can use the TABLE menu in ASK mode to evaluate a function. See Section A.6 for details. ▪

Figure A.5.3

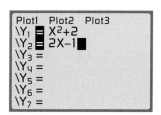

Selecting and Deselecting Functions in the Y= Editor   Move the cursor to the equal sign and press $\boxed{ENTER}$. When the equal sign is highlighted, the function is selected. When it is not highlighted, the function is not selected.

You can enter more than one function in the Y= Editor. Figure A.5.3 shows two functions entered in the Y= Editor.

Deleting Functions in the Y= Editor   Move the cursor to the right of the equal sign of the function you wish to delete. Press $\boxed{CLEAR}$.

A.6 Building a Table

To create a table of values for a function, use the commands $\boxed{2ND}$ $\boxed{GRAPH\ (TABLE)}$ and $\boxed{2ND}$ $\boxed{WINDOW\ (TBLSET)}$. The function *must* be entered using the Y= Editor.

Example **1** **Generating a Table Automatically**

Display a table of values for $f(x) = \frac{2}{3}x - 2$, $x = -5, -4, -3, \ldots$.

▶Solution

1. Press $\boxed{\text{Y=}}$ and enter the function as Y_1:

$\boxed{(}$ **2** $\boxed{÷}$ **3** $\boxed{)}$ $\boxed{\text{X, T, }\theta\text{, }n}$ $\boxed{-}$ **2**. See Figure A.6.1.

2. Press $\boxed{\text{2ND}}$ $\boxed{\text{WINDOW (TBLSET)}}$. Fill in the following:

▶ `Tblstart=` $\boxed{(-)}$ **5**; $\triangle$`Tbl=` **1**

▶ Highlight `Auto` for both `Indpnt:` and `Depend:` options. This sets the beginning X value and the change in each X value. The highlighted `Auto` option for the independent variable will automatically generate the X values.

▶ Press $\boxed{\text{2ND}}$ $\boxed{\text{GRAPH (TABLE)}}$. See Figure A.6.2.

Figure A.6.1

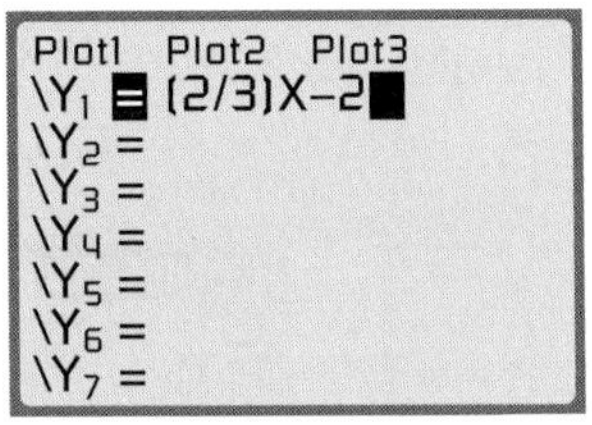

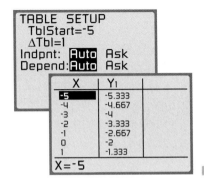

Figure A.6.2 Generating tables

Example 2 shows how to manually enter the values for the independent variable.

Figure A.6.3

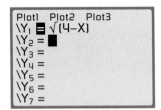

Figure A.6.4

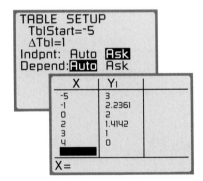

Example 2 **Generating a Table Manually**

Display a table of values for $f(x) = \sqrt{4 - x}$, $x = 4, 3, 2, 0, -1, -5$.

▶Solution

1. In the Y= Editor, enter the function as Y_1. See Figure A.6.3.

$\boxed{\text{2ND}}$ $\boxed{x^2(\sqrt{\ \ })}$ **4** $\boxed{-}$ $\boxed{\text{X, T, }\theta\text{, }n}$ $\boxed{)}$

2. Press $\boxed{\text{2ND}}$ $\boxed{\text{WINDOW (TBLSET)}}$. Fill in the following:

▶ `Tblstart=` $\boxed{(-)}$ **5**; $\triangle$`Tbl=` **1**

▶ Highlight `Ask` for `Indpnt` and `Auto` for `Depend`. The `Ask` option for the independent variable is highlighted to manually generate the X values. See Figure A.6.4.

3. Press $\boxed{\text{2ND}}$ $\boxed{\text{GRAPH (TABLE)}}$. The table is displayed with no entries. Move the cursor to the first entry position in the X input column to enter the X values. Press

$\boxed{(-)}$ **5** $\boxed{\text{ENTER}}$ $\boxed{(-)}$ **1** $\boxed{\text{ENTER}}$ **0** $\boxed{\text{ENTER}}$ **2** $\boxed{\text{ENTER}}$ **3** $\boxed{\text{ENTER}}$ **4** $\boxed{\text{ENTER}}$.

As each input value is entered, the corresponding output value appears in the Y_1 column. See Figure A.6.4. ■

A.7 Graphing Linear, Quadratic, and Piecewise-Defined Functions

To graph a function on a calculator, first enter the function in the Y= Editor. Then choose a window setting by specifying the minimum and maximum values of x and y. You can use a table of values to help you determine these values. Window settings are abbreviated as [Xmax, Xmin] (Xscl) by [Ymax, Ymin] (Yscl). Xscl and Yscl define the distance between the tick marks on the x- and y-axes, respectively. If Xscl $= 1$ or Yscl $= 1$, these values will be omitted from the keystroke sequence.

Example 1 Graphing a Function

Graph $y = \frac{2}{3}x - 2$ by using a table of values to choose an appropriate window.

▶Solution

Figure A.7.1

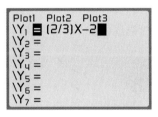

1. Enter the function. Press $\boxed{Y=}$ $\boxed{(}$ **2** $\boxed{\div}$ **3** $\boxed{)}$ $\boxed{X, T, \theta, n}$ $\boxed{-}$ **2** to enter the function as Y₁ in the equation editor. See Figure A.7.1.

2. Generate the table. Press $\boxed{2ND}$ $\boxed{WINDOW (TBLSET)}$ and fill in the following: Tblstart= -5; $\triangle$Tbl= **1**; highlight Auto for both Indpnt: and Depend: and press $\boxed{2ND}$ $\boxed{GRAPH (TABLE)}$. See Figure A.7.2.

3. Scroll through the table. The function crosses the x-axis at $(3, 0)$ and the y-axis at $(0, -2)$. These values must be displayed in the window. A window size of $[-7, 7]$ by $[-7, 5]$ will show the x- and y-intercepts and also give a good view of the graph. Other choices are also possible.

4. Enter the window dimensions by pressing $\boxed{WINDOW}$. Then set Xmin= $\boxed{(-)}$ 7, Xmax= 7, Xscl= **1** and Ymin= $\boxed{(-)}$ 7, Ymax= 5, Yscl= **1**. Press $\boxed{GRAPH}$. The graph in Figure A.7.3 is displayed.

Figure A.7.2   **Figure A.7.3**

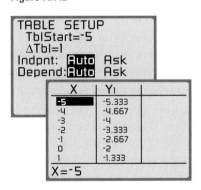   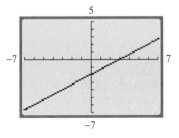

You can also graph a function by using built-in window settings accessed from the $\boxed{ZOOM}$ menu. This will be discussed in the following section.

Example 2 Built-in Window Settings

Graph the lines $y = -\frac{2}{3}x + \frac{1}{3}$ and $y = \frac{3}{2}x - 7$ using the decimal, standard, and square window settings.

▶Solution

1. Press $\boxed{Y=}$ and enter $\boxed{(-)}$ **2** $\boxed{\div}$ **3** $\boxed{X, T, \theta, n}$ $\boxed{+}$ **1** $\boxed{\div}$ **3** $\boxed{ENTER}$ for Y₁.

2. For Y₂, enter **3** $\boxed{\div}$ **2** $\boxed{X, T, \theta, n}$ $\boxed{-}$ **7** $\boxed{ENTER}$.

3. The keystrokes for each type of window are summarized in the following table.

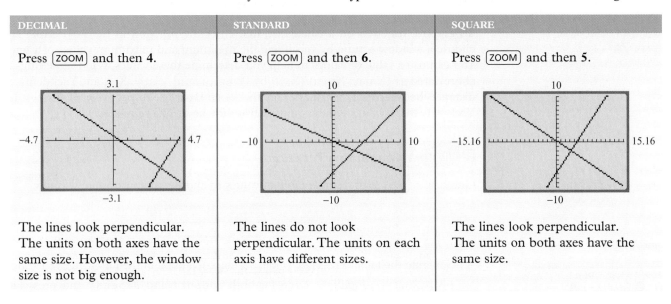

DECIMAL	STANDARD	SQUARE
Press ZOOM and then **4**.	Press ZOOM and then **6**.	Press ZOOM and then **5**.

The lines look perpendicular. The units on both axes have the same size. However, the window size is not big enough.

The lines do not look perpendicular. The units on each axis have different sizes.

The lines look perpendicular. The units on both axes have the same size.

To see the true shape of a circle, an ellipse, or any other figure, you will need to use a decimal or a square window. ▪

Example 3 Graphing a Quadratic Function

Graph the function $f(x) = x^2 + x - 12$.

▶Solution

1. Enter the function as Y₁ in the equation editor by pressing

$$\boxed{Y=}\ \boxed{X, T, \theta, n}\ \boxed{x^2}\ \boxed{+}\ \boxed{X, T, \theta, n}\ \boxed{-}\ \textbf{12}.$$

If you simply press ZOOM **6** to graph in a standard window, you will not obtain a complete picture. Part of the parabola will be cut off. See Figure A.7.4. Since the graph of this function is a parabola, the vertex should be visible on your graph.

Figure A.7.4

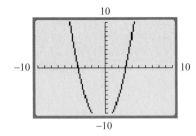

2. Generate a table of values to help you locate the vertex. Press 2ND WINDOW (TBLSET) and fill in the following: TblStart= (−)**10**, △Tbl= **1**, highlight Auto for both Indpnt: and Depend:. Press 2ND GRAPH (TABLE). Refer to Section A.6 for more details. Scrolling through the table, you can see that the Y₁ values decrease until X = 0 and then start to increase. The vertex is near the point (0, −12). See Figure A.7.5.

3. Press WINDOW and enter window settings such as [−10, 10] by [−14, 10] or something similar. See Figure A.7.6.

4. Press GRAPH and a complete graph is displayed. See Figure A.7.7.

Figure A.7.5   **Figure A.7.6**   **Figure A.7.7**

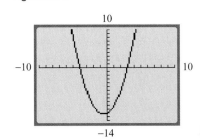

Figure A.7.8

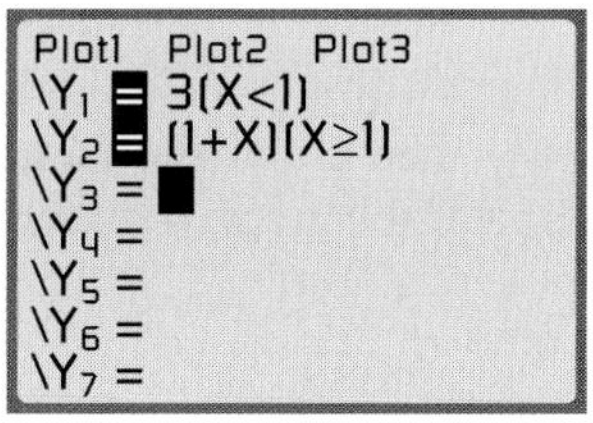

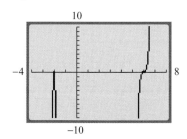

Example 4 Graphing a Piecewise-Defined Function

Graph the piecewise-defined function $H(x) = \begin{cases} 3, & x < 1 \\ 1 + x, & x \geq 1 \end{cases}$.

▶Solution

1. Assign a different name to each piece of the function in the Y= Editor and state the x values for which each piece is defined. In the Y= Editor, enter

$$Y_1 = 3 \;(\!(\; \boxed{\text{X, T, }\theta\text{, }n} \; \boxed{\text{2ND}} \; \boxed{\text{MATH (TEST)}} .$$

Press **5** to choose 5:<. Then enter **1** $\boxed{)}$.

Enter $Y_2 = \boxed{(}\; 1 \;\boxed{+}\; \boxed{\text{X, T, }\theta\text{, }n}\; \boxed{)}\; \boxed{(}\; \boxed{\text{X, T, }\theta\text{, }n}\; \boxed{\text{2ND}}\; \boxed{\text{MATH (TEST)}}$. Press **4** for 4:≥. Then enter **1** $\boxed{)}$. See Figure A.7.8.

2. Press $\boxed{\text{ZOOM}}$ **6** to get the graph. See Figure A.7.9.

3. At $x = 1$, the function jumps from 3 to 2, and the screen may show a line connecting these values. This is not part of the actual graph. To keep the calculator from connecting across the jump, press $\boxed{\text{MODE}}$ and set the mode to DOT. Now press $\boxed{\text{GRAPH}}$. See Figures A.7.10 and A.7.11.

Figure A.7.9

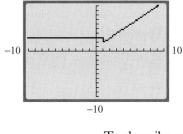

Figure A.7.10

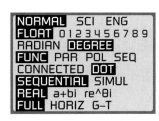

Figure A.7.11

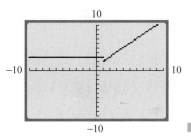

To describe an interval such as $1 \leq x \leq 3$ for a piecewise-defined function, you must rewrite the interval as $x \geq 1$ *and* $x \leq 3$. To enter *and*, press $\boxed{\text{2ND}}$ $\boxed{\text{MATH (TEST)}}$ and select the LOGIC submenu. Press **1** for 1:and.

A.8 Graphing Polynomials, Rational Functions, and Inequalities

For more complicated functions such as polynomials, you may have to try a few window settings before you get a reasonable view of the graph.

Figure A.8.1

Example 1 Graphing a Polynomial Function

Graph the function $f(x) = (x - 6)^3(x + 2)^2$.

▶Solution

1. In the Y= Editor, enter

$$\boxed{(}\; \boxed{\text{X, T, }\theta\text{, }n}\; \boxed{-}\; 6\; \boxed{)}\; \boxed{\wedge}\; 3\; \boxed{(}\; \boxed{\text{X, T, }\theta\text{, }n}\; \boxed{+}\; 2\; \boxed{)}\; \boxed{\wedge}\; 2$$

2. Since this polynomial has zeros at $x = 6$ and $x = -2$, choose Xmin and Xmax so that the zeros lie between them. Press $\boxed{\text{WINDOW}}$ and choose the settings $[-4, 8]$ by $[-10, 10]$. Press $\boxed{\text{GRAPH}}$. Note that a part of the graph is cut off. See Figure A.8.1.

3. Press (ZOOM) and press **0** for **0:Zoom Fit** to fit the *y* values. Now you can see the zeros and the end behavior of the function. See Figure A.8.2.

4. To get a better view of the middle portion of the graph, press (WINDOW) and set **Ymin= (-) 2000, Ymax= 2000, Yscl= 500** and press (GRAPH). The graph shown in Figure A.8.3 is displayed. The zeros, end behavior, and shape of the graph between the zeros are now visible on the screen.

Figure A.8.2

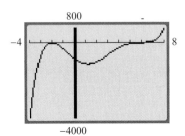

Figure A.8.3

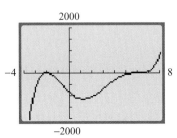

Care is required when using a graphing calculator to graph a rational function because of the presence of vertical asymptotes. By default, the calculator will graph in CONNECTED mode. This can give strange views when vertical asymptotes are present. This situation is illustrated in Example 2.

Example 2 Graphing a Rational Function

Graph $f(x) = \dfrac{1}{x-1}$ in a standard window. Change to DOT mode and graph again in the standard window.

▶Solution

1. Enter $Y_1 = 1/(X - 1)$ in the Y= Editor. Note that you *must* enclose the denominator X − 1 in parentheses. Press (ZOOM) and then **6** for **6:ZStandard** to get the graph in Figure A.8.4 in the standard window.

Figure A.8.4

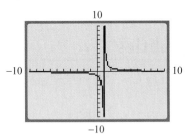

2. The vertical line that connects the negative and positive portions of the graph is *not* the vertical asymptote. To get around this, you can set the mode to DOT by pressing (MODE), selecting **DOT**, and then pressing (ENTER). Press (GRAPH) to display the function again in the standard window. See Figure A.8.5. You will no longer see the vertical line. However, the collection of dots results in poor resolution. Make sure you reset the mode to CONNECTED for graphing other types of functions.

Because the vertical asymptote is at $x = 1$, we can also use a decimal window in CONNECTED mode to display the graph. Press (ZOOM) and then **4**. Because the *x* values increase by 0.1 in a decimal window, the calculator will simply omit the value $x = 1$ and will not connect the positive and negative values. See Figure A.8.6.

Figure A.8.5

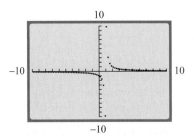

Figure A.8.6

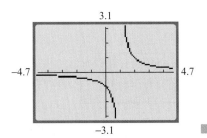

If you know that a vertical asymptote occurs at a number that is an integer multiple of 0.1, then using a decimal window in CONNECTED mode gives the best picture. For any other window, use the DOT mode.

Graphing Inequalities

You can shade the area above or below a function entered in the Y= Editor by using the marker setting to the left of the function.

| Example | 3 | Graphing a System of Inequalities |

Graph the following system of inequalities.

$$\begin{cases} y \geq x \\ y \leq -x \end{cases}$$

▶Solution To satisfy this system of inequalities, we must shade the area above $y = x$ and below $y = -x$.

1. In the Y= Editor, enter $\boxed{\text{X, T, }\theta\text{, }n}$ in Y_1 and then use the $\boxed{\triangleleft}$ key to move to the leftmost end of the screen. Press $\boxed{\text{ENTER}}$ $\boxed{\text{ENTER}}$ to activate the "shade above" command. See Figure A.8.7.

2. In the Y= Editor, enter $\boxed{(-)}$ $\boxed{\text{X, T, }\theta\text{, }n}$ in Y_2 and then use the $\boxed{\triangleleft}$ key to move to the leftmost end of the screen. Press $\boxed{\text{ENTER}}$ $\boxed{\text{ENTER}}$ $\boxed{\text{ENTER}}$ to activate the "shade below" command. See Figure A.8.7.

3. Press $\boxed{\text{ZOOM}}$ **6** to graph. The region in the xy plane satisfying both inequalities is shaded, along with both the horizontal and vertical lines. See Figure A.8.8.

Figure A.8.7

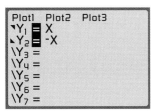

Figure A.8.8

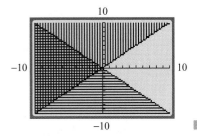

A.9 Solving Equations

There are two ways to solve an equation on a graphing calculator. One is by calculating the zero(s) of the corresponding functions and the other is by finding the intersection of the graphs of the two functions.

Solving an Equation Using the ZERO Feature

To solve an equation using the ZERO feature, you first must write the equation in the form $f(x) = 0$. You then proceed to find the zero(s).

| Example | 1 | Solving a Quadratic Equation Using the ZERO Feature |

Find all real number solutions of $3x^2 - 6x - 1 = 0$.

Figure A.9.1

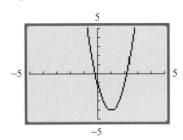

▶**Solution** Solving the equation is the same as finding the real zeros of $f(x) = 3x^2 - 6x - 1$.

1. In the Y= Editor, enter Y₁ = **3** $\boxed{\text{X, T, }\theta\text{, }n}$ $\boxed{x^2}$ $\boxed{-}$ **6** $\boxed{\text{X, T, }\theta\text{, }n}$ $\boxed{-}$ **1**.

2. Press $\boxed{\text{WINDOW}}$ and use a window setting of $[-5, 5]$ by $[-5, 5]$. If needed, refer to Section A.7 for more details. See Figure A.9.1.

3. Press $\boxed{\text{2ND}}$ $\boxed{\text{TRACE (CALC)}}$ to display the CALCULATE menu.

4. Press **2** for **2:zero** to find a real zero. You are prompted for a left bound. Move the arrow key to the left of one of the zeros and press $\boxed{\text{ENTER}}$. See Figure A.9.2.

5. You are now prompted for a right bound. Use $\boxed{▷}$ to move to the right of the zero on the graph and press $\boxed{\text{ENTER}}$. See Figure A.9.3.

Figure A.9.2   **Figure A.9.3**

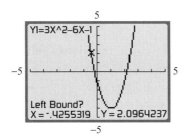   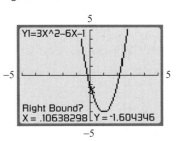

6. Now you will need a guess for the zero. See Figure A.9.4. Move the cursor very near to the zero on the graph and press $\boxed{\text{ENTER}}$. The zero is $x \approx -0.1547005$. See Figure A.9.5. The other zero, at $x \approx 2.1547005$, can be found similarly.

Figure A.9.4   **Figure A.9.5**

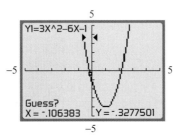   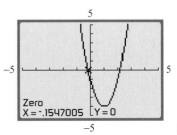

Solving an Equation Using the INTERSECT Feature

You can also solve an equation by using the **intersect** option from the CALCULATE menu.

Example **2** **Solving an Equation Using the INTERSECT Feature**

Solve the equation $2x + 1 = -3x + 11$.

▶**Solution** Solving this equation is the same as finding the intersection of the lines $y_1 = 2x + 1$ and $y_2 = -3x + 11$.

1. In the Y= Editor, enter Y₁ = $2x + 1$ and Y₂ = $-3x + 11$. Use the $\boxed{(-)}$ key to enter the negative sign. Press $\boxed{\text{ZOOM}}$ **6** to graph in the standard window. See Section A.7 for details.

2. Press (2ND) (TRACE (CALC)) to display the CALCULATE menu.

3. Press **5** for 5:intersect to find the intersection point(s).

4. Press (ENTER) when asked for the first and second curves. See Figure A.9.6.

5. Now you will need a guess for the intersection point. See Figure A.9.7. Move the cursor very near to the intersection point on the graph and press (ENTER). The intersection is at $x = 2$, $y = 5$, or (2, 5). See Figure A.9.8.

Figure A.9.6

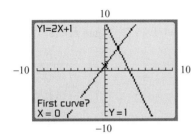

Figure A.9.7

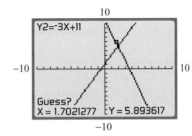

Figure A.9.8

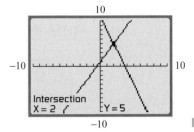

A.10 Finding the Maximum and Minimum of a Function

To obtain the maximum and minimum values of a function, also referred to as turning points or extrema, use the CALCULATE menu.

Example **1** **Determining Local Extrema**

Find the local maximum and minimum values of $f(x) = -2x^4 + 8x^2$.

▶Solution

1. Enter $Y_1 = -2x^4 + 8x^2$ in the Y= Editor and graph the function in the standard window. See Section A.7 for basic graphing details.

2. To get a better view of the locations of the maxima and minima, adjust the window size to $[-5, 5]$ by $[-10, 10]$. For other problems, you may have to adjust the window size first just to see the maxima and minima. See Figure A.10.1.

3. Each extremum must be computed separately. To calculate the maximum in the first quadrant, press (2ND) (TRACE (CALC)) and then **4** for 4:maximum. You are prompted for Left Bound. See Figure A.10.2.

4. Use the left or right arrow keys to move to the *left* of this maximum. Press (ENTER).

Figure A.10.1

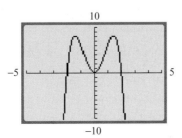

Figure A.10.2

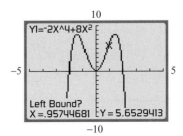

5. You are next prompted for **Right Bound**. Use the left or right arrow keys to move to the *right* of this maximum. Press (ENTER). Note that ▶ and ◀ show the interval on which the calculator program looks for a maximum. See Figure A.10.3.

6. You are now prompted for **Guess**. See Figure A.10.4. Move the cursor very close to the maximum on the graph and press (ENTER). The maximum is at $x \approx 1.414214$, $y = 8$. See Figure A.10.5. Similarly, the other maximum can be found at $x \approx -1.414214$, $y = 8$. You can also find that the minimum is at $x = 0$, $y = 0$.

Figure A.10.3

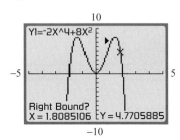

Figure A.10.4

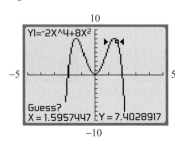

Figure A.10.5

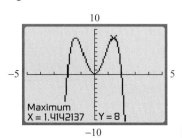

A.11 Complex Numbers

Figure A.11.1

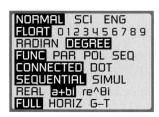

To enter complex numbers into the calculator, you first must set the mode to $a + bi$ by accessing the MODE menu. Move the arrow keys until $a + bi$ is highlighted. Press (ENTER). See Figure A.11.1. Press (2ND) (MODE (QUIT)) to exit the menu.

The imaginary number $i = \sqrt{-1}$ is accessed by pressing (2ND) (.(i)), on the bottom row of the keyboard. For instance, to enter $2 + 3i$ from the Home Screen, press **2** (+) **3** (2ND) (.(i)).

| Example **1** | **Operations with Complex Numbers** |

(a) Subtract: $(1 + i) - (2 - i)$

(b) Multiply: $(1 + 3i)(2 - 4i)$

(c) Find the conjugate of $1 + 2i$.

Figure A.11.2

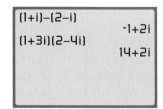

(d) Divide: $\dfrac{2}{-3 + 2i}$

▶Solution

All of the following operations are entered in the Home Screen.

(a) Enter (() **1** (+) (2ND) (.(i)) ()) (−) (() **2** (−) (2ND) (.(i)) ()) (ENTER). The result is $-1 + 2i$. See Figure A.11.2.

(b) Enter (() **1** (+) **3** (2ND) (.(i)) ()) (() **2** (−) **4** (2ND) (.(i)) ()) (ENTER). The result is $14 + 2i$. See Figure A.11.2.

(c) Press [MATH] [▷][▷] to highlight CPX. Press **1** for **1:conj(** and enter **1** [+] **2** [2ND] [.(i)] [)] [ENTER]. The result is $1 - 2i$. See Figures A.11.3 and Figure A.11.4.

(d) Enter **2** [÷] [(][(−)]**3** [+] **2** [2ND] [.(i)][)] [ENTER]. Press [MATH] **1** to access ▶**1:Frac** and press [ENTER]. The result is $\dfrac{-6}{13} - \dfrac{4}{13}i$. See Figure A.11.5.

Figure A.11.3   **Figure A.11.4**   **Figure A.11.5**

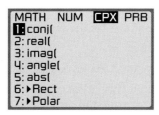   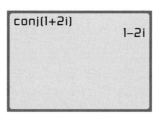   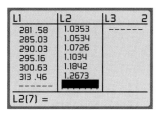

A.12 Fitting Curves to Data (Regression)

You can use the graphing calculator to fit a line or curve through a set of data points. This procedure is referred to as **curve-fitting** or **regression.**

Example 1 Modeling the Relation Between Body Weight and Organ Weight

Body Weight (grams)	Heart Weight (grams)
281.58	1.0353
285.03	1.0534
290.03	1.0726
295.16	1.1034
300.63	1.1842
313.46	1.2673

The table at the left gives the body weights of laboratory rats and the corresponding weights of their hearts, in grams. All data points are given to five significant digits. (*Source:* NASA Life Sciences Data Archive, 2005)

Find an expression for the *linear* function that best fits the given data points, and graph the function.

▶**Solution**

1. Enter the data into the calculator as follows.

    (a) Press [STAT] **1** to display the list editor. Clear any existing data from each list by pressing [△] [CLEAR] [ENTER].

    (b) Highlight the first entry position in L1. Enter the values for the independent variable, body weight, here. Enter **281.58** [ENTER] **285.03** [ENTER] **290.03** [ENTER] **295.16** [ENTER] **300.63** [ENTER] **313.46** [ENTER].

    (c) Use the arrow keys to highlight the first position in L2. Enter **1.0353** [ENTER] **1.0534** [ENTER] **1.0726** [ENTER] **1.1034** [ENTER] **1.1842** [ENTER] **1.2673** [ENTER]. The table in Figure A.12.1 is displayed.

2. Construct the graph.

    (a) Press [Y=] and turn off or clear any functions.

    (b) Press [2ND] [Y= (STAT PLOT)] **1** to select **1:Plot1.**

Figure A.12.1

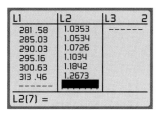

Figure A.12.2

(c) Highlight 0n and press (ENTER) to turn Plot 1 on. See Figure A.12.2.

(d) Highlight the following selections and press (ENTER). See Figure A.12.2.

Type:	Scatter Plot
	(the first icon)
Xlist:	L₁
Ylist:	L₂
Mark:	∘

Figure A.12.3

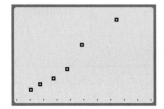

(e) Press (ZOOM) 9 for 9:ZoomStat for the scatter plot of the data. See Figure A.12.3.

3. Find the linear function of best fit.

(a) Press (STAT) (▷) to display the CALCULATE menu.

(b) Press 4 to select regression type 4:LinReg(ax+b) and then press (ENTER). The regression coefficients and equation are displayed. See Figure A.12.4. The linear equation of best fit is

$$y \approx 0.0075854x - 1.1131.$$

4. To plot the line of best fit, you must first copy the equation into the Y= Editor and then graph.

(a) Press (Y=) to display the equation editor. Move to an empty space.

(b) Press (VARS) 5 to access 5:Statistics. Press (▷)(▷) to move to EQ and then press 1 for 1:RegEQ. The current regression equation will be copied to the Y= Editor.

(c) Press (GRAPH). See Figure A.12.5.

Figure A.12.4

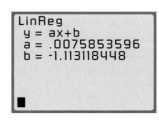

Figure A.12.5

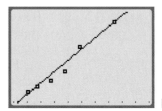

To fit functions other than linear functions to a data set, follow the same steps as in the previous example, but use the corresponding number for the desired regression type in Step 3(b). For example, you can find a quadratic function of best fit by pressing 5 in Step 3(b).

A.13 Matrices

To enter matrices, access the matrix menu by pressing (2ND) (x⁻¹ (MATRIX)). (On the TI-83, enter (2nd) (x⁻¹ (MATRX)).)

Example 1 Entering Matrices

Enter the matrix $\begin{bmatrix} 3 & 1 & -10 & -8 \\ 1 & 1 & -2 & -4 \\ -2 & 0 & 9 & 5 \end{bmatrix}$ as matrix A.

▶Solution

1. Press [2ND] [x⁻¹ (MATRIX)] [▷] [▷] to access the EDIT submenu.
2. Press **1** to select **1:[A]** and press [ENTER]. The row dimension is highlighted. Now press **3** [ENTER]. Next, the column dimension is highlighted. Press **4** [ENTER]. You have requested a 3×4 matrix. See Figure A.13.1.

Figure A.13.1

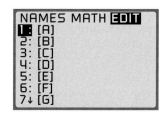 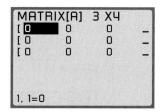

Figure A.13.2

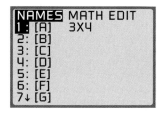

3. The first position is highlighted. Enter the individual elements of the matrix as follows.
   (a) Press **3** [ENTER].
   (b) The second position in row 1 is highlighted. Press **1** [ENTER].
   (c) The third position in row 1 is highlighted. Press [(−)] **10** [ENTER].

   Continue in this manner until all the elements have been entered. The entire matrix does not fit on one screen. See Figure A.13.2. Press [▷] to view the last column of the matrix.

4. Press [2ND] [MODE (QUIT)] to save the matrix. To access the matrix A from the Home Screen, press [2ND] [x⁻¹ (MATRIX)]. Select the NAMES submenu. Press **1** to select **1:[A]** and then press [ENTER]. You can also just press [ENTER] [ENTER] since **1:** is already selected. The matrix A now appears on the Home Screen. See Figure A.13.3.

Figure A.13.3

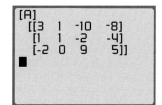

Gauss-Jordan Elimination

Your calculator will compute the reduced row echelon form, abbreviated rref, of a matrix using Gauss-Jordan elimination.

Example 2 Gauss-Jordan Elimination

Solve the following system of equations.

$$3x + y - 10z = -8$$
$$x + y - 2z = -4$$
$$-2x + 9z = 5$$

▶Solution

1. The augmented matrix for this system is the matrix from the previous example. Enter it as matrix A if you have not already done so.

2. From the Home Screen, Press [2ND] [x⁻¹ (MATRIX)]. Press [▷] to highlight MATH. Press [ALPHA] **B** to access the command B:rref. You will see rref on the Home Screen. See Figure A.13.4.

3. Input the name of the matrix by pressing [2ND] [x⁻¹ (MATRIX)]. Select the NAMES submenu. Press **1** to select 1:[A]. Close the parentheses by pressing [)] [ENTER].

4. The matrix in Figure A.13.5 is displayed. You can read off the answers: $x = 2$, $y = -4$, and $z = 1$.

Figure A.13.4

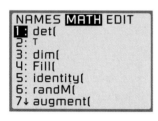

Figure A.13.5

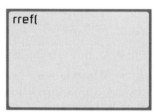

Matrix Arithmetic and Inverses

To perform arithmetic with matrices on the calculator, first enter and store the matrices in the MATRIX menu, as shown in Example 1. The arithmetic operations are performed on the Home Screen, with the matrices pasted in from the MATRIX ▶ NAMES submenu.

Example 3 **Matrix Arithmetic**

Let A, B, and C be given as follows.

$$A = \begin{bmatrix} 1 & -1 & 0 \\ 2 & -5 & 1 \end{bmatrix}, \quad B = \begin{bmatrix} 2 & -2 \\ -1 & 4.2 \\ 2 & -6 \end{bmatrix}, \quad C = \begin{bmatrix} 0 & -4 \\ 2 & 6 \\ 0 & 1 \end{bmatrix}$$

Find the following.

(a) B + C   (b) AC

▶Solution Enter the matrices A, B, and C using the technique described in Example 1. Then do the following.

(a) On the Home Screen, press [2ND] [x⁻¹ (MATRIX)]. Choose B in the NAMES submenu by pressing **2**. Press [+]. Press [2ND] [x⁻¹ (MATRIX)] and press **3** to choose C in the NAMES submenu. Press [ENTER] to perform the computation. See Figure A.13.6. An error message will be displayed if you try to add or subtract matrices of unequal dimensions.

Figure A.13.6

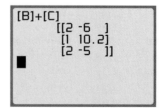

(b) On the Home Screen, press $\boxed{\text{2ND}}$ $\boxed{\text{x}^{-1} \text{ (MATRIX)}}$. Choose A in the NAMES submenu by pressing **1**. Press the $\boxed{\times}$ sign. Press $\boxed{\text{2ND}}$ $\boxed{\text{x}^{-1} \text{ (MATRIX)}}$ and press **3** to choose C in the NAMES submenu. Press $\boxed{\text{ENTER}}$ to perform the computation. See Figure A.13.7.

Figure A.13.7

```
[A]*[C]
      [[ -2 -10]
       [ -10 -37]]
```

Matrix Inverses and Determinants

Example 4 **Finding a Matrix Inverse and a Determinant**

Find the inverse and determinant of the square matrix

$$A = \begin{bmatrix} 3 & 3 & 9 \\ 1 & 0 & 2 \\ -2 & 3 & 0 \end{bmatrix}.$$

▶**Solution**

1. Enter the matrix A as directed in Example 1. On the Home Screen, display A by pressing $\boxed{\text{2ND}}$ $\boxed{\text{x}^{-1} \text{ (MATRIX)}}$ and then choose A by pressing **1**. See Figure A.13.8.

2. To find the inverse, press $\boxed{\text{x}^{-1} \text{ (MATRIX)}}$ $\boxed{\text{ENTER}}$. Only part of the matrix shows on the screen. Use the left and right arrow keys to scroll through the rest of the matrix.

3. To change the decimals to fractions, press $\boxed{\text{MATH}}$ **1** for ▶1:Frac and then press $\boxed{\text{ENTER}}$. See Figure A.13.9.

4. To check your work, store the inverse in B as follows. Press $\boxed{\text{2ND}}$ $\boxed{(-) \text{ (ANS)}}$ $\boxed{\text{STO} \blacktriangleright}$ $\boxed{\text{2ND}}$ $\boxed{\text{x}^{-1} \text{ (MATRIX)}}$. Press **2** to store in 2:[B] and press $\boxed{\text{ENTER}}$.

   Multiply matrix A by its inverse B as instructed in the section on matrix arithmetic. You should see the identity matrix.

5. To compute det(A), press $\boxed{\text{2ND}}$ $\boxed{\text{x}^{-1} \text{ (MATRIX)}}$. Press $\boxed{\triangleright}$ to highlight MATH. Then press **1** for 1:det(. Display the matrix name A as in Step 1, and press $\boxed{)}$ $\boxed{\text{ENTER}}$. The answer is -3. See Figure A.13.10.

Figure A.13.8

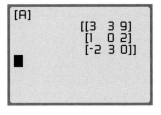

```
[A]
      [[3  3 9]
       [1  0 2]
       [-2 3 0]]
■
```

Figure A.13.9

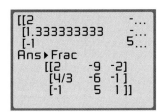

```
[[2              -...
 [1.333333333    -...
 [-1             5...
Ans▶Frac
      [[2   -9 -2]
       [4/3 -6 -1]
       [-1   5  1]]
```

Figure A.13.10

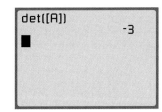

```
det([A])
                -3
■
```

A.14 Sequences and Series

To plot sequences, you must change a setting in the MODE menu:

1. Press MODE ▽ ▽ ▽ ▷ ▷ ▷ to highlight Seq.
2. Press ENTER 2ND MODE (QUIT). Remember to change back to FUNCTION mode once you finish working with sequences.

Figure A.14.1

Example 1 Entering and Graphing a Sequence

Graph the sequence defined by $u(n) = 100 + 5n$, $n = 0, 1, 2, 3, \ldots, 30$.

▶Solution Make sure you have set the calculator to SEQUENCE mode as outlined in the beginning of this section.

1. In the Y= Editor, set nMin to 0. Then enter $100 + 5n$ for $u(n)$. The variable n is entered by pressing X,T,θ,n. Since $n = 0$ is the minimum n value, enter $u(nMin) = 100$. See Figure A.14.1.
2. Press 2ND GRAPH (TABLE) to see a table of values, as shown in Figure A.14.2.
3. The plot will start at $n = 0$ and end at $n = 30$. From the table, the $u(n)$ values range from 100 to 250. Enter the data as shown in the following screens and press GRAPH. See Figure A.14.3.

Figure A.14.2

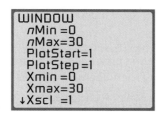

Figure A.14.3

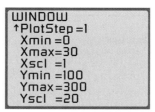

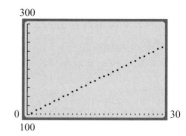

Example 2 Summing the Terms of a Sequence

Find the sum of the first 11 terms of the sequence defined by $a_j = 2 + 3j$, $j = 0, 1, 2 \ldots$.

▶Solution From the Home Screen, press 2ND STAT (LIST) ▷ ▷ 5 to choose the sum function. Then press 2ND STAT (LIST) ▷ 5 to choose the seq function. Enter

2 + 3 X,T,θ,n , X,T,θ,n , 0 , 10)) ENTER.

The answer is 187. See Figure A.14.4.

Figure A.14.4

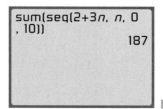

Answers to Check It Out Exercises

Chapter P, Section 1

Check It Out 1 (page 3)

Distributive property of multiplication over addition

Check It Out 2 (page 5)

All real numbers greater than or equal to 3 and less than or equal to 5

Check It Out 3 (page 6)

$[-5, 4]$

Check It Out 4 (page 6)

-1

Check It Out 5 (page 7)

9

Check It Out 6 (page 8)

18

Check It Out 7 (page 9)

-6

Chapter P, Section 2

Check It Out 1 (page 12)

460; In 2011, there will be 460 students in the elementary school.

Check It Out 2 (page 13)

$\dfrac{y^3}{x^2}$

Check It Out 3 (page 14)

$\dfrac{256x^2}{y^{10}}$

Check It Out 4 (page 15)

3.15×10^{-2}

Check It Out 5 (page 15)

0.000705

Check It Out 6 (page 16)

198 people per square kilometer

Check It Out 7 (page 17)

Four

Check It Out 8 (page 17)

110 miles

Chapter P, Section 3

Check It Out 1 (page 20)

3

Check It Out 2 (page 21)

$\dfrac{5\sqrt{2}}{2}$

Check It Out 3 (page 22)

$3\sqrt{5}; \; xy^2\sqrt[3]{x^2}$

Check It Out 4 (page 22)

$-2 + 2\sqrt{3}$

Check It Out 5 (page 23)

$(1 + x)\sqrt{x}$

Check It Out 6 (page 23)

-2

Check It Out 7 (page 24)

(a) 4 **(b)** 64 **(c)** $\dfrac{1}{4}$

Check It Out 8 (page 25)

(a) 32 **(b)** $\dfrac{512s^5}{t^{1/2}}$

Chapter P, Section 4

Check It Out 1 (page 28)

$4x^5 - 3x^2 + 7$; degree: 5; terms: $4x^5, -3x^2, 7$; coefficients: $a_5 = 4, a_2 = -3, a_0 = 7$; constant term: $a_0 = 7$

Check It Out 2 (page 29)

$x^4 + x^3 + 3x^2 + 5x - 4$

Check It Out 3 (page 29)

$-12x^9$

Check It Out 4 (page 30)

$2x^2 + 5x - 12$

Check It Out 5 (page 30)

$6x^2 - 13x - 5$

Check It Out 6 (page 31)

$3y^3 + 3y^2 - 13y + 15$

Chapter P, Section 5

Check It Out 1 (page 34)

$5y(1 + 2y - 5y^2)$

Check It Out 2 (page 34)

$(x + 4)(x + 3)$

Check It Out 3 (page 36)

$2(x + 5)(x - 1)$

Check It Out 4 (page 37)

$(2x - 3)(x + 2)$

Check It Out 5 (page 38)

$4(y + 5)(y - 5)$

Chapter P, Section 6 _____

Check It Out 1 (page 41)

$x \neq 1, -1$

Check It Out 2 (page 41)

$\dfrac{x - 2}{x + 3}$

Check It Out 3 (page 42)

$\dfrac{(x + 1)(x + 2)}{x - 2}$

Check It Out 4 (page 43)

$\dfrac{x + 3}{x}$

Check It Out 5 (page 44)

$\dfrac{-2x + 1}{x^2 - 4}$

Check It Out 6 (page 45)

$\dfrac{x}{y(x - y)}$

Check It Out 7 (page 45)

$\dfrac{2y + 1}{x - 2y}$

Chapter P, Section 7 _____

Check It Out 1 (page 49)

$36 + \dfrac{9}{2}\pi \approx 50.137\,\text{square inches}$

Check It Out 2 (page 50)

$V = \dfrac{1}{3}\pi(1.5)^2(5) \approx 11.781\,\text{cubic inches}$

Check It Out 3 (page 51)

$\sqrt{2}$

Check It Out 4 (page 51)

$2\sqrt{13}\,\text{feet} \approx 7.21\,\text{feet}$

Chapter P, Section 8 _____

Check It Out 1 (page 54)

$x = 6$

Check It Out 2 (page 54)

$x = 5$

Check It Out 3 (page 55)

$x = \dfrac{2}{3}$

Check It Out 4 (page 55)

$2l + 2w = 20;\; w = 10 - l$

Chapter 1, Section 1 _____

Check It Out 1 (page 69)

(a) 165 miles

(b) $d(8.25) = 20(8.25) = 165$

Check It Out 2 (page 70)

(a) This is not a function.

(b) This is a function.

Check It Out 3 (page 71)

This table represents a function because each input value has only one corresponding output value.

Check It Out 4 (page 72)

(a) 1

(b) $-a^2 - 2a + 1$

(c) $-x^4 + 2$

Check It Out 5 (page 72)

$-\dfrac{3}{5}$

Check It Out 6 (page 73)

The rate is $1.06.

Check It Out 7 (page 74)

$T(\text{August}) \approx 80°\text{F}$

Check It Out 8 (page 75)

(a) $(-\infty, \infty)$

(b) $[4, \infty)$

(c) $\left(-\infty, -\dfrac{1}{2}\right) \cup \left(-\dfrac{1}{2}, \infty\right)$

(d) $(4, \infty)$

Chapter 1, Section 2 _____

Check It Out 1 (page 80)

1980

Check It Out 2 (page 81)

This set of points does not define a function.

Check It Out 3 (page 82)

The domain of the function is $(-\infty, \infty)$. The range of the function is $(-\infty, \infty)$.

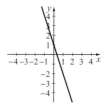

Check It Out 4 (page 83)

The domain of the function is $(-\infty, \infty)$. The range of the function is $[-2, \infty)$.

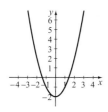

Check It Out 5 (page 84)

The domain of the function is $[4, \infty)$. The range of the function is $[0, \infty)$.

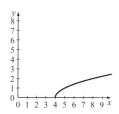

Check It Out 6 (page 85)

No. The graph does not pass the vertical line test.

Check It Out 7 (page 86)

$f(1) \approx 1\frac{1}{2}$ and $f(-3) = 1$

Chapter 1, Section 3

Check It Out 1 (page 91)

(a) Let the variables be defined as follows:
Input variable: x (amount of sales generated in one week, in dollars)
Output variable: $P(x)$ (pay for that week, in dollars)
(b) Eduardo's pay for a given week consists of a fixed portion, $500, plus a commission based on the amount of sales generated that week. Since he receives 15% of the sales generated, the commission portion of his pay is given by $0.15x$. Hence his total pay for the week is given by

Pay = fixed portion + commission portion

$P(x) = 500 + 0.15x$.

(c)

Sales, x	Pay, $P(x)$
0	500
500	575
1000	650
1500	725
2000	800

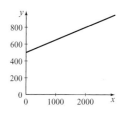

Check It Out 2 (page 91)

(a) $g(x) = -2x + \frac{1}{3}$ is a linear function, where $m = -2$ and $b = \frac{1}{3}$.

(b) $f(x) = x^2 + 4$ is not a linear function.

(c) $H(x) = 3x$ is a linear function, where $m = 3$ and $b = 0$.

Check It Out 3 (page 93)

$m = \frac{1}{14}$

Check It Out 4 (page 95)

The equation is $y = 2x - \frac{1}{3}$. The point $\left(\frac{1}{6}, 0\right)$ lies on the line because $0 = 2\left(\frac{1}{6}\right) - \frac{1}{3}$.

Check It Out 5 (page 96)

$y - 1 = \frac{1}{2}\left(x + \frac{3}{2}\right)$

Check It Out 6 (page 97)

$y = \frac{1}{2}x + \frac{7}{4}$

Check It Out 7 (page 97)

$y = \frac{3}{4}x - \frac{5}{4}$

Check It Out 8 (page 98)

$y = -1$

Check It Out 9 (page 99)

$x = 4$

Check It Out 10 (page 100)

$y = 4x + 6$

Check It Out 11 (page 101)

$y = \frac{1}{2}x - 2$

Chapter 1, Section 4 _____

Check It Out 1 (page 107)
(a) $B(t) = 1400 + 300t$
(b) $3800
(c) $t = 6$
(d) The slope of the line is 300. It signifies that Jocelyn's bonus will increase by $300 for each year that she works for the company. The y-intercept is $(0, 1400)$. It signifies that Jocelyn will receive a bonus of $1400 to start, at time $t = 0$. The graphical interpretation is shown below.

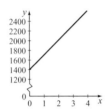

Check It Out 2 (page 108)
$f(x) = 2x + 2.5; f(-1) = 0.5$

Check It Out 3 (page 109)
(a) The input variable, t, is the number of years after purchase of the car. The output variable, v, is the value of the car after t years.
(b) $v(t) = -2000t + 14,000$
(c) $v(0) = 14,000$
(d) $t = 7$

Check It Out 4 (page 110)
(a) The graph of the function is shown below.

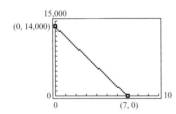

The values of t are greater than or equal to zero because the function represents the number of years since 2002.
(b) The y-intercept is $(0, 14,000)$. The original purchase price of the car is $14,000.
(c) $t = 7$
(d) At time $t = 7$, the value of the car is $0.

Check It Out 5 (page 112)
$h(275) = 0.97289$ gram

Check It Out 6 (page 113)
$k = \dfrac{7}{9}$

Check It Out 7 (page 114)
$k = 12,500$

Chapter 1, Section 5 _____

Check It Out 1 (page 120)
$t = 262.5$ minutes

Check It Out 2 (page 121)
$(1, -1)$

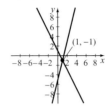

Check It Out 3 (page 123)
In interval notation, these values are $(2, \infty)$.

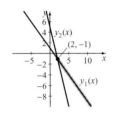

Check It Out 4 (page 124)
$(-2, \infty)$

Check It Out 5 (page 125)
$\left[-\dfrac{3}{11}, \infty \right)$

Check It Out 6 (page 126)
$p \leq 1111.11$

Check It Out 7 (page 127)
$x \geq 95.0°F$

Check It Out 8 (page 128)
$q > \dfrac{250}{3}$. Fewer pounds need to be sold when the price per pound is higher for revenue to exceed cost.

Chapter 2, Section 1 _____

Check It Out 1 (page 143)
(a) $5\sqrt{2}$

(b) $\left(-\dfrac{3}{2}, \dfrac{9}{2} \right)$

Check It Out 2 (page 144)
$(x - 4)^2 + (y + 1)^2 = 9$

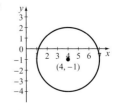

Check It Out 3 (page 144)

$(x - 3)^2 + (y + 1)^2 = 18$

Check It Out 4 (page 146)

$(x + 1)^2 + (y - 3)^2 = 16$. The center is $(-1, 3)$ and the radius is 4.

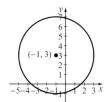

Check It Out 5 (page 147)

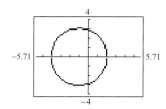

Chapter 2, Section 2 _____

Check It Out 1 (page 151)

$P(t) = 8 + 0.7t$

Check It Out 2 (page 153)

(a) $(f + g)(x) = \dfrac{2x^2 + 5x + 3}{x + 2}$

(b) $\left(\dfrac{f}{g}\right)(x) = \dfrac{1}{(2x + 1)(x + 2)}$

Check It Out 3 (page 154)

(a) 8

(b) $\dfrac{1}{3}$

Check It Out 4 (page 155)

$105

Check It Out 5 (page 157)

$(f \circ g)(-1) = 7$; $(g \circ f)(-1) = -3$

Check It Out 6 (page 158)

$f \circ g = \sqrt{x + 1}$. The domain of $f \circ g$ is $[-1, \infty)$.

Check It Out 7 (page 158)

A possible set of functions is $g(x) = x^3 + 9$ and $f(x) = x^5$.

Check It Out 8 (page 159)

$\dfrac{f(x + h) - f(x)}{h} = -2x - h$

Chapter 2, Section 3 _____

Check It Out 1 (page 164)

The domain of $f(x)$ and $g(x)$ is $(-\infty, \infty)$. The range of $f(x)$ is $[0, \infty)$, and the range of $g(x)$ is $[3, \infty)$.

x	$f(x)$	$g(x)$
-3	3	6
-2	2	5
-1	1	4
0	0	3
1	1	4
2	2	5
3	3	6

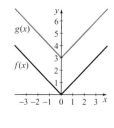

Check It Out 2 (page 165)

The domain of $f(x)$ and $g(x)$ is $(-\infty, \infty)$. The range of $f(x)$ and $g(x)$ is $[0, \infty)$.

x	$f(x)$	$g(x)$
-3	3	4
-2	2	3
-1	1	2
0	0	1
1	1	0
2	2	1
3	3	2

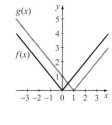

Check It Out 3 (page 167)

The graph is that of $g(x) = \sqrt{x}$ moved to the right by 2 and upward by 1.

Check It Out 4 (page 169)

(a) The graph is that of $f(x) = x^2$ stretched vertically away from the x-axis by a factor by 3 and reflected across the x-axis.

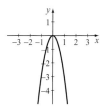

(b) The graph is that of $f(x) = |x|$ moved to the right by 1 unit, reflected across the x-axis, and moved upward by 2 units.

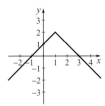

Check It Out 5 (page 170)

$g(x) = 3\sqrt{x + 2} + 1$

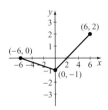

Check It Out 6 (page 173)

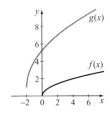

Chapter 2, Section 4

Check It Out 1 (page 181)

(a) Even **(b)** Neither odd nor even

Check It Out 2 (page 182)

f is decreasing on the intervals $(-2, 1)$ and $(2, 5)$.

Check It Out 3 (page 183)

14

Check It Out 4 (page 184)

$$\frac{\text{Increase in distance traveled}}{\text{Increase in amount of gasoline used}} = \frac{200 - 100}{10 - 5}$$

$$= \frac{100}{5}$$

$$= \frac{20 \text{ miles}}{\text{gallon}}$$

Chapter 2, Section 5

Check It Out 1 (page 189)

The solution set is $\left\{\frac{14}{5}, -2\right\}$.

Check It Out 2 (page 191)

$\left[-\frac{7}{2}, \frac{5}{2}\right]$

Check It Out 3 (page 192)

$|x - 6| \geq 3$

Check It Out 4 (page 193)

$|x - 49| > 5$

Chapter 2, Section 6

Check It Out 1 (page 196)

2.50; At 9 P.M., the fare is $2.50.

Check It Out 2 (page 196)

(a) $H(4) = -3$

(b) $H(-3) = 1$

Check It Out 3 (page 197)

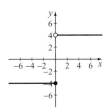

Check It Out 4 (page 198)

$$C(t) = \begin{cases} 40, & \text{if } x \leq 700 \\ 40 + 0.1(x - 700), & \text{if } x > 700 \end{cases}$$

Check It Out 5 (page 199)

$f(-2.5) = -3$

Chapter 3, Section 1

Check It Out 1 (page 214)

$A = 60l - l^2$

Check It Out 2 (page 216)

The domain of both functions is $(-\infty, \infty)$, and the range of both functions is $[0, \infty)$. Both graphs open upward, and the graph of $g(x)$ is scaled by a factor of 0.5 compared with the graph of $f(x)$.

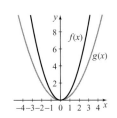

Check It Out 3 (page 217)

The lowest point on the graph is $(2, -1)$.

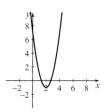

Check It Out 4 (page 218)

$f(x) = 3(x - 2)^2 - 5$. The vertex of the parabola is the point $(2, -5)$, which is a minimum point.

Check It Out 5 (page 220)

(a) The vertex of the function is $(1, 1)$.
(b) The axis of symmetry is $t = 1$.
(c) Two additional points on the graph are $(0, -2)$ and $(2, -2)$.
(d) The function is increasing on the interval $(-\infty, 1)$ and decreasing on the interval $(1, \infty)$. The range of the function is $(-\infty, 1]$.

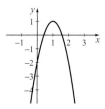

Check It Out 6 (page 221)

30 feet × 30 feet produces a maximum area of 900 square feet.

Check It Out 7 (page 222)

$c(12) = 0.0428(12)^2 - 0.5229(12) + 16.364 \approx 16.252\%$

Check It Out 8 (page 224)

The values produced by the model are off by less than 6% from the actual values.

Chapter 3, Section 2

Check It Out 1 (page 231)

$x = -\dfrac{2}{5}$ and $x = 1$

Check It Out 2 (page 233)

The x-intercepts are $\left(-\dfrac{1}{2}, 0\right)$ and $(2, 0)$. The zeros are $x = -\dfrac{1}{2}$ and $x = 2$.

Check It Out 3 (page 233)

$f(x) = x^2 + 2x - 8$

Check It Out 4 (page 234)

$x = \sqrt{5}$ and $x = -\sqrt{5}$

Check It Out 5 (page 235)

The zeros are $x = 1 + \dfrac{\sqrt{6}}{2}$ and $x = 1 - \dfrac{\sqrt{6}}{2}$. The x-intercepts are $\left(1 + \dfrac{\sqrt{6}}{2}, 0\right)$ and $\left(1 - \dfrac{\sqrt{6}}{2}, 0\right)$.

Check It Out 6 (page 237)

The zeros are $x = 1 + \dfrac{\sqrt{6}}{2}$ and $x = 1 - \dfrac{\sqrt{6}}{2}$. The x-intercepts are $\left(1 + \dfrac{\sqrt{6}}{2}, 0\right)$ and $\left(1 - \dfrac{\sqrt{6}}{2}, 0\right)$.

Check It Out 7 (page 238)

$x = 2$

Check It Out 8 (page 239)

There are no real solutions. The value of the discriminant is -11.

Check It Out 9 (page 240)

$t \approx 3.27$ seconds

Check It Out 10 (page 240)

Sometime during 1992 and sometime during 1998

Chapter 3, Section 3

Check It Out 1 (page 246)

$x = \pm 3i$

Check It Out 2 (page 247)

$5i, 6i\sqrt{3}, \dfrac{2}{3}i$

Check It Out 3 (page 247)

$-\dfrac{1}{2} \pm \dfrac{\sqrt{15}}{2}i$

Check It Out 4 (page 248)

$\sqrt[3]{5} + 0i, (-1 + \sqrt{2}) + 0i, 0 + \dfrac{1}{4}i$

Check It Out 5 (page 248)

The real part is $\sqrt[3]{5}$ and the imaginary part is 0; the real part is $(-1 + \sqrt{2})$ and the imaginary part is 0; the real part is 0 and the imaginary part is $\dfrac{1}{4}$.

Check It Out 6 (page 249)

(a) $-1 + 3i$ **(b)** $-2 + i$ **(c)** i

Check It Out 7 (page 250)

$-7 + 26i$

Check It Out 8 (page 251)

$-3 + 7i$

Check It Out 9 (page 251)

58; a real number

Check It Out 10 (page 251)

$-\dfrac{6}{29} + \dfrac{14}{29}i$

Check It Out 11 (page 252)

$s = \dfrac{1}{3} \pm \dfrac{\sqrt{2}}{3}i$; no x-intercepts

Check It Out 12 (page 253)

$t = \dfrac{3}{4} \pm \dfrac{\sqrt{7}}{4}i$

Chapter 3, Section 4

Check It Out 1 (page 257)

(a) $-2 \le x \le 3$, or $[-2, 3]$
(b) $x < -2$ or $x > 3$, or $(-\infty, -2) \cup (3, \infty)$

Check It Out 2 (page 259)

The solution set is $x \le -7$ or $x \ge 1$, or $(-\infty, -7] \cup [1, \infty)$.

Check It Out 3 (page 260)

The solution set is $x > \dfrac{3}{2}$ or $x < 1$, or $(-\infty, 1) \cup \left(\dfrac{3}{2}, \infty\right)$.

Check It Out 4 (page 261)

The solution set is $-\dfrac{3}{4} - \dfrac{\sqrt{17}}{4} \le x \le -\dfrac{3}{4} + \dfrac{\sqrt{17}}{4}$, or

$\left[-\dfrac{3}{4} - \dfrac{\sqrt{17}}{4}, -\dfrac{3}{4} + \dfrac{\sqrt{17}}{4}\right]$.

Check It Out 5 (page 262)

The solution set is $-\infty < x < \infty$, or $(-\infty, \infty)$.

Check It Out 6 (page 263)

(a) The revenue equation remains

$$R(q) = -0.1q^2 + 200q.$$

(b) The cost equation becomes

$$C(q) = 25,000 + 15q.$$

(c) Now that the profit equation is

$$P(q) = -0.1q^2 + 185q - 25,000$$

the values of q such that $-0.1q^2 + 185q - 25,000 > 0$ are $147 \le q \le 1703$ (to produce between 147 and 1703 units).

Chapter 3, Section 5

Check It Out 1 (page 266)

$x = \pm 1$ and $x = \pm \dfrac{\sqrt{6}}{2}i$

Check It Out 2 (page 267)

$t = \dfrac{\sqrt[3]{9}}{3}$ and $t = \sqrt[3]{-2}$

Check It Out 3 (page 268)

$x = 1$ and $x = -2$

Check It Out 4 (page 269)

$x = 1$ and $x = -1$

Check It Out 5 (page 270)

$x = 1$

Check It Out 6 (page 270)

$$\left(\dfrac{350 - 4x}{x}\right)(x + 10) = 350$$

$$\left(\dfrac{350 - 4(25)}{25}\right)(25 + 10) \overset{?}{=} 350$$

$$\left(\dfrac{250}{25}\right)(35) \overset{?}{=} 350$$

$$(10)(35) = 350$$

Check It Out 7 (page 272)

$x = 8.6524$ kilometers along the river from point B

Chapter 4, Section 1

Check It Out 1 (page 285)

$4x^3 - 32x^2 + 64x$

Check It Out 2 (page 285)

(a) Polynomial function of degree 0; $a_0 = 6$
(b) Polynomial function of degree 2; $a_2 = 1$
(c) Not a polynomial function

Check It Out 3 (page 288)

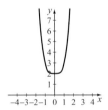

Check It Out 4 (page 291)

(a) Behaves like $y = -3x^3$: $h(x) \to -\infty$ as $x \to +\infty$; $h(x) \to +\infty$ as $x \to -\infty$.
(b) Behaves like $y = 2x^2$: $s(x) \to +\infty$ as $x \to +\infty$; $s(x) \to +\infty$ as $x \to -\infty$.

Check It Out 5 (page 291)

$x = -2, 0, 2; (-2, 0), (0, 0), (2, 0)$

Check It Out 6 (page 293)

$f(x) \to -\infty$ as $x \to +\infty$; $f(x) \to +\infty$ as $x \to -\infty$. $(0, 0)$,
$(-3, 0), (3, 0)$

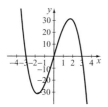

Check It Out 7 (page 294)

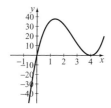

Check It Out 8 (page 295)

1,752,000

Chapter 4, Section 2 _____

Check It Out 1 (page 300)

$x = 0$ has multiplicity 2, so the graph will touch the x-axis at
$(0, 0)$. $x = 5$ has multiplicity 2, so the graph will touch the
x-axis at $(5, 0)$.

Check It Out 2 (page 301)

(a) Neither
(b) Odd

Check It Out 3 (page 302)

Local minimum at $(0.7746, -0.1859)$
Local maximum at $(-0.7746, 0.1859)$

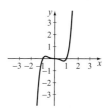

Check It Out 4 (page 303)

$f(x) = x^3 - 2x^2 - x + 2$

Check It Out 5 (page 305)

75.5 million people

Chapter 4, Section 3 _____

Check It Out 1 (page 309)

Quotient: $x + 3$; remainder: 0

Check It Out 2 (page 311)

Quotient: $3x^2 + x - 1$; remainder: 0

Check It Out 3 (page 312)

$3x^2 - 2x + 2 + \dfrac{-3}{x + 1}$

Check It Out 4 (page 313)

$-2x^2 + x + 1$

Check It Out 5 (page 313)

37

Check It Out 6 (page 314)

Yes; $p(1) = 0$

Chapter 4, Section 4 _____

Check It Out 1 (page 316)

$p(-2) = -16 + 4 + 10 + 2 = 0$
$p(x) = (x - 1)(2x - 1)(x + 2)$

Check It Out 2 (page 317)

Function	Zero	x-Intercept	Factor
$p(x)$	-6	$(-6, 0)$	$x + 6$
$h(x)$	4	$(4, 0)$	$x - 4$
$g(x)$	2	$(2, 0)$	$x - 2$

Check It Out 3 (page 319)

$x = -1, \pm\dfrac{1}{2}$

Check It Out 4 (page 319)

$(-2, 0), (-0.3028, 0), (3.3028, 0)$

Check It Out 5 (page 321)

$x = -2, 1, -\dfrac{3}{2}$

Check It Out 6 (page 322)

The number of positive zeros is 3 or 1. The number of negative zeros is 1.

Chapter 4, Section 5 _____

Check It Out 1 (page 326)

2

Check It Out 2 (page 326)

$h(t) = (t - 6)(t^2 + 5)$

Check It Out 3 (page 327)

$h(t) = (t - 6)(t + i\sqrt{5})(t - i\sqrt{5})$

Check It Out 4 (page 328)

$p(x) = -2(x + 2)^2(x - 1)^2(x - 4)$

Chapter 4, Section 6 _____

Check It Out 1 (page 331)

$A(x) = \dfrac{50}{x}$, where x is the number of miles driven per day.

Check It Out 2 (page 333)

$(-\infty, 0) \cup (0, \infty)$

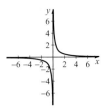

Check It Out 3 (page 335)

$x = \pm 3$

Check It Out 4 (page 337)

$y = 0$

Check It Out 5 (page 339)

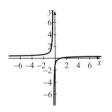

Check It Out 6 (page 341)

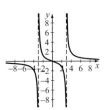

Check It Out 7 (page 342)

$y = x + 2$

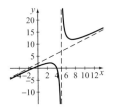

Check It Out 8 (page 343)

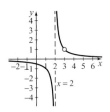

Chapter 4, Section 7 _____

Check It Out 1 (page 349)

$[-2, 0] \cup [2, \infty)$

Check It Out 2 (page 349)

$[5, \infty)$

Check It Out 3 (page 350)

$(-2, 1)$

Check It Out 4 (page 351)

$\left(-3, \dfrac{1}{2}\right]$

Chapter 5, Section 1 _____

Check It Out 1 (page 365)

$(G \circ L)(x) = \dfrac{(3.785x)}{3.785} = x$. Because gallons of fuel are converted to liters of fuel and then back to gallons, the input is the same as the output, where x represents gallons of fuel.

Check It Out 2 (page 366)

$$f(g(x)) = 2\left(\frac{x}{2} + \frac{9}{2}\right) - 9 = x + 9 - 9 = x$$

$$g(f(x)) = \frac{(2x - 9)}{2} + \frac{9}{2} = x - \frac{9}{2} + \frac{9}{2} = x$$

Check It Out 3 (page 367)

$$f^{-1}(x) = \frac{x}{4} + \frac{5}{4}$$

$$(f \circ f^{-1})(x) = f(f^{-1}(x)) = 4\left(\frac{x}{4} + \frac{5}{4}\right) - 5 = x + 5 - 5 = x$$

$$(f^{-1} \circ f)(x) = f^{-1}(f(x)) = \frac{(4x - 5)}{4} + \frac{5}{4} = x - \frac{5}{4} + \frac{5}{4} = x$$

Since $(f \circ f^{-1})(x) = (f^{-1} \circ f)(x) = x$, the inverse checks.

Check It Out 4 (page 369)

Assuming $f(a) = f(b)$, we must show that $a = b$. Substituting for $f(a) = f(b)$, $4a - 6 = 4b - 6$. Adding 6 to both sides, $4a = 4b$. Dividing both sides by 4, $a = b$. So, $f(x) = 4x - 6$ is one-to-one.

Check It Out 5 (page 370)

The graph of $f^{-1}(x) = \frac{x}{3} + \frac{2}{3}$ is a reflection of the graph of $f(x) = 3x - 2$ about the line $y = x$.

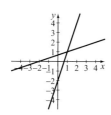

Check It Out 6 (page 371)

(a) To show that f is one-to-one, assume $f(a) = f(b)$ and determine if $a = b$. Assuming $-a^3 + 2 = -b^3 + 2$, subtract 2 from both sides. Then, multiply both sides of $-a^3 = -b^3$ by -1. Take the cube root of both sides of $a^3 = b^3$ to get $a = b$.
(b) $f^{-1}(x) = (2 - x)^{1/3}$

(c)

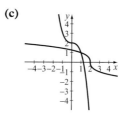

Check It Out 7 (page 372)

Due to the domain restriction $x \geq 0$, the function f passes the horizontal line test. The function is thus one-to-one and therefore has an inverse.
$f^{-1}(x) = \sqrt[4]{x} = x^{1/4}, x \geq 0.$

Chapter 5, Section 2 _____

Check It Out 1 (page 376)

$P(9) = 512$. This means that there are 512 bacteria present after 9 hours.

Check It Out 2 (page 378)

As $x \to +\infty$, the value of $g(x)$ gets very large. As $x \to -\infty$, the value of $g(x)$ gets extremely small, but never reaches zero. The range is $(0, \infty)$.

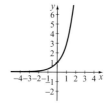

x	$g(x)$
-10	$\frac{1}{3^{10}} \approx 0.0000169$
-5	$\frac{1}{3^5} \approx 0.00412$
-2	$\frac{1}{3^2} \approx 0.1111$
-1	$\frac{1}{3^1} \approx 0.3333$
0	$3^0 = 1$
1	$3^1 = 3$
2	$3^2 = 9$
5	$3^5 = 243$
10	$3^{10} = 59,049$

Check It Out 3 (page 379)

As $x \to -\infty$, the value of $g(x)$ gets very large. As $x \to +\infty$, the value of $g(x)$ gets extremely small, but never reaches zero. The range is $(0, \infty)$.

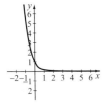

x	g(x)
−10	$\frac{1}{6^{-10}} = 60{,}466{,}176$
−5	$\frac{1}{6^{-5}} = 7776$
−2	$\frac{1}{6^{-2}} = 36$
−1	$\frac{1}{6^{-1}} = 6$
0	$6^0 = 1$
1	$6^{-1} \approx 0.1667$
2	$6^{-2} \approx 0.02778$
5	$6^{-5} \approx 0.000129$
10	$6^{-10} \approx 0.0000000165$

Check It Out 4 (page 380)

As $s \to +\infty$, the value of $h(s)$ continues to decrease. As $s \to -\infty$, the value of $h(s)$ gets increasingly close to zero, but never actually reaches zero. The domain is all real numbers. The range is $(-\infty, 0)$.

s	h(s)
−10	$-\frac{1}{3^{10}} \approx -0.0000169$
−5	$-\frac{1}{3^5} \approx -0.00412$
−2	$-\frac{1}{3^2} \approx -0.1111$
−1	$-\frac{1}{3^1} \approx -0.3333$
0	$-3^0 = -1$
1	$-3^1 = -3$
2	$-3^2 = -9$
5	$-3^5 = -243$
10	$-3^{10} = -59{,}049$

Check It Out 5 (page 382)

The y-intercept is $(0, 1)$. To determine the range of h, note that e raised to any power is positive, and every positive number can be expressed as e raised to a power. The domain is all real numbers and the range is all positive real numbers. As $x \to -\infty$, the value of $h(x)$ gets increasingly large. As $x \to +\infty$, the value of $h(x)$ gets increasingly close to zero, but never actually reaches zero.

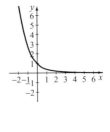

x	h(x)
−10	$e^{10} \approx 22{,}026.47$
−5	$e^5 \approx 148.413$
−2	$e^2 \approx 7.389$
−1	$e^1 \approx 2.7182$
0	$e^0 = 1$
1	$e^{-1} \approx 0.3679$
2	$e^{-2} \approx 0.1353$
5	$e^{-5} \approx 0.00674$
10	$e^{-10} \approx 0.0000454$

Check It Out 6 (page 383)

$v(t) = 12{,}000(0.9)^t$

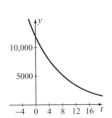

Check It Out 7 (page 384)

$162.89

Check It Out 8 (page 385)

(a) $3644.01
(b) $3645.93

Check It Out 9 (page 386)

$t \approx 11.58$ years

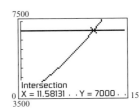

Chapter 5, Section 3 _____

Check It Out 1 (page 391)

$t = 6$

Check It Out 2 (page 393)

Logarithmic Statement	Exponential Statement
$\log_3 27 = 3$	$3^3 = 27$
$\log_4 \dfrac{1}{4} = -1$	$4^{-1} = \dfrac{1}{4}$

Check It Out 3 (page 393)

Exponential Statement	Logarithmic Statement
$4^3 = 64$	$\log_4 64 = 3$
$10^{1/2} = \sqrt{10}$	$\log_{10} \sqrt{10} = \dfrac{1}{2}$

Check It Out 4 (page 394)

(a) $\log_6 36 = 2$

(b) $\log_b b^{1/3} = \dfrac{1}{3}$

(c) $10^{\log_{10} 9} = 9$

Check It Out 5 (page 394)

$x = 125$

Check It Out 6 (page 395)

$\log 10^{2/3} = \dfrac{2}{3}$ and $\ln e^{4/3} = \dfrac{4}{3}$

Check It Out 7 (page 396)

Approximating the x-value corresponding to $f(x) = 8$ on a graph of $f(x) = e^x$ will provide an accurate value for $\ln 8$, although with less precision than the value produced by a calculator, which is $\ln 8 \approx 2.0794$.

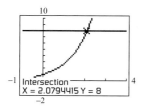

Check It Out 8 (page 397)

$\log_6 15 = \dfrac{\ln 15}{\ln 6} \approx 1.511$

Check It Out 9 (page 400)

The domain of $g(x)$ is $(0, \infty)$ and the range is $(-\infty, \infty)$. The vertical asymptote is $x = 0$.

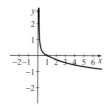

Check It Out 10 (page 401)

$I \approx 15,848,932 I_0$

Check It Out 11 (page 402)

$\log(9.3 \times 10^8) \approx 8.9685$

Chapter 5, Section 4 _____

Check It Out 1 (page 408)

$\log 2500 \approx 3.3979$. This is a close approximation to four decimal places of the value given by a calculator.

Check It Out 2 (page 409)

$\log 4 - \dfrac{1}{3} \log x + \dfrac{1}{2} \log y$

Check It Out 3 (page 410)

$\dfrac{1}{2} \log(x - 1) - \log(x^2 + 4)$

Check It Out 4 (page 411)

$\log_a \dfrac{x^3}{x^2 + 1}$

Check It Out 5 (page 412)

7.495

Chapter 5, Section 5

Check It Out 1 (page 415)

$t = 9$

Check It Out 2 (page 416)

$$10^{2x-1} = 3^x$$
$$\log_3 10^{2x-1} = \log_3 3^x$$
$$(2x - 1)(\log_3 10) = x$$
$$2x \log_3 10 - \log_3 10 = x$$
$$2x \log_3 10 - x = \log_3 10$$
$$x(2 \log_3 10 - 1) = \log_3 10$$
$$x = \frac{\log_3 10}{2 \log_3 10 - 1} \approx 0.6567$$

Check It Out 3 (page 416)

$t \approx 0.7324$

Check It Out 4 (page 417)

$t \approx 6.73$ years

Check It Out 5 (page 418)

$C(9) \approx 0.014$ cent per megabyte

Check It Out 6 (page 419)

$t \approx 23.22$ hours

Check It Out 7 (page 419)

$x = 32$

Check It Out 8 (page 420)

$x = 5$

Check It Out 9 (page 420)

$x \approx 3.164$

Check It Out 10 (page 421)

$x \approx 1.998$ weeks

Chapter 5, Section 6

Check It Out 1 (page 427)

$A(t) = A_0 e^{-0.11552t}$

Check It Out 2 (page 428)

Approximately 321 million people

Check It Out 3 (page 429)

$18,290 billion

Check It Out 4 (page 430)

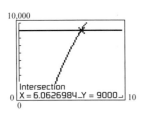

Check It Out 5 (page 431)

Compared to the projection of 484 million given by the model, the U.S. Census Bureau projection of 468 million differs by less than 4%.

Chapter 6, Section 1

Check It Out 1 (page 447)

No. $3(15) + 8(15) = 165 < 180$

Check It Out 2 (page 448)

$x = 5, y = 25$

Check It Out 3 (page 449)

$$y = -2x + 5$$
$$\underline{-(y = -2x + 2)}$$
$$0 = 3 \qquad \text{False statement}$$

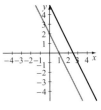

Check It Out 4 (page 451)

$(5, 15), (8, 12), (-10, 30), (20, 0)$

Check It Out 5 (page 453)

$(22, 50)$

Check It Out 6 (page 454)

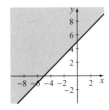

Check It Out 7 (page 456)

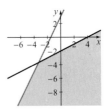

Check It Out 8 (page 457)

$$\begin{cases} x + y \le 30 \\ x \ge 8 \\ y \le 15 \\ x \ge 0 \\ y \ge 0 \end{cases}$$

Check It Out 9 (page 458)

$(6, 24)$; 210 calories

Chapter 6, Section 2 _____

Check It Out 1 (page 465)

No. These allocations do not satisfy $2x + 4y + 5z = 3.5$.

Check It Out 2 (page 466)

$x = 10, y = -1, z = -1$

Check It Out 3 (page 469)

$x = 1, y = 0, z = -1$

Check It Out 4 (page 470)

$x = -1, y = 3, z = -1$

Check It Out 5 (page 471)

$y = -10 - 3z, x = -13 - 4z$

Check It Out 6 (page 472)

No solution

Chapter 6, Section 3 _____

Check It Out 1 (page 478)

$$\left[\begin{array}{ccc|c} 0 & 0 & -2 & -1 \\ -2 & 4 & 1 & 0 \\ 3 & 4 & 2 & 5 \end{array}\right]$$

Check It Out 2 (page 480)

$x = 1, y = 0, z = -1$

Check It Out 3 (page 481)

$x = -2, y = 1, z = -\frac{5}{3}; x = 2, y = 3, z = \frac{1}{3}.$

Answers may vary.

Check It Out 4 (page 482)

No solution

Check It Out 5 (page 484)

$x = 1, y = 1, z = 0$

Check It Out 6 (page 485)

$x = 3 + 2u, y = -4 - u, z = 2u, u$ any real number

Check It Out 7 (page 487)

$w = 4, p = 6, b = 6$

Chapter 6, Section 4 _____

Check It Out 1 (page 492)

$$\begin{bmatrix} 5.0 & 4.4 \\ 4.9 & 4.6 \end{bmatrix}$$

Check It Out 2 (page 493)

3×3; 4; 0

Check It Out 3 (page 494)

(a) $\begin{bmatrix} 0 & 2 \\ 3.1 & 4.5 \\ -0.5 & 1 \end{bmatrix}$ (b) $\begin{bmatrix} 6 & -6 \\ -1.1 & 0.5 \\ -0.5 & -7 \end{bmatrix}$ (c) $\begin{bmatrix} -6 & 6 \\ 1.1 & -0.5 \\ 0.5 & 7 \end{bmatrix}$

Check It Out 4 (page 495)

$$\begin{array}{c} 2002 \\ 2003 \\ 2004 \end{array}\begin{array}{cc} \text{Baltimore} & \text{Annapolis} \\ \begin{bmatrix} 102,000 & 78,000 \\ 108,000 & 90,000 \\ 114,000 & 96,000 \end{bmatrix} \end{array}$$

Check It Out 5 (page 496)

6.38 million

Check It Out 6 (page 498)

9.5

Check It Out 7 (page 499)

$$\begin{bmatrix} 0 & 11 \\ 6 & -5 \end{bmatrix}$$

Check It Out 8 (page 500)

$$\begin{bmatrix} -8 & 20 & -4 \\ 14 & -32 & 6 \\ 2 & -5 & 1 \end{bmatrix}$$

Check It Out 9 (page 501)

Outfit	Total Cost of Fabric
1	$21.20
2	$24.60
3	$24.90
4	$28.70

Check It Out 10 (page 503)

$$\begin{bmatrix} 23 \\ -44 \\ -26 \\ -23 \end{bmatrix}\begin{bmatrix} 13 \\ -5 \\ -1 \\ -13 \end{bmatrix}$$

Chapter 6, Section 5 _____

Check It Out 1 (page 507)

$$AI = \begin{bmatrix} -2 & 3 \\ 4 & 7 \end{bmatrix}\begin{bmatrix} 1 & 0 \\ 0 & 1 \end{bmatrix} = \begin{bmatrix} -2 & 3 \\ 4 & 7 \end{bmatrix}$$

Check It Out 2 (page 510)

$$\begin{bmatrix} -2 & 5 \\ -1 & 3 \end{bmatrix}$$

Check It Out 3 (page 510)

$$\begin{bmatrix} -1 & 0 & -2 \\ 1 & 1 & 1 \\ 1 & 1 & 0 \end{bmatrix} \begin{bmatrix} -1 & -2 & 2 \\ 1 & 2 & -1 \\ 0 & 1 & -1 \end{bmatrix} = \begin{bmatrix} 1 & 0 & 0 \\ 0 & 1 & 0 \\ 0 & 0 & 1 \end{bmatrix}$$

Check It Out 4 (page 512)

$$\begin{bmatrix} -5 & -2 & 2 \\ 5 & 2 & -1 \\ 2 & 1 & -1 \end{bmatrix}$$

Check It Out 5 (page 513)

$$x = \frac{2}{15}, y = -\frac{17}{15}, z = \frac{4}{5}$$

Check It Out 6 (page 514)

$$\begin{bmatrix} 6 \\ 9 \\ 18 \\ 5 \end{bmatrix}; \text{FIRE}$$

Chapter 6, Section 6 _____

Check It Out 1 (page 519)

1

Check It Out 2 (page 520)

$-4, 4$

Check It Out 3 (page 521)

17

Check It Out 4 (page 523)

$$\left(-\frac{1}{2}, -\frac{3}{2} \right)$$

Check It Out 5 (page 523)

Yes. $D \neq 0$

Check It Out 6 (page 525)

$$x = \frac{4}{5}, y = \frac{8}{5}, z = -\frac{3}{5}$$

Chapter 6, Section 7 _____

Check It Out 1 (page 530)

$$\frac{1}{x} - \frac{2}{x-1}$$

Check It Out 2 (page 531)

$$-\frac{2}{x^2} + \frac{1}{x+3}$$

Check It Out 3 (page 532)

$b^2 - 4ac = (1)^2 - 4(2)(-3) > 0.$ Reducible

Check It Out 4 (page 533)

$$\frac{3}{x^2+1} + \frac{1}{x-3}$$

Check It Out 5 (page 534)

$$\frac{-2}{x^2+1} + \frac{3}{(x^2+1)^2}$$

Chapter 6, Section 8 _____

Check It Out 1 (page 537)

$(1, -1), (-1, 1)$

Check It Out 2 (page 538)

$(-1.517, 1.303), (1.517, 1.303)$

Check It Out 3 (page 539)

$(-2.128, -0.470), (-0.202, -4.959), (2.330, 0.429)$

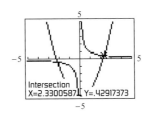

Check It Out 4 (page 540)

6 in., 4 in.

Chapter 7, Section 1 _____

Check It Out 1 (page 558)

The focus is at $(0, -2)$, the equation of the directrix is $y = 2$, and the equation of the axis of symmetry is $x = 0$.

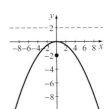

Check It Out 2 (page 559)

$y^2 = 8x$

Check It Out 3 (page 561)

$(x + 2)^2 = -8(y - 7)$

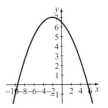

Check It Out 4 (page 563)

Vertex: $(2, -1)$; focus: $(2, 0)$; directrix: $y = -2$

Check It Out 5 (page 563)

(a) $y^2 = \dfrac{8}{3}x$

(b) The coordinates of the focus are $\left(\dfrac{2}{3}, 0\right)$, so the bulb should be placed $\dfrac{2}{3}$ of an inch away from the vertex.

Chapter 7, Section 2 _____

Check It Out 1 (page 572)

Vertices: $(0, 3)$ and $(0, -3)$; foci: $\left(0, \sqrt{5}\right)$ and $\left(0, -\sqrt{5}\right)$; major axis lies on the y-axis

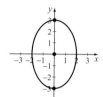

Check It Out 2 (page 572)

$\dfrac{x^2}{25} + \dfrac{y^2}{9} = 1$

Check It Out 3 (page 573)

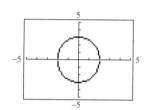

Check It Out 4 (page 575)

$\dfrac{(x - 3)^2}{5} + \dfrac{y^2}{9} = 1$

Check It Out 5 (page 577)

$\dfrac{(x - 1)^2}{3} + \dfrac{y^2}{12} = 1$

Check It Out 6 (page 577)

Approximately 17.43 AU

Chapter 7, Section 3 _____

Check It Out 1 (page 585)

Vertices: $(2, 0)$ and $(-2, 0)$; foci: $\left(2\sqrt{2}, 0\right)$ and $\left(-2\sqrt{2}, 0\right)$; asymptotes: $y = x$ and $y = -x$

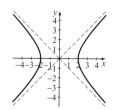

Check It Out 2 (page 586)

$\dfrac{y^2}{9} - \dfrac{x^2}{7} = 1$

Check It Out 3 (page 587)

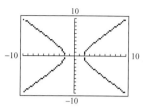

Check It Out 4 (page 589)

$\dfrac{x^2}{4} - \dfrac{(y - 3)^2}{12} = 1$

Check It Out 5 (page 590)

$\dfrac{(x + 1)^2}{9} - \dfrac{(y + 2)^2}{9} = 1$

Check It Out 6 (page 592)

$40\sqrt{2}$ inches

Chapter 8, Section 1 _____

Check It Out 1 (page 602)

Year	Interest ($)	Total value ($)
0	0	10,000
1	500	10,500
2	500	11,000
3	500	11,500
4	500	12,000
5	500	12,500

Check It Out 2 (page 603)

$f(n) = 10{,}000 + 500n$

Check It Out 3 (page 604)

$a_0 = -1 - 2(0) = -1$, $a_1 = -1 - 2(1) = -3$, $a_2 = -1 - 2(2) = -5$, $a_3 = -1 - 2(3) = -7$

Check It Out 4 (page 605)

(a) $a_8 = -22 + 32 = 10$; $a_{12} = -22 + 48 = 26$

(b) $a_n = 42 - 3n$

Check It Out 5 (page 606)

$a_n = 10{,}000(1.05)^n$; $10{,}000(1.05)^{20} \approx \$26{,}532.98$

Check It Out 6 (page 607)

(a) $a_2 = 3(2)^2 = 12$; $a_3 = 3(2)^3 = 24$ **(b)** $a_n = 2(3)^n$

Check It Out 7 (page 608)

(a) $a_2 = 128\left(\dfrac{1}{2}\right)^2 = 32$; $a_5 = 128\left(\dfrac{1}{2}\right)^5 = 4$

(b) $a_n = \dfrac{1}{9}(3)^n$

Check It Out 8 (page 609)

(a) Geometric
(b) 560
(c) $f_n = 35(2)^{n-1}$, $n = 1, 2, 3, \ldots$

Check It Out 9 (page 610)

9 cycles

Chapter 8, Section 2

Check It Out 1 (page 615)

120 gifts

Check It Out 2 (page 617)

420

Check It Out 3 (page 618)

675

Check It Out 4 (page 618)

$$\sum_{i=3}^{7} 4i = 12 + 16 + 20 + 24 + 28$$

Check It Out 5 (page 619)

$$\sum_{i=1}^{9} 4i; \ 180$$

Check It Out 6 (page 620)

1365

Check It Out 7 (page 620)

200

Check It Out 8 (page 621)

$\dfrac{8}{3}$

Check It Out 9 (page 622)

340 seats

Check It Out 10 (page 623)

$26,972.70

Chapter 8, Section 3

Check It Out 1 (page 627)

1, 3, 9, 19

Check It Out 2 (page 628)

$a_n = \dfrac{1}{n^3}$, $n = 1, 2, 3, \ldots$

Check It Out 3 (page 628)

$a_n = (-1)^{n+1}$, $n = 0, 1, 2, \ldots$

Check It Out 4 (page 629)

n	1	2	3	4	5
$f(n)$	-1	$\dfrac{1}{2}$	$-\dfrac{1}{3}$	$\dfrac{1}{4}$	$-\dfrac{1}{5}$

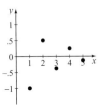

Check It Out 5 (page 630)

(a) 1, 0, 1, 0, 1
(b) Range of $f(n)$: $\{0, 1\}$

Check It Out 6 (page 630)

8, 13, 21, 34

Check It Out 7 (page 631)

(a) 21
(b) 35

Check It Out 8 (page 632)

(a)

n	a_n, Amount in Bloodstream (mg)
0	40
1	28
2	25.6
3	25.12
4	25.024
5	25.0048
6	25.0096

(b)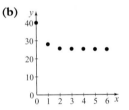

(c) $a_n = 0.20a_{n-1} + 20$

Chapter 8, Section 4

Check It Out 1 (page 636)
1320

Check It Out 2 (page 637)
900

Check It Out 3 (page 637)
120

Check It Out 4 (page 638)
16!

Check It Out 5 (page 638)
210

Check It Out 6 (page 639)
6720

Check It Out 7 (page 640)
720

Check It Out 8 (page 641)
6

Check It Out 9 (page 642)
455

Check It Out 10 (page 643)
560

Chapter 8, Section 5

Check It Out 1 (page 646)
$\frac{1}{8}$

Check It Out 2 (page 647)
$\{2, 3, 4, 5\}$

Check It Out 3 (page 648)
(a) $\{$HHHH, HHHT, HHTH, HTHH, THHH, HHTT, HTHT, HTTH, THHT, THTH, TTHH, HTTT, THTT, TTHT, TTTH, TTTT$\}$
(b) $\{$HTTT, THTT, TTHT, TTTH$\}$
(c) $\{$HHTT, HTHT, HTTH, THHT, THTH, TTHH, HTTT, THTT, TTHT, TTTH, TTTT$\}$

Check It Out 4 (page 649)
$\frac{1}{12}$

Check It Out 5 (page 649)
$\frac{15}{16}$

Check It Out 6 (page 649)
(a) Not mutually exclusive (b) Mutually exclusive

Check It Out 7 (page 650)
$\frac{1}{4}$

Check It Out 8 (page 651)
$\frac{1}{4}$

Check It Out 9 (page 651)
0.2

Check It Out 10 (page 652)
$\frac{11}{16}$

Check It Out 11 (page 653)
(a) 0.035 (b) 0.956

Chapter 8, Section 6

Check It Out 1 (page 658)
(a) $x^5, -x^4y, x^3y^2, -x^2y^3, xy^4, -y^5$ (b) 5

Check It Out 2 (page 659)
15

Check It Out 3 (page 659)
$16y^4 - 32y^3x + 24y^2x^2 - 8yx^3 + x^4$

Check It Out 4 (page 660)
$243u^5 + 810u^4v + 1080u^3v^2 + 720u^2v^3 + 240uv^4 + 32v^5$

Check It Out 5 (page 660)
$7290x^5$

Check It Out 6 (page 661)
$\binom{5}{2} = \frac{5!}{(5-2)!\,2!} = \frac{5!}{3!\,(5-3)!} = \binom{5}{3}$

Chapter 8, Section 7

Check It Out 1 (page 664)
$2(k+1)(k+3)$

Check It Out 2 (page 665)
$P_1: 2(1) - 1 = (1)^2$
Assume $P_k: 1 + 3 + 5 + \cdots + (2k - 1) = k^2$.
Prove $P_{k+1}: 1 + 3 + 5 + \cdots + (2k - 1) + (2(k+1) - 1)$
$$= (k+1)^2.$$
$1 + 3 + 5 + \cdots + (2k - 1) + (2(k+1) - 1)$
$$= k^2 + (2(k+1) - 1)$$
$$= k^2 + 2k + 1$$
$$= (k+1)^2$$

Check It Out 3 (page 666)

P_1: $3(1) - 1 = \dfrac{1}{2}(1)(3(1) + 1)$

Assume P_k: $2 + 5 + 8 + \cdots + (3k - 1) = \dfrac{1}{2}k(3k + 1)$.

Prove P_{k+1}: $2 + 5 + 8 + \cdots + (3k - 1) + (3(k + 1) - 1)$
$$= \frac{1}{2}(k + 1)(3(k + 1) + 1).$$

$2 + 5 + 8 + \cdots + (3k - 1) + (3(k + 1) - 1)$
$$= \frac{1}{2}k(3k + 1) + \frac{1}{2}(6k + 4)$$
$$= \frac{1}{2}(k(3k + 1) + (6k + 4))$$
$$= \frac{1}{2}(k + 1)(3(k + 1) + 1)$$

Check It Out 4 (page 666)

P_1: $4^{1-1} = \dfrac{4^1 - 1}{3}$

Assume P_k: $1 + 4 + 4^2 + \cdots + 4^{k-1} = \dfrac{4^k - 1}{3}$.

Prove P_{k+1}: $1 + 4 + 4^2 + \cdots + 4^{k-1} + 4^{(k+1)-1} = \dfrac{4^{(k+1)} - 1}{3}$.

$1 + 4 + 4^2 + \cdots + 4^{k-1} + 4^{(k+1)-1} = \dfrac{4^k - 1}{3} + 4^k$
$$= \frac{4^k - 1 + 3(4^k)}{3}$$
$$= \frac{4^k + 3(4^k) - 1}{3}$$
$$= \frac{4^{k+1} - 1}{3}$$

Answers to Odd-Numbered Exercises

Chapter P Exercises

Exercise Set P.1 (page 9)

1. $-1, 0, 10, 40$ **3.** $-1.67, -1, 0, 0.5, \frac{4}{5}, 10, 40$

5. $0, 10, 40$ **7.** $-1, 0$

9. Associative property of multiplication

11. Distributive property of multiplication over addition

13. Commutative property of multiplication

15.   **17.**

19.   **21.**

23.   **25.** $[-3, 4]$ **27.** $[0, \infty)$

29. $(-2, 4)$ **31.** $(1, \infty)$

33. $[4, 10]$ **35.** $[-3, 0)$

37. $2 < x \le 12; (2, 12]$ **39.** 3.2 **41.** 253 **43.** $\frac{5}{4}$ **45.** $\frac{4}{3}$

47. 6 **49.** -4.5 **51.** 7 **53.** -3 **55.** 6 **57.** 9 **59.** 4.5

61. 12.2 **63.** 3 **65.** $\frac{17}{15}$ **67.** -5 **69.** 4 **71.** 14

73. -240 **75.** $\frac{-6}{11}$ **77.** -9 **79.** $-\frac{43}{3}$ **81.** $\frac{1}{4}$ **83.** $\frac{19}{5}$

85. $5, 6, 7, 8, 9$ **87.** $[25, 36]$ **89.** $20{,}602$ feet

91. No; it can be zero, which is neither negative nor positive.

93. $-7, 1$

Exercise Set P.2 (page 18)

1. 13 **3.** 10 **5.** 25 **7.** -5 **9.** -9 **11.** $\frac{1}{64}$ **13.** -1

15. $\frac{16}{9}$ **17.** $-16x^4y^8$ **19.** $\frac{1}{4x^{10}}$ **21.** $9a^4b^6$ **23.** $-2a^4b^{10}$

25. $\frac{1}{16x^2y^4}$ **27.** $\frac{4x^2y^2}{9}$ **29.** $\frac{x^3y^2}{3}$ **31.** $\frac{4}{y}$ **33.** $\frac{4}{x^6y^2}$

35. $\frac{4}{x^5yz^2}$ **37.** $\frac{9t^6}{s^4}$ **39.** 5.1×10^{-3} **41.** 5.6×10^3

43. 5.67×10^{-5} **45.** 1.76×10^6 **47.** 3.1605×10

49. 2.8×10^8 **51.** 371 **53.** 0.028 **55.** $596{,}000$

57. $43{,}670{,}000$ **59.** 0.008673 **61.** 0.00000465

63. 6×10^2 **65.** 9.03×10^7 **67.** 2×10^2 **69.** 2×10^{-1}

71. 3 **73.** 4 **75.** 4 **77.** 12 **79.** 30 **81.** 29

83. 3.68×10^4 **85.** 5×10^{-3}

87. 82.8 people per square mile   **89.** $31{,}100$ U.S. dollars

91. 129 square feet   **93.** No. $2 \ne \frac{1}{2}$

95. $-\infty < x < \infty, y \ne 0$

Exercise Set P.3 (page 26)

1. 7 **3.** $\frac{1}{2}$ **5.** 343 **7.** $\frac{2}{5}$ **9.** $4\sqrt{2}$ **11.** $5\sqrt[3]{2}$ **13.** 16

15. $15\sqrt{2}$ **17.** $\frac{-2\sqrt{2}}{5}$ **19.** $\frac{5\sqrt{6}}{21}$ **21.** $\frac{\sqrt{15}}{5}$ **23.** $\frac{\sqrt[3]{21}}{3}$

25. $\frac{-2\sqrt[3]{2}}{3}$ **27.** $xy^2\sqrt{x}$ **29.** $x^2y\sqrt[3]{x^2y}$ **31.** $3xy^2\sqrt{5x}$

33. $6y^2z^3$ **35.** $2xy\sqrt[3]{3x^2}$ **37.** $19\sqrt{2}$ **39.** $-2\sqrt{6} + \sqrt{3}$

41. -4 **43.** $-2\sqrt{3} + \sqrt{2} + 1 - 2\sqrt{6}$

45. $\sqrt{30} + \sqrt{15} - 2\sqrt{3} - \sqrt{6}$ **47.** $-1 - \sqrt{5}$

49. $\sqrt{3} + \sqrt{2}$ **51.** $\frac{1}{3^{2/3}}$ **53.** 1 **55.** $\frac{1}{7^{3/4}}$ **57.** $\frac{1}{4^{7/12}}$

59. $\frac{1}{xy^{3/2}}$ **61.** $\frac{1}{x^{7/6}}$ **63.** $\frac{4r}{s^{4/3}}$ **65.** $\frac{y^{10/3}}{x^{2/3}}$ **67.** $10\sqrt{2}$ inches

69. About 816 species   **71.** Yes. $11.163 < 16.464$

73. $a = 36, b = 8, c = 4$ **75.** $(-\infty, \infty)$

77. If $a = \sqrt{10}$, then $a^2 = 10$. Since $3^2 = 9, a > 3$.

Exercise Set P.4 (page 31)

1. $5y + 26; 1$ **3.** $3t^2 - 2t + 5; 2$ **5.** $-4s^2 - 6s + 7; 2$

7. $-4v^3 - 3; 3$ **9.** $-10z^5 - 3z^4 + 7z^3 + 1; 5$

11. $6z + 14$ **13.** $-5y^2 + 5y + 3$

15. $-9x^3 + 7x^2 - 25x - 3$ **17.** $x^5 + 4x^4 + 6x^3 - 3x^2 + 1$

19. $-3x^5 - 3x^4 + 3x^3 - x - 1$

21. $-9t^5 + 2t^4 + 7t^3 - 3t^2 - 4$ **23.** $-9v^2 + 17v + 6$

25. $9t^3 - 21t^2 + 21t + 23$ **27.** $2s^2 + s$ **29.** $-18z^3 - 15z$

31. $7t^3 - 3t^2 - 9t$ **33.** $7z^4 + 63z^3 - 56z^2$

35. $y^2 + 11y + 30$ **37.** $-v^2 - 9v + 36$ **39.** $7t^2 + 52t - 32$

41. $-28v^2 - 59v - 30$ **43.** $u^3 + 3u^2 - 9u - 27$

45. $x^2 + 8x + 16$ **47.** $s^2 + 12s + 36$ **49.** $25t^2 + 40t + 16$

51. $36v^2 - 36v + 9$ **53.** $9z^2 - 6z + 1$

55. $25t^2 - 60t + 36$ **57.** $v^2 - 81$ **59.** $-81s^2 + 49$

61. $v^4 - 9$ **63.** $25y^4 - 16$ **65.** $16z^4 - 25$ **67.** $x^2 - 4z^2$

69. $25s^2 - 16t^2$ **71.** $-t^3 - 11t^2 - 29t + 6$

73. $28z^3 + 5z^2 + 23z - 20$ **75.** $x^3 + x^2 - 13x + 14$

77. $-20u^4 + 59u^3 - 14u^2 - 85u + 63$

79. $24x$ **81. a.** $4s$ **b.** $8s$

83. a. $1000r^2 + 2000r + 1000$ **b.** $\$1102.50$

85. $(x - 3)^2 - 9 + 6x = x^2$

$x^2 - 6x + 9 - 9 + 6x = x^2$

87. Not if the coefficients of the third degree terms are additive inverses

89. 0 **91.** $a = 0$

Exercise Set P.5 (page 39)

1. $x^2 + 12x + 36$ **3.** $72x^2 + 168x + 98$ **5.** $9y^2 - 100$

7. $2x(x^2 + 3x - 4)$ **9.** $-3(y^3 - 2y + 3)$

11. $-2t^2(t^4 + 2t^3 - 5)$ **13.** $-5x^3(x^4 - 2x^2 + 3)$

15. $(3 + x)(x + 1)$ **17.** $(s - 3)(s + 3)(s - 5)$

19. $(2u - 1)(2u + 1)(3u + 1)$ **21.** $(x + 1)(x + 3)$

23. $(x - 8)(x + 2)$ **25.** $3(s + 4)(s + 1)$

27. $-6(t-6)(t+2)$ **29.** $-5(z+6)(z-2)$
31. $(3x+4)(x-3)$ **33.** $(4z+1)(z-6)$
35. $(x+4)(x-4)$ **37.** $(3x+2)(3x-2)$
39. $(y+\sqrt{3})(y-\sqrt{3})$ **41.** $3(x+2)(x-2)$ **43.** $(x+3)^2$
45. $(y-7)^2$ **47.** $(2x+1)^2$ **49.** $(3x-1)^2$ **51.** $3(2x+1)^2$
53. $(y+4)(y^2-4y+16)$ **55.** $(u-5)(u^2+5u+25)$
57. $2(x-2)(x^2+2x+4)$ **59.** $(2y+1)(4y^2-2y+1)$
61. $(z+6)(z+7)$ **63.** $(x+6)^2$ **65.** $-(y-2)^2$
67. $(z-8)^2$ **69.** $-(2y-1)(y-3)$ **71.** $(3y+2)^2$
73. $(3z-1)^2$ **75.** $(2z-5)^2$ **77.** $3(2z+3)(z-2)$
79. $-5(3t-1)(t+5)$ **81.** $-5(2u+1)(u+4)$
83. $(7-s)(7+s)$ **85.** $(v+2)(v-2)$
87. $(-5t+2)(5t+2)$ **89.** $(3z+1)(3z-1)$ **91.** $t^2(t-16)$
93. $4u(3u-5)(u+2)$ **95.** $-5t(2t-3)(t+1)$
97. $-5z(3z+4)(z-1)$ **99.** $(y-2)(y+2)(2y+3)$
101. $4x^2(x+2)(x+3)$ **103.** $3y^2(y+4)(y+2)$
105. $-x^2(x-3)(x+2)$ **107.** $7x^3(x-3)(x+3)$
109. $5y^3(y-2)(y+2)$ **111.** $8(x+2)(x^2-2x+4)$
113. $(1-2y)(1+2y+4y^2)$ **115.** $4x^4=(2x^2)(2x^2)$
117. $x^2+6x+9=(x+3)^2$
119. $(2x+3)(2x-3)(4x^2+9)$

Exercise Set P.6 (page 46)

1. $2x(x-7)$ **3.** $(x+9)(x-9)$ **5.** $-(x-3)^2$ **7.** $\dfrac{19}{8}$

9. $\dfrac{x-2}{6}, x\neq -2$ **11.** $\dfrac{x+2}{x+3}, x\neq 3,-3$

13. $x^2(x-1), x\neq -1$ **15.** $\dfrac{x^2+x+1}{x+1}$ **17.** $\dfrac{1}{xy}$

19. $\dfrac{1}{(x-3)(x+2)}$ **21.** $\dfrac{6}{x+6}$ **23.** $\dfrac{2(x-2)}{x(x+2)(x+5)}$

25. $\dfrac{(x^2-x+1)(2x+1)}{x+2}$ **27.** $\dfrac{5}{(x-4)(x+1)}$

29. $\dfrac{(x-2)(x-1)}{x+1}$ **31.** $\dfrac{x^2+2x+4}{(x+2)(x-2)}$ **33.** $\dfrac{2x+3}{x^2}$

35. $\dfrac{3x-4}{x^2}$ **37.** $\dfrac{5x+3}{x^2-1}$ **39.** $\dfrac{3(x-5)}{(x+4)(2x-1)}$

41. $\dfrac{2x(x^2-3x+6)}{3(x+3)(x-3)}$ **43.** $\dfrac{2z^2+z+5}{5(z-2)^2}$ **45.** $\dfrac{2(x+2)}{(x+1)(x-1)}$

47. $\dfrac{x(x-3)(x-1)}{(x+4)(x-4)}$ **49.** $\dfrac{z^2-8z+3}{3(z-5)^2}$ **51.** $\dfrac{1}{1-x}$

53. $\dfrac{2x-11}{(x+2)(x-2)}$ **55.** $\dfrac{-8x-7}{(x-3)(x+2)}$ **57.** $\dfrac{x}{x-1}$

59. $\dfrac{y(y+x)}{x-2y^2}$ **61.** $\dfrac{rst}{st+rt+rs}$ **63.** $\dfrac{x}{1-x}$

65. $\dfrac{-2(x+1)}{(x-3)(5x-1)}$ **67.** $-\dfrac{1}{x(x+h)}$

69. $\dfrac{x+4}{4(x-2)}$ **71.** $\dfrac{a+ba-b^2}{a+b}$

73. 3.50: Average cost per book for 100 booklets

75. $t=\dfrac{12}{7}$ hr   **77.** $x=1, y=1$ **79.** Not true for $x=0$

Exercise Set P.7 (page 51)

1. 24 inches   **3.** 37.699 inches   **5.** 22 centimeters

7. 15 square centimeters   **9.** 28.274 square feet
11. 197.920 cubic inches   **13.** 268.083 cubic inches
15. 226.195 cubic centimeters   **17.** 113.097 square inches
19. 62.832 square inches   **21.** 140.883 square centimeters
23. $(36+9\pi)$ square inches ≈ 64.274 square inches
25. 72 square inches   **27.** 5   **29.** 26   **31.** 29   **33.** $\sqrt{34}$
35. 30 feet, 50 square feet   **37.** 25.133 cubic feet
39. 3.606 inches   **41.** 18.850 feet   **43.** 56.549 cubic inches
45. $r>0$ **47.** 4

Exercise Set P.8 (page 56)

1. $x=1$ **3.** $x=-3$ **5.** $x=-3$ **7.** $x=-9$ **9.** $x=\dfrac{19}{2}$

11. $x=-\dfrac{18}{13}$ **13.** $x=2$ **15.** $x=9$ **17.** $x=2$

19. $x=\dfrac{4}{5}$ **21.** $x=6$ **23.** $x=\dfrac{60}{19}$ **25.** $x=10$

27. $x=\dfrac{40}{49}$ **29.** $x=\dfrac{1}{\pi}$ **31.** $y=5-x$ **33.** $y=3+2x$

35. $y=\dfrac{5}{2}-\dfrac{5}{4}x$ **37.** $y=5-4x$

39. $40x-200=800; x=25$ DVD players   **41.** 7 inches
43. 12.5 feet   **45.** 18 inches   **47.** 2.5 feet
49. No. The variable cancels on both sides of the equation, leaving the false statement $2=0$.
51. Step 2 should be $(x+1)+4=16; x=11$.

Chapter P Review Exercises (page 63)

Section P.1 (page 63)

1. $-5, 3, 8$ **3.** $3, 8$ **5.** Associative property of addition

7.   **9.**   **11.** 10

13. -3.7 **15.** -5 **17.** -22

Section P.2 (page 64)

19. 2   **21.** $\dfrac{6y^4}{x^2}$ **23.** $\dfrac{4x^4}{y^3}$ **25.** $\dfrac{16x^{12}}{y^6}$ **27.** 4.67×10^6

29. 30,010   **31.** 6.4×10^2 **33.** 2.0

Section P.3 (page 64)

35. $5\sqrt[3]{3}$ **37.** $5x\sqrt{2x}$ **39.** $\dfrac{5\sqrt{2}}{6}$ **41.** $-\sqrt{x}+4$

43. $\dfrac{15+5\sqrt{2}}{7}$ **45.** -4 **47.** 512   **49.** $36x^{7/12}$ **51.** $3x^{1/6}$

Section P.4 (page 64)

53. $6y^3+13y^2+24y-12$ **55.** $9t^5+3t^4-10$
57. $-12u^2-43u-10$ **59.** $-24z^2+91z-72$
61. $-6z^3+13z^2-29z+40$ **63.** $9x^2-4$
65. $25-10x+x^2$

Section P.5 (page 64)

67. $4z^2(2z+1)$ **69.** $(y+4)(y+7)$ **71.** $(3x+5)(x-4)$
73. $(5x+2)(x-2)$ **75.** $(3u+7)(3u-7)$ **77.** $z(z^2+8)$
79. $2(x+1)^2$ **81.** $4(x+2)(x^2-2x+4)$

Section P.6 (page 65)

83. $x + 3, x \neq 3$ **85.** $\dfrac{(x - 2)(x + 1)}{x - 1}$ **87.** $\dfrac{(x + 2)^2}{x(x - 2)}$

89. $\dfrac{5x + 1}{(x + 1)(x - 3)}$ **91.** $\dfrac{-2x^4 + 11x^2 + 7x + 78}{(x + 3)(x - 3)}$ **93.** $\dfrac{a + b}{a}$

Section P.7 (page 65)

95. 12 square inches   **97.** 615.752 cubic centimeters
99. 113.097 square inches   **101.** $2\sqrt{13}$

Section P.8 (page 65)

103. $x = 3$ **105.** $x = -\dfrac{1}{2}$ **107.** $x = 1$ **109.** $y = -3x + 5$

Applications (page 65)

111. 1.6×10^{-2} gram   **113.** $\dfrac{5\sqrt{2}}{4}$ seconds

115. 156π cubic inches

Chapter P Test (page 66)

1. $-1, 1.55, 41$ **2.** Distributive property

3. $\overset{}{\underset{-5\,-4\,-3\,-2\,-1\ \ 0\ \ 1\ \ 2\ \ 3}{\longleftrightarrow}}$ **4.** 10.3   **5.** $\dfrac{9}{7}$ **6.** -25

7. 8.903×10^6 **8.** $-36x^4y^{10}$ **9.** $\dfrac{64x^{27}}{y^{12}}$ **10.** $4x\sqrt{3x}$

11. $-30x^{8/15}$ **12.** $\dfrac{9x^{1/3}}{y^3}$ **13.** $\sqrt[3]{2}$ **14.** 25   **15.** $-\dfrac{7 - 7\sqrt{5}}{4}$

16. $(5 + 7y)(5 - 7y)$ **17.** $(2x + 5)^2$ **18.** $(3x - 5)(2x + 1)$
19. $x(2x + 3)(2x - 3)$ **20.** $(3x - 7)(x + 5)$

21. $2(x + 2)(x^2 - 2x + 4)$ **22.** $\dfrac{2(2x + 1)}{(x + 3)(x - 2)}$

23. $\dfrac{x - 1}{(x + 2)(x - 2)}$ **24.** $\dfrac{19 - 7x}{x^2 - 4}$

25. $\dfrac{3x^2 + 6x - 4}{(2x + 1)(x - 1)(x + 3)}$ **26.** $\dfrac{8x - 6}{-3x - 2}$

27. 25π square inches   **28.** 360π cubic centimeters

29. $x = \dfrac{11}{4}$ **30.** 2.85×10^{-5} grams   **31.** 140π cubic inches

Chapter 1 Exercises

Exercise Set 1.1 (page 76)

1. False   **3.** $f(3) = 18; f(-1) = -2; f(0) = 3$

5. $f(3) = \dfrac{-17}{2}; f(-1) = \dfrac{11}{2}; f(0) = 2$

7. $f(3) = 11; f(-1) = 3; f(0) = 2$
9. $f(3) = -36; f(-1) = -4; f(0) = -6$
11. $f(3) = \sqrt{13}; f(-1) = 1; f(0) = 2$

13. $f(3) = \dfrac{4}{3}; f(-1) = 0; f(0) = -\dfrac{1}{3}$

15. $f(a) = 4a + 3; f(a + 1) = 4a + 7; f\left(\dfrac{1}{2}\right) = 5$

17. $f(a) = -a^2 + 4; f(a + 1) = -a^2 - 2a + 3; f\left(\dfrac{1}{2}\right) = \dfrac{15}{4}$

19. $f(a) = \sqrt{3a - 1}; f(a + 1) = \sqrt{3a + 2}; f\left(\dfrac{1}{2}\right) = \dfrac{\sqrt{2}}{2}$

21. $f(a) = \dfrac{1}{a + 1}; f(a + 1) = \dfrac{1}{a + 2}; f\left(\dfrac{1}{2}\right) = \dfrac{2}{3}$

23. $g(-x) = \sqrt{6}; g(2x) = \sqrt{6}; g(a + h) = \sqrt{6}$
25. $g(-x) = -2x - 3; g(2x) = 4x - 3;$
$g(a + h) = 2a + 2h - 3$
27. $g(-x) = 3x^2; g(2x) = 12x^2; g(a + h) = 3a^2 + 6ah + 3h^2$

29. $g(-x) = -\dfrac{1}{x}; g(2x) = \dfrac{1}{2x}; g(a + h) = \dfrac{1}{a + h}$

31. $g(-x) = -x^2 + 3x + 5; g(2x) = -4x^2 - 6x + 5;$
$g(a + h) = -a^2 - 2ah - h^2 - 3a - 3h + 5$
33. a. $g(5) = -1$ **b.** $g(0) = 2$
c. $g(3)$ is not defined, because the table does not include a value for $g(t)$ at $t = 3$.
35. Yes   **37.** No   **39.** Yes   **41.** Yes   **43.** $(-\infty, \infty)$
45. $(-\infty, -1) \cup (-1, \infty)$ **47.** $(-\infty, 3) \cup (3, \infty)$
49. $(-\infty, -2) \cup (-2, 2) \cup (2, \infty)$ **51.** $(-\infty, 2]$
53. $(-\infty, \infty)$ **55.** $(-7, \infty)$ **57.** $V(3) = 36\pi$
59. $S(30) = 1600.$ This means that the salesperson receives a commission of \$1600 when 30 items are sold.
61. a. $d(t) = 45t$
b. The car travels 90 miles in 2 hours. $d(2) = 90$
c. The domain of the function is the set of all real numbers greater than or equal to zero. The range of the function is the set of all real numbers greater than or equal to zero.
63. a. $D(\text{\textit{Finding Nemo}}) \approx 350$ million dollars. This means the dollar amount grossed by *Finding Nemo* is about 350 million dollars.
b. The domain is the set of movies consisting of *Titanic, Star Wars, Spiderman, Shrek,* and *Finding Nemo.*
c. No. As in this case, the domain can be a set of objects.
65. $A(w) = 3w^2$
67. a. $h(0) = 100.$ This means that the height of the ball 0 seconds after it is dropped—that is, at its starting position—is 100 feet.
b. $h(2) = 36.$ This means that the height of the ball 2 seconds after it is dropped is 36 feet.
69. a. This table represents a function because for each year (input), there is only one value for per capita consumption (output).
b. $S(2001) = 62.6$ pounds. This means that in 2001, the per capita consumption of high-fructose corn syrup was 62.6 pounds.
c. This is reflected in the table by a steady increase over three decades.
71. $D(g) = 26g$
73. a. $L(t) = 10t$ **b.** $M(t) = 8.2t$ **c.** 14.4 tons
75. $c = 7$ **77.** Domain: $[-1, \infty)$, range: $[1, \infty)$

Exercise Set 1.2 (page 87)

1. 2   **3.** 4   **5.** 4
7. No   **9.** No   **11.** No

13.

x	-4	-2	0	2	4
$f(x) = -\dfrac{1}{2}x - 4$	-2	-3	-4	-5	-6

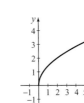

15.

x	0	2	$\dfrac{9}{2}$	8	18
$f(x) = \sqrt{2x}$	0	2	3	4	6

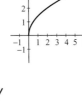

17. Domain: $(-\infty, \infty)$, range: $(-\infty, \infty)$
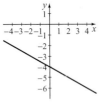

19. Domain: $(-\infty, \infty)$, range: $(-\infty, \infty)$
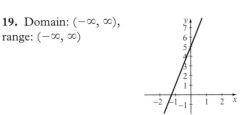

21. Domain: $(-\infty, \infty)$, range: $(-\infty, \infty)$
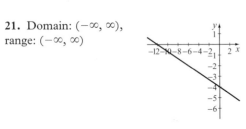

23. Domain: $(-\infty, \infty)$, range: $(-\infty, \infty)$
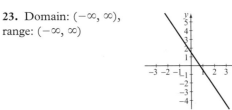

25. Domain: $(-\infty, \infty)$, range: $\{4\}$
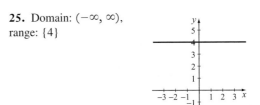

27. Domain: $(-\infty, \infty)$, range: $[0, \infty)$
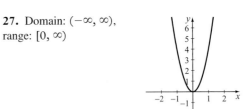

29. Domain: $(-\infty, \infty)$, range: $(-\infty, 4]$
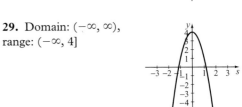

31. Domain: $[-4, \infty)$, range: $[0, \infty)$
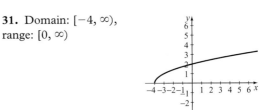

33. Domain: $[0, \infty)$, range: $[0, \infty)$
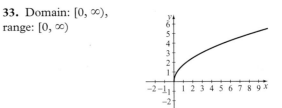

35. Domain: $[0, \infty)$, range: $[0, \infty)$
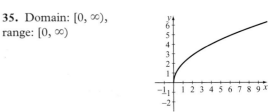

37. Domain: $(-\infty, \infty)$, range: $[0, \infty)$
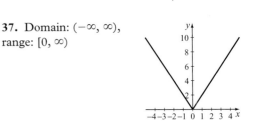

39. Domain: $(-\infty, \infty)$, range: $(-\infty, 0]$

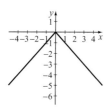

41. Domain: $(-\infty, \infty)$, range: $[0, \infty)$

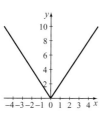

43. Domain: $(-\infty, 0]$, range: $[0, \infty)$

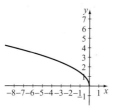

45. Domain: $(-\infty, \infty)$, range: $(-\infty, \infty)$

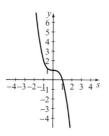

47. This graph does not depict a function. There are values in the domain for which there is more than one corresponding value in the range.

49. This graph depicts a function. There are no values in the domain for which there is more than one corresponding value in the range.

51. Domain: $(-\infty, \infty)$, range: $(-\infty, 1]$

53. Domain: $(-\infty, \infty)$, range: $[-1, 2]$

55.

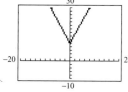

57.

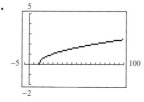

59.

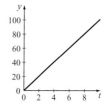

61. a. $f(-1) = 1.5, f(0) = 1, f(2) = 0$
b. Domain: $(-\infty, \infty)$
c. x-intercept: $(2, 0)$; y-intercept: $(0, 1)$
63. a. $f(-1) = 1, f(0) = 3, f(2) = 3$
b. Domain: $(-\infty, \infty)$
c. No x-intercept; y-intercep: $(0, 3)$
65. a. $f(-1) = 1, f(0) = 2, f(2) = 0$
b. Domain: $(-\infty, \infty)$
c. x-intercept: $(-2, 0)$ and $(2, 0)$; y-intercept: $(0, 2)$
67. a. $f(-1) \approx -1.7, f(0) = -2, f(2) \approx -2.4$
b. Domain: $[-4, \infty)$
c. x-intercept: $(-4, 0)$; y-intercept: $(0, -2)$
69. a. 2006   **b.** About $1 billion
71. Radius is the measure of the distance from the center of a sphere to any point on the sphere, and distance is a positive number.

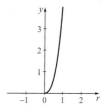

73. $D(t) = 55t$. The values of t must be greater than or equal to zero.

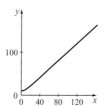

75. $d(x) = \sqrt{x^2 + 100}$

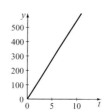

77. a. $F(t) = 10.1t$. The units of the input variable are time in years and the units of the output variable are tons of greenhouse gases.

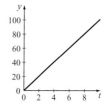

b. $O(t) = 8.1t$

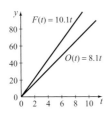

c. As the value of t increases, the difference between the corresponding values of the functions increases.

79. $h(x)$ decreases as x increases, whereas $g(x)$ increases as x increases.

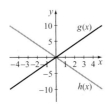

81. The lowest value of $f(x)$ is 0, whereas the lowest value of $g(x)$ is 1.

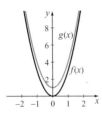

83. No. Note that neither the domains nor the ranges are the same for these two functions.

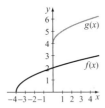

85. The range of $f(x)$ is the set of all real numbers greater than or equal to 4. The range of $g(x)$ is the set of all real numbers greater than or equal to −4.

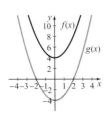

87. The range of $f(x)$ is the set of all real numbers. The range of $g(x)$ is the set of all real numbers.

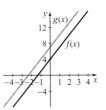

Exercise Set 1.3 (page 102)

1. $y = -6$ **3.** $y = 2$ **5.** $y = 20$
7. a. Yes, the equation fits the form $f(x) = mx + b$, where $m = 3$ and $b = 1$.
b. No, the equation does not fit the form $f(x) = mx + b$.
c. Yes, the equation fits the form $f(x) = mx + b$, where $m = -5$ and $b = 0$.
d. No, the equation does not fit the form $f(x) = mx + b$.

9. -7 **11.** 0 **13.** $\dfrac{1}{2}$ **15.** Undefined

17. Undefined **19.** 1 **21.** $\dfrac{2}{9}$ **23.** $m = \dfrac{5}{5} = 1$

25. The values of y do not change as x changes.

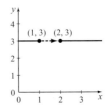

27. Yes, $(2, 1)$ lies on $y = -2x + 5$.
29. No, $(-1, 0)$ does not lie on $y = -2x + 5$.
31. $y = 6x - 1$; $f(x) = 6x - 1$. The slope is 6. The x-intercept is $\left(\dfrac{1}{6}, 0\right)$ and the y-intercept is $(0, -1)$.
33. $y = 2x + \dfrac{3}{5}$; $f(x) = 2x + \dfrac{3}{5}$. The slope is 2. The x-intercept is $\left(-\dfrac{3}{10}, 0\right)$ and the y-intercept is $\left(0, \dfrac{3}{5}\right)$.
35. $y = 3x - 11$; $f(x) = 3x - 11$. The slope is 3. The x-intercept is $\left(\dfrac{11}{3}, 0\right)$ and the y-intercept is $(0, -11)$.
37. $y = \dfrac{1}{2}x + 3$; $f(x) = \dfrac{1}{2}x + 3$. The slope is $\dfrac{1}{2}$. The x-intercept is $(-6, 0)$ and the y-intercept is $(0, 3)$.
39. $y = 2x + 13$; $f(x) = 2x + 13$. The slope is 2. The x-intercept is $\left(-\dfrac{13}{2}, 0\right)$ and the y-intercept is $(0, 13)$.
41. $y = \dfrac{2}{5}x - 2$; $f(x) = \dfrac{2}{5}x - 2$. The slope is $\dfrac{2}{5}$. The x-intercept is $(5, 0)$ and the y-intercept is $(0, -2)$.

43. $y = -2x + 6; f(x) = -2x + 6.$ The slope is $-2.$ The x-intercept is $(3, 0)$ and the y-intercept is $(0, 6).$

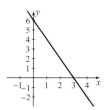

45. $y = \frac{4}{3}x + \frac{2}{3}; f(x) = \frac{4}{3}x + \frac{2}{3}.$ The slope is $\frac{4}{3}.$ The x-intercept is $\left(-\frac{1}{2}, 0\right)$ and the y-intercept is $\left(0, \frac{2}{3}\right).$

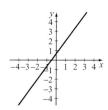

47. $y = -2x + 15; f(x) = -2x + 15.$ The slope is $-2.$ The x-intercept is $\left(\frac{15}{2}, 0\right)$ and the y-intercept is $(0, 15).$

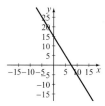

49. $y = \frac{5}{3}x + 3; f(x) = \frac{5}{3}x + 3.$ The slope is $\frac{5}{3}.$ The x-intercept is $\left(-\frac{9}{5}, 0\right)$ and the y-intercept is $(0, 3).$

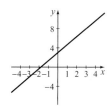

51. $y + 1 = -5(x - 2)$ **53.** $y + 4 = \frac{1}{3}(x + 1)$

55. $y - 5 = 0(x - 4)$ **57.** $y - 3.6 = 0.9(x - 1.5)$

59. $y + 1 = \frac{3}{5}\left(x - \frac{1}{2}\right)$ **61.** $y = x + 1; f(x) = x + 1$

63. $y = \frac{2}{5}x - \frac{6}{5}; f(x) = \frac{2}{5}x - \frac{6}{5}$

65. $x = 2.$ The equation cannot be expressed as a function because the line is not the graph of a function.

67. 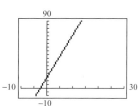 **69.**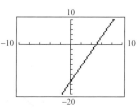

71.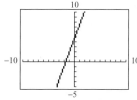

73. $y = -x + 2$ **75.** $y = \frac{1}{3}x + 1$ **77.** $y = -\frac{1}{2}x + \frac{5}{2}$

79. $y = -3x + 8$ **81.** $y = \frac{3}{4}x - \frac{3}{2}$ **83.** $y = -6x + 3$

85. $x = 4$ **87.** $y = -1$ **89.** $y = 0.5$ **91.** $y = -3x - 1$

93. $y = -\frac{1}{2}x + 1$ **95.** $y = -\frac{2}{3}x - 1$ **97.** $y = -\frac{3}{2}x - 1$

99. $y = -\frac{1}{2}x$ **101.** $y = 2x - 3$ **103.** $x = 0$ **105.** $x = -2$

107. a. $f(x) = 50x + 650$
b. $m = 50.$ This represents the number of additional dollars earned for each additional computer sold. $b = 650.$ This represents the amount earned if no computers are sold.
109. a. 173.4 million
b. The y-intercept of 167 million represents the total number of moviegoers in 2003.
111. a. 980 handbags
b. 500 handbags. This value represents the number of handbags sold in 2003.
c. 2007
113. $f(t) = 25.50 + 1.5t$ **115.** $C(S) = 1.06S + 500$
117. a. $C(x) = 5x + 5000$
b. The domain is the set of all real numbers greater than or equal to zero. The range is the set of all real numbers greater than or equal to 5000.
c. Slope: 5, y-intercept: $(0, 5000).$ The slope represents the increase in cost for each additional watch manufactured. The y-intercept represents the fixed cost.
d. $11,250

e.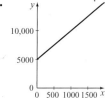

119. The graph does not appear in the viewing window.

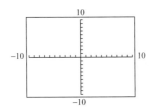

The maximum and minimum values of x and y shown in the viewing window must be manually changed to accommodate the values of this function.

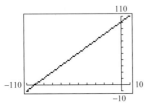

121. $y = -\dfrac{4}{5}x - 1$

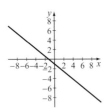

123. $y = \dfrac{5}{3}x + \dfrac{28}{3}$

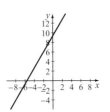

125. The function cannot be linear because the ratio of the change in $g(t)$ to the change in t is not consistent between consecutive points.
127. $m = 0$; $b = 4$; $g(t) = 4$

Exercise Set 1.4 (page 115)
1. 5   **3.** 5.15   **5.** 88   **7.** $f(x) = 4x - 7$
9. $f(x) = 2.4x - 9.94$ **11.** $f(x) = 15x + 150$
13. $k = 4$, $y = 4x$ **15.** $k = 10$, $y = \dfrac{10}{x}$
17. $k = 3.5$, $y = 3.5x$ **19.** $k = 15$, $y = \dfrac{15}{x}$
21. $k = \dfrac{1}{6}$, $y = \dfrac{x}{6}$ **23.** $k = 6$, $y = \dfrac{6}{x}$ **25.** $k = 5$, $y = 5x$
27. $k = 98$, $y = \dfrac{98}{x}$ **29.** $c(n) = 4.50 + 0.07n$

31. $F = \dfrac{9}{5}C + 32$

33. $k = \dfrac{5}{4}$. The volume at 60°C is 75 cc.

35. 12.5 feet   **37.** $\dfrac{10}{3}$ days

39. a. $C(m) = 19.95 + 0.99m$
b. The slope is 0.99 and the y-intercept is 19.95. The slope refers to the increment of change for each additional mile driven. The y-intercept corresponds to the cost of the rental for one day when 0 miles are driven.
c. $75.39

41. a.

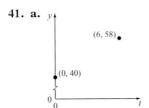

b. The slope is 3.
c. The slope represents the number of additional visitors each month, in millions.
d. $v(t) = 40 + 3t$
e. 70 million visitors
43. a. $f(t) = 203 + 50t$
b. The slope is 50. It signifies the number of additional cases sold each year, in millions.
c. The y-intercept is 203. It signifies the number of cases sold in 2002, in millions.
45. a. $P(t) = 3t + 20$
b. 38%
c. 2010
d. No, because the model predicts a percentage of consumers over 100% in 2030.
e. This model cannot accurately predict the buying habits over a long period of time.
47. a. Traffic increased by 6250 vehicles per year. Mathematically, this is the slope of a linear model of the traffic.
b. $T(t) = 6250t + 175,000$; $T(6) = 212,500$
49. a. The percentage is decreasing over time.
b. $m(t) = -0.114t + 20.34$
51. a. The number of college students is increasing.
b. $P(x) = 0.38x + 15.24$
c. 19.04 million
53. a. As the price increases, the number of rings sold decreases.
b. $s(x) = -1.55x + 1744$
c. 194
55. No, because direct variation refers to an equation of the form $y = kx$, not $y = kx + b$, $b \neq 0$.
57. Yes, because there is a constant change in y for each incremental change in x. This can be modeled by the equation $y = 6x$.

Exercise Set 1.5 (page 129)
1. slope; *y*-intercept **3.** True

5. 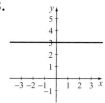 **7.**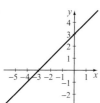

9. 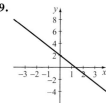 **11.**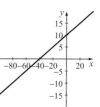

13. (1, 2)
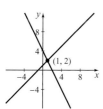

15. (3, 1) 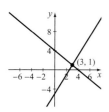 **17.** (3, 3)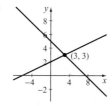

19. (−1, 3) **21.** (−1, 1) **23.** (−4, −2) **25.** $\left(8, -\dfrac{1}{3}\right)$

27. (−6, −11) **29.** $\left(-\dfrac{200}{11}, \dfrac{16}{11}\right)$ **31.** (7, 9)

33. (−2, 0) **35.** (−1, 4) **37.** $x \geq 4$
39. No **41.** Yes

43. $[-1, \infty)$
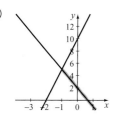

45. $[4, \infty)$

47. $(10, \infty)$
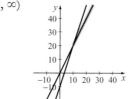

49. $\left[-\dfrac{3}{2}, \infty\right)$ **51.** $(2, \infty)$ **53.** $(-\infty, 1]$ **55.** $\left(-\infty, -\dfrac{13}{5}\right]$

57. $[3, \infty)$ **59.** $(-4, \infty)$ **61.** $\left[\dfrac{18}{7}, \infty\right)$ **63.** $\left[-\dfrac{3}{2}, 1\right]$

65. (1, 5) **67.** (−3, 3)
69. Break-even point: $q = 5$. Values for which revenue exceeds cost: $q > 5$
71. Break-even point: $q = 40$. Values for which revenue exceeds cost: $q > 40$
73. $70 \leq x \leq 89.1$ **75.** 21.25 million tickets
77. $3.54 **79.** At least 62.5 minutes
81. The score on the fifth exam must be at least 89.
83. a. $A(m) = 45$
b. $B(m) = 25 + 0.25m$
c. Driving 80 miles in a car rented from either company would cost $45.

d.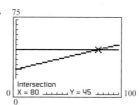

85. a. The ratio of students to computers is $S(t):1$, where $S(t) = -0.75t + 8$.
b. In or after 2006

c.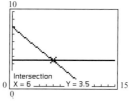

87. The lines do not intersect. They have the same slope and different y-intercepts. When graphed, these functions are parallel lines.

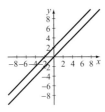

89. The intersection is all coordinate pairs that satisfy either equation. The lines intersect at all points on either line.

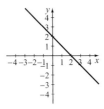

Chapter 1 Review Exercises (page 136)

Section 1.1 (page 136)

1. a. $f(4) = 11$ **b.** $f(-2) = -7$ **c.** $f(a) = 3a - 1$
d. $f(a + 1) = 3a + 2$

3. a. $f(4) = \dfrac{1}{17}$ **b.** $f(-2) = \dfrac{1}{5}$ **c.** $f(a) = \dfrac{1}{a^2 + 1}$

d. $f(a + 1) = \dfrac{1}{a^2 + 2a + 2}$

5. a. $f(4) = 9$ **b.** $f(-2) = 3$ **c.** $f(a) = |2a + 1|$
d. $f(a + 1) = |2a + 3|$

7. Domain: $(-\infty, \infty)$

9. Domain: $(-\infty, 2) \cup (2, \infty)$

11. Domain: $(-\infty, \infty)$

13. Domain: $(-\infty, \infty)$

Section 1.2 (page 137)

15. Domain: $(-\infty, \infty)$,
range: $\{3\}$

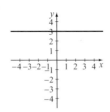

17. Domain: $(-\infty, \infty)$,
range: $(-\infty, \infty)$

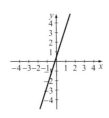

19. Domain: $(-\infty, \infty)$,
range: $(-\infty, 3]$

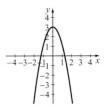

21. Domain: $(-\infty, \infty)$,
range: $(-\infty, 0]$

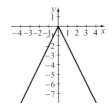

23. Yes

25. Graphs (b) and (d) are functions because they pass the vertical line test.

27. $f(x)$ does not change as the values of x change, whereas $g(x)$ changes as the values of x change.

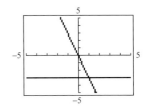

29. The graph of $h(w)$ opens upward, whereas the graph of $f(w)$ opens downward.

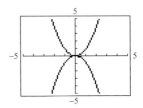

Section 1.3 (page 137)

31. Yes. This is a linear function because it is of the form $ax + b = 0$, where $a = \dfrac{3}{4}$ and $b = 0$.

33. No. This is not a linear function because it cannot be written in the form $ax + b = 0$.

35. $\dfrac{5}{2}$ **37.** $-\dfrac{12}{7}$ **39.** 0

41. $f(x) = -2x + 7$

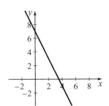

43. $f(x) = \dfrac{3}{2}x + 3$

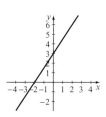

45. $f(x) = -\dfrac{1}{5}x - \dfrac{23}{5}$

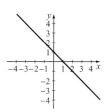

47. $f(x) = -x + 1$

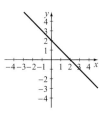

49. $f(x) = -x + 2$

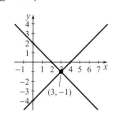

Section 1.4 (page 138)

51. $f(x) = \dfrac{2}{3}x + 4$ **53.** $y = 2.5x$ **55.** $y = \dfrac{54}{x}$

Section 1.5 (page 138)

57. $(3, -1)$

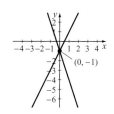

59. $(0, -1)$

61. $(1, 2)$ **63.** $\left(-\dfrac{6}{7}, -\dfrac{13}{7}\right)$ **65.** $\left(-\infty, \dfrac{5}{12}\right)$

67. $\left(-\infty, -\dfrac{15}{11}\right]$ **69.** $\left[5, \dfrac{19}{2}\right]$

71. $x \le 0.943$

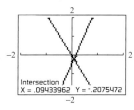

73. $-6 \le x \le -1.3333$

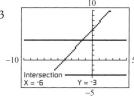

Applications (page 138)

75. a. About 51%   **b.** 1992
77. $S(20,000) = 2800.$ This value represents the earnings when $20,000 worth of sales are generated.
79. a. 640 pens
b. The y-intercept is 400. It represents the number of pens sold in 2003.
c. 2012
81. 600 units
83. The break-even point is $q = 60$ items.
85. a. They are generally decreasing.
b. $f(x) = -48.8x + 925$ **c.** 583.4 million

Chapter 1 Test (page 139)

1. a. -8 **b.** $-a^2 + 4a - 3$ **c.** 3   **d.** 4
2. $(-\infty, \infty)$ **3.** $(-\infty, 5) \cup (5, \infty)$

4. Domain: $(-\infty, \infty)$

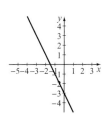

5. Domain: $[-3, \infty)$

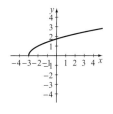

6. Domain: $(-\infty, \infty)$,
range: $(-\infty, 4]$

7. No **8.** Yes

9. The variable x is raised to the -1 power.

10. a. $\dfrac{7}{2}$ **b.** Undefined **11.** $y = -4x - 1$

12. $y = -x + 3$ **13.** $y = -2x + 6$

14. $y = -4x - 12$ **15.** $y = -5$ **16.** $x = 7$

17. $k = \dfrac{9}{2}$, $y = \dfrac{9}{2}x$ **18.** $k = 70$, $y = \dfrac{70}{x}$

19. $(1, 3)$

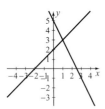

20. $\left[-\dfrac{17}{2}, \infty \right)$ **21.** $\left[-\dfrac{3}{5}, \dfrac{9}{5} \right)$

22. a. $y = 300,000 + 15,000t$ **b.** In 8 years

23. 100 minutes **24.** 400 units

Chapter 2 Exercises

Exercise Set 2.1 (page 148)

1. hypotenuse **3.** 3 **5.** $4x^2 - 20x + 25$

7. Distance: $7\sqrt{5}$; midpoint: $\left(-1, \dfrac{15}{2} \right)$

9. Distance: $6\sqrt{2}$; midpoint: $(-7, 17)$

11. Distance: $2\sqrt{13}$; midpoint: $(3, 2)$

13. Distance: 14; midpoint: $(6, 4)$

15. Distance: $\dfrac{\sqrt{17}}{4}$; midpoint: $\left(-\dfrac{3}{8}, \dfrac{1}{2} \right)$

17. Distance: $\sqrt{(b_1 - a_1)^2 + (b_2 - a_2)^2}$;

midpoint: $\left(\dfrac{a_1 + b_1}{2}, \dfrac{a_2 + b_2}{2} \right)$

19. $x^2 + y^2 = 25$ **21.** $(x + 1)^2 + y^2 = 9$

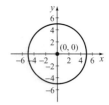

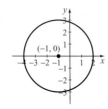

23. $(x - 3)^2 + (y + 1)^2 = 25$ **25.** $(x - 1)^2 + y^2 = \dfrac{9}{4}$

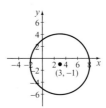

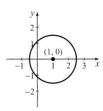

27. $(x - 1)^2 + (y - 1)^2 = 3$ **29.** $x^2 + y^2 = 10$

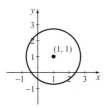

31. $(x - 2)^2 + y^2 = 25$ **33.** $(x - 1)^2 + (y + 2)^2 = 25$

35. $\left(x - \dfrac{1}{2} \right)^2 + y^2 = \dfrac{37}{4}$ **37.** 36 **39.** $\dfrac{25}{4}$ **41.** $\dfrac{9}{4}$

43. Center: $(0, 0)$; radius: 6 **45.** Center: $(1, -2)$; radius: 6

47. Center: $(8, 0)$; radius: $\dfrac{1}{2}$ **49.** Center: $(3, -2)$; radius: 4

51. Center: $(1, -1)$; radius: 3

53. Center: $(3, 2)$; radius: $3\sqrt{2}$

55. Center: $\left(\dfrac{1}{2}, 0 \right)$; radius: $\dfrac{3}{2}$ **57.** $x^2 + y^2 = 4$

59. $(x + 1)^2 + (y + 1)^2 = 4$

61. Center: $(0, 0)$; radius: 2.5

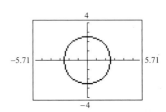

63. Center: $(3.5, 0)$; radius:

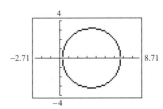

65. a. $x^2 + y^2 = 196$ **b.** 28π or about 88 plants

67. $x^2 + y^2 = 625$ **69.** 2.5 miles

71. 4.5 miles

73. a. A: $(0, 0)$; B: $(5, 0)$; C: $(10, 0)$; D: $(15, 0)$; E: $(20, 0)$; F: $(15, 12)$; G: $(10, 12)$; H: $(5, 12)$ **b.** 118 feet

Answers to Odd-Numbered Exercises ▪ Chapter 2   **A13**

75. $(x + 1)^2 + (y - 3)^2 = 5$

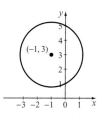

77. $(x + 5)^2 + (y + 1)^2 = 64$
79. $(0, -2 - 3\sqrt{7})$ and $(0, -2 + 3\sqrt{7})$

Exercise Set 2.2 (page 159)

1. rational expression; zero **3.** $(-\infty, \infty)$ **5.** $(-\infty, -3] \cup [3, \infty)$
7. a. $(f + g)(x) = 2x - 2$. The domain is $(-\infty, \infty)$.
b. $(f - g)(x) = 4x - 8$. The domain is $(-\infty, \infty)$.
c. $(fg)(x) = -3x^2 + 14x - 15$. The domain is $(-\infty, \infty)$.
d. $\left(\dfrac{f}{g}\right)(x) = \dfrac{3x - 5}{-x + 3}$. The domain is $(-\infty, 3) \cup (3, \infty)$.
9. a. $(f + g)(x) = x^2 + x - 2$. The domain is $(-\infty, \infty)$.
b. $(f - g)(x) = -x^2 + x - 4$. The domain is $(-\infty, \infty)$.
c. $(fg)(x) = x^3 - 3x^2 + x - 3$. The domain is $(-\infty, \infty)$.
d. $\left(\dfrac{f}{g}\right)(x) = \dfrac{x - 3}{x^2 + 1}$. The domain is $(-\infty, \infty)$.
11. a. $(f + g)(x) = \dfrac{3x - 1}{2x^2 - x}$. The domain is
$(-\infty, 0) \cup \left(0, \dfrac{1}{2}\right) \cup \left(\dfrac{1}{2}, \infty\right)$.
b. $(f - g)(x) = \dfrac{x - 1}{2x^2 - x}$. The domain is
$(-\infty, 0) \cup \left(0, \dfrac{1}{2}\right) \cup \left(\dfrac{1}{2}, \infty\right)$.
c. $(fg)(x) = \dfrac{1}{2x^2 - x}$. The domain is
$(-\infty, 0) \cup \left(0, \dfrac{1}{2}\right) \cup \left(\dfrac{1}{2}, \infty\right)$.
d. $\left(\dfrac{f}{g}\right)(x) = \dfrac{2x - 1}{x}$. The domain is $(-\infty, 0) \cup (0, \infty)$.
13. a. $(f + g)(x) = \sqrt{x} - x + 1$. The domain is $[0, \infty)$.
b. $(f - g)(x) = \sqrt{x} + x - 1$. The domain is $[0, \infty)$.
c. $(fg)(x) = -x\sqrt{x} + \sqrt{x}$. The domain is $[0, \infty)$.
d. $\left(\dfrac{f}{g}\right)(x) = \dfrac{\sqrt{x}}{-x + 1}$. The domain is $[0, 1) \cup (1, \infty)$.
15. a. $(f + g)(x) = |x| + \dfrac{1}{2x + 5}$. The domain is
$\left(-\infty, -\dfrac{5}{2}\right) \cup \left(-\dfrac{5}{2}, \infty\right)$.
b. $(f - g)(x) = |x| - \dfrac{1}{2x + 5}$. The domain is
$\left(-\infty, -\dfrac{5}{2}\right) \cup \left(-\dfrac{5}{2}, \infty\right)$.

c. $(fg)(x) = \dfrac{|x|}{2x + 5}$. The domain is $\left(-\infty, -\dfrac{5}{2}\right) \cup \left(-\dfrac{5}{2}, \infty\right)$.
d. $\left(\dfrac{f}{g}\right)(x) = |x|(2x + 5)$. The domain is $(-\infty, \infty)$.

17. 1 **19.** 3 **21.** -1 **23.** $-\dfrac{8}{3}$ **25.** -7 **27.** 5 **29.** -3

31. -7 **33.** -30 **35.** 3 **37.** $-\dfrac{1}{10}$ **39.** 0 **41.** -2

43. 3 **45.** 0 **47.** Undefined because $g(3)$ is undefined
49. 210 **51.** 42 **53.** -6 **55.** 3 **57.** 6 **59.** $\sqrt{6}$

61. $\sqrt{6}$ **63.** -18 **65.** $-\dfrac{9}{4}$

67. $(f \circ g)(x) = -x^2 - 2x$
$(g \circ f)(x) = -x^2 + 2$
The domain of $f \circ g$ is $(-\infty, \infty)$.
The domain of $g \circ f$ is $(-\infty, \infty)$.
69. $(f \circ g)(x) = x$
$(g \circ f)(x) = x$
The domain of $f \circ g$ is $(-\infty, \infty)$.
The domain of $g \circ f$ is $(-\infty, \infty)$.
71. $(f \circ g)(x) = 3x^2 + 16x + 12$
$(g \circ f)(x) = 3x^2 + 4x + 2$
The domain of $f \circ g$ is $(-\infty, \infty)$.
The domain of $g \circ f$ is $(-\infty, \infty)$.
73. $(f \circ g)(x) = \dfrac{1}{2x + 5}$
$(g \circ f)(x) = \dfrac{2}{x} + 5$
The domain of $f \circ g$ is $\left(-\infty, -\dfrac{5}{2}\right) \cup \left(-\dfrac{5}{2}, \infty\right)$.
The domain of $g \circ f$ is $(-\infty, 0) \cup (0, \infty)$.
75. $(f \circ g)(x) = \dfrac{3}{4x^2 + 1}$
$(g \circ f)(x) = \dfrac{18}{4x^2 + 4x + 1}$
The domain of $f \circ g$ is $(-\infty, \infty)$.
The domain of $g \circ f$ is $\left(-\infty, -\dfrac{1}{2}\right) \cup \left(-\dfrac{1}{2}, \infty\right)$.
77. $(f \circ g)(x) = \sqrt{-3x - 3}$
$(g \circ f)(x) = -3\sqrt{x + 1} - 4$
The domain of $f \circ g$ is $(-\infty, -1]$.
The domain of $g \circ f$ is $[-1, \infty)$.
79. $(f \circ g)(x) = \left|\dfrac{2x}{x - 1}\right|$
$(g \circ f)(x) = \dfrac{2|x|}{|x| - 1}$
The domain of $f \circ g$ is $(-\infty, 1) \cup (1, \infty)$.
The domain of $g \circ f$ is $(-\infty, -1) \cup (-1, 1) \cup (1, \infty)$.
81. $(f \circ g)(x) = x^2$
$(g \circ f)(x) = x^2 - 2x + 2$
The domain of $f \circ g$ is $(-\infty, \infty)$.
The domain of $g \circ f$ is $(-\infty, \infty)$.

83. $(f \circ g)(x) = \dfrac{|x|^2 + 1}{|x|^2 - 1}$

$(g \circ f)(x) = \left| \dfrac{x^2 + 1}{x^2 - 1} \right|$

The domain of $f \circ g$ is $(-\infty, -1) \cup (-1, 1) \cup (1, \infty)$.
The domain of $g \circ f$ is $(-\infty, -1) \cup (-1, 1) \cup (1, \infty)$.

85. $(f \circ g)(x) = \dfrac{9x^2 - 6x + 1}{13x^2 - 2x + 2}$

$(g \circ f)(x) = \dfrac{3 + x^2}{2 - x^2}$

The domain of $f \circ g$ is $(-\infty, \infty)$.
The domain of $g \circ f$ is $(-\infty, -\sqrt{2}) \cup (-\sqrt{2}, \sqrt{2}) \cup (\sqrt{2}, \infty)$.
87. $f(x) = x^2$, $g(x) = 3x - 1$ **89.** $f(x) = \sqrt[3]{x}$, $g(x) = 4x^2 - 1$

91. $f(x) = \dfrac{1}{x}$, $g(x) = 2x + 5$

93. $f(x) = \sqrt{x} + 5$, $g(x) = x^2 + 1$
95. $f(x) = 4x^5 - x^8$, $g(x) = 2x + 9$
97. $(f \circ f)(-1) = -1$
99. $(f \circ f)(t) = -t^4$. The domain is $(-\infty, \infty)$.
101. $(f \circ f)(2) = 22$
103. $(f \circ f)(t) = 9t + 4$. The domain is $(-\infty, \infty)$.

105. 3   **107.** $-2x - h + 1$ **109.** $-\dfrac{1}{x^2 - 6x + 9 + xh - 3h}$

111. $P(t) = 75 - 20t$
113.

Week, x	Difference in Hours Billed, $(f - g)(x)$
1	1
2	-7
3	3
4	3

$(f - g)(x)$ represents how many more hours Employee 1 billed for than Employee 2.

115. $\dfrac{n(t)}{p(t)}$ represents the number of students with whom each tutor worked during a given week.
117. $(f \circ R)(t) = 0.82(40 + 2t) = 32.8 + 1.64t$. This function represents the GlobalEx revenue in euros.
119. $(C \circ A)(r) = 25.8064\pi r^2$. This function represents the surface area of a sphere in square centimeters based on its radius in inches.
121. While there are some functions f and g for which $(fg)(x) = (f \circ g)(x)$ (for example, $f(x) = 2x$ and $g(x) = 3x$), the statement is not true for all functions f and g. For example, it is not true for $f(x) = x + 5$ and $g(x) = 2x$.
123. $(f + g)(x) = ax + b + (cx + d) = (a + c)x + (b + d)$. Because $a, b, c,$ and d are constants, $a + c$ and $b + d$ are constants, and so this result is of the form $ax + b$.
$(f - g)(x) = ax + b - (cx + d) = (a - c)x + (b - d)$. Because $a, b, c,$ and d are constants, $a - c$ and $b - d$ are constants, and so this result is of the form $ax + b$.

Exercise Set 2.3 (page 174)
1. The graph of $g(t)$ is the graph of $f(t) = t^2$ moved up by 1 unit.

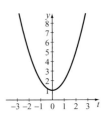

3. The graph of $f(x)$ is the graph of $g(x) = \sqrt{x}$ moved down by 2 units.

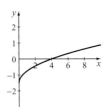

5. The graph of $h(x)$ is the graph of $f(x) = |x|$ moved to the right by 2 units.

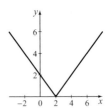

7. The graph of $F(s)$ is the graph of $f(s) = s^2$ moved to the left by 5 units.

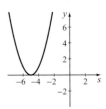

9. The graph of $f(x)$ is the graph of $g(x) = \sqrt{x}$ moved to the right by 4 units.

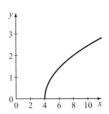

11. The graph of $H(x)$ is the graph of $g(x) = |x|$ moved to the right by 2 units and up by 1 unit.

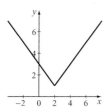

13. The graph of $S(x)$ is the graph of $g(x) = x^2$ moved to the left by 3 units and down by 1 unit.

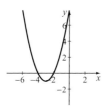

15. The graph of $H(t)$ is the graph of $g(t) = t^2$ stretched vertically away from the x-axis by a factor of 3.

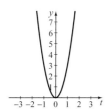

17. The graph of $S(x)$ is the graph of $f(x) = |x|$ reflected in the x-axis and stretched vertically away from the x-axis by a factor of 4.

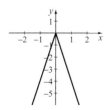

19. The graph of $H(s)$ is the graph of $f(s) = |s|$ reflected in the x-axis and moved down by 3 units.

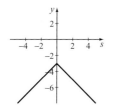

21. The graph of $h(x)$ is the graph of $f(x) = |x|$ reflected in the x-axis, compressed vertically toward the x-axis by a factor of $\frac{1}{2}$, moved to the left by 1 unit, and moved down by 3 units.

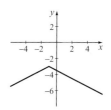

23. The graph of $g(x)$ is the graph of $f(x) = x^2$ reflected in the x-axis, stretched vertically away from the x-axis by a factor of 3, moved to the left by 2 units, and moved down by 4 units.

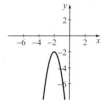

25. The graph of $f(x)$ is the graph of $g(x) = |x|$ compressed horizontally toward the y-axis and scaled by a factor of $\frac{1}{2}$.

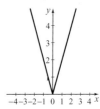

27. The graph of $f(x)$ is the graph of $g(x) = x^2$ compressed horizontally toward the y-axis and scaled by a factor of $\frac{1}{2}$.

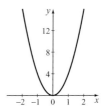

29. The graph of $g(x)$ is the graph of $f(x) = \sqrt{x}$ compressed horizontally toward the y-axis and scaled by a factor of $\frac{1}{3}$.

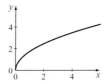

31. The graph is translated to the left by 1 unit. The function is $g(x) = |x + 1|$.

33. The graph is reflected in the x-axis and moved up by 1 unit. The function is $g(x) = -|x| + 1$.

35. The graph is moved to the right by 2 units. The function is $g(x) = (x - 2)^2$.

37. The graph is moved to the right by 1 unit and down by 2 units. The function is $g(x) = (x - 1)^2 - 2$.

39. $g(t) = |t + 4| - 3$ **41.** $g(t) = -3(t - 1)^2$

43. $k(t) = \sqrt{-t} + 3$ **45.** $h(t) = \dfrac{1}{2}|t| + 4$

47.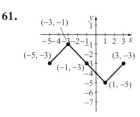

49.

51.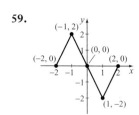

53.

55.

57.

59.

61.

63. $k = -2$

65. $k = 1$

67.

x	$f(x)$	$g(x) = f(x) - 3$
-2	36	33
-1	25	22
0	16	13
1	9	6
2	4	1

69. The graph of $f(x)$ is a translation of $h(x)$ to the left by 3.5 units, whereas the graph of $g(x)$ is a translation of $h(x)$ up by 3.5 units.

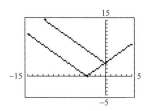

71. The graph of $f(x - 4.5)$ is a translation of the graph of $f(x)$ to the right by 4.5 units.

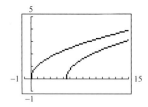

73. The graphs are different. The graph of $-2f(x)$ is a vertical scaling of the graph of $f(x)$ by a factor of 2 away from the x-axis, and then a reflection in the x-axis. The graph of $f(-2x)$ is a horizontal scaling of the graph of $f(x)$ by a factor of $\dfrac{1}{2}$ toward the y-axis, and then a reflection in the y-axis.

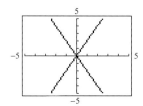

75. The graph of g is the graph of f moved to the right by 7 units.

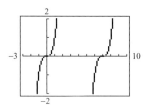

77. $T(x) = 0.06P(x)$

79. a. $C(x) = 450 + 3x$

b. The graph of the decreased cost function has a lower y-intercept but the same slope as the graph of the original cost function.

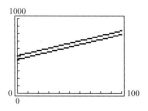

81. a. 14 meters **b.** 34 meters
c. The value of $s(t)$ is increased by h meters when $h > 0$, and decreased by $|h|$ meters when $h < 0$.
83. $g(x) = (x + 5)^2$ is $f(x) = x^2$ shifted to the left by 5 units.
85. a. $g(x) = 2|x| + 3$ **b.** $g(x) = 2|x| + 6$
c. In part (b), the y-intercept is 3 units higher than in part (a) because scaling by 2 after adding 3 units effectively scales the original equation by 2 and then moves it up by 6 units, as we can see from the equation $g(x) = 2(|x| + 3) = 2|x| + 6$.
In part (a), the function is moved up by 3 units *after* it is scaled by 2, as we can see from the equation $g(x) = 2|x| + 3$.

Exercise Set 2.4 (page 185)
1. Even **3.** Neither **5.** Odd **7.** Even **9.** Odd
11. $[-2, 2]$ **13.** $(0, 4)$ **15.** $(0, 1)$ **17.** Even
19. Decreasing on $(-3, -2)$ and $(3, 4)$ **21.** $(0, 1)$ **23.** $\dfrac{1}{2}$
25. 3 **27.** Decreasing on $(0, 1)$ **29.** -4 **31.** Neither
33. Odd because $f(-x) = -2(-x) = 2x = -(-2x) = -f(x)$
35. Neither, because $f(x) \neq f(-x)$ and $f(-x) \neq -f(x)$
37. Neither, because $f(x) \neq f(-x)$ and $f(-x) \neq -f(x)$
39. Odd because $f(-x) = 2(-x) = -(2x) = -f(x)$
41. Odd because $f(-x) = (-x)^5 - 2(-x) = -x^5 + 2x = -(x^5 - 2x) = -f(x)$
43. Even because $f(-x) = ((-x)^2 - 3)((-x)^2 - 4) = (x^2 - 3)(x^2 - 4) = f(x)$
45. 15 **47.** -8 **49.** 10 **51.** -22 **53.** -1 **55.** $1 - \dfrac{\sqrt{2}}{2}$

57. The function is neither odd nor even.

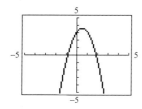

59. The function is neither odd nor even.

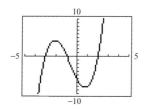

61. The function is even.

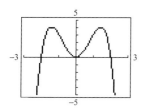

63. This is an increasing function because as x increases, $R(x)$ increases. That is, if $x_1 > x_2$, then $R(x_1) > R(x_2)$.
65. The average rate of change is 0.05 dollar per year, which means that over the interval $[0, 4]$, the value of the stamp increases at an average rate of 0.05 dollar each year.
67. As the endpoints of the interval get closer together, the average rate of change over the interval appears to get closer and closer to a single value, in this case, 6.

Interval	Average Rate of Change
$[1, 2]$	9
$[1, 1.1]$	6.3
$[1, 1.05]$	6.15
$[1, 1.01]$	6.03
$[1, 1.001]$	6.003

69. If f is constant on an interval, it is modeled by the function $f(x) = c$, and the average rate of change on $[a, b]$ is
$$\frac{f(b) - f(a)}{b - a} = \frac{c - c}{b - a} = \frac{0}{b - a} = 0.$$
71. If $[c, d]$ is within a decreasing interval, then the function is also decreasing on that subinterval, so that for $d > c$, $f(d) < f(c)$. The average rate of change on the interval $[c, d]$ is $\dfrac{f(d) - f(c)}{d - c}$. Because $d > c$, $d - c$ is positive. Because $f(d) < f(c)$, $f(d) - f(c)$ is negative. The quotient of a negative and a positive is negative, so $\dfrac{f(d) - f(c)}{d - c}$ is negative.

Exercise Set 2.5 (page 193)
1. 3 **3.** 4 **5.** $x > -4$ **7.** $-5 \leq x \leq 4$ **9.** $\{2, -10\}$
11. $\{-2, 6\}$ **13.** $\{1, -5\}$ **15.** $\left\{\dfrac{23}{4}, -\dfrac{17}{4}\right\}$ **17.** $\{-1, -9\}$
19. No solution **21.** $\{8, 2\}$ **23.** $\{5, -3\}$ **25.** $\left\{\dfrac{3}{2}, \dfrac{7}{2}\right\}$
27. $\{\sqrt{7}, -\sqrt{7}, 3, -3\}$ **29.** No **31.** Yes
33.
```
    ○────
-4-3-2-1 0 1 2 3
```
35.
```
      ●───●
-4-3-2-1 0 1 2 3
```
37.
```
     ●───●
-3-2-1 0 1 2 3
```
39.
```
  ○───────○
-8 -4  0  4  8
```
41. $(-\infty, -4) \cup (4, \infty)$ **43.** $[-7, 1]$
```
    ○───────○
-6-4-2 0 2 4 6
```
```
  ●───────────●
-8-6-4-2 0 2
```
45. $(-\infty, 4) \cup (16, \infty)$ **47.** $(-\infty, 2) \cup (5, \infty)$
```
       ○─────○
0  4  8 12 16
```
```
      ○───○
-2 0 2 4 6
```
49. $\left[-\dfrac{8}{3}, 4\right]$ **51.** $[-22, -2]$
```
   ●───────●
-4-2 0 2 4 6
```
```
  ●─────────●
-25-20-15-10-5 0
```

53. $(-37, 23)$

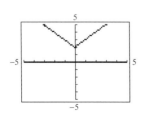

55. $(-\infty, -4] \cup [12, \infty)$

57. $\left(-\dfrac{17}{3}, 1\right)$

59. $\varnothing$

61. $(3.999, 4.001)$

63. $|x + 7| = 3$

65. $|x - 8| < 5$
67. $|x + 6.5| > 8$
69. a. $x = 1$

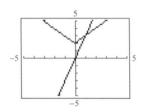

b. There are no solutions when $k = 1$.

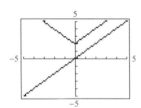

c. There are no solutions when $k = -\dfrac{1}{2}$.

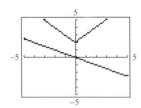

d. There are no solutions when $k = 0$.

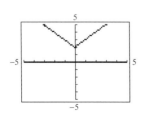

71. The solution set is $\left\{-4, \dfrac{4}{3}\right\}$.

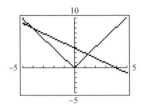

73. The solution set is $\left\{\dfrac{1}{2}, \dfrac{9}{2}\right\}$.

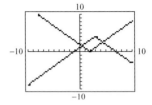

75. The solution set is $[-4, 4]$.

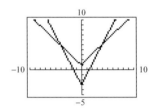

77. $-30 < T < 10$. This means the temperature is always less than 10 degrees Fahrenheit but greater than -30 degrees Fahrenheit.
79. If Omaha is assumed to be the origin, then all points within 30 miles north or south of the center of Omaha would be given by the inequality $|x| < 30$, where x is a distance directly north of Omaha (points due south would be represented by a negative distance).
81. $|T - 68| \le 1.5$
83. These expressions are not the same because $|-3(x + 2)|$ will produce strictly nonnegative values for all real x, whereas $-3|x + 2|$ will produce strictly nonpositive values for all real x.
85. By definition, $|x - k| = x - k$ or $k - x$. Similarly, $|k - x| = k - x$ or $x - k$. Therefore, these are identical values.
87. Any equation for which $|x| = a$, where a is any negative number, has no solution.

Exercise Set 2.6 (page 200)
1. $f(-2) = 1; f(0) = 1; f(1) = 1$
3. $f(-2)$ is undefined.
$f(0) = \dfrac{1}{2}; f(1) = \dfrac{1}{2}$
5. $f(-2) = -2; f(0) = 1; f(1) = 1$
7. $f(-2) = -1; f(0) = 2; f(1) = 2$
9. $f(-2)$ is undefined.
$f(0) = 0; f(1) = 1$
11. 3 **13.** 3 **15.** -4

17. 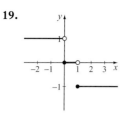 **19.**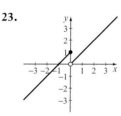

43. a. $S(4) = 20.75$, which means that 20.75 million portable CD players were sold in 2004. **b.** 32 million
45. a. $1.06 **b.** $1.75 **c.** $1.52
47. a. Because any call lasting 20 minutes or less will cost $1; the cost is not proportional to the length of the call in this case.

b. $C(t) = \begin{cases} 1, & \text{if } 0 < t \le 20 \\ 1 + 0.7(t - 20), & \text{if } t > 20 \end{cases}$

c. A 5-minute call will cost $1, a 20-minute call will cost $1, and a 30-minute call will cost $1.70.

21. 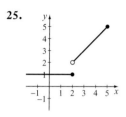 **23.**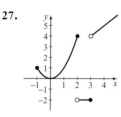

49. 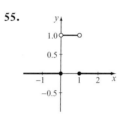 **51.** 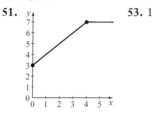 **53.** 1

25. 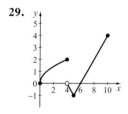 **27.**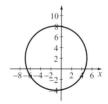

55.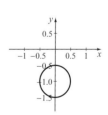

29.

Chapter 2 Review Exercises (page 207)

Section 2.1 (page 207)

1. Distance: $\sqrt{34}$; midpoint: $\left(-\dfrac{3}{2}, \dfrac{1}{2}\right)$

3. Distance: $\sqrt{41}$; midpoint: $\left(\dfrac{17}{2}, 11\right)$

31. $f(-1) = 2; f(0) = 2; f(2)$ is undefined.
33. Domain: $(-\infty, \infty)$
Range: $\{2\} \cup [3, \infty)$
x-intercept(s): none
y-intercept: $(0, 3)$
$f(x) = \begin{cases} 2, & \text{whenever } x < 0 \\ x^2 + 3, & \text{whenever } x \ge 0 \end{cases}$
35. Domain: $(-\infty, \infty)$
Range: $(-\infty, 1]$
x-intercept(s): $(-2, 0)$ and $(2, 0)$
y-intercept: $(0, 1)$
$f(x) = \begin{cases} x + 2, & \text{whenever } x \le -1 \\ 1, & \text{whenever } -1 < x < 1 \\ -x + 2, & \text{whenever } x \ge 1 \end{cases}$

5. $(x + 1)^2 + (y - 2)^2 = 36$ **7.** $x^2 + (y + 1)^2 = \dfrac{1}{4}$

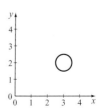

37. 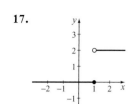 **39.**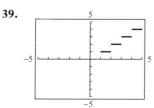

9. The center is $(3, 2)$ and the radius is $\dfrac{1}{2}$.

41. a. $20 **b.** $70

11. The center is $(4, -1)$ and the radius is $\sqrt{22}$.

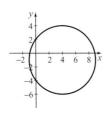

13. The center is $(1, -1)$ and the radius is 3.

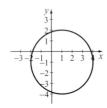

15. The center is $(0, 2.3)$ and the radius is $\sqrt{17}$.

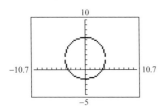

Section 2.2 (page 208)

17. a. $(f + g)(x) = 4x^2 + x + 2$. The domain is $(-\infty, \infty)$.
b. $(f - g)(x) = 4x^2 - x$. The domain is $(-\infty, \infty)$.
c. $(fg)(x) = 4x^3 + 4x^2 + x + 1$. The domain is $(-\infty, \infty)$.
d. $\left(\dfrac{f}{g}\right)(x) = \dfrac{4x^2 + 1}{x + 1}$. The domain is $(-\infty, -1) \cup (-1, \infty)$.

19. a. $(f + g)(x) = \dfrac{x^2 + 2x + 1}{2x^3 + 2x}$. The domain is
$(-\infty, 0) \cup (0, \infty)$.
b. $(f - g)(x) = \dfrac{x^2 - 2x + 1}{2x^3 + 2x}$. The domain is
$(-\infty, 0) \cup (0, \infty)$.
c. $(fg)(x) = \dfrac{1}{2x^3 + 2x}$. The domain is $(-\infty, 0) \cup (0, \infty)$.
d. $\left(\dfrac{f}{g}\right)(x) = \dfrac{x^2 + 1}{2x}$. The domain is $(-\infty, 0) \cup (0, \infty)$.

21. a. $(f + g)(x) = \dfrac{3x^3 - 12x^2 + 2}{x - 4}$. The domain is
$(-\infty, 4) \cup (4, \infty)$.
b. $(f - g)(x) = \dfrac{-3x^3 + 12x^2 + 2}{x - 4}$. The domain is
$(-\infty, 4) \cup (4, \infty)$.
c. $(fg)(x) = \dfrac{6x^2}{x - 4}$. The domain is $(-\infty, 4) \cup (4, \infty)$.

d. $\left(\dfrac{f}{g}\right)(x) = \dfrac{2}{3x^2(x - 4)}$. The domain is
$(-\infty, 0) \cup (0, 4) \cup (4, \infty)$.

23. 0 **25.** 5 **27.** 7 **29.** $\dfrac{5}{8}$ **31.** -13 **33.** -19

35. $(f \circ g)(x) = -x^2 + 4x$
$(g \circ f)(x) = -x^2 + 2$
The domain of $f \circ g$ is $(-\infty, \infty)$.
The domain of $g \circ f$ is $(-\infty, \infty)$.
37. $(f \circ g)(x) = -x^2 + 9x - 18$
$(g \circ f)(x) = -x^2 + 3x - 3$
The domain of $f \circ g$ is $(-\infty, \infty)$.
The domain of $g \circ f$ is $(-\infty, \infty)$.

39. $(f \circ g)(x) = \dfrac{1}{x^2 + x - 2}$

$(g \circ f)(x) = \dfrac{x - 1}{(x - 2)^2}$

The domain of $f \circ g$ is $(-\infty, -2) \cup (-2, 1) \cup (1, \infty)$.
The domain of $g \circ f$ is $(-\infty, 2) \cup (2, \infty)$.

41. $(f \circ g)(x) = \dfrac{|x|}{|x| + 3}$

$(g \circ f)(x) = \left|\dfrac{x}{x + 3}\right|$

The domain of $f \circ g$ is $(-\infty, \infty)$.
The domain of $g \circ f$ is $(-\infty, -3) \cup (-3, \infty)$.
43. 4 **45.** $4x + 2h - 3$

Section 2.3 (page 209)

47.

49.

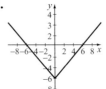

51.

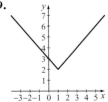

53.

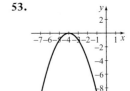

55.

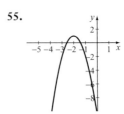

57.

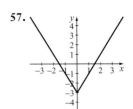

59. $g(x) = |x - 3| + 1$ **61.** $g(x) = 2(x + 1)^2$

63. 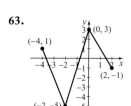   **65.**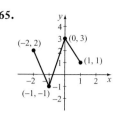

67. The first function moves $f(x) = x^2$ to the right by 1.5 units, and the second function moves $f(x) = x^2$ down by 1.5 units.

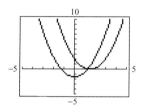

Section 2.4 (page 209)

69. f is increasing on the intervals $(-2, 2)$ and $(3, 4)$.
71. f is constant on the interval $(4, 5)$.
73. 1   **75.** Odd   **77.** Even   **79.** Odd
81. This function is neither even nor odd.

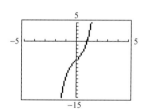

83. -2 **85.** -1

Section 2.5 (page 210)

87. $\{-1, 11\}$ **89.** $\left\{-\dfrac{15}{4}, \dfrac{17}{4}\right\}$ **91.** $\{2, 3\}$

93. $(-\infty, -5) \cup \left(-\dfrac{5}{3}, \infty\right)$ **95.** $[-22, -2]$

97. $\left(-\dfrac{8}{3}, \dfrac{4}{3}\right)$ **99.** $[-1, 3]$

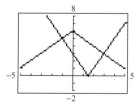

Section 2.6 (page 210)

101. 0   **103.** 3
105. 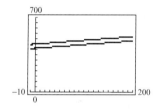   **107.**

Applications (page 210)

109. $x^2 + y^2 = 484$
111. $P(t) = 15 + 2.6t$
113. a. $C(x) = 410 + 0.35x$
b. $S(x) = 18 + 0.014x$
c. Graph of original cost function shifted downward (Answers may vary.)

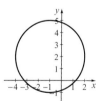

115. a. \$3   **b.** $C(x) = \begin{cases} 0.25x, & \text{if } x \le 10 \\ 2.50 + 0.10x, & \text{if } x > 10 \end{cases}$

Chapter 2 Test (page 211)

1. $\sqrt{26}; \left(-\dfrac{3}{2}, \dfrac{5}{2}\right)$ **2.** $(x - 2)^2 + (y - 5)^2 = 36$
3. Center: $(-1, 2)$
Radius: 3

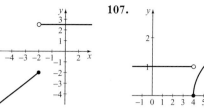

4. 11   **5.** -2 **6.** 3   **7.** 0   **8.** -1 **9.** -3
10. $\left(\dfrac{f}{g}\right)(x) = \dfrac{x^2 + 2x}{2x - 1}, x \ne \dfrac{1}{2}$ **11.** $2x + h + 2$
12. $\dfrac{1}{2x^2 - 2}, x \ne -1, 1$
13. 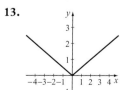   **14.**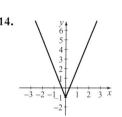

15. 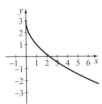 **16.**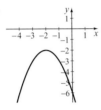

17. $g(x) = |x + 2| + 1$ **18.** $g(x) = (2x)^2 - 1$
19. Neither **20.** Even **21.** Odd

22. a. $(0, 2)$ **b.** $(-5, 0)$ **c.** $(2, 5)$ **23.** $-\dfrac{1}{6}$

24. 20 **25.** $\left\{ \dfrac{11}{18}, -\dfrac{19}{18} \right\}$ **26.** $\left\{ 4, -\dfrac{1}{2} \right\}$

27. $\left[-\dfrac{28}{5}, 4 \right]$ **28.** $(-\infty, 1] \cup [4, \infty)$

29. $\left(-\dfrac{9}{5}, 1 \right)$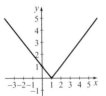

30. a. 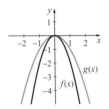 **b.** $f(x) = |x - 1|$

31. $|x| \geq 53$; Use the center of St. Louis as the origin.
32. a. $260 million **b.** $P(t) = 6.5t - 15$
c. In approximately 3.85 years
33. $c(x) = \begin{cases} 3.50x, & 0 < x \leq 50 \\ 175 + 3(x - 50), & x > 50 \end{cases}$

Chapter 3 Exercises

Exercise Set 3.1 (page 225)

1. up **3.** to the left **5.** $(x - 8)^2$ **7.** $(8x + 3)^2$
9. The domain for both functions is $(-\infty, \infty)$. The range for both functions is $(-\infty, 0]$.

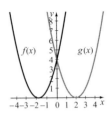

11. The domain for both functions is $(-\infty, \infty)$. The range for both functions is $[0, \infty)$.

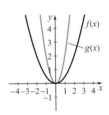

13. The domain for both functions is $(-\infty, \infty)$. The range for $f(x)$ is $[1, \infty)$ and the range for $g(x)$ is $[-1, \infty)$.

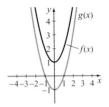

15. The domain for both functions is $(-\infty, \infty)$. The range for both functions is $[0, \infty)$.

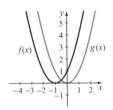

17. The domain for both functions is $(-\infty, \infty)$. The range for both functions is $[0, \infty)$.

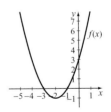

19. b **21.** a **23.** e **25.** f
27. The vertex is $(-2, -1)$.

29. The vertex is $(-1, -1)$.

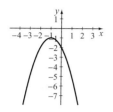

31. The vertex is $(-4, -2)$.

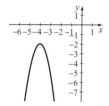

33. The function can be written as $g(x) = (x + 1)^2 + 4$. The vertex is $(-1, 4)$ and it is a minimum point.
35. The function can be written as $w(x) = -(x - 3)^2 + 13$. The vertex is $(3, 13)$ and it is a maximum point.
37. The function can be written as $h(x) = \left(x + \frac{1}{2}\right)^2 - \frac{13}{4}$.
The vertex is $\left(-\frac{1}{2}, -\frac{13}{4}\right)$ and it is a minimum point.
39. The function can be written as $f(x) = 3(x + 1)^2 - 7$. The vertex is $(-1, -7)$ and it is a minimum point.
41. The vertex is $(1, 1)$ and the axis of symmetry is $x = 1$.
Points on the parabola include $(0, -1)$ and $(2, -1)$.

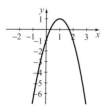

43. The vertex is $(2, 1)$ and the axis of symmetry is $x = 2$.
Points on the parabola include $(0, -3)$ and $(4, -3)$.

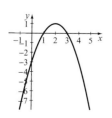

45. The vertex is $\left(\frac{3}{2}, \frac{11}{4}\right)$ and the axis of symmetry is $x = \frac{3}{2}$.
Points on the parabola include $(0, 5)$ and $(3, 5)$.

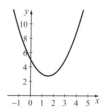

47. The vertex is $(0, 100)$ and the axis of symmetry is $t = 0$.
Points on the parabola include $(2, 36)$ and $(-2, 36)$.

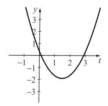

49. The vertex is $\left(\frac{3}{2}, -\frac{23}{12}\right)$ and the axis of symmetry is
$t = \frac{3}{2}$. Points on the parabola include $\left(0, \frac{1}{3}\right)$ and $\left(3, \frac{1}{3}\right)$.

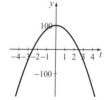

51. The vertex is $(5, 17)$ and the axis of symmetry is $x = 5$.
The function is increasing on the interval $(-\infty, 5)$ and decreasing on the interval $(5, \infty)$. The range of the function is $(-\infty, 17]$.

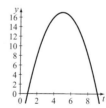

53. The vertex is $(-1, -5)$ and the axis of symmetry is
$x = -1$. The function is increasing on the interval $(-1, \infty)$ and decreasing on the interval $(-\infty, -1)$. The range of the function is $[-5, \infty)$.

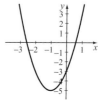

55. The vertex is $(1, -2)$ and the axis of symmetry is $x = 1$. The function is increasing on the interval $(-\infty, 1)$ and decreasing on the interval $(1, \infty)$. The range of the function is $(-\infty, -2]$.

57. The vertex is $\left(-1, -\dfrac{9}{4}\right)$ and the axis of symmetry is $x = -1$. The function is increasing on the interval $(-1, \infty)$ and decreasing on the interval $(-\infty, -1)$. The range of the function is $\left[-\dfrac{9}{4}, \infty\right)$.

59. a. $(-\infty, \infty)$ **b.** $(-\infty, 0]$ **c.** Maximum of 0
d. $x = 1$ **e.** $(-\infty, 1)$ **f.** $(1, \infty)$
61. a. $(-\infty, \infty)$ **b.** $[1, \infty)$ **c.** Minimum of 1
d. $x = -3$ **e.** $(-3, \infty)$ **f.** $(-\infty, -3)$
63. The vertex is $(0, 20)$.

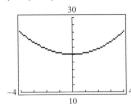

65. The vertex is $(-0.3536, 0.8232)$.

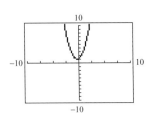

67. The vertex is $(1.25, 145)$.

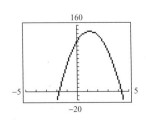

69. a. The axis of symmetry is $x = 2$.
b. By symmetry, a third point is $(1, -1)$.

c.

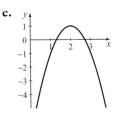

71. $b = 4$ **73.** 250,000 square feet
75. a. 39.0625 feet **b.** 19.53125 feet
77. 83.8% at 49°F, 2.2% at 47°F
79. a. model estimate: 8.9534 million; prediction within $\dfrac{1}{2}\%$ of actual value
b. Predicts 21.335 million
c. Vertex is $(7.99, 7.1875)$. The function predicts attendance decreasing from 10.31 million in 1981 to 7.1875 million in 1988, and then increasing after that.
d. 2025 is too many years past data used to create the model.
e. $0 \le t \le 20$ is a possible answer.

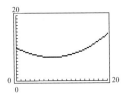

81. $15

83. a. $\dfrac{\pi x}{2}$

b. $x + 2y + \dfrac{\pi x}{2} = 24$. In terms of y, this is $y = 12 - \dfrac{\pi x}{4} - \dfrac{x}{2}$.

c. The area of the window is

$$A(x) = x\left(12 - \frac{\pi x}{4} - \frac{x}{2}\right) + \frac{\pi\left(\dfrac{x}{2}\right)^2}{2}.$$

d. $x = \dfrac{48}{\pi + 4} \approx 6.72$, $y = \dfrac{24}{\pi + 4} \approx 3.36$;

$A = \dfrac{1152 + 288\pi}{(\pi + 4)^2} \approx 40.327$ square feet

85. a. $d(t) = a(t - 3)^2 + 144$ **b.** $a = -16$
c. $d(t) = -16(t - 3)^2 + 144$
d. Yes. $144 = -16(3 - 3)^2 + 144$ and $0 = -16(0 - 3)^2 + 144$.

87. $f(x) = -\dfrac{20}{9}\left(x - \dfrac{3}{2}\right)^2 + 5$

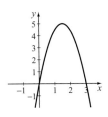

89. a.

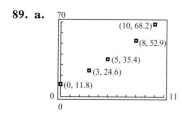

70
(10, 68.2)■
■(8, 52.9)
■(5, 35.4)
■(3, 24.6)
■(0, 11.8)
0 ⌊_____⌋ 11
0

b. $f(x) = 0.1954x^2 + 3.6454x + 11.8844$

c.

Year	1995	1998	2000	2003	2005
Actual	11.8	24.6	35.4	52.9	68.2
Predicted	11.88	24.58	35.0	53.56	67.88

d. 60.52   **e.** Error of about 1% in prediction
91. A linear function has a constant rate of change (slope) but a quadratic function does not. A linear function increases (or decreases) over its entire domain, but a quadratic function decreases and then increases (or increases and then decreases) over its domain.
93. a. $(-6, 0)$ **b.** $f(x) = -\dfrac{2}{25}(x + 1)^2 + 2$

c.

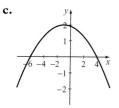

95. a. $h = 2, k = 8$ **b.** $f(x) = a(x - 2)^2 + 8$ **c.** $a = -2$
d. $f(x) = -2(x - 2)^2 + 8$ **e.**

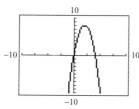

10
-10 ⌊_____⌋ 10
-10

Exercise Set 3.2 (page 241)

1. $(x - 8)(x - 5)$ **3.** $(x - 1)^2$ **5.** 9   **7.** $(x - 4)^2$
9. $x = 5$ and $x = -5$ **11.** $x = 4$ and $x = 3$

13. $x = 2$ and $x = -2$ **15.** $x = \dfrac{2}{3}$ and $x = -\dfrac{1}{2}$ **17.** $x = \dfrac{1}{2}$

19. The x-intercepts are $(3, 0)$ and $(-3, 0)$, and the zeros are $x = 3$ and $x = -3$.
21. The s-intercept is $(1, 0)$, and the zero is $s = 1$.
23. The x-intercepts are $\left(\dfrac{1}{2}, 0\right)$ and $(-3, 0)$, and the zeros are $x = \dfrac{1}{2}$ and $x = -3$.
25. The t-intercepts are $\left(\dfrac{3}{2}, 0\right)$ and $(-1, 0)$, and the zeros are $t = \dfrac{3}{2}$ and $t = -1$.

27. $f(x) = x^2 + 2x - 3$ or, more generally, any equation of the form $f(x) = n(x^2 + 2x - 3), n \neq 0$
29. $f(x) = x^2 + 3x$ or, more generally, any equation of the form $f(x) = n(x^2 + 3x), n \neq 0$
31. $f(x) = 2x^2 - 7x + 3$ or, more generally, any equation of the form $f(x) = n(2x^2 - 7x + 3), n \neq 0$
33. $x = -1$ and $x = -3$ **35.** $x = 1 + \sqrt{5}$ and $x = 1 - \sqrt{5}$
37. $x = 1$ and $x = -2$
39. $x = -2 + \dfrac{3\sqrt{2}}{2}$ and $x = -2 - \dfrac{3\sqrt{2}}{2}$
41. $x = -1 + \sqrt{2}$ and $x = -1 - \sqrt{2}$
43. $x = \dfrac{1}{2} + \dfrac{\sqrt{3}}{2}$ and $x = \dfrac{1}{2} - \dfrac{\sqrt{3}}{2}$
45. $x = -\dfrac{1}{2} + \dfrac{\sqrt{13}}{2}$ and $x = -\dfrac{1}{2} - \dfrac{\sqrt{13}}{2}$
47. No real solution
49. $l = 20 + 10\sqrt{3}$ and $l = 20 - 10\sqrt{3}$
51. $t = 4 + \sqrt{22}$ and $t = 4 - \sqrt{22}$
53. $x = -\dfrac{4}{3} + \dfrac{2\sqrt{10}}{3}$ and $x = -\dfrac{4}{3} - \dfrac{2\sqrt{10}}{3}$
55. $x = 2$ and $x = -2$ **57.** $x = 1$
59. $x = -1$ and $x = -\dfrac{1}{2}$
61. $x = 1 + \sqrt{10}$ and $x = 1 - \sqrt{10}$
63. $x = -\dfrac{1}{2} + \dfrac{\sqrt{13}}{2}$ and $x = -\dfrac{1}{2} - \dfrac{\sqrt{13}}{2}$
65. The discriminant is 8, so the graph of f has two x-intercepts and the equation $f(x) = 0$ has two real solutions.
67. The discriminant is -7, so the graph of f has no x-intercepts and the equation $f(x) = 0$ has no real solutions.
69. The discriminant is 0, so the graph of f has one x-intercept and the equation $f(x) = 0$ has one real solution.
71. The vertex is $(3, -4)$. The axis of symmetry is $x = 3$. The x-intercepts are $(1, 0)$ and $(5, 0)$.

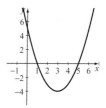

73. The vertex is $\left(\dfrac{3}{4}, \dfrac{17}{8}\right)$. The axis of symmetry is $s = \dfrac{3}{4}$.
The s-intercepts are $\left(\dfrac{3}{4} + \dfrac{\sqrt{17}}{4}, 0\right)$ and $\left(\dfrac{3}{4} - \dfrac{\sqrt{17}}{4}, 0\right)$.

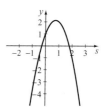

75. The vertex is $\left(-\frac{1}{2}, \frac{3}{4}\right)$. The axis of symmetry is $t = -\frac{1}{2}$. The function has no t-intercepts.

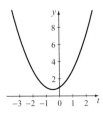

77. a.

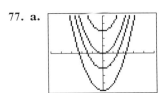

b. f will have two x-intercepts if $c < 0$. Because the parabola opens upward ($a > 0$), the vertex has to be below the x-axis for the graph of f to intersect the x-axis.
c. f will have two real zeros if $c < 0$. A zero represents the x-coordinate of an x-intercept, so the same reasoning as in part (b) applies.
d. f will have one x-intercept if $c = 0$. The vertex has to be on the x-axis for the graph to intersect the x-axis in only one place. For the same reason, f will have one real zero if $c = 0$.
e. f will have no x-intercepts if $c > 0$. Because the parabola opens upward ($a > 0$), the vertex has to be above the x-axis for the graph of f not to intersect the x-axis. For the same reason, f will have no real zeros if $c > 0$.
79. a. 1.58 seconds   **b.** 2.5 seconds   **c.** $0 \le t \le 2.5$
81. x in the diagram is either $5 + \sqrt{10}$ or $5 - \sqrt{10}$.
83. 5 feet × 14 feet or 7 feet × 10 feet
85. Near the end of 1998
87. $x \approx 3.79$ feet
89. a. $4x + 2y = 500$ **b.** $y = 250 - 2x$
c. $A(x) = -2x^2 + 250x$ **d.** 62.5 feet × 125 feet
91. a. (300, 0) and (−300, 0). These points represent the points at which the arch intersects the ground.
b. The vertex represents the highest point on the arch.
c. $y = -\dfrac{x^2}{144} + 625$
d. $y = 555.\overline{5}$
93. a. The rate increases by an increasing amount each year (with the exception of the last year).
b. Linear: $y = 54.836 + 2.814x$;
quadratic: $y = 0.236x^2 + 0.649x + 57.386$
c. The quadratic model, because it more closely models the data set.
d. During 1999
95. $f(x) = x^2 + 4x + 3$
97. $f(l) = -l^2 + 40l$ and $y(l) = 100$
99. $f(x) = x^2 - 5x + 6$, or any multiple of this function

Exercise Set 3.3 (page 253)

1. $x^2 + x - 6$ **3.** $x^2 - 16$ **5.** $\dfrac{5 - \sqrt{13}}{2}$ and $\dfrac{5 + \sqrt{13}}{2}$

7. $\dfrac{3 - \sqrt{17}}{2}$ and $\dfrac{3 + \sqrt{17}}{2}$ **9.** $4i$ **11.** $2i\sqrt{3}$

13. $\dfrac{2}{5}i$ **15.** $\pm 4i$ **17.** $\pm 2i\sqrt{2}$ **19.** $\pm i\sqrt{10}$

21. The real part is 2 and the imaginary part is 0.
23. The real part is 0 and the imaginary part is $-\pi$.
25. The real part is $1 + \sqrt{5}$ and the imaginary part is 0.
27. The real part is 1 and the imaginary part is $\sqrt{5}$.
29. -2 **31.** $-1 - i$ **33.** $3 + \sqrt{2}$ **35.** -1

37. $2 + 2i$; $-2 + 4i$; $3 + 6i$; $-\dfrac{3}{5} + \dfrac{6}{5}i$

39. $-1 + 2i$; $-5 + 8i$; $9 + 19i$; $-\dfrac{21}{13} + \dfrac{1}{13}i$

41. $7 - 3i$; $1 - 7i$; $22 - 7i$; $\dfrac{2}{13} - \dfrac{23}{13}i$

43. $\dfrac{7}{10} - \dfrac{5}{3}i$; $\dfrac{3}{10} - \dfrac{13}{3}i$; $\dfrac{41}{10} + \dfrac{1}{15}i$; $-\dfrac{1755}{818} - \dfrac{285}{409}i$

45. $-\dfrac{5}{6} - \sqrt{5}i$; $\dfrac{1}{6} + 3\sqrt{5}i$; $\dfrac{61}{6} + \dfrac{\sqrt{5}}{6}i$; $-\dfrac{118}{243} - \dfrac{14\sqrt{5}}{243}i$

47. $-\dfrac{5}{2} + 2i$; $-\dfrac{7}{2}$; $-\dfrac{5}{2} - \dfrac{5}{2}i$; $-\dfrac{2}{5} + \dfrac{14}{5}i$

49. $x = \dfrac{3\sqrt{2}}{2}i$ and $x = -\dfrac{3\sqrt{2}}{2}i$

51. $x = \dfrac{\sqrt{30}}{3}i$ and $x = -\dfrac{\sqrt{30}}{3}i$

53. $x = -\dfrac{1}{2} + \dfrac{\sqrt{3}}{2}i$ and $x = -\dfrac{1}{2} - \dfrac{\sqrt{3}}{2}i$

55. $t = \dfrac{1 + 2\sqrt{7}}{3}$ and $t = \dfrac{1 - 2\sqrt{7}}{3}$

57. $x = \dfrac{-1 + \sqrt{23}}{2}$ and $x = \dfrac{-1 - \sqrt{23}}{2}$

59. $x = -4$ and $x = \dfrac{4}{3}$

61. The zeros of the function $f(x) = x^2 + 2x + 3$ are $-1 + \sqrt{2}i$ and $-1 - \sqrt{2}i$.
63. The zeros of the function $f(x) = -3x^2 + 2x - 4$ are $\dfrac{1}{3} + \dfrac{\sqrt{11}}{3}i$ and $\dfrac{1}{3} - \dfrac{\sqrt{11}}{3}i$.
65. The zeros of the function $f(x) = 5x^2 - 2x + 3$ are $\dfrac{1}{5} + \dfrac{\sqrt{14}}{5}i$ and $\dfrac{1}{5} - \dfrac{\sqrt{14}}{5}i$.
67. The zeros of the function $f(x) = 5x^2 + 2x + 3$ are $-\dfrac{1}{5} + \dfrac{\sqrt{14}}{5}i$ and $-\dfrac{1}{5} - \dfrac{\sqrt{14}}{5}i$.
69. The zeros of the function $f(x) = -3x^2 + 8x - 16$ are $\dfrac{4}{3} + \dfrac{4\sqrt{2}}{3}i$ and $\dfrac{4}{3} - \dfrac{4\sqrt{2}}{3}i$.
71. The zeros of the function $f(t) = -4t^2 + t - \dfrac{1}{2}$ are $\dfrac{1}{8} + \dfrac{\sqrt{7}}{8}i$ and $\dfrac{1}{8} - \dfrac{\sqrt{7}}{8}i$.

73. The zeros of the function $f(x) = \frac{2}{3}x^2 + x + 1$ are $-\frac{3}{4} + \frac{\sqrt{15}}{4}i$ and $-\frac{3}{4} - \frac{\sqrt{15}}{4}i$.

75. The zeros of the function $f(x) = x^2 + 2x + 26$ are $-1 + 5i$ and $-1 - 5i$.

77. $z + \bar{z} = (a + bi) + (a - bi) = (a + a) + (b - b)i = 2a$ and $z - \bar{z} = (a + bi) - (a - bi) = (a - a) + (b + b)i = 2bi$.

79. If $z = a + bi$ and $\bar{z} = a - bi$, then $\frac{z + \bar{z}}{2} = \frac{(a + bi) + (a - bi)}{2} = \frac{2a}{2} = a$. This is equal to the real part of z.

81. $4 - 4a$ **83.** $a = 1$

85. $x \approx 2.280 - 2.191i$ and $x \approx 2.280 + 2.191i$

87. $x \approx 0.333 + 1.1583i$ and $x \approx 0.333 - 1.1583i$

89. a. The vertex is above the x-axis and the parabola opens upward. Graphically, the parabola does not intersect the x-axis and so there are no real zeros of the function graphed.
b. The vertex form of a parabola is $f(x) = a(x - h)^2 + k$. If the vertex is at $(0, 4)$, the equation would be $f(x) = ax^2 + 4$. If $a > 0$, this parabola would open up. Thus $f(x) = 2x^2 + 4$ is a possible equation.
c. $x = \pm i\sqrt{2}$

91. a. $x^2 + 1$ **b.** $\pm i$
c. $(x + i)(x - i)$ represents the factorization of $x^2 + 1$, and the solutions of $(x + i)(x - i) = 0$ provide the zeros of $f(x) = x^2 + 1$.
d. $(x + 3i)(x - 3i)$ **e.** $(x + ci)(x - ci)$

93. a. The line of symmetry for the associated parabola is $x = 0$ because $f(x) = f(-x)$, and so $(x, f(x))$ and $(-x, f(-x))$ are the same distance from $(0, f(x))$ for all x.
b. The minimum value of the function is 1, which occurs at $x = 0$.

c.

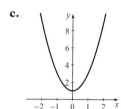

d. The function has two nonreal zeros, since its graph doesn't intersect the x-axis.

Exercise Set 3.4 (page 264)

1. The solution set is $-1 \le x \le 2$, or $[-1, 2]$.
3. The solution set is $-1 < x < 2$, or $(-1, 2)$.
5. The solution set is $t \le -2$ or $t \ge 0$, or $(-\infty, -2] \cup [0, \infty)$.
7. The solution set is $t < -2$ or $t > 0$, or $(-\infty, -2) \cup (0, \infty)$.
9. The solution set is $t \le -4$ or $t \ge 0$, or $(-\infty, -4] \cup [0, \infty)$.
11. The solution set is $-0.5 \le x \le 2$, or $[-0.5, 2]$.
13. The solution set is $0 \le t \le 1$, or $[0, 1]$.
15. The solution set is $-1 \le x \le 1$, or $[-1, 1]$.

17. The solution set is $x \ge 1$ or $x \le -\frac{5}{2}$, or $\left(-\infty, -\frac{5}{2}\right] \cup [1, \infty)$.

19. The solution set is $x \ge 1$ or $x \le -\frac{2}{3}$, or $\left(-\infty, -\frac{2}{3}\right] \cup [1, \infty)$.

21. The solution set is $-\frac{1}{2} < x < 1$, or $\left(-\frac{1}{2}, 1\right)$.

23. The solution set is $x \ge 2$ or $x \le -\frac{2}{5}$, or $\left(-\infty, -\frac{2}{5}\right] \cup [2, \infty)$.

25. The solution set is $-\frac{3}{2} \le x \le \frac{1}{5}$, or $\left[-\frac{3}{2}, \frac{1}{5}\right]$.

27. The solution set is $x < 1$ or $x > 1$, or $(-\infty, 1) \cup (1, \infty)$.

29. The solution set is $\frac{3}{4} - \frac{\sqrt{17}}{4} < x < \frac{3}{4} + \frac{\sqrt{17}}{4}$, or $\left(\frac{3}{4} - \frac{\sqrt{17}}{4}, \frac{3}{4} + \frac{\sqrt{17}}{4}\right)$.

31. The solution set is $x \le \frac{1}{2} - \frac{\sqrt{17}}{2}$ or $x \ge \frac{1}{2} + \frac{\sqrt{17}}{2}$, or $\left(-\infty, \frac{1}{2} - \frac{\sqrt{17}}{2}\right] \cup \left[\frac{1}{2} + \frac{\sqrt{17}}{2}, \infty\right)$.

33. The solution set is $x \le \frac{1}{6} - \frac{\sqrt{13}}{6}$ or $x \ge \frac{1}{6} + \frac{\sqrt{13}}{6}$, or $\left(-\infty, \frac{1}{6} - \frac{\sqrt{13}}{6}\right] \cup \left[\frac{1}{6} + \frac{\sqrt{13}}{6}, \infty\right)$.

35. The solution set is $\varnothing$.
37. The solution set is all real values of x.
39. $0 < w \le 20$ or $30 \le w < 50$, or $(0, 20] \cup [30, 50)$
41. $0 < x \le 5 - \sqrt{10}$ or $5 + \sqrt{10} \le x < 10$, or $(0, 1.838] \cup [8.162, 10)$
43. Between 1981 and 1984 or between 1993 and 2000
45. $n \ge 9$
47. If x is a real number, $x^2 \ge 0$ for all real x, so $ax^2 \le 0$ only if $a \le 0$.
49. The only solution of $(x + 1)^2 \le 0$ is $x = -1$. Thus $(x + 1)^2 < 0$ has no solution, because any real number squared cannot be negative.

Exercise Set 3.5 (page 273)

1. True **3.** 3 **5.** $2x - 10$
7. $x = \pm\sqrt{7}$ and $x = \pm i\sqrt{7}$
9. $x = \pm\sqrt{7}$ and $x = \pm\sqrt{3}$
11. $s = \pm\frac{\sqrt{6}}{3}$ and $s = \pm\frac{i\sqrt{2}}{2}$
13. $x = \pm\sqrt{2}$ and $x = \pm\frac{i}{2}$
15. $x = -1$ and $x = \sqrt[3]{5}$ **17.** $t = \sqrt[3]{-\frac{2}{3}}$ and $t = \sqrt[3]{-4}$
19. $x = \frac{30}{19}$ **21.** $x = \frac{4}{3}$ **23.** $x = -\frac{1}{2}$ and $x = \frac{1}{5}$
25. $x = 3$ **27.** $x = \frac{13 + \sqrt{249}}{8}$ and $x = \frac{13 - \sqrt{249}}{8}$
29. $x = 0$ **31.** No solution **33.** $x = 0$ and $x = 3$
35. $x = 22$ **37.** $x = \pm 4$ **39.** $x = -8$ and $x = 2$
41. $x = \frac{5 + \sqrt{13}}{2}$ **43.** $x = 122$ **45.** $x = 17$

47. $x = 3$ and $x = 11$ **49.** $x = 6$ **51.** $x = 1$ and $x = 9$

53. $x = -1$ and $x = \dfrac{1}{27}$

55. $x \approx 1.4475$

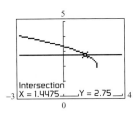

57. $x \approx 1.0699$

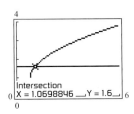

59. $k = \dfrac{1}{2}$ has one solution, $x = 0.866$; $k = 1$ has two solutions, $x = 0$ and -1; and $k = 2$ has no solutions.

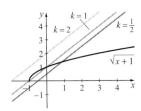

61. $120
63. 12 hours
65. 4.681 kilometers east of the starting point
67. 9.196 kilometers east of the starting point
69. $\sqrt{x+1} - 2 = 0$
$(x + 1) + 4 = 0$
The left-hand side of the equation is squared incorrectly (there should be a middle term).
The 2 should be moved to the other side first: $\sqrt{x+1} = 2$, then square both sides: $x + 1 = 4$, then solve: $x = 3$.
71. Four zeros: $x = \pm 1$, $x = \pm i$; two x-intercepts: $(1, 0)$ and $(-1, 0)$

Chapter 3 Review Exercises (page 280)

Section 3.1 (page 280)
1. The domain of $f(x)$ is $(-\infty, \infty)$, and the range of $f(x)$ is $[0, \infty)$. The domain of $g(x)$ is $(-\infty, \infty)$, and the range of $g(x)$ is $[0, \infty)$.

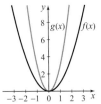

3. The vertex is $(-3, -1)$.

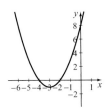

5. The vertex is $(0, -1)$.

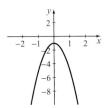

7. An equivalent function is $f(x) = (x - 2)^2 - 1$. The vertex is $(2, -1)$ and it is a minimum point.
9. An equivalent function is $f(x) = 4(x + 1)^2 - 5$. The vertex is $(-1, -5)$ and it is a minimum point.
11. The vertex is $(-1, -6)$ and the axis of symmetry is $x = -1$. Points on the parabola include $(0, -3)$ and $(-2, -3)$.

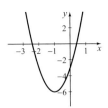

13. The vertex is $\left(-\dfrac{3}{2}, \dfrac{13}{4}\right)$ and the axis of symmetry is $s = -\dfrac{3}{2}$. Points on the parabola include $(0, 1)$ and $(-3, 1)$.

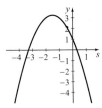

15. The vertex is $\left(-\dfrac{3}{4}, -\dfrac{27}{8}\right)$ and the axis of symmetry is $x = -\dfrac{3}{4}$. Points on the parabola include $\left(\dfrac{3}{2}, 0\right)$ and $(-3, 0)$.

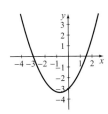

17. The vertex is $(1, 0)$ and the axis of symmetry is $x = 1$. The function is increasing on the interval $(1, \infty)$ and decreasing on the interval $(-\infty, 1)$. The range of the function is $[0, \infty)$.

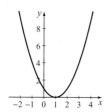

19. The vertex is $(2, 3)$ and the axis of symmetry is $x = 2$. The function is decreasing on the interval $(-\infty, 2)$ and increasing on the interval $(2, \infty)$. The range of the function is $[3, \infty)$.

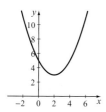

Section 3.2 (page 280)

21. $x = 3$ and $x = -3$ **23.** $x = 5$ and $x = 4$

25. $x = \dfrac{3}{2}$ and $x = -\dfrac{4}{3}$

27. The zeros of the function are $x = 1$ and $x = -\dfrac{5}{2}$, and the x-intercepts of the function are $(1, 0)$ and $\left(-\dfrac{5}{2}, 0\right)$.

29. The zeros of the function are $x = \dfrac{1}{3}$ and $x = -2$, and the x-intercepts of the function are $\left(\dfrac{1}{3}, 0\right)$ and $(-2, 0)$.

31. $x = 2 + \sqrt{6}$ and $x = 2 - \sqrt{6}$

33. $x = -\dfrac{3}{2} + \dfrac{\sqrt{37}}{2}$ and $x = -\dfrac{3}{2} - \dfrac{\sqrt{37}}{2}$

35. $x = \dfrac{-1 + \sqrt{37}}{6}$ and $x = \dfrac{-1 - \sqrt{37}}{6}$

37. No real solutions

39. $t = \dfrac{-1 + \sqrt{13}}{3}$ and $t = \dfrac{-1 - \sqrt{13}}{3}$

41. $s = \dfrac{-\sqrt{2} + 2}{2}$ and $s = \dfrac{-\sqrt{2} - 2}{2}$

43. Because $b^2 - 4ac = (-6)^2 - 4(1)(4) = 20 > 0$, the function has two x-intercepts and two real solutions.

45. Because $b^2 - 4ac = (-6)^2 - 4(1)(9) = 0$, the function has one x-intercept and one real solution.

Section 3.3 (page 280)

47. The real part is $\sqrt{3}$ and the imaginary part is 0.

49. The real part is 7 and the imaginary part is -2.

51. $\dfrac{1}{2}$

53. $4 - i$

55. $x + y = 3 + i$, $x - y = -1 + 7i$, $xy = 14 + 5i$, and $\dfrac{x}{y} = -\dfrac{10}{13} + \dfrac{11}{13}i$

57. $x + y = 0.5 - i$, $x - y = 2.5 - 5i$, $xy = 6i + 4.5$, and $\dfrac{x}{y} = -1.5$

59. $x + y = \dfrac{1}{2} + \dfrac{3}{2}i$, $x - y = -\dfrac{5}{2} - \dfrac{1}{2}i$, $xy = -2 - \dfrac{1}{4}i$, and $\dfrac{x}{y} = -\dfrac{4}{13} + \dfrac{7}{13}i$

61. The solutions are $x = \dfrac{1}{2} + \dfrac{\sqrt{11}}{2}i$ and $x = \dfrac{1}{2} - \dfrac{\sqrt{11}}{2}i$.
Possible answer for $f(x)$: These values of x are also the zeros of the function $f(x) = -x^2 + x - 3$.

63. The solutions are $t = -\dfrac{2}{5} + \dfrac{\sqrt{21}}{5}i$ and $t = -\dfrac{2}{5} - \dfrac{\sqrt{21}}{5}i$.
Possible answer for $f(t)$: These values of t are also the zeros of the function $f(t) = t^2 + \dfrac{4}{5}t + 1$.

Section 3.4 (page 281)

65. $(-\infty, -2] \cup [1, \infty)$

67. $(-\infty, -2) \cup (1, \infty)$

69. $x \le -2$ or $x \ge 2$; $(-\infty, -2] \cup [2, \infty)$

71. $-5 \le x \le -\dfrac{1}{4}$; $\left[-5, -\dfrac{1}{4}\right]$

73. $\dfrac{3}{2} \le x \le 4$; $\left[\dfrac{3}{2}, 4\right]$

75. $\dfrac{1}{2} < x < 3$; $\left(\dfrac{1}{2}, 3\right)$

77. $x \le \dfrac{3 - \sqrt{13}}{2}$ or $x \ge \dfrac{3 + \sqrt{13}}{2}$;
$\left(-\infty, \dfrac{3 - \sqrt{13}}{2}\right] \cup \left[\dfrac{3 + \sqrt{13}}{2}, \infty\right)$

Section 3.5 (page 281)

79. $x = \pm\sqrt{3}$ and $x = \pm 2\sqrt{2}$ **81.** $x = \pm 2$ and $x = \pm\dfrac{\sqrt{3}}{3}i$

83. $x = 2$ and $x = 1$ **85.** $x = 0$ and $x = 2$ **87.** $x = 4$

89. $x = 3$ and $x = 7$

Applications (page 281)

91. 30 feet × 60 feet

93. a. The y-intercept is 2.90, and it represents $f(0) = 2.9\%$ of the total budget in 1980.
b. There is no year between 1980 and 2000 during which the expenditure for airline food was 2% of the total operating expenses. Over this interval, the values are always greater than or equal to 2.9%.
c. The model is not a reliable long-term indicator because as t increases, the value of $f(t)$ decreases, eventually passing 0 and becoming negative.

Chapter 3 Test (page 282)

1. $f(x) = 2(x - 1)^2 - 1$; vertex: $(1, -1)$; minimum point
2. Vertex: $(1, 2)$; axis of symmetry: $x = 1$; range: $(-\infty, 2]$

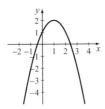

3. Vertex: $(-2, -2)$; axis of symmetry: $x = -2$; range: $[-2, \infty)$

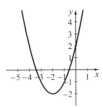

4. Vertex: $(2, 4)$; axis of symmetry: $x = 2$; range: $(-\infty, 4]$

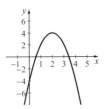

5. Vertex: $(-1, -3)$; axis of symmetry: $x = -1$;
decreasing on $(-\infty, -1)$; increasing on $(-1, \infty)$

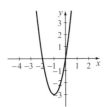

6. $(6, 0)$, $(-1, 0)$; zeros: $x = 6$, $x = -1$
7. $x = 1 + \frac{1}{2}\sqrt{10}, 1 - \frac{1}{2}\sqrt{10}$
8. $x = -\frac{1}{6} + \frac{\sqrt{13}}{6}, -\frac{1}{6} - \frac{\sqrt{13}}{6}$ **9.** $x = \frac{1}{2} + \frac{\sqrt{7}}{2}, \frac{1}{2} - \frac{\sqrt{7}}{2}$
10. $x = \frac{4}{3}, -1$ **11.** No real solutions
12. $x = -\frac{1}{2} + \frac{1}{2}\sqrt{11}, -\frac{1}{2} - \frac{1}{2}\sqrt{11}$
13. Real part: 4; imaginary part: $-\sqrt{2}$
14. $-1 + 2i$ **15.** $-2 + 11i$ **16.** $\frac{4}{13} + \frac{7}{13}i$
17. $x = -1 + i\sqrt{2}, -1 - i\sqrt{2}$
18. $x = \frac{1}{4} + \frac{1}{4}i\sqrt{7}, \frac{1}{4} - \frac{1}{4}i\sqrt{7}$ **19.** $\left(-\frac{5}{3}, 3\right)$

20. $-2 < b < 2$ **21.** $x = i\frac{\sqrt{2}}{2}, -i\frac{\sqrt{2}}{2}, \frac{2\sqrt{3}}{3}, -\frac{2\sqrt{3}}{3}$
22. $x = \frac{4}{7}$ **23.** $x = 5$
24. 22 feet by 9 feet or 18 feet by 11 feet
25. a. $h(0) = 256$ feet. It signifies the initial height of the ball.
b. $t = 4$ seconds **c.** $0 \le t \le 2$

Chapter 4 Exercises

Exercise Set 4.1 (page 296)
1. degree **3.** zeros **5.** 9 **7.** $x(x - 4)(x + 1)$ **9.** up
11. Yes; no breaks, no corners, end behavior like a polynomial
13. No; has a corner **15.** Yes; 3
17. No. This function cannot be written as a sum of terms in which t is raised to a nonnegative integer power.
19. Yes; 0 **21.** Yes; 3
23. $f(t) \to \infty$ as $t \to \infty$; $f(t) \to -\infty$ as $t \to -\infty$
25. $f(x) \to \infty$ as $x \to -\infty$; $f(x) \to -\infty$ as $x \to \infty$
27. $H(x) \to -\infty$ as $x \to \infty$; $H(x) \to -\infty$ as $x \to -\infty$
29. $g(x) \to \infty$ as $x \to -\infty$; $g(x) \to -\infty$ as $x \to \infty$
31. $f(s) \to \infty$ as $s \to \infty$; $f(s) \to -\infty$ as $s \to -\infty$
33. 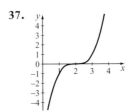 **35.**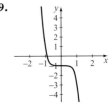
37. 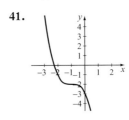 **39.**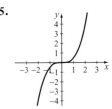
41.   **43.**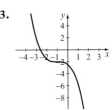
45. $h(x) = -5x^3$

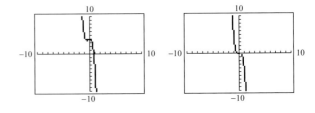

47. $g(x) = 1.5x^5$

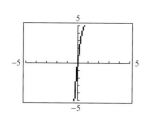

49. a. $g(x) = -2x^3$ **b.** $(0, 0), (2, 0), (-2, 0)$
c. $(-\infty, -2) \cup (0, 2)$ **d.** $(-2, 0) \cup (2, \infty)$
e.

51. a. $h(x) = x^3$ **b.** $(3, 0), (-4, 0), (1, 0), (0, 12)$
c. $(-4, 1) \cup (3, \infty)$ **d.** $(-\infty, -4) \cup (1, 3)$
e.

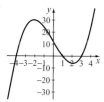

53. a. $g(x) = -\dfrac{1}{2}x^4$ **b.** $(0, -2), (-2, 0), (2, 0), (1, 0), (-1, 0)$
c. $(-2, -1) \cup (1, 2)$ **d.** $(-\infty, -2) \cup (-1, 1) \cup (2, \infty)$
e.

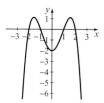

55. a. $g(x) = x^3$ **b.** $(0, 0), (3, 0), (-1, 0)$
c. $(-1, 0) \cup (3, \infty)$ **d.** $(-\infty, -1) \cup (0, 3)$

e.

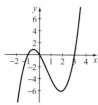

57. a. $g(x) = -2x^3$ **b.** $(0, 0), \left(-\dfrac{1}{2}, 0\right), (3, 0)$

c. $\left(-\infty, -\dfrac{1}{2}\right) \cup (0, 3)$ **d.** $\left(-\dfrac{1}{2}, 0\right) \cup (3, \infty)$

e.

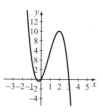

59. a. $g(x) = -x^4$ **b.** $(0, -6), (-3, 0), (-1, 0), (1, 0), (2, 0)$
c. $(-3, -1) \cup (1, 2)$ **d.** $(-\infty, -3) \cup (-1, 1) \cup (2, \infty)$
e.

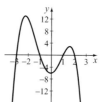

61. a. $h(x) = 2x^3$ **b.** $(0, 0), (-3, 0)$ **c.** $(-3, 0) \cup (0, \infty)$
d. $(-\infty, -3)$ **e.**

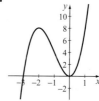

63. a. $g(x) = 2x^4$ **b.** $(0, -3), (3, 0), \left(-\dfrac{1}{2}, 0\right)$

c. $\left(-\infty, -\dfrac{1}{2}\right) \cup (3, \infty)$ **d.** $\left(-\dfrac{1}{2}, 3\right)$ **e.**

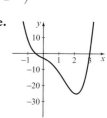

65. a. $(-1, 0)$ **b.** $(0, 3)$ **c.** Odd **d.** Positive
67. a. $(0, 0), (-1, 0), (1, 0)$ **b.** $(0, 0)$ **c.** Odd **d.** Negative
69. a. The graph appears parabolic in the standard window.

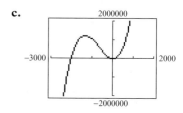

b.

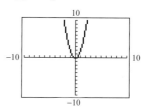

c.

71. a. $(0, 3{,}073{,}000)$; the number of burglaries in 1985
b. $3{,}103{,}353$; the number of burglaries in 1991
c. $1{,}848{,}285$ burglaries
d. The end behavior of the function makes it inaccurate for large values of x.
73. a. $V(x) = x(10 - 2x)(15 - 2x)$
b. $0 < x < 5$, to produce nonnegative values of measure
75. a. $h = 20 - 4s$ **b.** $V = (20 - 4s)s^2$
c. $0 < s < 5$, so that the terms have positive values
77. The y-intercept of all polynomial functions is $(0, a_0)$. No. For example, $f(x) = x^2 + 1$ has no x-intercept.
79. The end behavior is determined by the leading coefficient, ax^n. Assuming n is odd, as $x \to \infty$, $f(x) \to \infty$ if $a > 0$ and $f(x) \to -\infty$ if $a < 0$. Also, as $x \to -\infty$, $f(x) \to -\infty$ if $a > 0$ and $f(x) \to \infty$ if $a < 0$.

Exercise Set 4.2 (page 305)
1. y-axis; even **3.** Even **5.** Odd
7. $x = 2$ has multiplicity 2 and the graph touches the x-axis; $x = -5$ has multiplicity 5 and the graph passes through the x-axis.
9. $t = 0$ has multiplicity 2 and the graph touches the t-axis; $t = 1$ has multiplicity 1 and the graph passes through the t-axis; $t = -2$ has multiplicity 1 and the graph passes through the t-axis.
11. $x = -1$ has multiplicity 2 and the graph touches the x-axis.
13. $s = 0$ has multiplicity 1 and the graph passes through the s-axis; $s = -1$ has multiplicity 2 and the graph touches the s-axis.
15. Symmetric with respect to the y-axis; even
17. No symmetry; neither
19. Symmetric with respect to the origin; odd
21. Symmetric with respect to the y-axis; even
23. $x = -1$, odd multiplicity; $x = 0$, odd multiplicity; $x = \dfrac{3}{2}$, odd multiplicity
25. $x = 0$, even multiplicity; $x = 4$, even multiplicity
27. a. $f(x) \to \infty$ as $x \to \infty$; $f(x) \to -\infty$ as $x \to -\infty$
b. $(0, 0)$
c. $x = 0$, multiplicity 2; $x = 1$, multiplicity 1
d. No symmetry
e. Positive: $(1, \infty)$; negative: $(-\infty, 0) \cup (0, 1)$

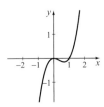

29. a. $f(x) \to \infty$ as $x \to \infty$; $f(x) \to -\infty$ as $x \to -\infty$
b. $(0, 8)$
c. $x = 2$, multiplicity 2; $x = -2$, multiplicity 1
d. No symmetry

e. Positive: $(-2, 2) \cup (2, \infty)$; negative: $(-\infty, -2)$

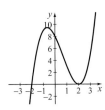

31. a. $g(x) \to \infty$ as $x \to \infty$; $g(x) \to \infty$ as $x \to -\infty$
b. $(0, -6)$
c. $x = -1$, multiplicity 2; $x = -3$, multiplicity 1; $x = 2$, multiplicity 1
d. No symmetry
e. Positive: $(-\infty, -3) \cup (2, \infty)$; negative: $(-3, -1) \cup (-1, 2)$

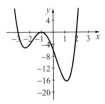

33. a. $g(x) \to -\infty$ as $x \to \infty$; $g(x) \to -\infty$ as $x \to -\infty$
b. $(0, -18)$
c. $x = -1$, multiplicity 2; $x = 3$, multiplicity 2
d. No symmetry
e. Negative: $(-\infty, -1) \cup (-1, 3) \cup (3, \infty)$

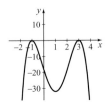

35. a. $f(x) \to \infty$ as $x \to \infty$; $f(x) \to -\infty$ as $x \to -\infty$
b. $(0, 0)$
c. $x = 0$, multiplicity 1; $x = -2$, multiplicity 2
d. No symmetry
e. Positive: $(0, \infty)$; negative: $(-\infty, -2) \cup (-2, 0)$

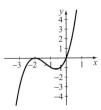

37. a. $h(x) \to -\infty$ as $x \to \infty$; $h(x) \to -\infty$ as $x \to -\infty$
b. $(0, 0)$
c. $x = 0$, multiplicity 2; $x = 1 - \sqrt{2}$, multiplicity 1; $x = 1 + \sqrt{2}$, multiplicity 1
d. No symmetry

e. Positive: $(1 - \sqrt{2}, 0) \cup (0, 1 + \sqrt{2})$;
negative: $(-\infty, 1 - \sqrt{2}) \cup (1 + \sqrt{2}, \infty)$

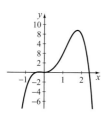

39. $f(x) = (x + 2)(x - 5)(x - 6)$ **41.** $f(x) = (x - 2)^2(x - 4)^2$
43. $f(x) = (x - 2)(x + 3)^2$
45. $f(x) = (x + 2)(x + 1)(x - 5)^3$
47. a. $(-1.5321, 0), (-0.3473, 0), (1.8794, 0)$
b. Positive: $(-\infty, -1.5321) \cup (-0.3473, 1.8794)$;
negative: $(-1.5321, -0.3473) \cup (1.8794, \infty)$
c. Local maximum: $(1, 3)$; local minimum: $(-1, -1)$
d. No symmetry

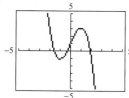

49. a. $(0.7167, 0), (-2.1069, 0)$
b. Positive: $(-\infty, -2.1069) \cup (0.7167, \infty)$;
negative: $(-2.1069, 0.7167)$
c. Local minimum: $(-1.5, -2.6875)$
d. No symmetry

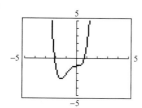

51. a. $V = h(h + 3)^2$ **b.**   **c.** $h > 0$

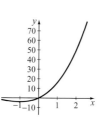

53. a. $5778, $9378, $4402
b. 2002 is too far outside the range of the data used to model
the equation.

c. 1943

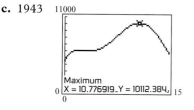

55. a.

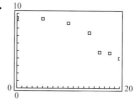

b. $f(t) = 0.0008928t^3 - 0.04556t^2 + 0.2775t + 9.2971$

c.   **d.** 1.7096 million tons

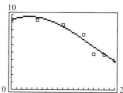

e. No. Based on the actual data for 1996–2000, the rate at
which the decrease occurs slows.
f. Yes. The values decrease markedly after 1990.
57. $f(x) = x(x - 1)(x + 1)$. No. Any function of the form
$f(x) = ax^i(x - 1)^j(x + 1)^k, a, i, j, k \neq 0$, will satisfy these
conditions.
59. a. Possible answer: **b.** Answers may vary.

c. Possible answer:
$f(x) = (x + 1)(x - 3)$

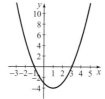

d.

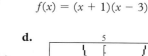

61. a. Possible answer:

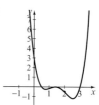

b. Answers may vary.

c. Possible answer: $h(x) = (x - 3)\left(x - \dfrac{1}{2}\right)(x - \sqrt{2})^2$

d.

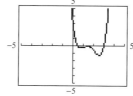

Exercise Set 4.3 (page 314)

1. $2x + 3$ **3.** $x^2 - 4$ **5.** $x^2 - 5x + 12 + \dfrac{-28}{x + 2}$

7. $-x^3 - \dfrac{x^2}{3} + \dfrac{2x}{9} + \dfrac{2}{27} - \dfrac{52}{27(3x - 1)}$

9. $x^5 - x^4 + x^3 - x^2 + x - 1 + \dfrac{2}{x + 1}$

11. $x + 2 + \dfrac{-1 + 2x}{x^2 - 2}$ **13.** $x^2 - x + 1 + \dfrac{3x^2 - 3x + 2}{x^3 + x - 1}$

15. $x^2 + x + 1 = (x + 1)x + 1$
17. $3x^3 + 2x - 8 = (x - 4)(3x^2 + 12x + 50) + 192$
19. $x^6 - 3x^5 + x^4 - 2x^2 - 5x + 6 =$
$(x^2 + 2)(x^4 - 3x^3 - x^2 + 6x) + (6 + 17x)$

21. $11; 95$ **23.** $21; -108$ **25.** $201; -4$ **27.** $\dfrac{9}{16}$

29. No. $\dfrac{p(x)}{q(x)}$ has a nonzero remainder.

31. Yes. $\dfrac{p(x)}{q(x)}$ has no remainder.

33. No. $\dfrac{p(x)}{q(x)}$ has a nonzero remainder.

35. No. $\dfrac{p(x)}{q(x)}$ has a nonzero remainder.

37. Yes. $\dfrac{p(x)}{q(x)}$ has no remainder.

39. $x - 1$ **41.** 8 **43.** 0 **45.** 3

Exercise Set 4.4 (page 323)

1. integers   **3.** True   **5.** $x = 10$ **7.** Neither   **9.** $x = \sqrt{3}$
11. $p(2) = (2)^3 - 5(2)^2 + 8(2) - 4 = 0;$
$p(x) = (x - 2)^2(x - 1)$
13. $p(1) = -(1)^4 - (1)^3 + 18(1)^2 + 16(1) - 32 = 0;$
$p(x) = -(x - 1)(x - 4)(x + 4)(x + 2)$

15. $p\left(\dfrac{2}{3}\right) = 3\left(\dfrac{2}{3}\right)^3 - 2\left(\dfrac{2}{3}\right)^2 + 3\left(\dfrac{2}{3}\right) - 2 = 0;$

$p(x) = (3x - 2)(x^2 + 1)$

17. $p\left(-\dfrac{1}{3}\right) = 3\left(-\dfrac{1}{3}\right)^3 + \left(-\dfrac{1}{3}\right)^2 + 24\left(-\dfrac{1}{3}\right) + 8 = 0;$

$p(x) = (3x + 1)(x^2 + 8)$

19. $-2; x + 2$ **21.** $(-4, 0); x + 4$ **23.** $x = -1, -3, 2$

25. $x = -3, -1, 0, 4$ **27.** $s = -2, -\dfrac{3}{2}, \dfrac{3}{2}, 2$ **29.** $x = -3, \dfrac{1}{4}$

31. $x = -2, -\dfrac{2}{7}, 2$ **33.** $x = -4, 3$ **35.** $x = -1$

37. $x = -1, 3, 4$ **39.** $x = -2, \dfrac{1}{2}, 3$ **41.** $x = -1, 1$

43. Positive zeros: 3 or 1; negative zeros: 1
45. Positive zeros: 3 or 1; negative zeros: 0
47. Positive zeros: 2 or 0; negative zeros: 2 or 0
49. Positive zeros: 2 or 0; negative zeros: 3 or 1
51. Positive zeros: 2 or 0; negative zeros: 2 or 0

53. $x = 4$

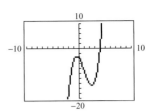

55. $x = -\dfrac{1}{2}, \dfrac{3}{2}, 2$

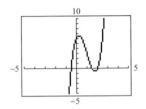

57. $x = -\dfrac{1}{2}, 1, 3$

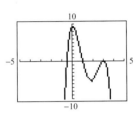

59. $x \approx -3.8265$

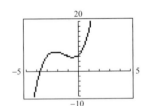

61. 2 inches × 12 inches × 7 inches
63. Radius ≈ 5.5047 inches; height ≈ 10.5047 inches
65. $y = (x + 1)^3; y = -(x + 1)(x^2 + 1)$

Exercise Set 4.5 (page 329)

1. complex   **3.** $2 - 3i$ **5.** $-3i$
7. $x = 1$, multiplicity 3; $x = 4$, multiplicity 5
9. $s = \pi$, multiplicity 10; $s = -\pi$, multiplicity 3

11. $x = 1, \dfrac{3}{2}; p(x) = (2x - 3)(x - 1)$

13. $x = -i\sqrt{5}, 0, i\sqrt{5}; p(x) = x(x - i\sqrt{5})(x + i\sqrt{5})$
15. $x = \pi, -\pi; p(x) = (x - \pi)(x + \pi)$
17. $x = -\sqrt[4]{3}, \sqrt[4]{3}; p(x) = (x - \sqrt[4]{3})(x + \sqrt[4]{3})$
19. $x = 3i, -3i; p(x) = (x - 3i)(x + 3i)$
21. $x = \sqrt{3}, -\sqrt{3}, i\sqrt{3}, -i\sqrt{3};$
$p(x) = (x - \sqrt{3})(x + \sqrt{3})(x - i\sqrt{3})(x + i\sqrt{3})$
23. $(x - 2)(x^2 + 1)$
25. $(x - 5)(2x - 3)(x + 2)$
27. $(x - 3)(x - 2)(x^2 + 1)$
29. $(x + 5)(x - 1)(x + 2i)(x - 2i)$

31. $(x - 2)(x + i)(x - i)$
33. $(x - 5)(x + 2)(2x - 3)$
35. $(x - 3)(x - 2)(x + i)(x - i)$
37. $(x + 5)(x - 1)(x + 2i)(x - 2i)$
39. $p(x) = x^2 - x - 2$
41. $p(x) = x^3 - 2x^2 + x$
43. $p(x) = 9x^4 - 24x^3 + 22x^2 - 8x + 1$
45. $1 - i$
47. Each of these zeros would have to be of even multiplicity, and so the function would have to be at least a quartic function.
49. 3; $x = 1$ is a zero of even multiplicity, $x = -1$ is a zero of odd multiplicity; $p(x) = x^3 - x^2 - x + 1$

Exercise Set 4.6 (page 344)

1. polynomials
3. $-\dfrac{x - 1}{x + 2}$
5. $\dfrac{x - 1}{x - 3}$
7. $(-\infty, -6) \cup (-6, \infty)$; vertical asymptote: $x = -6$; horizontal asymptote: $y = 0$
9. $(-\infty, -2) \cup (-2, 2) \cup (2, \infty)$; vertical asymptotes: $x = \pm 2$; horizontal asymptote: $y = 0$
11. $(-\infty, -2) \cup (-2, 2) \cup (2, \infty)$; vertical asymptotes: $x = \pm 2$; horizontal asymptote: $y = \dfrac{1}{2}$
13. $(-\infty, 2) \cup (2, \infty)$; vertical asymptote: $x = 2$; horizontal asymptote: $y = 0$
15. $(-\infty, -1) \cup (-1, \infty)$; vertical asymptote: $x = -1$; horizontal asymptote: none
17. $(-\infty, -3) \cup \left(-3, \dfrac{1}{2}\right) \cup \left(\dfrac{1}{2}, \infty\right)$; vertical asymptotes: $x = -3, \dfrac{1}{2}$; horizontal asymptote: $y = 0$
19. $(-\infty, \infty)$; vertical asymptote: none; horizontal asymptote: $y = 0$
21. $(-\infty, 2) \cup (2, \infty)$; vertical asymptote: $x = 2$; horizontal asymptote: $y = 3$; x- and y-intercept: $(0, 0)$
23. $(-\infty, -2) \cup (-2, 1) \cup (1, \infty)$; vertical asymptotes: $x = 1, -2$; horizontal asymptote: $y = 0$; x-intercept: $(-3, 0)$; y-intercept: $\left(0, -\dfrac{3}{2}\right)$
25. $(-\infty, -1) \cup (-1, 2) \cup (2, \infty)$; vertical asymptotes: $x = -1, 2$; horizontal asymptote: $y = 0$; x-intercept: $(4, 0)$; y-intercept: $(0, 2)$
27. a. $f(x) \to -\infty$ as $x \to -1$ from the left; $f(x) \to \infty$ as $x \to -1$ from the right.

x	-1.5	-1.1	-1.01	-0.99	-0.9	-0.5
$f(x)$	-4	-20	-200	200	20	4

b. $f(x)$ gets close to zero.

x	10	50	100	1000
$f(x)$	$\dfrac{2}{11}$	$\dfrac{2}{51}$	$\dfrac{2}{101}$	$\dfrac{2}{1001}$

c. $f(x)$ gets close to zero.

x	-1000	-100	-50	-10
$f(x)$	$-\dfrac{2}{999}$	$-\dfrac{2}{99}$	$-\dfrac{2}{49}$	$-\dfrac{2}{9}$

29. a. $f(x) \to -\infty$ as $x \to 0$ from the left; $f(x) \to -\infty$ as $x \to 0$ from the right.

x	-0.5	-0.1	-0.01	0.01	0.1	0.5
$f(x)$	-2	-98	-9998	-9998	-98	-2

b. $f(x) \to 2$ as $x \to \infty$.

x	10	50	100	1000
$f(x)$	1.99	1.9996	1.9999	1.999999

c. $f(x) \to 2$ as $x \to -\infty$.

x	-1000	-100	-50	-10
$f(x)$	1.999999	1.9999	1.9996	1.99

31. Vertical asymptote: $x = 2$; horizontal asymptote: $y = 0$; $\left(0, -\dfrac{1}{2}\right)$

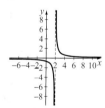

33. Vertical asymptote: $x = -6$; horizontal asymptote: $y = 0$; $(0, -2)$

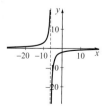

35. Vertical asymptote: $x = 3$; horizontal asymptote: $y = 0$; $(0, 4)$

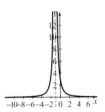

37. Vertical asymptote: $x = -1$; horizontal asymptote: $y = 0$; $(0, 3)$

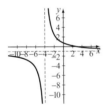

39. Vertical asymptote: $x = -4$; horizontal asymptote: $y = -1$; $\left(0, \frac{3}{4}\right)$, $(3, 0)$

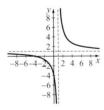

41. Vertical asymptote: $x = 1$; horizontal asymptote: $y = 1$; $(-4, 0)$; $(0, -4)$

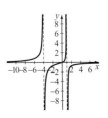

43. Vertical asymptotes: $x = -4, 1$; horizontal asymptote: $y = 0$; $(0, 0)$

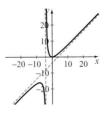

45. Vertical asymptotes: $x = -1, 2$; horizontal asymptote: $y = 3$; $(0, 0)$

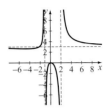

47. Vertical asymptotes: $x = -\frac{1}{2}, 3$; horizontal asymptote: $y = 0$; $(1, 0)$, $\left(0, \frac{1}{3}\right)$

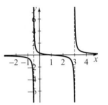

49. Vertical asymptotes: $x = \pm 1$; horizontal asymptote: $y = 1$; $(0, 6)$, $(-3, 0)$, $(2, 0)$

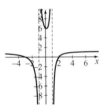

51. Vertical asymptotes: none; horizontal asymptote: $y = 0$; $(0, 1)$

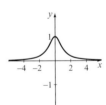

53. $y = x - 4$; $(0, 0)$

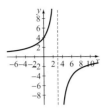

55. $y = -3x - 15$; $(0, 0)$

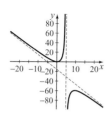

57. $y = -x$; $(2, 0), (-2, 0)$

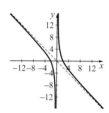

59. $y = x + 2$; $(0, -1)$

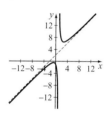

61. $y = 3x + 2$; $\left(\dfrac{1}{3}, 0\right), (-2, 0)$; $(0, -2)$

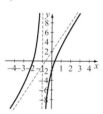

63. $y = x - 3$; $(-1, 0)$

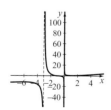

65. Asymptotes: $x = 3$; $y = 0$; y-intercept: $(0, -1)$; no x-intercept

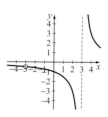

67. Asymptotes: $x = -3, y = 1$; y-intercept: $\left(0, \dfrac{2}{3}\right)$;
x-intercept: $(-2, 0)$

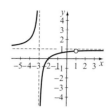

69. Asymptotes: none; y-intercept: $(0, 5)$;
x-intercept: $(-5, 0)$

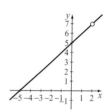

71. a. ≈ 1.2308 milligrams per liter
b. $C(t) = 0$; the concentration will near zero as time increases.
73. $C(x) = \dfrac{15}{x} + 0.25$; $0.55

75. a. $C(x) = \dfrac{30}{x}, \ 0 < x \le 250$

b. $C(x) = \dfrac{30}{x} + \dfrac{0.60(x - 250)}{x}, \ x \ge 250$

c. $0.40
77. 1.75 ounces

79. a. $V(x) = x(5 - 2x)(3 - 2x)$; $0 < x < \dfrac{3}{2}$

b. $S(x) = (3 - 2x)(5 - 2x) + 2x(3 - 2x) + 2x(5 - 2x)$

c. $r(x) = \dfrac{x(5 - 2x)(3 - 2x)}{(3 - 2x)(5 - 2x) + 2x(3 - 2x) + 2x(5 - 2x)}$

d.

x	0.2	0.4	0.6	0.8	1.0	1.2	1.4
$r(x)$	0.1612	0.2574	0.3027	0.3061	0.2727	0.2026	0.0860

e. The ratio increases as x increases and then decreases as x increases; 0.7
f. $x \approx 0.7170$

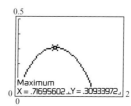

81. $r(x) = \dfrac{-2x(x - 2)}{(x - 1)^2}$

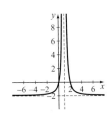

83. $r(x) = \dfrac{1}{x}$

85. It doesn't show the behavior of the graph near the vertical asymptote $x = 10$.

Exercise Set 4.7 (page 352)

1. $[-2, 2]$ **3.** $[-3, 2]$ **5.** $(-\infty, -1] \cup \left[\dfrac{5}{3}, 0\right)$

7. $[-5, 0] \cup [3, \infty)$ **9.** $(-\infty, -4) \cup (0, 4)$ **11.** $[4, \infty)$
13. $(-\infty, -4) \cup (-1, 0)$ **15.** $(-\infty, -2)$ **17.** $[-2, \infty)$
19. $(-\infty, -1.5175) \cup (1.5175, \infty)$ **21.** $(-\infty, -2] \cup [-1, 2]$
23. $(-\infty, -2] \cup [0, 2]$ **25.** $(-1, 0) \cup (3, \infty)$
27. $[-2, 1)$ **29.** $(-\infty, -2] \cup [2, 3)$
31. $(-\infty, -1] \cup [0, \infty)$ **33.** $(-\infty, -2) \cup (1, 2)$

35. $(-\infty, 0) \cup \left(\dfrac{1}{2}, 1\right]$ **37.** $(-\infty, 1) \cup \left[\dfrac{5}{2}, \infty\right)$

39. $(-\infty, -2) \cup [1, \infty)$ **41.** $\left(-\infty, -\dfrac{1}{2}\right)$

43. $(-\infty, -3) \cup (-1, 3)$ **45.** $(-\infty, -4) \cup \left[\dfrac{1}{5}, 3\right)$

47. $x \geq 4$ inches   **49.** $(1, 3)$
51. Each term doesn't necessarily have to be less than 2 for the product of the terms to be less than 2. The student should expand the product and bring the lone constant term to the left-hand side before attempting to factor to find the zeros.
53. Example: $p(x) = x(x - 1)(x - 3)$

Chapter 4 Review Exercises (page 359)

Section 4.1 (page 359)

1. Yes; degree 3; $-1, -6, 5; -1$ **3.** No
5. $f(x) \to -\infty$ as $x \to \infty$; $f(x) \to \infty$ as $x \to -\infty$
7. $H(s) \to -\infty$ as $s \to \infty$; $H(s) \to -\infty$ as $s \to -\infty$
9. $h(s) \to \infty$ as $s \to \infty$; $h(s) \to -\infty$ as $s \to -\infty$
11. a. $f(x) \to -\infty$ as $x \to \infty$; $f(x) \to \infty$ as $x \to -\infty$
b. $(1, 0), (-2, 0), (-4, 0), (0, 8)$
c. $(-\infty, -4) \cup (-2, 1)$
d. $(-4, -2) \cup (1, \infty)$

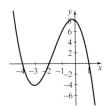

13. a. $f(t) \to \infty$ as $t \to \infty$; $f(t) \to -\infty$ as $t \to -\infty$

b. $(0, 0), \left(\dfrac{1}{3}, 0\right), (-4, 0)$ **c.** $(-4, 0) \cup \left(\dfrac{1}{3}, \infty\right)$

d. $(-\infty, -4) \cup \left(0, \dfrac{1}{3}\right)$

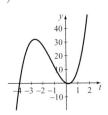

15. a. $f(x) \to \infty$ as $x \to \infty$; $f(x) \to -\infty$ as $x \to -\infty$

b. $(0, 0), \left(\dfrac{1}{2}, 0\right), (-1, 0)$

c. $(-1, 0) \cup \left(\dfrac{1}{2}, \infty\right)$

d. $(-\infty, -1) \cup \left(0, \dfrac{1}{2}\right)$

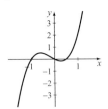

Section 4.2 (page 359)

17. $x = -2$, multiplicity 3, crosses the x-axis; $x = -7$, multiplicity 2, touches the x-axis
19. $t = -1$, multiplicity 1, crosses the t-axis; $t = 0$, multiplicity 2, touches the t-axis; $t = 2$, multiplicity 1, crosses the t-axis
21. $x = 0$, multiplicity 1, crosses the x-axis; $x = -1$, multiplicity 2, touches the x-axis
23. a. $(0, 0), \left(-\dfrac{1}{2}, 0\right)$

b. $x = 0$, multiplicity 2; $x = -\dfrac{1}{2}$, multiplicity 1

c. $f(x) \to \infty$ as $x \to \infty$; $f(x) \to -\infty$ as $x \to -\infty$

d. Positive: $\left(-\dfrac{1}{2}, 0\right) \cup (0, \infty)$; negative: $\left(-\infty, -\dfrac{1}{2}\right)$

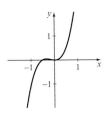

25. a. $(0, -1)$, $\left(\dfrac{1}{2}, 0\right)$, $(4, 0)$

b. $x = \dfrac{1}{2}$, multiplicity 2; $x = 4$, multiplicity 1

c. $f(x) \to \infty$ as $x \to \infty$; $f(x) \to -\infty$ as $x \to -\infty$

d. Positive: $(4, \infty)$; negative: $\left(-\infty, \dfrac{1}{2}\right) \cup \left(\dfrac{1}{2}, 4\right)$

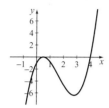

27. a. $(0, -2)$, $(1, 0)$, $(-2, 0)$
b. $t = 1$, multiplicity 1; $t = -2$, multiplicity 1
c. $f(t) \to \infty$ as $t \to \infty$; $f(t) \to \infty$ as $t \to -\infty$
d. Positive: $(-\infty, -2) \cup (1, \infty)$; negative: $(-2, 1)$

29. a. $(0, 0)$, $(6, 0)$, $(-3, 0)$
b. $x = 0$, multiplicity 2; $x = -3$, multiplicity 1; $x = 6$, multiplicity 1
c. $g(x) \to \infty$ as $x \to \infty$; $g(x) \to \infty$ as $x \to -\infty$
d. Positive: $(-\infty, -3) \cup (6, \infty)$; negative: $(-3, 0) \cup (0, 6)$

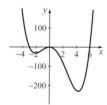

Section 4.3 (page 360)

31. $-4x - 15 + \dfrac{-67}{x - 4}$ **33.** $x^3 - x^2 - 3x + 4 + \dfrac{6x - 11}{x^2 + 3}$

35. 0; yes, no remainder **37.** 0; yes, no remainder

Section 4.4 (page 360)

39. $p(2) = (2)^3 - 6(2)^2 + 3(2) + 10 = 0$;
$p(x) = (x - 5)(x - 2)(x + 1)$
41. $p(3) = -(3)^4 + 3^3 + 4(3)^2 + 5(3) + 3 = 0$;
$p(x) = -(x - 3)(x + 1)(x^2 + x + 1)$

43. $x = \pm 1, \dfrac{3}{2}$ **45.** $x = 3$ **47.** $x = -5, -\dfrac{1}{2}, 1$

49. $x = 1, 2, 4$
51. Positive zeros: 2 or 0; negative zeros: 2 or 0

Section 4.5 (page 360)
53. $x = 0$; ± 5; $p(x) = x(x + 5)(x - 5)$
55. $x = \pm i$, ± 3; $p(x) = (x + 3)(x - 3)(x + i)(x - i)$

Section 4.6 (page 360)
57. a. $f(x) \to \infty$ as $x \to -1$ from the left and right.

x	-1.5	-1.1	-1.01	-0.99	-0.9	-0.5
$f(x)$	4	100	10,000	10,000	100	4

b. $f(x) \to 0$ as $x \to \infty$

x	10	50	100	1000
$f(x)$	0.008264	0.0003845	0.00009803	0.0000009980

c. $f(x) \to 0$ as $x \to -\infty$

x	-1000	-100	-50	-10
$f(x)$	0.000001002	0.0001020	0.0004165	0.01235

59. Vertical asymptote: $x = 1$; horizontal asymptote: $y = 0$; $(0, -2)$

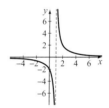

61. Vertical asymptotes: $x = \pm 2$;

horizontal asymptote: $y = 0$; $\left(0, -\dfrac{1}{4}\right)$

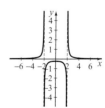

63. Vertical asymptotes: $x = -1, 3$;

horizontal asymptote: $y = 0$; $(2, 0)$, $\left(0, \dfrac{2}{3}\right)$

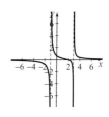

65. Vertical asymptote: $x = -2$; horizontal asymptote: none;
slant asymptote: $y = x - 2$; $(1, 0)$, $(-1, 0)$, $\left(0, -\dfrac{1}{2}\right)$

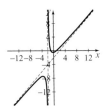

Section 4.7 (page 361)

67. $(-3, -1) \cup (0, 3)$ **69.** $(-\infty, -3] \cup [-2, 1]$

71. $(-\infty, -1) \cup (-1, 1]$ **73.** $\left[0, \dfrac{1}{3}\right)$

Applications (page 361)

75. a. $V(x) = x(11 - 2x)(8 - 2x)$ **b.** $0 < x < 4$
c. Approx. 1.5252 inches

77. $h(h + 3)(h + 4) = 60$; $h = 2$ **79.** $s^2(s - 1) \geq 48$; $s \geq 4$

Chapter 4 Test (page 362)

1. Degree: 5; coefficients: 3, 4, -1, 7; leading coefficient: 3
2. As $x \to -\infty$, $p(x) \to -\infty$; as $x \to \infty$, $p(x) \to -\infty$.
3. 0, multiplicity 2; 3, -3, each with multiplicity 1; touches
x-axis at $(0, 0)$ and crosses it at $(-3, 0)$ and $(3, 0)$
4. a. x-intercepts: $(0, 0)$, $(2, 0)$, $(-1, 0)$; y-intercept: $(0, 0)$
b. 0, -1, and 2, each of multiplicity 1
c. As $x \to -\infty$, $f(x) \to \infty$; as $x \to \infty$, $f(x) \to -\infty$
d. Positive: $(-\infty, -1)$, $(0, 2)$;
negative: $(-1, 0)$, $(2, \infty)$

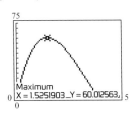

5. a. x-intercepts: $(-1, 0)$, $(2, 0)$; y-intercept: $(0, 4)$
b. -1, multiplicity 1; 2, multiplicity 2
c. As $x \to -\infty$, $f(x) \to -\infty$; as $x \to \infty$, $f(x) \to \infty$
d. Positive: $(-1, 2)$, $(2, \infty)$;
negative: $(-\infty, -1)$

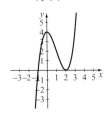

6. a. x-intercepts: $(-1, 0)$, $(0, 0)$; y-intercept: $(0, 0)$
b. -1, multiplicity 2; 0, multiplicity 1
c. As $x \to -\infty$, $f(x) \to \infty$; as $x \to \infty$, $f(x) \to -\infty$
d. Positive: $(-\infty, -1)$, $(-1, 0)$; negative: $(0, \infty)$

7. a. x-intercepts: $(-2, 0)$, $\left(-\dfrac{1}{2}, 0\right)$, $(0, 0)$; y-intercept: $(0, 0)$

b. 0, multiplicity 2; -2 and $-\dfrac{1}{2}$, each of multiplicity 1

c. As $x \to -\infty$, $f(x) \to \infty$; as $x \to \infty$, $f(x) \to \infty$

d. Positive: $(-\infty, -2)$, $\left(-\dfrac{1}{2}, 0\right)$; negative: $\left(-2, -\dfrac{1}{2}\right)$, $(0, \infty)$

8. $p(x) = (x^2 + 1)(3x^2 - 9) + x + 8$
9. $p(x) = (x - 1)(-2x^4 - x^3 - x^2 - 5x - 5) - 2$
10. Remainder is 0. Yes, by the Factor Theorem, since
$p(2) = 0$.
11. $p(x) = (x - 1)(x - 3)(x^2 + x + 1)$ **12.** 2, $-\sqrt{3}$, $\sqrt{3}$

13. $-\dfrac{1}{2}$, -2, -1 **14.** $\dfrac{1}{2}$, -1, -2 **15.** 1, -1, 3

16. Positive zeros: two or none; negative zeros: two or none
17. Zeros: -2, 0, 2, $-2i$, $2i$;
$p(x) = x(x - 2)(x + 2)(x - 2i)(x + 2i)$
18. Zeros: -2, 3, $2i$, $-2i$; $p(x) = (x + 2)(x - 3)(x - 2i)(x + 2i)$
19. Asymptotes: $x = -3$, $y = 0$;
y-intercept: $(0, -1)$

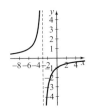

20. Asymptotes: $x = 2$, $y = -2$;
y-intercept: $(0, 0)$;
x-intercept: $(0, 0)$

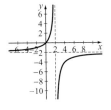

21. Asymptotes: $x = -\dfrac{1}{2}$, 2 and $y = 0$; y-intercept: $(0, -1)$

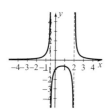

22. $(-\infty, -3], [-2, 2]$ **23.** $(-2, -1) \cup (5, \infty)$

24. $\left(-\dfrac{1}{3}, \dfrac{1}{3}\right]$ **25.** Height: 4 inches

26. a. $C(x) = 0.25 + \dfrac{50}{x}$

b. \$0.45 per mile

Chapter 5 Exercises _____

Exercise Set 5.1 (page 373)

1. a   **3.** $(f \circ g)(x) = x + 3$ **5.** x **7.** x

9. $f(g(x)) = -(-x - 3) - 3 = x$
$g(f(x)) = -(-x - 3) - 3 = x$

11. $f(g(x)) = 6\left(\dfrac{1}{6}x\right) = x$

$g(f(x)) = \dfrac{1}{6}(6x) = x$

13. $f(g(x)) = -3\left(-\dfrac{1}{3}x + \dfrac{8}{3}\right) + 8 = x - 8 + 8 = x$

$g(f(x)) = -\dfrac{1}{3}(-3x + 8) + \dfrac{8}{3} = x - \dfrac{8}{3} + \dfrac{8}{3} = x$

15. $f(g(x)) = (\sqrt[3]{x - 2})^3 + 2 = x - 2 + 2 = x$
$g(f(x)) = \sqrt[3]{x^3 + 2 - 2} = \sqrt[3]{x^3} = x$

17. $f(g(x)) = (\sqrt{x - 3})^2 + 3 = x - 3 + 3 = x$
$g(f(x)) = \sqrt{x^2 + 3 - 3} = \sqrt{x^2} = x$

19. The function is one-to-one because there are no values of a and b in the chart such that $f(a) = f(b)$, where $a \neq b$.

21. The function is not one-to-one because there are values of a and b in the chart such that $f(a) = f(b)$, where $a \neq b$. Here, $f(0) = f(2) = 9$.

23. Not one-to-one   **25.** One-to-one   **27.** Not one-to-one

29. One-to-one   **31.** Not one-to-one   **33.** One-to-one

35. $f^{-1}(x) = -\dfrac{3}{2}x$ **37.** $f^{-1}(x) = -\dfrac{1}{4}x + \dfrac{1}{20}$

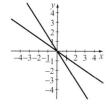   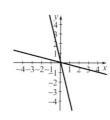

39. $f^{-1}(x) = \sqrt[3]{x + 6}$

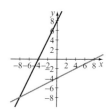

41. $f^{-1}(x) = 2x + 8$

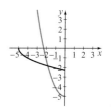

43. $g^{-1}(x) = \sqrt{8 - x}$

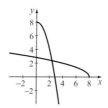

45. $g^{-1}(x) = -\sqrt{x + 5}$

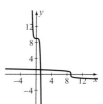

47. $f^{-1}(x) = \sqrt[3]{\dfrac{7 - x}{2}}$

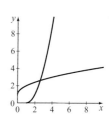

49. $f^{-1}(x) = \sqrt[5]{\dfrac{9 - x}{4}}$

51. $f^{-1}(x) = \dfrac{1}{x}$

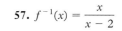

53. $g^{-1}(x) = \sqrt{x} + 1, x \geq 0$

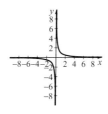   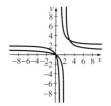

55. $f^{-1}(x) = x^2 - 3, x \geq 0$ **57.** $f^{-1}(x) = \dfrac{x}{x - 2}$

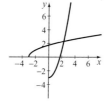

59. Domain of f: $[-3, 3]$
Range of f: $[0, 4]$
Domain of f^{-1}: $[0, 4]$
Range of f^{-1}: $[-3, 3]$

61. Domain of f: $[-3, 2]$
Range of f: $[-5, 5]$
Domain of f^{-1}: $[-5, 5]$
Range of f^{-1}: $[-3, 2]$

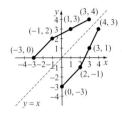

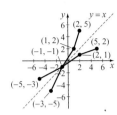

63. -2 **65.** -1 **67.** 0 **69.** -1.5
71. $f(x) = 4x$, where x is the number of gallons and $f(x)$ is the number of quarts in x gallons. The inverse, $f^{-1}(x) = \frac{1}{4}x$, gives the number of gallons, where x is the number of quarts.
73. Solve the equation for q: $q = 1000 - 10p$.
75. a. The range of f is even integers in the interval $[32, 54]$.
b. $f^{-1}(s) = s - 30$; f^{-1} gives American size when French size is input.
77. a. $x = -2$ **b.** $g(2) = 1$ **c.**

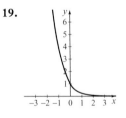

79. No. Linear functions that are horizontal lines are not one-to-one, so they do not have inverses.
81. Quadrant II
83. The function $f(x) = x^5 + x^3 - x$ is an odd function that is not one-to-one.
85. One possible answer is $x \leq 0$.

Exercise Set 5.2 (page 387)

1. 125 **3.** $\frac{1}{4}$ **5.** 18 **7.** 1.2806 **9.** 9.1896

11. 4.7288 **13.** 20.0855 **15.** 0.0821

17.

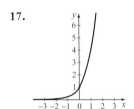

19.

21.

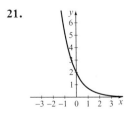

23.

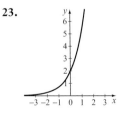

25.

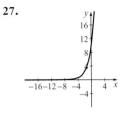

27.

29.

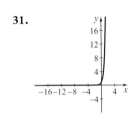

31.

33.

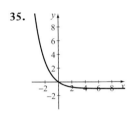

35.

37.

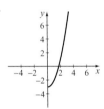

a. $(0, -1)$
b. The domain is all real numbers, and the range is $(-\infty, 0)$.
c. The horizontal asymptote is $y = 0$.
d. As $x \to \infty$, $f(x)$ approaches $-\infty$. As $x \to -\infty$, $f(x)$ approaches 0.

39.

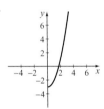

a. $(0, 2)$
b. The domain is all real numbers, and the range is $(-\infty, 3)$.
c. The horizontal asymptote is $y = 3$.
d. As $x \to -\infty$, $f(x)$ approaches 3. As $x \to \infty$, $f(x)$ approaches $-\infty$.

41.

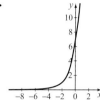

a. (0, 7)
b. The domain is all real numbers, and the range is (0, ∞).
c. The horizontal asymptote is $y = 0$.
d. As $x \to \infty$, $f(x)$ approaches ∞. As $x \to -\infty$, $f(x)$ approaches 0.

43.

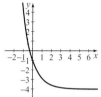

a. (0, −1)
b. The domain is all real numbers, and the range is (−4, ∞).
c. The horizontal asymptote is $y = -4$.
d. As $x \to -\infty$, $f(x)$ approaches ∞. As $x \to \infty$, $f(x)$ approaches −4.
45. This graph does not represent an exponential function, as it does not include a horizontal asymptote.
47. This graph does not represent an exponential function, as it includes a vertical asymptote.
49. d **51.** a
53. $x \approx 1.46497$ **55.** $x \approx -3.32193$

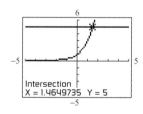 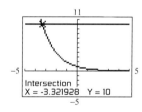

57. $x \approx 11.5525$

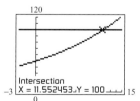

59. a.

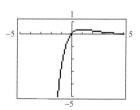

b. The domain is all real numbers, and the range is approximately (−∞, 0.3679].
c. The function intersects the axes at the origin, (0, 0).
d. As $x \to \infty$, $f(x) \to 0$. As $x \to -\infty$, $f(x) \to -\infty$.
61. $2007.34 **63.** $2023.28 **65.** $1795.83 **67.** $1793.58

69.

Years at Work	Annual Salary
0	$10,000.00
1	$10,500.00
2	$11,025.00
3	$11,576.25
4	$12,155.06

$S(t) = 10,000(1.05)^t$

71.

Years Since Purchase	Value
0	$20,000
1	$18,000
2	$16,200
3	$14,580
4	$13,122

$V(t) = 20,000(0.90)^t$

73.

Years in the Future	Value
1	$12,600.00
2	$8820.00
3	$6174.00
4	$4321.80
5	$3025.26

$V(t) = 18,000(0.70)^t$

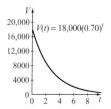

75. $1348.35. If the bonds continued paying interest, their value would have no upper bound. Thus it would be financially onerous for the government to guarantee this rate over a long period of time. For instance, after 80 years, a bond purchased for $1000 would be worth nearly 11 times its purchase price.

77. a. $W(t) = 17.48(1.027)^t$

b.

Year	Hourly Wage
2000	$17.48
2001	$17.95
2002	$18.44
2003	$18.93
2004	$19.45
2005	$19.97
2006	$20.51
2007	$21.06

c. The predicted value is within 1% of the actual value.
79. a. 625 feet
b. 587.66 feet
c. $x \approx -243.06, 243.06$

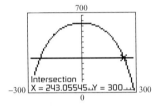

81. a. 2
b. You could observe the graph to determine whether the function is asymptotic to the line $y = 2$.

83. a.

x	$f(x) = 2x$	$g(x) = 2^x$
-1	-2	0.5
-0.5	-1	$\dfrac{\sqrt{2}}{2}$
0	0	1
0.5	1	$\sqrt{2}$
1	2	2
1.5	3	$2\sqrt{2}$
2	4	4
2.5	5	$4\sqrt{2}$
3	6	8
3.5	7	$8\sqrt{2}$
4	8	16

b. $(-\infty, 1) \cup (2, \infty)$
c. $(1, 2)$
d. When x increases by 1 unit, $f(x)$ increases by 2 units.
e. When x increases by 1 unit, $g(x)$ doubles.
f. For values above $x = 2$, doubling produces a greater value than adding 2 units, so the value of $g(x)$ increases much faster than the value of $f(x)$.

Exercise Set 5.3 (page 403)
1. $3^{1/2}$ **3.** $10^{1/3}$ **5.** True   **7.** True   **9.** 8.45×10^6

11.

Logarithmic Statement	Exponential Statement
$\log_3 1 = 0$	$3^0 = 1$
$\log 10 = 1$	$10^1 = 10$
$\log_5 \dfrac{1}{5} = -1$	$5^{-1} = \dfrac{1}{5}$
$\log_a x = b,\ a > 0$	$a^b = x$

13.

Exponential Statement	Logarithmic Statement
$3^4 = 81$	$\log_3 81 = 4$
$5^{1/3} = \sqrt[3]{5}$	$\log_5 \sqrt[3]{5} = \dfrac{1}{3}$
$6^{-1} = \dfrac{1}{6}$	$\log_6 \dfrac{1}{6} = -1$
$a^v = u,\ a > 0$	$\log_a u = v,\ a > 0$

15. 4   **17.** $\dfrac{1}{3}$ **19.** 2   **21.** $\dfrac{1}{3}$ **23.** $x + y$ **25.** k **27.** $\dfrac{1}{2}$
29. -4 **31.** -2 **33.** $x^2 + 1$ **35.** ≈ 1.2041
37. ≈ 0.3466 **39.** ≈ 3.1461 **41.** ≈ -1.3979
43. 0.2031 **45.** -0.4307 **47.** 3.5850 **49.** 2.5750
51. $x = 8$ **53.** $x = \sqrt[3]{3}$ **55.** $x = 6$
57. The domain is $(0, \infty)$. The asymptote is $x = 0$.
The x-intercept is $(1, 0)$.

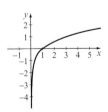

59. The domain is $(0, \infty)$. The asymptote is $x = 0$.
The x-intercept is $(1, 0)$.

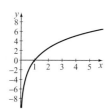

61. The domain is $(0, \infty)$. The asymptote is $x = 0$.
The x-intercept is $(1000, 0)$.

63. The domain is $(-1, \infty)$. The asymptote is $x = -1$.
The x-intercept is $(0, 0)$.

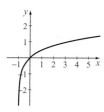

65. The domain is $(-4, \infty)$. The asymptote is $x = -4$.
The x-intercept is $(-3, 0)$.

67. The domain is $(1, \infty)$. The asymptote is $x = 1$.
The x-intercept is $(2, 0)$.

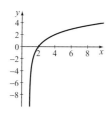

69. The domain is $(0, \infty)$. The asymptote is $t = 0$.
The t-intercept is $(1, 0)$.

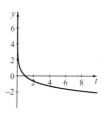

71. The domain is $(-\infty, 0) \cup (0, \infty)$. The asymptote is $x = 0$.
The x-intercepts are $(1, 0)$ and $(-1, 0)$.

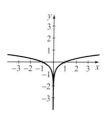

73. Approximately 0.85. The graph corresponds
to $f(x) = 10^x$, so $\log f(x) = x$. Here, when
$f(x) = 7$, $x \approx 0.85$.
75. c
77. b
79. $\log (7) = t$ has the solution $t = 0.8451$.

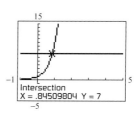

81. $\log (5) = x$ has the solution $x \approx 0.69897$.

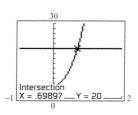

83. 4
85. $63{,}095{,}734.45 I_0$
87. Approximately 794.33:1
89. 4

91. a. To sort 100 items, the bubble sort requires 10,000 operations, while the heap sort requires 200 operations.
b. The corresponding increase would be 300 operations.

n	Operations
5	25
10	100
15	225
20	400

c. The corresponding increase would be 16.02 operations.

n	Operations
5	3.49
10	10
15	17.64
20	26.02

d. The heap sort is more efficient because n^2 grows faster than $n \log n$.
e. n^2 grows faster than $n \log n$.

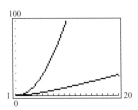

93. Because $10^2 = 100$ and $10^3 = 1000$, the value of x for which $10^x = 400$ is between 2 and 3, as is the value of log 400.
95. log 1000 = x where $f(x) = 10^x$ and $f(x) = 1000$, so $x = 3$.
97. log 0.5 = x where $f(x) = 10^x$ and $f(x) = 0.5$, so $x = -0.3010$.
99. $f(x) = 3 \log x$

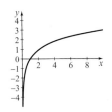

101. The graphs of the two functions are identical.

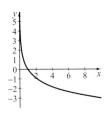

Exercise Set 5.4 (page 413)
1. $x^{1/5}$ **3.** $x^{3/5}$ **5.** True **7.** 1.5441 **9.** -0.3980
11. 0.1505 **13.** 2.097 **15.** $\log x + 3 \log y$
17. $\frac{1}{3} \log x + \frac{1}{4} \log y$ **19.** $\frac{1}{4} \log x + \log y$
21. $2 \log x + 5 \log y - 1$ **23.** $\frac{2}{3} \ln x - 2$
25. $\frac{1}{2} \log_a(x^2 + y) - 3$ **27.** $3 \log_a x - \frac{3}{2} \log_a y - \frac{5}{2} \log_a z$
29. $\frac{1}{3} \log_a x + \log_a y - \frac{5}{3} \log_a z$ **31.** $\log 2.1$ **33.** $\log 3x\sqrt{y}$
35. $\log \dfrac{x^3\sqrt{y}}{z}$ **37.** $\log 80$ **39.** $\ln y^2 e^3$ **41.** $\log_3 3y^5$
43. $\ln(x + 1)$ **45.** $\log\left(\dfrac{\sqrt[3]{x + 3}}{x}\right)$ **47.** $b + 1$ **49.** $3b$ **51.** $-b$
53. $\sqrt{2}$ **55.** $\sqrt{3}$ **57.** $5x$ **59.** $3x + 1$ **61.** 3 **63.** $\frac{2}{5}$
65. $f(x) = g(x) + 1$

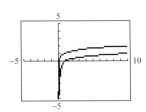

67. $f(x) = g(x) + 2$

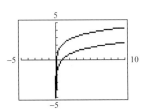

69. 10,000
71. 7.2
73. $10^{-3.4}$
75. 13.01 dB

77. a.

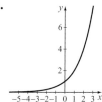

b. The domain of $f(x)$ is $(-\infty, \infty)$. The range of $f(x)$ is $(0, \infty)$.

c.

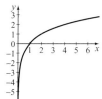

d. The domain of $f^{-1}(x)$ is $(0, \infty)$. The range of $f^{-1}(x)$ is $(-\infty, \infty)$.

79.

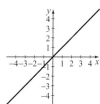

a. The domain of $f(x)$ is $(0, \infty)$. The domain of $g(x)$ is $(-\infty, \infty)$.
b. For all real values of $x > 0$
81. No, because there is no real value of y for which a^y is equal to a negative number, in this case $x = -3$.

Exercise Set 5.5 (page 421)
1. True **3.** False **5.** $x = 3$ **7.** $x = 3$
9. $x = -2$ **11.** $x \approx 2.197$ **13.** $x \approx 2.322$
15. $x \approx 1.947$ **17.** $x \approx 0.926$ **19.** $x \approx -0.380$
21. $x \approx 17.329$ **23.** $x \approx 1.610$ **25.** $x \approx -7.640$
27. $x \approx \pm 0.781$ **29.** $x \approx \pm 1.716$
31. $x \approx 1.302$ **33.** $x \approx 0.524$

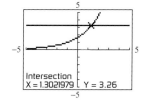 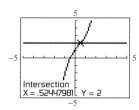

35. $x = 1$ **37.** $x = e^2 + 1$ **39.** $x = 8$ **41.** $x = 5$
43. $x = \sqrt{2}$ **45.** $x = 2$ **47.** $x = 4$
49. No solutions **51.** $x = 0, x = 3$
53. $x = \dfrac{-7 + \sqrt{89}}{4}$ **55.** $x = 5$

57. $x \approx 3.186$ **59.** $x \approx 5.105, x \approx -3.105$

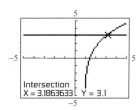 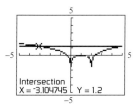

61. $t \approx 11.55$ years **63.** $t \approx 12.05$ years **65.** $t \approx 9.24$ years
67. $r \approx 5.75\%$ **69.** $r \approx 5.07\%$ **71.** $r \approx 3.25\%$
73. $k \approx 0.0578$; $t = 24$ hours; $t = 36$ hours
75. a. 2297.1 transistors
b. 2.09 years
77. a. \$43,173 **b.** 20% **c.** 3 years
79. $f(t) = 10e^{-0.0000285t}$; $t \approx 56,472$ years
81. $10^{-1.5}$ mole per liter **83.** $P_1 \approx 88$ W
85. a. $A(t) = 5e^{-0.087t}$
b. $B(t) = 5e^{-0.231t}$
c. 5.87 days; 2.21 days
d. The solution produced with water that has a pH of 6.0 should be used because more malathion will remain after several days than will remain in the solution produced with water that has a pH of 7.0.
e.

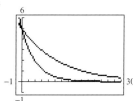

87. A little under two years **89.** $V(t) = 5000e^{0.06t}$; $r = 6\%$
91. This equation has no solution because there is no value of x for which the left side of the equation will equal a nonpositive value.
93. The step assumes that the logarithms of both sides of the equation have been taken, but this can be done only if the logarithms have the same base, which is not the case in this equation.
95. $x = 10$ **97.** $x \approx 2.859$; $x \approx 0.141$

Exercise Set 5.6 (page 432)
1. growth **3.** ∞ **5.** 0 **7.** 10 **9.** $t \approx 0.6931$
11. $4e \approx 10.8731$ **13.** $\ln 2 \approx 0.6931$ **15.** $\dfrac{10}{3}$
17. 9.0944 **19.** -1 **21.** $x = e^2 \approx 7.3891$
23. c **25.** a
27. a. $A(t) = A_0 e^{-0.0001216t}$ **b.** About 9035 years
29. Approximately 2900 years old
31. a. $P(t) = 16e^{0.02097t}$ **b.** 19.7 million
33. $P(t) = 123,000e^{0.04153t}$
35. a. $P(t) = 23,024e^{-0.293956t}$ **b.** $\approx \$9532$
37. a. 20 deer **b.** About 32 deer

39. a. $f(x) = 2.8265(1.1975)^x$

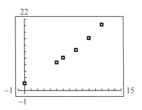

b. $f(18) = \$72.4784$ billion
c. This is not a realistic model over the long term because as x increases, the function produces values for $f(x)$ that represent a greater revenue than the global population can provide.
41. a. A logistic function fits this data well because there is an upper limit to the speed that a land vehicle can attain under the constraints of the laws of physics.
b. $f(x) = \dfrac{330}{1 + 2.23e^{-0.0232x}}$

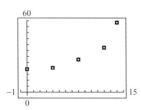

c. c is the limiting speed of the car. **d.** 240 mph
43. a. $P(t) = 16.76(1.079)^t$

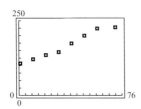

b. About $61.04 per barrel
45. a. The annual cost increases as the arsenic concentration is lowered.
b. $C(x) = 774.6092(0.8786)^x$

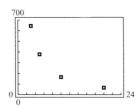

c. a must be less than 1 because the cost is decreasing as the concentration is increasing.
d. $163.8993 million
e. $C(2) \approx 598$. This means it would cost nearly $600 million a year to keep the arsenic concentration to 2 micrograms per liter.

47. If the quantity of strontium-90 is half its original amount after the time deemed its "half-life" has passed, then after another half-life, the remaining quantity will be further reduced by half. So it will take two half-lives to reach the value $\dfrac{A_0}{4}$. Because the half-life is 29 years, it would take 58 years for the initial amount A_0 to be reduced to $\dfrac{A_0}{4}$.

49. The function $f(t) = e^{(1/2)t}$ cannot model exponential decay because as $t \to \infty$, $f(t)$ increases. To model decay, as $t \to \infty$, the function $f(t)$ must decrease. This would occur if e were raised to a negative exponent, as opposed to a fractional positive exponent.

Chapter 5 Review Exercises (page 441)

Section 5.1 (page 441)

1. $f(g(x)) = 2\left(\dfrac{x-7}{2}\right) + 7 = x - 7 + 7 = x$

$g(f(x)) = \dfrac{(2x+7) - 7}{2} = \dfrac{2x}{2} = x$

3. $f(g(x)) = 8\left(\dfrac{\sqrt[3]{x}}{2}\right)^3 = \dfrac{8x}{8} = x$

$g(f(x)) = \dfrac{\sqrt[3]{8x^3}}{2} = \dfrac{2x}{2} = x$

5. $g(x) = -\dfrac{5}{4}x$

7. $g(x) = \dfrac{6-x}{3}$

9. $g(x) = \sqrt[3]{x-8}$
11. $f(x) = \sqrt{8-x}, \ x \le 8$
13. $g(x) = -x - 7$ **15.** $g(x) = \sqrt[3]{1-x}$

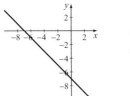 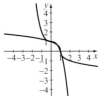

Section 5.2 (page 441)
17. The y-intercept is $(0, -1)$. Other points include $(1, -4)$ and $(2, -16)$. As $x \to \infty$, $f(x) \to -\infty$. As $x \to -\infty$, $f(x) \to 0$. Domain: $(-\infty, \infty)$; range: $(-\infty, 0)$

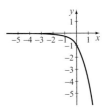

19. The y-intercept is $(0, 1)$. Other points include $\left(1, \frac{2}{3}\right)$ and $\left(-1, \frac{3}{2}\right)$. As $x \to \infty$, $g(x) \to 0$. As $x \to -\infty$, $g(x) \to \infty$. Domain: $(-\infty, \infty)$; range: $(0, \infty)$

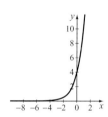

21. The y-intercept is $(0, 4)$. Other points include $(1, 4e)$ and $\left(-1, \frac{4}{e}\right)$. As $x \to \infty$, $f(x) \to \infty$. As $x \to -\infty$, $f(x) \to 0$. Domain: $(-\infty, \infty)$; range: $(0, \infty)$

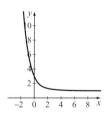

23. The y-intercept is $(0, 3)$. Other points include $\left(1, \frac{2}{e} + 1\right)$ and $(-1, 2e + 1)$. As $x \to \infty$, $g(x) \to 1$. As $x \to -\infty$, $g(x) \to \infty$. Domain: $(-\infty, \infty)$; range: $(1, \infty)$

Section 5.3 (page 441)

25.

Logarithmic Statement	Exponential Statement
$\log_3 9 = 2$	$3^2 = 9$
$\log 0.1 = -1$	$10^{-1} = 0.1$
$\log_5 \dfrac{1}{25} = -2$	$5^{-2} = \dfrac{1}{25}$

27. 4 **29.** 2 **31.** $\dfrac{1}{2}$ **33.** $\dfrac{1}{3}$

35. $x + 2$ **37.** 1.2041 **39.** 1.0397
41. 1.3277 **43.** -0.1606
45. The domain of the function is $(0, \infty)$. The asymptote is $x = 0$. The x-intercept is $(10^6, 0)$.

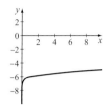

47. The domain of the function is $(0, \infty)$. The asymptote is $x = 0$. The x-intercept is $(1, 0)$.

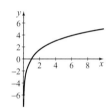

Section 5.4 (page 442)
49. 1.3222 **51.** 0.2219
53. $\dfrac{1}{4} \log x + \dfrac{1}{3} \log x$
55. $3 \log_a x - \dfrac{3}{2} \log_a y - \dfrac{5}{2} \log_a z$
57. $\dfrac{1}{3} \ln x + \ln y - \dfrac{5}{3} \ln z$ **59.** $\ln x$
61. $\log x^3 \sqrt[4]{x - 1}$ **63.** $\log_3 x^{23/6}$

Section 5.5 (page 442)
65. $x = 4$ **67.** $x = -2$ **69.** $x = 25 \ln 4$
71. $x = -\dfrac{1}{2}$ **73.** $x = \dfrac{\ln 4 - 1}{2}$ **75.** $x = 1$
77. $x = e^4$ **79.** $x = \dfrac{5}{2}$ **81.** $x = 0, 3$ **83.** $x = 1$

Section 5.6 (page 442)
85. 4; 0.0022
87. 14.8629; 33.1888
89. 20; 31.2965

Applications (page 442)
91. \$2021.03
93. \$2065.69

95. $v(t) = 17{,}000(0.75)^t$

Number of Years After Purchase	Value
1	$12,750.00
2	$9562.50
3	$7171.88
4	$5378.91
5	$4034.18

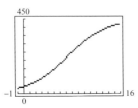

97. 1

99. a. 45 trout

b. About 409 trout

c. As t increases, $N(t)$ increases, but at an increasingly slower rate. The function is asymptotic to $N(t) = 450$.

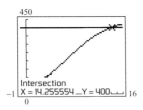

d. Approximately 14.3 months

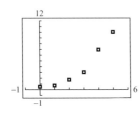

101. a. $f(x) = 0.4627(1.8887)^x$

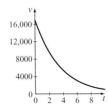

b. 39.668 million

c. 2006

Chapter 5 Test (page 444)

1. $f(g(x)) = x$ and $g(f(x)) = x$, so the functions are inverses.

2. $f^{-1}(x) = \sqrt[3]{\dfrac{x + 1}{4}}$

3. $f^{-1}(x) = \sqrt{x + 2}$

4. As $x \to -\infty$, $f(x) \to 1$; as $x \to \infty$, $f(x) \to -\infty$.

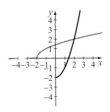

5. As $x \to -\infty$, $f(x) \to \infty$; as $x \to \infty$, $f(x) \to -3$.

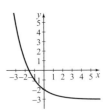

6. As $x \to -\infty$, $f(x) \to \infty$; as $x \to \infty$, $f(x) \to 0$.

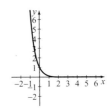

7. $6^{-3} = \dfrac{1}{216}$ **8.** $\log_2 32 = 5$ **9.** -2 **10.** 3.2   **11.** 0.8178

12. Vertical asymptote: $x = -2$;
x-intercept: $(-1, 0)$;
y-intercept: $(0, \ln 2)$

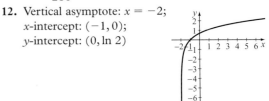

13. $\dfrac{2}{3} \log x + \dfrac{4}{3} \log y$ **14.** $2 + 2 \ln x + \ln y$

15. $\ln(x(x + 2))$ **16.** $\log_2 x^2$ **17.** $x = \dfrac{1}{2}$ **18.** $x = 1.4139$

19. $x = -0.6137$ **20.** $t = 6.9315$ **21.** $x = 0$ **22.** $x = 2$
23. \$4042.05 **24.** \$4934.71 **25.** $f(t) = 900(0.6)^t$
26. $10^{6.2}I_0$, or about $(1.585 \times 10^6)I_0$
27. a. 30 students **b.** Approximately 114 students
28. $P(t) = 28,000e^{0.06677t}$

Chapter 6 Exercises

Exercise Set 6.1 (page 459)
1. $(-1, 2)$ **3.** $(3, 5)$ **5.** $(4, 2)$ **7.** $\begin{cases} 4 + 5(-2) = -6 \\ -4 + 2(-2) = -8 \end{cases}$

9. $\begin{cases} 2(0) - 3(2) = -6 \\ -0 + 2(2) = 4 \end{cases}$ **11.** $\begin{cases} a - (a - 5) = 5 \\ -2a + 2(a - 5) = -10 \end{cases}$

13. $x = 1, y = 3$ **15.** $x = -1, y = 2$ **17.** No solution
19. $x = 2, y = -3$ **21.** Dependent system

23. $x = -\dfrac{16}{3}, y = -3$ **25.** $x = 6, y = -4$ **27.** $x = \dfrac{1}{2}, y = 1$

29. $x = -2, y = -3$ **31.** $\begin{cases} 3x + 3y = 1 \\ -2x + y = 2 \end{cases}; x = -\dfrac{5}{9}, y = \dfrac{8}{9}$

33. $\begin{cases} x + 4y = -5 \\ 2x + 3y = 6 \end{cases}; x = \dfrac{39}{5}, y = \dfrac{-16}{5}$

35. $\begin{cases} 3x + 2y = 5 \\ 3x + y = 7 \end{cases}; x = 3, y = -2$

37. $x = -3.75, y = 2.875$

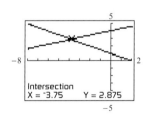

39. $x = -0.75, y = 2.75$

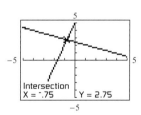

41. $\begin{cases} x - y = 0 \\ -2x - y = -3 \end{cases}$ **43.**

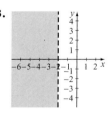

45.  **47.**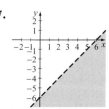

49. 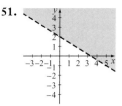 **51.**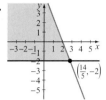

53. 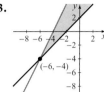 **55.**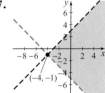

57. 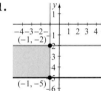 **59.**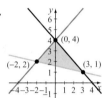

61.   **63.**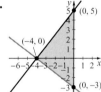

65.   **67.**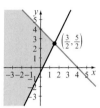

69. 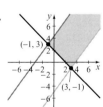 **71.** $P = 170$ **73.** $P = 180$

75. Tickets: \$600 million; merchandise: \$2.4 billion
77. a. Running: 26 minutes; walking: 14 minutes
b. Running: 38 minutes; walking: -18 minutes. No. Time
must be nonnegative.
79. \$380: 53 tickets; \$700: 27 tickets
81. a. $x + y = 10$ **b.** $0.10x + 0.25y = 1.5$
c. $x = \dfrac{20}{3}$ gallons, $y = \dfrac{10}{3}$ gallons; $\dfrac{20}{3}$ gallons of 10%

solution and $\dfrac{10}{3}$ gallons of 25% solution are needed for a

15% solution.

d. No. The acidity of the less acidic of the two solutions used to make the mixture is greater than the acidity desired.

83. a. $\begin{cases} x + y \le 10{,}000 \\ x \ge 6000 \\ y \le 4000 \\ y \ge 0 \end{cases}$ **b.**

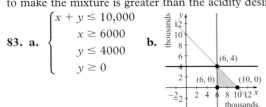

85. a. $\begin{cases} x + y \le 200 \\ x \ge 40 \\ 6x + 8y \ge 960 \\ y \ge 0 \end{cases}$ **b.**

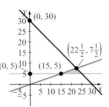

87. a. $\begin{cases} x + y \le 30 \\ x \ge 3y \\ y \ge 5 \\ x \ge 0 \end{cases}$ **b.**

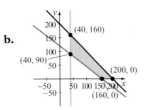

89. 1000 of Model 120; 0 of Model 140 **91.** 10 days
93. 50 drivers, 30 putters
95. 0 acres of cucumbers, 50 acres of peanuts
97. 80 mini pizzas, 20 mini quiches **99.** $a = 0.15, b = 10$
101. a. **b.** No

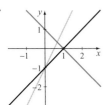

c. No. There is no point of intersection.
103. No. The resulting equations form a dependent system.

Exercise Set 6.2 (page 473)

1. $\begin{cases} 3(1) - (0) = 3 \\ 2(1) + (-2) - 2(0) = 0 \\ 3(1) - 2(-2) + (0) = 7 \end{cases}$

3. $\begin{cases} 5(0) - (1) + 3(-4) = -13 \\ (0) - (1) + 2(-4) = -9 \\ 4(0) - (1) + (-4) = -5 \end{cases}$

5. $\begin{cases} (-9 - 11z) + 2(5 + 7z) - 3z = 1 \\ 2(-9 - 11z) + 3(5 + 7z) + z = -3 \end{cases}$

7. $x = 4, y = 1, z = -2$ **9.** $u = 7, v = 4, w = 3$
11. $x = 2, y = 0, z = 2$ **13.** $x = -4, y = 2, z = -1$
15. $x = 0, y = 0, z = -2$ **17.** $x = -6, y = -9, z = 11$
19. $x = \dfrac{2}{3} + \dfrac{1}{6}y, z = -\dfrac{1}{3} - \dfrac{5}{6}y$ **21.** $x = 3, y = \dfrac{1}{2}, z = -\dfrac{3}{2}$

23. $x = 0, y = -\dfrac{4}{3}, z = \dfrac{2}{3}$ **25.** $x = 2, y = -3, z = -5$
27. $u = 4, v = 2, w = -5$ **29.** $r = 3, s = 11, t = 8$
31. $x = -1 - 5z, y = -4 - z$
33. $r = 9 - 8t, s = -4 + 4t$ **35.** $y = 4, z = -15 - 3x$
37. Inconsistent
39. Mutual fund: 50%, high-yield bond: 20%, CD: 30%
41. A: 20 ohms, B: 35 ohms, C: 45 ohms
43. Level A: $45, Level B: $35, Level C: $30
45. Florida: 23.3%, New York: 23.3%, California: 22.1%
47. a. $a + b + c = 1; 4a - 2b + c = -8$
b. $a = -1, b = 2, c = 0$ **c.** $f(x) = -x^2 + 2x$
49. $f(x) = 5x^2 - 10x + 6$

Exercise Set 6.3 (page 488)

1. $\begin{bmatrix} 4 & 1 & -2 & | & 6 \\ -1 & -1 & 1 & | & -2 \\ 3 & 0 & -1 & | & 4 \end{bmatrix}$ **3.** $\begin{bmatrix} 3 & -2 & 1 & | & -1 \\ 1 & 1 & -4 & | & 3 \\ -2 & -1 & 3 & | & 0 \end{bmatrix}$

5. $\begin{bmatrix} 6 & -2 & 1 & | & 0 \\ -5 & 1 & -3 & | & -2 \\ 2 & -3 & 5 & | & 7 \end{bmatrix}$ **7.** $\begin{bmatrix} 1 & 1 & 2 & | & -3 \\ -3 & 2 & 1 & | & 1 \end{bmatrix}$

9. $\begin{bmatrix} 1 & -2 & 0 & | & -1 \\ 1 & -4 & -1 & | & \frac{1}{2} \\ 3 & 5 & 1 & | & 2 \end{bmatrix}$ **11.** $\begin{bmatrix} 2 & -8 & -2 & | & 1 \\ 1 & -2 & 0 & | & -1 \\ 3 & 5 & 1 & | & 2 \end{bmatrix}$

13. $\begin{bmatrix} 1 & -2 & 0 & | & -1 \\ 4 & -12 & -2 & | & -1 \\ 3 & 5 & 1 & | & 2 \end{bmatrix}$ **15.** $\begin{cases} 1x + 0y = -7 \\ 0x + 1y = 3 \end{cases}$
$x = -7, y = 3$

17. $\begin{cases} 2x + 0y = 6 \\ 0x + 1y = 5 \end{cases}$ **19.** $\begin{cases} x + 3z = 5 \\ y - 2z = -2 \end{cases}$
$x = 3, y = 5$ $\quad x = 5 - 3z, y = -2 + 2z$

21. $\begin{cases} x - 5u = 2 \\ y - 2u = -3 \\ z + 3u = 5 \end{cases}$
$x = 2 + 5u, y = -3 + 2u, z = 5 - 3u$

23. $x = -14, y = 8$ **25.** $x = \dfrac{1}{2}, y = -4$ **27.** $x = 2, y = 4$

29. $x = 3, y = -1$ **31.** $y = -\dfrac{1}{3} - \dfrac{5}{3}x, x$ is a real number

33. Inconsistent system **35.** $x = -3, y = 2, z = 4$
37. $x = 1, y = 1, z = 0$ **39.** $x = 2, y = -2, z = 5$
41. $x = -y, z = y - 2, y$ is a real number
43. Inconsistent system **45.** $x = 4y + 24, z = 10$
47. $r = 10, s = 2t - 4$ **49.** Inconsistent system
51. $x = -1, y = 3, z = -6$
53. $6\dfrac{2}{3}$ pounds of Colombian, $2\dfrac{2}{9}$ pounds of Java,

$1\dfrac{1}{9}$ pounds of Kona
55. 5 cheese pizzas, 3 pepperoni pizzas
57. 10 boxes of Brand A, 6 boxes of Brand B
59. Mutual fund: 50%, bond: 12.5%, CD: 37.5%
61. $A = 40$ ohms, $B = 20$ ohms, $C = 80$ ohms

63. Infinitely many solutions: $a = 0$. One solution: $a \neq 0$. If $a = 0$, this is a dependent system. If $a \neq 0$, the solution is $(-2, 5, 0)$.

65.

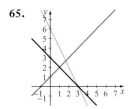

67. a. The first equation is the second equation multiplied by 3. The second equation is the first multiplied by $\frac{1}{3}$.

b. Because the equations are multiples of each other, the values that satisfy one will satisfy the other.

c. $w = 2u + 2v + 1$

d. $(1, 1, 5), (-1, 1, 1)$

Exercise Set 6.4 (page 503)

1. -1 **3.** 3×4 **5.** π

7. False. For addition, matrices must have the same dimensions so that corresponding entries can be added.

9. True. The product of an $m \times n$ matrix and an $n \times p$ matrix is an $m \times p$ matrix.

11. $\begin{bmatrix} 4 & 5 \\ 3 & -1 \\ 0 & 1 \end{bmatrix}$ **13.** $\begin{bmatrix} 12 & 5 \\ 6 & -3 \\ 2 & -5 \end{bmatrix}$ **15.** $\begin{bmatrix} 20 & -15 \\ 3 & -5 \\ 8 & -27 \end{bmatrix}$

17. Not defined. The dimensions do not allow for addition.

19. $\begin{bmatrix} -\frac{1}{3} & 4 \\ 36 & 12 \end{bmatrix}$

21. Not defined. The number of columns of the first matrix does not match the number of rows of the second.

23. $\begin{bmatrix} \frac{1}{2} & -\frac{3}{2} & \frac{1}{6} \\ \frac{5}{2} & 0 & -1 \end{bmatrix}$ **25.** $\begin{bmatrix} -5 & \frac{25}{3} \\ 20 & 23 \end{bmatrix}$ **27.** $\begin{bmatrix} \frac{544}{3} & 44 \\ 36 & 12 \\ \frac{758}{3} & 76 \end{bmatrix}$

29. a. $\begin{bmatrix} -10 \\ -1 \end{bmatrix}$ **b.** The dimensions do not allow for addition.

c. The dimensions do not allow for multiplication in the given order.

31. a. $\begin{bmatrix} 8 & -4 \\ -4 & 35 \end{bmatrix}$ **b.** $\begin{bmatrix} 9 & -29 \\ 18 & -17 \end{bmatrix}$ **c.** $\begin{bmatrix} 42 & -37 \\ 6 & 1 \end{bmatrix}$

33. a. $\begin{bmatrix} -6 & -12 \\ 13 & -17 \\ 13 & -13 \end{bmatrix}$

b. The dimensions do not allow for addition.

c. The dimensions do not allow for multiplication in the given order.

35. $\begin{bmatrix} 28 \\ 27 \end{bmatrix}$ **37.** $\begin{bmatrix} -5 & 3 \\ 4 & 52 \end{bmatrix}$ **39.** $\begin{bmatrix} 10 & -2 \\ 23 & -7 \end{bmatrix}$

41. $\begin{bmatrix} -18 & -4 \\ 20 & -2 \\ -10 & 2 \end{bmatrix}$ **43.** $ag + bj$; $ei + fl$ **45.** $eh + fk$; $ci + dl$

47. $\begin{bmatrix} 3 & -2 \\ 2 & -1 \end{bmatrix}, \begin{bmatrix} 4 & -3 \\ 3 & -2 \end{bmatrix}$ **49.** $\begin{bmatrix} 16 & 0 \\ 0 & 9 \end{bmatrix}, \begin{bmatrix} -64 & 0 \\ 0 & 27 \end{bmatrix}$

51. $\begin{bmatrix} 9 & 0 & 0 \\ -4 & 2 & 1 \\ -12 & 1 & 1 \end{bmatrix}, \begin{bmatrix} 27 & 0 & 0 \\ -16 & 3 & 2 \\ -40 & 2 & 1 \end{bmatrix}$

53. $a = -1, 1$ **55.** $a = 0; b = 4$

57. $a = 2, 1; b = -2$

59. Regular: \$28.80; high-octane: \$31.80

61. a. $\begin{bmatrix} 1854.86 & 834.687 & 129.8402 \\ 1964.92 & 884.214 & 137.5444 \\ 2016.44 & 907.398 & 141.1508 \end{bmatrix}$

b. It represents the breakdown of the contribution from each of these three sectors per year.

c. Yes. It represents the total contribution of these three sectors over the given years.

63. Sofa: \$122; loveseat: \$94; chair: \$48.50

65. Keith: \$107.70; Sam; \$102.75; Cody: \$98.75

67. $\begin{bmatrix} 10 \\ -19 \\ -7 \\ -10 \end{bmatrix}, \begin{bmatrix} -9 \\ 8 \\ 75 \\ -10 \end{bmatrix}, \begin{bmatrix} 34 \\ -127 \\ -68 \\ -34 \end{bmatrix}$

69. $AB = \begin{bmatrix} 0 & 0 \\ 0 & 0 \end{bmatrix}$. No, as demonstrated by the given matrices.

71. $AI = \begin{bmatrix} 2 & -1 \\ 1 & 0 \end{bmatrix}\begin{bmatrix} 1 & 0 \\ 0 & 1 \end{bmatrix} = \begin{bmatrix} 2 & -1 \\ 1 & 0 \end{bmatrix}$

$IA = \begin{bmatrix} 1 & 0 \\ 0 & 1 \end{bmatrix}\begin{bmatrix} 2 & -1 \\ 1 & 0 \end{bmatrix} = \begin{bmatrix} 2 & -1 \\ 1 & 0 \end{bmatrix}$

$AI = IA$

73. $AB = \begin{bmatrix} 2 & 3 \\ 4 & 7 \end{bmatrix}$; $BA = \begin{bmatrix} 3 & 4 \\ 4 & 6 \end{bmatrix}$. No. The corresponding entries of the resulting matrices are computed using different rows and columns.

Exercise Set 6.5 (page 515)

1. Yes   **3.** Yes   **5.** Yes

7. $\begin{bmatrix} -1 & 3 \\ 1 & -2 \end{bmatrix}$ **9.** $\begin{bmatrix} -4 & 3 \\ -1 & 1 \end{bmatrix}$ **11.** $\begin{bmatrix} 2 & -3 \\ -3 & 5 \end{bmatrix}$

13. $\begin{bmatrix} 4 & 0 & -5 \\ -18 & 1 & 24 \\ -3 & 0 & 4 \end{bmatrix}$ **15.** $\begin{bmatrix} -4 & -1 & -1 \\ 0 & 0 & 1 \\ 3 & 1 & 0 \end{bmatrix}$

17. $\begin{bmatrix} 0 & -\frac{2}{5} & \frac{1}{5} \\ -\frac{2}{5} & -\frac{3}{5} & \frac{2}{5} \\ \frac{1}{5} & \frac{2}{5} & 0 \end{bmatrix}$ **19.** $\begin{bmatrix} -1 & 1 & 1 \\ 1 & -1 & 1 \\ 1 & 1 & -1 \end{bmatrix}$

21. $\begin{bmatrix} 7 & 0 & 2 & -1 \\ 6 & 3 & 2 & 2 \\ 3 & 1 & 1 & 1 \\ 0 & 1 & 0 & 1 \end{bmatrix}$ **23.** $x = 4, y = 18, z = 44$

25. $x = -\dfrac{3}{2}, y = -\dfrac{1}{2}, z = \dfrac{1}{2}$ **27.** $x = -3, y = -1$

29. $x = -\dfrac{9}{2}, y = \dfrac{5}{2}$ **31.** $x = 1, y = -2$ **33.** $x = 4, y = 3$

35. $x = -3, y = 3$ **37.** $x = 36, y = 9, z = -5$
39. $x = -47, y = -35, z = 17$ **41.** $x = 4, y = 2, z = -3$
43. $x = -49, y = -28, z = 5.5$
45. $x = -3, y = 0, z = -5, w = 0$

47. $\begin{bmatrix} 1 & -2 \\ 0 & 1 \end{bmatrix}, \begin{bmatrix} 1 & -3 \\ 0 & 1 \end{bmatrix}$ **49.** $\begin{bmatrix} \frac{1}{4} & -\frac{1}{4} \\ 0 & 1 \end{bmatrix} \begin{bmatrix} \frac{1}{8} & \frac{3}{8} \\ 0 & -1 \end{bmatrix}$

51. $\begin{bmatrix} \frac{1}{4} & 0 & 0 \\ 0 & 1 & -4 \\ 0 & 0 & 1 \end{bmatrix}, \begin{bmatrix} \frac{1}{8} & 0 & 0 \\ 0 & 1 & -6 \\ 0 & 0 & 1 \end{bmatrix}$ **53.** Adult: \$12; child: \$5

55. Cheese: 200 calories; Meaty Delite: 270 calories; Veggie
Delite: 150 calories
57. Red: \$72; white: \$54; blue: \$81

59. $\begin{bmatrix} 3 & 7 \\ -2 & 5 \end{bmatrix}$ **61.** $\begin{bmatrix} 1.5 & -1.5 & 0.5 & 4.5 \\ 0.5 & -2.5 & 1.5 & 8.5 \\ 0.5 & -0.5 & 0 & 1.5 \\ 0 & 1 & -0.5 & -3 \end{bmatrix}$

63. $\begin{bmatrix} 16 \\ 9 \\ 18 \end{bmatrix}$ PIR, $\begin{bmatrix} 1 \\ 20 \\ 5 \end{bmatrix}$ ATE, $\begin{bmatrix} 19 \\ 0 \\ 0 \end{bmatrix}$ S   **65.** $\begin{bmatrix} 19 \\ 14 \\ 5 \end{bmatrix}, \begin{bmatrix} 5 \\ 26 \\ 25 \end{bmatrix}$; SNEEZY

67. $\begin{bmatrix} \frac{1}{a} & -1 & -1 \\ 0 & 1 & 0 \\ 0 & 0 & 1 \end{bmatrix}$. If $a = 1$, $\begin{bmatrix} 1 & -1 & -1 \\ 0 & 1 & 0 \\ 0 & 0 & 1 \end{bmatrix}$.

69. $(A^2)^{-1} = (A^{-1})^2 = \begin{bmatrix} 11 & 8 \\ 4 & 3 \end{bmatrix}$

71. $(A^3)^{-1} = (A^{-1})^3 = \begin{bmatrix} -13 & -17 \\ 68 & 89 \end{bmatrix}$

73. $(A^2)^{-1} = \begin{bmatrix} 4 & -5 \\ -15 & 19 \end{bmatrix}$; $(A^3)^{-1} = \begin{bmatrix} 19 & -24 \\ -72 & 91 \end{bmatrix}$

75. a. $\begin{bmatrix} -1 \\ 2 \end{bmatrix}$

b. The coordinates were transposed from (a, b) to (b, a).

c. Multiply the product by A.

1. -14 **3.** -9 **5.** 35 **7.** 11 **9.** $49; 49$ **11.** $10; -10$
13. -85 **15.** 62 **17.** 0 **19.** 0 **21.** 6 **23.** $x = 2$
25. $x = -4$ **27.** $x = -1$ **29.** $x = 7, y = -26$

31. $x = -\dfrac{5}{2}, y = -\dfrac{17}{2}$ **33.** $x = 0, y = -2$

35. $x = -2, y = 1$ **37.** $x \approx -6.9286, y = 4.85$

39. $x = \dfrac{1}{2}, y = \dfrac{1}{2}, z = \dfrac{1}{2}$ **41.** $x = 0, y = -2, z = 1$

43. $x = \dfrac{25}{32}, y = \dfrac{133}{32}, z = \dfrac{71}{16}$ **45.** $x - \dfrac{1}{2}, y = -\dfrac{3}{2}, z = 1$

47. The entries of the second row are all zero, so the sum of
the products of these entries and their cofactors will be zero.

49. $\begin{cases} (1) + 2(2) = & 5 \\ 4(1) + (2) - (0) = & 6 \\ -2(1) - 4(2) = -10 \end{cases}$ $D = \begin{vmatrix} 1 & 2 & 0 \\ 4 & 1 & -1 \\ -2 & -4 & 0 \end{vmatrix} = 0$.

Cramer's Rule applies only for $D \neq 0$.

1. $\dfrac{Ax + B}{x^2 - x - 3}$ **3.** $\dfrac{A}{x + 5} + \dfrac{B}{(x + 5)^2}$ **5.** $\dfrac{Ax + B}{x^2 + 2} + \dfrac{C}{2x + 1}$

7. $\dfrac{A}{x - 2} + \dfrac{B}{x + 2} + \dfrac{Cx + D}{x^2 + 4}$ **9.** $\dfrac{A}{3x} + \dfrac{B}{x + 1} + \dfrac{C}{(x + 1)^2}$

11. Irreducible   **13.** Irreducible   **15.** Reducible

17. $\dfrac{-1}{x + 4} + \dfrac{1}{x - 4}$ **19.** $\dfrac{-2}{x} + \dfrac{4}{2x - 1}$ **21.** $\dfrac{-2}{x + 2} + \dfrac{3}{x + 3}$

23. $\dfrac{-2}{x} + \dfrac{1}{x + 1} + \dfrac{-2}{x - 1}$ **25.** $\dfrac{-2}{x - 1} + \dfrac{4}{(x - 1)^2}$

27. $\dfrac{2}{x^2} + \dfrac{-1}{x + 2}$ **29.** $\dfrac{-1}{x - 1} + \dfrac{-1}{x + 2} + \dfrac{2}{(x + 2)^2}$

31. $\dfrac{-1}{x + 2} + \dfrac{x + 2}{x^2 + 3}$ **33.** $\dfrac{3}{x + 2} + \dfrac{-3x}{x^2 + x + 1}$

35. $\dfrac{-1}{x + 1} + \dfrac{1}{x - 1} + \dfrac{-1}{x^2 + 1}$ **37.** $\dfrac{-1}{x^2 + 2} + \dfrac{-2x}{(x^2 + 2)^2}$

39. $\dfrac{3}{2(x + 1)} + \dfrac{-3}{2(x - 1)} + \dfrac{x}{x^2 + 1}$

41. $\dfrac{1}{3(t + 5)} + \dfrac{2t - 5}{3(t^2 - 5t + 25)}$

43. $\dfrac{2}{s} + \dfrac{-2s}{s^2 + 1}$ **45.** $\dfrac{1}{x(x^2 + 2x - 3)} = \dfrac{A}{x} + \dfrac{B}{x + 3} + \dfrac{C}{x - 1}$

47. $\dfrac{1}{(x - c)^2}$

1. $(2, 3), (-3, -2)$ **3.** $(1, 2), (-1, -2)$

5. $\left(-\dfrac{1}{2}, -\dfrac{1}{2}\right), (3, 17)$ **7.** $\left(1, -\dfrac{5}{4}\right), \left(-\dfrac{2}{3}, 0\right)$

9. $(5, 7), \left(-\dfrac{5}{3}, \dfrac{1}{3}\right)$ **11.** $(-2, -3)$ **13.** $(2, -2), (-2, 2)$

15. $\left(2\sqrt{2}, -1\right), \left(-2\sqrt{2}, -1\right), (-2.598, 1.5), (2.598, 1.5)$

17. $\left(-\sqrt{2}, 0\right), \left(\sqrt{2}, 0\right)$ **19.** $(-2, -1), (2, -1), (5, 6), (-5, 6)$

21. No solution   **23.** $(4, 2), (4, -2)$ **25.** $(0, -2)$

27. $(4, 0), (-4, 0), \left(\sqrt{7}, 3\right), \left(-\sqrt{7}, 3\right)$

29. $(-2, 3), \left(\dfrac{11}{2}, -\dfrac{39}{22}\right)$

31. $(2, \sqrt{10}), (-2, \sqrt{10}), (2, -\sqrt{10}), (-2, -\sqrt{10})$

33. $\left(20, \dfrac{1}{4}\right)$ **35.** $(3, 3)$

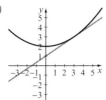

37. $(3, -1), (7, -5)$

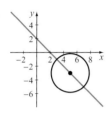

39. $(1, -3), (5, -3)$

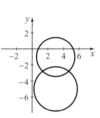

41. $(0.343, -1.646), (-0.834, 0.088)$

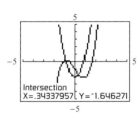

43. $(0.789, 2.378), (-1.686, 0.157)$

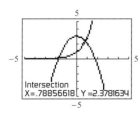

45. No solution

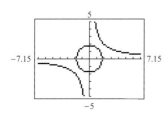

47. $(-1.123, -3.695), (-1.542, 1.895), (1.123, -3.695),$
$(1.542, 1.895)$

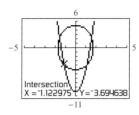

49. 28 feet $\times$ 12 feet

51. 8 inches, 5 inches

53. $r = \dfrac{3}{2}$ inches, $h = 3$ inches

55. Radius = 3 inches, height = 12 inches

57. 6 and 7

59. Property: 30 feet $\times$ 10 feet; pool: 15 feet $\times$ 5 feet; area of paved portion: 225 square feet

61. $b < \dfrac{9}{4}$

63. If $x, y \neq 0$, then $x^2, y^2 \neq 0$. If $xy = 18$, then $2xy = 36$. Expanding the first equation, we get $x^2 + 2xy + y^2 = 36$. Because $2xy = 36$, we can subtract these terms from both sides of the equation. Then $x^2 + y^2 = 0$, which contradicts the assumption that $x, y \neq 0$.

65. a. $h = \pm 2r$

b. $0 < h < 2r$ or $-2r < h < 0$

c. $h = 0$

d. $h > 2r, h < -2r$

Chapter 6 Review Exercises (page 548)

Section 6.1 (page 548)

1. $x = 4, y = 3$ **3.** $y = 5 - x$, x a real number

5.

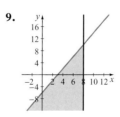

7.

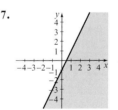

9.

11. $P = 105$

Section 6.2 (page 549)

13. $x = 1, y = -1, z = -1$

15. $x = 1 - z, y = 3 - z$

Section 6.3 (page 549)

17. $\begin{bmatrix} 4 & 1 & 1 & | & 0 \\ 0 & -1 & 2 & | & -1 \\ 1 & 0 & 1 & | & 3 \end{bmatrix}$ **19.** $x = 16, y = -6$

21. $x = 1, y = -3, z = -2$

23. $\begin{cases} x - 2z = 3 \\ y + z = 5 \end{cases}$
$x = 3 + 2z, y = 5 - z$

Section 6.4 (page 549)

25. 0 **27.** 3×4 **29.** $\begin{bmatrix} -1 & 6 \\ 3 & -4 \\ 4 & -9 \end{bmatrix}$ **31.** $\begin{bmatrix} -5 & 7 \\ 5 & -7 \\ 6 & -15 \end{bmatrix}$

33. $\begin{bmatrix} -17 & 4 \\ 4 & -3 \end{bmatrix}$

Section 6.5 (page 550)

35. $\begin{bmatrix} -1 & 5 \\ 1 & -4 \end{bmatrix}$ **37.** $\begin{bmatrix} -1 & 1 \\ -3 & 4 \end{bmatrix}$ **39.** $\begin{bmatrix} 5 & 1 & 1 \\ 4 & 1 & 1 \\ 10 & 3 & 2 \end{bmatrix}$

41. $x = -6, y = 1$

Section 6.6 (page 550)

43. 10 **45.** -42 **47.** $x = -2, y = 4$

49. $x = 2, y = \dfrac{3}{4}, z = \dfrac{13}{4}$

Section 6.7 (page 550)

51. $\dfrac{14}{x + 2} - \dfrac{17}{x + 3}$ **53.** $-\dfrac{1}{x + 2} + \dfrac{x - 2}{x^2 + 5}$

Section 6.8 (page 550)

55. $(1.56, 2.56), (-2.56, -1.56)$ **57.** $(0, 0), (1, 1)$
59. $100; 350$ **61.** Soft taco: 190; tostada: 200; rice: 210
63. 15 inches, 18 inches

Chapter 6 Test (page 551)

1. $x = -2, y = 3$ **2.** $x = 1, y = -2$

3.

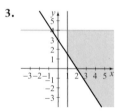

4. $P = 62.5$ at $(3.5, 1)$ **5.** $x = -3, y = -1, z = 2$

6. $x = \dfrac{4}{3} + \dfrac{1}{3}z, y = \dfrac{2}{3} + \dfrac{2}{3}z, z$ any real number

7. $\begin{bmatrix} -19 & -2 \\ 11 & -8 \\ -3 & 1 \end{bmatrix}$ **8.** $\begin{bmatrix} 46 & -12 & 33 \\ -20 & -20 & -3 \\ 21 & 8 & 9 \end{bmatrix}$

9. $\begin{bmatrix} 2 & -5 \\ -1 & 3 \end{bmatrix}$ **10.** $\begin{bmatrix} 1 & 0 & -2 \\ 2 & 1 & -4 \\ -4 & 2 & 9 \end{bmatrix}$

11. $x = 13, y = -4, z = 1$ **12.** 36

13. $x = 0, y = 2$ **14.** $x = -\dfrac{7}{4}, y = -\dfrac{5}{4}, z = \dfrac{3}{4}$

15. $-\dfrac{3}{x + 1} + \dfrac{1}{(x + 1)^2} + \dfrac{2}{x}$ **16.** $\dfrac{2x - 1}{x^2 + 1} + \dfrac{3}{x - 3}$

17. $\left(\dfrac{\sqrt{10}}{5}, \dfrac{3\sqrt{10}}{5} \right), \left(-\dfrac{\sqrt{10}}{5}, -\dfrac{3\sqrt{10}}{5} \right)$ **18.** $(2, 1), (-4, 1)$

19. 600 CX100 models; 100 FX100 models;
maximum profit: $100,000
20. 4 first class; 2 business class; 4 coach

21.

	Total
Customer 1	$40
Customer 2	$40
Customer 3	$30

Chapter 7 Exercises

Exercise Set 7.1 (page 564)

1. True   **3.** 10 **5.** $f(x) = (x - 1)^2$ **7.** True
9. $x^2 + 6x + 9$ **11.** c   **13.** b
15. Vertex: $(0, 0)$; focus: $(3, 0)$; directrix: $x = -3$

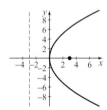

17. Vertex: $(0, 0)$; focus: $(2, 0)$; directrix: $x = -2$

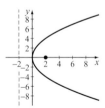

19. Vertex: $(0, 0)$; focus: $(0, -1)$; directrix: $y = 1$

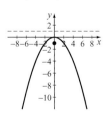

21. Vertex: $(0, 0)$; focus: $\left(\dfrac{2}{3}, 0\right)$; directrix: $x = -\dfrac{2}{3}$

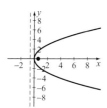

23. Vertex: $(0, 0)$; focus: $\left(\dfrac{1}{5}, 0\right)$; directrix: $x = -\dfrac{1}{5}$

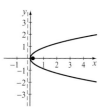

25. Vertex: $(2, 0)$; focus: $(-1, 0)$; directrix: $x = 5$

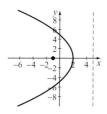

27. Vertex: $(5, -1)$; focus: $(5, -2)$; directrix: $y = 0$

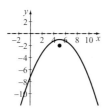

29. Vertex: $(1, 4)$; focus: $\left(\dfrac{3}{4}, 4\right)$; directrix: $x = \dfrac{5}{4}$

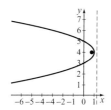

31. Vertex: $(1, 2)$; focus: $(0, 2)$; directrix: $x = 2$

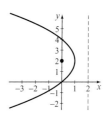

33. Vertex: $(1, 1)$; focus: $\left(1, \dfrac{3}{4}\right)$; directrix: $y = \dfrac{5}{4}$

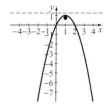

35. Vertex: $(-3, -4)$; focus: $(-3, -5)$; directrix: $y = -3$

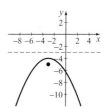

37. Vertex: $(4, -3)$; focus: $\left(4, -\dfrac{7}{4}\right)$; directrix: $y = -\dfrac{17}{4}$

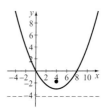

39. $y^2 = 4x$ **41.** $x^2 = 12y$ **43.** $x^2 = 8y$ **45.** $y^2 = -8x$
47. $x^2 = 16y$ **49.** $y^2 = -20x$ **51.** $(x + 2)^2 = 12(y - 1)$
53. $x^2 = 16y$ **55.** $y^2 = 9x$ **57.** $(x - 4)^2 = -(y - 3)$
59. $(x - 4)^2 = 8(y - 1)$ **61.** $(x - 5)^2 = -3(y - 3)$
63. 5.0625 inches from the vertex
65. a. 128 inches **b.** 12 inches from the ends **67.** 3.2 feet
69. a. $y = 4 + x - \dfrac{x^2}{16}$ or $(x - 8)^2 = -16(y - 8)$

b. Different. Under conditions of weaker gravity, the object travels higher and farther.
71. 5 **73.** 4 **75.** $y^2 = -8x$ **77.** $(2, -2)$

Exercise Set 7.2 (page 578)
1. True **3.** $2\sqrt{10}$ **5.** 36 **7.** c **9.** d
11. Center: $(0,0)$; vertices: $(5,0)$ and $(-5,0)$;
foci: $(3,0)$ and $(-3,0)$

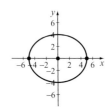

13. Center: $(0,0)$; vertices: $(0,5)$ and $(0,-5)$;
foci: $(0,3)$ and $(0,-3)$

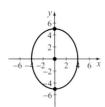

15. Center: $(0,0)$; vertices: $(3,0)$ and $(-3,0)$;
foci: $\left(\sqrt{5},0\right)$ and $\left(-\sqrt{5},0\right)$

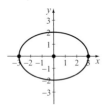

17. Center: $(0,0)$; vertices: $\left(\frac{3}{2},0\right)$ and $\left(-\frac{3}{2},0\right)$;
foci: $\left(\frac{\sqrt{17}}{6},0\right)$ and $\left(-\frac{\sqrt{17}}{6},0\right)$

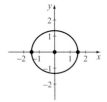

19. Center: $(-3,-1)$; vertices: $(-3,-5)$ and $(-3,3)$;
foci: $\left(-3,-1+\sqrt{7}\right)$ and $\left(-3,-1-\sqrt{7}\right)$

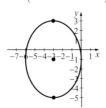

21. Center: $(1,-1)$; vertices: $(11,-1)$ and $(-9,-1)$;
foci: $(-7,-1)$ and $(9,-1)$

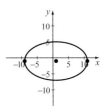

23. Center: $(1,-2)$; vertices: $(-3,-2)$ and $(5,-2)$;
foci: $\left(1-\sqrt{7},-2\right)$ and $\left(1+\sqrt{7},-2\right)$

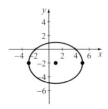

25. Center: $(0,0)$; vertices: $(-2,0)$ and $(2,0)$;
foci: $(-1,0)$ and $(1,0)$

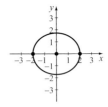

27. Center: $(0,0)$; vertices: $\left(0,\sqrt{5}\right)$ and $\left(0,-\sqrt{5}\right)$;
foci: $\left(0,\sqrt{3}\right)$ and $\left(0,-\sqrt{3}\right)$

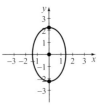

29. Center: $(3,4)$; vertices: $(3,6)$ and $(3,2)$;
foci: $\left(3,4-\sqrt{3}\right)$ and $\left(3,4+\sqrt{3}\right)$

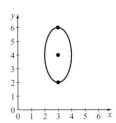

31. Center: $(2, -3)$; vertices: $(5, -3)$ and $(-1, -3)$;
foci: $(4, -3)$ and $(0, -3)$

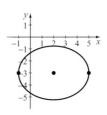

33. Center: $(4, -3)$; vertices: $(4, -4.75)$ and $(4, -1.25)$;
foci: $(4, -4.05)$ and $(4, -1.95)$

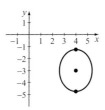

35. 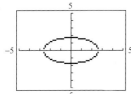   **37.**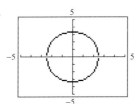

39.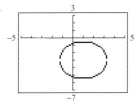

41. $\dfrac{x^2}{36} + \dfrac{y^2}{27} = 1$ **43.** $\dfrac{x^2}{64} + \dfrac{y^2}{39} = 1$ **45.** $\dfrac{4x^2}{49} + \dfrac{4y^2}{81} = 1$

47. $\dfrac{x^2}{4} + \dfrac{y^2}{81} = 1$ **49.** $\dfrac{(x+2)^2}{16} + \dfrac{(y-4)^2}{7} = 1$

51. $\dfrac{(x+3)^2}{5} + \dfrac{(y-1)^2}{9} = 1$ **53.** $\dfrac{(x-4)^2}{12} + \dfrac{(y+1)^2}{16} = 1$

55. $\dfrac{(x-7)^2}{36} + \dfrac{(y+8)^2}{16} = 1$

57. $\dfrac{4x^2}{81} + \dfrac{4y^2}{25} = 1$; $\dfrac{4x^2}{25} + \dfrac{4y^2}{81} = 1$

59. $\dfrac{(x-5)^2}{36} + \dfrac{(y-4)^2}{16} = 1$; $\dfrac{(x-9)^2}{16} + \dfrac{y^2}{36} = 1$

61. $\left(\sqrt{29.75}, 0\right), \left(-\sqrt{29.75}, 0\right)$
63. Minimum distance: 226,335 miles;
maximum distance: 251,401 miles

65. $4\sqrt{34}$ feet   **67.** $4\sqrt{13}$ feet   **69.** 5 inches   **71.** 13
73. $\dfrac{x^2}{11} + \dfrac{y^2}{36} = 1$ **75.** $\dfrac{4}{5}$; $\dfrac{4}{5} > \dfrac{3}{5}$

Exercise Set 7.3 (page 592)
1. $d = \sqrt{(c-w)^2 + (d-v)^2}$ **3.** 3   **5.** $x^2 + 22x + 121$
7. d   **9.** c
11. Center: $(0, 0)$; vertices: $(4, 0)$ and $(-4, 0)$;
foci: $(-5, 0)$ and $(5, 0)$; asymptotes: $y = \dfrac{3}{4}x$ and $y = -\dfrac{3}{4}x$

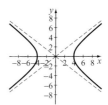

13. Center: $(0, 0)$; vertices: $(0, 3)$ and $(0, -3)$;
foci: $(0, -5)$ and $(0, 5)$; asymptotes: $y = \dfrac{3}{4}x$ and $y = -\dfrac{3}{4}x$

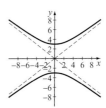

15. Center: $(0, 0)$; vertices: $(0, 8)$ and $(0, -8)$;
foci: $(0, -10)$ and $(0, 10)$; asymptotes: $y = \dfrac{4}{3}x$ and $y = -\dfrac{4}{3}x$

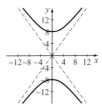

17. Center: $(0, 0)$; vertices: $(3, 0)$ and $(-3, 0)$;
foci: $\left(-3\sqrt{2}, 0\right)$ and $\left(3\sqrt{2}, 0\right)$; asymptotes: $y = x$ and $y = -x$

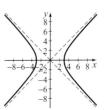

19. Center: $(-3, -1)$; vertices: $(-7, -1)$ and $(1, -1)$;
foci: $(2, -1)$ and $(-8, -1)$;

asymptotes: $y = \dfrac{3}{4}x + \dfrac{5}{4}$ and $y = -\dfrac{3}{4}x - \dfrac{13}{4}$

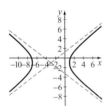

21. Center: $(-5, -1)$; vertices: $(-10, -1)$ and $(0, -1)$;
foci: $\left(-5 + \sqrt{41}, -1\right)$ and $\left(-5 - \sqrt{41}, -1\right)$;

asymptotes: $y = \dfrac{4}{5}x + 3$ and $y = -\dfrac{4}{5}x - 5$

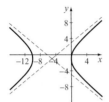

23. Center: $(-4, 0)$; vertices: $(-4, 1)$ and $(-4, -1)$;
foci: $\left(-4, \sqrt{2}\right)$ and $\left(-4, -\sqrt{2}\right)$;
asymptotes: $y = x + 4$ and $y = -x - 4$

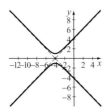

25. Center: $(-1, 3)$; vertices: $(-1, 0)$ and $(-1, 6)$;
foci: $\left(-1, 3 + \sqrt{34}\right)$ and $\left(-1, 3 - \sqrt{34}\right)$;

asymptotes: $y = \dfrac{3}{5}x + \dfrac{18}{5}$ and $y = -\dfrac{3}{5}x + \dfrac{12}{5}$

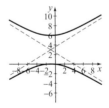

27. Center: $(-2, 3)$; vertices: $(10, 3)$ and $(-14, 3)$;
foci: $(-15, 3)$ and $(11, 3)$;

asymptotes: $y = \dfrac{5}{12}x + \dfrac{23}{6}$ and $y = -\dfrac{5}{12}x + \dfrac{13}{6}$

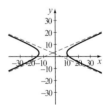

29. Center: $(0, 0)$; vertices: $\left(\dfrac{5}{2}, 0\right)$ and $\left(-\dfrac{5}{2}, 0\right)$;

foci: $\left(-\dfrac{5\sqrt{13}}{4}, 0\right)$ and $\left(\dfrac{5\sqrt{13}}{4}, 0\right)$;

asymptotes: $y = \dfrac{3}{2}x$ and $y = -\dfrac{3}{2}x$

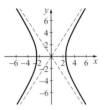

31. Center: $(-3, 0)$; vertices: $(-6, 0)$ and $(0, 0)$;
foci: $\left(-3 - 3\sqrt{10}, 0\right)$ and $\left(-3 + 3\sqrt{10}, 0\right)$;
asymptotes: $y = 3x + 9$ and $y = -3x - 9$

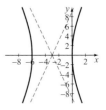

33. Center: $(2, -3)$;
vertices: $\left(2 + 2\sqrt{2}, -3\right)$ and $\left(2 - 2\sqrt{2}, -3\right)$;
foci: $\left(2 - 6\sqrt{2}, -3\right)$ and $\left(2 + 6\sqrt{2}, -3\right)$;
asymptotes: $y = 2\sqrt{2}x - 4\sqrt{2} - 3$ and
$y = -2\sqrt{2}x + 4\sqrt{2} - 3$

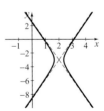

35. $\dfrac{x^2}{9} - \dfrac{y^2}{7} = 1$ **37.** $\dfrac{5x^2}{16} - \dfrac{5y^2}{64} = 1$ **39.** $\dfrac{x^2}{1} - \dfrac{y^2}{3} = 1$

41. $\dfrac{x^2}{16} - \dfrac{3y^2}{4} = 1$ **43.** $\dfrac{(y+4)^2}{2} - \dfrac{(x+3)^2}{2} = 1$

45. $\dfrac{(x-3)^2}{4} - \dfrac{(y+2)^2}{25} = 1$ **47.** $\dfrac{(x-1)^2}{25} - \dfrac{(y+4)^2}{39} = 1$

49. 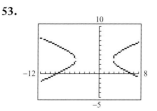 **51.**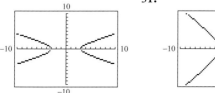

53.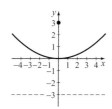

55. $\dfrac{x^2}{36} - \dfrac{y^2}{576} = 1$; $\dfrac{y^2}{576} - \dfrac{x^2}{36} = 1$

57. $\dfrac{(y-4)^2}{9} - \dfrac{(x+1)^2}{40} = 1$; $\dfrac{(y+2)^2}{9} - \dfrac{(x+1)^2}{40} = 1$

59. 20 feet apart **61.** $4\sqrt{2}$ inches

63. $\dfrac{x^2}{25{,}000{,}000} - \dfrac{y^2}{144{,}000{,}000} = 1$ **65.** $x \geq 3$ or $x \leq -3$

67. $2c$ **69.** $\dfrac{(y+3)^2}{9} - \dfrac{(x-3)^2}{16} = 1$

71. The hyperbola $xy = 10$ lies in the first and third quadrants. The hyperbola $xy = -10$ lies in the second and fourth quadrants.

Chapter 7 Review Exercises (page 598)

Section 7.1 (page 598)

1. Vertex: $(0, 0)$; focus: $(0, 3)$; directrix: $y = -3$

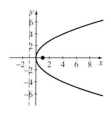

3. Vertex: $(0, 0)$; focus: $(1, 0)$; directrix: $x = -1$

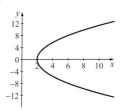

5. Vertex: $(3, -9)$; focus: $(3, -8.75)$; directrix: $y = -9.25$

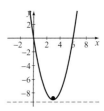

7. Vertex: $(5, -5)$; focus: $(3.75, -5)$; directrix: $x = 6.25$

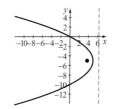

9. Vertex: $(0, 0)$; focus: $\left(\dfrac{1}{28}, 0\right)$; directrix: $x = -\dfrac{1}{28}$

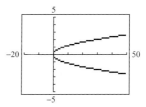

11. $x^2 = -\dfrac{2}{3}y$

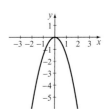

13. $x^2 = -20y$

15. $y^2 = 16(x - 2)$

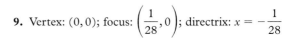

Section 7.2 (page 598)

17. Vertices: $(2, 0)$ and $(-2, 0)$; foci: $\left(-\sqrt{3}, 0\right)$ and $\left(\sqrt{3}, 0\right)$

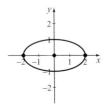

19. Vertices: $(3, 1)$ and $(-7, 1)$; foci: $(1, 1)$ and $(-5, 1)$

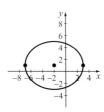

21. Vertices: $(-2, 4)$ and $(-2, -4)$;
foci: $\left(-2, -\sqrt{7}\right)$ and $\left(-2, \sqrt{7}\right)$

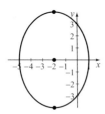

23. Vertices: $\left(0, \sqrt{7}\right)$ and $\left(0, -\sqrt{7}\right)$;

foci: $\left(0, \dfrac{\sqrt{42}}{3}\right)$ and $\left(0, -\dfrac{\sqrt{42}}{3}\right)$

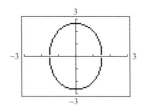

25. Vertices: $\left(2 + 2\sqrt{6}, 0\right)$ and $\left(2 - 2\sqrt{6}, 0\right)$
foci: $\left(2 + 2\sqrt{5}, 0\right)$ and $\left(2 - 2\sqrt{5}, 0\right)$

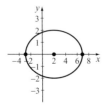

27. Vertices: $\left(-3, 3 - \sqrt{7}\right)$ and $\left(-3, 3 + \sqrt{7}\right)$;
foci: $\left(-3, 3 - \sqrt{3}\right)$ and $\left(-3, 3 + \sqrt{3}\right)$

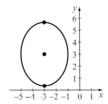

29. $\dfrac{x^2}{16} + \dfrac{y^2}{25} = 1$ **31.** $\dfrac{x^2}{20} + \dfrac{y^2}{36} = 1$ **33.** $\dfrac{x^2}{25} + \dfrac{y^2}{16} = 1$

Section 7.3 (page 599)

35. Center: $(0, 0)$; vertices: $(0, -6)$ and $(0, 6)$;

foci: $\left(0, -\sqrt{61}\right)$ and $\left(0, \sqrt{61}\right)$; asymptotes: $y = \pm \dfrac{6}{5}x$

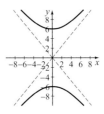

37. Center: $(1, -2)$; vertices: $(1, -4)$ and $(1, 0)$;
foci: $\left(1, -2 - 2\sqrt{2}\right)$ and $\left(1, -2 + 2\sqrt{2}\right)$;
asymptotes: $y = x - 3$ and $y = -x - 1$

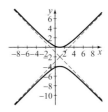

39. Center: $(0, -2)$; vertices: $(-2, -2)$ and $(2, -2)$;
foci: $\left(-\sqrt{13}, -2\right)$ and $\left(\sqrt{13}, -2\right)$;

asymptotes: $y = \dfrac{3}{2}x - 2$ and $y = -\dfrac{3}{2}x - 2$

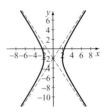

41. Center: $(-2, 0)$; vertices: $(-2, -1)$ and $(-2, 1)$;
foci: $\left(-2, -\sqrt{17}\right)$ and $\left(-2, \sqrt{17}\right)$;
asymptotes: $y = \dfrac{1}{4}x + \dfrac{1}{2}$ and $y = -\dfrac{1}{4}x - \dfrac{1}{2}$

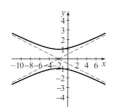

43. Center: $(0, 0)$; vertices: $\left(-2\sqrt{2}, 0\right)$ and $\left(2\sqrt{2}, 0\right)$;
foci: $\left(-6\sqrt{2}, 0\right)$ and $\left(6\sqrt{2}, 0\right)$;
asymptotes: $y = 2\sqrt{2}x$ and $y = -2\sqrt{2}x$

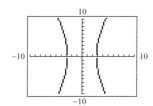

45. $\dfrac{y^2}{4} - \dfrac{x^2}{32} = 1$ **47.** $\dfrac{(y + 4)^2}{9} - \dfrac{(x - 3)^2}{16} = 1$
49. $\dfrac{(x + 4)^2}{16} - \dfrac{(y - 2)^2}{20} = 1$

Applications (page 599)
51. $\dfrac{5\sqrt{3}}{3}$ feet **53.** $20\sqrt{15}$ AU **55.** $2\sqrt{113}$ feet

Chapter 7 Test (page 599)
1. $(y + 3)^2 = -12(x + 1)$
Vertex: $(-1, -3)$
Focus: $(-4, -3)$
Directrix: $x = 2$

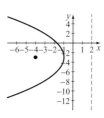

2. $y^2 - 4y + 4x = 0$
Vertex: $(1, 2)$
Focus: $(0, 2)$
Directrix: $x = 2$

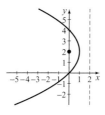

3. $x^2 = 8(y + 2)$ **4.** $y^2 = 12x$

5. $\dfrac{(x + 2)^2}{16} + \dfrac{(y - 3)^2}{25} = 1$
Vertices: $(-2, -2), (-2, 8)$
Foci: $(-2, 0), (-2, 6)$

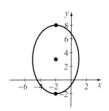

6. $\dfrac{x^2}{9} + \dfrac{(y + 1)^2}{4} = 1$
Vertices: $(-3, -1), (3, -1)$
Foci: $(-\sqrt{5}, -1), (\sqrt{5}, -1)$

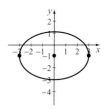

7. $\dfrac{(x + 1)^2}{1} + \dfrac{y^2}{4} = 1$ **8.** $\dfrac{x^2}{16} + \dfrac{y^2}{25} = 1$
9. $\dfrac{y^2}{4} - x^2 = 1$
Center: $(0, 0)$
Vertices: $(0, 2), (0, -2)$
Foci: $(0, \sqrt{5}), (0, -\sqrt{5})$
Asymptotes: $y = 2x, y = -2x$

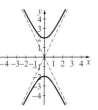

10. $\dfrac{(x + 2)^2}{9} - \dfrac{y^2}{16} = 1$
Center: $(-2, 0)$
Vertices: $(-5, 0), (1, 0)$
Foci: $(-7, 0), (3, 0)$
Asymptotes: $y = \dfrac{4}{3}(x + 2)$,

$y = -\dfrac{4}{3}(x + 2)$

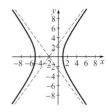

11. $\dfrac{x^2}{1} - \dfrac{y^2}{8} = 1$ **12.** $\dfrac{y^2}{1} - \dfrac{9x^2}{1} = 1$
13. 16 feet from the center
14. $\dfrac{1}{16}$ inch from the vertex

Chapter 8 Exercises _____

Exercise Set 8.1 (page 610)
1. True **3.** \$108.24 **5.** 4, 10, 16 **7.** $-5, -2, 1$
9. $-4, -8, -12$ **11.** 8, 6, 4 **13.** 7, 28, 112 **15.** 5, 15, 45
17. $-2, -6, -18$ **19.** $-3, -15, -75$ **21.** $a_n = -3 + 6n$
23. $g(n) = -16 + 4n$ **25.** $b_n = -20 + 3n$
27. $a_n = 10 + 5n$ **29.** $a_n = 21 - 2n$ **31.** $b_n = 10 + \dfrac{1}{2}n$
33. $a_n = 3(2)^n$ **35.** $h(n) = 2(4)^n$ **37.** $b_n = 3\left(\dfrac{1}{2}\right)^n$
39. $a_n = 4(2)^n$ **41.** $b_n = 4\left(\dfrac{3}{2}\right)^n$ **43.** $a_n = 4(5)^n$

45.

n	0	1	2	3
a_n	2	5	8	11

47.

n	0	1	2	3
a_n	-2	2	6	10

49.

n	0	1	2	3
a_n	5	3	1	-1

51.

n	0	1	2	3
a_n	1	3	9	27

53.

n	0	1	2	3
a_n	2	10	50	250

55.

n	0	1	2	3
a_n	$\frac{1}{2}$	$\frac{1}{4}$	$\frac{1}{8}$	$\frac{1}{16}$

57. Arithmetic **59.** Geometric **61.** Arithmetic
63. Geometric **65.** Arithmetic **67.** Geometric

69. a.

Year	Total Value ($)
1	2120
2	2240
3	2360
4	2480

b. $V = 2000 + 120n$

71. a. Arithmetic **b.** $a_n = 100 + 4(n - 1)$ **c.** 18 rows
73. a. Geometric **b.** 440 Hz, 880 Hz **c.** $a_n = 55(2)^{n-1}$
75. Geometric; $50,613
77. a. 160,000 **b.** Geometric **c.** 7 hours
79. a. Arithmetic; each term is increased by 4.
   b. Neither; it is a sequence of squared terms.
81. a. $f(n) = 550.1 + 28.11n$ **b.** $775 billion
83. $a_0, a_0, a_0, \ldots$

85. a. $a_n = -3 + 4n$ **b.** $a_n = \frac{1}{3}(3)^n$

c. $a_4 = 13, a_5 = 17, a_6 = 21, a_7 = 25, a_8 = 29$
d. $a_4 = 27, a_5 = 81, a_6 = 243, a_7 = 729, a_8 = 2187$
e. The sequence in part (b) because it is geometric
87. $(1 + x)^4$, $(1 + x)^5$; $r = (1 + x)$
89. A geometric sequence; each term can be rewritten in the
form $a_n^3 = a_0^3(r^3)^n$.

Exercise Set 8.2 (page 624)
1. False **3.** $1040.60 **5.** 315 **7.** 371 **9.** 483 **11.** 765

13. $11\frac{61}{64}$ **15.** $98\frac{33}{64}$ **17.** 1395 **19.** 520 **21.** $\frac{n}{2}(5n + 8)$

23. 97,656 **25.** 2186 **27.** 3965 **29.** 3195 **31.** 4288

33. 2632 **35.** 42 **37.** 165 **39.** $1\frac{15}{16}$ **41.** 968

43. a. $\sum_{i=0}^{19} 2 + 2i$ **b.** 420 **45. a.** $\sum_{i=1}^{60} ia$ **b.** $1830a$

47. a. $\sum_{i=1}^{10} 2^i$ **b.** 2046 **49. a.** $\sum_{i=1}^{40} a^i$ **b.** $a\left(\dfrac{1 - a^{40}}{1 - a}\right)$

51. a. $\sum_{i=0}^{24} 2.5i$ **b.** 750 **53. a.** $\sum_{i=0}^{44} (0.5)^i$ **b.** 2

55. 49, arithmetic, $d = 2$ **57.** 105, arithmetic, $d = 0.5$

59. $\dfrac{127}{2}$, geometric, $r = 2$ **61.** $\dfrac{63}{64}$, geometric, $r = 0.5$

63. 1.9375, geometric, $r = 0.5$ **65.** 108, arithmetic, $d = 6$

67. 12 **69.** $\dfrac{27}{2}$ **71.** No **73.** 4 **75.** $\dfrac{15}{4}$

77. a. 10 **b.** The next two possible arrangements can be
made with 66 and 78 boxes. **c.** 11; 4
79. $50,639.99 **81.** 36.25 feet

83. a. $\sum_{i=0}^{4} 7^i$; geometric **b.** 2801

85. $\sum_{i=0}^{7} 4 + 4i$; 144 **87.** 52 **89.** $\dfrac{48}{31}$

Exercise Set 8.3 (page 633)

1. $6, 2, -2, -6, -10$ **3.** $-4, -\dfrac{4}{3}, -\dfrac{4}{9}, -\dfrac{4}{27}, -\dfrac{4}{81}$

5. $4, 5, 8, 13, 20$ **7.** $\dfrac{1}{3}, \dfrac{2}{9}, \dfrac{3}{19}, \dfrac{4}{33}, \dfrac{5}{51}$ **9.** $e, -e^2, e^3, -e^4, e^5$

11. $\dfrac{3}{2}, 1, \dfrac{9}{10}, 1, \dfrac{33}{26}$ **13.** $2, \sqrt{6}, \sqrt{8}, \sqrt{10}, \sqrt{12}$

15. $2, 3, 6, 11, 18$ **17.** $0, 2, 16, 54, 128$
19. $a_n = -2 - 4n, n = 0, 1, 2, 3, \ldots$

21. $a_n = \dfrac{1}{2^n}, n = 0, 1, 2, 3, \ldots$ **23.** $a_n = \sqrt{n}, n = 1, 2, 3, \ldots$

25. $a_n = (0.4)^n, n = 0, 1, 2, 3, \ldots$

27. $a_n = \sqrt{3n}, n = 1, 2, 3, \ldots$ **29.** $6, 4, 2, 0$ **31.** $4, 1, \dfrac{1}{4}, \dfrac{1}{16}$

33. $-1, 0, 2, 5$
35. $\sqrt{3}, \sqrt{\sqrt{3} + 3}, \sqrt{\sqrt{\sqrt{3} + 3} + 3}, \sqrt{\sqrt{\sqrt{\sqrt{3} + 3} + 3} + 3}$
37. $-4, -2, 0, 2; a_n = -4 + 2n, n = 0, 1, 2, 3, \ldots$

39. $6, 3, \dfrac{3}{2}, \dfrac{3}{4}; b_n = 6\left(\dfrac{1}{2}\right)^n, n = 0, 1, 2, 3, \ldots$

41. a. $p_1 = 200, p_n = 1.04(p_{n-1})$ **b.** $p_n = 200(1.04)^n$
c. $224.97 **d.** 6 years **e.** 11 years
43. a. $P_n = 2(36) + 2(18 + 0.5n)$ **b.** $A_n = (36)(18 + 0.5n)$
c. 36 inches × 20.5 inches, 113 inches, 738 square inches
d. $3330
45. a. $0.06 **b.** $7.56 **c.** $9.66 **47.** $n = 5$
49. 4 **51.** $a_0 < 0$

Exercise Set 8.4 (page 643)
1. 24 **3.** 60 **5.** 30 **7.** 24 **9.** 2520 **11.** 6720 **13.** 4
15. 56 **17.** 28 **19.** 1 **21.** 100
23. BCZ, BZC, CZB, CBZ, ZBC, ZCB

25. John, Maria, Susan; Maria, Susan, Angelo;
John, Susan, Angelo; John, Maria, Angelo
27. 24   **29.** 12   **31.** 220   **33.** 45   **35.** 4845   **37.** 336
39. 38,760   **41.** 175,560   **43.** 1350   **45.** 2730   **47.** 35
49. 17,576,000   **51.** 1,010,129,120   **53.** 635,013,559,600
55. 240 (if the bride and groom can switch positions)
57. 56,800,235,584; 37,029,625,920
59. a. 6   **b.** 6   **c.** 36   **d.** 66
61. $(n - 1) + (n - 2) + \cdots + 1$
63. 5; 20; $2(n - 3) + (n - 4) + (n - 5) + \cdots + 1$
65. 34,650

Exercise Set 8.5 (page 654)

1. $\{HH, HT, TH, TT\}$ **3.** $\{TT\}$ **5.** $\{1, 3, 5\}$ **7.** $\dfrac{1}{3}$

9. {2 of spades, 3 of spades, 4 of spades, 5 of spades,
6 of spades, 7 of spades, 8 of spades, 9 of spades, 10 of spades,
jack of spades, queen of spades, king of spades, ace of spades}

11. $\dfrac{1}{52}$ **13.** {quarter, dime, nickel, penny}   **15.** $\dfrac{1}{4}$

17. False   **19.** True   **21.** True

23.

Number of Heads	Probability
0	$\dfrac{1}{16}$
1	$\dfrac{1}{4}$
2	$\dfrac{3}{8}$
3	$\dfrac{1}{4}$
4	$\dfrac{1}{16}$

If n = number of heads, $P(n) = P(4 - n)$. No.

25. $\dfrac{1}{13}$ **27.** $\dfrac{1}{52}$ **29.** $\dfrac{3}{13}$ **31.** $\dfrac{1}{6}$ **33.** $\dfrac{2}{9}$ **35.** 10,000

37. 0.504   **39.** $\dfrac{1}{9}$ **41.** $\dfrac{1}{16}$ **43.** $\dfrac{1}{4}$

45. This answer overcounts a card, since there is one spade
that is a king. $\dfrac{4}{13}$

47. $\dfrac{36}{10,000}$ **49.** Neither

51. a. {YYYY, YYYN, YYNY, YNYY, NYYY, NNYY,
YYNN, NYNY, YNYN, NYYN, YNNY, YNNN, NYNN,
NNYN, NNNY, NNNN}

b. $\dfrac{3}{8}$ **c.** $\dfrac{1}{16}$

53. a. 28

b. $\dfrac{5}{28}$

Exercise Set 8.6 (page 662)

1. $a^5, a^4b, a^3b^2, a^2b^3, ab^4, b^5$
3. $x^7, x^6y, x^5y^2, x^4y^3, x^3y^4, x^2y^5, xy^6, y^7$ **5.** 24   **7.** 3   **9.** 15
11. 21   **13.** 1   **15.** 1   **17.** $x^4 + 8x^3 + 24x^2 + 32x + 16$
19. $8x^3 - 12x^2 + 6x - 1$
21. $243 + 405y + 270y^2 + 90y^3 + 15y^4 + y^5$
23. $x^4 - 12x^3z + 54x^2z^2 - 108xz^3 + 81z^4$
25. $x^6 + 3x^4 + 3x^2 + 1$
27. $y^4 - 8y^3x + 24y^2x^2 - 32yx^3 + 16x^4$ **29.** 160
31. 2916   **33.** 28   **35.** $240x^4$ **37.** $1024y^5$ **39.** $2160x^2$
41. $-10,240x^3$

43. $\dbinom{n}{r} = \dfrac{n!}{(n - r)!\, r!} = \dfrac{n!}{(n - r)!(n - (n - r))!} = \dbinom{n}{n - r}$

45. 1

Exercise Set 8.7 (page 667)

1. $(k + 1)(k + 2)(k + 3)$ **3.** $\dfrac{k + 1}{k + 2}$

5. P_1: $2(1) + 1 = 1(1 + 2)$
Assume P_k: $3 + 5 + \cdots + (2k + 1) = k(k + 2)$.
Prove P_{k+1}: $3 + 5 + \cdots + (2k + 1) + (2(k + 1) + 1)$
$$= (k + 1)(k + 3).$$
$3 + 5 + \cdots + (2k + 1) + (2(k + 1) + 1)$
$$= k(k + 2) + (2(k + 1) + 1)$$
$$= k^2 + 2k + 2k + 2 + 1$$
$$= k^2 + 4k + 3$$
$$= (k + 1)(k + 3)$$

7. P_1: $3(1) - 2 = \dfrac{1(3(1) - 1)}{2}$

Assume P_k: $1 + 4 + 7 + \cdots + (3k - 2) = \dfrac{k(3k - 1)}{2}$.

Prove P_{k+1}: $1 + 4 + 7 + \cdots + (3k - 2) + (3(k + 1) - 2)$
$$= \dfrac{(k + 1)(3(k + 1) - 1)}{2}.$$

$1 + 4 + 7 + \cdots + (3k - 2) + (3(k + 1) - 2)$
$$= \dfrac{k(3k - 1)}{2} + (3(k + 1) - 2)$$
$$= \dfrac{k(3k - 1)}{2} + 3k + 1$$
$$= \dfrac{3k^2 + 5k + 2}{2}$$
$$= \dfrac{(k + 1)(3k + 2)}{2}$$
$$= \dfrac{(k + 1)(3(k + 1) - 1)}{2}$$

9. P_1: $9 - 2(1) = -(1)^2 + 8(1)$
Assume P_k: $7 + 5 + 3 + \cdots + (9 - 2k) = -k^2 + 8k$.
Prove P_{k+1}: $7 + 5 + 3 + \cdots + (9 - 2k) + (9 - 2(k + 1))$
$$= -(k + 1)^2 + 8(k + 1).$$
$7 + 5 + 3 + \cdots + (9 - 2k) + (9 - 2(k + 1))$
$$= -k^2 + 8k + (9 - 2(k + 1))$$
$$= -k^2 + 6k + 7$$
$$= -(k^2 + 2k + 1) + 8k + 8$$
$$= -(k + 1)^2 + 8(k + 1)$$

11. P_1: $3(1) - 1 = \dfrac{1}{2}(1)(3(1) + 1)$

Assume P_k: $2 + 5 + 8 + \cdots + (3k - 1) = \dfrac{1}{2}k(3k + 1)$.

Prove P_{k+1}: $2 + 5 + 8 + \cdots + (3k - 1) + (3(k + 1) - 1)$
$$= \frac{1}{2}(k + 1)(3(k + 1) + 1).$$

$2 + 5 + 8 + \cdots + (3k - 1) + (3(k + 1) - 1)$
$$= \frac{1}{2}k(3k + 1) + (3(k + 1) - 1)$$
$$= \frac{1}{2}(3k^2 + k) + (3k + 2)$$
$$= \frac{1}{2}(3k^2 + k) + \frac{1}{2}(6k + 4)$$
$$= \frac{1}{2}(3k^2 + k + 6k + 4)$$
$$= \frac{1}{2}(3k^2 + 7k + 4)$$
$$= \frac{1}{2}(3k + 4)(k + 1)$$
$$= \frac{1}{2}(k + 1)(3(k + 1) + 1)$$

13. P_1: $(1)^3 = \dfrac{1^2((1) + 1)^2}{4}$

Assume P_k: $1^3 + 2^3 + \cdots + k^3 = \dfrac{k^2(k + 1)^2}{4}$.

Prove P_{k+1}: $1^3 + 2^3 + \cdots + k^3 + (k + 1)^3 = \dfrac{(k + 1)^2(k + 2)^2}{4}$.

$1^3 + 2^3 + \cdots + k^3 + (k + 1)^3 = \dfrac{k^2(k + 1)^2}{4} + (k + 1)^3$
$$= \frac{k^2(k + 1)^2 + 4(k + 1)^3}{4}$$
$$= \frac{(k + 1)^2(k^2 + 4(k + 1))}{4}$$
$$= \frac{(k + 1)^2(k^2 + 4k + 4)}{4}$$
$$= \frac{(k + 1)^2(k + 2)^2}{4}$$

15. P_1: $(2(1) - 1)^2 = \dfrac{1(2(1) - 1)(2(1) + 1)}{3}$

Assume P_k: $1^2 + 3^2 + \cdots + (2k - 1)^2 = \dfrac{k(2k - 1)(2k + 1)}{3}$.

Prove P_{k+1}: $1^2 + 3^2 + \cdots + (2k - 1)^2 + (2(k + 1) - 1)^2$
$$= \frac{(k + 1)(2(k + 1) - 1)(2(k + 1) + 1)}{3}.$$

$1^2 + 3^2 + \cdots + (2k - 1)^2 + (2(k + 1) - 1)^2$
$$= \frac{k(2k - 1)(2k + 1)}{3} + (2(k + 1) - 1)^2$$
$$= \frac{k(4k^2 - 1)}{3} + \frac{3(2k + 1)^2}{3}$$
$$= \frac{4k^3 - k + 3(4k^2 + 4k + 1)}{3}$$

$$= \frac{4k^3 + 12k^2 + 11k + 3}{3}$$
$$= \frac{(k + 1)(2k + 1)(2k + 3)}{3}$$
$$= \frac{(k + 1)(2(k + 1) - 1)(2(k + 1) + 1)}{3}$$

17. P_1: $1(1 + 1) = \dfrac{1(1 + 1)(1 + 2)}{3}$

Assume P_k: $1 \cdot 2 + 2 \cdot 3 + 3 \cdot 4 + \cdots + k(k + 1)$
$$= \frac{k(k + 1)(k + 2)}{3}.$$

Prove P_{k+1}:
$1 \cdot 2 + 2 \cdot 3 + 3 \cdot 4 + \cdots + k(k + 1) + (k + 1)(k + 2)$
$$= \frac{(k + 1)(k + 2)(k + 3)}{3}.$$

$1 \cdot 2 + 2 \cdot 3 + 3 \cdot 4 + \cdots + k(k + 1) + (k + 1)(k + 2)$
$$= \frac{k(k + 1)(k + 2)}{3} + (k + 1)(k + 2)$$
$$= \frac{k(k + 1)(k + 2) + 3(k + 1)(k + 2)}{3}$$
$$= \frac{(k + 1)(k + 2)(k + 3)}{3}$$

19. P_1: $5^{1-1} = \dfrac{5^1 - 1}{4}$

Assume P_k: $1 + 5 + 5^2 + \cdots + 5^{k-1} = \dfrac{5^k - 1}{4}$.

Prove P_{k+1}: $1 + 5 + 5^2 + \cdots + 5^{k-1} + 5^k = \dfrac{5^{k+1} - 1}{4}$.

$1 + 5 + 5^2 + \cdots + 5^{k-1} + 5^k = \dfrac{5^k - 1}{4} + 5^k$
$$= \frac{5^k - 1 + 4(5^k)}{4}$$
$$= \frac{5^k + 4(5^k) - 1}{4}$$
$$= \frac{5^{k+1} - 1}{4}$$

21. P_1: $3^1 - 1$ is divisible by 2.
Assume P_k: $3^k - 1$ is divisible by 2.
Prove P_{k+1}: $3^{k+1} - 1$ is divisible by 2.
$3^{k+1} - 1 = 3(3^k) - 1$
$\qquad\quad = 3(3^k - 1) - 1 + 3$
$\qquad\quad = 3(3^k - 1) + 2$
Because $3^k - 1$ is divisible by 2, $3(3^k - 1)$ is divisible by 2,
and so $3(3^k - 1) + 2$ is divisible by 2.
23. P_1: $1^2 + 3(1)$ is divisible by 2.
Assume P_k: $k^2 + 3k$ is divisible by 2.
Prove P_{k+1}: $(k + 1)^2 + 3(k + 1)$ is divisible by 2.
$(k + 1)^2 + 3(k + 1) = k^2 + 2k + 1 + 3k + 3$
$\qquad\qquad\qquad = (k^2 + 3k) + (2k + 1 + 3) = (k^2 + 3k) + 2(k + 2)$
Both $(k^2 + 3k)$ and $2(k + 2)$ are divisible by 2, so
$(k^2 + 3k) + 2(k + 2)$ is divisible by 2.
25. P_1: $2^1 > 1$
Assume P_k: $2^k > k$.
Prove P_{k+1}: $2^{k+1} > k + 1$.
$2^{k+1} = 2(2^k) > 2(k) = k + k \geq k + 1$

27. $a^3 - b^3 = (a - b)(a^2 + ab + b^2)$. This product is divisible by $a - b$.
29. For $1 + 4 + 4^2 + \cdots + 4^{n-1}$, $a_0 = 1$, $r = 4$, and
$a_0\left(\dfrac{1 - r^n}{1 - r}\right) = \left(\dfrac{1 - 4^n}{1 - 4}\right) = \dfrac{4^n - 1}{3}$.

Chapter 8 Review Exercises (page 673)

Section 8.1 (page 673)
1. $a_n = 7 + 2n$ **3.** $a_n = 1 + 4n$ **5.** $a_n = 5(2)^n$
7. $a_n = 2(3)^n$

Section 8.2 (page 673)
9. 496   **11.** 1030   **13.** 508   **15.** $\dfrac{63}{32}$ **17. a.** $\displaystyle\sum_{i=0}^{24} 3i$ **b.** 900
19. a. $\displaystyle\sum_{i=0}^{7} 2^i$ **b.** 255   **21.** $\dfrac{3}{2}$ **23.** No sum

Section 8.3 (page 673)
25. $5, 3, 1, -1, -3$ **27.** $-3, -\dfrac{3}{2}, -\dfrac{3}{4}, -\dfrac{3}{8}, -\dfrac{3}{16}$
29. $1, 0, -3, -8, -15$ **31.** $\dfrac{1}{2}, \dfrac{2}{5}, \dfrac{3}{10}, \dfrac{4}{17}, \dfrac{5}{26}$
33. $-\dfrac{1}{3}, \dfrac{1}{9}, -\dfrac{1}{27}, \dfrac{1}{81}, -\dfrac{1}{243}$ **35.** $4, 7, 10, 13$
37. $-1, 1, 5, 11$ **39.** 28

Section 8.4 (page 674)
41. a. 12   **b.** 120   **c.** 24   **d.** 120   **43.** 120   **45.** 144
47. 120

Section 8.5 (page 674)
49. 3 heads; 2 heads, 1 tail; 1 head, 2 tails; 3 tails
51. {HHH, HHT, HTH, THH, TTH, THT, HTT, TTT}
53. {TTT}   **55.** $\dfrac{2}{9}$

Section 8.6 (page 674)
57. $x^4 + 12x^3 + 54x^2 + 108x + 81$
59. $27x^3 + 27x^2y + 9xy^2 + y^3$

Section 8.7 (page 675)
61. 15   **63.** $6x^2y$
65. P_1: $1 + \dfrac{1}{2^1} = 2 - \dfrac{1}{2^1}$

Assume P_k: $1 + \dfrac{1}{2} + \dfrac{1}{4} + \cdots + \dfrac{1}{2^k} = 2 - \dfrac{1}{2^k}$.

Prove P_{k+1}: $1 + \dfrac{1}{2} + \dfrac{1}{4} + \cdots + \dfrac{1}{2^k} + \dfrac{1}{2^{k+1}} = 2 - \dfrac{1}{2^{k+1}}$.

$$1 + \dfrac{1}{2} + \dfrac{1}{4} + \cdots + \dfrac{1}{2^k} + \dfrac{1}{2^{k+1}} = 2 - \dfrac{1}{2^k} + \dfrac{1}{2^{k+1}}$$
$$= 2 - \dfrac{2}{2^{k+1}} + \dfrac{1}{2^{k+1}}$$
$$= 2 - \dfrac{1}{2^{k+1}}$$

67. P_1: $(1 + 1)^2 + 1$ is odd.
Assume P_k: $(k + 1)^2 + k$ is odd.
Prove P_{k+1}: $(k + 2)^2 + (k + 1)$ is odd.
$$(k + 2)^2 + (k + 1) = k^2 + 4k + 4 + (k + 1)$$
$$= k^2 + 5k + 5$$
$$= k^2 + 2k + 1 + k + 2k + 4$$
$$= [(k + 1)^2 + k] + 2(k + 2)$$
The sum of an odd number and an even number is odd.

69. P_1: $3(1) = \dfrac{3}{2}(1)(1 + 1)$

Assume P_k: $3 + 6 + 9 + \cdots + 3k = \dfrac{3}{2}k(k + 1)$.

Prove P_{k+1}: $3 + 6 + 9 + \cdots + 3k + 3(k + 1)$
$$= \dfrac{3}{2}(k + 1)(k + 2).$$

$$3 + 6 + 9 + \cdots + 3k + 3(k + 1) = \dfrac{3}{2}k(k + 1) + 3(k + 1)$$
$$= (k + 1)\left(\dfrac{3}{2}k + 3\right)$$
$$= \dfrac{3}{2}(k + 1)(k + 2)$$

Applications (page 675)
71. \$43,500   **73.** 14   **75.** 30   **77.** $\dfrac{253}{9996}$

Chapter 8 Test (page 675)
1. $a_n = 8 + 3n$ **2.** $a_n = 7 + 5n$ **3.** $a_n = 15\left(\dfrac{1}{3}\right)^n$
4. $a_n = \dfrac{9}{2}\left(\dfrac{2}{3}\right)^n$ **5.** 1853   **6.** 1680   **7.** $\displaystyle\sum_{i=0}^{9} 5i = 225$
8. $a_0 = 2, a_1 = 0, a_2 = -14, a_3 = -52$
9. $a_0 = 1, a_1 = -\dfrac{1}{4}, a_2 = \dfrac{1}{16}, a_3 = -\dfrac{1}{64}, a_4 = \dfrac{1}{256}$
10. $a_0 = 4, a_1 = 3, a_2 = 1, a_3 = -2, a_4 = -6$ **11.** 26
12. a. 60   **b.** 15   **c.** 7   **d.** 360   **13.** 220   **14.** 720
15. 1728   **16.** {red, blue, white}   **17.** $\dfrac{2}{11}$ **18.** $\dfrac{6}{11}$ **19.** $\dfrac{1}{26}$
20. $81x^4 + 216x^3 + 216x^2 + 96x + 16$
21. P_1 is true: $2 = 2(1)^2$. Assume P_k is true. Then
P_{k+1}: $2 + 6 + 10 + \cdots + 4k - 2 + 4(k + 1) - 2$
$$= 2k^2 + 4(k + 1) - 2 = 2k^2 + 4k + 2 = 2(k + 1)^2.$$
Since P_{k+1} is of the form $2(k + 1)^2$, the proposition holds true for all k.
22. 24   **23.** 327,600

INDEX

Library of Functions

Linear Function

$f(x) = mx + b$, where m and b are constants

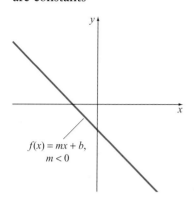

$f(x) = mx + b$,
$m < 0$

Constant Function

$f(x) = k$

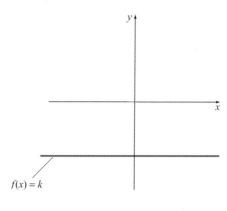

$f(x) = k$

Absolute Value Function

$f(x) = |x|$

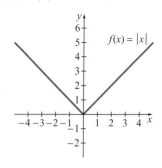

$f(x) = |x|$

Greatest Integer Function

$f(x) = [\![x]\!]$ (the largest integer greater than or equal to x)

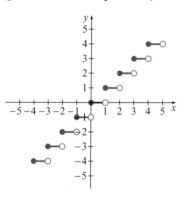

Square Root Function

$f(x) = \sqrt{x}$

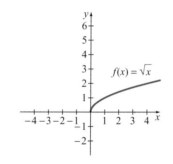

$f(x) = \sqrt{x}$

Quadratic Function

$f(x) = ax^2 + bx + c$, where a, b, and c are real numbers and $a \neq 0$

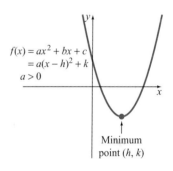

$f(x) = ax^2 + bx + c$
$= a(x - h)^2 + k$
$a > 0$

Minimum point (h, k)

Cubic Function

$f(x) = x^3$

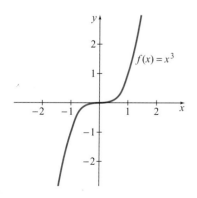

$f(x) = x^3$

A Simple Rational Function

$f(x) = \dfrac{1}{x - b}$

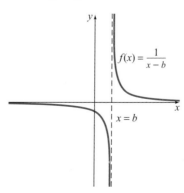

$f(x) = \dfrac{1}{x - b}$

$x = b$

Logistic Function

$f(x) = \dfrac{c}{1 + ae^{-bx}}$

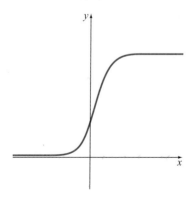

continued

continued

Exponential Function

$f(x) = Ca^x$, where a and C are constants such that $a > 0$, $a \neq 0$, and $C \neq 0$.

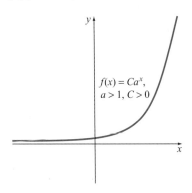

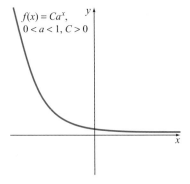

Logarithmic Function

$f(x) = \log_a x, a > 1$

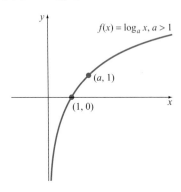

$f(x) = \log_a x, 0 < a < 1$

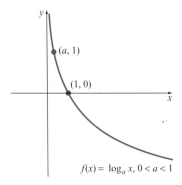

Concepts of Functions

A **function** establishes a correspondence between a set of input values and a set of output values in such a way that for each allowable input value, there is *exactly one* corresponding output value.

Domain

The **domain** of a function is the set of all input values for which the function will produce a real number.

Range

The **range** of a function is the set of all output values that are possible for the given domain of the function.

Even Function

A function is **symmetric with respect to the y-axis** if

$$f(x) = f(-x)$$ for each x in the domain of f.

Functions having this property are called **even functions**.

Odd Function

A function is **symmetric with respect to the origin** if

$$f(-x) = -f(x)$$ for each x in the domain of f.

Functions having this property are called **odd functions**.

Increasing, Decreasing, and Constant Functions

▶ A function f is **increasing** on an open interval I if, for any a, b in the interval, $f(a) < f(b)$ for $a < b$.

▶ A function f is **decreasing** on an open interval I if, for any a, b in the interval, $f(a) > f(b)$ for $a < b$.

▶ A function f is **constant** on an open interval I if, for any a, b in the interval, $f(a) = f(b)$.

Properties

Properties of Exponents

1. $a^m \cdot a^n = a^{m+n}$
2. $(a^m)^n = a^{mn}$
3. $(ab)^m = a^m b^m$
4. $\left(\dfrac{a}{b}\right)^m = \dfrac{a^m}{b^m}, b \neq 0$
5. $\dfrac{a^r}{a^s} = a^{r-s}, a \neq 0$
6. $a^1 = a$
7. $a^0 = 1, a \neq 0$
8. $a^{-s} = \dfrac{1}{a^s}, a \neq 0$

Properties of Radicals

1. $\sqrt[n]{a} \cdot \sqrt[n]{b} = \sqrt[n]{ab}$
2. $\dfrac{\sqrt[n]{a}}{\sqrt[n]{b}} = \sqrt[n]{\dfrac{a}{b}}, b \neq 0$
3. $a^{m/n} = \sqrt[n]{a^m} = (\sqrt[n]{a})^m$
4. If n is odd, then $\sqrt[n]{a^n} = a, a \neq 0$
   If n is even, then $\sqrt[n]{a^n} = |a|, a \neq 0$

Properties of Logarithms

1. $y = \log_a x$ if and only if $x = a^y$.
2. $y = \log x$ if and only if $x = 10^y$.
3. $y = \ln x$ if and only if $x = e^y$.
4. $\log_a(xy) = \log_a x + \log_a y$.
5. $\log_a x^k = k \log_a x$.
6. $\log_a \dfrac{x}{y} = \log_a x - \log_a y$.
7. $\log_a a = 1$
8. $\log_a 1 = 0$

Formulas

Geometric Formulas

Rectangle	Square	Circle	Triangle	Parallelogram
Length: l Width: w	Length: s Width: s	Radius: r	Base: b Height: h	Base: b Height: h Side: s
Perimeter $P = 2l + 2w$	Perimeter $P = 4s$	Circumference $C = 2\pi r$	Perimeter $P = a + b + c$	Perimeter $P = 2b + 2s$
Area $A = lw$	Area $A = s^2$	Area $A = \pi r^2$	Area $A = \frac{1}{2}bh$	Area $A = bh$

Rectangular Solid	Right Circular Cylinder	Sphere	Right Circular Cone
Length: l Width: w Height: h	Radius: r Height: h	Radius: r	Radius: r Height: h
Surface Area $S = 2(wh + lw + lh)$	Surface Area $S = 2\pi rh + 2\pi r^2$	Surface Area $S = 4\pi r^2$	Surface Area $S = \pi r(r^2 + h^2)^{1/2} + \pi r^2$
Volume $V = lwh$	Volume $V = \pi r^2 h$	Volume $V = \frac{4}{3}\pi r^3$	Volume $V = \frac{1}{3}\pi r^2 h$

continued

Algebraic Formulas

Definition of a Slope

The **slope** of a line containing the points (x_1, y_1) and (x_2, y_2) is given by

$$m = \frac{y_2 - y_1}{x_2 - x_1},$$

where $x_1 \neq x_2$.

Average Rate of Change

The **average rate of change** of a function f on an interval $[x_1, x_2]$ is given by

$$\text{Average rate of change} = \frac{f(x_2) - f(x_1)}{x_2 - x_1}.$$

Distance Formula

The distance d between the points (x_1, y_1) and (x_2, y_2) is given by

$$d = \sqrt{(x_2 - x_1)^2 + (y_2 - y_1)^2}.$$

Slope-Intercept Form of the Equation of a Line

The **slope-intercept form** of the equation of a line with slope m and y-intercept $(0, b)$ is given by

$$y = mx + b.$$

Equation of a Line in Point-Slope Form

The equation of a line with slope m and containing the point (x_1, y_1) is given by

$$y - y_1 = m(x - x_1)$$

or, equivalently,

$$y = m(x - x_1) + y_1.$$

The Quadratic Formula

The solutions of $ax^2 + bx + c = 0$, with $a \neq 0$, are given by the **quadratic formula**

$$x = \frac{-b \pm \sqrt{b^2 - 4ac}}{2a}.$$

Theorems

The Pythagorean Theorem

$c^2 = a^2 + b^2$

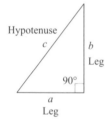

The Remainder Theorem

When a polynomial $p(x)$ is divided by $x - c$, the remainder is equal to the value of $p(c)$.

The Factor Theorem

The term $x - c$ is a *factor* of a polynomial $p(x)$ if and only if $p(c) = 0$.

The Fundamental Theorem of Algebra

Every nonconstant polynomial function with real or complex coefficients has at least one complex zero.

Binomial Theorem

Let n be a positive integer. Then

$(a + b)^n$

$$= \sum_{i=0}^{n} \binom{n}{i} a^{n-i} b^i$$

$$= \binom{n}{0} a^n + \binom{n}{1} a^{n-1} b + \binom{n}{2} a^{n-2} b^2 + \cdots + \binom{n}{n} b^n.$$